CUMULATIVE SUBJECT INDEX

VOLUMES 1–75 (including Revised Series Volumes 1–31)

PART II: F–N

HANDBOOK OF CLINICAL NEUROLOGY

Editors

PIERRE J. VINKEN   GEORGE W. BRUYN

Executive Editor

KENNETH ELLISON DAVIS

VOLUME 77

ELSEVIER

AMSTERDAM • LONDON • NEW YORK • OXFORD • PARIS •
SHANNON • TOKYO

CUMULATIVE SUBJECT INDEX

VOLUMES 1–75 (including Revised Series Volumes 1–31)

PART II: F–N

Editors

PIERRE J. VINKEN GEORGE W. BRUYN

In collaboration with

WILLEKE VAN OCKENBURG

REVISED SERIES 33

ELSEVIER

AMSTERDAM • LONDON • NEW YORK • OXFORD • PARIS •
SHANNON • TOKYO

ELSEVIER SCIENCE B.V.
Sara Burgerhartstraat 25
P.O. Box 211, 1000 AE Amsterdam, The Netherlands

First edition 2002

Library of Congress Cataloging in Publication Data
A catalog record from the Library of Congress has been applied for.

ISBN: 0-444-50919-4 (Part I: A–E)
0-444-50918-6 (Part II: F–N)
0-444-50917-8 (Part III: O–Z)
ISSN: 0072-9752

∞ The paper used in this publication meets the requirements of ANSI/NISO Z39.48-1992 (Permanence of Paper).
Printed in The Netherlands.

Handbook of Clinical Neurology, Vol. 77 (33): Cumulative Subject Index, Part II: F–N

Cumulative Subject Index to Volumes 1–75 (including Revised Series 1–31) Part II: F–N

In collaboration with W. van Ockenburg

F protein
  measles virus, 56/423
F protein antibody
  subacute sclerosing panencephalitis, 56/427
F syndrome
  hypertelorism, 30/249
  skeletal malformation, 30/249
F wave
  chronic inflammatory demyelinating polyradiculoneuropathy, 51/536
  direct muscle response, 7/142
  motor nerve conduction velocity, 1/57
F wave latency
  Guillain-Barré syndrome, 51/248
  thoracic outlet syndrome, 51/126
Fab fragment
  chemistry, 9/502
  viral infection, 56/51
Fc fragment
  chemistry, 9/502
  Ig, *see* Ig Fc fragment
  viral infection, 56/51
Fabry-Anderson disease, *see* Fabry disease
Fabry corneal dystrophy
  amiodarone, 66/228
  chloroquine, 66/228
Fabry disease, 66/225-241
  amniocentesis, 29/363
  anhidrosis, 60/172
  arterial hypertension, 60/172
  ataxia, 10/543
  autonomic dysfunction, 51/377
  autonomic nervous system, 42/427, 75/21
  autonomic polyneuropathy, 51/476
  Bean syndrome, 14/106
  biochemistry, 10/301, 544
  bowel, 75/650
  brain hemorrhage, 55/457
  brain infarction, 53/33, 165, 55/457
  brain ischemia, 53/165
  brain vessel, 11/464
  burning pain, 60/172
  cardiomegaly, 51/377
  cataract, 51/377, 60/172
  ceramide dihexoside, 10/300
  ceramide polyhexoside, 10/300
  ceramide trihexosidase, 10/544
  ceramide trihexoside, 10/300, 344, 544, 579
  cerebrovascular disease, 53/28, 33, 55/456
  chemical defect, 60/172
  chemistry, 10/300
  choreoathetosis, 10/543
  chronic meningitis, 56/644
  clinical features, 10/343, 542, 29/360-363, 51/376
  cornea verticillata, 60/172
  corneal opacity, 42/427, 51/376
  crisis, 51/376
  dementia, 55/457
  diabetes mellitus, 60/172
  diagnosis, 51/379, 66/239
  dialysis, 66/240
  digalactosyl ceramide, 42/428
  EEG, 42/427
  enzyme defect, 29/350

enzyme deficiency polyneuropathy, 51/369, 376
exanthema, 60/172
features, 10/300-302, 343-345, 542-545, 579-583
fever attack, 60/172
fucosidosis, 42/550
α-galactosidase, 42/427-429
β-galactosidase, 10/302
α-galactosidase A, 55/456, 60/172, 66/234
gastrointestinal disease, 39/464
gene site, 60/172
genetics, 10/544, 579-581, 66/239
globoside, 10/544, 581
glomerular endothelium, 10/468
heterozygote, 66/229, 239
heterozygote detection, 42/428
histochemistry, 40/37
histology, 55/457
histopathology, 10/344
iatrogenic brain infarction, 53/165
internuclear ophthalmoplegia, 55/457
lactosylceramide, 10/544
lesion, 66/237
lymphocyte vacuolation, 10/283
metabolic defect, 66/226
mutation, 66/75
nerve conduction, 51/376
neurocutaneous syndrome, 14/780
neuropathology, 10/301, 51/377, 60/173, 66/22
orthostatic hypotension, 63/156
pain, 51/376, 66/226
pathology, 10/543
periodic fever, 60/172
polyneuropathy, 42/427, 60/171
prenatal diagnosis, 42/429
prevalence, 42/428
proteinuria, 60/172
psychiatric disorder, 42/428
psychosis, 60/172
Refsum disease, 21/216
renal insufficiency, 51/377
seizure, 42/427, 55/457, 60/172
sex linked transmission, 55/456
shooting pain, 60/172
shooting pain origin, 60/174
skeletal deformity, 42/428
skin lesion, 55/456
sural nerve biopsy, 51/378
symptom, 66/226, 229
systemic brain infarction, 11/464
transient ischemic attack, 55/457
treatment, 51/379, 55/457
trihexosyl ceramide galactosidase, 21/51
ultrastructure, 10/300, 51/378
X-linked anomaly, 21/51
X-linked recessive disorder, 51/376
Xq21.33-Xq22, 60/172
Fabry disease type I
muscle fiber feature, 62/17
Face
acrocephalosyndactyly type I, 31/231
acrocephalosyndactyly type II, 31/235
ataxia, 1/314
Beckwith-Wiedemann syndrome, 31/330
bird like, *see* Bird like face
brain arteriovenous malformation, 12/229
Cockayne-Neill-Dingwall disease, 31/236
Coffin-Lowry syndrome, 31/244
Coffin-Siris syndrome, 31/239
craniofacial dysostosis (Crouzon), 31/233
craniofacial injury, 57/309
cri du chat syndrome, 31/563
cyclopia, 30/442
De Lange syndrome, 31/280
Down syndrome, 31/413
frozen, *see* Frozen face
Goldenhar syndrome, 31/246
Hallermann-Streiff syndrome, 31/248
holoprosencephaly, 50/232
hypoesthesia, 12/4
leonine, *see* Leontiasis
Lowe syndrome, 31/302
mask like, *see* Mask like face
microcephaly, 50/272
moon, *see* Moon face
multiple nevoid basal cell carcinoma syndrome, 14/456
penetrating head injury, 57/309
poker, *see* Poker face
Potter, *see* Potter face
progressive hemifacial atrophy, 31/252
Rubinstein-Taybi syndrome, 31/282, 43/234, 50/276
Turner syndrome, 31/499
Face-brain correlation, *see* Brain-face correlation
Face-hand test
sensory extinction, 3/190
Facet
cervical vertebral, *see* Cervical vertebral facet
Facet fracture
cervical vertebral column injury, 61/29
thoracolumbar vertebral column injury, 61/36
vertebral column injury, 61/523
Facet interlocking
bilateral cervical, *see* Bilateral cervical facet interlocking
bony fusion, 25/340

instability, 25/340
spinal injury, 25/136, 329-340
thoracolumbar spine injury, 26/296
thoracolumbar vertebral column injury, 61/93
unilateral cervical, *see* Unilateral cervical facet interlocking
Facet joint
intervertebral, *see* Intervertebral facet joint
Facet joint plane
cervical vertebral column, 61/8
lumbar vertebral column, 61/8
thoracic vertebral column, 61/8
Facet subluxation
cervical, *see* Cervical facet subluxation
cervical vertebral column injury, 61/62
Facial anastomosis
facio, *see* Faciofacial anastomosis
hypoglossal, *see* Hypoglossal facial anastomosis
Facial anesthesia
congenital, *see* Congenital facial anesthesia
Facial angiofibroma
tuberous sclerosis complex, 68/292
Facial angioma
glaucoma, 14/61
Sturge-Weber syndrome, 14/66
Facial apraxia
aphasia, 4/93
developmental dyspraxia, 4/458
Facial asymmetry
acrocephalosyndactyly type I, 43/318
cardiofacial syndrome, 42/307
congenital torticollis, 43/146
facial paralysis, 42/313
mental deficiency, 43/296
platysma sign, 1/183
Saethre-Chotzen syndrome, 43/322
unilateral facial paresis, 42/313
Wildervanck syndrome, 43/343
Facial burning
lateral medullary infarction, 53/381
Facial cleft syndrome
median, *see* Median facial cleft syndrome
Facial clonus
see also Facial myoclonia
clinical picture, 8/274
myokymia, 1/280, 285, 2/62
spasm, 1/285
taxonomy, 1/278
terminology, 1/418
Facial contracture
facial paralysis, 8/270, 274
postparalytic, 1/269
terminology, 1/418
Facial diplegia
congenital, *see* Möbius syndrome
congenital acetylcholine receptor deficiency/short channel opentime syndrome, 62/437
Guillain-Barré syndrome, 51/245
hypoglossal nerve agenesis, 50/220
peripheral neurolymphomatosis, 56/181
rabies, 56/388
trigeminal nerve, 30/404
Facial dysmorphia
adult GM1 gangliosidosis, 60/666
ARG syndrome, 50/582
dihydroxycholestanemia, 60/666
infantile Refsum disease, 60/666
juvenile metachromatic leukodystrophy, 60/666
mevalonate kinase deficiency, 60/666
neonatal adrenoleukodystrophy, 60/666
peroxisomal acetyl-coenzyme A acyltransferase deficiency, 66/518
recessive spinocerebellar ataxia, 60/666
trihydroxycholestanemia, 60/666
trisomy 8, 50/563
Facial dysmorphism
Salla disease, 59/360
Facial dystonia
progressive supranuclear palsy, 49/240
Facial edema
cavernous sinus thrombosis, 52/173
echinoidea intoxication, 37/67
ichthyohepatoxic fish intoxication, 37/87
Melkersson-Rosenthal syndrome, 8/208-211, 14/790, 16/222, 42/322
Facial expression
echinoidea intoxication, 37/68
emotional disorder, 3/356, 358
kuru, 56/556
Lafora progressive myoclonus epilepsy, 27/173
ontogeny, 45/221
pathologic laughing and crying, 3/356, 45/221
Facial flushing
Balaenoptera borealis intoxication, 37/94
cluster headache, 48/222
glyceryl trinitrate intoxication, 37/454
ichthyohepatoxic fish intoxication, 37/87
papaverine intoxication, 37/456
Facial granuloma
idiopathic, *see* Midline granuloma
Facial grimacing
Bolivian hemorrhagic fever, 56/369
Brissaud-Sicard syndrome, 2/240
hereditary paroxysmal dystonic choreoathetotis, 49/349
Facial growth

craniosynostosis, 50/124
Facial hemihypertrophy
François syndrome, 14/627, 635, 638
Sachsalber syndrome, 14/398
Sturge-Weber syndrome, 22/549
Facial hemispasm, *see* Brissaud-Sicard syndrome
Facial injury
facial nerve injury, 24/107
ice hockey injury, 23/583
optic nerve injury, 24/36
skiing injury, 23/581
Facial mycosis fungoides, *see* Midline granuloma
Facial myoclonia
see also Facial clonus
type, 2/62
Facial myokymia
definition, 1/280
Joseph-Machado disease, 42/262
multiple sclerosis, 9/169, 41/300
Facial nerve
abnormality, *see* Möbius syndrome
anastomosis, 8/245
anatomy, 48/487, 74/427
auricularis magnus nerve, 5/286
autonomic nervous system, 2/110
buccopharyngeal herpes, 7/483
conduction, 7/165
crocodile tear syndrome, 2/69, 123
diabetes mellitus, 27/118
embryology, 30/397
geniculate ganglion, 48/488
hemophilia, 38/63
leprosy, 33/436
leprous neuritis, 51/220
lesion site assay, 2/59-72
main component, 8/253
myoclonus, 8/274
neuralgia, *see* Intermedius neuralgia
neurinoma, 68/481, 538
paralysis, *see* Facial paralysis
parasympathetic fiber, 2/110
pathway, 8/243
petrosal nerve, 48/488
phylogenesis, 8/246
physiology, 8/247
pontine infarction, 12/16
regeneration, 8/254
sensory fiber, 5/337
somatic afferent, 48/489
sphenoidal sinus syndrome, 2/296
subarachnoid hemorrhage, 55/18
synkinesis, 1/416
temporal arteritis, 55/342
trichloroethylene polyneuropathy, 51/280
tympanic plexus, 48/488
Uhrmacher tic, 7/334
vascularization, 8/246, 251
vestibular neurinoma, 68/441, 443
visceral efferent, 48/487
Facial nerve agenesis
bilateral form, 50/215
Poland-Möbius syndrome, 50/215
trigeminal nerve agenesis, 50/214
Facial nerve injury
accessory facial anastomosis, 24/113
acoustic neuroma, 24/109
auriculotemporal syndrome, 1/457, 8/276
birth incidence, 24/109
clinical examination, 57/137
combat, 24/107
conservative treatment, 24/112, 114
delayed, 57/138
diagnosis, 24/109
ear surgery, 24/108
electrodiagnosis, 24/111
extratemporal treatment, 24/114
facial injury, 24/107
faciofacial anastomosis, 24/115
hypoglossal facial anastomosis, 24/113
immediate, 24/107, 57/137
intratemporal treatment, 24/114
localization, 24/109, 125
optic nerve injury, 24/36
paralysis, *see* Facial paralysis
parotid gland, 24/109
plastic surgery, 24/115
prevention, 24/112
saliva production, 24/111
stapedius muscle, 24/111
surgical, 24/107, 112-114
taste, 24/111
tear secretion, 24/110
temporal bone fracture, 24/105, 124
tumor, 24/109
variety, 24/105
Facial nerve neuralgia, *see* Intermedius neuralgia
Facial nerve palsy, *see* Facial paralysis
Facial neuralgia, *see* Intermedius neuralgia
Facial neurinoma
skull base, 68/466
Facial pain
see also Cranial neuralgia *and* Trigeminal neuralgia
atypical, 8/349
facial paralysis, 8/260
headache, 5/209, 48/2

lateral medullary infarction, 53/381
multiple sclerosis, 5/28
osteitis deformans (Paget), 70/13
paratrigeminal syndrome, 48/336
posttraumatic, 5/401
superior petrosal sinus thrombophlebitis, 33/180
surgery, 5/401
Facial paralysis
à frigore, 8/241
annual incidence, 42/313
anterior inferior cerebellar artery syndrome, 53/396, 55/90
anticoagulant overdose, 65/461
apraxia, 45/431
arthropod envenomation, 65/195
assessment, 2/61
auriculotemporal syndrome, 8/270
bacterial endocarditis, 52/300
Bannwarth syndrome, 51/200
barium, 51/274
bilateral, *see* Bilateral facial paresis
blink reflex, 2/65
brain lacunar infarction, 53/210
branchio-otodysplasia, 42/359
bulbopontine paralysis, 42/96
cardiofacial syndrome, 42/307
cat scratch disease, 52/130
cause, 2/66
central core myopathy, 43/80
cerebellar hemorrhage, 54/314
cerebellar infarction, 53/383
cerebellopontine angle syndrome, 2/318
chloroma, 63/343
chronaximetry, 8/267
congenital heart disease, 63/11
congenital muscular dystrophy, 43/93
congenital unilateral, 42/313
Coxsackie virus infection, 34/139
craniometaphyseal dysplasia, 31/255, 43/362
crocodile tear syndrome, 2/69, 123, 8/270, 275
cryptogenic, 8/241
cryptogenic associated movement, 8/279
cytomegalovirus, 56/270
depressor anguli oris muscle, 42/313
diabetes mellitus, 8/269, 27/118
diabetic neuropathy, 27/118
diagnosis, 8/263-265
diaphyseal dysplasia, 43/376
differential diagnosis, 40/304
dysosteosclerosis, 43/394
echinoidea intoxication, 37/68
Edwards syndrome, 50/561
electrical injury, 61/195
electrodiagnosis, 2/64, 8/265
electrogustometry, 8/265
electrotherapy, 8/286
EMG, 8/268, 281
enterovirus, 56/351
enterovirus 70, 56/351
Epstein-Barr virus, 56/253
Epstein-Barr virus infection, 56/252
expression, 1/421
eye movement, 1/613
eyelid, 1/613
facial asymmetry, 42/313
facial contracture, 8/270, 274
facial pain, 8/260
facial spasm, 42/314
facial tic, 6/801, 8/270
facioplegic migraine, 5/75-77
familial, 21/12
faulty regeneration, 8/270
Fazio-Londe disease, 42/94
Freeman-Sheldon syndrome, 43/354
frozen face, 8/276
heart disease, 42/313
Heerfordt syndrome, 40/306
hemiplegic, 1/178
hemophilia A, 42/738
hereditary, *see* Hereditary facial nerve palsy
hereditary amyloid polyneuropathy, 42/523
hereditary cranial nerve palsy, 60/41
herpes zoster, 2/71, 34/170, 172
herpes zoster ganglionitis, 51/181
herpes zoster oticus, 8/282, 56/118
herpetic geniculate neuralgia, 5/210
histiocytosis X, 42/442
hypertension, 8/283
idiopathic, *see* Idiopathic facial paralysis
infection, 8/282
infectious mononucleosis, 56/252
intensity duration curve, 8/267
ischemic, 8/241
Kawasaki syndrome, 51/453
Kearns-Sayre-Daroff-Shy syndrome, 43/142
Köhlmeier-Degos disease, 39/436, 51/454
Krabbe muscular hypoplasia, 43/110
lesion site, 2/66
Lyme disease, 52/262
lymphatic leukemia, 18/250
lymphomatoid granulomatosis, 51/451
Melkersson-Rosenthal syndrome, 2/72, 8/215, 279, 344, 14/790, 16/222, 42/322
meningococcal meningitis, 52/26
methyldopa intoxication, 37/441
migraine, 48/156

Millard-Gubler syndrome, 32/76
Möbius syndrome, 42/324
multiple myeloma, 63/393
multiple sclerosis, 9/167, 47/56
mumps, 7/486, 56/430
myotonic dystrophy, 43/152
myotubular myopathy, 43/113
nemaline myopathy, 43/122
neonatal myasthenia gravis, 41/100
neuroborreliosis, 51/199
neurosarcoidosis, 63/422
nonrecurrent, 60/41
occupational, 7/335
oculopharyngeal muscular dystrophy, 43/101
ophthalmoplegia, 43/138, 145
osteitis deformans (Paget), 17/76, 38/363, 43/451, 70/13
osteopetrosis, 43/456
otitic, 8/280
pain, 42/313
pathology, 8/259
platysma sign, 1/183
poker face, 8/276
posterior inferior cerebellar artery occlusion, 55/90
postherpetic neuralgia, 5/324
postparalytic symptom, 8/270
Prader-Labhart-Willi syndrome, 43/463
prognosis, 2/64, 8/289
pseudobulbar paralysis, 2/241
recurrent, 21/12, 60/41
refrigeratory, 8/241
regeneration, 8/270
rheumatic, 8/205, 241
salivation, 8/265
sarcoid neuropathy, 51/195
sarcoidosis, 71/485
sarcotubular myopathy, 43/129
scarlet fever, 7/487
sclerosteosis, 31/258, 43/475
sea snake intoxication, 37/92
sensory loss, 42/313
skeletal deformity, 42/313
spastic eyelid closure, 1/617
stapedius reflex, 8/263
surgery, 8/286, 24/107, 112-114
taste test, 2/64
testing, 8/290
thalidomide syndrome, 42/660
tick bite intoxication, 37/111
tick venom, 65/195
total ophthalmoplegia, 43/145
Toxopneustes pileolus intoxication, 37/68
traumatic, 8/280
treatment, 8/283
tumor, 8/282
unilateral, 42/313
Van Buchem disease, 31/258, 38/408, 43/410
varicella zoster virus, 56/229
vascular hypertension, 8/283
Waldenström macroglobulinemia, 63/396
watershed infarction, 53/210
weakness, 40/304, 422
Wegener granulomatosis, 51/451
Facial paresthesia
brain lacunar infarction, 53/395
lateral medullary infarction, 53/381
pontine infarction, 53/390
thalamic infarction, 53/395
vertebrobasilar system syndrome, 53/390, 394
Facial spasm
areflexia, 42/314
Argyll Robertson pupil, 42/314
clonic, *see* Facial clonus
diplopia, 42/314
eyelid, 1/618
facial paralysis, 42/314
hemi, *see* Hemifacial spasm
hyporeflexia, 42/314
myoclonus, 42/220
nerve conduction velocity, 42/315
nystagmus, 42/315
polyneuropathy, 42/314-316
primary, 8/277
secondary, 8/274
tic, 6/801
tremor, 42/315
trophic ulcer, 42/315
Facial structure
embryology, 30/434-438
Facial sweating
see also Forehead sweating
headache, 48/335
paratrigeminal syndrome, 2/113, 48/329
Facial sympathalgia, *see* Sphenopalatine neuralgia
Facial tic
facial paralysis, 6/801, 8/270
Facial weakness
abnormal myomuscular junction myopathy, 62/334
cap disease, 62/334
central core myopathy, 62/334
congenital fiber type disproportion, 62/334
congenital myasthenic syndrome, 62/334
congenital myopathy, 62/334
cytoplasmic inclusion body myopathy, 62/334

differential diagnosis, 40/304
Fazio-Londe disease, 40/306
herpes zoster ganglionitis, 56/231
Mallory body myopathy, 62/334
mitochondria jagged Z line myopathy, 62/334
multicore myopathy, 62/334
muscle weakness, 40/304, 422
nemaline myopathy, 62/334
olivopontocerebellar atrophy (Wadia-Swami), 60/494
Pyle disease, 22/493, 60/774
trilaminar myopathy, 62/334
tubular aggregate myopathy, 62/334
tubulomembranous inclusion myopathy, 62/334
uniform muscle fiber type I myopathy, 62/334
Facilitating myasthenic syndrome, *see* Eaton-Lambert myasthenic syndrome
Facilitation
aphasia, 46/616
cerebellar function, 2/411
definition, 1/50
postactivation, *see* Postactivation facilitation
reticular formation, 2/533
Facioacoustic oxalosis
ethylene glycol intoxication, 64/124
Faciocephalalgia
autonomic, *see* Sphenopalatine neuralgia
Faciodigital genital syndrome, *see* Aarskog syndrome
Faciofacial anastomosis
facial nerve injury, 24/115
Facio-oculoacousticorenal syndrome
cataract, 43/400
heterochromia iridis, 43/400
hypertelorism, 43/399
hypoplasia iridis, 43/400
proteinuria, 43/399
sella turcica, 43/400
sensorineural deafness, 43/399
swayback nose, 43/399
Facioplegic migraine
facial paralysis, 5/75-77
hereditary, 42/748
schema, 48/156
Facioscapulohumeral muscular atrophy
fasciculation, 42/76
genetics, 41/441
progressive amyotrophy, 21/29
Facioscapulohumeral muscular dystrophy
age at onset, 40/460, 62/163
asymptomatic case, 40/460, 62/164
birth incidence, 43/99
clinical features, 41/485, 62/162
Coats disease, 62/167
corticosteroid, 40/423
course, 22/58
creatine kinase, 40/420, 62/166
diabetes mellitus, 43/99
differential diagnosis, 40/494, 41/3, 58
distal muscle weakness, 40/481
Duchenne muscular dystrophy, 40/419
Emery-Dreifuss muscular dystrophy, 62/155
EMG, 40/420, 463, 62/166
genetics, 62/163
hearing loss, 40/420
heart disease, 40/419
histochemistry, 40/50
histopathology, 62/166
inflammatory myopathy, 40/421
laboratory features, 62/166
lactate dehydrogenase, 62/166
lactic acidosis, 43/99
limb girdle syndrome, 40/438, 441
lordosis, 40/418
mitochondria, 43/99
Möbius syndrome, 40/419, 422, 62/162
muscle computerized assisted tomography, 62/162
muscle fiber type I, 43/99
muscle necrosis, 40/252, 420
myalgia, 40/460
myoglobinuria, 62/556
myotonic dystrophy, 62/240
myotubular myopathy, 62/168
neurogenic muscular atrophy, 22/75
neuromuscular disease, 41/418
pathology, 40/252, 420, 463
penetrance, 62/165
poikiloderma atrophicans vasculare, 40/420
polymyositis, 41/59, 43/99
prevalence, 43/98, 62/165
ptosis, 43/98
retinopathy, 62/168
scapulohumeral distribution, 62/170
scapuloperoneal muscular dystrophy, 62/169
scapulothoracic fixation, 40/423
scoliosis, 40/422
serum alanine aminotransferase, 62/166
survey, 40/460
treatment, 62/168
winged scapula, 40/417, 62/162
Facioscapulohumeral spinal muscular atrophy
clinical features, 59/42
differential diagnosis, 59/375
electrophysiologic study, 59/42
inappropriate name, 59/41
muscle biopsy, 59/42

myopathic features, 59/42
neurogenic features, 59/42
Facioscapulohumeral syndrome, 40/415-423
central core myopathy, 62/168
classification, 40/280, 416
clinical features, 40/302, 305, 308, 313, 322, 342, 441
Coats disease, 62/167
differentiation, 62/161
inflammatory form, 41/80
mitochondrial abnormality, 41/218
mitochondrial myopathy, 62/168
muscle computerized assisted tomography, 62/162
myasthenia gravis, 62/168
myopathy, 40/280
nemaline myopathy, 62/168
neuralgic amyotrophy, 51/175
neurogenic, 40/423
pathologic reaction, 40/252, 415
polymyositis, 62/167
scapuloperoneal syndrome, 40/416, 418, 423
spinal muscular atrophy, 40/423
survey, 40/460-466
Factor
growth, *see* Growth factor
Hageman, *see* Hageman factor
hereditary, *see* Hereditary factor
intrinsic, *see* Intrinsic factor
motility, *see* Motility factor
myelinotoxic, *see* Myelinotoxic factor
nerve growth, *see* Nerve growth factor
neurovirulence, 56/37
platelet activating, *see* Platelet activating factor
serotonin releasing, *see* Serotonin releasing factor
transcription, *see* Transcription factor
transfer, *see* Transfer factor
Fahr disease
intracerebral calcification, *see* Striatopallidodentate calcification
Fakirism
theta rhythm, 45/240
Falcine meningioma
spastic paraplegia, 19/86, 59/428
Fallon triad
intraspinal neurenteric cyst, 32/425, 431
Fallopian canal
neurinoma, 68/538
Fallopius, G., 2/10
Fallot tetralogy
brain abscess, 52/145, 63/5
cerebrovascular disease, 53/31
congenital heart disease, 63/2
holoprosencephaly, 50/237
infectious endocarditis, 63/112
r(13) syndrome, 50/585
False localization sign
ependymal cyst, 18/135
lateral ventricle tumor, 17/597
occipital lobe tumor, 2/672, 17/332
False porencephaly
epilepsy, 72/109
False transmitter
myasthenia gravis, 41/113
Falx cerebri
absence, 42/18, 39
anatomy, 1/569, 30/417
calcification, 14/114, 463
cerebellar agenesis, 42/18
embryology, 30/416
herniation, 1/569
lissencephaly, 30/485, 42/39
meningioma, 17/542
physiologic striatopallidodentate calcification, 49/417
Falx cerebri hypoplasia
Arnold-Chiari malformation type II, 50/405, 408
Familial akinetorigid syndrome, *see* Pure akinesia
Familial amaurotic idiocy, 10/212-228
adult neuronal ceroid lipofuscinosis, 10/218, 225, 588, 622, 21/62
Batten disease, 10/215
Bielschowsky-Jansky type, 10/215
brain biopsy, 10/680, 685
cherry red spot, 10/338
chorioretinal degeneration, 13/39
classification, 10/15, 225
congenital type, 10/299
Creutzfeldt-Jakob disease, 6/748
Dollinger-Bielschowsky type, 10/216
epilepsy, 15/423
eponymic classification, 10/217
glycosaminoglycan, 10/343
history, 10/212-215, 226-228, 332
juvenile, 10/589
leukodystrophy, 10/53
main clinical feature, 10/338-340
mental deficiency, 46/56
metachromatic leukodystrophy, 10/53
neurochemistry, 10/332-341
neuronal ceroid lipofuscinosis, 10/215
Norman-Wood type, 10/218
nosology, 10/222
optic atrophy, 13/39
pigment variation, 6/609, 616, 10/550-552
Schaffer-Spielmeyer cell change, 10/213
spongy degeneration, 10/203

terminology, 10/588-592
transitional type, 10/53
type, 10/299, 333
ultrastructure, 10/340
unitarian concept, 10/216, 220
unitarian pitfall, 10/218
Familial amyotrophic chorea-acanthocytosis, *see* Chorea-acanthocytosis
Familial amyotrophic dystonic paraplegia
symptomatic dystonia, 6/556
Familial amyotrophic lateral sclerosis, 59/241-249
age at onset, 22/329, 59/243
amorphous hyaline, 59/246
anterior horn, 21/30
autosomal dominant, 22/132
autosomal dominant inheritance, 59/241
autosomal recessive inheritance, 59/242
bulbar type, 22/132, 319
cerebellum, 59/246
chromosome 21, 59/249
Clarke column, 21/30
clinical features, 59/236, 243-245
deafness, 22/138
dementia, 59/236, 245
etiology, 59/241-243
features, 21/26
Guam, 22/129, 132, 319
incidence, 22/321, 59/245
lower limb, 59/243
medial longitudinal fasciculus, 59/246
molecular genetic study, 59/248
molecular genetics, 59/248
neuropathology, 22/133, 342, 59/233, 236, 245-248
oculomotor nucleus, 59/246
Onufrowicz nucleus, 59/246
optic atrophy, 22/138
Parkinson dementia, 59/236
parkinsonism, 22/319, 59/236, 245
posterior column, 21/30, 22/323
Purkinje cell loss, 59/246
sensory sign, 59/244
sex ratio, 59/180, 243
spinocerebellar tract, 21/30
sporadic type, 22/320, 340
substantia nigra depigmentation, 59/246
variability, 59/246
Familial autonomic dysfunction
intestinal pseudo-obstruction, 51/493
Familial basal ganglion calcification, *see* Striatopallidodentate calcification
Familial Bell palsy
features, 21/12
Familial benign tremor, *see* Benign essential tremor
Familial betalipoproteinemia, *see* Hyperlipoproteinemia type IIA
Familial brucellosis
benign intracranial hypertension, 67/111
Mollaret meningitis, 34/549, 56/629
primary amyloidosis, 51/415
Familial bulbopontine paralysis
deafness, *see* Brown-Vialetto-Van Laere syndrome
Fazio-Londe disease, 22/108
Familial choreoathetosis
progressive, *see* Progressive familial choreoathetosis
Familial colloid body, *see* Hutchinson-Tay choroidopathy
Familial combined hyperlipidemia
apolipoprotein A-I/C-III/A-IV gene complex, 66/543
defect, 66/543
Familial cortical arteriolosclerosis
intracerebral hematoma, 11/672
Familial craniometaphyseal dysplasia, *see* Pyle disease
Familial defective apolipoprotein B-100
apolipoprotein B, 66/545
Familial dementia
spastic paraparesis, 46/401
Familial drusen, *see* Hutchinson-Tay choroidopathy
Familial dysautonomia, *see* Hereditary sensory and autonomic neuropathy type III
Familial dysmyelination
Alexander disease, 10/101
Familial ectodermal dysplasia
anodontia, 14/112
computer assisted tomography, 60/671
deafness, 60/657
dwarfism, 60/666
hypogonadism, 60/668
mental deficiency, 60/664
Familial essential tremor, *see* Essential tremor
Familial exercise intolerance
phosphohexose isomerase deficiency, 27/236
Familial gliomatosis-glioblastomatosis
case report, 14/496
dysraphia, 14/498
genetics, 14/496-500
twin study, 14/499
Familial glycinuria
renal glycosuria, 29/187
Familial hemiplegic migraine, 48/141-151
autosomal dominant, 5/83, 42/748
brain blood flow, 48/27, 70
case series, 5/266

classification, 48/6
clinical picture, 48/142-148
diagnosis, 48/147-149
literature review, 5/82, 48/142-147
pathogenesis, 48/149
schema, 48/156
secondary pigmentary retinal degeneration, 48/27, 60/737
treatment, 48/151
Familial hepatocerebral degeneration, *see* Wilson disease
Familial holotopistic striatal necrosis, *see* Hereditary striatal necrosis
Familial hydrocephalus
anencephaly, 50/289
encephalocele, 50/289
spina bifida, 50/289
Familial hyperargininemia
clinical features, 59/358
hyperammonemia, 59/358
tetraplegia, 59/358
treatment, 59/359
Familial hyperbetalipoproteinemia, *see* Hyperlipoproteinemia type II
Familial hyperbeta-prebetalipoproteinemia, *see* Hyperlipoproteinemia type III
Familial hypercholesterolemia, *see* Hyperlipoproteinemia type II
Familial hypercholesterolemic xanthomatosis, *see* Hyperlipoproteinemia type II
Familial hyperchylomicronemia with hyperprebetalipoproteinemia, *see* Hyperlipoproteinemia type V
Familial hyperprebetalipoproteinemia, *see* Hyperlipoproteinemia type IV
Familial hypobetalipoproteinemia, 29/391-398
ataxia, 13/427
autosomal dominant, 13/427, 29/395
Babinski sign, 13/427
Bassen-Kornzweig syndrome, 29/391, 60/132
clinical features, 60/133
dysesthesia, 13/427
lower extremity, 13/427
metabolic polyneuropathy, 51/395
neuroacanthocytosis, 63/271
neurologic dysfunction, 29/394
progressive weakness, 13/427
secondary pigmentary retinal degeneration, 60/728
Familial hypoglycemia
mental deficiency, 46/19
Familial hypokalemic periodic paralysis
EMG, 62/69
Familial iminoglycinuria
biochemistry, 29/186
clinical symptom, 29/185
diagnosis, 29/186
genetics, 29/187
history, 29/185
Familial incidence
amyotrophic lateral sclerosis, 59/245
brain aneurysm, 12/86
celiac disease, 28/222
Hirayama disease, 59/107
juvenile parkinsonism, 49/159
Marcus Gunn phenomenon, 30/403, 42/320
multiple neuronal system degeneration, 59/143
multiple sclerosis, 9/94, 47/297, 306
myelinoclastic diffuse sclerosis, 9/481
optic nerve aplasia, 30/400
porencephaly, 30/685, 42/49, 50/358
spinal arachnoid cyst, 32/395
true porencephaly, 30/685
Turner syndrome, 31/497
Familial infantile cerebral sclerosis
history, 10/4
Familial infantile myasthenia
age improvement, 62/427
apnea, 62/427
compound muscle action potential, 62/428
congenital myasthenic syndrome, 62/427
congenital myopathy, 62/332
EMG, 62/427
exacerbation, 62/427
muscle biopsy, 62/428
neuromuscular junction disease, 62/392
ptosis, 62/427
single fiber electromyography, 62/427
treatment, 62/429
Familial insomnia
fatal, *see* Fatal familial insomnia
sleep, 74/549
Familial intermittent ophthalmoplegia
myositis, 22/190
Familial inverted choreoathetosis
Huntington chorea, 49/293
Familial juvenile epithelial dystrophy
primary pigmentary retinal degeneration, 13/248
Familial late onset ataxia
dystonia, 60/664
Familial leukodystrophy
elfin face syndrome, 10/54
Norman microcephalic, *see* Norman microcephalic familial leukodystrophy
Familial limb girdle myasthenia
congenital, *see* Congenital familial limb girdle

myasthenia
proximal muscle weakness, 43/156
Familial lysis muscle fiber type I
congenital myopathy, 41/21
Familial metaphyseal dysplasia
blindness, 46/88
fibrous dysplasia, 14/170
Familial multilocular cystic jaw disease, *see* Cherubism
Familial multiple dentigerous cyst, *see* Multiple nevoid basal cell carcinoma syndrome
Familial myoclonus epilepsy, *see* Progressive myoclonus epilepsy
Familial myopathy
multiple mitochondrial DNA deletion, 66/429
Familial neurovisceral lipidosis, *see* GM1 gangliosidosis
Familial orthochromatic leukodystrophy with diffuse leptomeningeal angiomatosis, *see* Divry-Van Bogaert syndrome
Familial osteopetrosis, *see* Hereditary osteopetrosis
Familial pallidoluysionigral degeneration
reticuloendothelial disorder, 6/665, 49/483
Familial pallidonigral degeneration
Huntington chorea, 49/293
pallidonigral necrosis, 49/466
reticuloendothelial disorder, 6/665, 49/483
Familial pancytopenia, *see* Fanconi syndrome
Familial paroxysmal ataxia
nuclear magnetic resonance, 60/673
Familial paroxysmal choreoathetosis
see also Hereditary exertional paroxysmal dystonic choreoathetosis, Hereditary paroxysmal dystonic choreoathetotis *and* Hereditary paroxysmal kinesigenic choreoathetosis
ballismus, 42/205
choreoathetoid movement, 6/355, 458, 42/205
classification, 49/349
striatal convulsion, 6/169
symptomatic dystonia, 6/556, 42/205
terminology, 42/205, 49/349
Familial periodic paralysis
hyperkalemic periodic paralysis, 62/457
hypokalemic periodic paralysis, 62/457
migraine, 5/47
normokalemic periodic paralysis, 28/597, 41/162, 62/457
progressive spinal muscular atrophy, 22/32
Familial pes cavus, 21/263-268
clinical features, 21/267
diagnosis, 21/268
Friedreich ataxia, 21/263
GM1 gangliosidosis, 10/223
group classification, 21/263
heredodegenerative disease, 21/263
idiopathic, 21/264
lymphedema, 21/264
mechanism, 21/265-267
muscle factor, 21/265
prognosis, 21/268
treatment, 21/263
Familial polyradiculopathy syndrome
dysarthria, 42/97
dysphagia, 42/97
Familial progressive myoclonus epilepsy (Unverricht-Lundborg), *see* Unverricht-Lundborg progressive myoclonus epilepsy
Familial protein intolerance
hyperdibasic aminoaciduria, 29/105, 210
lysinuric, *see* Lysinuric protein intolerance
periodic ataxia, 21/578
Familial recessive tremor
Parkinson disease, 49/108
Familial recurrent myoglobinuria
early onset, 62/558
multiple mitochondrial DNA deletion, 66/429
Familial relapsing ophthalmoplegia
orbital myositis, 62/302
Familial spastic paraplegia, 22/421-430, 59/301-313
adrenomyeloneuropathy, 59/308
amyotrophic lateral sclerosis, 59/306
amyotrophy, 21/16, 463, 22/427, 445-465, 469
arthritis, 22/425
associated disease, 22/422
ataxia, 22/425, 458, 59/305
athetosis, 22/453
autosomal dominant, 59/308
autosomal recessive, 59/309
baclofen, 59/313
Bardet-Biedl syndrome, 59/330-332
Bassen-Kornzweig syndrome, 59/327
Behr disease, 59/334
benzodiazepine, 59/313
bulbar paralysis, 22/139
cardiac disorder, 22/425, 427
cassava intoxication, 59/306
central retinal degeneration, 22/469
cerebral palsy, 59/306
cervical myelopathy, 59/306
chorea, 22/453
clinical features, 22/424, 59/304-306
Cockayne-Neill-Dingwall disease, 59/326, 332
complicated type, 22/425, 59/306
congenital retinal blindness, 22/454, 59/334

cranial nerve, 59/305
CSF homocarnosine, 59/312
CSF protein, 59/312
dantrolene, 59/313
deafness, 22/471
dementia, 22/424, 426, 433, 469
diagnostic criteria, 59/306
differential diagnosis, 59/306-308, 334
dysarthria, 22/435
dystonia, 22/433, 456
ECG abnormality, 60/670
electrophysiologic study, 59/311
epidemiology, 21/19, 59/302-304
epidural lipomatosis, 59/306
epilepsy, 15/327, 22/426, 428
extrapyramidal disorder, *see* Hereditary dystonic paraplegia
extrapyramidal sign, *see* Ferguson-Critchley syndrome
Ferguson-Critchley syndrome, 59/319-323
foot deformity correction, 59/313
Friedreich ataxia, 21/19, 22/422, 425
gaze paralysis, 22/434, 440
genetics, 22/422
globoid cell leukodystrophy, 59/352
Gordon-Capute-Konigsmark syndrome, 59/326, 333
gyrate atrophy, 59/327
H-reflex study, 59/312
Hallervorden-Spatz syndrome, 59/326, 333
hereditary, 22/422
hereditary dystonic paraplegia, 59/346
hereditary motor and sensory neuropathy type I, 22/422, 425
hereditary motor and sensory neuropathy type II, 22/422, 425
hereditary motor and sensory neuropathy variant, 60/244
hereditary olivopontocerebellar atrophy (Menzel), 21/444
hereditary sensory and autonomic neuropathy type I, 21/12
hereditary spinal muscular atrophy, 59/26
history, 22/421, 59/301
homocarnosinosis, 59/308, 326, 333
human T-lymphotropic virus type I, 59/306
Huntington chorea, 22/430, 455
ichthyosis, 22/427
inheritance mode, 59/308
Kearns-Sayre-Daroff-Shy syndrome, 59/334
Kjellin syndrome, 59/326
laboratory data, 22/429
lathyrism, 59/306
Laurence-Moon syndrome, 59/330-332
Leber optic atrophy, 13/97, 22/454
leukodystrophy, 22/433, 441
linkage study, 59/310
macular degeneration, 22/468
mannosidosis, 59/308
mental deficiency, 59/312, 374
metabolic disorder, 59/351-361
multiple sclerosis, 59/306
muscular dystrophy, 22/433
myelopathy, 59/306
Nasu-Hakola disease, 59/308
nerve conduction velocity, 59/311
neuropathology, 22/423, 438, 462, 59/303
neurosyphilis, 59/306
nosology, 22/461
nystagmus, 22/427, 458
obesity, 22/425
olivopontocerebellar atrophy, 21/400
ophthalmologic examination, 59/326, 328, 332
ophthalmoplegia, *see* Ferguson-Critchley syndrome
optic atrophy, 21/16, 22/425, 427, 472, 42/409
parkinsonism, 22/434
progabide, 59/313
progressive external ophthalmoplegia classification, 22/179, 62/290
pseudodeficiency state, 59/352
pure Strümpell type, 22/422, 424
Refsum disease, 59/327, 334
retinal disease, 59/325-335
retinal pigmentation, 59/312
rubella, 59/327
secondary pigmentary retinal degeneration, 22/467-473, 60/733
sensory evoked potential, 59/312
sensory impairment, 22/434, 59/305
sensory loss, 60/244
sensory neuropathy, 21/16
Sjögren-Larsson syndrome, 59/326-328
spastic tetraplegia, 42/168-170
sphincter disturbance, 59/305
spinal cord compression, 59/306
spinal cord segmental narrowing, 59/306
spinocerebellar degeneration, 21/15
spinocerebellar degeneration classification, 60/281
striatal necrosis, 22/455
subacute combined spinal cord degeneration, 59/306
syndactyly, 22/468
syringomyelia, 59/306
tertiary syphilis, 59/327

thioridazine intoxication, 59/327
threonine, 59/313
tizanidine, 59/313
treatment, 22/429, 59/312
tropical ataxic neuropathy, 59/306
tropical spastic paraplegia, 59/306
Troyer syndrome, 59/374
visual evoked response, 59/311
X-linked recessive, 59/310
Familial spherocytosis
chromosome 8, 63/263
hereditary spinal muscular atrophy, 59/25
Familial spinal arachnoiditis
autosomal dominant, 33/269
neurofibromatosis type I, 42/107
opticochiasmatic arachnoiditis, 42/107
pain, 33/269, 42/107
spastic paraplegia, 42/107
Familial striatopallidodentate calcification
ataxia, 6/707, 49/420
dementia, 46/401
pigmentary macular degeneration, 6/707, 49/420
Familial syringomyelia
hereditary sensory and autonomic neuropathy, 60/5
hereditary sensory and autonomic neuropathy type I, 60/5
Familial tremor
essential, *see* Essential tremor
Familial tremor syndrome
Parkinson disease, 49/131
Familial ulceromutilating acropathy, *see* Hereditary sensory and autonomic neuropathy type I
Familial visceral neuropathy
autonomic nervous system, 75/641
classification, 75/642
type, 70/328
Family life
sexual function, 26/453
Fanconi anemia, *see* Fanconi syndrome
Fanconi syndrome
birth incidence, 43/17
café au lait spot, 43/17
carnitine deficiency, 62/493
chromosomal aberration, 43/17
cystinosis, 29/173
fructose intolerance, 29/172, 256
galactosemia, 29/172
glycogen storage disease, 29/172
glycogen storage disease type I, 29/172
Goltz-Gorlin syndrome, 14/113
heavy metal intoxication, 29/172
hereditary tyrosinemia, 29/172
heterozygote detection, 43/18
heterozygote frequency, 43/17
hyperpigmentation, 43/17
hypertelorism, 30/249
hypogonadism, 43/16
leukemia, 31/319
lipid metabolic disorder, 62/493
Lowe syndrome, 29/172
mental deficiency, 31/320, 43/16
microcephaly, 30/508, 31/321, 43/16
microphthalmia, 43/16
multiple myeloma, 29/172
prenatal diagnosis, 43/18
renal abnormality, 43/16
secondary systemic carnitine deficiency, 62/493
skeletal malformation, 30/249
spastic paraplegia, 59/439
strabismus, 43/16
thrombocytopenia, 43/16
tyrosinemia, 29/172
Wilson disease, 27/387, 29/172
Far eastern tick-borne encephalitis
acute viral encephalitis, 56/134, 139
bulbar paralysis, 56/140
fatality rate, 56/140
headache, 56/139
Koshevnikoff epilepsy, 56/140
lower motoneuron, 56/140
photophobia, 56/139
vomiting, 56/139
Farber disease, 66/211-221
acid ceramidase, 66/60
acid ceramidase deficiency, 42/592-594, 51/379
acylsphingosine deacylase, 66/219
amyotrophy, 51/379
autosomal recessive, 51/379
biochemistry, 66/217
ceramide, 66/211, 216
cherry red spot, 60/166, 66/19
classic phenotype, 66/213, 215
contracture, 42/593
CSF, 42/593
enzyme deficiency polyneuropathy, 51/369, 379
epilepsy, 72/222
foam cell, 10/353, 545
genetics, 66/218
glycosaminoglycanosis, 10/431
GM2 gangliosidosis type II, 66/215
granuloma, 42/593
history, 66/211
myoclonic epilepsy, 60/166
nerve conduction velocity, 51/379
pathogenesis, 66/219

pathology, 66/215
phenotype, 66/212
polyneuropathy, 60/166
prenatal diagnosis, 42/594
Schwann cell vacuole, 51/379
sphingosine, 66/216
treatment, 66/220
Farnoquinone, *see* Vitamin K2
Fascia lata syndrome
nerve intermittent claudication, 20/795
Fasciculation
amyotrophic lateral sclerosis, 22/292, 42/65, 59/185, 192
Aran-Duchenne disease, 42/86
ataxia telangiectasia, 60/659
carbamate intoxication, 64/186, 190
clofibrate, 41/300
dominant ataxia, 60/659
facioscapulohumeral muscular atrophy, 42/76
hyperparathyroidism, 27/293, 41/247
infantile spinal muscular atrophy, 42/88
intermediate spinal muscular atrophy, 42/90
Isaacs syndrome, 62/273
Joseph-Machado disease, 60/659
Kjellin syndrome, 42/173
late onset ataxia, 60/659
lathyrism, 65/3
Machado disease, 42/155
motoneuron disease, 41/300
motor unit, 62/50
motor unit hyperactivity, 41/299
multiple neuronal system degeneration, 59/138
muscular atrophy, 42/78, 82-84
myoclonus, 42/240
myokymia, 41/299
myopathy, 40/330, 62/50
neuropathy, 60/659
nystagmus, 42/240
olivopontocerebellar atrophy (Dejerine-Thomas), 60/659
organophosphate intoxication, 64/168
pellagra, 7/573
progressive dysautonomia, 59/138
progressive muscular atrophy, 59/13
pyrethroid intoxication, 64/219
renal insufficiency, 27/324-326, 336
rigidity, 60/659
Ryukyan spinal muscular atrophy, 42/93
scapuloperoneal syndrome, 42/99
spastic paraplegia, 42/172
striatonigral degeneration, 42/262
tick bite intoxication, 37/111
tongue paralysis, 30/411
toxic myopathy, 62/596
Wohlfart-Kugelberg-Welander disease, 22/70, 42/91, 59/85
xeroderma pigmentosum, 60/659
Fasciculation discharge
features, 62/50
motor unit, 62/50
myopathy, 62/50
Fasciculation syndrome
differential diagnosis, 59/412
muscular pain, *see* Muscular pain fasciculation syndrome
Fasciculus
lenticularis, *see* Lenticular fasciculus
longitudinal, *see* Longitudinal fasciculus
medial longitudinal, *see* Medial longitudinal fasciculus
Fasciculus formation
nerve, 51/3
Fasciculus retroflexus
Meynert, *see* Meynert fasciculus retroflexus
Fasciculus subcallosus
anatomy, 6/16
Huntington chorea, 6/404
Fasciculus thalamicus
anatomy, 6/5, 20
Fasciitis
eosinophilia myalgia syndrome, 64/259
eosinophilic, *see* Eosinophilic fasciitis
hypereosinophilic syndrome, 41/383
muscle tumor, 41/383
Fasciotomy
Duchenne muscular dystrophy, 41/470
Fast pain, *see* First pain
Fastigial nucleus
cerebellar ablation, 2/404
lesion, 2/404
orthostatic hypotension, 63/232
Fastigiovestibular system
dentatorubropallidoluysian atrophy, 49/442
Fasting
carnitine palmitoyltransferase deficiency, 62/494
glutaric aciduria type II, 62/495
headache, 48/36
migraine, 48/120, 122
multiple acyl-coenzyme A dehydrogenase deficiency, 62/495
painful dominant myotonia, 62/266
Fat
brain embolism, 11/397
Fat absence
subcutaneous, *see* Subcutaneous fat absence
Fat embolism, 38/563-572

anemia, 23/635
anoxic ischemic leukoencephalopathy, 47/538
brain, *see* Brain fat embolism
brain embolism, 53/161
brain hemorrhage, 9/588
brain infarction, 53/161, 55/188
brain injury, 23/631
brain ischemia, 53/161
brain microembolism, 9/588
cardiac surgery, 53/161
chemical manifestation, 38/564-566
deafness, 55/132
diagnosis, 23/633
experimental injury, 23/634
fracture, 53/161
head injury, 57/139, 236
history, 23/631, 38/563
iatrogenic neurological disease, 63/178
lung, 23/633
microembolism, 9/588
parturition, 11/460
pathogenesis, 23/631, 38/569
pathology, 38/566-569
pathophysiology, 23/632, 38/569-671
pontine infarction, 12/44
retinal artery embolism, 55/178
sickle cell anemia, 38/42-44
symptom, 11/397-400
systemic brain infarction, 11/438
thrombocytopenia, 23/635
treatment, 23/635, 38/571
Fat metabolism
aging, 74/237
Fatal familial insomnia
autonomic nervous system, 75/21, 408
molecular biology, 75/408
Fatal infantile encephalomyopathy
complex III deficiency, 66/423
Fatal infantile myopathy/cardiopathy
creatine kinase, 62/510
mitochondrial disease, 62/510
mitochondrial DNA point mutation, 62/509
ragged red fiber, 62/510
Fatality rate
acute mercury intoxication, 64/375
bacterial meningitis, 55/416
congenital fiber type disproportion, 62/356
diamorphine intoxication, 65/355
far eastern tick-borne encephalitis, 56/140
heterocyclic antidepressant intoxication, 65/324
Japanese B encephalitis, 56/138
marine toxin intoxication, 65/142
Minamata disease, 46/391, 64/415
Murray Valley encephalitis, 34/76, 56/139
myotubular myopathy, 62/348
opiate intoxication, 65/355
St. Louis encephalitis, 56/138
tricyclic antidepressant intoxication, 65/324
Fatigue
acrylamide intoxication, 64/68
ACTH induced myopathy, 62/537
adrenal insufficiency myopathy, 62/536
antihypertensive agent, 63/87
atrial myxoma, 63/97
auditory, *see* Auditory fatigue
barotrauma, 63/416
brain edema, 63/416
brain hypoxia, 63/416
buspirone intoxication, 65/344
cardiovascular agent intoxication, 37/426
chronic, *see* Chronic fatigue
corticosteroid withdrawal myopathy, 62/536
dermatomyositis, 62/371
dipeptidyl carboxypeptidase I inhibitor intoxication, 65/445
disulfiram intoxication, 37/321
dysbarism, 63/416
dysphrenia hemicranica, 5/79
endocrine myopathy, 62/527
eosinophilia myalgia syndrome, 63/377
epilepsy, 15/480
eye, 5/35, 204, 206
glycoside intoxication, 37/426
hypereosinophilic syndrome, 63/371
hypokalemia, 63/557
hyponatremia, 63/545
iron deficiency anemia, 63/253
lead intoxication, 64/435
Lyme disease, 51/204
methyldopa, 65/448
migraine, 48/122
mitochondrial myopathy, 43/119
motion sickness, 74/357
mountain sickness, 63/416
multiple sclerosis, 47/59
myasthenia gravis, 43/156
myopathy, 40/330
myophosphorylase deficiency, 62/485
Nelson syndrome, 62/537
nerve conduction, 7/132
neuroborreliosis, 51/204
paralysis periodica paramyotonica, 43/167
Parkinson disease, 49/90
pellagra, 28/86, 46/336
pheochromocytoma, 39/501, 42/764
postural tremor, 6/818

prazosin intoxication, 65/447
primary amyloidosis, 51/415
Shapiro syndrome, 50/167
stress, 74/325
temporal arteritis, 55/342
tick bite intoxication, 37/111
traumatic psychosyndrome, 24/552, 686
tremor, 6/818, 62/52
uremia, 63/504
uremic encephalopathy, 63/504
Waldenström macroglobulinemia, 63/396
Fatty acid
adrenoleukodystrophy, 66/96
beta oxidation, 66/505
brain injury, 23/127
carnitine, 66/400
chorea-acanthocytosis, 63/285
CSF, 16/371
diabetes mellitus, 70/153
diet, 9/399
essential, *see* Essential fatty acid
free, *see* Free fatty acid
gray matter, 9/5
hepatic coma, 49/215
hepatic encephalopathy, 27/351
metabolism, 66/40
monounsaturated 18 C, 9/6
monounsaturated 20 C, 9/6
multiple sclerosis, 9/321, 399
myelin, 7/45, 9/6
myelin lipid, 10/235
nomenclature, 66/35
oxidation, 41/206
plaque, 9/314
polyunsaturated, *see* Polyunsaturated fatty acid
Refsum disease, 8/23, 10/345, 13/314, 21/50, 186, 196, 202, 27/519, 36/347, 40/514, 41/433, 66/485, 491
saturation, 9/5, 10/671
serum, 9/321
short chain, 27/351
sphingomyelin, 10/284
structure, 66/35
terminology, 66/34
unsaturated, *see* Unsaturated fatty acid
very long chain, *see* Very long chain fatty acid
vitamin B_{12}, 70/381
vitamin B_{12} deficiency, 70/381
white matter, 9/5
Fatty acid biosynthesis
disorder, 66/656
malonyl-coenzyme A decarboxylase deficiency, 66/656
Fatty acid oxidation
carnitine cycle, 66/399, 402
childhood myoglobinuria, 62/566
defect, 66/399
hypoglycin, 37/521
hypoglycin intoxication, 65/85
lipid metabolic disorder, 62/491
metabolic myopathy, 62/491
metabolic pathway, 66/399
mitochondrial disease, 66/394, 398
mitochondrial myopathy, 66/394
muscle carnitine deficiency, 66/404
myoglobinuria, 41/277, 62/558
primary carnitine deficiency, 66/401
Fatty acid oxidation disorder
acute metabolic encephalopathy, 66/406
carnitine deficiency, 66/406
mitochondrial matrix, 66/410
vitamin B_t, 66/399
Fatty acyl-carnitine
lipid metabolic disorder, 41/195
Fatty acyl-coenzyme A
lipid metabolic disorder, 41/194
Fatty alcohol NAD oxidoreductase, *see* Fatty alcohol oxidoreductase
Fatty alcohol oxidoreductase
Sjögren-Larsson syndrome, 66/617, 619
Fatty aldehyde
myelin, 7/45
Fatty aldehyde dehydrogenase
Sjögren-Larsson syndrome, 66/617
Faulk-Epstein-Jones syndrome
vertebral abnormality, 30/100
Favism, *see* Glucose-6-phosphate dehydrogenase deficiency
Favre disease, *see* Hyaloidoretinal degeneration (Wagner)
Fay, T., 1/422
Fazio-Londe disease, 22/103-109, 59/121-130
age at onset, 22/107, 128
brain stem encephalitis, 22/107
brain stem glioma, 22/108
Brown-Vialetto-Van Laere syndrome, 59/129
characteristics, 22/523
clinical features, 22/107, 59/125-128
course, 22/107
cranial nerve, 42/95
differential diagnosis, 22/107, 59/129, 224
dysarthria, 42/94
external ophthalmoplegia, 22/140
facial paralysis, 42/94
facial weakness, 40/306
familial bulbopontine paralysis, 22/108

genetics, 41/440
history, 22/103, 59/121
infantile spinal muscular atrophy, 22/103, 105, 108, 59/128
main features, 21/25, 29
nemaline myopathy, 22/108
neuropathology, 22/105, 59/124, 128
nosology, 22/103, 59/128
oculomotor nucleus, 22/181
ophthalmoplegia, 42/95
presenting symptom, 22/107
pseudobulbar paralysis, 22/107
slow progression, 59/129
spinocerebellar degeneration, 22/104
tongue fibrillation, 42/95
unverified case, 59/123
vocal cord paralysis, 42/95
Wohlfart-Kugelberg-Welander disease, 22/109
Fc receptor
multiple sclerosis, 47/250
viral infection, 56/60
FDH syndrome, *see* Goltz-Gorlin syndrome
Fear, *see* Anxiety
Febrile convulsion
child, 74/456
clinical features, 15/247
diphtheria pertussis tetanus vaccine, 52/243
EEG, 15/254-259, 72/310
epilepsy, 15/259
etiology, 15/248
familial history, 15/249
gene defect, 72/137
genetics, 15/250, 436
Harvey syndrome, 42/223
hereditary sensory and autonomic neuropathy type IV, 60/12
incidence, 15/246-248
infectious, 15/252
mortality, 15/259
prevalence, 15/246
prevention, 15/260
sex ratio, 15/251
treatment, 15/260
Febrile seizure
abnormal theta wave, 42/667
aminoaciduria, 42/517
classification, 73/309
definition, 72/16
differential diagnosis, 73/312
epidemiology, 73/309
epilepsy, 75/351
etiology, 73/309
management, 73/312
natural history, 73/310
prevalence, 42/694
survey, 73/309
Febrile status epilepticus
cryptogenic, 15/179
encephalitis, 15/179
meningitis, 15/179
Fecal incontinence
anal disorder, 75/637
caudal agenesis, 42/50
caudal aplasia, 42/50
Charlevoix-Saguenay spastic ataxia, 60/454, 664
dentatorubropallidoluysian atrophy, 60/664
hereditary spastic ataxia, 60/664
hydrocephalic dementia, 46/328
lathyrism, 65/3
multiple sclerosis, 9/171
neuronal intranuclear hyaline inclusion disease, 60/664
orthostatic hypotension, 43/66
Shy-Drager syndrome, 1/474
Feedback
emotion, 3/317, 322, 325
negative, 6/92
occupational neurosis, 1/290
positive, 1/290, 6/92
restoration of higher cortical function, 3/385, 394
sensory, *see* Sensory feedback
spinal segmental, 6/94
Feeding behavior
brain cortex, 46/587
lateral hypothalamic syndrome, 46/587
Feeding disorder
see also Aphagia *and* Dysphagia
argininosuccinic aciduria, 42/524
citrullinuria, 42/538
CNS spongy degeneration, 42/506
congenital muscular dystrophy, 43/93
glycogen storage disease, 43/178
hyperammonemia, 42/560
hyperglycinemia, 42/565
isovaleric acidemia, 42/579
myotonic dystrophy, 43/152
nemaline myopathy, 43/122
Robin syndrome, 43/471
Russell-Silver syndrome, 43/477, 46/21
thyroid gland dysgenesis, 42/632
trichopoliodystrophy, 42/584
Feeding reflex
newborn behavior, 4/345
rooting reflex, 4/345
Feer disease, *see* Acrodynia
Feet syndrome

burning, *see* Burning feet syndrome
painful, *see* Burning feet syndrome
Felbamate
antiepileptic agent, 65/496
epilepsy, 73/360
Felbamate intoxication
ataxia, 65/512
diplopia, 65/512
headache, 65/512
nystagmus, 65/512
vertigo, 65/512
Feline ataxia
animal viral disease, 34/296
panleukopenia virus, 34/296
Felinosis, *see* Cat scratch disease
Female sexual function
diabetes mellitus, 75/602
excitation phase, 61/332
functional capability, 61/336
orgasm, 61/333
paraplegia, 61/331
physiology, 61/332
plateau phase, 61/333
rehabilitation, 61/331
resolution, 61/334
spinal cord injury, 61/331
tetraplegia, 61/331
transverse spinal cord lesion, 61/331
Femoral angiography
brain tumor, 16/641
Femoral artery
temporal arteritis, 48/316
Femoral fracture
peroneal nerve, 70/38
sciatic nerve, 70/38
Femoral nerve
anatomy, 8/304, 305
compression neuropathy, 51/107
conduction, 7/164
cutaneous, *see* Cutaneous femoral nerve
hip fracture, 70/35
lateral cutaneous, *see* Lateral femoral cutaneous nerve
posterior cutaneous, *see* Posterior cutaneous femoral nerve
synovial cyst, 71/19
topographical diagnosis, 2/40
vascularization, 8/309
Femoral neuropathy, 8/303-310
see also Lumbosacral plexus neuritis
anatomy, 8/303
anticoagulant, 8/13
aorta aneurysm, 63/49
aorta surgery, 63/58
hemophilia, 8/13
heparin, 63/49
iatrogenic neurological disease, 63/49, 177, 530
idiopathic, 8/307
iliopsoas muscle hematoma, 63/49
intra-aortic balloon, 63/178
metabolism, 8/307
plexopathy, 51/166
postoperative hematoma, 63/530
pressure, 8/306
renal transplantation, 63/530
rheumatoid arthritis, 71/18
symptomatology, 8/303
trauma, 8/306
Femoxetine
migraine, 48/196
Fenamic acid
migraine, 48/203
Fencing
rehabilitative sport, 26/532
Fenclofenac
headache, 48/174
migraine, 48/174
Fenclofos
organophosphate intoxication, 64/155
Fenestrae parietalis symmetricae, *see* Foramina parietalia permagna
Fenestration
basilar artery, 32/47
septum pellucidum, *see* Septum pellucidum fenestration
Fenfluramine
amphetamine intoxication, 65/256
drug addiction, 65/256
myophosphorylase deficiency, 41/190
toxic myopathy, 62/601
Fenofibrate
toxic myopathy, 62/601
Fenoldopam
classification, 74/149
Fenoprofen
headache, 48/174
migraine, 48/174
Fenoterol
myotonia, 62/276
Fenpropathrin
neurotoxin, 64/213
pyrethroid, 64/213
pyrethroid intoxication, 64/213
Fentanyl
designer drug, 65/349
opiate, 65/350

Fentanyl intoxication
fatality, 65/358
forgotten respiration, 1/654
Fenvalerate
neurotoxin, 64/212
pyrethroid, 64/212
pyrethroid intoxication, 64/212
Ferguson-Critchley like syndrome
Ferguson-Critchley syndrome, 59/322
without ataxia, 59/322
Ferguson-Critchley syndrome, 22/433-442
β-*N*-acetylhexosaminidase A deficiency, 59/322
all-system atrophy, 59/322
ataxia, 42/142
bladder function, 42/142
cerebellopallidoluysionigral atrophy, 59/322
clinical features, 22/434
consanguinity, 22/435
dentatorubropallidoluysian atrophy, 59/322
diagnosis, 22/439-441
diplopia, 42/142
distal amyotrophy, 59/321
external ophthalmoplegia, 59/319
extrapyramidal symptom, 22/427, 436
eyelid retraction, 59/319
familial spastic paraplegia, 59/319-323
features, 21/16
Ferguson-Critchley like syndrome, 59/322
genetics, 22/434, 436
hereditary cerebellar ataxia, 59/319-323
history, 22/433
laboratory data, 22/436
metachromasia, 22/441
multiple sclerosis, 22/442
neuropathology, 22/436
nystagmus, 42/142, 59/319
ocular symptom, 22/436
olivopontocerebellar atrophy, 59/322
ophthalmoplegia, 22/191, 471
optic atrophy, 42/142, 59/319
parkinsonism, 59/321
posterior column degeneration, 22/437
progressive external ophthalmoplegia, 62/304
Shy-Drager syndrome, 59/322
speech disorder, 42/142
striatonigral degeneration, 59/322
urinary incontinence, 59/319
without ataxia, 59/322
Woods-Schaumburg syndrome, 59/321
Fernel, J., 2/7, 8
Ferraro disease
history, 10/13
Ferric hydroxamate method, 9/31
Ferrier, D., 1/6
Ferritin H
muscle tissue culture, 62/92
Ferrochelatase
lead, 64/434
porphyria variegata, 42/620
Fertility
paraplegia, 61/313
spinal cord injury, 61/313, 316
tetraplegia, 61/313
Festination
hastening phenomenon, 49/68
Parkinson disease, 49/68
Fetal alcohol syndrome
alcoholism, 30/119-121, 43/197, 46/81, 70/355
arhinencephaly, 50/242
brain cortex heterotopia, 50/40, 279
CNS, 30/121
corpus callosum agenesis, 50/153
Dandy-Walker syndrome, 50/333
epidemiology, 30/121, 43/197
growth retardation, 30/120, 43/197
hydrocephalus, 50/279
malformation, 30/120
mental deficiency, 30/120, 46/81
microcephaly, 30/120, 50/279
pathology, 30/121
philtrum absence, 50/279
ptosis, 30/120, 50/279
pyruvate decarboxylase deficiency, 62/502
upper lip hypoplasia, 50/279
Fetal face syndrome
hypertelorism, 30/248, 43/400
hypogonadism, 43/400
macrocephaly, 30/100
mental deficiency, 43/400
micrognathia, 43/400
short stature, 43/400
skeletal malformation, 30/248
vertebral abnormality, 43/400
Fetal hemoglobin
thalassemia major, 42/629
Fetal hydantoin syndrome
antiepileptic agent, 65/514
features, 65/514
mental deficiency, 30/122, 50/279
microcephaly, 30/508, 50/279
prevalence, 42/641
Fetal Minamata disease
athetosis, 64/418, 421
blindness, 64/418
blood level, 64/419
brain level, 64/419

cerebral palsy, 64/418
dose response, 64/419
epilepsy, 64/418, 421
hair level, 64/419
infantile hypotonia, 64/420
intoxication, 64/418
mental deficiency, 64/418
microcephaly, 64/419
myoclonia, 64/418
neuronal migration disorder, 64/419
neuropathology, 64/419
spastic paraplegia, 64/418
spastic tetraplegia, 64/418, 421
symptom, 64/418
Fetal movement
arthrogryposis multiplex congenita, 42/73
infantile spinal muscular atrophy, 42/88
Fetal trimethadione syndrome
antiepileptic agent, 65/514
features, 65/514
mental deficiency, 30/122
Fetal warfarin syndrome
congenital malformation, 46/80
deformity, 46/80
mental deficiency, 30/123, 46/80
stippled epiphysis, 46/50
α-Fetoprotein
amniotic fluid, 32/563
anencephaly, 30/178, 42/13, 50/86
ataxia telangiectasia, 60/353
chemistry, 32/562
CSF, 32/563
Down syndrome, 50/531
encephalomeningocele, 42/28
function, 32/562
hydrocephalus, 30/179
iniencephaly, 30/179
maternal serum, 32/563, 567
prenatal diagnosis, 32/562, 564-572
serum, *see* Serum α-fetoprotein
spina bifida, 30/179, 42/56
synthesis, 32/562
teratoma, 31/60
Fetoscopy
holoprosencephaly, 42/34
prenatal diagnosis, 32/570
Fetuin
galactose cleavage, 10/479
Fetus
Campylobacter, *see* Campylobacter fetus
harlequin, *see* Congenital ichthyosis
heterotopia, 30/499
Fever
Abt-Letterer-Siwe disease, 38/96, 42/441
acetylcholine, 74/449
adrenal insufficiency myopathy, 62/536
adrenogenital syndrome, 43/68
Alexander disease, 42/484
allergic granulomatous angiitis, 63/383
anterior cerebral artery syndrome, 53/349
anterior hypothalamus, 74/448
antipyresis, 74/451
antipyretic agent, 74/452
Argentinian hemorrhagic, *see* Argentinian hemorrhagic fever
atrial myxoma, 63/97
autonomic nervous system, 43/68, 74/453, 75/543
basofrontal syndrome, 53/349
Bolivian hemorrhagic, *see* Bolivian hemorrhagic fever
brain cortex, 74/453
brain fat embolism, 55/178
cat scratch disease, 52/127
chemical connection, 74/448
Colorado tick, *see* Colorado tick fever
congenital hyperphosphatasia, 31/258
convulsion, *see* Febrile convulsion
corticosteroid withdrawal myopathy, 62/536
critical illness polyneuropathy, 51/576, 578, 581, 584
cytokine, 74/449
dengue, *see* Dengue fever
dermatomyositis, 62/373
efferent pathway, 74/450
endocarditis, 63/111
engineering model, 74/440
eosinophilia myalgia syndrome, 63/375, 64/252
etiocholanolone, 43/68
frontal lobe syndrome, 53/349
head injury, 57/238
headache, 48/6
hemorrhagic, *see* Argentinian hemorrhagic fever *and* Bolivian hemorrhagic fever
hereditary sensory and autonomic neuropathy type III, 21/109, 60/26
histiocytosis X, 42/441
hydralazine neuropathy, 42/647
hypothalamic neuron, 74/449
ichthyohepatoxic fish intoxication, 37/87
inclusion body myositis, 62/373
infection, 74/457
initial process, 74/444
intracranial pressure, 74/454
Kawasaki syndrome, 56/638
Lassa, *see* Lassa fever
leprosy, 51/216

loxosceles intoxication, 37/112
Lyme disease, 51/204
lymphedema, 42/726
mackeral intoxication, 37/87
mechanism, 74/437
Mediterranean, *see* Brucellosis
mephenytoin intoxication, 37/202
metabolic study, 74/438
metastatic cardiac tumor, 63/96
methyl salicylate intoxication, 37/417
migraine, 5/46, 42/746, 48/156, 160
monoamine, 74/449
Muckle-Wells syndrome, 42/396
multiple myeloma, 20/11
neuroborreliosis, 51/204
neuroendocrinology, 74/451
neurologic intensive care, 55/221
Omsk hemorrhagic, *see* Omsk hemorrhagic fever
periodic, 43/67-69
polyarteritis nodosa, 8/125, 39/295, 55/354, 396
polymyositis, 38/485, 62/373
preoptic area, 74/449
primary cardiac tumor, 63/96
progressive myositis ossificans, 43/191
prostaglandin, 74/449
pyrogen, 74/438
pyrogenic cytokine, 74/446
relapsing, *see* Relapsing fever
renal transplantation, 63/531
rheumatic, *see* Rheumatic fever
Rift Valley, *see* Rift Valley fever
Rocky Mountain spotted, *see* Rocky Mountain spotted fever
salicylic acid intoxication, 37/417
scarlet, *see* Scarlet fever
sodium salicylate intoxication, 37/417
South African tick-bite, *see* South African tick-bite fever
steroid induced, 1/448
stress, 74/455
subacute necrotizing encephalomyelopathy, 42/625
subarachnoid hemorrhage, 55/15
survey, 74/437
survival value, 74/439
swine, *see* Swine fever
Takayasu disease, 55/337
temporal arteritis, 55/342
thermoregulation, 1/447, 74/439
toxic oil syndrome, 63/380
transverse sinus thrombosis, 54/397
treatment, 74/458
typhoid, *see* Typhoid fever
venous sinus occlusion, 54/429
yellow, *see* Yellow fever
zinc intoxication, 64/362
FG syndrome
contracture, 43/250
corpus callosum agenesis, 30/100, 43/250, 50/163
cryptorchidism, 43/250
digital abnormality, 43/250
heart disease, 43/250
heterozygote detection, 43/250
imperforate anus, 43/250
mental deficiency, 43/250
pectus excavatum, 43/250
syndactyly, 43/250
Fiber, *see* Muscle fiber *and* Nerve fiber
Fiber branching
nerve, 51/4
Fiber electromyography
single, *see* Single fiber electromyography
Fiber loss
myelinated, *see* Myelinated fiber loss
nerve, *see* Nerve fiber loss
Fiber type disproportion
congenital, *see* Congenital fiber type disproportion
Fiber vacuole
muscle, *see* Rimmed muscle fiber vacuole
Fibers
muscle, *see* Muscle fiber
myelinated, *see* Myelinated nerve fiber
postganglionic nerve, *see* Postganglionic nerve fiber
ragged red, *see* Ragged red fiber
Remak, *see* Unmyelinated nerve fiber
unmyelinated, *see* Unmyelinated nerve fiber
Fibril
amyloid, *see* Amyloid fibril
intervertebral disc, 20/532
Fibrillary astrocytoma
angiography, 18/25
classification, 18/2
histogenesis, 67/10
low grade, *see* Low grade fibrillary astrocytoma
microscopy, 18/9
nomenclature, 18/2
optic chiasm compression, 68/75
tissue culture, 18/13
Fibrillary chorea
Morvan, *see* Morvan fibrillary chorea
Fibrillation
atrial, *see* Atrial fibrillation
cardiac dysrhythmia, 38/144
EMG, 19/276, 40/143
motor unit hyperactivity, 41/299

muscle, *see* Muscle fibrillation
organophosphate induced delayed polyneuropathy, 64/177
paroxysmal ventricular, *see* Paroxysmal ventricular fibrillation
tongue, *see* Tongue fibrillation
ventricular, *see* Ventricular fibrillation
Fibrillation potential
features, 62/61
motor unit, 62/50, 61, 67
myopathy, 62/50, 61, 67
Fibrin
hemostasis, 63/303
Fibrinoid arteriolar degeneration, *see* Arteriolar fibrinohyalinoid degeneration
Fibrinoid arteriolar necrosis, *see* Arteriolar fibrinohyalinoid degeneration
Fibrinoid arteritis, *see* Arteriolar fibrinohyalinoid degeneration
Fibrinoid artery degeneration, *see* Arteriolar fibrinohyalinoid degeneration
Fibrinoid degeneration
brain hemorrhage, 11/637
Fibrinolysis
brain infarction, 53/426
head injury, 57/238
heat stroke, 23/672, 677
hemostasis, 63/305
intravascular consumption coagulopathy, 11/456
spinal cord injury, 26/392
Fibrinolytic agent
anticoagulant, 63/308
brain infarction, 11/355
cardiovascular agent, 65/433
hemostasis, 63/308
iatrogenic neurological disease, 65/461
middle cerebral artery syndrome, 53/368
Fibrinolytic agent overdose
brain hemorrhage, 65/462
neurologic adverse effect, 65/461
subarachnoid hemorrhage, 65/462
Fibrinolytic system
hemostasis, 63/301, 304
intravascular consumption coagulopathy, 55/494
Fibrinolytic treatment
brain hemorrhage, 54/291
carotid artery, 54/182
complication, 54/182
contraindication, 54/182
local intra-arterial, 54/182
neurologic intensive care, 55/211
pontine infarction, 12/33
Fibroblast
α-*N*-acetylgalactosaminidase deficiency, 66/346
fibroma, 18/321
muscle tissue culture, 62/87
perineural, *see* Perineural fibroblast
Fibroblastic endocarditis
Loeffler, *see* Idiopathic hypereosinophilic syndrome
Fibrocystic bone disease, *see* Aneurysmal bone cyst
Fibrocystic dysplasia
polyostotic, *see* Polyostotic fibrocystic dysplasia
Fibrodysplasia elastica, *see* Ehlers-Danlos syndrome
Fibrodysplasia ossificans progressiva, *see* Progressive myositis ossificans
Fibroelastosis
endocardial, *see* Endocardial fibroelastosis
Fibrolipoma
epidural spinal lipoma, 20/405
intradural spinal lipoma, 20/396
Fibroma
fibroblast, 18/321
gingival, *see* Gingival fibroma
Koenen, *see* Koenen fibroma
nasopharyngeal, *see* Nasopharyngeal fibroma
nerve, 8/460
orbital tumor, 17/177
ossifying, *see* Ossifying fibroma
periungual, *see* Periungual fibroma
primary cardiac tumor, 63/93
skull base tumor, 17/147, 177
subungual, *see* Subungual fibroma
supratentorial brain tumor, 18/321
tuberous sclerosis, 43/49
Fibroma molluscum
neurofibromatosis type I, 14/135, 149
Fibromuscular dysplasia, 11/366-383, 55/283-290
adventional type, 55/284
age, 11/368, 53/39, 163, 55/284, 288
angiographic diagnosis, 55/285
angiography, 11/368, 55/283
arteriosclerosis, 11/381, 55/288
bead string, 55/285
benign nature, 55/288
brain aneurysm, 55/283
brain angiography, 11/372, 381
brain embolism, 53/163
brain infarction, 11/369, 371, 53/163, 55/283
brain ischemia, 11/383, 53/163, 55/283
carotid angiography, 11/372
carotid artery, 55/283
carotid artery bifurcation, 55/284
carotid bruit, 11/371, 55/289
carotid cavernous fistula, 55/288
carotid dissecting aneurysm, 54/271, 273, 275,

55/283, 288
carotid involvement level, 55/284
carotid system syndrome, 53/308
cerebrovascular disease, 11/371, 53/28
clinical diagnosis, 11/369
clinical features, 55/287
computer assisted tomography, 54/64
definition, 55/283
differential diagnosis, 55/286
etiology, 55/286
focal ischemic symptom, 55/288
frequency, 55/287
headache, 11/372, 48/273, 286
histology, 11/367, 55/283
history, 11/366-368
intimal type, 55/284
intracranial pressure, 48/285
literature, 11/368
medial type, 55/284
natural history, 55/288
nonrecurrent nonhereditary multiple cranial neuropathy, 51/571
pathology, 11/374, 55/284
pericarotid syndrome, 48/337
sex ratio, 11/368, 55/283, 288
subarachnoid hemorrhage, 11/368, 371, 55/2, 29, 284
subclavian artery stenosis, 53/372
systemic brain infarction, 11/454
systemic disease, 55/283
transient ischemic attack, 11/369, 371, 53/163, 266
treatment, 55/288
vertebral artery, 11/373, 53/378, 55/283, 286
vertebral artery angiography, 11/373
vertebral dissecting aneurysm, 54/277
vertebrobasilar system syndrome, 53/378
Fibromyolipoma
renal lesion, 14/50
Fibromyoma
uterus, 14/51
Fibroplasia
retrolental, *see* Retrolental fibroplasia
Fibrosarcoma
Bean syndrome, 14/106
brain sarcoma, 16/20
cerebral, *see* Cerebral fibrosarcoma
cranial vault tumor, 17/117
dura mater, 16/20
misdiagnosis, 14/30
nervous system, 68/380
skull base, 68/34
skull base tumor, 17/168
spinal cord compression, 19/368
survey, 68/386
Fibrosing alveolitis
chronic axonal neuropathy, 51/531
Fibrosing myositis
myosclerosis, 41/65
Fibrosis
actinomycosis, 35/383
cardiac interstitial, *see* Cardiac interstitial fibrosis
carotid sheath, 48/338
choroid plexus, *see* Choroid plexus fibrosis
cystic, *see* Cystic fibrosis
endoneurial, *see* Endoneurial fibrosis
interstitial, *see* Interstitial fibrosis
multiple sclerosis, 9/292
muscle, *see* Muscle fibrosis
plaque, 9/292
pulmonary, *see* Pulmonary fibrosis
pulmonary interstitial, *see* Pulmonary interstitial fibrosis
retroperitoneal, *see* Retroperitoneal fibrosis
serosal, *see* Serosal fibrosis
vascular, *see* Vascular fibrosis
Fibrositis
myalgia, 40/325, 41/385
tenderness, 40/325
Fibrotic disorder
migraine, 5/56
Fibrous band
cervical, *see* Cervical fibrous band
Fibrous dysplasia, 14/163-205, 38/381-389
see also Albright syndrome
acromegaly, 14/199
age distribution, 38/382
alkaline phosphatase, 14/180
anosmia, 14/190
basilar impression, 50/399
bone, *see* Albright syndrome
Charles Ruppe disease, 14/196
cherubism, 14/170
Chiari basal hyperostosis, 14/199
classification, 14/169, 38/382
clinical features, 14/179
concentric hyperostosis, 14/199
course, 38/387
cranial vault tumor, 17/118
craniosynostosis, 14/199
deafness, 14/170
diabetes mellitus, 14/186
differential diagnosis, 14/193-200
elephantiasis, 14/185
endocrinology, 14/183-186
etiology, 38/387

familial metaphyseal dysplasia, 14/170
glycosaminoglycanosis type I, 14/170
growing skull fracture, 14/198
Halliday hyperostosis, 14/199
Hand-Schüller-Christian disease, 14/198
hemangioma, 14/197
hematologic disease, 14/199
history, 14/164, 38/381
hyperparathyroidism, 14/163, 186, 199, 38/385
hyperphosphatasemia, 14/170, 199
ivory vertebra, 19/172
laboratory finding, 14/180, 38/386
lesion, 38/382
lesion distribution, 14/167
malignant degeneration, 14/159
melorheostosis, 14/199
mental deficiency, 14/187
metastasis, 14/198
microscopy, 14/171-175
monostotic, *see* Monostotic fibrous dysplasia
Morgagni-Stewart-Morel syndrome, 14/199
nature, 14/200
neurofibromatosis, 14/83, 177, 182, 185, 187, 193, 198
neurofibromatosis type I, 14/83, 134, 165, 492
neurologic manifestation, 38/383
neurology, 14/187-190
neuropathology, 14/83
occipital pug, 14/168
Ollier disease, 14/192
optic atrophy, 14/170
ossifying fibroma, 14/176
osteitis deformans (Paget), 14/193, 199
osteogenesis imperfecta, 14/199
osteopetrosis, 14/198
osteosarcoma, 14/198
pathology, 14/165-179
pigmentation, 14/180-182
precocious puberty, 14/183
prevalence, 14/165
progressive hereditary craniodiaphyseal dysplasia, 14/198
Pyle disease, 13/83, 14/198, 46/88
radiologic features, 20/28, 38/384
roentgenology, 14/190-193
shepherd crook, 14/169
skull, 14/193, 46/88
skull base tumor, 17/173
Sturge-Weber syndrome, 14/199, 201
treatment, 14/203, 38/388
tuberous sclerosis, 14/199
vertebral scalloping, 14/193
Fibrous histiocytoma
malignant, *see* Malignant fibrous histiocytoma
Fibrous histiosarcoma
malignant, *see* Malignant fibrous histiosarcoma
Fibrous meningitis, *see* Spinal arachnoiditis
Fibrous osteodysplasia, *see* Fibrous dysplasia
Fibular aplasia
craniosynostosis, *see* Lowry syndrome
Fick principle
brain blood flow, 11/120
Fickler-Winkler olivopontocerebellar atrophy, *see* Olivopontocerebellar atrophy (Fickler-Winkler)
Field fever, *see* Leptospirosis
Fièvre boutonneuse
tick-borne typhus, 34/657
Fifth phakomatosis, *see* Multiple nevoid basal cell carcinoma syndrome
Fifth ventricle
striatopallidodentate calcification, 6/712
Fight
autonomic nervous system, 74/1
Filamentous body
muscle, *see* Muscle filamentous body
Filaria
helminthiasis, 52/513-519
nematode, 52/513
Filariasis, 35/161-172
apathy, 52/516
carbamate intoxication, 64/187
choreoathetosis, 52/516
clinical features, 35/165, 52/515
CNS, 35/163-168
coma, 35/168, 52/516
CSF, 35/167, 52/514, 516, 518
diagnosis, 35/169, 52/517
diethylcarbamazine, 35/163, 170, 52/519
Dipetalonema, 35/161, 52/514
Dipetalonema perstans, 35/161, 163, 52/513
dipsomania, 52/516
dracunculiasis, 35/161, 52/514
Dracunculus medinensis, 35/161, 167, 52/513
dystonia, 52/517
EEG, 35/168, 52/517
elephantiasis, 35/169, 52/513
encephalopathy, 35/165, 167, 170, 52/514, 517
eosinophilia, 35/163, 52/516
epidemiology, 35/11, 161-163, 212
epidural empyema, 52/516
epilepsy, 35/168, 52/514, 517
granulomatous CNS vasculitis, 55/388
headache, 35/167, 52/515, 517
helminthiasis, 35/212
immunology, 52/518
intracranial hypertension, 35/167, 170, 52/515

Loa loa, 52/513
loiasis, 35/161, 163, 52/514, 517
myelomalacia, 52/516
neurosis, 35/168, 52/517
Onchocerca volvulus, 35/161, 52/513
onchocerciasis, 35/161
paraplegia, 35/167, 52/516
parasitic disease, 35/11-15
parasitology, 35/161-163, 52/513
parkinsonism, 52/516
psychiatric symptom, 35/163, 52/516
psychosis, 35/168
reactive encephalopathy, 52/519
spastic paraplegia, 59/436
spinal cord compression, 35/163, 167, 170, 52/514, 516
treatment, 35/170-172, 52/517, 519
tropical myeloneuropathy, 56/526
Wuchereria bancrofti, 35/161, 168, 52/513
wuchereriasis, 52/513
Filix mas intoxication
optic atrophy, 13/63
Filobasidiella neoformans, *see* Cryptococcus neoformans
Filum reduplication
diastematomyelia, 50/438
diplomyelia, 50/438
Filum terminale
anatomy, 19/78
diastematomyelia, 50/438
diplomyelia, 50/438
spina bifida, 50/496, 501
spinal ependymoma, 19/359, 20/356, 363, 375
Filum terminale ependymoma
monomelic spinal muscular atrophy, 59/376
Final autonomic motor pathway
autonomic nervous system, 74/21
Final autonomic pathway
autonomic nervous system, 74/20
Final common path
definition, 1/50
Fincher syndrome
cauda equina tumor, 20/367
headache, 20/367
sciatica, 20/367
Finger
clubbing, 1/488
Dawson, *see* Dawson finger
Finger agnosia
see also Gerstmann syndrome
Alzheimer disease, 46/251
autotopagnosia, 45/381
body scheme disorder, 2/668, 4/219-222
Gerstmann syndrome, 2/604, 686
number aphasia, 2/600
schizophrenia, 46/490
Wernicke aphasia, 4/98
Finger aphasia
body scheme disorder, 4/220
Finger apraxia
body scheme disorder, 4/220
Finger deformity
Patau syndrome, 14/121, 50/558
Finger to ear test
cerebellar ataxia, 1/324, 333
dysmetria, 1/324, 333
hypermetria, 1/324, 333
Finger flexor reflex
diagnostic value, 1/244
Finger to nose test
cerebellar ataxia, 1/324, 333
dysmetria, 1/324, 333
hypermetria, 1/324, 333
prehension, 4/447
Fingerprint body
lipid, 21/52
muscle fiber, 40/39, 43, 62/44
Fingerprint body myopathy
classification, 40/288
congenital myopathy, 41/20, 62/332
differential diagnosis, 62/347
histochemistry, 40/47
mental deficiency, 62/347
mitochondria, 43/82
muscle fiber size, 43/82
muscle fiber type I, 43/82
nonspecific, 40/39
origin, 40/113
pathologic reaction, 40/263
sex ratio, 43/82
Fingerprints, *see* Dermatoglyphics
Finkel spinal muscular atrophy
adult onset, 59/43
autosomal dominant inheritance, 59/43
cramp, 59/43
inheritance mode, 59/43
suffocation, 59/43
First pain
pain type, 1/85
First rib syndrome
see also Cervical rib syndrome
compression syndrome, 7/430, 443
First and second branchial arch syndrome, *see* Goldenhar syndrome
Fish intoxication, 37/78-84
ciguatoxin, 37/82-84

classification, 37/78
tetrodotoxin, 37/79-82
Fish roe intoxication
chest pain, 37/86
coma, 37/86
convulsion, 37/86
diarrhea, 37/86
dyspnea, 37/86
nausea, 37/86
syncope, 37/86
tinnitus, 37/86
vomiting, 37/86
Fisher syndrome
acute cerebellar ataxia, 34/629
Guillain-Barré syndrome, 2/296
ophthalmic finding, 74/430
ophthalmoplegia, 22/180
Fissure syndrome
Down syndrome, 50/526
olfactory, *see* Olfactory fissure syndrome
sphenoidal, *see* Superior orbital fissure syndrome
superior orbital, *see* Superior orbital fissure syndrome
Fissures
interhemispheric, *see* Interhemispheric fissure
medullary, *see* Medullary fissure
transverse basilar, *see* Transverse basilar fissure
Fist-ring test
premotor cortex syndrome, 2/731
Fistula
arteriovenous, *see* Arteriovenous fistula
brain arteriovenous malformation, 5/145-149
carotid cavernous, *see* Carotid cavernous fistula
congenital arteriovenous, *see* Congenital arteriovenous fistula
CSF, *see* CSF fistula
gastrocolic, *see* Gastrocolic fistula
spontaneous CSF, *see* Spontaneous CSF fistula
tracheoesophageal, *see* Tracheoesophageal fistula
traumatic carotid cavernous sinus, *see* Traumatic carotid cavernous sinus fistula
Fitness
space, 74/295
Fixations
eye, *see* Eye fixation
formalin, *see* Formalin fixation
gaze, *see* Gaze fixation
joint, *see* Joint fixation
ocular, *see* Eye fixation
scapulothoracic, *see* Scapulothoracic fixation
visual, *see* Visual fixation
Fixed dystonia
dystonia musculorum deformans, 49/522
Fixed heart rate
acute pandysautonomia, 51/475
FK-506, *see* Tsukubaenolide
FK-506 intoxication, *see* Tsukubaenolide intoxication
Flaccid paralysis
acute intermittent porphyria, 42/618
botulinum toxin, 65/245
cataplexy, 42/711
cephalopoda intoxication, 37/62
chronaximetry, 1/224
chronaxy, 1/224
CNS spongy degeneration, 42/506
hapalochlaena lunulata intoxication, 37/62
hapalochlaena maculosa intoxication, 37/62
hyperpipecolatemia, 29/222
metabolic encephalopathy, 69/399
octopoda intoxication, 37/62
pseudo-Babinski sign, 19/37
puffer fish intoxication, 37/80
restoration of higher cortical function, 3/389, 395
spinal lipoma, 20/400
tetrodotoxin intoxication, 65/156
tick bite intoxication, 37/111
total transverse spinal cord syndrome, 12/498
unilateral pontine hemorrhage, 12/40
Flaccid paraplegia
aorta rupture, 61/116
Burkitt lymphoma, 63/352
clinical aspect, 1/206
EMG, 1/224
hemiplegia, 1/176
Flapping tremor, *see* Asterixis
Flatfoot
diastematomyelia, 50/436
diplomyelia, 50/436
Flavine adenine dinucleotide
electron transfer flavoprotein, 66/409
mitochondrial matrix system, 66/409
Flavivirus
acute viral encephalitis, 56/137
basal ganglion, 56/135
brain stem, 56/135
dengue fever, 56/12
headache, 56/12
Japanese B encephalitis, 56/12
Murray Valley encephalitis, 34/76, 56/12
neurotropic virus, 56/12
rocio virus, 56/12
spinal cord, 56/135
St. Louis encephalitis, 56/12
thalamus, 56/135
tick-borne encephalitis, 56/12

viral meningitis, 56/12
West Nile virus, 56/12
yellow fever, 56/12
Flavobacterium
gram-negative bacillary meningitis, 52/117
Flavobacterium meningosepticum
gram-negative bacillary meningitis, 52/104, 120
infantile enteric bacillary meningitis, 33/63
meningitis, 33/103
Flavoprotein
electron transfer, *see* Electron transfer flavoprotein
Flea-borne typhus
rickettsial infection, 34/646
Flecainide
cardiac pharmacotherapy, 63/192
epilepsy, 63/192
iatrogenic neurological disease, 63/192, 65/455
myopathy, 63/192
tremor, 63/192
Flecainide intoxication
encainide, 65/455
headache, 65/456
paresthesia, 65/456
polyneuropathy, 65/456
sensory polyneuropathy, 65/456
tremor, 65/456
vertigo, 65/456
visual impairment, 65/456
Fleck retina, *see* Kandori disease
Flecked retina syndrome, *see* Fundus flavimaculatus
Flexibilitas cerea, *see* Catatonia *and* Plastic rigidity
Flexion adduction sign
neuralgic amyotrophy, 51/172
Flexion contracture
Addison disease, 39/480
claw hand, 20/827
exercise, 20/828
osteoma, 20/827
Parkinson dementia, 49/173
talipes, 20/827
treatment, 20/827
Flexion dystonia
cerebellar ablation, 21/521
pallidal syndrome, 21/526
Parkinson disease, 6/140, 142-144
Flexor reflex
afferent, 1/57
ataxia telangiectasia, 60/364
clinical aspect, 1/172
finger, *see* Finger flexor reflex
nociception, 1/69
scratch reflex inhibition, 1/50
spinal automatism, 1/251
spinal lesion, 2/208
Flexor spasm
epilepsy, 14/347
multiple sclerosis, 9/412
taxonomy, 1/279
treatment, 9/412
tuberous sclerosis, 14/347
Flick sign
carpal tunnel syndrome, 51/94
Flicker fusion frequency
A wave, 13/19
B wave, 13/19
ERG, 13/19
Flight
autonomic nervous system, 74/1
Flip sign
interosseous nerve syndrome, 2/36
FLM syndrome, *see* Medial longitudinal fasciculus
Floating-Harbor syndrome
craniolacunia, 50/139
Floccular artery
radioanatomy, 11/93
topography, 11/93
Flocculonodular lobe
archeocerebellum, 1/329
cerebellar ablation, 2/399
Flocculonodular lobe syndrome
ataxia, 21/569
Floppy baby syndrome
congenital ophthalmoplegia, 29/232
Floppy infant, *see* Infantile hypotonia
Flourens, P., 1/7
Flucytosine
aspergillosis, 52/381
Candida meningitis, 52/404
candidiasis, 52/403
cladosporiosis, 52/484
coccidioidomycosis, 35/452, 52/419
cryptococcosis, 35/490, 52/433
drechsleriasis, 52/486
granulomatous amebic encephalitis, 52/330
sporotrichosis, 52/495
Fludarabine
adverse effect, 69/488
antineoplastic agent, 65/528
cranial neuropathy, 65/528
encephalopathy, 65/528
neuropathy, 69/461
neurotoxin, 65/528
Fludarabine intoxication
leukoencephalopathy, 65/534
Fludrocortisone
autonomic polyneuropathy, 51/487

orthostatic hypotension, 75/723
Fluent agraphia
definition, 45/459
paragraphia, 45/460
transcortical sensory aphasia, 45/460
Wernicke aphasia, 45/460
Flufenamic acid
headache, 48/174
migraine, 48/174, 203
Fluid balance
brain injury, 23/109
head injury, 57/227
Flumazenil
benzodiazepine intoxication, 65/334
Flunarizine
hereditary periodic ataxia, 60/442
migraine, 48/200
postlumbar puncture syndrome, 61/156
Flunitrazepam
toxic myopathy, 62/601
Fluorescein angiography
carbon disulfide intoxication, 64/26
colloid body, 13/75
optic atrophy, 13/48
primary pigmentary retinal degeneration, 13/225-231
retinal, *see* Retinal fluorescence angiography
retinal hemangioblastoma, 14/645
Fluorescein photography
brain scanning, 11/250-252
retinal, *see* Retinal fluorescence photography
Fluorescent treponemal antibody absorption test
parainfectious amyotrophy, 22/21
syphilis, 33/343, 52/276
Fluoride intoxication
acute, 36/431-433
chronic, 36/431
spastic paraplegia, 59/438
Fluorochrome
myelin, 9/26
Fluorocortisone
progressive dysautonomia, 59/140
5-Fluorocytosine, *see* Flucytosine
2-Fluoro-2-deoxyglucose F 18, *see* 2-Deoxy-2-fluoroglucose F 18
6-Fluorodopamine
Parkinson disease, 49/39
9α-Fluorohydrocortisone
see also Corticosteroid
normokalemic periodic paralysis, 28/598, 41/163
thyrotoxic periodic paralysis, 41/242
6-Fluorolevodopa F 18
Parkinson disease, 49/99
Fluorosis
foramen magnum, 50/396
ivory vertebra, 19/171
skeletal, *see* Skeletal fluorosis
Fluorouracil
adverse effect, 69/485
antimetabolite, 39/104
antineoplastic agent, 65/528
astrocytoma, 18/39
cerebellar encephalopathy, 64/5
chemotherapy, 39/104
encephalopathy, 46/600, 65/528, 67/363
epilepsy, 65/528
late cerebellar atrophy, 60/586
leukemia, 39/17
neurologic toxicity, 39/104
neurotoxin, 39/17, 94, 99, 104, 119, 46/600, 60/586, 64/5, 65/528, 532, 67/363
toxic encephalopathy, 64/5
Fluorouracil intoxication
ataxia, 65/532
cerebellar syndrome, 65/532
confusion, 65/532
dysarthria, 65/532
epilepsy, 65/532
headache, 65/532
incidence, 65/532
leukoencephalopathy, 65/532
nystagmus, 65/532
symptom, 64/5
treatment, 65/532
vitamin B1, 65/532
Fluorouridine
brain tumor, 18/491
radiosensitizer, 18/491
Fluoxetine
neuroleptic akathisia, 65/289
obstructive sleep apnea, 63/460
selective serotonin reuptake inhibitor, 65/312
Fluoxetine intoxication
headache, 65/320
insomnia, 65/320
tremor, 65/320
Fluphenazine
chemical formula, 65/275
neuroleptic agent, 65/275
phenothiazine, 65/275
toxic myopathy, 62/601
Flurazepam
half-life, 37/358
hypnotic agent, 37/355
Fluroxene intoxication
convulsion, 37/412

death, 37/412
diarrhea, 37/412
dog, 37/412
experimental, 37/412
inhalation anesthetics, 37/412
seizure, 37/412
vomiting, 37/412
Flushing
alcohol intoxication, 64/112
bradykinin, 48/97
cluster headache, 48/9
facial, *see* Facial flushing
gustatory, 75/126
hydralazine intoxication, 37/431
hyperserotonemia, 29/225
hypothalamic syndrome, 2/450
pheochromocytoma, 14/252, 39/498
proctalgia, 43/69
Fluvalinate
neurotoxin, 64/212
pyrethroid, 64/212
pyrethroid intoxication, 64/212
Fluvoxamine intoxication
headache, 65/320
insomnia, 65/320
tremor, 65/320
Flying Squad
spinal cord injury, 26/297
Flynn-Aird syndrome
aphasia, 14/112, 42/327
ataxia, 14/112, 42/327
cataract, 14/112, 22/519, 42/327, 60/653
Cockayne-Neill-Dingwall disease, 14/113
Cowden syndrome, 14/113
deafness, 14/112, 60/657
differential diagnosis, 14/113
Eldridge-Berlin-Money-McKusick syndrome, 14/113
endocrine disorder, 42/328
epilepsy, 14/112
hereditary hearing loss, 22/519
kyphoscoliosis, 22/519, 42/327
mental deficiency, 42/327
muscular atrophy, 42/327
myopia, 14/112, 22/519, 42/327
neurocutaneous syndrome, 14/101
neuropathy, 22/519, 42/327
osteoporosis, 42/327
paresthesia, 42/327
Refsum disease, 21/217, 22/520, 60/235, 66/497
secondary pigmentary retinal degeneration, 14/112, 22/519, 42/327, 60/730
sensorineural deafness, 42/327
skeletal lesion, 22/519
skin lesion, 22/519
Usher syndrome, 22/520
Von Gräfe-Sjögren syndrome, 14/113
Flynn phenomenon
mechanism, 74/416
Foam cell
Alport syndrome, 42/376
characteristics, 10/88
cytoside lipidosis, 10/545
Farber disease, 10/353, 545
GM1 gangliosidosis, 10/299
GM1 gangliosidosis type I, 42/431
GM1 gangliosidosis type II, 42/432
Goldberg syndrome, 42/440
hyperphosphatasia, 42/568
juvenile dystonic lipidosis, 42/447
Lemieux-Neemeh syndrome, 42/371
leprous neuritis, 51/226
mannosidosis, 42/597
mucolipidosis type I, 42/475
Niemann-Pick disease, 10/282, 488, 498, 66/21
Niemann-Pick disease type A, 42/469
Niemann-Pick disease type C, 42/472
Tangier disease, 10/548, 42/627
Wolman disease, 10/504, 546
Foamy spheroid body
Guam amyotrophic lateral sclerosis, 59/286-291
Macaca irus, 59/291-294
twilight state, 59/291
FOAR syndrome, *see* Facio-oculoacousticorenal syndrome
Focal athetosis
axial dystonia, 49/383
blepharospasm, 49/382
definition, 49/382
dystonia, 49/382
orofacial dyskinesia, 49/383
oromandibular dyskinesia, 49/383
posthemiplegic athetosis, 49/383
spasmodic torticollis, 49/382
writers cramp, 49/383
Focal brain cortex dysplasia, *see* Brain cortex focal dysplasia
Focal brain damage
brain contusion, 57/44
corpus callosum, 57/51
cranial nerve, 57/51
hypothalamus, 57/51
neurologic sign, 57/134
pathology, 57/44
pontomedullary rent, 57/51
vascular lesion, 57/51

Focal cerebellar cortical panatrophy
neuropathology, 21/500
nomenclature, 21/498
Focal cortical cerebellar sclerosis, *see* Focal cerebellar cortical panatrophy
Focal dermal dysplasia
microcephaly, 14/788
trigeminal nerve agenesis, 50/214
Focal dermal hypoplasia, *see* Goltz-Gorlin syndrome
Focal dermal hypoplasia syndrome, *see* Goltz-Gorlin syndrome
Focal disorder
consciousness, *see* Consciousness focal disorder
Focal dysplasia
brain cortex, *see* Brain cortex focal dysplasia
epilepsy, 72/109
Focal dystonia
brain lacunar infarction, 54/247
clinical type, 49/521
exogenous, 49/541
tardive dyskinesia, 49/186
Focal dystonia musculorum deformans
epidemiology, 49/520
hereditary, 49/520
Focal epilepsy, *see* Epileptic focus
Focal myoclonus
action, 49/612
clinical features, 49/611
reflex, 49/612
spontaneous, 49/612
Focal myopathy
incomplete dystrophin deficiency, 62/135
needle, 62/611
toxic myopathy, 62/611
Focal neuropathy
diabetic polyneuropathy, 51/501
Focal neuropsychological syndrome
dementia, 46/208
Focal sensory epilepsy, *see* Epileptic focus
Fog, M., 1/12
Fog rule
plaque, 9/259
Foix-Alajouanine disease
see also Angiodysgenetic necrotizing myelomalacia
age, 9/454
angiography, 55/103
angiography failure, 32/490
arterial involvement, 9/462
carcinoma, 38/669-671
clinical features, 55/103
clinical type, 12/135
course, 9/454
CSF, 9/455
decompression myelopathy, 61/223
demyelination, 10/15
histopathology, 9/458
history, 9/452
idiopathic nonvascular form, 9/452
intraspinal angioma, 20/481
lateral pontine syndrome, 12/17
malignancy, 38/669-671
necropsy data, 9/457
neuropathology, 55/103
nomenclature, 9/453, 461-467
paraneoplastic syndrome, 71/696
pathogenesis, 9/460
prodromic, 9/454
race, 9/454
sex ratio, 9/454
space occupying lesion, 19/378
spastic paraplegia, 59/437
spinal automatism, 19/38
spinal cord compression, 19/378
syringobulbia, 9/462
syringomyelia, 9/462
vascular, 9/461-467
Foix-Chavany-Marie syndrome, *see* Operculum syndrome
Foix-Jefferson syndrome
skull base, 67/146
skull base tumor, 17/181
Foix syndrome
ataxia, 2/277
chorea, 2/277
nucleus ruber lesion, 2/277
Foix-Thévenard tonic postural reflex, *see* Tonic postural reflex (Foix-Thévenard)
Folate, *see* Vitamin B_c
Folch-Lees proteolipid protein
amino acid composition, 7/56
Folic acid, *see* Vitamin B_c
Folic acid deficiency, *see* Vitamin B_c deficiency
Folinic acid
methotrexate intoxication, 65/532
Folium
tuber vermis, 2/402
Follicle stimulating hormone, *see* Human menopausal gonadotropin
Follicular keratosis, *see* Darier disease
Follitropin/luteinizing hormone secreting tumor
pituitary tumor, 68/344
Fonsecaea
fungal CNS disease, 35/565, 52/487
Fonsecaeasis

brain abscess, 35/566, 52/487
epilepsy, 35/566, 52/487
lymphocytic meningitis, 35/566, 52/487
meningitis, 35/566, 52/487
multiple cranial neuropathy, 35/566, 52/487
Food
anthozoa intoxication, 37/54
headache, 48/20
instant, *see* Instant food
manganese intoxication, 64/303
migraine, 48/20, 36, 60, 121
snack, *see* Snack food
staple, *see* Staple food
trepang intoxication, 37/66
Food allergy
multiple sclerosis, 47/154
Food-borne botulism
biology, 51/188
Food intoxication
neuropathy, 7/520
Staphylococcus, 37/87
Fool plaque
white matter, 9/587
Foot
club, *see* Talipes
flat, *see* Flatfoot
hypotrophic, *see* Hypotrophic foot
immersion, *see* Immersion foot
Mendel-Von Bechterew reflex, 1/248
osteoporosis, 70/20
rockerbottom, *see* Rockerbottom foot
Rossolimo sign, 1/248
sweating, 28/517
trench, *see* Trench foot
Foot and mouth disease virus
picornavirus, 34/5, 140, 56/36, 309
Foot deformity
Charlevoix-Saguenay spastic ataxia, 60/453
equinovarus, *see* Talipes equinovarus
methotrexate syndrome, 30/122
neurogenic acro-osteolysis, 42/294
Parkinson dementia, 49/172
scapuloperoneal spinal muscular atrophy, 59/45
spina bifida, 32/522, 545, 552
Foot drop
dialysis, 63/530
infantile distal myopathy, 43/114, 62/201
lateral popliteal nerve palsy, 8/153
lead polyneuropathy, 64/436
Foot injury
neurolysis, 70/42
tarsal tunnel syndrome, 70/42
Foot sweat test
skin temperature, 75/57
Foot ulcer
hereditary perforating, *see* Hereditary perforating foot ulcer
perforating, *see* Perforating foot ulcer
Football injury
atlantoaxial dislocation, 24/157
brain concussion, 23/571
carotid artery thrombosis, 23/574
cervicomedullary injury, 24/157-160, 174
epidural hematoma, 23/572, 574
head injury, 23/572, 24/160
intracerebral hematoma, 23/574
pontine lesion, 23/574
skull fracture, 23/573
statistic, 23/571, 573
subdural hematoma, 23/572, 574
traumatic carotid thrombosis, 23/574
vertebral artery, 24/157-160
Footballer migraine
case report, 48/146
definition, 48/124
literature, 48/388
Foramen
condyloid emissary, *see* Condyloid emissary foramen
intervertebral, *see* Intervertebral foramen
jugular, *see* Jugular foramen
mastoid emissary, *see* Mastoid emissary foramen
optic, *see* Optic foramen
parietal emissary, *see* Parietal emissary foramen
parietal permagna, *see* Foramina parietalia permagna
Foramen arcuale
atlas, 32/38
atlas dysplasia, 50/393
frequency, 32/38
Foramen cecum
sincipital cephalocele, 50/102
Foramen lacerum syndrome (Jefferson)
optic atrophy, 13/82
Foramen of Luschka-Magendie
atresia, *see* Dandy-Walker syndrome
Foramen magnum
achondroplasia, 43/316, 50/395
Arnold-Chiari malformation type I, 42/15
Arnold-Chiari malformation type II, 50/408
asymmetry, 32/6
basilar impression, 32/18, 42/17
bony masses on anterior rim, 32/31
craniodiaphyseal dysplasia, 43/356
craniometaphyseal dysplasia, 43/362-365
enlarged, *see* Enlarged foramen magnum

fluorosis, 50/396
frontometaphyseal dysplasia, 31/257, 43/402
iniencephaly, 50/129
labia, 32/34
meningioma, 68/413
meroanencephaly, 50/77
occipital dysplasia, 50/395
osteitis deformans (Paget), 38/361, 365, 50/396
skull base tumor, 68/469
tonsillar herniation, 1/565-567, 16/120
Foramen magnum decompression
syringomyelia, 50/459
Foramen magnum herniation, *see* Tonsillar herniation
Foramen magnum stenosis
asymmetry, 32/6
hydromyelia, 50/428
Foramen magnum tumor
age at onset, 17/721
atlanto-occipital synostosis, 20/184
cervical cord, 2/208
clinical features, 17/721-723, 725
clinical symptomatology, 2/208
cold sensation, 17/724
compression, 2/208
cranial nerve, 17/725
cranial nerve lesion, 19/39
CSF, 17/728
epilepsy, 20/199
hourglass tumor, 20/184, 291
incidence, 17/719, 720
literature, 17/719
location, 17/720
meningioma, 17/721, 19/67-69
motor sign, 17/722
multiple tumors, 17/721
myelography, 17/726, 19/193
neurofibroma, 17/721
papilledema, 19/60
paresthesia, 19/58
pathology, 17/723
psychosis, 20/199
radiology, 17/728
remote atrophy, 17/722
sensory deficit, 19/58
sensory symptom, 17/724
spastic paraplegia, 59/431
spinal meningioma, 19/68, 20/184
spinal neurinoma, 20/253
Foramen of Monro cyst, *see* Ependymal cyst
Foramen ovale patency
decompression myelopathy, 61/224
Foramen syndrome
jugular, *see* Jugular foramen syndrome
Foramina atresia
lateral cerebellar, *see* Lateral cerebellar foramina atresia
Luschka, *see* Luschka foramina atresia
Magendie, *see* Magendie foramina atresia
Foramina parietalia permagna
see also Parietal emissary foramen
associated abnormality, 50/142
autosomal dominant inheritance, 50/142
cleidocranial hypoplasia, 50/142
clinical features, 30/274, 50/140
congenital heart malformation, 50/142
craniofacial dysostosis (Crouzon), 50/142
deafness, 50/142
dermatoglyphics, 30/278
embryopathogenesis, 50/137
embryopathology, 50/141
emissary vein, 42/28
epidemiology, 30/155, 276, 50/141
etiology, 50/142
genetics, 30/155, 274, 277
headache, 42/29
heart disease, 42/29
hydrocephalus, 30/277, 42/29
incidence, 50/60, 141
Klinefelter syndrome, 50/142
location, 30/276
mental deficiency, 30/277
microcephaly, 42/29
monozygotic twins, 30/277, 50/142
parietal emissary foramen, 30/276, 50/141
pathology, 30/277
Saethre-Chotzen syndrome, 50/142
seizure, 30/275, 278, 42/29
size, 30/277
synonym, 30/274
treatment, 30/278, 50/142
Foraminal impaction, *see* Tonsillar herniation
Foraminotomy
spondylotic radiculopathy, 26/107
Forbes limit dextrinosis, *see* Debrancher deficiency
Forced laughing, *see* Pathologic laughing
Forced laughing and crying, *see* Pathologic laughing and crying
Forearm intermittent claudication
thoracic outlet syndrome, 51/124
Forebrain
autonomic nervous system, 74/152
basal, *see* Basal forebrain
Foregut cyst, *see* Intraspinal neurenteric cyst
Foregut mediastinal cyst, *see* Intraspinal neurenteric cyst

Hochwart, L.
Fraser syndrome
hereditary cerebellar ataxia (Marie), 21/369
hypertelorism, 30/247
Freckling
neurofibromatosis type I, 50/366
pseudoprogeria Hallermann-Streiff syndrome, 43/405
Free erythrocyte protoporphyrin
lead intoxication, 36/23
Free fatty acid
acquired hepatocerebral degeneration, 49/216
brain fat embolism, 55/180
brain metabolism, 57/82
Huntington chorea, 49/261
migraine, 48/95
plasma, 48/95
platelet serotonin, 48/96
Reye syndrome, 49/218, 56/238, 65/118
wine, 48/96
Free floating anxiety
phobic anxiety, 45/266
stress, 45/249-251
Free radical
action, 64/18
antiepileptic agent, 65/516
brain edema, 67/81
brain infarction, 53/130
brain ischemia, 53/130
brain metabolism, 57/81
brain microcirculation, 53/81
enzyme, 64/18
manganese deficiency, 64/305
manganese intoxication, 64/305
1-methyl-4-phenyl-1,2,3,6-tetrahydropyridine, 65/382, 387
1-methyl-4-phenyl-1,2,3,6-tetrahydropyridine intoxication, 65/382, 387
neuron death, 63/212
neurotoxicology, 64/18
neurotoxin, 64/18
oxygen, *see* Oxygen free radical
Parkinson disease, 49/129
respiratory encephalopathy, 63/416
type, 64/18
Free radical scavengers
brain infarction, 53/429
catalase, 64/19
ceruloplasmin, 64/19
glutathione, 64/19
glutathione peroxidase, 64/19
superoxide dismutase, 64/19
transferrin, 64/19
vitamin C, 64/19
Freeman-Sheldon syndrome, 43/352-354
anterior cranial fossa, 38/413
contracture, 43/353
digital abnormality, 38/413, 43/353
equinovarus, 38/413
facial paralysis, 43/354
growth retardation, 43/353
hypertelorism, 30/247, 43/353
kyphoscoliosis, 43/353
mental deficiency, 43/353
micrognathia, 43/353
microstomia, 38/413, 43/353
muscular atrophy, 38/413
ocular defect, 30/247
short stature, 38/413
talipes, 38/413, 43/353
Freeze fracture
axon, 47/9-11, 15, 17, 51/47
muscular dystrophy, 40/371
Ranvier node, 47/9, 11, 51/14
Freezing
akinesia, 49/67
Freezing gait
threo-3,4-dihydroxyphenylserine, 49/69
EMG, 49/69
gait apraxia, 49/69
paradoxical kinesia, 49/69
Freezing phenomenon
Parkinson disease, 49/68
pathophysiology, 49/68
pure akinesia, 49/68
Frenula abnormality
orofaciodigital syndrome type I, 50/167
Freon
distal axonopathy, 64/13
neurotoxin, 64/13
toxic neuropathy, 64/13
Freud, S., 1/5
Freund adjuvant
experimental allergic encephalomyelitis, 9/517, 522, 47/430, 451
experimental allergic neuritis, 9/506, 47/454
glomerulonephritis, 9/506
immunosuppression, 47/191
orchitis, 9/506
thyroiditis, 9/506
uveitis, 9/506
Frey syndrome, *see* Auriculotemporal syndrome
Fried-Emery classification
infantile spinal muscular atrophy, 59/51
Fried-Emery syndrome, *see* Intermediate spinal muscular atrophy

upper type, 2/316
Fowler test, *see* Loudness recruitment
Fox intoxication
arsenic intoxication, 37/96
brain tumor, 37/96
diarrhea, 37/96
diplopia, 37/96
drowsiness, 37/96
headache, 37/96
ingestion, 37/95
irritability, 37/96
papilledema, 37/96
scurvy, 37/96
vitamin A intoxication, 37/96
vomiting, 37/96
Fraccaro-Parenti achondrogenesis, *see* Achondrogenesis (Fraccaro-Parenti)
Fracture
ankle, *see* Ankle fracture
ankylosing spondylitis, 70/3
anterior cranial fossa, *see* Anterior cranial fossa fracture
blowout, *see* Blowout fracture
brain injury mechanism, 57/31
burst, *see* Burst fracture
cervical compression, *see* Cervical compression fracture
cervical tilting, *see* Cervical tilting fracture
compound skull, *see* Compound skull fracture
compression, *see* Compression fracture
dens, *see* Odontoid fracture
depressed skull, *see* Depressed skull fracture
Duchenne muscular dystrophy, 41/474
ethmoid bone, *see* Ethmoid bone fracture
facet, *see* Facet fracture
fat embolism, 53/161
femoral, *see* Femoral fracture
freeze, *see* Freeze fracture
hangman, *see* Hangman fracture
hip, *see* Hip fracture
humeral, *see* Humeral fracture
Jefferson, *see* Jefferson fracture
knee, *see* Knee fracture
Le Fort, *see* Le Fort fracture
limb, *see* Limb fracture
mandible, *see* Mandible fracture
maxillary, *see* Maxillary fracture
middle fossa, *see* Middle fossa fracture
nasomaxillary, *see* Nasomaxillary fracture
naso-orbital, *see* Naso-orbital fracture
nerve lesion, 70/26
odontoid, *see* Odontoid fracture
optic canal, *see* Optic canal fracture
optic foramen, 13/69
orbit, *see* Orbital fracture
pelvic, *see* Pelvic fracture
pillar, *see* Pillar fracture
rehabilitation, 41/474
rib, *see* Rib fracture
skull, *see* Skull fracture
skull base, *see* Skull base fracture
slice, *see* Slice fracture
spina bifida, 50/495
spinal injury, 26/200
spinous process, 25/366
sternum, *see* Sternum fracture
styloid process, 48/502
superior orbital, *see* Superior orbital fracture
surgery, 26/226-232
teardrop, *see* Vertebral teardrop fracture
temporal bone, *see* Temporal bone fracture
thoracic spinal cord injury, 26/265
treatment, 26/226-232
ulnar, *see* Ulnar fracture
vertebral, *see* Vertebral fracture
vertebral wedge compression, *see* Vertebral wedge compression fracture
wrist, *see* Wrist fracture
zygomatic bone, *see* Zygomatic bone fracture
Fragile chromosome X, *see* Sex chromosome fragility
Fragilitas ossium, *see* Osteogenesis imperfecta
Frameless system
stereotaxy, 67/243
France
neurology, 1/7
Franceschetti cranial dysostosis, *see* Cranial dysostosis (Franceschetti)
Francesconi index
basilar impression, 32/19
François syndrome
buphthalmos, 14/635
dwarfism, 13/436
facial hemihypertrophy, 14/627, 635, 638
hydrophthalmia, 14/627, 635, 638
Maffucci syndrome, 14/119
neurofibromatosis type I, 14/627, 635, 638
plexiform neuroma, 14/627, 635, 638
radiology, 14/627
Frankel Scale
spinal cord injury, 61/422
spinal cord injury recovery score, 61/423
transverse spinal cord lesion, 61/422
Frankfurt, Germany
neurology, 1/9
Frankl-Hochwart, L. Von, *see* Von Frankl-

Forsius-Eriksson syndrome
dyschromatopsia, 42/397
foveal hypoplasia, 42/397
genetic linkage, 42/398
myopia, 42/397
Nettleship-Falls syndrome, 42/398
nystagmus, 42/397
oculocutaneous albinism, 42/398
Forssmann carotid syndrome, *see* Forssmann-Skoog syndrome
Forssmann-Skoog syndrome
cerebellar syndrome, 2/345
experimental animal, 2/345
Forstenon intoxication
chemical classification, 37/545
organophosphorus compound, 37/545
Förster, O., 1/1, 9, 19, 22, 28, 29, 31, 36, 38
Förster syndrome
see also Astasia abasia
astasia abasia, 1/175
ataxia, 42/202
Babinski sign, 42/202
basal ganglion degeneration, 42/202
brain atrophy, 42/202
epilepsy, 42/202
hyperreflexia, 42/202
hypotonic, 1/175
mental deficiency, 1/175, 42/202
psychomotor retardation, 42/202
speech disorder in children, 42/202
strabismus, 42/202
striatal necrosis, 49/509
Fortification spectrum
hallucination classification, 45/56
Foscarnet
neurotoxicity, 71/368
side effect, 71/381
Fossa cribrosa
radiography, 19/161
Foster frame
decubitus, 20/820
Foster Kennedy, 1/5
Foster Kennedy syndrome
false localization, 2/570
frontal lobe tumor, 2/570, 13/69, 17/260
intracranial pressure, 2/53
meningioma, 2/570
nonrecurrent nonhereditary multiple cranial neuropathy, 51/571
olfactory nerve, 2/53, 570
optic atrophy, 2/53, 13/58, 68, 82
papilledema, 2/53
skull base, 67/146
X-linked neuromuscular dystrophy, 62/117
Fou rire prodromique
emotional disorder, 3/358, 362
pathologic laughing and crying, 3/358, 45/221
Founder effect
Ellis-Van Creveld syndrome, 43/348
GM2 gangliosidosis, 42/434
Foundry fever, *see* Zinc intoxication
Fourth ventricle
Arnold-Chiari malformation type II, 50/406
striatopallidodentate calcification, 6/712
unilateral pontine hemorrhage, 12/40
Fourth ventricle cyst
Dandy-Walker syndrome, 30/624, 50/328
Fovea
pseudo, *see* Pseudofovea
Foveal dystrophy
butterfly shaped pigmentary, *see* Pigmentary foveal dystrophy
vitelliform, *see* Vitelliform foveal dystrophy
Foveal dystrophy (Behr)
hereditary adult, *see* Hereditary adult foveal dystrophy (Behr)
Foveal hypoplasia
Forsius-Eriksson syndrome, 42/397
Foveal reflex
yellow mutant albinism, 43/7
Foville-Millard-Gubler syndrome, *see* Foville syndrome
Foville pontine syndrome, *see* Foville syndrome
Foville syndrome
see also Cestan-Chenais syndrome *and* Pontine syndrome
brain stem infarction, 12/16
caudal tegmental syndrome, 12/20
Cestan-Chenais syndrome, 1/188, 2/239
diplopia, 2/278
eponymic classification, 2/264, 12/13
Horner syndrome, 2/316
inferior pontine, 1/188, 2/239
lesion site, 2/301
localization, 2/301, 304, 315-318
midbrain, 1/187
Millard-Gubler syndrome, 2/242, 317
nuclear lesion, 2/316-318
nystagmus, 2/316
oculomotor disorder, 2/301, 304, 315-318
peduncular type, 2/304, 316
pontine infarction, 2/242, 261, 12/13
root lesion, 2/315
superior pontine, 1/188
superior type, 2/239
tegmental pontine syndrome, 12/20

Forehead anhidrosis
paratrigeminal syndrome, 48/334, 336
pericarotid syndrome, 48/336
Forehead sweating
see also Facial sweating
chronic paroxysmal hemicrania, 48/262
cluster headache, 48/236
Foreign antigen
experimental allergic neuritis, 9/505
hypersensitivity, 9/504
lymphocytic choriomeningitis, 9/536, 544
measles, 9/545
mumps, 9/545
Forel field H
anatomy, 6/5
Forel field H1
anatomy, 6/20
constituent fiber, 2/472
Forel field H2
anatomy, 6/20
constituent fiber, 2/472
Forgetfulness
depressive pseudodementia, 46/207
head injury outcome, 57/405
migraine, 48/162
Forgotten respiration
autonomic nervous system, 74/583
bilateral cervical cordotomy, 63/489
bilateral medullary infarction, 53/382
brain stem compression, 63/490
brain stem hemorrhage, 1/654
brain stem infarction, 1/654, 63/441, 488
central sleep apnea, 63/461
congenital megacolon, 63/489
cordotomy, 63/489
fentanyl intoxication, 1/654
hypoventilation, 74/583
lateral medullary infarction, 53/382, 63/441
morphine intoxication, 1/654
multiple neuronal system degeneration, 63/489
myotonic dystrophy, 63/490
neuroanatomy, 63/487
olivopontocerebellar atrophy, 63/488
poliomyelitis, 1/654, 63/441, 488
sleep apnea syndrome, 63/488
syringobulbia, 63/441, 490
transverse myelitis, 63/489
Forhead plaque
tuberous sclerosis complex, 68/292
Forking
aqueduct, *see* Aqueduct forking
muscle fiber, 40/206
Formalin fixation
myelin, 10/34
pseudomicrogyria, 50/254
Formations
fasciculus, *see* Fasciculus formation
mesencephalic reticular, *see* Mesencephalic reticular formation
midbrain reticular, *see* Midbrain reticular formation
onion bulb, *see* Onion bulb formation
pontine paramedian reticular, *see* Pontine paramedian reticular formation
reticular, *see* Reticular formation
Forme fruste
ataxia telangiectasia, 14/311
cluster headache, 48/221
dystonia musculorum deformans, 6/530, 49/520
Friedreich ataxia, 21/351
holoprosencephaly, 50/240
Klippel-Trénaunay syndrome, 14/116, 395
Krabbe-Bartels disease, 14/412
neurocutaneous melanosis, 14/424, 43/33
neurofibromatosis, 14/490
neurofibromatosis type I, 14/134, 143, 145, 151, 490
progressive bulbar palsy, 22/137
pseudopseudohypoparathyroidism, 42/578
Refsum disease, 21/189
Sydenham chorea, 6/427
tuberous sclerosis, 14/349-352
Von Hippel-Lindau disease, 14/245
Formiminoglutamic acid
formiminotransferase deficiency, 42/546
Formiminotetrahydrofolate cyclodeaminase deficiency
neurologic abnormality, 42/547
Formiminotransferase deficiency
brain atrophy, 42/546
EEG, 42/546
formiminoglutamic acid, 42/546
mental deficiency, 42/546
short stature, 42/546
ventricular dilatation, 42/546
Forney-Robinson-Pascoe syndrome, *see* Forney syndrome
Forney syndrome
LEOPARD syndrome, 14/117
vertebral abnormality, 30/101
Fornix
anatomy, 2/770
memory, 2/771
section, 2/771
Wernicke encephalopathy, 2/772
Wernicke-Korsakoff syndrome, 2/772

Friedländer arteritis proliferans, *see* Thromboangiitis obliterans
Friedman-Roy syndrome
convulsion, 43/251
CSF, 43/251
EEG, 43/251
hyperreflexia, 43/251
intracerebral calcification, 43/251
mental deficiency, 43/251
seizure, 43/251
strabismus, 43/251
talipes, 43/251
Friedreich ataxia, 21/319-359, 60/299-326
abnormality, 21/338, 348
abortive form, 60/322
adiposogenital syndrome, 21/348
age at onset, 21/327, 60/307
β-alanine, 42/144
amino acid metabolism, 60/305
amyotrophic lateral sclerosis, 21/354
amyotrophy, 21/320, 333, 355
anhidrotic ectodermal dysplasia, 14/788
arachnodactyly, 21/354
areflexia, 42/143, 60/660
ataxia, 60/311
ataxia telangiectasia, 14/75, 307-313, 316, 21/21, 60/383
atypical form, 60/322
auditory evoked potential, 60/321
autonomic change, 21/334
autonomic nervous sign, 60/312
autosomal dominant, 21/321
autosomal recessive, 21/321, 51/388
Babinski sign, 21/332, 42/143, 51/388
Bassen-Kornzweig syndrome, 10/548, 13/413, 418, 21/20, 357
Behr disease, 13/91
Betz cell, 21/324
bilirubin, 42/144, 60/307
biochemistry, 6/128
bladder dysfunction, 60/326
blood group, 21/322
brain embolism, 55/163
brain hemisphere, 60/303
brain infarction, 55/163
brain stem, 60/302
carbohydrate metabolism, 60/304
cardiac interstitial fibrosis, 21/344, 346
cardiac lesion, 21/325, 342-345
cardiac manifestation, 40/508, 60/316
cardiomyopathy, 21/342-346, 40/508, 51/388, 60/318, 667
cataract, 21/347, 60/310, 653
catecholamine, 21/326
cause, 21/320-322
cerebellum, 60/303
cerebrovascular disease, 55/163
chorea, 6/435, 21/330
choreiform instability, 1/319
chorioretinal degeneration, 13/39
choroid degeneration, 13/301
choroid sclerosis, 21/348
chromosome 9, 60/322
Clarke column, 42/144
claw hand, 21/341
clinical features, 21/327-349, 60/307-319
clinical rating scale, 60/312
color blindness, 21/347
combined form, 21/352
computer assisted tomography, 60/320, 671
consanguinity, 21/322
course, 21/357, 60/323
cranial nerve, 60/302, 309
CSF protein, 21/350, 60/320
cutaneous reflex, 21/332
deafness, 21/336, 347, 22/517, 51/388
dementia, 46/401
dentate nucleus, 42/144
dentatorubral atrophy, 21/530
dentatorubropallidoluysian atrophy, 49/442, 60/613
diabetes mellitus, 21/349, 42/144, 51/388, 60/319, 325, 674
diagnosis, 21/355-357
diagnostic criteria, 60/299
differential diagnosis, 22/15
distal amyotrophy, 60/658
dysarthria, 21/337, 51/388
dysgraphia, 21/329
dysphagia, 21/348
dysraphia, 2/326
dyssynergia cerebellaris myoclonica, 21/20, 351, 354, 530, 42/211
dyssynergia cerebellaris progressiva, 42/213
dystonia, 6/555
ECG, 21/326, 343-346, 356
ECG abnormality, 60/316, 670
echoCG, 60/316
EEG, 21/349, 60/320
EMG, 21/350, 51/388, 60/320
enzyme deficiency polyneuropathy, 51/388
enzyme study, 21/326
epidemiology, 21/3, 15, 60/307
epilepsy, 15/327, 21/347
eye movement disorder, 40/302
familial pes cavus, 21/263

familial spastic paraplegia, 21/19, 22/422, 425
forme fruste, 21/351
gait, 21/328
gaze limitation, 40/302
gaze paretic nystagmus, 60/310
genetic counseling, 60/324
genetics, 21/320-322
glutamate dehydrogenase deficiency, 51/389
glutamic acid, 51/389
Hallervorden-Spatz syndrome, 6/623, 21/354
Hallgren syndrome, 13/301
hearing loss, 51/388, 60/326
heart disease, 42/143
hereditary cerebellar ataxia (Marie), 21/434
hereditary olivopontocerebellar atrophy (Menzel), 21/444
hereditary sensory and autonomic neuropathy type II, 21/11
hereditary spastic ataxia, 21/369
hereditary spinal muscular atrophy, 59/24
history, 21/319
Hutchinson-Laurence-Moon syndrome, 13/381
hydrolytic enzyme, 60/307
hypertension, 60/667
hypertrophic interstitial neuropathy, 21/147
hypoesthesia, 60/661
hypogonadal cerebellar ataxia, 21/348, 476, 42/129
hypogonadism, 21/348
incidence, 60/307
infantilism, 21/348
initial symptom, 60/308
insulin resistance, 60/319
intermediate form, 21/352
iris coloboma, 21/347
kyphoscoliosis, 21/320, 341, 60/669
laboratory investigation, 60/319-322
Lafora progressive myoclonus epilepsy, 21/509
Laurence-Moon-Biedl syndrome, 21/348
Laurence-Moon syndrome, 13/300
Lichtenstein-Knorr disease, 22/515
lipid metabolism, 60/306
lipoamine dehydrogenase, 51/388
Machado disease, 21/357
macular degeneration, 21/347, 60/310
magnetic stimulation, 60/322
main features, 21/16
mental change, 60/308
mental deficiency, 21/337, 46/90
mirror movement, 21/330, 42/233
mitochondrial malic enzyme, 51/389
mixed genetic form, 21/321
motor system, 60/310
muscle tone, 21/333
muscular atrophy, 42/145
myatrophic ataxia, 21/355
myopathy, 21/355
myotatic reflex, 21/331
myotonic dystrophy, 21/355, 40/508
nature of defect, 21/325-327
nerve biopsy, 21/81, 325
nerve conduction study, 60/320
nerve conduction velocity, 42/144, 51/388, 60/670
neurocardiac disease, 21/344
neurochemistry, 60/304
neuropathology, 21/322-325, 60/301-304
neuropathy, 60/659
nigrospinodentate degeneration, 22/174
nongenetic cause, 21/322
nuclear magnetic resonance, 60/320
nystagmus, 21/335, 348, 42/143, 60/309
olivopontocerebellar atrophy, 21/21, 353, 22/172
olivopontocerebellar atrophy (Dejerine-Thomas), 60/531
ophthalmoplegia, 2/326, 13/304, 21/347, 355, 22/191
optic atrophy, 13/39, 82, 21/325, 334, 42/143, 409, 51/388, 60/309, 654
opticocochleodentate degeneration, 60/757
palmoplantar keratosis, 21/348
paresis, 21/330
parkinsonism, 21/354, 42/252
peripheral nerve, 60/303
personality change, 21/338
pes cavus, 21/338-341, 40/321, 42/143, 60/315, 325, 669
polydactyly, 21/348
polyneuropathy, 21/355
presenting sign, 21/328, 60/308
prevalence, 21/22, 327, 42/144
prognosis, 21/357, 60/323
progressive external ophthalmoplegia, 21/347, 22/191, 43/136, 60/310, 62/304
progressive spinal muscular atrophy, 22/13
psychotherapy, 60/324
ptosis, 21/347, 60/310
pulmonary function, 60/318
pupillary reaction, 21/347
pyramidal tract absence, 42/144
pyramidal tract degeneration, 42/144
pyruvate carboxylase, 51/389
pyruvate oxidation, 42/144
pyruvic acid, 51/388
Ramsay Hunt syndrome, 42/211, 49/616, 60/597
rebound nystagmus, 60/310
recurrent polyneuropathy, 21/355

Refsum disease, 8/23, 21/199, 60/235, 66/490, 497
rehabilitation, 60/325
retinitis punctata albescens, 21/347
Richards-Rundle syndrome, 21/359, 43/264
Roussy-Lévy syndrome, 21/12, 172-174, 351, 42/108
scoliosis, 51/388, 60/313, 325
secondary pigmentary retinal degeneration, 13/297-304, 21/347, 60/310, 654, 734
sensory change, 60/311
sensory loss, 21/320, 330, 42/143
sensory neuropathy, 8/184
sensory polyneuropathy, 51/388
serum transaminase, 60/319
sex linked recessive, 21/322
sex ratio, 21/327
skeletal deformity, 60/313
somatosensory evoked potential, 60/321
spastic paraplegia, 13/307, 21/19
speech, 21/329, 337
speech disorder, 60/309
speech disorder in children, 42/143
spina bifida, 21/348
spinal cord, 60/301
spinal cord lesion, 21/322-324
spinal nerve root, 60/303
spinocerebellar degeneration, 42/144
spinocerebellar degeneration classification, 60/275, 277
spinopontine degeneration, 21/397, 400
Stargardt disease, 13/132
sudden death, 21/346
Sylvester disease, 22/505
symptomatic dystonia, 6/555
talipes, 21/320
taurine, 42/144
thalamus degeneration, 21/598, 600
treatment, 21/358, 60/323
tremor, 21/330, 60/661
trophic lesion, 21/334
Tunbridge-Paley disease, 22/507
vestibular disorder, 21/337, 387
vestibular function, 60/658
visual evoked response, 60/321
xeroderma pigmentosum, 14/13, 315, 21/348
Friedreich ataxia like Sylvester syndrome
optic atrophy, 60/654
Friedreich disease, *see* Hemifacial hypertrophy
Friedreich, N.
history, 1/8, 60/300
Friedreich paramyoclonus multiplex, *see* Hereditary essential myoclonus
Friedreich tabes, *see* Friedreich ataxia
Fritsch, G., 1/8
Fröhlich syndrome
aqueduct stenosis, 30/619
obesity, 2/449
pituitary tumor, 2/454
Frohse arcade
posterior interosseous nerve syndrome, 51/100
supinator syndrome, 51/100
Froin syndrome, *see* Nonne-Froin syndrome
Froment sign
ulnar paralysis, 2/37
Frontal acalculia
secondary acalculia, 4/189
Frontal alexia
definition, 45/441
synonym, 45/441
Frontal aphasia
interpretation, 45/36
Frontal artery
inferior, *see* Inferior frontal artery
posterior, *see* Posterior frontal artery
Frontal bone
development, 30/215
methotrexate syndrome, 30/122
Frontal bossing
Coffin-Lowry syndrome, 43/238
frontometaphyseal dysplasia, 43/401
hereditary sensory and autonomic neuropathy type III, 60/18
Lowe syndrome, 42/606
multiple nevoid basal cell carcinoma syndrome, 43/31
progeria, 43/465
13q partial monosomy, 43/526
7q partial trisomy, 43/506
recombinant chromosome 3 syndrome, 43/495
Rothmund-Thomson syndrome, 43/460
Rubinstein-Taybi syndrome, 43/234
Schwartz-Lelek syndrome, 31/257
sclerosteosis, 43/474
Frontal cortex
depression, 46/428
emotion, 3/324, 329
Huntington chorea, 49/317
orbital, *see* Orbital frontal cortex
Frontal cortex nodular dysgenesia
striatopallidodentate calcification, 49/424
Frontal dementia
Alzheimer disease, 45/30
frontal lobe lesion, 45/30
Pick disease, 45/30
Frontal diploic vein
anatomy, 11/55

Frontal gyrus
  Marchiafava-Bignami disease, 9/653
  superior, *see* Superior frontal gyrus
Frontal headache
  cervical vertebral column injury, 61/33
Frontal lobe
  akinesia, 45/162
  anatomy, 2/725, 45/23
  apathy, 55/137
  association zone, 2/727
  cerebral dominance, 4/2
  cortex morphology, 2/727
  cytoarchitecture, 45/24
  Down syndrome, 50/531
  emotion, 45/273
  function, 3/17
  intrinsic zone, 2/727
  limbic encephalitis, 8/135
  movement plan, 3/34
  neuropsychology, 45/516
  oligodendroglioma, 18/82
  ontogeny, 2/725-727
  projection zone, 2/726
  schizophrenia, 46/458, 509
  size, 50/530
  subcortical dementia, 46/313
  transcortical motor aphasia, 45/19
  traumatic psychosis, 24/531
  traumatic psychosyndrome, 24/558, 560
  vestibular system, 16/331
  Wernicke-Korsakoff syndrome, 45/197
Frontal lobe ataxia
  clinical features, 1/340-342
  frontal lobe tumor, 17/251
Frontal lobe atrophy
  Pick disease, 27/489, 42/285, 46/239, 242, 424
  total, *see* Total frontal atrophy (Lüers-Spatz)
Frontal lobe epilepsy
  autosomal dominant nocturnal, *see* Autosomal dominant nocturnal frontal lobe epilepsy
  benign, *see* Benign frontal lobe epilepsy
  characteristics, 73/40
  classification, 72/10
  differential diagnosis, 73/44
  EEG, 73/45
  generalized, 73/38
  historic development, 73/37
  neuropathology, 73/46
  pathology, 72/117
  primary motor cortex seizure, 73/41
  seizure, 73/39
  treatment, 73/46
Frontal lobe injury
  head injury, 46/626
Frontal lobe lesion
  abulia, 2/741
  acalculia, 4/189
  affect, 45/31
  agrammatism, 45/35
  alien hand syndrome, 45/32
  amusia, 45/31
  anatomy, 45/24-36
  animal experiment, 2/737
  apraxia, 4/54-57
  attention disorder, 3/148-150, 189
  Balint syndrome, 45/410
  buccofacial apraxia, 4/53
  constructional apraxia, 45/500
  epilepsy, 2/752
  frontal dementia, 45/30
  gait apraxia, 1/341, 4/53, 17/252
  hallucination, 45/32
  incentive, 2/738
  memory, 45/29
  memory disorder, 3/284-286
  motor initiation, 45/32
  motor neglect, 45/32
  neglect, 45/31
  neuropsychology, 45/27-29
  perseverative agraphia, 45/467
  personality, 45/29
  psychometry, 45/27-29
  purposive behavior, 2/737
  site, 2/751
  size, 2/751
  speech, 45/32-36
  stereotyped behavior, 2/737
  symptom, 45/29-36
  vocal amusia, 45/488
Frontal lobe syndrome, 2/725-753, 45/23-38
  see also Anterior cerebral artery syndrome *and* Prefrontal syndrome
  abstraction, 55/138
  abulia, 53/349, 55/138
  aggression, 55/138
  akinesia, 55/138
  akinetic mutism, 55/138
  alogia, 45/24
  Alzheimer disease, 46/432
  amnesia, 53/348
  anatomy, 45/23
  anosodiaphoria, 53/349
  apathy, 2/739, 53/349, 55/138
  behavior disorder, 53/348
  brain infarction, 55/138
  clinical study, 45/24-26

confabulation, 53/349
decorum loss, 53/349
dementia, 45/30, 46/124
diabetes insipidus, 53/349
euphoria, 2/738, 45/24-26, 53/349
fever, 53/349
frontal lobe tumor, 46/432
frontal lobectomy, 45/27
general paresis, 46/432
grasp reflex, 55/138
head injury outcome, 57/407
Huntington chorea, 46/432
inappropriate behavior, 53/349
indifference, 55/138
initiative loss, 53/349
lobotomy, 45/26
morphologic data, 2/725-729
motor cortex, 2/725, 727
organic personality syndrome, 46/431
paratonic rigidity, 55/138
perseveration, 55/138
personality change, 45/29
Pick disease, 46/239, 432
prefrontal leukotomy, 45/25-27
psychological test, 45/27-29
stimulus boundedness, 55/138
subarachnoid hemorrhage, 46/432
thalamic lesion, 21/590
thalamic syndrome (Dejerine-Roussy), 2/484
urinary incontinence, 55/138
Wernicke-Korsakoff syndrome, 45/29

Frontal lobe tumor
agraphia, 17/256
angiography, 17/265
anomia, 17/255, 272
anosmia, 17/260
aphasia, 17/253, 271, 67/150
apraxia, 17/252, 67/149
astrocytoma, 17/238
autonomic disturbance, 67/151
autonomic dysfunction, 17/257
brain scanning, 17/264
cerebral dominance, 17/235
cognitive dysfunction, 67/148
comprehension, 67/150
cranial nerve, 17/260
defecation, 17/258
diagnosis, 17/236, 261
echoEG, 17/263
EEG, 17/263
epilepsy, 17/247, 270
euphoria, 46/429
extrapyramidal disorder, 17/257
extrapyramidal sign, 17/257
features, 67/147
Foster Kennedy syndrome, 2/570, 13/69, 17/260
frontal lobe ataxia, 17/251
frontal lobe syndrome, 46/432
frontocallosal, 17/237
gait, 67/149
gait apraxia, 4/53, 17/252
gastrointestinal disease, 17/258
gaze apraxia, 17/253
hallucination, 17/259
headache, 17/246
hyperosmia, 17/260
instrumental apraxia, 45/486
lateralization, 17/235
literature, 17/234
lumbar puncture, 17/261
meningioma, 17/238
mental disorder, 17/241, 244
micturition, 17/258
motor disorder, 17/256, 270
motor weakness, 17/270
muscle rigidity, 17/257
naming, 67/150
occurrence, 17/235
papilledema, 17/260
paresthesia, 17/259
personality, 45/29
postcentral gyrus, 17/272
posture, 67/149
precentral area, 17/270, 272
prefrontal, 17/239
pseudohemianopia, 17/260
psychiatric disorder, 17/240, 242, 244
reading, 67/150
reflex disorder, 17/248
repetition, 67/150
rubbing nose sign, 17/259
sensory disorder, 17/259
site, 17/235
sleep, 17/259
somnolence, 17/259
spatial orientation disorder, 17/259
speech disorder, 17/253, 271, 67/150
sphenoidal ridge meningioma, 17/238
sweating, 17/259
treatment, 17/269
tremor, 17/257
unilateral exophthalmos, 18/327
ventriculography, 17/267
writing, 67/150

Frontal lobectomy
frontal lobe syndrome, 45/27

Frontal sinus
absence, 43/332
arthrodento-osteodysplasia, 43/332
Frontal vein
inferior, *see* Inferior frontal vein
middle, *see* Middle frontal vein
superior, *see* Superior frontal vein
Frontalis muscle tension
headache, 48/360
Frontobasal brain injury
traumatic psychosis, 24/531
Frontobasal syndrome
traumatic, *see* Traumatic frontobasal syndrome
Frontobasilar artery
radioanatomy, 11/78
topography, 11/78
Frontometaphyseal dysplasia
autosomal dominant, 31/256
brain atrophy, 38/410
contracture, 43/401
dental abnormality, 31/257, 43/402
diplopia, 38/410
foramen magnum, 31/257, 43/402
frontal bossing, 43/401
headache, 38/410
hearing loss, 31/256, 38/410
Hershey kiss, 31/257
mental deficiency, 31/256
muscular atrophy, 31/256, 43/401
psychomotor retardation, 38/410
sensorineural deafness, 43/401
spasticity, 38/410
vertebral abnormality, 31/257, 43/402
visual impairment, 38/410
Frontonasal duct
headache, 5/214
Fronto-orbital artery
medial, *see* Inferior frontal artery
Frontopolar artery
radioanatomy, 11/79
topography, 11/79
Frontopolar vein
anatomy, 11/47
Frontopontine bundle
double athetosis, 6/448
Frontotemporal atrophy
glutaryl-coenzyme A dehydrogenase deficiency, 66/647
organic acid metabolism, 66/641
Frostbite
immersion foot, 8/342
trench foot, 8/342
Frozen face
facial paralysis, 8/276
Fructose
CSF pressure, 16/361
diabetic polyneuropathy, 51/507
intracranial pressure, 23/212
Fructose-bisphosphate aldolase
brain glycolysis, 27/9
Duchenne muscular dystrophy, 62/118
eosinophilia myalgia syndrome, 63/377
fructose intolerance, 42/548
GM1 gangliosidosis, 10/467, 470
GM2 gangliosidosis, 10/289, 291, 335, 393, 560
hypothyroid myopathy, 62/533
limb girdle syndrome, 62/185
muscular dystrophy, 40/384
polymyositis, 22/25
thyrotoxic myopathy, 62/529
wallerian degeneration, 7/217
Wohlfart-Kugelberg-Welander disease, 22/71
Fructose-1,6-diphosphatase deficiency
autosomal recessive inheritance, 29/257
epilepsy, 72/222
lactic acidosis, 41/208, 42/586
seizure, 42/586
Fructose-diphosphate aldolase deficiency
fructose intolerance, 29/255, 257
Fructose intolerance
aminoaciduria, 42/548
attack, 42/548
diet, 42/549
Fanconi syndrome, 29/172, 256
fructose-bisphosphate aldolase, 42/548
fructose-diphosphate aldolase deficiency, 29/255, 257
hereditary, *see* Hereditary fructose intolerance
hypoglycemia, 42/548
jaundice, 42/548
mental deficiency, 46/19
proteinuria, 42/548
Fructose-1-phosphate aldolase, *see* Fructose-bisphosphate aldolase
Frusemide, *see* Furosemide
FTA test, *see* Fluorescent treponemal antibody absorption test
5-FU, *see* Fluorouracil
Fuchs gyrate atrophy, *see* Gyrate atrophy (Fuchs)
Fuchs heterochromic iridocyclitis
hemiatrophy, 59/480
Fuchs sign, 2/319
Fuchs syndrome, *see* Gyrate atrophy (Fuchs)
Fucose
multiple sclerosis, 9/321
serum, 9/321

Fucosidase
α-L-fucosidase, 42/549
α-galactosidase, 42/550
α-glucosidase, 42/550
α-L-Fucosidase
enzyme deficiency polyneuropathy, 51/370, 381
fucosidase, 42/549
fucosidosis, 51/381
gene localization, 42/550
mucolipidosis type II, 42/448
mucolipidosis type III, 42/449
pseudodeficiency, 66/58
α-L-Fucosidase deficiency
secondary pigmentary retinal degeneration, 60/726
transplant, 66/88
Fucosidosis
see also Mucolipidosis type II
alpha, *see* Alpha fucosidosis
biochemical diagnosis, 66/334
clinical features, 66/334
clinical spectrum, 66/334
demyelination, 42/550
differential diagnosis, 66/336
enzyme, 29/350, 66/335
enzyme deficiency polyneuropathy, 51/370, 381
epilepsy, 72/222
excessive sweating, 51/381
Fabry disease, 42/550
α-L-fucosidase, 51/381
genetics, 66/335
glycoprotein degradation, 66/62
glycosaminoglycan, 27/157
H antigen, 29/369
heart disease, 42/550
hepatosplenomegaly, 42/550
mental deficiency, 42/550, 66/330
neuropathology, 66/334
original description, 10/431
psychomotor retardation, 42/550
Schwann cell vacuole, 51/381
skeletal lesion, 10/466
sphingolipid, 29/350
survey, 66/329
ultrastructure, 51/381
vacuolated lymphocyte, 42/550
vertebral abnormality, 42/550
β-xylosidase, 42/550
Fugu intoxication, *see* Puffer fish intoxication *and* Tetrodotoxin intoxication
Fukuyama cerebromuscular dystrophy, *see* Fukuyama syndrome
Fukuyama congenital muscular atrophy, *see* Fukuyama syndrome
Fukuyama syndrome
see also Congenital muscular dystrophy, HARD syndrome *and* HARDE syndrome
clinical features, 41/36
clinodactyly, 43/92
CNS, 41/38
contracture, 43/91
creatine kinase, 43/91
dental abnormality, 43/92
EEG, 43/92
epilepsy, 43/91, 72/129
gene, 72/135
genetics, 41/36
HARD syndrome, 50/261
HARDE syndrome, 50/261
hypotonia, 41/36
laboratory finding, 41/37
malignant hyperthermia, 62/570
mental deficiency, 41/37, 43/91
muscular dystrophy, 43/91
nerve conduction velocity, 43/91
pathogenesis, 41/41
pathology, 41/37
polymicrogyria, 41/38
prevalence, 41/35
pseudohypertrophy, 43/91
seizure, 43/92
sex ratio, 43/92
ventricular dilatation, 43/92
Fulminant meningococcemia
see also Waterhouse-Friderichsen syndrome
meningococcal meningitis, 52/24
Fulton, J.F., 1/29
Fumarase, *see* Fumarate hydratase
Fumarase deficiency, *see* Fumarate hydratase deficiency
Fumarate hydratase
wallerian degeneration, 7/217
Fumarate hydratase deficiency
amaurosis, 66/420
dystonia, 66/420
enzyme defect, 66/420
enzyme deficiency, 66/420
fumarate hydratase gene, 66/420
hepatomegaly, 66/420
infant, 66/655
infantile spasm, 66/420
mitochondrial disease, 62/497
neutropenia, 66/420
organic acid metabolism, 66/639
oxoglutarate dehydrogenase deficiency, 62/503
polyhydramnios, 66/420

tricarboxylic acid cycle, 66/396, 420
urinary organic acid, 66/420
vomiting, 66/420
Fumarate hydratase gene
fumarate hydratase deficiency, 66/420
Fumaric acid
hepatic coma, 49/215
Fumaric aciduria, *see* Fumarate hydratase deficiency
Functional deficit zone
definition, 73/387
Functional evaluation
Duchenne muscular dystrophy, 41/494
schema, 41/458
Functional hallucination
hallucinosis, 46/562
Functional hyperinsulinism
hypoglycemia, 27/69
Functional psychosis, 46/443-463
affective psychosis, 46/444-446
brain blood flow, 46/453
classification, 46/444-446
clinical features, 46/455-457
computer assisted tomography, 46/451-453
delirium, 46/532
EEG, 46/460
historic note, 46/443
neurochemistry, 46/446-450
neuropathology, 46/453-455
neuropsychologic assessment, 46/457-460
neuroradiology, 46/450-452
psychopharmacology, 46/446-450
schizophrenia, 46/444
Functional spinal unit
definition, 61/4
load, 61/4
lumbar segment, 61/16, 18
rotation, 61/4
stretch test, 61/10
thoracic segment, 61/13
three column concept, 61/90
x axis, 61/5
y axis, 61/5
z axis, 61/5
Fundus albipunctatus, *see* Retinitis punctata albescens
Fundus camera
involuntary eye movement, 13/7
Fundus flavimaculatus, 13/263-266
chorea, 13/263
differential diagnosis, 13/133
primary pigmentary epithelium degeneration, 13/30
Stargardt disease, 13/28, 129, 133, 136, 42/420
Fungal CNS disease
Allescheria boydii, 35/558-560
Alternaria, 35/561, 52/480
Cephalosporium, 35/562, 52/479, 481
Cladosporium, 35/563, 52/480, 482
Cryptococcus, 52/496
Cryptococcus neoformans, 35/570, 52/480, 496
Curvularia, 52/479, 484
dematiomycosis, 52/480
Diplorhinotrichum, 35/565
Drechslera, 35/565, 52/485
epilepsy, 72/160
Fonsecaea, 35/565, 52/487
Fusarium, 52/488
Madurella, 35/566, 52/479, 488
mycetoma, 52/479
Nocardia, 52/479
nomenclature, 52/479
Paecilomyces, 35/567, 52/489
Penicillium, 35/567, 52/489
Pseudoallescheria, 52/490
Pseudoallescheria boydii, 52/479
Rhodotorula, 52/492
Sepedonium, 52/493
Sporotrichum, 35/568-570, 52/493
Streptomyces, 35/570, 52/479
Trichophyton, 35/571, 52/496
Trichosporon, 52/497
Ustilago, 35/572, 52/498
Fungal endocarditis
infectious endocarditis, 52/289
postoperative, *see* Postoperative fungal endocarditis
Fungal meningitis
brain vasculitis, 55/416, 434
CSF, 55/433
dementia, 46/387
neurologic symptom, 55/432
treatment, 55/433, 436
Fungus
brain, *see* Brain fungus
cancer, 69/438
CNS infection, 69/435, 438, 446
Funicular myelitis, *see* Subacute combined spinal cord degeneration
Funicular myelosis, *see* Subacute combined spinal cord degeneration
Funicular pain
intramedullary spinal tumor, 19/69
spinal meningioma, 20/198, 201
spinal tumor, 19/55
Funicular spinal disease, *see* Subacute combined spinal cord degeneration

Funktionswandel, 2/659, 3/224, 4/25, 30, 37
apraxia, 4/60
constructional apraxia, 4/60
Fur nevus
neurofibromatosis type I, 14/102
Furaltadone
polyneuropathy, 7/537
Furan
neuropathy, 7/517
Furor bulborum
extrapyramidal disorder, 2/344
opsoclonus, 2/344
Furor transitorius
migrainous acute confusional state, 48/165
Furosemide
brain aneurysm, 55/53
brain aneurysm neurosurgery, 55/53
brain edema, 16/204
deafness, 65/440
diuretic agent, 65/437
hydrocephalus, 50/297
hypercalcemia, 63/562
neurologic intensive care, 55/213
subarachnoid hemorrhage, 55/53
thallium intoxication, 64/327
tinnitus, 65/440
vertigo, 65/440
Furukara disease
ataxic neuropathy, 42/334
Fusariosis
brain abscess, 52/488
Fusarium
brain abscess, 52/151
fungal CNS disease, 52/488
Fusions
bony, *see* Bony fusion
spinal, *see* Spinal fusion
vertebral, *see* Vertebral fusion
Fusobacterium
brain abscess, 52/149
Fusobacterium necrophorum
spinal epidural empyema, 52/187

G band
chromosome, 31/344, 350
G-protein, *see* Guanine nucleotide binding protein
G syndrome
brain defect, 30/247
hypertelorism, 30/247
neurologic disorder, 30/247
GA2 glycolipid
degradation, 66/59
GABA
Alzheimer disease, 46/267
Amanita muscaria intoxication, 29/507, 36/534
amino acid neurotransmitter, 29/485-507
analogue, 29/507
anxiety, 45/267
autonomic nervous system, 74/56
ballismus, 49/372
basal ganglion, 49/21, 33
basal ganglion tier III, 49/22
biochemistry, 29/486-491
brain, 9/16
brain amino acid metabolism, 29/19
brain metabolism, 27/17
cerebellar tremor, 49/588
chemistry, 10/258
CSF, *see* CSF GABA
dentatorubropallidoluysian atrophy, 49/443
dystonia, 29/506
epilepsy, 15/62, 29/503-505, 72/54, 58, 65, 84, 87, 89, 73/348
galactosemia, 29/485-507
GM2 gangliosidosis, 10/296
hemiballismus, 49/372
historic consideration, 29/485
homocarnosinase deficiency, 42/558
Huntington chorea, 29/505, 42/227, 49/74, 257
mesolimbic noradrenalin system, 49/51
neostriatum, 49/47
nervous system tumor, 27/504, 510
neurofibrillary tangle, 46/269
neurotransmitter, 29/493
pallidonigral fiber, 49/8
pallidosubthalamic projection, 49/7
paraplegia, 61/401
Parkinson disease, 29/505, 42/246, 49/123
pathway, 29/19, 498-503
pedunculopontine tegmental nucleus, 49/12, 51
reduced neostriatal, *see* Reduced neostriatal GABA
regional distribution, 29/491
release, 29/496
spastic paraplegia, 61/367
spinal cord injury, 61/367, 401
striatonigral fiber, 49/8
striatopallidal fiber, 49/6
subcellular distribution, 29/492
substantia nigra, 49/8
subthalamopallidal fiber, 49/6
tetanus, 29/506
tetraplegia, 61/401
transverse spinal cord lesion, 61/367, 401
tricarboxylic acid cycle, 27/15, 17
turning behavior, 45/167

uptake, 29/497
uremic encephalopathy, 63/514
uremic polyneuropathy, 63/514
vitamin B6, 28/113
GABA metabolism
biochemistry, 28/113
succinate semialdehyde dehydrogenase deficiency, 66/654
GABA receptor
alcohol intoxication, 64/114
characteristics, 29/495
Huntington chorea, 49/259
organochlorine insecticide intoxication, 64/202
GABA receptor stimulating agent
ibotenic acid, 65/39
muscimol, 65/39
GABA synthesis rate
Huntington chorea, 49/290
Gabapentin
epilepsy, 73/360
Gabapentin intoxication
ataxia, 65/512
diplopia, 65/512
headache, 65/512
nystagmus, 65/512
tremor, 65/512
vertigo, 65/512
Gag reflex
brain stem death, 57/473
hereditary sensory and autonomic neuropathy type III, 21/109, 60/27
pseudobulbar paralysis, 2/780
Gain control
basal ganglion function, 6/92
Gait
coordination, 1/308
dentatorubropallidoluysian atrophy, 21/519
diaphyseal dysplasia, 43/376
disturbance, 46/327
Down syndrome, 50/528
dromedary, *see* Dromedary gait
freezing, *see* Freezing gait
Friedreich ataxia, 21/328
frontal lobe tumor, 67/149
headache, 48/50
hemiplegia, 1/420
hereditary cerebello-olivary atrophy (Holmes), 21/403, 409, 60/569
hereditary motor and sensory neuropathy type I, 21/281
hereditary motor and sensory neuropathy type II, 21/281
hereditary olivopontocerebellar atrophy (Menzel), 21/441
hereditary sensory and autonomic neuropathy type III, 43/59, 60/26, 29
hereditary spastic ataxia, 21/366
hydrocephalic dementia, 46/327
juvenile hereditary benign chorea, 49/346
MASA syndrome, 43/255
olivopontocerebellar atrophy (Dejerine-Thomas), 21/424
Parkinson dementia, 49/172
parkinsonism, 1/346
peacock, *see* Peacock gait
Roussy-Lévy syndrome, 21/175, 42/108
spastic, *see* Spastic gait
spina bifida, 50/493
steppage, *see* Steppage gait
vestibular system, 16/315
Wilson disease, 49/228
Gait apraxia
aqueduct stenosis, 50/311
astasia abasia, 1/346
Binswanger disease, 54/222
cerebellar hemorrhage, 54/314
freezing gait, 49/69
frontal lobe lesion, 1/341, 4/53, 17/252
frontal lobe tumor, 4/53, 17/252
progressive supranuclear palsy, 22/219, 49/240
Rett syndrome, 60/637
test, 45/423
Gait ataxia
Behçet syndrome, 56/598
cerebellar component, 1/323-325
clinical features, 1/312
error source, 1/310
erythrokeratodermia ataxia syndrome, 21/216
Gerstmann-Sträussler-Scheinker disease, 60/622
Huntington chorea, 49/280
labyrinthine disease, 1/336
Minamata disease, 64/413
pyroglutamic aciduria, 29/230
Gait disturbance
acanthocytosis, 13/417
acquired immune deficiency syndrome dementia, 56/491
aqueduct stenosis, 50/311
archicerebellar syndrome, 2/418
bismuth intoxication, 64/338
brain lacunar infarction, 1/199
camptocormia, 1/291
cerebellopontine angle syndrome, 2/94
concentric sclerosis, 47/414
corpus callosum ataxia, 1/343, 17/508
Divry-Van Bogaert syndrome, 55/319

ethylene oxide polyneuropathy, 51/275
hydrocephalic dementia, 46/327
intraspinal tumor, 19/24
Joseph-Machado disease, 42/261
kuru, 56/556
kyphoscoliosis, 13/414
late cerebellar atrophy, 42/135
Lindenov-Hallgren syndrome, 13/454
neuromyelitis optica, 47/401
osteitis deformans (Paget), 38/362, 366, 43/451, 46/400
pallidoluysionigral degeneration, 6/665
Parkinson disease, 6/189
parkinsonism, 1/346
phenylketonuria, 42/611, 59/75
Von Hippel-Lindau disease, 14/115
Galactoceramidase deficiency
globoid cell leukodystrophy, 59/355
Galactocerebrosidase deficiency, *see* Galactosylceramidase deficiency
Galactocerebrosidase pseudodeficiency
globoid cell leukodystrophy, 59/356
Galactocerebroside
demyelination, 47/447
globoid cell, 10/87
multiple sclerosis, 47/109, 114
Galactocerebroside-β-galactosidase
globoid cell leukodystrophy, 10/147, 47/598
Galactocerebroside-β-galactosidase deficiency, *see* Globoid cell leukodystrophy
Galactocerebroside sulfotransferase
globoid cell leukodystrophy, 10/146
Galactokinase deficiency
clinical aspect, 29/259
course, 29/259
definition, 29/258
diagnosis, 29/260
genetics, 29/258
incidence, 29/258
inheritance, 29/259
pathophysiology, 29/260
sign, 29/259
symptom, 29/259
treatment, 29/260
Galactolipid
chemical structure, 66/198
metabolism, 66/3, 199
Galactorrhea
methyldopa intoxication, 37/440
nonpurpural type, 2/456
pituitary adenoma, 17/423
Galactosamine
chemistry, 10/293
β-Galactosaminidase deficiency
GM2 gangliosidosis type III, 10/298, 414
Galactose
mucopolysaccharide cleavage, 10/479
multiple sclerosis, 9/402
Galactosemia
aminoaciduria, 42/552
benign intracranial hypertension, 16/160, 42/551
birth incidence, 42/551
brain edema, 42/551
diet, 42/552
Fanconi syndrome, 29/172
GABA, 29/485-507
galactose-1-phosphate uridylyltransferase, 29/260, 46/53
genetics, 29/260
hypoglycemia, 27/72, 42/551
incidence, 29/260
jaundice, 42/551
lenticular opacity, 42/551
liver disease, 42/552
mental deficiency, 42/551, 46/19, 52
Moro reflex, 42/551
pathophysiology, 29/261
prenatal diagnosis, 42/552
proteinuria, 42/552
Purkinje cell loss, 42/551
space occupying lesion, 16/242
treatment, 29/261
Galactose-1-phosphate uridylyltransferase
galactosemia, 29/260, 46/53
9p partial monosomy, 50/579
Galactose-1-phosphate uridylyltransferase deficiency, *see* Galactosemia
Galactose-6-sulfatase, *see* Galactose-6-sulfurylase
Galactose-6-sulfurylase
N-acetylgalactosamine-6-sulfatase, 66/64
glycosaminoglycanosis type IVA, 66/64
Galactosialidase deficiency
motor conduction velocity, 51/380
neurogenic muscular atrophy, 51/380
Schwann cell vacuole, 51/380
spinal muscular atrophy, 51/380
Galactosialidosis
clinical finding, 66/360
defect, 66/359
enzyme deficiency polyneuropathy, 51/370
gene, 66/360
genetics, 66/271, 361
GM1 gangliosidosis, 66/249
infantile type, 66/361
laboratory finding, 66/359
neuropathology, 66/22

protein, 66/57
sialic acid, 66/354
treatment, 66/362
α-Galactosidase
Fabry disease, 42/427-429
fucosidase, 42/550
GM1 gangliosidosis, 10/477
myelin, 9/34
pseudodeficiency, 66/58
β-Galactosidase
chemistry, 10/258
enzyme deficiency polyneuropathy, 51/369
Fabry disease, 10/302
GLB1 gene, 66/282
glycoprotein, 10/480
glycosaminoglycanosis type I, 10/223, 42/450
glycosaminoglycanosis type III, 10/430
glycosaminoglycanosis type IV, 27/150, 42/457
glycosaminoglycanosis type IVB, 66/64, 282, 306
GM1 gangliosidosis, 10/223, 299, 467, 27/155, 66/60
GM1 gangliosidosis type I, 10/223, 42/431
GM1 gangliosidosis type II, 42/432
Goldberg syndrome, 42/440
metachromatic leukodystrophy, 10/48
mucolipidosis type II, 42/488
mucolipidosis type III, 29/367, 42/449
organ activity, 10/474-479
vertical supranuclear ophthalmoplegia, 42/606
wallerian degeneration, 7/215, 9/36
α-Galactosidase A
enzyme deficiency polyneuropathy, 51/369, 376
Fabry disease, 55/456, 60/172, 66/234
gene, 66/237
genetics, 66/235
β-Galactosidase A
gene localization, 42/432
α-Galactosidase A deficiency, *see* Fabry disease
α-Galactosidase deficiency, *see* Fabry disease
β-Galactosidase deficiency, *see* Glycosaminoglycanosis type IVB *and* GM1 gangliosidosis
β-Galactosidase deficiency type I, *see* GM1 gangliosidosis type I
Galactosidase gene
phosphoglycerate kinase, 66/119
Galactosylceramidase
globoid cell leukodystrophy, 66/60
pseudodeficiency, 66/58
Galactosylceramidase deficiency
bone marrow transplantation, 66/89
Galactosylceramide
globoid cell leukodystrophy, 42/489
lipidosis, *see* Globoid cell leukodystrophy
Galactosylceramide β-galactosidase
enzyme deficiency polyneuropathy, 51/369, 373
Galactosylceramide lipidosis, *see* Globoid cell leukodystrophy
Galactosyl-3-sulfate ceramide
metachromatic leukodystrophy, 42/494
Galactosylsulfatide
enzyme deficiency polyneuropathy, 51/367
Galen, C., 1/3, 2/6
Galen classification
epilepsy, 15/1
Galen vein
adulthood, 31/195
anatomy, 11/51, 30/418
arteriovenous shunt, 12/103
brain aneurysm, 12/103
brain arteriovenous malformation, 2/257, 12/103, 236, 18/179-284, 54/43
brain phlebothrombosis, 54/399
cervical angioma, 32/486
child, 31/195
clinical features, 31/193-196
cystic encephalomalacia, 30/665
Dandy-Walker syndrome, 30/638
diagnosis, 31/196
displacement, 30/550
drainage, 30/421, 32/486
drainage territory, 9/579
dura mater, 30/421
hydranencephaly, 30/665
hydrocephalus, 30/543, 72/210
infant, 31/193-195
malformation, 12/103, 31/191-202, 72/210
newborn period, 31/193
prognosis, 31/196
radioanatomy, 11/108
thrombophlebitis, 33/172
treatment, 31/196
variation, 30/422
venous malformation, 12/236
venous system, 11/49, 63
Galen vein aneurysm
aqueduct stenosis, 18/282, 50/308
brain arteriovenous malformation, 12/260
features, 31/191-202
hemorrhage, 18/282
hydrocephalus, 18/282
neurosurgery, 12/260
posterior cerebral artery, 18/279
Galen vein thrombosis
basal ganglion infarction, 54/399
thalamic infarction, 54/399

Gallamine
  malignant hyperthermia, 41/268
Gallbladder
  diabetes mellitus, 75/594
  dystopic, *see* Dystopic gallbladder
Galloway-Mowat syndrome
  brain atrophy, 43/431
  diaphragm hernia, 43/431
  edema, 43/431
  glomerulonephritis, 43/431
  microcephaly, 43/431
  nephrotic syndrome, 43/431
Gallstone
  Wilson disease, 49/224
GALNS gene
  N-acetylgalactosamine-6-sulfatase, 66/282
  glycosaminoglycanosis type IVA, 66/282
Galvanic skin response
  experimental aspect, 2/408
  hyperactivity, 46/183
  orthostatic hypotension, 63/160
  stimulation, 2/408
Galveston orientation and amnesia test
  coma, 57/383
  head injury outcome, 57/398
Gamma efferent system
  cerebellar ablation, 2/396
Gamma encephalography
  astrocytoma, 18/29-33
Gamma fiber
  muscle tone, 1/258
  physiology, 1/65
Gamma heavy chain disease
  classification, 69/290
  paraproteinemia, 69/290, 307
Gamma loop
  gamma system, 1/302
  muscle spindle, 6/95
Gamma myelography
  cervical tumor, 19/248
Gamma rigidity
  see also Decerebration
  cerebellar ablation, 2/399
Gamma system
  cerebellar, 1/332
  gamma loop, 1/302
  muscle tone, 1/258
  physiology, 1/65
  rehabilitation, 12/472
  spasticity, 1/67
  stimulation, 6/97-99
Gammaglobulin, *see* Ig
Gammaherpes virus
  Epstein-Barr virus, 56/8
  Marek disease virus, 56/8
  saimiri-ateles virus, 56/8
Gammopathy
  benign monoclonal, *see* Benign monoclonal gammopathy
  biclonal, *see* Biclonal gammopathy
  IgM monoclonal, *see* IgM monoclonal gammopathy
  immunocompromised host, 56/469
  monoclonal, *see* Monoclonal gammopathy
  neuropathic, *see* Neuropathic gammopathy
  undetermined significant monoclonal, *see* Undetermined significant monoclonal gammopathy
Gamper, R., 1/10
Gamstorp-Wohlfart syndrome, *see* Isaacs syndrome
Ganciclovir
  cytomegalovirus infection, 56/274
  neurotoxicity, 71/368
  side effect, 71/381
Gangliocytoma, *see* Neuroastrocytoma
Ganglioglioma, *see* Neuroastrocytoma
Gangliomyelopathy
  varicella zoster virus myelitis, 56/236
Ganglion
  autonomic, *see* Autonomic ganglion
  basal, *see* Basal ganglion
  biliary system, 75/615
  carpal tunnel syndrome, 7/293
  cerebrospinal, *see* Cerebrospinal ganglion
  ciliary, *see* Ciliary ganglion
  dorsal spinal nerve root, *see* Dorsal spinal root ganglion
  function, 74/39
  gasserian, *see* Gasserian ganglion
  geniculate, *see* Geniculate ganglion
  histology, 1/432-437
  Langley, *see* Langley ganglion
  median nerve, 7/293
  middle cervical, *see* Middle cervical ganglion
  myenteric, *see* Myenteric ganglion
  nodosum, *see* Nodosum ganglion
  otic, *see* Otic ganglion
  pancreas, 75/615
  parasympathetic, *see* Parasympathetic ganglion
  petrosum, *see* Petrosum ganglion
  prevertebral, *see* Prevertebral ganglion
  Scarpa, *see* Scarpa ganglion
  sphenopalatine, *see* Sphenopalatine ganglion
  spinal, *see* Spinal ganglion
  stellate, *see* Stellate ganglion
  sympathetic, *see* Sympathetic ganglion

trigeminal, *see* Trigeminal ganglion
vertebral joint, *see* Vertebral joint ganglion
Ganglion blockade
pharmacology, 74/169
Valsalva maneuver, 74/169
Ganglion caroticum, *see* Carotid body
Ganglion cell tumor
biopsy, 67/227
Ganglion cyst
epidural, *see* Epidural ganglion cyst
Ganglion degeneration
dorsal spinal nerve root, *see* Dorsal spinal root ganglion degeneration
Ganglion intercaroticticum, *see* Carotid body
Ganglion intercaroticum, *see* Carotid body
Ganglion pterygopalatinum, *see* Pterygopalatine ganglion
Ganglionectomy
geniculate, *see* Geniculate ganglionectomy
trigeminal, *see* Trigeminal ganglionectomy
Ganglioneuroma, *see* Neuroastrocytoma
Ganglioneuropathy
human immunodeficiency virus neuropathy, 71/354
Ganglionic transmission
chemical, 1/48
Ganglionitis
dorsal spinal nerve root, *see* Dorsal spinal nerve root ganglionitis
herpes zoster, *see* Herpes zoster ganglionitis
Kawasaki syndrome, 52/266
Ganglioradiculopathy
paraneoplastic polyneuropathy, 51/466
vitamin B6 intoxication, 65/572
Ganglioside
amaurotic idiocy, 10/220
astrocytoma, 67/297
brain metabolism, 57/84
chemistry, 10/249, 251, 415, 66/37
composition, 10/293
definition, 66/247
degeneration, 10/257, 310
diabetes mellitus, 70/153
disialosyl, *see* Disialosyl ganglioside
G1-4, 9/17
glia, 9/4, 17
glycosaminoglycanosis type I, 9/4
GM1, *see* GM1 ganglioside
GM2, *see* GM2 ganglioside
GM2 gangliosidosis, 10/220, 390
GM3, *see* GM3 ganglioside
gray matter, 9/4
head injury, 57/84
leukodystrophy, 10/24
lysosomal degradation, 66/59
malnutrition, 9/6, 29/6
metabolism, 10/257, 66/251
metachromatic leukodystrophy, 9/4
mucolipidosis type IV, 66/384
myelin, 9/4
nervous system tumor, 27/504
neuropathy, 69/298
nomenclature, 10/253
paraproteinemia, 69/297
puromycin, 9/9
sphingolipid, 10/327
subacute sclerosing panencephalitis, 9/4
synapse, 29/6
synthesis blockade, 9/9
undetermined significant monoclonal gammopathy, 63/402
white matter, 9/4
Ganglioside sialidase deficiency, *see* Mucolipidosis type IV
Gangliosidosis
adult GM1, *see* Adult GM1 gangliosidosis
classification, 10/222
GD3, *see* GD3 gangliosidosis
generalized, *see* GM1 gangliosidosis
genetic, *see* Genetic gangliosidosis
GM1, *see* GM1 gangliosidosis
GM2, *see* GM2 gangliosidosis
GM3, *see* Gm3 gangliosidosis
infantile GM2, *see* Infantile GM2 gangliosidosis
metabolic ataxia, 21/575
neurolipidosis, 66/19
nosologic evolution, 10/224
sialic acid, 66/250
survey, 66/247
Gangrenous osteomyelitis, *see* Midline granuloma
Ganser dorsal commissure
accessory optic bundle, 2/529
Ganser syndrome
consciousness, 3/126
delirium, 46/552
dementia, 46/124
Gap substance
Ranvier node, 47/10, 51/14, 16
Garcin, R., 1/24, 30
Garcin syndrome
aneurysm, 2/313
angioma, 2/313
basal meningitis, 2/313
cavernous sinus thrombosis, 2/313
classification, 2/101
clinical features, 2/313

cranial nerve, 2/313
Gradenigo syndrome, 2/91
hemorrhage, 2/313
multiple cranial neuropathy, 51/570
nasopharyngeal tumor, 2/313
polyneuritis, 2/313
skull base fracture, 2/313
skull base tumor, 2/103, 17/182
trochlear nerve, 2/308
Gardner disease
bilateral acoustic neurinoma, 22/517
hereditary hearing loss, 22/517
neurofibromatosis type I, 22/517
secondary pigmentary retinal degeneration, 60/737
Gardner theory
hydromyelia, 50/428
Garel-Bernfeld syndrome, *see* Stylohyoid syndrome
Garel syndrome, *see* Stylohyoid syndrome
Gargoylism, *see* Glycosaminoglycanosis type IH
Gari, *see* Cassava
Garin-Bujadoux-Bannwarth syndrome, *see* Bannwarth syndrome
Gas
blood, *see* Blood gas
industrial, *see* Industrial gas
Gas myelography
spinal angioma, 32/470
spinal cord atrophy, 19/181
spinal cord tumor, 19/181
syringomyelia, 19/181
technique, 19/184
Gaskoyen syndrome, *see* Bean syndrome
Gasoline
leukoencephalopathy, 66/721
neuropathy, 7/521
organic solvent intoxication, 64/40
toxic myopathy, 62/601
Gasoline intoxication
toxic myopathy, 62/613
Gasp reflex
pure autonomic failure, 22/234
Gasperini syndrome
case history, 2/222
conjugate deviation, 2/317
deafness, 2/317
definition, 2/240
eponymic classification, 2/264
gaze paralysis, 2/317
medulla oblongata syndrome, 2/222
motor deficiency, 2/317
Gasping respiration
brain stem encephalitis, 63/487
cerebellar hemorrhage, 63/487
clinical significance, 1/662-665
definition, 63/487
lateral medullary tegmentum, 63/487
lesion site, 1/653, 63/487
prognostic significance, 1/663
site, 63/478
sobbing syncope, 15/827
toxic oil syndrome, 63/480
traumatic vegetative syndrome, 24/586
Gasserian anastomosis
carotid, *see* Carotid gasserian anastomosis
Gasserian ganglion
carotid artery, 5/273, 335
compression, 5/397
decompression, 5/397
electrocoagulation, 5/306
gangliectomy, 5/301, 304
gangliolysis, 5/397
injection, 5/388, 390
Kirschner coagulation, 5/306
meningioma, 5/273
neurocysticercosis, 5/273
pain, 45/229
pathology, 5/273
surgery history, 5/300-302
trigeminal neuralgia, 5/273, 42/352
trigeminal pain, 5/273
vitamin B_6 intoxication, 65/572
Gasserian ganglion syndrome
skull base, 67/146
Gastaut classification
epilepsy, 15/26
Gastrectomy
amyotrophic lateral sclerosis, 22/29
amyotrophy, 22/29
progressive spinal muscular atrophy, 22/29
restless legs syndrome, 8/315, 51/545
vitamin B_{12} deficiency, 51/339
vitamin B_c deficiency, 51/336
Gastric aperistalsis
sensory radicular neuropathy, 21/82
Gastric bypass
deficiency neuropathy, 51/328
Gastric dysrhythmia
pathogenesis, 75/631
Gastric pacemaker dysfunction
pathogenesis, 75/631
Gastrin
cluster headache, 48/240
progressive dysautonomia, 59/154
Gastrin releasing peptide
headache, 48/109

Gastrocolic fistula
deficiency neuropathy, 7/618
intestinal malabsorption, 7/618
Gastrocytoma, *see* Intraspinal neurenteric cyst
Gastrogenous polyneuropathy
incidence, 7/600
malabsorption syndrome, 7/598, 600
symptomatology, 7/600
weight loss, 7/699
Gastrointestinal bleeding
brain aneurysm neurosurgery, 55/54
corticosteroid, 24/632
liver cirrhosis, 27/351
subarachnoid hemorrhage, 55/54
Gastrointestinal disease
APUD cell, 39/454-456
cholecystokinin, 39/452
endorphin, 39/453
enkephalin, 39/453
Fabry disease, 39/464
frontal lobe tumor, 17/258
intracranial hypertension, 16/140
multiple sclerosis, 39/456-459
neurology, 39/449-464
neurotensin, 39/452
somatostatin, 39/451
substance P, 39/453
vasoactive intestinal polypeptide, 39/451
Whipple disease, 39/459-464
Gastrointestinal disorder
see also Enteric nervous system
acetazolamide intoxication, 37/202
autonomic nervous system, 75/613
Behçet syndrome, 34/482
disulfiram intoxication, 37/321
drug induced, 75/651
endocrine communication, 75/627
enteric neuron, 75/621
hereditary amyloid polyneuropathy type 1, 8/9, 368
lissencephaly syndrome, 42/40
motility disorder, 75/627
neuroimmune communication, 75/627
Parkinson disease, 6/195
phenacemide intoxication, 37/202
propranolol intoxication, 37/447
psychiatry, 75/651
reserpine intoxication, 37/437
thalidomide syndrome, 42/660
Gastrointestinal disturbance
autonomic nervous system, 75/571
Gastrointestinal dysfunction
diabetes mellitus, 75/591
spinal cord injury, 75/577
Gastrointestinal neuropathy
diabetes mellitus, 70/141
Gastrointestinal reflex
survey, 75/623
Gastrointestinal system, *see* Digestive system
Gastrointestinal tract
autonomic nervous system, 74/14, 116
congestin intoxication, 37/37
constipation, 74/116
contraction, 75/622
hereditary amyloid polyneuropathy, 8/368, 21/123, 60/105
hereditary amyloid polyneuropathy type 1, 60/95
hereditary amyloid polyneuropathy type 2, 60/98
hereditary angioedema, 43/60
neuromuscular disease, 70/325
pellagra, 6/748, 7/572
Peutz-Jeghers syndrome, 14/12, 121, 526, 43/41
sleep, 74/540
Smith-Lemli-Opitz syndrome, 66/586
Gastrointestinal tract disorder
survey, 70/224
Gastropleuroschisis
anencephaly, 50/84
Gastropoda intoxication, *see* Conidae intoxication
Gate control system
causalgia, 22/255
pain, 45/231
Gaucher cell
birefringency, 10/307
cell ultrastructure, 10/509-512
globoid cell leukodystrophy, 10/69, 42/437
Gaucher disease, 10/509-529, 66/123-130
acid phosphatase, 10/308, 42/437
acute infantile type, 10/513
acute neurology, 10/513
acute neuropathic, 29/351
acute neuropathology, 10/514
adolescent adult type, 10/564
adult, *see* Adult Gaucher disease
amniocentesis, 29/353
associated neurological disease, 10/527
atypical case, 10/514, 521
Babinski sign, 42/438
biochemistry, 29/353
birth incidence, 42/437
bone marrow transplantation, 66/129
brain injury, 10/525
brain pathology, 10/525
ceramide galactose, 10/509
ceramide glucose, 10/307, 509, 512, 525, 564, 21/50

cerebral cell, 10/519
cerebroside, 10/308
chemical pathology, 10/308, 329
chorioretinal degeneration, 13/39
chromosome 1, 60/154
classification, 60/153
clinical features, 10/307, 329, 29/351
clinical phenotype, 66/123
cutaneous pigmentation, 10/307
cytoside, 10/308, 544
definition, 60/147
dementia, 46/402
detection, 29/353
differential diagnosis, 10/527
early confusion, 10/219
enzyme, 66/61
enzyme defect, 29/350, 352
enzyme deficiency polyneuropathy, 51/369, 379
enzyme replacement therapy, 66/127
epidemiology, 10/566
epilepsy, 15/426, 42/438, 72/222
etiology, 10/509
features, 10/328
gene frequency, 42/437
gene therapy, 66/129
genetic mutation, 66/126
genetics, 10/309, 566
globoid cell, 10/89
globoid cell leukodystrophy, 10/528
β-glucosidase deficiency, 10/306, 309, 526, 42/437-439, 46/55
glucosylceramidase, 10/330, 51/379, 66/60
glucosylceramidase cDNA, 66/119
glucosylcerebroside, 10/544
hematoside, 10/308
heterozygote detection, 42/438
heterozygote frequency, 42/437
histopathology, 10/307
hyperreflexia, 42/438
infantile, *see* Infantile Gaucher disease
juvenile, *see* Juvenile Gaucher disease
malignant type, 10/513
mental deficiency, 42/438, 46/55
metabolic defect, 66/125
neuropathology, 10/512, 515-518, 51/379, 66/23
noninfantile neuronopathic, *see* Noninfantile neuronopathic Gaucher disease
optic atrophy, 13/39
pain, 42/437
PAS positive microglia, 10/308
plasma, 10/308
polyneuropathy, 21/65
prenatal diagnosis, 42/438
primary defect, 10/309
prognosis, 10/528
N-stearylglucocerebrosidase, 10/565
strabismus, 42/438
supranuclear ophthalmoplegia, 60/656
survey, 66/123
treatment, 10/528, 29/353, 66/127
type I, *see* Adult Gaucher disease
type II, *see* Infantile Gaucher disease
type III, *see* Juvenile Gaucher disease
type IIIA, *see* Juvenile Gaucher disease type A
type IIIB, *see* Juvenile Gaucher disease type B
ultrastructure, 21/50
visceral chemical pathology, 10/512
Gaze
attention, 3/23
automatic, 2/652
cerebellum, 2/345
conjugate, *see* Conjugate gaze
cortical influence, 2/274
lateral, *see* Lateral gaze
occipital lesion, 2/652
optokinetic nystagmus, 2/653
vision, 2/666
volitional, 2/652
Gaze apraxia
ataxia telangiectasia, 21/581
Balint syndrome, 45/406, 55/144
bilateral occipital lobe infarction, 53/413
blinking, 45/406
eye convergence, 45/406
frontal lobe tumor, 17/253
Huntington chorea, 6/322
opsoclonus, 45/406
posterior cerebral artery syndrome, 53/413
Gaze center
cortical, *see* Cortical gaze center
pontine, *see* Pontine gaze center
Gaze defect
alexia, 45/439
visuospatial agnosia, 45/170
Gaze deviation
brain infarction, 55/122
Gaze fixation
test technique, 2/333
Gaze movement
see also Eye movement
command movement, 2/331
compensating movement, 2/331
cortical area, 2/331
following movement, 2/331
guiding movement, 2/331
type, 2/331

Gaze paralysis
ataxia telangiectasia, 14/652
Benedikt syndrome, 2/298
brain amyloid angiopathy, 54/337
brain lacunar infarction, 54/245
brain stem arteriovenous malformation, 2/258
cardiac arrest, 63/215
cavernous sinus thrombosis, 52/173
cerebrovascular disease, 55/121
Cestan-Chenais syndrome, 2/239
climbing eye movement, 2/334
dentatorubropallidoluysian atrophy, 49/440
diagnostic principle, 2/347
downward paralysis, 1/604
ethylene glycol intoxication, 64/123
examination, 2/332
examination method, 2/332-340
familial spastic paraplegia, 22/434, 440
frontal gaze palsy, 2/334
Gasperini syndrome, 2/317
gaze spasm, 2/336
hepatic polyneuropathy, 51/326
hereditary olivopontocerebellar atrophy (Menzel), 21/443
horizontal, *see* Horizontal gaze paralysis
Huntington chorea, 6/322, 358
lateral, *see* Lateral gaze palsy
leukemia, 63/342
Lyle syndrome, 2/302
multiple sclerosis, 9/167
nystagmus, 2/335
occipital gaze center, 2/339
occipital gaze palsy, 2/339
occipital lobe ischemia, 2/672
oculomotor apraxia, 2/332
pontine, *see* Pontine gaze paralysis
pontine hemorrhage, 2/254
pontine tuberculoma, 2/261
progressive supranuclear palsy, 22/217, 219
psychic, *see* Psychic gaze paralysis
spinopontine degeneration, 21/371, 390
table, 2/346
temporal arteritis, 55/344
thalamic syndrome (Dejerine-Roussy), 2/485
type, 1/601
upward, *see* Upward gaze paralysis
vertical, *see* Vertical gaze palsy
Gaze spasm
see also Oculogyric crisis
bradykinesia, 2/344
extrapyramidal disorder, 2/344
gaze paralysis, 2/336
GD3 gangliosidosis
Norman-Wood type, 10/225
Gee disease, *see* Celiac disease
Gee-Herter disease, *see* Celiac disease
Gegendruckphänomen
spasmodic torticollis, 6/573
Gegenhalten
Parkinson disease, 49/94
spastic paraplegia, 59/429
Gehirnpathologie
critique, 1/28
definition, 1/24
task, 1/38
Geisbock disease
see also Polycythemia vera
definition, 55/467
Gelastic epilepsy, *see* Gelolepsy
Gelatinoid myxoglioma
definition, 14/32
neurofibromatosis type I, 14/32
Gellé syndrome
deafness, 2/241
dizziness, 2/241
eponymic classification, 2/264
pain, 2/241
pontine syndrome, 2/241
statoacoustic nerve, 2/241
tinnitus, 2/241
Gelolepsy
EEG, 45/509
GM2 gangliosidosis, 10/392
pathologic laughing, 45/509
pathologic laughing and crying, 55/142
Gelsemium
trigeminal neuralgia, 5/299
Gelsolin variant
hereditary amyloid polyneuropathy type 3, 60/108
Gemfibrozil
acquired myoglobinuria, 62/576
toxic myopathy, 62/601
Gemistocytic astrocytoma
classification, 18/2
histology, 18/8-10
tissue culture, 17/56
Gemonil, *see* Metharbital
Gene
ABH secretor, *see* ABH secretor gene
α-*N*-acetylgalactosaminidase, 66/347
adrenoleukodystrophy, 66/96
adult neuronal ceroid lipofuscinosis, 72/135
apolipoprotein B-100, *see* Apolipoprotein B-100 gene
apolipoprotein C-II, *see* Apolipoprotein C-II gene
apolipoprotein E, *see* Apolipoprotein E gene

ARSB, *see* ARSB gene
biological rhythm, 74/467
cancer, *see* Cancer gene
cerebroside sulfatase, *see* Cerebroside sulfatase gene
cerebrotendinous xanthomatosis, 66/607
cholesterol ester transfer protein, *see* Cholesterol ester transfer protein gene
CNS malformation, 30/126-129
congenital bilateral perisylvian syndrome, 72/138
definition, 30/95
eukaryotic, *see* Eukaryotic gene
Fukuyama syndrome, 72/135
fumarate hydratase, *see* Fumarate hydratase gene
galactosialidosis, 66/360
galactosidase, *see* Galactosidase gene
α-galactosidase A, 66/237
GALNS, *see* GALNS gene
GLB1, *see* GLB1 gene
glucosylceramidase, *see* Glucosylceramidase gene
β-glucuronidase, *see* β-Glucuronidase gene
GM2 gangliosidosis, 66/252
GNAT, *see* GNAT gene
G6S, *see* G6S gene
GUSB, *see* GUSB gene
haptoglobin1, *see* Haptoglobin1 gene
HARD syndrome, 72/135
HARDE syndrome, 72/135
HSS, *see* HSS gene
IDS, *see* IDS gene
IDUA, *see* IDUA gene
immune response, *see* Immune response gene
infantile neuronal ceroid lipofuscinosis, 72/134
late infantile neuronal ceroid lipofuscinosis, 72/134
lipoprotein lipase, *see* Lipoprotein lipase gene
lissencephaly, 72/132
long chain acyl-coenzyme A dehydrogenase, *see* Long chain acyl-coenzyme A dehydrogenase gene
MERRF syndrome, 72/136
mitochondrial DNA, *see* Mitochondrial DNA gene
mitochondrial DNA defect, 66/397
myelin, *see* Myelin gene
NAGLU, *see* NAGLU gene
neurofibromatosis type I, 67/32
neurofibromatosis type II, 67/33
Northern epilepsy syndrome, 72/135
organolead intoxication, 64/129
phosphatidylcholine sterol acyltransferase, *see* Phosphatidylcholine sterol acyltransferase gene
proteolipid protein, *see* Proteolipid protein gene
proteolipid protein DM-20, *see* Proteolipid protein DM-20 gene
Rb, *see* Rb gene
Sturge-Weber syndrome, 72/137
suicide, *see* Suicide gene
suppressor, *see* Suppressor gene
TP53, *see* TP53 gene
tricarboxylic acid cycle deficiency, 66/419
tumor suppressor, *see* Tumor suppressor gene
very long chain acyl-coenzyme A dehydrogenase, *see* Very long chain acyl-coenzyme A dehydrogenase gene
X-linked mental deficiency, 72/136
Gene delivery method
brain tumor, 67/299
Gene delivery vector
gene therapy, 66/113
Gene frequency
Bassen-Kornzweig syndrome, 21/580
cerebro-oculofacioskeletal syndrome, 43/342
congenital deafness, 42/365
deafness, 43/43
Gaucher disease, 42/437
globoid cell leukodystrophy, 42/490
glucose-6-phosphate dehydrogenase deficiency, 42/644
GM2 gangliosidosis, 10/560
infantile spinal muscular atrophy, 42/89
Klein-Waardenburg syndrome, 43/53
muscular atrophy, 42/85
Niemann-Pick disease, 10/569
Pena-Shokeir syndrome type I, 43/439
Pendred syndrome, 42/369
piebald pigmentation, 43/43
sclerosteosis, 43/475
sensorineural deafness, 43/43
Sjögren-Larsson syndrome, 43/307
spastic paraplegia, 42/170
Tangier disease, 42/627
Troyer syndrome, 42/194
Usher syndrome, 42/392
Wilson disease, 42/271
Gene instability
myotonic dystrophy, 62/213
Gene linkage, *see* Genetic linkage
Gene localization
β-*N*-acetylhexosaminidase A, 42/434
β-*N*-acetylhexosaminidase B, 42/435
argininosuccinase, 42/525
cri du chat syndrome, 31/549, 568
α-L-fucosidase, 42/550
β-galactosidase A, 42/432
hemochromatosis, 42/553
hemophilia A, 42/738

hemophilia B, 42/738
hereditary olivopontocerebellar atrophy (Menzel), 42/162
human, 30/91
Lesch-Nyhan syndrome, 42/154
phosphoribosyl pyrophosphate synthetase, 42/573
poliovirus, 42/653
poliovirus receptor, 42/653
retinoblastoma, 42/468
Gene locus
benign neonatal familial convulsion, 72/129
epilepsy, 72/129
metachromatic leukodystrophy, 47/590
Gene penetrance
dystonia musculorum deformans, 6/519
Klinefelter syndrome, 13/25
multiple sclerosis, 47/299
Gene product
myotonic dystrophy, 62/213
Gene therapy
adenovirus, 66/114
adenovirus associated vector, 66/116
adenovirus vector, 66/115
asialoglycoprotein receptor, 66/114
astrocytoma, 67/291
brain tumor, 67/291, 299
Gaucher disease, 66/129
gene delivery vector, 66/113
glycosaminoglycanosis type VI, 66/309
herpes simplex virus, 66/114, 116
history, 66/113
immunotherapy, 67/291
liver cell, 66/114
lysosomal storage disease, 66/113
metachromatic leukodystrophy, 66/176
Moloney murine leukemia virus, 66/114
plasmid DNA, 66/114
polycationic liposome, 66/114
retrovirus, 66/114
Gene transcription
core promotor element, 66/70
DNA, 66/70
modification, 66/72
regulation, 66/72
steroid responsive element, 66/72
upstream promotor element, 66/71
Gene transfer
brain, 66/118
hematopoietic cell, 66/117
General adaptation syndrome
adrenal cortex hypertrophy, 45/252
adrenal medulla, 45/253
histopathology, 45/252
intestinal ulcer, 45/253
strain, 45/247
stress, 45/247
thymicolymphatic involution, 45/252
General anesthesia
Huntington chorea, 49/284
lumbosacral plexus, 51/161
malignant hyperthermia, 38/550, 62/567
multiple sclerosis, 47/151
myoglobinuria, 40/339
sensory extinction, 3/193
traumatic intracranial hematoma, 57/275
General anesthetic agent
anesthetic agent intoxication, 37/408-413
isoflurane, 37/412
General paresis
depression, 46/427
euphoria, 46/429
frontal lobe syndrome, 46/432
mania, 46/429
General paresis of the insane
see also Neurosyphilis
apathy, 52/279
Argyll Robertson pupil, 52/279
brain atrophy, 52/279
computer assisted tomography, 52/279
CSF, 52/279
dementia, 52/279
hydrocephalus, 52/279
neurosyphilis, 33/358-362, 46/390, 52/279
penicillin, 52/279
psychiatric symptom, 52/279
tremor, 52/279
Generalized chorea
regressive, *see* Regressive generalized chorea
stationary, *see* Stationary generalized chorea
Generalized epilepsy, 15/107-128, 73/235
clonic seizure, 15/121
myoclonic epilepsy, 15/121-128
pathology, 72/117
tonic clonic epilepsy, 15/108-116
tonic seizure, 15/116-120
Generalized gangliosidosis, *see* GM1 gangliosidosis
Generalized hyperkeratosis, *see* Urbach-Wiethe disease
Generalized lipodystrophy, *see* Berardinelli-Seip syndrome
Generalized myoclonic status
status epilepticus, 15/160
Generalized myoclonus, *see* Myoclonic epilepsy
Generalized myotonia (Becker)
creatine kinase, 43/164
heterozygote detection, 43/164

muscle hypertrophy, 43/163
pes cavus, 43/163
prevalence, 43/164
Generalized peroxisomal disorder, 66/519
pathogenesis, 66/520
peroxisomal β-oxidation, 66/520
Generalized tonic clonic seizure, 72/3
awake-sleep cycle, 73/175
clinical picture, 15/108, 111
grand mal on awakening, 73/175
grand mal on sleeping, 73/177
juvenile myoclonic epilepsy, 73/160
Lennox-Gastaut syndrome, 72/7
precipitating factor, 73/177
random epilepsy, 73/177
sleep-wake cycle, 73/177
survey, 73/175
Generation
tremor, *see* Tremor generation
Genetic code, 30/94
Genetic counseling
branchio-otodysplasia, 42/360
branchio-otorenal dysplasia, 42/363
carrier detection, 41/415, 484, 489
congenital spondyloepiphyseal dysplasia, 43/480
craniosynostosis, 42/21
Friedreich ataxia, 60/324
neuromuscular disease, 41/484
pseudoxanthoma elasticum, 43/47, 55/454
rehabilitation, 41/484-490
SBLA syndrome, 42/771
Genetic drift
GM2 gangliosidosis type I, 42/434
Genetic epilepsy
animal model, 72/63
Genetic gangliosidosis
metabolism, 66/248
Genetic hypothesis test
deletion mapping, 30/111
linkage analysis, 30/111
segregation analysis, 30/111
somatic cell hybridization, 30/111
Genetic linkage
congenital adrenal hyperplasia, 42/515
deutan color blindness, 40/352, 388, 43/88, 131
Emery-Dreifuss muscular dystrophy, 43/88
Forsius-Eriksson syndrome, 42/398
glucose-6-phosphate dehydrogenase deficiency, 40/352, 388
hereditary olivopontocerebellar atrophy (Menzel), 42/162
HLA antigen, 42/128, 163, 515
Kell blood group, 42/611
Lutheran blood group, 40/522, 43/153
multiple malformation, 30/6
multiple sclerosis, 42/497, 47/307-309
muscular dystrophy, 43/458
myotonic dystrophy, 40/521, 43/153
Nettleship-Falls albinism, 42/405
olivopontocerebellar atrophy (Schut-Haymaker), 42/163
Pelger-Huët abnormality, 43/108, 458
phenylketonuria, 42/611
phosphoglucomutase-1, 42/611
proximal muscular dystrophy, 43/108
scapuloperoneal myopathy, 43/131
Schut family ataxia, 60/487
secretor locus, 40/522, 43/153
spinocerebellar ataxia, 42/128
Xg blood group, 42/398, 405
Genetics
see also Inheritance *and* Mendel law
α-*N*-acetylgalactosaminidase, 66/347
α-*N*-acetylgalactosaminidase deficiency, 66/342
achondroplasia, 31/269
acid maltase deficiency, 41/180, 426
acoustic nerve agenesis, 50/218
acrocephalosyndactyly type I, 31/231
acrocephalosyndactyly type II, 31/235
acrodysostosis, 31/228
adrenergic receptor, 75/242
adrenoleukodystrophy, 47/597
Aicardi syndrome, 31/226
alpha adrenergic receptor, 75/242
Alzheimer disease, 46/272-274
anencephaly, 30/193
anophthalmia, 30/470
anterior sacral meningocele, 32/206
aqueduct stenosis, 30/157, 609, 618, 50/306
arachnoid cyst, 31/89
argininosuccinic aciduria, 29/100
Arnold-Chiari malformation type I, 32/103
arthrogryposis multiplex congenita, 30/165, 32/512
aspartylglucosaminuria, 29/228
ataxia telangiectasia, 14/274, 515-517, 31/24, 60/358
autism, 46/191
autonomic dysfunction, 75/409
autonomic nervous system, 74/218
autosomal recessive inheritance, 13/25
basilar impression, 30/161, 32/71
Bassen-Kornzweig syndrome, 10/549, 13/425, 29/405, 410
Batten disease, 10/576, 623
Becker muscular dystrophy, 40/387, 41/411, 486

Beckwith-Wiedemann syndrome, 31/329
benign essential tremor, 49/568, 586
benign partial epilepsy with somatosensory evoked potential, 73/26
beta adrenergic receptor, 75/243
beta mannosidosis, 66/334
bipolar affective disorder, 43/209, 46/472
blastomatosis, 16/30
Bonnet-Dechaume-Blanc syndrome, 14/514
Börjeson-Forssman-Lehmann syndrome, 31/311
brain aneurysm, 14/517-520
brain arteriovenous malformation, 30/160
brain tumor, 14/500, 16/28-30
brancher deficiency, 41/185, 427
carbohydrate deficient glycoprotein syndrome, 66/627
carnitine deficiency, 41/198, 200, 427
carnosinase deficiency, 29/198
caudal aplasia, 32/348
central core myopathy, 41/430
cerebellar hypoplasia, 30/387
cerebrohepatorenal syndrome, 31/305
cervico-oculoacusticus syndrome, 32/127-129
chondrodystrophic myotonia, 31/279, 41/424
chorea-acanthocytosis, 49/327, 63/283
citrullinemia, 29/97
CNS malformation, 50/50
Cockayne-Neill-Dingwall disease, 31/236
Coffin-Lowry syndrome, 31/244
congenital aneurysm, 31/139
congenital fiber type disproportion, 41/431, 62/356
congenital muscular dystrophy, 41/420
congenital Pelizaeus-Merzbacher disease, 10/162
congenital ptosis, 30/100, 401
congenital retinal blindness, 22/539, 542, 59/342
Conradi-Hünermann syndrome, 31/275, 38/401
contingency method, 15/430
corpus callosum agenesis, 30/155, 285
cranial cephalocele, 30/209
craniocervical region, 32/7
craniofacial dysostosis (Crouzon), 31/233
craniometaphyseal dysplasia, 38/393
craniosynostosis, 30/154, 221
cystinosis, 29/178
cystinuria, 29/188
Dandy-Walker syndrome, 30/158, 630, 50/332
debrancher deficiency, 41/183, 427
dentatorubropallidoluysian atrophy, 49/440
depression, 46/471
dermal sinus and dermoid, 32/455
developmental dyslexia, 46/118
distal muscular dystrophy, 41/421
Divry-Van Bogaert syndrome, 14/512-514
dopamine receptor, 75/244
double athetosis, 49/384
Down syndrome, 31/372, 391-393
Duchenne muscular dystrophy, 40/352-357, 41/487, 489, 62/125
dyslexia, 46/130
dystonia musculorum deformans, 49/520
EEG, 15/429, 522
Emery-Dreifuss muscular dystrophy, 62/156
Engelmann disease, 38/400
enteric nervous system, 75/614
epidemiology, 30/142
epilepsy, 15/429, 72/70, 125, 340, 75/351
epileptic focus, 15/435
Fabry disease, 10/544, 579-581, 66/239
facioscapulohumeral muscular atrophy, 41/441
facioscapulohumeral muscular dystrophy, 62/163
familial gliomatosis-glioblastomatosis, 14/496-500
familial iminoglycinuria, 29/187
familial spastic paraplegia, 22/422
Farber disease, 66/218
Fazio-Londe disease, 41/440
febrile convulsion, 15/250, 436
Ferguson-Critchley syndrome, 22/434, 436
foramina parietalia permagna, 30/155, 274, 277
Friedreich ataxia, 21/320-322
fucosidosis, 66/335
Fukuyama syndrome, 41/36
galactokinase deficiency, 29/258
galactosemia, 29/260
galactosialidosis, 66/271, 361
α-galactosidase A, 66/235
Gaucher disease, 10/309, 566
Gerhardt syndrome, 50/219
Gerstmann-Sträussler-Scheinker disease, 56/554
giant axonal neuropathy, 60/82
glioblastomatosis, 14/496
globoid cell leukodystrophy, 66/204
glutamate dehydrogenase, 60/564
glycosaminoglycanosis, 10/456, 60/176
glycosaminoglycanosis type IVB, 66/268
GM1 gangliosidosis, 10/466, 480
GM2 gangliosidosis, 10/297, 388
Goldenhar syndrome, 31/246
Goltz-Gorlin syndrome, 31/288
grand mal on awakening, 73/178
Gruber syndrome, 31/242
Guam amyotrophic lateral sclerosis, 49/170, 59/262
Guam Parkinson dementia, 49/170
Hallervorden-Spatz syndrome, 66/713

Hartnup disease, 7/577, 29/153
headache, 5/258
hereditary amyloid polyneuropathy, 21/129, 41/435, 60/106
hereditary cerebello-olivary atrophy (Holmes), 21/405
hereditary cylindric spirals myopathy, 62/348
hereditary deafness, 50/218
hereditary dystonic paraplegia, 59/346
hereditary essential myoclonus, 42/235
hereditary fructose intolerance, 29/255
hereditary hemorrhagic telangiectasia (Osler), 38/258
hereditary motor and sensory neuropathy type I, 21/272
hereditary motor and sensory neuropathy type II, 21/272
hereditary multiple recurrent mononeuropathy, 51/559, 60/66
hereditary neuropathy, 60/265
hereditary olivopontocerebellar atrophy (Menzel), 21/435
hereditary orotic aciduria, 29/245
hereditary paroxysmal dystonic choreoathetotis, 49/351
hereditary paroxysmal kinesigenic choreoathetosis, 49/354
hereditary progressive diurnal fluctuating dystonia, 49/536
hereditary sensory and autonomic neuropathy type II, 60/11
hereditary sensory and autonomic neuropathy type III, 21/108, 60/24, 75/148
hereditary sensory and autonomic neuropathy type IV, 60/13
hereditary tyrosinemia, 29/216
heterogeneity, 30/89
heterotopia, 30/503
histidinemia, 29/197
holoprosencephaly, 30/156, 458, 42/33
Horner syndrome, 30/100
Huntington chorea, 6/380, 49/273
hydrocephalus, 50/289
hydromyelia, 50/427
hyperglycinemia, 29/203
hyperkalemic periodic paralysis, 41/160, 165, 425, 62/463
hypertension, 75/246
hypertrophic interstitial neuropathy, 21/147
hypokalemic periodic paralysis, 41/151, 425
hypophosphatasia, 31/315
idiopathic partial epilepsy, 73/7
idiopathic scoliosis, 32/150, 42/52
Ig, 56/55
incontinentia pigmenti, 14/215, 31/241
infantile autism, 46/191
infantile Gaucher disease, 10/309
infantile spinal muscular atrophy, 22/82, 89, 91, 41/437, 59/54
iniencephaly, 30/266
intelligence, 3/303, 313
intracranial aneurysm, 30/160
Joseph-Machado disease, 42/263, 60/471
juvenile Gaucher disease, 10/309, 522
juvenile hereditary benign chorea, 49/346
juvenile myoclonic epilepsy, 73/165
Kearns-Sayre-Daroff-Shy syndrome, 62/312
Klein-Waardenburg syndrome, 43/53, 50/218
Klippel-Feil syndrome, 30/150, 162, 32/112-115
Klippel-Trénaunay syndrome, 14/520-524
Krabbe-Bartels disease, 14/528
kyphosis, 32/147
learning disability, 46/130
Lesch-Nyhan syndrome, 29/270-273
limb girdle syndrome, 41/419, 486, 62/181
lissencephaly, 30/486
Lowe syndrome, 31/301
Löwenberg-Hill leukodystrophy, 10/176
lysinuric protein intolerance, 29/211
macrencephaly, 30/158, 658
malignant hyperthermia, 38/550, 41/429
mania, 46/471
α-mannosidase, 66/332
maple syrup urine disease, 29/70
Marinesco-Sjögren syndrome, 21/557
membranous labyrinth, 30/408
mental deficiency, 14/528
metachromatic leukodystrophy, 10/46, 314, 568, 29/358, 42/491, 47/592, 60/125
microcephaly, 30/156, 507, 512
microphthalmia, 30/470
migraine, 5/42, 55, 258-268, 48/23, 27
mitochondrial disease, 62/498
mitochondrial myopathy, 41/431
Möbius syndrome, 30/156, 50/215
molecular, *see* Molecular genetics
motoneuron disease, 41/441
mucolipidosis, 66/384
mucolipidosis type I, 66/358
mucolipidosis type II, 66/384
multiple sclerosis, 9/85, 97, 47/289-313
myophosphorylase deficiency, 41/189, 427
myotonia congenita, 41/421
myotonic dystrophy, 41/421, 486, 62/209
myotubular myopathy, 41/431
nemaline myopathy, 41/432, 62/342

neurocutaneous melanosis, 14/524-526
neurofibromatosis, 30/150, 158, 50/366
neurofibromatosis type I, 14/488, 16/30, 30/150, 158, 31/10, 32/141, 68/287
neuronal ceroid lipofuscinosis, 72/134
Niemann-Pick disease, 10/282, 485, 66/134
Niemann-Pick disease type C, 66/143
noradrenalin, 75/245
normokalemic periodic paralysis, 41/162, 425
Norrie disease, 31/292
nuclear DNA defect, 66/395
oculopharyngeal muscular dystrophy, 62/296
olivopontocerebellar atrophy variant, 21/454
opticocochleodentate degeneration, 21/535
ornithine carbamoyltransferase deficiency, 29/93
orofaciodigital syndrome type I, 31/250
orthochromatic leukodystrophy, 10/109, 113, 42/501
osteitis deformans (Paget), 43/452
pain, 75/325
pallidal degeneration, 49/446
pallidoluysian atrophy, 6/636
pallidoluysiodentate degeneration, 49/446
pallidoluysionigral degeneration, 49/446
pallidonigral degeneration, 49/446
paramyotonia congenita, 41/423
Parkinson dementia, 49/170
Parkinson disease, 49/129
Pelizaeus-Merzbacher disease, 10/154, 162, 176
periodic arthralgia, 42/301
periodic paralysis, 41/425
Peutz-Jeghers syndrome, 14/526-528, 43/41
phakomatosis, 14/6-8, 50, 482, 30/158-160, 31/30
phenylketonuria, 29/29
pheochromocytoma, 14/492
phosphofructokinase deficiency, 41/191, 427, 62/488
Pick disease, 46/243
platybasia, 30/151
pleiotropy, 13/25, 30/89
porencephaly, 50/358
porphyria, 60/119
progressive diaphyseal dysplasia, 38/400
progressive external ophthalmoplegia, 22/185, 41/420, 43/137, 62/295, 312
progressive myoclonus epilepsy, 27/173
propionic acidemia, 29/209
protective protein deficiency, 66/271
psychomotor disorder, 4/445
pyruvate dehydrogenase complex, 66/413
Refsum disease, 21/184-187, 41/434, 60/233, 66/487, 494
renal glycosuria, 29/184
retinoblastoma, 42/468, 50/586
Rett syndrome, 63/437
Riley-Smith syndrome, 14/76, 483
Salla disease, 66/364
scapuloperoneal spinal muscular atrophy, 22/75
schizoidism, 43/209
schizophrenia, 46/498
scoliosis, 30/162
septo-optic dysplasia, 30/321
sex chromatin, 13/24
Shy-Drager syndrome, 38/234, 75/245
Sjögren-Larsson syndrome, 13/470, 22/475
Smith-Lemli-Opitz syndrome, 66/592
sphingolipidosis, 10/556-587
spina bifida, 30/163, 32/533, 547, 548
spinal muscular atrophy, 41/440
spongy cerebral degeneration, 10/574, 66/664
Stilling-Türk-Duane syndrome, 50/213
striatopallidodentate calcification, 42/534
Sturge-Weber syndrome, 14/238, 507, 30/159, 31/19, 50/377
syringomyelia, 30/164, 32/258, 50/427, 451
telangiectasia, 14/520
temporal lobe epilepsy, 73/68
thanatophoric dwarfism, 31/274
thyrotoxic periodic paralysis, 62/461
trichopoliodystrophy, 29/279
trigeminal neuralgia, 5/297
tuberous sclerosis, 14/493-496, 16/30, 30/159, 31/5, 67/6
twin method, 15/431
tyrosyluria, 29/216
unipolar affective disorder, 43/209, 46/472
Unverricht-Lundborg progressive myoclonus epilepsy, 27/173
vagus nerve agenesis, 50/219
Van Buchem disease, 38/410
vertebral canal stenosis, 50/474
vestibular neurinoma, 68/435
Von Hippel-Lindau disease, 14/242, 502-504, 16/31, 30/159, 31/16
Wilson disease, 27/406-408, 49/235
Wohlfart-Kugelberg-Welander disease, 22/69, 75, 41/439, 59/82-85
Wolf-Hirschhorn syndrome, 30/508
xeroderma pigmentosum, 14/529, 530, 31/291
Geniculate ganglion
ageusia, 24/15
anatomy, 48/487
crocodile tear syndrome, 2/70, 123
facial nerve, 48/488
intermedius neuralgia, 5/338
Geniculate ganglionectomy

intermedius neuralgia, 48/491
Geniculate herpes
Ramsay Hunt syndrome, 5/209
Geniculate neuralgia, *see* Intermedius neuralgia
Geniculate nucleus
lateral, *see* Lateral geniculate nucleus
medical, *see* Medial geniculate nucleus
Geniculate otalgia, *see* Intermedius neuralgia
Geniculocalcarine tract
see also Optic tract
bundle, 2/557
bundle course, 2/534
cerebrovascular disease, 55/117
injuries, *see* Optic tract injury
macular representation, 2/558
Meyer-Archambault loop, 2/534, 557
optic neurofibromatosis, 68/79
pupil, 1/621
retinotopic organization, 2/557
temporal lobectomy, 2/558
visual fiber, 2/557
Geniculocalcarine tract injury
brain injury, 24/46
homonymous hemianopia, 24/46
partial lesion, 24/46
penetrating head injury, 24/46
Genital abnormality
Saldino-Noonan syndrome, 31/243
WARG syndrome, 50/581
Genital malformation
caudal aplasia, 50/511
Smith-Lemli-Opitz syndrome, 66/586
Genital tract
nerve, 74/113
Genital tract tumor
brain metastasis, 18/205
Genitofemoral nerve
compression neuropathy, 51/108
entrapment, 7/314
topographical diagnosis, 2/39
Genitourinary apparatus
trisomy 8, 50/563
Gennari-Bamberger stripe
visual cortex, 2/538
Genodermatosis
neuroaxonal leukodystrophy, 49/406
Genotype
Duchenne muscular dystrophy, 62/126
oligosaccharidosis, 66/329
pyruvate dehydrogenase complex deficiency, 66/416
Gentamicin
acoustic neuropathy, 64/9
bacterial meningitis, 52/10, 12
deafness, 65/483
drug induced polyneuropathy, 51/295
gram-negative bacillary meningitis, 52/106
neurobrucellosis, 52/596
neurotoxin, 50/219
staphylococcal meningitis, 33/72-74
toxic neuropathy, 64/9
ventriculitis, 52/120
Gentamicin intoxication
symptom, 64/9
Gentamycin, *see* Gentamicin
Genu valgum
acrocephalosyndactyly type II, 38/422
congenital spondyloepiphyseal dysplasia, 43/478
glycosaminoglycanosis type IVB, 66/306
homocystinuria, 55/329
neurofibromatosis type I, 50/369
Genu varum
congenital spondyloepiphyseal dysplasia, 43/478
neurofibromatosis type I, 50/369
Schwartz-Lelek syndrome, 31/257
Gerhardt syndrome
see also Laryngeal abductor paralysis
bilateral medial rectus palsy, 2/301
genetics, 50/219
mental deficiency, 50/219
recurrent laryngeal nerve palsy, 50/219
stridor, 50/219
Gerlach, J., 1/46
Germ cell
neural tube, 50/5
Germ cell cyst
organic acid metabolism, 66/641
Germ cell tumor
chemotherapy, 68/250
classification, 68/230
diagnosis, 68/239
epidemiology, 68/231
histogenesis, 68/230
leptomeningeal metastasis, 71/662
location, 68/232
metastasis, 71/662
nongerminomatous, *see* Nongerminomatous germ cell tumor
pathology, 68/234
staging, 68/243
surgery, 68/243
survey, 68/229
Germ layer
phakomatosis, 14/102
Germ line therapy
lysosomal storage disease, 66/113

Germany
neurology, 1/8, 13
neuropsychiatry, 1/8
Germine acetate
Eaton-Lambert myasthenic syndrome, 41/360
myasthenia gravis, 41/132
Germinoma
brain biopsy, 16/723
choroid plexus carcinoma, 67/177
hypothalamic tumor, 68/71
imaging, 67/178
incidence, 68/232
location, 68/232
management, 68/251
optic chiasm compression, 68/75
pathology, 68/235
radiotherapy, 68/245-247
survey, 68/235
thalamic tumor, 68/65
Gerstmann-Sträussler disease, *see* Gerstmann-Sträussler-Scheinker disease
Gerstmann-Sträussler-Scheinker disease, 60/619-631
age, 56/553
amyloid, 60/629
amyloid plaque, 56/554
amyotrophy, 21/553
animal model, 56/585
animal transmission, 56/555
areflexia, 56/553, 60/660
ataxia, 56/553
atypical, 60/623
brain amyloid angiopathy, 56/555
cerebellar ataxia, 56/553
choreoathetosis, 60/663
classification question, 60/630
clinical features, 56/553
computer assisted tomography, 56/554, 60/671
Creutzfeldt-Jakob disease, 46/295
CSF, 56/554
deafness, 56/553
dementia, 21/553, 56/553, 60/622
dysarthria, 56/553
dysphagia, 56/553
EMG, 56/554
epidemiology, 56/554
extrapyramidal rigidity, 56/553
extrapyramidal sign, 60/622
family, 60/522-627
features, 21/455
gait ataxia, 60/622
genetics, 56/554
history, 60/619
hypoesthesia, 60/661
hyporeflexia, 60/622
immunochemistry, 60/629
immunology, 56/556
kuru, 56/554
laboratory study, 60/627
leukoencephalopathy, 56/555
myoclonus, 56/553, 60/662
neurofibrillary tangle, 56/555, 60/623
neuronal argyrophilic dystrophy, 21/64
neuropathology, 56/554
nuclear magnetic resonance, 60/673
ocular dysmetria, 56/553
onset, 60/622
optic atrophy, 56/553
other prion disease, 60/628
pallidal degeneration, 49/454
pallidoluysian degeneration, 49/454
pallidoluysionigral degeneration, 49/454, 459
prion disease, 56/543, 553
prion protein gene, 60/629
pseudobulbar sign, 60/622
pyramidal sign, 60/622
Refsum disease, 60/661
senile plaque, 21/553, 56/554, 569
sensory disturbance, 56/553
spastic paraplegia, 59/436
spasticity, 60/661
spongiform encephalopathy, 56/555
transmissibility, 60/628
tremor, 60/661
upper eyelid retraction, 60/656
white matter demyelination, 60/623
without cerebellar involvement, 60/623
Gerstmann syndrome
see also Finger agnosia
acalculia, 1/107, 2/686, 4/189, 222, 45/70, 474
agraphia, 1/107, 2/686, 4/95, 146, 222, 45/70, 461
alexia, 4/128, 45/70, 461
aphasia, 45/383
apraxia, 2/691, 4/53
associative feature, 2/692
autotopagnosia, 4/222
body scheme disorder, 1/107, 2/668, 4/222, 297, 398, 401, 45/383
calculation defect, 45/70
central alexia, 45/440
clumsiness, 46/160
constructional apraxia, 2/690-692, 4/78, 45/500
definition, 2/588
developmental dyslexia, 4/401, 46/117-119
dyslexia, 2/691, 694, 46/127-129, 131
fiction, 2/691

finger agnosia, 2/604, 686
hemianopic inattention, 2/691, 695
left hemisphere, 2/687
mental deficiency, 4/389
middle cerebral artery syndrome, 53/363
occipital lobe tumor, 17/325
pain agnosia, 2/691
parietal lobe syndrome, 2/686, 45/69
parietal lobe tumor, 17/302
right left disorientation, 1/107, 2/687, 45/70, 383
symptom, 53/363
Turner syndrome, 50/545
visual object agnosia, 2/656
Geschwind syndrome
temporal lobe epilepsy, 73/84
Gestagen
headache, 48/437
migraine, 48/437
Gestalt theory
agnosia, 3/12, 4/27
Gestaltzerfall
visual agnosia, 2/655
visual perception, 2/662
Geste antagonistique, *see* Counter pressure phenomenon
Gesture
alexia, 4/96
cerebral dominance, 4/5
motor activity, 4/445
Geyelin-Penfield calcification
striatopallidodentate calcification, 6/720
Ghent, Belgium
neurology, 1/10
Giant axonal change
carbon disulfide polyneuropathy, 51/279
Giant axonal neuropathy, 60/75-85
bulbar paralysis, 60/76
chronic axonal neuropathy, 51/531
clinical features, 60/75
definition, 60/75
differential diagnosis, 60/82
electrophysiologic study, 60/77, 84
epidemiology, 60/75
genetics, 60/82
hexacarbon neuropathy, 51/278, 60/82
intermediate neurofilament, 60/84
intra-axonal granular mass, 60/85
multiple neuronal system degeneration, 60/82
muscle biopsy, 60/79
nerve biopsy, 60/79
neuropathology, 60/82
paracrystalline body, 60/85
pathogenetic consideration, 60/84
spinocerebellar degeneration, 60/82
toxic polyneuropathy, 60/82
trichopoliodystrophy, 60/82
Giant axonal swelling
acetonylacetone intoxication, 64/85, 88
chorea-acanthocytosis, 63/283
Giant brain aneurysm
headache, 5/147
Giant cell arteritis
see also Polymyalgia rheumatica *and* Temporal arteritis
aortitis, 63/52
blindness, 71/194
classification, 71/4, 199
CSF, 54/198
diagnosis, 71/198
headache, 71/193
hypothyroidism, 70/101
jaw claudication, 71/193
laboratory, 71/197
polymyalgia rheumatica, 55/346, 71/193
prognosis, 71/199
retinal artery occlusion, 55/112
survey, 71/191
symptom, 71/192
treatment, 71/199
vasculitis, 71/8
Giant cell astrocytoma
giant cell tumor, 14/46
neurofibromatosis, 68/293
subependymal, *see* Subependymal giant cell astrocytoma
tuberous sclerosis, 14/46
tuberous sclerosis complex, 68/293
Giant cell glioblastoma
tissue culture, 17/63
Giant cell infiltration
angiopathic polyneuropathy, 51/445
Giant cell osteoma, *see* Benign osteoblastoma
Giant cell polymyositis
myasthenia gravis, 62/402
Giant cell tumor
see also Tumeur à myéloplaxe
aneurysmal bone cyst, 19/301
central, *see* Aneurysmal bone cyst
cranial vault tumor, 17/109
giant cell astrocytoma, 14/46
glioblastoma, 17/63
incidence, 68/530
nonrecurrent nonhereditary multiple cranial neuropathy, 51/571
pituitary adenoma, 17/413
sacrum, 67/195

skull base tumor, 17/160
spinal cord tumor, 68/516
spinal epidural tumor, 20/103, 128
subependymal, *see* Subependymal giant cell tumor
vertebral column, 20/443
Giant follicular lymphadenopathy
malignant change, 8/144
polyneuropathy, 8/144
Giant osteoid osteoma, *see* Benign osteoblastoma
Giant plasmatic glial cell
subacute sclerosing panencephalitis, 10/7
Giant pyramidal cell
Meynert, *see* Meynert giant pyramidal cell
Giant tumor (Elsberg)
spinal ependymoma, 20/376
Gibbs classification
epilepsy, 15/7
Gibson, E.H., 8/188
Giddiness, *see* Dizziness
Gifford progeria
Cockayne-Neill-Dingwall disease, 13/323
Gigantism
acromegaly, 2/454
adenohypophyseal syndrome, 2/454
brain, *see* Cerebral gigantism
cerebral, *see* Cerebral gigantism
Klinefelter variant XXYY, 31/484
partial, *see* Partial gigantism
Gignoux syndrome
Schmidt syndrome, 2/100
Gilchrist disease, *see* North American blastomycosis
Gilles de la Tourette syndrome
see also Jumpers of the Maine *and* Tic
agraphia, 45/466
Babinski sign, 49/631
behavior, 49/627
case series, 6/789
childhood chronic multiple tic, 49/627, 630, 635
clinical features, 49/628
coprolalia, 1/291, 6/788, 42/221, 45/468, 49/627, 629
copropraxia, 49/627
definition, 49/627
differential diagnosis, 49/635
dopamine β-mono-oxygenase, 42/222
echolalia, 6/788, 42/221, 49/627, 629
echopraxia, 6/788, 49/627
EEG, 6/790, 49/631
epidemiology, 49/632
erythrocyte choline level, 49/633
haloperidol, 49/632
history, 49/628
homovanillic acid, 49/633
hyperactivity, 46/181
hyperfunction, 49/632
hypoxanthine phosphoribosyltransferase, 42/222
incoordination, 49/631
intelligence quotient, 49/629
involuntary eye movement, 49/632
latah, 49/629, 635
mesencephalic gray matter, 49/632
metenkephalin, 49/633
migraine, 49/631
motor tic, 6/787, 49/628
multiple tics, 1/291
myriachit, 49/629, 635
neuropsychology, 49/630
palilalia, 6/788
pathogenesis, 49/632
pathophysiology, 49/632
phenothiazine derivative, 6/791
readiness potential, 49/74, 632
self-mutilation, 42/221
somnambulism, 49/631
startle myoclonus, 49/636
suppression, 49/631
thermoregulation, 49/631
tic, 42/221
treatment, 6/792, 49/634
vocalization, 6/788, 49/629
Gillespie syndrome
aniridia, 60/655
computer assisted tomography, 60/671
hereditary congenital cerebellar atrophy, 60/289
mental deficiency, 42/123
pulmonary stenosis, 60/668
pupillary disorder, 60/655
Gilmour disease, *see* Temporal arteritis
Ginger Jake paralysis, *see* Tri-*o*-cresyl phosphate intoxication
Gingiva hyperplasia
phenytoin intoxication, 42/641
Gingival fibroma
tuberous sclerosis complex, 68/292
Gingivitis
Melkersson-Rosenthal syndrome, 8/213
Gingivostomatitis
herpes simplex virus, 56/8
Giroux-Barbeau syndrome, *see* Erythrokeratodermia ataxia syndrome
Giving way of legs
astasia, 1/347
cataplexy, 1/347
diagnostic problem, 1/347
drop attack, 1/347
Gland

adrenal, *see* Adrenal gland
lacrimal, *see* Lacrimal gland
parathyroid, *see* Parathyroid gland
pineal, *see* Pineal gland
pituitary, *see* Pituitary gland
sweat, *see* Sweat gland
thymus, *see* Thymus gland
thyroid, *see* Thyroid gland
Glandula carotica, *see* Carotid body
Glandular fever, *see* Infectious mononucleosis
Glandular lipoma, *see* Hibernoma
Glandular tympanica
glomus jugulare, 18/435
Glanzmann disease, *see* Glanzmann thrombasthenia
Glanzmann thrombasthenia
brain hemorrhage, 63/325
coagulopathy, 63/325
epidural hematoma, 63/325
Glasgow Coma Scale
brain embolism, 53/209
brain infarction, 53/209
brain injury, 57/127
brain injury prognosis, 24/677
cerebrovascular disease, 53/209
child, 57/127
coma, 57/381
component, 53/209
consciousness, 57/132
eye response, 57/126
Glasgow Outcome Scale, 57/382
good recovery, 57/115
head injury, 45/188, 57/123, 132, 330
head injury outcome, 57/382
intracranial hemorrhage, 53/209
motor response, 57/126
pediatric head injury, 57/330
prognostic significance, 46/611
recovery, 57/382
stroke, 53/210
sum score, 57/126, 382
transient ischemic attack, 53/209
traumatic intracranial hematoma outcome, 57/284
verbal response, 57/126
Glasgow Outcome Scale
5 categories, 57/368
8 categories, 57/368
coma, 57/370
death, 57/132
Glasgow Coma Scale, 57/382
good recovery, 57/132
head injury, 57/132, 368
head injury outcome, 57/114, 132, 374
interrater agreement, 57/368
moderate disability, 57/115, 132
persistent vegetative state, 57/115
severe disability, 57/115, 132
subjective complaint, 57/374
vegetative state, 57/132
Glasgow, United Kingdom
neurology, 1/7
Glauber salt, *see* Sulfate sodium
Glaucoma
acrocephalosyndactyly type I, 43/318
altitudinal hemianopia, 2/563
altitudinal scotoma, 2/563
ARG syndrome, 50/581
buphthalmos, 14/398
carotid cavernous fistula, 24/413, 416
cerebrohepatorenal syndrome, 43/339
cluster headache, 48/226
congenital, *see* Congenital glaucoma
facial angioma, 14/61
glyceryl trinitrate intoxication, 37/453
glycosaminoglycanosis type IVB, 66/307
glycosaminoglycanosis type VI, 66/308
guanethidine, 37/436
headache, 5/206, 48/6, 34
hemihypertrophy, 14/398
hereditary amyloid polyneuropathy, 42/523
hereditary vitreoretinal degeneration, 13/274
homocystinuria, 42/555
hyaloidoretinal degeneration (Wagner), 13/37, 274
Klein-Waardenburg syndrome, 13/462
Lowe syndrome, 42/606
multiple neurofibromatosis, 14/398
neurofibromatosis, 14/398, 635
neurofibromatosis type I, 14/635
pontine, *see* Pontine glaucoma
primary pigmentary retinal degeneration, 13/178
pseudoprogeria Hallermann-Streiff syndrome, 43/405
Rieger syndrome, 43/470
ring scotoma, 2/565
Robin syndrome, 43/472
Seidel scotoma, 2/565
stage, 14/647
Sturge-Weber syndrome, 14/223, 227, 234, 644, 43/48, 50/376
temporal arteritis, 5/120
Von Hippel-Lindau disease, 14/647, 50/375
GLB1 gene
β-galactosidase, 66/282
glycosaminoglycanosis type IVB, 66/282
Glia
Alzheimer type 1, *see* Alzheimer type 1 glia
Alzheimer type 2, *see* Alzheimer type 2 glia

carbonate dehydratase, 9/14
cerebroside, 9/6
cholesterol, 9/6
ganglioside, 9/4, 17
lipid, 9/6
liver, *see* Liver glia
mercury intoxication, 64/381
neuronal ceroid lipofuscinosis, 10/659
phosphatidylethanolamine, 9/6
phosphatidylinositol, 9/6
phosphatidylserine, 9/6
plasmalogen, 9/6
porphyrin metabolism, 9/14
potassium, 28/473
white matter, 9/6
Glia cell
brain maturation, 30/71
Creutzfeldt-Jakob disease, 6/750
giant plasmatic, *see* Giant plasmatic glial cell
multiple sclerosis plaque, 9/37
Glia regulation
axon, 51/43
Glial dystrophy
age, 21/58
Alexander disease, 21/67
Alpers disease, 21/68
astroglial type, 21/66
classification, 21/61, 65
combined type, 21/68
Creutzfeldt-Jakob disease, 21/67
demyelinating type, 21/66
gliofibrillary type, 21/67
glioneuronal type, 21/67
infantile spongy dystrophy, 21/68
main senile morphology type, 21/54
metachromatic leukodystrophy, 21/65
morphology, 21/47
oligodendroglial dystrophy, 21/65
Pelizaeus-Merzbacher disease, 21/66
presenile, *see* Presenile glial dystrophy
Rosenthal fiber, 21/67
senile, *see* Senile glial dystrophy
spongy, *see* Spongy glial dystrophy
spongy cerebral degeneration, 21/68
vasoglial type, 21/68
Wilson disease, 21/68
Glial heterotopia
diffuse glioblastosis, 14/145
neurofibromatosis, 14/37
parietal encephalocele, 50/110
Glial membrane
lipid, 9/18
Glial pigment
brown, *see* Brown glial pigment
neuroaxonal dystrophy, 49/392
Glial protein
CSF, *see* CSF glial protein
Joseph-Machado disease, 60/472
Glial septum
aqueduct stenosis, 50/302
Glioblastoma
age, 68/98
apraxia, 45/426
biopsy, 67/226
brachytherapy, 68/107
bromouridine, 18/491
capillary, 67/73
chemotherapy, 18/65
citrovorum factor, 18/65
distribution, 68/90
doughnut sign, 18/55
family, 68/88
giant cell, *see* Giant cell glioblastoma
giant cell tumor, 17/63
hemorrhage, 68/95
histology, 68/93
infant, 42/733
laboratory diagnosis, 68/97
metastasis, 68/92
monstrocellular sarcoma, 18/51
multiple neurofibromatosis, 43/35
oligodendroglioma, 72/357
optic, 42/733
parietal lobe tumor, 17/297
pontine, 42/332
posterior fossa tumor, 18/397
prognosis, 68/97
spinal, *see* Spinal glioblastoma
spinal cord compression, 19/361
spinal glioma, 19/359, 361, 20/335
surgery, 68/101
survival, 68/5, 102, 112
terminology, 67/225
therapy, 68/99
Glioblastoma multiforme, 18/49-68
see also Astrocytoma grade IV *and* High grade glioma
age, 18/50, 53
brain angiography, 18/59
brain biopsy, 16/713
brain edema, 18/53
brain scanning, 16/685
brain scintigraphy, 18/55
brain stem, 18/54, 397
bromouridine, 18/65
carmustine, 18/65

cerebral fibrosarcoma, 42/730
chemotherapy, 18/65
child, 18/54
chromosome, 67/50
chromosome abnormality, 67/45
classification, 18/3, 20/335
clinical features, 31/48
congenital, 31/47-49
corpus callosum, 18/51
corpus callosum tumor, 17/499
corticosteroid, 18/65
CSF, 18/54
CSF cytology, 16/392
cytostatic treatment, 18/65
diagnosis, 18/54
dura mater, 18/51
EEG, 18/58
epilepsy, 18/20
headache, 18/53
histology, 18/52
hyperbaric oxygen, 18/65
hyperthermia, 18/64
hypervascular, 68/270
hypoglycorrhachia, 18/54
hypothalamic tumor, 68/71
hypothermia, 18/64
immunotherapy, 18/66
incidence, 68/87
location, 31/47, 48
macroscopic aspect, 18/50
malignant astrocytoma, 18/15
metastasis, 18/51
methotrexate, 18/65
microscopy, 16/16
multicentric, 18/52
needle biopsy, 18/60
neuropathology, 18/50
nonhereditary, 67/43
optic nerve tumor, 17/353
parietal lobe tumor, 17/297
PEG, 18/60
preference site, 18/50
prevalence, 18/50
psychiatric disorder, 18/53
radiotherapy, 18/63, 489
recurrence, 18/67
secondary, 18/11
sensitization, 18/64
sex ratio, 18/50, 53
spinal, *see* Spinal glioblastoma
subarachnoid hemorrhage, 12/119
surgical treatment, 18/61
survey, 68/87, 270
survival, 18/62, 66, 68/113
symptomatology, 18/52-56
terminology, 18/1
thalamic tumor, 68/65
tissue culture, 17/63
treatment, 18/60, 31/48
twins, 18/52
Glioblastomatosis
familial gliomatosis, *see* Familial gliomatosis-glioblastomatosis
genetics, 14/496
temporal lobe, 14/496
Glioblastosis
diffuse, *see* Diffuse glioblastosis
Glioblastosis cerebri, *see* Diffuse glioblastosis
Gliocytoma embryonale, *see* Cerebellar astrocytoma
Gliodystrophic encephalopathy
hepatogenic, *see* Hepatogenic gliodystrophic encephalopathy
Glioependymoma, *see* Ependymoma
Glioepithelioma, *see* Ependymoma
Gliofibrillogenesis
astrocyte, 9/617, 625
subacute combined degeneration, 9/617, 625
Gliofibroma
histopathology, 68/379
infratentorial, 68/382
spinal cord tumor, 68/383
supratentorial, 68/382
Gliogenesis
brain, 67/10
Glioma, *see* Astrocytoma
Glioma antigen sensitized cytotoxic T-lymphocyte, 67/295
Glioma durum, *see* Astrocytoma
Glioma muqueux, *see* Spongioblastoma
Glioma polyposis syndrome
autosomal recessive, 42/735
cerebellar medulloblastoma, 42/742
chromosome, 67/55
CNS tumor, 30/101, 42/735, 67/54
ependymoma, 42/735
neurofibromatosis, 68/299
secondary pigmentary retinal degeneration, 60/737
spinal medulloblastoma, 42/735
Gliomatosis
diffuse leptomeningeal, *see* Diffuse leptomeningeal gliomatosis
familial, *see* Familial gliomatosis-glioblastomatosis
meningeal, *see* Meningeal gliomatosis
meningoencephalic, *see* Meningoencephalic

gliomatosis
neurofibromatosis, 14/35, 102
Gliomatosis cerebri
dementia, 46/386
neurofibromatosis type I, 14/143, 151
Gliomatous hypertrophy, *see* Diffuse glioblastosis
Gliomyxoma, *see* Spongioblastoma
Gliosarcoma
radiation induced, 68/393
Gliosis, *see* Astrocytosis
Gliosome, 18/84
Gliotoxicity
antibody, 9/519
Glisson sling
cervical spondylosis, 7/464
cervical vertebral column injury, 25/291
Global amnesia
hippocampus, 45/44
migraine, 48/162
migrainous transient, *see* Migrainous transient global amnesia
rehabilitation, 46/622
transient, *see* Transient global amnesia
Global aphasia
agraphia, 45/460
brain embolism, 53/215
brain infarction, 53/215
cerebrovascular disease, 53/215
computer assisted tomography, 45/312
hemispherectomy, 45/312
intracranial hemorrhage, 53/215
perisylvian region, 45/20
prognosis, 45/323
radioisotope diagnosis, 45/298
transient ischemic attack, 53/215
Wernicke aphasia, 4/98
Global associated movement, *see* Synkinesis
Globefish intoxication, *see* Puffer fish intoxication
Globoid cell
acid phosphatase, 10/79, 303
ceramide glucose, 10/87
diffuse sclerosis, 10/68-70, 74
enzyme histochemistry, 10/79, 303
galactocerebroside, 10/87
Gaucher disease, 10/89
histochemistry, 10/79-84
histogenesis, 10/81
lipid hexosamine, 10/89
multinucleated, *see* Multinucleated globoid cell
origin, 10/70
pathology, 66/195
review, 10/67-90
ultrastructure, 10/84
Globoid cell leukodystrophy, 10/67-90, 42/489-491, 66/187-205
acid glycolipid, 10/79, 84
acid phosphatase, 10/303
adult form, 59/355, 66/191
anatomic pathology, 10/73
animal experiment, 10/351
animal model, 47/599, 66/202-204
astrocyte, 10/76
atypical case, 10/72
autosomal recessive, 29/356, 51/373
biochemistry, 10/71, 79, 85, 145, 304-306, 350, 29/356
birth incidence, 42/490
blood vessel, 10/77
bone marrow transplantation, 66/87, 91
brain biopsy, 10/685
cerebron, 10/304
cerebroside, 10/71, 304, 327, 29/356
cerebroside-sulfatide ratio, 10/71, 84
cerebroside sulfotransferase, 10/21, 87, 306, 29/356
chromosome, 10/71
chromosome 18 abnormality, 10/71
chronic inflammatory demyelinating polyradiculoneuropathy, 51/530
classification, 9/472, 10/18-21, 28, 37, 66/5
clinical features, 10/71-73, 349, 51/373
clinical presentation, 66/188
clinical variant, 66/189
congenital fiber type disproportion, 62/355
CSF, 10/73, 350, 42/489
CSF protein, 10/304, 59/355
cytoside, 10/304
deafness, 51/373
demyelinating neuropathy, 47/622
demyelination, 10/20, 303
diagnosis, 10/304, 66/204
dog, 10/147
enzyme, 10/146, 29/350
enzyme change, 10/77-81
enzyme deficiency polyneuropathy, 51/369
epilepsy, 15/374, 425, 72/222
epithelioid cell, 10/69, 74, 79-84
experimental, 10/71, 87
familial spastic paraplegia, 59/352
free body, 10/75
galactoceramidase deficiency, 59/355
galactocerebrosidase pseudodeficiency, 59/356
galactocerebroside-β-galactosidase, 10/147, 47/598
galactocerebroside sulfotransferase, 10/146
galactosylceramidase, 66/60

galactosylceramide, 42/489
Gaucher cell, 10/69, 42/437
Gaucher disease, 10/528
gene frequency, 42/490
genetics, 66/204
globoid body, 10/74, 29/356
glucose-6-phosphate dehydrogenase, 10/303
β-glucosidase deficiency, 10/306
β-glucuronidase, 10/21, 86, 147, 304, 306
glycolipid, 10/79, 84
gray matter change, 10/85
heterozygote detection, 42/490
heterozygote frequency, 42/490
hexosamine, 10/85
histology, 10/350
histopathology, 10/74
history, 10/4, 8, 67-69
hypertonia, 10/302
hypotonia, 40/336
infantile, 66/190
infantile optic atrophy, 42/408
infantile spasm, 51/373
juvenile, 66/191
kerasin, 10/304
lactate dehydrogenase, 10/303
leptomeninges, 10/77
leukocytosis, 10/73
leukoencephalopathy, 47/598
lipid concentration, 10/145
lipid metabolic disorder, 10/302-306
lipid metabolism, 10/84-87
macrophage inclusion, 51/373
mental deficiency, 46/55
microglia, 10/77
multinucleated globoid cell, 66/10
muscle weakness, 40/336
myelin, 10/145
myelin body, 10/69, 74
myelin change, 10/74
myoclonia, 10/72
myoclonic jerk, 10/303
nerve cell, 10/77
nerve conduction, 51/373
nerve conduction velocity, 42/489
neuropathology, 10/37, 21/50, 51/373
nuclear magnetic resonance, 59/355, 66/193
oligodendrocyte, 10/77
opisthotonos, 10/303, 29/356, 42/489
optic atrophy, 10/302, 350, 13/83, 29/356
pathogenesis, 10/88
pathology, 10/73, 84, 14/405, 66/194-198
pathophysiology, 66/201
Pelizaeus-Merzbacher disease, 10/4
peripheral disease, 10/77, 351
phospholipid, 10/85
poikilothermia, 10/303
prenatal diagnosis, 42/491
prognosis, 10/306
progressive bulbar palsy, 59/375
segmental demyelination, 51/71, 373
serum protein change, 10/304
sex ratio, 10/70
spasticity, 10/303
sphingolipid, 10/89, 351, 29/350
spinocerebellar syndrome, 51/373
spongy glial dystrophy, 21/66
startle reaction, 10/72, 303, 51/373
sulfatide, 10/21, 304
survey, 66/187
temperature spiking, 10/73, 29/356
treatment, 10/306, 66/205
ultrastructure, 10/81, 89
uninucleated epithelioid cell, 66/10
urinary finding, 10/73
visceral involvement, 10/78
Von Hippel-Lindau disease, 14/7
white matter, 10/85
white matter degeneration, 42/490
Globoside
Fabry disease, 10/544, 581
GM2 gangliosidosis, 10/414, 417
GM2 gangliosidosis type III, 10/298
Globular neuropathy, *see* Hereditary multiple recurrent mononeuropathy
Globulin
antilymphocytic, *see* Lymphocyte antibody
antithymocyte, *see* Thymocyte antibody
α-Globulin
Alzheimer disease, 46/265
α-2-Globulin
GM2 gangliosidosis, 10/389
Niemann-Pick disease, 10/284
Sydenham chorea, 6/426
temporal arteritis, 55/342
β-2-Globulin
CSF, *see* CSF β-2-globulin
γ-Globulin, *see* Ig
γ-2-Globulin
myelinotoxic factor, 9/143
Globus hystericus
stylohyoid syndrome, 48/502
symptomatology, 46/574
Globus pallidus
see also Paleostriatum
afferent connection, 49/5
akinesia, 49/68

anatomy, 6/19-30
anterior choroidal artery, 49/467
bilirubin encephalopathy, 27/420
carbon monoxide encephalopathy, 49/472
carbon monoxide intoxication, 64/33
cell, 6/28
chorea-acanthocytosis, 49/331
cytology, 49/5
demyelinating type, 6/714
dorsal, *see* Dorsal globus pallidus
epilepsy, *see* Infantile spasm
exogenous lesion, 6/633
external, *see* External pallidum
function, 6/109, 634
glutamic acid, 49/5
Hallervorden-Spatz syndrome, 49/396, 66/713
histochemistry, 6/28
Huntington chorea, 6/3, 337, 49/319
infantile neuroaxonal dystrophy, 49/396
kernicterus, 6/497
lenticulostriate artery, 49/467
lesion, *see* Pallidal necrosis, Pallidonigral necrosis *and* Pallidostriatal necrosis
manganese intoxication, 64/307
mesolimbic dopamine system, 49/50
motor function, 1/160
pallidal lesion, 6/633
Parkinson disease, 6/212
physiology, 49/22
rigidity, 49/71
survey, 6/632
vascular supply, 49/467
venous drainage, 49/467
ventral, *see* Ventral globus pallidus
Globus pallidus externus
hyperkinesia, 21/520, 526
lesion, 21/520, 526
Globus pallidus infarction
cyanide intoxication, 64/232
Globus pallidus internus
hypokinesia, 21/520, 526
lesion, 21/520, 526
rigidity, 21/520, 526
Glomangioma
tissue culture, 17/70
Glomerocytoma, *see* Glomus jugulare tumor
Glomerular filtration rate
renal insufficiency, 63/503
Glomeruli arteriosi intercarotici, *see* Carotid body
Glomerulonephritis
Behçet syndrome, 56/595
Freund adjuvant, 9/506
Galloway-Mowat syndrome, 43/431
hydrocephalus, 30/596
hypertensive encephalopathy, 54/212
Glomerulopathy
progressive, *see* Progressive glomerulopathy
Glomus
carotid, *see* Carotid body
Glomus caroticum, *see* Carotid body
Glomus jugulare
anatomy, 18/436
definition, 18/435
glandular tympanica, 18/435
history, 18/435
paraganglioma, 18/435, 42/762
synonym, 18/435
Glomus jugulare tumor, 18/435-452
Bean syndrome, 14/106
bulbar syndrome, 18/444
carotid body tumor, 18/438
cerebellar syndrome, 18/445
cerebral syndrome, 18/445
clinical features, 18/442
course, 18/448
cranial nerve, 18/443
definition, 18/435
diagnosis, 18/448
differential diagnosis, 18/448
early symptom, 18/442
extension, 18/441
histology, 18/439
history, 18/435
malignancy, 18/449
misdiagnosis, 18/437, 440
multiplicity, 18/438
neuroradiology, 18/446
neurosurgery, 18/450
noradrenalin, 18/442
paraganglioma, 18/435
paraganglioma nodosum, 18/493
parapharyngeal space syndrome, 18/445
pheochromocytoma, 18/440
prognosis, 18/448
radiotherapy, 18/451
remote symptom, 18/444
site, 18/437
skull base, 68/482
synonym, 18/435
treatment, 18/450
Glomus tumor
nerve tumor, 8/455-458
nonrecurrent nonhereditary multiple cranial neuropathy, 51/571
pain, 8/456
phakomatosis, 14/10

Glossal spasm
athetosis, 49/382
Glossina, *see* Tsetse fly
Glossitis granulomatosa
Melkersson-Rosenthal syndrome, 8/213
Glossodynia
tabetic, *see* Tabetic glossodynia
Glossolaryngeal hemiplegia, *see* Tapia syndrome
Glossopharyngeal breathing
respiratory dysfunction, 26/346
Glossopharyngeal nerve
abnormality, 30/410
anatomy, 2/76, 5/350
autonomic nervous system, 2/110
buccopharyngeal herpes, 7/483
congenital larynx stridor, 30/410
embryology, 30/398-400
injury, 24/179
neurinoma, 68/539
respiration, 1/655
subarachnoid hemorrhage, 55/18
vocal cord paralysis, 30/411
Glossopharyngeal nerve agenesis
cranial nerve agenesis, 50/219
dysphagia, 50/219
hereditary sensory and autonomic neuropathy type III, 50/219, 60/30
stridor, 50/219
Glossopharyngeal nerve palsy
carotid dissecting aneurysm, 54/272
petrobasilar suture syndrome, 2/313
Glossopharyngeal neuralgia, 5/350-359, 48/459-468
age, 48/460
autonomic nervous system, 75/298
blood pressure, 48/465, 467
bradycardia, 48/465, 467
carbamazepine, 5/357
cervical neurotomy, 5/357
cluster headache, 48/468
cranial neuralgia, 5/312, 316
differential diagnosis, 5/316, 48/468
dizziness, 48/466
drug treatment, 5/383
dysautonomia, 22/254
ECG, 48/466
elongated styloid process, 48/464
foramen lacerum, 48/464
frequency ratio, 48/460
history, 48/459
idiopathic form, 5/351
incidence, 5/272
intermedius neuralgia, 48/468
Jacobson nerve, 5/351
lateral, 48/460
limb clonus, 48/465
metabolic, 48/465
neoplasm, 48/464
neurinoma, 48/464, 68/540
neurofibroma, 48/464
neuropathy, 48/467
onset of attack, 48/461
otalgia, 5/209
otalgic type, 48/462
pain distribution, 48/462
paroxysm, 48/461
pathogenesis, 48/466
pathophysiology, 5/275
pharyngeal form, 48/462
pulse rate, 48/467
salivation, 48/466
sex ratio, 48/460
Sjögren syndrome, 71/74
sphenopalatine neuralgia, 48/468, 478
superior laryngeal neuralgia, 48/468
surgical treatment, 5/358, 361, 400
symptomatic form, 5/355
tabetic glossodynia, 48/468
therapy, 48/468
tonsillar fossa, 48/461
trauma, 48/464
treatment, 5/357, 361
trigger area, 48/461
tympanic form, 48/462
tympanic plexus neuralgia, 48/490
vagal neuralgia, 22/253
variant, 48/465
vascular lesion, 48/464
wisdom tooth, 48/464
Glossoptosis
Robin syndrome, 43/471
Glucagon
accidental injury, 74/324
clonidine, 74/167
myophosphorylase deficiency, 41/190
osteitis deformans (Paget), 43/452
pharmacology, 74/167
progressive dysautonomia, 59/154, 157
Reye syndrome, 56/152
Glucagon deficiency
hypoglycemia, 27/71
α-1,4-Glucan-6-glucosyltransferase deficiency, *see* Brancher deficiency
Glucantransferase
debrancher deficiency, 43/179
Glucocerebrosidase, *see* Glucosylceramidase
Glucocerebrosidase gene, *see* Glucosylceramidase

gene
Glucocerebroside, *see* Ceramide glucose
Glucocorticoid
see also Corticosteroid *and* Steroid
brain edema, 16/203, 67/88
brain metastasis, 69/139, 71/625
brain tumor, 67/315
epidural metastasis, 69/208
head injury, 57/232
leprous neuritis, 51/233
listeriosis, 52/91
multiple sclerosis, 47/192, 353
muscle tissue culture, 62/90
neurologic intensive care, 55/209
pain, 69/46
primary CNS lymphoma, 68/262
radiation myelopathy, 61/210
spinal cord metastasis, 69/183
stress, 45/248
Glucocorticosteroid, *see* Glucocorticoid
Gluconeogenesis
brain, 27/13
pyruvate carboxylase, 66/418
Gluconeogenesis inhibition
hypoglycin, 65/99
hypoglycin intoxication, 65/97
mechanism, 65/99
4-pentenoic acid, 65/99
in vitro, 65/97
in vivo, 65/98
Glucosamine
heparin, 27/144
Glucosamine-6-phosphate deaminase
metachromatic leukodystrophy, 10/48
Glucosamine-6-phosphate synthesis
Huntington chorea, 49/290
Glucose
brain, 27/19
brain injury, 23/123
brain metabolism, 54/110, 112, 57/71
CSF, *see* CSF glucose
Jamaican vomiting sickness, 37/535
oxidation, 27/19
Parkinson disease, 75/181
positron emission tomography, 54/112, 72/371, 380
progressive dysautonomia, 59/156
transport into brain, 27/6
Glucose-galactose malabsorption
deficiency neuropathy, 7/596
Glucose intolerance
amyotrophic lateral sclerosis, 22/30
mental deficiency, 43/299
motoneuron disease, 22/30
progressive spinal muscular atrophy, 22/30
Glucose-6-phosphatase
glycogen storage disease type I, 27/222, 226
Glucose-6-phosphatase deficiency, *see* Glycogen storage disease type I
Glucose-6-phosphate dehydrogenase
globoid cell leukodystrophy, 10/303
myoglobinuria, 62/558
wallerian degeneration, 7/217
Glucose-6-phosphate dehydrogenase deficiency
anemia, 42/643
back pain, 42/644
bilirubin encephalopathy, 27/423
gene frequency, 42/644
genetic linkage, 40/352, 388
Heinz body, 42/644
heterozygote detection, 42/645
jaundice, 42/644
lyonization, 42/645
neurofibromatosis, 14/492
primaquine, 42/644
Glucose tolerance test
anorexia nervosa, 46/586
α-Glucosidase
acid maltase deficiency, 41/180
fucosidase, 42/550
pseudodeficiency, 66/58
α-1,4-Glucosidase
acid maltase deficiency, 51/383, 62/481
enzyme deficiency neuropathy, 51/371
enzyme deficiency polyneuropathy, 51/371
α-1,6-Glucosidase
acid maltase deficiency, 62/481
β-Glucosidase
cyanogenic glycoside intoxication, 65/27
infantile Gaucher disease, 10/309
wallerian degeneration, 9/36
Glucosidase deficiency
Isaacs syndrome, 62/263
α-1,4-Glucosidase deficiency, *see* Acid maltase deficiency
β-Glucosidase deficiency
Gaucher disease, 10/306, 309, 526, 42/437-439, 46/55
globoid cell leukodystrophy, 10/306
Glucosteroid
brain edema, 67/74
Glucosylceramidase
adult Gaucher disease, 66/126
degradation, 66/59
enzyme deficiency neuropathy, 51/369
enzyme deficiency polyneuropathy, 51/369

Gaucher disease, 10/330, 51/379, 66/60
juvenile Gaucher disease, 42/439
juvenile Gaucher disease type B, 66/128
Glucosylceramidase cDNA
Gaucher disease, 66/119
Glucosylceramidase gene
lysosomal storage disease, 66/118
Glucosylceramide
catabolism, 66/60
Glucosylceramide lipidosis, *see* Gaucher disease
Glucosylcerebroside
Gaucher disease, 10/544
Glucuronate pathway, *see* Polyol pathway
Glucuronate-2-sulfatase
heparan sulfate substrate, 66/65
Glucuronic acid
heparin, 27/144
mucoploysaccharide constituent, 10/432
β-Glucuronidase
CSF, *see* CSF β-glucuronidase
globoid cell leukodystrophy, 10/21, 86, 147, 304, 306
glycosaminoglycanosis type VII, 27/152, 42/459, 66/282
GUSB gene, 66/282
heparan sulfate substrate, 66/65
mucolipidosis type II, 42/448
myelin, 9/34
pseudodeficiency, 66/58
wallerian degeneration, 7/215, 9/36
β-Glucuronidase deficiency, *see* Glycosaminoglycanosis type VII
β-Glucuronidase gene
glycosaminoglycanosis type VII, 66/311
Glucuronosyltransferase
Crigler-Najjar syndrome, 6/492, 21/52
Glucuronylparagloboside sulfate
undetermined significant monoclonal gammopathy, 63/401
Glue sniffing
acetonylacetone intoxication, 64/82
brain infarction, 55/523
organic solvent intoxication, 64/41
Glutaconic aciduria
symptomatic dystonia, 49/543
Glutamate, *see* Glutamic acid
Glutamate aspartate
Pick disease, 46/242
Glutamate decarboxylase
basal ganglion, 49/21
epileptic focus, 15/68
Huntington chorea, 42/227, 49/74, 256
pallidosubthalamic projection, 49/7
Parkinson disease, 49/130
progressive supranuclear palsy, 49/251
reduced neostriatal, *see* Reduced neostriatal glutamic acid decarboxylase
regional distribution, 29/492
substantia nigra, 49/8
vitamin B6 dependency, 42/716
Glutamate dehydrogenase
genetics, 60/564
Huntington chorea, 49/261
olivopontocerebellar atrophy, 60/553
Parkinson disease, 49/130
particulate form, 60/553
progressive supranuclear palsy, 49/251
soluble form, 60/553
Glutamate dehydrogenase deficiency
biochemical study, 60/552, 559
Friedreich ataxia, 51/389
glutamic acid receptor, 60/563
neuropathologic study, 60/556-559
olivopontocerebellar atrophy, 49/108, 130, 209, 60/551-566
parkinsonism, 49/130
pathogenesis, 60/560
supranuclear palsy, 60/553
Glutamate formiminotransferase deficiency
epilepsy, 72/222
Glutamate oxaloacetate transaminase, *see* Aspartate aminotransferase
Glutamate pyruvate transaminase, *see* Alanine aminotransferase
Glutamic acid
Alzheimer disease, 46/267
amino acid neurotransmitter, 29/512-514
amygdalostriate fiber, 49/4
autonomic nervous system, 74/55
basal ganglion, 49/33
brain, 9/16
brain amino acid metabolism, 29/17-22
brain edema, 67/80, 84
brain ischemia, 63/211
brain metabolism, 57/80
Chinese restaurant syndrome, 75/502
corticostriate fiber, 49/4
corticosubthalamic fiber, 49/11
epilepsy, 72/84
excitotoxin, 64/20
Friedreich ataxia, 51/389
globus pallidus, 49/5
GM2 gangliosidosis, 10/296
head injury, 57/80
hippocampus, 46/267
Huntington chorea, 49/261

hyponatremia, 63/547
neostriatum, 49/47
neurofibrillary tangle, 46/269
neurologic intensive care, 55/211
neuron death, 63/211
neurotransmitter, 29/512-514
Parkinson disease, 49/123
respiratory encephalopathy, 63/416
spastic paraplegia, 61/367
spinal cord compression, 69/170
spinal cord injury, 61/367
thalamostriate fiber, 49/4
transverse spinal cord lesion, 61/367
Glutamic acid receptor
arthropod envenomation, 65/200
domoic acid, 65/150
excitotoxic mechanism, 64/20
glutamate dehydrogenase deficiency, 60/563
olivopontocerebellar atrophy, 60/563
3-oxalylaminoalanine, 65/3
Glutamic aciduria
nuclear magnetic resonance, 60/673
symptomatic dystonia, 49/543
tendon hypertrophy, 60/669
Glutamic oxalacetic transaminase
CSF, *see* CSF glutamic oxalacetic transaminase
GM2 gangliosidosis, 10/289
Glutamine
brain, 9/16
brain amino acid metabolism, 29/19
CSF, *see* CSF glutamine
hepatic coma, 49/215
hyponatremia, 63/547
osmoregulation, 63/555
Reye syndrome, 56/153
Glutamine synthetase
Huntington chorea, 49/262
manganese intoxication, 64/305
γ-Glutamyl transpeptidase deficiency
amino acid, 29/122
cysteine peptiduria, 29/123
mental deficiency, 29/122
urine, 29/122
Glutaric aciduria type I, *see* Glutaryl-coenzyme A dehydrogenase deficiency
Glutaric aciduria type II
see also Multiple acyl-coenzyme A dehydrogenase deficiency
coma, 62/495
epilepsy, 72/222
episodic encephalopathy, 62/495
exercise intolerance, 62/495
fasting, 62/495
hypoglycemia, 62/495
hypoglycin intoxication, 65/104
infantile hypotonia, 62/495
Jamaican vomiting sickness, 37/533-535, 65/104
lethargy, 62/495
2 main forms, 62/495
metabolic encephalopathy, 62/495
metabolic myopathy, 62/495
myalgia, 62/495
recurrent coma, 62/495
type, 66/412
Glutaryl coenzyme A
hypoglycin, 65/130
hypoglycin intoxication, 65/130
Glutaryl-coenzyme A dehydrogenase deficiency
basal ganglion, 66/647
computer assisted tomography, 66/643
dyskinesia, 66/647
dystonia, 66/647
epilepsy, 72/222
epileptic discharge, 66/647
frontotemporal atrophy, 66/647
head circumference, 66/647
hyperthermia, 66/647
insomnia, 66/647
lysine metabolism, 66/648
macrocephaly, 66/647
nuclear magnetic resonance, 66/644
organic acid metabolism, 66/639, 641
sweating, 66/647
symptomatic dystonia, 49/543
tetraplegia, 66/647
Glutaryl-coenzyme A oxidase deficiency
peroxisomal disorder, 66/510
Glutathione
acrylamide intoxication, 64/72
free radical scavengers, 64/19
mercury intoxication, 64/391
Glutathione deficiency
hemolytic anemia, 60/674
Glutathione peroxidase
free radical scavengers, 64/19
Glutathione synthetase
nitrofurantoin polyneuropathy, 51/296
Glutathione-cysteine-disulfide
Hallervorden-Spatz syndrome, 49/402
Gluteal nerve
topographical diagnosis, 2/40, 43
Trendelenburg sign, 2/43
Gluten
multiple sclerosis, 47/155
Gluten induced enteropathy, *see* Celiac disease
Glutethimide

neuropathy, 7/543
toxic myopathy, 62/601
Glutethimide intoxication
neuropathy, 7/519, 37/354
Gluteus maximus reflex
diagnostic value, 1/247
Glyceraldehyde-3-phosphate dehydrogenase
brain glycolysis, 27/9
hypoglycin intoxication, 65/130
rat, 13/202
retinal degeneration, 13/202
δ-Glyceric aciduria
epilepsy, 72/222
Glycerol
brain edema, 16/202, 57/215
brain infarction, 53/427
neurologic intensive care, 55/213
Glycerol-3-phosphate dehydrogenase
skeletal muscle, 40/5
wallerian degeneration, 7/217
Glycerol-ether-phosphatide
chemistry, 10/243
Glycerol kinase deficiency
characteristics, 62/127
Duchenne muscular dystrophy, 62/127
syndrome, 70/217
L-α-Glycerophosphate
biochemistry, 9/7
Glycerophosphate dehydrogenase
biochemistry, 9/7
white matter, 9/16
α-Glycerophosphate dehydrogenase, *see* Glycerol-3-phosphate dehydrogenase
Glycerophosphatide
brain development, 9/17
Pelizaeus-Merzbacher disease, 10/151
sphingolipid, 10/266
synthesis, 9/7
Glycerophospholipid
biochemistry, 66/36
biosynthesis, 10/237, 66/42
chemistry, 10/235
degradation, 10/239, 243
metabolism, 66/42
myelin, 9/2
N-containing type, 10/240, 243
N-free type, 10/237, 239
Glycerophosphoric acid
chemistry, 10/235
Glyceryl trinitrate
action mechanism, 37/453
brain blood flow, 37/454
chemical structure, 37/453
chronic migrainous neuralgia, 48/251
cluster headache, 37/454
dyspnea, 37/454
headache, 48/228, 281, 443, 75/285
hypertensive encephalopathy, 54/216
migraine, 75/295
myocardial infarction, 37/453
nitrate intoxication, 65/443
restless legs syndrome, 8/318, 42/261
vasodilator agent, 37/453
vasospasm, 37/454
Glyceryl trinitrate intoxication
brain blood flow, 65/443
brain ischemia, 65/443
facial flushing, 37/454
glaucoma, 37/453
headache, 37/453, 65/443
hypotension, 37/453
mania, 37/454
methemoglobinemia, 37/453
polyneuropathy, 37/453
psychosis, 37/453
Glyceryl trinitrate test
cluster headache, 48/228
Glycinamide phosphonucleotide
Down syndrome, 50/531
Glycinamide ribonucleotide synthetase, *see* Phosphoribosylamine glycine ligase
Glycine
amino acid neurotransmitter, 29/508-511
ballismus, 49/372
basal ganglion, 49/33
biochemistry, 29/508
brain, 9/16, 29/508
CNS, 29/508
CSF, 6/348
epilepsy, 72/95
familial spinal cord disorder, 29/511
GM2 gangliosidosis, 10/295
hemiballismus, 49/372
history, 29/508
Huntington chorea, 6/348
hyperglycinemia, 29/511
hypoglycin intoxication, 65/104
Jamaican vomiting sickness, 65/104
ketotic hyperammonemia, 29/106
neurotransmitter, 29/508
spastic paraplegia, 61/367
spinal cord, 29/508
spinal cord injury, 61/367
tetanus, 29/511
transverse spinal cord lesion, 61/367
uremic encephalopathy, 63/514

uremic polyneuropathy, 63/514
Glycine cleavage enzyme system
hyperglycinemia, 42/566
Glycinuria
familial, *see* Familial glycinuria
indolylacrolyl, *see* Indolylacrolyl glycinuria
Glycoconjugate
amniotic fluid, 66/56
glycoprotein, 66/61
Glycogen
see also Corpora amylacea *and* Progressive myoclonus epilepsy
anesthesia, 27/207
anoxia, 27/206
biogenic amine, 27/208-210
brain injury, 27/206
cytoplasmic excess, 40/36, 98, 241
disorder, 27/169-211
distribution, 27/169
drug effect, 27/208-210
histochemistry, 27/169, 40/36
hypoxia, 27/206
ischemia, 27/206
metabolism, 27/20-23, 170, 221-224
muscle fiber, 62/20
necrosis, 27/210
nerve section, 27/210
neuroleptic agent, 27/208
normally in muscle, 40/240
penetrating brain injury, 27/206
radiation, 27/207
radiation injury, 27/207
reactive astrocytosis, 27/205
REM sleep deprivation, 27/210
spongy cerebral degeneration, 27/203
status spongiosus, 27/203
temperature, 27/210
ultrastructure, 40/98
white matter, 9/15
β-Glycogen
acid maltase deficiency, 51/383
Glycogen debrancher enzyme deficiency, *see* Debrancher deficiency
Glycogen phosphorylase
phosphorolysis, 70/424
Glycogen storage disease, 27/221-237
acid maltase, 27/222
acid maltase deficiency, 40/53, 99, 153, 190, 255, 280, 317, 438, 62/480
areflexia, 43/180
arthrogryposis multiplex congenita, 43/78
brain, 62/490
brain embolism, 55/163
brain infarction, 55/163
brancher deficiency, 62/483
cardiomyopathy, 62/490
cat, 27/199
central core myopathy, 43/81
cerebrovascular disease, 55/163
childhood myoglobinuria, 62/564
classification, 27/222
corpus callosum agenesis, 42/8
debrancher deficiency, 27/222, 40/256, 41/181, 62/482
dementia, 43/132
diabetes mellitus, 42/543
diet, 43/178, 182
enzymopathy, 21/52
exercise intolerance, 43/181, 183
Fanconi syndrome, 29/172
feeding disorder, 43/178
glycogen synthetase, 27/222
glycogen synthetase deficiency, 27/222
growth retardation, 43/179
hepatomegaly, 43/179
hepatosplenomegaly, 43/180
hereditary ataxia, 27/199
histochemistry, 40/52
histology, 40/52
history, 62/479
hypokalemic periodic paralysis, 43/171
hyporeflexia, 43/178, 180
infantile acid maltase deficiency, 27/222
ischemic exercise test, 41/193
Klinefelter variant XXXY, 43/559
limb girdle syndrome, 62/190
limit dextrin type glycogen, *see* Debrancher deficiency
liver disease, 43/179
mental deficiency, 46/19, 62/490
metabolic myopathy, 41/175-194, 62/479
muscle cramp, 41/186, 190, 193, 297
myalgia, 43/181
myoglobinuria, 41/265, 62/558
myopathy, 22/189, 43/78, 132
myophosphorylase deficiency, 27/232, 40/4, 53, 189, 257, 317, 327, 339, 342, 438, 62/485
myotonic dystrophy, 43/152
neuromuscular disease, 41/189, 426
other glycogen storage disease, 41/192-194
pathologic reaction, 40/254
periodic paralysis, 41/170
phosphofructokinase deficiency, 27/235, 40/53, 190, 264, 327, 339, 342, 41/191, 62/487
phosphoglucomutase, 27/222
phosphoglycerate kinase deficiency, 62/489

phosphoglycerate mutase deficiency, 62/489
phosphohexose isomerase, 27/222
phosphorylase kinase, 27/222, 234
phosphorylase kinase deficiency, 62/486
progressive external ophthalmoplegia, 22/189, 43/136, 62/298
progressive external ophthalmoplegia classification, 22/179, 62/290
rabbit, 27/199
rimmed muscle fiber vacuole content, 62/18
rimmed muscle fiber vacuole type, 62/18
rimmed vacuole distal myopathy, 62/490
secondary neuroaxonal dystrophy, 49/408
spinal muscular atrophy, 59/371
type II, *see* Acid maltase deficiency
type III, *see* Debrancher deficiency
type IV, *see* Brancher deficiency
type V, *see* Myophosphorylase deficiency
type VII, *see* Phosphofructokinase deficiency
type VIII, *see* Phosphorylase kinase deficiency
type IX, *see* Phosphoglycerate kinase deficiency
type X, *see* Phosphoglycerate mutase deficiency
type XI, *see* Lactate dehydrogenase deficiency
ultrastructure, 40/98
unclassified disease, 27/198, 237
uridine diphosphoglucose-glycogen synthetase, 40/3
variant type, 62/490
vitamin B1, 7/570
Glycogen storage disease type I
bilirubin encephalopathy, 27/423
biochemistry, 27/221, 224
clinical features, 27/225
clinical laboratory examination, 27/225
differential diagnosis, 27/224
Fanconi syndrome, 29/172
glucose-6-phosphatase, 27/222, 226
hypoglycemia, 27/71, 225
lactic acidosis, 41/208, 42/586
pathology, 27/225
seizure, 42/586
treatment, 27/226
xanthoma, 27/225
xanthomatosis, 42/586
Glycogen synthetase
glycogen storage disease, 27/222
Glycogen synthetase deficiency
clinical features, 27/222
glycogen storage disease, 27/222
hypoglycemia, 27/72
Glycogenosis, *see* Glycogen storage disease
Glycolipid
GA2, *see* GA2 glycolipid
globoid cell leukodystrophy, 10/79, 84
history, 10/21
IgM monoclonal gammopathy, 69/297
orcinol reaction, 9/26
paraproteinemia, 69/297
resorcinol reaction, 9/26
Glycolipidosis
mental deficiency, 10/552
Minkowitz disease, 10/552
Glycolysis
anaerobic, *see* Anaerobic glycolysis
brain, 27/6-13
brain infarction, 54/122
brain metabolism, 57/75
ethyl iodoacetate, 13/210
infarction, 27/31
ischemia, 27/31
mercury intoxication, 64/391
Glycolytic enzyme
brain tumor, 27/508
CSF, *see* CSF glycolytic enzyme
nervous system tumor, 27/508
Glycoprotein
acid sialidase, 66/377
carbohydrate deficient, *see* Carbohydrate deficient glycoprotein
carbohydrate deficient glycoprotein syndrome, 66/631
cephalopoda intoxication, 37/62
chemistry, 27/153
deficient degradation, 66/61
envelope, *see* Envelope glycoprotein
β-galactosidase, 10/480
glycoconjugate, 66/61
human immunodeficiency virus, 56/490
inborn error of metabolism, 27/153-155
intervertebral disc, 20/526, 538
multiple sclerosis, 9/321, 47/109
myelin, 9/31
myelin associated, *see* Myelin associated glycoprotein
octopoda intoxication, 37/62
oligosaccharidosis, 66/329
senile plaque, 46/253
Glycoprotein degradation
α-*N*-acetylgalactosaminidase deficiency, 66/62
aspartylglycosaminuria, 66/62
fucosidosis, 66/62
glycoprotein storage disease, 66/62
glycoproteinosis, 66/62
GM2 gangliosidosis, 66/62
heritable, 66/62
oligosaccharidosis, 66/62

Glycoprotein lamp 1, *see* Lysosome associated membrane protein 1
Glycoprotein storage disease
  glycoprotein degradation, 66/62
Glycoproteinosis
  degradation, 66/63
  enzyme defect, 66/63
  glycoprotein degradation, 66/62
  inclusion cell disease, 21/48
  survey, 66/26
  urinary oligosaccharide, 66/63
Glycosaminoglycan
  biological action, 10/432
  chemistry, 27/144
  congenital spondyloepiphyseal dysplasia, 43/480
  degradation, 66/63
  familial amaurotic idiocy, 10/343
  fucosidosis, 27/157
  glycosaminoglycanosis, 66/281
  glycosaminoglycanosis type VI, 66/308
  GM1 gangliosidosis, 10/473
  heparan sulfate, 10/327
  heparin, 10/327
  Huntington chorea, 6/350
  Hurler-Hunter disease, 10/341
  inborn error of metabolism, 27/145-153
  intervertebral disc, 20/526, 533, 535
  myelin, 9/31
  phenotype, 66/282
  striatopallidodentate calcification, 6/715
  ultrastructure, 10/377
  undetermined significant monoclonal gammopathy, 63/402
Glycosaminoglycan excretion
  glycosaminoglycanosis type IV, 66/303
  glycosaminoglycanosis type VII, 66/313
Glycosaminoglycan metabolism
  glycosaminoglycanosis, 66/281
  glycosaminoglycanosis type IH, 66/281
  glycosaminoglycanosis type II, 66/281
Glycosaminoglycanosis, 10/427-458, 27/143-152
  see also Hurler-Hunter disease
  acid phosphatase, 10/377
  age at onset, 10/434
  benign intracranial hypertension, 67/111
  biochemistry, 10/431
  blood vessel, 10/452
  bone, 10/455
  brain, 10/47
  brown metachromasia, 10/52
  carbohydrate metabolism, 27/162
  carotid system syndrome, 53/309
  cattle, 10/457
  cerebellum, 10/444
  chemistry, 10/341-343
  chondroitin, 10/327
  chondroitin sulfate, 66/281
  chondroitin sulfate A, 10/327, 432
  chondroitin sulfate C, 10/327, 432
  chorioretinal degeneration, 13/40
  classification, 60/176
  clinical features, 10/434-438, 60/176
  deafness, 60/176
  dementia, 60/176
  dermatan sulfate, 66/281
  enzyme defect, 60/176
  enzyme deficiency polyneuropathy, 51/370, 379
  extrinsic nerve cell, 10/446
  eye, 10/452
  Farber disease, 10/431
  genetics, 10/456, 60/176
  glycosaminoglycan, 66/281
  glycosaminoglycan metabolism, 66/281
  heart, 10/450
  heart ultrastructure, 10/451
  heparan sulfate, 10/429, 66/281
  histochemistry, 10/434, 443-445
  history, 10/427-429
  hyaluronic acid, 10/432
  keratan, 10/327, 432
  keratan sulfate, 66/281
  kidney, 10/454
  laboratory test, 10/433
  liver, 10/448
  liver ultrastructure, 10/449
  macrencephaly, 50/263
  marginal case, 10/430
  mental deficiency, 46/19, 53-55
  metachromasia, 10/454
  neurology, 10/441
  neuropathology, 10/442
  osseous change, 10/438-440
  pituitary gland, 10/449
  polyneuropathy, 60/175
  prenatal diagnosis, 27/163
  prognosis, 10/434
  secondary pigmentary retinal degeneration, 13/332
  skin, 10/454
  sleep apnea, 66/293
  sphingolipidosis, 29/372
  spinal cord, 10/445
  spleen, 10/454
  symptomatic dystonia, 49/543
  systemic dystonia, 49/543
  tryptophan, 10/432

type, 10/299, 341, 60/175, 66/26
ultrastructure, 10/445-454
white blood cell ultrastructure, 10/455
white matter hyperdensity, 60/175
zebra body, 10/379, 382
Glycosaminoglycanosis type I
see also Hurler-Hunter disease
age at onset, 10/434
atlantoaxial subluxation, 50/392
basilar impression, 50/399
birth incidence, 42/451
brain embolism, 55/163
brain infarction, 55/163
cerebrovascular disease, 55/163
corneal clouding, 10/435
corneal opacity, 10/436, 42/450, 60/177
corpus callosum agenesis, 50/151
cranial nerve, 42/450
deafness, 10/441
dementia, 10/441, 60/175
dermatan sulfate, 27/146, 46/54
diagnostic pitfalls, 10/219
dwarfism, 10/435
epilepsy, 15/426
fibrous dysplasia, 14/170
β-galactosidase, 10/223, 42/450
ganglioside, 9/4
GM2 gangliosidosis, 10/431
Hallervorden-Spatz syndrome, 6/616
heparan sulfate, 27/146, 46/54
heparin, 46/54
hepatosplenomegaly, 10/436, 42/450
heterozygote detection, 42/452
hydrocephalus, 10/441
hypertelorism, 42/450
α-L-iduronidase, 27/146, 42/452
inborn error of carbohydrate metabolism, 27/162
intracranial hypertension, 60/175
macrencephaly, 10/441
mannosidosis, 42/597
membranous cytoplasmic inclusion body, 21/52
mental deficiency, 10/435, 441, 27/147, 42/450, 46/53
metachromasia, 10/53
metachromatic leukodystrophy, 10/53
microcephaly, 42/451
mucolipidosis type II, 42/448
neurofibromatosis type I, 14/492
optic atrophy, 13/40
prenatal diagnosis, 27/162, 42/452
prognosis, 10/435
progressive external ophthalmoplegia, 22/212
psychomotor retardation, 42/450
retinal degeneration, 10/441
secondary pigmentary retinal degeneration, 10/436, 42/450-452
sella turcica, 42/451
skeletal deformity, 10/435, 42/450
spastic paraplegia, 60/175
Glycosaminoglycanosis type IH
anesthesia, 66/291, 297
bone marrow transplantation, 66/87, 99
brain biopsy, 10/685
carpal tunnel syndrome, 66/285, 291
cervical spinal cord compression, 66/297
clinical presentation, 66/283
corneal opacity, 60/177
corneal transplantation, 66/291
deafness, 60/177
dementia, 60/175
dermatan sulfate, 10/429
enzymatic diagnosis, 66/314
enzyme replacement therapy, 66/291
glycosaminoglycan metabolism, 66/281
glycosaminoglycanosis type IS, 66/289
glycosaminoglycanosis type VII, 66/310
hearing aid, 66/297
heart disease, 66/297
hydrocephalus, 66/297
IDUA gene, 66/282, 289
α-L-iduronidase, 66/65, 282, 289
intracranial hypertension, 60/175
lymphocyte vacuolation, 10/335, 342
management, 66/291
α-mannosidase, 10/430
membranous cytoplasmic inclusion body, 10/445-447
neurolipidosis, 66/19
nuclear magnetic resonance, 66/99
respiratory system, 66/294
sea blue histiocyte, 60/674
secondary pigmentary retinal degeneration, 60/176
skeletal system, 66/294
sleep, 66/297
spastic paraplegia, 60/175
torpedo, 10/444
tracheostomy, 66/297
vertebral subluxation, 66/285
viscus, 66/294
white matter atrophy, 66/26
Glycosaminoglycanosis type IH-S
carpal tunnel syndrome, 66/292
clinical presentation, 66/283
enzymatic diagnosis, 66/314
hearing impairment, 66/292

IDUA gene, 66/282, 289
α-L-iduronidase, 66/282
phenotype, 66/287
Glycosaminoglycanosis type IS
cardiac valve replacement, 66/292
carpal tunnel syndrome, 10/441, 42/453, 66/292
clinical presentation, 66/283
contracture, 42/453
corneal clouding, 10/437
corneal opacity, 42/452, 60/177
corneal transplantation, 66/292
dermatan sulfate, 10/429
enzymatic diagnosis, 66/314
glycosaminoglycanosis type IH, 66/289
heterozygote detection, 42/453
hypertrichosis, 10/437
IDUA gene, 66/282
α-L-iduronidase, 27/151, 42/453, 66/282
optic atrophy, 13/40
physical finding, 66/286
secondary pigmentary retinal degeneration, 42/452-455
skeletal deformity, 42/453
spastic paraplegia, 60/175
Glycosaminoglycanosis type II
age at onset, 10/434
bone marrow transplantation, 66/298
carpal tunnel syndrome, 66/295
Caucasian patient, 42/454
claw hand, 42/454
corneal clouding, 10/435, 42/454, 66/295
corneal opacity, 60/177
deafness, 10/436, 441, 42/454, 60/177
dementia, 60/175
dental care, 66/298
dermatan sulfate, 10/429, 42/454
dwarfism, 10/435, 42/454
endocrine system, 66/295
enzymatic diagnosis, 66/314
enzyme defect, 66/296
female, 66/296
gene location, 66/296
genetic feature, 10/456
glycosaminoglycan metabolism, 66/281
hearing loss, 66/295
heparan sulfate, 42/454
hepatosplenomegaly, 10/436, 42/454
histochemistry, 10/443
hydrocephalus, 42/454
hypertrichosis, 42/454
IDS gene, 66/282, 296
iduronate sulfatase, 42/454, 66/65
iduronate-2-sulfatase, 66/282
intracranial hypertension, 60/175
kyphoscoliosis, 42/454, 66/298
membranous cytoplasmic inclusion body, 10/445-447
mental deficiency, 10/435, 441, 42/454, 46/54
multiple sulfatase deficiency, 66/314
mutation analysis, 66/296
optic atrophy, 13/40
papilledema, 42/454
pathology, 10/442
phenotype, 66/293
prognosis, 10/435, 66/295
secondary pigmentary retinal degeneration, 10/436, 42/454
selective IgA deficiency, 66/294
sensorineural deafness, 42/454
skeletal deformity, 10/435
spastic paraplegia, 60/175
sulfatase deficiency, 66/314
symptom, 66/293
zebra body, 10/445
Glycosaminoglycanosis type IIA
secondary pigmentary retinal degeneration, 60/176
Glycosaminoglycanosis type IIB
papilledema, 60/176
Glycosaminoglycanosis type III
acetyl-coenzyme A α-glucosaminide N-acetyltransferase, 66/65
acetyl coenzyme A:α-glycosaminedisulfate-*N*-acetyltransferase, 42/456
N-acetylglucosamine-6-sulfatase, 66/65
α-*N*-acetylglucosaminidase, 66/65
N-acetyl-α-D-glucosaminidase, 42/456
birth incidence, 42/456
brain cortex atrophy, 66/301
carpal tunnel syndrome, 66/302
contracture, 42/455
dementia, 60/175, 66/302
diarrhea, 66/301
dyssomnia, 66/299
endocrine system, 66/300
enzyme, 66/301
β-galactosidase, 10/430
hearing loss, 66/299
heart involvement, 66/299
heparan *N*-sulfatase, 66/65
heparan sulfate, 66/302
heparan sulfate sulfatase, 42/456
hepatosplenomegaly, 10/342, 42/455, 66/300
hirsutism, 66/300
hydrocephalus, 66/302
hypertrichosis, 66/300

intracranial hypertension, 60/175
mental deficiency, 10/342, 42/455, 46/54, 66/299, 302
mental deterioration, 66/299
ocular involvement, 66/300
optic atrophy, 10/342, 13/40
otitis media, 66/302
pathology, 66/301
pneumonia, 66/299
prenatal diagnosis, 42/456
prognosis, 66/301
respiratory infection, 66/299
Salam-Idriss syndrome, 10/343
seizure, 66/301
sensorineural deafness, 42/455
skeletal system, 66/300
sleep disorder, 42/456
spastic paraplegia, 60/175
speech, 66/300
speech delay, 66/298
speech disorder in children, 42/455
splenomegaly, 66/300
type, 66/299, 302
type a, 27/150
type b, 27/150
vertebral abnormality, 42/456
Glycosaminoglycanosis type IIIA
enzymatic diagnosis, 66/314
epilepsy, 72/222
heparan *N*-sulfatase, 66/282, 302
HSS gene, 66/282
secondary pigmentary retinal degeneration, 60/176
Glycosaminoglycanosis type IIIB
α-*N*-acetylglucosaminidase, 66/282, 302
enzymatic diagnosis, 66/315
NAGLU gene, 66/282
Glycosaminoglycanosis type IIIC
acetyl-coenzyme A α-glucosaminide N-acetyltransferase, 66/282, 302
enzymatic diagnosis, 66/315
GNAT gene, 66/282
Glycosaminoglycanosis type IIID
N-acetylglucosamine-6-sulfatase, 66/64, 282, 302
enzymatic diagnosis, 66/315
G6S gene, 66/282
Glycosaminoglycanosis type IV
N-acetylgalactosamine-6-sulfatase, 42/457, 66/303
arylsulfatase B deficiency, 42/458
atlantoaxial dislocation, 42/458
atlantoaxial subluxation, 50/392, 60/175
barrel chest, 66/304
birth incidence, 42/457
corneal opacity, 42/457, 60/177
coxa valga, 66/304
deafness, 60/177
β-galactosidase, 27/150, 42/457
glycosaminoglycan excretion, 66/303
hyperextensible joint, 42/457
inguinal hernia, 66/304
intelligence, 66/304
juvenile spinal cord injury, 61/233
juvenile vertebral column injury, 61/233
keratan, 10/342
keratan sulfate, 66/303, 305
kyphoscoliosis, 66/304
maxilla, 66/303
mental deficiency, 42/457
nose, 66/303
odontoid aplasia, 66/304
osteoporosis, 66/303
pathology, 66/305
platyspondyly, 66/304
prognosis, 66/305
skeletal system, 66/304
spinal cord compression, 66/304
spondyloepiphyseal dysplasia, 42/457
survey, 66/247
Glycosaminoglycanosis type IVA
N-acetylgalactosamine-6-sulfatase, 66/282
enzymatic diagnosis, 66/315
galactose-6-sulfurylase, 66/64
GALNS gene, 66/282
gene location, 66/305
genotype-phenotype correlation, 66/248
lysosomal inclusion, 66/305
Glycosaminoglycanosis type IVB
acid β-galactosidase, 66/247, 269
bone marrow transplantation, 66/306
enzymatic diagnosis, 66/315
enzyme, 66/57
β-galactosidase, 66/64, 282, 306
gene location, 66/306
genetics, 66/268
genotype-phenotype correlation, 66/248
genu valgum, 66/306
glaucoma, 66/307
GLB1 gene, 66/282
GM1 gangliosidosis, 66/269, 306
heart disease, 66/307
keratan sulfate, 66/57
mRNA, 66/269
mutation, 66/306
skeletal deformity, 66/265
spinal cord compression, 66/306

survey, 66/247
therapy, 66/307
wrist, 66/307
Glycosaminoglycanosis type V, *see* Glycosaminoglycanosis type IS
Glycosaminoglycanosis type VI
N-acetylgalactosamine-4-sulfatase, 66/282, 309
ARSB gene, 66/282
arylsulfatase B deficiency, 27/152, 42/458
atlantoaxial dislocation, 42/458
atlantoaxial subluxation, 50/392, 66/308
bone marrow transplantation, 66/309
cardiomyopathy, 66/307
carpal tunnel syndrome, 42/458, 66/307
cervical myelopathy, 66/309
congestive heart failure, 66/309
corneal dystrophy, 66/309
corneal opacity, 42/458, 60/177, 66/308
dermatan sulfate, 10/429
dwarfism, 42/458
enzymatic diagnosis, 66/315
enzyme defect, 66/308
gene location, 66/309
gene therapy, 66/309
glaucoma, 66/308
glycosaminoglycan, 66/308
growth retardation, 66/307
hepatosplenomegaly, 42/458
hydrocephalus, 42/458, 66/309
inguinal hernia, 66/308
kyphoscoliosis, 42/458
Legg-Perthes disease, 66/307
leukocyte, 66/308
liver, 66/308
mental deficiency, 66/308
multiple deficiency syndrome, 66/315
mutation analysis, 66/309
odontoid process, 66/308
respiratory system, 66/307
retinopathy, 66/308
skeletal deformity, 42/458
sternum malformation, 10/437
tracheostomy, 66/307
Glycosaminoglycanosis type VII
amniocentesis, 29/369
ascites, 66/311
Berry spot test, 66/313
biochemical aspect, 29/368
bone marrow transplantation, 66/312
carrier detection, 66/312
cetylpyridiniumchloride-carbazole test, 66/313
cornea, 66/310
dementia, 60/175
detection, 29/369
diagnostic procedure, 66/312
dwarfism, 42/459
electrophoresis, 66/313
enzymatic diagnosis, 66/316
enzyme, 66/312
β-glucuronidase, 27/152, 42/459, 66/282
β-glucuronidase gene, 66/311
glycosaminoglycan excretion, 66/313
glycosaminoglycanosis type IH, 66/310
GUSB gene, 66/282
hematologic finding, 66/311
hepatosplenomegaly, 42/459, 66/310
hypertelorism, 42/459, 66/310
infantile, 66/310
intracranial hypertension, 60/175
jaundice, 66/311
juvenile, 66/310
kyphoscoliosis, 42/459
kyphosis, 66/310
Legg-Perthes disease, 66/311
liver, 66/311
macrocephaly, 66/310
mutation, 66/311
nasal bridge, 66/310
neonatal, 66/310
newborn screening, 66/313
optic nerve, 66/311
pectus excavatum, 66/310
prenatal diagnosis, 66/312
respiratory complication, 66/310
scoliosis, 66/310
sella turcica, 42/459
spastic paraplegia, 60/175
spleen, 66/311
treatment, 29/369
urine analysis, 66/313
Glycosaminoglycanosis type VIII
N-acetylglucosamine sulfate sulfatase, 42/460
hepatosplenomegaly, 42/459
mental deficiency, 42/460
skeletal deformity, 42/460
Glycoside
cardiac, *see* Cardiac glycoside
Glycoside intoxication, 37/425-430
cardiac dysrhythmia, 37/426
color vision, 37/426, 429
fatigue, 37/426
gynecomastia, 37/426
headache, 37/426
nausea, 37/426
scotoma, 37/426, 429
seizure, 37/426

syncope, 37/426
transient global amnesia, 37/426
trigeminal neuralgia, 37/429
visual acuity, 37/426, 429
vomiting, 37/426
weakness, 37/426
Glycosphingolipid
N-acetylneuraminic acid, 10/248
GM3 hematoside lipodystrophy, 29/370
lysosomal degradation, 66/59
nervous system tumor, 27/504, 508
Glycosphingolipid lipidosis, *see* Gangliosidosis, Gaucher disease *and* Metachromatic leukodystrophy
Glycosphingolipidosis, *see* Gangliosidosis, Gaucher disease *and* Metachromatic leukodystrophy
Glycosuria
hypothalamus, 2/440
renal, *see* Renal glycosuria
Wilson disease, 27/387, 49/233
Glycosylasparaginase, *see* *N*4-(β-*N*-Acetylglucosaminyl)asparaginase
Glycosylation
endoplasmic reticulum, 66/82
Glycosylcerebroside lipidosis, *see* Gaucher disease
Glycyrrhiza
Glycyrrhiza glabra, 62/601, 603, 64/17
hypokalemic periodic paralysis, 64/17
myasthenic syndrome, 64/17
toxic myopathy, 62/601, 603, 64/17
Glycyrrhiza intoxication
symptom, 64/17
Glyoxalate aminotransferase, *see* Alanine glyoxylate aminotransferase
Glyoxylic acid metabolism
peroxisome, 66/507
GM1 galactosidase
Hallervorden-Spatz syndrome, 66/714
GM1 ganglioside
catabolism, 66/60
cervical spinal cord injury, 61/61
GM1 ganglioside antibody
multiple sclerosis, 47/109
GM1 gangliosidosis, 10/462-482
N-acetyl-β-D-galactosaminidase, 10/559, 561
N-acetyl-β-D-glucosaminidase, 10/477, 481, 561
N-acetylglucosaminyltransferase, 66/83
acid phosphatase, 10/410
Alder-Reilly granule, 10/430
amaurotic idiocy, 10/390
amniocentesis, 29/369
atypical Derry-Suzuki case, 10/480
biochemical aspect, 29/368
biochemistry, 10/470, 29/368
bone deformity, 10/463-466
carbohydrate metabolism, 27/163
carrier detection, 10/481
cherry red spot, 10/528
clinical features, 10/462
detection, 29/369
differential diagnosis, 10/468, 470
enzyme defect, 10/475, 29/367
enzyme deficiency polyneuropathy, 51/369, 374
enzymopathy, 21/51
epilepsy, 72/222
familial pes cavus, 10/223
foam cell, 10/299
fructose-bisphosphate aldolase, 10/467, 470
galactosialidosis, 66/249
α-galactosidase, 10/477
β-galactosidase, 10/223, 299, 467, 27/155, 66/60
genetics, 10/466, 480
glycosaminoglycan, 10/473
glycosaminoglycanosis type IVB, 66/269, 306
Goldberg syndrome, 42/440
histochemistry, 10/470
inborn error of carbohydrate metabolism, 27/162
infantile, 66/264
keratan, 10/223, 473
keratan sulfate, 10/479, 27/155
laboratory test, 10/466
late onset form, 66/264
lymphocyte vacuolation, 10/466
α-mannosidase, 10/477
metabolic block, 10/475
metabolic scheme, 10/475
mucolipidosis type I, 10/466, 470
mucolipidosis type II, 10/466, 470
neuronal lipidosis, 10/467
nomenclature, 10/473
pathology, 10/467
prenatal diagnosis, 27/162
prevention, 10/481
renal epithelial vacuolation, 10/467
survey, 66/247
symptomatic dystonia, 49/543
synonym, 10/222
treatment, 10/481, 29/369
type, 10/224, 298
visceral histiocyte, 10/467
visceral keratan sulfate, 10/223
GM1 gangliosidosis type I
foam cell, 42/431
β-galactosidase, 10/223, 42/431
heterozygote detection, 42/432
mental deficiency, 42/431

prenatal diagnosis, 42/432
psychomotor retardation, 42/431
sella turcica, 42/431
skeletal deformity, 42/431
GM1 gangliosidosis type II
age at onset, 10/223
ataxia, 42/432
foam cell, 42/432
β-galactosidase, 42/432
heterozygote detection, 42/433
mental deficiency, 42/432
prenatal diagnosis, 42/433
psychomotor retardation, 42/432
seizure, 42/433
GM2 ganglioside
catabolism, 66/60
chemistry, 10/416, 472
degradation, 66/59
GD1a ganglioside, 10/471
survey, 66/247
Wohlfart-Kugelberg-Welander disease, 59/93
GM2 ganglioside activator protein
degradation, 66/59
GM2 ganglioside activator protein deficiency
clinical phenotype, 66/262
GM2 gangliosidosis, 10/385-417
β-*N*-acetylhexosaminidase, 66/257
β-*N*-acetylhexosaminidase A, 46/56, 66/58, 60, 64
N-acetyl-β-D-hexosaminidase A, 10/297, 417
N-acetyl-β-D-hexosaminidase A deficiency, 10/287
N-acetyl-α-D-hexosaminidase deficiency, 27/156
acid phosphatase, 10/291
acute infantile type, 66/252
adenosine triphosphatase, 10/291
adult, *see* Adult Tay-Sachs disease
alanine, 10/295
amniocentesis, 29/366
Ashkenazim, 10/562
aspartic acid, 10/296
astrocytosis, 10/291, 398
athetosis, 6/457
atypical case, 10/411, 414
biochemistry, 10/221, 335, 415
birth incidence, 42/434
blood, 10/409
brain, 10/385
brain ganglioside, 66/260
carbohydrate metabolism, 27/163
cerebellar atrophy, 10/291, 394
cerebellum, 10/399
cherry red spot, 10/334, 385, 387, 558, 13/326, 42/433, 66/19
classic, 29/365
classification, 29/363-365
clinical features, 10/287, 391-393
congenital Epstein-Norman-Wood type, 10/414
cortical neuron, 66/255
cystic leukoencephalopathy, 10/394
cytosome, 10/365
demyelination, 10/397
detection, 29/366
diphosphopyridine nucleotide diaphorase, 10/291
dystonia, 60/664
ECG, 10/392
EEG, 10/334, 391
endoplasmic reticulum, 10/417
enzyme, 66/61
enzyme deficiency polyneuropathy, 51/369, 375
enzyme histochemistry, 10/402-404
enzymology, 66/256
epidemiology, 10/297, 558-562
epilepsy, 60/666, 72/222
eponymic classification, 10/217
ethnic predominance, 10/388
founder effect, 42/434
fructose-bisphosphate aldolase, 10/289, 291, 335, 393, 560
GABA, 10/296
ganglioside, 10/220, 390
gelagoplegia, 10/334
gelolepsy, 10/392
gene, 66/252
gene frequency, 10/560
genetic basis, 66/251
genetic defect, 66/256
genetics, 10/297, 388
globoside, 10/414, 417
α-2-globulin, 10/389
glutamic acid, 10/296
glutamic oxalacetic transaminase, 10/289
glycine, 10/295
glycoprotein degradation, 66/62
glycosaminoglycanosis type I, 10/431
GM2 gangliosidosis type II, 66/249, 260
hepatosplenomegaly, 60/668
hereditary congenital cerebellar atrophy, 60/289
hexosaminidase, 51/375
histochemistry, 10/396, 401
historic aspect, 29/363
history, 10/385-389
inborn error of carbohydrate metabolism, 27/162
inclusion body, 51/375
juvenile, *see* Juvenile Tay-Sachs disease
lactate dehydrogenase, 10/288-291
lactate dehydrogenase isoenzyme 1, 10/289

lactate dehydrogenase isoenzyme 2, 10/289
lactate dehydrogenase isoenzyme 3, 10/289
lactate dehydrogenase isoenzyme 5, 10/289
Lafora body, 10/411
late infantile, 29/365
late onset form, 66/253
leptomeninx, 10/395
leucine, 10/295
liver, 10/412
liver cell, 10/409
lymphocyte, 10/335, 409
lymphocyte vacuolation, 10/335, 409
macrencephaly, 10/394
malate dehydrogenase, 10/291
megalocephaly, 10/206
membranous cytoplasmic inclusion body, 10/291, 295, 365-372, 408-410, 446, 467
mental deficiency, 46/56
micrencephaly, 10/415
microglia, 10/398
mucolipid, 10/390
muscle, 10/410
myenteric plexus, 10/406
myoclonic epilepsy, 49/617
Negro, 10/563
neurolipidosis, 66/20
neuronal ceroid lipofuscinosis, 10/221, 390
neuronal storage dystrophy, 21/61
neuropathology, 10/385-406
nosology, 10/224
optic atrophy, 13/83, 42/433
pathology, 66/18
phenylalanine, 10/295
pituitary gland, 10/406
prenatal diagnosis, 27/162, 42/434
retina, 10/405
Schaffer-Spielmeyer cell change, 10/216, 386
secondary neuroaxonal dystrophy, 49/408
Sephardim, 10/562
serine, 10/295
sialic acid, 10/292
spinal cord, 10/400
spinal muscular atrophy, 51/375
startle reaction, 10/288, 391, 42/433
strandin, 10/390
survey, 66/247
symptomatic dystonia, 49/543
terminology, 10/588-590
thalamus, 2/487
torpedo, 10/397, 410
treatment, 29/367
triphosphopyridine nucleotide diaphorase, 10/291
tyrosine, 10/296
ultrastructure, 10/291, 362, 407, 410, 412
valine, 10/296
variant, 66/252

GM2 gangliosidosis type I
β-*N*-acetylhexosaminidase A, 42/433
β-*N*-acetylhexosaminidase B, 27/156, 42/434
amniocentesis, 29/366
blindness, 22/537, 42/433
cherry red spot, 22/537, 42/433
genetic drift, 42/434
heterozygote detection, 42/434
heterozygote frequency, 42/434
hyperacusis, 42/433
hypothalamic dysfunction, 42/433
nosology, 10/224
optic atrophy, 42/433
opticocochleodentate degeneration, 21/550
prenatal diagnosis, 42/434
reproductive fitness, 42/434
selective advantage, 42/434
startle reaction, 42/433

GM2 gangliosidosis type II
β-*N*-acetylhexosaminidase A, 27/156, 29/365, 42/435, 66/60, 64
β-*N*-acetylhexosaminidase A deficiency, 59/402
β-*N*-acetylhexosaminidase B, 27/156, 29/365, 42/435, 66/60, 64
β-*N*-acetylhexosaminidase B deficiency, 27/154, 59/402
blindness, 42/435
carbohydrate metabolism, 27/163
cherry red spot, 42/435
dystonia, 60/664
Farber disease, 66/215
GM2 gangliosidosis, 66/249, 260
heterozygote detection, 42/436
hyperacusis, 42/435
inborn error of carbohydrate metabolism, 27/162
nosology, 10/224
prenatal diagnosis, 27/163, 42/436
seizure, 42/435
startle reaction, 42/435

GM2 gangliosidosis type III
β-*N*-acetylhexosaminidase A, 42/437
asialoganglioside, 10/298
biochemistry, 10/221
ceramide trihexoside, 10/298
computer assisted tomography, 60/671
β-galactosaminidase deficiency, 10/298, 414
globoside, 10/298
heterozygote detection, 42/436
mental deficiency, 42/436
optic atrophy, 42/436

prenatal diagnosis, 42/436
psychomotor retardation, 42/436
seizure, 42/436
spasticity, 60/661
unclassified disease, 10/505
GM3 ganglioside
catabolism, 66/60
white matter lipid, 9/4
GM3 gangliosidosis
epilepsy, 72/222
GM3 hematoside lipodystrophy
enzyme, 29/350
glycosphingolipid, 29/370
sphingolipid, 29/350
GNAT gene
acetyl-coenzyme A α-glucosaminide N-acetyltransferase, 66/282
glycosaminoglycanosis type IIIC, 66/282
Gnathanacanthidae intoxication
family, 37/78
venomous spine, 37/78
Gnathanacanthus intoxication
family, 37/78
venomous spine, 37/78
Gnathostomiasis
see also Eosinophilic encephalomyelitis
angiostrongyliasis, 35/329
differential diagnosis, 35/329
eosinophilic encephalomyelitis, 35/329
helminthiasis, 35/212
Gnosis
agnosia, 4/26
higher nervous activity disorder, 3/8
time sense, 3/229
Goatfish intoxication
CNS, 37/86
hallucination, 37/86
Godfried-Prick-Carol-Prakken syndrome
mental deficiency, 14/113
neurocutaneous syndrome, 14/101
Godin syndrome
differential diagnosis, 5/330
headache, 5/330
hypotension, 5/330
vasoconstriction, 5/330
Goiter
epiphysis, 42/375
exophthalmos, 2/320
multiple sclerosis, 9/112
Pendred syndrome, 42/368-370
sensorineural deafness, 42/375
Gökay-Tükel syndrome
hereditary cerebellar ataxia, 21/575
intention tremor, 21/575
Gold
chronic axonal neuropathy, 51/531
distal axonopathy, 64/13
neurologic complication, 38/497
neuropathy, 7/532
neurotoxicity, 51/265, 294, 305
neurotoxin, 7/532, 36/332, 38/497, 46/392, 51/265, 274, 294, 305, 531, 62/597, 64/4, 10, 13, 353
toxic encephalopathy, 64/4
toxic myopathy, 62/597
toxic neuropathy, 64/10, 13
toxic polyneuropathy, 51/274
Gold encephalopathy
epilepsy, 51/305
hallucination, 51/305
ophthalmoplegia, 51/307
Gold intoxication
action site, 64/10
agranulocytosis, 64/356
aplastic anemia, 64/356
auranofin, 64/355
aurothiomalate, 64/355
chorea, 49/559
dementia, 46/392
diarrhea, 64/356
elimination, 64/356
EMG, 64/356
exfoliative dermatitis, 64/356
half-life, 64/355
metal intoxication, 64/353
mortality, 64/356
Morvan fibrillary chorea, 49/293, 559
myokymia, 51/306, 64/13
neuropathy, 36/332
organ distribution, 64/355
polyneuropathy, 64/356
prevention, 64/356
psychiatric manifestation, 36/333
sensorimotor neuropathy, 64/356
sensorimotor polyneuropathy, 64/356
symptom, 64/4, 356
toxic encephalitis, 64/356
treatment, 64/356
Gold polyneuropathy
amyotrophy, 51/306
ataxia, 51/307
CSF, 51/306
meningism, 51/305
motor neuropathy, 51/306
neuralgic attack, 51/306
Gold sodium thiomalate, *see* Aurothiomalate

Gold therapy, *see* Chrysotherapy
Goldberg syndrome
cherry red spot, 42/439
dwarfism, 42/439
foam cell, 42/440
β-galactosidase, 42/440
GM1 gangliosidosis, 42/440
hyperreflexia, 42/439
kyphoscoliosis, 42/439
mental deficiency, 42/439
seizure, 42/439
sella turcica, 42/439
sensorineural deafness, 42/439
Goldenhar-Gorlin syndrome, *see* Goldenhar syndrome
Goldenhar syndrome
autosomal recessive, 38/412
axial mesodermal dysplasia spectrum, 50/516
birth incidence, 43/443
caudal aplasia, 50/516
CNS, 31/247
cri du chat syndrome, 31/566
differential diagnosis, 31/248
dysmorphic features, 50/214
ear malformation, 31/247
epibulbar dermoid, 32/127, 38/410, 43/442
eye lesion, 31/247
eyelid coloboma, 38/411, 43/442
face, 31/246
genetics, 31/246
heart disease, 43/443
hemifacial hypoplasia, 38/411
hemivertebra, 43/442
hypertelorism, 30/243
mental deficiency, 31/247, 46/46
oculovertebral syndrome, 31/248
oral lesion, 31/247
progressive hemifacial atrophy, 31/254, 59/479
skeletal abnormality, 31/247
trigeminal nerve agenesis, 38/411, 50/214
VACTERL syndrome, 50/514
vertebral abnormality, 43/443
Goldmann-Favre disease, *see* Hyaloidotapetoretinal degeneration (Goldmann-Favre)
Goldmann perimeter
retinal function, 13/12
Goldstein heredofamilial angiomatosis, *see* Hereditary hemorrhagic telangiectasia (Osler)
Goldstein, K., 1/9, 28
Golgi apparatus
muscle cell, 40/212
muscle pathology, 40/67
organolead intoxication, 64/132
organotin intoxication, 64/140
Golgi, C., 1/10, 27
Golgi cell
action, 2/416
cerebellar stimulation, 2/409, 411
Golgi-Mazzoni corpuscle
proprioception, 1/93
sensation, 7/82
Golgi route
nerve myelin, 51/30
Golgi system
muscle fiber, 62/21
Golgi tendon organ
muscle tone, 1/259
physiology, 1/65
Goltz-Gorlin syndrome
Aicardi syndrome, 31/227
alopecia, 43/19
brachycephaly, 14/113
CNS, 31/288
congenital dyskeratosis, 14/113
craniosynostosis, 14/788
cutaneous hypoplasia, 43/19
diastasis recti, 14/788
differential diagnosis, 14/113, 31/288
dwarfism, 14/113
Fanconi syndrome, 14/113
genetics, 31/288
growth retardation, 14/788
hemiparesis, 14/788
Hoffmann-Zurhelle syndrome, 14/113
hyperextensible joint, 43/19
hyperpigmentation, 43/19
incontinentia pigmenti, 14/108
iris coloboma, 14/787, 43/20
kyphoscoliosis, 43/19
mental deficiency, 14/787, 31/288, 43/20, 46/61
microcephaly, 14/113, 788, 30/508, 31/288
microphthalmia, 14/787, 43/20
mucocutaneous abnormality, 31/286
Naegeli syndrome, 14/113
neurocutaneous syndrome, 14/101
nevus lipomatosus cutaneous superficialis, 31/288
nystagmus, 14/113, 43/20
optic atrophy, 13/82, 14/113
optic nerve coloboma, 43/20
Pasini-Pierini syndrome, 14/113
poikiloderma, 46/61
polydactyly, 14/113
Rothmund-Thomson syndrome, 14/112, 31/288
scoliosis, 14/788
secondary pigmentary retinal degeneration, 14/113
sex ratio, 14/786, 43/20

skeletal abnormality, 14/787, 31/287
spina bifida, 14/788, 43/19
strabismus, 43/20
syndactyly, 43/19
telangiectasia, 43/198
Goltz-Peterson-Gorlin-Ravits syndrome, *see* Goltz-Gorlin syndrome
Goltz syndrome, *see* Goltz-Gorlin syndrome
Gonadal dysfunction
pineal tumor, 17/658
Gonadal dysgenesis
see also Noonan-Ehmke syndrome, Noonan syndrome *and* Turner syndrome
ataxia, 21/467
mixed, 31/503
monosomy X, 43/550-552
pseudohypoparathyroidism, 6/706, 49/420
pure, 31/503
Turner syndrome, 14/109, 43/550
Ullrich-Turner syndrome, 14/109
Gonadal function
sexual function, 26/451
Gonadal hypoplasia
xeroderma pigmentosum, 51/399
Gonadoblastoma
ARG syndrome, 50/582
Gonadorelin
Huntington chorea, 49/263
Gonadotropin
human menopausal, *see* Human menopausal gonadotropin
hypogonadal cerebellar ataxia, 21/468
pituitary adenoma, 17/393, 68/358
pituitary dwarfism type II, 42/613
Gonadotropin releasing hormone, *see* Gonadorelin
Gonda sign
diagnostic value, 1/250
Goniodysgenesis
Rieger syndrome, 43/470
Gonococcal bacteremia
meningitis rash, 52/25
Gonococcal meningitis
epidemiology, 33/103
penicillin, 33/103
Gonyalgia paresthetica
terminology, 1/108
Gonyaulax genus, *see* Alexandrium genus
Goodpasture syndrome
penicillamine intoxication, 49/235
vasculitis, 71/8
Gordon-Capute-Konigsmark syndrome
familial spastic paraplegia, 59/326, 333
secondary pigmentary retinal degeneration, 60/721
Gordon-Hay sign
juvenile hereditary benign chorea, 49/345
Gordon phenomenon
hereditary periodic ataxia, 21/569
Gordon reflex
Babinski sign, 1/184, 250
Gordon test
handedness, 4/255
Gorlin-Chaudry-Moss syndrome
antimongoloid eyes, 43/369
craniosynostosis, 43/369
dental abnormality, 43/369
hirsutism, 43/369
microphthalmia, 43/369
midface hypoplasia, 43/369
Gorlin-Goltz-Ward syndrome, *see* Multiple nevoid basal cell carcinoma syndrome
Gorlin LEOPARD syndrome, *see* LEOPARD syndrome
Gorlin syndrome, *see* Multiple nevoid basal cell carcinoma syndrome
Gossypol
hypokalemic periodic paralysis, 64/16
Malvaceae, 64/16
Gottron rash
dermatomyositis, 62/371
Gougerot-Sjögren syndrome, *see* Sjögren syndrome
Gout
carpal tunnel syndrome, 7/295, 42/309
lead intoxication, 64/435
migraine, 5/47
mummy, 74/188
neuropathy, 8/3
Gowers bundle, *see* Anterolateral funiculus
Gowers classification
epilepsy, 15/7
Gowers distal myopathy
clinical features, 40/306, 319, 322, 473
differential diagnosis, 40/426, 472
histochemistry, 40/50
history, 40/471
neuropathy, 40/471-481
non-Swedish type, 40/473
pathologic reaction, 40/253
Swedish type, 40/473
Gowers hemoglobin
Patau syndrome, 43/528
Gowers sign
Duchenne muscular dystrophy, 40/314, 359, 43/105, 62/117
late infantile acid maltase deficiency, 27/228
lumbar intervertebral disc prolapse, 20/587

Gowers, W.R., 1/4, 6
Goyer syndrome
  dominant ichthyosis, 22/490
  extremity involvement, 22/490
  hearing loss, 22/490
  hereditary progressive cochleovestibular atrophy, 60/772
  prolinuria, 22/490
  renal disease, 22/490
G6PD, *see* Glucose-6-phosphate dehydrogenase
GPI, *see* General paresis of the insane
Gracile nucleus
  Bassen-Kornzweig syndrome, 63/278
  neuroaxonal dystrophy, 49/392
Gradenigo-Lannois syndrome, *see* Gradenigo syndrome
Gradenigo postotitic inferior petrous sinus thrombophlebitis, *see* Gradenigo syndrome
Gradenigo syndrome
  abducent nerve, 48/302
  abducent nerve paralysis, 2/91, 313, 16/210
  apical petrositis, 33/181
  cranial epidural empyema, 52/168
  Garcin syndrome, 2/91
  headache, 5/210, 48/6
  Horner syndrome, 2/313
  mastoiditis, 2/313
  neuralgia, 2/91, 16/210
  nonrecurrent nonhereditary multiple cranial neuropathy, 51/571
  ophthalmic nerve, 2/91, 313
  orbital pain, 2/313
  otitis media, 2/91, 313, 16/210
  pain, 2/91, 16/210
  paralysis, 2/91, 16/210
  posterior fossa tumor, 18/408
  skull base, 67/146
  skull base tumor, 17/181, 68/468
  transverse sinus thrombosis, 52/176
  trigeminal nerve, 2/91
  trigeminal neuralgia, 2/313, 16/210, 48/302
Gram-negative bacillary meningitis
  Acinetobacter, 52/110, 117
  Acinetobacter calcoaceticus, 52/104
  adult, 33/97-105, 52/103
  aminoglycoside, 52/105
  antibiotic agent, 52/105
  antibiotic agent choice, 52/121
  azlocillin, 52/108
  aztreonam, 52/108
  bacterial meningitis, 52/3
  Bacteroides, 52/117, 121
  blood-brain barrier, 52/105
  blood clotting test, 52/104
  brain abscess, 52/120
  carbenicillin, 52/108
  cause, 52/104
  cefamandole, 52/107
  cefoperazone, 52/107
  cefoxitin, 52/107
  cefsulodin, 52/108
  ceftazidime, 52/108
  ceftizoxime, 52/107
  ceftriaxone, 52/107
  cefuroxime, 52/107
  cephalosporin, 52/105, 107
  chloramphenicol, 52/105, 108
  Citrobacter, 52/104, 117
  clindamycin, 52/121
  clinical features, 52/104, 118
  coma, 52/119
  complement deficiency, 52/105
  complication, 52/119
  co-trimoxazole, 52/109
  counterimmunoelectrophoresis, 52/104
  CSF, 33/103, 52/104, 119
  CSF culture, 52/105
  CSF endotoxin, 52/105
  CSF glucose, 52/104
  CSF lactic acid, 52/119
  deafness, 52/123
  diagnosis, 52/104
  Enterobacter, 52/104
  Enterobacter sakazakii, 52/118
  epilepsy, 52/122
  erythromycin, 52/121
  Escherichia coli, 52/104, 117
  Flavobacterium, 52/117
  Flavobacterium meningosepticum, 52/104, 120
  gentamicin, 52/106
  group B streptococci, 52/117
  head injury, 52/103
  Hemophilus influenzae, 33/53, 52/117
  host defence, 52/104
  hyponatremia, 52/122
  imipenem, 52/108
  incidence, 52/103, 117
  infant, 52/117
  intracranial hypertension, 52/119
  intrathecal treatment, 52/106
  irritability, 52/119
  jaundice, 52/119
  Klebsiella, 52/117
  Klebsiella pneumoniae, 52/104
  Klebsiella type 26, 52/118
  latex fixation test, 52/104

lethargy, 52/119
limulus lysate assay, 52/105
Listeria monocytogenes meningitis, 52/110
management, 52/109
mastoiditis, 52/103
mental status, 52/104
metronidazole, 52/121
mezlocillin, 52/108
Moraxella, 52/104
mortality, 52/103
moxalactam, 52/110
netilmicin, 52/106
neurosurgery, 52/103
nosocomial infection, 52/117
opsonin, 52/105
otitis media, 52/103
Paracolobactrum, 52/118
Paracolobactrum aeroginoides, 52/118
Pasteurella multocida, 33/103, 52/104
penicillin, 33/103, 52/105, 108
piperacillin, 52/108
predisposing factor, 52/103
Proteus, 52/104, 117
Providencia, 52/104
Pseudomonas aeruginosa, 52/104
rifampicin, 52/121
Salmon-Rickham reservoir, 52/120
Salmonella, 52/104, 117, 121
Serratia marcescens, 52/104
sex ratio, 52/117
Shigella, 52/121
spinal anesthesia, 52/103
Staphylococcus epidermidis, 52/110
Streptococcus pneumoniae, 52/110
suppurative foci removal, 52/110
ticarcillin, 52/108
tobramycin, 52/106
treatment, 52/105, 121
treatment senario, 52/109
vancomycin, 52/110, 121
ventriculitis, 52/119
vomiting, 52/119
X-linked recessive immunodeficiency, 52/117
Gram-negative bacilli
bacterial meningitis, 52/3
Gram-positive meningitis, *see* Staphylococcal meningitis *and* Streptococcal meningitis
Grand mal on awakening
age at onset, 73/178
EEG, 73/179
generalized tonic clonic seizure, 73/175
genetics, 73/178
prevalence, 73/178
seizure type, 73/176
Grand mal epilepsy
age at onset, 15/462
EEG, 15/468
hyperbeta-alaninemia, 29/230
hypoparathyroidism, 27/304
infantile spasm, 14/347
juvenile Gaucher disease, 42/438
lesion location, 15/463
postictal symptom, 15/112
preictal symptom, 15/108
psychology, 15/465
reading epilepsy, 2/621
sleep, 15/463
terminology, 15/3, 22, 108
theta rhythm, 42/667
Grand mal on sleeping
generalized tonic clonic seizure, 73/177
physiology, 73/180
Grand mal status
EEG, 15/150
prognosis, 15/151
status epilepticus, 15/146, 150, 462
Grande aphthose de Touraine, *see* Behçet syndrome
Granular brain cortex atrophy
brain thromboangiitis obliterans, 55/310
Divry-Van Bogaert syndrome, 55/318
Granular cell hypertrophy
cerebellar, *see* Cerebellar granular cell hypertrophy
Granular cell layer
mercury polyneuropathy, 51/272
Granular cortical atrophy
Binswanger disease, 54/224
brain thromboangiitis obliterans, 12/385
Granular layer
cerebellar external, *see* Cerebellar external granular layer
état glacé, 21/502
subpial, *see* Subpial granular layer
Granular layer atrophy
cerebellar, *see* Cerebellar granular layer atrophy
late cortical cerebellar atrophy, 21/500
Granular layer hypoplasia
cerebellar, *see* Cerebellar granular layer hypoplasia
Granule
Alder-Reilly, *see* Alder-Reilly granule
Birbeck, *see* Birbeck granule
eosinophil, 63/370
Reich, *see* Schwann cell
Reich μ, *see* Reich μ-granule
Reich π, *see* Reich π-granule

Tuffstein, *see* Tuffstein granule
Granule cell
cerebellar stimulation, 2/409, 411
epilepsy, 72/108
Granulocyte colony stimulating factor
Pearson syndrome, 66/427
Granulocytic leukemia, *see* Myeloid leukemia
Granulocytopenia, *see* Agranulocytosis
Granulofilamentous myopathy
congenital myopathy, 62/332
histopathology, 62/345
Granuloma
angiopathic polyneuropathy, 51/445
brain, *see* Brain granuloma
brucellosis, 52/585
Candida, *see* Candida granuloma
cerebrovascular disease, 55/432
coccidioidomycosis, 52/411
congenital cerebral, *see* Congenital brain granuloma
cysticercotic, *see* Cysticercotic granuloma
eosinophilic, *see* Eosinophilic granuloma
extradural nocardia, *see* Extradural Nocardia granuloma
Farber disease, 42/593
gangraenescens, *see* Midline granuloma
hypernatremia, 2/446
idiopathic face, *see* Midline granuloma
intraventricular, *see* Intraventricular granuloma
malignant, *see* Midline granuloma
neurobrucellosis, 52/585
nocardiosis, 35/524, 52/449
pituitary tumor, 17/427
spinal, *see* Spinal granuloma
spinal cord, *see* Spinal cord granuloma
spinal eosinophilic, *see* Spinal eosinophilic granuloma
subdural, *see* Subdural granuloma
subependymal, *see* Subependymal granuloma
thalamic tumor, 68/65
tuberculoid, *see* Tuberculoid granuloma
tuberculous meningitis, 52/200
Granulomatosis
brain, 16/229, 233
cerebral lymphomatoid, *see* Cerebral lymphomatoid granulomatosis
chronic, *see* Chronic granulomatosis
headache, 48/285
lipoid, *see* Lipoid granulomatosis
lymphomatoid, *see* Lymphomatoid granulomatosis
pathergic, *see* Pathergic granulomatosis
recurrent, 8/205, 222, 224
uranitis, *see* Uranitis granulomatosis
Wegener, *see* Wegener granulomatosis
Granulomatosis infantiseptica, *see* Listeria monocytogenes meningitis
Granulomatous amebic encephalitis
Acanthamoeba, 52/314, 316
Acanthamoeba astronyxis, 52/317
Acanthamoeba castellani, 52/317
Acanthamoeba culbertsoni, 52/317
Acanthamoeba polyphaga, 52/317
Acanthamoeba rhysodes, 52/317
ataxia, 52/317
bizarre behavior, 52/317
brain edema, 52/317
brain herniation, 52/317
brain tissue, 52/316
cerebellitis, 52/327
clinical features, 52/316
CNS invasion, 52/325
CSF, 52/316, 319
demyelination, 52/325
diagnostic test, 52/316
differential diagnosis, 52/316
encephalopathy, 52/317
entry portal, 52/316
flucytosine, 52/330
headache, 52/317
histopathology, 52/316
immunity, 52/325
immunocompromised host, 52/314
intracranial hypertension, 52/317
invasion route, 52/316
laboratory finding, 52/317
meningitis, 52/317, 327
multifocal demyelination, 52/325
neuroamebiasis, 52/309
olfactory nerve, 52/325
pathogenesis, 52/325
pathology, 52/316
phospholipase, 52/325
polymyxin B, 52/330
prognosis, 52/329
psychiatric symptom, 52/317
serologic test, 52/320
sex ratio, 52/316
sulfadiazine, 52/329
treatment, 52/316, 329
Granulomatous angiitis, *see* Allergic granulomatous angiitis
Granulomatous arteritis
allergic, *see* Allergic granulomatous angiitis
aorta, 48/316
cerebrovascular disease, 53/41

temporal arteritis, 48/320
temporal artery, 48/309
varicella zoster virus, 56/237
Granulomatous basal meningitis
Tolosa-Hunt syndrome, 48/303
Granulomatous cerebral arteritis
primary, *see* Primary granulomatous cerebral arteritis
Granulomatous CNS vasculitis, 55/387-397
see also Brain vasculitis
age, 55/389
allergic granulomatous angiitis, 55/396
amyloid angiopathy, 55/388
angiography, 55/395
asthma, 55/388
ataxia, 55/390
azathioprine, 55/397
brain angiography, 55/391
brain biopsy, 55/394
brain hemorrhage, 55/390
brain infarction, 55/389
cat scratch disease, 55/387
chronic meningitis, 55/391
chronic myelogenous leukemia, 55/388
chronic pyelonephritis, 55/388
clinical features, 55/388
corticosteroid, 55/387, 397
CSF, 55/390
cyclophosphamide, 55/387, 397
cytomegalovirus infection, 55/387
dementia, 55/390
diabetes mellitus, 55/388
differential diagnosis, 55/395
disease duration, 55/392-394
diverticulitis, 55/388
drug abuse, 55/396
EEG, 55/391
filariasis, 55/388
headache, 55/390
herpes zoster, 55/388
herpes zoster infection, 55/396
Hodgkin lymphoma, 55/388
human T-lymphotropic virus type III, 55/388
hypertension, 55/388
intracranial hypertension, 55/390
iridocyclitis, 55/388
leptomeningeal biopsy, 55/395
lymphomatoid granulomatosis, 55/396
meningeal carcinomatosis, 55/391
misdiagnosis, 55/390
mortality, 55/392-394
multifocal encephalopathy, 55/389
multiple sclerosis, 55/390
myelopathy, 55/390
nephritis, 55/388
neurosarcoidosis, 55/388
non-Hodgkin lymphoma, 55/388
pathogenesis, 55/387
pathology, 55/388
polyarteritis nodosa, 51/446, 55/396
relapsing myelopathy, 55/390
renal transplant recipient, 55/388
saccular aneurysm, 55/388
sarcoidosis, 55/396
schistosomiasis, 55/388
sex ratio, 55/389
Sneddon syndrome, 55/408
spinal tumor, 55/390
Takayasu disease, 55/395
temporal arteritis, 55/346
treatment, 55/397
Wegener granulomatosis, 55/396
Granulomatous disease
chronic, *see* Chronic granulomatous disease
polymyositis, 62/373
progressive multifocal leukoencephalopathy, 9/491, 34/307
Granulomatous encephalitis
Chagas disease, 52/346
history, 18/235
space occupying lesion, 16/231
Granulomatous inflammation
neurosarcoidosis, 63/422
reticulosarcomatosis, 18/241
Vogt-Koyanagi-Harada syndrome, 56/618
Granulomatous myositis
see also Inflammatory myopathy *and* Sarcoidosis
inflammatory myopathy, 41/85
Granulomatous vasculitis, *see* Allergic granulomatous angiitis
Granulovacuolar degeneration
Alzheimer disease, 22/399, 46/259, 49/250
experimental induction, 46/272
Guam amyotrophic lateral sclerosis, 22/399, 59/257
Guam amyotrophic lateral sclerosis-Parkinson dementia complex, 65/22
Guam motoneuron disease, 65/22
hippocampus, 46/259
Parkinson dementia, 49/177
Parkinson dementia complex, 22/399, 49/250
Pick disease, 49/250
progressive supranuclear palsy, 22/223, 49/248, 250
puromycin, 46/259, 272
Granulovacuolar lobular myopathy

congenital myopathy, 62/332
rimmed muscle fiber vacuole variant, 62/352
Granulovacuolar neuronal degeneration
description, 21/57
Down syndrome, 50/530
progressive supranuclear palsy, 22/223, 49/250
Graphanesthesia
terminology, 1/89, 104
Graphesthesia
tactile agnosia, 45/66
terminology, 1/104
Graphic dyskinesia, *see* Writers cramp
Graphospasm, *see* Writers cramp
Grasp reflex
Alzheimer disease, 46/252
brain death, 24/718
brain injury, 24/718
cingulate gyrus, 1/71
definition, 1/70
elicitation, 1/253
features, 1/253
frontal lobe syndrome, 55/138
infant, 4/346
Klüver-Bucy syndrome, 45/258
lesion site, 1/70
pellagra, 28/85, 46/337
Pick disease, 46/234
tonic, *see* Tonic grasp reflex
Grasping
oral, *see* Oral grasping
Grass pea, *see* Lathyrus sativus
Graves-Basedow disease, *see* Basedow disease
Graves disease, *see* Hyperthyroidism
Graves ophthalmopathy, *see* Thyroid associated ophthalmopathy
Graves ophthalmoplegia, *see* Thyroid associated ophthalmopathy
Gravitational stress
sympathetic nervous system, 74/653
Gravity
see also Space
autonomic nervous system, 74/274, 609
brain blood flow, 53/65
brain blood flow autoregulation, 53/65
cardiovascular function, 74/277
standing, 74/609
Gray matter
brain stem reticular formation, 3/64
central, *see* Central gray matter
central hemorrhagic necrosis, 12/582
central ischemic infarction, 12/586
composition, 66/39
fatty acid, 9/5
ganglioside, 9/4
mesencephalic, *see* Mesencephalic gray matter
periaqueductal, *see* Periaqueductal gray matter
Gray matter heterotopia
Arnold-Chiari malformation type II, 50/407
corpus callosum agenesis, 50/165
porencephaly, 50/356
Gray matter necrosis
anoxic leukomalacia, 47/526-528
Great anterior cerebellar vein, *see* Petrosal vein
Great auricular nerve, *see* Auricularis magnus nerve
Great cerebral vein, *see* Galen vein
Greater occipital nerve
anatomy, 2/134
Greater superficial petrosal neuralgia, *see* Cluster headache *and* Sphenopalatine neuralgia
Grebe-Myle-Löwenthal syndrome
Krabbe-Bartels disease, 14/407
Green seaweed intoxication
anesthetic effect, 37/97
caulerpicin, 37/97
depression, 37/97
respiratory complication, 37/97
vertigo, 37/97
Greenfield classification
diffuse sclerosis, 9/472, 10/15-17
hereditary olivopontocerebellar atrophy (Menzel), 21/416
nerve disease, 7/201
olivopontocerebellar atrophy, 21/397, 42/161
Greeting movement, *see* Infantile spasm
Gregg syndrome, *see* Postrubella embryopathy
Greig syndrome, *see* Hypertelorism
Grenet syndrome
definition, 2/239
Griesinger sign
transverse sinus thrombophlebitis, 33/174
transverse sinus thrombosis, 52/176, 54/397
venous sinus thrombosis, 54/431
Grimacing
dentatorubropallidoluysian atrophy, 21/519
facial, *see* Facial grimacing
hereditary olivopontocerebellar atrophy (Menzel), 21/441
neuroleptic dystonia, 65/281
olivopontocerebellar atrophy variant, 21/452-454, 456
progressive pallidal atrophy (Hunt-Van Bogaert), 6/457, 21/531
Rubinstein-Taybi syndrome, 31/282, 43/234
Grinker meylinopathy
carbon monoxide intoxication, 64/35
Grip

milking, *see* Milking grip
milkmaid, *see* Milking grip
tenodesis, *see* Tenodesis grip
Grisel syndrome
atlantoaxial dislocation, 5/373, 32/83
juvenile spinal cord injury, 61/233
juvenile vertebral column injury, 61/233
Griseofulvin
ataxia, 65/472
headache, 65/472
neurotoxin, 65/472
Groll-Hirschowitz syndrome
autonomic nervous system, 75/642
Grönblad-Strandberg syndrome
angioid streak, 13/141, 14/114
clinical features, 14/114
cutaneous dysplasia, 14/101
drusen, 13/251
Laurence-Moon-Bardet-Biedl syndrome, 14/114
neurocutaneous syndrome, 14/101
pseudoxanthoma elasticum, 14/114, 55/452
Stargardt disease, 13/134
Urbach-Wiethe disease, 14/114
Grönblad-Strandberg-Touraine syndrome, *see* Grönblad-Strandberg syndrome
Grooming reflex
inhibition, 1/50
Groove sign
eosinophilia myalgia syndrome, 63/375
Groping reflex
diagnostic value, 1/253
Grouper intoxication
vitamin A intoxication, 37/87
Grouting compound
neurotoxicity, 51/265
Grouting process
acrylamide polyneuropathy, 51/281
Growing pain
see also Abdominal migraine
restless legs syndrome, 8/313, 51/545
Growth
brain, *see* Brain growth
ECHO virus, 34/139
facial, *see* Facial growth
JC virus, 71/414
lung, *see* Lung growth
nerve, 51/18
neurite, *see* Neurite growth
perineural, *see* Perineural growth
plaque, 9/242-252
Growth acceleration
intrauterine, *see* Intrauterine growth acceleration
Growth deficiency syndromes
classification, 31/269-292
Growth factor
amyotrophic lateral sclerosis, 59/232
brain metastasis, 69/109
epidermal, *see* Epidermal growth factor
meningioma, 68/403
nerve, *see* Nerve growth factor
Growth factor receptor
brain metastasis, 69/109
epidermal, *see* Epidermal growth factor receptor
Growth failure
Edwards syndrome, 50/560
Patau syndrome, 31/506, 43/527, 50/558
Growth hormone
anorexia nervosa, 46/586
benign intracranial hypertension, 67/111
brain injury, 23/122
cluster headache, 74/508
depression, 46/449
Huntington chorea, 49/260, 262, 290
infantile optic glioma, 42/734
microcephaly, 30/513
multiple neuronal system degeneration, 75/173
myotonic dystrophy, 62/230, 249
optic nerve glioma, 42/734
pituitary dwarfism type III, 42/614
progressive dysautonomia, 59/154
pure autonomic failure, 75/173
Reye syndrome, 56/152
Growth hormone deficiency
lymphocytic choriomeningitis, 56/35
pituitary dwarfism type II, 42/614
Growth hormone releasing factor receptor
tumor cell, 67/23
Growth hormone secreting tumor
pituitary tumor, 68/344
Growth hormone secretion
depression, 46/474
pituitary tumor, 68/351
Russell syndrome, 16/349
schizophrenia, 46/474
Growth retardation
see also Short stature
Aarskog syndrome, 43/313
anencephaly, 30/175
ataxia telangiectasia, 14/290, 42/120, 60/361, 373
Berlin syndrome, 43/9
Bloom syndrome, 14/775, 43/10
Bowen-Conradi syndrome, 43/335
cerebro-oculofacioskeletal syndrome, 43/341
cleidocranial dysplasia, 43/348
Coffin-Siris syndrome, 43/240
craniodigital syndrome, 43/243

De Lange syndrome, 43/246
Down syndrome, 50/519
Dubowitz syndrome, 43/247
dyskeratosis congenita, 43/392
elfin face syndrome, 31/318, 43/259
Ellis-Van Creveld syndrome, 43/347
fetal alcohol syndrome, 30/120, 43/197
Freeman-Sheldon syndrome, 43/353
glycogen storage disease, 43/179
glycosaminoglycanosis type VI, 66/307
Goltz-Gorlin syndrome, 14/788
Hartnup disease, 29/155
hereditary sensory and autonomic neuropathy type III, 21/111, 43/58, 60/28
intrauterine, *see* Intrauterine growth retardation
Klinefelter variant XXXXY, 31/483
LEOPARD syndrome, 43/27
Marden-Walker syndrome, 43/423
Marinesco-Sjögren syndrome, 42/184
mental deficiency, 43/262, 272
methotrexate syndrome, 30/122
microcephaly, 30/510, 43/427
Mietens-Weber syndrome, 43/304
mitochondrial myopathy, 43/188
monosomy 21, 43/540
N syndrome, 43/283
Neu syndrome, 43/430
4p partial monosomy, 43/497
4p partial trisomy, 43/500
10p partial trisomy, 43/519
11p partial trisomy, 43/525
Patau syndrome, 31/506, 43/527, 50/558
Pena-Shokeir syndrome type I, 43/437
phosphorylase kinase deficiency, 41/192, 62/486
progeria, 43/465
4q partial monosomy, 43/496
11q partial monosomy, 43/523
18q partial monosomy, 43/533
4q partial trisomy, 43/498
10q partial trisomy, 43/518
14q partial trisomy, 43/529
19q partial trisomy, 43/538
r(13) syndrome, 50/585
r(14) syndrome, 50/590
r(21) syndrome, 43/543
recombinant chromosome 3 syndrome, 43/495
Rett syndrome, 60/637
Rothmund-Thomson syndrome, 14/315, 46/61
Rowley-Rosenberg syndrome, 43/473
Rubinstein-Taybi syndrome, 31/282, 284, 43/234
Russell-Silver syndrome, 43/476, 46/21
trichopoliodystrophy, 14/784
trisomy 22, 43/548
Werner syndrome, 43/489
xeroderma pigmentosum, 43/12
Gruber petrosphenoid ligament
abducent nerve paralysis, 2/91
nerve pressure, 2/91
Gruber syndrome
abdominal organ, 31/243
acrocephaly, 14/13
anencephaly, 14/115, 505, 43/391
arhinencephaly, 50/242
birth incidence, 43/392
cerebrohepatorenal syndrome, 31/243
cheilopalatoschisis, 30/247
CNS, 31/243
computer assisted tomography, 60/672
congenital brain, 30/99
congenital heart disease, 14/505
corpus callosum agenesis, 50/163
craniorachischisis, 14/505
differential diagnosis, 31/243
encephalomeningocele, 31/242
exophthalmos, 14/115
genetics, 31/242
holoprosencephaly, 43/391, 50/236
hypertelorism, 14/115, 30/244, 247
hyperventilation syndrome, 63/438
hypospadias, 14/115
iris coloboma, 14/13
meningocele, 14/13
microcephaly, 14/115, 505, 30/508, 31/242, 43/391
microphthalmia, 14/505
myelomeningocele, 14/115, 43/391
occipital encephalocele, 14/505, 43/391
oral lesion, 31/243
palatoschisis, 14/505
polycystic kidney, 14/505, 31/243
polydactyly, 14/505, 43/391
prenatal diagnosis, 43/392
pseudohermaphroditism, 14/505
renal abnormality, 43/391
Saldino-Noonan syndrome, 31/243
skeletal abnormality, 31/243
Smith-Lemli-Opitz syndrome, 31/243
spina bifida, 14/115, 505, 43/391
syndactyly, 14/13, 114
Ullrich-Feichtiger syndrome, 14/115, 31/244
VACTERL syndrome, 50/514
Von Hippel-Lindau disease, 14/60, 252, 505
G6S gene
N-acetylglucosamine-6-sulfatase, 66/282
glycosaminoglycanosis type IIID, 66/282
GTCS, *see* Generalized tonic clonic seizure

GTP binding protein, *see* Guanine nucleotide binding protein
Guam amyotrophic lateral sclerosis, 22/339-345, 59/253-266
aluminosilicate, 59/265
aluminum, 59/264-266, 275
aluminum uptake, 59/266
2-amino-3-methylaminopropionic acid, 59/263, 274
amyloid β-protein, 59/259
brain cortex, 22/287, 343
calcium, 59/264, 275
calcium dietary deficiency, 59/266
Chamorro people, 22/341
chronic aluminum intoxication, 59/266
clinical features, 59/255
cycad, 22/130, 344
Cycas circinalis hypothesis, 59/263
Cycas circinalis neurotoxin, 59/274
cytoskeletal protein, 59/257
epidemiology, 21/34, 59/259-262
etiology, 22/343-345, 59/262-265
experimental model, 59/265
features, 21/26
foamy spheroid body, 59/286-291
genetics, 49/170, 59/262
geographic trends, 59/259
granulovacuolar degeneration, 22/399, 59/257
Guam motoneuron disease, 41/442, 42/70
Hallervorden-Spatz syndrome, 22/403, 59/287
hereditary motor and sensory neuropathy type I, 21/26
hydroxyapatite, 59/265
hyperparathyroidism, 59/281
immunochemistry, 59/258
infectious agent, 59/262
intracisternal intoxication, 59/266
investigation, 22/283
iron, 59/265
Kii Peninsula amyotrophic lateral sclerosis, 59/253
lead, 59/264
lytico, 59/255, 259
magnesium, 59/276
manganese, 59/264, 275
manganese miner, 59/276
metal, 59/275-281
methylazoxymethanol glucoside, 22/130, 344
migration study, 59/261
mineral, 59/264
multinucleated giant cell, 59/294
myotonic dystrophy, 21/27, 22/342
neuroaxonal dystrophy, 21/63
neurofibrillary degeneration, 22/343, 59/256
neurofibrillary tangle, 22/377-380, 59/257-259, 278, 282-284
neuronal argyrophilic dystrophy, 21/64
neuropathology, 59/233, 237, 256-259, 294-296
New Zealand white rabbit, 59/266
nonhuman primate, 59/266
3-oxalylaminoalanine, 59/274
paralytico, 59/255, 259
Parkinson dementia, 22/341, 42/70, 249-251, 59/237
pathology, 22/129, 59/184
plant neurotoxin, 59/263
pseudopolyneuritic type, 59/256
secondary hyperparathyroidism, 59/265
silicon, 59/264
spheroid body, 59/287-291
substantia nigra, 22/287, 59/287-291
temporal trends, 59/259
tissue aluminum, 59/276-281
tissue calcium, 59/276-280
tissue metal, 59/276-281
trace element, 59/263
type, 59/180
ultrastructure, 59/258
Guam amyotrophic lateral sclerosis-Parkinson dementia complex
see also Guam motoneuron disease
aluminum intoxication, 64/276
dementia, 65/22
granulovacuolar degeneration, 65/22
neurofibrillary tangle, 65/22
parkinsonism, 65/22
Guam motoneuron disease
see also Cycad intoxication *and* Guam amyotrophic lateral sclerosis-Parkinson dementia complex
age at onset, 65/22
amyotrophic lateral sclerosis, 65/21
clinical form, 65/22
dementia, 65/22
granulovacuolar degeneration, 65/22
Guam amyotrophic lateral sclerosis, 41/442, 42/70
Guam Parkinson dementia, 42/70
neurofibrillary tangle, 65/22
neuropathology, 65/22
neurotoxicology, 65/22
Parkinson disease, 65/22
parkinsonism, 65/22
pathogenic mechanism, 41/336
progressive supranuclear palsy, 65/21
Guam Parkinson dementia, 42/249-251
Alzheimer disease, 27/488

biochemistry, 27/486
features, 46/378
genetics, 49/170
geography, 27/488, 49/167
Guam motoneuron disease, 42/70
motoneuron disease, 42/70
neurofibrillary tangle, 27/486
Parkinson disease, 49/108
Guanethidine
action mechanism, 37/435
antihypertensive agent, 37/435
antihypertensive agent intoxication, 37/436
autonomic nervous system, 75/17
chemical structure, 37/435
contraindication, 37/436
glaucoma, 37/436
pheochromocytoma, 37/436
Guanethidine intoxication
asthma, 37/436
bradycardia, 37/436
diarrhea, 37/436
exacerbation, 37/436
hypersensitivity, 37/436
noradrenalin, 37/436
orthostatic hypotension, 37/436
Guanidine
botulism, 41/289
Eaton-Lambert myasthenic syndrome, 41/358, 62/426
motoneuron disease, 22/39
myasthenic syndrome, 41/356
neuromuscular transmission, 7/112
poliomyelitis, 22/24
uremic polyneuropathy, 63/514
Guanidinosuccinic acid
uremic encephalopathy, 63/514
uremic polyneuropathy, 63/514
Guanine nucleotide binding protein
brain metastasis, 69/109
tumor cell, 67/23
Guanosine triphosphate cyclohydrolase I
defect, 75/236
epilepsy, 72/222
Gubler swelling
nerve lesion, 2/27
Gudden-Meynert ventral commissure
accessory optic bundle, 2/529
Guérin-Stern syndrome
albinism, 14/105
mental deficiency, 14/105
Turner syndrome, 14/109
Guillain-Barré polyradiculoneuropathy, *see* Guillain-Barré syndrome
Guillain-Barré syndrome, 7/495-507
see also Acute inflammatory polyradiculoneuropathy, Chronic inflammatory demyelinating polyradiculoneuropathy *and* Polyradiculitis
acquired immune deficiency syndrome, 56/493
acrodynia, 64/374
acute cord transection, 19/38
acute demyelinating neuropathy, 47/605, 608-612
acute disseminated encephalomyelitis, 47/606
acute dysautonomia, 51/247
acute polyneuropathy, 51/529
adenovirus, 56/290
allergy, 8/87, 9/500
anhidrosis, 63/153, 75/685
animal model, 34/403, 51/243, 56/581
annual incidence, 42/465
antecedent event, 51/239
antibody mediated demyelination, 51/250
antiphospholipid antibody, 63/330
areflexia, 42/465
arrhythmia, 51/257
arsenic intoxication, 51/266
arsenic polyneuropathy, 51/266
assisted ventilation, 51/256
ataxia, 51/245
atelectasis, 51/255
autoimmune response, 47/612
autonomic dysfunction, 51/256, 63/148
autonomic nervous system, 42/645, 51/245, 74/336, 75/14, 681
autonomic neuropathy, 51/245
autonomic polyneuropathy, 51/478
axonal damage, 51/244
B-lymphocyte, 51/248
bladder, 75/685
blood pressure, 51/245
bradyarrhythmia, 51/245
bradycardia, 51/257
brain stem encephalitis, 51/246
Brucella, 51/185
Burkitt lymphoma, 63/352
C-reactive protein, 51/248
cardiac dysrhythmia, 51/255, 63/238
cardiac pacemaker, 51/257
cardiovascular function, 51/257
cause, 9/596
cell mediated immunity, 51/249
chickenpox, 34/168
chlamydia, 51/242
chronic inflammatory demyelinating polyneuropathy, 51/247
chronic inflammatory demyelinating

polyradiculoneuropathy, 51/537
chronic mercury intoxication, 64/371
circulating immune complex, 51/249
clinical classification, 7/500
clinical features, 7/497-499, 40/297, 303-308, 318, 326, 336, 41/287, 47/610, 51/245
clinical syndrome, 34/393-395
clinical variant, 71/553
course, 7/499, 51/529
cranial nerve, 7/497, 42/645
cranial nerve palsy, 51/245
creatine kinase, 51/245
critical illness polyneuropathy, 51/576
CSF, 7/495, 498, 42/645, 51/247
CSF examination, 47/610
cytomegalovirus infection, 51/240
demyelination, 9/543, 42/645
diagnosis, 71/554
diagnostic criteria, 7/498, 47/609
differential diagnosis, 41/289, 291, 51/246
diphtheria, 7/503, 51/183
disulfiram polyneuropathy, 51/309
dysphagia, 51/245
Eaton-Lambert myasthenic syndrome, 62/425
EMG, 1/638
endoneurial mononuclear cell, 51/243
enterovirus, 56/330
epidemiology, 34/393, 51/239, 71/551
Epstein-Barr virus, 47/608, 51/240, 56/8, 254
Epstein-Barr virus infection, 56/252
etiology, 71/554
experimental allergic neuritis, 47/452, 458, 489, 611, 51/243, 252, 254
F wave latency, 51/248
facial diplegia, 51/245
Fisher syndrome, 2/296
heparin, 51/257
hepatitis B virus infection, 56/296, 298
hereditary cranial nerve palsy, 60/42
herpes zoster, 51/182
hexane polyneuropathy, 51/277
2-hexanone polyneuropathy, 51/277
HLA antigen, 42/646, 51/248
Hodgkin disease, 39/56
human immunodeficiency virus neuropathy, 71/354
hypertrophic interstitial neuropathy, 21/145
IgG, 9/140
immune mediated syndrome related to virus infection, 34/392-403, 423-425
immunologic aspect, 34/401-403
immunology, 47/611
inappropriate antidiuretic hormone secretion, 28/499
infection, 47/608-610
infectious mononucleosis, 9/597, 34/188, 56/252
inflammatory infiltration, 47/611
influenza vaccination, 51/242, 56/168
influenza virus, 51/187, 56/168
malignant lymphoma, 39/56
management, 75/689
Marek disease, 47/609, 51/250, 56/190, 581
mesaxon, 51/251
Miller Fisher syndrome, 51/245
multiple cranial neuropathy, 51/245, 247
multiple myeloma, 39/153, 63/394
multiple sclerosis, 34/437
mycoplasma, 47/608
Mycoplasma pneumoniae, 51/240
myelin basic protein P2, 47/611
myelin lesion, 51/251
nerve biopsy, 51/244
neurobrucellosis, 33/316, 52/594
neurologic intensive care, 55/204, 219
neuropathy, 8/365, 40/297, 303, 308, 318, 326, 336
neurophysiologic test, 47/610
ophthalmoplegia, 22/180, 51/245
orthostatic hypotension, 51/257, 63/148, 151
pain, 7/495, 51/245
papilledema, 7/501, 21/151, 51/246
paralysis, 7/497, 42/645
paraproteinemia, 39/153
pathogenesis, 7/502, 41/346, 71/554
pathology, 34/396-399, 47/611, 51/243
Pena-Shokeir syndrome type I, 59/369
periodic ataxia, 21/570, 60/648
periodic dysarthria, 21/570
peripheral neurolymphomatosis, 56/179-181
perivascular cuff, 47/611
physiology, 34/400
plasmapheresis, 51/255, 258
pneumonitis, 51/255
polyneuritis, 42/645
polyneuropathy, 41/287-291, 346
postinfectious complication, 47/481-489
precipitating event, 47/608-610
pregnancy, 8/14
primary demyelinating neuropathy, 47/605-612
primary demyelination, 51/243
prodromal factor, 34/400
prognosis, 7/506
progressive bulbar palsy, 59/375
progressive spinal muscular atrophy, 59/406
protein P2, 47/484
proteinocytologic dissociation, 7/495, 498

rabies, 56/385, 388, 391
rabies paralysis, 56/387
rabies vaccine, 9/544, 47/609
recurrent nonhereditary multiple cranial neuropathy, 51/570
Refsum disease, 60/235, 66/487, 497
respiratory insufficiency, 7/499, 51/255
sarcoid neuropathy, 51/196
sciatic nerve, 51/254
segmental demyelination, 51/70, 243
serogenetic neuropathy, 8/100
serum nerve antibody, 51/253
smallpox, 7/487
spinal muscular atrophy, 59/370
spinal nerve root swelling, 51/243
steroid, 51/257
survey, 71/551, 75/681
sweat production, 75/686
sweating, 7/499, 51/245
swine influenza vaccination, 47/483
symptom, 71/552
systemic lupus erythematosus, 51/447
T-lymphocyte, 47/611, 51/248
tachycardia, 7/499, 51/245, 257
tissue culture study, 51/250
treatment, 7/507, 47/612, 51/255
tremor, 49/589
tuberculosis, 51/186
typhoid fever, 51/187
ultrastructure, 51/243, 251
undetermined significant monoclonal gammopathy, 63/399, 402
undetermined significant monoclonal gammopathy IgM, 63/402
uremic polyneuropathy, 51/356, 360
vaccination induced, 56/391
vaccinogenic, 8/88
varicella zoster virus, 56/236
venous thrombosis, 51/257
ventilatory failure, 51/246
viral disease, 51/242
viral infection, 47/481
Waldenström macroglobulinemia, 63/397
zimeldine, 51/304
Guillain, G., 1/7, 38
Guillain-Garcin syndrome, *see* Garcin syndrome
Guillain-Mollaret triangle
progressive bulbar palsy, 22/140
tic, 6/798
Gumma
brain, *see* Brain gumma
neurosyphilis, 33/367, 384, 52/274, 277, 282
spinal cord, *see* Spinal cord gumma
Gunn, M., 1/6
Gunshot injury
see also Missile injury
amnesia, 46/622
angiography, 57/303
antibiotic agent, 57/311
anticonvulsant, 57/311
brachial plexus, 51/144
brain abscess, 33/116
civilian, 57/299
classification, 25/49
computer assisted tomography, 57/302
congenital arteriovenous fistula, 31/200
corticosteroid, 57/311
epilepsy, 57/311
hemorrhagic necrosis, 25/50
intracerebral hematoma, 57/303
intracranial infection, 24/227
medulla oblongata, 25/50
neuropathology, 25/49-51
penetrating, *see* Penetrating gunshot injury
pneumocephalus, 57/312
prognosis, 46/611
spinal cord injury, 25/49
subdural hematoma, 57/303
Gunshot sequel
paraplegia, 61/255
Gunther disease, *see* Congenital erythropoietic porphyria
GUSB gene
β-glucuronidase, 66/282
glycosaminoglycanosis type VII, 66/282
Gustatory hallucination
agnosia, 4/14
parietal operculum, 45/357
temporal lobe tumor, 17/286
Gustatory nerve
anatomy, 5/351, 358
Gustatory piloerection
gustatory sweating, 2/122
Gustatory sweating, 75/126
auriculotemporal nerve, 8/216
auriculotemporal syndrome, 1/457, 2/122, 8/276
cervical plexus, 2/121
gustatory piloerection, 2/122
syringomyelia, 50/445
thermal salivation, 2/122
Gustolacrimal reflex, *see* Crocodile tear syndrome
Guttmann sign
paralysis, 2/190, 202
spinal cord lesion, 2/190, 202
vasoconstriction, 2/190, 202
Guttmann, Sir L., 26/410

Guyon canal
compression neuropathy, 51/99
rheumatoid arthritis, 38/483
Gymnapistes intoxication
convulsion, 37/76
death, 37/76
delirium, 37/76
diarrhea, 37/76
hypotension, 37/76
hypothermia, 37/76
ischemia, 37/76
local pain, 37/76
nausea, 37/76
pallor, 37/76
radiation, 37/76
respiratory distress, 37/76
vomiting, 37/76
Gymnuridae intoxication
elasmobranche, 37/70
Gynecomastia
ataxia, 60/668
bulbar paralysis, 22/137
bulbospinal muscular atrophy, 59/44
cardiovascular agent intoxication, 37/426
digitalis intoxication, 37/427
glycoside intoxication, 37/426
hypogonadism, 60/668
Klinefelter syndrome, 43/557, 50/546, 550
Klinefelter variant XXXY, 43/559
Klinefelter variant XXYY, 31/484, 43/563
Klippel-Trénaunay syndrome, 14/524
mental deficiency, 43/273-275
POEMS syndrome, 47/619, 51/430, 63/394
spinal muscular atrophy, 59/44
Wohlfart-Kugelberg-Welander disease, 59/88
XX male, 43/555
Gyral abnormality
Edwards syndrome, 43/536
Gyrate atrophy
Alder abnormality, 13/259
choroid, 13/256-260
familial spastic paraplegia, 59/327
primary pigmentary retinal degeneration, 13/247
retina, 13/256-260
Gyrate atrophy (Fuchs)
choroid atrophy, 13/35
Klein-Waardenburg syndrome, 14/116
optic atrophy, 13/82
Gyrectomy
causalgia, 1/110
phantom limb pain, 1/110
Gyromitra esculenta
mushroom intoxication, 36/536-538
mushroom toxin, 65/36
Gyromitrine
see also Mushroom toxin
ataxia, 64/3
brain edema, 64/3
chemistry, 65/36
epilepsy, 64/3
headache, 64/3
neurotoxin, 36/536, 64/3, 65/36
source, 64/3
toxic encephalopathy, 64/3
Gyromitrine intoxication
carcinogeneticity, 65/36
coma, 65/36
delirium, 65/36
diplopia, 65/36
epilepsy, 65/36
headache, 65/36
hepatotoxicity, 65/36
1-methyl-1-formylhydrazine, 65/36
mydriasis, 65/36
neurologic symptom, 65/36
nystagmus, 65/36
renal effect, 65/36
symptom, 64/3
vertigo, 65/36
Gyrus
anterior cingulate, *see* Anterior cingulate gyrus
arcuate, *see* Arcuate gyrus
cingulate, *see* Cingulate gyrus
dentate, *see* Dentate gyrus
frontal, *see* Frontal gyrus
Heschl, *see* Heschl gyrus
postcentral, *see* Postcentral gyrus
precentral, *see* Precentral gyrus
superior frontal, *see* Superior frontal gyrus
superior temporal, *see* Superior temporal gyrus

H protein
measles virus, 56/423
H protein antibody
subacute sclerosing panencephalitis, 56/427
H-reflex
acrylamide intoxication, 64/71
alpha activity, 1/67
amyotrophic lateral sclerosis, 22/304
critical illness polyneuropathy, 51/578
diabetic polyneuropathy, 51/504
direct muscle response, 7/142
dystonia musculorum deformans, 6/537
leprous neuritis, 51/222
method, 1/244
Parkinson disease, 49/70

progressive spinal muscular atrophy, 22/33, 38
rehabilitation, 12/473
schizophrenia, 46/505
spinal cord experimental injury, 25/13
Sydenham chorea, 6/421
vibration, 1/64
H-substance
causalgia, 8/328, 22/255
H zone
skeletal muscle, 40/63
H-2 locus
mouse, 42/662
Habenular commissure
embryo, 50/11
Habenular nucleus
anatomy, 2/473
connection, 2/473
lateral, *see* Lateral habenular nucleus
Hachinski ischemic score
multi-infarct dementia, 46/354, 55/142
Haferkamp syndrome
Bean syndrome, 14/106
Haff disease
acquired myoglobinuria, 62/577
myalgia, 41/272
myoglobinuria, 40/339, 41/272
toxic myopathy, 62/601, 612
unknown origin, 41/271
Hageman factor
coagulation, 1/129
Hair
argininosuccinic aciduria, 42/525
arsenic intoxication, 36/206
brittle, *see* Brittle hair
Coffin-Siris syndrome, 43/240
hair brain syndrome, 43/297
lead intoxication, 64/445
manganese intoxication, 64/309
trichopoliodystrophy, 29/288, 42/584
twisted, *see* Pili torti
white, *see* White hair
white forelock, *see* White hair
Hair abundance
9p partial monosomy, 50/579
Hair brain syndrome
brittle hair, 43/297-299
hair, 43/297
hair protein, 43/298
mental deficiency, 43/297
short stature, 43/297-299
trichorrhexis nodosa, 43/297
Hair loss, *see* Alopecia
Hair protein
hair brain syndrome, 43/298
Hair tuft
diastematomyelia, 50/436
diplomyelia, 50/436
Hairy cell leukemia
brain tumor, 18/235
immunocompromised host, 56/470
Hajdu-Cheney syndrome, *see* Arthrodento-osteodysplasia
Haldane rule
Becker muscular dystrophy, 41/416
Duchenne muscular dystrophy, 40/353, 41/416
mutation rate, 40/353, 41/416
Half base syndrome, *see* Garcin syndrome
Half skull base syndrome, *see* Garcin syndrome
Halichoerus grypus intoxication
vitamin A, 37/96
Haliotis intoxication
ingestion, 37/64
Hall, M., 1/6
Hallermann-Streiff syndrome
bird like face, 43/403
cataract, 43/403
CNS, 31/249
cutaneous atrophy, 43/403
dental abnormality, 43/403
differential diagnosis, 31/250
eye lesion, 31/248
face, 31/248
hydrocephalus, 31/248
hypotrichosis, 43/403
mental deficiency, 43/404, 46/93
microphthalmia, 43/403
oral lesion, 31/249
pseudoprogeria, *see* Pseudoprogeria Hallermann-Streiff syndrome
short stature, 43/403
skin lesion, 31/249
spina bifida, 43/403
Hallervorden classification
diffuse sclerosis, 10/15, 24
Hallervorden-Spatz syndrome, 6/604-628, 66/711-718
acanthocyte, 49/332
acanthocytosis, 63/271, 290
β-*N*-acetylhexosaminidase A, 66/714
adult onset, 66/716
age at onset, 49/399
Alzheimer disease, 49/401
Alzheimer neurofibrillary change, 6/623
amyotrophic lateral sclerosis, 6/623, 22/323, 383, 397, 59/410
aplastic anemia, 6/625

athetosis, 6/61, 155, 454, 49/386
atom bomb victim, 6/625
autosomal recessive inheritance, 49/399
basal ganglion, 22/403
basal ganglion tier III, 49/24
behavior disorder, 42/257
behavior disorder in children, 42/257
biochemistry, 6/128
brain cortex, 49/396
brain stem nucleus, 49/396
caudate nucleus atrophy, 49/396
caudatolenticular bridge, 49/396
ceroid lipofuscin, 66/713
ceruloplasmin, 66/714
childhood, 66/715
choreoathetosis, 42/257, 49/385, 399, 66/712, 714
chorioretinal degeneration, 13/39
classification, 66/715
clinical features, 49/399, 66/712
cutaneous pigmentation, 6/605, 610
cysteine dioxygenase, 66/713
cystine, 49/402, 66/713
dementia, 42/257, 46/130, 401, 49/399, 59/410
dentatorubropallidoluysian atrophy, 21/520
diagnosis, 66/714
double athetosis, 6/155, 49/386
duration, 49/399
dwarfism, 6/604, 607
dysmyelinic state, 6/661
dystonia, 6/523, 551-554, 49/399, 66/712, 714
dystonia musculorum deformans, 49/524
early onset, 66/715
epidemiology, 66/713
epilepsy, 15/337
extrapyramidal disorder, 66/714
extrapyramidal system, 22/403
familial form, 66/712
familial spastic paraplegia, 59/326, 333
Friedreich ataxia, 6/623, 21/354
genetics, 66/713
globus pallidus, 49/396, 66/713
glutathione-cysteine-disulfide, 49/402
glycosaminoglycanosis type I, 6/616
GM1 galactosidase, 66/714
Guam amyotrophic lateral sclerosis, 22/403, 59/287
hereditary case, 6/605-614
histochemistry, 6/622
history, 66/711
Hodgkin disease, 6/625
Huntington chorea, 6/337, 49/293, 66/714
hyperkinesia, 22/403, 42/257
hyperpigmentation, 14/424
infantile neuroaxonal dystrophy, 42/230
infantile type, 6/609
intracerebral calcification, 49/417
iron, 49/35
juvenile type, 6/617
Kii Peninsula amyotrophic lateral sclerosis, 22/403
laboratory evaluation, 66/716
late cortical cerebellar atrophy, 21/477
late infantile neuroaxonal dystrophy, 42/257, 49/399
late infantile type, 6/609, 617
late onset, 49/130, 399
Leber optic atrophy, 13/99
Lewy body, 6/624, 49/401
Luys nucleus, 66/713
management, 66/717
muscular atrophy, 6/605, 626
myoclonus, 6/166, 618, 21/514, 49/618
neuroacanthocytosis, 49/401, 60/673, 63/271
neuroaxonal dystrophy, 6/608, 21/63, 49/395, 397
neurocutaneous melanosis, 14/425
neurofibromatosis, 6/625, 49/401
neuroleptic malignant syndrome, 66/715
neuropathology, 49/400, 66/22
nontypical, 49/398
nosology, 6/604, 627
nuclear magnetic resonance, 66/716
nyctalopia, 49/401
olivopontocerebellar atrophy, 49/401
opisthotonos, 66/715
optic atrophy, 6/626, 13/39, 49/401, 66/712
pallidal degeneration, 6/650
pallidal necrosis, 49/465, 481
pallidoluysionigral degeneration, 49/458
pallidonigral necrosis, 49/465
pallidostriatal necrosis, 49/465
Parkinson dementia, 49/401
Parkinson dementia complex, 6/623, 22/383, 397
Parkinson disease, 49/130
parkinsonism, 6/624, 628, 22/403, 49/130
Pick disease, 6/614
primary neuroaxonal dystrophy, 49/397
progressive, 66/715
progressive bulbar palsy, 59/225
progressive pallidal degeneration, 42/257
pyramidal system dysfunction, 49/399
rheumatic encephalitis, 6/625
rigidity, 49/399, 66/712, 714
sea blue histiocyte, 49/401
secondary pigmentary retinal degeneration, 6/606, 610, 13/336, 60/731
solitary case, 6/614-624

spastic paraplegia, 59/439
spasticity, 66/712
spheroid body, 6/605, 22/403
striatal syndrome, 2/500
striatopallidodentate calcification, 49/417
substantia nigra, 22/403, 49/395
substantia nigra degeneration, 42/257
symptomatic dystonia, 6/551, 554
therapy, 66/716
tremor, 6/606, 626, 66/712
typical familial, 49/398
typical nonfamilial, 49/398
Hallgren syndrome
chorioretinal degeneration, 13/40
deafness, 13/242, 338
Friedreich ataxia, 13/301
optic atrophy, 13/40
primary pigmentary retinal degeneration, 13/242, 298, 320
Hallipre test
topographical diagnosis, 2/28
Hallpike technique
description, 1/577
Hallucination, 4/327-336
acrylamide intoxication, 64/68
acrylamide polyneuropathy, 51/281
agnosia, 4/14, 45/352
alcohol intoxication, 4/329, 331
alcoholism, 43/197
Alice in Wonderland syndrome, 5/244
aluminum neurotoxicity, 63/525
Amanita muscaria intoxication, 37/329
Amanita pantherina intoxication, 36/534
amantadine, 46/594
amphetamine intoxication, 65/50
antihypertensive agent, 63/87
aprindine intoxication, 65/455
auditory, *see* Auditory hallucination
autoscopic, *see* Autoscopia
barbiturate intoxication, 65/332
Behçet syndrome, 56/598
beta adrenergic receptor blocking agent intoxication, 65/436
bismuth intoxication, 64/337
body scheme disorder, 3/31, 254
brain cortex stimulation, 4/336
brain edema, 63/417
brain embolism, 53/206
brain hypoxia, 63/417
brain infarction, 53/206
bufotenine intoxication, 65/45
calcium antagonist intoxication, 65/442
cannabis, 4/332
cerebrovascular disease, 53/206
chlormethine intoxication, 65/534
ciclosporin intoxication, 65/553
ciguatoxin intoxication, 65/163
cocaine intoxication, 4/331, 65/49, 255
consciousness, 3/116, 45/117
cortical blindness, 2/670
damsel fish intoxication, 37/86
deafferentation, 4/213
definition, 4/327, 45/351
delirium, 46/525-527
dextropropoxyphene intoxication, 65/358
dialysis encephalopathy, 64/277
diffuse Lewy body disease, 75/189
digoxin intoxication, 65/450
disulfiram intoxication, 37/321
drug induced, 4/331-336
dysphrenia hemicranica, 5/79
elementary, *see* Elementary hallucination
epilepsy, 4/330, 15/95, 45/508
extracampine, *see* Extracampine hallucination
frontal lobe lesion, 45/32
frontal lobe tumor, 17/259
functional, *see* Functional hallucination
general paralysis, 4/330
goatfish intoxication, 37/86
gold encephalopathy, 51/305
gustatory, *see* Gustatory hallucination
hallucinogenic agent, 3/125, 4/330
hallucinosis, 46/561
Hartnup disease, 29/155
headache, 5/244, 250, 48/361
hereditary acute porphyria, 51/391
hippocampus, 45/355
historic survey, 4/327
hyoscyamine intoxication, 65/48
hyperosmolar hyperglycemic nonketotic diabetic coma, 27/90
hypnagogic, *see* Hypnagogic hallucination
hypoparathyroidism, 27/301
ictal, 4/330
imagery, 45/351-369
intracranial hemorrhage, 53/206
kinesthetic, *see* Kinesthetic hallucination
Lafora progressive myoclonus epilepsy, 15/385, 27/172
lead encephalopathy, 64/436
lead intoxication, 4/330
levodopa, 46/593
lilliputian, *see* Lilliputian hallucination
limbic system, 45/355
lymphocytic choriomeningitis, 9/197
lysergide intoxication, 4/334, 37/329, 332, 65/44

macrosomatognosia, *see* Body scheme disorder
manganese intoxication, 64/308
manic depressive psychosis, 43/208
mescaline, 4/333, 45/365
mescaline intoxication, 65/45
metabolic deprivation, 4/329
metamorphopsia, 45/361-363
methamphetamine, 65/366
3-methoxy-4,5-methylenedioxyamphetamine intoxication, 65/50
3,4-methylenedioxyamphetamine, 65/366
3,4-methylenedioxyamphetamine intoxication, 65/50
3,4-methylenedioxyethamphetamine intoxication, 65/50
3,4-methylenedioxymethamphetamine intoxication, 65/50
1-methyl-4-phenyl-1,2,3,6-tetrahydropyridine intoxication, 65/370
migraine, 5/244, 250, 42/746, 48/161, 166, 361
mountain sickness, 63/417
muscimol intoxication, 65/40
musical, *see* Musical hallucination
myoclonic epilepsy, 42/702
nalorphine, 37/366
narcolepsy, 2/448, 45/354, 511
negative, *see* Negative hallucination
nifedipine intoxication, 65/442
nimodipine intoxication, 65/442
norpethidine intoxication, 65/357
occipital lobe, 45/364-368
occipital lobe lesion, 4/330
occipital lobe syndrome, 45/56-58
occipital lobe tumor, 67/154
olfactory, *see* Olfactory hallucination
organic solvent intoxication, 64/42
organolead intoxication, 64/133
palinacousia, 45/367
parietal lobe, 45/358-364
parietal lobe epilepsy, 73/101
pathogenesis, 4/329-331
peduncular, *see* Peduncular hallucination *and* Peduncular hallucinosis
pellagra, 7/571, 38/648, 46/336
pentazocine intoxication, 65/357
perception, 45/351-369
pethidine intoxication, 65/357
phencyclidine intoxication, 65/52, 367
phenethylamine intoxication, 65/50
phenytoin intoxication, 65/500
prazosin intoxication, 65/447
procainamide intoxication, 37/451
progressive myoclonus epilepsy, 42/702
propranolol intoxication, 37/447
pseudo, *see* Pseudohallucination
psilocin, 65/40
psilocybine, 4/333, 65/40
psychomotor, *see* Psychomotor hallucination
psychosis, *see* Peduncular hallucinosis
psychotropic substance, 4/330
Reye syndrome, 29/333, 65/116
rudderfish intoxication, 37/86
schizophrenia, 43/212, 45/355, 46/419
scopolamine intoxication, 65/48
sensory deprivation, 4/329
sleep deprivation, 4/329
spiromustine intoxication, 65/534
steroid function, 4/330
striatopallidodentate calcification, 49/423
tabernathe iboga intoxication, 37/329
television, *see* Television hallucination
temporal arteritis, 55/344
temporal lobe, 45/355-358
temporal lobe epilepsy, 2/706, 42/685
temporal lobe lesion, 4/331
temporal lobe tumor, 17/286, 46/568
terminology, 4/328
tetrahydrocannabinol intoxication, 65/47
thallium intoxication, 64/325
tocainide intoxication, 65/455
transient ischemic attack, 53/206
tricyclic antidepressant intoxication, 65/321
3,4,5-trimethoxyamphetamine intoxication, 65/50
tryptamine derivative intoxication, 37/329, 331
uremic encephalopathy, 63/504
verbal, *see* Verbal hallucination
verbal anosognosia, 45/378
vincristine, 63/359
visual, *see* Visual hallucination
visual perseveration, 45/364-366
Wernicke-Korsakoff syndrome, 4/331, 45/194
Hallucination classification
allesthesia, 45/56
chromatopsia, 45/56
cinematographic vision, 45/56
dysmorphopsia, 45/56
fortification spectrum, 45/56
macropsia, 45/56
metachromatopsia, 45/56
metamorphopsia, 45/56
micropsia, 45/56
mosaic vision, 45/56
phosphene, 45/56
polyopia, 45/56
pseudohallucination, 4/328
scintillation, 45/56

scotoma, 45/56
teichopsia, 45/56
teleopsia, 45/56
visual perseveration, 45/56
zoom vision, 45/56
Hallucinatory fish intoxication
classification, 37/79
CNS, 37/86
damsel fish, 37/86
goatfish, 37/86
rudderfish, 37/86
Hallucinogen, *see* Hallucinogenic agent
Hallucinogenic agent
see also Amphetamine, Brolamfetamine, Bufotenine, Cocaine, *N,N*-Dimethyltryptamine, Hyoscyamine, Lysergide, 3-Methoxy-4,5-methylenedioxyamphetamine, 4-Methyl-2,5-dimethoxyamphetamine, 3,4-Methylenedioxyamphetamine, 3,4-Methylenedioxyethamphetamine, 3,4-Methylenedioxymethamphetamine, Phencyclidine, Scopolamine, Tetrahydrocannabinol *and* 3,4,5-Trimethoxyamphetamine
Amanita muscaria, 37/329
amphetamine, 65/41
behavior disorder, 46/594
brolamfetamine, 65/42
bufotenine, 65/42
cannabinol, 65/41
classification, 65/42
cocaine, 65/41
consciousness, 3/125
N,N-dimethyltryptamine, 65/42
epilepsy, 72/234
hallucination, 3/125, 4/330
hyoscyamine, 65/42
ibogaine, 65/43
ibotenic acid, 65/42
indole alkylamine, 65/41, 43
levodopa, 37/235
lysergide, 37/329, 65/42
mescaline, 65/42
4-methoxyamphetamine, 65/42
5-methoxy-*N,N*-dimethyltryptamine, 65/42
3-methoxy-4,5-methylenedioxyamphetamine, 65/42
4-methyl-2,5-dimethoxyamphetamine, 65/42
3,4-methylenedioxyamphetamine, 65/42
3,4-methylenedioxyethamphetamine, 65/42
3,4-methylenedioxymethamphetamine, 65/42
muscimol, 65/42
phencyclidine, 65/41
phenethylamine, 65/41
psilocin, 65/42
psilocybine, 64/3, 65/42
psychosis, 46/594
schizophrenia, 46/448
scopolamine, 65/42
structure, 65/42
tabernathe iboga, 37/329
tetrahydrocannabinol, 65/42
2,4,5-trimethoxyamphetamine, 65/42
3,4,5-trimethoxyamphetamine, 65/42
tropane alkaloid, 65/41
tryptamine derivative, 37/329
Hallucinogenic agent intoxication, 37/329-342
Amanita muscaria, 37/329
cannabis, 37/329
chemistry, 37/334
clinical value, 37/340-342
ibogaine, 37/329
lysergide, 37/329, 332-342
mescaline, 37/330
pharmacology, 37/336-340
symptom, 37/329
tabernathe iboga, 37/329
tryptamine, 37/329, 331
Hallucinosis, 46/561-568
alcoholic, *see* Alcoholic hallucinosis
anatomoclinical correlation, 46/567
autophonia, 46/562
autoscopia, 2/714, 46/562
brain stimulation, 46/566
cerebral localization, 46/566
chronic tactile, *see* Chronic tactile hallucinosis
definition, 46/563
dysphrenia hemicranica, 5/79
extracampine hallucination, 46/562
functional hallucination, 46/562
hallucination, 46/561
hallucination like experience, 46/562
lilliputian hallucination, 46/562
musical hallucination, 46/568
negative hallucination, 46/562
palinacousia, 46/562
pathogenesis, 46/595
peduncular, *see* Peduncular hallucinosis
polyarteritis nodosa, 55/356
pseudohallucination, 46/563
psychomotor behavior, 46/562
psychomotor hallucination, 46/562
sound induced photism, 46/562
specific syndrome, 46/564-566
systemic lupus erythematosus, 55/376
television hallucination, 46/567

temporal lobe, 46/566
temporal lobe epilepsy, 46/566
visual perseveration, 46/562
Haloperidol
chemical formula, 65/276
chemical structure, 65/398
classification, 74/149
Gilles de la Tourette syndrome, 49/632
1-methyl-4-phenylpyridinium, 65/397
neuroleptic agent, 65/276
pathochemistry, 65/398
toxic myopathy, 62/601
Halothane
malignant hyperthermia, 38/552, 41/268, 270, 43/118, 62/567
muscle, 40/546
succinylcholine, 41/268, 270
Halothane intoxication
animal experiment, 37/411
convulsion, 37/411
inhalation anesthetics, 37/410
in vitro study, 37/410
in vivo, 37/411
Haloxon
opiate, 65/350
Halpern, L., 1/14, 28
Halstead Category Test
epilepsy, 15/565
Halstead-Reitan test battery
neuropsychologic defect, 45/525
Halsted, W.S., 1/17
Halter traction
cervical vertebral column injury, 25/291
Haltia-Hagberg-Santavuori syndrome, *see* Infantile neuronal ceroid lipofuscinosis
Haltia-Santavuori syndrome, *see* Infantile neuronal ceroid lipofuscinosis
Hamartie, *see* Phakos
Hamartoblastoma, *see* Phakoblastoma
Hamartoblastomatosis, *see* Phakoblastomatosis
Hamartoma
see also Phakoma
benign, *see* Benign hamartoma
brain, 12/104
choroid, *see* Choroid hamartoma
definition, 14/53
hypothalamic, *see* Hypothalamic hamartoma
hypothalamic tumor, 68/71
hypothalamus tumor, 18/366
multiple visceral hamartoma, 14/72
neurofibromatosis, 14/36, 68/291
neurofibromatosis type I, 14/36, 136
optic chiasm compression, 68/75
precocious puberty, 18/367
primary cardiac tumor, 63/93
retinal, *see* Retinal hamartoma
spinal lipoma, 20/397
Sturge-Weber syndrome, 18/276
subarachnoid hemorrhage, 12/104
survey, 68/333
thalamic tumor, 68/65
tongue, *see* Tongue hamartoma
tuberous sclerosis, 18/300
Hamartomelanosis
leptomeningeal, *see* Leptomeningeal melanosis
Hamartosis, *see* Phakomatosis
Hamburg, Germany
neurology, 1/9
Hammer toe
diastematomyelia, 50/436
diplomyelia, 50/436
Refsum disease, 60/231
Hammersmith, United Kingdom
neurology, 1/7
Hamstring reflex
internal, *see* Internal hamstring reflex
Hamycin
North American blastomycosis, 35/407, 52/392
Hand-ear-eye test
head, *see* Head's hand-ear-eye test
Hand-eye coordination
corpus callosum agenesis, 50/161
Hand flapping
autism, 43/203
Hand malformation
arthrogrypotic like, *see* Arthrogrypotic like hand malformation
claw, *see* Claw hand
Parkinson dementia, 49/172
Hand-Schüller-Christian disease
see also Histiocytosis X
brain biopsy, 10/685
cerebral histiocytosis, 39/531
cholesterol, 10/343, 38/96, 42/442
cranial lesion, 38/95
cranial vault tumor, 17/115
diabetes insipidus, 38/96, 42/442
exophthalmos, 38/95
fibrous dysplasia, 14/198
histiocytosis X, 38/95, 42/441
history, 38/95
hypernatremia, 2/446
Langerhans cell histiocytosis, 70/56
polyuria, 38/95
skull base, 18/322
soft tissue, 38/95

space occupying lesion, 16/231
xanthoma, 27/248
xanthomatosis, 10/343
Hand sign
hollow, *see* Hollow hand sign
Hand turning test
hypermetria, 1/324
Hand writhing
Rett syndrome, 29/305, 312, 60/636, 664, 63/437
Handedness, 4/248-270, 273-289, 291-309, 312-324
see also Cerebral dominance, Dominant hemisphere *and* Manual dexterity
acquired factor, 4/252
agraphia, 4/150
ambidexterity, 4/316, 348, 466
ambilevous, 4/316
amphiocularity, 4/258
anatomic basis, 4/253, 269
André-Thomas extensibility test, 4/255
animal, 4/250, 307
aphasia, 4/6, 88, 264, 316-318, 45/303
aphasia in children, 4/265
apraxia, 45/429
audition, 4/320-323
behavior disorder in children, 4/465-469
brain maturation, 4/255
classification, 4/257-259
Corominas test, 4/255
cultural factor, 4/250
degree, 4/256-259, 315, 466
developmental deprivation, 4/468
developmental dyslexia, 4/380, 383, 46/117
developmental dyspraxia, 4/465-469
dichotic listening, 4/320-323
dynamometer test, 4/315
dyslexia, 46/129, 131
earedness, 4/259, 320-323, 466
EEG, 4/259, 260
examination, 4/254-259, 269, 313
eyedness, 4/258, 323, 466
footedness, 4/466
genotypical factor, 4/251-253
Gordon test, 4/255
higher nervous activity, 4/465-469
historic survey, 4/248, 291
historic time, 4/249
human body asymmetry, 4/250
inheritance, 4/251, 269, 316
language disorder, 4/261
language function, 4/249
lateralization pattern, 4/466
left, 45/429
manual dexterity, 4/449, 466
manual preference, 4/249-251, 347
manual superiority, 4/249-261
mixed preference, 4/258, 312, 315
nervous control, 4/249
ontogeny, 4/250, 312
Ozeretski test, 4/255
pathologic handedness, 4/254
phenotypical factor, 4/251-253
prevalence, 43/222
reading disability, 4/261, 320
restoration of higher cortical function, 3/379
right and left side, 4/251
schizophrenia, 46/456, 505
sidedness, 4/249-251
skilled performance, 4/249
social factor, 4/250
somatic sensation, 4/324
speech, 43/223
speech brainedness, 4/321
stuttering, 4/261
synkinesis, 4/255
tachistoscope study, 4/323
test battery, 4/255-259, 313
twins, 4/252
unilateral dominance alpha rhythm, 4/259
vision, 4/323
Handgrip
orthostatic hypotension, 63/157
Handwriting, *see* Writing
Hangman fracture
C2 spondylolisthesis, 61/28
cauda equina injury, 25/450
cervical vertebral column injury, 61/28
mechanics, 61/29
radiology, 61/517
spinal injury, 25/32, 134, 326-329
surgical treatment, 61/64
vertebral column injury, 61/519
Hangover
headache, 48/5, 446
Hansen disease, *see* Leprosy
Hapalochlaena lunulata intoxication
artificial respiration, 37/62
convulsion, 37/62
fatal, 37/62
flaccid paralysis, 37/62
paralysis, 37/62
Hapalochlaena maculosa intoxication
artificial respiration, 37/62
convulsion, 37/62
fatal, 37/62
flaccid paralysis, 37/62
paralysis, 37/62

Haploidy
definition, 50/540
Happy puppet syndrome
ataxia, 30/515, 31/308
brachycephaly, 31/309
convulsion, 31/309
corticotropin, 31/309
hypsarrhythmia, 31/309
mental deficiency, 30/515, 31/309, 46/24, 94
microcephaly, 30/509, 515, 31/309
muscular hypotonia, 31/309
PEG, 31/309
seizure, 31/309
treatment, 31/309
ventricular system, 31/309
Haptene
multiple sclerosis, 9/145
Haptoglobin
multiple sclerosis, 9/99, 320
spinal cord injury, 26/380, 398
Haptoglobin1 gene
Alzheimer disease, 46/273
Harada disease, *see* Vogt-Koyanagi-Harada syndrome
Harboyan syndrome
corneal dystrophy, 42/367
corneal transplantation, 42/368
sensorineural deafness, 42/367
vision loss, 42/368
HARD syndrome
see also Fukuyama syndrome
computer assisted tomography, 60/672
Dandy-Walker syndrome, 50/593
definition, 50/260
Fukuyama syndrome, 50/261
gene, 72/135
lissencephaly, 50/593
HARDE syndrome
see also Fukuyama syndrome
Dandy-Walker syndrome, 50/593
definition, 50/260
Fukuyama syndrome, 50/261
gene, 72/135
lissencephaly, 50/593
Hardy-Weinberg equation
genotype distribution, 30/143
Harelip
hereditary sensory and autonomic neuropathy type I, 60/9
iniencephaly, 50/131
Harlequin fetus, *see* Congenital ichthyosis
Harlequin syndrome
ophthalmic finding, 74/430
sympathetic innervation, 75/125
HARP syndrome
acanthocytosis, 63/290
Harper dwarf
microcephaly, 30/100
Harrington-Flocks test
retinal function, 13/11
Harrington rod
kyphoscoliosis, 25/187
traumatic spine deformity, 25/187, 26/168
Harris neuralgia, *see* Chronic migrainous neuralgia
Hartmannella
Amoebae, 35/33
culture, 52/318
direct observation, 52/318
isolation, 52/318
limax form, 52/312
microscopy, 52/312
neuroamebiasis, 52/312
primary amebic meningoencephalitis, 35/25
Hartnup disease, 29/149-168
amino acid intestinal absorption, 29/163, 167
aminoaciduria, 7/577, 14/315, 29/148-163, 42/146
ataxia, 7/577, 21/574, 579, 29/155
ataxia telangiectasia, 14/315
atypical, 29/153
autosomal recessive, 21/564
biochemical finding, 7/577
blue diaper syndrome, 29/167
cerebellar ataxia, 14/315, 42/146
clinical features, 29/154-156
deficiency neuropathy, 7/577
dermatitis, 42/146
development, 29/155
diarrhea, 29/155
diplopia, 42/146
dwarfism, 60/666
dystonia, 60/664
genetics, 7/577, 29/153
growth retardation, 29/155
hallucination, 29/155
Hartnup family, 29/149
headache, 21/564
hereditary periodic ataxia, 21/564
history, 29/149
hypotonia, 7/577
incidence, 29/149-153
indolylacrolyl glycinuria, 29/231
intention tremor, 7/577
jejunointestinal mucosa, 21/564
mental deficiency, 42/146, 46/59
mental symptom, 29/155
metabolic ataxia, 21/574, 579

neurologic manifestation, 29/155
nystagmus, 7/577, 21/564
peptide intestinal absorption, 29/163, 167
photosensitivity, 21/579, 42/146, 60/667
precipitating factor, 29/167
psychiatric disorder, 42/146
renal tubular function, 21/564
skin lesion, 14/315, 29/154
spastic paraplegia, 21/564
symptomatic dystonia, 49/543
syncope, 21/564
tryptophan, 7/577, 21/564, 29/163
tryptophan metabolism, 29/156
tryptophanuria, 42/146
variant, 29/156
vitamin B3 deficiency, 7/577
vitamin PP deficiency, 21/564
Harvest fever, *see* Leptospirosis
Harvey syndrome
areflexia, 42/222
athetosis, 42/222
Babinski sign, 42/222
EEG, 42/223
febrile convulsion, 42/223
kyphoscoliosis, 42/223
mental deficiency, 42/223
psychomotor retardation, 42/223
talipes, 42/223
Hashimoto disease, *see* Hashimoto thyroiditis
Hashimoto encephalopathy
see also Hashimoto thyroiditis
hypothyroidism, 70/101
Hashimoto-Pritzker syndrome
Langerhans cell histiocytosis, 70/56
Hashimoto thyroiditis
see also Hashimoto encephalopathy
autoimmune disease, 41/77, 240
Down syndrome, 50/532
hypothyroidism, 70/101
myasthenia gravis, 62/534
polymyositis, 62/373
thyroid associated ophthalmopathy, 62/530
Turner syndrome, 50/544
Hashish, *see* Cannabis
Haskovec, M.L., 1/11, 21
Hastening phenomenon
festination, 49/68
Huntington chorea, 49/68
normal pressure hydrocephalus, 49/68
olivopontocerebellar atrophy, 49/68
Parkinson disease, 49/68
pathophysiology, 49/68
pure akinesia, 49/68
striatonigral degeneration, 49/68
Hatahata intoxication
vitamin A intoxication, 37/87
Hatfield-Turner disk
heat transfer, 1/463
Hausmanowa-Petrusewicz classification
infantile spinal muscular atrophy, 59/51
Haw River syndrome
computer assisted tomography, 60/671
dementia, 60/665
mental deficiency, 60/664
myoclonus, 60/662
Haymaker, W., 1/4, 28
Head abnormality
anencephaly, 50/83
Head banging, *see* Jactatio capitis nocturna
Head circumference
Down syndrome, 50/525
glutaryl-coenzyme A dehydrogenase deficiency, 66/647
intracranial pressure, 57/132
microcephaly, 30/507, 519, 50/270
Head erythromelalgia, *see* Cluster headache
Head flexion test, *see* Hyndham sign
Head, H., 1/6, 80
Head injury
see also Brain concussion, Brain injury, Commotio cerebri, Craniofacial injury, Scalp injury *and* Skull injury
A wave, 57/218
abducent nerve, 24/146
abducent nerve injury, 24/68
accident room attender, 57/9
acetylcholine metabolism, 57/79
acute subdural hematoma, 24/278
admission rate, 57/9
age, 23/26, 46/612
age specific death rate, 57/8
aggression, 46/626
airway obstruction, 57/434
ambulance, 57/433
amnesia, 3/293-295
angiography, 57/165
anosmia, 57/135
anoxic ischemic leukoencephalopathy, 47/541
antibiotic agent, 57/234
anticoagulant, 55/476
anticonvulsant, 57/230
aphasia, 45/297, 46/150, 57/404
apnea, 57/65
aqueduct stenosis, 50/308
armed forces, 24/455
arterial pressure, 57/227, 230

assault, 23/27
astrocytoma, 18/14
atlantoaxial dislocation, 24/143
attention defect, 46/624
attention deficit, 46/624
audiological examination, 24/134, 137
audit, 57/437
auditing effectiveness, 57/429
auditory evoked potential, 57/194
autonomic dysfunction, 57/238
autopsy, 57/431
autoregulation, 57/70
B wave, 57/218
battered child syndrome, 23/603
behavior disorder, 46/625-629
biomechanics, 23/67
blood pressure, 57/226
blunt, 23/36
boxing injury, 23/527
brain abscess, 52/145, 57/238
brain aneurysm, 12/99
brain angiography, 57/212
brain atrophy, 57/164
brain blood flow, 54/150, 57/211
brain concussion, 57/23
brain contusion, 57/33, 104
brain death leukoencephalopathy, 9/587
brain edema, 57/69, 105, 213
brain hypoxia, 23/137, 141
brain injury, 57/4
brain injury model, 57/28
brain ischemia, 23/43
brain parenchyma lesion, 57/104
brain perfusion pressure, 57/226
brain stem auditory evoked potential, 57/193
brain swelling, 57/58, 105, 159
bronchoscopy, 57/235
bullet tract, 57/148
calcium channel, 57/80
calcium channel blocking agent, 57/77
cardiac dysrhythmia, 63/236
care organization, 57/433
carotid artery occlusion, 26/43
carotid artery thrombosis, 57/374
carotid cavernous fistula, 24/399
cause, 23/27, 57/10
cause of death, 57/431
cell ischemia, 57/86
cerebral vasodilation, 57/231
cerebromalacia, 57/164
cervical spine, 57/147
child, *see* Pediatric head injury
Children Coma Score, 57/330
chronic cachexia, 57/234
chronic subdural hematoma, 57/48, 154
classification, 23/1, 11, 24/684, 57/101-116
clinical examination, 57/123-140, 430
clinical finding, 57/106
clinical symptom, 57/430
closed head injury, 57/102
coagulopathy, 57/238
cognitive impairment, 57/397
coma, 57/228
coma duration, 57/109
Coma Scale, 57/107
comminuted fracture, 57/102
computer assisted tomography, 57/224, 272, 274, 381
concomitant spinal injury, 24/141-178
conscious level, 57/273
consciousness level, 57/107
controlled ventilation, 57/227
cortical electrical function, 57/70
cranial epidural empyema, 52/168
cranial nerve palsy, 57/135
cranial subdural empyema, 52/170
critical care, 57/215
cruciate paralysis, 24/149
CSF fistula, 24/186, 57/311
death, 57/7, 433
definition, 23/2
delirium, 46/545
dementia, 46/394
demyelination, 9/587
dens hypoplasia, 24/143-147
depressed fracture, 57/102
depressed skull fracture, 24/225-227
depression, 46/461
diabetes insipidus, 46/395, 57/237
diagnostic procedure, 23/28
diffuse axonal injury, 57/52, 151, 153, 380
diffuse brain injury, 57/34, 105
diffuse brain lesion, 57/105
diffuse brain swelling, 57/58
diffuse vascular injury, 57/59
dizziness, 24/121, 126
Doppler ultrasonography, 57/226
driving, 46/619
dynamic change, 57/65
dysphasia, 46/614
early epilepsy, 57/192
early nutrition, 57/234
ECG, 57/226
EEG, 23/79, 324, 333, 344, 57/181-193, 225
EEG frequency analysis, 23/79
electrolyte abnormality, 57/236

electrophysiology, 57/181-199
β-endorphin, 57/83
endotracheal intubation, 57/216
ENG, 24/127
epidural hematoma, 24/264, 57/47, 103, 335
epilepsy of late onset, 57/130, 134
epistaxis, 57/132
ERG, 57/198
evoked potential, 57/112, 193-199, 385
evolving pattern, 57/169
excitatory aminoacid, 57/79
experimental, *see* Experimental head injury
experimental brain concussion, 57/23
experimental model, 23/70, 76
fat embolism, 57/139, 236
fatal nonmissile, 57/43
fever, 57/238
fibrinolysis, 57/238
fluid abnormality, 57/236
fluid balance, 57/227
fluid maintenance, 57/228
fluid percussion model, 57/30, 86
focal brain injury, 57/32
football injury, 23/572, 24/160
frontal lobe injury, 46/626
function restoration, 3/370
future, 57/239
ganglioside, 57/84
Glasgow Coma Scale, 45/188, 57/123, 132, 330
Glasgow Outcome Scale, 57/132, 368
glucocorticoid, 57/232
glutamic acid, 57/80
gram-negative bacillary meningitis, 52/103
head injury outcome, 57/114
headache, 46/629, 48/32
hearing loss, 24/124, 133, 57/138
hemiathetosis, 49/384
hemidystonia, 49/542
hemophilia, 63/310
hemotympanum, 57/132
high dose barbiturate, 57/232
history, 23/1, 1-7, 57/101
Horner syndrome, 24/148
hospital admission, 23/28, 57/8, 433
hypotension, 57/67, 217
hypothalamic dysfunction, 46/626
hypovolemic shock, 57/67
hypoxia, 23/117, 141, 57/216, 434
hypoxic brain damage, 57/56
hypoxic cortical damage, 57/57
impact, 23/98
inappropriate antidiuretic hormone secretion, 57/237
industrial, 23/27
infection, 57/139
infectious complication, 57/317
initial examination, 57/125
intermittent mandatory ventilation, 57/229
intermittent pressure stocking, 57/236
internal carotid artery thrombosis, 57/52
intracerebral hematoma, 11/663, 57/48, 156
intracranial arterial spasm, 57/57, 166
intracranial hemorrhage, 57/47
intracranial hypertension, 24/608, 55/215
intracranial hypertension correction, 57/231
intracranial infection, 24/215
intracranial pressure, 57/49, 67, 114, 132, 160, 227
intrauterine, *see* Intrauterine head injury
intravascular consumption coagulopathy, 57/223, 238, 63/315
intraventricular hemorrhage, 57/158
intubation, 57/216
isotope, 57/165
Kohlu injury, 23/597
law, 24/829
legal aspect, 24/829, 833
linear fracture, 57/102
lipid metabolism, 57/81
localization, 23/27
luxury perfusion syndrome, 57/70
management algorhythm, 57/273
management guideline, 57/123
mannitol, 57/227, 232
mechanical ventilation, 57/229
medical complication, 57/234
medical management, 57/207
medical management basic principle, 57/227
meningitis, 57/51, 237
mental symptom, 46/419
metabolism, 57/233
N-methylaspartic acid receptor, 57/80
midline shift, 57/217
migraine, 48/146, 150
mild, 57/3, 373, 389, 434
minor, *see* Minor head injury
missile injury, 24/455
model, 57/380
moderate, 57/3, 373, 389
morphologic lesion, 57/102
mortality, 57/113
multimodality evoked potential, 57/198
multiple injury, 57/106, 386
multiple major extracranial injury, 57/433
multiple organ failure, 57/235
myelomalacia, 55/100
nasotracheal tube, 57/230

nerve growth factor, 57/84
neurochemistry, 57/385
neurologic intensive care, 55/212, 215
neuro-otology, 24/119-140
neuropathology, 23/35, 57/43-61
neuropsychology, 45/9
neurotransmitter, 57/78
nimodipine, 57/77
nodular headache, 5/157
nonhospitalized, 57/9
nonrecurrent nonhereditary multiple cranial neuropathy, 51/571
normothermia, 57/227, 231
nuclear magnetic resonance, 57/167, 222, 381
nutrition, 57/233
nystagmus, 24/122, 125, 127, 135
obstructive hydrocephalus, 57/164
occipitoatlantal dislocation, 24/141
oculomotor nerve injury, 24/61
open head injury, 57/102
optic canal fracture, 24/39
orbital injury, 24/87
otologic examination, 24/127
pallidal necrosis, 49/470
parietal lobe epilepsy, 73/104
pathology, 57/378
patient position, 57/230
penetrating, *see* Penetrating head injury
persistent vegetative state, 57/341
personality, 46/625-629
phenytoin, 57/231
plateau wave, 57/218
pneumocephalus, 57/139, 312
pneumonia, 57/234
porencephaly, 23/307, 42/48
positive end expiratory pressure, 57/229
postconcussional syndrome, 23/454, 57/375
posterior fossa hematoma, 57/157
postresuscitation syndrome, 57/228
posttraumatic amnesia, 23/430, 24/488, 684, 27/468, 45/185-189, 57/373, 398
posttraumatic brain abscess, 57/321
posttraumatic dementia, 57/52
posttraumatic epilepsy, 46/395, 57/192
posttraumatic headache, 24/501
posttraumatic meningitis, 57/317
postural tremor, 49/589
prehospital care, 57/216
prehospital treatment, 57/216
pressure volume curve, 57/50
prevention, 23/32, 24/835
previous head injury, 57/378
primary brain injury mechanism, 57/31
primary brain stem hemorrhage, 57/60
process, 57/437
prophylactic antibiotic agent, 57/237
prospective study, 23/6
protection, 23/74
providing service, 57/429
provision, 57/437
psychological aspect, 57/397-408
psychosocial aspect, 23/1, 14
pulmonary artery pressure, 57/236
pulmonary care, 57/228
pulmonary edema, 57/235
pulmonary embolism, 57/236
pulse oximetry, 57/226
recovery, 57/367, 436
rehabilitation, 46/610, 617-620, 624-629, 57/388, 436
respiration, 57/125
respiratory distress syndrome, 57/235
restoration of higher cortical function, 3/370, 373
retrograde amnesia, 45/185
road accident prevention, 57/13
scalp, 57/131
scalp bruising, 57/131
schizophrenia, 46/461
schizophreniform psychosis, 46/420
season, 23/26
severe, 24/833, 57/3, 370, 378, 388
severe brain injury, 57/107
sex ratio, 23/26
sexual dysfunction, 75/100
shock, 57/66
skull fracture, 57/102, 147
skull radiation guideline, 57/271
social outcome, 57/410
social sequel, 57/408-415
somatosensory evoked potential, 23/80, 82, 57/196
spatial disorder, 46/617-620
spinal cord injury, 61/545
spinal injury, 26/188, 224
sport injury, 23/27
statistic, 23/26-30
status epilepticus, 57/192
stress wave, 23/94, 98
subarachnoid hemorrhage, 12/74, 120, 57/157
subdural empyema, 52/73, 57/237, 323
subdural hematoma, 57/34, 48, 103, 335
supracallosal hernia, 57/50
surgery indication, 57/273
swelling, 57/131
symptomatic dystonia, 49/542
terminology, 23/6, 23

tinnitus, 57/138
tonsillar herniation, 57/50
traffic injury, 23/27, 32
transcranial Doppler, 57/225
transient global amnesia, 45/206, 55/141
translation, 57/25
transtentorial herniation, 57/50
traumatic, 57/123-140
traumatic aneurysm, 24/381
traumatic brain injury, 57/43
traumatic carotid cavernous sinus fistula, 57/345
traumatic hydrocephalus, 24/231
traumatic intracerebral hematoma, 57/104
traumatic intracranial hematoma, 57/249
traumatic intracranial hematoma outcome, 57/282
treatment, 23/28, 32, 57/387
tremor, 49/604
trochlear nerve injury, 24/65
ultrasound, 57/165
urban, 23/26
vasogenic edema, 57/213
vasomotor paralysis, 57/231
vasospasm, 57/76, 225
venous sinus thrombosis, 24/369
ventilation, 57/229
ventricular drainage, 57/227, 232
ventriculitis, 57/237
vertebral artery, 24/145, 156, 57/52
vertigo, 24/121, 126
vestibular examination, 24/127, 135
vestibulocochlear nerve injury, 57/138
vibration, 23/95
visual disorder, 46/617-620
visual evoked response, 57/197
visual loss, 57/373
volume pressure relationship, 57/220
whiplash injury, 23/76, 93
Head injury epidemiology, 23/23-33, 57/1-14
accident room attender, 57/9
admission rate, 57/9
age specific death rate, 57/8
assault, 23/27, 57/11
bicycle riding, 57/11
boxing, 57/11
cause, 57/10
data, 57/5
death, 57/7
fall, 23/26, 57/10
firearm, 57/12
head injury incidence, 57/7
homicide, 57/12
hospital admission, 23/27, 57/8
household head injury, 57/7
ICD classification, 57/2
industrial, 23/26
initial clinical features, 57/3
International Bank of Severe Head Injuries, 57/5
minor head injury, 57/9
mortality statistics, 57/2
National Head and Spinal Cord Survey, 57/6
nonhospitalized head injury, 57/9
Olmsted County Survey, 57/6
pedestrian, 57/12
pediatric head injury, 23/446
population based study, 57/5
posttraumatic amnesia, 57/4
road accident, 23/26, 57/10
road accident prevention, 23/32, 57/13
San Diego County Survey, 57/6
severe, 57/2
sport, 23/26, 57/11
suicide, 57/12
traffic accident death, 57/2
United Kingdom, 57/5
United States, 57/6, 396
US Hospital Discharge Survey, 57/6
US National Coma Data Bank, 57/5
Head injury management
avoidable factor, 57/431-433
comprehensive service, 57/429
ethics, 57/436
guideline, 57/435
mild injury, 57/389, 434
moderate, 57/373, 389
organised care, 57/430
prehospital care, 57/434
severe brain injury, 57/435
severe head injury, 57/370, 388
temporal pattern, 57/430
triage, 57/431
Head injury mechanism
experimental model, 57/86
head injury outcome, 57/372, 376
Head injury outcome, 57/367-390
accident neurosis, 57/413
age, 57/111, 377
anomic aphasia, 57/404
anterograde amnesia, 57/398
anxiety, 57/407
aphasia, 57/404
attention, 57/399
attention deficit, 57/405
attention lapse, 57/401
attention recovery, 57/405
audit, 57/438
blow, 57/372

Bourdon-Wiersma test, 57/401
brain blood flow, 57/384
brain hyperemia, 57/384, 387
brain ischemia, 57/370, 386
brain stem auditory evoked potential, 57/195
brain stem reflex, 57/383
child, 57/375
choice reaction time, 57/401
cognitive impairment, 57/389, 397
cognitive impairment recovery, 57/405
cognitive problem, 57/376
cognitive-social behavior relationship, 57/415
coma duration, 57/110
coma level, 57/110
complication, 57/4
compressed spectral analysis, 57/382
computer assisted tomography, 57/111
concentration, 57/406
course, 57/414
creatine kinase BB, 57/385
CSF lactic acidosis, 57/386
daily life activity, 57/388
death, 57/368
depressed mood, 57/413
depression, 57/407, 413
deterioration, 57/376
determinant factor, 57/376
diffuse brain swelling, 57/380
digit span test, 57/399
Disability Rating Scale, 57/369
disinhibition, 57/407
divided attention, 57/400
divided attention deficit, 57/406
domestic situation, 57/408
dysgraphia, 57/404
early nutrition, 57/234
emotional reaction, 57/411
epilepsy of late onset, 57/193
etiology, 57/414
evoked potential, 57/385
extracranial insult, 57/125
fall, 57/372, 377
family problem, 57/408
fear in traffic, 57/411
fibrin degradation product, 57/385
financial burden, 57/389
focal neurologic deficit, 57/375
focussed attention, 57/399
forgetfulness, 57/405
friendship, 57/410
frontal lobe syndrome, 57/407
Galveston orientation and amnesia test, 57/398
Glasgow Assessment Schedule, 57/369
Glasgow Coma Scale, 57/382
Glasgow-Liège Coma Scale, 57/385
Glasgow Outcome Scale, 57/114, 132, 374
good recovery, 57/115, 368
head injury, 57/114
head injury mechanism, 57/372, 376
helicopter transport, 57/387
hypercapnia, 57/387
hypotension, 57/387
hypoxemia, 57/387
iatrogenic factor, 57/412
information processing, 57/406
injury susceptibility, 57/377
insight loss, 57/407
intellectual handicap, 57/403
intelligence, 57/403
intelligence recovery, 57/406
intracranial pressure, 57/114, 379, 383
Katz Adjustment Scales, 57/409
language, 57/404
language recovery, 57/406
loneliness, 57/410
management, 57/4
marietal relationship, 57/409
memory disorder, 57/397
memory loss, 57/397
memory recovery, 57/405
mental slowness, 57/400
metabolic demand, 57/386
mild head injury, 57/373
Mill Hill Scale, 57/403
moderate, 57/373
moderate disability, 57/115, 368
mortality rate, 57/377
multimodality evoked potential, 57/114
multiple injury, 57/386
neurobehavior test, 57/375
neurochemistry, 57/385
nuclear magnetic resonance, 57/379
outcome scale, 57/369
Paced Auditory Serial Addition Task, 57/400, 406
paraphasia, 57/404
pathology, 57/378
pediatric head injury, 57/377
persistent vegetative state, 57/115, 368, 372
personality change, 57/407
postconcussional syndrome, 57/375
posttraumatic amnesia, 45/189, 57/373, 382, 398
posttraumatic epilepsy, 57/373, 375
posttraumatic neurosis, 57/412
posttraumatic stress disorder, 57/413
posttraumatic syndrome, 57/376
prediction, 57/110, 132

pressure volume index, 57/383
pretraumatic personality, 57/407
previous head injury, 57/378
primary death, 57/370
psychogenic factor, 57/411
psychological impairment, 57/414
psychology, 57/397-408
pupillary reflex, 57/383
Raven Progressive Matrices, 57/403
recovery evaluation, 57/367
rehabilitation program, 57/388
retrieval slowness, 57/415
retrograde amnesia, 57/397
return to work, 57/410
road accident, 57/372, 376
secondary death, 57/370
severe disability, 57/115, 368
severe head injury, 57/370
shock, 57/66
skull fracture, 57/376
sleep disorder, 57/413
smoking, 57/378
social, 57/410
social contact, 57/409
social isolation, 57/409
social reintegration, 57/410
social sequel, 57/408-415
sociogenic factor, 57/411
Stroop effect, 57/400
study, 57/370
subdural hematoma, 57/378
subjective complaint, 57/374
subjective symptom, 57/376
supervisory attentional control, 57/402
survey of series, 57/371
survival limit, 57/110
sustained attention, 57/401
systemic complication, 57/378
talk and deteriorate, 57/438
talk and die, 57/431, 438
time on task effect, 57/401
Token test, 57/405
Traumatic Coma Data Bank, 57/371
traumatic intracerebral hematoma, 57/379
treatment, 57/387
verbal expansiveness, 57/409
verbal impoverishment, 57/404
verbosity, 57/405
visual field defect, 57/374
visual loss, 57/373
Wechsler Adult Intelligence Scale, 57/403
word fluency, 57/405
work ability, 57/115

Head injury prognosis
  age, 23/450, 57/113
  brain stem auditory evoked potential, 57/195
  computer assisted tomography, 57/164
  intracerebral lesion, 57/113
  massive diffuse injury, 57/379
  multimodality evoked potential, 57/198
  rehabilitation, 46/611
  severe head injury, 24/834, 46/611
  subdural hematoma, 57/379
  uncontrollable intracranial hypertension, 57/379
Head paradoxical reflex
  autonomic nervous system, 74/574
  vagus nerve block, 1/494
Head perfusion
  isolated, *see* Isolated head perfusion
Head retraction reflex
  technique, 1/241
Head retroflexion
  infantile Gaucher disease, 10/307, 565, 66/124
  iniencephaly, 50/129
Head tilting test, *see* Doll head eye phenomenon
Head trauma
  epilepsy, 72/191
  neurologic intensive care, 55/204
  paraballism, 49/374
  staphylococcal meningitis, 33/69
  stress, 74/318
  Von Willebrand disease, 63/312
Head tremor
  Bassen-Kornzweig syndrome, 51/394
  pallidoluysian degeneration, 49/455
  Parkinson disease, 49/96
Head turning reflex
  anatomy, 2/275
Head up tilt maneuver
  autonomic failure, 75/717
  hypotension, 75/717
  orthostatic tolerance, 75/717
Head up tilt test
  autonomic nervous system, 74/615
  blood pressure, 74/614
  heart rate, 74/614
Headache
  see also Migraine
  abdominal pain, 48/159
  abscess, 48/6
  accelerated hypertension, 48/6
  acetic acid, 48/174
  acetonylacetone intoxication, 64/87
  acetylsalicylic acid, 48/39, 174
  acetylsalicylic acid intoxication, 65/460
  acidosis, 48/422

acoustic nerve tumor, 17/670
acquired immune deficiency syndrome, 56/492
acquired toxoplasmosis, 52/354
acromegaly, 48/423
acute amebic meningoencephalitis, 52/317
acute hypertensive encephalopathy, 5/154, 48/282
acute muscle contraction, *see* Acute muscle contraction headache
acute paroxysmal hypertension, 48/282
acute petrositis, 5/210
acute posttraumatic, 48/6
acute raised pressure, 48/6
acute subdural hematoma, 24/279
Addison disease, 48/6
adrenal gland, 5/230
adrenalin, 48/91
African trypanosomiasis, 52/341
aggravation, 48/47
alcohol, 48/6, 122, 445
alcohol intoxication, 48/420, 64/112
aldosterone, 5/231
algie meningée postérieure, 5/187
Alice in Wonderland syndrome, 5/244
allergic, *see* Allergic headache
allergic granulomatous angiitis, 5/149
altitude illness, 48/395
amine, 48/6
amitriptyline, 48/364
amnesic shellfish poisoning, 65/167
amphetamine addiction, 48/420, 55/519
amphotericin, 65/472
amyloid angiopathy, 48/278
analgesic abuse, 5/34
analgesic agent, 48/6, 39
aneurysm, 48/6, 275
angiography, 48/287
angiostrongyliasis, 35/323, 52/553, 556
angiotensin, 48/109
aniline intoxication, 48/420
anisocoria, 75/283
anorexia, 48/32
anterior cerebral artery, 5/130
anterior ethmoidal nerve syndrome, 5/215
anterior uveitis, 5/206
anthranilic acid, 48/174
anxiety, 5/254, 48/6, 20, 354, 359
anxiolytic agent, 48/203
aorta, 48/316
apical periodontitis, 5/224, 48/391
apicitis, 48/6
aprindine intoxication, 65/455
Argentinian hemorrhagic fever, 56/364
arsenic intoxication, 48/420, 64/284
arterial dissection, 48/286
arterial pressure, 5/5, 48/278
arteriovenous fistula, 5/145, 148, 31/182
arteriovenous malformation, 48/273, 277
arteritis, 5/119, 48/278
arthrodento-osteodysplasia, 43/332
aspergillosis, 35/397, 52/378
associated symptom, 48/47
astigmatism, 48/34
astrocytoma, 48/33
atherothrombotic cerebrovascular disease, 48/279
atropine intoxication, 48/420
attack, 48/44
atypical facial, *see* Sphenopalatine neuralgia
aura, 48/5
autohypnosis, 48/39
autonomic abnormality, 75/282
autonomic crisis, 26/494
autonomic nervous system, 75/281
azapropazone, 48/174
bacterial endocarditis, 52/291, 294, 296, 298
Balaenoptera borealis intoxication, 37/94
barbiturate, 5/212
barium intoxication, 64/354
barogenic, 48/395-403
barotrauma, 63/416
basal ganglion, 5/134, 138
basal ganglion hemorrhage, 5/138
basilar artery, 5/134, 138
basilar artery aneurysm, 53/387
basilar artery thrombosis, 48/274
basilar dissecting aneurysm, 54/280
basilar ischemia, 12/6
basophil adenoma, 48/424
Behçet syndrome, 56/598
benign intracranial hypertension, 5/155, 42/740, 48/6, 33
benorilate, 48/174
benzodiazepine withdrawal, 65/340
biofeedback, 48/39, 203, 350
birgus latro intoxication, 37/58
blackout, 48/48
blindness, 48/280
blood-brain barrier, 48/64, 79
blood flow, 75/285
blood hyperviscosity, 55/484
blood pressure, 48/77, 282, 418, 75/282, 287
body mass, 5/231
Bonnet-Dechaume-Blanc syndrome, 43/36
borate intoxication, 48/420
Bornholm disease, 56/198
bradykinin, 48/97
brain abscess, 5/220, 52/151

brain amyloid angiopathy, 54/337
brain aneurysm, 5/141, 143, 42/723
brain angioma, 5/145
brain arteriovenous malformation, 5/145-149, 12/248, 31/182, 53/35, 54/376
brain arteritis, 5/119, 149
brain artery system, 5/125
brain blood flow, 48/78
brain edema, 48/6, 418, 63/416
brain embolism, 5/126
brain hematoma, 48/278
brain hemorrhage, 5/138, 48/273, 277, 53/205, 54/302
brain hypoxia, 63/416
brain infarction, 53/205
brain injury, 5/178, 23/12
brain ischemia, 48/280
brain lacunar infarction, 5/137
brain metastasis, 18/216
brain phlebothrombosis, 5/154
brain thrombosis, 5/126
brain tumor, *see* Brain tumor headache
brain vasculitis, 48/287
brain vasospasm, 48/64, 287
bromate intoxication, 48/420
bromide intoxication, 36/300
brucellosis, 48/31
bulbar paralysis, 65/178
bullous myringitis, 5/210
buspirone intoxication, 65/344
caffeine, 48/6
calcium, 48/111
calcium antagonist intoxication, 65/441
camphor intoxication, 48/420
cannabis intoxication, 48/420
capsaicin, 48/111
captopril intoxication, 65/445
carbamate intoxication, 64/186
carbamazepine intoxication, 65/505
carbon disulfide intoxication, 48/420, 64/24
carbon monoxide intoxication, 48/417, 420, 64/31
carbon tetrachloride intoxication, 48/420
carboxylic acid, 48/174
cardiovascular agent intoxication, 37/426
cardiovascular system, 75/288
carotid angiography, 5/66
carotid artery, 5/128, 48/77
carotid body tumor, 48/337
carotid cavernous fistula, 5/149, 24/414, 48/374
carotid dissecting aneurysm, 53/40, 54/272
carotid dissection, 48/286
carotidynia, 5/375, 48/6
cat scratch disease, 52/127, 130
catecholamine, 48/437
cavernous sinus thrombosis, 5/154, 220, 48/283, 52/173, 54/398
cerebellar hemorrhage, 5/140, 48/273, 278, 54/314
cerebellar infarction, 53/383
cerebellar tumor, 17/710
cerebral amebiasis, 52/315
cerebral malaria, 35/145, 149
cerebral toxoplasmosis, 52/354
cerebrovascular disease, 5/124, 48/273, 280, 288
cerebrovascular dissection, 48/286
cervical, *see* Cervical headache
cervical arthrosis, 5/198
cervical injury, 5/180, 200
cervical spine, 48/406, 411
cervical spondylosis, 7/450
cervical syndrome, 5/194
cervicogenic, *see* Cervicogenic headache
Chagas disease, 52/347
chelonia intoxication, 37/91
chemical, 48/441, 443
child, 5/239, 48/3, 31-40, 48
Chinese restaurant syndrome, 48/6, 419
chlorate intoxication, 48/420
chlordecone intoxication, 48/420
chlorinated cyclodiene, 64/200
chlorinated cyclodiene intoxication, 64/200
chlormethine intoxication, 64/233
chloroquine intoxication, 65/483
cholecystokinin octapeptide, 48/109
chromophobe adenoma, 48/423
chronic, *see* Chronic headache
chronic cluster, *see* Chronic migrainous neuralgia
chronic enterovirus infection, 56/330
chronic mountain sickness, 48/397
chronic muscle contraction, *see* Chronic muscle contraction headache
chronic paroxysmal hemicrania, 48/6
chronic posttraumatic, 48/6
chronic recurrent, 48/277
chronobiology, 75/286
ciclosporin, 63/179, 537
ciclosporin intoxication, 63/537, 65/552
ciguatoxin intoxication, 65/162
cinchonism, 52/372, 65/452, 485
ciprofloxacin intoxication, 65/486
classic, 48/5
classification, 5/11-14, 48/1-10, 55
climate, 48/123
clinical features, 48/37
cluster, *see* Cluster headache
cocaine addiction, 55/521
cocaine intoxication, 65/255

coccidioidomycosis, 52/413
coitus, *see* Exertional headache
cold, 5/188
Colorado tick fever, 56/141
combined, 48/6
common, 48/5
common carotid artery, 5/128
compensation neurosis, 5/183
compression, 48/6
computer assisted tomography, 48/277-279
concentric sclerosis, 47/414
confusion, 48/284
congenital heart disease, 38/133, 63/4
conversion, *see* Conversion headache
copper intoxication, 48/420
coprine intoxication, 65/39
corpus callosum syndrome, 2/761
corticosteroid, 48/325
Costen syndrome, 5/345, 48/6, 226
cough, *see* Cough headache
couvade syndrome, 48/358
crab intoxication, 37/58
cranial epidural empyema, 52/168
cranial nerve, 48/6, 50
cranial subdural empyema, 52/170
cranial vasculitis, 48/273, 284
craniodiaphyseal dysplasia, 43/356
craniometaphyseal dysplasia, 43/362
craniopharyngioma, 18/541
cryptococcal meningitis, 52/431
Cryptococcus neoformans, 63/532
CSF flow, 48/32
CSF hypertension, 5/155
cubomedusae intoxication, 37/43
Cushing reflex, 75/289
Cushing syndrome, 5/230
cyanide intoxication, 65/28
cyanogenic glycoside intoxication, 65/28
cycloserine, 65/481
cycloserine intoxication, 65/481
decaborane intoxication, 64/360
decompression illness, 61/221
definition, 5/2, 48/1-10
delusion, 5/249, 48/5, 361
dental, *see* Dental headache
dental causalgia, 5/226
depression, 48/6, 34, 359
diabetes mellitus, 5/230
diagnostic process, 48/52
diagnostic value, 48/287
dialysis disequilibrium syndrome, 63/523
diborane intoxication, 64/359
diclofenac, 48/174
dietary factor, 48/36
differential diagnosis, 5/25, 48/43-58
digitalis intoxication, 48/420
dimercaprol intoxication, 48/420
dipeptidyl carboxypeptidase I inhibitor, 65/445
dipeptidyl carboxypeptidase I inhibitor intoxication, 65/445
dissecting aorta aneurysm, 63/47
disulfiram intoxication, 37/321, 48/419
diving, 48/401
drug, 48/6
drug holiday, 48/6
drug induced, 69/482
drug treatment, 48/173-204
dry socket, 5/226
duration, 48/46
dysbarism, 63/416
ear, 48/6
ear nose and throat disease, 5/35, 208-210, 212, 48/6
early morning, 5/150
ECG, 75/282, 286
echinococcosis, 52/524
edetic acid intoxication, 48/420
EEG, 5/33, 43, 48/39
embolic cerebrovascular disease, 48/279
embolic disease, 48/280
EMG, 48/349
emphysema, 48/6
empty sella syndrome, 48/432
encainide intoxication, 65/456
encephalitis, 48/6
endocrine, 5/229, 48/422, 431-438
enolic acid, 48/174
eosinophilic granuloma, 5/229
ependymoma, 18/126
epidural lymphoma, 63/349
epilepsy, 5/28
episodic, *see* Episodic headache
Epstein-Barr virus, 56/253
ergot intoxication, 65/67
ergotamine, 48/6
estrogen, 5/232, 48/433, 437
ethionamide, 65/481
ethionamide intoxication, 65/481
ethmoid cell, 5/214
ethosuximide intoxication, 65/507
ethylene glycol intoxication, 64/123
examination, 48/49
exercise, *see* Exertional headache
exertional, *see* Exertional headache
experimental, 5/4
extracranial, 48/6, 384

eye, 5/35, 204, 206, 48/6
eye movement, 48/50
facial pain, 5/209, 48/2
facial sweating, 48/335
facial thermography, 48/77
familial history, 48/25
far eastern tick-borne encephalitis, 56/139
fasting, 48/36
felbamate intoxication, 65/512
female-male ratio, 48/18
fenclofenac, 48/174
fenoprofen, 48/174
fever, 48/6
fibromuscular dysplasia, 11/372, 48/273, 286
filariasis, 35/167, 52/515, 517
Fincher syndrome, 20/367
flavivirus, 56/12
flecainide intoxication, 65/456
flufenamic acid, 48/174
fluorouracil intoxication, 65/532
fluoxetine intoxication, 65/320
fluvoxamine intoxication, 65/320
food, 48/20
foramina parietalia permagna, 42/29
fox intoxication, 37/96
frequency, 48/14, 46
freudian explanation, 48/357
frontal, *see* Frontal headache
frontal lobe tumor, 17/246
frontalis muscle tension, 48/360
frontometaphyseal dysplasia, 38/410
frontonasal duct, 5/214
gabapentin intoxication, 65/512
gait, 48/50
gastrin releasing peptide, 48/109
general disease, 5/34
genetics, 5/258
gestagen, 48/437
giant brain aneurysm, 5/147
giant cell arteritis, 71/193
glaucoma, 5/206, 48/6, 34
glioblastoma multiforme, 18/53
glossopharyngeal, 48/6
glyceryl trinitrate, 48/228, 281, 443, 75/285
glyceryl trinitrate intoxication, 37/453, 65/443
glycoside intoxication, 37/426
Godin syndrome, 5/330
Gradenigo syndrome, 5/210, 48/6
granulomatosis, 48/285
granulomatous amebic encephalitis, 52/317
granulomatous CNS vasculitis, 55/390
griseofulvin, 65/472
guilt, 48/354
gyromitrine, 64/3
gyromitrine intoxication, 65/36
hallucination, 5/244, 250, 48/361
hangover, 48/5, 446
Hartnup disease, 21/564
head injury, 46/629, 48/32
heart rate, 75/282, 286
hematoma, 48/5
hemiplegic, 48/5
hemodialysis, 48/419
hemolytic uremic syndrome, 63/324
heptachlor intoxication, 48/420
hereditary hemorrhagic telangiectasia (Osler), 42/739
hereditary Meniere disease, 42/402
herpes simplex virus encephalitis, 48/32, 56/214
herpes zoster, 5/207
herpes zoster oticus, 5/207
heterocyclic acetic acid, 48/174
heterocyclic antidepressant, 65/316, 320
heterophoria, 5/205
high-pressure nervous syndrome, 48/402
histamine, *see* Cluster headache
histoplasmosis, 52/438
history, 5/15, 48/44
Hodgkin lymphoma, 63/349
Holmes-Adie syndrome, 74/411
homocystinuria, 55/328
Horner syndrome, 48/286, 335, 75/284
Horton, *see* Chronic migrainous neuralgia *and* Cluster headache
hostility, 48/354
hot dog, *see* Hot dog headache
hydralazine intoxication, 37/431
hydralazine neuropathy, 42/647
hydrocephalus, 48/32, 34
hydrogen sulfide intoxication, 48/420
5-hydroxyindoleacetic acid, 48/90
hyperbaric, *see* Hyperbaric headache
hypercapnia, 48/6, 422
hyperextension neck injury, 48/364
hypermetropia, 48/34
hyperparathyroidism, 27/288, 48/425
hyperprolactinemia, 48/424
hypertension, 48/5, 19, 282
hypertensive encephalopathy, 11/552, 48/418, 54/212
hyperthyroidism, 48/378, 70/92
hyperventilation, 75/282
hyperventilation syndrome, 63/433
hyperviscosity syndrome, 55/484
hypnotic agent, 48/39
hypocalcemia, 48/6, 34, 63/561

hypochondriacal, 48/5
hypoglycemia, 48/6, 34
hyponatremia, 63/545
hypophyseal tumor, 5/229
hypotensive, 5/175
hypoxia, 48/6
hysteria, 43/207, 48/356, 359
ibuprofen, 48/174
ice cream, *see* Ice cream headache
ice pick, *see* Ice pick headache
ichthyohepatoxic fish intoxication, 37/87
IgE, 48/36
imipramine intoxication, 48/420
indoleacetic acid, 48/174
indometacin responsive, *see* Indometacin responsive headache
indurative, 5/157
infectious endocarditis, 63/23, 112, 116
inflammation, 48/6, 281, 284
intermittent, 48/3
internal auditory artery, 48/274
intracranial, 48/6
intracranial aneurysm, 48/277
intracranial hemorrhage, 48/32
intracranial hypertension, 5/172, 16/128
intracranial hypotension, 5/175
intracranial infection, 48/31
intracranial inflammation, 5/21
intracranial pressure, 48/33, 283-285, 67/141
intracranial source, 5/229
intracranial thrombophlebitis, 48/287
intracranial vasodilatation, 48/419
intracranial venous occlusion, 48/283
iodine intoxication, 48/420
iridocyclitis, 5/206
ischemic, 48/72
isolated severe, 48/276
jaw, 48/392
kerosine intoxication, 48/420
ketoprofen, 48/174
Köhlmeier-Degos disease, 39/437, 55/276
kuru, 56/556
lacrimation, 75/284
lamotrigine intoxication, 65/512
lateral medullary infarction, 48/281, 53/381
lateral medullary syndrome, 5/132
laughing, 48/367
lead encephalopathy, 64/435
lead intoxication, 48/420, 64/435
learning, 48/34
legionellosis, 52/254
lethargy, 48/32
leukemia, 48/33, 63/344
lindane intoxication, 65/479
β-lipoprotein, 48/100
Listeria monocytogenes, 63/531
Listeria monocytogenes encephalitis, 33/87
Listeria monocytogenes meningitis, 33/85, 52/94, 63/531
lithium intoxication, 48/420
lobar hemorrhage, 48/278
long chain alcohol intoxication, 48/420
lower face, *see* Sphenopalatine neuralgia
lower half, *see* Sphenopalatine neuralgia
luetic, 5/28
lumbar puncture, 5/34, 175, 33/6
Lyme disease, 51/204, 208
lymphocytic choriomeningitis, 9/197, 56/359
lymphoma, 48/285
lymphomatoid granulomatosis, 48/284
mackeral intoxication, 37/87
major illness, 48/52
malignant hypertension, 48/6
MAO, 48/93, 437
MAO inhibitor, 48/6, 420
marine toxin intoxication, 37/33, 65/153
mastication, 48/392
masticatory muscle, 48/414
mastoiditis, 5/210
maxillary antrum, 48/393
maxillary sinus, 5/213
maxillary sinusitis, 5/207, 213, 218
mechanism, 5/15
meclofenamic acid, 48/174
mediastinal obstruction, 48/6
mefenamic acid, 48/174
mefloquine intoxication, 65/485
Melkersson-Rosenthal syndrome, 14/790, 42/322
Meniere disease, 5/208
meningeal, 48/6
meningeal irritation, 48/284
meningeal leukemia, 63/341
meningeal lymphoma, 63/347
meningeal pain, 1/540
meningitis, 48/31
meningococcal meningitis, 33/23, 52/24
menopause, 48/433
menstrual, *see* Catamenial migraine
mercury intoxication, 48/420
merostomata intoxication, 37/58
metabolic, 48/6, 417-428
metabolic factor, 48/419
methanol intoxication, 36/353, 48/420, 64/98
methotrexate intoxication, 65/529
methyl bromide intoxication, 48/420
methyl chloride intoxication, 48/420

methyldopa, 65/448
methyldopa intoxication, 37/440
methyl iodide intoxication, 48/420
methyl salicylate intoxication, 37/417
mexiletine intoxication, 65/455
mianserin intoxication, 65/320
middle cerebral artery, 5/129, 48/109
middle ear tumor, 5/211
migraine, 48/77, 144, 201
migrainous, *see* Migrainous headache
Minnesota Multiphasic Personality Inventory, 48/345-347, 359
miosis, 48/330, 75/281, 283
mixed, *see* Mixed headache
monoamine, 48/79
morning, *see* Morning headache
mountain sickness, 48/397, 419, 63/416
moyamoya disease, 12/360, 53/32, 55/293
mucormycosis, 35/544, 52/471
multiple myeloma, 18/255
multiple sclerosis, 9/182
mumps, 56/430
muscarine intoxication, 65/39
muscle contraction, *see* Muscle contraction headache
muscle spasm, 5/254
muscle tension, 48/34
mycotic brain aneurysm, 52/294, 296
myoglobinuria, 40/338, 65/178
nalidixic acid intoxication, 65/486
naphthalene intoxication, 48/420
naproxen, 48/174
nasal, *see* Nasal headache
nasal secretion, 75/284
nasal septum, 5/212
nasal vasomotor reaction, 48/5
nausea, 75/286
neck muscle, 5/20
neck pain, 48/406
neptunea intoxication, 37/64
neuralgia, 48/5
neuritis, 48/5
neuroborreliosis, 51/204, 208
neurobrucellosis, 33/309, 312, 52/587
neurocysticercosis, 16/225, 52/530
neurokinin, 48/97
neuroradiology, 5/33
neurosyphilis, 52/277
nicotine intoxication, 36/427, 48/420, 65/262
nitrate intoxication, 48/6, 419, 65/443
nitrite intoxication, 48/6, 419, 444
nitrite-nitrate, 48/78
nitrogen oxide intoxication, 48/420
nodular, *see* Nodular headache
nomenclature, 48/1
non-Hodgkin lymphoma, 63/349
nonsteroid anti-inflammatory agent intoxication, 65/416, 563
noradrenalin, 48/91
North American blastomycosis, 35/404, 406, 52/389
obsessional, 48/354
obstructive sleep apnea, 63/454
occipital, *see* Occipital headache
occipital nerve, 5/110
ocular, 48/5
oculomotor nerve, 48/275
oculosympathetic paralysis, 48/337
oculosympathetic pathway, 48/331
OKT3, 63/536
ophthalmoplegia, 65/178
oral contraceptive agent, 48/201, 435
organochlorine insecticide intoxication, 64/200
organolead intoxication, 64/133
organophosphate intoxication, 37/555, 64/167, 226
organophosphorus compound intoxication, 48/420
organotin intoxication, 64/138, 141
orgasm, 48/6, 378
osteitis deformans (Paget), 38/362, 43/451, 48/6
ostrea gigas intoxication, 37/64
otitic hydrocephalus, 5/155, 48/6
otitis externa, 5/209
otitis media, 5/210
otologic disease, 5/35, 208, 212, 214, 330
oxicam derivative, 48/174
pain, 26/494, 48/1, 358
pain sensitive structure, 5/229
pain site, 48/46
papilledema, 48/33
paralytic shellfish poisoning, 37/33, 65/153
paraplegic autonomic crisis, 26/494
parathion intoxication, 48/420
paratrigeminal syndrome, 48/6
paresthesia, 48/443
paroxetine intoxication, 65/320
paroxysmal, 48/3, 18
paroxysmal hypertension, 26/494
pathophysiology, 5/2, 37, 48/406
patient examination, 5/2, 25, 32-34
pellagra, 7/573, 28/86
pentaborane intoxication, 64/359
pentazocine intoxication, 65/357
pericarotid sympathetic plexus, 48/336
periodic, 48/3, 46
periosteal, 48/6

peritoneal dialysis, 27/323
perivascular nerve, 48/79
perivascular nerve fiber, 75/290
pethidine, 5/98
phenacetin, 5/34
phenethylamine, 48/97
phenylacetic acid, 48/174
phenylbutazone, 48/174
phenylpropanolamine, 55/520
phenytoin intoxication, 65/500
pheochromocytoma, 5/230, 39/498, 500, 42/764, 48/6, 427
phonophobia, 75/286
photophobia, 75/286
physiologic, 48/55
piroxicam, 48/174
pituitary adenoma, 17/390
pituitary apoplexy, 5/141, 48/424
pituitary disorder, 48/432
pituitary gland, 5/141, 229
pituitary hemorrhage, 48/279
pituitary tumor, 48/422-424
pizotifen, 48/39
platelet, 48/281
platelet MAO, 48/93
polar bear intoxication, 37/96
pollen, 5/237
polyarteritis nodosa, 8/125, 39/300, 48/284, 55/354
polycythemia, 63/250
polycythemia vera, 55/469, 63/250
pontine, 48/278
pontine hemorrhage, 5/139, 48/279
pontine infarction, 2/241, 12/15
postconcussional syndrome, 23/454, 57/412, 424
postconvulsive, 48/6
postcraniotomy, 48/6
posterior cerebral artery, 5/130
posterior inferior cerebellar artery, 5/137
postheart transplantation, 63/179
postherpetic, 48/6
postlumbar puncture, 5/175, 9/324, 48/5
postlumbar puncture syndrome, 61/152
postoperative, 48/392
posttraumatic, *see* Posttraumatic headache
posttraumatic syndrome, 24/501, 536, 686, 48/385, 387
potassium, 48/111
precipitation, 48/46
pregnancy, 48/424, 433
premenstrual, *see* Premenstrual headache
premenstrual syndrome, 48/426
preorgasmic, 48/380
prevalence, 48/13-15, 31
primary granulomatous cerebral arteritis, 48/284
primary pontine hemorrhage, 12/39
progesterone, 5/231, 48/434
prognosis, 48/39
progressive multifocal leukoencephalopathy, 9/495
prolactin, 48/436
prophylactic treatment, 48/39
propionic acid, 48/174
propranolol, 48/39
prostacyclin, 48/97, 281
prostaglandin, 48/97
prostaglandin induced, 48/78
psychiatric, *see* Psychiatric headache
psychogenesis, 48/357
psychogenic, *see* Psychogenic headache
psychological test, 48/359
psychotherapy, 5/256, 48/39, 360
ptosis, 48/330, 65/178, 75/282
pulpitis, 5/222, 48/391
pupil, 75/281
putaminal hemorrhage, 5/138, 48/278
pyrazolones, 48/174
pyrethroid intoxication, 64/219
pyrethrum intoxication, 48/420
quality, 48/47
quantity, 48/47
quinidine intoxication, 37/448
quinolone intoxication, 65/486
railway spine, 48/385
recurrent abdominal pain, *see* Abdominal migraine
referred, 48/6
reflex, 48/6
refractive disorder, 5/204
registration, 48/53
rejection encephalopathy, 63/530
relief, 48/47
renal disease, 5/28
renal insufficiency, 27/323
renal transplantation, 63/531
reserpine, 48/72, 78
reserpine intoxication, 37/438
respiratory encephalopathy, 63/414
respiratory paralysis, 65/178
reversible angiopathy, 48/287
rheumatic disease, 5/28, 157
rhodotorulosis, 52/493
rugby football injury, 23/570
running, 48/124
sagittal sinus thrombosis, 5/155
salicylic acid, 48/174

salicylic acid intoxication, 37/417
salivation, 75/284
salt and water balance, 5/231
scalp injury, 5/180
scalp muscle contraction, 5/181
schistosomiasis, 52/538
schizophrenia, 48/361
school absence, 48/34
school refusal, 48/40
schoolchild, 48/407
scintillating scotoma, 48/442
sclerosteosis, 31/258, 38/412, 43/475
screening test, 48/55
seasonal pattern, 5/236
selective serotonin reuptake inhibitor, 65/316, 320
sensitive vessel, 48/281
sepsis, 48/284
serotonin, 48/88, 90, 97, 446
serotonin releasing factor, 48/89
sertraline intoxication, 65/320
sex hormone, 48/425
sexual intercourse, 48/10, 373-381
sickle cell anemia, 63/259
sinus, 48/6
sinusitis, 5/207, 218, 330, 48/33
skull, 48/50
skull fracture, 5/180
skull osteomyelitis, 5/219
sleep, 74/550
snake envenomation, 65/178
sneezing, 48/367
social history, 48/48
sodium, 48/436
sodium glutamate, 48/6, 420, 444
sodium salicylate intoxication, 37/417
somatostatin, 48/109
space occupying, 48/6
speech disorder, 48/118
spinal lesion, 26/494
sporotrichosis, 35/569, 52/495
sport, 48/124
spreading depression, 48/72
squint, 5/205
statistic, 5/6, 25
steroid, 48/434, 437
stooping, 48/367
subarachnoid hemorrhage, 12/183, 48/6, 33, 273, 275, 277, 286, 53/205, 240, 55/11, 19
subclavian steal syndrome, 5/149, 53/373
subdural CSF collection, 5/182
subdural lymphoma, 63/349
substance P, 48/111, 274, 75/291
sulfamethoxazole, 65/472
sulindac, 48/174
superior longitudinal sinus thrombosis, 48/283, 52/178
supratrochlear artery, 75/285
surgical treatment, 5/104
sweating, 75/282, 284
symbolic aspect, 5/249
Symonds, *see* Symonds headache
sympathectomy, 5/105
sympathetic nervous system, 48/112
sympathetic pathway, 48/331
symptom, 48/40
syncope, 48/443
syphilis, 5/28, 52/274
systemic infection, 48/5
systemic lupus erythematosus, 39/286, 48/284, 55/374, 378, 71/44
Takayasu disease, 5/149, 48/284, 55/337, 71/202
tempo, 48/51
temporal arteritis, 5/149, 39/316-319, 48/6, 77, 284, 310, 316, 55/341, 342
temporal lobe tumor, 17/291, 67/152
temporomandibular cephalalgia, 48/413
temporomandibular joint dysfunction, 48/34, 47
tension, *see* Muscle contraction headache
tension vascular, *see* Tension vascular headache
testosterone, 48/427, 433
tetracycline, 48/34
tetraplegia, 26/494
thalamic hemorrhage, 5/139, 48/278
thalamic syndrome (Dejerine-Roussy), 42/771
thalassemia, 55/466
thiotepa intoxication, 65/534
throbbing, 5/153
thrombocythemia, 55/473
thrombocytopenia, 55/472, 63/322
thrombotic thrombocytopenic purpura, 55/472, 63/324
thromboxane, 48/281
thrombus, 48/281
thunderclap, *see* Thunderclap headache
thyroid gland, 5/230
tolfenamic acid, 48/174
tolmetin, 48/174
Tolosa-Hunt syndrome, 48/6
toxic, 48/6, 417-428
toxic encephalopathy, 52/296
toxic oil syndrome, 63/380
toxic shock syndrome, 52/259
traction, 5/21, 48/5
transient global amnesia, 45/209
transient ischemic attack, 12/6, 53/263
transverse sinus thrombosis, 5/154, 48/283,

52/176
traumatic, *see* Traumatic headache
traumatic vegetative syndrome, 24/580
trazodone intoxication, 65/320
treatment, 48/52
tricyclic antidepressant, 65/316
trigeminal, 48/6
trigeminal ganglion, 48/108-110
trigeminal ganglionectomy, 48/109
trigeminal nerve, 48/107, 274
trigeminal pain, 48/281
trigeminopupillary reflex, 75/283
trigeminovascular connection, 48/107
trigeminovascular system, 48/107
trimethoprim, 65/472
tsukubaenolide intoxication, 65/558
tuberculous meningitis, 48/32
tumor, 48/5
tumor localization, 5/30
tumor necrosis factor, 65/562
tyramine, 48/97, 445
unexplained recurrent, 48/34
unilateral, 48/3, 9, 118
urticaria pigmentosa, 14/790
uveitis, 5/206
vacuum, *see* Vacuum headache
Van Buchem disease, 31/258, 38/409, 43/410
varicella zoster virus meningoencephalitis, 56/235
vascular, *see* Vascular headache
vascular dilatation, 48/418
vascular disease, 48/274
vascular hypertension, 5/138, 149, 153
vascular hypotension, 5/154
vascular malformation, 48/287
vasculitis, 48/285
vasoactive intestinal polypeptide, 48/109
vasoconstriction, 5/190, 48/72
venerupis semidecussata intoxication, 37/64
Venezuelan equine encephalitis, 56/137
venous pressure, 48/6
venous sinus occlusion, 54/429
venous sinus thrombosis, 5/154, 54/395, 429
ventricular extension, 48/278
vertebral artery, 5/132
vertebral artery arteriosclerosis, 53/376
vertebral dissecting aneurysm, 54/279
vertebrobasilar disease, 48/279
vertebrobasilar system syndrome, 53/373, 376
vessel calibre, 75/285
vestibular neurinoma, 68/425, 446
vigabatrin intoxication, 65/511
viral meningitis, 56/125
vitamin A intoxication, 37/96, 65/568
vitamin B1 intoxication, 65/572
vomiting, 48/32, 75/286
Von Hippel-Lindau disease, 14/115
Waldenström macroglobulinemia, 55/486, 63/396, 398
water balance, 5/231
water intoxication, 64/241
water retention, 48/436
weather, 48/123
weekend, 74/508
Wegener granulomatosis, 48/284
weight lifting, 48/124
whiplash injury, 5/200
Whipple disease, 52/138
Willis, *see* Willis headache
withdrawal, 48/421
wolf intoxication, 37/96
zinc intoxication, 64/362
Head's hand-ear-eye test
body scheme disorder, 4/214, 220
Head's zone
dermatome, 2/159
enterotome, 2/160
hyperalgesia, 2/161
referred pain, 2/174
Hearing loss
see also Deafness *and* Neuro-otology
α-amino-n-butyric acid, 13/339
ascertainment, 4/420
ataxia, 60/659
auditory imperception, 4/428
branchio-otodysplasia, 42/359
cause, 4/420
central deafness, 4/428
chloroquine intoxication, 65/483
cinchonism, 65/452, 485
cisplatin intoxication, 64/358
clinical examination, 57/138
Coats disease, 62/167
complex I deficiency, 62/504, 66/422
conduction test, 24/133, 137
congenital hyperphosphatasia, 31/258
craniometaphyseal dysplasia, 43/364
CSF hypotension, 61/165
decompression illness, 61/219
developmental dysphasia, 46/141
dialysis, 63/530
distal amyotrophy, 60/659
dominant low frequency, *see* Dominant low frequency hearing loss
dominant midfrequency, *see* Dominant midfrequency hearing loss
Down syndrome, 50/526

Ekman-Lobstein disease, 22/494
electrical injury, 61/195
examination, 4/420
facioscapulohumeral muscular dystrophy, 40/420
Friedreich ataxia, 51/388, 60/326
frontometaphyseal dysplasia, 31/256, 38/410
glycosaminoglycanosis type II, 66/295
glycosaminoglycanosis type III, 66/299
Goyer syndrome, 22/490
head injury, 24/124, 133, 57/138
Helweg-Larsen-Ludvigsen syndrome, 22/491
hereditary, *see* Hereditary hearing loss
hereditary Meniere disease, 22/486, 42/402
hereditary sensory and autonomic neuropathy type I, 60/8
high frequency, *see* High frequency hearing loss
high tone, *see* High frequency hearing loss
Huntington chorea, 6/316
iatrogenic factor, 61/165
iatrogenic neurological disease, 61/165
infantile spinal muscular atrophy, 59/70
Klippel-Feil syndrome, 32/120
lumbar puncture, 61/165
management, 4/421
mental deficiency, 60/659
MERRF syndrome, 62/509
methyl salicylate intoxication, 37/417
midfrequency, *see* Midfrequency hearing loss
midtone, *see* Midfrequency hearing loss
multiple mitochondrial DNA deletion, 62/510
multiple sclerosis, 9/168
myelography, 61/165
myotonic dystrophy, 62/234
neurosensory, *see* Neurosensory hearing loss
nonsteroid anti-inflammatory agent intoxication, 65/416
osteitis deformans (Paget), 22/493, 38/362, 43/451, 50/396, 60/773
Pyle disease, 60/773
quinidine intoxication, 37/448
quinine intoxication, 13/224
radiation, 7/397
recessive Meniere disease, 22/486
recessive progressive sensorineural, *see* Recessive progressive sensorineural hearing loss
Refsum disease, 8/22, 21/196, 22/501, 36/348, 42/148, 66/490
renal acidosis, 22/488
salicylic acid intoxication, 37/417
Schimke syndrome, 62/304
secondary speech disorder, 4/420
sex linked progressive sensorineural, *see* Sex linked progressive sensorineural hearing loss
sodium salicylate intoxication, 37/417
speech discrimination, 24/134, 137
speech disorder in children, 4/420
spinal anesthesia, 61/165
temporal bone fracture, 24/123, 125, 185
temporal lobe tumor, 17/289
treatment, 60/326
uremic polyneuropathy, 63/512
Waldenström macroglobulinemia, 63/396
Wegener granulomatosis, 71/177
Heart
acetylcholine, 74/150
autonomic nervous system, 74/14
cortical influence, 1/459, 461
electrical injury, 74/321
glycosaminoglycanosis, 10/450
hereditary amyloid polyneuropathy, 60/105
hereditary amyloid polyneuropathy type 2, 60/98
hereditary amyloid polyneuropathy type 3, 60/100
hypothalamus, 1/459, 461
primary carnitine deficiency, 66/402
Refsum disease, 8/22, 10/345, 21/246, 36/347, 41/434, 60/231, 66/497
rheumatic disease, 11/435
Smith-Lemli-Opitz syndrome, 66/585
Heart atrioventricular block
congenital heart disease, 63/5
Huntington chorea, 49/284
Heart block
congenital heart disease, 63/5
Kearns-Sayre-Daroff-Shy syndrome, 38/224, 62/508
progressive external ophthalmoplegia, 13/311, 22/193, 204, 211
syncope, 22/193
Heart death
brain death, 55/257
brain injury, 24/811, 823
sleep, 74/552
Heart disease
see also Cardiomyopathy
acrocephalosyndactyly type I, 43/320
acrocephalosyndactyly type II, 43/324
amyloidosis, 71/508
anencephaly, 42/13
anoxic ischemic leukoencephalopathy, 47/525
aspartylglycosaminuria, 42/527
atrophoderma vermiculatum, 43/8
autonomic nervous system, 75/425
Baller-Gerold syndrome, 43/370
Becker muscular dystrophy, 43/86
Beckwith-Wiedemann syndrome, 43/398
Bowen syndrome, 43/335

brain infarction, 53/155
brain ischemia, 53/155
cardiofacial syndrome, 42/307
caudal aplasia, 42/50
cerebrovascular disease, 53/17
cervicothoracic syndrome, 7/455
Chagas disease, 75/393
congenital, *see* Congenital heart disease
congenital muscular dystrophy, 43/93
corpus callosum agenesis, 42/8
cyanotic, *see* Cyanotic heart disease
De Lange syndrome, 43/246
dementia, 46/397
Down syndrome, 43/545
Duchenne muscular dystrophy, 40/362, 43/106
dwarfism, 43/382, 435
Edwards syndrome, 43/536
elfin face syndrome, 43/259
Ellis-Van Creveld syndrome, 43/347
Emery-Dreifuss muscular dystrophy, 40/389, 43/88
epilepsy, 73/378
facial paralysis, 42/313
facioscapulohumeral muscular dystrophy, 40/419
FG syndrome, 43/250
foramina parietalia permagna, 42/29
Friedreich ataxia, 42/143
fucosidosis, 42/550
glycosaminoglycanosis type IH, 66/297
glycosaminoglycanosis type IVB, 66/307
Goldenhar syndrome, 43/443
hemivertebra, 42/31
hemochromatosis, 42/553
hereditary amyloid polyneuropathy, 42/518, 524
hereditary long QT syndrome, 42/718, 75/442
hypertelorism hypospadias syndrome, 43/413
hypertension, 8/283
hypokalemic periodic paralysis, 62/459
idiopathic long QT syndrome, 75/442
idiopathic scoliosis, 42/52
ischemic, *see* Ischemic heart disease
Klinefelter syndrome, 43/560
Klinefelter variant XXXXY, 43/560
Klippel-Feil syndrome, 42/37
LEOPARD syndrome, 43/27
lissencephaly syndrome, 42/40
Marden-Walker syndrome, 43/424
mental deficiency, 43/257, 262
middle cerebral artery syndrome, 53/356
muscular atrophy, 42/343-345
myotonic dystrophy, 40/507, 522, 62/226
ophthalmoplegia, 43/141
orthostatic hypotension, 63/145
osteogenesis imperfecta, 43/453
4p partial monosomy, 43/497
10p partial trisomy, 43/519
20p partial trisomy, 43/539
Patau syndrome, 43/528
polymyositis, 41/73
Potter syndrome, 43/469
progressive bulbar palsy, 22/114
pseudoxanthoma elasticum, 43/45
4q partial monosomy, 43/496
11q partial monosomy, 43/523
18q partial monosomy, 43/533
4q partial trisomy, 43/498
10q partial trisomy, 43/517
respiratory function, 75/425
rheumatic, *see* Rheumatic heart disease
Rubinstein-Taybi syndrome, 43/235, 46/92
severe congenital muscular dystrophy, 43/93
spastic paraplegia, 42/169
survey, 75/425
sweating, 1/458
thalidomide syndrome, 42/660
triploidy, 43/566
trisomy 22, 43/548, 50/564
tuberous sclerosis, 14/372
Turner syndrome, 43/550
unilateral facial paresis, 42/313
valvular, *see* Valvular heart disease
Heart dysplasia
acrocephalosyndactyly type II, 50/123
Heart dysrhythmia, *see* Cardiac dysrhythmia
Heart failure
autonomic nervous system, 74/173, 75/443
baroreflex, 75/444
brain arteriovenous malformation, 31/183-185
Chagas disease, 75/399
congestive, *see* Congestive heart failure
Duchenne muscular dystrophy, 62/134
infantile acid maltase deficiency, 62/480
malignant hyperthermia, 38/551
meningococcal meningitis, 33/24, 52/30
pharmacology, 74/173
primary amyloidosis, 51/415
scorpion sting intoxication, 37/112
Heart malformation
congenital, *see* Congenital heart malformation
trisomy 8, 50/563
Heart rate
aging, 74/228
arterial hypertension, 63/82
autonomic nervous system, 74/246, 581, 608, 611, 75/426
baroreflex, 75/467

blood pressure, 75/467
brain blood flow, 63/81
change, 74/581
Cheyne-Stokes respiration, 63/483
chronic hypoxia, 75/268
cold pressor test, 74/605
fixed, *see* Fixed heart rate
head up tilt test, 74/614
headache, 75/282, 286
hereditary sensory and autonomic neuropathy type III, 75/152
measurement, 74/600
migraine, 75/296
obstructive sleep apnea, 63/456
odor, 74/367
orthostatic hypotension, 63/144
pure autonomic failure, 22/232, 63/147
REM sleep, 74/538
respiration, 74/581, 603
response to standing, 74/597
sleep, 3/82
space, 74/296
spinal cord injury, 26/313-333, 61/439
squatting, 74/619
stress, 74/309
SUNCT syndrome, 75/299
tetraplegia, 61/439
transverse spinal cord lesion, 61/439
vagus nerve, 74/619
Valsalva maneuver, 74/615, 618
Heart rate control
autonomic polyneuropathy, 51/487
Heart rhabdomyoma
cerebrovascular disease, 53/43
tuberous sclerosis, 50/373, 63/94, 103
Heart transplantation
autonomic nervous system, 75/446
brachial plexus lesion, 63/181
brain abscess, 63/179
brain arteriolar dilatation, 63/175
brain blood flow, 63/175
brain capillary dilatation, 63/175
brain embolism, 63/175
brain fat embolism, 63/178
brain hemorrhage, 63/179
brain infarction, 63/179
brain lymphoma, 63/10
brain perfusion pressure, 63/188
brain reticulosarcoma, 63/10
cerebrovascular complication frequency, 63/178
CNS lymphoma, 63/181
congenital heart disease, 63/10
coronary bypass surgery, 63/181
cytomegalovirus encephalitis, 63/180
cytomegalovirus infection, 63/180
encephalopathy, 63/180
epilepsy, 63/178, 180
herpes simplex virus encephalitis, 63/180
iatrogenic neurological disease, 63/175
immunosuppressive therapy, 63/10
meningitis, 63/179
meningoencephalitis, 63/179
migraine, 63/179
mortality, 63/178
multiple organ failure, 63/180
neurologic complication, 63/10
neurologic complication prophylaxis, 63/188
peroneal nerve lesion, 63/181
phrenic nerve lesion, 63/181
prior brain infarction, 63/179
prior stroke significance, 63/188
psychosis, 63/180
recurrent laryngeal nerve lesion, 63/181
regional brain blood flow, 63/175
rhinocerebral infection, 63/179
ulnar nerve lesion, 63/181
vascular headache, 63/179
viral encephalitis, 63/180
Heart valve disease, *see* Cardiac valvular disease
Heat
migraine, 48/122
myoglobinuria, 41/266
stress, 74/320
Heat conservation
see also Thermoregulation
mechanism, 75/53
Heat cramp
heat stroke, 23/670
management, 75/63
Heat edema
management, 75/63
Heat exhaustion
heat stroke, 23/670
management, 75/63
Heat gain
thermoregulation, 26/361, 363
transverse spinal cord lesion, 26/358, 361
Heat injury
see also Thermal injury *and* Thermoregulation
anhidrosis, 8/342
urticaria, 8/342
Heat intolerance
heat stroke, 75/63
oxidation-phosphorylation coupling defect, 62/506
pheochromocytoma, 39/498

Heat loss
see also Thermoregulation
mechanism, 75/53
spinal cord injury, 61/280
thermoregulation, 26/359, 364
Heat stroke, 38/543-549
see also Thermoregulation
acquired myoglobinuria, 62/572
amphetamine addiction, 55/519
ataxia, 21/583
blood clotting, 23/672, 677
brain injury, 23/669
Cheyne-Stokes respiration, 23/674
chronic course, 23/678
classification, 23/670
clinical features, 23/674, 38/544-546
clinical picture, 71/587
coma, 71/588
convulsion, 71/588
cooling time, 71/591
CSF, 23/678
delirium, 46/546
diagnosis, 71/588
electrolyte disorder, 23/676, 679
epilepsy, 23/674, 679
fibrinolysis, 23/672, 677
heat cramp, 23/670
heat exhaustion, 23/670
heat intolerance, 75/63
heat pyrexia, 23/671
hemiplegia, 71/588
hemolysis, 23/672, 677
history, 38/543
hypothalamus, 2/447
intravascular consumption coagulopathy, 23/671, 677, 55/495
laboratory finding, 23/676
malignant hyperthermia, 38/550, 62/571, 573
management, 23/679, 75/63
metabolic encephalopathy, 62/573
myoclonus, 49/618
myoglobinuria, 40/339
paraventricular nucleus, 2/447
pathogenesis, 23/670, 38/546-548, 71/600
pathology, 23/672, 674, 38/548
prevention, 38/549, 71/591
regulation body heat, 23/669
renal insufficiency, 23/674, 679
subarachnoid hemorrhage, 12/122, 23/673
sunstroke, 23/669
survey, 71/585
symptom, 71/600
treatment, 38/549, 71/588
Heat syncope
see also Thermoregulation
classification, 15/824
Heat syndrome
pathogenesis, 71/600
symptom, 71/600
Heautoscopia, see Autoscopia
Heautoscopic hallucination, see Autoscopia
Heavy chain
alpha, 9/502
gamma, 9/502
pi, 9/502
type specific, 9/502
viral infection, 56/52
Heavy chain disease
gamma, see Gamma heavy chain disease
hyperviscosity syndrome, 55/484
immunocompromised host, 56/470
paraproteinemic polyneuropathy, 51/429
plasma cell dyscrasia, 63/391
Heavy duty below elbow splint
brachial plexus, 51/151
Heavy metal
late cortical cerebellar atrophy, 21/477
neurotoxicity, 51/265
Raynaud phenomenon, 8/337
Heavy metal intoxication
Fanconi syndrome, 29/172
neuropathy, 7/511
progressive muscular atrophy, 59/18
Heavy metal motoneuron disorder
WFN classification, 59/3
Hebephrenia
brain ventricle enlargement, 46/482
diffuse focal sign, 46/484
regional brain blood flow, 46/453
Hebephrenic anergic syndrome
diffuse focal sign, 46/485
HeCNU, see Elmustine
Hedgehog breathing
respiration, 1/674
Hedonia
amphetamine intoxication, 65/257
Heemstede Neurotoxicity Scale
epilepsy, 72/419
Heerfordt syndrome
facial paralysis, 40/306
idiopathic facial paralysis, 8/279
Heidelberg, Germany
neurology, 1/9, 19
Heidenhain disease, see Creutzfeldt-Jakob disease
Heidenhain-Nevin disease, see Creutzfeldt-Jakob disease

Heine-Medin disease, *see* Poliomyelitis
Heinz body
  glucose-6-phosphate dehydrogenase deficiency, 42/644
HeLa cell
  neural inductor, 30/46, 48
Held, H., 1/47
Heliotrope rash
  dermatomyositis, 41/57, 62/371
Helmet erythrocyte
  thrombotic thrombocytopenic purpura, 9/600, 607
Helminthiasis
  angiostrongyliasis, 52/510
  Angiostrongylus cantonensis, 52/506, 545
  antibody, 52/509
  ascariasis, 35/211
  biology, 52/505
  brain cyst, 52/511
  Brugia malayi, 52/506
  cestode, 52/505
  Dirofilaria immitis, 52/506
  Echinococcus, 35/217, 52/523
  Echinococcus granulosus, 52/506
  eosinophilia, 52/509, 511
  eosinophilic meningitis, 35/213
  epidemiology, 35/211-218, 52/507
  Filaria, 52/513-519
  filariasis, 35/212
  general features, 35/209-225
  gnathostomiasis, 35/212
  Hill-Sherman syndrome, 21/565
  history, 35/209-211
  hookworm disease, 35/211
  hydatidosis, 35/217
  immunologic aspect, 35/222-224
  immunology, 52/508
  larva migrans, 35/212
  nematode, 52/505, 509
  neurocysticercosis, 35/218, 52/529
  neurotoxin, 52/509
  organophosphate, 64/155
  paragonimiasis, 35/215
  Paragonimus, 52/511
  Paragonimus westermani, 52/506
  pathogenesis, 35/218-225
  Schistosoma, 52/511, 535-541
  Schistosoma haematobium, 52/506
  Schistosoma japonicum, 52/506
  Schistosoma mansoni, 52/506
  schistosomiasis, 35/216
  Strongyloides stercoralis, 52/506
  Taenia solium, 52/506, 511
  Toxocara canis, 52/506
  Toxocara cati, 52/506
  toxocariasis, 35/212, 52/510
  trematode, 52/505, 511
  Trichinella, 52/563
  Trichinella spiralis, 52/506
  trichinosis, 35/213
  Wuchereria bancrofti, 52/506
Helminthic infection
  epilepsy, 72/165
Helweg-Larsen-Ludvigsen syndrome
  dominant anhidrosis, 22/491
  hearing loss, 22/491
  hereditary progressive cochleovestibular atrophy, 60/772
Hemagglutinating encephalomyelitis virus
  pig, *see* Porcine hemagglutinating encephalomyelitis virus
Hemangiectasia hypertrophica (Weber), *see* Klippel-Trénaunay syndrome
Hemangiectatic hypertrophy, *see* Klippel-Trénaunay syndrome
Hemangioblastoma
  see also Angioreticuloma
  brain, 14/60, 244
  cerebellar, *see* Cerebellar hemangioblastoma
  chromosome abnormality, 67/45
  clinical features, 68/272
  collagen fiber, 14/55
  congenital spinal cord tumor, 32/355-386
  cyst transformation, 14/56
  diagnosis, 68/272
  distribution, 67/130
  epilepsy, 72/180
  fibrous transformation, 14/55
  hemangiopericytoma, 14/52
  histologic differentiation, 18/124
  histology, 14/648
  hypervascular, 68/270
  imaging, 67/179, 68/272
  mastocyte, 14/55
  medulla oblongata, 14/244, 249
  optic chiasm compression, 68/75
  optic pathway tumor, 68/74
  pathology, 14/52, 54
  posterior fossa tumor, 18/401
  pseudoxanthoma cell, 14/244
  radiotherapy, 18/510
  reticulin fiber, 14/55
  retinal, *see* Retinal hemangioblastoma
  site preference, 14/60
  spinal, *see* Spinal hemangioblastoma
  spinal angioma, 20/448
  spinal cord, *see* Spinal cord hemangioblastoma

spinal cord compression, 19/362
spinal tumor, 19/70
subarachnoid hemorrhage, 12/119
survey, 68/269
symptom, 14/102
syringomyelia, 14/60
treatment, 68/273
Von Hippel-Lindau disease, 14/242, 18/401, 20/449, 68/294
xanthomatous degeneration, 14/55
Hemangioendothelioblastoma (Mallory), *see* Hemangiopericytoma
Hemangioendothelioma
see also Cerebellar hemangioblastoma
anastomosis, 18/288
bleeding tendency, 18/288
blindness, 18/288
CNS tumor, 68/280
congenital spinal cord tumor, 32/355-386
diagnostic features, 68/281
epidemiology, 68/281
hypervascular, 68/270
malignant, *see* Malignant hemangioendothelioma
management, 68/281
pathology, 68/281
reticulin fiber, 18/288
sella turcica, 18/288
spinal cord, 18/288
survey, 68/270
Hemangiolipoma
Klippel-Feil syndrome, 20/440
Hemangioma
brain, *see* Brain hemangioma
capillary, *see* Nevus flammeus
cavernous, *see* Cavernous hemangioma
cerebellar, *see* Cerebellar hemangioblastoma
cerebellar tumor, 17/715
cranial vault tumor, 17/110
cryptic, *see* Cryptic hemangioma
fibrous dysplasia, 14/197
incidence, 68/530
intravascular consumption coagulopathy, 63/318
macrencephaly, 42/41
macrocephaly, 14/480, 768
Maffucci-Kast syndrome, 43/29
multiple capillary, *see* Multiple capillary hemangioma
multiple cutaneous, *see* Multiple cutaneous hemangioma
muscle tumor, 41/382
nerve, 8/459
optic nerve tumor, 17/365
port wine, *see* Nevus flammeus
Rubinstein-Taybi syndrome, 31/284
scalp tumor, 17/124
skull base tumor, 17/163
spinal cord, *see* Spinal cord hemangioma
spinal tumor, 68/515
third ventricle tumor, 17/444
vertebral, *see* Vertebral hemangioma
Hemangiomatosis
Bean syndrome, 14/106
leptomeningeal, *see* Meningeal angiomatosis
Hemangiomatous nevus
diastematomyelia, 50/436
diplomyelia, 50/436
Hemangiopericytoma
brain, 18/293-297
diagnostic features, 68/276
epidemiology, 68/275
hemangioblastoma, 14/52
hypervascular, 68/270
pathology, 68/276
spinal cord tumor, 68/576
spinal epidural tumor, 20/103
survey, 68/270
treatment, 68/277
vertebra, 20/443
Hemangiosarcoma
Bean syndrome, 14/106
brain, 18/288-292
Hematemesis
dermatomyositis, 62/374
inclusion body myositis, 62/374
polymyositis, 62/374
Hematin
chemistry, 41/259
Hematocephalus, *see* Intracerebral hematoma
Hematocephalus externus
brain hemisphere, 12/159
subarachnoid hemorrhage, 12/159
Hematocrit
spinal cord injury, 26/379, 398
Hematoma
acute subdural, *see* Acute subdural hematoma
bilateral periorbital, *see* Bilateral periorbital hematoma
brain, *see* Brain hematoma
brain hemorrhage, 54/89
brain stem, *see* Brain stem hematoma
cerebellar, *see* Cerebellar hematoma
chronic epidural, *see* Chronic epidural hematoma
chronic subdural, *see* Chronic subdural hematoma
epidural, *see* Epidural hematoma
headache, 48/5
hyperaldosteronism, 39/496

intracerebellar, *see* Intracerebellar hematoma
intracerebral, *see* Intracerebral hematoma
intracranial, *see* Intracranial hematoma
intracranial epidural, *see* Intracranial epidural hematoma
intracranial subdural, *see* Intracranial subdural hematoma
intraventricular, *see* Intraventricular hematoma
local anesthetic agent, 65/421
nasal septum, 5/218
nerve, *see* Nerve hematoma
nerve injury, 51/134
organic acid metabolism, 66/641
parasagittal, *see* Parasagittal hematoma
periorbital, *see* Periorbital hematoma
pontine, *see* Pontine hematoma
posterior fossa, *see* Posterior fossa hematoma
posterior fossa epidural, *see* Posterior fossa epidural hematoma
posttraumatic amnesia, 24/489
posttraumatic intracranial, *see* Posttraumatic intracranial hematoma
psoas, *see* Psoas hematoma
scalp injury, 24/655
spinal epidural, *see* Spinal epidural hematoma
spinal subdural, *see* Spinal subdural hematoma
subarachnoid hemorrhage, 12/71
subdural, *see* Subdural hematoma
subgaleal, *see* Subgaleal hematoma
traumatic intracerebral, *see* Traumatic intracerebral hematoma
traumatic intracranial, *see* Traumatic intracranial hematoma
Hematomyelia
see also Hemorrhagic necrosis
angiodysgenetic necrotizing myelomalacia, 20/487
cause, 12/581
central hemorrhagic necrosis, 25/272
central spinal syndrome, 2/211
definition, 25/47
hemophilia, 38/60
hypoprothrombinemia, 38/77-79
posttraumatic exception, 61/378
posttraumatic syringomyelia, 61/378
progressive posttraumatic, 25/266
sickle cell anemia, 38/60
spinal cord injury, 61/523
spinal cord vascular disease, 12/557, 580
spinal injury, 12/557, 580
subarachnoid hemorrhage, 12/173
thrombocytopenia, 38/69
traumatic, 12/581
Hematopoietic cell
gene transfer, 66/117
Hematoside
Gaucher disease, 10/308
neonatal brain, 10/305
Hematoside lipidosis, *see* Farber disease
Hematoside lipodystrophy
GM3, *see* GM3 hematoside lipodystrophy
Hematothorax
spinal injury, 26/194, 225
thoracic spinal column injury, 26/224
Hematoxylin body
systemic lupus erythematosus, 11/139
Hematuria
Lesch-Nyhan syndrome, 6/456, 29/265, 42/153
subarachnoid hemorrhage, 12/151
Wilson disease, 49/229
Heme induced nephropathy
myoglobinuria, 62/556
Heme moiety
myoglobin property, 41/261
Heme synthesis
acute intermittent porphyria, 51/390
lead, 64/434
Heme synthetase
lead polyneuropathy, 51/270
Hemeralopia
Refsum disease, 21/199, 27/522, 60/728
Stargardt disease, 13/262
trimethadione intoxication, 37/202
Hemeralopic retinosis, *see* Primary pigmentary retinal degeneration *and* Secondary pigmentary retinal degeneration
Hemi-3 syndrome
hemihypertrophy, 59/483
syringomyelia, 50/453
Hemiachromatopsia
see also Monochromatism
hemiamblyopia, 2/583
hemianopia, 2/583
homonymous, *see* Homonymous hemiachromatopsia
neuropsychology, 45/15
Hemiagnosia
color, *see* Color hemiagnosia
parietal lobe syndrome, 45/70
Hemiakinesia
neglect, 45/16
Hemialexia
corpus callosum, 4/117
corpus callosum defect, 45/444
reading, 45/448
Hemiamblyopia

hemiachromatopsia, 2/583
Hemianalgesia
hysteria, 1/111
Hemianesthesia
Avellis syndrome, 1/189
Benedikt syndrome, 2/298
brain lacunar infarction, 46/355
hemi-inattention, 45/153
parietal, *see* Parietal hemianesthesia
transient ischemic attack, 12/5
Hemianesthesia cruciata
lateral pontine syndrome, 12/17
Hemianhidrosis
survey, 75/67
Hemianopia
altitudinal, *see* Altitudinal hemianopia
anosognosia, 4/216
Balint syndrome, 1/599
binasal, *see* Binasal hemianopia
bitemporal, *see* Bitemporal hemianopia
brain arteriovenous malformation, 53/36
brain embolism, 53/207
brain infarction, 53/207
capsula interna lesion, 2/585
cerebrovascular disease, 53/207
dementia, 42/278
Divry-Van Bogaert syndrome, 14/108
hemiachromatopsia, 2/583
hemi-inattention, 45/153
homonymous, *see* Homonymous hemianopia
intracranial hemorrhage, 53/207
Köhlmeier-Degos disease, 39/437
migraine, 48/131, 143
occipital lobe tumor, 17/314-316, 319, 337
parietal lobe syndrome, 45/69
pituitary adenoma, 17/387
porencephaly, 42/49
pseudo, *see* Pseudohemianopia
radiation injury, 23/653
temporal lobe tumor, 17/290
transient ischemic attack, 53/207
uncinate fit, 2/586
unilateral, 2/563, 565, 567
visuospatial agnosia, 45/168
Hemianopic inattention
Gerstmann syndrome, 2/691, 695
Hemianopic pupillary phenomenon
bitemporal hemianopia, 2/575, 577
Hemiaplasia
cerebellar, *see* Cerebellar hemiaplasia
cerebellar agenesis, 50/181
lateral, *see* Lateral hemiaplasia
Hemiasomatognosia
amorphosynthesis, 45/375
attention disorder, 3/194
body scheme disorder, 2/478, 4/22, 214, 297, 45/374-376
conscious, *see* Conscious hemiasomatognosia
dressing apraxia, 4/53, 215
epilepsy, 45/509
hemiplegia, 45/375
higher nervous activity disorder, 3/29
nonconscious, *see* Hemidepersonalization
parietal lobe, 45/375
parietal lobe syndrome, 45/69
spatial agnosia, 45/375
Hemiataxia
brain arteriovenous malformation, 53/36
posterolateral thalamic syndrome, 2/485
thalamic syndrome, 2/481
Hemiathetosis
acute infantile hemiplegia, 53/31
brain arteriovenous malformation, 53/36
cause, 6/446
cerebrovascular disease, 53/31
definition, 49/382
head injury, 49/384
infantile hemiplegia, 49/384
Hemiatrophia cerebri, *see* Cerebral hemiatrophy
Hemiatrophy, 22/545-553
cerebral, *see* Cerebral hemiatrophy
classification, 59/475
congenital general fibromatosis, 59/480
differential diagnosis, 59/475, 480
facial, *see* Hemifacial atrophy
Fuchs heterochromic iridocyclitis, 59/480
infantile spastic hemiplegia, 59/480
ipsilateral benign hemiparkinsonism, 59/480
linear scleroderma, 59/480
lipodystrophy, 59/480
poliomyelitis, 59/480
primary cerebral, *see* Primary cerebral hemiatrophy
syringobulbia, 59/480
toxic phocomelia, 59/480
Hemiballismus
amiodarone intoxication, 65/458
anatomy, 49/371
arteriovenous malformation, 49/373
Behçet syndrome, 49/373
brain embolism, 53/211
brain infarction, 53/211
brain injury, 49/374
brain lacunar infarction, 54/246
brain metastasis, 49/373
carotid artery occlusion, 49/373

cerebellar atrophy, 42/224
cerebrovascular disease, 53/211
choreoathetosis, 42/224
clinical features, 49/370
contracture, 42/224
cyst, 49/373
deanol, 49/376
diazepam, 49/376
disease duration, 49/371
etiology, 49/373
GABA, 49/372
glycine, 49/372
hereditary olivopontocerebellar atrophy (Menzel), 21/441, 445
hereditary pallidoluysiodentate degeneration, 49/451
heredodegeneration, 42/224
hyperglycemia, 49/374
infarction, 49/373
intracranial hemorrhage, 53/211
lateral pallidal segment, 49/372
levodopa induced dyskinesia, 49/374
medial pallidal segment, 49/372
multiple sclerosis, 49/374
neostriatum, 49/370
neuroleptic agent, 6/261, 49/375
neuropharmacology, 49/58
outer pallidal segment, 49/370
pallidoluysian atrophy, 42/224
pallidoluysiodentate degeneration, 49/456
pallidothalamic pathway, 49/370
pathophysiology, 49/371
phenytoin intoxication, 49/374
postcentral gyrus, 49/370
posthemiplegic chorea, 6/437
precentral gyrus, 49/370
progabide, 49/376
prognosis, 49/371
Reye syndrome, 49/373
scleroderma, 49/373
speech disorder, 42/224
stereotaxic operation, 49/374
subclavian steal syndrome, 49/373
subthalamic nucleus, 42/224, 49/369
superior frontal gyrus, 49/370
systemic lupus erythematosus, 49/373, 55/377
tetrabenazine, 49/375
thalamotomy, 1/288
thalamus, 49/369, 373
therapy, 49/374
transient ischemic attack, 53/211
tuberculoma, 49/373
tuberculous meningitis, 49/373
valproic acid, 49/376
vertebrobasilar insufficiency, 49/373
Hemibase syndrome, *see* Garcin syndrome
Hemicerebellopyramidal syndrome
brain lacunar infarction, 46/355
Hemicerebral atrophy
Sturge-Weber syndrome, 18/276
Hemicholinium
acetylcholine, 41/105, 111
experimental myopathy, 40/150
neuromuscular transmission, 40/143
synaptic vesicle, 41/105, 111
Hemichordata intoxication
chordata, 37/68
Hemichorea
brain embolism, 53/211
brain infarction, 53/211
brain lacunar infarction, 54/246
cerebrovascular disease, 53/211
diphtheria, 33/486
ECHO virus, 56/330
intracranial hemorrhage, 53/211
putaminal hemorrhage, 54/305
transient ischemic attack, 53/211
vertebrobasilar system syndrome, 53/397
Hemicord
diastematomyelia, 50/438
diplomyelia, 50/438
Hemicrania
chronic paroxysmal, *see* Chronic paroxysmal hemicrania
migraine, 48/3
Hemicrania angioparalytica, *see* Cluster headache
Hemicrania angioparalytica sive neuroparalytica, *see* Cluster headache
Hemicrania continua
chronic paroxysmal hemicrania, 48/260
SUNCT syndrome, 75/299
Hemicrania dysphrenica, *see* Dysphrenia hemicranica
Hemidepersonalization
body scheme disorder, 4/214, 45/375
consciousness, 3/56
parietal lobe syndrome, 45/69, 375
sensory extinction, 3/56
Hemidysplasia
brain atrophy, 43/416
congenital heart disease, 43/416
cranial nerve, 43/416
dystonia musculorum deformans, 6/528
ichthyosis, 43/416
pectoralis muscle agenesis, 43/416
Hemidystonia

see also Symptomatic dystonia
head injury, 49/542
Hemifacial atrophy, 43/407-409
congenital, *see* Congenital hemifacial atrophy
congenital progressive, *see* Congenital progressive hemifacial atrophy
Klippel-Trénaunay syndrome, 14/524
progressive, *see* Progressive hemifacial atrophy
Hemifacial hypertrophy
see also Sachsalber syndrome
Bencze syndrome, 59/482
Bitter syndrome, 14/107
hemimegalencephaly, 59/482
hydrophthalmia, 14/635
Klippel-Trénaunay syndrome, 14/397-399
macrodactyly, 22/550
mental deficiency, 59/482
neurofibromatosis, 14/635
nevus unius lateris, 59/482
polydactyly, 22/550
Sturge-Weber syndrome, 14/67
syndactyly, 22/550
variant type, 59/481
Hemifacial hypoplasia
Goldenhar syndrome, 38/411
mental deficiency, 46/46
Hemifacial microsomia, *see* Goldenhar syndrome
Hemifacial spasm
botulinum toxin A, 65/213
intermedius neuralgia, 48/492
jerk, 49/610
Joubert syndrome, 63/437
motor unit hyperactivity, 41/302
tuberculosis, 51/186
Hemihypertonia postapoplectica, *see* Hemirigor apoplecticus
Hemihypertrophy, 22/545-553
adrenogenital syndrome, 22/550
Bannayan-Zonana syndrome, 59/482
cerebrovascular anomaly, 59/482
classification, 59/475
congenital, *see* Congenital hemihypertrophy
definition, 59/475
differential diagnosis, 59/482
facial, *see* Facial hemihypertrophy
glaucoma, 14/398
hemi-3 syndrome, 59/483
hemimegalencephaly, 59/482
Ito hypomelanosis, 59/483
lipodystrophy, 59/483
Maffucci-Kast syndrome, 59/483
nerve entrapment syndrome, 59/481
neurofibromatosis, 59/483
neurofibromatosis type I, 22/549
nevus unius lateris, 59/482
partial, 59/482
Proteus syndrome, 59/482
Russell-Silver syndrome, 22/550, 59/483
tuberous sclerosis, 22/549
variant type, 59/480-483
visceral malignancy, 59/480
Von Hippel-Lindau disease, 22/549
Hemihypesthesia
concentric sclerosis, 47/416
Hemi-inattention
Balint syndrome, 45/410
body scheme, 45/154
brain infarction, 55/145
cerebrovascular disease, 55/145
diagnosis, 45/73
hemianesthesia, 45/153
hemianopia, 45/153
inferior parietal lobule, 45/157
middle cerebral artery syndrome, 53/363
neglect, 45/16
neglect syndrome, 45/153-158
pathophysiology, 45/154
response mode, 45/154
superior temporal sulcus, 45/157
syndrome component, 45/73
testing, 45/154
Hemikeratosis
mental deficiency, 46/95
Hemimegalencephaly
epilepsy, 50/263, 72/342, 73/393
hemifacial hypertrophy, 59/482
hemihypertrophy, 59/482
neuronal migration disorder, 73/400
polymicrogyria, 50/263
Hemimyelocele, *see* Spina bifida
Hemineglect syndrome
diagnosis, 55/145
Hemiparesis
acute subdural hematoma, 24/280
ataxic, *see* Ataxic hemiparesis
Behçet syndrome, 56/598
brachiocrural, *see* Brachiocrural hemiparesis
brachiofacial, *see* Brachiofacial hemiparesis
brain amyloid angiopathy, 54/337
brain arteriovenous malformation, 53/36
brain embolism, 53/206
brain infarction, 53/206
Brissaud-Sicard syndrome, 2/240
cerebellar, *see* Marie-Foix-Alajouanine disease
cerebellum, *see* Marie-Foix-Alajouanine disease
cerebrovascular disease, 53/206

Cestan-Chenais syndrome, 2/318
diphtheria, 33/485
Goltz-Gorlin syndrome, 14/788
hemolytic uremic syndrome, 63/324
hyperosmolar hyperglycemic nonketotic diabetic coma, 27/90
intracranial hemorrhage, 53/206
lead encephalopathy, 64/436
Listeria monocytogenes meningitis, 63/531
Lyme disease, 51/205
migraine, 48/142
multiple sclerosis, 47/56
neuroborreliosis, 51/205
occipital lobe tumor, 17/335
periodic ataxia, 21/571
pontine infarction, 12/15, 23
spreading depression, 48/68
thrombotic thrombocytopenic purpura, 63/324
transient ischemic attack, 53/206
uremic encephalopathy, 63/505
water intoxication, 64/241
Hemiplegia
acute infantile, *see* Acute infantile hemiplegia
air embolism, 61/216
alternating, *see* Alternating hemiplegia
alternating infantile, *see* Alternating infantile hemiplegia
amyloidosis, 63/137
angioneurotic edema, 43/61
anosognosia, *see* Anton-Babinski syndrome
arthropathy, 1/183
arthropod envenomation, 65/197
ascending spinal, *see* Mills syndrome
ataxic, *see* Ataxic hemiparesis
athetotic, *see* Athetotic hemiplegia
attitude, 6/138
Avellis syndrome, 1/189
Babinski sign, 1/190
Binswanger disease, 54/222
biofeedback, 46/621
brachiofacial, *see* Brachiofacial hemiplegia
brain abscess, 1/194
brain infarction, 1/192, 63/345
brain lacunar infarction, 46/355
brain metastasis, 18/216
brain stem, 1/186
brain tumor, 1/193
Brissaud-Sicard syndrome, 2/240
capsular, 1/186, 2/481
caudal tegmental syndrome, 12/20
cavernous angioma, 42/724
central motor disorder, 1/178-185
centrofacial lentiginosis, 14/782
Chagas disease, 63/135
comatous, 1/178
common migraine, 48/144, 149
complicated migraine, 5/63, 48/145
concentric sclerosis, 47/416
congestive heart failure, 63/132
contracture, 2/481
corpus callosum agenesis, 50/162
cortical, 1/185
crossed, *see* Crossed hemiplegia
cruciate paralysis, 1/189
decompression myelopathy, 61/217
differential diagnosis, 1/190
endocarditis, 63/111
essential thrombocythemia, 63/331
etiology, 1/191
flaccid paraplegia, 1/176
gait, 1/420
hand contracture, 2/481
heat stroke, 71/588
hemiasomatognosia, 45/375
hereditary hemorrhagic telangiectasia (Osler), 1/193
hypercalcemia, 63/561
infantile, *see* Infantile hemiplegia
infantile spastic, *see* Infantile spastic hemiplegia
infectious disease, 1/194
intracerebral hematoma, 1/192
intracranial hypertension, 16/134
leukemia, 63/342, 345
Listeria monocytogenes meningitis, 63/531
mental disorder, 4/219
middle cerebral artery occlusion, 53/122
misoplegia, 4/219
moyamoya disease, 12/360, 42/749, 53/33
multiple myeloma, 18/255
multiple sclerosis, 47/56
neglect, 2/695
neurogenic osteoarthropathy, 1/183
nevus unius lateris, 43/40
occipital lobe tumor, 17/335
Opalski syndrome, 1/189
pallidonigral degeneration, 42/242
paracentral lobule, 1/186
paralysed hatred, 4/219
paralysed limb personification, 4/219
polyarteritis nodosa, 39/301
pontine, *see* Pontine hemiplegia
pontine infarction, 12/15, 23
porencephaly, 42/49
pronator sign, 1/184
pulmonary overpressure, 61/216
pure motor, *see* Pure motor hemiplegia

Raimiste sign, 1/184
recurrent infantile, *see* Recurrent infantile hemiplegia
somatoparaphrenia, 4/219
spastic, *see* Spastic hemiplegia
spasticity, 1/180
striatal, *see* Striatal hemiplegia
striatal necrosis, 49/501
Sturge-Weber syndrome, 43/48
subacute bacterial endocarditis, 63/111
subarachnoid hemorrhage, 12/131
synkinesis, 1/182
systemic lupus erythematosus, 39/279, 43/420
thalamic lesion, 2/486
thalamic syndrome (Dejerine-Roussy), 42/771
thalassemia major, 42/629
tongue, 1/180
transient infantile, *see* Transient infantile hemiplegia
transient ischemic attack, 12/5
transtentorial herniation, 1/560
traumatic, 1/191
trophic disorder, 1/487
tumor, 1/192
unilateral pontine hemorrhage, 12/40
vascular, 1/192
venous sinus occlusion, 54/429
Waldenström macroglobulinemia, 63/396
Hemiplegia inflection
pallidal necrosis, 49/470
Hemiplegic dystonia
basal ganglion disease, 6/145-149
characteristics, 6/145-149
spastic dystonia, 6/146, 149
Wilson disease, 6/651
Hemiplegic migraine
angioneurotic edema, 48/148
child, 6/37
classification, 48/6
clinical symptom, 48/26
clinical type, 5/63
familial, *see* Familial hemiplegic migraine
features, 48/141-144
history taking, 48/157
pathomechanism, 48/141-144
sensory disturbance, 5/46, 6/131
transient infantile hemiplegia, 12/348
vascular impairment, 5/19
Hemiplegie à bascule
operculum syndrome, 2/781
Hemirigor apoplecticus
plastic rigidity, 6/673
Hemiskull base syndrome, *see* Garcin syndrome
Hemispace
visuospatial agnosia, 45/169
Hemispasm
facial, *see* Brissaud-Sicard syndrome
Hemispatial agnosia, *see* Visuospatial agnosia
Hemispatial attention deficit
visuospatial agnosia, 45/170
Hemispatial neglect, *see* Visuospatial agnosia
Hemisphere
asymmetry, *see* Cerebral dominance
brain, *see* Brain hemisphere
cerebellar atrophy, 2/419
dominant, *see* Dominant hemisphere
left, *see* Left hemisphere
right, *see* Right hemisphere
Hemisphere syndrome
minor, *see* Minor hemisphere syndrome
Hemispherectomy
aphasia, 46/152
body scheme, 4/211
cerebral dominance, 4/265
chemical, 4/266
child, 45/303
epilepsy, 73/390, 406
global aphasia, 45/312
intelligence, 4/9
language, 45/303
sensory extinction, 3/193
Hemispheric atrophy
epilepsy, 72/109
Hemispheric difference, *see* Cerebral dominance
Hemispheric disconnection, *see* Disconnection syndrome
Hemivertebra
see also Vertebral abnormality
cheilopalatoschisis, 42/31
craniolacunia, 32/187
diastematomyelia, 42/24, 30
Goldenhar syndrome, 43/442
heart disease, 42/31
ichthyosis, 43/416
imperforate anus, 42/31
incontinentia pigmenti, 43/23
Klippel-Feil syndrome, 42/37
mental deficiency, 43/299
prenatal diagnosis, 42/31
prevalence, 42/31
scoliosis, 42/31
spina bifida, 32/545, 42/30, 55
Sprengel deformity, 42/30, 43/483
trisomy 8, 50/563
Hemochromatosis
brain embolism, 55/163

brain infarction, 55/163
cerebrovascular disease, 55/163
cutaneous pigmentation, 42/553
diabetes mellitus, 42/553
encephalopathy, 42/553
gene localization, 42/553
heart disease, 42/553
hepatosplenomegaly, 42/553
HLA antigen, 42/553
hypogonadism, 42/553
idiopathic, *see* Idiopathic hemochromatosis
jaundice, 42/553
listeriosis, 52/91
liver disease, 42/553
neuropathy, 8/11-13, 42/553
polyarthritis, 42/553
prevalence, 42/552
sex ratio, 42/553

Hemodialysis
see also Dialysis *and* Uremic polyneuropathy
alcohol intoxication, 64/116
brain edema, 16/179
brain embolism, 55/166
brain hemorrhage, 54/291
carnitine deficiency, 62/494
cerebrovascular disease, 55/166
dementia, 46/398, 542
disequilibrium syndrome, 28/501
ethylene glycol intoxication, 64/125
ethylene oxide polyneuropathy, 51/275
headache, 48/419
intermittent, *see* Intermittent hemodialysis
mercury intoxication, 64/396
methanol intoxication, 36/357, 64/102
Minamata disease, 64/416
neurologic complication, 27/341
secondary systemic carnitine deficiency, 62/494
uremic polyneuropathy, 63/511

Hemoglobin
Bart, *see* Bart hemoglobin
brain injury, 23/127
CSF, 12/145
fetal, *see* Fetal hemoglobin
Gowers, *see* Gowers hemoglobin
Patau syndrome, 31/514, 43/528
spinal cord injury, 26/379, 397
temporal arteritis, 48/320

Hemoglobin C
sickle cell anemia, 42/624

Hemoglobin F, *see* Fetal hemoglobin

Hemoglobin H
thalassemia intermedia, 42/630

Hemoglobin S
sickle cell anemia, 42/624, 55/503

Hemoglobinopathy
see also Sickle cell anemia, Sickle cell trait *and* Thalassemia
classification, 63/256
hemoneurological disease, 63/249
hyperviscosity syndrome, 55/484
type, 55/503

Hemolysis
CSF, 12/144
heat stroke, 23/672, 677
loxosceles intoxication, 37/112
phosphofructokinase deficiency, 41/191
sickle cell anemia, 55/503
Streptococcus, 52/78
Wilson disease, 49/223

Hemolytic anemia
autoimmune disease, 41/77
congenital erythropoietic porphyria, 27/433
cryoglobulinemia, 39/181
glutathione deficiency, 60/674
methyldopa intoxication, 37/440
organic acid metabolism, 66/641
paxillus involutus intoxication, 36/542
phosphofructokinase deficiency, 41/191, 43/183, 62/487
phosphoglycerate kinase deficiency, 62/489
spinocerebellar degeneration, 60/674
vitamin K3 intoxication, 65/570

Hemolytic uremic syndrome
ataxia, 63/324
brain infarction, 63/324
cerebrovascular disease, 53/34
coma, 63/324
cranial neuropathy, 63/324
epilepsy, 63/325
headache, 63/324
hemiparesis, 63/324
papilledema, 63/324
retinal hemorrhage, 63/325
symptom, 63/324

Hemoneurological disease
acanthocytosis, 63/271
anemia, 63/249
hemoglobinopathy, 63/249
polycythemia, 63/249
survey, 63/249

Hemopexin
Duchenne carrier, 40/356, 385

Hemophilia
brain hemorrhage, 53/34, 54/71, 293, 63/310
cerebrovascular disease, 55/475
cranial nerve, 38/63

defect, 38/54-56
epidural hematoma, 26/4, 19, 63/310
facial nerve, 38/63
femoral nerve paralysis, 55/475, 63/311
femoral neuropathy, 8/13
head injury, 63/310
hematomatous nerve impression, 63/311
hematomyelia, 38/60
iliopsoas hematoma, 63/311
intracranial hemorrhage, 7/665, 38/55-60
intramuscular hemorrhage, 63/311
intraventricular hemorrhage, 63/310
lumbosacral plexus, 51/164
median nerve palsy, 63/311
nerve, 38/61-63
nerve hematoma, 55/475
nerve lesion, 7/666
neuropathy, 7/664, 8/13
peripheral nerve impression, 63/311
radial nerve palsy, 63/311
spinal cord, 38/60
spinal hemorrhage, 7/665
spinal subdural hemorrhage, 55/475
subdural hematoma, 26/33, 63/310
treatment, 38/63, 64
ulnar nerve palsy, 63/311
Hemophilia A
birth incidence, 42/738
blood clotting factor VIII, 42/737
classic, 42/737-739
coagulopathy, 63/309
facial paralysis, 42/738
gene localization, 42/738
lyonization, 42/738
paralysis, 42/738
polyneuropathy, 42/738
seizure, 42/738
sensory loss, 42/738
Hemophilia B
gene localization, 42/738
Hemophilus
bacterial endocarditis, 52/302
Hemophilus aphrophilus
brain abscess, 52/150
Hemophilus influenzae
bacterial meningitis, 52/2
brain abscess, 52/150
brain metastasis, 69/116
cranial subdural empyema, 52/170
gram-negative bacillary meningitis, 33/53, 52/117
incidence, 52/59
microbiology, 52/59
spinal meningitis, 52/190
superior longitudinal sinus thrombosis, 52/179
type b, 33/53
Hemophilus influenzae meningitis
age distribution, 33/54
ampicillin, 33/56
antibiotic agent, 52/64
antibody response, 33/58
brain edema, 52/61
carbenicillin, 33/56
chloramphenicol, 33/56-58
clinical features, 52/62
clinical symptom, 33/55
complication, 33/58
computer assisted tomography, 52/61
CSF, 33/55, 52/63
culture, 33/55
deafness, 52/63, 66
diagnosis, 33/55
epidemiology, 33/54, 52/59
etiology, 33/53
genetic predisposition, 52/62
history, 33/53
host immunity, 52/61
inappropriate antidiuretic hormone secretion, 52/63
intracranial pressure, 52/61
laboratory finding, 52/62
limulus lysate test, 33/55
mental deficiency, 46/19
mortality, 33/54
pathogenesis, 33/53, 52/60
pathology, 52/60
pathophysiology, 52/60
prevention, 52/66
prognosis, 52/66
relapse, 33/58
seasonal pattern, 33/55
septic shock, 52/62
streptomycin, 33/56
subdural effusion, 33/58
sulfonamide, 33/56
supportive care, 52/65
tetracycline, 33/56
toxic encephalopathy, 52/61
treatment, 33/55-58, 52/64
upper respiratory infection, 52/60, 62
uridine monophosphate kinase, 52/62
vaccine, 52/67
Hemophilus influenzae type B
bacterial meningitis, 52/3, 59, 55/416
capsule, 50/60
Hemophilus parainfluenzae
brain abscess, 52/150

spinal cord abscess, 52/191
Hemophilus paraphrophilus
brain abscess, 52/150
Hemorrhage
a-alphalipoproteinemia, 51/396
Abt-Letterer-Siwe disease, 42/441
aqueduct stenosis, 50/306
astrocytoma, 68/95
basal ganglion, *see* Basal ganglion hemorrhage
Benedikt syndrome, 2/298
bilateral pontine, *see* Bilateral pontine hemorrhage
brain, *see* Brain hemorrhage
brain arteriosclerosis, 53/95
brain arteriovenous malformation, 54/91, 373
brain infarction, 53/365
brain injury, 23/49
brain matrix, *see* Brain matrix hemorrhage
brain stem, *see* Brain stem hemorrhage
brain tumor, *see* Brain tumor hemorrhage
brain ventricle, *see* Brain ventricle hemorrhage
capillary apoplexy, 11/583
carotid cavernous fistula, 24/420
cavernous, *see* Cavernous hemorrhage
central cord necrosis, 61/378
cerebellar, *see* Cerebellar hemorrhage
cerebral lymphomatoid granulomatosis, 39/524
ependymoma, 68/95
epilepsy, 72/345, 73/254
Galen vein aneurysm, 18/282
Garcin syndrome, 2/313
glioblastoma, 68/95
hereditary brain, *see* Hereditary brain hemorrhage
hypernatremia, 2/446
idiopathic multiple hemorrhagic sarcoma, 18/288
intervertebral disc, *see* Intervertebral disc hemorrhage
intracerebellar, *see* Intracerebellar hemorrhage
intracerebral, *see* Brain hemorrhage
intracranial, *see* Intracranial hemorrhage
intraventricular, *see* Intraventricular hemorrhage
Jürgens-Von Willebrand hereditary thrombocytopenia, 9/598
labyrinthine, *see* Labyrinthine hemorrhage
leptomeningeal venous angioma, 18/272-274
leukoencephalopathy, 9/597
medulloblastoma, 68/95
oligodendroglioma, 68/95
optic nerve injury, 24/41
orbital, *see* Orbital hemorrhage
orbital injury, 24/87
parenchymal, *see* Parenchymal hemorrhage
pituitary, *see* Pituitary hemorrhage
pituitary gland, 5/141
pituitary tumor, 68/345
plaque, 12/26
pontine, *see* Pontine hemorrhage
pontine infarction, 2/250, 12/26
prepontine, *see* Prepontine hemorrhage
primary pontine, *see* Primary pontine hemorrhage
pseudoinflammatory foveal dystrophy (Sorsby), 13/138
putaminal, *see* Putaminal hemorrhage
retinal, *see* Retinal hemorrhage
retroperitoneal, *see* Retroperitoneal hemorrhage
scurvy, 38/71
Sheehan syndrome, 2/455
spinal cord, *see* Spinal cord hemorrhage
spinal cord compression, 19/372, 69/170
spinal cord injury, 25/266
spinal leukemia, 20/112
spinal subarachnoid, *see* Spinal subarachnoid hemorrhage
spinal subdural, *see* Spinal subdural hemorrhage
spinal tumor, 19/56
spontaneous cerebellar, *see* Spontaneous cerebellar hemorrhage
stingray intoxication, 37/70
striatal lesion, 6/673
subarachnoid, *see* Subarachnoid hemorrhage
subependymal, *see* Subependymal hemorrhage
tegmental pontine, *see* Tegmental pontine hemorrhage
thalamus, *see* Thalamic hemorrhage
unilateral pontine, *see* Unilateral pontine hemorrhage
vasa nervorum disease, 12/655
ventricular, *see* Ventricular hemorrhage
white matter, *see* White matter hemorrhage
yellow jacket intoxication, 37/107
Hemorrhagic amaurosis, *see* Blindness
Hemorrhagic conjunctivitis
acute, *see* Acute hemorrhagic conjunctivitis
Hemorrhagic diathesis, *see* Bleeding tendency
Hemorrhagic encephalitis
acute, *see* Hyperacute disseminated encephalomyelitis
acute amebic meningoencephalitis, 52/326
aspergillosis, 52/379
multifocal, 52/379
varicella zoster virus meningoencephalitis, 56/235
Hemorrhagic encephalopathy
ethylene glycol intoxication, 64/124
lead, 51/264
methanol intoxication, 64/100
organolead intoxication, 64/133
Hemorrhagic ependymitis

pneumococcal meningitis, 33/37, 52/43
Hemorrhagic fever, *see* Arenavirus infection, Argentinian hemorrhagic fever, Bolivian hemorrhagic fever *and* Omsk hemorrhagic fever
Hemorrhagic lesion
hydrozoa intoxication, 37/40
lumbosacral plexus, 51/164
Hemorrhagic leukoencephalitis
acute, *see* Acute hemorrhagic leukoencephalitis
acute infantile hemiplegia, 53/32
acute necrotizing, *see* Acute necrotizing hemorrhagic leukoencephalitis
cerebrovascular disease, 53/32
influenza A virus, 56/15
Hemorrhagic leukoencephalopathy
cerebral malaria, 52/368
Hemorrhagic necrosis
see also Hematomyelia
brain infarction, 54/438
brain phlebothrombosis, 54/438
central, *see* Central hemorrhagic necrosis
cervical spinal cord, 25/266
gunshot injury, 25/50
histopathology, 25/73, 77
neuropathology, 25/64
pathology specimen, 25/30-42
spinal cord contusion, 25/47
spinal cord experimental injury, 25/21
spinal cord injury, 25/30-42, 80, 84, 266
stab wound, 25/52
thoracolumbar spinal cord, 25/451
Hemorrhagic nephritis
sensorineural deafness, 42/376
Hemorrhagic osteomyelitis, *see* Aneurysmal bone cyst
Hemorrhagic polioleukoencephalopathy
methanol intoxication, 64/100
Hemorrhagic shock
pallidal necrosis, 49/466
Hemorrhagic telangiectasia, *see* Hereditary hemorrhagic telangiectasia (Osler)
Hemorrhagic thrombocythemia
primary, *see* Primary hemorrhagic thrombocythemia
Hemosiderin
epileptogenesis, 72/177
idiopathic multiple hemorrhagic sarcoma, 18/288
nuclear magnetic resonance, 54/88
Hemostasis
acetylsalicylic acid, 63/306
adenyl cyclase, 63/303
antithrombotic agent, 63/306
Bernard-Soulier syndrome, 63/302
bleeding time, 63/306
brain embolism, 11/417
calcium, 63/303
coagulation inhibitor, 63/305
coagulation system, 63/304
D-Dimer test, 63/306
endothelium, 63/301, 305
euglobulin lysis test, 63/306
extrinsic factor, 63/303
fibrin, 63/303
fibrin degradation product, 63/306
fibrinolysis, 63/305
fibrinolytic agent, 63/308
fibrinolytic system, 63/301, 304
fibronectin, 63/301
final common pathway, 63/303
fluid phase, 63/301
glycoprotein Ib, 63/301
heparin, 63/306
intravascular consumption coagulopathy, 63/306
intrinsic factor, 63/303
kaolin cephalin test, 63/306
kininogen, 63/303
laboratory test, 63/306
pharmacology, 63/306
platelet, 63/301
platelet activating factor, 63/302
platelet adhesion, 63/301
platelet aggregation, 63/302
platelet alpha granule, 63/303
prekallikrein, 63/303
prostacyclin, 63/303
prostaglandin endoperoxide, 63/302
protein C, 63/305
protein S, 63/305
prothrombin, 63/303
prothrombin time, 63/306
subarachnoid hemorrhage, 12/176
thrombin, 63/303
thrombin time, 63/306
thromboplastin, 63/303
thromboxane A_2, 63/302
tissue plasminogen activator, 63/305
tissue plasminogen activator inhibitor, 63/305
Von Willebrand disease, 63/302
Von Willebrand factor, 63/301
warfarin, 63/306
warfarin action, 63/307
Hemotympanum
clinical examination, 57/132
head injury, 57/132
temporal bone fracture, 24/120
Henle fiber layer

visual sensory system, 2/510
Hennebert syndrome
temporomandibular joint dysfunction, 5/345
Henneman size principle
tremor, 49/573
Henner, K., 1/11, 14, 17, 31, 35
Henoch-Bergeron chorea, *see* Chorea electrica
Henoch chorea, *see* Chorea electrica
Henoch-Schönlein purpura
angiopathic polyneuropathy, 51/446
leukoencephalopathy, 9/599, 607
neurologic symptom, 38/71
polyarteritis nodosa, 55/359
segmental fibrinoid necrosis, 9/607
subarachnoid hemorrhage, 55/3
systemic brain infarction, 11/453
urticaria, 38/71
vasculitis, 71/8
Henschen, S., 1/12
Hensen node
autoneuralization, 30/49
experiment, 30/47
neural induction, 30/49, 50/22
in vitro, 30/47
Hepadnavirus
hepatitis B like virus, 56/10
hepatitis B virus, 56/295
human, 56/2
taxonomic place, 56/2
Heparan *N*-sulfatase
glycosaminoglycanosis type III, 66/65
glycosaminoglycanosis type IIIA, 66/282, 302
HSS gene, 66/282
Heparan sulfate
chemistry, 10/432, 27/144
glycosaminoglycan, 10/327
glycosaminoglycanosis, 10/429, 66/281
glycosaminoglycanosis type I, 27/146, 46/54
glycosaminoglycanosis type II, 42/454
glycosaminoglycanosis type III, 66/302
intervertebral disc, 20/526
muscle fiber, 62/31
Heparan sulfate sulfatase
glycosaminoglycanosis type III, 42/456
Heparin
acid glycosaminoglycan, 10/432
anticoagulant overdose, 65/460
brain embolism, 63/306
brain fat embolism, 55/191
brain infarction, 53/425, 63/306
cardiac arrest, 55/210
cavernous sinus thrombosis, 52/175
femoral neuropathy, 63/49
glucosamine, 27/144
glucuronic acid, 27/144
glycosaminoglycan, 10/327
glycosaminoglycanosis type I, 46/54
Guillain-Barré syndrome, 51/257
hemostasis, 63/306
internal carotid artery syndrome, 53/325
multiple sclerosis, 9/400
neurologic intensive care, 55/210
nonbacterial thrombotic endocarditis, 63/121
radiation myelopathy, 61/210
transient ischemic attack, 53/325
venous sinus thrombosis, 63/306
Heparitin sulfate, *see* Heparan sulfate
Hepatic arteritis
temporal arteritis, 48/317
Hepatic carnitine palmitoyltransferase deficiency
hypoglycemic encephalopathy, 62/494
Hepatic cirrhosis, *see* Liver cirrhosis
Hepatic coma
see also Hepatic encephalopathy
acid base disturbance, 27/354
acquired hepatocerebral degeneration, 49/213
action tremor, 49/213
acute, 27/361-364
Alzheimer type I astrocyte, 9/635, 49/214
Alzheimer type II astrocyte, 49/214
amino acid, 27/360, 49/215
ammonia, 27/355-360
asterixis, 49/214
blood ammonia, 27/335-358, 368
carbohydrate metabolism, 27/353
cause, 7/617
central neurogenic hyperventilation, 63/441
chronic, 27/364-366
citric acid, 49/215
clinical chemistry, 27/368
clinical features, 27/349, 49/213
course, 27/350
CSF ammonia, 27/357, 368
CSF cerebral edema, 27/350
CSF 2-oxoglutaramic acid, 27/349, 359, 364
dementia, 49/213
demyelination, 9/634
differential diagnosis, 27/370
EEG, 27/349, 370, 49/214
electrolyte disorder, 27/354
epilepsy, 27/362
fatty acid, 49/215
fumaric acid, 49/215
glutamine, 49/215
historic aspect, 27/349
hyperammonemia, 49/215

hypoglycemia, 27/353, 371
intermittent, 27/365
laboratory diagnosis, 27/368-370
lactic acid, 49/215
leukoencephalopathy, 9/634-636
levodopa, 27/360, 372
mental change, 49/213
mercaptan, 27/361, 49/215
neuropathology, 27/351, 49/214
oxaloacetic acid, 49/215
oxidative metabolism, 27/353
2-oxoglutaramic acid, 49/215
pathogenesis, 27/353-361, 49/214
pathophysiology, 27/350
prognosis, 27/370
pyruvic acid, 49/215
short chain fatty acid, 27/361
survey, 70/249
terminal liver cirrhosis, 6/281
toxic factor, 27/365
treatment, 27/371, 49/217
Hepatic disease, *see* Liver disease
Hepatic disorder
nitrogen metabolism, 70/250
Hepatic encephalopathy
see also Hepatic coma, Hepatogenic gliodystrophic encephalopathy, Posticteric encephalopathy *and* Wilson disease
acquired, *see* Acquired hepatocerebral degeneration
alcoholism, 70/356
ammonia, 27/355-360
bilirubin encephalopathy, 49/218
carnitine deficiency, 62/493
Creutzfeldt-Jakob disease, 6/748
CSF glutamine, 27/349, 369
CSF 2-oxoglutaramic acid, 27/349, 359, 364
delirium, 46/542
dementia, 46/128, 398
demyelination, 9/634
EEG, 27/349, 370
fatty acid, 27/351
hyperammonemia, 21/576
hypoglycemia, 27/353
influenza A virus, 56/149
influenza B virus, 56/149
levodopa, 27/360, 372
lipid metabolic disorder, 62/493
liver cirrhosis, 27/351, 49/384
mercaptan, 27/361
metabolic encephalopathy, 69/400
respiratory failure, 63/477
Reye syndrome, 27/361-363, 363, 49/217
survey, 70/249
syncope, 72/263
systemic carnitine deficiency, 62/493
treatment, 49/217
varicella zoster virus, 56/149
Hepatic insufficiency polyneuropathy
see also Deficiency neuropathy
cholesterol ester, 51/363
hyperlipidemia, 51/363
primary biliary cirrhosis, 51/362
sensory polyneuropathy, 51/362
xanthomatous deposit, 51/363
Hepatic myelopathy
malabsorption syndrome, 70/232
Hepatic necrosis
acute, *see* Acute hepatic necrosis
valproic acid induced, *see* Valproic acid induced hepatic necrosis
Hepatic polyneuropathy
ataxia, 51/326
axonal dystrophy, 51/326
cause, 7/608-613
deficiency neuropathy, 51/324, 326
gaze paralysis, 51/326
liver cirrhosis, 7/609
metabolic neuropathy, 51/68
viral hepatitis, 7/486
Hepatic porphyria, *see* Acute intermittent porphyria, δ-Aminolevulinic acid dehydratase deficiency porphyria, Hereditary coproporphyria *and* Porphyria variegata
Hepatic stupor
terminal liver cirrhosis, 6/281
Hepatitis
cytomegalic inclusion body disease, 56/267
cytomegalovirus infection, 34/219
diamorphine, 37/393
drug abuse, 37/393, 55/517
epidemic, *see* Epidemic hepatitis
nerve conduction, 7/609
nerve lesion, 7/486
neuropathy, 7/609
polyradiculitis, 7/486
serum, *see* Serum hepatitis
viral, *see* Viral hepatitis
Hepatitis A virus
picornavirus, 56/11
Hepatitis B virus
acute viral myositis, 56/195
chronic relapsing polyneuropathy, 56/300
core antigen, 56/295
cryoglobulinemia, 63/403
DNA virus, 56/295

epidemiology, 56/295
hepadnavirus, 56/295
liver cell carcinoma, 56/10
polyarteritis nodosa, 56/10
polyneuropathy, 56/10
replication, 56/10
reverse transcriptase, 56/295
serum hepatitis, 56/295
surface antigen, 56/295
systemic vasculitis, 56/301
transmission, 56/295
viral infection, 56/295
Hepatitis B virus infection, 56/295-303
case report, 56/298
clinical pattern, 56/296
course, 56/296
cryoglobulinemia, 56/302
CSF, 56/297, 302
dermatomyositis, 56/303
Guillain-Barré syndrome, 56/296, 298
mononeuritis multiplex, 56/300
neurologic complication, 56/295
pathogenesis, 56/296
polyarteritis nodosa, 39/296, 55/359, 56/302
polymyositis, 56/303
polyneuropathy, 56/297
retrobulbar neuritis, 56/303
systemic vasculitis, 56/297
viral marker, 56/296
Hepatitis C
antiphospholipid antibody, 63/330
Hepatitis virus
deafness, 56/107
neurotropic mouse, *see* JHM virus
vertigo, 56/107
Hepatobiliary disorder
neurology, 70/249
survey, 70/249
Hepatobiliary transplantation
see also Liver transplantation
autopsy, 70/251
sex distribution, 70/251
Hepatoblastoma
Beckwith-Wiedemann syndrome, 43/398
Hepatocerebral degeneration, *see* Wilson disease
Hepatocerebral syndrome
acquired, *see* Acquired hepatocerebral degeneration
Hepatocuprein
liver specific copper protein, 6/119
Hepatogenic gliodystrophic encephalopathy
see also Hepatic encephalopathy
Crigler-Najjar syndrome, 21/52
Hepatolenticular degeneration, *see* Wilson disease
Hepatomegaly
see also Hepatosplenomegaly
Abt-Letterer-Siwe disease, 38/97, 42/441
Berardinelli-Seip syndrome, 42/591
carnitine deficiency, 43/175
cerebrohepatorenal syndrome, 43/339
chelonia intoxication, 37/91
debrancher deficiency, 62/482
fumarate hydratase deficiency, 66/420
glycogen storage disease, 43/179
human T-lymphotropic virus type I associated myelopathy, 56/527
infantile Gaucher disease, 10/565
lipodystrophy, 8/346
long chain acyl-coenzyme A dehydrogenase deficiency, 62/495
medium chain acyl-coenzyme A dehydrogenase deficiency, 62/495
methylmalonic acidemia, 29/205, 42/601
mitochondrial DNA depletion, 66/430
mixed connective tissue disease, 71/122
Mulibrey dwarfism, 43/435
neonatal adrenoleukodystrophy, 60/668
phosphorylase kinase deficiency, 27/222, 235, 41/192
pickwickian syndrome, 38/300
POEMS syndrome, 47/619, 51/430, 63/394
primary amyloidosis, 51/416
pyruvate carboxylase deficiency, 62/503
Reye syndrome, 29/334, 31/168, 56/150
Saldino-Noonan syndrome, 31/243
Hepatopathy, *see* Liver disease
Hepatophosphorylase deficiency
biochemical aspect, 27/234
clinical features, 27/222, 234
hypoglycemia, 27/72
liver phosphorylase, 27/222, 41/188
treatment, 27/234
Hepatophosphorylase kinase deficiency, *see* Phosphorylase kinase deficiency
Hepatorenal tubular syndrome
tyrosinuria, *see* Tyrosinosis
Hepatosplenomegaly
see also Hepatomegaly
adult Gaucher disease, 66/123
brancher deficiency, 62/483
Chédiak-Higashi syndrome, 14/783, 42/536, 60/668
fucosidosis, 42/550
glycogen storage disease, 43/180
glycosaminoglycanosis type I, 10/436, 42/450
glycosaminoglycanosis type II, 10/436, 42/454

glycosaminoglycanosis type III, 10/342, 42/455, 66/300
glycosaminoglycanosis type VI, 42/458
glycosaminoglycanosis type VII, 42/459, 66/310
glycosaminoglycanosis type VIII, 42/459
GM2 gangliosidosis, 60/668
hemochromatosis, 42/553
hereditary amyloid polyneuropathy, 42/521
Hurler-Hunter disease, 10/436
hyperlipoproteinemia type I, 42/443
hyperlipoproteinemia type IV, 10/276
infantile Gaucher disease, 10/307, 66/124
juvenile Gaucher disease, 10/307, 29/352, 42/438
juvenile Gaucher disease type B, 66/125, 128
juvenile metachromatic leukodystrophy, 60/668
leprosy, 51/216
lipoprotein lipase gene, 66/541
mannosidosis, 27/157, 42/597
mevalonate kinase deficiency, 60/668
multiple sulfatase deficiency, 42/493
Niemann-Pick disease type A, 42/468
Niemann-Pick disease type C, 60/668
oligosaccharidosis, 66/329
organic acid metabolism, 66/641
osteopetrosis, 43/456
pallidoluysionigral degeneration, 6/665
pernicious anemia, 42/609
Salla disease, 59/360
sea blue histiocyte disease, 42/472
Tangier disease, 10/278, 42/627
thalassemia major, 42/629
vertical supranuclear ophthalmoplegia, 42/605
Wilson disease, 42/270
Wolman disease, 10/504, 546

Hepatotoxicity
diamorphine intoxication, 65/355
gyromitrine intoxication, 65/36
mephenytoin intoxication, 37/202
opiate intoxication, 37/393, 65/355
phenacemide intoxication, 37/202

Heptachlor
chlorinated cyclodiene, 64/200
neurotoxin, 36/392, 417, 48/420, 64/197, 201
organochlorine, 36/393, 64/197
organochlorine insecticide intoxication, 36/417, 64/197
toxic encephalopathy, 64/201

Heptachlor intoxication
animal study, 36/417
headache, 48/420

Herbicide
neurotoxicity, 51/265
toxic neuropathy, 51/283

Herbicide intoxication
Parkinson disease, 7/523, 49/136

Hereditary acute porphyria
abdominal pain, 51/391
anxiety, 51/391
autonomic polyneuropathy, 51/391
axonal degeneration, 51/392
bathing trunk hypalgesia, 51/391
bladder function, 51/391
CSF protein, 51/392
diarrhea, 51/391
hallucination, 51/391
laboratory test, 51/392
metabolic polyneuropathy, 51/389
nerve biopsy, 51/391
pathogenesis, 51/392
photosensitivity, 51/391
prevalence, 51/391
seizure, 51/391
tachycardia, 51/391
treatment, 51/392

Hereditary adult foveal dystrophy (Behr)
Behr disease, 13/88
clinical course, 13/88
color vision, 13/89
differential diagnosis, 13/91
EOG, 13/90
ERG, 13/89
hereditary, 13/90
histology, 13/90
visual adaptation, 13/89
visual function, 13/89

Hereditary amyloid polyneuropathy, 21/119-141, 60/89-110
abdominal organ, 60/105
amyloid disposition pattern, 60/104
Andrade, *see* Hereditary amyloid polyneuropathy type 1
autonomic nervous system, 60/104
biochemistry, 60/106
bladder, 75/672
blood vessel, 60/102
brain, 60/102
burning pain, 8/368
cardiovascular system, 21/124
chromosome 18, 60/107
classification, 60/91
clinical features, 21/121, 41/373
corneal lattice dystrophy, *see* Hereditary amyloid polyneuropathy type 3
course, 21/121
cranial nerve, 42/523
CSF protein, 21/127

cutaneous atrophy, 42/524
definition, 21/120
diabetic polyneuropathy, 21/129
diagnosis, 21/127
diagnostic guideline, 60/109
differential diagnosis, 21/127, 60/19
distal muscle weakness, 42/518-521
DNA analysis, 60/108
dorsal spinal nerve root ganglion, 60/102
ECG, 21/125
EMG, 21/126
epidemiology, 21/8
experimental model, 60/109
facial paralysis, 42/523
Finnish, *see* Hereditary amyloid polyneuropathy type 3
gastrointestinal tract, 8/368, 21/123, 60/105
genetics, 21/129, 41/435, 60/106
geography, 21/130
glaucoma, 42/523
heart, 60/105
heart disease, 42/518, 524
hepatosplenomegaly, 42/521
hereditary motor and sensory neuropathy type I, 21/128
hereditary sensory and autonomic neuropathy type I, 21/128
hereditary sensory and autonomic neuropathy type II, 21/128
history, 60/89
Indiana, *see* Hereditary amyloid polyneuropathy type 2
Iowa, *see* Hereditary amyloid polyneuropathy type 1
Japanese (Kiushu), 21/8, 119, 130
leprosy, 21/128
lower limb, *see* Hereditary amyloid polyneuropathy type 1
lymph node, 60/106
Meretoja, *see* Hereditary amyloid polyneuropathy type 3
monoclonal gammopathy, 71/446
motor neuropathy, 21/122
multiple myeloma, 63/393
muscle biopsy, 60/104
muscular atrophy, 42/518, 521
nerve biopsy, 21/135-137
neuropathology, 21/134-139, 60/100
onset, 21/121
orthostatic hypotension, 42/518, 63/156
pain, 8/368, 42/518, 521
paresthesia, 8/368, 42/518, 521
pathology, 21/131-134
peripheral nerve, 60/102
peristalsis, 21/124
Portuguese, *see* Hereditary amyloid polyneuropathy type 1
prealbumin mutation, 60/91, 106-108
reflex, 21/123
respiratory system, 60/106
Rukavina, *see* Hereditary amyloid polyneuropathy type 2
sensory loss, 42/518, 521
sensory neuropathy, 21/121
sexual disorder, 8/368, 21/123
Sjögren syndrome, 21/125
skeletal system, 60/106
skin, 60/100, 106
specific amyloid stain, 60/100
sphincter function, 8/368, 21/123
spleen, 60/106
Swiss type, 21/131
testis, 60/106
treatment, 21/141, 60/109
trophic lesion, 21/123
trophic ulcer, 42/518
ultrastructural finding, 60/104
ultrastructure, 21/138-140
upper limb, *see* Hereditary amyloid polyneuropathy type 2
Van Allen, *see* Hereditary amyloid polyneuropathy type 1
8 varieties, 21/12
Hereditary amyloid polyneuropathy type 1, 42/518-520
anhidrosis, 42/518, 51/421
apolipoprotein A, 60/108
autonomic nervous system, 42/518, 522
autonomic polyneuropathy, 51/420, 479
autonomic system, 60/92
cardiomyopathy, 60/92
cataract, 51/424
clinical features, 60/91-97
course, 60/96
CSF protein, 51/424, 60/92
deafness, 51/424
dissociated sensory loss, 51/420
electrophysiologic study, 60/92
epidemiology, 21/130, 60/96
eye, 60/95
features, 21/8
gastrointestinal disorder, 8/9, 368
gastrointestinal tract, 60/95
geography, 60/96
history, 21/119
hypertrophic interstitial neuropathy, 21/128

laboratory test, 51/421
muscular atrophy, 51/420
neuropathology, 51/422, 424
non-Portuguese, 51/422
ocular abnormality, 51/421
orthostatic hypotension, 51/421
pain, 8/368, 51/420
peptic ulcer, 42/522
prealbumin, 60/106-108, 71/512-514
renal disease, 42/522, 60/95
respiratory tract, 60/95
sensorimotor polyneuropathy, 51/424
sexual disorder, 8/368
sexual impotence, 8/9, 368
shooting pain, 51/424
skin, 60/95
sphincter function, 8/368
steppage gait, 51/420
symptom, 51/420
vitreous opacity, 42/518, 51/424
Hereditary amyloid polyneuropathy type 2
age at onset, 51/423
autonomic system, 60/98
cardiovascular system, 60/98
carpal tunnel syndrome, 42/520, 51/423
clinical features, 60/97
clinical presentation, 60/97
course, 60/98
electrophysiologic study, 60/98
epidemiology, 21/130, 60/98
eye, 60/98
features, 21/8
gastrointestinal tract, 60/98
heart, 60/98
history, 21/120
kidney, 60/98
liver, 60/98
neuropathology, 51/423
prealbumin, 60/106-108, 71/512-514
sex ratio, 51/423
vitreous opacity, 42/520, 51/423
Hereditary amyloid polyneuropathy type 3
axonal neuropathy, 51/425
bilateral facial paresis, 51/424
clinical features, 60/99
corneal dystrophy, 51/424
cystine metabolism, 42/524
dermatopathology, 60/99
epidemiology, 21/8, 60/99
features, 21/8
gelsolin variant, 60/108
heart, 60/100
kidney, 60/100
lung, 60/100
lymph node, 60/100
male genital system, 60/100
meninges, 60/100
parotid gland, 60/100
presenting symptom, 60/99
prevalence, 42/523
pruritus, 60/100
renal biopsy, 60/100
skin biopsy, 60/100
thyroid, 60/100
Hereditary amyloidosis
hereditary motor and sensory neuropathy type I, 60/200
hereditary motor and sensory neuropathy type II, 60/200
incidence, 71/512
orthostatic hypotension, 63/149
Hereditary amyotrophic lateral sclerosis, *see* Familial amyotrophic lateral sclerosis
Hereditary angioedema
gastrointestinal tract, 43/60
swelling, 43/60
treatment, 43/61
Hereditary areflexic dystaxia, *see* Roussy-Lévy syndrome
Hereditary ataxia
see also Hereditary cerebellar ataxia
areflexia, 60/660
chorioretinal degeneration, 13/39
chronic axonal neuropathy, 51/531
dementia, 46/401
epilepsy, 15/326
glycogen storage disease, 27/199
myoclonus, *see* Hereditary myoclonus ataxia syndrome
ocular change, 13/298, 300
ocular pursuit, 60/657
optic atrophy, 13/39, 300, 305
progressive external ophthalmoplegia, 43/136, 62/304
progressive external ophthalmoplegia classification, 22/179, 62/290
secondary pigmentary retinal degeneration, 13/297-307
severe distal amyotrophy, 60/660
spastic, *see* Hereditary spastic ataxia
spastic paraplegia, 22/434
spheroid body, 6/626
Stevens type, 22/204, 207
tremor, 60/660
Hereditary autosomal recessive congenital optic atrophy

ERG, 13/91
tapetoretinal dysplasia, 13/91
Hereditary benign chorea
juvenile, *see* Juvenile hereditary benign chorea
speech disorder in children, 42/203
ventricular dilatation, 42/203
Hereditary benign myoclonus
involuntary movement, 1/282
Hereditary brachial neuropathy
classification, 21/12
Hereditary brachial plexus neuropathy, 60/71-73
autosomal dominant inheritance, 60/71
clinical features, 60/71
definition, 60/71
genetic study, 60/73
hereditary multiple recurrent mononeuropathy, 60/61, 68, 72
hypotelorism, 60/71
muscle biopsy, 60/72
nerve biopsy, 60/72
neurophysiologic study, 60/72
pregnancy, 60/71
tomacula absence, 60/72
treatment, 60/72
Hereditary brain hemorrhage
primary amyloidosis, 51/415
Hereditary centrolobar sclerosis, *see* Pelizaeus-Merzbacher disease
Hereditary cerebellar ataxia
see also Hereditary ataxia
Africa, 21/36
amyotrophic lateral sclerosis, 22/142
benign late, *see* Benign late hereditary cerebellar ataxia
bulbar paralysis, 22/142
chorea, 21/369
choreoathetosis, 21/575
deafness, 13/298
Ferguson-Critchley syndrome, 59/319-323
Gökay-Tükel syndrome, 21/575
intermittent, *see* Hill-Sherman syndrome
late onset, 59/319-323
Leber optic atrophy, 13/97
macular degeneration, 21/386
nystagmus, 21/369-371
ophthalmoplegia, 21/575, 22/191
optic atrophy, 42/117
pallidoluysian atrophy, 6/675
polyneuropathy, 42/117
prevalence, 21/21
progressive external ophthalmoplegia, 22/191, 208
Stevens type, 22/204, 207
upbeating nystagmus, 60/656
Hereditary cerebellar ataxia (Marie)
see also Cerebellar ataxia (Marie)
Behr disease, 13/91
ERG, 21/386
Fraser syndrome, 21/369
Friedreich ataxia, 21/434
hereditary cerebello-olivary atrophy (Holmes), 21/404
hereditary olivopontocerebellar atrophy (Menzel), 21/434
hereditary spastic ataxia, 21/370
Klippel-Durante family, 21/370
main features, 21/17
Nonne family, 21/369
optic atrophy, 13/83
prevalence, 21/22
spinopontine degeneration, 21/397, 400
Hereditary cerebellar palsy
choreoathetosis, 60/663
Hereditary cerebello-olivary atrophy (Holmes), 21/403-413
age at onset, 21/408, 60/569
anterograde degeneration, 21/404
brain atrophy, 21/412
brain stem, 21/407
Carpenter-Schumacher family, 21/413
Carter-Sukavajana syndrome, 21/411
catagenesis, 21/405
cerebellum, 21/406, 60/569
cerebrum, 21/407
clinical features, 21/410
course, 21/409
CSF protein, 21/410
dementia, 21/404, 409, 412
diagnosis, 21/411
differential diagnosis, 21/411
dominant late onset, 21/405
dysarthria, 21/403, 409, 60/569
epilepsy, 21/410
gait, 21/403, 409, 60/569
genetics, 21/405
hereditary cerebellar ataxia (Marie), 21/404
hereditary olivopontocerebellar atrophy (Menzel), 21/411, 434, 447
hypogonadism, 21/403, 60/569
Marie-Foix-Alajouanine disease, 21/411, 413
neuropathology, 21/406
nodding tremor, 21/403, 409, 60/569
nosology, 21/412
nystagmus, 21/403, 409, 60/569
olivary nucleus, 21/407
olivocerebellar atrophy, 21/18

olivopontocerebellar atrophy (Wadia-Swami), 60/497
parkinsonism, 21/410, 412
PEG, 21/410
prevalence, 21/406
pyramidal sign, 21/410
recessive late onset, 21/405
sex linked recessive, 21/412
sex ratio, 21/408
spasticity, 21/410
sphincter function, 21/410
spinal cord, 21/408
transsynaptic degeneration, 21/404
treatment, 21/411
tremor, 21/403, 409
Hereditary cerebral amyloid angiopathy
Dutch type, 54/341
English type, 59/439
Iceland, 54/341
migraine, 54/341
spastic paraplegia, 59/439
Hereditary chin tremor
nystagmus, 42/266
resting tremor, 49/585
Hereditary congenital cerebellar atrophy, 60/285-294
ataxia telangiectasia, 60/291
autosomal dominant early onset cerebellar vermis atrophy, 60/288
COACH syndrome, 60/292
congenital olivopontocerebellar atrophy, 60/291
Gillespie syndrome, 60/289
GM2 gangliosidosis, 60/289
infantile Gaucher disease, 60/289
infantile spinal muscular atrophy, 60/289
Joubert syndrome, 60/291
Kvistad-Dahl-Skre syndrome, 60/294
Malamud-Cohen family, 60/294
Marinesco-Sjögren syndrome, 60/288
Paine syndrome, 60/293
pontoneocerebellar atrophy (Brun), 60/290
primary granular cell layer atrophy, 60/287
trichopoliodystrophy, 60/289
Wolfslast-Blumel-Subrahmanyam syndrome, 60/293
X-linked recessive spastic ataxic syndrome, 60/293
Hereditary congenital optic atrophy
congenital retinal blindness, 13/96, 22/536
Hereditary connective tissue disease
Ehlers-Danlos syndrome, 39/381-389
Marfan syndrome, 39/389-393
neurologic manifestation, 39/379-406
osteogenesis imperfecta, 39/393-400
Rothmund-Thomson syndrome, 39/405
Werner syndrome, 39/400-405
Hereditary coproporphyria
aminolevulinic acid, 27/438
coproporphyrin III, 42/539
coproporphyrinogen oxidase, 42/539
features, 27/438
neuropathy, 60/120
Hereditary cranial nerve palsy
Albers-Schönberg disease, 60/43
cranial hyperostosis, 60/43
definition, 60/39
diabetes mellitus, 60/41
differential diagnosis, 60/42
dysglobulinemia, 60/42
facial paralysis, 60/41
Guillain-Barré syndrome, 60/42
hereditary facial nerve palsy, 60/39-42
hereditary multiple recurrent mononeuropathy, 60/43
Melkersson-Rosenthal syndrome, 60/39, 42
ophthalmoplegia, 60/43
ophthalmoplegic migraine, 60/42
rheumatoid serology, 60/41
sarcoidosis, 60/42
sclerosteosis, 60/43
Sjögren syndrome, 60/42
Tolosa-Hunt syndrome, 60/42
Van Buchem disease, 60/43
Hereditary cranial neuropathy
features, 21/10
Hereditary craniodiaphyseal dysplasia
progressive, *see* Progressive hereditary craniodiaphyseal dysplasia
Hereditary cutaneomandibular polyoncosis, *see* Multiple nevoid basal cell carcinoma syndrome
Hereditary cylindric spirals myopathy
congenital myopathy, 62/332
genetics, 62/348
muscle tissue culture, 62/108
percussion myotonia, 62/348
T tubule, 62/348
Hereditary deafness, 22/499-524
see also Hereditary hearing loss
antibiotic agent, 50/219
cannabis intoxication, 50/219
cortisone intoxication, 50/219
dementia, 22/516
diverticulitis, 22/524
dominant albinism, 50/218
epilepsy, 22/510-512
etacrynic acid intoxication, 50/219

genetics, 50/218
Huntington chorea, 49/291
Jervell-Lange-Nielsen syndrome, 50/218
keratopachydermia, 50/218
Klein-Waardenburg syndrome, 30/408, 50/218
LEOPARD syndrome, 50/218
lysergide intoxication, 50/219
myopia, 50/218
onychodystrophy, 50/218
optic atrophy, 50/218
polyneuropathy, 22/504
quinine intoxication, 13/63, 50/219
recessive atopic dermatitis, 50/218
Refsum disease, 50/218
salicylic acid derivative, 50/219
sensory radicular neuropathy, 22/524
thalidomide intoxication, 50/219
Usher syndrome, 13/441, 30/409, 50/218
X-linked pigmentary abnormality, 50/218
Hereditary diabetes mellitus
hereditary progressive cochleovestibular atrophy, 60/775
sensorineural deafness, 60/775
Hereditary diffuse leukoencephalopathy with spheroids, *see* Adult neuroaxonal dystrophy
Hereditary diurnal fluctuating dystonia
chorea, 49/42
juvenile, *see* Juvenile hereditary diurnal fluctuating dystonia
Hereditary dystonic paraplegia
amyotrophy, 22/460, 59/345
ataxia, 22/458
athetosis, 22/453
chorea, 22/453
chromosome 4, 59/346
chromosome 9, 59/346
clinical features, 22/446, 59/345
congenital retinal blindness, 59/346
dementia, 22/445-465
dorsal spinal nerve root ganglion, 22/452
dystonia, 22/456
familial spastic paraplegia, 59/346
genetics, 59/346
Huntington chorea, 59/346
mental deficiency, 59/345
muscle change, 22/447, 453
neuropathology, 22/449, 462, 59/345
nosology, 22/461
nystagmus, 22/447, 458
pallidopyramidal degeneration, 22/457
posterior column degeneration, 22/451
striatal necrosis, 22/455
tremor, 22/458
Hereditary episodic vertigo and deafness, *see* Hereditary Meniere disease
Hereditary essential myoclonus
benign essential tremor, 49/619
clinical features, 6/767, 49/618
EMG, 49/619
genetics, 42/235
juvenile hereditary benign chorea, 49/345
stimulus bound, 38/576
Hereditary essential tremor
brain atrophy, 42/269
ventricular dilatation, 42/269
Hereditary exertional paroxysmal dystonic choreoathetosis
see also Familial paroxysmal choreoathetosis
age at onset, 49/350
attack duration, 49/350
attack frequency, 49/350
EEG, 49/350
inheritance, 49/350
provocative factor, 49/350
sex ratio, 49/350
therapy, 49/350
trigger, 49/350
Hereditary facial nerve palsy
hereditary cranial nerve palsy, 60/39-42
Hereditary factor
neuroleptic parkinsonism, 65/295
scoliosis, 50/418
Hereditary fructose intolerance, 29/255-261
course, 29/255
definition, 29/255
diagnosis, 29/257
epilepsy, 72/222
genetics, 29/255
hypoglycemia, 27/72
incidence, 29/255
pathophysiology, 29/257
sign, 29/255
symptom, 29/255
treatment, 29/257
Hereditary hearing loss
see also Hereditary deafness
Alström-Hallgren syndrome, 22/502
ataxia, 22/500-510, 513-517
Cockayne-Neill-Dingwall disease, 13/242, 22/503
epilepsy, 22/510-513
Flynn-Aird syndrome, 22/519
Gardner disease, 22/517
Herrmann disease, 22/510
Jeune-Tommasi disease, 22/516
Latham-Munro syndrome, 22/512
Lemieux-Neemeh syndrome, 22/520

Lichtenstein-Knorr disease, 22/515
May-White syndrome, 22/510, 42/698
neural disease, 22/517-522
Norrie disease, 22/508
optic atrophy, 22/499, 504-508
opticocochleodentate degeneration, 22/508
Refsum disease, 22/501
retinal disease, 22/499, 508-510
Richards-Rundle syndrome, 22/516
Rosenberg-Bergström disease, 22/514
Rosenberg-Chutorian syndrome, 22/505, 42/418
secondary pigmentary retinal degeneration, 22/499-503
sensory radicular neuropathy, 22/518
Small disease, 22/509
Sylvester disease, 22/505
Telfer syndrome, 22/513
Tunbridge-Paley disease, 22/507
Usher syndrome, 22/499

Hereditary hemorrhagic telangiectasia (Osler), 38/257-277
anemia, 42/739
autosomal dominant, 55/448
Bean syndrome, 14/106
brain abscess, 55/449
brain aneurysm, 42/739, 55/450
brain angiography, 55/451
brain angioma, 55/450
brain arteriovenous malformation, 55/450
brain embolism, 55/449
brain infarction, 55/449
carotid cavernous fistula, 55/450
cerebral vascular malformation, 38/271-275
cerebrovascular disease, 53/31, 55/448
clinical features, 14/120
CNS vascular malformation, 38/270
computer assisted tomography, 55/451
CRST syndrome, 14/120
cryptic hemangioma, 12/104
CSF, 55/451
cutaneous dysplasia, 14/101
demyelination, 9/607
diagnosis, 55/451
general features, 38/259-263, 259-277
genetics, 38/258
headache, 42/739
hemiplegia, 1/193
hemoptysis, 55/450
history, 38/257
Huntington chorea, 49/291
leukoencephalopathy, 9/607
neurocutaneous hamartoma, 14/772
neurocutaneous syndrome, 14/101
neurofibromatosis, 68/297
neurologic manifestation, 38/263-277
neurologic symptom, 55/449
pathogenesis, 38/258
pathology, 38/258
polycythemia, 42/739, 55/449
prevalence, 42/739
pulmonary angiography, 55/451
pulmonary arteriovenous fistula, 42/739
pulmonary hypertension, 42/739
secondary pigmentary retinal degeneration, 13/251
septic embolism, 53/31
spinal angioma, 32/467, 474, 482, 484
spinal arteriovenous malformation, 38/275-277
spinal cord arteriovenous malformation, 55/450
stroke, 42/739
subarachnoid hemorrhage, 12/104, 42/739, 55/450
systemic brain infarction, 11/465
thrombocythemia, 55/449
treatment, 55/451
Ullmann syndrome, 14/72, 448

Hereditary hyperammonemia
metabolic ataxia, 21/574
periodic ataxia, 21/573

Hereditary hyperkalemic periodic paralysis
areflexia, 43/151
hyporeflexia, 43/151
paresthesia, 43/152
serum potassium, 43/151

Hereditary hyperlipidemia
arteriosclerosis, 11/462
ischemic heart disease, 11/462
xanthoma, 11/462

Hereditary hypophosphatemic osteomalacia
hereditary progressive cochleovestibular atrophy, 60/771
sensorineural deafness, 60/771

Hereditary interstitial nephritis
deafness, 27/335

Hereditary juvenile macular degeneration, *see* Stargardt disease

Hereditary leukodystrophy
animal experiment, 10/116
jimpy mouse, 10/116

Hereditary long QT syndrome
clonic seizure, 42/718
convulsion, 42/718
ECG QT interval, 63/238
epilepsy, 63/238
heart disease, 42/718, 75/442
syncope, 42/718, 63/238
tonic seizure, 42/718

Hereditary Meniere disease
episodic vertigo, 22/486, 42/402
headache, 42/402
hearing loss, 22/486, 42/402
hereditary progressive cochleovestibular atrophy, 60/768
Hereditary metaphyseal dysplasia, *see* Familial metaphyseal dysplasia
Hereditary mononeuritis multiplex
brachial plexus, 2/138
review, 21/87-89
Hereditary mononeuropathy
features, 21/10
recurrent, 21/87
Hereditary motoneuron disease
WFN classification, 59/1-4
Hereditary motor and sensory neuropathy plus syndrome
hereditary motor and sensory neuropathy variant, 60/243
Hereditary motor and sensory neuropathy type I
see also Hereditary motor and sensory neuropathy type II *and* Hypertrophic interstitial neuropathy
achalasia cardiae, 21/286
Africa, 21/36
amyotrophic lateral sclerosis, 22/326
amyotrophy, 21/274-279, 303
anosmia, 21/283, 308
anterior horn syndrome, 42/340
Aran-Duchenne amyotrophy, 21/279
Argyll Robertson pupil, 21/285
arm, 21/274
associated disorder, 60/191
ataxia, 42/117
atypical symptom, 21/283
autonomic change, 21/282
autonomic nervous system, 21/271-309, 75/24, 700
autonomic polyneuropathy, 51/479
autosomal dominant, 21/272
autosomal recessive, 21/272
Babinski sign, 21/283, 42/339
basal ganglion, 21/302
Behr disease, 21/308
benign essential tremor, 49/567, 569, 587
Brossard syndrome, 22/57
calf hypertrophy, 60/247
cardiac lesion, 21/286
cauda equina tumor, 19/87
cause, 21/272
cerebellar atrophy, 21/302
cerebellar peduncle, 21/302
cerebellar sign, 21/283
Chamorro people, 22/342
chorea-acanthocytosis, 63/280
chorioretinal degeneration, 13/39, 21/285, 308
chromosome 1, 60/200
chromosome 17, 60/202
classification, 60/185
claw hand, 21/279
clinical features, 21/274, 40/319, 322, 331, 60/188-193
congenital retinal blindness, 21/284
corticosteroid, 60/199
course, 21/286
cranial nerve, 21/283
CSF protein, 60/199
cutaneous reflex, 21/282
deafness, 21/284, 308
demyelinating polyneuropathy, 60/199
demyelination, 59/369
difference between type I and type II, 60/190
differential diagnosis, 19/87, 22/15, 40/472, 60/199
distal amyotrophy, 21/303
distal muscle weakness, 42/338
distal myopathy, 22/59
dorsal spinal root ganglion degeneration, 42/340
EMG, 1/641, 21/287
endocrine disorder, 21/286
epidemiology, 21/3, 5, 9, 60/187
epilepsy, 15/325
familial spastic paraplegia, 22/422, 425
fetal muscle fiber, 21/290
gait, 21/281
genetics, 21/272
Guam amyotrophic lateral sclerosis, 21/26
hereditary amyloid polyneuropathy, 21/128
hereditary amyloidosis, 60/200
hereditary motor and sensory neuropathy variant, 60/244, 247
hereditary multiple recurrent mononeuropathy, 60/65
hereditary nephritis, 21/286
hereditary sensory and autonomic neuropathy type I, 21/12, 78, 307, 60/8, 200
histochemistry, 40/44
history, 21/271, 60/185
hypertrophic interstitial neuropathy, 21/146, 148, 152, 304
hypertrophic type, 21/287, 305
hyporeflexia, 42/339
hypothermia, 21/282
inheritance, 21/12, 60/187
laboratory data, 21/288
Leber optic atrophy, 21/284

leg, 21/274
Lemieux-Neemeh syndrome, 22/521
medulla oblongata, 21/302
megaesophagus, 21/286
mental deficiency, 21/285
molecular genetics, 60/200, 202
motor end plate, 21/294
muscle lesion, 21/289
muscle spindle, 21/294
muscle ultrastructure, 21/290-296
muscular atrophy, 42/339
muscular interstitial tissue, 21/293
myopathic muscle change, 21/290
myopathy, 21/303
myotatic reflex, 21/282
myotonic dystrophy, 62/234
nemaline myopathy, 60/247
nerve biopsy, 60/195-199
nerve conduction, 7/179
nerve conduction velocity, 21/287, 42/339
nerve histology, 21/296-299
nerve ultrastructure, 21/296
neuropathology, 21/288-303
neuropathy, 40/319, 472, 60/659
neurophysiologic features, 60/193-195
nosology, 21/303-308
nystagmus, 21/284, 308
ocular myopathy, 21/285
ocular sign, 21/284
oculomotor disorder, 21/285
onion bulb formation, 42/340
ophthalmoplegia, 21/285, 308, 42/339
optic atrophy, 13/39, 81, 21/284, 308, 22/507, 42/117, 339, 409
optic tract, 21/303
orthostatic hypotension, 63/148
Papua New Guinea, 22/350
parkinsonism, 21/285, 308, 42/251-253
pathogenesis, 60/202
pes cavus, 21/278, 42/339
polyneuropathy, 41/432
pontine lesion, 21/302
posture, 21/281
presenting symptom, 21/273
prevalence, 21/13, 22, 42/340
progressive spinal muscular atrophy, 21/29, 22/13
pupillary disorder, 21/285
pyramidal features, 60/244
pyramidal sign, 21/283
Refsum disease, 21/197, 60/235, 66/489, 497
retrobulbar neuritis, 21/284
rockerbottom foot, 21/277
Roussy-Lévy syndrome, 21/172-174, 307, 42/108
scapulohumeral spinal muscular atrophy, 21/304, 42/108
scapuloperoneal muscular dystrophy, 40/425
sensorineural deafness, 42/370
sensory loss, 21/282, 42/339
sex chromosomes, 60/202
sex linked dominant, 21/273
sex linked recessive, 21/273
simian hand, 21/279
spina bifida, 21/286
spinal ganglion, 21/299
spinal nerve root, 21/299
spinocerebellar degeneration, 21/306, 60/245
sporadic case, 60/199
syringomyelia, 32/294
talipes, 21/278, 42/339
treatment, 21/309, 60/203
trophic ulcer, 42/339
type I molecular genetics, 60/200
variant, 21/308
vestibular disorder, 21/284, 308
vulnerability to other neuropathies, 60/193
X-linked hereditary motor and sensory neuropathy, 60/202
X-linked neuromuscular dystrophy, 62/117
Hereditary motor and sensory neuropathy type II
see also Hereditary motor and sensory neuropathy type I
achalasia cardiae, 21/286
Africa, 21/36
amyotrophic lateral sclerosis, 22/326
amyotrophy, 21/274-279, 303
anosmia, 21/283, 308
anterior horn syndrome, 42/340
Aran-Duchenne amyotrophy, 21/279
Argyll Robertson pupil, 21/285
arm, 21/274
associated disorder, 60/191
ataxia, 42/117
atypical symptom, 21/283-286
autonomic change, 21/282
autonomic nervous system, 75/24
autonomic polyneuropathy, 51/479
autosomal dominant, 21/272
autosomal dominant inherited motor neuropathy, 60/200
autosomal recessive, 21/272
Babinski sign, 21/283, 42/339
basal ganglion, 21/302
Behr disease, 21/308
benign essential tremor, 49/567, 569, 587
Brossard syndrome, 22/57
calf hypertrophy, 60/247

cardiac lesion, 21/286
cauda equina tumor, 19/87
cause, 21/272
cerebellar atrophy, 21/302
cerebellar peduncle, 21/302
cerebellar sign, 21/283
Chamorro people, 22/342
chorea-acanthocytosis, 63/280
chorioretinal degeneration, 13/39, 21/285, 308
classification, 60/185
claw hand, 21/279
clinical features, 21/274, 40/319, 322, 331, 60/188-193
course, 21/286
cranial nerve, 21/283
CSF protein, 60/199
cutaneous reflex, 21/282
deafness, 21/284, 308
difference between type I and type II, 60/190
differential diagnosis, 19/87, 22/15, 40/472, 60/199
distal amyotrophy, 21/303
distal muscle weakness, 42/338
distal myopathy, 22/59
EMG, 1/641, 21/287
endocrine disorder, 21/286
epidemiology, 21/3, 5, 9, 60/187
epilepsy, 15/325
familial spastic paraplegia, 22/422, 425
fetal muscle fiber, 21/290
gait, 21/281
genetics, 21/272
hereditary amyloidosis, 60/200
hereditary motor and sensory neuropathy variant, 60/244, 247
hereditary nephritis, 21/286
hereditary sensory and autonomic neuropathy, 60/200
hereditary sensory and autonomic neuropathy type I, 21/12, 73-86, 307, 60/8, 200
histochemistry, 40/44
history, 21/271, 60/185
hypertrophic interstitial neuropathy, 21/146, 148, 152
hyporeflexia, 42/339
hypothermia, 21/282
inheritance, 21/12, 60/187
laboratory data, 21/288
Leber optic atrophy, 21/284
leg, 21/274
medulla oblongata, 21/302
megaesophagus, 21/286
mental deficiency, 21/285
molecular genetics, 60/200, 202
motor end plate, 21/294
muscle lesion, 21/289
muscle spindle, 21/294
muscle ultrastructure, 21/290-296
muscular atrophy, 42/339
muscular interstitial tissue, 21/293
myopathic muscle change, 21/290
myopathy, 21/303
myotatic reflex, 21/282
myotonic dystrophy, 62/234
nerve biopsy, 60/195-199
nerve conduction, 7/179
nerve conduction velocity, 21/287, 42/339
nerve histology, 21/296-299
nerve ultrastructure, 21/296
neuronal type, 21/287, 305, 22/15
neuropathology, 21/288-303
neuropathy, 40/319, 472, 60/659
neurophysiologic features, 60/193-195
nosology, 21/303-308
nystagmus, 21/284, 308
ocular myopathy, 21/285
ocular sign, 21/284
oculomotor disorder, 21/285
ophthalmoplegia, 21/285, 308, 42/339
optic atrophy, 13/39, 81, 21/284, 308, 22/507, 42/117, 339, 409
optic tract, 21/303
orthostatic hypotension, 63/148
Papua New Guinea, 22/350
parkinsonism, 21/285, 308, 42/251-253
pathogenesis, 60/202
Pena-Shokeir syndrome type I, 59/369
pes cavus, 21/278, 42/339
polyneuropathy, 41/432
pontine lesion, 21/302
posture, 21/281
presenting symptom, 21/273
prevalence, 21/13, 22, 42/340
progressive spinal muscular atrophy, 21/29, 22/13, 59/406
pupillary disorder, 21/285
pyramidal features, 60/244
pyramidal sign, 21/283
retrobulbar neuritis, 21/284
rockerbottom foot, 21/277
Roussy-Lévy syndrome, 21/172-174, 307, 42/108
scapulohumeral spinal muscular atrophy, 21/304
scapuloperoneal muscular dystrophy, 40/425
scapuloperoneal spinal muscular atrophy, 21/304, 22/60, 59/44, 62/170
sensory loss, 21/282, 42/339

sex chromosomes, 60/202
sex linked dominant, 21/273
sex linked recessive, 21/273
simian hand, 21/279
spina bifida, 21/286
spinal ganglion, 21/299
spinal muscular atrophy, 59/373, 60/200
spinal nerve root, 21/299
spinal type, 21/288, 299, 305
spinocerebellar degeneration, 21/306, 60/245
sporadic case, 60/199
syringomyelia, 32/294
talipes, 21/278, 42/339
treatment, 21/309, 60/203
tremor, 60/661
trophic ulcer, 42/339
type II molecular genetics, 60/202
variant, 21/308
vestibular disorder, 21/284, 308
X-linked hereditary motor and sensory neuropathy, 60/202
X-linked neuromuscular dystrophy, 62/117
Hereditary motor and sensory neuropathy type III, *see* Hypertrophic interstitial neuropathy
Hereditary motor and sensory neuropathy type IV, *see* Refsum disease
Hereditary motor and sensory neuropathy variant, 60/243-249
amyotrophy, 60/244
ataxia, 60/245
chance association, 60/243
chromosome 19, 60/247
complex syndrome, 60/247-249
distal amyotrophy, 60/244
familial spastic paraplegia, 60/244
hereditary motor and sensory neuropathy plus syndrome, 60/243
hereditary motor and sensory neuropathy type I, 60/244, 247
hereditary motor and sensory neuropathy type II, 60/244, 247
infantile spinal muscular atrophy, 60/247
Leber optic atrophy, 60/246
myotonia, 60/247
nemaline myopathy, 60/247
neurofibromatosis type I, 60/247
optic atrophy, 60/245
paraplegia, 60/244
progressive external ophthalmoplegia, 60/247
scapuloperoneal spinal muscular atrophy, 60/247
secondary pigmentary retinal degeneration, 60/246
sensorineural deafness, 60/246
sensory loss, 60/244
spinocerebellar degeneration, 60/245
tremor, 60/245
Hereditary motor neuropathy
Pena-Shokeir syndrome type I, 59/369
Hereditary motor sensory neuropathy
ataxia telangiectasia, 60/386
demyelinating neuropathy, 47/621-623
Hereditary motor sensory neuropathy type I
axonal degeneration, 51/264
chronic inflammatory demyelinating polyradiculoneuropathy, 51/530
demyelinating neuropathy, 47/622
segmental demyelination, 51/264
Hereditary motor sensory neuropathy type II
chronic axonal neuropathy, 51/531
chronic inflammatory demyelinating polyradiculoneuropathy, 51/530
demyelinating neuropathy, 47/622
Hereditary motor sensory neuropathy type III
demyelinating neuropathy, 47/622
Hereditary multiple exostosis
see also Osteochondroma
Chamorro people, 22/342, 42/326
diagnosis, 19/317
malignancy, 19/317
osteochondroma, 42/326
prevalence, 19/313
radiology, 19/316
symptomatology, 19/314
treatment, 19/317
vertebral canal stenosis, 20/637
vertebral column, 19/313-319
Hereditary multiple recurrent mononeuropathy, 60/61-69
see also Pressure palsy
age, 51/553, 555
areflexia, 42/316
auditory evoked potential, 51/556
autosomal dominant, 51/551
axonal degeneration, 51/552
carpal tunnel syndrome, 51/560
clinical features, 51/551, 60/62
compound muscle action potential, 51/552
crossed legs, 51/554
definition, 60/61
differential diagnosis, 51/559, 60/66
Ehlers-Danlos syndrome, 55/456
electron microscopy, 60/63
electrophysiologic study, 60/62
EMG, 51/556
family report, 51/552
fiber group atrophy, 51/552

genetics, 51/559, 60/66
hereditary brachial plexus neuropathy, 60/61, 68, 72
hereditary cranial nerve palsy, 60/43
hereditary motor and sensory neuropathy type I, 60/65
hereditary neuralgic amyotrophy, 51/559
hypermyelination, 51/558
interval, 51/553
kneeling, 51/554
kyphoscoliosis, 60/62
large fiber involvement, 51/556
muscular atrophy, 42/316
myelin edema, 51/558
myelin lamella, 51/556
nerve action potential, 51/552
nerve biopsy, 51/552, 60/63
nerve compression, 51/559
nerve conduction, 42/316, 51/553
nerve conduction velocity, 42/316
occupational palsy, 51/560
paranodal area, 51/555
paranodal tomacula, 51/559
pathology, 51/556
pes cavus, 60/62
phospholipid, 42/316
pressure palsy, 51/551
prognosis, 51/554, 559
redundant myelin loop, 51/557
Schwann cell disorder, 51/558
second mesaxon, 51/558
segmental demyelination, 51/551
sensory loss, 42/316
sensory nerve conduction, 51/555
somatosensory evoked potential, 51/556
squatting, 51/554
survey, 51/553
tomacula, 51/551, 555, 558, 60/64
tremor, 51/552
trigeminal nerve, 51/554
ultrastructure, 51/558
Hereditary multiple sclerosis
pathology, 19/316
predisposing factor, 9/110
Hereditary muscular atrophy
ataxia, 13/306
chorioretinal degeneration, 13/39
diabetes mellitus, 13/306
optic atrophy, 13/39
primary pigmentary retinal degeneration, 13/306
Hereditary myoclonic dystonia
features, 6/556
symptomatic dystonia, 6/556
Hereditary myoclonus
clinical type, 38/576-579
dyssynergia cerebellaris myoclonica, 60/599
Hereditary myoclonus ataxia syndrome
absence, 21/511
autosomal dominant, 21/511
dyssynergia cerebellaris myoclonica, 21/510
progressive myoclonus epilepsy, 21/511
uric acid, 21/511
Hereditary myokymia
calf hypertrophy, 60/659
periodic ataxia, 60/659
Hereditary myotonic syndrome
animal model, 62/272
Hereditary nephritis
hereditary motor and sensory neuropathy type I, 21/286
hereditary motor and sensory neuropathy type II, 21/286
Hereditary nerve pressure palsy
Ehlers-Danlos syndrome, 55/456
Hereditary neuralgic amyotrophy
hereditary multiple recurrent mononeuropathy, 51/559
immune system, 51/174
Hereditary neurogenic muscular atrophy, *see* Hereditary motor and sensory neuropathy type I *and* Hereditary motor and sensory neuropathy type II
Hereditary neuropathy
areflexia, 60/256
autonomic features, 60/256
autonomic function measurement, 60/261
axonal degeneration, 60/258
chromosome 1, 60/265
chromosome 17, 60/265
classification, 21/5
clinical category, 60/254
clinical course, 60/258
clinical features, 60/255, 263
clinical work-up, 60/253-265
deformity, 60/257
demyelination, 60/258
differential diagnosis, 60/264
disability scale, 60/261
electrophysiologic study, 60/258
etiology, 60/262
features, 21/7
force measurement, 60/259
genetics, 60/265
involvement pattern, 60/257
liability to pressure palsy, *see* Hereditary multiple recurrent mononeuropathy

mononeuropathy, 60/254
motor sign, 60/255
motor symptom, 60/255
multifocal neuropathy, 60/254
multiple mitochondrial DNA deletion, 66/429
nerve biopsy, 60/259
nerve conduction, 7/179
nerve thickening, 60/257
neuromuscular disease, 41/432
neuropathy quantification, 60/259
pili torti, 60/263
plexopathy, 60/254
prolonged course, 60/263
secondary pigmentary retinal degeneration, 60/265, 733
sensation measurement, 60/260
sensory loss, 60/256
sensory symptom, 60/255
sex chromosomes, 60/265
skeletal deformity, 60/263
spinocerebellar degeneration, 60/733
sweating measurement, 60/261
symmetrical polyneuropathy, 60/254
Tangier disease, 21/8
trophic change, 60/257
Valsalva maneuver, 60/261
WFN classification, 60/1-3
yellow orange tonsil, 60/265

Hereditary nystagmus
albinism, 14/106

Hereditary oligophrenic cerebellolental degeneration, *see* Marinesco-Sjögren syndrome

Hereditary olivopontocerebellar atrophy
see also Joseph-Machado disease, Kjellin syndrome, Olivopontocerebellar atrophy (Schut-Haymaker) *and* Olivopontocerebellar atrophy (Wadia-Swami)
basal cell carcinoma, 60/647
bulbar paralysis, 60/658
HLA antigen, 42/162
myoclonus, 60/662
olivopontocerebellar atrophy (Wadia-Swami), 21/443
photosensitivity, 60/667

Hereditary olivopontocerebellar atrophy (Menzel), 21/433-447
age at onset, 21/440
amyotrophy, 21/443
athetosis, 21/441
autosomal dominant, 21/435
Babinski sign, 21/442
ballismus, 21/441, 445
basal ganglion, 21/438
brain cortex, 21/438
brain stem, 21/438
Carpenter-Schumacher family, 21/413
Carter-Sukavajana syndrome, 21/412, 447
cataract, 21/445
cause, 21/435
cerebellar ataxia, 21/444
cerebellum, 21/437
chorea, 21/441, 444
clinical features, 21/440
cranial neuropathy, 21/443
CSF, 21/445
dementia, 21/443, 445
diagnosis, 21/445
disease duration, 21/440
dysarthria, 21/441
dystonia musculorum deformans, 21/442
EEG, 21/445
Essick cell band, 21/435
familial spastic paraplegia, 21/444
familial type, 21/444
Friedreich ataxia, 21/444
gait, 21/441
gaze paralysis, 21/443
gene localization, 42/162
genetic linkage, 42/162
genetics, 21/435
Greenfield classification, 21/416
grimacing, 21/441
hemiballismus, 21/441, 445
hereditary cerebellar ataxia (Marie), 21/434
hereditary cerebello-olivary atrophy (Holmes), 21/411, 434, 447
hereditary spinocerebellar degeneration, 21/18
history, 21/435
HLA antigen, 42/162
intention tremor, 21/441
jactatio capitis nocturna, 21/441
kuru, 21/436
late onset, 21/445
mesencephalon, 21/438
neuropathology, 21/436
nosology, 21/446
nystagmus, 21/443
oculomotor paralysis, 21/443
optic atrophy, 21/443
parkinsonism, 21/412, 442
pathogenesis, 21/435
PEG, 21/445
pes cavus, 21/444
prevalence, 21/439
princeps case, 21/433
pyramidal sign, 21/443

Schut family, 21/435
sensory deficit, 21/443
sex linked recessive, 21/412
sex ratio, 21/440
spasmodic torticollis, 21/441
sphincter disturbance, 21/444
spinal cord, 21/439
spinopontine degeneration, 21/446
terminology, 21/434
thalamus, 21/438
treatment, 21/445
Hereditary optic atrophy, *see* Congenital retinal blindness
Hereditary orotic aciduria
clinical features, 29/245-247
diagnosis, 29/247
enzyme assay, 29/247
frequency, 29/245
genetics, 29/245
molecular defect, 29/250
pathogenesis, 29/249
treatment, 29/247-249
Hereditary osteodystrophy
Albright, *see* Pseudopseudohypoparathyroidism
Hereditary osteopetrosis
striatopallidodentate calcification, 49/424
Hereditary pallidal degeneration
nosologic group, 49/445-460
Hereditary pallidoluysiodentate degeneration
hemiballismus, 49/451
Hereditary pallidoluysionigral degeneration
nosologic group, 49/445-460
Hereditary pallidopyramidal degeneration
nosologic group, 49/445-460
Hereditary palmoplantar keratosis
nosology, 22/478
Hereditary paroxysmal dystonic choreoathetotis
see also Familial paroxysmal choreoathetosis *and* Hereditary paroxysmal kinesigenic choreoathetosis
age at onset, 49/350
anarthria, 49/349
attack duration, 49/350
attack frequency, 49/350
aura, 49/351
ballismus, 49/349
chorea, 49/349
computer assisted tomography, 49/351
CSF protein, 49/351
differential diagnosis, 49/352
dystonia, 49/349
EEG, 49/350, 352
facial grimacing, 49/349
genetics, 49/351
inheritance, 49/350
pathology, 49/352
provocative factor, 49/350
sex ratio, 49/350
slurred speech, 49/349
therapy, 49/350, 352
trigger, 49/350
Hereditary paroxysmal kinesigenic choreoathetosis, 42/207-209
see also Familial paroxysmal choreoathetosis, Hereditary paroxysmal dystonic choreoathetotis *and* Symptomatic paroxysmal kinesigenic choreoathetosis
age at onset, 42/207, 49/350, 353
attack duration, 49/350, 353
attack frequency, 42/207, 49/350, 353
attack suppression, 49/353
aura, 49/353
autosomal dominant, 42/208
computer assisted tomography, 49/353
diabetes mellitus, 49/355
dystonia, 42/205, 207, 49/352
dystonia musculorum deformans, 49/525
EEG, 42/206, 49/350, 353
genetics, 49/354
Huntington chorea, 6/355, 49/293
hypernatremia, 49/355
hyperthyroidism, 49/355
hypnogenic paroxysmal dystonia, 49/356
idiopathic hypoparathyroidism, 49/355
inheritance, 49/350
interictal chorea, 49/353, 355
mental deficiency, 42/206
multiple sclerosis, 49/355
newborn asphyxia, 49/355
Parkinson disease, 49/355
pathology, 49/354
pathophysiology, 49/355
phenytoin, 42/208
provocative factor, 49/350, 353
pyramidal tract sign, 49/353
readiness potential, 49/354
reflex epilepsy, 42/205, 207, 49/356
reflex myoclonus, 49/356
sex ratio, 42/206, 208, 49/350
somatosensory evoked potential, 49/354
speech disorder in children, 42/205
striatal convulsion, 6/169, 556
symptom, 49/385
symptomatic paroxysmal kinesigenic choreoathetosis, 49/355
thalamic infarction, 49/355

treatment, 42/708, 49/350, 354
trigger, 49/350
variant type, 49/355
visual evoked response, 49/354
Hereditary perforating foot ulcer
analgesia, 8/201
deafness, 8/201
dorsal spinal root ganglion degeneration, 8/201
hereditary sensory and autonomic neuropathy type I, 60/6
sensory radicular neuropathy, 8/201, 21/73, 76
shooting pain, 8/201
Hereditary periodic ataxia, 21/563-571, 60/433-442
acetazolamide, 60/442
age at onset, 60/434-437
alanine, 21/569
alprazolam, 60/442
aminoaciduria, 21/564
ascariasis, 21/569
attack duration, 60/433-437
autosomal dominant, 21/564, 569
cerebellar vermis atrophy, 60/441
chorea, 21/569
choreoathetosis, 21/564
clinical features, 21/563, 60/433-437
clonazepam, 60/442
course, 60/433-437
differential diagnosis, 21/569
EEG, 60/438-441
EMG, 60/438-441
ENG, 60/441
eosinophil count, 21/568
evoked potential, 60/438-441
family, 21/563
flunarizine, 60/442
Gordon phenomenon, 21/569
Hartnup disease, 21/564
indicanuria, 21/568
inheritance mode, 60/434-437
intention tremor, 21/565
interictal ataxia, 60/433
interictal finding, 60/433-437
kinesigenic ataxia, 60/433
laboratory study, 60/438-441
lactic acid, 21/565
main features, 21/17
muscle biopsy, 60/441
myokymia, 60/433
Negro family, 21/565
nerve conduction velocity, 60/439
nuclear magnetic resonance, 60/441
optic atrophy, 21/569
pathogenesis, 21/20, 60/441
pellagra like rash, 21/564
periodic nystagmus, 21/569
phenytoin, 60/442
photosensitivity, 21/564
precipitant, 60/433-437
progressive cerebellar ataxia, 21/564
protein intolerance, 21/578
pyruvic acid, 21/565, 569
subacute necrotizing encephalomyelopathy, 21/564
treatment, 60/434-437, 442
vitamin B1, 21/569
vomiting, 21/564
Hereditary polyneuropathy
tremor, 49/588
Hereditary posterior column ataxia, *see* Biemond syndrome
Hereditary primary amyloidosis
carpal tunnel syndrome, 7/296
Hereditary progressive arthro-ophthalmopathy
autosomal dominant inheritance, 13/41
blindness, 13/41
myopia, 13/41
retinal detachment, 13/41
Hereditary progressive cochleovestibular atrophy, 22/481-496
Alport syndrome, 60/768-770
branchio-otorenal dysplasia, 60/775
classification, 60/762
dominant low frequency hearing loss, 22/482, 60/766
dominant Meniere disease, 22/486
dominant midfrequency hearing loss, 22/482, 60/765
dominant progressive sensorineural deafness, 22/481
Ekman-Lobstein disease, 60/774
Goyer syndrome, 60/772
Helweg-Larsen-Ludvigsen syndrome, 60/772
hereditary diabetes mellitus, 60/775
hereditary hypophosphatemic osteomalacia, 60/771
hereditary Meniere disease, 60/768
hereditary renal tubular acidosis, 60/770
high frequency hearing loss, 60/764
inheritance mode, 60/761
integumentary disease, 22/489
Meige syndrome, 60/776
Muckle-Wells syndrome, 60/771
nonsyndromal atrophy, 60/761-768
osteitis deformans (Paget), 22/492, 60/772
prevalence, 60/761
Pyle disease, 60/773

recessive progressive sensorineural hearing loss, 60/767
renal disease, 22/487
sensorineural deafness, 60/770, 775
sex linked progressive sensorineural hearing loss, 60/767
syndromal atrophy, 60/768-776
syndromal *vs* nonsyndromal atrophy, 60/761
Van Buchem disease, 60/774
Hereditary progressive cone dystrophy
central scotoma, 13/134
dominant inheritance, 13/131, 134
function loss, 13/131
photophobia, 13/134
Hereditary progressive diurnal fluctuating dystonia
age at onset, 42/213, 49/529
case, 49/530
diagnosis, 49/536
dysarthria, 49/530
dystonia musculorum progressiva, 49/536
EMG, 49/533
genetics, 49/536
homoprotocatechuic acid, 42/214
homovanillic acid, 42/214
hypertonia, 49/530
juvenile parkinsonism, 42/213, 49/536
limb dystonia, 49/530
lumbar lordosis, 49/530
mask like face, 49/530
pallidal posture, 49/532
paradoxical contraction (Westphal), 49/533
parkinsonism, 49/533
pathology, 49/537
pathophysiology, 49/537
postural tremor, 49/530
pseudo-Babinski posture, 49/532
sporadic case, 49/531
treatment, 42/214, 49/535
vanilmandelic acid, 42/214
Hereditary progressive leukodystrophy
history, 10/12
Hereditary progressive myoclonus epilepsy, *see* Unverricht-Lundborg progressive myoclonus epilepsy
Hereditary progressive perceptive deafness, *see* Dominant progressive sensorineural deafness
Hereditary progressive spinal muscular atrophy
main features, 21/25
Hereditary protein intolerance, *see* Familial protein intolerance
Hereditary putaminal necrosis, *see* Hereditary striatal necrosis
Hereditary recurrent cranial nerve palsy
Tolosa-Hunt syndrome, 48/303
Hereditary recurrent neuropathy, 21/87-103
case report, 21/89-99
nerve biopsy, 21/90-98
nerve conduction velocity, 21/103
nerve ischemia, 21/102
pressure palsy, 21/89
segmental demyelination neuropathy, 21/90-98
Hereditary recurrent polyneuropathy, 21/87-103
classification, 21/87
Ig, 21/88
large family, 21/100
nerve conduction velocity, 21/103
nerve ischemia, 21/102
ultrastructure, 21/100
Hereditary renal tubular acidosis
hereditary progressive cochleovestibular atrophy, 60/770
sensorineural deafness, 60/770
Hereditary retinal aplasia, *see* Congenital retinal blindness
Hereditary retinal detachment
vitreoretinal degeneration, 13/37
Hereditary retinal dysplasia, *see* Congenital retinal blindness
Hereditary secondary dystonia, 59/339-346
autosomal dominant primary dystonia, 59/346
congenital retinal blindness, 59/339-345
corticospinal tract abnormality, 59/346
hereditary striatal necrosis, 59/343
mitochondrial DNA defect, 59/345
nuclear magnetic resonance, 59/344
nuclear magnetic resonance spectroscopy, 59/345
subacute necrotizing encephalomyelopathy, 59/343
Hereditary sensory and autonomic neuropathy
amyloidosis, 60/19
congenital analgesia, 60/19
congenital indifference to pain, 60/19
differential diagnosis, 60/18-20
dominant inherited type I, 60/6
dorsal spinal nerve root ganglion, 60/5
familial syringomyelia, 60/5
hereditary motor and sensory neuropathy type II, 60/200
leprous neuropathy, 60/19
Lesch-Nyhan syndrome, 60/19
management, 60/20
organophosphate ingestion, 60/19
particular features, 60/5
prevalence, 60/5
X-linked, *see* X-linked hereditary sensory and autonomic neuropathy

Hereditary sensory and autonomic neuropathy type I, 21/73-82
see also Congenital analgesia
amyloid deposit, 60/9
amyloidosis, 60/19
amyotrophy, 21/77
anhidrosis, 60/6
autonomic nervous system, 75/23
autonomic symptom, 60/7
autosomal dominant, 21/73, 79, 42/351, 60/7
axonal degeneration, 60/9
burning feet syndrome, 60/8
cataract, 60/8
cause, 21/78
clinical features, 21/74, 60/6
cochlear spiral ganglion, 60/9
compound action potential, 60/9
congenital abnormality, 21/73, 60/9
Corti organ, 60/9
course, 60/8
CSF, 42/348, 60/9
deafness, 21/73, 77
differential diagnosis, 21/74, 60/18
dissociated sensory loss, 60/8
dorsal spinal nerve root ganglion, 60/5, 9
electrophysiologic study, 60/9
epilepsy, 60/9
familial spastic paraplegia, 21/12
familial syringomyelia, 60/5
features, 21/8, 60/7
harelip, 60/9
hearing loss, 60/8
hereditary amyloid polyneuropathy, 21/128
hereditary motor and sensory neuropathy type I, 21/12, 78, 307, 60/8, 200
hereditary motor and sensory neuropathy type II, 21/12, 73-86, 307, 60/8, 200
hereditary perforating foot ulcer, 60/6
Hicks-Denny-Brown family, 21/74, 60/6
historic aspect, 60/5
history, 8/180
Huntington chorea, 49/291
hyperhidrosis, 21/73
hypertrophic polyneuropathy, 21/77
IgA, 21/82, 60/9
IgG, 60/9
inheritance mode, 60/7
leprous neuropathy, 60/19
lightning pain, 21/73, 77, 60/6
meningocele, 60/9
multiple neurofibromatosis, 14/398
Nageotte nodule, 60/9
nerve biopsy, 21/81
nerve conduction, 60/9
neurogenic acro-osteolysis, 42/293-295
neuropathology, 21/75, 80-82
nucleus ruber, 60/9
olivopontocerebellar atrophy, 21/11
onset, 21/73
optic atrophy, 13/42
palatognathoschisis, 21/77
palatoschisis, 60/9
paretoamyotrophic form, 21/78
pathology, 60/9
pes cavus, 21/77
polyneuropathy, 41/434
posterior column degeneration, 21/75, 60/9
pyramidal syndrome, 21/77
Roussy-Lévy syndrome, 21/77
scoliosis, 21/77
shooting pain, 60/8
small fiber loss, 60/10
spastic paraplegia, 21/77
spasticity, 60/9
spina bifida, 21/77, 60/9
spinal nerve root atrophy, 21/75
sural nerve biopsy, 60/7
syringomyelia, 21/76, 32/259, 294
thalamus, 60/9
ulceromutilating acropathy, 21/76, 60/6
vestibular nerve atrophy, 60/9
wallerian degeneration, 21/75
X-linked recessive, 60/16
Hereditary sensory and autonomic neuropathy type II
acropathie ulcéromutilante, 8/180
anhidrosis, 60/10
anosmia, 60/11
areflexia, 60/10
autonomic manifestation, 60/11
autonomic nervous system, 75/23
autonomic symptom, 60/7
autosomal recessive, 21/73, 51/565, 60/7, 11
bone dysplasia, 60/17
cardinal features, 60/10
childhood, 42/349
classification, 42/349
clinical features, 60/10, 17, 18
clumsiness, 60/11
congenital indifference to pain, 8/369, 21/5, 78, 60/10, 14, 27
congenital pain insensitivity, 51/564, 60/10, 14, 27
consanguinity, 21/80, 60/7
corneal reflex, 60/11
CSF, 42/350
deafness, 60/11

differential diagnosis, 42/350, 60/18
electrophysiologic study, 60/11, 17
features, 60/7
Friedreich ataxia, 21/11
genetics, 60/11
hereditary amyloid polyneuropathy, 21/128
hereditary sensory and autonomic neuropathy type III, 60/17
history, 8/180
Holmes-Adie syndrome, 60/18
hypotonia, 60/11
inheritance mode, 60/7
lesion site, 1/108
lingual fungiform papillae, 60/11
muscle stretch reflex, 60/11
Navajo indian, 60/18, 33
nerve biopsy, 60/11, 17, 18
nerve conduction, 7/187, 60/11
neurogenic osteoarthropathy, 42/349
neuropathology, 60/11, 17
neuroradiology, 60/17
neurotrophic keratitis, 60/17
ophthalmic finding, 74/430
orthostatic hypotension, 63/148
Pena-Shokeir syndrome type I, 59/369
perforating foot ulcer, 8/180
periodic fever, 60/11
posttraumatic syringomyelia, 21/74
secondary amyloidosis, 60/17
secondary pigmentary retinal degeneration, 60/11
sensory radicular neuropathy, 21/74
sexual impotence, 60/11
spastic paraplegia, 60/17
sporadic case, 21/78, 60/15
sural nerve biopsy, 60/7, 11
syringomyelia, 50/452, 60/6
tactile sensation, 60/11
tonic pupil, 60/18
ulceromutilating acropathy, 60/6, 10
vacuolated cell, 60/11
visceral pain, 60/11

Hereditary sensory and autonomic neuropathy type III
see also Congenital analgesia
abdominal colic, 63/147
acetylcholine, 51/565, 60/35
acetylcholinesterase, 21/114, 60/31
acrocyanosis, 43/58, 60
alacrimia, 1/475, 43/58, 60/28
arterial hypotension, 21/109, 60/7, 27
ataxia, 43/59, 60/27
autonomic nervous system, 1/475-480, 8/345, 42/300, 43/59, 60/23, 26, 28, 35, 75/23, 154, 700
autonomic symptom, 60/7, 26
biochemical study, 60/28
bird like face, 1/478
birth incidence, 43/59, 60/24
birth weight, 21/109, 60/26
bone dysplasia, 60/17
breath holding, 21/109, 60/27
cardiac arrest, 21/109, 60/27
cardiovascular regulation, 75/146, 156
catecholamine, 21/109, 115, 43/59, 60/27
central dysfunction, 75/147, 157
central sleep apnea, 63/462
circulatory symptom, 60/27
clinical features, 21/108, 60/26, 75/144
clinical subdivision, 60/24
cold face test, 75/152
cold pressor test, 75/152
congenital analgesia, 60/19
congenital indifference to pain, 21/110, 41/435, 60/7, 19, 27, 33
congenital pain insensitivity, 51/565, 60/27
cricopharyngeous muscle, 21/109, 60/26
definition, 60/23
demyelination, 8/345, 21/112, 60/30, 33
diagnostic criteria, 21/107, 60/24
diagnostic study, 75/153
differential diagnosis, 60/18
dysesthesia, 43/59
dysphagia, 21/109, 60/26
electrophysiologic study, 60/28
EMG, 21/110, 60/28
emotional disorder, 1/475, 43/58, 60/28
epidemiology, 10/562, 21/5, 11, 60/24
epilepsy, 15/350, 21/110, 60/27
evoked potential, 75/151
features, 60/7, 26
feeding difficulty, 60/26
fever, 21/109, 60/26
frontal bossing, 60/18
gag reflex, 21/109, 60/27
gait, 43/59, 60/26, 29
gastrointestinal dysmotility, 75/156
gastrointestinal motility, 75/146
genetics, 21/108, 60/24, 75/148
glossopharyngeal nerve agenesis, 50/219, 60/30
growth retardation, 21/111, 43/58, 60/28
heart rate, 75/152
hereditary sensory and autonomic neuropathy type II, 60/17
histamine, 21/107-116, 60/23-35, 75/150
histamine flare, 21/115, 60/7, 24
Holmes-Adie syndrome, 21/115, 60/18
homovanillic acid, 1/478, 21/78, 109, 60/25, 28

hyperhidrosis, 1/475, 21/114, 60/7, 28, 63/147
hypertelorism, 30/248
hyperthermia, 1/477, 21/111, 60/26
hypolacrima, 21/111, 60/28
hypotonia, 60/27, 75/144
indifference to pain, 1/475, 479, 8/199, 41/435, 42/300, 60/7, 27
inheritance mode, 60/7
initial symptom, 60/26
kyphoscoliosis, 43/59, 60/27
lacrimal gland, 60/32
lacrimation, 1/475, 21/109, 111, 51/565, 60/7, 28
lingual fungiform papillae, 21/109, 114, 60/24, 32
lingual vallate papillae, 21/114, 60/32
megaesophagus, 21/113
mental deficiency, 1/477, 30/248, 46/59, 60/27
mental function, 21/111
methacholine, 1/478, 21/108, 111, 115, 60/24, 28, 35
miosis, 43/58, 60/28
muscle biopsy, 60/31
muscular hypotonia, 21/110, 60/27
myenteric plexus, 21/113, 60/30
Navajo child, 60/18, 33
nerve biopsy, 21/113, 60/31
nerve conduction velocity, 75/151
neurogenic osteoarthropathy, 21/110, 60/27
neuropathology, 21/111-114, 60/29-31
neurotrophic keratitis, 60/17, 33
noradrenalin, 1/478, 21/115, 60/27, 35
nutrition, 75/144
ocular dysfunction, 60/27
ophthalmic finding, 74/430
ophthalmologic features, 75/145
orthostatic hypotension, 1/478, 21/109, 43/58, 60/7, 27
pain insensitivity, 60/7, 27
pathogenesis, 60/25
pathology, 75/154
primary, *see* Primary familial dysautonomia
pupillary pharmacology, 60/35
rectal biopsy, 60/31
respiratory control, 75/146
respiratory symptom, 60/27
reticular formation, 8/345, 21/111, 60/30
salivation, 21/109, 60/26
sensory deficit, 75/145
sensory problem, 75/155
sensory syndrome, 8/198
skin biopsy, 60/30-32
skin manifestation, 30/248, 60/28
skin response, 75/151
sleep, 74/546
speech, 21/110, 60/27
speech disorder in children, 1/477, 43/59, 60/27
substance P, 45/230, 60/29
sural nerve biopsy, 60/7, 31
survey, 75/143
sweating excess, 43/58, 60/28
sympathetic nervous system, 21/113, 42/300, 60/23, 25, 29, 31
symptom, 63/147
taste absence, 21/114, 60/27
thermotest, 75/150
tilt table test, 75/151
tongue, 75/149
treatment, 1/480, 21/111, 60/29, 75/155
upper lip, 75/145
vanilmandelic acid, 1/478, 21/109, 60/25, 28
vibrameter, 75/150
vomiting, 1/475, 21/109, 60/26

Hereditary sensory and autonomic neuropathy type IV
analgesia, 60/12
anhidrosis, 21/7, 78, 42/298, 346-348, 51/564, 60/12
autonomic nervous system, 75/24
autonomic symptom, 60/7, 12
autosomal recessive inheritance, 60/13, 17
axon reflex, 60/13
cardinal features, 60/11
case report, 60/13
child, 51/565, 60/13
clinical features, 60/12
congenital analgesia, 60/19
congenital indifference to pain, 21/5, 60/11, 17, 19
CSF, 60/13
differential diagnosis, 60/17, 19
electrophysiologic study, 60/13
features, 60/7
febrile convulsion, 60/12
genetics, 60/13
hyperplastic myelinopathy, 60/17
inheritance mode, 60/7, 13
intradermal histamine, 60/13
Lesch-Nyhan syndrome, 60/19
Lissauer tract absence, 8/199, 42/346, 60/13
mental deficiency, 60/12
methacholine pupillary response, 60/13
nerve biopsy, 60/13
nerve conduction, 60/13
nerve conduction velocity, 42/300
neurogenic osteoarthropathy, 42/346, 60/12
neuropathology, 60/13
organophosphate ingestion, 60/19
osteomyelitis, 60/12

periodic fever, 60/12
proprioceptive sensation, 60/12
self-mutilation, 60/11, 12, 17
septicemia, 60/12
skin biopsy, 60/13
spinal deformity, 60/12
spinal ganglion, 60/13
sural nerve biopsy, 60/7, 14
sweat test, 60/13
treatment, 60/13, 20
urinary catecholamine excretion, 60/13
Hereditary sensory and autonomic neuropathy type V
autonomic nervous system, 75/24
autonomic symptom, 60/7, 15
cardinal features, 60/7, 14
case report, 60/15
clinical features, 60/14
congenital analgesia, 60/19
congenital indifference to pain, 60/19
differential diagnosis, 60/19
electrophysiologic study, 60/15
inheritance mode, 60/7
intradermal histamine test, 60/15
keratitis, 60/17
laboratory study, 60/15
nerve biopsy, 60/15
neuroradiology, 60/15
pupillary reaction, 60/15
sural nerve biopsy, 60/7, 15
Hereditary sensory neuropathy, *see* Hereditary sensory and autonomic neuropathy type II
Hereditary sensory neuropathy type II, *see* Hereditary sensory and autonomic neuropathy type I
Hereditary sensory radicular neuropathy, *see* Hereditary sensory and autonomic neuropathy type I
Hereditary spastic ataxia, 21/365-374, 60/461-465
age at onset, 21/366, 60/461
autosomal dominant, 21/366
Behr disease, 21/372
bulbar paralysis, 60/658
cerebellar cortical atrophy, 60/463
chorea, 21/367, 383, 60/462, 663
clinical features, 21/366, 371, 60/461
congenital cataract, 60/464
congenital retinal blindness, 21/372
course, 21/368, 60/461
dementia, 60/462
differential diagnosis, 21/372
distal amyotrophy, 60/659
dysarthria, 21/367
dysphagia, 21/367, 60/462
extrapyramidal sign, 21/16
family, 21/366
fecal incontinence, 60/664
Friedreich ataxia, 21/369
gait, 21/366
hereditary cerebellar ataxia (Marie), 21/370
hereditary spinocerebellar degeneration, 21/372
inheritance, 60/461
Leber optic atrophy, 21/372
macular corneal dystrophy, 60/464
miosis, 60/464
multiple sclerosis, 21/372
myopia, 60/464
neuromyelitis optica, 21/372
neuropathology, 21/368, 60/462
nosology, 60/464
olivopontocerebellar atrophy, 21/369, 60/463
ophthalmoplegia, 21/383
optic atrophy, 21/365, 367, 60/462, 654
parkinsonian features, 60/463
posterior column degeneration, 60/463
presenting symptom, 21/366
ptosis, 21/366, 60/462, 655
pupillary reaction, 60/464
retinal degeneration, 60/463
retinovestibular, *see* Retinovestibular hereditary spastic ataxia
review, 21/365
rigidity, 60/661
secondary pigmentary retinal degeneration, 21/383-389, 60/734
spasticity, 60/461
spinopontine degeneration, 21/399, 60/463
tilted optic disc, 60/464
urinary incontinence, 60/664
vestibular disorder, 21/383-389
Hereditary spastic paraplegia, *see* Familial spastic paraplegia
Hereditary spinal ataxia, *see* Friedreich ataxia
Hereditary spinal muscular atrophy
β-*N*-acetylhexosaminidase A deficiency, 59/25
acid maltase deficiency, 59/26
differential diagnosis, 59/22-24
familial muscle cramp, 59/25
familial spastic paraplegia, 59/26
familial spherocytosis, 59/25
Friedreich ataxia, 59/24
Joseph-Machado disease, 59/26
neural Charcot-Marie-Tooth disease, 59/22
neuroacanthocytosis, 59/25
neurofibromatosis, 59/26
neuropathic gammopathy, 59/26

progressive bulbospinal muscular atrophy, 59/23
scapulohumeral muscular atrophy, 59/24
Sjögren-Larsson syndrome, 59/26
spinal muscular atrophy, 59/22
Wohlfart-Kugelberg-Welander disease, 59/24
Hereditary spinocerebellar ataxia
dementia, 21/553
Hereditary spinocerebellar degeneration
Bassen-Kornzweig syndrome, 21/14
epidemiology, 21/14-18
epilepsy, 15/327
hereditary olivopontocerebellar atrophy (Menzel), 21/18
hereditary spastic ataxia, 21/372
olivopontocerebellar atrophy (Fickler-Winkler), 21/18, 42/162
olivopontocerebellar atrophy (Schut-Haymaker), 21/18
slow eye movement, 21/443
Hereditary spinodentatonigro-oculomotor degeneration
case series, 22/157-175
Hereditary striatal necrosis
age at onset, 49/494
amaurosis, 49/495
apathy, 49/495
athetosis, 49/495
blindness, 49/495
computer assisted tomography, 49/495
course, 49/494
dysarthria, 49/495
dysphagia, 49/495
EEG, 49/496
epilepsy, 49/495
focal, 6/632
hereditary secondary dystonia, 59/343
history, 49/493
nuclear magnetic resonance, 49/496
pathology, 22/455, 49/493
psychomotor retardation, 49/495
psychosis, 49/495
rigidity, 49/495
spastic paraparesis, 22/455
spasticity, 49/495
tremor, 49/495
Hereditary telangiectasia
brain abscess, 52/145
Hereditary telangiectatic angiomatosis, *see* Hereditary hemorrhagic telangiectasia (Osler)
Hereditary thrombocytopenia
Jürgens-Von Willebrand, *see* Jürgens-Von Willebrand hereditary thrombocytopenia
Hereditary torsion dystonia, *see* Dystonia musculorum deformans
Hereditary tremor, *see* Essential tremor
Hereditary trophedema
chronic, *see* Milroy disease
Hereditary tyrosinemia, 29/213-221
biochemical finding, 29/215-217, 220
clinical features, 29/214-217, 220
diagnosis, 29/215, 220
enzyme defect, 29/215, 42/634
Fanconi syndrome, 29/172
genetics, 29/216
hepatorenal tubular dysfunction, 29/214-217
hypermethioninemia, 29/119
pathology, 29/215
polyneuropathy, 60/165
treatment, 29/217, 219, 221
Hereditary ulceromutilating acropathy, *see* Hereditary sensory and autonomic neuropathy type I
Hereditary vitreoretinal degeneration
acanthocyte, 49/332
autosomal dominant inheritance, 13/273
case history, 13/274-276
cataract, 13/274
choroid atrophy, 13/274
choroid sclerosis, 13/274
glaucoma, 13/274
hyaloidotapetoretinal degeneration (Goldmann-Favre), 13/178, 276
opacity, 13/178
pigmentary abnormality, 13/274
primary pigmentary retinal degeneration, 13/178, 274
strabismus, 13/274
Hereditary whispering dysphonia, *see* Focal dystonia musculorum deformans
Heredoataxia, *see* Hereditary cerebellar ataxia
Heredoataxia hemeralopica polyneuritiformis, *see* Refsum disease
Heredoataxia (Marie), *see* Hereditary cerebellar ataxia (Marie)
Heredoataxia (Stephens)
polyneuropathy, 22/207
progressive external ophthalmoplegia, 22/204, 207
Heredodegeneration
cerebellar, *see* Cerebellar heredodegeneration
hemiballismus, 42/224
optic atrophy, 13/68
Heredodegenerative disease
epidemiology, 21/3-42
familial pes cavus, 21/263
neuropathology, 21/43-71

nosology, 21/4
sign, 21/3
substrate, 21/3, 4
syndrome, 21/3
Heredofamilial myosclerosis
Emery-Dreifuss muscular dystrophy, 62/173
genetic heterogeneity, 62/173
rigid spine syndrome, 62/173
Heredopathia atactica polyneuritiformis, *see* Refsum disease
Heredopathia ophthalmo-otoencephalica
brain atrophy, 42/150
cataract, 42/150, 60/652
cerebellar ataxia, 42/150
cholesterol metabolism, 42/150
cranial nerve, 42/150
deafness, 60/657
dementia, 42/150
nystagmus, 42/150
psychiatric disorder, 42/150
sensorineural deafness, 42/150-152
speech disorder, 42/150
tremor, 42/150
Heredoretinopathia congenitalis, *see* Congenital retinal blindness
Heredoretinopathia congenitalis monohybridia recessiva autosomalis, *see* Congenital retinal blindness
Hering-Breuer deflation reflex
autonomic nervous system, 74/572
respiration, 74/572
Hering-Breuer inflation reflex
autonomic nervous system, 74/572
respiration, 74/572
Hering-Breuer reflex
anatomy, 1/652
apneustic breathing, 1/677
physiology, 1/493
pneumotaxic center, 63/479
respiratory center, 63/431
respiratory control, 63/431
Hering nerve
anatomy, 11/533
respiratory center, 63/431
Herman syndrome
Divry-Van Bogaert syndrome, 10/127
Hermansky-Pudlak syndrome
bleeding tendency, 43/5
oculocutaneous albinism, 43/5
squamous carcinoma, 43/6
ulcerative colitis, 43/6
Hernia
cervical disc, *see* Cervical intervertebral disc prolapse
diaphragm, *see* Diaphragm hernia
inguinal, *see* Inguinal hernia
lumbar disc, *see* Lumbar intervertebral disc prolapse
muscle, *see* Muscle hernia
thoracic disc, *see* Thoracic intervertebral disc prolapse
umbilical, *see* Umbilical hernia
Hernia nuclei pulposi, *see* Intervertebral disc prolapse
Herniation
brain, *see* Brain herniation
brain infarction, 54/404
brain stem, *see* Brain stem herniation
cerebellar, *see* Cerebellar herniation
cerebellar tonsil, *see* Tonsillar herniation
cingulate, *see* Cingulate herniation
clivus ridge syndrome, 2/297
cribriform plate, 1/571
depressed skull fracture, 24/225
falx cerebri, 1/569
foraminal, 1/565
intracranial hypertension, 16/95, 97
pituitary fossa, 1/571
sphenoid wing, 1/571
spinal nerve root, *see* Spinal nerve root herniation
tentorial, *see* Transtentorial herniation
tonsillar, *see* Tonsillar herniation
transtentorial, *see* Transtentorial herniation
Heroin, *see* Diamorphine
Herophilus
historic review, 2/5
Herpangina
Coxsackie virus, 34/139
Herpes
auricular, *see* Auricular herpes
buccopharyngeal, *see* Buccopharyngeal herpes
geniculate, *see* Geniculate herpes
herpes simplex virus, 56/8
Herpes genitalis
meningitis, 71/267
myelitis, 71/267
radiculitis, 71/267
Herpes labialis
meningococcal meningitis, 52/24, 27
Herpes neonatorum
see also Herpes simplex virus encephalitis
herpes simplex virus type 1, 56/214
herpes simplex virus type 2, 56/214
Herpes simplex, 34/145-156
acquired immune deficiency syndrome, 56/495
acquired myoglobinuria, 62/577

acute cerebellar ataxia, 34/622
cluster headache, 48/220
CNS infection, 69/447
cytarabine, 34/154
differential diagnosis, 53/237
history, 34/145
iododeoxyuridine, 34/154
meningitis, 34/85, 156
meningitis serosa, 34/85
meningococcal meningitis, 33/23
multiple sclerosis, 9/126, 34/437, 47/329-331
polyradiculitis, 34/156
radiculitis, 34/156
vidarabine, 34/154
virus-host relationship, 34/146
Herpes simplex virus, 56/207-221
see also Herpes virus hominis
acquired immune deficiency syndrome, 56/478, 514
acute polymyositis, 56/200
acute viral encephalitis, 56/126, 131
acute viral myositis, 56/195
agammaglobulinemia, 56/211
animal model, 56/211
antibody, 56/210
ascending myelitis, 56/478
biological property, 34/145
biology, 34/145-147
brain tumor, 67/300
C-3b complement, 56/208
capsid, 56/208
composition, 56/207
Cowdry A inclusion body, 56/221
cytotoxic T-lymphocyte, 56/210
deafness, 56/107
definition, 34/8
DNA, 56/212
enhancer, 56/37
envelope glycoprotein, 56/208
gene therapy, 66/114, 116
gingivostomatitis, 56/8
herpes, 56/8
hydranencephaly, 50/341
immune response, 56/209
immunocompromise, 56/478
immunocompromised host, 56/477
interferon, 56/209
intra-axonal spread, 56/211
keratitis, 56/8
latency, 56/211
meningoencephalitis, 56/478
morphology, 56/207
mRNA, 56/212
natural killer cell, 56/209
necrotizing encephalitis, 71/266
neurology, 34/145-147
nonstructural protein, 56/208
nucleoprotein, 56/208
persistence, 56/211
polyradiculitis, 56/221
reactivation, 56/211
replication, 56/7
replication cycle, 56/209
Reye syndrome, 29/332, 49/217, 56/150
structural protein, 56/208
structure, 56/207
tegument, 56/208
teratogenic agent, 34/377
type 1, 56/8, 207
type 2, 56/8, 207
ultrastructure, 56/7
vertigo, 56/107
virus-host relationship, 34/146
virus replication, 56/7
Wiskott-Aldrich syndrome, 56/211
Herpes simplex virus antigen
CSF, *see* CSF herpes simplex virus antigen
Herpes simplex virus 1 antigen
CSF, *see* CSF herpes simplex virus 1 antigen
Herpes simplex virus encephalitis
see also Herpes neonatorum
aciclovir, 56/219
acquired immune deficiency syndrome, 56/478
blood-brain barrier, 56/85
brain abscess, 56/218
brain biopsy, 10/685, 56/215, 217
brain blood flow, 56/87
brain hemorrhagic infarction, 54/72
brain tumor, 56/218
5-(2-bromovinyl)-2'-deoxyuridine, 56/219
child, 34/155, 56/219
clinical features, 34/150-153, 56/214
coma, 56/214
computer assisted tomography, 54/72, 56/215
consciousness, 45/120
course, 34/154
cranial nerve palsy, 56/214
Creutzfeldt-Jakob disease, 56/546
CSF, 34/151
CSF antibody index, 56/216
CSF cell hybridization, 56/218
CSF herpes simplex virus antigen, 56/218
CSF IgG, 56/215
cytarabine, 34/154
diagnosis, 34/153, 56/215
differential diagnosis, 56/217

EEG, 56/215
epidemiology, 34/149
epilepsy, 56/214
headache, 48/32, 56/214
heart transplantation, 63/180
history, 56/207
Hodgkin disease, 39/45
inappropriate antidiuretic hormone secretion, 28/499
iododeoxyuridine, 34/154
Klüver-Bucy syndrome, 45/260
malignant lymphoma, 39/45
Melkersson-Rosenthal syndrome, 8/210
mucous membrane, 56/8
multiple sclerosis, 9/126
neonate, 56/219
neuropsychology, 45/9
newborn, 56/207
nuclear magnetic resonance, 56/216
pathology, 34/149
personality change, 56/214
prognosis, 34/154
skin malformation, 56/8
symptom, 16/218
temporal bone tumor, 56/214
treatment, 34/154, 56/219
vidarabine, 34/154, 56/219
Wernicke-Korsakoff syndrome, 45/199

Herpes simplex virus meningitis
causative agent, 56/8
clinical features, 56/220
CSF, 56/126
epidemiology, 56/128
frequency, 56/126
newborn, 56/207
subfamily, 56/8

Herpes simplex virus myelitis
acquired immune deficiency syndrome, 56/221
blood-brain barrier, 56/85
clinical features, 56/220
lifetime persistence, 56/8
newborn, 56/207

Herpes simplex virus radiculitis
clinical features, 56/220

Herpes virus
avian, *see* Avian herpes virus
beta subfamily, 56/8
cytomegalovirus, 56/8
epilepsy, 72/155
Epstein-Barr virus, 56/7
features, 56/6
gamma subfamily, *see* Gammaherpes virus
human, 56/2
immunocompromised host, 56/476
Marek disease virus, 56/189, 581
nervous system, 71/261
neurotropic virus, 56/6
varicella zoster virus, 56/8

Herpes virus hominis
see also Herpes simplex virus
chronic meningitis, 56/646

Herpes virus hominis syndrome
clinical features, 31/216
hydrocephalus, 31/212
mental deficiency, 31/212, 216
microcephaly, 30/508, 510, 31/212
seizure, 31/212, 216
socioeconomic factor, 31/216
treatment, 31/216
venereal transmission, 31/216

Herpes zoster
see also Chickenpox, Postherpetic neuralgia *and* Varicella zoster
acquired immune deficiency syndrome, 56/498, 71/268
auditory symptom, 2/71
auricular, 2/71, 7/483
autonomic nervous system, 75/21
cephalic, 34/171-173
classification, 34/161
cranial nerve palsy, 56/231
CSF, 34/174
demyelinating polyneuropathy, 51/182
dermatomal pain, 56/233
EMG, 51/181
encephalitis, 34/173-175, 56/230, 71/268
epidemiology, 56/230
extremity involvement, 34/173
facial paralysis, 2/71, 34/170, 172
granulomatous CNS vasculitis, 55/388
Guillain-Barré syndrome, 51/182
headache, 5/207
herpes sine herpete, 56/233
history, 34/161
Hodgkin disease, 20/121, 39/44, 69/279
Hodgkin lymphoma, 63/353
immunosuppression, 56/230
incidence, 34/170
intractable hiccup, 63/491
lymphoma, 63/353
malignant lymphoma, 39/44
motor radiculitis, 51/181
myelitis, 34/173-175, 71/268
nerve lesion, 7/483
neuralgia, 5/207, 209
neuritis, *see* Herpes zoster ganglionitis

neurologic complication, 34/161-177
neuromyelitis optica, 47/406
non-Hodgkin lymphoma, 69/279
nonrecurrent nonhereditary multiple cranial neuropathy, 51/571
ophthalmoplegia, 56/231
pain sequence, 56/233
papillitis, 56/231
paraneoplastic infection, 18/250
paratrigeminal syndrome, 48/337
pathology, 34/175-177
polyneuritis, 34/173-175
polyradiculitis, 7/483
postherpetic neuralgia, 5/316, 56/230, 233
primary cerebral reticulosarcoma, 18/237
radiculitis, 71/268
radiculoneuropathy, 56/231
Ramsay Hunt syndrome, 34/171-173, 56/231
risk factor, 34/170
sclerotomal pain, 56/233
survey, 34/170-177
temporal arteritis, 55/346
teratogenic agent, 34/378
thoracolumbar spinal root, 56/232
trigeminal nerve, 34/171, 56/232
truncal involvement, 34/173
tympanic membrane, 48/489
varicella zoster virus, 56/229
Herpes zoster ganglionitis
see also Herpes zoster ophthalmicus
aciclovir, 51/181
cranial neuritis, 51/181
CSF lymphocytosis, 51/181
CSF protein, 51/181
facial paralysis, 51/181
facial weakness, 56/231
multiple cranial nerve, 51/181
postherpetic neuralgia, 51/181
trigeminal nerve, 51/181
virus sequestration, 51/179
Herpes zoster neuritis, *see* Herpes zoster ganglionitis
Herpes zoster ophthalmicus
see also Herpes zoster ganglionitis
brain infarction, 53/161
brain ischemia, 53/161
Herpes zoster oticus
see also Ramsay Hunt syndrome
clinical features, 56/119
diagnosis, 56/119
ear ache, 5/209
epidemiology, 34/171-173
facial paralysis, 8/282, 56/118
headache, 5/207
history, 2/70
management, 56/119
myoclonus, 38/578
neuralgia, 5/509
otalgia, 56/118
Ramsay Hunt syndrome, 2/360, 34/172
temporal bone pathology, 56/119
varicella zoster virus, 56/118
vertigo, 2/360
Herpes zoster virus
see also Varicella zoster virus
acute viral myositis, 56/195
Reye syndrome, 34/169
Herpetic geniculate neuralgia
criticism, 48/478
facial paralysis, 5/210
Herpetic meningoencephalitis
speech therapy, 46/616
Herpetic stomatitis
childhood myoglobinuria, 62/562
Herring body
neuroaxonal dystrophy, 49/395
Herrmann disease
deafness, 22/510
diabetes mellitus, 22/510
hereditary hearing loss, 22/510
progressive myoclonus epilepsy, 22/510
renal disease, 22/510
Herrmann-Opitz syndrome
brachycephaly, 43/325
brachydactyly, 43/325
craniosynostosis, 43/325
micrognathia, 43/325
syndactyly, 43/325
Hers disease, *see* Hepatophosphorylase deficiency
Hershey kiss
frontometaphyseal dysplasia, 31/257
Hertwig-Magendie symptom, *see* Skew deviation
Herxheimer reaction
corticosteroid, 52/285
neurosyphilis, 52/279
tabes dorsalis, 52/280
trichinosis, 52/574
tuberculous meningitis, 52/211
Heschl gyrus
lesion, 2/704-706
stimulation, 2/704-706
Hesitancy
urinary, *see* Urinary hesitancy
Hess chart
orbital fracture, 24/93
Hess screen

description, 2/288-290
eye movement disorder, 2/288-290
Heterochromatin
chromosome, 30/94
Heterochromatism
Alzheimer disease, 46/273
Heterochromia iridis
angioma, 14/643
Bonnet-Dechaume-Blanc syndrome, 14/263
dysraphia, 2/325
epilepsy, 42/688
facio-oculoacousticorenal syndrome, 43/400
Horner syndrome, 22/256, 43/65
Klein-Waardenburg syndrome, 14/115, 30/408, 43/51, 50/218
neurofibromatosis type I, 14/492
oculocutaneous albinism, 42/404
phakomatosis, 14/639
Porot-Filiu syndrome, 42/688
progressive hemifacial atrophy, 31/253, 43/408
seizure, 42/688
sensorineural deafness, 42/688
Stargardt disease, 13/133
Sturge-Weber syndrome, 14/512
Heterocyclic acetic acid
headache, 48/174
migraine, 48/174
Heterocyclic antidepressant
adrenergic receptor affinity, 65/315
amoxapine, 65/312
amoxapine metabolite, 65/319
ataxia, 65/316
chemical structure, 65/314
chorea, 65/319
delirium, 65/316
dopamine reuptake, 65/315
dopaminergic receptor affinity, 65/315
dystonia, 65/316
elimination half-life, 65/314
epilepsy, 65/316, 319
extrapyramidal effect, 65/316
headache, 65/316, 320
histaminic receptor affinity, 65/315
intoxication treatment, 65/323
malignant neuroleptic syndrome, 65/319
maprotiline, 65/312
metabolism, 65/313
mianserin, 65/312
muscarinic receptor affinity, 65/315
myoclonus, 65/319
neuropathy, 65/316
neurotoxin, 65/311
noradrenalin reuptake, 65/315
oculogyric crisis, 65/319
orthostatic hypotension, 65/316
pharmacokinetics, 65/313
polyneuropathy, 65/316
prolactinemia, 65/319
receptor affinity, 65/315
serotonin reuptake, 65/315
serotonin reuptake inhibition, 65/313
steady state concentration, 65/314
tardive dyskinesia, 65/319
trazodone, 65/312
tremor, 65/316, 320
vertigo, 65/320
visual accommodation, 65/316
Heterocyclic antidepressant intoxication
coma, 65/323
epilepsy, 65/323
fatality rate, 65/324
relative toxicity, 65/324
Heterodontus francisci intoxication
elasmobranche, 37/70
local pain, 37/70
muscle weakness, 37/70
spinal injury, 37/70
Heterodontus portusjacksoni intoxication
elasmobranche, 37/70
Heterogeneity
allelic, *see* Allelic heterogeneity
cerebrotendinous xanthomatosis, 14/779
definition, 30/89
distal myopathy, 62/197
environmental, 30/89
etiologic, *see* Etiologic heterogeneity
genetics, 30/89
Huntington chorea, 49/274-276
infantile spinal muscular atrophy, 59/97
neurofibromatosis, 50/365
pallidodentate degeneration, 21/531
pallidoluysian atrophy, 21/531
progressive myoclonus epilepsy, 1/282
quadriceps myopathy, 62/174
Heterogeneous system degeneration, *see* Progressive supranuclear palsy
Heterophenia
tuberous sclerosis, 14/351
Heterophoria
differential diagnosis, 2/291
diplopia, 2/290
headache, 5/205
paralytic strabismus, 2/291
Heterotopia
Arnold-Chiari malformation type II, 50/405
ataxia telangiectasia, 14/74

brain cortex, *see* Brain cortex heterotopia
cell rest, 30/499-503
cerebellar, *see* Cerebellar heterotopia
cerebellar cortex, *see* Cerebellar cortex heterotopia
cerebellar dysgenesis, 30/377
cerebellar vermis, *see* Cerebellar vermis heterotopia
cerebral, *see* Cerebral heterotopia
cortical, *see* Cortical heterotopia
definition, 30/492
epilepsy, 72/109, 340, 342
fetus, 30/499
genetics, 30/503
glial, *see* Glial heterotopia
gray matter, *see* Gray matter heterotopia
hippocampus, *see* Hippocampus heterotopia
inferior olivary nucleus, *see* Inferior olivary nucleus heterotopia
Krabbe-Bartels disease, 14/410
lissencephaly, 30/485
malformation syndrome, 30/499-502
microgyria, 30/484
nerve tissue, *see* Nerve tissue heterotopia
neurofibromatosis, 14/145
neurofibromatosis type I, 14/145
neuronal, *see* Neuronal heterotopia
nodular, *see* Nodular heterotopia
normal population, 30/499, 503
occipital encephalocele, 42/26
pathogenesis, 30/503
periventricular, *see* Periventricular heterotopia
polymicrogyria, 50/40
premature, 30/499
Purkinje cell, *see* Purkinje cell heterotopia
spina bifida, 32/525
subcortical, *see* Subcortical heterotopia
subependymal, *see* Subependymal heterotopia
testicular, *see* Testicular heterotopia
testis, *see* Testicular heterotopia
thalamic tumor, 68/65
tuberous sclerosis, 14/45, 47, 354, 360, 366
Heterotopic brain tissue
anencephaly, 50/84
Heterotopic endometrium
vertebral column, 20/443
Heterotopic hemopoietic tissue
spinal cord, 20/443
Heterotopic ossification, *see* Neurogenic osteoarthropathy
Heterotropia, *see* Concomitant strabismus
Heterozygote detection
acid maltase deficiency, 43/178
argininosuccinic aciduria, 42/525
aspartylglycosaminuria, 42/528
Becker muscular dystrophy, 43/87
brancher deficiency, 43/180
Chédiak-Higashi syndrome, 42/537
debrancher deficiency, 43/179
Duchenne muscular dystrophy, 43/106
Emery-Dreifuss muscular dystrophy, 43/88
Fabry disease, 42/428
Fanconi syndrome, 43/18
FG syndrome, 43/250
Gaucher disease, 42/438
generalized myotonia (Becker), 43/164
globoid cell leukodystrophy, 42/490
glucose-6-phosphate dehydrogenase deficiency, 42/645
glycosaminoglycanosis type I, 42/452
glycosaminoglycanosis type IS, 42/453
GM1 gangliosidosis type I, 42/432
GM1 gangliosidosis type II, 42/433
GM2 gangliosidosis type I, 42/434
GM2 gangliosidosis type II, 42/436
GM2 gangliosidosis type III, 42/436
homocystinuria, 42/557
hyperargininemia, 42/563
isovaleric acidemia, 42/580
Lesch-Nyhan syndrome, 42/154
mannosidosis, 42/598
maple syrup urine disease, 29/69
megalotestes, 43/287
mental deficiency, 43/287
metachromatic leukodystrophy, 29/359, 42/495
myophosphorylase deficiency, 43/181
Niemann-Pick disease type A, 29/355
Refsum disease, 21/185, 189, 27/520, 41/434, 42/149, 66/495
scapulohumeroperoneal muscular atrophy, 42/343
Wilson disease, 42/271
xeroderma pigmentosum, 43/13
Heterozygote frequency
ataxia telangiectasia, 42/121
cerebrocostomandibular syndrome, 43/237
cerebro-oculofacioskeletal syndrome, 43/342
corpus callosum agenesis, 42/103
dystonia musculorum deformans, 42/217
Fanconi syndrome, 43/17
Gaucher disease, 42/437
globoid cell leukodystrophy, 42/490
GM2 gangliosidosis type I, 42/434
Huntington chorea, 42/226
infantile spinal muscular atrophy, 42/89
muscular atrophy, 42/84
Niemann-Pick disease type A, 42/469

Pena-Shokeir syndrome type I, 43/439
sensorimotor neuropathy, 42/103
sickle cell anemia, 42/624
Sjögren-Larsson syndrome, 43/307
Smith-Lemli-Opitz syndrome, 43/308
suxamethonium sensitivity, 42/655
thalassemia major, 42/629
Troyer syndrome, 42/194
Werner syndrome, 43/489
Wilson disease, 42/271
Heubner arteritis
neurosyphilis, 52/278
Heubner artery
anatomy, 11/5, 7-10
radioanatomy, 11/76
supply area, 2/585
topography, 11/76
Hexacarbon
chronic axonal neuropathy, 51/531
distal axonopathy, 64/11
neurotoxin, 51/531, 64/10
toxic neuropathy, 64/10
Hexacarbon intoxication
action site, 64/10
pathomechanism, 64/11
Hexacarbon neuropathy
acrylamide, 51/278
carbon disulfide, 51/278
giant axonal neuropathy, 51/278, 60/82
3,3'-iminodipropionitrile intoxication, 51/278
Hexacarbon polyneuropathy
acrylamide intoxication, 64/63
Hexachlorobenzene
organochlorine, 64/197
organochlorine insecticide intoxication, 64/197
Hexachlorocyclohexane
benign intracranial hypertension, 67/111
chemical formula, 64/203
organochlorine, 64/197, 202
organochlorine insecticide intoxication, 36/405-408, 64/197, 202
Hexachlorocyclohexane intoxication
acute, 36/405, 407
chronic, 36/407
clinical features, 36/405
diagnosis, 36/407
treatment, 36/408
Hexachlorophene
iatrogenic neurological disease, 65/471
neurotoxin, 37/479, 51/68, 64/10, 14, 65/471
segmental demyelination, 64/14
spongiform leukoencephalopathy, 65/471
toxic encephalopathy, 65/471
toxic neuropathy, 64/10, 14, 65/471
toxic polyneuropathy, 51/68
Hexachlorophene intoxication, 37/479-509
action site, 64/10, 14
biochemical finding, 37/491
birth weight, 37/499
blindness, 37/486, 65/476
brain edema, 37/488
brain level, 65/478
brain ultrastructure, 65/474
cat, 37/485
cattle, 37/485
circulatory disturbance, 37/486
CNS, 37/488, 499
coma, 37/486, 65/476
convulsion, 37/486
cramp, 37/486
diarrhea, 37/486
disorientation, 37/486
dog, 37/485
EEG, 65/478
encephalopathy, 65/478
epidemic, 65/478
epilepsy, 65/478
excretion, 37/482
experimental animal, 37/484-486
fatal case, 65/476
laboratory diagnosis, 37/483
metabolism, 37/482
monkey, 37/486
mouse, 37/484
muscle spasm, 65/476
muscle weakness, 37/486
muscular twitching, 37/486
myelin ultrastructure, 65/475
nausea, 37/486
neuropathology, 37/499, 65/473, 476
neuropathy, 65/472
newborn, 37/479
nystagmus, 37/486, 65/476
optic nerve kinking, 65/476
papilledema, 37/486, 65/476, 478
peripheral nerve, 37/479
pontine change, 65/477
premature infant, 37/479
pupillary reflex, 37/486
rat, 37/484
retinal change, 65/476
retinopathy, 37/479, 65/472
serum level, 37/486, 65/473, 478
sheep, 37/485
skin contact, 37/481, 486
somnolence, 37/486, 65/476

spongiform leukoencephalopathy, 37/479, 488, 47/565, 65/471
spongiform myelopathy, 37/499
swine, 37/485
visual acuity, 37/486
vomiting, 37/486, 65/476
Hexadactyly
Patau syndrome, 14/121, 31/504, 50/556
r(13) syndrome, 50/585
Hexadecanol
plasmalogen, 10/243
Hexamethylmelamine, *see* Altretamine
Hexamethylmelamine intoxication, *see* Altretamine intoxication
Hexane
biotransformation, 64/83
chemical formula, 64/82
distal axonopathy, 64/12
element data, 64/82
industrial use, 64/81
neuropathy, 7/522
neurotoxicity, 51/276
neurotoxin, 7/522, 36/361, 51/68, 276, 64/12, 81
organic solvent, 64/82
pyrroles, 64/83
toxic encephalopathy, 64/12
toxic polyneuropathy, 51/68
Hexane intoxication
clinical features, 36/362
course, 36/365
history, 36/361
laboratory finding, 36/363
pathology, 36/363
symptom, 64/12
Hexane polyneuropathy
anhidrosis, 51/277
autonomic neuropathy, 51/277
axonal swelling, 51/277
2-butanone potentiation, 51/276
Guillain-Barré syndrome, 51/277
hyperhidrosis, 51/277
neurofilament, 51/277
pathologic change, 51/277
rapid ascending polyneuropathy, 51/277
sensorimotor polyneuropathy, 51/276
sexual impotence, 51/277
2,5-Hexanedione, *see* Acetonylacetone
2,5-Hexanedione intoxication, *see* Acetonylacetone intoxication
2-Hexanone
biotransformation, 64/83
chemical formula, 64/82
distal axonopathy, 64/12
metabolite, 64/91
neurotoxicity, 51/276
neurotoxin, 36/365, 51/68, 276, 64/12, 49, 81, 91
organic solvent, 64/82
pyrroles, 64/83
TLV, 64/91
toluene intoxication, 64/50
toxic encephalopathy, 64/12
toxic polyneuropathy, 51/68
2-Hexanone intoxication
symptom, 64/12
2-Hexanone polyneuropathy
anhidrosis, 51/277
autonomic neuropathy, 51/277
axonal swelling, 51/277
2-butanone potentiation, 51/276
Guillain-Barré syndrome, 51/277
hyperhidrosis, 51/277
neurofilament, 51/277
pathologic change, 51/277
rapid ascending polyneuropathy, 51/277
sensorimotor polyneuropathy, 51/276
sexual impotence, 51/277
Hexokinase
brain glycolysis, 27/7
Hexomonophosphate shunt
biochemistry, 27/13
CNS, 27/13, 79
hyperglycemia, 27/79
peripheral nervous system, 27/79
Hexosamine
globoid cell leukodystrophy, 10/85
multiple sclerosis, 9/321
myelin, 9/31
serum, 9/321
Hexosaminidase
GM2 gangliosidosis, 51/375
Hexosaminidase A, *see* β-*N*-Acetylhexosaminidase A
Hexosaminidase A deficiency, *see* β-*N*-Acetylhexosaminidase A deficiency
Hexosaminidase B, *see* β-*N*-Acetylhexosaminidase B
Hexosaminidase B deficiency, *see* β-*N*-Acetylhexosaminidase B deficiency
Hexose
multiple sclerosis, 9/321
serum, 9/321
Hexose phosphate isomerase
brain glycolysis, 27/9
Hexoside sulfate
hypertrophic interstitial neuropathy, 21/148
HHE syndrome

Krabbe-Bartels disease, 14/529
5-HIAA, *see* 5-Hydroxyindoleacetic acid
Hibernoma, 18/195-199
clinical features, 18/196
diagnosis, 18/196
sex and age, 18/196
treatment, 18/199
Hiccough, *see* Hiccup
Hiccup
Addison disease, 39/479
anatomic classification, 1/279
brain lacunar infarction, 54/253
brain stem tumor, 63/440
cat scratch disease, 52/130
cause, 1/284
corticosteroid intoxication, 65/555
diborane intoxication, 64/359
frequency, 63/490
glottal closure, 63/490
intractable, *see* Intractable hiccup
Köhlmeier-Degos disease, 39/437
medulla oblongata, 2/228, 235
pentaborane intoxication, 64/359
physiology, 63/490
posterior inferior cerebellar artery occlusion, 55/90
respiration reflex, 63/490
Hidrotic ectodermal dysplasia
striatopallidodentate calcification, 49/424
High altitude
see also Hypoxia
autonomic nervous system, 74/588, 75/259
central sleep apnea, 63/463
Cheyne-Stokes respiration, 63/482
dysbarism, 63/416
hypoxia, 74/588
pregnancy, 74/334
pulmonary edema, 63/416
secondary polycythemia, 63/250
syncope, 75/216
High altitude brain edema, 63/416
survey, 75/271
High arched palate
Pena-Shokeir syndrome type I, 43/438
Rubinstein-Taybi syndrome, 31/282, 43/234
Saethre-Chotzen syndrome, 43/322
trisomy 8, 50/563
High conductance fast channel syndrome
congenital, *see* Congenital high conductance fast channel syndrome
High density lipoprotein
a-alphalipoproteinemia, 51/396
antihypertensive agent, 63/83
chemistry, 10/268
familial, 66/549
reverse cholesterol transport, 66/549
Tangier disease, 10/269, 278, 547, 42/627, 66/549
High density lipoprotein deficiency, *see* Tangier disease
High frequency hearing loss
hereditary progressive cochleovestibular atrophy, 60/764
hypogonadism, 42/377
inverse secondary pigmentary retinal degeneration, 42/377
Lemieux-Neemeh syndrome, 42/371
Pendred syndrome, 42/368
High grade glioma
see also Anaplastic astrocytoma *and* Glioblastoma multiforme
cancer family syndrome, 68/88
carmustine, 68/111
dissemination, 68/91
histology, 68/93
morphology, 68/90
pesticide, 68/89
predilection site, 68/89
High pitched voice
Dubowitz syndrome, 43/249
Peter Pan syndrome, 18/545
High-pressure nervous syndrome
drowsiness, 48/402
dysbarism, 23/623
EEG, 48/402
headache, 48/402
tremor, 48/402
High voltage treatment
pontine tumor, 17/703
Higher nervous activity
cerebral dominance, 3/27-33, 4/465-469
cerebral localization, 3/3
conjugate eye movement, 3/23-26
development, 4/340-376
developmental diagnosis, 4/365-373
developmental disorder, 4/340-469
future, 1/38
handedness, 4/465-469
introduction, 1/2
newborn behavior, 4/344-347
tactile agnosia, 4/14
tectoreticular system, 3/24
Higher nervous activity disorder
behavior development, 4/342-347
brain injury, 4/360-365, 466
brain maturation, 4/364
cerebral localization, 3/22-38

constructional apraxia, 3/30
developmental dysarthria, 4/465
developmental dyscalculia, 4/465
dysmegalopsia, 3/29
gnosis, 3/8
hemiasomatognosia, 3/29
hyperkinetic child syndrome, 4/363, 465
idiopathic developmental dyslexia, 4/378-383
introduction, 3/2
macropsia, 3/29
memory disorder, 3/37
metamorphognosia, 3/29
metamorphotaxia, 3/31
micropsia, 3/29
newborn behavior, 4/344-347
phantom limb, 3/29
porropsia, 3/29
tonic neck reflex, 4/346
Hill-Sherman syndrome
helminthiasis, 21/565
hyperalaninemia, 21/578
hyperpyruvic acidemia, 21/578
Hip arthroplasty
lumbosacral plexus, 51/161
Hip dip
proximal muscle weakness, 40/314
Hip dislocation
congenital, *see* Congenital hip dislocation
Hip fracture
femoral nerve, 70/35
sciatic nerve, 70/37
Hippel-Lindau disease, *see* Von Hippel-Lindau disease
Hippocampal sclerosis
epilepsy, 72/112, 114
temporal lobe epilepsy, 73/69
Hippocampal vein
anatomy, 11/51
Hippocampectomy
schizophrenia, 46/497
Hippocampus
alcohol intoxication, 64/114
Alzheimer disease, 46/249
amnesic shellfish poisoning, 65/166
amygdala, 73/61
anatomy, 73/62
apnea, 63/481
arousal disorder, 45/109
arthropod envenomation, 65/200
calcification, 14/777
cardiac dysrhythmia, 63/239
epilepsy, 15/99
functional anatomy, 73/61
global amnesia, 45/44
glutamic acid, 46/267
granulovacuolar degeneration, 46/259
hallucination, 45/355
iniencephaly, 50/131
kernicterus, 6/502
lead, 64/434
limbic encephalitis, 8/135
limbic system, 73/60
loop, 73/63
neurofibrillary tangle, 46/258
Norman microcephalic familial leukodystrophy, 10/103
opticocochleodentate degeneration, 21/548
organolead intoxication, 64/131, 145
organophosphate intoxication, 37/555, 64/228
organotin intoxication, 64/140, 145
pathway, 73/65
pertussis encephalopathy, 9/551
Pick disease, 46/238, 242
posttraumatic amnesia, 45/185
respiration, 63/481
schizophrenia, 46/506-508
spreading depression, 48/65, 68
strain, 45/249
stress, 45/251
temporal lobe epilepsy, 72/86, 116, 73/59
Hippocampus commissure
corpus callosum agenesis, 50/150
Hippocampus heterotopia
Edwards syndrome, 50/275
Hippocampus lesion
memory disorder, 2/713, 46/357
Hippocrates
historic review, 2/5
Hippuran excretion test
CSF, 30/586
Hippus
pupil, 74/404
Hirano body
amyotrophic lateral sclerosis, 22/131, 287
description, 21/57
Parkinson dementia, 49/177
Hirayama disease, 59/107-118
age at onset, 59/107
amyotrophic lateral sclerosis, 59/113
anterior tephromalacia, 59/113
autonomic disturbance, 59/108
autonomic nervous system, 59/111
carpal tunnel syndrome, 59/114
cervical rib, 59/114
cervical spondylosis, 59/113
cold paresis, 59/107

computer assisted tomography, 59/111
course, 59/110
CSF, 59/110
EMG, 59/110
familial incidence, 59/107
fascicular twitching, 59/108
mechanism, 59/117
monomelic spinal muscular atrophy, 59/376
muscle biopsy, 59/110
myelography, 59/111
myelopathy, 59/113
negative sign, 59/108
negative symptom, 59/108
neuropathology, 59/114
nuclear magnetic resonance, 59/111
oblique amyotrophy, 59/108
onset, 59/107
pathogenesis, 59/117
poliomyelitis, 59/114
radiologic study, 59/111
sex ratio, 59/107
spinal cord tumor, 59/114
syringomyelia, 59/113
therapy, 59/117
traumatic myelopathy, 59/114
tremulous movement, 59/108
unilateral predominance, 59/108
Wohlfart-Kugelberg-Welander disease, 59/93
Hirch-Peiffer test
metachromatic leukodystrophy, 8/16, 10/44
Hirnpathologie, *see* Gehirnpathologie
Hirnwarten, *see* Status verrucosus
Hirschsprung disease, *see* Congenital megacolon
Hirsutism
craniodigital syndrome, 43/243
De Lange syndrome, 43/246
glycosaminoglycanosis type III, 66/300
Gorlin-Chaudry-Moss syndrome, 43/369
holoprosencephaly, 50/237
Huntington chorea, 49/291
Hurler-Hunter disease, 10/436
incontinentia pigmenti, 14/12
leprechaunism, 42/589
lipodystrophy, 8/346
multiple hamartoma syndrome, 42/754
porphyria variegata, 42/620
r(9) syndrome, 50/580
Rubinstein-Taybi syndrome, 31/284, 43/234
Histamine
Balaenoptera borealis intoxication, 37/95
basal ganglion, 49/33
bee sting intoxication, 37/107
brain edema, 67/85
brain infarction, 53/421
brain microcirculation, 53/80
brain vasospasm, 11/125
cephalalgia, *see* Cluster headache
cephalopoda intoxication, 37/62
chronic paroxysmal hemicrania, 48/260, 265
classic migraine, 5/17
cluster headache, 5/4, 116, 48/232
eosinophil, 63/370
epilepsy, 72/94
headache, *see* Cluster headache
hereditary sensory and autonomic neuropathy type III, 21/107-116, 60/23-35, 75/150
migraine, 5/39, 48/92, 113
multiple sclerosis, 9/391
neptunea intoxication, 37/64
octopoda intoxication, 37/62
pain mechanism, 1/142
scombroid intoxication, 37/85
smooth muscle function, 75/624
substance P, 48/98
triple response, 1/485
urticaria pigmentosa, 14/789
Histamine cephalalgia, *see* Cluster headache
Histamine desensitization
cluster headache, 48/230
Histamine flare
hereditary sensory and autonomic neuropathy type III, 21/115, 60/7, 24
Histamine flare response
spinal nerve root injury, 25/420
Histamine H1 receptor antagonist
cluster headache, 48/230
neuroleptic agent, 65/280
Histamine H2 receptor antagonist
cluster headache, 48/230
Histamine headache, *see* Cluster headache
Histamine skin test
cluster headache, 48/228, 233
description, 2/173
segmental innervation, 2/173
Histaminic cephalgia, *see* Cluster headache
Histidase deficiency
histidinemia, 42/554
Histidine
homocarnosinase deficiency, 42/558
Histidinemia
biochemical finding, 29/196
birth incidence, 42/554
clinical features, 29/195
CSF, 42/554
diagnosis, 29/196
diet, 42/555

enzyme defect, 29/196
epilepsy, 42/554
genetics, 29/197
histidase deficiency, 42/554
mental deficiency, 42/554
speech disorder in children, 42/554
treatment, 29/197
Histiocyte
Langerhans, *see* Langerhans histiocyte
RES brain tumor, 18/233
sea blue, *see* Sea blue histiocyte
Histiocytic sarcoma
nomenclature, 18/234
Histiocytoid cardiomyopathy
infantile, *see* Infantile histiocytoid cardiomyopathy
Histiocytoma
malignant fibrous, *see* Malignant fibrous histiocytoma
Histiocytosis
cerebral, *see* Cerebral histiocytosis
hypothalamic tumor, 68/71
Langerhans cell, *see* Langerhans cell histiocytosis
optic chiasm compression, 68/75
Histiocytosis X, 38/93-113
see also Abt-Letterer-Siwe disease, Hand-Schüller-Christian disease *and* Lipoid granulomatosis
Abt-Letterer-Siwe disease, 38/96
Birbeck granule, 38/97-99
brain, 16/231
clinical features, 38/109-112
CSF pleocytosis, 63/422
CSF protein, 63/422
dermatitis, 42/441
diabetes insipidus, 42/442, 63/422
diagnosis, 38/109-112
eosinophilic granuloma, 38/97
facial paralysis, 42/442
fever, 42/441
Hand-Schüller-Christian disease, 38/95, 42/441
history, 38/93-95
hypothalamic involvement, 38/100-102
Langerhans cell, 38/97-99
Langerhans histiocyte, 63/422
lipoid granulomatosis, 38/95
neurologic involvement, 38/99-102
orbital, 18/340
pathology, 38/102-109
prognosis, 38/111
sex ratio, 63/422
skull base, 18/322
spinal, *see* Spinal histiocytosis X
spinal cord compression, 19/365
thrombocytopenia, 42/441
treatment, 38/112
xanthoma, 27/248
xanthomatosis, 10/539, 27/248
xanthomatous transformation, 27/248
Histiosarcoma
malignant fibrous, *see* Malignant fibrous histiosarcoma
Histocompatibility complex, *see* Major histocompatibility complex
Histologic anaplasia
prognosis, 67/6
Histone
nuclear, *see* Nuclear histone
Histoplasma capsulatum
see also Histoplasma meningitis *and* Histoplasmosis
antigen, 52/439
bacteriology, 52/437
brain abscess, 52/151
cerebrovascular disease, 55/432
chronic meningitis, 56/645
complement fixation, 52/439
meningitis, 55/432
Histoplasma duboisii
see also African histoplasmosis
African histoplasmosis, 35/513
Histoplasma meningitis
see also Histoplasma capsulatum
amphotericin B, 52/441
basal form, 52/437
CSF, 52/438
dementia, 46/388
diagnosis, 52/441
ketoconazole, 52/442
serial computer assisted tomography, 52/442
treatment, 52/441
Histoplasmosis, 35/503-513
see also Histoplasma capsulatum
acquired immune deficiency syndrome, 56/515, 71/297
African, *see* African histoplasmosis
amphotericin B, 35/511-513, 52/440
ataxia, 52/438
brain angiography, 52/439
brain granuloma, 11/230, 52/437
chorioretinitis, 35/510
chronic meningitis, 56/645
CNS infection, 69/447
computer assisted tomography, 52/439
cranial nerve palsy, 52/438
CSF, 52/438, 71/297

CSF culture, 52/440
EEG, 52/439
epilepsy, 52/438
etiology, 35/503
headache, 52/438
histoplasmin skin test, 52/441
histoplasmoma, 52/438, 441
history, 35/503
ketoconazole, 52/441
meningoencephalitis, 52/437
mental symptom, 52/438
mycosis, 35/376-378
neurologic involvement rate, 52/437
neurologic symptom, 52/438
neurosurgery, 52/441
serology, 52/439
treatment, 52/440
Histopteridae intoxication
family, 37/78
venomous spine, 37/78
History of neurology, 1/1-38
Kaiser Wilhelm Institute, 1/37
Max Planck Gesellschaft, 1/26, 37
Waldeyer neuron theory, 1/46
Histotoxic anoxia
brain anoxia, 46/358
Histotoxic hypoxia
complex IV, 9/621
structural damage, 9/621
tricarboxylic acid cycle, 9/621
Hitzig, E., 1/8
HIV, *see* Human immunodeficiency virus
HLA antigen
amyotrophic lateral sclerosis, 42/66
arteritis, 48/310
chronic inflammatory demyelinating polyradiculoneuropathy, 51/538
congenital adrenal hyperplasia, 42/515
CSF, *see* CSF HLA antigen
diabetes mellitus, 42/544
disease association, 47/304
ethnic variation, 47/305
genetic linkage, 42/128, 163, 515
geographic variation, 47/305
Guillain-Barré syndrome, 42/646, 51/248
hemochromatosis, 42/553
hereditary olivopontocerebellar atrophy, 42/162
hereditary olivopontocerebellar atrophy (Menzel), 42/162
juvenile optic atrophy, 42/410
leprosy, 42/650
manic depressive psychosis, 43/209
multiple sclerosis, 42/497, 47/97, 189, 299-313, 371
muscle tissue culture, 62/92
myasthenia gravis, 41/101, 43/158
olivopontocerebellar atrophy (Schut-Haymaker), 42/162
Parkinson disease, 49/133
poliomyelitis, 42/653
spinocerebellar ataxia, 42/128
Sydenham chorea, 49/362
HLA-B7 antigen
lymphocytotoxic antibody, 47/369
multiple sclerosis, 47/340
HLA-B40 antigen
lymphocytotoxic antibody, 47/369
HLA-DR antigen
eosinophilia myalgia syndrome, 64/253
HLA-DR2 antigen
interferon, 47/354
lymphocytotoxic antibody, 47/369
HLA-DR4 antigen
temporal arteritis, 55/346
HLA-Dw2 antigen
interferon, 47/354
HLA haplotype
thyrotoxic periodic paralysis, 62/532
HLA histocompatibility antigen system
cluster headache, 48/233
HLA loci
ankylosing spondylitis, 41/77
inflammatory myopathy, 41/78
myasthenia gravis, 41/101
HLA system
see also Major histocompatibility complex
Mycobacterium leprae, 51/215
viral infection, 56/34, 66
H-M interval discrimination
diabetic polyneuropathy, 51/504
HMC syndrome, *see* Hypertelorism microtia clefting syndrome
Hoarse voice
primary amyloidosis, 51/416
vitamin B1 deficiency, 51/332
Hoarseness
lateral medullary infarction, 53/381
lymphoma, 63/350
pontine infarction, 53/389
Hobara focus
Kii Peninsula amyotrophic lateral sclerosis, 22/360, 373
Hodgkin antigen
paraneoplastic syndrome, 69/330
Hodgkin disease
see also Hodgkin lymphoma, Malignant

lymphoma *and* RES brain tumor
acquired toxoplasmosis, 35/124, 52/353
angioimmunoblastic lymphadenopathy, 39/530
ataxia telangiectasia, 46/58
autonomic neuropathy, 75/535
brachial neuritis, 39/56
brain, 18/262
brain tumor, 18/234
bulbar encephalitis, 39/54
carcinoma, 38/687
carcinomatous cerebellar degeneration, 21/495
cause, 20/117
cerebral lymphomatoid granulomatosis, 39/522-524
classification, 39/27, 69/263
clinical features, 18/260
CNS infection, 39/41-45
CNS involvement, 18/259-261
CNS localization, 69/274
cranial vault tumor, 17/123
Cryptococcus, 39/43
CSF, 20/121
dermatomyositis, 39/56
differential diagnosis, 20/119
Eaton-Lambert myasthenic syndrome, 39/56
epidural disease, 69/276
etiology, 20/117
Guillain-Barré syndrome, 39/56
Hallervorden-Spatz syndrome, 6/625
herpes simplex virus encephalitis, 39/45
herpes zoster, 20/121, 39/44, 69/279
histologic form, 18/259
history, 20/113
immunocompromised host, 56/469
intracranial metastasis, 39/29-32
intraspinal metastasis, 39/32-38
ivory vertebra, 19/171
late cerebellar atrophy, 60/585
late cortical cerebellar atrophy, 21/495
leptomeningeal metastasis, 39/36-38
leukoencephalopathy, 9/594
limbic encephalitis, 39/54
Listeria monocytogenes meningitis, 33/90
lymphangiography, 39/49
lymphnode histology *vs* cat scratch disaese, 52/128
lymphogranulomatous meningoencephalopathy, 18/261
lymphoma, 63/345
malnutrition, 39/38-40
metabolic disorder, 39/38-40
metastasis, 39/28-38
multiple myeloma, 71/440
myasthenic syndrome, 39/56
neuromyopathy, 39/56
neuropathology, 18/262
neuropathy, 6/143
nonmetastatic complication, 39/38-48
optic neuritis, 39/54
paraneoplastic polyneuropathy, 51/468
paraneoplastic syndrome, 39/53-56, 69/337, 71/689
pathology, 20/114
peripheral nerve, 39/38
plexus, 39/38
polymyositis, 39/56
radiation injury, 39/50-53
radiotherapy, 39/50-53
Reed-Sternberg cell, 20/114
RES tumor, 18/235, 259-263
skull base tumor, 17/200
spinal, *see* Spinal lymphogranulomatosis
spinal cord compression, 19/364, 366
spinal epidural tumor, 20/113-128
splenectomy, 39/50
stage, 20/117
subacute cerebellar degeneration, 39/54
subacute malignant cerebellar degeneration, 39/54
toxic disorder, 39/38-40
toxoplasmosis, 39/43
treatment, 18/262
type, 20/114
vascular disorder, 39/45-48
Hodgkin-Huxley model
membrane current, 7/65
myotonia, 40/554
strength-duration curve, 7/67
transmembrane model, 40/551
Hodgkin lymphoma
see also Hodgkin disease
brain tumor, 63/348
cauda equina compression, 63/346
chlormethine, 63/358
chronic meningitis, 56/645
corticosteroid, 63/348
diffuse brain infiltration, 63/348
epidural lymphoma, 63/349
epilepsy, 63/349
granulomatous CNS vasculitis, 55/388
headache, 63/349
herpes zoster, 63/353
irradiation myelopathy, 63/356
leukemia, 63/354
Listeria monocytogenes, 63/353
measles encephalitis, 63/353
neurologic complication, 63/345

nitrosourea, 63/358
progressive multifocal leukoencephalopathy, 63/354
radiotherapy, 63/348
spinal cord compression, 63/346, 349
spinal nerve root compression, 63/346
subdural lymphoma, 63/349
vincristine, 63/358
Hoffmann, C., 1/9
Hoffmann disease, *see* Hypertrophic interstitial neuropathy
Hoffmann reflex, *see* H-reflex
Hoffmann sign
elicitation, 1/184
Hoffmann syndrome
see also Kocher-Debré-Sémélaigne syndrome
hypothyroid myopathy, 41/243, 43/84, 62/533
hypothyroidism, 70/97
Hoffmann-Zurhelle syndrome
Goltz-Gorlin syndrome, 14/113
Hog cholera virus
animal viral disease, 34/298
cerebellar defect, 34/298
cholera, 34/298
Hollenhorst plaque
amaurosis fugax, 55/108
displacement, 55/109
retinal artery aneurysm, 55/108
Hollow foot, *see* Pes cavus
Hollow hand sign
athetosis, 1/184
chorea, 1/184
metacarpal, 1/184
muscle weakness, 1/184
Holmes-Adie syndrome
anhidrosis, 8/344
anisocoria, 74/411
areflexia, 43/72
Argyll Robertson pupil, 1/621
autonomic nervous system, 43/72, 75/125
autonomic polyneuropathy, 51/475
ciliary ganglion, 43/72
dysautonomia, 22/259
headache, 74/411
hereditary sensory and autonomic neuropathy type II, 60/18
hereditary sensory and autonomic neuropathy type III, 21/115, 60/18
hypertrophic interstitial neuropathy, 21/151
hyporeflexia, 43/72
miosis, 43/72
orthostatic hypotension, 22/262, 63/156
photophobia, 74/411
pupil, 74/411, 413
pure autonomic failure, 22/236
Ross syndrome, 75/125
Sjögren syndrome, 51/449
tonic pupil, 2/109, 22/259
Holmes cerebellar ataxia
brief survey, 42/127
Holmes, G., 1/6
Holmes hereditary cerebello-olivary atrophy, *see* Hereditary cerebello-olivary atrophy (Holmes)
Holoacrania, *see* Anencephaly
Holoanencephaly
anencephaly, 30/175, 50/72
Holocarboxylase
epilepsy, 72/222
Holocardiac twins
acardiac monster, 30/51
neural induction, 50/25
Holocyclotoxin
neurotoxin, 65/194
Holoprosencephaly, 30/431-472, 42/32-34
see also Arhinencephaly, Cebocephaly, Cyclopia, Ethmocephaly, Median facial cleft syndrome *and* Otocephaly
abnormal karyotype, 50/240
accessory spleen, 50/237
agyria, 50/237
alobar, *see* Alobar holoprosencephaly
anal atresia, 50/237
anatomy, 30/438-441
anencephaly, 42/12, 50/235, 237
animal, 30/457
anophthalmia, 30/470
atrial septal defect, 50/237
birth incidence, 42/32
borderline syndrome, 50/242
brain cortex disorganization, 50/249
brain-face correlation, 30/450, 50/235
cardiac ventricular septum defect, 50/237
cebocephaly, 30/446, 42/32, 50/233
cerebellar heterotopia, 50/251
chromosome 18, 42/33
classification, 30/438, 50/230
clinical features, 30/442-451, 50/232
clitorimegaly, 50/237
cocaine intoxication, 65/255
computer assisted tomography, 50/237
congenital heart disease, 63/11
corpus callosum agenesis, 30/455, 50/155
course, 30/457, 50/239
cryptorchidism, 50/237
cyclopia, 42/32, 50/232
cystic kidney, 50/237

dextroposed aorta, 50/237
diagnosis, 30/468
diagnostic criteria, 50/241
diastasis recti, 50/237
double ureter, 50/237
double vagina, 50/237
Down syndrome, 50/531
dystopic gallbladder, 50/237
Edwards syndrome, 31/521, 43/537, 50/557, 562
EEG, 30/452, 456, 459, 50/238
embryology, 30/434-438, 50/226
encephalocele, 50/235-237
endocrine disorder, 30/453, 456, 50/237
epidemiology, 30/156, 457
epilepsy, 30/459, 72/109, 344
ethmocephaly, 30/445, 42/32, 50/233
etiology, 30/467, 50/241
experimental, 30/468
extracephalic abnormality, 50/236
face, 50/232
Fallot tetralogy, 50/237
fetoscopy, 42/34
forme fruste, 50/240
genetics, 30/156, 458, 42/33
gross anatomy, 50/230
gross morphology, 50/231
Gruber syndrome, 43/391, 50/236
hirsutism, 50/237
history, 30/431-434, 50/225
homeostasis, 50/238
hydranencephaly, 50/347
hydrocephalus, 42/32, 50/237
hydroureter, 50/237
hypospadias, 50/237
hypotelorism, 30/431, 448-450, 42/32, 50/233
incidence, 50/55, 61, 63
inheritance, 50/239
iniencephaly, 50/131
intermaxillary rudiment, 50/234
Kallmann syndrome, 30/471
karyotype, 30/459-466, 42/33
lobar, *see* Lobar holoprosencephaly
lung abnormality, 50/237
malformation, 30/451-453
management, 30/468, 469, 50/241
Meckel diverticulum, 50/237
median cleft lip, 42/32, 50/234
median cleft lip face with hypotelorism, 30/447
mental deficiency, 30/459, 42/32, 50/239
mental disorder, 50/239
metabolism, 50/238
metopic suture, 50/233
microcephaly, 30/451, 42/32
microphthalmia, 30/470
neural induction, 50/25
neurologic manifestation, 30/451-453
nomenclature, 30/438, 50/230
normal face, 30/450
orbital hypotelorism, 30/431, 446-450, 42/32, 50/233
otocephaly, 30/437, 50/233
18p partial monosomy, 43/534
18p syndrome, 31/582
parietal cephalocele, 50/108
parietal encephalocele, 50/108
Patau syndrome, 31/504, 507, 512, 43/527, 50/235, 237, 557-559
phenotype-karyotype correlation, 50/240
pituitary gland aplasia, 30/470
poikilothermia, 50/239
polydactyly, 50/237
polyploidy, 50/555
prenatal diagnosis, 50/240
proboscis, 50/232
prognosis, 30/457, 50/239
prosoposchisis, 50/233
13q partial monosomy, 43/526
r(13) syndrome, 50/585
rachischisis, 50/235
radiologic appearance, 30/453-456
seizure, 30/452, 456, 459, 42/32, 50/238, 240
semilobar, *see* Semilobar holoprosencephaly
sex ratio, 50/239
sincipital encephalocele, 50/103
spina bifida, 50/237
syndactyly, 50/237
syndrome, 30/469-472
talipes, 50/237
teratologic series definition, 30/431
thumb agenesis, 50/237
trigonocephaly, 30/451, 43/486, 50/235
triploidy, 43/566
trisomy 13-15, 14/774
truncus arteriosus communication, 50/237
umbilical hernia, 50/237
uterus bicornis, 50/237
Holotelencephaly, *see* Holoprosencephaly
Holothurian intoxication
holothurioidea, 37/66
Holothurioidea intoxication
sea cucumber, 37/66
sea slug, 37/66
Holt-Oram syndrome
Patau syndrome, 14/121
Holthouse-Batten disease, *see* Hutchinson-Tay choroidopathy

Homatropine
mydriasis, 2/108
Homeostasis
autonomic nervous system, 74/15
cholesterol, *see* Cholesterol homeostasis
copper, 29/279-286
dysautonomia, 22/245
holoprosencephaly, 50/238
salt appetite, 28/497-499
thermoregulation, 61/275
Homer Wright rosette
cerebellar medulloblastoma, 16/18, 18/174, 177
ependymoma, 18/116, 177
spinal ependymoma, 20/359
sympathicoblastoma, 18/177
Homicide
head injury epidemiology, 57/12
migraine, 48/162
Homidium bromide
neurotoxin, 47/569, 64/10, 17
oligodendrocyte inclusion body, 64/17
spongiform leukoencephalopathy, 47/569
toxic neuropathy, 64/10
Homidium bromide intoxication
action site, 64/10
mitochondrial inhibition, 64/17
oligodendrocyte inclusion body, 64/17
Homocarnosinase deficiency, 42/557-559
brain atrophy, 42/558
CNS, 42/558
CSF, 42/558
GABA, 42/558
histidine, 42/558
homocarnosine, 42/558
intellectual impairment, 42/577
retinal pigmentation, 42/558
spastic paraplegia, 42/557
urine, 42/558
visual acuity, 42/558
Homocarnosinase metabolism
spastic paraplegia, 42/179
Homocarnosine
CSF, *see* CSF homocarnosine
homocarnosinase deficiency, 42/558
homocarnosinosis, 59/359
Homocarnosinosis
brain atrophy, 42/558
CSF, 42/557-559
dementia, 42/557, 59/359
familial spastic paraplegia, 59/308, 326, 333
homocarnosine, 59/359
progressive paraparesis, 59/359
retinal degeneration, 59/359
secondary pigmentary retinal degeneration, 42/557-559, 60/727
spastic paraplegia, 42/557
Homocitrullinemia
epilepsy, 72/222
Homocitrinullinuria
hyperornithinemia, 29/103
Homocysteine
endothelial loss, 55/326
homocystinuria, 55/326
platelet, 55/326
vessel fibrosis, 55/326
Homocystine
leukoencephalopathy, 47/585
Homocystinuria, 55/325-332
arachnodactyly, 55/329
autosomal dominant hypoalphalipoproteinemia, 60/136
autosomal recessive, 55/326
6-azauridine, 55/325
betaine, 55/325, 332
birth incidence, 42/556
brain embolism, 55/328
brain infarction, 53/33, 165, 55/328
brain ischemia, 53/165
brain thrombosis, 55/328
carotid dissecting aneurysm, 54/282
carotid system syndrome, 53/309
cerebrovascular disease, 53/28, 33, 43
clinical features, 29/114, 55/327
CNS spongy degeneration, 42/556
collagen disease, 42/556
corpus callosum hypoplasia, 55/330
CSF, 55/328
cystathionine synthetase, 29/113, 115
cystathionine synthetase deficiency, 42/555-557, 46/52
ectopia lentis, 39/392, 42/555, 55/327
EEG, 55/328
enzyme defect, 29/111, 115
epilepsy, 55/328, 72/222
genu valgum, 55/329
glaucoma, 42/555
headache, 55/328
heterozygote detection, 42/557
heterozygous form, 55/327, 329
high dose pyridoxine responding cystathionine synthetase deficiency, 29/116
history, 55/327
homocysteine, 55/326
hypermethioninemia, 29/119, 55/325
iatrogenic brain infarction, 53/165
iridodonesis, 55/327

kyphoscoliosis, 42/555
lethargy, 55/329
low dose pyridoxine responding cystathionine synthetase deficiency, 29/116
mental deficiency, 42/556, 46/51, 55/327-329
metabolic pathway, 55/325
methionine, 55/325
methylenetetrahydrofolate reductase deficiency, 29/116
methylfolate-H4-methyltransferase deficiency, 29/116
methylmalonic aciduria, 55/327, 329
microgyria, 55/330
myopia, 55/327
neuropathology, 55/329
occlusive vasculopathy, 55/328
olfactory bulb, 55/330
olfactory bulb hypoplasia, 55/330
optic atrophy, 13/42
osteoporosis, 42/555
pectus excavatum, 55/329
plasma methionine, 55/329
platelet abnormality, 29/115
polygyria, 55/330
polymicrogyria, 55/330
prenatal diagnosis, 42/557
prevalence, 55/325
prognathism, 55/329
pyridoxal-5-phosphate, 28/120-122, 29/115
scoliosis, 50/418
serine, 55/325
sex ratio, 55/325
skeletal deformity, 55/329
stroke, 42/555
subacute combined spinal cord degeneration, 63/255
symptom, 29/112-116
symptomatic dystonia, 49/543
systemic brain infarction, 11/462
treatment, 29/115, 55/331
vertebral dissecting aneurysm, 54/282
vitamin B_6, 28/120-123, 29/115, 55/332
vitamin B_6 intoxication, 65/572
vitamin B_{12}, 29/115, 55/331
vitamin B_{12} deficiency, 55/325
vitamin B_c, 55/325
vitamin B_c deficiency, 51/336, 55/325
Homodimer
dihydrolipoamide dehydrogenase, 66/415, 420
Homolateral ataxia crural paresis syndrome
brain lacunar infarction, 54/57
computer assisted tomography, 54/57
Homonymous hemiachromatopsia
cortical blindness, 2/650
occipital lobe syndrome, 2/650
Homonymous hemianopia
alexia, 2/607, 609
bilateral, 2/593
brain injury, 24/45, 57/134
cerebrovascular disease, 55/117
clinical examination, 57/134
compensation mechanism, 2/673
completion phenomenon, 2/584
critical flicker fusion, 2/584
development, 2/582
geniculocalcarine tract injury, 24/46
hyperosmolar hyperglycemic nonketotic diabetic coma, 27/90
incongruity, 2/577
lateral geniculate body, 2/480
lesion site, 2/572
migraine, 48/135, 161, 166
occipital lobe lesion, 2/672
occipital lobe tumor, 2/672
occipital lobectomy, 2/671
optic tract injury, 24/45
parietal lobe lesion, 2/587
Sturge-Weber syndrome, 50/376
temporal lobe, 2/708
temporal lobe lesion, 2/586
transtentorial herniation, 24/64
visual adaptation, 2/584
visual cortex injury, 24/47
Homonymous paracentral scotoma
reading disorder, 2/585
visual field defect, 2/585
Homoprotocatechuic acid
hereditary progressive diurnal fluctuating dystonia, 42/214
Parkinson disease, 49/123
urinary, *see* Urinary homoprotocatechuic acid
Homosexual man
acquired immune deficiency syndrome, 56/490
cell number, 74/500
melatonin, 74/501
sexual dimorphic nucleus of preoptic area, 74/500
suprachiasmatic nucleus, 74/500
vasopressin neuron, 74/501
Homosexuality
prevalence, 43/215
sex ratio, 43/215
Homovanillic acid
akinesia, 49/67
Alzheimer disease, 46/266
amyotrophic lateral sclerosis, 22/30
catabolism, 8/478

CSF, *see* CSF homovanillic acid
daytime hypersomnia, 45/136
dystonia musculorum deformans, 49/524, 526
Gilles de la Tourette syndrome, 49/633
hereditary progressive diurnal fluctuating dystonia, 42/214
hereditary sensory and autonomic neuropathy type III, 1/478, 21/78, 109, 60/25, 28
hyperactivity, 46/182
Parkinson dementia, 49/174
Parkinson disease, 6/180, 49/120
pharmacology, 74/144
progressive supranuclear palsy, 49/251
Reye syndrome, 56/153
rigid Huntington chorea, 49/285, 295
rigidity, 49/70
striatonigral degeneration, 42/262
tremor, 49/73
Homoveratrylamine
Parkinson disease, 49/120
Honeycomb dystrophy (Doyne), *see* Hutchinson-Tay choroidopathy
Honeycomb myopathy
clinical features, 62/353
congenital myopathy, 62/332
zebra body, 62/353
Honeycomb structure
muscle fiber, 40/106, 224
oligodendroglioma, 16/13, 22, 20/333
transverse tubule, 40/106, 224
Honeymoon palsy
compression neuropathy, 51/92
Hooft-Bruens syndrome
epileptic seizure, 42/508
hyperreflexia, 42/508
mental deficiency, 42/508
optic atrophy, 42/508
spasticity, 42/508
tetraplegia, 42/508
Hooft disease
chorioretinal degeneration, 13/41
optic atrophy, 13/41
Hookworm disease, 35/361-364
clinical features, 35/363
epidemiology, 35/211, 362
geographic distribution, 35/362
helminthiasis, 35/211
morphology, 35/361
parasitology, 35/361
pathogenesis, 35/364
pathology, 35/364
prophylaxis, 35/364
treatment, 35/364
Hopping reaction
significance, 1/69
Horizontal gaze
brain lacunar infarction, 54/246
complex dysfunction, 54/246
Horizontal gaze paralysis
cerebellar hemorrhage, 54/314
conjugate gaze, 1/602
differential feature, 2/335
juvenile Gaucher disease type A, 66/125
juvenile Gaucher disease type B, 66/125
mental deficiency, 42/333
myokymia, 42/333
myopia, 42/333
pontine infarction, 53/389
scoliosis, 42/332
tremor, 42/333
Hormone
accidental injury, 74/324
adenohypophysis, 2/442
brain tumor carcinogenesis, 16/40
epilepsy, 72/232
growth, *see* Growth hormone
luteinizing, *see* Luteinizing hormone
metabolic, *see* Metabolic hormone
migraine, 48/18, 60, 86, 100, 119
parathyroid, *see* Parathyroid hormone
progressive dysautonomia, 59/153
sex, *see* Sex hormone
sexual function, 26/451
sexual impotence, 75/94
steroid, *see* Steroid hormone
suprachiasmatic nucleus, 74/487
thyroid, *see* Thyroid hormone
toluene intoxication, 64/50
vampire, *see* Vampire hormone
Horn syndrome
anterior, *see* Anterior horn syndrome
Horner syndrome
anterior inferior cerebellar artery, 2/244
anterior inferior cerebellar artery syndrome, 11/31, 53/396
aorta aneurysm, 63/50
autonomic nervous system, 43/64, 74/406
Babinski-Nageotte syndrome, 2/218
basilar artery system, 11/31
brachial paralysis, 43/64
brachial plexus, 43/64, 51/144
brachial plexus injury, 7/410
brachial plexus paralysis, 2/136
brain embolism, 53/213
brain infarction, 53/213
Budge ciliospinal center, 2/202

C8 to T1 root, 2/174
C8-T1 symptom, 2/205
carotid dissecting aneurysm, 54/272
central pontine myelinolysis, 63/549
cerebellar artery syndrome, 55/90
cerebrovascular disease, 53/213
cervical root lesion, 2/171
cervicomedullary injury, 24/148
chronic migrainous neuralgia, 48/251
ciliospinal center, 19/59, 20/199
cluster headache, 5/113-115, 48/9, 222
definition, 2/111
differentiation, 2/318
dysautonomia, 22/256
dysraphia, 2/318, 325
enophthalmos, 43/64
epidural anesthesia, 61/150
exophthalmos, 19/59
extramedullary spinal tumor, 19/59
Foville syndrome, 2/316
genetics, 30/100
Gradenigo syndrome, 2/313
head injury, 24/148
headache, 48/286, 335, 75/284
heterochromia iridis, 22/256, 43/65
intracranial hemorrhage, 53/213
intramedullary spinal tumor, 19/32, 59
lateral medullary infarction, 2/234, 53/382
lateral medullary syndrome, 2/232
lesion site, 1/624
miosis, 19/59, 43/64
mydriasis, 19/59
myopathy, 40/302
nasociliary nerve, 2/58
neuroblastoma, 43/64
neuromyelitis optica, 9/428
ophthalmoplegia, 22/209
orthostatic hypotension, 63/145
pain, 48/330
Pancoast tumor, 2/143
paratrigeminal syndrome, 1/514, 48/329, 75/293
partial, *see* Partial Horner syndrome
pontine glioma, 18/398
pontine tegmental hemorrhage, 2/254
progressive hemifacial atrophy, 8/347, 14/777, 31/253, 43/408
ptosis, 43/64
pupil, 74/408
scalenus anticus syndrome, 2/150
spinal cord injury, 25/383, 415, 417
spinal cord lesion, 2/209
spinal cord tumor, 19/59
spinal nerve root injury, 25/415, 417
spinal sympathicoblastoma, 20/112
spinal tumor, 19/35
stress, 74/317
superior cerebellar artery syndrome, 2/240, 53/397
sweating, 1/455
sympathetic outflow, 75/124
symptom, 2/113
syringobulbia, 50/445
syringomyelia, 32/261, 268, 50/445
thalamus lesion, 2/482
transient ischemic attack, 53/213
vertebrobasilar system syndrome, 53/397
Villaret syndrome, 2/100
Wegener granulomatosis, 71/179
Hornet intoxication, *see* Hymenoptera intoxication
Hornet venom
myasthenic syndrome, 64/16
myoglobinuria, 62/575, 64/16
rhabdomyolysis, 64/16
toxic myopathy, 62/601, 612, 64/16
Horseradish peroxidase
middle cerebral artery, 48/108
neurotropic virus, 56/26
trigeminal ganglion, 48/108
Horseshoe crab intoxication, *see* Xiphosura intoxication
Horseshoe kidney
Edwards syndrome, 43/536, 50/275
Horsley, V., 1/6
Horton cephalalgia, *see* Cluster headache
Horton headache, *see* Chronic migrainous neuralgia *and* Cluster headache
Horton-Magath-Brown syndrome, *see* Temporal arteritis
Horton syndrome, *see* Chronic migrainous neuralgia *and* Cluster headache
Hospitalization time
cervical spinal cord injury, 61/66
determinant factor, 61/66
Host immune response
slow virus disease, 56/19
viral infection, 56/51
Host reaction
CNS infection, 55/417
immunodeficient category, 55/417
Hot dog headache
classification, 48/6
definition, 48/10
Hot water epilepsy
clinical features, 73/190
pathogenic mechanism, 73/190
Hourglass neurinoma
intervertebral foramen, 20/230, 242

Hourglass tumor, 20/177-312
clinical features, 20/295
CSF, 20/307
diagnosis, 20/298
differential diagnosis, 20/308
foramen magnum tumor, 20/184, 291
histogenesis, 20/289
histology, 20/292
history, 20/289
intervertebral foramen, 8/429, 19/161
lesion site, 20/290
malignant degeneration, 20/294
metastasis, 20/294
multiple tumors, 8/429, 20/294
neurinoma, 19/161
neurofibroma, 8/440, 19/161
neurofibromatosis, 20/289
neurofibromatosis type I, 14/139, 154, 20/289
prevalence, 20/290
pseudopsoas shadow, 20/297
radiology, 20/300
segmental lesion distribution, 20/291
spinal cord, 8/429
spinal cord compression, 19/161, 20/295
spinal meningioma, 19/358, 20/184
spinal neurinoma, 19/358, 20/242
terminology, 8/429
treatment, 20/310
vertebral column syndrome, 20/298
Howship-Romberg sign
obturator nerve, 2/40
HRR plate
neuroretinal degeneration, 13/15
HSS gene
glycosaminoglycanosis type IIIA, 66/282
heparan *N*-sulfatase, 66/282
5-HT1A receptor, *see* Serotonin 1A receptor
HTLV-1, *see* Human T-lymphotropic virus type I
Hu antibody
paraneoplastic syndrome, 69/353, 355
Hu family
paraneoplastic syndrome, 69/330
Hubbard tank
hydrotherapy, 8/384
Huffer neuropathy
acetonylacetone intoxication, 64/82
organic solvent intoxication, 36/374
Hultkranz line
basilar impression, 32/18
Human immunodeficiency virus
acute viral myositis, 56/195
antiphospholipid antibody, 63/330
autonomic nervous system, 74/335, 75/18, 412
CD4 antigen, 56/490
glycoprotein, 56/490
Kaposi sarcoma, 56/489
lentivirus, 56/460, 584
retrovirus, 56/490
seroconversion, 56/492
Sjögren syndrome, 71/74
stress, 74/335
T-helper lymphocyte, 56/490
T-lymphocyte lymphoma, 56/489
transmission, 56/490
visna-maedi virus, 56/584
visna virus, 56/463
Human immunodeficiency virus dementia
progressive multifocal leukoencephalopathy, 71/277
Human immunodeficiency virus infection
see also Acquired immune deficiency syndrome *and* Acquired immune deficiency syndrome treatment
amphotericin B, 71/384
antineoplastic agent, 71/385
autonomic dysfunction, 75/412
brain atrophy, 75/413
carbamate intoxication, 64/187
cerebrovascular disease, 71/335, 337
ciprofloxacin, 71/385
cytarabine, 71/386
dapsone, 71/384
diamorphine intoxication, 65/356
embolus, 71/339
encephalitis, 71/268
epidemiology, 71/353
epilepsy, 72/145, 155
Kaposi sarcoma, 71/340
lymphoma, 71/341, 343
methotrexate, 71/386
neoplasm, 71/340
noninfectious disorder, 71/335
opiate intoxication, 65/356
primary CNS lymphoma, 71/343
survey, 71/353
systemic lymphoma, 71/341
thalidomide, 71/388
thrombosis, 71/337
treatment, 71/265, 367
vinca alkaloid, 71/385
zidovudine, 71/377
Human immunodeficiency virus myopathy
acquired immune deficiency syndrome, 71/374
diagnosis, 71/380
polymyositis, 71/374
Human immunodeficiency virus neuropathy

autonomic neuropathy, 71/357
brachial neuritis, 71/354
cytomegalovirus polyradiculopathy, 71/358
demyelinating, 71/355
distal symmetric polyneuropathy, 71/356
epidemiology, 71/353
ganglioneuropathy, 71/354
Guillain-Barré syndrome, 71/354
mononeuropathy multiplex, 71/356
peripheral facial paralysis, 71/354
polyneuropathy, 71/354
Human immunodeficiency virus treatment
neuropathy, 71/362
Human immunodeficiency virus type 1
polymyositis, 56/528, 62/381
retrovirus, 56/528
Human immunodeficiency virus type 1 infection
acquired immune deficiency syndrome dementia complex, 71/238
Human immunodeficiency virus type 2
human T-lymphotropic virus type I associated myelopathy, 56/529
retrovirus, 56/528
Human leukocyte antigen, *see* HLA antigen
Human menopausal gonadotropin
adenohypophysis, 2/442
anorexia nervosa, 46/586
Human neurolymphomatosis, *see* Neurolymphomatosis
Human rights
law, 24/792, 795
Human T-lymphotropic virus type I
adult T-cell leukemia, 56/525
autonomic nervous system, 75/414
blood transfusion, 56/530
familial spastic paraplegia, 59/306
genome, 56/530
husband to wife, 56/530
incubation period, 56/530
mother to child, 56/530
multiple sclerosis, 56/18
myelopathy, 59/306, 65/356, 75/414
neurolymphomatosis, 56/179
polymyositis, 56/528, 62/382
retrovirus, 56/18, 525, 528
spastic paraplegia, 59/436
structure, 56/530
transmission, 56/530
tropical spastic paraplegia, 56/525
Human T-lymphotropic virus type I associated myelopathy, 59/306, 447-453
see also Chronic myelitis *and* Tropical spastic paraplegia
adult T-cell leukemia flower appearance, 59/450
age, 56/534, 59/452
amyotrophic lateral sclerosis, 59/451-453
arthropathy, 59/450
brain stem auditory evoked potential, 56/538
cause, 56/529
chronic myelitis, 56/529
clinical features, 56/535, 59/448-450
clinical presentation, 59/452
cryoglobulinemia, 59/450
CSF, 56/538
CSF pleocytosis, 59/450
CSF protein, 59/450
cystometry, 56/537
definition, 56/529
diagnostic guideline, 59/451
electrophysiologic study, 59/453
epidemiology, 56/533
evoked potential, 56/537
familial aggregation, 56/535
female preponderance, 56/527
flower lymphocyte, 56/538
geography, 56/534
hepatomegaly, 56/527
history, 56/535
human immunodeficiency virus type 2, 56/529
human T-lymphotropic virus type I DNA, 59/450
hypercalcemia, 56/527
ichthyosis, 59/450
IgG antibody intrathecal synthesis, 59/450
IgM monoclonal gammopathy, 59/450
incidence, 56/531
initial symptom, 59/452
laboratory examination, 56/538
laboratory study, 59/450, 453
late onset, 56/527
leukocyte polymerase chain reaction, 59/453
lymphadenopathy, 56/527
lymphocyte, 59/450
lymphocytic meningitis, 56/538
monoclonal gammopathy, 56/525
murine neurotropic retrovirus, 59/447
mycosis fungoides, 56/527
nerve conduction, 56/537
neuroepidemiology, 56/531
neurologic sign, 59/452
neuropathologic study, 59/453
neuropathology, 56/531-533, 536
neurophysiologic study, 59/450
OKT4/OKT8 ratio, 59/450
onset, 56/535
pathogenesis, 56/531
peripheral blood, 59/450

prevalence, 56/531
prevention, 56/538
primary lateral sclerosis, 56/527
progression, 56/535
pseudoamyotrophic lateral sclerosis, 59/451
pulmonary alveolitis, 59/450
race, 56/535, 59/452
secondary pigmentary retinal degeneration, 60/736
segmental demyelination, 56/531
serum human T-lymphotropic virus type I antibody, 59/453
serum IgG, 59/450
sex ratio, 56/531, 59/452
Sjögren syndrome, 59/450
spastic paraplegia, 56/525
splenomegaly, 56/527
treatment, 56/538
Treponema pallidum, 56/538
uveitis, 59/450
vasculitis, 59/450
viral meningitis, 56/538
xerosis, 59/450
Human T-lymphotropic virus type I DNA
human T-lymphotropic virus type I associated myelopathy, 59/450
Human T-lymphotropic virus type Ib
retrovirus, 56/529
Human T-lymphotropic virus type II
autonomic nervous system, 75/415
myelopathy, 65/356
retrovirus, 56/18, 529
Human T-lymphotropic virus type III
granulomatous CNS vasculitis, 55/388
retrovirus, 56/18
spastic paraplegia, 59/436
Human T-lymphotropic virus type V
retrovirus, 56/529
Humeral dislocation
brachial plexus injury, 7/406
Humeral fracture
brachial plexus, 70/31
radial nerve, 70/31
Humeroscapular paralysis
acute, *see* Brachial neuralgia
Humidity
olfaction, 74/363
Humoral immune response
CSF, *see* CSF humoral immunity
multiple sclerosis, 34/435-440, 47/363-372
subacute sclerosing panencephalitis, 56/425
viral infection, 56/51
Hungary
neurology, 1/11
Hunt geniculate neuralgia, *see* Intermedius neuralgia
Hunt, J.R., 1/4, 5/337
Hunt neuralgia, *see* Intermedius neuralgia
Hunt-Van Bogaert disease, *see* Hereditary pallidal degeneration
Hunt-Van Bogaert progressive pallidal atrophy, *see* Progressive pallidal atrophy (Hunt-Van Bogaert)
Hunter disease, *see* Glycosaminoglycanosis type II, Glycosaminoglycanosis type IIA *and* Glycosaminoglycanosis type IIB
Hunter-Russell syndrome
Minamata disease, 36/74-85, 103
neurotoxicology, 64/8
organic mercury intoxication, 36/73-75
Huntington chorea
acetylcholine, 29/447-450
acetylcholine receptor, 49/259
acetylcholinesterase, 49/257
adenocarcinoma, 49/291
age distribution, 6/389
age pattern, 6/379
agraphia, 45/466
allelic heterogeneity, 49/276
Alzheimer disease, 49/291
amino acid metabolism, 6/348
amyotrophic lateral sclerosis, 22/138, 59/410
amyotrophy, 49/291
animal model, 49/261
annual incidence, 42/226
anteposition, 6/306
apathy, 46/306
aspartic acid, 49/261
aspiny neuron, 49/256, 260
associated disease, 49/291
astrocyte loss, 49/319
astrocytosis, 49/318
ataxia, 42/226, 49/280
Australia, 49/269
autoimmune disease, 6/382-384
autonomic dysfunction, 49/280
autonomic vegetative symptom, 6/314
basal ganglion degeneration, 42/226
basal ganglion tier II, 49/24
benzodiazepine receptor, 49/259
biochemical aspect, 27/492-496
biochemistry, 6/129, 342-353
blink reflex, 49/290
bradycardia, 49/284
brain atrophy, 6/331, 42/226, 46/305
brain biopsy, 10/685
brain cortex, 49/318
brain stem, 49/320

brain stem auditory evoked potential, 49/290
brain weight, 49/316
cachexia, 49/284
Canada, 49/269
capillary increase, 49/319
captopril intoxication, 65/445
Caribbean, 49/269
catecholamine, 6/351-353
caudate nucleus, 49/318
caudate nucleus atrophy, 60/671
cause, 6/379
cerebellum, 6/340, 49/320
chemical pathology, 49/255
cholecystokinin, 49/258, 298
choline acetyltransferase, 42/227, 49/75, 258
chorea-acanthocytosis, 49/292
chorea electrica, 6/301, 49/293
chorée variable des dégénérés, 49/294
choreoathetosis, 49/385
choriopathy, 49/281
choriophrenia, 49/281
chronic psychotic choreoathetosis, 6/355
classification, 46/205
clinical features, 6/307-314, 49/277
computer assisted tomography, 49/288
constructional dyspraxia, 46/308
contravoluntary movement, 49/279
corpus callosum dyspraxia, 49/280
cortical architecture, 49/318
corticostriate fiber, 6/399, 403
Creutzfeldt-Jakob disease, 6/746, 49/293
CSF, 6/348
cytochrome oxidase aa1, 49/262
deafness, 49/291
delusion, 49/283
dementia, 27/492-496, 42/225, 46/130, 305-309, 49/281-283, 59/410
dentatorubropallidoluysian atrophy, 49/437, 60/614
dermatoglyphics, 6/300, 49/284
differential diagnosis, 6/353-356, 49/292
dipeptidyl carboxypeptidase I inhibitor, 65/445
dipeptidyl carboxypeptidase I inhibitor intoxication, 65/445
disease duration, 6/305, 49/274, 284
disease progression phase, 6/380
dopamine, 49/75, 257
dopamine receptor, 49/259, 295
dopamine receptor type D1, 49/260
dopamine receptor type D2, 49/260
dorsal paramedian nucleus, 6/331
dysarthria, 49/277
dysphagia, 49/280
dystonia, 6/533, 555, 49/280
dystonia musculorum deformans, 42/215, 49/524
EEG, 6/314, 49/288
electron microscopy, 6/337
emaciation, 6/305
EMG, 49/73
enalapril intoxication, 65/446
endocrine disorder, 42/225
endocrine parameter, 49/290
enzyme metabolism, 6/350
EOG, 49/289
epidemiology, 6/303, 49/268
epilepsy, 15/339, 72/129, 131
etiologic classification, 46/205
etiology, 6/302
event related potential, 49/290
evoked potential, 49/289
eyelid apraxia, 49/279
familial biotype, 6/306
familial inverted choreoathetosis, 49/293
familial pallidonigral degeneration, 49/293
familial spastic paraplegia, 22/430, 455
fasciculus subcallosus, 6/404
fiber change, 6/399-408
forbidden clone, 6/381
free fatty acid, 49/261
frontal cortex, 49/317
frontal lobe syndrome, 46/432
GABA, 29/505, 42/227, 49/74, 257
GABA receptor, 49/259
GABA synthesis rate, 49/290
gait ataxia, 49/280
gaze apraxia, 6/322
gaze paralysis, 6/322, 358
gene transmission, 6/323
general anesthesia, 49/284
genetics, 6/380, 49/273
global saccadic palsy, 55/123
globus pallidus, 6/3, 337, 49/319
glucosamine-6-phosphate synthesis, 49/290
glutamate decarboxylase, 42/227, 49/74, 256
glutamate dehydrogenase, 49/261
glutamic acid, 49/261
glutamine synthetase, 49/262
glycine, 6/348
glycosaminoglycan, 6/350
gonadorelin, 49/263
grading system, 49/315
growth control, 6/382-386
growth hormone, 49/260, 262, 290
Hallervorden-Spatz syndrome, 6/337, 49/293, 66/714
hastening phenomenon, 49/68

hearing loss, 6/316
heart atrioventricular block, 49/284
hereditary, 6/302, 304
hereditary benign juvenile chorea, 49/292
hereditary deafness, 49/291
hereditary dystonic paraplegia, 59/346
hereditary hemorrhagic telangiectasia (Osler), 49/291
hereditary paroxysmal kinesigenic choreoathetosis, 6/355, 49/293
hereditary sensory and autonomic neuropathy type I, 49/291
heterogeneity, 49/274-276
heterogeneous disease, 6/391-394
heterozygote frequency, 42/226
hirsutism, 49/291
history, 6/298-302, 49/267
hydrocephalus, 42/226
hyperreflexia, 42/225
hypotonia, 49/71
ibotenic acid, 49/58, 261
ichthyosis, 49/291
India, 49/269
inferior olivary nucleus, 49/320, 323
intelligence, 46/307
involuntary movement, 2/504
Japan, 49/269
Jews, 49/270
juvenile, *see* Juvenile Huntington chorea
juvenile hereditary benign chorea, 49/292, 335
juvenile hereditary diurnal fluctuating dystonia, 49/293
kainic acid, 49/58, 261
kainic acid neurotoxicity, 49/297
karyorrhexis, 49/325
kinesimetry, 49/289
Klüver-Bucy syndrome, 49/291
Lake Maracaibo, 49/270
language, 46/308
learning, 46/307
leprosy, 49/291
leptomeninges, 49/316
levodopa provocation test, 49/290
lipid metabolism, 6/350
α-2-macroglobulin, 6/380, 395
Madelung deformity, 49/291
marginal gliosis, 6/334
medium spiny neuron, 49/256
Melanesia, 49/269
memory, 46/307
mental deficiency, 46/45
mental manifestation, 49/281
metal metabolism, 6/342-347
metenkephalin, 49/258, 298
N-methyl-D-aspartic acid, 49/58, 261
m-*O*-methylation, 49/285
p-*O*-methylation, 49/285
methyldopa, 37/440
migraine, 5/80, 49/291
milking grip, 49/280
mimical apraxia, 6/328, 49/280
modifying gene, 6/395
Morvan fibrillary chorea, 49/293
motor symptom, 49/277
movement type, 6/157-159
muscarinic acetylcholine receptor, 49/295
muscle contraction, 49/277
mutation, 49/292
mutation frequency, 49/273
myoclonus, 21/514, 49/618, 60/662
natural history, 6/304-306
Negro, 49/269
neurofibromatosis, 6/315, 14/492, 49/291
neurogenic muscular atrophy, 6/315
neuromedin B, 49/296
neuromedin K, 49/296
neuron-astrocyte ratio, 49/319
neuronal loss, 49/318
neuronal nucleus membrane, 49/323
neuropathology, 6/329-342, 49/256, 315
neuropeptide Y, 49/256, 260, 319
neuropeptide Y neuron, 49/298
neuropsychologic test, 46/307-309, 49/283
neurosis, 46/306
neurotensin, 49/258
neurotransmitter, 42/227
nicotinate nucleotide pyrophosphorylase (carboxylating), 49/298
nonhereditary chorea, 6/306
nuclear magnetic resonance, 49/288
nucleus accumbens, 49/318, 323
oculomotor apraxia, 6/308, 49/280
olivopontocerebellar atrophy, 49/293
orthostatic hypotension, 63/147
osteitis deformans (Paget), 6/316, 49/291
ouabain sensitive adenosine triphosphatase, 49/261
pallidal atrophy, 6/645-647
pallidoluysionigral degeneration, 49/459
pallidonigral degeneration, 6/356, 49/293
pallidopyramidal degeneration, 6/356
paradoxical contraction (Westphal), 6/310, 49/280
paranoia, 49/282
parathyroid adenoma, 49/291
parkinsonism, 42/226
patchy depletion, 49/318

pathology, 6/57-59
pathophysiology, 6/139
PEG, 6/314, 49/288
Pelizaeus-Merzbacher disease, 49/293
personality, 46/306
personality change, 49/282
pharmacologic model, 49/294
phosphofructokinase, 49/261
Pick disease, 49/291, 293
picolinic acid, 49/298
platelet dopamine uptake, 49/290
platelet 5-hydroxytryptamine uptake, 49/290
poikilotonia, 49/280
polycythemia vera, 6/316
positron emission tomography, 49/42, 262, 288
prevalence, 6/302, 42/226, 49/270-272
progeric senescence, 49/284
progressive muscular dystrophy, 49/291
prolactin, 49/263
prolactin secretion, 49/290
protein metabolism, 6/349
protirelin, 49/258, 263
psychiatric disorder, 42/225
psychiatric symptom, 6/311-314
psychopathic behavior, 49/283
psychopathology, 46/306
Purkinje cell, 49/320
putamen, 49/318
pyruvate decarboxylase, 49/262
pyruvic acid, 6/349
quantitative study, 49/315
quinolinic acid, 49/58, 261, 298
random initiating event, 6/380
readiness potential, 49/74
reduced neostriatal GABA, 49/297
reduced neostriatal glutamic acid decarboxylase, 49/297
reduced nicotinamide adenine dinucleotide phosphate dehydrogenase, 49/260, 298, 319
reproductive fitness, 42/226
Rett syndrome, 49/293
ribosomal system, 49/325
rigid, *see* Rigid Huntington chorea
rigidity, 49/69
schizophrenia, 46/306, 49/282
scleroderma, 49/291
Scotland, 49/270
seizure, 42/226
self-negligence, 49/283
senile chorea, 6/159
serotonin receptor, 49/295
Simola test, 6/348
small-large neuron ratio, 49/318
somatic cell study, 49/291
somatic mutation, 6/380
somatosensory evoked potential, 49/289
somatostatin, 49/75, 256, 260
somatostatin like immunoreactivity, 49/258
somatostatin neuron, 49/298
South Africa, 49/269
South America, 49/269
spastic paraplegia, 49/291
speech disorder in children, 42/225
spinal cord, 6/340-342, 49/320
spinal muscular atrophy, 49/291
statokinesimetry, 49/289
status epilepticus, 15/339
status subchoreaticus, 49/287
stereoencephalotomy, 6/360
stereotypy, 49/277
stimulus myoclonus, 6/153
striatal cell depletion, 6/335-337
striatal fiber, 6/399
striatal necrosis, 49/509, 511
striatal syndrome, 6/322
striatonigral degeneration, 6/355, 697, 699, 49/206, 293
striatopallidocorticodentate calcification, 49/293
striatopallidodentate calcification, 49/293
striosome, 49/257
Stroop test, 46/309
subcortical dementia, 46/312-314, 378
substance K, 49/296, 298
substance P, 49/75, 258, 260, 296
substantia nigra, 6/407, 49/320
subthalamic nucleus, 49/320
suicide, 6/313
supranuclear ophthalmoplegia, 60/656
Sydenham chorea, 6/354
symptomatic dystonia, 6/555
syphilis, 6/315
syringomyelia, 6/315, 49/291
thalamic microneuron, 49/319
thalamus, 2/491, 6/338, 49/319
thalamus degeneration, 21/598
threonine, 6/348
treatment, 6/356-360, 49/294
tremor, 6/328, 42/226, 49/293
tremor dysarthria ataxia syndrome, 49/293
tyrosine hydroxylase, 49/75
ultrastructure, 49/315, 322
USSR, 49/269
variability, 49/274
vasoactive intestinal polypeptide, 49/258
vasopressin, 49/258
ventricular dilatation, 42/226

vertical gaze palsy, 49/280
visual evoked response, 49/289
vitamin PP, 49/298
voice, 49/277
weight loss, 49/261
Wernicke-Korsakoff syndrome, 45/199
Westphal type, *see* Rigid Huntington chorea
Huntington disease, *see* Huntington chorea
Hurler disease, *see* Glycosaminoglycanosis type IH
Hurler-Hunter disease
see also Glycosaminoglycanosis *and* Glycosaminoglycanosis type I
biochemical abnormality, 10/341
claw hand, 10/436
clinical features, 10/341, 434-442
corneal opacity, 10/341, 436
deafness, 10/436
dolichocephaly, 10/436
dwarfism, 10/341, 436
glycosaminoglycan, 10/341
hepatosplenomegaly, 10/436
hirsutism, 10/436
hypertelorism, 10/436
hypertrichosis, 10/436
inguinal hernia, 10/436
kyphosis, 10/436
mental deficiency, 10/341
short neck, 10/436
trunk, 10/436
umbilical hernia, 10/436
Hurler polydystrophy
pseudo, *see* Mucolipidosis type III
Hurler-Scheie disease, *see* Glycosaminoglycanosis type IH-S
Hurler syndrome variant, *see* GM1 gangliosidosis
Hurst disease, *see* Acute hemorrhagic leukoencephalitis
Hutchinson-Gilford syndrome, *see* Progeria
Hutchinson-Laurence-Moon syndrome
Friedreich ataxia, 13/381
spastic paraplegia, 13/381, 59/439
Hutchinson-Tay choroiditis, *see* Hutchinson-Tay choroidopathy
Hutchinson-Tay choroidopathy
primary pigmentary epithelium degeneration, 13/30, 138
Stargardt disease, 13/133
Hutchinson teeth
congenital syphilis, 33/371
Treponema pallidum, 33/371
HVA, *see* Homovanillic acid
HY antigen
XX male, 43/556
Hyaline
collagen type, *see* Arteriolar fibrinohyalinoid degeneration
hypertensive fibrinoid arteritis, 11/136
Hyaline angionecrosis, *see* Arteriolar fibrinohyalinoid degeneration
Hyaline arterionecrosis, *see* Arteriolar fibrinohyalinoid degeneration
Hyaline body, *see* Neurofibrillary tangle
Hyaline degeneration, *see* Arteriolar fibrinohyalinoid degeneration
Hyaline macular dystrophy, *see* Hutchinson-Tay choroidopathy
Hyaline vessel wall degeneration, *see* Arteriolar fibrinohyalinoid degeneration
Hyalinosis, *see* Arteriolar fibrinohyalinoid degeneration
Hyalinosis cutis et mucosae, *see* Urbach-Wiethe disease
Hyaloid artery
persistent, *see* Persistent hyaloid artery
Hyaloidoretinal degeneration (Wagner)
astigmatism, 13/37
autosomal dominant inheritance, 13/37, 273
case history, 13/274-276
cataract, 13/274
choroid atrophy, 13/274
choroid sclerosis, 13/274
divergent strabismus, 13/274
EOG, 13/274
ERG, 13/274
glaucoma, 13/37, 274
myopia, 13/37, 274
primary pigmentary retinal degeneration, 13/37, 178
ring scotoma, 13/274
visual acuity, 13/274
vitreoretinal degeneration, 13/37
vitreous body, 13/273
vitreous opacity, 13/178
Hyaloidotapetoretinal degeneration (Goldmann-Favre)
autosomal recessive inheritance, 13/37
cataract, 13/37, 276
central scotoma, 13/276
hereditary vitreoretinal degeneration, 13/178, 276
macular degeneration, 13/37, 276
nyctalopia, 13/37, 276
primary pigmentary retinal degeneration, 13/37, 178, 276
retinoschisis, 13/276
ring scotoma, 13/276
visual adaptation, 13/37

vitreous body, 13/276
vitreous opacity, 13/178
Hyaluronic acid
chemistry, 27/144
CNS myelin, 9/31
connective tissue, 10/432
glycosaminoglycanosis, 10/432
intervertebral disc, 20/526
Hyaluronidase
arthropod envenomation, 65/198
infantile acid maltase deficiency, 41/180
tuberculous meningitis, 52/213
Hycodan, *see* Hydrocodone
Hydantoin
epilepsy, 15/627, 670
serum level, 15/684, 686, 690
side effect, 15/711
Hydantoin intoxication
anticonvulsant, 37/200
Hydantoin syndrome
fetal, *see* Fetal hydantoin syndrome
fetal intoxication, *see* Fetal hydantoin syndrome
Hydatid cyst
cerebral, *see* Cerebral hydatid cyst
classification, 68/310
spinal epidural cyst, 20/164
Hydatid cyst disease, 35/175-207
see also Echinococcosis
cerebral, 35/183-193
spinal, 35/193-202
vertebral, 35/194-198
Hydatid disease
cranial, *see* Cranial hydatid disease
Echinococcus, 52/523
vertebral, *see* Vertebral hydatid disease
Hydatidosis
epilepsy, 72/168
helminthiasis, 35/217
tropical myeloneuropathy, 56/526
Hydergin, *see* Co-dergocrine
Hydralazine
action mechanism, 37/430
antihypertensive agent, 37/430, 63/83
antihypertensive agent intoxication, 37/431
antiphospholipid antibody, 63/330
arterial hypertension, 54/209, 63/74
brain infarction, 63/74
chemical structure, 37/430
drug induced systemic lupus erythematosus, 16/222
hypertensive encephalopathy, 54/216
iatrogenic neurological disease, 65/446
migraine, 5/56
neurotoxin, 37/430, 42/647, 51/68, 63/74, 83, 87, 65/446
slow acetylator, 63/87
toxic polyneuropathy, 51/68
vitamin B6 deficiency, 63/87, 70/430
Hydralazine intoxication
anxiety, 37/431
dizziness, 37/431
flushing, 37/431
headache, 37/431
neurologic adverse effect, 65/447
palpitation, 37/431
polyneuropathy, 7/519, 37/432, 65/447
serum sickness, 37/431
sodium retention, 37/431
systemic lupus erythematosus like syndrome, 37/431
vitamin B6 deficiency neuropathy, 65/447
water retention, 37/431
Hydralazine neuropathy
fever, 42/647
headache, 42/647
sensory symptom, 7/532
systemic lupus erythematosus, 42/647
vitamin B6 deficiency, 42/467, 51/297, 334
Hydralazine polyneuropathy
chemistry, 51/297
vitamin B6 deficiency, 63/87
Hydramnios
iniencephaly, 50/135
myotonic dystrophy, 40/490
Pena-Shokeir syndrome type I, 59/71
Hydranencephaly, 30/661-677
Akabane virus, 34/301, 50/340
anencephaly, 50/347
animal viral disease, 34/301
arthrogryposis, 50/340, 345
bacterial meningitis, 50/337
basal ganglion, 30/673, 50/347
birth incidence, 42/34
bluetongue virus, 50/340
brain stem tegmentum, 30/673, 675, 50/347
capillary proliferation, 50/345
cerebellar hemorrhage, 50/340
cerebellum, 30/671, 50/347
cerebral membrane, 30/673, 50/347
classification, 50/345, 347
clinical features, 30/675, 50/349
computer assisted tomography, 50/351
corpus callosum agenesis, 50/159
Coxsackie virus B, 50/340
cranial transilluminability, 30/674, 50/346, 351
cyclopia, 50/347

cytomegalic inclusion body disease, 56/268
cytomegalovirus, 56/268
Dandy-Walker syndrome, 50/344
decerebrate rigidity, 30/675, 50/349
destructive process, 50/342
diagnosis, 50/350
epidemiology, 30/661
Escherichia coli meningitis, 50/342
etiology, 50/337
familial type, 50/345
Galen vein, 30/665
genetic factor, 30/669, 50/342
glia reaction, 50/338
herpes simplex virus, 50/341
holoprosencephaly, 50/347
hydrocephalus, 30/668, 50/341
hypoxic ischemia, 50/338
incidence, 50/53, 63
infantile hypotonia, 50/349
internal carotid artery, 50/338
intraventricular hemorrhage, 50/337
lissencephaly, 30/667, 50/337, 340, 342, 347
mental deficiency, 46/11
microcephaly, 50/345
microgyria, 50/337, 342
microhydrocephaly, 50/345
microphthalmia, 30/676, 50/342
Moro reflex, 30/675, 50/349
multiple cystic encephalopathy, 30/665, 50/341, 346
myoclonus, 30/675, 50/349
nosologic place, 30/670, 50/345
occurrence time, 30/669, 50/345
onset timing, 50/337
pathogenesis, 30/662, 50/338
pathogenetic mechanism, 50/340
pathologic finding, 50/339
periventricular necrosis, 50/347
placental abruption, 50/338
polyploidy, 50/555
porencephaly, 30/682, 684, 50/340, 346, 356, 358
prenatal vascular disease, 53/27
respiratory failure, 30/675, 50/349
Rift Valley fever, 50/340
schizencephaly, 30/665-667, 50/198, 342, 347
seizure, 30/675, 50/349
sex difference, 50/56
spinal cord, 30/671, 50/347
subcortical heterotopia, 50/337, 342
survival, 50/349
terminology, 30/661
thalamus, 30/670, 50/346
Togavirus, 50/340
toxoplasmosis, 50/337
true porencephaly, 30/684
ultrasound scan, 50/351
in utero viral infection, 50/337, 340
vascular lesion, 30/662, 50/338
vascular system, 50/338
ventriculoatrial shunt, 50/351
viral encephalitis, 50/337
Hydratropic acid
migraine, 48/175
Hydrazine
chemical formula, 64/82
neurotoxin, 36/380, 64/3, 82
organic solvent, 64/82
toxic encephalopathy, 64/3
vitamin B6 metabolism, 70/430
Hydrazine intoxication
clinical features, 36/381
prognosis, 36/382
treatment, 36/382
Hydrencephaly, *see* Hydrocephalus
Hydrencephaly syndrome
hydrocephalus, 42/35
Hydrocarbon
brain tumor carcinogenesis, 17/6-8
Hydrocephalic dementia, 46/323-331
akinetic mutism, 46/328
Alzheimer disease, 46/326, 329
arrested hydrocephalus, 46/324
ataxia, 46/327
brain cyst, 46/326
clinical features, 46/326-328
computer assisted tomography, 46/328
CSF shunt, 46/329
diagnosis, 46/328
EEG, 46/328
fecal incontinence, 46/328
gait, 46/327
gait disturbance, 46/327
hydrocephalus ex vacuo, 46/326
intracranial pressure, 46/329
intracranial pressure monitoring, 46/329
isotope cisternography, 46/328
lumboperitoneal shunt, 46/329
memory, 46/327
mental change, 46/327
multi-infarct dementia, 46/326
normal pressure hydrocephalus, 46/325
obstructive hydrocephalus, 46/324
pathogenesis, 46/323-326
PEG, 46/328
Pick disease, 46/326
prognosis, 46/330

reflex disorder, 46/327
subarachnoid hemorrhage, 46/330
subcortical, 46/205
treatment, 46/329
urinary incontinence, 46/328
ventriculoatrial shunt, 46/329
Hydrocephalus, 30/661-677, 50/285-298
see also Spina bifida
abducent nerve paralysis, 50/291
acetazolamide, 50/296
achondroplasia, 50/289, 395
acquired obstruction, 50/286
acrocephalosyndactyly type I, 43/320, 50/122
acrocephalosyndactyly type II, 50/123
acrodysostosis, 30/100, 31/230, 43/291
air embolism, 30/579
Alexander disease, 10/209
aminopterin syndrome, 30/122
anencephaly, 30/181, 42/36
animal model, 30/662, 50/289, 427
antenatal diagnosis, 50/297
aqueduct forking, 30/538, 609, 613, 32/532, 540, 50/287
aqueduct membranous occlusion, 30/538
aqueduct obstruction, 32/529
aqueduct stenosis, 30/538, 609-622, 42/36, 56-58, 50/285-287, 306, 309
arachnoid cyst, 30/541
argininosuccinic aciduria, 50/289
Arnold-Chiari malformation type I, 30/162, 538, 32/103, 42/15, 50/254
Arnold-Chiari malformation type II, 50/405, 407
arousal disorder, 45/112
arrested, *see* Arrested hydrocephalus
astrocytosis, 50/290
athetotic hemiplegia, 42/200
atlanto-occipital synostosis, 50/393
autosomal recessive, 50/289
axial mesodermal dysplasia spectrum, 50/516
axonal degeneration, 50/290
B-mode sonography, 54/36
bacterial meningitis, 52/13
basal ganglion, 30/673
basilar impression, 32/63
benzodiazepine, 50/297
Bickers-Adams syndrome, 30/610
birth incidence, 42/35
birth order, 50/288
Bonnet-Dechaume-Blanc syndrome, 14/262, 43/36
brain aneurysm neurosurgery, 55/55
brain blood flow, 54/147
brain death, 24/744
brain edema, 47/535
brain remnant, 30/674
brain scintigraphy, 23/294
brain stem, 30/673, 32/531
brain tumor, 31/40, 46/324-326
Burkitt lymphoma, 63/351
Candida meningitis, 52/400
cardiac catheter obstruction, 30/586
cardiac tamponade, 30/597
cardiotonic steroid agent, 50/297
carotid angiography, 30/549-550
cat, 32/550
catheter detachment, 30/586
cavum vergae, 2/767
cephalocele, 30/212
cerebellar hypoplasia, 42/18
cerebellar medulloblastoma, 42/742
cerebellum, 30/671
chlormethine intoxication, 64/233
choroid plexus extirpation, 30/569
choroid plexus fibrosis, 50/290
choroid plexus papilloma, 17/570, 30/540, 50/286
choroid plexus pulsation, 32/533
choroid plexus radiation, 30/566
classification, 30/536-538, 670, 46/323, 50/285
clinical features, 30/547, 549, 675-677, 50/290
cloverleaf skull syndrome, 38/422
CNS, 30/600
CNS malformation, 50/57
Cobb syndrome, 14/438
coccidioidomycosis, 52/412, 415
Coffin-Lowry syndrome, 31/245, 43/239
colloid cyst, 30/541, 543
complication, 30/579-600
computer assisted tomography, 30/555, 50/293
congenital ichthyosis, 43/406
congenital obstruction, 50/285
congenital rubella, 50/278
congenital toxoplasmosis, 35/130, 42/662, 52/358
contrast radiography, 30/584
convulsion, 30/555
Cordis-Hakim valve, 30/575-578
corpus callosum agenesis, 2/764, 30/287, 42/7, 50/159, 161
corpus callosum perforation, 30/568
cracked pot sign, 18/545, 50/292
craniolacunia, 30/271, 50/139
craniosynostosis, 30/228
cri du chat syndrome, 31/566, 43/501
cryptococcal meningitis, 52/432
CSF, 30/556, 32/528, 531, 42/35
CSF fistula, 24/186
CSF oversecretion, 50/286

CSF pressure, 30/532-534
CSF shunt, 30/573-575
cytomegalovirus infection, 31/313, 315, 56/266
Dandy-Walker syndrome, 30/539, 623-646, 31/110-112, 32/59, 42/21, 36, 50/286, 324
decreasing CSF production, 30/567
demyelination, 50/290
dental abnormality, 42/691
developmental pathology, 32/533, 544
diagnosis, 30/545-548, 32/550-552, 50/292
diastematomyelia, 50/441
diencephalic tumor, 67/156
diplomyelia, 50/441
Divry-Van Bogaert syndrome, 14/513
drug treatment, 50/296
dwarfism, 43/385
dye ventriculography, 30/553
dyssynergia cerebellaris progressiva, 42/212
echoEG, 30/548
education, 30/601
EEG, 30/558
Ekbom other syndrome, 63/85
Ellis-Van Creveld syndrome, 43/348
embryology, 30/662
EMG, 32/548-550
encephalocele, 50/286
encephalomeningocele, 42/27
environmental factor, 30/535
epidemiology, 30/149-151, 157, 163, 534, 661
epilepsy, 30/600, 42/691
etiology, 50/287
exertional headache, 48/374
experimental, 30/534, 32/532
familial, *see* Familial hydrocephalus
fetal alcohol syndrome, 50/279
α-fetoprotein, 30/179
fontanelle, 32/541
foramina parietalia permagna, 30/277, 42/29
furosemide, 50/297
Galen vein, 30/543, 72/210
Galen vein aneurysm, 18/282
gas ventriculography, 30/552
general paresis of the insane, 52/279
genetics, 50/289
glomerulonephritis, 30/596
glycosaminoglycanosis type I, 10/441
glycosaminoglycanosis type IH, 66/297
glycosaminoglycanosis type II, 42/454
glycosaminoglycanosis type III, 66/302
glycosaminoglycanosis type VI, 42/458, 66/309
Hallermann-Streiff syndrome, 31/248
headache, 48/32, 34
hemodynamic factor, 30/578
herpes virus hominis syndrome, 31/212
history, 30/661, 32/528
holoprosencephaly, 42/32, 50/237
Holter shunt system, 30/576
Huntington chorea, 42/226
hydranencephaly, 30/668, 50/341
hydrencephaly syndrome, 42/35
hydromyelia, 50/425
inadequate valve function, 30/580, 585
inappropriate antidiuretic hormone secretion, 28/499
incidence, 50/53, 62, 287
incontinentia pigmenti, 14/85, 30/100
increased intracranial pressure, 50/291
infected shunt, 30/595
infective organism, 30/594
iniencephaly, 30/258, 50/131
intelligence, 30/600
intracranial cyst, 31/97, 100, 105, 117, 119
intracranial hypertension, 67/144
intracranial infection, 24/229
intracranial nervous hypertension, 30/539
intracranial pressure, 30/547, 556-558
kaolin, 50/427
laryngeal abductor paralysis, 42/319
latent, 16/242
Laurence-Moon-Bardet-Biedl syndrome, 31/328
lead encephalopathy, 64/437
lissencephalocele, 50/286
lissencephaly, 30/485, 50/258
Listeria monocytogenes encephalitis, 52/95
Lowe syndrome, 42/607
lowering ventricular CSF pressure, 30/566
lumboperitoneal drainage, 30/571
lymphocytic choriomeningitis, 9/198, 34/198, 50/308, 56/360
macrencephaly, 30/658, 42/41
management, 30/677
median cleft lip hypotelorism syndrome, 50/236
meningeal leukemia, 63/341
meningitis, 50/288
meningococcal meningitis, 33/23, 52/28
mental deficiency, 30/600, 32/552, 46/8, 50/291
microcephaly, 30/510, 50/278
Mohr syndrome, 43/450
multiple cystic encephalopathy, 30/665
multiple neurofibromatosis, 43/34
multiple nevoid basal cell carcinoma syndrome, 14/79, 464, 31/26
mumps virus, 9/551
neurocutaneous melanosis, 14/119, 414, 418, 31/26, 43/33
neurofibromatosis type I, 14/146

neurosarcoidosis, 38/530, 63/422
neurosyphilis, 52/277
normal development, 30/677
normal pressure, *see* Normal pressure hydrocephalus
North American blastomycosis, 35/403, 52/388
nuclear magnetic resonance, 50/292
obstructive, *see* Obstructive hydrocephalus
occipital cephalocele, 50/101
occult communicating internal, 16/245
Ommaya reservoir, 30/552
optic atrophy, 13/69
optic function, 30/676
orofaciodigital syndrome type I, 30/100, 31/252, 43/448
orthostatic hypotension, 22/262, 63/156
osteitis deformans (Paget), 38/366, 46/400, 70/14
osteogenesis imperfecta, 30/100, 43/454
otitic, *see* Otitic hydrocephalus
otogenic, *see* Otogenic hydrocephalus
4p partial monosomy, 43/497
4p partial trisomy, 43/500
4p syndrome, 31/557
18p syndrome, 31/579, 582
parietal cephalocele, 50/108
parietal encephalocele, 50/108
partial, 30/677
pathogenesis, 30/662-669, 32/529-533
pathology, 30/534, 663, 670-675, 32/531, 544, 50/289
platybasia, 50/286
pneumococcal meningitis, 52/51
polyploidy, 50/555
porencephaly, 30/158, 684, 42/48
Portnoy valve, 30/576
positive contrast ventriculography, 30/554
posterior fossa hematoma, 57/157, 254
posterior groove occlusion, 30/538
posthemorrhagic, 30/543-545
postmeningitic, 30/545
Potter syndrome, 43/469
premature craniostenosis, 30/596
prenatal infection, 50/288
prognosis, 32/552-554, 50/296
progressive, 30/525-560
progressive cerebellar dyssynergia, 42/211
Pudenz-Heyer drain, 30/577
pyruvate dehydrogenase complex deficiency, 66/416
13q partial monosomy, 43/526
r(13) syndrome, 50/585
r(21) partial syndrome, 43/543
radiation exposure, 50/288
radioisotope ventriculography, 30/553
radiologic appearance, 30/547, 32/550
recurrence risk, 30/536
renal agenesis syndrome, 50/513
rheumatoid arthritis, 71/29
Roberts syndrome, 30/100
Robin syndrome, 43/472
rubella syndrome, 31/212
sarcoidosis, 38/530, 71/476, 488, 490
schizencephaly, 30/339, 345, 662
seizure, 30/555, 42/691, 50/291
septum pellucidum agenesis, 30/314-321, 326, 328
septum pellucidum fenestration, 50/290
setting sun sign, 30/546
sex linked recessive aqueduct stenosis, 50/62
sex ratio, 50/62
shunt choice, 50/294
shunt dependent, 30/598
shunt device, 50/293
shunt infection, 50/295
shunt revision, 50/296
shunted patient, 30/583-585
sincipital encephalocele, 50/103
skin flap decubitus, 30/579
skull, 30/600
spina bifida, 30/157, 535, 32/528-533, 535, 42/35, 50/287, 482-485, 492
spinal cord, 30/671-673
spinal lipoma, 20/398
spongy edema, 50/290
streptococcal meningitis, 52/81
subarachnoid hemorrhage, 12/152, 179, 55/5, 8, 55
subdural hematoma, 30/583
superior longitudinal sinus, 12/427
supratentorial midline tumor, 16/112
surgical infection, 30/579, 593-595
surgical treatment, 30/569, 573, 579, 590
suture, 32/541
syringomyelia, 50/446
technical aid, 30/552
temporal lobe agenesis syndrome, 31/105, 50/289
teratogenic agent, 34/383
thalamus, 30/670
thanatophoric dwarfism, 31/274, 43/388
third ventriculostomy, 30/568
time at onset, 30/669
toxic, *see* Toxic hydrocephalus
Toxoplasma gondii syndrome, 31/212
toxoplasmosis, 30/545, 35/130, 50/277, 52/358
transillumination, 30/548, 50/292
transverse sinus thrombosis, 52/176

traumatic, *see* Traumatic hydrocephalus
treatment, 30/565-607, 50/293
triploidy syndrome, 50/289
trisomy 8 mosaicism, 43/511
trisomy 8 syndrome, 31/528
trisomy 9, 50/564
trisomy 22, 50/565
tuberculous meningitis, 30/545, 33/206, 52/200, 202, 207, 214, 55/431
tuberous sclerosis, 31/2, 6
type, 50/285
unblocking CSF pathway, 30/567
VACTERL syndrome, 50/514
valve system, 30/573-578
valve system overfunction, 30/581-583
vascular malformation, 30/543, 31/137, 145, 177, 193
vena cava thrombosis, 30/587-590
venous sinus thrombosis, 12/427
ventricular catheter obstruction, 30/590-593
ventricular diverticulum, 50/290
ventricular hemorrhage, 30/579
ventriculocisternostomy, 30/568
ventriculography, 30/550-555
ventriculolumbar ureterostomy, 30/573
ventriculoperitoneal drainage, 30/570
ventriculopleural drainage, 30/569
viral infection, 50/289
virus induced, 34/383
Whipple disease, 52/138
white matter atrophy, 50/290
X-linked, *see* X-linked hydrocephalus
yellow teeth, 42/691
Hydrocephalus ex vacuo
hydrocephalic dementia, 46/326
Hydrochlorothiazide
antihypertensive agent, 63/87
brain edema, 16/204
Hydrocodeine, *see* Dihydrocodeine
Hydrocodone
chemistry, 37/365
opiate, 65/350
Hydrocortisone
biological rhythm, 74/506
brain injury, 23/120, 122
brain tumor, 16/344
cluster headache, 48/240
Cushing syndrome, 70/206
depression, 46/449, 74/506
migraine, 5/380
plasma, *see* Plasma hydrocortisone
progressive dysautonomia, 59/154
Reye syndrome, 56/152
Hydrocortisone half-life
anorexia nervosa, 46/586
Hydrocyanic acid
trigeminal neuralgia, 5/299
Hydrodynamic theory
hydromyelia, 50/428
Hydrogen sulfide intoxication
headache, 48/420
Hydroids intoxication, *see* Coelenterata intoxication
Hydrolagus intoxication
family, 37/78
venomous spine, 37/78
Hydrolase
acid maltase deficiency, 41/179
Hydromania
pellagra, 7/573
Hydromorphone
chemistry, 37/365
opiate, 65/350
Hydromyelia, 32/231-236, 50/425-431
see also Syringobulbia *and* Syringomyelia
acrocephalosyndactyly type I, 50/428
age at onset, 50/429
amyotrophy, 50/429
animal model, 50/427
areflexia, 50/429
Arnold-Chiari malformation type I, 32/104, 231, 50/425, 428
Arnold-Chiari malformation type II, 50/407
ataxia, 50/429
cerebellar heterotopia, 50/425
clinical features, 32/236, 50/429
communicating syringomyelia, 50/425
complicated, *see* Complicated hydromyelia
computer assisted tomography, 50/430
cranial nerve palsy, 50/429
CSF protein, 50/428
Dandy-Walker syndrome, 42/23, 50/425
definition, 32/231, 50/425
diagnosis, 50/430
dissociated sensory loss, 50/429
embryology, 32/232-234, 50/426
encephalocele, 50/426
encephalomeningocele, 42/26
epidemiology, 30/150
etiology, 50/427
exomphalos, 50/427
foramen magnum stenosis, 50/428
Gardner theory, 50/428
genetics, 50/427
history, 32/231, 50/425
hydrocephalus, 50/425
hydrodynamic theory, 50/428

kaolin, 50/427
kyphoscoliosis, 50/425, 429
limb atrophy, 50/429
malformation, 32/231, 234
monomelic spinal muscular atrophy, 59/376
myelocystocele, 32/235, 50/427
neural tube closure failure, 50/426
neurotropic agent, 50/427
nuclear magnetic resonance, 50/430
obex plugging, 50/431
pain, 50/429
paraplegia, 50/429
patent central canal, 50/427
pathogenesis, 32/232, 50/427
pathology, 32/234-236, 50/427
progressive, *see* Progressive hydromyelia
shunt procedure, 50/431
spina bifida, 50/486
spina bifida cystica, 50/425
spina bifida occulta, 50/425
stage, 50/426
surgical decompression, 50/431
syringomyelia, 32/235, 256, 42/60, 50/425, 457
syringomyelocele, 50/427
treatment, 32/236, 50/430
valve mechanism, 50/428
viral infection, 50/427
Hydronephrosis
anencephaly, 50/84
aneuploidy, 50/553
Patau syndrome, 31/508, 50/558
trisomy 9, 50/564
Hydrophiidae intoxication
fatality, 37/91
Hydrophis cyanocintus intoxication
antivenene, 37/94
Hydrophobia
rabies, 34/235, 253, 56/387, 75/418
Hydrophthalmia
François syndrome, 14/627, 635, 638
hemifacial hypertrophy, 14/635
phakomatosis, 14/643
Hydrops meningeus, *see* Spinal arachnoid cyst
Hydrotherapy
Hubbard tank, 8/384
Hydroureter
anencephaly, 50/84
holoprosencephaly, 50/237
Patau syndrome, 31/508, 50/558
Hydroxocobalamin, *see* Vitamin B12a
Hydroxyapatite
Guam amyotrophic lateral sclerosis, 59/265
striatopallidodentate calcification, 49/426
α-Hydroxybutyric acid
dihydrolipoamide dehydrogenase, 66/417
methionine malabsorption syndrome, 29/127, 42/600
4-Hydroxybutyric aciduria, *see* Succinate semialdehyde dehydrogenase deficiency
Hydroxycobalamin, *see* Vitamin B12a
4-Hydroxydebrisoquine
Parkinson disease, 49/133
6-Hydroxydopamine, *see* Oxidopamine
11-Hydroxydronabinol
cannabis, 65/46
tetrahydrocannabinol, 65/46
D-2-Hydroxyglutaric aciduria
clinical finding, 66/655
nuclear magnetic resonance, 66/655
organic acid metabolism, 66/639, 641
origin, 66/657
L-2-Hydroxyglutaric aciduria
clinical finding, 66/655
nuclear magnetic resonance, 66/655
organic acid metabolism, 66/639, 641
origin, 66/656
5-Hydroxyindole acetaldehyde
migraine, 48/86
5-Hydroxyindoleacetic acid
Alzheimer disease, 46/266
CSF, 45/136
daytime hypersomnia, 45/136
dystonia musculorum deformans, 49/524
headache, 48/90
migraine, 48/86, 90
Parkinson dementia, 49/174
3-Hydroxyisobutyric aciduria
valine metabolism, 66/653
α-Hydroxyisovaleric acid
dihydrolipoamide dehydrogenase, 66/417
β-Hydroxyisovaleric aciduria, *see* 3-Hydroxy-3-methylbutyric aciduria
Hydroxykynureninuria
tryptophan metabolism, 29/223
Hydroxyl apatite
pseudocalcium, 6/715
Hydroxylamine reaction
lipid ester, 9/25
11-Hydroxylase deficiency
adrenal disorder, 70/214
congenital adrenal hyperplasia, 42/514
genetic, 70/214
survey, 70/214
17-Hydroxylase deficiency
adrenal disorder, 70/215
congenital adrenal hyperplasia, 42/514

21-Hydroxylase deficiency
adrenal disorder, 70/213
childhood, 70/213
congenital adrenal hyperplasia, 42/514, 70/213
genetic, 70/214
late onset, 70/213
prenatal, 70/214
simple virilizing type, 70/213
Hydroxylignoceric acid
metachromatic leukodystrophy, 10/49
Hydroxylysinemia
biochemistry, 29/223
epilepsy, 29/223
mental deficiency, 29/223
screening test, 29/223
Hydroxylysinuria
mental deficiency, 42/704
myoclonic epilepsy, 42/704
3-Hydroxy-3-methylbutyric aciduria
spinal muscular atrophy, 59/370
3-Hydroxy-3-methylglutaric aciduria
organic acid metabolism, 66/641
Hydroxymethyl-pyridine, *see* Nicotinyl alcohol
4-Hydroxyphenylpyruvic acid
Parkinson disease, 49/120
4-Hydroxyphenylpyruvic acid oxidase
tyrosinemia, 42/635
4-Hydroxyphenylpyruvic acid oxidase deficiency
epilepsy, 72/222
Hydroxyproline
iminoacidemia, 29/129, 131
metabolism, 29/131-133
muscular dystrophy, 40/380
osteitis deformans (Paget), 38/361, 43/451
Hydroxyprolinemia
see also Iminoacidemia
biochemistry, 29/142
clinical features, 29/141
history, 29/141
incidence, 29/142
inheritance, 29/143
treatment, 29/143
3β-Hydroxysteroid dehydrogenase deficiency
adrenal disorder, 70/215
congenital adrenal hyperplasia, 42/514
steroidogenesis, 70/215
Hydroxystilbamidine
North American blastomycosis, 35/407, 52/392
2-Hydroxytetracosanoic acid
cerebroside, 9/6
sulfatide, 9/6
11-Hydroxy-Δ9-tetrahydrocannabinol, *see* 11-Hydroxydronabinol
5-Hydroxytryptamine, *see* Serotonin
5-Hydroxytryptophan, *see* Oxitriptan
Hydroxyurea
astrocytoma, 18/39
chemotherapy, 39/117
hypereosinophilic syndrome, 63/374
neurologic toxicity, 39/117
neurotoxin, 67/364
Hydrozoa intoxication
ataxia, 37/40
convulsion, 37/40
crying, 37/38
cyanosis, 37/40
death, 37/40
defecation, 37/40
delirium, 37/40
disorientation, 37/40
distress, 37/40
dyspnea, 37/38, 40
emaciation, 37/38
experimental animal, 37/40
eye injury, 37/40
flaccid, 37/40
hemorrhagic lesion, 37/40
local pain, 37/40
meiosis, 37/40
mouse, 37/40
myalgia, 37/38
necrotizing, 37/40
painful, 37/38
respiratory distress, 37/40
scar, 37/40
sequela, 37/38
shock, 37/38
skin manifestation, 37/40
ulceration, 37/38, 40
Hygroma
subdural, *see* Subdural hygroma
Hymenoptera intoxication
envenomation, 37/107-111
Hyndham sign
lumbar intervertebral disc prolapse, 20/586
Hyoid bone
obstructive sleep apnea, 63/454
snoring, 63/450
Hyoscyamine
see also Hallucinogenic agent
anticholinergic activity, 65/48
belladonna alkaloid, 65/48
biokinetic, 65/48
Datura stramonium, 65/48
hallucinogenic agent, 65/42
Hyoscyamine intoxication

hallucination, 65/48
mydriasis, 65/48
symptom, 65/48
treatment, 65/49
Hypalgesia
anterior cervical cord syndrome, 25/270
bathing trunk, *see* Bathing trunk hypalgesia
definition, 1/86
pellagra, 28/90
pure sensory stroke, 54/250
Hyperabduction maneuver
diagnosis, 2/150
hyperabduction syndrome, 2/150
thoracic outlet syndrome, 51/124
Hyperabduction syndrome
brachial plexus, 7/441
compression syndrome, 7/430, 432, 434, 441, 8/335
costoclavicular maneuver, 2/150
hyperabduction maneuver, 2/150
paresthesia, 7/441
sign, 2/148, 150
treatment, 2/148, 8/337
Hyperactive carotid sinus reflex
cardioinhibitory type, 11/538
cerebral type, 11/540
clinical aspect, 11/535
clinical features, 11/537
type identification, 11/539
vasodepressor type, 11/539
Hyperactive motoneuron disorder
WFN classification, 59/6
Hyperactivity
aggression, 46/177-180
auditory evoked potential, 46/184
brain injury, 46/182
child, 46/175-184
classification, 46/176-181
clinical features, 46/175
developmental dyslexia, 46/117
EEG, 46/183
galvanic skin response, 46/183
Gilles de la Tourette syndrome, 46/181
homovanillic acid, 46/182
hyperkinetic child syndrome, 3/195, 4/341, 363, 46/176
inferior oblique muscle, 2/308
intelligence, 46/180
language, 46/181
learning disability, 46/181
manganese intoxication, 64/306
motor unit, *see* Motor unit hyperactivity
neurologic sign, 46/182
neurophysiology, 46/183
neurotransmitter, 46/182
pathogenesis, 46/182-184
phenylketonuria, 42/611
physical abnormality, 46/183
schizophrenia, 46/485
speech, 46/181
symptom definition, 46/175
temporal lobe, 46/182
temporal lobe epilepsy, 46/182
tic, 46/181
visual evoked response, 46/184
Hyperacusis
GM2 gangliosidosis type I, 42/433
GM2 gangliosidosis type II, 42/435
Melkersson-Rosenthal syndrome, 14/790, 42/322
myasthenia gravis, 41/98
phencyclidine intoxication, 65/368
sarcoid neuropathy, 51/195
sulfite oxidase deficiency, 29/121
Hyperacute disseminated encephalomyelitis
Bordetella pertussis, 9/526
experimental allergic encephalomyelitis, 9/526-529
Hyperadrenocorticism
vitamin B2 deficiency, 51/332
Hyperagammaglobulinemic purpura
polymyositis, 62/373
Hyperalaninemia
chemistry, 29/211
Hill-Sherman syndrome, 21/578
metabolic ataxia, 21/574, 578
pyruvate carboxylase deficiency, 29/212, 42/516
pyruvate dehydrogenase deficiency, 29/212, 42/516
Hyperaldosteronism
adrenal adenoma, 39/492
adrenal gland, 39/492-496
adrenocortical hyperplasia, 39/493
brain hemorrhage, 39/496
brain infarction, 39/496
CSF, 39/496
dexamethasone, 39/493
EEG, 39/496
hematoma, 39/496
hypertension, 39/494, 496
laboratory examination, 39/493
mental symptom, 39/493
periodic paralysis, 62/457
primary, 75/475
renal artery stenosis, 39/492
sex ratio, 39/493
sign, 39/493

skeletal muscle, 40/438
spirolactone, 39/494
symptom, 39/493
transient ischemic attack, 39/496
weakness, 41/250
Hyperaldosteronism myopathy
creatine kinase, 62/536
EMG, 39/494, 62/536
hypokalemic periodic paralysis, 39/494, 41/152, 62/536
muscle biopsy, 62/536
periodic weakness, 39/494, 62/536
potassium depletion, 39/494, 62/536
Hyperalgesia
bulbopontine, 2/477
definition, 1/106
Head's zone, 2/161
malignant radiculopathy, 69/73
neuralgic amyotrophy, 51/172
neuronal mechanism, 61/125
pontine, *see* Pontine hyperpathia
reflex sympathetic dystrophy, 61/121
serotonin, 48/194
terminology, 1/86
thalamic, 2/476
Hyperalgesic diabetic neuropathy
painful, 70/136
Hyperalimentation
anorexia nervosa, 46/589
hypophosphatemia, 63/564
Hyperammonemia
acquired hepatocerebral degeneration, 6/282, 285
Alzheimer type II astrocyte, 22/577
argininosuccinic aciduria, 22/577, 42/524-526
ataxia, 42/560
brain atrophy, *see* Rett syndrome
congenital, *see* Congenital hyperammonemia
convulsion, 42/560
cyanide intoxication, 64/231
diet, 42/561
epilepsy, 72/222
familial hyperargininemia, 59/358
feeding disorder, 42/560
hepatic coma, 49/215
hepatic encephalopathy, 21/576
hereditary, *see* Hereditary hyperammonemia
hyperdibasic aminoaciduria, 42/564
ketotic, *see* Ketotic hyperammonemia
mental deficiency, 42/560
methylmalonic aciduria, 29/105
Moro reflex, 42/560
organic acid metabolism, 66/641
β-oxothiolase deficiency, 29/106
propionic acidemia, 29/105
respiratory distress, 42/560
Rett syndrome, 29/305, 325, 60/637
Reye syndrome, 21/578, 27/361, 363, 29/331, 335, 338, 34/169, 49/218
spastic diplegia, 42/560
trichorrhexis nodosa, 21/577
valproic acid induced hepatic necrosis, 65/125
Wilson disease, 6/292
Hyperanteflexion
disruptive, *see* Disruptive hyperanteflexion
injury, *see* Cervical hyperanteflexion injury
sprain, *see* Cervical hyperanteflexion sprain
Hyperapobetalipoproteinemia
familial, 66/544
molecular basis, 66/544
Hyperargininemia
ataxia, 42/562
Babinski sign, 42/562
brain atrophy, 42/562
diet, 29/102, 42/563
epileptic seizure, 29/102
familial, *see* Familial hyperargininemia
heterozygote detection, 42/563
hyperreflexia, 42/562
mental deficiency, 42/562
psychomotor retardation, 29/102, 42/562
seizure, 29/102, 42/562
short stature, 42/562
spasticity, 42/562
ventricular dilatation, 42/562
Hyperbaric headache
ama, 48/401
Hyperbaric oxygenation
brain edema, 16/204, 24/618
intracranial pressure, 23/215
multiple sclerosis, 47/160, 188
neurologic intensive care, 55/210
pallidal necrosis, 6/660
subacute myelo-optic neuropathy, 37/135
Hyperbeta-alaninemia
amino acid metabolism, 29/230
grand mal epilepsy, 29/230
lethargy, 29/230
screening test, 29/230
tuberculosis, 29/230
Hyperbetalipoproteinemia
acute intermittent porphyria, 42/618
familial, *see* Hyperlipoproteinemia type II
type, 10/269-278
Hyperbeta-prebetalipoproteinemia
familial, *see* Hyperlipoproteinemia type III
Hyperbilirubinemia

acholuric jaundice, 6/492
cerebellar agenesis, 50/180
congenital deafness, 42/364
double athetosis, 49/384
subarachnoid hemorrhage, 12/113
valproic acid induced hepatic necrosis, 65/126
vitamin K3 intoxication, 65/570
Hyperbradykininemia
orthostatic hypotension, 43/67
Hypercalcemia
asterixis, 63/561
basilar artery vasoconstriction, 53/388
brain vasospasm, 63/561
calcitonin, 63/562
carcinoma, 38/652-654
confusion, 63/561
confusional state, 63/561
convulsion, 63/561
delirium, 46/543, 63/561
dialysis, 63/562
diphosphonate, 63/562
electrolyte disorder, 63/561
elfin face syndrome, 31/317
furosemide, 63/562
hemiplegia, 63/561
human T-lymphotropic virus type I associated myelopathy, 56/527
hyperparathyroidism, 27/288-290, 63/561
hypophosphatasia, 42/578
infantile idiopathic, *see* Elfin face syndrome
juvenile spinal cord injury, 61/243
malignancy, 38/652-654, 63/561
manic depression, 63/561
manic status, 63/561
mental depression, 63/561
metabolic encephalopathy, 69/401
mithramycin, 63/562
multiple myeloma, 20/11, 14, 39/138, 142, 63/392
myoclonus, 63/561
myoglobinuria, 62/556
neurologic manifestation, 28/536-539
obtundation, 63/561
paranoia, 63/561
rigidity, 63/561
Sipple syndrome, 42/752
symptom, 70/111
transient hemiparesis, 63/561
transient ischemic attack, 63/561
treatment, 27/289
vitamin A intoxication, 65/569
vitamin D3 intoxication, 65/569
Williams syndrome, 43/260, 46/92
Hypercalciuria
Wilson disease, 49/229
Hypercapnia
brain edema, 16/180
brain infarction, 53/422
brain injury, 57/212
central sleep apnea, 63/461
cluster headache, 48/242
congenital central alveolar hypoventilation, 63/464
head injury outcome, 57/387
headache, 48/6, 422
hypoventilation, 63/413
obstructive sleep apnea, 63/455
pickwickian syndrome, 38/299, 63/455
Hypercarbia, *see* Hypercapnia
Hypercholesterolemia
acute intermittent porphyria, 42/618
familial, *see* Hyperlipoproteinemia type II
neuropathy, 8/18
Hypercitrullinemia
epilepsy, 72/222
Hypercorticism, *see* Hyperadrenocorticism
Hypercortisolism
adrenal disorder, 70/209
alcoholism, 70/209
benign intracranial hypertension, 70/206
classification, 70/207
depression, 70/207
imaging, 70/209
myopathy, 70/206
perineural lipomatosis, 70/207
psychiatry, 70/207, 209
spinal lipomatosis, 70/207
Hyperdeflexion
anterior cervical cord syndrome, 25/359
disruptive, *see* Disruptive hyperdeflexion
injury, *see* Cervical hyperdeflexion injury
sprain, *see* Cervical hyperdeflexion sprain
Hyperdibasic aminoaciduria
diet, 42/564
familial protein intolerance, 29/105, 210
hyperammonemia, 42/564
leukopenia, 42/564
mental deficiency, 42/564
short stature, 42/563
Hyperdiploidy
Alzheimer disease, 46/273
Hyperemesis
neuropathy, 8/14
Hyperemia
brain, *see* Brain hyperemia
reactive, *see* Reactive hyperemia
Hyperencephalus

encephalocele, 50/73
Hypereosinophilia
see also Idiopathic hypereosinophilic syndrome
brain embolism, 63/138
encephalopathy, 63/138
endocarditis, 63/111
endomyocarditis, 63/123
neuropathy, 63/138
Hypereosinophilic syndrome
amnesia, 63/372
anticoagulation, 63/373
ataxia, 63/372
axonal degeneration, 63/373
Balint syndrome, 41/65
behavior change, 63/372
brain atrophy, 63/372
brain embolism, 63/372
computer assisted tomography, 63/372
confusion, 63/372
confusional state, 63/372
corticospinal tract, 63/372
corticosteroid, 63/372, 374
CSF, 63/372
dementia, 63/372
dysesthesia, 63/373
EEG, 63/372
EMG, 63/373
encephalopathy, 63/372
endocardial fibrosis, 63/373
eosinophil cationic protein, 63/374
eosinophilia, 63/371
eosinophilia myalgia syndrome, 64/261
eosinophilic polymyositis, 41/63, 65
epilepsy, 63/372
fasciitis, 41/383
fatigue, 63/371
hydroxyurea, 63/374
idiopathic, *see* Idiopathic hypereosinophilic syndrome
leukoencephalopathy, 63/372
liver, 63/372
lung, 63/372
mononeuritis multiplex, 63/373
muscle biopsy, 63/374
muscle fiber type II atrophy, 63/374
myalgia, 63/371, 374
nerve biopsy, 63/373
neurologic involvement, 63/372
neuropathy, 63/372
nuclear magnetic resonance, 63/372
paresthesia, 63/373
pathogenesis, 63/374
polymyositis, 41/65, 62/373
polyneuropathy, 63/373
pyramidal tract syndrome, 63/372
radiculopathy, 63/373
restrictive infiltrative cardiomyopathy, 63/372
sensory neuropathy, 63/373
skin, 63/372
symptom, 63/371
transient ischemic attack, 63/372
treatment, 63/374
vasculitis, 63/373
Hyperesthesia
caffeine intoxication, 65/260
eosinophilia myalgia syndrome, 63/375, 64/254
organochlorine insecticide intoxication, 64/199
pellagra, 28/90
terminology, 1/89
thalamic syndrome (Dejerine-Roussy), 42/771
Hyperexplexia
see also Startle epilepsy *and* Syncinésie sursaut
brain stem excitability, 42/228
brain stem reflex, 42/228
epilepsy, 42/228, 72/255
involuntary movement, 6/301
myoclonus, 38/577
startle syndrome, 49/620
Hyperextensible joint
Ehlers-Danlos syndrome, 43/14
glycosaminoglycanosis type IV, 42/457
Goltz-Gorlin syndrome, 43/19
LEOPARD syndrome, 43/28
Stickler syndrome, 43/484
Hyperextensible skin
Coffin-Lowry syndrome, 43/239
Ehlers-Danlos syndrome, 43/14
Hyperextensions, *see* Cervical hyperdeflexion injury *and* Cervical hyperdeflexion sprain
Hyperfibrinogenemia
hyperviscosity syndrome, 55/484
Hyperflexion, *see* Cervical hyperanteflexion injury, Cervical hyperanteflexion sprain *and* Spinal hyperflexion injury
Hypergammaglobulinemia
diamorphine addiction, 55/519
multiple sclerosis, 9/326, 354
Hyperglycemia
see also Diabetes mellitus
ataxia, 21/578
brain infarction, 53/128, 167
brain injury, 23/122
brain ischemia, 53/128, 167
diabetic neuropathy, 27/92-94
diazoxide intoxication, 37/433
effect, 27/79-95

hemiballismus, 49/374
hexomonophosphate shunt, 27/79
hypothalamus, 2/440
lipodystrophy, 8/346
metabolic ataxia, 21/578
multiple myeloma polyneuropathy, 51/430
nervous system, 27/79-95
nonketotic, *see* Nonketotic hyperglycemia
pheochromocytoma, 39/498
POEMS syndrome, 51/430
polyol pathway, 27/79-83, 86
sorbitol, 27/79-83
vascular disease, 27/94

Hyperglycinemia, 29/199-210
see also Methylmalonic acidemia, β-Oxothiolase deficiency *and* Propionic acidemia
biochemical finding, 29/201-203
clinical features, 29/200
convulsion, 42/565
CSF, 42/565
diagnosis, 29/203
familial spinal cord disorder, 29/203, 511
feeding disorder, 42/565
genetics, 29/203
glycine, 29/511
glycine cleavage enzyme system, 42/566
hyperreflexia, 42/565
hypsarrhythmia, 42/565
isovaleric acidemia, 29/204
ketotic, *see* Propionic acidemia
mental deficiency, 42/565
metabolic study, 29/201-203
methylmalonyl CoA carbamyl mutase deficiency, 29/204
neutropenia, 42/565
nonketotic, *see* Nonketotic hyperglycinemia
β-oxothiolase deficiency, 29/204
pathology, 29/200
propionic acidemia, 28/520, 29/204
seizure, 42/565
spastic paraplegia, 42/170
thrombocytopenia, 42/565
treatment, 29/203

Hypergonadotropic hypogonadism
Klinefelter syndrome, 50/546, 549
Turner syndrome, 50/544
XXXX syndrome, 50/555

Hypergraphia
agraphia, 45/468
schizophrenia, 45/469

Hyperhidrosis
acrylamide intoxication, 64/67
acrylamide polyneuropathy, 51/281
alcoholic polyneuropathy, 51/316
arsenic intoxication, 64/287
autonomic polyneuropathy, 51/486
cause, 75/70
cluster headache, 48/9
diphtheria, 51/183
dysautonomia, 22/260
dyskeratosis congenita, 14/784, 43/392
essential, 8/344
hereditary sensory and autonomic neuropathy type I, 21/73
hereditary sensory and autonomic neuropathy type III, 1/475, 21/114, 60/7, 28, 63/147
hexane polyneuropathy, 51/277
2-hexanone polyneuropathy, 51/277
Isaacs syndrome, 43/160
leprosy, 51/218
malignancy, 75/529
mechanism, 75/70
mercury polyneuropathy, 51/272
myokymia, 43/160
myotonia, 43/160
palmar, *see* Palmar hyperhidrosis
pathophysiology, 75/70
POEMS syndrome, 63/394
Shapiro syndrome, 50/167
survey, 75/67
syringomyelia, 50/445
treatment, 75/71
tumor, 75/530
Wilson disease, 49/227

Hyperinsulin neuropathy
nerve conduction, 7/166

Hyperinsulinemia
dementia, 46/397

Hyperinsulinism
functional, *see* Functional hyperinsulinism
hypoglycemia, 27/55
insulinoma, 27/68
islet cell adenoma, 27/68
myotonic dystrophy, 62/230

Hyperkalemia
acute muscle weakness, 41/287
areflexia, 63/558
Burkitt lymphoma, 63/352
delirium, 46/543
electrolyte disorder, 63/558
flaccid quadriplegia, 63/558
generalized weakness, 63/558
malignant hyperthermia, 38/551, 43/117
muscle weakness, 63/558
nerve conduction, 7/169
paresthesia, 63/558

periodic paralysis, 63/558
salbutamol, 41/161
sea snake intoxication, 37/92
spinal cord injury, 61/286
tetraplegia, 63/558
transverse spinal cord lesion, 61/286
Hyperkalemic periodic paralysis, 28/592-597
see also Paramyotonia congenita
acetazolamide, 62/470, 472
age, 62/462
age at onset, 28/595
areflexia, 43/151
attack, 62/461
attack duration, 62/462
attack frequency, 62/462
attack provocation, 62/462
case history, 40/562
channel α-subunit gene, 62/463
chromosome 17, 62/463
Chvostek sign, 62/462
classification, 40/537
clinical features, 28/593-595, 40/317, 325, 335, 544, 41/158
differential diagnosis, 28/592
ECG, 43/152
EMG, 28/595
familial periodic paralysis, 62/457
genetics, 41/160, 165, 425, 62/463
hereditary, *see* Hereditary hyperkalemic periodic paralysis
hyporeflexia, 43/151
inheritance, 28/595
laboratory study, 28/595
morphology, 40/547
myotonia, *see* Myotonic hyperkalemic periodic paralysis
myotonic syndrome, 40/544
orciprenaline, 62/472
paramyotonia, 40/317, 325, 335, 537, 544, 547, 557, 562
paresthesia, 43/152
pathology, 28/596, 41/160
permanent muscular weakness, 62/462
progressive supranuclear palsy, 49/246
salbutamol, 62/472
serum potassium, 43/151
sex distribution, 28/595
symptom, 62/461
treatment, 28/596, 40/557, 41/161
Hyperkeratosis
arsenic intoxication, 64/287
congenital ichthyosis, 43/405
generalized, *see* Urbach-Wiethe disease
incontinentia pigmenti, 43/23
linear nevus sebaceous syndrome, 43/38
pellagra, 28/89
Richner-Hanhart syndrome, 42/581
sensorineural deafness, 42/395
tyrosinemia, 42/635
Werner syndrome, 43/489
Hyperkinesia
see also Extrapyramidal disorder
barbiturate intoxication, 65/509
Börjeson-Forssman-Lehmann syndrome, 42/530
boxing injury, 23/543
brain lacunar infarction, 54/247
cerebellar catalepsy (Babinski), 21/521
cervical cordotomy, 6/867
congenital Pelizaeus-Merzbacher disease, 42/503
drug induced extrapyramidal syndrome, 6/251
Dubowitz syndrome, 43/247
extrapyramidal, *see* Extrapyramidal hyperkinesia
globus pallidus externus, 21/520, 526
Hallervorden-Spatz syndrome, 22/403, 42/257
history, 6/851
hypervalinemia, 42/574
infantile neuroaxonal dystrophy, 6/627, 42/229
infantile optic glioma, 42/733
neuroleptic syndrome, 6/251
opticocochleodentate degeneration, 21/540
pathogenesis, 6/845-847
Pelizaeus-Merzbacher disease, 42/503
pellagra, 6/748, 7/571
rigidity, 6/851
Russell syndrome, 16/349
Stargardt disease, 13/132, 42/420
surgery, 6/851
vigabatrin intoxication, 65/511
Hyperkinetic agraphia
cerebellar tremor, 45/466
parkinsonism, 45/465
postural tremor, 45/465
tremor, 45/465-467
Hyperkinetic child syndrome
see also Minimal brain dysfunction
aggression, 4/341, 43/201
amphetamine, 3/196, 4/363, 46/177
attention disorder, 3/195-198, 43/200-202
barbiturate, 3/197
behavior disorder in children, 4/363
brain injury, 3/195, 197, 46/177
cerebral dominance, 4/465
classification, 3/197
clinical features, 3/195
EEG, 3/196, 4/363, 43/201
emotional disorder, 43/201

epilepsy, 3/196, 4/363, 46/177
higher nervous activity disorder, 4/363, 465
hyperactivity, 3/195, 4/341, 363, 46/176
methylphenidate, 3/197
mirror movement, 43/201
neurologic abnormality, 3/195
pathogenesis, 3/197
poor school performance, 3/195, 198
prevalence, 43/200
reticular formation, 3/197
sphincter disturbance, 43/201
tranquilizing agent, 3/197
Hyperkinetic mutism
brain lacunar infarction, 54/247
Hyperlactic acidemia
chemistry, 29/211
ketotic hyperammonemia, 29/106
Hyperlaxity
joint, *see* Joint hyperlaxity
Hyperlexia syndrome
mental deficiency, 46/29
Hyperlipemic neuropathy, 29/429-432
biochemistry, 29/431
history, 29/429
pathology, 29/430
treatment, 29/432
Hyperlipidemia
alcohol intoxication, 64/111
arterial hypertension, 63/77
cerebrovascular disease, 53/34
familial combined, *see* Familial combined hyperlipidemia
hepatic insufficiency polyneuropathy, 51/363
hereditary, *see* Hereditary hyperlipidemia
polyneuropathy, 60/167
spinal muscular atrophy, 59/103
systemic brain infarction, 11/462
Wolman disease, 60/165
Hyperlipidemia type II
mucolipidosis type IV, 51/380
Hyperlipidemic neuropathy
cerebrotendinous xanthomatosis, 60/167
differential diagnosis, 60/167
Hyperlipoproteinemia
Wohlfart-Kugelberg-Welander disease, 59/372
Hyperlipoproteinemia type I
classification, 66/540
cranial nerve, 42/443
hepatosplenomegaly, 42/443
lipoprotein lipase, 42/443
pain, 42/443
polyneuritis, 42/443
xanthomatosis, 42/443
Hyperlipoproteinemia type II
brain arteriosclerosis, 53/98
cerebrotendinous xanthomatosis, 10/537, 539
cutaneotendinous xanthoma, 10/272
β-lipoprotein, 10/272
lipoprotein metabolism, 66/545
low density lipoprotein, 10/272, 42/445
neuropathy, 42/445
prevalence, 42/445
stroke, 42/445
xanthoma, 10/272-274
xanthomatosis, 10/272, 42/444
Hyperlipoproteinemia type IIA
amyotrophic lateral sclerosis, 22/46
amyotrophy, 22/45
Hyperlipoproteinemia type III
brain arteriosclerosis, 53/98
cerebrotendinous xanthomatosis, 10/537, 539
hyperuricemia, 10/275
very low density lipoprotein, 66/541
Hyperlipoproteinemia type IV
amyotrophy, 22/45
hepatosplenomegaly, 10/276
ischemic heart disease, 10/275
prebetalipoprotein, 10/275
xanthoma, 10/276
Hyperlipoproteinemia type V
abdominal pain, 10/277
chylomicron, 10/277
xanthoma, 10/277
Hyperlordosis
Bassen-Kornzweig syndrome, 51/394
diastematomyelia, 50/436
diplomyelia, 50/436
Hyperlysinemia
diet, 42/567
epilepsy, 72/222
mental deficiency, 29/221, 42/567
microcephaly, 30/509
saccharopine dehydrogenase, 42/567
saccharopine oxidoreductase, 42/567
saccharopinuria, 29/221, 42/567
Hypermagnesemia
acetylcholine release, 63/563
areflexia, 63/563
cardiac arrest, 63/563
clinical features, 28/569, 570
delirium, 46/543
Eaton-Lambert myasthenic syndrome, 63/564
electrolyte disorder, 63/563
magnesium imbalance, 28/568-570
muscle weakness, 63/563
myasthenia gravis, 63/564

neonatal, 28/569
neurologic symptom, 28/568
periodic paralysis, 62/457
symptom, 70/114
tetraplegia, 63/563
treatment, 28/570
weakness, 63/563
Hypermelanosis
Albright syndrome, 1/484
brain, 14/418
neurofibromatosis, 1/484
Hypermethioninemia
hereditary tyrosinemia, 29/119
homocystinuria, 29/119, 55/325
tyrosinemia, 29/119
Hypermetria
see also Dysmetria
animal experiment, 2/403
cerebellar ataxia, 1/323
definition, 2/394
finger to ear test, 1/324, 333
finger to nose test, 1/324, 333
hand turning test, 1/324
lobulus ansoparamedianus, 2/403
Hypermotility
esophageal spasm, 75/629
Hypermyelination
hereditary multiple recurrent mononeuropathy, 51/558
status marmoratus, 6/676
Hypernatremia
aging, 74/235
agitation, 63/554
brain atrophy, 63/554
brain hemorrhage, 63/554
brain infarction, 63/554
brain venous infarction, 63/554
cardiac glycoside, 28/447
cerebellar hemorrhage, 63/554
cerebral lesion, 2/446, 28/455-457
chorea, 63/554
clinical features, 28/450
coma, 63/554
confusion, 63/554
CSF, 28/452
delirium, 46/543, 63/554
diabetes insipidus, 63/554
EEG, 28/452, 63/554
electrolyte disorder, 63/554
encephalopathy, 63/554
epilepsy, 28/451, 63/554
essential, *see* Essential hypernatremia
excessive salt intake, 63/554
granuloma, 2/446
Hand-Schüller-Christian disease, 2/446
hemorrhage, 2/446
hereditary paroxysmal kinesigenic choreoathetosis, 49/355
hypernatremic dehydration, 28/453
hypothalamus tumor, 2/446
infectious disease, 2/446
laboratory aspect, 28/451
metabolic encephalopathy, 63/554, 69/402
myelinolysis, 63/554
myoclonic jerk, 63/554
myoclonus, 63/554
myoglobinuria, 41/274
neurologic manifestation, 28/443-458
oligodendroglial shrinkage, 63/554
opisthotonos, 63/554
pathology, 28/447
pathophysiology, 28/448-450
prevention, 28/457
prognosis, 28/452
rigidity, 63/554
salt administration, 28/454
salt ingestion, 28/454
sodium retention, 28/457
status epilepticus, 63/554
stupor, 63/554
subarachnoid hemorrhage, 12/113, 63/554
subdural hematoma, 63/554
syndrome, 28/455-457
systemic brain infarction, 11/463
thirst, 63/554
treatment, 28/457
tremor, 63/554
venous sinus thrombosis, 63/554
Hypernephroma
CSF cytology, 16/407
secondary polycythemia, 63/250
tuberous sclerosis, 14/686
vertebral metastasis, 20/423
Von Hippel-Lindau disease, 14/242, 245, 249, 695, 16/31
Hyperornithinemia
homocitrinullinuria, 29/103
metabolic ataxia, 21/574, 577
ocular manifestation, 29/104
periodic ataxia, 21/577
Hyperosexia
attention disorder, 3/139
Hyperosmia
frontal lobe tumor, 17/260
olfactory illusion, 73/72
Hyperosmolality

diabetes mellitus, 70/196
Hyperosmolar coma
systemic brain infarction, 11/462
Hyperosmolar hyperglycemic nonketotic diabetic coma
age, 27/89
Babinski sign, 27/90
dysphagia, 27/90
hallucination, 27/90
hemiparesis, 27/90
hemisensory deficit, 27/90
homonymous hemianopia, 27/90
hyperreflexia, 27/90
hyponatremia, 27/91
nystagmus, 27/90
photophobia, 27/90
polyol pathway, 27/91
prodromal symptom, 27/131
seizure, 27/89
sorbitol, 27/91
treatment, 27/92
Hyperosmolar hyperglycemic nonketotic syndrome
syncope, 72/262
Hyperosmolarity syndrome
acute pancreatitis, 27/456
Hyperostosis
congenital hyperphosphatasia, 31/258
cranial, *see* Cranial hyperostosis
Schwartz-Lelek syndrome, 31/257
sclerosteosis, 31/258
Hyperostosis corticalis deformans
bone fragility, 43/409
juvenile, 43/409
sensorineural deafness, 43/409
vertebral abnormality, 43/409
Hyperostosis corticalis generalisata familiaris, *see* Van Buchem disease
Hyperostosis cranii, *see* Cranial hyperostosis
Hyperostosis frontalis interna, *see* Morgagni-Stewart-Morel syndrome
Hyperoxaluria
primary, *see* Primary hyperoxaluria
Hyperparathyroid myopathy
EMG, 62/540
incidence, 62/540
muscular atrophy, 62/540
neuropathy, 62/540
osteomalacia, 62/538
pathogenesis, 62/540
proximal limb weakness, 62/540
rickets, 62/538
treatment, 62/541
waddling gait, 62/540
Hyperparathyroidism
amyotrophic lateral sclerosis, 59/202
ataxia, 27/292
basilar impression, 50/399
calcitonin, 27/290
CSF, 27/290
CSF protein, 27/291
deafness, 27/293
delirium, 46/544
dementia, 46/396
depression, 46/427
EEG, 27/290
endocrine myopathy, 41/246, 324, 338
fasciculation, 27/293, 41/247
fibrous dysplasia, 14/163, 186, 199, 38/385
Guam amyotrophic lateral sclerosis, 59/281
headache, 27/288, 48/425
history, 27/284
hypercalcemia, 27/288-290, 63/561
hypertension, 27/286
ivory vertebra, 19/171
magnesium, 27/291
mental change, 27/286-288
multiple endocrine adenomatosis, 27/286
muscle intermittent claudication, 20/799
myeloma, 20/12
myelopathy, 27/293, 70/120
myopathy, 27/293-296
neuromuscular disease, 70/118
neuropathy, 27/293
parkinsonism, 27/292
pathogenesis, 27/284-286
prevalence, 27/284
primary myopathy, 28/538
proximal muscle weakness, 40/317
psychiatric symptom, 27/286-288
psychiatry, 70/118
secondary, 70/121
skeletal manifestation, 27/288
skeletal muscle, 40/438
sleep disorder, 27/288
striatopallidodentate calcification, 27/292
symptom, 70/121
systemic brain infarction, 11/458
uremic encephalopathy, 63/515
uremic polyneuropathy, 63/515
vitamin D, 41/247
Hyperpathia, *see* Hyperalgesia
Hyperphagia
meningeal leukemia, 63/341
Hyperphenylalaninemia
see also Phenylketonuria
biochemistry, 29/31

dermatitis, 42/611
diet, 29/37
differential diagnosis, 29/38-42
enamel hypoplasia, 42/611
epilepsy, 42/611
maternal, *see* Maternal hyperphenylalaninemia
mental deficiency, 42/611
phenylketonuria, 29/44, 59/75
Hyperphenylalaninemic embryopathy syndrome
see also Phenylketonuria
diagnostic criteria, 31/299-301
Hyperphosphatasemia
see also Van Buchem disease
chronica tarda, 14/170
fibrous dysplasia, 14/170, 199
Hyperphosphatasia
congenital, *see* Congenital hyperphosphatasia
foam cell, 42/568
hypoparathyroidism, 42/577
macrocephaly, 30/100
mental deficiency, 42/568
microcephaly, 42/578
osteoectasia, 43/409
seizure, 42/568
Hyperphosphatemia
Burkitt lymphoma, 63/352
lipocalcinogranulomatosis, 43/185, 187
myoglobinuria, 62/556
pseudohypoparathyroidism, 42/622, 46/85
symptom, 70/116
Hyperphosphaturia
Wilson disease, 49/229
Hyperpigmentation
ACTH induced myopathy, 62/537
congenital dyskeratosis, 43/392
congenital torticollis, 43/146
Fanconi syndrome, 43/17
Goltz-Gorlin syndrome, 43/19
Hallervorden-Spatz syndrome, 14/424
keloid, 43/146
multiple myeloma polyneuropathy, 51/430
Nelson syndrome, 62/537
pituitary adenoma, 17/423
POEMS syndrome, 47/619, 51/430, 63/394
Hyperpipecolatemia
adrenomyeloneuropathy, 60/170
case, 29/222
CSF pipecolic acid, 29/222
demyelination, 29/222
extremity involvement, 29/222
flaccid paralysis, 29/222
hypotonia, 29/222
intention tremor, 29/222
liver, 29/222
mental deficiency, 29/222
nystagmus, 29/222
pathology, 29/222
peroxisomal disease, 51/383
peroxisomal disorder, 66/510
secondary pigmentary retinal degeneration, 13/326
vomiting, 29/222
Hyperpipecolic acidemia, *see* Hyperpipecolatemia
Hyperpituitarism
pituitary adenoma, 17/393, 423
Hyperplasia
adrenocortical, *see* Adrenocortical hyperplasia
C cell, *see* C cell hyperplasia
congenital adrenal, *see* Congenital adrenal hyperplasia
diaphyseal, *see* Diaphyseal hyperplasia
gingiva, *see* Gingiva hyperplasia
leptomeningeal melanoblastic, *see* Leptomeningeal melanoblastic hyperplasia
oligodendrocyte, *see* Oligodendrocyte hyperplasia
pancreatic islet cell, *see* Pancreatic islet cell hyperplasia
pedicular arch, *see* Pedicular arch hyperplasia
scrotum, *see* Scrotum hyperplasia
thymus, *see* Thymus hyperplasia
tongue, *see* Tongue hyperplasia
Hyperplastic myelinopathy
autosomal dominant pain insensitivity, 60/17
hereditary sensory and autonomic neuropathy type IV, 60/17
Hyperpnea
see also Hyperventilation syndrome *and* Respiration
brain stem tumor, 63/440
central neurogenic hyperventilation, 63/436, 440
Cheyne-Stokes respiration, 1/497
classification, 1/657
continuous, 63/436
episodic, *see* Episodic hyperpnea
frontal lobe meningioma, 63/440
Joubert syndrome, 60/667, 63/437
malignant hyperthermia, 63/440
neurologic disease, 63/436
pontine glioma, 63/440
primary CNS lymphoma, 63/440
pyruvate dehydrogenase deficiency, 63/439
Rett syndrome, 63/437
Reye syndrome, 31/168, 63/438
Smith-Lemli-Opitz syndrome, 60/667, 63/438
vitamin Bw dependent multiple carboxylase deficiency, 63/439

Hyperprebetalipoproteinemia
familial, *see* Hyperlipoproteinemia type IV
familial hyperchylomicronemia, *see* Hyperlipoproteinemia type V
Hyperprolactinemia
cocaine intoxication, 65/253, 256
diagnosis, 68/350
headache, 48/424
iatrogenic, 68/348
Hyperprolinemia type I
see also Iminoacidemia
biochemical finding, 29/136
clinical finding, 29/133-135
convulsion, 29/134
deafness, 29/133
EEG, 29/133
inheritance, 29/137
mental deficiency, 29/133, 42/569
proline oxidase, 42/569
renal disease, 29/133
seizure, 42/569
treatment, 29/137
urinary tract defect, 29/133
Hyperprolinemia type II
see also Iminoacidemia
biochemical finding, 29/138-140
clinical finding, 29/138
convulsion, 29/138
EEG, 29/138
inheritance, 29/140
mental deficiency, 29/138, 42/569
proline oxidase, 42/569
1-pyrroline-5-carboxylate dehydrogenase, 42/569
seizure, 29/138, 42/569
treatment, 29/141
Hyperpyrexia, *see* Hyperthermia
Hyperpyruvic acidemia
chemistry, 29/211
Hill-Sherman syndrome, 21/578
ketotic hyperammonemia, 29/106
metabolic ataxia, 21/574
subacute necrotizing encephalomyelopathy, 29/211-213
Hyperreflexia
adrenoleukodystrophy, 42/483
amyotrophic dystonic paraplegia, 42/199
amyotrophic lateral sclerosis, 42/65
aqueduct stenosis, 42/57
ataxic diplegia, 42/121
autonomic, *see* Autonomic hyperreflexia
bulbopontine paralysis, 42/96
CNS spongy degeneration, 42/506
congenital retinal blindness, 42/401
craniodiaphyseal dysplasia, 43/356
craniometaphyseal dysplasia, 43/362
dystonia, 42/219
elfin face syndrome, 31/318
Förster syndrome, 42/202
Friedman-Roy syndrome, 43/251
Gaucher disease, 42/438
Goldberg syndrome, 42/439
Hooft-Bruens syndrome, 42/508
Huntington chorea, 42/225
hyperargininemia, 42/562
hyperglycinemia, 42/565
hyperosmolar hyperglycemic nonketotic diabetic coma, 27/90
hypoparathyroidism, 42/577
infantile Gaucher disease, 42/438, 66/124
Lafora progressive myoclonus epilepsy, 27/173
Leber hereditary optic neuropathy, 62/509, 66/428
Leber optic atrophy, 42/401
Lesch-Nyhan syndrome, 42/153
lysergide, 65/45
Marinesco-Sjögren syndrome, 42/184
Mast syndrome, 42/282
mental deficiency, 43/268, 277
microcephaly, 43/429
multiple neuronal system degeneration, 75/162
multiple sclerosis, 42/496
myelopathy, 42/156
myopathy, 43/123-125
nystagmus myoclonus, 42/240
oculocerebrocutaneous syndrome, 43/444
opticocochleodentate degeneration, 42/241
organolead intoxication, 64/133
Paine syndrome, 43/433
pallidopyramidal degeneration, 42/244
Parkinson dementia complex, 42/249
phenylketonuria, 42/610
phenytoin intoxication, 42/641
presenile dementia, 42/288
respiratory encephalopathy, 63/414
Salla disease, 43/306
Sjögren-Larsson syndrome, 43/307
Sjögren syndrome, 71/75
spastic ataxia, 42/165
spastic paraplegia, 42/172, 180
spinocerebellar degeneration, 42/190
spinopontine degeneration, 42/192
Stargardt disease, 42/420
subsarcolemmal myofibril, 42/219
trigeminal nerve agenesis, 50/214
Troyer syndrome, 42/193
unilateral pontine hemorrhage, 12/40
Von Hippel-Lindau disease, 42/156

W syndrome, 43/271
Hyperrotation injury
cervical spine, 25/259
Hypersalivation
chronic mercury intoxication, 64/371
Lafora progressive myoclonus epilepsy, 27/173
Minamata disease, 36/105, 123
neophocaena phocaenoides intoxication, 37/95
pallidal degeneration, 49/455
porpoise intoxication, 37/95
Hypersarcosinemia
autosomal recessive, 29/230
psychomotor retardation, 29/228
screening test, 29/229
Hypersensitivity
angiitis, *see* Allergic granulomatous angiitis
antidepressant intoxication, 37/319
autoantigen, 9/504
carotid sinus, *see* Carotid sinus hypersensitivity
CNS, 9/504
delayed, *see* Delayed hypersensitivity
denervation, *see* Denervation hypersensitivity
diamorphine addiction, 55/519
dopamine, *see* Dopamine hypersensitivity
foreign antigen, 9/504
guanethidine intoxication, 37/436
immunity, 9/503
Koch reaction, 9/503
lymphocytic choriomeningitis, 9/501, 544
model, 9/504
nitrate antigen, 9/504
platinum intoxication, 64/357
radiation, *see* Radiation hypersensitivity
tricyclic antidepressant intoxication, 37/319
ultraviolet light, 13/432
Hypersensitivity angiitis, *see* Allergic granulomatous angiitis
Hypersensitivity vasculitis, *see* Allergic granulomatous angiitis
Hyperserotonemia
ataxia, 29/225
emotional lability, 29/225
flushing, 29/225
infantile autism, 46/191
mental deficiency, 29/225
slurred speech, 29/225
tongue, 29/225
Hyperserotoninemia, *see* Hyperserotonemia
Hypersexuality
temporal lobe epilepsy, 46/433
Hypersomnia
see also Lethargy *and* Sleep
African trypanosomiasis, 52/341
amphetamine intoxication, 65/258
brain death, 24/701, 711, 713
brain injury, 24/593, 701
brain lacunar infarction, 53/395
CNS stimulant, 46/592
daytime, *see* Daytime hypersomnia
hypothalamus disorder, 46/418
idiopathic CNS, *see* Idiopathic CNS hypersomnia
intermittent daytime, *see* Intermittent daytime hypersomnia
multiple sclerosis, 3/96
myotonic dystrophy, 40/517, 62/217, 235
narcolepsy, 3/96, 15/838
paroxysmal disorder, 45/511
pickwickian syndrome, 38/299, 63/455
postencephalitic, *see* Postencephalitic hypersomnia
respiratory encephalopathy, 63/414
sleep disorder, 3/67, 96
thalamic infarction, 53/395
thalamic lesion, 2/482
vertebrobasilar system syndrome, 53/395
Hypersomnolence, *see* Hypersomnia
Hypertelorism
Aarskog syndrome, 30/249, 43/313
acanthosis nigricans, 30/248
achondroplasia, 30/248
acrocephalosyndactyly type I, 30/246, 43/318
acrocephalosyndactyly type II, 30/246, 38/422
acrocephalosyndactyly type V, 30/246, 43/320, 50/123
Albers-Schönberg disease, 30/246
albinism, 30/248
basal encephalocele, 50/106
basal meningoencephalocele, 50/106
Bonham-Carter syndrome, 30/248
cat eye syndrome, 30/250
cerebral gigantism, 30/247, 43/336
cerebrohepatorenal syndrome, 30/247, 43/338
cleidocranial dysostosis, 30/248
Coffin-Lowry syndrome, 30/248, 43/238
congenital heart disease, 63/11
Conradi-Hünermann syndrome, 30/248
corpus callosum agenesis, 42/101, 43/417, 50/162
craniodiaphyseal dysplasia, 30/248
craniofacial dysostosis (Crouzon), 30/246, 43/359-361
craniometaphyseal dysplasia, 30/246, 31/255, 43/362-364
craniosynostosis, 50/125
cri du chat syndrome, 30/249, 43/501
De Lange syndrome, 30/248
Dubowitz syndrome, 30/248

Ehlers-Danlos syndrome, 14/111
elfin face syndrome, 31/318
encephalomeningocele, 42/27
epicanthus, 42/459
F syndrome, 30/249
facio-oculoacousticorenal syndrome, 43/399
Fanconi syndrome, 30/249
fetal face syndrome, 30/248, 43/400
Fraser syndrome, 30/247
Freeman-Sheldon syndrome, 30/247, 43/353
G syndrome, 30/247
glycosaminoglycanosis type I, 42/450
glycosaminoglycanosis type VII, 42/459, 66/310
Goldenhar syndrome, 30/243
Gruber syndrome, 14/115, 30/244, 247
hereditary sensory and autonomic neuropathy type III, 30/248
Hurler-Hunter disease, 10/436
hypohidrotic ectodermal dysplasia, 30/248
hypospadias, 43/413
iris dysplasia, 43/417
Juberg-Hayward syndrome, 30/247
Klein-Waardenburg syndrome, 14/116, 30/408, 50/218
Klinefelter syndrome, 43/560
Klinefelter variant XXXXY, 30/250, 31/483, 43/560
Klippel-Feil syndrome, 30/249
Larsen syndrome, 30/249
LEOPARD syndrome, 14/117, 30/248, 43/27
leprechaunism, 42/590
median facial cleft syndrome, 30/247, 46/89
Melnick-Needles syndrome, 30/249, 43/452
mental deficiency, 43/257, 277, 417, 439
methotrexate syndrome, 30/122
Minkowitz disease, 10/552
Mohr syndrome, 30/247
multiple nevoid basal cell carcinoma syndrome, 30/248, 43/31
Neu syndrome, 43/430
neurocutaneous melanosis, 14/600
neurogenic acro-osteolysis, 42/294
Nielsen syndrome, 30/249
Noonan syndrome, 30/249, 46/75
orbital, *see* Orbital hypertelorism
orofaciodigital syndrome type I, 30/247
otopalatodigital syndrome, 30/248
9p partial trisomy, 43/515
10p partial trisomy, 43/519
11p partial trisomy, 43/525
Patau syndrome, 30/249, 31/504, 43/528
Pena-Shokeir syndrome type I, 43/438, 59/71
Pinsky-DiGeorge-Harley-Bird syndrome, 30/247
Potter syndrome, 30/250
Pyle disease, 13/83, 14/198
13q interstitial deletion, 50/587
2q partial trisomy, 43/493
4q partial trisomy, 43/498
6q partial trisomy, 43/503
11q partial trisomy, 43/523
r(15) syndrome, 43/530
recombinant chromosome 3, 43/495
Rieger syndrome, 30/247
Roberts syndrome, 30/247
Rubinstein-Taybi syndrome, 31/282, 43/234, 46/91
Russell-Silver syndrome, 30/248
Saethre-Chotzen syndrome, 30/246, 43/322
sclerosteosis, 31/258
Seckel dwarf, 30/248
Seckel syndrome, 43/379
self-mutilation encephalopathy, 42/141
sensorineural deafness, 43/417
sincipital encephalocele, 50/104
Sjögren-Larsson syndrome, 13/474, 30/248, 66/615
Sprengel deformity, 30/248
Stickler syndrome, 30/247
swayback nose, 43/439
trigeminal nerve agenesis, 50/214
trisomy 8, 50/563
trisomy 8 mosaicism, 43/510
Turner syndrome, 14/109, 30/248-250, 43/550, 50/543
ventricular dilatation, 43/417
Vogt-Waardenburg syndrome, 30/246
Wildervanck syndrome, 30/247
Williams syndrome, 30/250
Wolf-Hirschhorn syndrome, 30/249, 43/497
XXXX syndrome, 30/250, 43/554
XXXXX syndrome, 30/250, 43/554
Hypertelorism hypospadias syndrome
cheilopalatoschisis, 43/413
cryptorchidism, 43/413
dermatoglyphics, 43/413
heart disease, 43/413
imperforate anus, 43/413
mental deficiency, 43/413
telecanthus, 43/413
Hypertelorism microtia clefting syndrome, 43/414-416
mental deficiency, 43/415
syndactyly, 43/415
Hypertension
see also Blood pressure
acute intermittent porphyria, 41/434, 42/617

acute paroxysmal, *see* Acute paroxysmal hypertension
adrenalin, 75/471
alcohol, 55/523
allergic granulomatous angiitis, 63/383
amphetamine addiction, 55/519
anoxic ischemic leukoencephalopathy, 9/589
arterial, *see* Arterial hypertension
arteriolar fibrinohyalinoid degeneration, 11/601
arteriosclerosis, 75/479
autonomic nervous system, 74/118, 170, 75/246, 453-455
baroreflex, 75/476
benign intracranial, *see* Benign intracranial hypertension
Binswanger disease, 46/318, 359
brain arteriosclerosis, 63/77
brain blood flow, 54/144
brain hemorrhage, 11/587-601, 54/288
brain infarction, 53/464
burning feet syndrome, 28/12
carbon disulfide intoxication, 64/27
cause, 75/455
cerebellar hemorrhage, 54/291, 55/91
cerebrovascular disease, 12/449, 453, 53/17, 75/368
Charcot-Bouchard aneurysm, 12/447
chronic, *see* Chronic hypertension
ciclosporin, 75/476
congenital torticollis, 43/146
CSF, *see* CSF hypertension
diamorphine addiction, 55/517
dissecting aorta aneurysm, 63/46
distal, *see* Distal hypertension
endothelin 1, 75/473
ephedrine intoxication, 65/260
facial paralysis, 8/283
Friedreich ataxia, 60/667
genetic influence, 75/474
genetics, 75/246
granulomatous CNS vasculitis, 55/388
headache, 48/5, 19, 282
heart disease, 8/283
hyperaldosteronism, 39/494, 496
hyperparathyroidism, 27/286
insulin resistance, 75/480
intracranial, *see* Intracranial hypertension
intracranial nervous, *see* Intracranial nervous hypertension
keloid, 43/146
left ventricular hypertrophy, 75/479
leukoaraiosis, 63/76
malignant, *see* Malignant hypertension
malignant hyperthermia, 43/117
migraine, 48/19, 85, 156
morning headache, 5/150
moyamoya disease, 55/295
neurogenic arterial, *see* Neurogenic arterial hypertension
nitric oxide, 75/465
noradrenalin, 75/459
obesity, 75/477
organophosphate intoxication, 64/228
papaverine intoxication, 37/456
paroxysmal, *see* Paroxysmal hypertension
Parry disease, 42/467, 60/667
pharmacology, 74/170
phencyclidine addiction, 55/522
phenylpropanolamine intoxication, 65/259
pheochromocytoma, 11/458, 14/252, 702, 750, 39/498, 42/764, 48/427, 75/476
pinealoma, 42/767
polyarteritis nodosa, 8/125, 39/296, 308, 55/354, 356, 396
propranolol, 37/444
pseudoxanthoma elasticum, 11/465, 14/775, 43/45, 55/451-453
pulmonary, *see* Pulmonary hypertension
recumbent, 75/115
renal, *see* Renal hypertension
renal insufficiency, 75/475
renin-angiotensin system, 75/472
renovascular, *see* Renovascular hypertension
retinal artery occlusion, 55/112
scorpion sting intoxication, 37/112
secondary, 75/474
sensorineural deafness, 42/389
sleep apnea, 75/477
Sneddon syndrome, 55/404
spontaneous cerebellar hemorrhage, 12/54
stroke, 75/368
subarachnoid hemorrhage, 55/2
survey, 75/453
sympathetic nerve activity, 75/456
sympathoactivation, 75/463
transient global amnesia, 45/211
vascular, *see* Vascular hypertension
ventricular arrhythmia, 75/480
yohimbine, 74/168

Hypertensive encephalopathy, 11/552-574
acute, *see* Acute hypertensive encephalopathy
acute headache, 5/154
age, 54/211
aneurysm, 11/562
anoxic ischemic leukoencephalopathy, 47/536-538
aorta coarctation, 54/212

aphasia, 54/212
arterial hypertension, 63/72
arteriolar fibrinohyalinoid degeneration, 11/560-562, 47/536, 54/213
autoregulation failure, 63/72
blood-brain barrier, 54/214
brain blood flow, 11/571, 53/51, 54/213
brain blood flow increase, 63/72
brain edema, 11/564, 16/195, 54/213, 63/72
brain hemorrhage, 11/587
brain infarction, 11/559, 53/158, 54/215
brain ischemia, 53/158
brain lacunar infarction, 11/559, 47/537, 54/213
cerebrovascular disease, 53/39, 41
chronic hypertension, 47/538
clonidine, 54/217
collagen vascular disease, 54/212
coma, 11/555, 54/212
computer assisted tomography, 54/60, 212
cranial nerve palsy, 11/558
CSF, 11/557, 54/212
definition, 54/211
diazoxide, 54/216
diuretic agent, 54/217
ECG, 54/213
eclampsia, 54/212
EEG, 11/559, 54/213
epilepsy, 11/553, 54/212
essential hypertension, 54/212
glomerulonephritis, 54/212
glyceryl trinitrate, 54/216
headache, 11/552, 48/418, 54/212
hydralazine, 54/216
interstitial nephritis, 54/212
isoelectric EEG, 54/215
labetalol, 54/216
leukoencephalopathy, 9/613
malignant hypertension, 9/613, 54/211
mechanism, 53/158
mental confusion, 53/237
mental disorder, 11/557
methyldopa, 54/217
microinfarction, 54/213
nifedipine, 54/217
papilledema, 11/557, 54/212, 63/72
pathogenesis, 54/213
pathology, 54/213
petechial hemorrhage, 54/213
phencyclidine addiction, 55/522
phentolamine, 54/216
phenylpropanolamine intoxication, 65/259
pheochromocytoma, 39/501, 54/212
pontine hemorrhage, 12/45, 49
pontine infarction, 12/24
pregnancy, 63/73
prevalence, 54/212
prognosis, 54/217
pseudouremia, 54/211
renal hypertension, 54/212
renal insufficiency, 11/554
reserpine, 54/217
retinopathy, 47/536, 54/212
sodium nitroprusside, 54/216
space occupying lesion, 16/239
stupor, 54/212
subarachnoid hemorrhage, 55/20
symptom, 54/211
systemic sign, 54/212
Takayasu disease, 55/337
treatment, 11/573, 54/215
trimetaphan, 54/216
ultrastructure, 11/565
uremia, 63/504
uremic encephalopathy, 54/211
urine, 54/213
vascular change, 47/536
vascular permeability, 54/214
vascular spasm, 11/566, 571
visual disorder, 11/555
Hypertensive fibrinoid arteritis
azocarmine, 11/136
basal ganglion, 11/136
brain hemorrhage, 11/139
cerebrovascular disease, 11/136-139
general pathology, 11/136-139
hyaline, 11/136
inflammatory response, 11/137
phosphotungstic acid hematoxylin, 11/136
pons, 11/136
staining, 11/136
Hyperthermia
alcohol, 75/62, 495
amphetamine, 75/64
amphetamine intoxication, 65/257
anencephaly, 50/80
autonomic lesion, 1/438
beta adrenergic receptor blocking agent, 75/62
brain embolism, 53/207
brain infarction, 53/207
cerebrovascular disease, 53/207
disorder, 75/62
drug, 74/457, 75/62
endocrine disorder, 75/62
glioblastoma multiforme, 18/64
glutaryl-coenzyme A dehydrogenase deficiency, 66/647

hereditary sensory and autonomic neuropathy type III, 1/477, 21/111, 60/26
hypothalamic lesion, 2/447, 75/64
infantile Gaucher disease, 42/438
intracranial hemorrhage, 53/207
late cortical cerebellar atrophy, 21/477
lithium intoxication, 64/295
lysergide intoxication, 75/64
malignant, *see* Malignant hyperthermia
memory disorder, 27/463
microcephaly, 50/279
midbrain syndrome, 24/585
migraine coma, 60/651
multiple sclerosis, 9/120, 412
oxidation-phosphorylation coupling defect, 62/506
primary pontine hemorrhage, 2/253
spinal cord, *see* Spinal cord hyperthermia
spinal cord injury, 26/369, 372, 61/283
stress, 74/320, 455
survey, 71/585
tetraplegia, 61/275
thalamic syndrome (Dejerine-Roussy), 2/482
thermoregulation, 26/369, 372, 75/62
transient ischemic attack, 53/207
Hyperthyroid chorea, 70/87
clinical features, 27/279
CSF homovanillic acid, 27/280
dopamine, 27/280
predisposition, 49/365
restless, 27/268
terminology, 27/268
treatment, 27/279
Hyperthyroid polyneuropathy
see also Endocrine polyneuropathy
Basedow paraplegia, 51/516
carpal tunnel syndrome, 51/517
clinical features, 51/516
incidence, 51/517
pathogenesis, 51/517
pathology, 51/518
sensorimotor polyneuropathy, 51/518
subclinical polyneuropathy, 51/516
treatment, 51/518
type, 51/516
Hyperthyroidism
see also Basedow disease *and* Thyroid disease
amyotrophic lateral sclerosis, 59/202
anxiety, 46/431
autoimmune disease, 41/235-240, 335
brain embolism, 70/92
chorea, *see* Hyperthyroid chorea
choreoathetosis, 49/385
clinical features, 40/303, 317, 332
corticospinal tract disease, 70/86
Dalrymple sign, 2/318
delirium, 46/544
dementia, 46/396
demyelination, 9/627
differential diagnosis, 40/438, 41/209, 335, 337, 62/528
Eaton-Lambert myasthenic syndrome, 62/421
EMG, 1/642
endocrine myopathy, 62/528
exertional headache, 48/375
headache, 48/378, 70/92
hereditary paroxysmal kinesigenic choreoathetosis, 49/355
leukoencephalopathy, 9/627
Luft syndrome, 41/209
mania, 46/450
maternal, *see* Maternal hyperthyroidism
movement disorder, 70/87
myasthenia gravis, 41/101, 43/157, 62/402, 534, 70/85
myopathy, *see* Thyrotoxic myopathy
neurologic manifestation, 27/260
neuropathy, 70/86
ocular change, 27/269-271
oligosymptomatic, *see* Oligosymptomatic hyperthyroidism
osteitis fibrosa generalisata, 14/163
Parkinson disease, 49/96
penis erection, 75/89
periodic paralysis, 27/268, 28/597, 40/285, 41/165, 241, 62/457, 468, 70/84
propranolol, 37/446
pseudopseudohypoparathyroidism, 14/185
psychosis, 46/462
spasmodic torticollis, 42/264
thyroid associated ophthalmopathy, 41/240
thyrotoxic tremor, 6/818
tremor, 70/87
vitamin B1 deficiency, 51/331
Von Gräfe sign, 2/318
Hypertonia
Alzheimer disease, 46/252
astasia abasia, 1/346
cri du chat syndrome, 50/275
differential diagnosis, 1/268
drug induced extrapyramidal syndrome, 6/251
extrapyramidal, *see* Rigidity
globoid cell leukodystrophy, 10/302
hereditary progressive diurnal fluctuating dystonia, 49/530
intention, *see* Intention hypertonia

Mast syndrome, 42/282
muscle tone, 1/264-269
muscular, *see* Muscular hypertonia
pyramidal, *see* Spasticity
pyramidal sign, *see* Spasticity
spasticity, 6/851
spinal type, 1/268
stiff-man syndrome, 1/268

Hypertonia musculorum vera
myalgia, 43/84

Hypertonic urea
brain edema, 16/198

Hypertrichosis
De Lange syndrome, 14/784
diastematomyelia, 50/498
glycosaminoglycanosis type IS, 10/437
glycosaminoglycanosis type II, 42/454
glycosaminoglycanosis type III, 66/300
Hurler-Hunter disease, 10/436
intraspinal neurenteric cyst, 20/65, 67, 71
Miescher syndrome type II, 14/120
multiple myeloma polyneuropathy, 51/430
phakomatosis, 14/10
POEMS syndrome, 47/619, 51/430, 63/394
spinal arachnoid cyst, 20/82
spinal intradural cyst, 20/78

Hypertriglyceridemia
neuroacanthocytosis, 60/674
organic acid metabolism, 66/641
Wohlfart-Kugelberg-Welander disease, 59/372

Hypertrophia musculorum vera
areflexia, 43/84
creatine kinase, 43/84
Kocher-Debré-Sémélaigne syndrome, 43/84
lactate dehydrogenase, 43/84
masseter muscle, 43/84
muscle fiber size, 43/84
muscle hypertrophy, 43/83-85
paresthesia, 43/84

Hypertrophic cardiomyopathy
age, 63/136
atrial fibrillation, 63/136
brain embolism, 63/136
brain infarction, 63/137
infectious endocarditis, 63/137
sex ratio, 63/136
symptom, 63/136
syncope, 63/136
transient ischemic attack, 63/137

Hypertrophic interstitial neuropathy, 21/145-165, 60/213-221
see also Hereditary motor and sensory neuropathy type I
acromegaly, 21/145, 51/519
age at onset, 21/149
amyloidosis, 21/145
areflexia, 42/317
Argyll Robertson pupil, 21/151
ataxia, 42/317
autonomic change, 21/152
autosomal dominant, 21/147, 152, 41/432
autosomal recessive, 21/147, 152
axon size, 21/158
biochemistry, 21/161
ceramide hexose, 21/12
ceramide hexoside, 21/148, 161
classification, 7/199
clinical features, 21/148, 60/214-216
clinical variability, 21/146
congenital, 60/214, 217
course, 21/152
cranial nerve, 21/151
CSF protein, 21/161, 60/216
dementia, 42/317
epidemiology, 21/3
epilepsy, 42/317
essential, 21/9
experimental allergic neuritis, 21/163
external ophthalmoplegia, 21/152
Friedreich ataxia, 21/147
genetics, 21/147
π-granule, 21/159
Guillain-Barré syndrome, 21/145
hereditary amyloid polyneuropathy type 1, 21/128
hereditary motor and sensory neuropathy type I, 21/146, 148, 152, 304
hereditary motor and sensory neuropathy type II, 21/146, 148, 152
hexoside sulfate, 21/148
history, 8/169, 21/145, 60/213
Holmes-Adie syndrome, 21/151
hypertrophic nerve, 21/146
incoordination, 21/150
infantile, 60/215, 218
juvenile, 60/216, 219
kyphoscoliosis, 21/152, 42/317
laboratory study, 60/216
leprosy, 21/145
megadolichocolon, 21/152
motor symptom, 21/149
muscular atrophy, 42/317
Nattrass type, 21/146
nerve biopsy, 60/214
nerve enlargement, 60/214
nerve hypertrophy, 21/149
nerve lipid, 21/162

neuropathology, 21/154-160, 51/63
neurophysiology, 21/160
nosology, 60/213
nystagmus, 42/317
onion bulb formation, 42/317, 51/71
optic atrophy, 13/82, 42/317
papilledema, 21/151
pathogenesis, 21/162, 60/219
peripheral pseudotabes, 1/318
pes cavus, 21/152
Pierre Marie-Boveri type, 21/146, 177
polyneuropathy, *see* Hypertrophic polyneuropathy
prevalence, 21/148
pyruvic acid, 21/161
recurrent polyneuropathy, 21/146
reflex, 21/150
Refsum disease, 7/181, 8/23, 21/9, 146, 199, 41/434, 51/385, 60/235, 66/497
remyelination, 51/72
Roussy-Cornil type, 21/146
Roussy-Lévy syndrome, 21/148, 177, 60/220
scapuloperoneal spinal muscular atrophy, 62/170
Schwann cell proliferation, 51/72
secondary pigmentary retinal degeneration, 13/316, 21/151
segmental demyelination, 21/162, 51/72
sensorineural deafness, 21/152
sensory loss, 42/317
sensory symptom, 21/151
transketolase, 21/161
treatment, 21/164
tremor, 42/317
type classification, 21/146, 148
wallerian degeneration, 21/162
Hypertrophic mononeuropathy
monomelic spinal muscular atrophy, 59/376
Hypertrophic nerve
hypertrophic interstitial neuropathy, 21/146
leprous neuritis, 51/225, 231
neurolymphomatosis, 56/179
scapuloperoneal spinal muscular atrophy, 62/170
Hypertrophic neuropathy, *see* Hypertrophic interstitial neuropathy
Hypertrophic osteoarthropathy, *see* Bamberger-Marie syndrome
Hypertrophic pachymeningitis
Brucella, 51/185
decompressive laminectomy, 33/190
dura mater, 33/190
radicular pain, 33/190
spinal cord compression, 33/190
Hypertrophic polyneuropathy
Behr disease, 13/91
cranial nerve, 56/183
hereditary sensory and autonomic neuropathy type I, 21/77
Leber optic atrophy, 13/97
nerve conduction, 7/180
neurolymphomatosis, 56/179
optic atrophy, 13/82
pachydermia, 14/791
secondary pigmentary retinal degeneration, 13/316
spinal tumor, 19/87
Hypertrophic sensory neuropathy
Sjögren syndrome, 51/449
Hypertrophied muscle, *see* Muscle hypertrophy
Hypertrophy
adrenal cortex, *see* Adrenal cortex hypertrophy
bladder wall, *see* Bladder wall hypertrophy
calf, *see* Calf hypertrophy
cerebellar cortex, *see* Cerebellar cortex hypertrophy
cerebellar granular cell, *see* Cerebellar granular cell hypertrophy
cranial bone, *see* Cranial bone hypertrophy
diffuse cerebellar, *see* Diffuse cerebellar hypertrophy
hemangiectatic, *see* Klippel-Trénaunay syndrome
hemifacial, *see* Hemifacial hypertrophy
idiopathic esophagus, *see* Idiopathic esophagus hypertrophy
inferior olivary, *see* Inferior olivary hypertrophy
ligamentum flavum, *see* Ligamentum flavum hypertrophy
muscle, *see* Muscle hypertrophy
muscle fiber, 62/15
muscle fiber type II, *see* Muscle fiber type II hypertrophy
pickwickian syndrome, 38/299
posterior arch, *see* Posterior arch hypertrophy
pseudo, *see* Pseudohypertrophy
sternocleidomastoid muscle, *see* Sternocleidomastoid muscle hypertrophy
tendon, *see* Tendon hypertrophy
Hyperuricemia
see also Lesch-Nyhan syndrome
ataxia, 42/571
Burkitt lymphoma, 63/352
diazoxide intoxication, 37/433
Down syndrome, 50/531
hyperlipoproteinemia type III, 10/275
Lesch-Nyhan syndrome, 6/456, 22/514, 29/265, 42/153, 46/194, 49/385, 59/360, 60/19
Machado disease, 42/155
muscular atrophy, 42/571

organic acid metabolism, 66/641
phosphoribosyl pyrophosphate synthetase, 42/572
sensorineural deafness, 42/571
Wolff-Parkinson-White syndrome, 42/571
Hypervalinemia, 42/573-575
diet, 42/574
EEG, 42/574
hyperkinesia, 42/574
maple syrup urine disease, 29/55
mental deficiency, 42/574
nystagmus, 42/574
valine transamination, 42/574
vomiting, 42/574
Hyperventilation
autonomic nervous system, 74/586, 75/261
brain infarction, 53/422
brain injury, 24/605, 611
brain stem glioma, 63/440
central neurogenic, *see* Central neurogenic hyperventilation
cerebellar hypoplasia, 42/19
cerebellar vermis agenesis, 42/4
Cheyne-Stokes respiration, 1/497
cluster headache, 75/289
cyanide intoxication, 65/28
cyanogenic glycoside intoxication, 65/28
EEG, 23/356, 359
headache, 75/282
hyperventilation syndrome, 63/432
hypoxia, 75/261
intracranial hypertension, 24/611
intracranial pressure, 23/214
juvenile myoclonic epilepsy, 73/165
malignant hyperthermia, 38/556, 43/117, 63/439
neurogenic, 1/496
neurologic intensive care, 55/214
pheochromocytoma, 39/501, 42/764
pontine glioma, 63/440
postoperative brain swelling, 57/280
pyruvate dehydrogenase deficiency, 60/667
respiration, 1/659, 675
Rett syndrome, 29/315, 60/636, 667, 63/437
Reye syndrome, 29/333, 34/169, 49/218, 56/151, 238, 63/438, 65/116
severe brain injury, 24/612
spinal cord injury, 61/263, 445
tetraplegia, 61/445
transverse spinal cord lesion, 61/445
valproic acid induced hepatic necrosis, 65/125
Hyperventilation syndrome, 38/309-354
see also Hyperpnea
absent alveolar carbon dioxide rise, 63/433
acroparesthesia, 63/432
adrenalin, 63/434
age, 38/327, 63/434
alveolar carbon dioxide, 38/333, 63/432
ammonium chloride, 63/429
antidepressant, 38/329, 45/254
anxiety, 38/320, 63/436
arthralgia, 38/319, 63/435
beta adrenergic receptor blocking agent, 38/329, 63/442
blood pH, 38/315, 63/434
brain blood flow, 38/318, 63/435
brain vasoconstriction, 38/318, 63/435
capnography, 63/432
carbon dioxide administration, 63/442
carpopedal spasm, 38/319, 326, 63/435
cause, 38/320-322
cell calcium, 63/435
chest pain, 63/436
Chvostek sign, 38/326, 63/432
clinical features, 38/318-320
criteria, 63/432
diagnosis, 38/322-327, 63/432
diagnostic test, 63/433
differential diagnosis, 38/323, 63/441
dysphagia, 38/320, 63/436
ECG, 38/319, 63/436
EEG, 38/318, 63/435
globus, 38/320, 63/436
Gruber syndrome, 63/438
headache, 63/433
history, 38/309-315, 63/429
hyperventilation, 63/432
hypnosis, 38/330, 63/443
incidence, 38/327
metabolic alkalosis, 63/435
muscle cramp, 63/435
muscle spasm, 63/435
myalgia, 38/319, 63/435
myoclonus, 63/435
occupational therapy, 38/330, 63/443
palpitation, 63/436
paresthesia, 38/318, 63/433
pathogenesis, 63/429
pathophysiology, 63/433
predisposing factor, 38/327, 63/433
prevalence, 63/433
pulse rate, 63/436
rebreathing, 38/328, 63/442
recovery phase, 63/433
respiratory alkalosis, 38/315-317, 63/434
respiratory exercise, 38/328, 63/442
respiratory frequency, 38/333, 63/433
respiratory tetany, 38/315, 326, 328, 63/429

sequela, 63/441
sex ratio, 38/327, 63/433
Smith-Lemli-Opitz syndrome, 63/438
symptom, 63/432
symptom frequency, 63/433
syncope, 63/441
tachycardia, 38/319, 63/436
tetany, 38/315, 318, 63/435
treatment, 38/327, 63/442
tremor, 38/319, 63/435
Trousseau sign, 38/326, 63/432
vertigo, 63/433

Hyperviscosity syndrome
see also Blood hyperviscosity, Blood viscosity *and* Brain blood flow
ataxia, 55/484
brain infarction, 53/165, 55/489
brain ischemia, 53/165
Burr cell formation, 55/484
cerebrovascular disease, 55/470, 483
CNS circulation, 71/439
confusional state, 55/470
corpuscular type, 55/483
cryofibrinogen, 55/490
cryoglobulinemia, 55/484, 490, 71/439
dementia, 55/487
epilepsy, 55/484
erythrocytosis, 55/484
headache, 55/484
heavy chain disease, 55/484
hemoglobinopathy, 55/484
hyperfibrinogenemia, 55/484
leukemia, 55/471, 484
lymphoma, 55/484
lymphosarcoma, 55/487
macroglobulinemia, 55/484
malignancy, 55/485
malignant lymphoma, 55/487
multiple myeloma, 55/484, 486
nonbacterial thrombotic endocarditis, 55/489
paraproteinemia, 55/484, 71/439
plasmatic type, 55/483
polycythemia vera, 55/467, 484
pyruvate kinase deficiency, 55/484
Raynaud phenomenon, 55/485
reticulosarcoma, 55/487
retinopathy, 55/484
rheumatoid arthritis, 55/485
sickle cell anemia, 55/484
Sjögren syndrome, 55/485
spherocytosis, 55/484
symptom, 55/470, 484
syphilis, 55/485
systemic lupus erythematosus, 55/485
transient ischemic attack, 55/485
vertigo, 55/484
Waldenström macroglobulinemia, 55/485, 63/397
Waldenström macroglobulinemia polyneuropathy, 39/532

Hypervitaminosis A, *see* Vitamin A intoxication

Hypholoma fasciculare
mushroom intoxication, 36/542

Hypnagogic hallucination
definition, 45/352
diagnosis, 45/352
dreamy state, 45/355
narcolepsy, 2/448, 15/845, 45/148, 354
narcoleptic syndrome, 42/712
obstructive sleep apnea, 63/453
sleep, 3/86, 90
sleep paralysis, 42/712

Hypnagogic imagery
definition, 45/352
psychosis, 45/353

Hypnic headache syndrome
biological rhythm, 74/507
lithium, 74/509
suprachiasmatic nucleus, 74/507

Hypnogenic paroxysmal dystonia
hereditary paroxysmal kinesigenic choreoathetosis, 49/356

Hypnosis
consciousness, 3/117, 126
hyperventilation syndrome, 38/330, 63/443
muscle contraction headache, 5/163
pain, 45/239
phantom limb, 4/227-230, 45/398
posttraumatic amnesia, 23/423

Hypnotic agent
adverse effect, 65/329
barbiturate, 37/349-353
benzodiazepine, 37/355-360
carbamate, 37/353
chronic muscle contraction headache, 48/350
delirium, 46/539
EEG, 37/351
flurazepam, 37/355
headache, 48/39
history, 37/347
iatrogenic neurological disease, 65/329
memory disorder, 27/463
methyprylon, 37/354
nitrazepam, 37/355
pain, 69/47
progressive dysautonomia, 59/160
quinazolinone, 37/355

sleep disorder, 3/91
Hypnotic agent withdrawal delirium
atypical features, 46/541
Hypnotic blindness
consciousness, 3/117
Hypnotoxin intoxication
physalia, 37/37
Hypoadrenocorticism
pituitary adenoma, 17/393
Hypoalbuminemia
brain fat embolism, 55/179
pressure sore, 61/350
Hypoalphalipoproteinemia
autosomal dominant, *see* Autosomal dominant hypoalphalipoproteinemia
Hypobetalipoproteinemia
acanthocyte, 49/327, 332
apolipoprotein B, 63/275
apolipoprotein B-100 gene, 66/546
autosomal dominant, 63/275
familial, *see* Familial hypobetalipoproteinemia
homozygous, 63/275
neuroacanthocytosis, 29/394, 60/673, 63/271
neurologic dysfunction, 29/394
secondary pigmentary retinal degeneration, 60/653
Hypocalcemia
Burkitt lymphoma, 63/352
calcium, 28/434
calcium imbalance, 28/536-539
central manifestation, 28/534-536
Chvostek sign, 40/330
coma, 63/561
confusion, 63/558, 561
confusional state, 63/558, 561
delirium, 46/543
dementia, 63/561
dystonia, 63/561
EEG, 28/534
electrolyte disorder, 63/558
epilepsy, 63/558
ethylene glycol intoxication, 64/122
headache, 48/6, 34, 63/561
hypoparathyroid myopathy, 62/541
hypoparathyroidism, 27/296-298, 303, 42/577, 63/558, 561
intracranial calcification, 28/536
intracranial hypertension, 63/558, 561
intracranial pressure, 63/561
isovaleric acidemia, 42/579
mental change, 28/536
metabolic encephalopathy, 63/558
multiple sclerosis, 47/39
myoglobinuria, 62/556
neonatal, 70/123
nerve conduction, 47/39
newborn epilepsy, 15/200
obtundation, 63/558
opisthotonos, 63/559
papilledema, 63/561
peripheral manifestation, 28/532-534
personality change, 63/561
pertussis, 33/278
pseudohypoparathyroidism, 42/621, 46/85, 63/558, 561
pseudopseudohypoparathyroidism, 63/561
psychiatry, 70/122
rickets, 63/558
spinal reflex change, 28/535
striatopallidodentate calcification, 63/561
symptom, 70/112
syncope, 72/262
tetany, 27/273, 28/532-536, 33/278, 63/558
Hypocapnia
autonomic nervous system, 75/261
brain injury, 57/212
enflurane intoxication, 37/412
neurologic intensive care, 55/214
vinyl ether intoxication, 37/412
Hypocarbia, *see* Hypocapnia
Hypochlorite Schiff staining
protein, 9/23
Hypocholesterolemia
Bassen-Kornzweig syndrome, 42/511
Hypochondriasis
atlas syndrome, 5/373
body scheme disorder, 45/380
dominant hemisphere, 46/581
hysteria, 46/580
Hypochromic microcytic anemia, *see* Iron deficiency anemia
Hypocomplementemia
polymyositis, 41/78
Hypocortisolism
adrenal disorder, 70/211
classification, 70/211
Hypocupremia
trichopoliodystrophy, 46/61
Hypodipsia
hypothalamic lesion, 1/500
Hypoesthesia
anterior cervical cord syndrome, 25/270
benign late hereditary cerebellar ataxia, 60/661
Biemond syndrome, 60/661
blennorrhagic polyneuritis, 7/478
Charlevoix-Saguenay spastic ataxia, 60/661

deafness, 60/661
dentatorubropallidoluysian atrophy, 60/661
Ekbom syndrome type II, 60/661
face, 12/4
Friedreich ataxia, 60/661
Gerstmann-Sträussler-Scheinker disease, 60/661
Joseph-Machado disease, 60/661
olivopontocerebellar atrophy (Dejerine-Thomas), 60/661
olivopontocerebellar atrophy (Schut-Haymaker), 60/661
olivopontocerebellar atrophy (Wadia-Swami), 60/661
optic atrophy, 60/661
parietal lobe tumor, 17/300
pharyngeal, *see* Pharyngeal hypoesthesia
polyneuropathy, 60/661
sensation, 1/89
terminology, 1/89
Hypofibrinogenemia
intravascular consumption coagulopathy, 55/496
Hypofrontality
schizophrenia, 46/485, 487
Hypogammaglobulinemia
ataxia telangiectasia, 14/269, 42/120
X-linked, *see* X-linked hypogammaglobulinemia
Hypogastric nerve
anatomy, 1/361-363
physiology, 1/369
Hypogastric plexus
anatomy, 1/362
physiology, 1/369
Hypogenitalism, *see* Hypogonadism
Hypogeusia
hyponatremia, 63/545
Sjögren syndrome, 71/74
zinc intoxication, 36/335
Hypoglossal artery
anatomy, 11/1
primitive, *see* Primitive hypoglossal artery
Hypoglossal facial anastomosis
facial nerve injury, 24/113
Hypoglossal nerve
abnormality, 30/411
anatomy, 2/80
embryology, 30/398-400
injury, 24/179
Jackson syndrome, 1/189
neurinoma, 14/152, 68/542
subarachnoid hemorrhage, 55/18
tongue paralysis, 30/411
Hypoglossal nerve abnormality
trigeminal nerve agenesis, 50/213
Hypoglossal nerve agenesis
clinical features, 50/220
cranial nerve agenesis, 50/220
external ophthalmoplegia, 50/220
facial diplegia, 50/220
pathology, 50/220
Hypoglossal nerve injury
radiation induced, 67/343
Hypoglossal paralysis
carotid dissecting aneurysm, 54/272
cephalic tetanus, 7/476
tetanus, 7/476
Hypoglossal vein
anatomy, 11/61
Hypoglycemia
acyl-coenzyme A dehydrogenase, 37/527
adrenal insufficiency, 27/70
adult, 27/53
amyotrophy, 22/29
Beckwith-Wiedemann syndrome, 43/398
brain development, 27/65
brain energy metabolism, 27/23, 56-67
brain infarction, 53/167
brain injury, 70/184
brain ischemia, 53/167
Burkitt lymphoma, 63/352
carnitine deficiency, 43/176
child, 27/54-56
chronic axonal neuropathy, 51/531
classification, 27/55, 70/175
CNS, 27/53-73, 70/171
debrancher deficiency, 27/72, 43/179
definition, 70/171
delirium, 46/542
demyelination, 9/626
diabetes mellitus, 27/131
diagnosis, 27/53
endocrine disorder, 27/70
epidemiology, 70/174
epilepsy, 70/185, 72/222
ethylmalonic aciduria, 37/535
experimental, 27/56-67
familial, *see* Familial hypoglycemia
focal deficit, 70/179
food deprived, 70/176
fructose intolerance, 42/548
functional hyperinsulinemia, 27/69
functional hyperinsulinism, 27/69
galactosemia, 27/72, 42/551
glucagon deficiency, 27/71
glutaric aciduria type II, 62/495
glycogen storage disease type I, 27/71, 225
glycogen synthetase deficiency, 27/72

headache, 48/6, 34
hepatic coma, 27/353, 371
hepatic encephalopathy, 27/353
hepatophosphorylase deficiency, 27/72
hereditary fructose intolerance, 27/72
hereditary hepatic enzyme deficiency, 27/71
hyperinsulinism, 27/55
hypoglycin intoxication, 65/79, 82
hypopituitarism, 27/70
hypothyroidism, 27/70
idiopathic, *see* Idiopathic hypoglycemia
infant, 27/54-56
insulin induced, 27/57-64
ischemia, 70/185
islet cell carcinoma, 27/68
Jamaican vomiting sickness, 37/511, 532, 65/79
ketotic, *see* Ketotic hypoglycemia
late cortical cerebellar atrophy, 21/477
leukoencephalopathy, 9/626
lipid metabolic disorder, 62/494
liver disease, 27/71
malaria, 52/369
maple syrup urine disease, 28/517, 29/64, 42/599
metabolic encephalopathy, 69/404
methylmalonic acidemia, 29/205, 42/601
migraine, 48/36, 120
multiple acyl-coenzyme A dehydrogenase deficiency, 62/495
neuropathy, 8/52, 70/157
newborn epilepsy, 15/200, 203
organic acid metabolism, 66/641
orthostatic hypotension, 63/148
pallidostriatal necrosis, 49/466
pancreatic polypeptide, 74/624
pertussis encephalopathy, 33/277
pituitary dwarfism type I, 42/612
pituitary dwarfism type IV, 42/616
pituitary dwarfism type V, 42/616
pituitary gland agenesis, 42/11
primary carnitine deficiency, 66/402
progressive dysautonomia, 59/155
progressive spinal muscular atrophy, 22/29
pyruvate carboxylase deficiency, 29/212, 42/516
Reye syndrome, 9/549, 27/363, 29/335, 34/80, 169, 56/152, 65/118
seizure, 72/261
sequela, 70/180
severe, 70/178
spinal cord injury, 61/446
spontaneous, 27/55, 67-73
striatal necrosis, 49/500
surgically induced, 27/56
survey, 70/131
symptom, 70/173
syncope, 72/261
systemic brain infarction, 11/457
tetraplegia, 61/446
transient neonatal, *see* Transient neonatal hypoglycemia
transverse spinal cord lesion, 61/446
valproic acid induced hepatic necrosis, 65/126
vitamin B12 intoxication, 28/212
Hypoglycemia unawareness
pathogenesis, 70/182
survey, 70/180
Hypoglycemic coma
differential diagnosis, 12/44
primary pontine hemorrhage, 12/44
Hypoglycemic encephalopathy
carnitine deficiency, 62/493
childhood myoglobinuria, 62/563, 566
hepatic carnitine palmitoyltransferase deficiency, 62/494
long chain 3-hydroxyacyl-coenzyme A dehydrogenase deficiency, 62/495
medium chain acyl-coenzyme A dehydrogenase deficiency, 62/495
primary carnitine deficiency, 66/403
primary systemic carnitine deficiency, 62/493
striatal lesion, 6/675
striatal necrosis, 49/503
Hypoglycemic shock
pallidal necrosis, 49/479
Hypoglycin
Acer pseudoplatanus, 65/81
acyl-coenzyme A dehydrogenase, 37/523-526, 65/130
amino acid metabolism, 37/521
analogue, 65/82
fatty acid oxidation, 37/521
gluconeogenesis inhibition, 65/99
glutaryl coenzyme A, 65/130
isovaleryl coenzyme A, 65/130
leucine similarity, 65/83
Litchi chinensis, 65/81
metabolism, 65/83
α-methylbutyryl coenzyme A, 65/130
3-methylenecyclopropylacetic acid, 65/83
neurotoxin, 37/511, 49/217, 65/79, 130
pharmacology, 65/84
toxic encephalopathy, 49/217
toxicity, 65/84
valproic acid induced hepatic necrosis, 65/126
Hypoglycin A intoxication
Jamaican vomiting sickness, 37/511-539
Hypoglycin intoxication

see also Jamaican vomiting sickness
acer tree, 65/81
acidosis, 65/80
acyl glycinuria, 65/93
animal experiment, 65/85
Blighia sapida seed, 65/82
branched chain amino acid, 65/95
2 clinical types, 65/80
clofibrate feeding, 65/96
epilepsy, 65/79
fatty acid oxidation, 65/85
fatty acid β-oxidation, 65/83, 87, 89
fatty liver degeneration, 65/85
fulminant course, 65/80
gluconeogenesis inhibition, 65/97
glutaric aciduria type II, 65/104
glutaryl coenzyme A, 65/130
glyceraldehyde-3-phosphate dehydrogenase, 65/130
glycine, 65/104
hypoglycemia, 65/79, 82
hypoglycin A, 37/514, 65/80
hypoglycin B, 37/514, 65/80
hypoglycin chemistry, 65/80
intravenous glucose, 65/105
isomer, 65/81
isovaleric acidemia, 65/131
isovaleryl coenzyme A, 65/130
litchi fruit, 65/81
malate shuttle, 65/130
α-methylbutyryl coenzyme A, 65/130
3-methylenecyclopropylacetic acid, 37/517, 65/82
methylenecyclopropylglycine, 65/91
mitochondria, 65/86
mortality, 65/80
organic acidemia, 65/84, 94
organic aciduria, 37/529, 65/93
β-oxidation inhibition, 65/90
palmitoylkarnitine oxidation, 65/92
4-pentenoic acid, 65/85, 92
peroxisome, 65/96
plasma free fatty acid, 65/84
Reye syndrome, 37/514, 65/80
salicylic acid like syndrome, 65/80
short chain fatty acid, 65/131
spiropentaneacetic acid, 65/92
stereochemistry, 65/81
toxic encephalopathy, 65/79
treatment, 65/104
valproic acid like syndrome, 65/80
vitamin B2, 65/105
Hypoglycorrhachia
bacterial meningitis, 33/7
glioblastoma multiforme, 18/54
neurocutaneous melanosis, 43/33
neurocysticercosis, 52/531
staphylococcal meningitis, 33/71
streptococcal meningitis, 33/71
varicella zoster virus meningoencephalitis, 56/236
Hypogonadal cerebellar ataxia, 21/467-476
arachnodactyly, 21/467
autosomal recessive, 21/475
cause, 21/475
clinical features, 21/467-469
dementia, 21/469
diabetes mellitus, 21/476
Friedreich ataxia, 21/348, 476, 42/129
gonadotropin, 21/468
neuropathology, 21/469
optic atrophy, 21/468, 476
pathology, 21/469-475
pes cavus, 21/468
retinal degeneration, 21/468
Richards-Rundle syndrome, 21/468, 42/129
Roussy-Lévy syndrome, 21/476
Hypogonadism
acrocephalosyndactyly type II, 38/422
acrodysostosis, 31/230
adrenomyeloneuropathy, 60/170
alopecia, 42/374
Alström-Hallgren syndrome, 22/502
anencephaly, 42/13
anosmia, 60/652
ataxia, *see* Matthews-Rundle syndrome
ataxia telangiectasia, 14/274, 291, 517, 42/120, 60/374
Bardet-Biedl syndrome, 13/389, 462, 22/508, 43/233
Behr disease, 60/669
Biemond syndrome type II, 43/334
Börjeson-Forssman-Lehmann syndrome, 42/530
bulbospinal muscular atrophy, 59/44
caudal aplasia, 42/50
cerebellar ataxia, *see* Hypogonadal cerebellar ataxia
congenital muscular dystrophy, 43/89
cryptophthalmos syndrome, 43/374
Down syndrome, 43/545, 50/532
familial ectodermal dysplasia, 60/668
Fanconi syndrome, 43/16
fetal face syndrome, 43/400
Friedreich ataxia, 21/348
gynecomastia, 60/668
hemochromatosis, 42/553
hereditary cerebello-olivary atrophy (Holmes), 21/403, 60/569

high frequency hearing loss, 42/377
hypergonadotropic, *see* Hypergonadotropic hypogonadism
hypogonadotrophic, *see* Hypogonadotrophic hypogonadism
Kallmann syndrome, 43/418
Klinefelter syndrome, 43/558, 560, 60/668
Klinefelter variant XXXY, 43/559
Klinefelter variant XXYY, 43/563
Klippel-Feil syndrome, 42/37
Laurence-Moon syndrome, 43/253
Marinesco-Sjögren syndrome, 42/184, 60/343, 668
masseter muscle, 43/138
Matthews-Rundle syndrome, 60/578, 652, 668
mental deficiency, 43/123-125, 272-275
multiple nevoid basal cell carcinoma syndrome, 14/466, 43/32
muscular dystrophy, 43/89, 139
myopathy, 43/123-125
myotonic dystrophy, 43/152
neuroaxonal dystrophy, 60/669
ophthalmoplegia, 43/138-140
pallidocerebello-olivary degeneration, 60/668
POEMS syndrome, 51/430
Prader-Labhart-Willi syndrome, 31/322, 324, 40/336, 43/463, 464
ptosis, 43/138
18q partial monosomy, 43/533
6q partial trisomy, 43/503
Richards-Rundle syndrome, 21/468, 22/516, 43/264, 60/668
Rothmund-Thomson syndrome, 14/777, 39/405, 43/460, 46/61
Rud syndrome, 13/321, 43/284, 51/398, 60/721
secondary pigmentary retinal degeneration, 60/722
sensorineural deafness, 42/374, 377, 389, 43/272
spinal muscular atrophy, 59/44
trisomy 22, 43/548
Tunbridge-Paley disease, 22/508
Turner syndrome, 43/550, 50/544
Weiss-Alström syndrome, 13/463
xeroderma pigmentosum, 60/668
XX male, 43/555

Hypogonadotrophic hypogonadism
Möbius syndrome, 59/375

Hypohidrosis
thermoregulation, 75/64

Hypohidrotic ectodermal dysplasia
hypertelorism, 30/248
mental deficiency, 30/248
skin manifestation, 30/248

Hypokalemia
areflexia, 63/557
barium carbonate intoxication, 63/557
barium intoxication, 64/354
Bartter syndrome, 42/528
clinical features, 41/287
cotton seed intoxication, 63/557
delirium, 46/543
diuretic agent, 65/438
diuretic agent intoxication, 65/440
electrolyte disorder, 63/557
endemic periodic paralysis, 63/557
fatigue, 63/557
generalized weakness, 63/557
hypoparathyroidism, 27/298
malaise, 63/557
muscle necrosis, 41/274
muscle weakness, 63/557
nerve conduction, 7/169
pancreatic encephalopathy, 27/456
pancreatitis, 27/456
quadriparesis, 63/557
restless legs, 63/557
rhabdomyolysis, 63/557
tetany, 63/558
tetraplegia, 63/557
toxic myopathy, 62/603

Hypokalemic periodic paralysis, 28/582-593, 41/149-158
abortive attack, 41/149, 62/458
age at onset, 28/583
anesthesia, 62/460
associated disease, 41/152
attack, 41/149, 62/457
attack frequency, 28/583, 41/150
attack provocation, 41/150, 62/458
attack rate, 62/458
barium, 64/16
clinical features, 28/582, 584, 592, 41/149
differential diagnosis, 28/592
EMG, 28/583, 62/459
familial, *see* Familial hypokalemic periodic paralysis
familial periodic paralysis, 62/457
fatality, 28/592, 41/151
female nonpenetrance, 62/460
genetics, 41/151, 425
geographic distribution, 28/583, 41/153
glycogen storage disease, 43/171
glycyrrhiza, 64/17
gossypol, 64/16
heart disease, 62/459
histopathology, 28/587-591, 41/153

hyperaldosteronism myopathy, 39/494, 41/152, 62/536
inheritance, 28/582
inheritance mode, 41/152, 62/460
laboratory examination, 28/583
life expectancy, 41/151, 62/459
male preponderance, 62/460
metabolic alteration, 28/584-586
muscle biochemical change, 28/586
myopathy, 43/170
myotonia, 41/150, 43/170, 62/458
pain, 43/170
pathogenesis, 28/591
permanent weakness, 41/150, 62/458
prevalence, 43/170
prognosis, 28/592
racial distribution, 28/583, 41/153
serum potassium, 43/170
sex distribution, 28/583
Sjögren syndrome, 71/74, 89
sodium channel α-subunit gene, 62/460
sporadic case, 41/152, 62/461
surgery, 62/460
symptom, 62/457
treatment, 28/591, 41/155-158, 62/469
Hypokinesia
brain blood flow, 73/477
drug induced extrapyramidal syndrome, 6/250
features, 6/135-139
globus pallidus internus, 21/520, 526
juvenile parkinsonism, 49/155
Mast syndrome, 42/282
motor apraxia, 6/135
neuroleptic syndrome, 6/250
Parkinson dementia complex, 42/249
Parkinson disease, 6/180-182, 42/245
progressive pallidal atrophy (Hunt-Van Bogaert), 42/247
schizophrenia, 73/477
Hypokinetic agraphia, *see* Micrographia
Hypokinetic rigid Huntington chorea, *see* Rigid Huntington chorea
Hypolipidemia syndrome, *see* Hooft disease
Hypomagnesemia
alcohol, 63/562
alcoholism, 63/562
aminoglycoside, 63/562
ataxia, 63/562
athetosis, 63/562
chorea, 63/562
Chvostek sign, 63/562
ciclosporin, 63/10
cisplatin, 63/562
coma, 63/562
confusion, 63/562
confusional state, 63/562
convulsion, 63/562
Crohn disease, 63/563
CSF magnesium, 28/435, 552
delirium, 46/543, 63/562
diagnosis, 28/565
diuretic agent, 63/562
dysphagia, 63/562
EEG, 28/549
EMG, 28/547
ENG, 28/548
epilepsy, 63/562
etiology, 28/554
hypoparathyroidism, 27/297
magnesium imbalance, 28/546-568
malabsorption, 63/562
muscle cramp, 63/562
myoclonus, 63/562
natural history, 28/553
neurologic form, 28/557-565
neurologic symptom, 28/546
nystagmus, 63/562
pathophysiology, 28/555
renal tubular acidosis, 63/562
starvation, 63/562
stupor, 63/562
symptom, 70/114
syncope, 72/262
tetany, 63/559, 562
treatment, 28/566-568
tremor, 63/562
Trousseau phenomenon, 63/562
vertical nystagmus, 63/562
Hypomelanosis
Ito, *see* Ito hypomelanosis
Hypometria
definition, 2/394
Hypometric saccade
progressive supranuclear palsy, 49/241, 243
Hyponatremia
Addison disease, 28/499
ageusia, 63/545
aging, 74/235
apathy, 63/545
aspartic acid, 63/547
asymptomatic, *see* Asymptomatic hyponatremia
biphasic course, 63/548
brain, 28/495-497
brain amino acid, 63/547
brain computer assisted tomography, 63/546
brain edema, 63/546

brain potassium, 63/547
brain taurine, 63/547
carbamazepine intoxication, 65/507
central pontine myelinolysis, 47/586, 63/548
clinical features, 28/499-503
CNS, 28/430-432
CNS sodium content, 28/432
coma, 63/545
confusional state, 63/545
congestive heart failure, 28/499
consciousness, 63/545
creatine, 63/547
delirium, 46/543, 63/545
dipeptidyl carboxypeptidase I inhibitor intoxication, 65/445
disequilibrium syndrome, 28/501
diuretic agent, 65/438
dizziness, 63/545
dopamine β-mono-oxygenase deficiency, 59/163
EEG, 63/546
electrolyte disorder, 63/545
encephalopathy, 63/547
epilepsy, 63/545
extra pontine myelinolysis, 63/548
fatigue, 63/545
glutamic acid, 63/547
glutamine, 63/547
gram-negative bacillary meningitis, 52/122
headache, 63/545
hyperosmolar hyperglycemic nonketotic diabetic coma, 27/91
hypogeusia, 63/545
hypo-osmolar, *see* Hypo-osmolar hyponatremia
hypothalamic lesion, 2/446
inappropriate antidiuretic hormone secretion, 28/499, 63/545
inositol, 63/547
legionellosis, 52/254
lethargy, 63/545
leukoencephalopathy, 47/586
liver disease, 28/499
metabolic encephalopathy, 63/547
muscle cramp, 63/545
myelinolysis, 63/548
nausea, 63/545
nephrotic syndrome, 28/499
neurologic manifestation, 28/503
opisthotonos, 63/545
oxcarbazepine intoxication, 65/507
pathophysiology, 28/496
potassium, 28/430
potassium imbalance, 28/482
sign, 28/500
spinal cord disease, 75/581
status epilepticus, 63/545
stupor, 63/545
symptom, 28/500
syncope, 72/261
taste sensation, 63/545
thiazide induction, 63/547
traumatic intracranial hematoma, 57/281
treatment, 28/502
tremor, 63/545
trismus, 63/545
tuberculous meningitis, 28/499
water intoxication, 28/482, 63/545, 64/239

Hypo-osmolar hyponatremia
syncope, 72/261

Hypoparathyroid myopathy
creatine kinase, 62/541
hypocalcemia, 62/541
Kearns-Sayre-Daroff-Shy syndrome, 62/541
paresthesia, 62/541
tetany, 62/541

Hypoparathyroidism
see also Myxedema
alcohol intoxication, 64/113
anxiety, 46/431
athetosis, 27/308
basal ganglion, 70/122
benign intracranial hypertension, 67/111
cataract, 42/577
chorea, 27/308
Chvostek sign, 27/299, 41/248
congenital anosmia, 42/357
convulsion, 27/300, 303-305
delirium, 46/544
dementia, 46/396
DiGeorge syndrome, 70/121
dystonia, 27/308
EEG, 27/304, 42/577
EMG, 1/642, 27/300
enamel defect, 42/577
endocrine candidiasis syndrome, 42/642
endocrine myopathy, 41/215, 247
epilepsy, 27/303-305, 70/123
Erb sign, 27/300
extrapyramidal motor abnormality, 27/305-310
familial idiopathic type, 6/707
grand mal epilepsy, 27/304
hallucination, 27/301
hyperphosphatasia, 42/577
hyperreflexia, 42/577
hypocalcemia, 27/296-298, 303, 42/577, 63/558, 561
hypokalemia, 27/298

hypomagnesemia, 27/297
idiopathic, *see* Idiopathic hypoparathyroidism
increased intracranial pressure, 27/310
intestinal pseudo-obstruction, 51/493
intracerebral calcification, 42/577
intracranial hypertension, 70/124
Kearns-Sayre-Daroff-Shy syndrome, 41/215, 43/142, 62/309, 508, 70/121
mental deficiency, 46/19
mental symptom, 27/301
mitochondrial abnormality, 41/248
movement disorder, 70/122
multiple mitochondrial DNA deletion, 62/510
myalgia, 42/577
myopathy, 27/312
neonatal tetany, 27/299
neuropathy, 27/312
papilledema, 27/310
paraplegia, 27/311
Parkinson disease, 27/308
parkinsonism, 27/308
peripheral nervous system manifestation, 27/298-300
postthyroidectomy type, 6/705
progressive external ophthalmoplegia, 62/309
pseudo, *see* Pseudohypoparathyroidism
pseudopseudo, *see* Pseudopseudohypoparathyroidism
psychiatric disorder, 42/577
psychiatric symptom, 27/301
renal dysplasia, 70/121
seizure, 27/300, 303-305, 42/577
sensorineural deafness, 42/383, 70/123
spasmodic torticollis, 27/308
striatopallidodentate calcification, 27/305-310, 49/418
symptom, 70/121
tetany, 27/298-300, 70/123
treatment, 27/312-314
Trousseau sign, 27/300, 41/248
Hypoperfusion
brain, *see* Brain hypoperfusion
migraine, 48/72
Hypophosphatasia
alkaline phosphatase, 42/578
birth incidence, 42/579
bone fragility, 42/578
convulsion, 42/578
craniosynostosis, 42/578
diet, 42/579
exophthalmos, 42/578
genetics, 31/315
hypercalcemia, 42/578
mental deficiency, 31/315
metabolic abnormality, 31/316
pathology, 31/316
phosphorylethanolamine, 42/578
prognosis, 31/317
radiologic appearance, 31/315
skeletal abnormality, 31/314
skeletal deformity, 42/578
treatment, 31/316
Hypophosphatemia
alcoholic polyneuropathy, 63/565
alcoholism, 63/564
apathy, 63/564
areflexia, 63/564
coma, 63/564
confusion, 63/564
dysarthria, 63/564
EMG, 63/565
encephalopathy, 63/564
epilepsy, 63/564
hyperalimentation, 63/564
iatrogenic neurological disease, 63/564
metabolic encephalopathy, 63/564
muscle weakness, 63/565
obtundation, 63/564
paresthesia, 63/565
perioral paresthesia, 63/565
rhabdomyolysis, 63/565
symptom, 70/115
tremor, 63/564
Hypophosphatemic osteomalacia
hereditary, *see* Hereditary hypophosphatemic osteomalacia
Hypophyseal adenoma
basophil adenoma, 18/353
brain tumor, 18/506
child, 18/352-355
chromophobe adenoma, 18/352
clinical features, 18/354
CSF cytology, 16/402
diagnosis, 18/355
differential diagnosis, 18/355
hypothalamus tumor, 18/366
pathology, 18/352-354
prognosis, 18/355
radiotherapy, 18/507
sella turcica, 18/352
sign, 18/354
subarachnoid hemorrhage, 12/119
symptom, 18/354
treatment, 18/355
Hypophyseal artery
inferior, *see* Inferior hypophyseal artery

Hypophyseal diabetes insipidus
paraventricular nucleus degeneration, 42/543
sex ratio, 42/543
supraoptic nucleus degeneration, 42/543
Hypophyseal diencephalic disturbance
brain injury, 23/11
Hypophyseal duct tumor, *see* Craniopharyngioma
Hypophyseal syndrome
clinical type, 2/453
hypothalamic, *see* Hypothalamic hypophyseal syndrome
Hypophyseal system
hypothalamic, *see* Hypothalamic hypophyseal system
Hypophyseal tumor
adenohypophyseal syndrome, 2/456-461
headache, 5/229
optic atrophy, 13/69
tuberculoma, 18/417
Hypophysectomy
adenohypophyseal syndrome, 2/461-463
Hypophysis, *see* Pituitary gland
Hypopigmentation
Bloom syndrome, 14/775, 43/10
iris, *see* Iris hypopigmentation
mental deficiency, 43/257
neurofibromatosis type I, 50/366
oculocerebrocutaneous syndrome, 43/444
olivocochlear tract, 43/43
skin, *see* Cutaneous hypopigmentation
Hypopituitarism
brain tumor, 16/347
empty sella syndrome, 17/431
hypoglycemia, 27/70
late cerebellar atrophy, 60/587
Möbius syndrome, 50/215
optic nerve agenesis, 50/211
pituitary adenoma, 17/393, 421
sexual impotence, 18/155
Sipple syndrome, 42/752
Hypoplasia
abducent nerve abnormality, 30/401
adrenal, *see* Adrenal hypoplasia
amyotrophic cerebellar, *see* Amyotrophic cerebellar hypoplasia
anterior cerebral artery, 11/176
arm, *see* Arm hypoplasia
auditory osscile, *see* Auditory ossicle hypoplasia
autosomal recessive cerebellar, *see* Autosomal recessive cerebellar hypoplasia
brain hemisphere, 42/12
brain stem, *see* Brain stem hypoplasia
carotid artery, 12/306
cerebellar, *see* Cerebellar hypoplasia
cerebellar agenesis, 50/181
cerebellar granular cell, *see* Cerebellar granular cell hypoplasia
cerebellar granular layer, *see* Cerebellar granular layer hypoplasia
cerebellar vermis, *see* Cerebellar vermis aplasia
clavicle, *see* Clavicle hypoplasia
cleidocranial, *see* Cleidocranial hypoplasia
cochlear, *see* Cochlear hypoplasia
condylar, *see* Condylar hypoplasia
congenital muscular, *see* Congenital muscular hypoplasia
corpus callosum, *see* Corpus callosum hypoplasia
cutaneous, *see* Cutaneous hypoplasia
definition, 30/4
dens, *see* Dens hypoplasia
enamel, *see* Enamel hypoplasia
falx cerebri, *see* Falx cerebri hypoplasia
foveal, *see* Foveal hypoplasia
gonadal, *see* Gonadal hypoplasia
hemifacial, *see* Hemifacial hypoplasia
internal carotid, *see* Internal carotid hypoplasia
internal carotid artery, 12/306
iris, *see* Hypoplasia iridis
iris mesenchymal, *see* Iris mesenchymal hypoplasia
Klinefelter variant XXXXY, 31/484
Krabbe muscular, *see* Krabbe muscular hypoplasia
lung, *see* Pulmonary hypoplasia
macular, *see* Macular hypoplasia
mandibular, *see* Mandibular hypoplasia
midface, *see* Midface hypoplasia
muscle fiber type II, *see* Muscle fiber type II hypoplasia
nail, *see* Nail hypoplasia
occipital bone shortening, 32/14
occipital condylus, 32/22
occipital dysplasia, 32/14, 20-22
olfactory bulb, *see* Olfactory bulb hypoplasia
operculum, *see* Operculum hypoplasia
optic disc, *see* Optic disc hypoplasia
patellar, *see* Patellar hypoplasia
pontocerebellar, *see* Pontocerebellar hypoplasia
pontoneocerebellar, *see* Pontoneocerebellar hypoplasia
pulmonary, *see* Pulmonary hypoplasia
pyramidal tract, *see* Pyramidal tract hypoplasia
retinal, *see* Retinal hypoplasia
rhinencephalic, *see* Rhinencephalic hypoplasia
Saldino-Noonan syndrome, 31/243
skin, *see* Cutaneous hypoplasia
superior cerebellar vermis, *see* Superior cerebellar

vermis hypoplasia
superior temporal lobe, *see* Superior temporal lobe hypoplasia
tentorial, *see* Tentorial hypoplasia
thumb, *see* Thumb hypoplasia
thymus, *see* Thymus hypoplasia
vertebral, *see* Vertebral hypoplasia
vertebral abnormality, *see* Vertebral hypoplasia
Hypoplasia iridis
facio-oculoacousticorenal syndrome, 43/400
megalocorneal mental deficiency syndrome, 43/270
mental deficiency, 43/268
Hypoplastic cerebellum, *see* Cerebellar hypoplasia
Hypoplastic left heart syndrome, *see* Congenital heart disease
Hypoplastic nail, *see* Nail hypoplasia
Hypoplastic pyramid, *see* Pyramidal hypoplasia
Hypoplastic upper lip, *see* Upper lip hypoplasia
Hypopnea
bradypnea, 1/666
classification, 1/657
tachypnea, 1/665
Hypoprothrombinemia, 38/73-80
clinical features, 38/74
etiology, 38/73, 74
hematomyelia, 38/77-79
intracranial hemorrhage, 38/75-77
pathology, 38/74
peripheral nerve involvement, 38/79
spinal cord hemorrhage, 38/77
subdural hematoma, 38/76
Hypopyon iritis
Behçet syndrome, 71/215
Hyporeflexia
see also Areflexia
anterior horn syndrome, 42/72
benign dominant myopathy, 43/111
cerebrohepatorenal syndrome, 43/338
chlormethine intoxication, 64/233
chondrodystrophic myotonia, 62/269
congenital ataxic diplegia, 42/121
craniodiaphyseal dysplasia, 43/356
CSF paraprotein, 42/604
dermatoleukodystrophy, 42/488
disequilibrium syndrome, 42/139
distal muscular dystrophy, 43/96
Duchenne muscular dystrophy, 43/106
dystonia, 42/253
Eaton-Lambert myasthenic syndrome, 62/426
facial spasm, 42/314
Gerstmann-Sträussler-Scheinker disease, 60/622
glycogen storage disease, 43/178, 180
hereditary hyperkalemic periodic paralysis, 43/151
hereditary motor and sensory neuropathy type I, 42/339
hereditary motor and sensory neuropathy type II, 42/339
Holmes-Adie syndrome, 43/72
hyperkalemic periodic paralysis, 43/151
infantile spinal muscular atrophy, 42/88
Lowe syndrome, 42/606
Machado disease, 42/155
metachromatic leukodystrophy, 8/16, 42/491
mitochondrial myopathy, 43/127
multicore myopathy, 43/120
muscular atrophy, 42/82, 84, 91
muscular dystrophy, 43/109
myopathy, 43/193
nemaline myopathy, 43/122
neurogenic acro-osteolysis, 42/293
oxoglutarate dehydrogenase deficiency, 62/503
10p partial trisomy, 43/519
parkinsonism, 42/253
pellagra, 28/90
Richards-Rundle syndrome, 43/264
Rosenberg-Chutorian syndrome, 22/506
Ryukyan spinal muscular atrophy, 42/93
sarcotubular myopathy, 43/129
scapulohumeroperoneal muscular atrophy, 42/343
scapuloilioperoneal muscular atrophy, 42/344
scapuloperoneal myopathy, 43/131
sensory radicular neuropathy, 42/351
serum paraprotein, 42/604
spinocerebellar ataxia, 42/182
thalidomide, 42/660
tonic pupil, 42/72
toxic oil syndrome, 63/381
xanthinuria, 43/193
xeroderma pigmentosum, 43/13
Hyposensitivity
pickwickian syndrome, 38/301
Hyposmia
thalamic lesion, 2/480
Hyposomnia, *see* Insomnia
Hypospadias
ARG syndrome, 50/582
axial mesodermal dysplasia spectrum, 50/516
caudal aplasia, 50/512
congenital adrenal hyperplasia, 42/514
cytomegalic inclusion body disease, 56/267
Edwards syndrome, 50/561
Gruber syndrome, 14/115
holoprosencephaly, 50/237
hypertelorism, 43/413

LEOPARD syndrome, 43/28
Patau syndrome, 31/508, 50/558
pyruvate dehydrogenase complex deficiency, 66/416
r(9) syndrome, 50/580
r(13) syndrome, 50/585
renal agenesis syndrome, 50/512
Smith-Lemli-Opitz syndrome, 43/308, 50/276
XX male, 43/555
Hypotelorism
brachial neuritis, 42/304
congenital heart disease, 63/11
hereditary brachial plexus neuropathy, 60/71
holoprosencephaly, 30/431, 448-450, 42/32, 50/233
median cleft lip face, 30/447
N syndrome, 43/283
neuralgic amyotrophy, 51/175
orbital, *see* Orbital hypotelorism
Patau syndrome, 31/504, 512, 50/556
14q partial trisomy, 43/529
Hypotelorism syndrome
median cleft lip, *see* Median cleft lip hypotelorism syndrome
Hypotension
aging, 74/229
anoxic ischemic leukoencephalopathy, 9/576
apistus intoxication, 37/76
Argentinian hemorrhagic fever, 56/364
arterial, *see* Arterial hypotension
autonomic nervous system, 74/229, 613
brain air embolism, 55/193
brain infarction, 53/158
brain injury, 57/210, 433
brain ischemia, 53/158, 57/210
brain metabolism, 57/67
central pontine myelinolysis, 28/293
cerebrovascular disease, 53/29
CSF, *see* CSF hypotension
drug induced, *see* Drug induced hypotension
exercise, 75/716
extracranial insult, 57/125
glyceryl trinitrate intoxication, 37/453
Godin syndrome, 5/330
gymnapistes intoxication, 37/76
head injury, 57/67, 217
head injury outcome, 57/387
head up tilt maneuver, 75/717
idiopathic orthostatic, *see* Pure autonomic failure
intracranial, *see* Intracranial hypotension
intravascular consumption coagulopathy, 55/493
MAO inhibitor, 74/145
neurogenic orthostatic, *see* Neurogenic orthostatic hypotension
nothesthes intoxication, 37/76
orthostatic, *see* Orthostatic hypotension
pharmacology, 74/176
postalcohol, *see* Postalcohol hypotension
postural, *see* Orthostatic hypotension
primary orthostatic, *see* Pure autonomic failure
procainamide intoxication, 37/451
pterois volitans intoxication, 37/75
puffer fish intoxication, 37/80
reserpine intoxication, 37/438
scorpaena intoxication, 37/76
scorpaenodes intoxication, 37/76
scorpaenopsis intoxication, 37/76
sebastapistes intoxication, 37/76
secondary insult, 57/67
sodium nitroprusside intoxication, 37/435
straining, 75/115
sympathetic nerve, 74/659
systemic, *see* Systemic hypotension
traumatic, *see* Traumatic hypotension
traumatic CSF, *see* Traumatic CSF hypotension
urticaria pigmentosa, 14/790
vascular, *see* Vascular hypotension
Hypotensive state
hypoxic, *see* Hypoxic hypotensive state
Hypothalamic dysfunction
encephalomeningocele, 42/27
GM2 gangliosidosis type I, 42/433
head injury, 46/626
intracranial hypertension, 16/134
neurosarcoidosis, 63/422
Hypothalamic glioma
hypothalamic hamartoma, 68/72
neuroimaging, 68/72
Hypothalamic hamartoma
epilepsy, 75/354
hypothalamic glioma, 68/72
Hypothalamic hypophyseal adrenal system
multiple sclerosis, 47/353
Hypothalamic hypophyseal syndrome
etiology, 2/444
precocious puberty, 2/444
Hypothalamic hypophyseal system
depression, 46/449
Hypothalamic hypophyseal tract
neuroaxonal dystrophy, 49/395
Hypothalamic nucleus
dorsomedial, *see* Dorsomedial hypothalamic nucleus
posterior, *see* Posterior hypothalamic nucleus
ventromedial, *see* Ventromedial hypothalamic nucleus

Hypothalamic phakoma
cheilopalatoschisis, 42/737
seizure, 42/737
Hypothalamic pituitary adrenal axis disorder
benign intracranial hypertension, 67/111
Hypothalamic syndrome
autonomic nervous system, 75/533
bradycardia, 2/450
clinical type, 2/445
flushing, 2/450
lacrimation, 2/450
lateral, *see* Lateral hypothalamic syndrome
lymphatic leukemia, 18/250
meningeal leukemia, 63/341
sweating, 2/450
tumor, 2/450
viral infection, 2/450
yawning, 2/450
Hypothalamic tumor
anaplastic astrocytoma, 68/71
astrocytoma, 68/71
craniopharyngioma, 68/71
epidermoid, 68/71
germinoma, 68/71
glioblastoma multiforme, 68/71
hamartoma, 68/71
histiocytosis, 68/71
lipoma, 68/71
lymphoma, 68/71
metastasis, 68/71
neurocytoma, 68/71
periodic paralysis, 62/457
sarcoid, 68/71
symptom, 68/69
teratoma, 68/71
toxoplasmosis, 68/71
treatment, 68/73
tuberculoma, 68/71
Hypothalamopituitary adrenal axis, *see* Hypothalamic hypophyseal adrenal system
Hypothalamopituitary axis, *see* Hypothalamic hypophyseal system
Hypothalamus
affective disorder, 46/473
afferent, 2/437
aggression, 2/440
akinesia, 45/165
akinetic mutism, 45/166
analgesia, 45/238-240
anorexia nervosa, 46/587
anxiety, 45/275
appetite, 1/500, 2/441
arousal disorder, 45/108, 111
autonomic nervous system, 74/2, 14, 59, 76, 152, 156
behavior, 2/441
biological rhythm, 74/469, 505
blood flow, 1/459
brain concussion, 23/465
brain granuloma, 16/230
brain injury, 23/59, 119
brain tumor, 16/343
carbohydrate metabolism, 2/440
cardiovascular control, 2/441
cardiovascular function, 1/461
centrencephalic system, 2/450
chronic migrainous neuralgia, 48/253
cluster headache, 48/242
connection, 1/460, 2/437
daily rhythm, 74/471
defecation, 2/441
depression, 74/504
development, 74/494
diencephalic epilepsy, 2/452
division, 2/434
efferent, 2/438
electrolyte balance, 1/500
emaciation, 2/449
emotion, 2/440, 715, 719, 3/319, 45/273-275, 277
emotional disorder, 74/505
β-endorphin, 45/238
feeding, 2/441
focal brain damage, 57/51
function, 2/438-442
gastric mucosal erosion, 2/441
gastrointestinal hemorrhage, 2/441
glycosuria, 2/440
heart, 1/459, 461
heat stroke, 2/447
hyperglycemia, 2/440
hypothermia, 75/59
leukemia, 39/9
light flight pattern, 2/441
malignant hyperthermia, 38/554
memory disorder, 2/450
micturition, 2/441
migraine, 74/508
neuroanatomy, 2/433
neurogenic arterial hypertension, 63/231
neurosecretion, 2/439
osmoreceptor, 2/438
pain, 45/236
pain control, 45/238-240
Parkinson disease, 49/110
pathologic laughing and crying, 3/362, 45/223
pharmacology, 74/156

precocious puberty, 18/367
preoptic, *see* Preoptic hypothalamus
radiation injury, 23/654, 67/330
retinohypothalamic tract, 74/478
Russell syndrome, 16/349
salt and water balance, 2/438
sex, 74/470
sexual dimorphic, 74/470
sexuality, 45/277
skin pigmentation, 2/441
sleep, 74/548
sleep regulation, 2/439
solitary tract, 74/153
stomach, 1/506
strain, 45/249
stress, 45/251, 74/504
suprachiasmatic nucleus, 74/487, 489
supraoptic, *see* Supraoptic hypothalamus
thermoregulation, 1/443, 2/438, 61/277, 71/570, 75/53, 59
toluene intoxication, 64/50
topography, 2/434
traumatic pathology, 23/466
vasoactive intestinal polypeptide, 74/482
visceral function, 2/441
water intake, 2/441
Wernicke encephalopathy, 2/450
Hypothalamus disorder
hypersomnia, 46/418
insomnia, 46/418
sleep disorder, 46/418
Hypothalamus-pituitary-adrenal axis
stress, 74/308
Hypothalamus tumor
anorexia nervosa, 46/417
appetite, 46/417
craniopharyngioma, 18/366
hamartoma, 18/366
hypernatremia, 2/446
hypophyseal adenoma, 18/366
meningioma, 18/366
pinealoma, 18/366
Hypothenar dimpling
muscle contraction, 41/303
Hypothermia
accidental, 71/572
alcohol, 75/60
anhidrosis, 42/346
apistus intoxication, 37/76
brain aneurysm, 12/211
brain aneurysm neurosurgery, 55/61
brain blood flow, 71/574
brain death, 55/263
brain function, 71/574
brain infarction, 53/417
brain stem death, 57/454, 471
brain vasospasm, 11/515
carbamate intoxication, 64/186
cardiac surgery, 71/578
cause, 1/446
clinical use, 71/576
congenital heart disease, 63/9
corpus callosum agenesis, 50/167
delirium, 46/546
disorder, 75/60
drug, 71/571, 75/60
episodic, *see* Episodic hypothermia
glioblastoma multiforme, 18/64
gymnapistes intoxication, 37/76
hereditary motor and sensory neuropathy type I, 21/282
hereditary motor and sensory neuropathy type II, 21/282
hypothalamus, 75/59
intracranial hypertension, 24/618
intracranial pressure, 23/215
ischemia, 71/577
metabolic encephalopathy, 69/400
metabolism, 71/575
migraine, 42/746
neurologic disease, 71/569
neurologic disorder, 71/571
neurologic intensive care, 55/217
neurotrauma, 71/579
nothesthes intoxication, 37/76
pial blood flow, 11/515
pontine hemorrhage, 12/50
pterois volitans intoxication, 37/75
scorpaena intoxication, 37/76
scorpaenodes intoxication, 37/76
scorpaenopsis intoxication, 37/76
sebastapistes intoxication, 37/76
sensory neuropathy, 42/346
Shapiro syndrome, 50/167
spinal cord injury, 26/368, 372, 61/282
stress, 74/319
subarachnoid hemorrhage, 12/211
survey, 71/569
symptom, 1/449
systemic disorder, 71/571
tetraplegia, 61/275
thermoregulation, 2/447, 26/368, 372, 71/569, 75/59
thyroid gland dysgenesis, 42/632
transtentorial herniation, 1/564
treatment, 71/574

trichopoliodystrophy, 42/584
Hypothyroid myopathy
aspartate aminotransferase, 62/533
atrophy fiber type II, 27/272
biochemistry, 62/533
creatine kinase, 62/533
EMG, 62/533
endocrine myopathy, 41/243-246, 298, 62/528
fructose-bisphosphate aldolase, 62/533
Hoffmann syndrome, 41/243, 43/84, 62/533
hypothyroid polyneuropathy, 62/533
Isaacs syndrome, 62/263
Kocher-Debré-Sémélaigne syndrome, 62/533
lactate dehydrogenase, 62/533
muscle biopsy, 62/533
muscle cramp, 41/243, 246, 62/532
muscle hypertrophy, 41/243
muscle stretch reflex time, 62/532
muscle ultrastructure, 62/533
myoedema, 62/532
precipitating factor, 62/532
pseudomyotonic, 62/532
treatment, 62/534
Hypothyroid polyneuropathy
see also Endocrine polyneuropathy
ataxia, 51/515
carpal tunnel syndrome, 51/516
clinical features, 51/514
coma, 51/515
deafness, 51/515
EMG, 51/515
hypothyroid myopathy, 62/533
mental deficiency, 51/515
mononeuropathy, 51/516
pathology, 51/515
symmetrical polyneuropathy, 51/515
type, 51/515
Hypothyroidism
see also Thyroid disease
acanthocyte, 49/332
acanthocytosis, 63/271, 290
acoustic nerve, 27/258
ataxia, 21/574, 27/260
basilar impression, 50/399
benign intracranial hypertension, 67/111
bowel dysfunction, 75/649
carpal tunnel syndrome, 41/244, 70/100
cerebellar ataxia, 70/95
clinical features, 40/324, 328
coma, 27/260, 70/93
compression neuropathy, 70/99
congenital, *see* Cretinism
cranial nerve disorder, 70/96
creatine kinase, 40/343
cretinism, 70/92
deafness, 27/258
delirium, 46/544
dementia, 46/205, 396
depression, 46/427
differential diagnosis, 40/438, 62/533
Down syndrome, 50/532
Eaton-Lambert myasthenic syndrome, 62/421
encephalopathy, 70/93
epilepsy, 27/260
giant cell arteritis, 70/101
Hashimoto encephalopathy, 70/101
Hashimoto thyroiditis, 70/101
Hoffmann syndrome, 70/97
hypoglycemia, 27/70
infantile neuroaxonal dystrophy, 59/75
Kocher-Debré-Sémélaigne syndrome, 40/324, 41/243
late cerebellar atrophy, 60/587
mental change, 70/94
metabolic ataxia, 21/575
metabolic neuropathy, 51/68
muscle hypertrophy, 43/84
muscle relaxation, 41/298
myasthenia gravis, 62/534, 70/101
myoedema, 40/330, 41/245
myopathy, 70/96
myotonia, 40/547
neuroleptic malignant syndrome, 70/101
neurology, 27/257-260
neuropathy, 70/99
papilledema, 42/577
Parkinson disease, 49/96
Pendred syndrome, 42/369
POEMS syndrome, 51/430
polymyalgia rheumatica, 70/101
polyneuropathy, 41/244, 70/99
pseudopseudo, *see* Pseudopseudohypothyroidism
psychosis, 27/259, 46/462
seizure, 70/93
sleep apnea syndrome, 45/132
sleep disorder, 70/95
snoring, 63/451
sodium nitroprusside intoxication, 37/435
special sense, 27/258-260
systemic brain infarction, 11/458
Hypotonia
see also Amyotonia congenita
amyotonia congenita, 1/271, 40/336, 43/79
arthrogryposis multiplex congenita, 41/32
ataxia telangiectasia, 60/361
benign congenital, *see* Amyotonia congenita

bilirubin encephalopathy, 27/421
cerebellar, *see* Cerebellar hypotonia
cerebellar component, 2/394
cerebellar pathophysiology, 2/403
childhood, *see* Childhood hypertonia
clinical features, 40/308, 332
complex I deficiency, 66/422
complex II deficiency, 66/422
complex IV deficiency, 66/424
congenital ataxic diplegia, 21/463, 42/121
congenital muscular dystrophy, 41/36
congenital myopathy, 41/1
congenital spondyloepiphyseal dysplasia, 43/478
differential diagnosis, 40/334
Ehlers-Danlos syndrome, 40/336
Fukuyama syndrome, 41/36
globoid cell leukodystrophy, 40/336
Hartnup disease, 7/577
hereditary sensory and autonomic neuropathy type II, 60/11
hereditary sensory and autonomic neuropathy type III, 60/27, 75/144
Huntington chorea, 49/71
hyperpipecolatemia, 29/222
infantile, *see* Infantile hypotonia
infantile myotonic dystrophy, 40/490
infantile neuronal ceroid lipofuscinosis, 66/689
ketotic hyperammonemia, 29/106
kwashiorkor, 51/322
Lafora progressive myoclonus epilepsy, 15/386, 27/173
late infantile metachromatic leukodystrophy, 29/357
long latency reflex, 49/70
metachromatic leukodystrophy, 8/16, 40/336
mitochondrial DNA depletion, 66/430
muscle tone, 1/263, 269-271
muscular, *see* Muscular hypotonia
neocerebellar syndrome, 2/419
nerve lesion, 1/219
organic acid metabolism, 66/641
osteogenesis imperfecta, 40/336
oxoglutarate dehydrogenase deficiency, 66/419
pathophysiology, 1/269-271, 49/71
pellagra, 6/748
peroxisomal acetyl-coenzyme A acyltransferase deficiency, 66/518
pertussis vaccine, 52/242
polymyositis, 40/335
Prader-Labhart-Willi syndrome, 40/336, 46/93
pyruvate dehydrogenase complex deficiency, 66/416
pyruvate dehydrogenase deficiency, 29/212
Rubinstein-Taybi syndrome, 31/283, 43/234
subacute necrotizing encephalomyelopathy, 10/209, 28/352
Sydenham chorea, 6/415, 49/71, 360
thalamic syndrome (Dejerine-Roussy), 2/481
thalamic ventral lateral nucleus, 49/71
type, 1/174
XYY syndrome, 50/551

Hypotonic bladder
acute pandysautonomia, 51/475

Hypotonic tetraparesis
critical illness polyneuropathy, 51/575, 581

Hypotrichosis
Berlin syndrome, 43/9
ectodermal dysplasia, 14/112
Hallermann-Streiff syndrome, 43/403
Rothmund-Thomson syndrome, 39/405

Hypotrophic foot
Rett syndrome, 60/669

Hypoventilation
alveolar, *see* Alveolar hypoventilation
alveolar carbon dioxide tension, 63/413
alveolar oxygen tension, 63/413
aminoglycoside intoxication, 65/483
arterial carbon dioxide tension, 63/413
autonomic nervous system, 1/427-515, 2/107-124, 74/583
brain stem infarction, 63/413
bulbar poliomyelitis, 63/413
carotid body sensitivity, 63/413
carotid endarterectomy, 63/413
central alveolar, *see* Central alveolar hypoventilation
chronic alveolar, *see* Chronic alveolar hypoventilation
chronic obstructive pulmonary disease, 63/413
chronic respiratory insufficiency, 63/413
classification, 63/413
congenital central alveolar, *see* Congenital central alveolar hypoventilation
congenital chemosensitive hypoventilation syndrome, 74/583
definition, 63/413
forgotten respiration, 74/583
hypercapnia, 63/413
hypercapnic acidosis, 63/413
hypoventilation syndrome, 63/413
hypoxemia, 63/413
myotonic dystrophy, 62/235
neurogenic, 1/498
obstructive apnea syndrome, 63/413
opiate use, 63/413
posttraumatic brain edema, 24/611

primary alveolar, *see* Primary alveolar hypoventilation
pulmonary, *see* Pulmonary hypoventilation
renal adaptation, 63/413
respiratory disease, 63/413
Hypoventilation syndrome
chronic obstructive pulmonary disease, 63/413
chronic respiratory insufficiency, 63/413
hypoventilation, 63/413
obesity, *see* Pickwickian syndrome
obstructive apnea syndrome, 63/413
respiratory disease, 63/413
Hypovolemia
orthostatic hypotension, 63/145
Hypoxanthine
brain edema, 67/81
CSF, 54/199
Hypoxanthine phosphoribosyltransferase
Gilles de la Tourette syndrome, 42/222
Lesch-Nyhan syndrome, 42/154, 59/360, 60/19
self-mutilation encephalopathy, 42/140
Hypoxanthine phosphoribosyltransferase deficiency
Lesch-Nyhan syndrome, 22/514, 29/268-270, 59/360
Hypoxemia
central sleep apnea, 63/461
cluster headache, 48/242
congenital central alveolar hypoventilation, 63/464
congenital heart disease, 63/3
dermatomyositis, 62/374
head injury outcome, 57/387
hypoventilation, 63/413
inclusion body myositis, 62/374
intravascular consumption coagulopathy, 55/493
myotonic dystrophy, 62/235
pallidal necrosis, 49/466
pickwickian syndrome, 38/300
polymyositis, 62/374
Hypoxia
see also Anoxia *and* High altitude
acute, *see* Acute hypoxia
adenosine triphosphate, 75/265
aging, 74/233
altitude illness, 48/397
anencephaly, 30/196
autonomic nervous system, 74/73, 108, 75/10, 259
blood pressure, 75/264
brain, *see* Brain hypoxia
brain fat embolism, 55/184
brain infarction, 53/167
brain injury, 23/117, 137, 141, 147
brain ischemia, 53/167
cell respiration, 9/621
cerebrovascular disease, 53/29
cervical spinal cord injury, 61/58
chronic, *see* Chronic hypoxia
congenital heart disease, 63/1
cortical blindness, 45/50
delirium, 46/541
diabetic polyneuropathy, 51/506
double athetosis, 49/384
edema, 16/180
EEG, 11/271
encephalopathy, 9/472
endoneurial, *see* Endoneurial hypoxia
epilepsy, 72/61
erythrocyte, 75/265
extracranial insult, 57/125
glycogen, 27/206
head injury, 23/117, 141, 57/216, 434
headache, 48/6
high altitude, 74/588
histotoxic, *see* Histotoxic hypoxia
hyperventilation, 75/261
hypobaric, 75/268
ischemic edema, 23/146
labyrinth concussion, 24/125
metabolic encephalopathy, 69/400
motion sickness, 74/357
myoclonus, 49/618
nerve conduction, 7/169
obstructive sleep apnea, 63/455
pallidal change, 6/59
perinatal, *see* Perinatal hypoxia
periventricular infarction, 53/29
pickwickian syndrome, 38/299
red blood cell, 75/265
respiratory encephalopathy, 63/416
sea snake intoxication, 37/92
structural damage, 9/621
vasodilatation, 75/264
Hypoxic hypotensive state
striatal necrosis, 49/502
Hypoxic ischemia
bee sting intoxication, 37/107
hydranencephaly, 50/338
perinatal, *see* Perinatal hypoxic ischemia
Hypoxic ischemic encephalopathy
brain death, 55/264
epilepsy, 73/254
iatrogenic neurological disease, 63/178
methanol intoxication, 64/102
Hypoxidosis
atmospheric, *see* Atmospheric hypoxidosis
Hypozoospermia

Down syndrome, 50/532
Hypsarrhythmia
Divry-Van Bogaert syndrome, 10/121
Down syndrome, 50/529
EEG, 42/686, 691-693, 72/303
happy puppet syndrome, 31/309
hyperglycinemia, 42/565
infantile spasm, 15/123, 219, 42/686, 73/200
linear nevus sebaceous syndrome, 43/39
pertussis vaccine, 52/242
phenylketonuria, 29/30
r(14) syndrome, 50/588
sleep, 15/471
tuberous sclerosis, 14/347, 43/49, 50/372
Hysteria, 46/573-582
abdominal pain, 48/357
agnosia, 4/23
agraphia, 45/468
amnesia, 43/207
analgesia, 1/111
attention disorder, 3/181
basilar artery migraine, 46/575
blindness, 24/51, 43/207
Briquet syndrome, 43/207
catatonia, 46/423
cognition, 46/579
consciousness, 3/126
definition, 46/573
dementia, 46/124
dysphrenia hemicranica, 5/79
dystonia, 6/561
headache, 43/207, 48/356, 359
hemianalgesia, 1/111
hypochondriasis, 46/580
memory, 46/621
migraine, 46/575, 48/356
multiple personality, 43/207, 46/576
multiple sclerosis, 9/177, 46/575, 47/72
neurophysiology, 46/577
neuropsychology, 46/580
oculomotor apraxia, 2/332
organic personality syndrome, 46/434
paralysis, 41/166, 43/207
personality, 46/578-580
pseudodementia, 46/224, 575
psychogenic pain, 43/207, 46/575
psychopathy, 46/581
psychophysiology, 46/577
ptosis, 1/617
schizophrenia, 46/576, 581
sex difference, 46/578-580
somnambulism, 43/207
status epilepticus, 46/576
symptomatic dystonia, 6/561
symptomatology, 46/573-576
Hysterical anosmia
nasolacrimal reflex, 2/54
Hysterical seizure
epilepsy, 46/575
Hysterical tremor
distinctive feature, 49/592

I band
calcavin, 40/157
dimethyl-*p*-phenylenediamine, 40/157
muscle fiber, 62/26
muscle ultrastructure, 40/212
plasmocid, 40/157
skeletal muscle, 40/63
I-cell disease, *see* Mucolipidosis type II
Iatrogenic brain infarction
angioendotheliomatosis, 53/165
brain angiography, 53/165
carotid endarterectomy, 53/165
carotid sinus compression, 53/165
Fabry disease, 53/165
homocystinuria, 53/165
hyperviscosity, 53/165
leukocytosis, 53/165
open heart surgery, 53/165
polycythemia, 53/165
pseudoxanthoma elasticum, 43/45, 53/165
Iatrogenic lesion
myelomalacia, 61/112
Iatrogenic meningocele, *see* Postoperative meningeal diverticulum
Iatrogenic nerve injury
brachial plexus, 51/144
general features, 7/262
ischemia, 51/159
lumbosacral plexus, 51/159, 161, 164
pallidal necrosis, 49/468
Iatrogenic neurological disease
acetrizoic acid, 63/44
acetylsalicylic acid intoxication, 65/460
air embolism, 55/192
amiodarone, 63/190, 65/457
angiography, 55/456
antiarrhythmic agent, 63/190
anticoagulant, 55/476, 65/461
antihypertensive treatment, 63/87
aorta surgery, 63/49, 55
aortography, 63/44
aprindine, 65/455
atrial thrombus, 63/177
beta adrenergic receptor blocking agent, 63/194,

65/435
brachial plexus lesion, 63/181
brain abscess, 52/144, 150
brain embolism, 55/161, 168, 63/175
brain fat embolism, 55/177, 63/182
brain hemorrhage, 61/164
brain perfusion pressure, 63/177
brain stem infarction, 63/177
bretylium, 63/192
bronchial arteriography, 63/44
calcium antagonist, 65/441
calcium channel blocking agent, 63/193
cardiac agent, 63/175
cardiac arrest, 61/165
cardiac catheterization, 55/161, 63/175
cardiac interventional procedure, 63/176
cardiac surgery, 55/168
cauda equina syndrome, 61/159
central pontine myelinolysis, 65/439
cerebellar infarction, 63/187
chloramphenicol, 65/486
chloroquine, 65/483
cinchonism, 65/451
ciprofloxacin, 65/486
CNS lymphoma, 63/181
coagulant overdose, 65/460
coma, 63/187
coronary bypass surgery, 63/181
cranial epidural empyema, 52/168
cranial subdural empyema, 52/170
cycloserine, 65/481
dapsone, 65/485
delirium, 61/165
diazoxide, 65/446
digoxin, 63/195, 65/449
diodone, 63/44
dipeptidyl carboxypeptidase I inhibitor, 65/445
discography complication, 61/147
disopyramide, 65/453
disopyramide phosphate, 63/192
diuretic agent, 65/440
encainide, 65/456
enoxacin, 65/486
epilepsy, 61/165
ergot intoxication, 65/68
ethambutol, 65/481
ethionamide, 65/481
facet joint denervation complication, 61/147
fat embolism, 63/178
femoral neuropathy, 63/49, 177, 530
fibrinolytic agent, 65/461
flecainide, 63/192, 65/455
hearing loss, 61/165
heart transplantation, 63/175
hexachlorophene, 65/471
hydralazine, 65/446
hypnotic agent, 65/329
hypophosphatemia, 63/564
hypoxic ischemic encephalopathy, 63/178
immunosuppression, 63/180
immunosuppressive agent, 63/189
intra-aortic balloon, 63/47, 177
intracranial hemorrhage, 54/38
intracranial subdural hematoma, 61/164
intrathecal chemotherapy, 61/147
intrathecal phenol, 55/101
iodoxyl, 63/44
isoniazid, 65/479
lidocaine, 63/190, 192, 65/454
lindane, 65/479
lumbar puncture, 55/476
lumbar puncture complication, 61/147
mefloquine, 65/485
meningitis, 63/179
meningoencephalitis, 63/179
mercury, 64/367
methyldopa, 65/448
metronidazole, 65/487
mexiletine, 63/192, 65/454
mitochondrial DNA depletion, 62/511
mitral stenosis, 63/177
mitral valvuloplasty, 63/177
myelography complication, 61/147
myelomalacia, 61/112, 63/178
nalidixic acid, 65/486
nitrofurantoin, 65/487
norfloxacin, 65/486
open heart surgery, 63/175
pefloxacin, 65/486
percutaneous transluminal coronary angioplasty, 63/177
percutaneous transluminal valvuloplasty, 63/177
peroneal nerve lesion, 63/181
phrenic nerve lesion, 63/181
pneumocephalus, 61/164
posterior spinal artery syndrome, 55/101
postsurgery coma, 63/187
prazosin, 65/447
procainamide, 63/190, 65/453
propafenone, 65/457
quinidine, 63/190
quinine, 65/485
radiation myelopathy, 61/199
recurrent laryngeal nerve lesion, 63/181
rhinocerebral infection, 63/179
sciatic neuropathy, 63/177

secondary systemic carnitine deficiency, 62/494
spinal angiography, 63/44
spinal cord abscess, 52/191
spinal epidural empyema, 52/186
spinal epidural hematoma, 61/137
spinal epidural hematoma incidence, 61/139
spinal epidural hematoma prophylaxis, 61/141
streptokinase, 65/461
subarachnoid hemorrhage, 61/164
suboccipital puncture complication, 61/147
tissue plasminogen activation, 65/461
tocainide, 63/192, 65/455
ulnar nerve lesion, 63/181
ventricular assist device, 63/178
Iatrogenic neuropathy
drug intoxication, 7/527-546
isoniazid neuropathy, 7/533
Iatrogenic paraplegia
spastic paraplegia, 59/437
thoracotomy, 26/69
Iberiotoxin
arthropod envenomation, 65/198
neurotoxin, 65/194, 198
Ibogaine
hallucinogenic agent, 65/43
hallucinogenic agent intoxication, 37/329
Ibotenate, *see* Ibotenic acid
Ibotenic acid
GABA receptor stimulating agent, 65/39
hallucinogenic agent, 65/42
Huntington chorea, 49/58, 261
mushroom toxin, 65/39
striatal necrosis, 49/511
Ibuprofen
brain metabolism, 57/82
chemical meningitis, 56/130
chronic paroxysmal hemicrania, 48/264
headache, 48/174
migraine, 48/174
ICAM-1, *see* Intercellular adhesion molecule 1
ICD classification
cerebrovascular disease, 53/2
epilepsy, 15/21
head injury epidemiology, 57/2
spinocerebellar degeneration, 60/272
ICD classification inconsistency
cerebrovascular disease, 53/2
Ice cream headache, 5/188-190, 48/441-446
classification, 48/6
cold induced pain, 48/441
definition, 48/10
duration, 48/441
experiment, 5/188
vascular response, 5/190
vasoconstriction, 5/190
Ice hockey injury
brain concussion, 23/583
facial injury, 23/583
prevention, 23/584
statistic, 23/583
Ice pick headache, 48/441-446
characteristics, 48/442
chronic paroxysmal hemicrania, 48/260
cluster headache, 48/260
location, 48/442
migraine, 48/442
Iceland disease, *see* Postviral fatigue syndrome
Ichthyohaemotix fish intoxication
abdominal pain, 37/87
fatal, 37/87
muscle weakness, 37/87
nausea, 37/87
paresthesia, 37/87
respiratory paralysis, 37/87
salivation, 37/87
urticaria, 37/87
vomiting, 37/87
Ichthyohepatoxic fish intoxication
alopecia, 37/87
arthralgia, 37/87
diarrhea, 37/87
facial edema, 37/87
facial flushing, 37/87
fever, 37/87
headache, 37/87
nausea, 37/87
vitamin A intoxication, 37/87
vomiting, 37/87
Ichthyoides
anencephaly, *see* Anencephalus ichthyoides
Ichthyo-otoxism intoxication, *see* Fish roe intoxication
Ichthyosiform erythroderma
autosomal recessive, 13/477
congenital, *see* Congenital ichthyosiform erythroderma
corpus callosum agenesis, 42/7
histology, 13/477
Netherton syndrome, 43/305
nevus unius lateris, 43/40
Sjögren-Larsson syndrome, 13/477, 43/307
Ichthyosiform erythromelia
unilateral hemimelic, *see* Unilateral hemimelic ichthyosiform erythromelia
Ichthyosis
Austin juvenile sulfatidosis, 60/667

autosomal dominant inheritance, 13/476
brain atrophy, 43/416
congenital, *see* Congenital ichthyosis
congenital heart disease, 43/416
congenital hemidysplasia, *see* Unilateral hemimelic ichthyosiform erythromelia
Conradi-Hünermann syndrome, 43/346
cranial nerve, 43/416
definition, 13/476
dominant, *see* Dominant ichthyosis
Down syndrome, 13/476
familial spastic paraplegia, 22/427
hemidysplasia, 43/416
hemivertebra, 43/416
histology, 13/476
human T-lymphotropic virus type I associated myelopathy, 59/450
Huntington chorea, 49/291
lipid storage myopathy, 43/184
mental deficiency, 13/476, 43/276
neuroaxonal leukodystrophy, 49/406
neurofibromatosis type I, 14/492
pectoralis muscle agenesis, 43/416
Refsum disease, 8/22, 13/479, 21/9, 202-205, 36/347, 41/434, 42/148, 60/228, 232, 667, 728, 66/491
Sjögren-Larsson syndrome, 13/468, 472, 476, 22/476, 59/358, 66/616
X-linked, *see* X-linked ichthyosis
Ichthyosis hystrix
nevus unius lateris, 43/40
Ichthyosis linguae, *see* Leukoplakia
Ichthyotic syndrome
Refsum disease, 38/9
Rud syndrome, 13/321, 479, 38/9, 43/284, 46/59, 51/398, 60/721
Sjögren-Larsson syndrome, 38/9
Icosisphingosine
biochemistry, 9/4
lipid metabolic disorder, 10/293, 295
Ictal aggression
epilepsy, 73/479
Ictal speech
paroxysmal aphasia, 45/322
Ictus
epilepsy, 73/473
Ideational apraxia
Alzheimer disease, 45/426
diagnosis, 45/424
parietal lobe syndrome, 45/74
Pick disease, 46/234
planning disturbance, 4/51
type, 2/601
Identical associated movement, *see* Synkinesis
Ideomotor apraxia
Alzheimer disease, 46/251
anterior cerebral artery syndrome, 53/346
brain infarction, 53/346
cerebellar catalepsy (Babinski), 21/526
classification, 45/425
constructional apraxia, 45/492, 498
middle cerebral artery syndrome, 53/363
parietal lobe syndrome, 45/74
perseveration, 4/50
rehabilitation, 46/621
spatial dyskinesia, 4/51, 60
supplementary motor area syndrome, 53/346
Idiocy
amaurotic, *see* Amaurotic idiocy
De Sanctis-Cacchione, *see* Xeroderma pigmentosum
familial amaurotic, *see* Familial amaurotic idiocy
juvenile amaurotic, *see* Juvenile amaurotic idiocy
leukodystrophy, 10/24
xerodermic, *see* Xeroderma pigmentosum
Idiomuscular contraction
reaction type, 1/221
Idiopathic autonomic neuropathy
infection, 74/336
Idiopathic cardiomyopathy
multiple mitochondrial DNA deletion, 66/429
Idiopathic CNS hypersomnia
daytime hypersomnia, 45/135
Idiopathic convulsion
classification, 15/5
Idiopathic demyelinating polyradiculoneuropathy
chronic, *see* Chronic idiopathic demyelinating polyradiculoneuropathy
Idiopathic developmental dyslexia
higher nervous activity disorder, 4/378-383
learning, 4/436
Idiopathic dystonia musculorum deformans
clinical features, 49/71
Idiopathic erythromelalgia
autonomic nervous system, 75/36
Idiopathic eosinophilic fasciitis, *see* Shulman syndrome
Idiopathic esophagus hypertrophy
amyotrophic lateral sclerosis, 22/143
Idiopathic face granuloma, *see* Midline granuloma
Idiopathic facial paralysis, 8/241-290
anatomy, 8/243-247
blink reflex, 8/263
clinical aspect, 8/259
clinical picture, 8/270-279
conduction block, 2/68

definition, 2/68
differential diagnosis, 8/279-283
electrophysiology, 8/266-270
etiology, 8/249-254
Heerfordt syndrome, 8/279
pathology, 8/258
physiology, 8/247-249
prognosis, 8/289
regeneration, 8/254-258
synonym, 8/241
treatment, 8/283-289
wallerian degeneration, 2/68
Idiopathic generalized myoclonic astatic epilepsy
seizure, 73/224
Idiopathic hemochromatosis
acanthocytosis, 63/290
Idiopathic hypercalcemia
infantile, *see* Elfin face syndrome
Idiopathic hypereosinophilic syndrome, 38/193-217
see also Hypereosinophilia
age, 63/124
brain embolism, 63/138
brain vasculitis, 63/124
cardiomyopathy, 63/138
cause, 63/123
classification, 38/213-215
clinical course, 63/125
course, 38/195
cranial neuropathy, 63/124
Davies disease, 38/212
encephalopathy, 63/138
eosinophilia, 63/371
eosinophilia myalgia syndrome, 64/257
epilepsy, 63/124
etiology, 38/215-217
features, 63/123
history, 38/193
intravascular consumption coagulopathy, 63/124
laboratory finding, 38/195
mental symptom, 63/124
mononeuritis multiplex, 63/124, 138
neurologic manifestation, 38/194-215
neurotrophic parietal endocarditis, 38/213
pathogenesis, 38/215-217
pathology, 38/195-197
polyneuropathy, 63/138
symptom, 63/124
treatment, 38/217, 63/125
Idiopathic hypoglycemia, 42/575-577
ataxia, 42/576
classification, 27/55
ketotic hypoglycemia, 27/72
leucine sensitivity, 27/73, 42/576
mental deficiency, 42/576
seizure, 42/576
Idiopathic hypoparathyroidism
hereditary paroxysmal kinesigenic choreoathetosis, 49/355
striatopallidodentate calcification, 6/705, 49/418
Idiopathic intracerebral calcinosis, *see* Familial striatopallidodentate calcification
Idiopathic long QT syndrome
heart disease, 75/442
Idiopathic medial aortopathy, *see* Takayasu disease
Idiopathic medial arteriopathy, *see* Takayasu disease
Idiopathic multiple hemorrhagic sarcoma, 18/288-293
brain vascular tumor, 18/288-292
case series, 18/290-293
endothelial proliferation, 18/288
fibroblastic proliferation, 18/288
hemorrhage, 18/288
hemosiderin, 18/288
inflammatory response, 18/288
lymphocyte, 18/288
malignant, 18/288
metastasis, 18/288
Idiopathic narcolepsy
cataplexy, 3/94
diplopia, 3/94
dreaming, 3/94
drowsiness, 3/94
drug treatment, 3/95
monotony, 3/93
paradoxical sleep, 3/94
paranoid psychosis, 3/95
paranoid symptom, 3/94
sleep paralysis, 3/94
Idiopathic orthostatic hypotension, *see* Pure autonomic failure
Idiopathic Parkinson disease
classification, 49/108
dementia, 46/378
etiologic critique, 6/174
Parkinson dementia, 46/378
Idiopathic partial epilepsy
age at onset, 73/8
benign infantile convulsion, 73/28
classification, 73/5
genetics, 73/7
incidence, 73/7
occipital paroxysm, 73/18
survey, 73/5
Idiopathic postural orthostatic tachycardia syndrome
viral infection, 74/327
weight loss, 74/327

Idiopathic retinoschisis
sex linked, *see* Sex linked idiopathic retinoschisis
Idiopathic scoliosis
birth incidence, 42/52
congenital hip dislocation, 42/52
epidemiology, 42/52
genetics, 32/150, 42/52
heart disease, 42/52
mental deficiency, 42/51
sex ratio, 32/132, 42/52
Idiopathic small fiber neuropathy
dysautonomia, 75/702
Idiopathic steatorrhea, *see* Celiac disease
Idiopathic vasculitis
polyarteritis nodosa, 71/154
Idiotype
CSF, *see* CSF idiotype
Idiotypic antibody
CSF, *see* CSF idiotypic antibody
multiple sclerosis, 47/111
Idiotypic IgG
multiple sclerosis, 47/362
Idoxuridine
progressive multifocal leukoencephalopathy, 47/519
IDPN intoxication, *see* 3,3'-Iminodipropionitrile intoxication
IDS gene
glycosaminoglycanosis type II, 66/282, 296
iduronate-2-sulfatase, 66/282
IDUA gene
glycosaminoglycanosis type IH, 66/282, 289
glycosaminoglycanosis type IH-S, 66/282, 289
glycosaminoglycanosis type IS, 66/282
α-L-iduronidase, 66/282
Iduronate sulfatase
glycosaminoglycanosis type II, 42/454, 66/65
Iduronate-2-sulfatase
glycosaminoglycanosis type II, 66/282
IDS gene, 66/282
Iduronic acid
connective tissue, 10/433
dermatan sulfate, 10/433
α-L-Iduronidase
glycosaminoglycanosis type I, 27/146, 42/452
glycosaminoglycanosis type IH, 66/65, 282, 289
glycosaminoglycanosis type IH-S, 66/282
glycosaminoglycanosis type IS, 27/151, 42/453, 66/282
IDUA gene, 66/282
mucolipidosis type II, 42/448
pseudodeficiency, 66/58
α-*L*-Iduronidase deficiency
bone marrow transplantation, 66/89
Ifosfamide
adverse effect, 69/492
antineoplastic agent, 65/528
cranial neuropathy, 65/528
epilepsy, 65/528
neuropathy, 69/461, 471
neurotoxin, 65/528
Ifosfamide intoxication
encephalopathy, 65/534
epilepsy, 65/534
Ig
Alzheimer disease, 46/265
antibody diversity, 56/55
Bloom syndrome, 43/10
chromosome, 56/55
class, 56/52
CSF, *see* CSF Ig
domain, 56/52
function, 56/54
genetics, 56/55
heavy chain gene, 56/55
hereditary recurrent polyneuropathy, 21/88
kappa light chain genes, 56/55
lambda light chain genes, 56/55
light chain, *see* Light chain Ig
molecular structure, 56/54
multiple sclerosis, 47/235-239, 309, 360
neuropathy, 8/72
physicochemical property, 56/53
polymyositis, 71/142
rabies, 34/266, 56/386
senile dementia, 46/284
structure, 56/52
tetanus, 65/222
undetermined significant monoclonal gammopathy, 63/403
vasculitis, 71/161
viral infection, 56/51
Ig binding lymphoid cell
multiple sclerosis, 47/81
Ig dyscrasia
progressive muscular atrophy, 59/16
Ig Fc fragment
multiple sclerosis, 47/365
Ig synthesis
multiple sclerosis, 47/329
Ig type
monoclonal gammopathy, 63/398
paraproteinemia, 63/398
IgA
ataxia, 21/581
ataxia telangiectasia, 14/75, 269-273, 60/375

characteristics, 9/502
chronic inflammatory demyelinating polyradiculoneuropathy, 51/537
CSF, *see* CSF IgA
deficiency, 40/422
demyelinating neuropathy, 47/616
function, 56/54
hereditary sensory and autonomic neuropathy type I, 21/82, 60/9
juvenile optic atrophy, 42/410
metabolic ataxia, 21/581
monoclonal, *see* Monoclonal IgA
multiple sclerosis, 47/339
myotonic dystrophy, 40/519
neurobrucellosis, 33/310
oculopharyngeal muscular dystrophy, 43/101
18q partial monosomy, 43/533
reticulosarcoma, 18/237
serum, *see* Serum IgA
undetermined significant monoclonal gammopathy, 63/402
IgA antibody
viral infection, 56/95
IgA deficiency
immunocompromised host, 56/468
polymyositis, 62/373
IgA increase
Parkinson dementia, 49/174
IgA paraproteinemia
paraproteinemic polyneuropathy, 51/436
IgD
characteristics, 9/502
CSF, *see* CSF IgD
function, 56/54
IgE
ataxia telangiectasia, 14/273, 60/375
characteristics, 9/502
CSF, *see* CSF IgE
function, 56/54
headache, 48/36
migraine, 48/122
IgE deficiency
ataxia telangiectasia, 60/351
IgG
atrial myxoma, 63/97
characteristics, 9/502
chronic inflammatory demyelinating polyradiculoneuropathy, 51/537
CSF, *see* CSF IgG
CSF oligoclonal, *see* CSF oligoclonal IgG
demyelinating antibody, 9/518
demyelinating neuropathy, 47/615
experimental allergic encephalomyelitis, 9/520
function, 56/54
Guillain-Barré syndrome, 9/140
hereditary sensory and autonomic neuropathy type I, 60/9
idiotypic, *see* Idiotypic IgG
juvenile optic atrophy, 42/410
mitochondrial myopathy, 43/119
monoclonal, *see* Monoclonal IgG
multiple sclerosis, 47/339
myotonic dystrophy, 43/153
neurobrucellosis, 33/310, 52/584
oculopharyngeal muscular dystrophy, 43/101
rabies postvaccinial encephalomyelitis, 9/140
serum, *see* Serum IgG
subacute sclerosing panencephalitis, 9/140, 349, 352
undetermined significant monoclonal gammopathy, 63/402
IgG allotype
multiple sclerosis, 47/87, 104-106
IgG antibody
Sydenham chorea, 49/362
viral infection, 56/95
IgG index
multiple sclerosis, 47/87
IgG-kappa paraproteinemia
multiple myeloma polyneuropathy, 51/430
IgG paraproteinemia
paraproteinemic polyneuropathy, 51/436
IgG subclass deficiency
immunocompromised host, 56/468
IgG synthesis
intrablood-brain barrier, *see* Intrablood-brain barrier IgG synthesis
intrathecal, *see* Intrathecal IgG synthesis
multiple sclerosis, 47/79-82
IgG2
ataxia telangiectasia, 60/375
IgG2 deficiency
meningococcal meningitis, 52/29
vasculitis, 56/468
IgM
ataxia telangiectasia, 14/75
B-lymphocyte lymphoma, 18/235
blood viscosity, 55/483
characteristics, 9/502
chronic inflammatory demyelinating polyradiculoneuropathy, 51/537
CSF, *see* CSF IgM
demyelinating neuropathy, 47/616-618
experimental allergic encephalomyelitis, 9/520
function, 56/54
juvenile optic atrophy, 42/410

monoclonal, *see* Monoclonal IgM
neurobrucellosis, 33/310, 52/584
oculopharyngeal muscular dystrophy, 43/101
plasma cell dyscrasia, 63/391
reticulosarcoma, 18/237
undetermined significant monoclonal gammopathy, 63/402
IgM antibody
viral infection, 56/95
IgM macroglobulin
Waldenström macroglobulinemia polyneuropathy, 51/432
IgM monoclonal gammopathy
cryoglobulinemia, 69/297
glycolipid, 69/297
human T-lymphotropic virus type I associated myelopathy, 59/450
paraproteinemia, 69/297
IgM paraproteinemic polyneuropathy
age, 51/437
ataxia, 51/437
demyelination, 51/437
EMG, 51/437
IgM-kappa, 51/437
IgM-lambda, 51/437
myelin associated glycoprotein, 51/437
nerve conduction velocity, 51/437
pathology, 51/437
sex ratio, 51/437
symptom, 51/436
tremor, 51/437
ultrastructure, 51/437
vibration neuropathy, 51/437
IgM reduction
Parkinson dementia, 49/174
Ileal receptor
vitamin B_{12} deficiency, 51/338
Ileum
cholinergic innervation, 74/43
Iliac crest abnormality
Dyggve-Melchior-Clausen syndrome, 43/266
Iliohypogastric nerve
entrapment, 7/313
topographical diagnosis, 2/39
Ilioinguinal nerve
compression neuropathy, 51/109
entrapment, 7/313
syndrome, 20/795
topographical diagnosis, 2/39
Illusion
auditory, *see* Auditory illusion
dysphrenia hemicranica, 5/79
epilepsy, 45/508
occipital lobe syndrome, 45/56-58
olfactory, *see* Olfactory illusion
parietal lobe, 45/361-363
parietal lobe epilepsy, 73/101
perception, 45/361
somatesthetic, *see* Somatesthetic illusion
temporal lobe tumor, 17/289
vertiginous, *see* Vertiginous illusion
Illusionary falsification
migraine, 48/162
Illusory visual speed
visual perseveration, 2/617
Imagery
definition, 45/351
hallucination, 45/351-369
hypnagogic, *see* Hypnagogic imagery
number-form, 43/224
type, 45/358-361
Wernicke-Korsakoff syndrome, 45/199
Imaging
acoustic neuroma, 67/189
astrocytoma, 67/169, 173
autonomic nervous system, 74/626
brain stem glioma, 67/179
brain tumor, 67/167, 173, 238, 313
cerebellar astrocytoma, 67/180
choroid plexus carcinoma, 67/177
choroid plexus papilloma, 67/177
choroid plexus tumor, 68/169
CNS, 74/626
craniopharyngioma, 67/188
embryonal cell carcinoma, 67/179
ependymoma, 67/176, 181, 68/168
germinoma, 67/178
hemangioblastoma, 67/179, 68/272
hypercortisolism, 70/209
infantile spasm, 72/379
Landau-Kleffner syndrome, 72/379
Lennox-Gastaut syndrome, 72/379
low grade glioma, 68/37
magnetoencephalography, 72/321
malignant astrocytoma, 67/169, 176
medulloblastoma, 67/183
meningioma, 67/182, 187, 189, 68/406
metastasis, 67/172, 179, 185
neurinoma, 67/183
neuroastrocytoma, 67/174
oligodendroglioma, 67/169, 172
pineal tumor, 67/177
pinealoblastoma, 67/179
pituitary adenoma, 67/186
pituitary microadenoma, 68/346
pleomorphic xanthoastrocytoma, 67/175

posttraumatic epilepsy, 72/202
primitive neuroectodermal tumor, 67/176, 182
sella turcica, 68/345
spinal cord hemangioma, 67/194
spinal cord metastasis, 67/204
spinal cord tumor, 68/500
teratoma, 67/179
vestibular neurinoma, 68/428
Wegener granulomatosis, 71/185
Imanenjana
involuntary movement, 6/301
Imbecillitas phenylpyruvica, *see* Phenylketonuria
Imidazole
myopathy, 40/150
Imidazole acidic acid riboside
chronic paroxysmal hemicrania, 48/261
Imidazole derivative
neurotoxin, 67/364
Iminoacidemia, 29/129-145
see also Hydroxyprolinemia, Hyperprolinemia type I *and* Hyperprolinemia type II
disorder, 29/144
hydroxyproline, 29/129, 131
neonate, 29/143
proline, 29/129-131
terminology, 29/129
Iminodibenzyl, *see* 10,11-Dihydro-5H-dibenz[*b*,*f*]azepine
3,3'-Iminodipropionitrile
acetonylacetone intoxication, 64/85
chemical formula, 64/82
neurotoxin, 64/10, 82
organic solvent, 64/82
toxic neuropathy, 64/10
3,3'-Iminodipropionitrile intoxication
action site, 64/10
bulbar paralysis, 22/127
demyelinating neuropathy, 51/264
demyelination, 51/264
exposure source, 51/278
hexacarbon neuropathy, 51/278
neuroaxonal dystrophy, 6/626
β,β'-Iminodipropionitrile intoxication, *see* 3,3'-Iminodipropionitrile intoxication
Iminoglycinuria
familial, *see* Familial iminoglycinuria
Iminostilbene intoxication, *see* Dibenzazepine intoxication
Imipramine
metabolism, 65/316
migraine, 5/101
myopathy, 40/169
myotonia, 62/276
narcolepsy, 3/95
neuropathy, 7/518, 533
neurotoxin, 7/518, 533, 46/394, 48/420, 65/311
tricyclic antidepressant, 65/312
Imipramine intoxication
headache, 48/420
Immaturity
developmental dyslexia, 46/114
Immersion foot
see also Trench foot
cold water, 75/697
cooling, 75/696
frostbite, 8/342
nerve injury, 51/136
vibration, 75/695
warm water, 75/699
Immobilization
compression neuropathy, 51/87
delirium, 46/535
Immune adherence
Newcastle disease, 47/358
Immune cell interaction
scheme, 65/550
Immune complex
allergic granulomatous angiitis, 63/383
angiopathic polyneuropathy, 51/445
antibody, 41/376
circulating, *see* Circulating immune complex
CSF, *see* CSF immune complex
diamorphine addiction, 55/519
multiple sclerosis, 47/114, 239, 370-372
Sydenham chorea, 49/362
systemic vasculitis, 56/297
toxic encephalopathy, 52/297
Immune complex myositis
dermatomyositis, 56/201
Immune complex nephritis
penicillamine intoxication, 49/235
Immune complex vasculitis, 71/8, 151
Kawasaki syndrome, 56/638
systemic lupus erythematosus, 71/39
Immune induced angiopathy
Sneddon syndrome, 55/408
Immune mediated encephalomyelitis
animal model, 34/421-423
chickenpox, 34/416-418
clinical syndrome, 34/404, 407
immune mediated syndrome related to virus infection, 34/403-425
influenza, 34/418
measles virus, 34/408-411
mumps, 34/411
pathology, 34/404-408

rabies, 34/414-416
rabies virus, 34/414-416
rubella encephalitis, 34/419-421
vaccinia, 34/412-414
varicella zoster, 34/416-418
Immune mediated polyneuritis
acute, *see* Acute immune mediated polyneuritis
Immune mediated syndrome related to virus infection, 34/391-425
acute immune mediated polyneuritis, 34/392-403, 423-425
acute paralytic brachial neuritis, 34/395
chronic relapsing polyneuritis, 34/396
Guillain-Barré syndrome, 34/392-403, 423-425
immune mediated encephalomyelitis, 34/403-425
recurrent polyneuropathy, 34/396
recurrent relapsing polyneuritis, 34/396
Immune response
abnormal, 56/20
brain thromboangiitis obliterans, 55/313
cell mediated, *see* Cell mediated immune response
experimental allergic encephalomyelitis, 47/437
herpes simplex virus, 56/209
host, *see* Host immune response
humoral, *see* Humoral immune response
JHM virus, 56/445
leprosy, 42/649
multiple sclerosis, 47/83
murine leukemia virus, 56/457
muscarinic acetylcholine, 75/553
progressive multifocal leukoencephalopathy, 34/307, 47/507
rabies, 34/263, 56/393
rabies postvaccinial encephalomyelitis, 56/394
rubella syndrome, 31/213
schistosomiasis, 52/539
slow virus disease, 56/20
sympathetic nervous system ablation, 75/555
toxoplasmosis, 42/662
Treponema pallidum syndrome, 31/218
viral infection, 34/39-44, 55, 56/71
Immune response gene
multiple sclerosis, 47/303
Immune serum
serogenetic neuropathy, 8/86, 95
Immune system
autonomic nervous system, 75/551
hereditary neuralgic amyotrophy, 51/174
multiple sclerosis, 47/337
sleep, 74/534
sympathetic nervous system, 75/557
Immunity
cell mediated, *see* Cell mediated immunity
cerebral amebiasis, 52/323
chemistry, 9/503
CSF humoral, *see* CSF humoral immunity
diphtheria, 52/228
granulomatous amebic encephalitis, 52/325
humoral, *see* Humoral immune response
hypersensitivity, 9/503
malaria, 35/144, 52/368
meningococcal meningitis, 33/22, 52/22
paraneoplastic syndrome, 69/340
pertussis, 52/234
poliomyelitis, 56/322
rabies, 34/249
toxoplasmosis, 35/118
Immunization
diphtheria, 52/228, 235
diphtheria pertussis tetanus, *see* Diphtheria pertussis tetanus immunization
multiple sclerosis, 47/150
neuropathy, 7/555
pertussis, 52/227, 235
serum paralysis, 7/555
streptococcal meningitis, 52/82
tetanus, 33/515-522, 52/227, 235, 65/220
Immunoaggression
antibody mediated, 9/622
cell mediated, 9/622
Immunocompromised host, 56/467-483
acquired toxoplasmosis, 35/128, 52/353
action mechanism, 56/471
acute cerebellar ataxia, 56/477
acute leukemia, 56/470
adenovirus, 56/474
agranulocytosis, 56/469
antilymphocyte serum, 56/471
antimetabolite, 56/469
aplastic anemia, 56/469
ascending myelitis, 56/477
aspergillosis, 52/378
Aspergillus fumigatus, 65/549
ataxia telangiectasia, 56/468
azathioprine, 56/471
bacterial meningitis, 52/1
benign monoclonal gammopathy, 56/470
brain abscess, 52/143
Burkitt lymphoma, 56/470
burn, 56/469, 474
candidiasis, 52/398
chromosomal abnormality, 56/469
chronic lymphocytic leukemia, 56/470
ciclosporin, 56/469, 471
CNS infection, 56/467, 65/548
coccidioidomycosis, 52/411

complement deficiency, 56/468
complication, 56/474
corticosteroid, 56/469
Coxsackie virus, 56/476
cryptococcosis, 52/430
Cryptococcus neoformans infection, 65/549
cyclophosphamide, 56/471
cytomegalovirus, 56/476
diabetes mellitus, 56/469, 473
Down syndrome, 56/469, 473
ECHO virus, 56/475
encephalitis, 56/474, 477
enterovirus, 56/475
Epstein-Barr virus, 56/477
exudative enteropathy, 56/469
gammopathy, 56/469
granulomatous amebic encephalitis, 52/314
hairy cell leukemia, 56/470
heavy chain disease, 56/470
herpes simplex virus, 56/477
herpes virus, 56/476
Hodgkin disease, 56/469
hyper-IgE, 56/468
IgA deficiency, 56/468
IgG subclass deficiency, 56/468
immunosuppression, 56/468
JC virus, 56/480
leukemia, 56/469
listeriosis, 52/91
lymphoma, 56/469
malignancy, 56/469
measles virus, 56/480
meningoencephalitis, 56/474
mucormycosis, 52/467
multiple myeloma, 56/470
non-Hodgkin lymphoma, 56/470
opportunistic infection, 56/472
organ transplantation, 56/472
papovavirus, 56/480
plasma cell leukemia, 56/470
poliomyelitis, 56/476
primary amebic meningoencephalitis, 52/314
progressive multifocal leukoencephalopathy, 56/480
radiation, 56/469, 471
sickle cell anemia, 56/469, 473
splenectomy, 56/469
surgery, 56/474
thymus hypoplasia, 56/468
Toxoplasma gondii infection, 65/549
transverse myelitis, 56/477
trauma, 56/474
varicella zoster virus, 56/478
viral infection, 56/467
viremia, 56/468
Waldenström macroglobulinemia, 56/470
Wiskott-Aldrich syndrome, 56/468
X-linked agammaglobulinemia, 56/468
Immunocytochemistry
leukemia, 69/241
muscle biopsy, 62/2
myasthenia gravis, 62/415
neurotropic virus, 56/26
Immunodeficiency
ataxia telangiectasia, 60/375
bacterial meningitis, 52/3
Chagas disease, 52/347, 75/390
Down syndrome, 50/530
enterovirus, 56/330
meningoencephalitis, 52/347
Nijmegen breakage syndrome, 60/426
Immunodeficiency syndrome
acquired, *see* Acquired immune deficiency syndrome
Immunoelectrophoresis
brain tumor, 16/364
multiple sclerosis, 47/87, 94, 111
pneumococcal meningitis, 33/40
Immunofixation, *see* Immunoelectrophoresis
Immunofluorescence test
lymphocytic choriomeningitis, 56/364
multiple sclerosis, 47/80, 236-239
neuroamebiasis, 52/318
North American blastomycosis, 52/390
progressive multifocal leukoencephalopathy, 47/510
protein, 9/24
rabies, 34/258-260, 56/396
viral infection, 56/96
Immunoglobulin, *see* Ig
Immunologic disorder
paraneoplastic polyneuropathy, 51/470
Immunology
acute amebic meningoencephalitis, 52/324
ataxia telangiectasia, 14/269, 60/348
autonomic nervous system, 75/551
Behçet syndrome, 56/603
brucellosis, 52/584
chronic inflammatory demyelinating polyradiculoneuropathy, 51/537
CNS, 9/501
Coxsackie virus infection, 34/134
Creutzfeldt-Jakob disease, 56/552
cytomegalovirus infection, 56/266
filariasis, 52/518
Gerstmann-Sträussler-Scheinker disease, 56/556

Guillain-Barré syndrome, 47/611
helminthiasis, 52/508
kuru, 56/558
leprosy, 33/426-428
meningococcal meningitis, 33/26, 52/23
multiple sclerosis, 9/128, 47/189, 337-380
neurobrucellosis, 52/584
North American blastomycosis, 35/408, 52/393
poliomyelitis, 56/323
prion, 56/565
prion disease, 56/556
syphilis, 33/347
trichinosis, 52/572
varicella zoster virus infection, 56/241
Vogt-Koyanagi-Harada syndrome, 56/623
Immunomodulating therapy
neurologic complication, 65/547
Immunopathology
brain vasculitis, 55/416
collagen vascular disease, 71/3
dermatomyositis, 62/379
inclusion body myositis, 62/379
Köhlmeier-Degos disease, 55/279
mechanism, 71/3, 5
multiple sclerosis, 47/232-235
poliomyelitis, 34/120
polyarteritis nodosa, 9/551
polymyositis, 62/379
tuberculosis, 55/423
undetermined significant monoclonal gammopathy, 63/400
Whipple disease, 52/136
Immunoperoxidase
multiple sclerosis, 47/235
Immunoregulation
Epstein-Barr virus infection, 56/250
multiple sclerosis, 47/368
Immunosuppression
adenovirus, 56/283
adenovirus encephalitis, 56/286
bacterial meningitis, 52/5
Behçet syndrome, 56/605
brain abscess, 52/148
brain embolism, 55/166
cerebrovascular disease, 55/166
copolymer I, 47/191-193
Eaton-Lambert myasthenic syndrome, 62/426
experimental autoimmune encephalomyelitis, 47/191
Freund adjuvant, 47/191
herpes zoster, 56/230
iatrogenic neurological disease, 63/180
immunocompromised host, 56/468
inflammatory myopathy, 41/80, 374
measles, 56/417
measles encephalitis, 56/417
mucormycosis, 35/544, 52/474
multiple sclerosis, 9/402, 47/202, 231
myasthenia gravis, 41/132
myelin basic protein, 47/191
neurocysticercosis, 52/533
nocardiosis, 52/451
primary CNS lymphoma, 56/255, 63/533
progressive multifocal leukoencephalopathy, 34/308, 47/503-506, 56/514, 71/400
progressive muscular atrophy, 59/16
renal transplantation, 63/530, 532
systemic lupus erythematosus, 55/381
thyroid associated ophthalmopathy, 62/532
visna infection, 56/462
Immunosuppressive agent
cardiac pharmacotherapy, 63/189
epilepsy, 72/232
iatrogenic neurological disease, 63/189
listeriosis, 52/91
myasthenia gravis, 41/96
temporal arteritis, 48/324
Immunosuppressive therapy
autoimmune disease, 65/550
azathioprine, 65/550
ciclosporin, 65/550
corticosteroid, 65/550
dermatomyositis, 62/385
Epstein-Barr virus infection, 65/549
heart transplantation, 63/10
inclusion body myositis, 62/385
Kaposi sarcoma, 65/549
lymphoma, 65/549
malignancy incidence, 65/549
neurologic complication, 65/547
non-Hodgkin lymphoma, 65/549
OKT3, 65/550
polymyositis, 62/385
squamous cell carcinoma, 65/549
undetermined significant monoclonal gammopathy, 63/403
Immunotherapy
acute disseminated encephalomyelitis, 71/550
astrocytoma, 67/291, 294
ataxia telangiectasia, 60/389
brain tumor, 67/291, 294
gene therapy, 67/291
glioblastoma multiforme, 18/66
leukemia, 18/251
quality of life, 67/399
vasculitis, 71/162

Immunothrombocytopenia
intraventricular hemorrhage, 63/327
leukoencephalopathy, 9/599
Impairment
cancer, 67/390
conjugate gaze, 1/599-601
intelligence, 4/6-9
memory, *see* Memory disorder
oxidative phosphorylation, 40/171
proprioceptive, *see* Proprioceptive impairment
quality of life, 67/390
visual fixation, 1/599
Impairments
motor, *see* Motor impairment
sensory, *see* Sensory impairment
Imperception
auditory, *see* Auditory imperception
visual agnosia, 2/601, 4/14
Imperforate anus
anencephaly, 42/13
anterior spinal meningocele, 42/43
birth incidence, 42/379
caudal aplasia, 42/50
corpus callosum agenesis, 42/7
Down syndrome, 50/274
FG syndrome, 43/250
hemivertebra, 42/31
hypertelorism hypospadias syndrome, 43/413
r(13) syndrome, 50/585
sensorineural deafness, 42/379
Townes-Brocks syndrome, 42/379
trigeminal nerve agenesis, 50/214
Impotence
sexual, *see* Sexual impotence
Impotentia coeundi
paraplegia, 61/315
spinal cord injury, 61/315
tetraplegia, 61/315
transverse spinal cord lesion, 61/315
Impression
basilar, *see* Basilar impression
Impulse conduction
axon, 51/44
nerve, 51/44
Impulsive grand mal, *see* Generalized tonic clonic seizure
Impulsive petit mal epilepsy (Janz), *see* Juvenile myoclonic epilepsy
Inability to open mouth
muscle shortening, 43/434
pseudocamptodactyly, 43/433
talipes, 43/434
Inactivation syndrome, *see* Motor ritardando
Inappropriate antidiuretic hormone secretion
brain aneurysm, 55/53
brain aneurysm neurosurgery, 55/53
brain tumor, 28/499
cause, 1/502
chronic meningitis, 56/644
cryptococcosis, 52/431
definition, 1/502
delirium, 46/543
epilepsy, 28/499
features, 1/502
Guillain-Barré syndrome, 28/499
head injury, 57/237
Hemophilus influenzae meningitis, 52/63
herpes simplex virus encephalitis, 28/499
hydrocephalus, 28/499
hyponatremia, 28/499, 63/545
meningitis, 28/499
myxedema, 28/499
neuroamebiasis, 52/320
recurrent meningitis, 52/52
sodium deficiency, 1/502
streptococcal meningitis, 52/82
subarachnoid hemorrhage, 55/9, 53
vincristine, 63/359
vincristine polyneuropathy, 51/299
Inappropriate laughing and crying, *see* Pathologic laughing and crying
Inattention
hemi, *see* Hemi-inattention
hemianopic, *see* Hemianopic inattention
sensory, *see* Sensory inattention
tactile, *see* Tactile inattention
visual, *see* Visual inattention
visuospatial, *see* Visuospatial inattention
Inborn error of carbohydrate metabolism
glycosaminoglycanosis type I, 27/162
GM_1 gangliosidosis, 27/162
GM_2 gangliosidosis, 27/162
GM_2 gangliosidosis type II, 27/162
mucolipidosis type II, 27/162
prenatal diagnosis, 27/163
Inborn error of metabolism, 72/221
classification, 46/17
dementia, 46/130
epilepsy, 72/223
general symptom, 72/223
glycoprotein, 27/153-155
glycosaminoglycan, 27/145-153
mental deficiency, 46/17
neonatal seizure, 73/255
Incentive
frontal lobe lesion, 2/738

Incertohypothalamic system
stress, 74/313
Incidences
autopsy, *see* Autopsy incidence
birth, *see* Birth incidence
Incipient myelomatosis, *see* Waldenström macroglobulinemia
Incisure
Lantermann-Schmidt, *see* Lantermann-Schmidt incisure
Schmidt-Lantermann, *see* Schmidt-Lantermann incisure
tentorial, *see* Tentorial incisure
Inclusion
argyrophilic, *see* Argyrophilic inclusion
basophilic granular, *see* Basophilic granular inclusion
eosinophilic cytoplasmic, *see* Eosinophilic cytoplasmic inclusion
leukocyte, *see* Leukocyte inclusion
macrophage, *see* Macrophage inclusion
paracrystalline, *see* Paracrystalline inclusion
Schwann cell, *see* Schwann cell inclusion
Inclusion body
amyotrophic lateral sclerosis, 22/323
atypical, *see* Atypical inclusion body
cap disease, 62/341
Cowdry A, *see* Cowdry A inclusion body
cytomegalovirus infection, *see* Cytomegalic inclusion body disease
cytoplasmic, *see* Cytoplasmic inclusion body
dermatomyositis, 40/264
eosinophilic, *see* Eosinophilic inclusion body
GM2 gangliosidosis, 51/375
lead intoxication, 64/436
membranous cytoplasmic, *see* Membranous cytoplasmic inclusion body
metachromatic leukodystrophy, 10/51
mitochondria, 40/91
muscle cell nucleus, 40/35, 90
Negri, *see* Negri inclusion body
nemaline myopathy, 62/342
neuronal ceroid lipofuscinosis, 51/382
Niemann-Pick disease type C, 51/375
oligodendrocyte, *see* Oligodendrocyte inclusion body
organolead intoxication, 64/132
organotin intoxication, 64/140
polymyositis, 40/264
Refsum disease, 51/385
Tomé-Fardeau body, 62/103
Inclusion body disease, *see* Mucolipidosis type II
Inclusion body encephalitis
brain biopsy, 10/685
Dawson, *see* Subacute sclerosing panencephalitis
leukodystrophy classification, 10/20
myoclonus, 6/166
Inclusion body myositis
adenovirus type 2, 56/290
age at onset, 62/204
amyloid deposit, 62/377
amyloid β-protein, 62/377
anti-Jo-1 antibody, 62/374
areflexia, 62/373
arthralgia, 62/374
aspiration pneumonia, 62/374
autoimmune antibody, 62/380
axonal neuropathy, 62/376
azathioprine, 62/385
basophilic granular inclusion, 62/377
calcinosis, 62/386
cardiac disorder, 62/374
cardiomyopathy, 62/374
ciclosporin, 62/386
classification, 40/284
clinical features, 40/322
collagen vascular disease, 62/373
congestive heart failure, 62/374
corticosteroid, 62/383
creatine kinase, 62/373, 375
cricopharyngeal myotomy, 62/386
cyclophosphamide, 62/385
diagnosis, 62/375
diagnostic criteria, 62/378
distal muscle, 62/373
dominant inheritance, 62/373
dysphagia, 62/206, 373
dyspnea, 62/374
electronmicroscopy, 62/377
EMG, 62/375
eosinophilic cytoplasmic inclusion, 62/377
extramuscular manifestation, 62/373
fever, 62/373
filamentous inclusion, 62/206
gastrointestinal symptom, 62/374
hematemesis, 62/374
histopathology, 62/206
hypoxemia, 62/374
immunopathology, 62/379
immunosuppressive therapy, 62/385
inflammatory myopathy, 40/52, 284, 322
intravenous Ig, 62/386
joint contracture, 62/374
lactate dehydrogenase, 62/375
leukapheresis, 62/386
leukoencephalopathy, 62/373

limb girdle syndrome, 62/187
lymphoid irradiation, 62/386
malaise, 62/373
malignancy, 62/374
melena, 62/374
membranolytic attack complex, 62/379
methotrexate, 62/385
mixed connective tissue disease, 62/374
multiple mitochondrial DNA deletion, 66/429
muscle biopsy, 62/373, 375
muscle tissue culture, 62/104
myocarditis, 62/374
nuclear magnetic resonance, 62/379
oculopharyngeal muscular dystrophy, 62/297
overlap syndrome, 62/374
pathogenesis, 62/379
physiotherapy, 62/386
picornavirus, 62/381
plasmapheresis, 62/386
pneumatosis intestinalis, 62/374
pneumonitis, 62/374
practical guideline, 62/386
primary inflammatory myositis, 62/369
pulmonary dysfunction, 62/374
Raynaud phenomenon, 62/374
rheumatoid arthritis, 62/374
rimmed muscle fiber vacuole, 62/376
rimmed muscle fiber vacuole content, 62/18
rimmed muscle fiber vacuole type, 62/18
scleroderma, 62/374, 71/109
serum alanine aminotransferase, 62/375
serum aspartate aminotransferase, 62/375
serum muscle enzyme, 62/375
sex ratio, 62/204
Sjögren syndrome, 62/374, 71/74
steroid myopathy, 62/384
systemic lupus erythematosus, 62/374
systemic sclerosis, 62/374
tachyarrhythmia, 62/374
treatment, 62/383
tubulofilament, 62/377
viral myositis, 62/381
viral origin, 56/201
virus isolation, 56/203
virus like particle, 40/52, 284, 322
weight loss, 62/374
Wohlfart-Kugelberg-Welander disease, 62/204
Inclusion cell disease
glycoproteinosis, 21/48
Incoherence of thought
migraine, 48/162
Incomplete cord lesion
cervical, 25/270-276, 283, 335, 359
child, 25/178
definition, 25/146
prognosis, 25/283, 26/308
sexual function, 26/449
Incomplete dystrophin deficiency
Becker muscular dystrophy, 62/135
cardiomyopathy, 62/135
exertional myalgia, 62/135
focal myopathy, 62/135
malignant hyperthermia, 62/135
myoglobinuria, 62/135
quadriceps myopathy, 62/135
Incomplete lesion
early burning pain, 26/490
physical rehabilitation, 61/473
spinal cord injury prognosis, 26/308
Incomplete penetrance
definition, 30/90
Incontinence
acquired immune deficiency syndrome dementia, 56/491
adrenoleukodystrophy, 60/169
Alzheimer disease, 46/328
brain lacunar infarction, 54/253
fecal, *see* Fecal incontinence
stress, *see* Stress incontinence
urinary, *see* Urinary incontinence
Wilson disease, 49/227
Incontinentia pigmenti, 14/213-221
alopecia, 14/214, 43/23
alopecia areata, 14/12
ataxia, 14/12, 108
ataxia telangiectasia, 14/75
blue sclerae, 14/108
Brandt syndrome, 14/108
case series, 14/216-220
cerebral palsy, 14/214
cheilopalatoschisis, 43/23
chromosomal aberration, 43/23
chronic meningitis, 56/644
clinical features, 14/107, 214
CNS, 31/242
congenital cataract, 14/12
convulsion, 14/85, 31/242, 43/23
dental abnormality, 14/214, 43/23
differential diagnosis, 14/108, 215, 31/242
EEG, 14/216, 43/23
epilepsy, 14/12, 85, 214, 46/68
eye finding, 31/241
eye lesion, 14/214
genetics, 14/215, 31/241
Goltz-Gorlin syndrome, 14/108
hemivertebra, 43/23

hereditary, 14/108
hirsutism, 14/12
history, 14/213
hydrocephalus, 14/85, 30/100
hyperkeratosis, 43/23
infantile spinal muscular atrophy, 59/75
Leschke-Ullmann syndrome, 14/117
Little disease, 14/108
mental deficiency, 14/12, 85, 31/242, 43/23, 46/67
microcephaly, 14/85, 216, 30/508, 43/23
microgyria, 14/214
Naegeli syndrome, 14/215, 31/242
neurocutaneous syndrome, 14/101
neuroectodermal dysplasia, 14/213
neurology, 14/214
neuropathology, 14/84, 220
optic atrophy, 13/42, 81, 14/12, 108, 43/23
oral lesion, 31/242
paresis, 14/85, 214
phakomatosis, 14/220, 31/26, 50/377
plagiocephaly, 14/216
porencephaly, 14/85
prevalence, 14/215
pseudoastrocytoma, 14/12, 43/23
psychiatric aspect, 14/768
related syndrome, 14/101
sex ratio, 43/23
short description, 14/12
Sjögren-Larsson syndrome, 13/474
skeletal abnormality, 31/242
skin lesion, 14/85, 214
spastic paraplegia, 43/23
spina bifida, 43/23
spinal muscular atrophy, 59/368, 370
strabismus, 43/23
symptom, 14/108
talipes, 43/23
treatment, 14/216

Incoordination
see also Ataxia *and* Coordination
astasia abasia, 1/345
Cockayne-Neill-Dingwall disease, 13/433
dentatorubrothalamic pathway, 2/481
direct dentatothalamic fiber, 2/481
Gilles de la Tourette syndrome, 49/631
hypertrophic interstitial neuropathy, 21/150
multiple sclerosis, 47/169
Sydenham chorea, 49/363
zinc intoxication, 36/338

Increased intracranial pressure
acrocephalosyndactyly type V, 50/123
brain abscess, 33/117
chronic respiratory insufficiency, 38/290, 298
elfin face syndrome, 31/318
hydrocephalus, 50/291
hypoparathyroidism, 27/310
pattern, 57/218

Increment sensitivity index
short, *see* Short increment sensitivity index

Indecisiveness
migraine, 48/162

Indeterminate leprosy
classification, 51/216

Indian club throwing
rehabilitative sport, 26/531

Indian tick-borne typhus
rickettsial infection, 34/657

Indicanuria
hereditary periodic ataxia, 21/568
metabolic ataxia, 21/579

Indifference
frontal lobe syndrome, 55/138
temporal lobe syndrome, 45/44

Indifference to pain
anesthesia, 8/199
aphasia, 8/196
body scheme disorder, 4/222, 45/383-385
congenital, *see* Congenital indifference to pain
epilepsy, 8/194
hereditary sensory and autonomic neuropathy type III, 1/475, 479, 8/199, 41/435, 42/300, 60/7, 27
lesion site, 8/197
Lissauer tract absence, 1/110, 8/199
neurogenic osteoarthropathy, 8/195
parietal lobe, 8/196-198
pathogenesis, 45/384
self-mutilation, 4/222
sensory syndrome, 8/199
smell, 8/195
sneezing, 8/195
taste, 8/195

Indifferent cell (Schaper), *see* Medulloblast

Indirect tumor sign, *see* Tumor sign

Indoleacetic acid
headache, 48/174
migraine, 48/174

Indoleamine-2,3-dioxygenase, *see* Indoleamine-pyrrole-2,3-dioxygenase

Indoleamine-pyrrole-2,3-dioxygenase
eosinophilia myalgia syndrome, 63/378

Indolence
migraine, 48/162

Indolylacrolyl glycinuria
bacterium, 29/231
Hartnup disease, 29/231
mental deficiency, 29/231

Indometacin
behavior disorder, 46/602
benign intracranial hypertension, 67/111
brain infarction, 53/429
chronic paroxysmal hemicrania, 48/263, 267
cluster headache, 48/230
cluster variant syndrome, 48/270
cough headache, 48/371
delirium, 46/540
migraine, 5/56, 102, 48/174, 177
multiple sclerosis, 47/355
myopathy, 40/343
postlumbar puncture syndrome, 61/156
progressive dysautonomia, 59/140, 158
Indometacin responsive headache
arthritis, 48/174
cluster variant syndrome, 48/227
Indonesian Weil disease, *see* Leptospirosis
Indoramin
migraine, 48/187
Indoxyl esterase
wallerian degeneration, 9/36
Industrial gas
neurotoxicity, 51/265
Industrial solvent
late cortical cerebellar atrophy, 21/497
neurotoxicity, 51/265
Industrial toxin
encephalopathy, 46/392
Inertia
thalamic dementia, 21/595
Infantile acid maltase deficiency
biochemical aspect, 27/195, 226
cardiomegaly, 27/195, 227, 62/480
clinical features, 27/195, 227
clinical laboratory examination, 27/227
diagnosis, 27/195
enzyme deficiency polyneuropathy, 51/371
glycogen storage disease, 27/222
heart failure, 62/480
histochemistry, 27/197
hyaluronidase, 41/180
infantile hypotonia, 27/195, 227, 62/480
late, *see* Late infantile acid maltase deficiency
limb girdle syndrome, 62/190
macroglossia, 62/480
muscle hypertrophy, 62/480
pulmonary failure, 62/480
ultrastructure, 27/197
Infantile amyotrophic lateral sclerosis
differential diagnosis, 59/377
Infantile attachment disorder
mental deficiency, 46/25
Infantile autism
behavior therapy, 46/192
computer assisted tomography, 46/191
criteria, 46/189-195
development, 46/190
differential diagnosis, 46/194
EEG, 46/191
epidemiology, 46/191
genetics, 46/191
hyperserotonemia, 46/191
mental deficiency, 46/25
motility, 46/190
mutism, 46/190
neuroleptic agent, 46/192
neurophysiology, 46/191
perception, 46/190
prognosis, 46/192
REM sleep, 46/191
social behavior, 46/190
speech, 46/190
speech therapy, 46/192
stereotypy, 46/190
treatment, 46/192
Infantile benign myoclonic epilepsy
clinical features, 73/137
definition, 72/5
differential diagnosis, 73/139
polygraphic recording, 73/138
Infantile cerebral diplegia, *see* Cerebral palsy
Infantile cerebral hemiplegia, *see* Cerebral palsy
Infantile cerebral paralysis, *see* Cerebral palsy
Infantile cerebrovascular disease
B-mode sonography, 54/35
cat scratch disease, 54/40
craniocerebral trauma, 54/40
extracorporeal membrane oxygenation, 54/38
fullterm infant, 54/37
hemorrhagic infarction, 54/40
intracranial hemorrhage, 54/35
ischemic infarction, 54/40
mental deficiency, 53/31
Mycoplasma infection, 54/40
periventricular hemorrhage, 54/36
periventricular leukomalacia, 54/40
preterm infant, 54/35
pyogenic meningitis, 54/40
subependymal hemorrhage, 54/36
tuberculous meningitis, 54/40
Infantile ceroid lipofuscinosis
late, *see* Late infantile ceroid lipofuscinosis
mental deficiency, 42/461, 463
Infantile diffuse sclerosis
Krabbe classification, 10/4

Infantile distal myopathy
calf hypertrophy, 62/201
clinical course, 62/201
foot drop, 43/114, 62/201
muscle fiber type I, 43/114, 62/201
pedigree, 62/201
taxonomy, 62/199
Infantile dominant optic atrophy, *see* Dominant infantile optic atrophy
Infantile enteric bacillary meningitis, 33/61-68
Bacteroides fragilis, 33/63
Citrobacter diversus, 33/62
counterimmunoelectrophoresis, 33/64
CSF, 33/64
Enterobacter, 33/62
Escherichia coli, 33/61-63
Flavobacterium meningosepticum, 33/63
Klebsiella, 33/62
limulus lysate test, 33/64
myelomeningocele, 33/61
Paracolobactrum, 33/62
Proteus, 33/62
Pseudomonas, 33/62
Salmonella Citrobacter, 33/62
Serratia, 33/62
Shigella sonnei, 33/63
subdural effusion, 33/67
Infantile epileptogenic encephalopathy
brain atrophy, 42/693
mental deficiency, 42/693
psychomotor retardation, 42/693
Infantile free sialicylic acid storage disease
defect, 66/364
sialic acid, 66/354
Infantile Gaucher disease
brain stem dysfunction, 10/307
ceramide glucose, 10/307, 66/124
classification, 29/351, 42/437
clinical features, 10/564, 60/156
CNS, 42/438, 66/124
CSF, 60/157
dysphagia, 42/438
electrophysiologic study, 60/157
extensor plantar response, 10/307, 42/438, 66/124
genetic mutation, 66/126
genetics, 10/309
β-glucosidase, 10/309
head retroflexion, 10/307, 565, 66/124
hepatomegaly, 10/565
hepatosplenomegaly, 10/307, 66/124
hereditary congenital cerebellar atrophy, 60/289
hyperreflexia, 42/438, 66/124
hyperthermia, 42/438
larynx stridor, 42/438
mental deficiency, 42/438
opisthotonos, 42/438
progressive bulbar palsy, 59/375
spasticity, 42/438, 66/124
splenomegaly, 10/565
strabismus, 10/307, 565
treatment, 66/128
trismus, 42/438, 66/124
Infantile GM1 gangliosidosis, *see* GM1 gangliosidosis type I
Infantile GM2 gangliosidosis
membranous cytoplasmic inclusion body, 66/21
Infantile hemiplegia
acute, *see* Acute infantile hemiplegia
alternating, *see* Alternating infantile hemiplegia
central motor disorder, 1/195
cerebral hemiatrophy, 50/206
cerebrovascular disease, 53/31
difference from aldult type, 1/195
hemiathetosis, 49/384
recurrent, *see* Recurrent infantile hemiplegia
transient, *see* Transient infantile hemiplegia
Infantile histiocytoid cardiomyopathy
complex III deficiency, 62/505
Infantile hypotonia
see also Congenital myopathy
aminoaciduria, 43/125-127
benign, 43/79
brancher deficiency, 62/484
carnitine deficiency, 62/493
cerebrohepatorenal syndrome, 43/338
complex I deficiency, 62/503, 66/422
complex III deficiency, 62/504, 66/423
congenital muscular dystrophy, 43/92
differential diagnosis, 59/368-373
fetal Minamata disease, 64/420
glutaric aciduria type II, 62/495
hydranencephaly, 50/349
infantile acid maltase deficiency, 27/195, 227, 62/480
infantile neuroaxonal dystrophy, 6/627, 49/402
long chain acyl-coenzyme A dehydrogenase deficiency, 62/495
Marfan syndrome, 40/336
mitochondrial myopathy, 43/127
multiple acyl-coenzyme A dehydrogenase deficiency, 62/495
nonprogressive cerebellar ataxia, 42/124
ophthalmoplegia, 43/125-127
oxoglutarate dehydrogenase deficiency, 62/503
pyruvate carboxylase deficiency, 41/211, 42/586
pyruvate decarboxylase deficiency, 62/502

severe congenital muscular dystrophy, 43/92
Infantile idiopathic hypercalcemia, *see* Elfin face syndrome
Infantile lactic acidosis
acid base metabolism, 28/516
pyruvate carboxylase deficiency, 41/211
pyruvate decarboxylase deficiency, 62/502
Infantile meningitis, *see* Infantile enteric bacillary meningitis
Infantile metachromatic leukodystrophy
late, *see* Late infantile metachromatic leukodystrophy
prismatic inclusion, 66/167
Infantile myoclonic encephalopathy
myoclonus, 38/583
Infantile myoclonic spasm
vitamin B6, 28/130
Infantile myoclonus
differential diagnosis, 73/242
neonatal myoclonus, 73/242
Infantile myopathy
distal muscle weakness, 43/114
mitochondrial DNA depletion, 66/430
Infantile myopathy/cardiopathy
fatal, *see* Fatal infantile myopathy/cardiopathy
Infantile myotonic dystrophy
hypotonia, 40/490
megacolon, 40/490
Infantile nemaline myopathy
muscle tissue culture, 62/99, 107
Infantile neuroastrocytoma
desmoplastic, *see* Desmoplastic infantile ganglioglioma
Infantile neuroaxonal dystrophy
arthrogryposis multiplex congenita, 49/402
autonomic nervous system, 49/402
blindness, 49/402
category, 6/617
cerebellar cortex, 49/396
clinical features, 49/402
convulsion, 49/402
deafness, 49/402
dementia, 42/229
diabetes insipidus, 59/75
EEG, 49/403
epilepsy, 59/75
globus pallidus, 49/396
Hallervorden-Spatz syndrome, 42/230
hyperkinesia, 6/627, 42/229
hypothalamic involvement, 59/75
hypothyroidism, 59/75
infantile hypotonia, 6/627, 49/402
infantile spinal muscular atrophy, 59/75
infundibular involvement, 59/75
lactate dehydrogenase, 42/229
late, *see* Late infantile neuroaxonal dystrophy
mental deficiency, 6/627, 59/75
neuropathology, 49/400, 403, 407, 66/22
nystagmus, 49/402
optic atrophy, 6/627, 13/83, 42/229, 49/402
pallidal degeneration, 6/650
Pelizaeus-Merzbacher disease, 10/151, 158, 160
plaque, 9/265
primary neuroaxonal dystrophy, 49/397
psychomotor retardation, 6/627, 49/402
recessive autosomal, 49/402
sensorineural deafness, 42/229
spasticity, 49/402
spinal muscular atrophy, 59/368, 370
status porosus, 42/230
substantia nigra, 49/396
Infantile neuronal ceroid lipofuscinosis
ataxia, 42/461, 66/689
blindness, 66/689
dementia, 60/665
differential diagnosis, 66/690
gene, 72/134
hypotonia, 66/689
late, 42/462-464
mental deficiency, 42/461, 463
microcephaly, 60/666, 66/689
myoclonic jerk, 66/689
prevalence, 42/462
psychomotor retardation, 66/689
secondary pigmentary retinal degeneration, 60/654
Infantile optic atrophy
autosomal dominant, *see* Autosomal dominant infantile optic atrophy
Behr disease, 42/408
centrocecal scotoma, 42/408
color perception, 42/408
dominant, *see* Dominant infantile optic atrophy
globoid cell leukodystrophy, 42/408
nystagmus, 42/408
recessive, *see* Recessive infantile optic atrophy
vision loss, 42/408
visual evoked response, 42/408
Infantile optic glioma
acromegaly, 42/733
behavior disorder in children, 42/733
CSF, 42/734
diencephalic syndrome, 42/733
growth hormone, 42/734
hyperkinesia, 42/733
nystagmus, 42/733

optic atrophy, 42/733
short stature, 42/733
subcutaneous fat absence, 42/733
Infantile paralysis, *see* Poliomyelitis
Infantile periarteritis nodosa, *see* Kawasaki syndrome
Infantile periventricular leukomalacia
topographical anatomy, 47/527
Infantile Pompe disease, *see* Infantile acid maltase deficiency
Infantile progressive bulbar paralysis, *see* Fazio-Londe disease
Infantile progressive poliodystrophy, *see* Alpers disease
Infantile Refsum disease
deafness, 60/657
epilepsy, 72/222
facial dysmorphia, 60/666
peroxisomal disease, 51/383
peroxisomal disorder, 66/510
peroxisome, 66/513
secondary pigmentary retinal degeneration, 60/654
Infantile sex linked cerebellar ataxia
Malamud-Cohen family, 21/464
Infantile sialic acid storage disease
Salla disease, 59/361
Infantile Sly disease, *see* Glycosaminoglycanosis type VII
Infantile spasm
Aicardi syndrome, 42/696
birth incidence, 42/687
Blitzkrämpf, 15/219
classification, 15/107, 73/200
clinical characteristic, 73/202
corpus callosum agenesis, 42/8, 696, 50/162
corticotropin, 73/200
diagnosis, 73/205
differential diagnosis, 15/220
diphtheria pertussis tetanus vaccine, 52/239
Down syndrome, 50/529
EEG, 15/223
encephalopathy, 73/200
epidemiology, 73/201
epilepsy, 42/682, 686, 692
etiology, 15/221, 73/203
familial occurrence, 73/201
follow-up, 15/230
fumarate hydratase deficiency, 66/420
globoid cell leukodystrophy, 51/373
grand mal epilepsy, 14/347
hypsarrhythmia, 15/123, 219, 42/686, 73/200
idiopathic type, 73/201
imaging, 72/379
incidence, 15/223
lethal case, 15/231
mental deficiency, 15/221, 42/687
mortality rate, 73/201
myoclonus, 38/579, 49/619
neurology, 15/221
neurophysiology, 73/202
Nick Krampf, 15/219
pathology, 15/226, 72/117
pathophysiology, 73/204
pertussis vaccine, 52/239
prevalence, 42/693
prognosis, 15/229, 788
sleep, 15/125
survey, 73/199
terminology, 15/219
treatment, 15/226, 73/200, 205
triad, 73/200
tuberous sclerosis, 14/347, 43/49
unilateral epileptic seizure, 15/241
Infantile spastic diplegia, *see* Symmetrical spastic cerebral palsy
Infantile spastic hemiplegia
hemiatrophy, 59/480
Infantile spinal muscular atrophy, 22/81-100, 59/51-76
acid maltase deficiency, 22/99
age at onset, 22/81
allelic nature, 59/41
anterior horn cell, 59/64
apoptosis, 59/60
areflexia, 42/88
arthrogryposis multiplex congenita, 59/71-74
autosomal recessive inheritance, 59/54
bell-shaped chest, 59/55
birth incidence, 42/89, 91
borderline form, 22/88
botulism, 59/74
bulbar paralysis, 22/135, 60/658
carnitine level, 59/56
cause, 22/92
cell death postnatal persistency, 59/60
cerebellar atrophy, 59/70
cerebellar hypoplasia, 30/386, 50/192
chest deformity, 59/54
childhood hypertonia, 1/271
chorea, 59/70
chromosome 5q allelic mutation, 59/54
Clarke column, 59/62
classification, 59/51-53, 367
clinical features, 22/82, 40/306, 308, 331, 334, 337, 342

CNS change, 41/28
computer assisted tomography, 59/57, 60/672
congenital fiber type disproportion, 22/99, 62/355
congenital myopathy, 41/1
cranial nerve deficit, 22/99
creatine kinase, 22/93
diaphragm, 59/62
differential diagnosis, 22/99, 59/70-75, 367-377
dominant transmission, 22/75
Dubowitz classification, 59/51
Duchenne muscular dystrophy, 22/77
electron microscopy, 59/61-63
electrophysiologic study, 59/56
EMG, 22/74, 93-95, 59/56
epidemiology, 59/53
extraocular motor nucleus, 59/62, 67
fasciculation, 42/88
Fazio-Londe disease, 22/103, 105, 108, 59/128
fetal movement, 42/88
Fried-Emery classification, 59/51
frog leg posture, 22/84
gene frequency, 42/89
genetic factor, 22/134
genetics, 22/82, 89, 91, 41/437, 59/54
Hausmanowa-Petrusewicz classification, 59/51
hearing loss, 59/70
hereditary congenital cerebellar atrophy, 60/289
hereditary motor and sensory neuropathy variant, 60/247
heterogeneity, 59/97
heterozygote frequency, 42/89
histochemistry, 22/95, 40/43
history, 22/81
hyporeflexia, 42/88
immunologic abnormality, 59/55
incidence, 59/53
incontinentia pigmenti, 59/75
infantile neuroaxonal dystrophy, 59/75
infantile neuronal degeneration, 59/70
intermediate form, 22/83
jug handle arm position, 22/84
late onset, 59/97
lethal congenital contracture syndrome, 59/72
lethal multiple pterygia syndrome, 59/72
leucine metabolism, 59/74
long chain dicarboxylic acid, 59/56
lung volume, 59/56
main features, 21/24, 27
management, 22/99
mental deficiency, 59/70
3-methylcrotonyl-coenzyme A carboxylase deficiency, 59/74
β-methylcrotonylglycinuria type I, 59/74
microcephaly, 59/69
mild form, 22/88
mortality, 22/82
motoneuron, 59/60
muscle biopsy, 59/58
muscle histology, 22/95-99
myocardium, 59/67
myosin isoform, 59/59
nerve conduction velocity, 22/95
nerve transitional zone, 51/12
neuropathology, 22/98, 59/60-63
neuropathy, 59/70
old taxonomy, 21/3
ophthalmoplegia, 22/180
optic atrophy, 59/70
paradoxical respiration, 59/66
pathognomonic EMG sign, 22/95
Pena-Shokeir syndrome type I, 59/71
phenylketonuria, 59/75
pontocerebellar hypoplasia, *see* Norman disease
progressive distal muscular atrophy, 59/70
progressive external ophthalmoplegia
classification, 22/179, 62/290
progressive muscular atrophy, 59/13
prolonged survival, 22/82
pseudomyotonic burst, 59/56
pulmonary function, 59/56
recessive form, 22/75, 77
reticular formation, 59/62
satellite cell, 59/59
scapuloperoneal form, 59/70
Schmidt-Lantermann incisure, 51/12
severe form, 22/82
skin, 59/56
spinal anterior root, 59/62
spinal muscular atrophy, 59/368
teeth, 59/54
terminology, 41/1
thalamus degeneration, 21/598
tongue atrophy, 22/85
tonsillar atrophy, 59/54
twins, 22/89
ultrasound imaging, 59/56
ultrasound muscle imaging classification, 59/57
Wohlfart-Kugelberg-Welander disease, 22/74, 82, 88, 91, 95
Infantile spongy dystrophy
glial dystrophy, 21/68
Infantilism
Friedreich ataxia, 21/348
Miescher syndrome type II, 14/120
Rud syndrome, 13/321, 479, 14/13, 38/9, 43/284
sexual, *see* Sexual infantilism

Infantus cerebro destituti, *see* Anencephaly
Infarction
anterior spinal artery, *see* Anterior spinal artery infarction
autonomic nervous system, 75/11
basal ganglion, *see* Basal ganglion infarction
basilar artery, 12/20
borderzone, *see* Borderzone infarction
brain, *see* Brain infarction
brain hemorrhagic, *see* Brain hemorrhagic infarction
brain lacunar, *see* Brain lacunar infarction
brain periventricular, *see* Brain periventricular infarction
brain stem, *see* Brain stem infarction
brain stem lacunar, *see* Brain stem lacunar infarction
brain venous, *see* Brain venous infarction
central ischemic, *see* Central ischemic infarction
cerebellar, *see* Cerebellar infarction
cerebellar arterial system, 12/20
corpus callosum, *see* Corpus callosum infarction
diabetic thigh muscle, *see* Diabetic thigh muscle infarction
epilepsy, 72/212
globus pallidus, *see* Globus pallidus infarction
glycolysis, 27/31
hemiballismus, 49/373
iatrogenic brain, *see* Iatrogenic brain infarction
ischemic, *see* Ischemic infarction
juvenile brain, *see* Juvenile brain infarction
lacunar, *see* Brain lacunar infarction
lateral medullary, *see* Lateral medullary infarction
local anesthetic agent, 65/421
mesencephalic, *see* Mesencephalic infarction
middle cerebral artery, *see* Middle cerebral artery infarction
middle cerebral artery syndrome, 11/304, 53/353
migraine, 48/135
multiple cerebral, *see* Multiple brain infarction
myocardial, *see* Myocardial infarction
occipital, *see* Occipital infarction
occipital lobe, *see* Occipital lobe infarction
optic nerve, *see* Optic nerve infarction
periventricular, *see* Periventricular infarction
pontine, *see* Pontine infarction
pontine lacunar, *see* Pontine lacunar infarction
pontine syndrome, 2/241
posterior cerebral artery, *see* Posterior cerebral artery infarction
retinal, *see* Retinal infarction
striatal lesion, 6/673
systemic brain, *see* Systemic brain infarction
temporal lobe, *see* Temporal lobe infarction
thalamic lacunar, *see* Thalamic lacunar infarction
thalamus, *see* Thalamic infarction
watershed, *see* Watershed infarction
Infection
acquired myoglobinuria, 62/577
adenovirus, *see* Adenovirus infection
anencephaly, 30/197
arenavirus, *see* Arenavirus infection
aspartylglucosaminuria, 29/225
Bartonella, *see* Bartonella infection
bladder, *see* Bladder infection
blood-brain barrier, 28/380
brain, *see* Brain infection
brain injury, 23/61
brain vasospasm, 23/173
Burkitt lymphoma, 63/352
Chagas disease, 75/389
childhood myoglobinuria, 62/561
chronic enterovirus, *see* Chronic enterovirus infection
Clostridium difficile, *see* Clostridium difficile infection
CNS, *see* CNS infection
coronavirus, *see* Coronavirus infection
Coxsackie virus, *see* Coxsackie virus infection
cranial dura, 33/149-162
critical illness polyneuropathy, 51/575
Cryptococcus neoformans, *see* Cryptococcus neoformans infection
CSF shunt, *see* CSF shunt infection
cytomegalovirus, *see* Cytomegalovirus infection
diamorphine intoxication, 65/355
ECHO virus, *see* ECHO virus infection
enterovirus, *see* Enterovirus infection
epilepsy, 72/145, 349
Epstein-Barr virus, *see* Epstein-Barr virus infection
facial paralysis, 8/282
fever, 74/457
Guillain-Barré syndrome, 47/608-610
head injury, 57/139
helminthic, *see* Helminthic infection
hepatitis B virus, *see* Hepatitis B virus infection
human immunodeficiency virus, *see* Human immunodeficiency virus infection
human immunodeficiency virus type 1, *see* Human immunodeficiency virus type 1 infection
idiopathic autonomic neuropathy, 74/336
intracranial, *see* Intracranial infection
intracranial pressure monitoring, 57/223
leukemia, 39/19
Listeria monocytogenes, *see* Listeria

monocytogenes infection
lung, *see* Pneumonitis
maternal, *see* Maternal infection
membranous labyrinth, 30/410
meningeal, *see* Meningeal infection
meningitis, 57/237
multiple sclerosis, 9/58, 117, 121, 47/149
mumps virus, *see* Mumps virus infection
mycoplasma, *see* Mycoplasma infection
mycotic aneurysm, 12/80
myoglobinuria, 41/275, 277, 62/557
nasal sinus, *see* Nasal sinus infection
neonatal seizure, 73/255
nonpolio enterovirus, *see* Nonpolio enterovirus infection
opiate intoxication, 65/355
opportunistic, *see* Opportunistic infection
pain, 69/35
paramyxovirus, *see* Paramyxovirus infection
Parkinson dementia, 49/178
penetrating head injury, 57/310
pertussis, 33/275-295
plaque, 9/295
pneumocephalus, 24/208
prenatal, *see* Prenatal infection
prophylactic antibiotic agent, 57/237
renal insufficiency, 27/343
respiratory, *see* Respiratory infection
respiratory virus, *see* Respiratory virus infection
retrovirus, *see* Retrovirus infection
rhinocerebral, *see* Rhinocerebral infection
rickettsial, *see* Rickettsial infection
sickle cell anemia, 38/40-42
sinopulmonary, *see* Sinopulmonary infection
sinus pain, 5/330
spastic paraplegia, 59/435
spinal cord, *see* Spinal cord infection
spinal cord injury, 25/61
status epilepticus, 72/145
subarachnoid hemorrhage, 55/32
subdural empyema, 57/237
sudden infant death, 63/465
systemic, *see* Systemic infection
Toxoplasma gondii, *see* Toxoplasma gondii infection
trigeminal neuralgia, 48/450
urinary tract, 26/415, 426-432
varicella zoster virus, *see* Varicella zoster virus infection
vasculitis, 71/157
venous sinus thrombosis, 12/431
ventriculitis, 57/237
vesicourethral neuropathy, 61/308
viral, *see* Viral infection
visna, *see* Visna infection
Infectious disease
acute benign ataxia, 21/563
acute pandysautonomia, 74/336
autonomic nervous system, 74/334
chorea, 49/553
CSF oligoclonal IgG, 47/96
epilepsy, 15/306
hemiplegia, 1/194
hypernatremia, 2/446
leprosy, 51/215
nerve, *see* Neuritis
neuralgic amyotrophy, 51/174
noninfectious sequelae, 71/547
Prague University, 1/17
pseudo-Meniere syndrome, 7/485
puerperal polyneuritis, 7/488
subarachnoid hemorrhage, 12/111
Infectious endocarditis
aged, 63/112
antibiotic agent, 63/24, 118
anticoagulation, 63/24
aorta stenosis, 63/112
aorta stenosis calcification, 63/23
aspergillosis, 52/302
bacteremia, 63/23
bacterial, *see* Bacterial endocarditis
bicuspid aortic valve, 63/112
blood culture, 63/112
brain abscess, 55/167, 63/23, 112, 115
brain angiography, 63/118
brain computer assisted tomography, 63/118
brain embolism, 55/166, 63/23, 114
brain hematoma, 63/23
brain hemorrhage, 55/167, 63/23, 115
brain infarction, 55/167, 63/23, 112, 114, 116
Candida albicans, 52/301
cardiac valvular disease, 63/22
cardiac ventricular septum defect, 63/112
cerebrovascular disease, 55/166
congenital heart defect, 63/23
congenital heart disease, 63/5, 112
CSF, 63/117
diagnosis, 63/117
discitis, 63/117
drug abuse, 55/517, 63/112
ECG, 63/112
echoCG, 63/117
encephalopathy, 63/23
epidemiology, 63/111
epilepsy, 63/23, 112, 116
Fallot tetralogy, 63/112

female sex, 63/114
fungal endocarditis, 52/289
headache, 63/23, 112, 116
hypertrophic cardiomyopathy, 63/137
low back pain, 63/117
lumbar discitis, 63/117
lupus erythematosus, 63/111
management, 63/24
meningitis, 55/167, 63/23, 112, 116
micro-organism type, 63/114
mitral annulus calcification, 63/23
mitral valve prolapse, 63/21, 23, 112
multiple brain microinfarction, 63/115
mycotic aneurysm, 63/23, 112
mycotic brain aneurysm, 55/167, 63/23, 112, 115
neurologic complication, 63/22, 112
neurologic complication frequency, 63/113
nuclear magnetic resonance, 63/118
pathophysiology, 63/114
prognosis, 63/118
prosthetic cardiac valve, 63/23, 112
Pseudomonas aeruginosa, 52/301
Psittacosis, 52/289, 302
rheumatic heart disease, 63/23
Rickettsia, 52/289, 302
spondylodiscitis, 63/117
Staphylococcus aureus, 63/114
subarachnoid hemorrhage, 63/23
toxic encephalopathy, 63/112, 117
transient ischemic attack, 63/116
treatment, 63/118
tricuspid valve, 63/112
Infectious leukoencephalomyelitis
animal viral disease, 34/294
Infectious mononucleosis
acute cerebellar ataxia, 34/623
autonomic neuropathy, 56/252
cerebellar ataxia, 56/252
clinical features, 56/251
cranial nerve palsy, 56/252
CSF, 34/185
deafness, 56/252
encephalitis, 34/188, 56/252
epilepsy, 56/252
Epstein-Barr virus, 34/185, 56/8, 249
Epstein-Barr virus infection, 34/185-189
facial paralysis, 56/252
Guillain-Barré syndrome, 9/597, 34/188, 56/252
meningitis, 56/252
meningitis serosa, 34/188
meningoencephalitis, 56/252
mononeuritis, 34/188
multiple sclerosis, 56/252
myelitis, 56/252
neuropathy, 7/555
organic brain syndrome, 56/252
polyneuritis, 34/188, 56/252
subacute sclerosing panencephalitis, 56/252
transverse myelitis, 56/252
Infectious striatopallidodentate calcification
abscess, 49/417
Addison disease, 49/417
cytomegalovirus infection, 49/417
encephalitis, 49/417
neurocysticercosis, 49/417
syphilis, 49/417
toxoplasmosis, 49/417
tuberculoma, 49/417
Inferior anastomotic vein, *see* Labbé vein
Inferior artery to dentate nucleus
anatomy, 11/26
Inferior cavernous sinus artery
carotid cavernous fistula, 12/272
radioanatomy, 11/69
Inferior cerebellar artery (Foix-Hillemand), *see* Posterior inferior cerebellar artery
Inferior cerebral vein
anatomy, 11/61
radioanatomy, 11/106
topography, 11/106
Inferior choroidal vein
anatomy, 11/51
Inferior colliculi
Parinaud syndrome, 2/339
Inferior frontal artery
radioanatomy, 11/76
topography, 11/76
Inferior frontal vein
anatomy, 11/47
Inferior hypophyseal artery
normal anatomy, 12/272
radioanatomy, 11/69
topography, 11/69
Inferior longitudinal sinus, *see* Inferior sagittal sinus
Inferior oblique muscle
hyperactivity, 2/308
pseudoparesis, *see* Brown syndrome
Inferior olivary hypertrophy
brain cortex focal dysplasia, 50/207
Inferior olivary nucleus
cerebellar agenesis, 50/188
Edwards syndrome, 50/250
Huntington chorea, 49/320, 323
olivopontocerebellar atrophy (Dejerine-Thomas), 21/418
pseudohypertrophy, 12/21

Inferior olivary nucleus heterotopia
cerebrohepatorenal syndrome, 50/250
Edwards syndrome, 50/250, 275
lissencephaly, 50/250, 259
Inferior olive
ataxia telangiectasia, 14/75
Marinesco-Sjögren syndrome, 21/560
mercury intoxication, 64/381
paramedian medullary syndrome, 2/227
Inferior ophthalmic vein
anatomy, 11/54
radioanatomy, 11/115
topography, 11/115
Inferior paravermal artery
anatomy, 11/26
Inferior parietal lobule
akinesia, 45/162
hemi-inattention, 45/157
Inferior petrosal sinus
anatomy, 11/54, 61
radioanatomy, 11/113
topography, 11/113
Inferior petrosal sinus thrombosis
abducent nerve paralysis, 54/398
Inferior posterior cerebellar vein
radioanatomy, 11/110
topography, 11/110
Inferior sagittal sinus
anatomy, 11/49
dura mater, 30/423
radioanatomy, 11/111
thrombophlebitis, 33/172
topography, 11/111
Inferior vena cava syndrome
metastatic cardiac tumor, 63/94
primary cardiac tumor, 63/94
Inferior ventricular vein
anatomy, 11/51
Inferior visual association cortex
monochromatism, 45/15
Infertility
epilepsy, 73/376
Turner syndrome, 50/543
Infiltration
corneal, *see* Corneal infiltration
giant cell, *see* Giant cell infiltration
Infiltrative ophthalmopathy, *see* Thyroid associated ophthalmopathy
Inflammation
demyelination, 47/43
dissecting aorta aneurysm, 63/45
experimental allergic neuritis, 9/505, 47/457
granulomatous, *see* Granulomatous inflammation
headache, 48/6, 281, 284
multiple sclerosis, 9/108, 47/80
nerve injury, 75/317
renal artery, *see* Renal artery inflammation
serous, *see* Serous inflammation
Tolosa-Hunt syndrome, 48/291
vasculitis, 71/153
ventral horn, *see* Ventral horn inflammation
Inflammatory angiopathy
angiopathic polyneuropathy, 51/445
Inflammatory cell
experimental allergic encephalomyelitis, 47/230
multiple sclerosis, 47/230
muscle, 62/42
scleroderma, 71/107
Inflammatory demyelinating neuropathy
acute, *see* Guillain-Barré syndrome
Inflammatory demyelinating polyneuropathy
acquired immune deficiency syndrome, 56/517
chronic, *see* Chronic inflammatory demyelinating polyneuropathy
Inflammatory demyelinating polyradiculo-neuropathy
chronic, *see* Chronic inflammatory demyelinating polyradiculoneuropathy
Inflammatory diffuse sclerosis
classification, 9/472, 10/6, 15, 24
concentric sclerosis, 47/411, 416
Inflammatory myopathy, 41/51-89
see also Collagen vascular disease, Dermatomyositis, Granulomatous myositis, Myositis *and* Polymyositis
antibody, 41/75
antibody formation, 40/264
associated condition, 62/372
blood vessel, 40/51, 164, 264
cell mediated immunity, 41/62, 74, 375
classification, 40/284
clinical features, 40/307, 313, 317, 323, 337, 339, 342, 41/370-372, 62/369
complement, 41/376
corticosteroid, 41/73, 79, 374
cysticercus, 62/373
differential diagnosis, 40/472, 41/373
experimental allergic myositis, 40/163-167
facioscapulohumeral muscular dystrophy, 40/421
granulomatous myositis, 41/85
histochemistry, 40/51
HLA loci, 41/78
immunosuppression, 41/80, 374
inclusion body myositis, 40/52, 284, 322
malignancy, 41/348-370
muscle necrosis, 40/264

necrosis, 40/264
paraneoplastic syndrome, 71/681
parasitic polymyositis, 62/373
pathogenesis, 41/375
pathologic reaction, 40/264
polyarteritis nodosa, 55/355
polymyalgia rheumatica, 41/83
polymyositis, 40/284, 41/54-83, 274, 62/369, 372
sarcoidosis, 40/52, 317
toxic myopathy, 62/604
Toxoplasma gondii, 62/373
treatment, 41/374
Trichinella, 62/373
trichinous myositis, 40/51, 325
Trypanosoma, 62/373
viral myositis, 41/53
Inflammatory neuropathy
chronic relapsing, *see* Chronic relapsing neuritis
Inflammatory polyradiculoneuropathy
acquired immune deficiency syndrome, 56/518
acute, *see* Acute inflammatory polyradiculoneuropathy
Inflammatory scapuloperoneal muscular dystrophy
facioscapulohumeral muscular dystrophy type, 62/170
Influenza
acquired myoglobinuria, 62/577
corpus callosum defect, 9/582
encephalitis lethargica, 34/452
immune mediated encephalomyelitis, 34/418
immune related encephalomyelitis, 34/418
nerve lesion, 7/478
polyradiculitis, 7/478
radiculoneuropathy, 7/478
respiratory virus infection, 31/219
Reye syndrome, 29/332, 34/80, 65/116, 70/288
rhinencephalic hypoplasia, 9/582
secondary pigmentary retinal degeneration, 13/214
subarachnoid hemorrhage, 12/111
teratogenic agent, 34/379
Influenza A virus
acute polymyositis, 56/199
acute viral myositis, 56/194
benign acute childhood myositis, 56/198
experimental acute viral myositis, 56/195
hemorrhagic leukoencephalitis, 56/15
hepatic encephalopathy, 56/149
neuralgic amyotrophy, 51/188
orthomyxovirus, 56/15
Reye syndrome, 34/169, 56/149, 161, 238
Influenza A2 virus
amantadine, 22/24
Influenza attack
zimeldine polyneuropathy, 51/304
Influenza B
acute encephalopathy, 47/479
postinfectious encephalopathy, 47/479
Influenza B virus
acute polymyositis, 56/199
acute viral myositis, 56/195
benign acute childhood myositis, 56/198
hepatic encephalopathy, 56/149
orthomyxovirus, 56/15
Reye syndrome, 29/332, 336, 34/169, 56/15, 149, 161, 238
Influenza B virus infection
Reye syndrome, 33/293, 34/80, 169, 49/217, 56/149, 63/438
Influenza C
multiple sclerosis, 47/329
Influenza vaccination
Guillain-Barré syndrome, 51/242, 56/168
swine, *see* Swine influenza vaccination
Influenza virus
acute encephalopathy, 56/168
cell surface receptor, 56/34
deafness, 56/107
encephalomyelitis, 56/168
Guillain-Barré syndrome, 51/187, 56/168
myeloradiculopathy, 51/187
myoglobinuria, 56/199
neuritis, 51/187
neurotropism, 56/34
postinfectious encephalomyelitis, 47/326
postviral encephalomyelitis, 56/168
Reye syndrome, 56/239
rhabdomyolysis, 56/199
vertigo, 56/107
viral protein HA, 56/34
Infraclinoid syndrome
multiple cranial neuropathy, 2/88
skull base tumor, 17/181
Infraorbital artery
radioanatomy, 11/66
topography, 11/66
Infraorbital nerve
avulsion, 5/392
injection, 5/388
Infratentorial syndrome
clinical examination, 57/135
Infratentorial tumor
syringomyelia, 50/456
Inguinal hernia
craniolacunia, 50/139
glycosaminoglycanosis type IV, 66/304

glycosaminoglycanosis type VI, 66/308
Hurler-Hunter disease, 10/436
Patau syndrome, 31/507, 43/528
r(9) syndrome, 50/580
INH, *see* Isoniazid
INH polyneuropathy, *see* Isoniazid
Inheritance
see also Genetics *and* Mendel law
acid maltase deficiency, 62/481
adrenoleukodystrophy, 47/594
allelic restriction, 30/106-109
autosomal, 30/96
autosomal dominant, *see* Autosomal dominant inheritance
autosomal recessive, *see* Autosomal recessive inheritance
autosomal recessive lissencephaly, 72/133
Bardet-Biedl syndrome, 13/396-398
Biemond syndrome, 60/448
brancher deficiency, 62/484
carnitine deficiency, 62/493
cerebrotendinous xanthomatosis, 10/547, 29/375
congenital retinal blindness, 13/27
cytoplasmic, 30/109
debrancher deficiency, 62/483
dentatorubropallidoluysian atrophy, 72/133
dominant, *see* Dominant inheritance
dystonia musculorum deformans, 6/519
galactokinase deficiency, 29/259
handedness, 4/251, 269, 316
hereditary exertional paroxysmal dystonic choreoathetosis, 49/350
hereditary motor and sensory neuropathy type I, 21/12, 60/187
hereditary motor and sensory neuropathy type II, 21/12, 60/187
hereditary paroxysmal dystonic choreoathetotis, 49/350
hereditary paroxysmal kinesigenic choreoathetosis, 49/350
hereditary spastic ataxia, 60/461
holoprosencephaly, 50/239
hydroxyprolinemia, 29/143
hyperkalemic periodic paralysis, 28/595
hyperprolinemia type I, 29/137
hyperprolinemia type II, 29/140
hypokalemic periodic paralysis, 28/582
intelligence, 3/303, 313
lactate dehydrogenase deficiency, 62/490
Lafora progressive myoclonus epilepsy, 72/133
law of independent assortment, 30/96
law of segregation, 30/95
lyonization, 30/97
malignant hyperthermia, 41/268, 62/568
Marinesco-Sjögren syndrome, 60/341
maternal, 30/110
medium chain acyl-coenzyme A dehydrogenase deficiency, 62/495
mitochondrial DNA depletion, 62/511
molecular basis, 30/94
multifactorial association, 30/127
multifactorial/threshold, 30/127
multiple mitochondrial DNA deletion, 62/510
multiple sclerosis, 47/300
myophosphorylase deficiency, 62/485
neuromyelitis optica, 47/400
normokalemic periodic paralysis, 41/162, 62/463
pattern, 30/95-97
phosphofructokinase deficiency, 62/488
phosphoglycerate kinase deficiency, 62/489
phosphoglycerate mutase deficiency, 62/489
Pick disease, 42/285, 46/243
polygenic, *see* Polygenic inheritance
primary pigmentary retinal degeneration, 13/150-156
recessive, *see* Recessive inheritance
Refsum disease, 10/345, 27/520
renal glycosuria, 29/184
sex influenced, 30/97
sex limited, 30/97
stuttering, 46/169
thyrotoxic periodic paralysis, 62/532
variable expressivity, 30/96
Wilson disease, 27/379
X-linked, *see* X-linked inheritance
X-linked recessive, *see* X-linked recessive inheritance
Inheritance mode
familial spastic paraplegia, 59/308
Finkel spinal muscular atrophy, 59/43
hereditary periodic ataxia, 60/434-437
hereditary progressive cochleovestibular atrophy, 60/761
hereditary sensory and autonomic neuropathy type I, 60/7
hereditary sensory and autonomic neuropathy type II, 60/7
hereditary sensory and autonomic neuropathy type III, 60/7
hereditary sensory and autonomic neuropathy type IV, 60/7, 13
hereditary sensory and autonomic neuropathy type V, 60/7
hypokalemic periodic paralysis, 41/152, 62/460
olivopontocerebellar atrophy (Wadia-Swami), 60/494

Pyle disease, 46/88
Sturge-Weber syndrome, 14/509-512
Inherited enzyme deficiency
autosomal recessive, 66/56
carrier, 66/56
neonatal detection, 66/56
protein abnormality, 66/57
Inherited metabolic disorder, *see* Inborn error of metabolism
Inherited spinocerebellar ataxia
autonomic nervous system, 75/35
Inhibition
central, *see* Central inhibition
cerebellar function, 2/408-412
chemistry, 1/51
epilepsy, 72/47
gluconeogenesis, *see* Gluconeogenesis inhibition
grooming reflex, 1/50
innervation, *see* Innervation inhibition
lateral geniculate nucleus, 2/532
myelination, *see* Myelination inhibition
nerve cell, 72/47
postsynaptic, *see* Postsynaptic inhibition
Renshaw cell, 1/52
somatomotor, *see* Somatomotor inhibition
strychnine, 1/51
tetanus toxin, 1/51
tyrosine hydroxylase, 75/239
Inhibition test
migration, *see* Migration inhibition test
Inhibitory postsynaptic potential
cerebellar stimulation, 2/413
organochlorine insecticide intoxication, 64/200
synapse action, 1/56
Iniencephalus apertus
definition, 30/257, 50/129
encephalocele, 50/129
Iniencephalus clausus
definition, 30/257, 50/129
Iniencephaly, 30/257-267
see also Cranium bifidum, Encephalocele *and* Rachischisis
anencephaly, 50/72, 131, 134
basilar bone, 50/134
brain stem, 50/131
cardinal features, 30/257
cardiovascular abnormality, 50/131
cerebellum, 50/131
cortex layer, 50/131
craniolacunia, 32/187
cranium bifidum, 50/129
cyclopia, 50/131
Dandy-Walker syndrome, 42/22
definition, 50/129
diagnosis, 50/135
diaphragm hernia, 50/131
differential diagnosis, 30/264
embryopathology, 30/264, 50/133
encephalocele, *see* Iniencephalus clausus
environmental factor, 30/266
epidemiology, 30/155, 265, 50/135
esophagus atresia, 50/131
etiology, 50/135
exencephaly, 50/71, 73
experimental, 30/266
external appearance, 30/257
α-fetoprotein, 30/179
foramen magnum, 50/129
genetics, 30/266
harelip, 50/131
head retroflexion, 50/129
heterotopic glial tissue, 50/131
hippocampus, 50/131
holoprosencephaly, 50/131
hydramnios, 50/135
hydrocephalus, 30/258, 50/131
incidence, 50/53, 60
Klippel-Feil syndrome, 30/257, 50/131
malformation, 30/258, 261-264
meroanencephaly, 50/134
microcephaly, 50/131
morphologic features, 50/129
nonschisis axial dysraphism, 32/187
other associated malformations, 50/131
palatoschisis, 50/131
pathology, 30/258
polymicrogyria, 50/131
prenatal diagnosis, 30/264
pulmonary disorder, 50/131
rachischisis, 50/129, 134
sex ratio, 50/60
situs inversus, 50/131
spina bifida complex, 50/134
spinal intradural cyst, 20/58
spine retroflexion, 50/129
streptonigrin, 50/135
superior cerebellar vermis hypoplasia, 50/131
tetracycline, 50/135
triparanol, 50/135
type, 50/129
urinary tract defect, 50/131
ventricular system atresia, 50/131
vinblastine, 50/135
without encephalocele, *see* Iniencephalus apertus
Inimicus didactylus intoxication
picture, 37/64

stonefish, 37/74
Injections
intrathecal, *see* Intrathecal injection
local, *see* Local injection
Injuries
abdominal, *see* Abdominal injury
acceleration, *see* Whiplash injury
accidental, *see* Accidental injury
ankle, *see* Ankle injury
birth, *see* Birth incidence
blast, *see* Blast injury
boxing, *see* Boxing injury
brachial plexus, *see* Brachial plexus injury
brain, *see* Brain injury
brain stem, *see* Brain stem injury
brain white matter, *see* Brain white matter injury
carotid, *see* Carotid injury
cauda equina, *see* Cauda equina injury
cerebral, *see* Brain injury
cervical, *see* Cervical injury
cervical hyperanteflexion, *see* Cervical hyperanteflexion injury
cervical hyperdeflexion, *see* Cervical hyperdeflexion injury
cervical spine, *see* Cervical vertebral column injury
cervicomedullary, *see* Cervicomedullary injury
chest, *see* Chest injury
clavicle, *see* Clavicle injury
cold, *see* Cold injury
compression, *see* Compression injury
craniofacial, *see* Craniofacial injury
craniospinal, *see* Craniospinal injury
decompression, *see* Decompression illness
diffuse axonal, *see* Diffuse axonal injury
electrical, *see* Electrical injury
experimental, *see* Experimental injury
experimental head, *see* Experimental head injury
eye, *see* Eye injury
facial, *see* Facial injury
facial nerve, *see* Facial nerve injury
foot, *see* Foot injury
football, *see* Football injury
frontal lobe, *see* Frontal lobe injury
frontobasal brain, *see* Frontobasal brain injury
geniculocalcarine tract, *see* Optic tract injury
gunshot, *see* Gunshot injury
head, *see* Head injury
heat, *see* Heat injury
hyperrotation, *see* Hyperrotation injury
hypoglossal nerve, *see* Hypoglossal nerve injury
iatrogenic nerve, *see* Iatrogenic nerve injury
ice hockey, *see* Ice hockey injury
intervertebral disc, *see* Intervertebral disc injury
intrauterine, *see* Intrauterine injury
intrauterine head, *see* Intrauterine head injury
Kohlu, *see* Kohlu injury
Laser, *see* Laser injury
lightning, *see* Lightning injury
lumbar spinal column, *see* Lumbar spinal column injury
lumbosacral nerve, *see* Lumbosacral nerve injury
lumbosacral trunk, *see* Lumbosacral trunk injury
maternal birth, *see* Maternal birth injury
mechanical, *see* Mechanical injury
midbrain, *see* Midbrain injury
minor head, *see* Minor head injury
missile, *see* Missile injury
neck, *see* Neck injury
nerve, *see* Nerve injury
obturator nerve, *see* Obturator nerve injury
occipitoaxial, *see* Occipitoaxial injury
ocular muscle, *see* Ocular muscle injury
oculomotor nerve, *see* Oculomotor nerve injury
optic chiasm, *see* Optic chiasm injury
optic nerve, *see* Optic nerve injury
optic radiation, *see* Geniculocalcarine tract injury
orbital, *see* Orbital injury
pediatric head, *see* Pediatric head injury
pelvic, *see* Pelvic injury
pencil, *see* Pencil injury
penetrating brain, *see* Penetrating brain injury
penetrating gunshot, *see* Penetrating gunshot injury
penetrating head, *see* Penetrating head injury
perinatal brain, *see* Perinatal brain injury
plexus, *see* Plexus injury
prenatal radiation, *see* Prenatal radiation injury
radiation, *see* Radiation injury
rotation, *see* Rotation injury
rugby football, *see* Rugby football injury
scalp, *see* Scalp injury
seat belt, *see* Seat belt injury
skiing, *see* Skiing injury
skull, *see* Skull injury
soccer, *see* Soccer injury
spinal, *see* Spinal injury
spinal cord, *see* Spinal cord injury
spinal cord experimental, *see* Spinal cord experimental injury
spinal cord missile, *see* Spinal cord missile injury
spinal hyperflexion, *see* Spinal hyperflexion injury
spinal nerve root, *see* Spinal nerve root injury
spine experimental, *see* Spine experimental injury
sport, *see* Sport injury
superior gluteal nerve, *see* Superior gluteal nerve

injury
thermal, *see* Thermal injury
thoracic, *see* Thoracic injury
thoracic spinal column, *see* Thoracic spinal column injury
thoracic spinal cord, *see* Thoracic spinal cord injury
thoracic vertebral column, *see* Thoracic vertebral column injury
thoracolumbar spine, *see* Thoracolumbar spine injury
thoracolumbar vertebral column, *see* Thoracolumbar vertebral column injury
traction, *see* Traction injury
traffic, *see* Traffic injury
trochlear nerve, *see* Trochlear nerve injury
vagus nerve, *see* Vagus nerve injury
vertebral artery, *see* Vertebral artery injury
vertebral column, *see* Vertebral column injury
vertebral end plate, *see* Vertebral end plate injury
vertebral experimental, *see* Vertebral experimental injury
vestibulocochlear nerve, *see* Vestibulocochlear nerve injury
visual cortex, *see* Visual cortex injury
whiplash, *see* Whiplash injury
white matter, *see* White matter injury
Inner ear
Edwards syndrome, 50/562
Inner mesaxon
Schwann cell, 51/9
Inner mitochondrial membrane system
acetyl-coenzyme A acyltransferase, 66/412
defect, 66/406
heart dysfunction, 66/406
long chain acyl-coenzyme A dehydrogenase, 66/407
long chain acyl-coenzyme A dehydrogenase cDNA, 66/407
long chain acyl-coenzyme A dehydrogenase protein, 66/407
long chain 3-hydroxyacyl-coenzyme A dehydrogenase, 66/406
long chain 3-hydroxyacyl-coenzyme A dehydrogenase coding, 66/408
long chain 3-hydroxyacyl-coenzyme A dehydrogenase deficiency, 66/408
medium chain acyl-coenzyme A dehydrogenase deficiency, 66/407
myopathy, 66/406
very long chain acyl-coenzyme A dehydrogenase, 66/406
very long chain acyl-coenzyme A dehydrogenase deficiency, 66/407
very long chain acyl-coenzyme A dehydrogenase gene, 66/407
very long chain acyl-coenzyme A dehydrogenase protein, 66/407
Innervation
apraxia, 4/50
autonomic, *see* Autonomic innervation
bladder, 1/361, 75/666
brain microcirculation, 53/79
cardiac pacemaker, 74/42
carotid sinus reflex, 11/535
chemoreceptor reflex, 75/112
colon, 75/634
cross, *see* Cross innervation
cutaneous, *see* Skin innervation
dura mater, 30/418
extraocular muscle, 30/397
intestinal pseudo-obstruction, 75/701
meninx, 1/537
micturition, 61/293
muscle, *see* Muscle innervation
penis erection, 61/314
reciprocal, *see* Reciprocal innervation
segmental, *see* Segmental innervation
skin, *see* Skin innervation
spinal epidural space, 26/7
sweating, 2/163
tentorium, 1/538
urethra, 75/669
Innervation inhibition
Sydenham chorea, 49/74
Innominate artery
anatomy, 53/371
pathology, 53/371
Innominate artery stenosis
frequency, 53/373
Inocybe
muscarine, 64/14
mushroom intoxication, 36/535
mushroom toxin, 65/39
Inorganic mercury
analysis, 64/369
mercury polyneuropathy, 51/271
Inorganic mercury intoxication, 36/148-167
see also Acrodynia
chorea, 49/556
late cortical cerebellar atrophy, 21/497
penicillamine, 36/183
Inorganic salt intoxication
anxiolytic agent, 37/348
behavior disorder, 37/348
Inosamine phosphatide

phosphatidylethanolamine lipidosis, 42/429
Inositol
diabetes mellitus, 70/154
diabetic polyneuropathy, 51/508
hyponatremia, 63/547
osmoregulation, 63/555
uremic encephalopathy, 63/514
uremic polyneuropathy, 51/361, 63/514
Inositol-CPM-phosphatidyltransferase
Pelizaeus-Merzbacher disease, 10/193
Inositol deficiency
diabetes mellitus, 70/150
Inositol nicotinate
restless legs syndrome, 8/318, 51/548
Inositol phospholipid
leukodystrophy, 10/30
Insect bite
death, 6/676
striatal necrosis, 6/676
Insect bite encephalopathy
striatal necrosis, 49/503
Insecticide intoxication
organochlorine, *see* Organochlorine insecticide intoxication
Insensitivity
congenital pain, *see* Congenital pain insensitivity
pain, *see* Pain insensitivity
universal, *see* Universal insensitivity
Insertional activity
features, 62/50
motor unit, 62/50
myopathy, 62/50
Insolation, *see* Heat stroke
Insomnia
acetonylacetone intoxication, 64/87
acrodynia, 64/374
African trypanosomiasis, 52/341
alcohol, 45/138
alcoholism, 45/138
amantadine, 46/594
amiodarone intoxication, 65/458
amphetamine intoxication, 65/257
anxiety, 45/139
barbiturate intoxication, 65/509
benzodiazepine intoxication, 65/335
beta adrenergic receptor blocking agent intoxication, 65/436
bismuth, 64/4
bismuth intoxication, 64/337
buspirone intoxication, 65/344
caffeine intoxication, 65/260
carbon disulfide intoxication, 64/25
carisoprodol, 5/168
central sleep apnea, 63/461
depression, 45/139
drug medication, 45/138
encainide intoxication, 65/456
eosinophilia myalgia syndrome, 63/375
etiology, 45/138
familial, *see* Familial insomnia
fatal familial, *see* Fatal familial insomnia
fluoxetine intoxication, 65/320
fluvoxamine intoxication, 65/320
glutaryl-coenzyme A dehydrogenase deficiency, 66/647
hypothalamus disorder, 46/418
lead encephalopathy, 64/443
locked-in syndrome, 49/245
mefloquine intoxication, 65/485
mexiletine intoxication, 65/455
Morvan fibrillary chorea, 6/355
nalorphine, 37/366
nocturnal myoclonus, 51/544
olivopontocerebellar atrophy, 49/245
organolead intoxication, 64/133
organophosphate intoxication, 64/167
organotin intoxication, 64/141
Parkinson disease, 49/97, 245
paroxetine intoxication, 65/320
pellagra, 28/86, 46/336
polychlorinated dibenzodioxin intoxication, 64/205
procainamide intoxication, 37/452
progressive supranuclear palsy, 46/301, 49/245
rebound, *see* Rebound insomnia
REM sleep, 45/138
restless legs syndrome, 8/313, 51/544
selective serotonin reuptake inhibitor, 65/320
senile dementia, 46/286
sertraline intoxication, 65/320
sleep disorder, 3/67, 81, 91-93
sleep medication, 45/138
subarachnoid hemorrhage, 46/418
2,3,7,8-tetrachlorodibenzo-*p*-dioxin, 64/205
tryptophan, 64/250
tsukubaenolide intoxication, 65/558
type, 45/137-139
uremic encephalopathy, 63/504
vitamin Bc deficiency, 63/255
vitamin Bc intoxication, 65/574
withdrawal syndrome, 45/138
Inspiration
nucleus tractus solitarii, 63/480
Inspiratory gasping
autonomic nervous system, 74/625
orthostatic hypotension, 63/146

sympathetic response, 74/607
Instability
atlantoaxial, 25/280
atlanto-occipital, *see* Atlanto-occipital instability
cervical spine, 25/293
cervical vertebral column, 61/10
cervical vertebral column injury, 25/293, 61/60
definition, 25/245
facet interlocking, 25/340
gene, *see* Gene instability
occipitoatlantoaxial complex, 61/6
odontoid fracture, 25/280
spinal injury, 25/245, 280, 293, 340, 447
thoracolumbar spine injury, 25/447
traumatic spine deformity, 26/162
vertebral fracture, *see* Vertebral fracture instability
Instant food
vitamin B1 deficiency, 51/331
Instinct
Pick disease, 46/235
Instrumental apraxia
amusia, 45/486
frontal lobe tumor, 45/486
Insufficiency polyneuropathy
hepatic, *see* Hepatic insufficiency polyneuropathy
renal, *see* Renal insufficiency polyneuropathy
Insula
bradycardia, 63/205
cardiac chronotropic center, 63/205
cardiac dysrhythmia, 63/205
limbic encephalitis, 8/135
tachycardia, 63/205
Insular sclerosis, *see* Multiple sclerosis
Insulin
aging, 74/233
brain infarction, 23/123
brain injury, 23/123
caudal aplasia, 32/347
depression, 46/474
epilepsy, 15/319
epileptic treatment, 15/731
muscle tissue culture, 62/88
myotonic dystrophy, 40/511
myotonic syndrome, 40/545
neuropathy, 8/53
progressive dysautonomia, 59/157
subarachnoid hemorrhage, 12/113
Insulin growth factor 1
muscle tissue culture, 62/90
Insulin resistance
Friedreich ataxia, 60/319
hypertension, 75/480
myotonic dystrophy, 62/230
Insulin resistant diabetes mellitus
ataxia telangiectasia, 60/361, 374
Insulinoma
anxiety, 46/431
hyperinsulinism, 27/68
pancreatic, *see* Pancreatic insulinoma
Insurance
epilepsy, 15/800, 811
Integrin
metastasis, 71/645
Integrin receptor
brain metastasis, 69/109, 111
Intellectual defect
agnosia, 4/27
traumatic, *see* Traumatic intellectual defect
Intelligence, 3/296-313
ablation experiment, 3/301
acquired component, 3/303-305
age, 3/310-312
aggression, 46/178
alexia, 4/97
aphasia, 3/7, 4/97, 105-111, 422, 45/294, 46/154
brain tumor, 16/729
cerebral localization, 3/300-303, 312, 4/8
Charlevoix-Saguenay spastic ataxia, 60/454
cognition, 3/298
cri du chat syndrome, 31/586
crystallized ability, 3/308
definition, 3/296, 313
development, 3/303, 306, 313, 4/9, 422
developmental dysphasia, 46/141
developmental dyspraxia, 4/460
developmental theory, 3/306
different primary mental ability theory, 3/305
Edwards syndrome, 50/562
environment, 3/303
epilepsy, 15/589
equipotentiality, 3/302
factor analysis, 3/299
fluid ability, 3/308
focal cerebral lesion, 3/309, 426
genetics, 3/303, 313
glycosaminoglycanosis type IV, 66/304
head injury outcome, 57/403
hemispherectomy, 4/9
hierarchic theory, 3/301
Huntington chorea, 46/307
hydrocephalus, 30/600
hyperactivity, 46/180
impairment, 4/6-9
inheritance, 3/303, 313
inherited component, 3/303-305
intelligence A and B, 3/307

intelligence quotient, 3/298
intuitive thought, 3/306
juvenile hereditary benign chorea, 49/346
Klinefelter syndrome, 50/549
learning, 3/302, 305, 312
localized lesion, 4/6-9
mass action, 3/302, 309, 313
mental quotient, 3/298
migraine, 48/19
neurofibromatosis, 14/776
neuropsychologic defect, 45/517
neuropsychologic theory, 3/307
nonverbal ability, 4/107
occipital lobe tumor, 17/334
operational thought, 3/307
Parkinson dementia, 46/375
Parkinson disease, 6/193
pediatric head injury, 23/451
phenytoin intoxication, 65/502
posttraumatic amnesia, 45/188, 46/613
preconceptual thought, 3/306
psychological aspect, 3/297-300
reasoning, 3/306
sampling theory, 3/306
sensorimotor intelligence, 3/306
sensory deprivation, 3/305
skill, 3/313, 4/6
spina bifida, 32/555-557, 50/494
spinal cord injury, 61/543
statistical study, 3/303
stereotaxic surgery, 46/380
theory, 3/305-308
theory of g, 3/298, 306
transient global amnesia, 45/212
Turner syndrome, 31/500, 50/544
twin study, 3/303
Two Factor Theory, 3/298
verbal ability, 4/107
Wechsler Intelligence Scale, *see* Wechsler test
Wernicke-Korsakoff syndrome, 45/199
Wolf-Hirschhorn syndrome, 31/559
Intelligence quotient
affective psychosis, 46/457
bone marrow transplantation, 66/97
congenital heart disease, 63/1
craniolacunia, 50/139
Down syndrome, 50/528
Gilles de la Tourette syndrome, 49/629
intelligence, 3/298
Klinefelter syndrome, 50/550
lead intoxication, 64/445
mental deficiency, 46/2-5
microcephaly, 30/519, 50/267
occipital cephalocele, 50/101
recurrent meningitis, 52/52
schizophrenia, 46/457
Intelligence test
ability level variation, 4/6, 8
aphasia, 4/106
child, 4/354-356, 372, 422
clumsiness, 46/163
conceptual defect, 4/106
learning disability, 46/125
neurology, 3/296-303
psychological test, 4/10
restoration of higher cortical function, 3/426
senile dementia, 3/311
Intensity duration curve
facial paralysis, 8/267
interpretation, 19/279
muscle innervation, 19/279
spinal cord tumor, 19/279
Intensive care
atracurium, 71/536
complication, 71/525, 539
critical illness polyneuropathy, 63/420
epilepsy, 73/378
multiple organ failure, 71/528
myopathy, 63/420
neurologic complication, 63/420
neurologic syndrome, 71/528
neurology, *see* Neurologic intensive care
pancuronium, 71/536
rocuronium, 71/536
survey, 71/528
vecuronium, 71/536
Intensive care complication
creatine kinase, 63/420
critical illness polyneuropathy, 63/420
myopathy, 63/420
rhabdomyolysis, 63/420
Intensive care syndrome
delirium, 46/534
muscular atrophy, 63/420
Intention
myoclonus, *see* Action myoclonus
neglect, 45/161-167
right hemisphere, 45/173
Intention hypertonia
thalamus lesion, 1/320
Intention myoclonus, *see* Action myoclonus
Intention tremor
see also Action tremor
acrylamide polyneuropathy, 51/281
action myoclonus, 21/529
amyotrophic lateral sclerosis, 59/235

anterolateral thalamic syndrome, 2/485
ataxia, 42/230, 49/590
Bassen-Kornzweig syndrome, 21/580, 51/394, 63/272
Behçet syndrome, 56/597
benign essential tremor, 49/568
Bolivian hemorrhagic fever, 56/369
bulbospinal muscular atrophy, 59/44
chronic mercury intoxication, 64/370
clinical features, 1/283
congenital ataxic diplegia, 42/121
definition, 2/394
dentatorubral atrophy, 21/529
dentatorubropallidoluysian atrophy, 21/528
diagnosis, 6/822
dysmetria, 49/590
dyssynergia cerebellaris myoclonica, 21/509, 513, 515, 60/597
EMG, 49/590
Gökay-Tükel syndrome, 21/575
Hartnup disease, 7/577
hereditary olivopontocerebellar atrophy (Menzel), 21/441
hereditary periodic ataxia, 21/565
hyperpipecolatemia, 29/222
Lafora progressive myoclonus epilepsy, 27/173
lipofuscinosis, 42/230
manganese intoxication, 64/306
Mast syndrome, 42/282
multiple hamartoma syndrome, 42/755
origin, 1/279
pallidoluysionigral degeneration, 49/456
pallidopyramidal degeneration, 49/456
periodic ataxia, 21/579
Refsum disease, 21/199, 60/231, 66/490
serial dysmetria, 49/590
spinopontine degeneration, 21/390
thalamus, 6/108
XYY syndrome, 50/551
zinc intoxication, 36/338
Intercanthal distance
normal value, 30/241
orbital hypertelorism, 30/239-241
Intercellular adhesion molecule 1
brain metastasis, 69/111
Intercostal artery aneurysm
congenital heart disease, 63/4
Intercostal nerve
compression neuropathy, 51/109
topographical diagnosis, 2/39
Intercrural perforating artery, *see* Paramedian pontine artery
Interferon
classification, 47/348
herpes simplex virus, 56/209
HLA-DR2 antigen, 47/354
HLA-Dw2 antigen, 47/354
multiple sclerosis, 47/204-206, 347, 350, 353-355
rabies, 56/398
varicella zoster virus, 56/231
viral infection, 56/73
Interferon α II, *see* α2-Interferon
α-Interferon
adverse effect, 69/496
Argentinian hemorrhagic fever, 56/367
brain tumor, 67/298
encephalopathy, 63/359
neurotoxicity, 63/359
Newcastle disease, 47/354
sensorimotor polyneuropathy, 63/359
subacute sclerosing panencephalitis, 56/421
α-Interferon intoxication
antineoplastic agent, 65/529
apathy, 65/562
aphasia, 65/562
catatonia, 65/538
dementia, 65/538
encephalopathy, 63/359, 65/529, 538, 562
epilepsy, 65/562
migraine, 65/562
motor aphasia, 65/562
neuralgic amyotrophy, 65/538
neurotoxin, 63/359, 65/529, 538, 562
polyneuropathy, 65/529, 538
α2-Interferon
neuropathy, 69/461, 472
β-Interferon
adverse effect, 69/496
brain tumor, 67/298
β-Interferon intoxication
antineoplastic agent, 65/529
encephalopathy, 65/529
mental depression, 65/562
neurotoxin, 65/529, 538, 562
polyneuropathy, 65/529
γ-Interferon
adverse effect, 69/496
muscle tissue culture, 62/92
γ-Interferon intoxication
apathy, 65/562
aphasia, 65/562
encephalopathy, 65/562
epilepsy, 65/562
migraine, 65/562
motor aphasia, 65/562
Interhemispheric cyst

corpus callosum agenesis, 50/154, 156
orofaciodigital syndrome type I, 50/168
Interhemispheric fissure
Arnold-Chiari malformation type II, 50/408
Interhemispheric relation, *see* Cerebral dominance
Interictal aggression
epilepsy, 73/481
Interictal ataxia
hereditary periodic ataxia, 60/433
Interictal chorea
hereditary paroxysmal kinesigenic choreoathetosis, 49/353, 355
Interleukin-1
adverse effect, 69/496-498
critical illness polyneuropathy, 51/584
eosinophilia myalgia syndrome, 64/265
lymphocyte, 56/69
multiple sclerosis, 47/189
Interleukin-2
adverse effect, 69/496-498
antineoplastic agent, 65/529
brain tumor, 67/298
encephalopathy, 65/529
eosinophilia myalgia syndrome, 64/265
lymphocyte, 56/69
multiple sclerosis, 47/189, 347, 375
neuropathy, 69/461, 472
neurotoxin, 65/529
Interleukin-2 intoxication
brachial plexus neuropathy, 65/539
delirium, 65/539
epilepsy, 65/539
leukoencephalopathy, 65/539
Interleukin-3
eosinophil, 63/369
Interleukin-5
eosinophil, 63/369
Interlockings
bilateral cervical facet, *see* Bilateral cervical facet interlocking
unilateral facet, *see* Unilateral cervical facet interlocking
Intermaxillary rudiment face with hypotelorism, *see* Holoprosencephaly
Intermediate latency response
direct muscle response, 7/142
Intermediate nerve
neurinoma, 68/538
Intermediate spinal muscular atrophy
age at onset, 42/89
birth incidence, 42/91
bulbar paralysis, 42/90
contracture, 42/90
cyproheptadine, 41/326
fasciculation, 42/90
methysergide, 41/326
scoliosis, 42/90
systemic neoplastic disease, 41/326
tremor, 42/90
Intermedius nerve
anatomy, 5/338, 351, 358, 8/245, 48/491
brain stem fiber, 5/339
localization, 2/59
Intermedius nerve neuralgia, *see* Intermedius neuralgia
Intermedius neuralgia, 5/337-343, 48/487-493
auriculotemporal syndrome, 48/490
chorda tympani, 48/487
cluster headache, 5/114
diagnosis, 5/341
drug treatment, 5/383
erythroprosopalgia, 5/115
geniculate ganglion, 5/338
geniculate ganglionectomy, 48/491
glossopharyngeal neuralgia, 48/468
hemifacial spasm, 48/492
herpetic, *see* Herpetic geniculate neuralgia
Jacobson neuralgia, 48/490
Jannetta mechanism, 48/492
otalgia, 48/487
otalgic type, 5/339
prosopalgic form, 5/340
surgery, 5/402
symptom, 5/339
symptomatology, 5/339-341
synonym, 5/337
treatment, 5/342, 383
trigeminal neuralgia, 48/487
vegetative symptom, 1/457
Intermedius thalamotomy
stereotaxic ventral, *see* Stereotaxic ventral intermedius thalamotomy
Intermittent ataxia, *see* Periodic ataxia
Intermittent claudication
cauda equina, *see* Cauda equina intermittent claudication
forearm, *see* Forearm intermittent claudication
mandibular, *see* Mandibular intermittent claudication
muscle, *see* Muscle intermittent claudication
nerve, *see* Nerve intermittent claudication
neurogenic, *see* Cauda equina intermittent claudication *and* Spinal cord intermittent claudication
spinal cord, *see* Spinal cord intermittent claudication

Intermittent daytime hypersomnia
Kleine-Levin syndrome, 2/449, 45/136
Intermittent familial cerebellar ataxia, *see* Hill-Sherman syndrome
Intermittent hemodialysis
uremic polyneuropathy, 51/355, 359
Intermittent heredoataxia, *see* Hill-Sherman syndrome
Intermittent light stimulation
epilepsy, 15/442
Intermittent nerve compression
compression neuropathy, 51/88
Intermittent ophthalmoplegia
familial, *see* Familial intermittent ophthalmoplegia
Intermittent porphyria
acute, *see* Acute intermittent porphyria
spinal muscular atrophy, 59/370
Intermittent spinal cord ischemia
see also Dejerine syndrome
angiography, 12/520
cauda equina, 12/507-547
cause, 12/524
clinical features, 12/511
course, 12/514
CSF pressure, 12/529
differential diagnosis, 12/530
discography, 12/520
examination, 12/516
history, 12/508
mechanical factor, 12/529
myelography, 12/518
ossovenography, 12/521
pathophysiology, 12/526
posture, 12/527
sex ratio, 12/515
surgical finding, 12/523
tomography, 12/523
treatment, 12/539
type, 12/511, 527
vascular factor, 12/528
Internal auditory artery
cerebrovascular disease, 55/129
embryology, 11/27
headache, 48/274
radioanatomy, 11/92
Internal auditory vein
anatomy, 11/54
Internal capsule, *see* Capsula interna
Internal carotid arteriosclerosis
amaurosis fugax, 55/107
Internal carotid artery
absence, 30/664
agenesis, 12/301
anastomosis, 53/296, 298
anatomy, 11/2, 53/293
aneurysm, 12/221
brain aneurysm, 12/89
brain infarction, 11/307
branch, 53/294, 296
carotid artery kinking, 11/173
carotid rete mirabile, 11/173, 12/308-311
cervical vertebral column injury, 61/33
coiling, 53/293
congenital aneurysm, 31/146
Doppler sonography, 54/6
dural branch, 11/69
embryology, 11/34
epidural branch, 11/69
hydranencephaly, 50/338
hypoplasia, 12/306
intradural branch, 11/69
kinking, 53/293
occlusion, 5/126
paratrigeminal syndrome, 48/332
partial absence, 12/306
pericarotid syndrome, 48/335
primitive, *see* Primitive internal carotid artery
radioanatomy, 11/67
radiologic segment, 11/68
rete mirabile, 53/294
segment, 53/293
supply area, 53/297
sympathetic plexus, 48/335
syndrome, 11/307
traumatic aneurysm, 24/384
tuberculous meningitis, 52/198
tympanic branch, 11/69
Internal carotid artery syndrome
acetylsalicylic acid, 53/324
amaurosis fugax, 53/315, 318
angioscopy, 53/315
anticoagulant, 53/325
asymptomatic carotid artery stenosis, 53/293, 320
bifurcation, 53/302
bilateral, 53/307, 318
bilateral carotid lesion, 53/320
brain infarction, 53/292
caroticotympanic artery, 11/308
carotid artery bifurcation, 53/303
carotid bruit, 53/321
carotid endarterectomy, 53/292, 322
collateral, 53/300
computer assisted tomography, 53/315
Doppler spectrum, 53/302
episodic limb shaking, 53/318

external internal carotid bypass, 53/323
frequency, 53/305
heparin, 53/325
infarction risk, 53/321
morbidity, 53/321
mortality, 53/307
natural history, 53/319
sex ratio, 53/305
stenosis degree, 53/305
surgical treatment, 53/322
transient ischemic attack, 53/307
treatment, 53/322
Internal carotid hypoplasia
moyamoya disease, 55/293
Internal cerebral vein
anatomy, 11/49
brain phlebothrombosis, 54/399
caudal migration, 11/63
radioanatomy, 11/107
Internal hamstring reflex
diagnostic value, 1/247
Internal jugular vein
anatomy, 11/54
Internal maxillary artery
anastomosis, 12/308
traumatic carotid cavernous sinus fistula, 57/353
Internal medicine
neurology, 1/16
Internal occipital vein, *see* Descending medial occipital vein
Internal pallidal segment
rigidity, 49/457
Internal parietal artery, *see* Posterior frontal artery
International Classification of Disease Coding, *see* ICD classification
International Classification of Diseases, *see* ICD classification
Internodal axolemma
nodal axolemma, 47/15
Ranvier node, 51/5, 15
Refsum disease, 51/15
Internode length
nerve, 51/8
Ranvier node, 51/11
Internode membrane
sodium channel, 51/47
Internuclear ophthalmoplegia
see also Medial longitudinal fasciculus *and* Unilateral elevator palsy
anterior type, 2/278, 328
Bassen-Kornzweig syndrome, 63/274
brain embolism, 53/212
brain infarction, 53/212
brain injury, 57/137
brain lacunar infarction, 54/246
brain stem infarction, 55/125
central pontine myelinolysis, 63/549
cerebrovascular disease, 53/212
clinical examination, 57/137
cocaine intoxication, 65/50
dissociated nystagmus, 2/367, 16/322
EOG, 54/164
Fabry disease, 55/457
features, 55/124
intracranial hemorrhage, 53/212
lymphomatoid granulomatosis, 51/451
median longitudinal fasciculus syndrome, 2/278
multiple sclerosis, 9/167, 47/55, 132, 140
nystagmus, 2/329
ocular dysmetria, 2/329
ocular imbalance, 24/75
olivopontocerebellar atrophy variant, 21/453
olivopontocerebellar atrophy with retinal degeneration, 60/507, 657
pontine infarction, 53/389, 394
posterior type, 2/278, 328
progressive supranuclear palsy, 49/242
superior cerebellar artery syndrome, 53/397
symptom, 2/328-330
systemic lupus erythematosus, 55/377
transient ischemic attack, 53/212
trigeminal nerve agenesis, 50/214
type, 2/278
vertebrobasilar system syndrome, 53/397
Interorbital distance
normal value, 30/240
orbital hypertelorism, 30/239
Interossei phenomenon
synkinesis, 1/184
Interosseous nerve syndrome
anterior, *see* Anterior interosseous nerve syndrome
flip sign, 2/36
posterior, *see* Posterior interosseous nerve syndrome
Interparietal syndrome
body scheme disorder, 4/233
parietal lobe lesion, 4/233
Interpedicular distance, *see* Elsberg sign
Interpedicular stenosis
lumbar vertebral canal stenosis, 20/774
Interpeduncular syndrome
oculomotor nerve injury, 2/298
Interpeduncular vein
anatomy, 11/51
Interpupillary distance
normal value, 30/241

orbital hypertelorism, 30/239, 241
Intersegmental artery
cervical, *see* Cervical intersegmental artery
proatlantal, *see* Proatlantal intersegmental artery
suboccipital, *see* Suboccipital intersegmental artery
Interstitial brachytherapy
brain metastasis, 69/144
Interstitial cystitis
migraine, 48/160
Interstitial fibrosis
myosclerosis, 41/65
pulmonary, *see* Pulmonary interstitial fibrosis
Interstitial keratitis
Cogan syndrome type II, 51/454
Interstitial lung disease
tuberous sclerosis, 50/371
Interstitial nephritis
Behçet syndrome, 56/595
hypertensive encephalopathy, 54/212
lead intoxication, 9/642
Interstitial neuropathy
adiposis dolorosa, 43/57
hypertrophic, *see* Hypertrophic interstitial neuropathy
Interstitial telangiectasia
multiple, *see* Multiple interstitial telangiectasia
Interventriculostomy
aqueduct stenosis, 50/315
Intervertebral disc
acid phosphatase, 20/538
age, 20/528, 536
anatomy, 20/525, 25/124, 61/14
annulus fibrosus, 20/525
annulus tear, 20/527
cell, 20/530
cervical vertebral column injury, 61/27
chondroitin sulfate A, 20/526, 535
chondroitin sulfate C, 20/526, 535
collagen fiber, 20/526
dermatan sulfate, 20/526, 535
development, 32/4
discitis, 19/208
fibril, 20/532
fluid transfer, 20/537
glycoprotein, 20/526, 538
glycosaminoglycan, 20/526, 533, 535
heparan sulfate, 20/526
histology, 20/529-532
hyaluronic acid, 20/526
interstitial fluid, 20/533
intradisc gap, 20/549
keratan sulfate, 20/526, 535
load, 20/540
lysosome, 20/538
matrix, 20/526
metabolism, 20/537
motor end plate, 20/525
movement, 20/527
noncollagen protein, 20/539
nucleus pulposus, 20/525
pH, 20/538
pressure, 20/540
radiology, *see* Discography
resorption, 25/505-508
scoliosis, 50/413
spine experimental injury, 25/1
stress, 20/528
vertebral column tumor, 20/21, 33
vertebral experimental injury, 25/1
Intervertebral disc calcification
plain X-ray, 20/566
spinal cord compression, 19/153
thoracic level, 20/566
Intervertebral disc degeneration, 20/525-544
biochemistry, 20/526-540
biophysics, 20/540-544
cervical disc, 20/549
concealed disc, 20/576
operant factor, 20/542
pathomechanism, 20/576
Intervertebral disc disruption, *see* Intervertebral disc prolapse
Intervertebral disc hemorrhage
burst fracture, 61/36
thoracolumbar vertebral column injury, 61/36
Intervertebral disc hernia, *see* Intervertebral disc prolapse
Intervertebral disc injury
classification, 25/484
definition, 25/482
disruption, 25/485
experimental, 25/1, 4, 6
incidence, 25/481
lumbosacral, 25/473
mechanism, 25/482
pathology, 25/27-42, 486
type, 25/484, 486
Intervertebral disc prolapse, 20/525-544, 25/493-505
see also Low back pain
cause, 25/487
cervical, *see* Cervical intervertebral disc prolapse
cervical radicular syndrome, 20/555
cervical vertebral column injury, 61/27, 31
clinical features, 25/488, 497

crossed Lasègue test, 20/577, 587
CSF, 25/499
definition, 25/486, 493
discography, 25/490, 500
dorsal, 20/574
EMG, 20/580, 25/500
epidurography, 25/500
force diagram, 20/543
fusion, 25/503
infraforaminal, 20/574
lateral, 20/574
lumbar, *see* Lumbar intervertebral disc prolapse
lumbar vertebral canal stenosis, 20/597
movement, 25/489
myelography, 19/196, 25/499, 506
neurobrucellosis, 33/316, 52/594
neurologic sign, 25/489
pathology, 25/494
posterior longitudinal ligament calcification, 19/154
progressive spinal cord deficit, 25/277
psychiatric sign, 25/489
radiology, 25/490, 499
site, 25/488
spinal phlebography, 19/239
spinal venography, 25/500
subdural, 20/575
surgery, 25/492, 501
teenage, *see* Teenage disc prolapse
thoracic, *see* Thoracic intervertebral disc prolapse
thoracic trauma, 61/80
thoracic vertebral column injury, 61/77
traumatic, 61/522
treatment, 25/491, 500
type, 20/574
vertebral column injury, 61/522
Intervertebral facet joint
epidural ganglion cyst, 20/605
Intervertebral foramen
anatomy, 20/550
functional anatomy, 25/238, 401
hourglass neurinoma, 20/230, 242
hourglass tumor, 8/429, 19/161
lumbosacral, 25/476
spinal nerve root, 25/400
Intervertebral foramen stenosis, 20/611-801
cauda equina intermittent claudication, 20/791
Intestinal aganglionosis
congenital, *see* Congenital megacolon
Intestinal autonomic neuropathy
paraneoplastic polyneuropathy, 51/494
Intestinal fistula
dorsal, *see* Neurenteric cyst
Intestinal function
aging, 74/237
drug treatment, 47/167-169
multiple sclerosis, 47/167
radiodiagnosis, 26/269
spina bifida, 50/494
spinal cord injury, 26/258
spinal shock, 26/258
thoracolumbar spine injury, 25/455
transverse spinal cord syndrome, 2/192-195
Intestinal lymphangiectasia
deficiency neuropathy, 51/326
Intestinal malabsorption
blind loop syndrome, 7/619
classification, 7/618
deficiency neuropathy, 7/603, 618
diverticulosis, 7/619
etiology, 7/630
gastrocolic fistula, 7/618
metabolic neuropathy, 51/68
operation, 7/619
small intestine, 7/618
treatment, 7/619
Intestinal neurinoma
neurofibromatosis type I, 14/604
Peutz-Jeghers syndrome, 14/604, 791
phakomatosis, 14/602
Intestinal peristalsis absence
acute pandysautonomia, 51/475
Intestinal pneumonitis
lymphoid, *see* Lymphoid interstitial pneumonitis
Intestinal pseudo-obstruction
see also Paralytic ileus
acetylcholine, 51/493
amyloidosis, 51/493
anthraquinone, 51/494
argyrophobic cell, 51/493
celiac disease, 51/493
Chagas disease, 51/493
chronic idiopathic type, 51/492
classification, 51/491, 70/320
congenital central alveolar hypoventilation, 63/415
congenital megacolon, 51/492
cytomegalovirus infection, 51/493
dermatomyositis, 51/493
diabetes mellitus, 51/493
diverticulosis, 70/320
Duchenne muscular dystrophy, 62/120
dysautonomia, 51/493
Ehlers-Danlos syndrome, 51/493
eosinophilic inclusion body, 51/493
esophageal manometry, 51/494

familial autonomic dysfunction, 51/493
familial type, 51/493, 70/324
hypoparathyroidism, 51/493
innervation, 75/701
Kearns-Sayre-Daroff-Shy syndrome, 62/313
lead neuropathy, 64/437
lead polyneuropathy, 64/437
mitochondrial myopathy, 62/313
MNGIE syndrome, 62/505
multiple sclerosis, 51/493
myenteric plexus, 51/491-493, 70/320
myopathy, 70/320
myotonic dystrophy, 51/493
neurology, 70/322
neuromuscular disorder, 70/320
neuronal intranuclear hyaline inclusion disease, 60/650, 668
oculogyric crisis, 60/650
paraneoplastic syndrome, 51/493
Parkinson disease, 51/494
parkinsonism, 51/493
pheochromocytoma, 51/493
postencephalitic syndrome, 51/493
primary cause, 51/492
primary type, 51/492
progressive external ophthalmoplegia, 62/313
progressive muscular dystrophy, 51/493
secondary cause, 51/493
Shy-Drager syndrome, 51/493
sporadic type, 51/493, 70/325
systemic disorder, 70/321
systemic lupus erythematosus, 51/493
systemic sclerosis, 51/493
Trypanosoma cruzi, 51/494
visceral myopathy, 51/492, 70/324
visceral neuropathy, 51/492, 70/324
Intestinal schwannoma, *see* Intestinal neurinoma
Intestinal tract tumor
brain metastasis, 18/204
Intestinal ulcer
Behçet syndrome, 56/594
general adaptation syndrome, 45/253
Intestine
see also Digestive system
neural control, 1/506
Intestinoma, *see* Intraspinal neurenteric cyst
Intoxications
Acanthaster planci, *see* Acanthaster planci intoxication
Acanthuridae, *see* Acanthuridae intoxication
Acanthurus, *see* Acanthurus intoxication
acetaldehyde, *see* Acetaldehyde intoxication
acetates, *see* Acetates intoxication
acetazolamide, *see* Acetazolamide intoxication
acetonylacetone, *see* Acetonylacetone intoxication
acetylcholine, *see* Acetylcholine intoxication
acetylsalicylic acid, *see* Acetylsalicylic acid intoxication
acetylurea, *see* Acetylurea intoxication
acivicin, *see* Acivicin intoxication
acorn worm, *see* Hemichordata intoxication
Acraniata, *see* Acraniata intoxication
acrylamide, *see* Acrylamide intoxication
actiniaria, *see* Actiniaria intoxication
aflatoxin, *see* Aflatoxin intoxication
agalgitakg, *see* Balaenoptera borealis intoxication
akee, *see* Hypoglycin A intoxication
alcohol, *see* Alcohol intoxication
aldrin, *see* Aldrin intoxication
alfadolone acetate with alfaxalone, *see* Alfadolone acetate with alfaxalone intoxication
alkylmercury, *see* Alkylmercury intoxication
alkyl tin, *see* Alkyl tin intoxication
almitrine, *see* Almitrine intoxication
alopex lagopus, *see* Alopex lagopus intoxication
althesin, *see* Alfadolone acetate with alfaxalone intoxication
altretamine, *see* Altretamine intoxication
aluminum, *see* Aluminum intoxication
alvodine, *see* Piminodine
Amanita muscaria, *see* Amanita muscaria intoxication
Amanita pantherina, *see* Amanita pantherina intoxication
Amanita phalloides, *see* Amanita phalloides intoxication
amanitin, *see* Amanitin intoxication
aminocaproic acid, *see* Aminocaproic acid intoxication
aminoglycoside, *see* Aminoglycoside intoxication
2-amino-3-methylaminopropionic acid, *see* 2-Amino-3-methylaminopropionic acid intoxication
amiodarone, *see* Amiodarone intoxication
ammonia, *see* Periodic ammonia intoxication
amoxapine, *see* Amoxapine intoxication
amphetamine, *see* Amphetamine intoxication
amphotericin B, *see* Amphotericin B intoxication
amsacrine, *see* Amsacrine intoxication
2-amyl-methylphosphonofluoridate, *see* 2-Amyl-methylphosphonofluoridate intoxication
amyl nitrite, *see* Amyl nitrite intoxication
Anemonia sulcata, *see* Anemonia sulcata intoxication
anesthetic agent, *see* Anesthetic agent intoxication

angatea, *see* Angatea intoxication
aniline, *see* Aniline intoxication
annelida, *see* Annelida intoxication
anthopleura elegantissima, *see* Anthopleura elegantissima intoxication
anthozoa, *see* Anthozoa intoxication
9-anthroic acid, *see* 9-Anthroic acid intoxication
antiarrhythmic agent, *see* Antiarrhythmic agent intoxication
anticonvulsant, *see* Anticonvulsant intoxication *and* Antiepileptic agent intoxication
antidepressant, *see* Antidepressant intoxication
antiepileptic agent, *see* Antiepileptic agent intoxication
antihypertensive agent, *see* Antihypertensive agent intoxication
antimony, *see* Antimony intoxication
antipsychotic agent, *see* Antipsychotic agent intoxication
apamin, *see* Apamin intoxication
apistus, *see* Apistus intoxication
aploactidae, *see* Aploactidae intoxication
aploactosoma, *see* Aploactosoma intoxication
aplysia, *see* Aplysia intoxication
aprindine, *see* Aprindine intoxication
araeosoma, *see* Araeosoma intoxication
arctic fox, *see* Fox intoxication
aromatic carboxylic acid, *see* Aromatic carboxylic acid intoxication
arsenic, *see* Arsenic intoxication
arsphenamine, *see* Arsphenamine intoxication
arthropoda, *see* Arthropoda intoxication
asparaginase, *see* Asparaginase intoxication
asteroidea, *see* Asteroidea intoxication
asthenosoma, *see* Asthenosoma intoxication
Atergatis floridus, *see* Atergatis floridus intoxication
Atopomycterus nicthemerus, *see* Puffer fish intoxication
atropine, *see* Atropine intoxication
azacitidine, *see* Azacitidine intoxication
azathioprine, *see* Azathioprine intoxication
azethion, *see* Azethion intoxication
azobenzene, *see* Azobenzene intoxication
bacterial food, *see* Bacterial food poisoning
BAL, *see* Dimercaprol intoxication
Balaenoptera borealis, *see* Balaenoptera borealis intoxication
balloon fish, *see* Puffer fish intoxication
barbiturate, *see* Barbiturate intoxication
barium, *see* Barium intoxication
barium carbonate, *see* Barium carbonate intoxication
bass, *see* Bass intoxication
bat ray, *see* Bat ray intoxication
Batrachoides, *see* Batrachoides intoxication
Batrachoididae, *see* Batrachoididae intoxication
batrachotoxin, *see* Batrachotoxin intoxication
Batrachus, *see* Batrachus intoxication
bee sting, *see* Bee sting intoxication
BEN, *see* Benzyl benzoate intoxication
benzene hexachlorine, *see* Hexachlorocyclohexane intoxication
benzodiazepine, *see* Benzodiazepine intoxication
benzomorphan, *see* Benzomorphan intoxication
benzyl benzoate, *see* Benzyl benzoate intoxication
beryllium, *see* Beryllium intoxication
beta adrenergic receptor blocking agent, *see* Beta adrenergic receptor blocking agent intoxication
bicuculline, *see* Bicuculline intoxication
birgus latro, *see* Birgus latro intoxication
bis(diethoxyphosphoryl)ethylamine, *see* Bis(diethoxyphosphoryl)ethylamine intoxication
bismuth, *see* Bismuth intoxication
bivalves, *see* Bivalves intoxication
black widow spider venom, *see* Black widow spider venom intoxication
Blighia sapida, *see* Hypoglycin A intoxication
blowfish, *see* Puffer fish intoxication
borate, *see* Borate intoxication
borax, *see* Borax intoxication
boric acid, *see* Boric acid intoxication
boron, *see* Boron intoxication
boron nitride, *see* Boron nitride intoxication
boron trifluoride, *see* Boron trifluoride intoxication
botulinum toxin, *see* Botulinum toxin intoxication
boxfish, *see* Boxfish intoxication
boxjelly, *see* Boxjelly intoxication
breast milk, *see* Subacute necrotizing encephalomyelopathy
brolamfetamine, *see* Brolamfetamine intoxication
bromate, *see* Bromate intoxication
bromide, *see* Bromide intoxication
bromoureide, *see* Bromoureide intoxication
broxuridine, *see* Broxuridine intoxication
buckthorn toxin, *see* Buckthorn toxin intoxication
bufotenine, *see* Bufotenine intoxication
α-bungarotoxin, *see* α-Bungarotoxin intoxication
β-bungarotoxin, *see* β-Bungarotoxin intoxication
bupivacaine, *see* Bupivacaine intoxication
buspirone, *see* Buspirone intoxication
O-*N*-butyl *O*-carbetoxymethyl ethylphosphonothioate, *see* *O*-*N*-Butyl *O*-carbetoxymethyl ethylphosphonothioate

intoxication
cadmium, *see* Cadmium intoxication
caffeine, *see* Caffeine intoxication
calcium antagonist, *see* Calcium antagonist intoxication
camphor, *see* Camphor intoxication
canis familiaris, *see* Canis familiaris intoxication
cannabis, *see* Cannabis intoxication
captopril, *see* Captopril intoxication
carbamate, *see* Carbamate intoxication
carbamazepine, *see* Carbamazepine intoxication
carbon dioxide, *see* Carbon dioxide intoxication
carbon disulfide, *see* Carbon disulfide intoxication
carbon monoxide, *see* Carbon monoxide intoxication
carbon tetrachloride, *see* Carbon tetrachloride intoxication
carboplatin, *see* Carboplatin intoxication
Carcinoscorpius rotundicauda, *see* Carcinoscorpius rotundicauda intoxication
cardiovascular agent, *see* Cardiovascular agent intoxication
caretta caretta, *see* Caretta caretta intoxication
Carpilius convexus, *see* Carpilius convexus intoxication
Carpilius noxius, *see* Carpilius noxius intoxication
Carukia barnesi, *see* Carukia barnesi intoxication
carybdea, *see* Carybdea intoxication
carybdeidae, *see* Carybdeidae intoxication
cassava, *see* Cassava intoxication
caulerpa, *see* Caulerpa intoxication
Caulerpa racemosa, *see* Caulerpa racemosa intoxication
caulerpicin, *see* Caulerpicin intoxication
Centaurea solstitialis, *see* Yellow star thistle intoxication
centropogon, *see* Centropogon intoxication
centrostephanus, *see* Centrostephanus intoxication
cephalochordata, *see* Cephalochordata intoxication
cephalopoda, *see* Cephalopoda intoxication
cetacea, *see* Cetacea intoxication
cheese, *see* Cheese reaction
chelidonichthys, *see* Chelidonichthys intoxication
chelonia, *see* Chelonia intoxication
chelonia mydas, *see* Chelonia mydas intoxication
chimaera, *see* Chimaera intoxication
chimaeridae, *see* Chimaeridae intoxication
chirodropus, *see* Chirodropus intoxication
chironex, *see* Chironex intoxication
chiropsalmus, *see* Chiropsalmus intoxication
chiropsoides, *see* Chiropsoides intoxication
chloral hydrate, *see* Chloral hydrate intoxication
chloramphenicol, *see* Chloramphenicol intoxication
chlorates, *see* Chlorate intoxication
chlordane, *see* Chlordane intoxication
chlordecone, *see* Chlordecone intoxication
chlorinated cyclodiene, *see* Chlorinated cyclodiene intoxication
chlormethine, *see* Chlormethine intoxication
chloroform, *see* Chloroform intoxication
chlorophyceae, *see* Chlorophyceae intoxication
chloroprocaine, *see* Chloroprocaine intoxication
chloroquine, *see* Chloroquine intoxication
chlorphenotane, *see* DDT intoxication
chlorpromazine, *see* Chlorpromazine intoxication
cholinergic receptor blocking agent, *see* Cholinergic receptor blocking agent intoxixation
chordata, *see* Chordata intoxication
chronic aluminum, *see* Chronic aluminum intoxication
chronic carbon monoxide, *see* Carbon monoxide encephalopathy
chronic lead, *see* Lead encephalopathy
chronic manganese, *see* Manganese encephalopathy
chrysaora, *see* Chrysaora intoxication
ciclosporin, *see* Ciclosporin intoxication
ciguatoxin, *see* Ciguatoxin intoxication
ciprofloxacin, *see* Ciprofloxacin intoxication
cisplatin, *see* Cisplatin intoxication
clam, *see* Paralytic shellfish poisoning
clioquinol, *see* Clioquinol intoxication
clofibrate, *see* Clofibrate intoxication
clomethiazole, *see* Clomethiazole intoxication
clonidine, *see* Clonidine intoxication
clupeotoxin, *see* Clupeotoxin intoxication
cnidarians, *see* Coelenterata intoxication
cnidoglanis, *see* Cnidoglanis intoxication
cobalt, *see* Cobalt intoxication
cobra snake venom, *see* Cobra snake venom intoxication
cocaine, *see* Cocaine intoxication
codeine, *see* Codeine intoxication
coelenterata, *see* Coelenterata intoxication
colchicine, *see* Colchicine intoxication
cone, *see* Cone intoxication
congestin, *see* Congestin intoxication
conidae, *see* Conidae intoxication
conus aulicus, *see* Conus aulicus intoxication
conus geographus, *see* Conus geographus intoxication
conus marmoreus, *see* Conus marmoreus intoxication
conus omaria, *see* Conus omaria intoxication

conus striatus, *see* Conus striatus intoxication
conus textile, *see* Conus textile intoxication
conus tulipa, *see* Conus tulipa intoxication
copper, *see* Copper intoxication
coprine, *see* Coprine intoxication
coral snake, *see* Coral snake intoxication
corticosteroid, *see* Corticosteroid intoxication
cortisone, *see* Cortisone intoxication
cotton seed, *see* Cotton seed intoxication
coumafos, *see* Coumafos intoxication
cow-nose ray, *see* Cow-nose ray intoxication
crab, *see* Crab intoxication
creseis, *see* Creseis intoxication
crinoidea, *see* Crinoidea intoxication
crotoxin, *see* Crotoxin intoxication
crustacea, *see* Crustacea intoxication
cryptotoxin, *see* Cryptotoxin intoxication
ctenochaetis, *see* Ctenochaetis intoxication
cubomedusae, *see* Cubomedusae intoxication
cubomedusan, *see* Cubomedusan intoxication
cuprizone, *see* Cuprizone intoxication
curare, *see* Curare intoxication
cyanea, *see* Cyanea intoxication
cyanide, *see* Cyanide intoxication
cyanogenic glycoside, *see* Cyanogenic glycoside intoxication
cycad, *see* Cycad intoxication
cyclopropane, *see* Cyclopropane intoxication
cycloserine, *see* Cycloserine intoxication
cyproterone acetate, *see* Cyproterone acetate intoxication
cytarabine, *see* Cytarabine intoxication
dactinomycin, *see* Dactinomycin intoxication
damsel fish, *see* Damsel fish intoxication
dapsone, *see* Dapsone intoxication
dasyatidae, *see* Dasyatidae intoxication
DDVP, *see* Dichlorvos intoxication
decaborane, *see* Decaborane intoxication
deferoxamine, *see* Deferoxamine intoxication
delphinapterus leucas, *see* Delphinapterus leucas intoxication
demeton, *see* Demeton intoxication
demeton-S, *see* Demeton-S intoxication
deoxybarbiturate, *see* Deoxybarbiturate intoxication
dermochelys coriacae, *see* Dermochelys coriacea intoxication
desferrioxamine, *see* Deferoxamine intoxication
designer drug, *see* Designer drug intoxication
devil ray, *see* Devil ray intoxication
dextropropoxyphene, *see* Dextropropoxyphene intoxication
DFP, *see* Dyflos intoxication
diadematidae, *see* Diadematidae intoxication
diamorphine, *see* Diamorphine intoxication
diazepam, *see* Diazepam intoxication
diazoxide, *see* Diazoxide intoxication
dibenzazepine, *see* Dibenzazepine intoxication
diborane, *see* Diborane intoxication
2,4-dichlorophenoxyacetic acid, *see* 2,4-Dichlorophenoxyacetic acid intoxication
dichlorvos, *see* Dichlorvos intoxication
dieldrin, *see* Dieldrin intoxication
diethyl bisdimethyl pyrophosphordiamide, *see* Diethyl bisdimethyl pyrophosphordiamide intoxication
diethylcarbethoxydichloromethylphosphonate, *see* Forstenon intoxication
O,*O*-diethyl *S*-(carbomethoxymethyl) phosphorodithioate, *see* Azethion intoxication
O,*O*-diethyl *O*-(3-chloro-4-methylumbelliferyl) phosphorothioate, *see* Coumafos intoxication
diethyl 4-chlorophenyl phosphate, *see* Diethyl 4-chlorophenyl phosphate intoxication
diethyl 2-chlorovinyl phosphate, *see* Diethyl 2-chlorovinyl phosphate intoxication
O,*O*-diethyl *S*-2-diethylmethylammoniumethyl phosphonothiolate methylsulfate, *see* *O*,*O*-Diethyl *S*-2-diethylmethylammoniumethyl phosphonothiolate methylsulfate intoxication
O,*O*-diethyl *S*-ethsulfonylmethyl phosphorodithioate, *see* Phorate sulfone intoxication
O,*O*-diethyl *S*-ethsulfonylmethyl phosphorothioate, *see* *O*,*O*-Diethyl *S*-ethsulfonylmethyl phosphorothioate intoxication
O,*O*-diethyl *S*-(2-eththioethyl) phosphorodithioate, *see* Disystox intoxication
O,*O*-diethyl *O*-(2-eththioethyl) phosphorothioate, *see* Demeton intoxication
O,*O*-diethyl *S*-(2-eththioethyl) phosphorothioate, *see* Demeton-S intoxication
O,*O*-diethyl *S*-eththiomethyl phosphorothioate, *see* *O*,*O*-Diethyl *S*-eththiomethyl phosphorothioate intoxication
O,*O*-diethyl *S*-(2-eththionylethyl) phosphorothioate, *see* Iso-Systox sulfoxide intoxication
O,*O*-diethyl *S*-eththionylmethyl phosphorothioate, *see* *O*,*O*-Diethyl *S*-eththionylmethyl phosphorothioate intoxication
O,*O*-diethyl *O*-(2-isopropyl-6-methyl-4-pyrimidyl) phosphorothioate, *see* Dimpylate intoxication
diethyl-3-methyl-5-pyrazolyl phosphate, *see*

Pyrazoxon intoxication
O,O-diethyl *O*-(4-methylumbelliferyl) phosphorothioate, *see* *O,O*-Diethyl *O*-(4-methylumbelliferyl) phosphorothioate intoxication
O,O-diethyl *O*-(4-nitrophenyl) phosphorothioate, *see* Parathion intoxication
O,O-diethyl s-(4-nitrophenyl) phosphorothioate, *see* *O,O*-Diethyl *S*-(4-nitrophenyl) phosphorothioate intoxication
O,S-diethyl *O*-(4-nitrophenyl) phosphorothioate, *see* *O,S*-Diethyl *O*-(4-nitrophenyl) phosphorothioate intoxication
diethyl phosphorocyanidate, *see* Diethyl phosphorocyanidate intoxication
diethyl phosphorofluoridate, *see* Diethyl phosphorofluoridate intoxication
O,O-diethyl *O*-(8-quinolyl) phosphorothioate, *see* *O,O*-Diethyl *O*-(8-quinolyl) phosphorothioate intoxication
N,N-diethyl-*m*-toluamide, *see* *N,N*-Diethyl-*m*-toluamide intoxication
diethyl trichloromethylphosphonate, *see* Diethyl trichloromethylphosphonate intoxication
O,O-diethyl-*S*-2-trimethylammoniumethyl phosphonothiolate iodide, *see* Ecothiopate iodide intoxication
di-2-fluoroethyl phosphorofluoridate, *see* Di-2-fluoroethyl phosphorofluoridate intoxication
digitalis, *see* Digitalis intoxication
digitoxin, *see* Digitoxin intoxication
digoxin, *see* Digoxin intoxication
diisopropyl phosphoroiodidate, *see* Diisopropyl phosphoroiodidate intoxication
dimefox, *see* Dimefox intoxication
dimercaprol, *see* Dimercaprol intoxication
2-dimethylaminoethyl dimethylphosphinate, *see* 2-Dimethylaminoethyl dimethylphosphinate intoxication
O,O-dimethyl *S*-(1,2-dicarbethoxyethyl) phosphorodithioate, *see* Malathion intoxication
dimethyl-2,2-dichlorovinyl phosphate, *see* Dichlorvos intoxication
O,O-dimethyl *O*-(2-ethylsulfonylethyl) phosphorothioate, *see* Methylsystox sulfone intoxication
O,O-dimethyl *S*-(4-nitrophenyl) phosphorothioate, *see* *O,O*-Dimethyl *S*-(4-nitrophenyl) phosphorothioate intoxication
dimethyl 1,2,2,2-tetrachloroethyl phosphate, *see* Dimethyl 1,2,2,2-tetrachloroethyl phosphate intoxication
N,N-dimethyltryptamine, *see* *N,N*-Dimethyltryptamine intoxication
dimpylate, *see* Dimpylate intoxication
dinitrophenol, *see* Dinitrophenol intoxication
dinoflagellates, *see* Dinoflagellates intoxication
dioxin, *see* Dioxin intoxication
dipeptidyl carboxypeptidase I inhibitor, *see* Dipeptidyl carboxypeptidase I inhibitor intoxication
diphenyl phosphorofluoridate, *see* Diphenyl phosphorofluoridate intoxication
diphenyl-2-trimethylammoniumethyl phosphate bromide, *see* Diphenyl-2-trimethylammoniumethyl phosphate bromide intoxication
diphtheria toxin, *see* Diphtheria toxin intoxication
di-*N*-propyl phosphorofluoridate, *see* Di-*N*-propyl phosphorofluoridate intoxication
disopyramide, *see* Disopyramide intoxication
disulfiram, *see* Disulfiram intoxication
disystox, *see* Disystox intoxication
diuretic agent, *see* Diuretic agent intoxication
docetaxel, *see* Taxotere intoxication
dogfish, *see* Squaliformes intoxication
dolphin, *see* Dolphin intoxication
dopaminergic, *see* Dopaminergic intoxication
doxorubicin, *see* Doxorubicin intoxication
DPDA, *see* Iso-Ompa intoxication
drug, *see* Drug intoxication
dyflos, *see* Dyflos intoxication
eagle ray, *see* Eagle ray intoxication
Echinodermata, *see* Echinodermata intoxication
echinoidea, *see* Echinoidea intoxication
echinothrix, *see* Echinothrix intoxication
echinothuridae, *see* Echinothuridae intoxication
ecothiopate iodide, *see* Ecothiopate iodide intoxication
edetic acid, *see* Edetic acid intoxication
EDTA, *see* Edetic acid intoxication
elapid, *see* Elapid intoxication
elasmobranche, *see* Elasmobranche intoxication
emerita analoga, *see* Emerita analoga intoxication
emetine, *see* Emetine intoxication
enalapril, *see* Enalapril intoxication
encainide, *see* Encainide intoxication
endosulfan, *see* Endosulfan intoxication
endrin, *see* Endrin intoxication
enflurane, *see* Enflurane intoxication
enhydrina schistosa, *see* Enhydrina schistosa intoxication
enkephalin, *see* Enkephalin intoxication
enoplosidae, *see* Enoplosidae intoxication
enoplosus, *see* Enoplosus intoxication

enoxacin, *see* Enoxacin intoxication
ephedrine, *see* Ephedrine intoxication
eretmochelys imbricata, *see* Eretmochelys imbricata intoxication
ergotamine, *see* Ergotamine intoxication
erignathus barbatus, *see* Erignathus barbatus intoxication
eriphia sebana, *see* Eriphia sebana intoxication
etacrynic acid, *see* Etacrynic acid intoxication
ethambutol, *see* Ethambutol intoxication
ethchlorvynol, *see* Ethchlorvynol intoxication
ethionamide, *see* Ethionamide intoxication
ethosuximide, *see* Ethosuximide intoxication
ethyl alcohol, *see* Alcohol intoxication
ethyl-*N*-diethyl phosphoramidocyanidate, *see* Ethyl-*N*-diethyl phosphoramidocyanidate intoxication
O-ethyl *S*-(2-dimethylaminoethyl) methylphosphonothioate, *see* *O*-Ethyl *S*-(2-dimethylaminoethyl) methylphosphonothioate intoxication
ethyl-*N*-dimethyl phosphoramidocyanidate, *see* Tabun intoxication
ethyl-*N*-dimethyl phosphoramidofluoridate, *see* Ethyl-*N*-dimethyl phosphoramidofluoridate intoxication
ethyl-*N*-dimethyl phosphoramidothiocyanidate, *see* Ethyl-*N*-dimethyl phosphoramidothiocyanidate intoxication
ethylene glycol, *see* Ethylene glycol intoxication
O-ethyl *S*-(2-ethylthioethyl) ethylphosphonothioate, *see* *O*-Ethyl *S*-(2-ethylthioethyl) ethylphosphonothioate intoxication
ethyl-4-nitrophenyl ethylphosphonate, *see* Ethyl-4-nitrophenyl ethylphosphonate intoxication
felbamate, *see* Felbamate intoxication
fentanyl, *see* Fentanyl intoxication
filix mas, *see* Filix mas intoxication
fish, *see* Fish intoxication
fish roe, *see* Fish roe intoxication
flecainide, *see* Flecainide intoxication
fludarabine, *see* Fludarabine intoxication
fluoride, *see* Fluoride intoxication
fluorouracil, *see* Fluorouracil intoxication
fluoxetine, *see* Fluoxetine intoxication
fluroxene, *see* Fluroxene intoxication
fluvoxamine, *see* Fluvoxamine intoxication
food, *see* Food intoxication
forstenon, *see* Forstenon intoxication
fox, *see* Fox intoxication
gabapentin, *see* Gabapentin intoxication
gasoline, *see* Gasoline intoxication
Gastropoda, *see* Conidae intoxication
gentamicin, *see* Gentamicin intoxication
globefish, *see* Puffer fish intoxication
glutethimide, *see* Glutethimide intoxication
glyceryl trinitrate, *see* Glyceryl trinitrate intoxication
glycoside, *see* Glycoside intoxication
glycyrrhiza, *see* Glycyrrhiza intoxication
gnathanacanthidae, *see* Gnathanacanthidae intoxication
gnathanacanthus, *see* Gnathanacanthus intoxication
goatfish, *see* Goatfish intoxication
gold, *see* Gold intoxication
green seaweed, *see* Green seaweed intoxication
grouper, *see* Grouper intoxication
guanethidine, *see* Guanethidine intoxication
gymnapistes, *see* Gymnapistes intoxication
gymnuridae, *see* Gymnuridae intoxication
gyromitrine, *see* Gyromitrine intoxication
halichoerus grypus, *see* Halichoerus grypus intoxication
haliotis, *see* Haliotis intoxication
hallucinatory fish, *see* Hallucinatory fish intoxication
hallucinogenic agent, *see* Hallucinogenic agent intoxication
halothane, *see* Halothane intoxication
Hapalochlaena lunulata, *see* Hapalochlaena lunulata intoxication
hapalochlaena maculosa, *see* Hapalochlaena maculosa intoxication
hatahata, *see* Hatahata intoxication
hemichordata, *see* Hemichordata intoxication
heptachlor, *see* Heptachlor intoxication
herbicide, *see* Herbicide intoxication
heroin, *see* Diamorphine
heterocyclic antidepressant, *see* Heterocyclic antidepressant intoxication
heterodontus francisci, *see* Heterodontus francisci intoxication
heterodontus portusjacksoni, *see* Heterodontus portusjacksoni intoxication
hexacarbon, *see* Hexacarbon intoxication
hexachlorocyclohexane, *see* Hexachlorocyclohexane intoxication
hexachlorophene, *see* Hexachlorophene intoxication
hexane, *see* Hexane intoxication
2-hexanone, *see* 2-Hexanone intoxication
histopteridae, *see* Histopteridae intoxication
holothurian, *see* Holothurian intoxication

holothurioidea, *see* Holothurioidea intoxication
homidium bromide, *see* Homidium bromide intoxication
hornet, *see* Hymenoptera intoxication
horseshoe crab, *see* Xiphosura intoxication
hydantoin, *see* Hydantoin intoxication
hydralazine, *see* Hydralazine intoxication
hydrazine, *see* Hydrazine intoxication
hydrogen sulfide, *see* Hydrogen sulfide intoxication
hydroids, *see* Coelenterata intoxication
hydrolagus, *see* Hydrolagus intoxication
Hydrophiidae, *see* Hydrophiidae intoxication
hydrophis cyanocintus, *see* Hydrophis cyanocintus intoxication
hydrozoa, *see* Hydrozoa intoxication
hynotoxin, *see* Hypnotoxin intoxication
hyoscyamine, *see* Hyoscyamine intoxication
hypoglycin, *see* Hypoglycin intoxication
ichthyohaemotix fish, *see* Ichthyohaemotix fish intoxication
ichthyohepatoxic fish, *see* Ichthyohepatoxic fish intoxication
ichthyo-otoxism, *see* Fish roe intoxication
ifosfamide, *see* Ifosfamide intoxication
3,3'-iminodipropionitrile, *see* 3,3'-Iminodipropionitrile intoxication
iminostilbene, *see* Dibenzazepine intoxication
imipramine, *see* Imipramine intoxication
inimicus didactylus, *see* Inimicus didactylus intoxication
inorganic mercury, *see* Inorganic mercury intoxication
inorganic salt, *see* Inorganic salt intoxication
α-interferon, *see* α-Interferon intoxication
β-interferon, *see* β-Interferon intoxication
γ-interferon, *see* γ-Interferon intoxication
interleukin-2, *see* Interleukin-2 intoxication
iodine, *see* Iodine intoxication
ipecac, *see* Ipecac intoxication
iproniazid, *see* Iproniazid intoxication
isobenz, *see* Isobenz intoxication
O-isobutyl *O*-carboxymethyl ethylphosphonothioate, *see* *O*-Isobutyl *O*-carboxymethyl ethylphosphonothioate intoxication
isodemeton, *see* Demeton-S intoxication
isoflurane, *see* Isoflurane intoxication
isoniazid, *see* Isoniazid intoxication
Iso-Ompa, *see* Iso-Ompa intoxication
isopropyl-N-dimethyl phosphoramidocyanidate, *see* Isopropyl-N-dimethyl phosphoramidocyanidate intoxication
isopropylmethylphosphonofluoridate, *see* Sarin intoxication
Iso-Systox sulfoxide, *see* Iso-Systox sulfoxide intoxication
iwashi-kujira, *see* Balaenoptera borealis intoxication
jackfish, *see* Jackfish intoxication
Japanese sandfish, *see* Japanese sandfish intoxication
jellyfish, *see* Scyphozoa intoxication
kaguo-kuzira, *see* Balaenoptera borealis intoxication
kainic acid, *see* Kainic acid intoxication
kanamycin, *see* Kanamycin intoxication
kepone, *see* Chlordecone intoxication
kerosine, *see* Kerosine intoxication
ketamine, *see* Ketamine intoxication
krait, *see* Krait intoxication
lamotrigine, *see* Lamotrigine intoxication
lapemis hardwicki, *see* Lapemis hardwicki intoxication
lapemis semifasciata, *see* Lapemis semifasciata intoxication
Lathyrus sativus, *see* Lathyrus sativus intoxication
α-latrotoxin, *see* α-Latrotoxin intoxication
lead, *see* Lead intoxication
leatherjacket, *see* Leatherjacket intoxication
levodopa, *see* Levodopa intoxication
lidocaine, *see* Lidocaine intoxication
lindane, *see* Lindane intoxication
lionfish, *see* Lionfish intoxication
lithium, *see* Lithium intoxication
livona, *see* Livona intoxication
Livona pica, *see* Livona pica intoxication
long chain alcohols, *see* Long chain alcohol intoxication
Loxosceles, *see* Loxosceles intoxication
lumbriconereis heteropoda, *see* Lumbriconereis heteropoda intoxication
lupin seed, *see* Lupin seed intoxication
lyngbya majuscula, *see* Lyngbya majuscula intoxication
lysergide, *see* Lysergide intoxication
mackeral, *see* Mackeral intoxication
magnesium, *see* Magnesium intoxication
malacostraca, *see* Malacostraca intoxication
malathion, *see* Malathion intoxication
mamba, *see* Mamba intoxication
mammalia, *see* Mammalia intoxication
manganese, *see* Manganese intoxication
manta ray, *see* Manta ray intoxication
MAO inhibitor, *see* MAO inhibitor intoxication
maprotiline, *see* Maprotiline intoxication

marine toxin, *see* Marine toxin intoxication
maternal drug use, *see* Maternal drug intoxication
meat, *see* Meat intoxication
mebaral, *see* Methylphenobarbital intoxication
meduso-congestin, *see* Meduso-congestin intoxication
mefloquine, *see* Mefloquine intoxication
meiacanthus, *see* Meiacanthus intoxication
mephenytoin, *see* Mephenytoin intoxication
mephobarbital, *see* Methylphenobarbital intoxication
mepivacaine, *see* Mepivacaine intoxication
meprobamate, *see* Meprobamate intoxication
mercury, *see* Mercury intoxication
merostomata, *see* Merostomata intoxication
mescaline, *see* Mescaline intoxication
mesuximide, *see* Mesuximide intoxication
metal, *see* Metal intoxication
metal fume, *see* Metal fume intoxication
methamphetamine, *see* Methamphetamine intoxication
methanol, *see* Methanol intoxication
metharbital, *see* Metharbital intoxication
methohexital, *see* Methohexital intoxication
methohexitone, *see* Methohexital intoxication
methotrexate, *see* Methotrexate intoxication
methoxychlor, *see* Methoxychlor intoxication
3-methoxy-4,5-methylenedioxyamphetamine, *see* 3-Methoxy-4,5-methylenedioxyamphetamine intoxication
methyl alcohol, *see* Methanol intoxication
methylazoxymethanol glucoside, *see* Methylazoxymethanol intoxication
methyl bromide, *see* Methyl bromide intoxication
methyl chloride, *see* Methyl chloride intoxication
4-methyl-2,5-dimethoxyamphetamine, *see* 4-Methyl-2,5-dimethoxyamphetamine intoxication
methyl-*N*-dimethyl phosphoramidocyanidate, *see* Methyl-*N*-dimethyl phosphoramidocyanidate intoxication
methyldopa, *see* Methyldopa intoxication
3,4-methylenedioxyamphetamine, *see* 3,4-Methylenedioxyamphetamine intoxication
3,4-methylenedioxyethamphetamine, *see* 3,4-Methylenedioxyethamphetamine intoxication
3,4-methylenedioxymethamphetamine, *see* 3,4-Methylenedioxymethamphetamine intoxication
methylenedioxyphenyl, *see* Methylenedioxyphenyl intoxication
α-methylfentanyl, *see* α-Methylfentanyl intoxication
methyl iodide, *see* Methyl iodide intoxication
methylmercury, *see* Minamata disease
methylphenobarbital, *see* Methylphenobarbital intoxication
1-methyl-4-phenyl-1,2,3,6-tetrahydropyridine, *see* 1-Methyl-4-phenyl-1,2,3,6-tetrahydropyridine intoxication *and* 1,2,3,6-Tetrahydroxy-1-methyl-4-phenylpyridine intoxication
methyl salicylate, *see* Methyl salicylate intoxication
methylsystox sulfone, *see* Methylsystox sulfone intoxication
metronidazole, *see* Metronidazole intoxication
mexiletine, *see* Mexiletine intoxication
mianserin, *see* Mianserin intoxication
mirex, *see* Mirex intoxication
misonidazole, *see* Misonidazole intoxication
mitotane, *see* Mitotane intoxication
mobulidae, *see* Mobulidae intoxication
mollusca, *see* Mollusca intoxication
molluscs, *see* Molluscs intoxication
monensin, *see* Monensin intoxication
moray, *see* Moray intoxication
morphine, *see* Morphine intoxication
mother milk, *see* Subacute necrotizing encephalomyelopathy
murex, *see* Murex intoxication
muscarine, *see* Muscarine intoxication
muscimol, *see* Muscimol intoxication
mushroom, *see* Mushroom intoxication
mussels, *see* Paralytic shellfish poisoning
myliobatidae, *see* Myliobatidae intoxication
nalidixic acid, *see* Nalidixic acid intoxication
nalline, *see* Nalorphine
naphthalene, *see* Naphthalene intoxication
neophocaena phocaenoides, *see* Neophocaena phocaenoides intoxication
Neosebastes nigropunctatus McCulloch, *see* Neosebastes nigropunctatus McCulloch intoxication
neptunea, *see* Neptunea intoxication
nereistoxin, *see* Nereistoxin intoxication
neuroleptic agent, *see* Neuroleptic agent intoxication
nicotine, *see* Nicotine intoxication
nifedipine, *see* Nifedipine intoxication
nimodipine, *see* Nimodipine intoxication
nitrate, *see* Nitrate intoxication
nitrite, *see* Nitrite intoxication
nitrofurantoin, *see* Nitrofurantoin intoxication
nitrogen oxide, *see* Nitrogen oxide intoxication

3-nitropropionic acid, *see* 3-Nitropropionic acid intoxication
nonsteroid anti-inflammatory agent, *see* Nonsteroid anti-inflammatory agent intoxication
norfloxacin, *see* Norfloxacin inoxication
norpethidine, *see* Norpethidine intoxication
notechis scutatus, *see* Notechis scutatus intoxication
nothesthes, *see* Nothesthes intoxication
octamethyl pyrophosphoramide, *see* Octamethyl pyrophosphoramide intoxication
octamethyl pyrophosphorotetramide, *see* Octamethyl pyrophosphoramide intoxication
octopoda, *see* Octopoda intoxication
octopus, *see* Octopus intoxication
OKT3, *see* OKT3 intoxication
OMPA, *see* Octamethyl pyrophosphoramide intoxication
opiate, *see* Opiate intoxication
opsanus, *see* Opsanus intoxication
orellanine, *see* Orellanine intoxication
organic mercury, *see* Organic mercury intoxication
organic solvent, *see* Organic solvent intoxication
organic thiocyanate, *see* Organic thiocyanate intoxication
organochlorine insecticide, *see* Organochlorine insecticide intoxication
organochlorine pesticide, *see* Organochlorine pesticide intoxication
organolead, *see* Organolead intoxication
organophosphate, *see* Organophosphate intoxication
organophosphorus compound, *see* Organophosphorus compound intoxication
organotin, *see* Organotin intoxication
ornate butterfly cod, *see* Ornate butterfly cod intoxication
ostracion lentiginosus, *see* Ostracion lentiginosus intoxication
ostracion meleagris Shaw, *see* Ostracion lentiginosus intoxication
ostraciontidae, *see* Ostraciontidae intoxication
ostrea gigas, *see* Ostrea gigas intoxication
ouabain, *see* Ouabain intoxication
3-oxalylaminoalanine, *see* 3-Oxalylaminoalanine intoxication
oxazepam, *see* Oxazepam intoxication
oxazolidinedione, *see* Oxazolidinedione intoxication
oxcarbazepine, *see* Oxcarbazepine intoxication
oxygen, *see* Oxygen intoxication
pahutoxic, *see* Pahutoxic intoxication
palfium, *see* Dextromoramide
Palythoa tuberculosa, *see* Palythoa tuberculosa intoxication
palytoxin, *see* Palytoxin intoxication
papaverine, *see* Papaverine intoxication
paraldehyde, *see* Paraldehyde intoxication
paraquat, *see* Paraquat intoxication
parathion, *see* Parathion intoxication
pargos, *see* Pargos intoxication
paroxetine, *see* Paroxetine intoxication
pefloxacin, *see* Pefloxacin intoxication
Pelamis platurus, *see* Pelamis platurus intoxication
pelochelys bibroni, *see* Pelochelys bibroni intoxication
penicillamine, *see* Penicillamine intoxication
pentaborane, *see* Pentaborane intoxication
pentaceropsis, *see* Pentaceropsis intoxication
pentazocine, *see* Pentazocine intoxication
perhexiline, *see* Perhexiline intoxication
permethrin, *see* Permethrin intoxication
pethidine, *see* Pethidine intoxication
phanerotoxin, *see* Phanerotoxin intoxication
phenacemide, *see* Phenacemide intoxication
phencyclidine, *see* Phencyclidine intoxication
phenethylamine, *see* Phenethylamine intoxication
phenobarbital, *see* Phenobarbital intoxication
phenothiazine, *see* Phenothiazine intoxication
O-phenyl-*N*-2-dimethylaminoethyl-4-phenylphosphonamidate methiodide, *see* *O*-Phenyl-*N*-2-dimethylaminoethyl-4-phenylphosphonamidate methiodide intoxication
phenylmercury derivative, *see* Phenylmercury derivative intoxication
phenylpropanolamine, *see* Phenylpropanolamine intoxication
phenytoin, *see* Phenytoin intoxication
phocca groenlandica, *see* Phocca groenlandica intoxication
phorate sulfone, *see* Phorate sulfone intoxication
phospholine, *see* Ecothiopate iodide intoxication
phosphorus, *see* Phosphorus intoxication
phthalate ester, *see* Phthalate ester intoxication
physalia, *see* Physalia intoxication
physeter catodon, *see* Physeter catodon intoxication
phytanic acid, *see* Phytanic acid intoxication
pinacolyl methylphosphonofluoridate, *see* Soman intoxication
piperidinedione, *see* Piperidinedione intoxication
pisces, *see* Pisces intoxication
plasmocid, *see* Plasmocid intoxication
platinum, *see* Platinum intoxication
platypodia granulosa, *see* Platypodia granulosa intoxication

Plectonema tetebrans, *see* Plectonema tetebrans intoxication
plotosidae, *see* Plotosidae intoxication
Plotosus, *see* Plotosus intoxication
polar bear, *see* Polar bear intoxication
polychaeta, *see* Polychaeta intoxication
polychlorinated biphenyl, *see* Polychlorinated biphenyl intoxication
polychlorinated dibenzodioxin, *see* Polychlorinated dibenzodioxin intoxication
porifera, *see* Porifera intoxication
porpoise, *see* Porpoise intoxication
potamotrygonidae, *see* Potamotrygonidae intoxication
potassium cyanide, *see* Potassium cyanide intoxication
prazosin, *see* Prazosin intoxication
prilocaine, *see* Prilocaine intoxication
primidone, *see* Primidone intoxication
prionurus, *see* Prionurus intoxication
procainamide, *see* Procainamide intoxication
procarbazine, *see* Procarbazine intoxication
promedol, *see* Trimeperidine
propafenone, *see* Propafenone intoxication
propanidid, *see* Propanidid intoxication
propranolol, *see* Propranolol intoxication
Proteus morganii, *see* Proteus morganii intoxication
protochordata, *see* Tunicata intoxication
protozoa, *see* Protozoa intoxication
psilocin, *see* Psilocin intoxication
psilocybine, *see* Psilocybine intoxication
pterois volitans, *see* Pterois volitans intoxication
puffer fish, *see* Puffer fish intoxication
pyrazoxon, *see* Pyrazoxon intoxication
pyrethroid, *see* Pyrethroid intoxication
pyrethrum, *see* Pyrethrum intoxication
pyridoxine, *see* Vitamin B6 intoxication
N-3-pyridylmethyl-*N'*-*p*-nitrophenyl urea, *see* Vacor intoxication
quinidine, *see* Quinidine intoxication
quiniform, *see* Quiniform intoxication
quinine, *see* Quinine intoxication
quinolone, *see* Quinolone intoxication
rauwolfia alkaloid, *see* Rauwolfia alkaloid intoxication
ray, *see* Ray intoxication
red fire fish, *see* Red fire fish intoxication
reserpine, *see* Reserpine intoxication
resitox, *see* Coumafos intoxication
rhinopteridae, *see* Rhinopteridae intoxication
rhizostomeae, *see* Rhizostomeae intoxication
rhodactis howesi, *see* Rhodactis howesi intoxication
river ray, *see* River ray intoxication
Ro 3-0658, *see* Diethyl trichloromethylphosphonate intoxication
rotenone, *see* Rotenone intoxication
round sting ray, *see* Round sting ray intoxication
rudderfish, *see* Rudderfish intoxication
ryania, *see* Ryania intoxication
salicylic acid, *see* Salicylic acid intoxication
salvarsan, *see* Arsphenamine intoxication
sarin, *see* Sarin intoxication
saxitoxin, *see* Saxitoxin intoxication
scatophagidae, *see* Scatophagidae intoxication
scombroid, *see* Scombroid intoxication
scopolamine, *see* Scopolamine intoxication
scorpaena, *see* Scorpaena intoxication
scorpaenidae, *see* Scorpaenidae intoxication
scorpaenodes, *see* Scorpaenodes intoxication
scorpaenopsis, *see* Scorpaenopsis intoxication
scorpion sting, *see* Scorpion sting intoxication
scyphozoa, *see* Scyphozoa intoxication
sea anemone, *see* Sea anemone intoxication
sea bass, *see* Sea bass intoxication
sea cucumber, *see* Sea cucumber intoxication
sea lillie, *see* Sea lillie intoxication
sea slug, *see* Sea slug intoxication
sea snake, *see* Sea snake intoxication
sea squirt, *see* Sea squirt intoxication
sea star, *see* Sea star intoxication
sea urchin, *see* Sea urchin intoxication
sea wasp, *see* Sea wasp intoxication
sebastapistes, *see* Sebastapistes intoxication
sebastapistes mcadamsi, *see* Sebastapistes mcadamsi intoxication
sei whale, *see* Balaenoptera borealis intoxication
seiwal, *see* Balaenoptera borealis intoxication
selective serotonin reuptake inhibitor, *see* Selective serotonin reuptake inhibitor intoxication
selenotoca, *see* Selenotoca intoxication
semaeostomeae, *see* Semaeostomeae intoxication
serax, *see* Oxazepam intoxication
serranid, *see* Serranid intoxication
sertraline, *see* Sertraline intoxication
shark, *see* Shark intoxication
shellfish, *see* Paralytic shellfish poisoning
short chain alkylmercury, *see* Short chain alkylmercury intoxication
siganidae, *see* Siganidae intoxication
siganus, *see* Siganus intoxication
siluridae, *see* Siluridae intoxication
silurus, *see* Silurus intoxication
snapper, *see* Snapper intoxication

sodium azide, *see* Sodium azide intoxication
sodium nitrate, *see* Sodium nitrate intoxication
sodium nitrite, *see* Sodium nitrite intoxication
sodium nitroprusside, *see* Sodium nitroprusside intoxication
sodium salicylate, *see* Sodium salicylate intoxication
solaster papposus, *see* Solaster papposus intoxication
soman, *see* Soman intoxication
sparfosic acid, *see* Sparfosic acid intoxication
sparid, *see* Sparid intoxication
Sphaeroides hamiltoni, *see* Puffer fish intoxication
spider bite, *see* Spider bite intoxication
spirogermanium, *see* Spirogermanium intoxication
spiromustine, *see* Spiromustine intoxication
sponge, *see* Porifera intoxication
squaliformes, *see* Squaliformes intoxication
squalus acanthias, *see* Squalus acanthias intoxication
squamata, *see* Squamata intoxication
squill, *see* Squill intoxication
starfish, *see* Starfish intoxication
stargazer, *see* Stargazer intoxication
stingray, *see* Stingray intoxication
stonefish, *see* Stonefish intoxication
streptomycin, *see* Streptomycin intoxication
strychnine, *see* Strychnine intoxication
sturgeon fish, *see* Sturgeon fish intoxication
sturgeon roe, *see* Sturgeon roe intoxication
styrene, *see* Styrene intoxication
succinimide, *see* Succinimide intoxication
succinylcholine, *see* Succinylcholine intoxication
sunfish, *see* Sunfish intoxication
suramin, *see* Suramin intoxication
swellfish, *see* Puffer fish intoxication
synanceja, *see* Synanceja intoxication
Synanceja trachynis Richardson, *see* Synanceja trachynis Richardson intoxication
tabernathe iboga, *see* Tabernathe iboga intoxication
tabun, *see* Tabun intoxication
tachypleus gigas, *see* Tachypleus gigas intoxication
tachypleus tridentatus, *see* Tachypleus tridentatus intoxication
tamoxifen, *see* Tamoxifen intoxication
tamoya, *see* Tamoya intoxication
taricha, *see* Taricha egg
taxol, *see* Taxol intoxication
tellurium, *see* Tellurium intoxication
TEPP, *see* Tetraethyl pyrophosphate intoxication
terrapin, *see* Terrapin intoxication
tetanospasmin, *see* Tetanospasmin intoxication
tetracaine, *see* Tetracaine intoxication
tetraethyllead, *see* Tetraethyllead intoxication
tetraethylphosphorodiamidic fluoride, *see* Tetraethylphosphorodiamidic fluoride intoxication
tetraethyl pyrophosphate, *see* Tetraethyl pyrophosphate intoxication
tetrahydrocannabinol, *see* Tetrahydrocannabinol intoxication
tetraisopropyl dithionopyrophosphate, *see* Tetraisopropyl dithionopyrophosphate intoxication
tetraisopropyl pyrophosphate, *see* Tetraisopropyl pyrophosphate intoxication
tetramethyllead, *see* Tetramethyllead intoxication
tetramethyl pyrophosphate, *see* Tetramethyl pyrophosphate intoxication
tetramonoisopropyl pyrophosphortetramide, *see* Iso-Ompa intoxication
tetraodontidae, *see* Tetraodontidae intoxication
tetraodontiformes, *see* Tetraodontiformes intoxication
thalidomide, *see* Thalidomide intoxication
thallium, *see* Thallium intoxication
tharlarctos maritimus, *see* Tharlarctos maritimus intoxication
thassalophryne, *see* Thassalophryne intoxication
thebaine, *see* Thebaine intoxication
thimet sulfone, *see* Phorate sulfone intoxication
thiodan, *see* Endosulfan intoxication
thiopental, *see* Thiopental intoxication
thiophene, *see* Thiophene intoxication
thioridazine, *see* Thioridazine intoxication
thiotepa, *see* Thiotepa intoxication
tick bite, *see* Tick bite intoxication
tick venom, *see* Tick venom intoxication
TIPP, *see* Tetraisopropyl pyrophosphate intoxication
tityustoxin, *see* Tityustoxin intoxication
toadfish, *see* Puffer fish intoxication
toados, *see* Puffer fish intoxication
tocainide, *see* Tocainide intoxication
toluene, *see* Toluene intoxication
toxaphene, *see* Toxaphene intoxication
toxic oil, *see* Toxic oil syndrome
Toxopneustes pileolus, *see* Toxopneustes pileolus intoxication
trachinidae, *see* Trachinidae intoxication
trachinus, *see* Trachinus intoxication
trazodone, *see* Trazodone intoxication
trepang, *see* Trepang intoxication

trichloroethanol, *see* Trichloroethanol intoxication
trichloroethylene, *see* Trichloroethylene intoxication
trichodontid, *see* Trichodontid intoxication
tri-*o*-cresyl phosphate, *see* Tri-*o*-cresyl phosphate intoxication
tricyclic antidepressant, *see* Tricyclic antidepressant intoxication
tridacna maxima, *see* Tridacna maxima intoxication
triethyltin, *see* Triethyltin intoxication
triglidae, *see* Triglidae intoxication
trimethadione, *see* Trimethadione intoxication
3,4,5-trimethoxyamphetamine, *see* 3,4,5-Trimethoxyamphetamine intoxication
2-trimethylammoniumethyl methylphosphonofluoridate iodide, *see* 2-Trimethylammoniumethyl methylphosphonofluoridate iodide intoxication
2-trimethylammonium-1-methylethyl methylphosphonofluoridate iodide, *see* 2-Trimethylammonium-1-methylethyl methylphosphonofluoridate iodide intoxication
3-trimethylammoniumpropyl methylphosphonofluoridate iodide, *see* 3-Trimethylammoniumpropyl methylphosphonofluoridate iodide intoxication
triperinol, *see* Triperinol intoxication
Tripneustes gratilla, *see* Tripneustes gratilla intoxication
tryptamine derivative, *see* Tryptamine derivative intoxication
tsukubaenolide, *see* Tsukubaenolide intoxication
tumor necrosis factor, *see* Tumor necrosis factor intoxication
tuna, *see* Tuna intoxication
tunicata, *see* Tunicata intoxication
turtle, *see* Turtle intoxication
uranoscopidae, *see* Uranoscopidae intoxication
uranoscopus, *see* Uranoscopus intoxication
Uranoscopus japonicus Houttyn, *see* Uranoscopus japonicus Houttyn intoxication
urilophidae, *see* Urilophidae intoxication
urochordata, *see* Tunicata intoxication
vacor, *see* Vacor intoxication
valproic acid, *see* Valproic acid intoxication
venerupis semidecussata, *see* Venerupis semidecussata intoxication
venomous fish, *see* Venomous fish intoxication
vertebrata, *see* Vertebrata intoxication
vidarabine, *see* Vidarabine intoxication
vigabatrin, *see* Vigabatrin intoxication
vinblastine, *see* Vinblastine intoxication
vincristine, *see* Vincristine intoxication
vindesine, *see* Vindesine intoxication
vinyl ether, *see* Vinyl ether intoxication
vitamin A, *see* Vitamin A intoxication
vitamin B1, *see* Vitamin B1 intoxication
vitamin B2, *see* Vitamin B2 intoxication
vitamin B3, *see* Vitamin B3 intoxication
vitamin B6, *see* Vitamin B6 intoxication
vitamin B12, *see* Vitamin B12 intoxication
vitamin Bc, *see* Vitamin Bc intoxication
vitamin E, *see* Vitamin E intoxication
water, *see* Water intoxication
whale, *see* Whale intoxication
whip ray, *see* Whip ray intoxication
wolf, *see* Wolf intoxication
wood alcohol, *see* Methanol intoxication
xanthodes reynaudi, *see* Xanthodes reynaudi intoxication
xiphosura, *see* Xiphosura intoxication
yellow jacket, *see* Yellow jacket intoxication
yellow star thistle, *see* Yellow star thistle intoxication
zinc, *see* Zinc intoxication
zozymus aeneus, *see* Zozymus aeneus intoxication
Intra-aortic balloon
femoral neuropathy, 63/178
iatrogenic neurological disease, 63/47, 177
myelomalacia, 63/178
sciatic neuropathy, 63/178
Intrablood-brain barrier IgG synthesis
see also Intrathecal IgG synthesis
CSF, *see* CSF intrablood-brain barrier IgG synthesis
CSF leukocyte count, 47/84
multiple sclerosis, 47/84, 97-101, 106
Intracerebellar hematoma
clinical diagnosis, 24/359
EEG, 11/280
posterior fossa hematoma, 24/346, 57/253
Intracerebellar hemorrhage
computer assisted tomography, 54/68
prevalence, 54/68
Intracerebral calcification, 42/534-536
see also Striatopallidodentate calcification
brain aneurysm, 49/417
brain angioma, 49/417
brain arteriovenous malformation, 54/93
calcium metabolism, 42/534
carbon monoxide intoxication, 49/417
cavernous angioma, 42/724
cerebro-oculofacioskeletal syndrome, 43/341
Cockayne-Neill-Dingwall disease, 43/350, 51/399
congenital toxoplasmosis, 52/359

corticostriatopallidodentate, *see* Striatopallidodentate calcification
endocrine candidiasis syndrome, 42/642
epidural hematoma, 49/417
epileptic seizure, 42/534
Fahr disease, *see* Striatopallidodentate calcification
Friedman-Roy syndrome, 43/251
Hallervorden-Spatz syndrome, 49/417
hypoparathyroidism, 42/577
intracerebral hematoma, 49/417
lead intoxication, 49/417
mental deficiency, 42/534
nuclear magnetic resonance, 54/93
oculodento-osseous dysplasia, 43/445
palilalia, 43/225
parathyroid hormone metabolism, 42/534
phakomatosis, 49/417
phosphorus, 42/534
pinealoma, 18/362, 42/767
poliodystrophy, 42/588
pseudohypoparathyroidism, 42/534, 621
pseudopseudohypoparathyroidism, 42/534
psychiatric disorder, 42/534
striatopallidodentate calcification, 6/703, 42/534
Sturge-Weber syndrome, 31/18-20, 43/47
subdural hematoma, 49/417
toxoplasmosis, 42/662, 50/277, 52/359
tuberous sclerosis, 43/50
Urbach-Wiethe disease, 42/595
Intracerebral hematoma, 11/660-697
see also Brain hemorrhage *and* Intracranial hematoma
acute type, 11/681
age, 11/680
air study, 24/360
aneurysm, 11/666
angiography, 11/690, 24/360
angioma, 11/668
anticoagulant, 24/365
arterial hypertension, 54/203
arteriolosclerosis, 11/671
arteriosclerosis, 11/671
arteritis, 11/672
basal ganglion, 11/671
basal ganglion hematoma, 57/49
bleeding tendency, 11/673
brain amyloid angiopathy, 54/337-340
brain aneurysm neurosurgery, 55/54
brain arteriovenous malformation, 54/382
brain blood flow, 54/147
brain infarction, 11/310-312
brain injury, 24/351
brain scintigraphy, 24/362
brain thrombophlebitis, 11/671
burr hole, 24/363
cause, 11/662
cavity, 54/89
chronic type, 11/684
classification, 11/680
clinical features, 11/681
computer assisted tomography, 24/362, 54/72
course, 11/682
craniotomy, 24/363
CSF, 11/684
delayed hematoma, 57/156
diagnosis, 24/359
differential diagnosis, 24/362
echoEG, 11/685, 24/361
EEG, 11/280, 686, 24/362
electrocorticography, 11/688
epilepsy, 24/360
familial cortical arteriolosclerosis, 11/672
football injury, 23/574
gunshot injury, 57/303
head injury, 11/663, 57/48, 156
hemiplegia, 1/192
history, 11/660
incapsulation, 11/678
intracerebral calcification, 49/417
localization, 11/679
lucid interval, 24/357
missile injury, 23/523
mortality, 24/365
neurosurgery, 12/211
nuclear magnetic resonance, 54/72, 89
onset, 24/357
outcome, 57/282
pallidal necrosis, 49/465
pallidonigral necrosis, 49/465
pallidostriatal necrosis, 49/465
papilledema, 24/360
pathogenesis, 11/676
pathology, 23/51, 57/48
PEG, 11/688
personal series, 24/365
polyarteritis nodosa, 11/672, 39/297
prevalence, 11/662
prognosis, 11/694-696, 24/364
pupil, 24/360
radiodiagnosis, 24/360, 57/156
recuperation period, 11/682
result, 24/364
sex ratio, 11/679
sign, 24/360
site, 24/358

subacute type, 11/682
subarachnoid hemorrhage, 12/70, 167, 55/54
surgery, 24/363, 656
thromboangiitis obliterans, 11/672
traumatic, *see* Traumatic intracerebral hematoma
traumatic aneurysm, 24/393
traumatic intracranial hematoma, 57/156, 253
traumatic psychosis, 24/532
treatment, 11/690, 24/363
tumor, 11/664
unknown origin, 11/674
vascular striatopallidodentate calcification, 49/417
Intracerebral hemorrhage, *see* Brain hemorrhage
Intracerebral metastasis, *see* Brain metastasis
Intracranial abscess
epilepsy, 72/158
Intracranial aneurysm
epidemiology, 30/160
genetics, 30/160
headache, 48/277
polycystic kidney, 14/519
Intracranial arterial occlusion
multiple, *see* Multiple intracranial arterial occlusion
Intracranial calcification
congenital toxoplasmosis, 35/131-134
cryptococcosis, 35/477
cytomegalic inclusion body disease, 56/110, 268
cytomegalovirus, 56/268
epilepsy, 15/537
hypocalcemia, 28/536
multiple nevoid basal cell carcinoma syndrome, 14/80, 457, 18/172
Sturge-Weber syndrome, 14/227
toxoplasmosis, 35/131-134
tuber histology, 14/364-366
tuberculous meningitis, 52/207
tuberous sclerosis, 14/356, 379
Urbach-Wiethe disease, 14/777, 38/5
Intracranial epidural hematoma
electrical injury, 61/192
Intracranial hematoma
see also Brain hemorrhage, Epidural hematoma, Intracerebral hematoma *and* Subdural hematoma
brain aneurysm neurosurgery, 55/54
brain concussion, 23/438
brain injury, 57/124
brain stab wound, 23/482
clinical features, 57/139
depressed skull fracture, 23/409
echoEG, 23/271, 274, 277
EEG, 23/344, 350
epilepsy, 72/211
intracranial pressure, 57/262
missile injury, 23/523
posttraumatic, *see* Posttraumatic intracranial hematoma
posttraumatic epilepsy, 24/448, 451
posttraumatic headache, 24/507
subarachnoid hemorrhage, 55/54
traumatic, *see* Traumatic intracranial hematoma
Intracranial hemorrhage
see also Brain hemorrhage
action myoclonus, 53/211
age, 53/200
akinetic mutism, 53/209
amnesia, 53/206
aphasia, 53/214
arteriosclerosis, 53/200
asterixis, 53/211
ataxia, 53/214
auditory symptom, 53/207
autonomic dysfunction, 53/207
Behr disease, 53/213
blepharospasm, 53/211
blindness, 53/207
blood pressure, 53/207
brain angiography, 53/228
brain arteriovenous malformation, 31/179-181
brain scintigraphy, 53/232
brain stem auditory evoked potential, 53/233
brain tumor, 69/422
Canadian Neurologic Scale, 53/209
cardiac disorder, 53/200
carotid bruit, 53/216
carotid sinus massage, 53/219
cause, 53/201
chronic vegetative state, 53/209
climate, 53/200
clinical course, 53/202
clinical evaluation, 53/199
cocaine intoxication, 65/255
coma, 53/208
computer assisted tomography, 53/220
conduction aphasia, 53/215
confusion, 53/206
conjugate gaze deviation, 53/212
consciousness loss, 53/206
correct diagnosis, 53/199
CSF, 53/224
CSF spectrophotometry, 53/224
decerebrate rigidity, 53/211
decortication, 53/211
diagnostic error, 53/220
differential diagnosis, 53/233

diplopia, 53/207
directional Doppler velocimetry, 53/225
duplex sonography, 53/225
duration, 53/203
dysphagia, 53/207
dysphonia, 53/207
dystonia, 53/211
ECG, 53/230
echoCG, 53/230
echoEG, 53/233
EEG, 53/232
emotional disorder, 53/206
eyelid nystagmus, 53/213
genetic predisposition, 53/199
Glasgow Coma Scale, 53/209
global aphasia, 53/215
hallucination, 53/206
head injury, 57/47
headache, 48/32
hematotachygraphy, 53/225
hemianopia, 53/207
hemiballismus, 53/211
hemichorea, 53/211
hemiparesis, 53/206
hemophilia, 7/665, 38/55-60
history, 53/199
Holter monitoring, 53/230
Horner syndrome, 53/213
hyperthermia, 53/207
hypoprothrombinemia, 38/75-77
iatrogenic neurological disease, 54/38
infantile cerebrovascular disease, 54/35
internuclear ophthalmoplegia, 53/212
intravenous digital subtraction angiography, 53/227
libido, 53/207
loxosceles intoxication, 37/112
mental function, 53/214
micturition, 53/207
missile injury, 23/521
monocular vision loss, 53/207
muscle tone, 53/211
nuchal rigidity, 53/207
nuclear magnetic resonance, 53/231, 54/88
nystagmus, 53/213
ocular bobbing, 53/213
oculoplethysmography, 53/225
oculopneumoplethysmography, 53/225
onset, 53/202
ophthalmodynamometry, 53/219
ophthalmoscopy, 53/218
oscillopsia, 53/207
palatal myoclonus, 53/211
Parinaud syndrome, 53/212
pathologic bruit, 53/217
phonoangiography, 53/218
physiologic bruit, 53/217
pontine hematoma, 53/207
positron emission tomography, 53/232
precipitating factor, 53/202
pruritus, 53/207
pseudoabducens nerve palsy, 53/213
race, 53/200
recovery, 53/203
respiration, 53/208
retinal artery embolism, 53/218
retraction nystagmus, 53/212
risk factor, 53/200
seizure, 53/206
sex ratio, 53/200
sexual impotence, 53/207
sickle cell anemia, 63/259
skew deviation, 53/213
snake envenomation, 65/179
speech disorder, 53/206
subclavian bruit, 53/216
systemic disease, 53/200
thermoregulation, 53/207
thrombocytopenia, 38/65-68
tinnitus, 53/207
transcortical aphasia, 53/215
traumatic, 23/49
traumatic carotid cavernous sinus fistula, 57/347
ultrasound diagnosis, 54/35
unilateral pulsating tinnitus, 53/207
vascular pulsation, 53/215
vertical gaze palsy, 53/212
vertical nystagmus, 53/213
vertical one and half syndrome, 53/212
vertigo, 53/207
visual evoked response, 53/232
visual hallucination, 53/207
visual manifestation, 53/207
vomiting, 53/206
xanthochromia, 53/224

Intracranial hypertension
see also CSF hypertension *and* Intracranial pressure
abducent nerve injury, 24/69, 86
acquired toxoplasmosis, 52/354
African trypanosomiasis, 52/342
anatomy, 16/95
angiostrongyliasis, 52/553
artificial brain tumor, 17/1
bacterial meningitis, 52/13
barbiturate, 55/215

Behçet syndrome, 71/224
benign, *see* Benign intracranial hypertension
blood flow disorder, 16/123
brain aneurysm neurosurgery, 55/55, 59
brain blood flow, 16/138, 54/147
brain compliance, 55/206
brain death, 24/742, 758, 760, 766
brain edema, 23/134, 137, 145, 63/417
brain hypoxia, 63/417
brain infarction, 53/159
brain injury, 23/43, 53, 145
brain ischemia, 53/159, 55/206
brain perfusion pressure, 53/159
brain tumor, 67/145
brain tumor headache, 5/173, 16/129
cause, 24/609
central herniation, 67/145
central pontine myelinolysis, 28/310
cerebellar herniation, 16/135
cerebellar tumor, 17/710
cerebellopontine angle syndrome, 2/93-97
child, 67/145
circadian rhythm, 55/219
clinical features, 16/126
coccidioidomycosis, 52/413
consciousness, 16/130, 136
continuous recording, 24/620
corpus callosum tumor, 17/509
corticosteroid, 24/618, 55/216
cranial subdural empyema, 52/170
craniotomy, 55/216
cryptococcosis, 52/431
CSF drainage, 55/216
CSF flow, 16/125
Cushing reflex, 16/137, 63/230
decerebration, 16/131, 134
dizziness, 16/129
echinococcosis, 52/524
EEG, 55/219
encephalitis, 55/215
epilepsy, 16/103
eye movement, 16/132
filariasis, 35/167, 170, 52/515
gastrointestinal disease, 16/140
glycosaminoglycanosis type I, 60/175
glycosaminoglycanosis type IH, 60/175
glycosaminoglycanosis type II, 60/175
glycosaminoglycanosis type III, 60/175
glycosaminoglycanosis type VII, 60/175
gram-negative bacillary meningitis, 52/119
granulomatous amebic encephalitis, 52/317
granulomatous CNS vasculitis, 55/390
head injury, 24/608, 55/215
head position, 55/206
headache, 5/172, 16/128
hemiplegia, 16/134
herniation, 16/95, 97
history, 16/90
hydrocephalus, 67/144
hyperventilation, 24/611
hypocalcemia, 63/558, 561
hypoparathyroidism, 70/124
hypothalamic dysfunction, 16/134
hypothermia, 24/618
isoelectric EEG, 55/219
laminaria, 17/1
lead encephalopathy, 64/436
lead intoxication, 64/436
mannitol, 24/616
meningococcal meningitis, 52/23, 25
mountain sickness, 63/417
nausea, 16/129
neurobrucellosis, 33/312, 52/587
neurocysticercosis, 35/300-302, 52/530, 533
neurogenic pulmonary edema, 63/441, 495
neurologic intensive care, 55/204
nuchal rigidity, 16/136
oculomotor nerve, 16/124, 132
organolead intoxication, 64/133
organotin intoxication, 64/138
osmolality correction, 63/547
osmotic dehydration, 24/617
pallidal necrosis, 49/465
pallidonigral necrosis, 49/465
pallidostriatal necrosis, 49/465
papilledema, 16/133, 42/740
paracoccidioidomycosis, 35/532, 534, 52/458
paresthesia, 16/135
pathogenesis, 24/609
pathology, 23/43, 46, 48
pineal tumor, 17/655
pinealoma, 42/767
pituitary adenoma, 17/408
plateau wave, 24/609
posterior fossa hematoma, 24/346
poststroke brain edema, 55/216
posttraumatic headache, 5/182
pressure reduction, 24/616
pulmonary edema, 16/139
pulse rate, 16/137
pupillary reflex, 16/132
respiratory dysfunction, 16/136
respiratory encephalopathy, 63/414
review, 16/89-139
Reye syndrome, 55/215, 56/161, 65/116
schistosomiasis, 52/538

skull base, 67/146
striatopallidodentate calcification, 49/422
subarachnoid hemorrhage, 12/177, 55/55, 59
subclavian herniation, 67/145
supratentorial brain tumor, 18/321
symptom, 67/143
systemic lupus erythematosus, 16/220
thermoregulation, 16/135
transverse sinus thrombosis, 52/177
traumatic, 23/43, 46, 48
traumatic hydrocephalus, 24/244, 248
treatment, 12/177, 24/609, 616
trichinosis, 52/566
tuberculous meningitis, 52/202, 215
venous pressure, 55/206
venous sinus occlusion, 54/429
venous sinus thrombosis, 54/395
vestibulo-ocular disorder, 16/133
visual field, 16/133
vitamin A intoxication, 65/568
Intracranial hypotension
chronic subdural hematoma, 24/299
headache, 5/175
Intracranial infection
see also Brain abscess, Brain infection, Meningioma, Meningitis *and* Subdural empyema
abducent nerve injury, 24/86
anterior fossa fracture, 24/222, 225
antibiotic agent, 24/224, 228
delirium, 46/544
depressed skull fracture, 24/224, 227
EEG, 57/187
epidural abscess, 24/219
gunshot injury, 24/227
head injury, 24/215
headache, 48/31
high velocity injury, 24/227
hydrocephalus, 24/229
incidence, 24/227
low velocity injury, 24/227
middle fossa fracture, 24/224
missile injury, 24/227, 465
osteomyelitis, 24/216, 657
posttraumatic, 57/187
scalp injury, 24/215
sequela, 24/229
skull, 24/215
skull base fracture, 24/220
skull vault fracture, 24/218
subdural empyema, 24/219
subgaleal abscess, 24/216
systemic infection, 24/228
thrombophlebitis, 24/218
vein, 24/218
Intracranial meningoencephalocele
neuropathology, 31/121-124
Intracranial neoplasm
recurrent nonhereditary multiple cranial neuropathy, 51/570
Intracranial nervous hypertension
hydrocephalus, 30/539
Intracranial pressure
see also CSF hypertension *and* Intracranial hypertension
A wave, 55/2206
acid base balance, 55/208
acrocephalosyndactyly type I, 43/319
actinomycosis, 35/385
age, 57/68
airway related pressure, 55/207
Alexander disease, 42/484
altitude illness, 48/6
angiography, 23/210
apnea, 1/659
aqueduct stenosis, 50/310
Arnold-Chiari malformation type I, 42/16
arterial hypertension, 55/207
autoregulation, 55/208
B wave, 23/202
barbiturate, 57/233
basal cistern, 57/162
basilar impression, 42/17
benign intracranial hypertension, 67/108
blood flow, 55/205
blood pressure, 55/208
bradypnea, 1/660
brain blood flow, 11/124, 23/201, 205, 55/207, 57/211, 384
brain blood flow autoregulation, 55/208
brain death, 24/758, 760
brain edema, 16/189
brain herniation, 23/204
brain injury prognosis, 24/674
brain ischemia, 55/206, 57/68
brain metabolism, 23/205, 57/67
brain perfusion pressure, 57/66, 130, 300, 384
brain swelling, 57/58
brain tumor headache, 67/141
brain ventricle drainage, 23/213
C wave, 23/202, 57/219
calcium imbalance, 28/536
cavernous angioma, 42/724
cerebrovascular resistance, 55/207
Cheyne-Stokes respiration, 63/483
clinical features, 23/206, 57/133

compartmentalization, 55/206
compensation mechanism, 16/93
compliance, 57/219
computer assisted tomography, 57/160, 276
corpus callosum syndrome, 2/761
cracked pot sign, 18/325, 329
CSF flow, 16/91-95
Cushing reflex, 57/130
Dandy-Walker syndrome, 42/22
dehydrating agent, 23/212, 215
delayed change, 23/201
dextrose, 23/212
diffuse brain swelling, 57/58
diuretic agent, 23/215
elevated, 57/68, 160, 176, 300, 431
epidural hematoma, 57/260
episodic pulmonary edema, 43/62
evaluation, 57/133
exertional headache, 5/148, 48/374
fever, 74/454
fibromuscular dysplasia, 48/285
focal tissue pressure, 57/67
fontanelle tension, 57/132
Foster Kennedy syndrome, 2/53
fructose, 23/212
general features, 16/91
global pressure, 57/67
head circumference, 57/132
head flexion, 57/230
head injury, 57/49, 67, 114, 132, 160, 227
head injury outcome, 57/114, 379, 383
head position, 55/206
headache, 48/33, 283-285, 67/141
Hemophilus influenzae meningitis, 52/61
high dose barbiturate, 57/232
hydrocephalic dementia, 46/329
hydrocephalus, 30/547, 556-558
hyperbaric oxygenation, 23/215
hyperventilation, 23/214
hypocalcemia, 63/561
hypothermia, 23/215
immediate change, 23/200
incidence, 23/206
increased, *see* Increased intracranial pressure *and* Intracranial hypertension
intracranial hematoma, 57/262
intracranial hypertension correction, 57/231
isosorbide, 23/212
loss of consciousness, 23/207
management, 23/212, 216
mannitol, 23/202, 212, 57/227, 232
mass lesion, 57/276
missile injury, 57/300
monitoring, *see* Intracranial pressure monitoring
mortality, 57/68
multiple myeloma, 18/255
North American blastomycosis, 35/404, 52/389
nuclear magnetic resonance, 57/176
outcome, 57/68
oxygen, 23/215
pallidal necrosis, 49/470
pathology, 23/210
patient position, 57/230
penetrating head injury, 57/300
physiology, 55/205
pinealoma, 18/362
plateau wave, 23/202, 55/206, 57/218
porencephaly, 42/49
postoperative elevation, 57/279
posttraumatic headache, 24/507, 48/384
pressure gradient, 23/200
pressure volume curve, 57/50
pressure volume index, 55/206
prognosis, 23/211, 217, 24/674, 57/379
radiodiagnosis, 23/209, 57/160
remote change, 23/203
respiratory encephalopathy, 38/290, 298
septum pellucidum tumor, 2/769
shock wave, 57/300
speech test, 16/336
steroid, 23/212
subarachnoid hemorrhage, 55/8
subdural hygroma, 57/286
sucrose, 23/212
supracallosal hernia, 57/50
supratentorial brain abscess, 33/117
supratentorial gunshot injury, 57/300
surgery, 23/213
systemic lupus erythematosus, 39/286
telemetry, 23/216
third ventricle, 57/162
tonsillar herniation, 57/50, 134
transducer, 23/216
transtentorial herniation, 55/206, 57/50, 133
traumatic hydrocephalus, 24/239, 244, 248
traumatic intracranial hematoma, 57/258, 275, 279
treatment, 57/68
type, 23/200
unconsciousness, 57/263
uncontrollable, 57/379
urea, 23/212
vegetative nervous function, 23/208
venous pressure, 55/206
ventricular drainage, 57/227, 232
volume pressure index, 57/68
volume pressure relationship, 57/220

volume pressure response, 23/217
Intracranial pressure monitoring
A wave, 57/218
B wave, 57/218
C wave, 57/219
compliance, 57/219
complication, 57/223
data display, 57/224
epidural monitor, 57/221
fiberoptic probe, 57/222
hydrocephalic dementia, 46/329
increased intracranial pressure pattern, 57/218
infection, 57/223
intravascular consumption coagulopathy, 57/223
method, 23/215
neurologic examination, 57/224
noninvasive method, 57/220
parenchymal monitor, 57/221
pediatric head injury, 57/329
safety, 57/220
subarachnoid bolt, 57/221
technique, 57/220
Traube-Hering-Mayer wave, 57/219
ventricular catheter, 57/221
volume pressure response, 57/219
Intracranial subdural hematoma
CSF hypotension, 61/164
discography, 61/164
iatrogenic neurological disease, 61/164
lumbar puncture, 61/164
myelography, 61/164
spinal anesthesia, 61/164
Intracranial thrombophlebitis
headache, 48/287
Intracranial tumor
cough headache, 5/185
Maffucci-Kast syndrome, 43/29
multiple nevoid basal cell carcinoma syndrome, 43/31
neurofibroma, 14/136
neurofibromatosis type I, 14/151
SBLA syndrome, 42/769
Intracranial venous occlusion
headache, 48/283
Intractable hiccup
auricular vagus branch, 63/491
brain abscess, 63/490
brain stem tumor, 63/490
carbamazepine, 63/491
diaphragmatic flutter, 63/491
encephalitis, 63/490
epilepsy, 63/490
herpes zoster, 63/491
multiple sclerosis, 63/490
nystagmus, 63/491
sarcoidosis, 63/490
sex ratio, 63/491
temporal lobe abscess, 63/490
temporal lobe epilepsy, 63/490
thoracic zoster, 63/491
Intractable pain
lobotomy, 45/27
Intradural spinal arachnoid cyst, *see* Spinal arachnoid cyst
Intradural spinal lipoma
fibrolipoma, 20/396
Intralamellar thalamic nuclear group
centrum medianum, 2/472
intralamellar nucleus, 2/472
limitans nucleus, 2/472
parafascicular nucleus, 2/472
Intralaminar nucleus
pain, 45/237
Intramedullary arterial system
spinal cord, 12/496
Intramedullary cyst
cystic necrosis, 25/80
Intramedullary hemorrhage, *see* Spinal cord hemorrhage
Intramedullary schwannosis
neurofibromatosis, 14/36
Intramedullary spinal abscess, *see* Spinal cord abscess
Intramedullary spinal tumor, 19/51-73
clinical picture, 19/32
duration, 19/54
funicular pain, 19/69
Horner syndrome, 19/32, 59
muscular atrophy, 19/32
myelography, 19/32, 191
papilledema, 19/60
reflex disorder, 19/32
sensory deficit, 19/58
sensory loss, 19/32
spinal cord compression, 19/351
syringomyelia, 50/455
type, 19/52
Intramedullary trigeminal tractotomy
trigeminal neuralgia, 5/397
Intranuclear inclusion body
nucleodegenerative myopathy, 62/351
Intraocular calcification
retinoblastoma, 42/768
Intraocular pressure
cluster headache, 48/224, 234, 241
papilledema, 16/275

Intraoperative monitoring
brain tumor, 67/250
motor evoked potential, 67/251
surgery, 67/253
Intrasacral meningocele
embryological development, 20/156
meningeal diverticulum, 20/153
nerve root compression, 20/156
physical examination, 20/157
reported cases, 20/158
sacral arachnoid cyst, 31/93, 95
terminology, 20/155
Intrasellar abscess
autopsy finding, 17/412
brain abscess, 52/152
myoblastoma, 17/426
Intrasellar aneurysm
brain aneurysm, 17/409
Intrasellar cyst
mucocele, 17/409, 428
pseudocyst, 17/430
Rathke cleft, 17/428
Intraspinal angioma
Foix-Alajouanine disease, 20/481
Intraspinal foregut cyst, *see* Intraspinal neurenteric cyst
Intraspinal lipoma
diastematomyelia, 50/439
diplomyelia, 50/439
Intraspinal neurenteric cyst, 32/420-445
age at diagnosis, 32/442
age at onset, 32/436
Arnold-Chiari malformation type I, 32/437
associated defect, 20/64-71
case series, 20/64-70
classification, 32/424
clinical criteria, 32/424
clinical features, 20/70, 32/437, 442
cutaneous manifestation, 32/442
definite enteric origin, 32/424-438
diastematomyelia, 32/433
embryology, 32/420-424, 439
enteric duplication, 32/425, 432, 434
enteric mediastinal cyst, 32/429-433
epidemiology, 32/437, 442
Fallon triad, 32/425, 431
histology, 32/435, 444
hypertrichosis, 20/65, 67, 71
Klippel-Feil syndrome, 32/429
lipoma, 20/65
localization, 32/441
malformation, 32/440
pathogenesis, 32/421-424
pathology, 20/62
possibly ependymal origin, 32/438-445
posterior dysraphia, 32/433
radiologic diagnosis, 32/445
sex chromatin, 20/70
sex chromosomes, 32/423
sex ratio, 20/71
spinal canal widening, 32/429
synonym, 32/420
treatment, 20/73, 32/445
vertebral abnormality, 32/426-429
vertebral body defect, 31/88, 32/426-429
widening, 19/157
Intraspinal tumor
gait disturbance, 19/24
Lhermitte sign, 19/35
neurofibroma, 14/136
neurofibromatosis type I, 14/153, 19/39
pseudoneuralgic prephase, 19/24
Intrathecal chemotherapy
alcohol, 61/152
arachnoiditis, 61/151
baclofen, 61/152
cauda equina syndrome, 61/159
central pontine myelinolysis, 61/151
cerebral mineralizing microangiopathy, 61/151
complication, 61/147, 151
corticosteroid, 61/151
cytarabine, 61/151
dementia, 61/151
disseminated necrotizing leukoencephalopathy, 63/355
encephalopathy, 61/151
iatrogenic neurological disease, 61/147
meningeal leukemia, 63/342, 69/247
meningeal lymphoma, 63/348
methotrexate, 39/100, 61/151
methylene blue, 61/152
morphine, 61/151
paraplegia, 61/151
phenol, 61/152
radiation myelopathy, 61/208
sodium chloride, 61/152
vincristine, 61/151
Intrathecal IgG synthesis
see also Intrablood-brain barrier IgG synthesis
multiple sclerosis, 47/81, 94, 104
Intrathecal injection
North American blastomycosis, 35/407, 52/391
Intrathecal phenol injection
spasticity, 9/413, 26/483
Intrathoracic gastrogenic cyst, *see* Intraspinal neurenteric cyst

Intrathoracic meningocele, 32/211-224
neurofibromatosis, 32/212-218, 220-222
neurofibromatosis type I, 14/38, 157
Intrathoracic pressure
micturition, 75/716
respiratory system, 26/339
Intrauterine growth acceleration
diabetes mellitus, 46/82
Marshall syndrome, 46/82
Weaver syndrome, 46/82
Intrauterine growth retardation
arthrogryposis multiplex congenita, 46/89
De Lange syndrome, 46/91
Russell-Silver syndrome, 46/82
trichomegaly, 46/94
Intrauterine head injury
biomechanics, 23/471
Intrauterine injury
spinal cord, 25/96
Intravascular coagulopathy, *see* Intravascular consumption coagulopathy
Intravascular consumption coagulopathy, 55/493-497
acidosis, 55/493
acute form, 55/495
amniotic fluid embolism, 63/315
antithrombin III, 55/494
Argentinian hemorrhagic fever, 56/367
blood clotting cascade, 55/493
blood clotting factor, 55/493
blood clotting inhibition, 55/494
blood clotting initiation, 55/493
blood clotting physiology, 55/493
brain computer assisted tomography, 63/317
brain embolism, 53/156
brain fat embolism, 55/180
brain hemorrhage, 38/72, 54/293, 55/497, 63/314, 344
brain infarction, 53/156, 166, 55/497, 63/316, 345
brain ischemia, 53/156, 166
brain nuclear magnetic resonance, 63/317
brain venous infarction, 63/314
burn, 55/495
cardiac valvular disease, 53/156
clinical features, 63/316
coagulopathy, 63/309
coma, 63/316
complement system, 55/495
confusional state, 63/316
CSF, 38/73
deep vein thrombosis, 55/496
delirium, 55/488
differential diagnosis, 63/318
dissecting aorta aneurysm, 63/47
endotoxin, 55/494
envenomation, 63/315
extrinsic pathway, 55/494
fibrinolysis, 11/456
fibrinolytic system, 55/494
head injury, 57/223, 238, 63/315
heat stroke, 23/671, 677, 55/495
hemangioma, 63/318
hemostasis, 63/306
hypofibrinogenemia, 55/496
hypotension, 55/493
hypoxemia, 55/493
idiopathic hypereosinophilic syndrome, 63/124
intracranial pressure monitoring, 57/223
intrinsic pathway, 55/493
Kasabach-Merritt syndrome, 63/318
laboratory diagnosis, 55/496
laboratory test, 63/316
lethargy, 63/316
leukemia, 39/8, 63/315, 345
liver disease, 63/314
loxosceles intoxication, 37/112
lymphoma, 63/350
major symptom, 55/496
malignancy, 55/493, 63/314
malignant hyperthermia, 38/551
malignant lymphoma, 55/488
meningitis, 63/314
meningococcal meningitis, 33/23, 25, 27, 52/25, 27, 30, 63/314
α-methylcyclohexane carboxylic acid, 63/318
microvascular occlusion, 55/496
neurologic complication, 55/497
neurologic manifestation, 38/73
nonarteriosclerotic brain infarction, 11/456
nonbacterial thrombotic endocarditis, 11/415, 55/165, 63/28, 111, 120
obstetric complication, 55/495
onset, 55/495
pancreatitis, 55/495
paracoagulation, 55/496
pathogenic factor, 63/315
pathophysiology, 55/494
properdine pathway, 55/495
protein C, 55/494
purpura fulminans, 55/495
sepsis, 55/493
septicemia, 63/314
snake bite, 63/315
snake venom, 63/315
spinal cord injury, 26/391
subarachnoid hemorrhage, 55/3, 497

thrombin, 55/494
thrombocytopenia, 55/496, 63/321
thrombophlebitis, 55/496
trauma, 55/493
traumatic intracerebral hematoma, 57/258
treatment, 38/73, 55/496, 63/317
venous sinus thrombosis, 63/345
Intravascular malignant lymphomatosis
lymphomatoid granulomatosis, 51/451
Intravenous pyelography
neurogenic bladder, 26/424
Intraventricular granuloma
cryptococcal meningitis, 52/431
Intraventricular hematoma
brain aneurysm neurosurgery, 55/54
resorption, 11/677
Intraventricular hemorrhage
aqueduct stenosis, 50/309
blood clotting factor V deficiency, 63/312
computer assisted tomography, 54/68, 57/148
electrical injury, 61/192
head injury, 57/158
hemophilia, 63/310
hydranencephaly, 50/337
immunothrombocytopenia, 63/327
newborn, 63/326
newborn alloimmunothrombocytopenia, 63/327
radiodiagnosis, 57/158
subarachnoid hemorrhage, 12/167, 55/54
traumatic intracranial hematoma, 57/148, 158
Intrinsic factor
hemostasis, 63/303
ophthalmoplegia, 2/293
pernicious anemia, 42/609
pressure sore, 26/463, 61/350
subacute combined spinal cord degeneration, 28/146
vitamin B12, 28/146
vitamin B12 deficiency, 51/337, 70/373
Intrinsic nerve
autonomic nervous system, 74/104
neurotransmitter, 74/104
Introspection
attention, 3/137
consciousness, 3/50
Invagination
basilar, *see* Basilar impression
Inverse primary pigmentary retinal degeneration
Stargardt disease, 13/28, 133
Inverse secondary pigmentary retinal degeneration
high frequency hearing loss, 42/377
sensorineural deafness, 42/377
Inversion of visual space, *see* Visual space
Involuntary eye movement
fundus camera, 13/7
Gilles de la Tourette syndrome, 49/632
Involuntary laughing, *see* Pathologic laughing
Involuntary movement, 1/277-291, 293-308
see also Athetosis, Chorea *and* Dystonia
akathisia, 1/279, 290
Behçet syndrome, 56/598
Benedikt syndrome, 2/276, 298
complicated migraine, 5/80
Critchley syndrome, 13/427
hereditary benign myoclonus, 1/282
history, 1/277
Huntington chorea, 2/504
hyperexplexia, 6/301
Imanenjana, 6/301
leaping ague, 6/301
mesencephalic lesion, 2/277
migraine, 5/80
neuroacanthocytosis, 42/209
Parry disease, 42/467
pontine hemorrhage, 2/253
pontine lesion, 2/251
Rett syndrome, 63/437
rhythmias, 1/290
rubral tremor, 2/481
startle epilepsy, 6/301
striatal syndrome, 2/499
Sydenham chorea, 2/501
syncinésie sursaut, 6/301
thalamic syndrome (Dejerine-Roussy), 2/481
tigretier, 6/301
tremor, 42/238
Iodamoeba bütschlii
Amoebae, 35/30
microbiology, 52/310
neuroamebiasis, 52/310
Iodine intoxication
headache, 48/420
Iodine starch test
sweat test, 2/17, 200
Iodoacetamide
muscle contraction, 41/297
Iodoacetate
creatine kinase inhibition, 41/264
retinal degeneration, 13/207
Iodoacetate sodium
animal experiment, 13/210
(3-Iodobenzyl)guanidine I 123
multiple neuronal system degeneration, 75/185
Iododeoxyuridine
herpes simplex, 34/154
herpes simplex virus encephalitis, 34/154

Iodomethamate, *see* Iodoxyl
Iodopyracet, *see* Diodone
Iodoxyl
  iatrogenic neurological disease, 63/44
Iofendylate polytomography
  acoustic nerve tumor, 17/682
Iohexol
  adverse effect, 61/149
Ionic current
  epilepsy, 72/47
  nerve cell, 72/47
Ionizing radiation
  nerve injury, 51/134
  nerve lesion, 7/388
  systemic brain infarction, 11/463
  teratogenic agent, 30/118
  vegetative dystonia syndrome, 74/322
Ionizing radiation neuropathy
  brachial plexus injury, 51/137
  carcinogenesis, 51/137
  cobalt therapy, 51/138
  cranial nerve, 51/137
  EMG, 51/138
  latency, 51/138
  lumbar plexus lesion, 51/137
  α-particle, 51/138
  radiation parameter, 51/138
  radioisotope, 51/138
  radiotherapy, 51/138
  sequence, 51/138
  spinal nerve root, 51/137
  topography, 51/138
Iopamidol
  adverse effect, 61/149
Ipecac intoxication
  cytoplasmic inclusion body myopathy, 62/345
Iprazochrome
  migraine, 48/191
Iproniazid intoxication
  liver toxicity, 37/321
IPSP, *see* Inhibitory postsynaptic potential
Irascibility
  migraine, 48/162
Iridis
  heterochromia, *see* Heterochromia iridis
Iridocyclitis
  ankylosing spondylitis, 26/176
  Behçet syndrome, 56/594
  Fuchs heterochromic, *see* Fuchs heterochromic iridocyclitis
  granulomatous CNS vasculitis, 55/388
  headache, 5/206
  Kawasaki syndrome, 56/638
  leprosy, 51/216
Iris, 74/404
  autonomic innervation, 74/419
  autonomic nervous system, 74/399
  hypoplasia, *see* Hypoplasia iridis
  nerve, 74/400
  neurofibromatosis, 14/634
  neurofibromatosis type I, 14/634
  pharmacology, 74/417
  sphincter, 74/400
  sphincter tear, 74/412
Iris coloboma
  Bardet-Biedl syndrome, 13/295
  Biemond syndrome type II, 13/295, 43/334
  cat eye syndrome, 31/525, 43/548
  congenital heart disease, 63/11
  Friedreich ataxia, 21/347
  Goltz-Gorlin syndrome, 14/787, 43/20
  Gruber syndrome, 14/13
  Klippel-Trénaunay syndrome, 14/524
  primary pigmentary retinal degeneration, 13/249
  11q partial monosomy, 43/523
  7q partial trisomy, 43/506
  Rubinstein-Taybi syndrome, 13/83, 31/232
  Sturge-Weber syndrome, 14/512
  thalidomide syndrome, 42/660
  trisomy 22, 43/548
  tuberous sclerosis, 14/496
Iris dysplasia
  hypertelorism, 43/417
  orbital hypertelorism, 43/417
  Rieger syndrome, 30/247, 43/471
Iris heterochromia, *see* Heterochromia iridis
Iris hypopigmentation
  tuberous sclerosis, 50/372
Iris mesenchymal hypoplasia
  Down syndrome, 50/526
Iris rubeosis
  temporal arteritis, 48/319
Iritis
  hypopyon, *see* Hypopyon iritis
  Sjögren syndrome, 71/86
Iron
  Alzheimer disease, 46/265
  basal ganglion, 49/35
  blindness, 13/224
  cerebrohepatorenal syndrome, 43/339
  CNS, 28/438
  epilepsy, 72/53
  Guam amyotrophic lateral sclerosis, 59/265
  Hallervorden-Spatz syndrome, 49/35
  lung, 26/345
  neuroaxonal dystrophy, 49/392

nyctalopia, 13/224
Parkinson disease, 49/36, 129
pseudocalcium, 6/714, 719
striatopallidodentate calcification, 6/716, 49/425
vascular mineralization, 11/141
Iron deficiency
congenital heart disease, 63/3
deficiency neuropathy, 51/327
manganese intoxication, 64/304
restless legs syndrome, 8/311, 315, 51/546
Iron deficiency anemia
aluminum intoxication, 64/274
benign intracranial hypertension, 67/111
cause, 63/252
cerebrovascular disease, 55/463
classification, 63/252
EEG dysrhythmia, 63/252
fatigue, 63/253
neurasthenia, 63/253
pagophagia, 55/464
papilledema, 63/252
pathophysiology, 63/252
pica syndrome, 63/252
posterior inferior cerebellar artery, 55/464
restless legs syndrome, 8/315, 63/252
retinopathy, 55/463
Rud syndrome, 38/9, 43/284
school performance, 63/253
thrombocytosis, 55/464
Irradiation
see also Radiotherapy
CNS, *see* CNS irradiation
dementia, 67/358
low grade glioma, 68/43
mineralizing microangiopathy, 67/361
necrotizing leukoencephalopathy, 67/358
polymyositis, 71/143
prophylactic cranial, *see* Prophylactic cranial irradiation
Irradiation encephalopathy
acute type, 63/354
brain edema, 63/355
chronic type, 63/356
disseminated necrotizing leukoencephalopathy, 63/355
late type, 63/354
leukemia, 63/354
mineralizing brain microangiopathy, 63/355
olivopontocerebellar atrophy (Dejerine-Thomas), 60/533
radiotherapy, 63/355
symptom, 63/354
Irradiation myelopathy
Hodgkin lymphoma, 63/356
leukemia, 63/356
radiotherapy, 63/356
Irregular sleep-wake syndrome
biological rhythm, 74/470
circadian pacemaker, 74/470
Irreversible coma, *see* Brain death
Irritability
cardiac arrest, 63/217
cocaine intoxication, 65/253
digoxin, 63/195
fox intoxication, 37/96
gram-negative bacillary meningitis, 52/119
mercury polyneuropathy, 51/272
migraine, 48/162
obstructive sleep apnea, 63/453
pellagra, 7/573, 28/86, 38/648, 46/336
phenylketonuria, 29/30
polar bear intoxication, 37/96
procainamide intoxication, 37/451
tick bite intoxication, 37/111
toxic oil syndrome, 63/380
vitamin A intoxication, 37/96
wolf intoxication, 37/96
Irritable heart, *see* Hyperventilation syndrome
Irritation
adversive seizure, 2/731
meningeal, *see* Meningeal irritation
premotor cortex, 2/731
spinal subarachnoid hemorrhage, 12/173
Irritative zone
definition, 73/386
Isaacs-Mertens syndrome, *see* Isaacs syndrome
Isaacs syndrome
cancer, 41/347
cervical radiculopathy, 62/274
characteristics, 62/273
chorea, 49/559
chronic muscle denervation, 62/263
clinical symptom, 62/273
cramp, 40/330, 546
curare, 40/546
2,4-dichlorophenoxyacetic acid, 41/301
differential diagnosis, 41/348, 59/376
distal muscle weakness, 43/160
EMG, 40/330
fasciculation, 62/273
gamma hyperactivity, 1/269
glucosidase deficiency, 62/263
hyperhidrosis, 43/160
hypothyroid myopathy, 62/263
malignancy, 41/347
motor unit hyperactivity, 40/330, 332, 41/296,

300, 303, 347
muscular atrophy, 43/160
myokymia, 42/237, 43/160
myokymic discharge, 62/263
myotonia, 43/160, 62/263, 273
myotonia paradoxa, 43/160
myotonic syndrome, 62/263, 273
neuromyotonic discharge, 62/263
ocular form, 41/301
periodic ataxia, 60/659
polyneuritis, 43/160
proximal muscle weakness, 43/160
repetitive electrical activity, 62/263
succinylcholine, 40/546
Ischemia
acquired myoglobinuria, 62/574
amyloid polyneuropathy, 51/419
angiopathic polyneuropathy, 51/445
apistus intoxication, 37/76
brain, *see* Brain ischemia
brain arteriovenous malformation, 12/232
brain stem, *see* Brain stem ischemia
centropogon intoxication, 37/76
cerebral, *see* Brain ischemia
cerebrovascular disease, *see* Brain ischemia
classic migraine, 48/118
compression neuropathy, 8/162, 51/88
demyelination, 9/573
diabetes mellitus, 70/142
epilepsy, 72/108, 210
glycogen, 27/206
glycolysis, 27/31
gymnapistes intoxication, 37/76
hypoglycemia, 70/185
hypothermia, 71/577
hypoxic, *see* Hypoxic ischemia
iatrogenic nerve injury, 51/159
intermittent spinal, *see* Intermittent spinal cord ischemia
labyrinthine, *see* Labyrinthine ischemia
leukoencephalopathy, 9/572
Levine experimental anoxia, *see* Levine experimental anoxia ischemia
lumbosacral plexus neuritis, 51/166
migraine, 48/73, 143, 150
muscle, *see* Muscle ischemia
myocardial, *see* Myocardial ischemia
myoglobinuria, 40/338, 41/270, 277, 62/557
nerve, *see* Nerve ischemia
nerve conduction, 7/132
nothesthes intoxication, 37/76
occipital, *see* Occipital ischemia
organophosphate intoxication, 64/228
papaverine, 37/455
paradoxical, *see* Paradoxical ischemia
perinatal hypoxic, *see* Perinatal hypoxic ischemia
pontine, *see* Pontine ischemia
pontine syndrome, 2/241
propranolol, 37/446
retinal, *see* Retinal ischemia
scorpaena intoxication, 37/76
scorpaenodes intoxication, 37/76
scorpaenopsis intoxication, 37/76
sebastapistes intoxication, 37/76
spinal cord compression, 69/170
spinal nerve root, *see* Spinal nerve root ischemia
syncope, 15/817
thalamus, *see* Thalamic ischemia
tourniquet, 70/40
vasa nervorum disease, 12/651
vertebral artery, 5/199
Ischemic attack
transient, *see* Transient ischemic attack
Ischemic cerebrovascular disease, *see* Brain ischemia
Ischemic encephalopathy
chronic mercury intoxication, 64/372
epilepsy, 72/210
hypoxic, *see* Hypoxic ischemic encephalopathy
symptomatic dystonia, 49/542
Ischemic exercise
acid maltase deficiency, 62/480
adenosine monophosphate deaminase deficiency, 62/512
carnitine palmitoyltransferase deficiency, 62/494
lactate dehydrogenase deficiency, 62/490
myophosphorylase deficiency, 27/233, 41/187, 62/485
phosphofructokinase deficiency, 27/235, 40/327, 339, 547, 41/190, 62/487
phosphoglycerate mutase deficiency, 62/489
phosphorylase kinase deficiency, 62/487
Ischemic exercise test
acid maltase deficiency, 41/176
alcoholic polyneuropathy, 41/273
carnitine palmitoyltransferase deficiency, 41/205
debrancher deficiency, 41/182
glycogen storage disease, 41/193
myophosphorylase deficiency, 27/233, 41/187, 62/485
Ischemic heart disease
arterial hypertension, 54/209
autonomic nervous system, 75/444
carbon disulfide intoxication, 64/27
cardiac valvular disease, 63/18
hereditary hyperlipidemia, 11/462

hyperlipoproteinemia type IV, 10/275
Kawasaki syndrome, 56/638
rheumatic heart disease, 63/18
snoring, 63/451
transient global amnesia, 45/211
transient ischemic attack, 53/291
Ischemic-hypoxic cycle
diabetes mellitus, 70/151
Ischemic infarction
central, *see* Central ischemic infarction
infantile cerebrovascular disease, 54/40
porencephaly, 42/48
Ischemic keratopathy
temporal arteritis, 48/319
Ischemic leukoencephalopathy
anoxic, *see* Anoxic ischemic leukoencephalopathy
chronic, *see* Chronic ischemic leukoencephalopathy
Ischemic myelopathy
see also Myelomalacia
abdominal trauma, 63/43
amphetamine, 63/55
aorta rupture, 63/43
aortography, 63/43
arteriosclerosis, 63/38
cardiac arrest, 63/40, 213
cervical spondylosis, 63/43
congenital heart disease, 63/4
dissecting aorta aneurysm, 61/115
epidural anesthesia, 61/137
exercise provocation, 63/40
lumbar puncture, 61/160
lumbosacral predilection, 63/40
paraplegia, 61/115, 160
polyarteritis nodosa, 63/54
postpneumonectomy, *see* Postpneumonectomy ischemic myelopathy
prevalence, 63/40
radiation myelopathy, 63/43
spastic paraplegia, 59/437
spinal artery thrombosis, 63/38
spinal cord intermittent claudication, 63/40
spinal cord vascular disease, 12/592
steal syndrome, 32/487
temporal arteritis, 55/342, 344
treatment, 63/44
vertebral subluxation, 63/43
Ischemic myopathy
blood vessel, 40/167-170
childhood dermatomyositis, 40/167
Duchenne muscular dystrophy, 40/167-170
experimental, *see* Experimental ischemic myopathy
serotonin, 40/167
Ischemic necrosis
dissecting aorta aneurysm, 39/248
optic nerve injury, 24/40
Ischemic neuropathy, 8/149-153, 154-163
amyloidosis, 8/161
anterior tibial syndrome, 8/152
arteriosclerosis, 8/161
cold injury, 8/162
compression, 8/162
cryoglobulinemia, 63/404
diabetes mellitus, 8/157-161
dialysis, 63/523, 529
experimental, 8/156
muscular atrophy, 8/152
peripheral nerve, 8/154-157
polyarteritis nodosa, 8/161
sensory neuropathy, 8/367
Ischemic optic neuropathy
acute, *see* Acute ischemic optic neuropathy
allergic granulomatous angiitis, 51/453
anterior, *see* Anterior ischemic optic neuropathy
antiphospholipid antibody, 63/330
carotid artery stenosis, 55/111
papilledema, 16/287
polyarteritis nodosa, 51/452
systemic lupus erythematosus, 51/447
temporal arteritis, 51/452
Wegener granulomatosis, 51/450
Ischemic papilledema, *see* Anterior ischemic optic neuropathy
Ischemic papillopathy
acute, *see* Acute ischemic papillopathy
temporal arteritis, 48/315
Ischemic paralysis
complication, 7/251
vascular occlusion, 7/251
Ischemic vasculitis
Sjögren syndrome, 51/449
Ischialgia
back pain, 69/57
cancer, 69/57
Ischuria paradoxa
spinal shock, 2/188
Islet of Calleja, *see* Calleja islet
Islet cell adenoma
hyperinsulinism, 27/68
Ismerlund-Grasbeck disease
vitamin B12 deficiency anemia, 63/254
Isobenz intoxication
clinical features, 36/411-414
EEG, 36/414-416
pathology, 36/417

prognosis, 36/416
treatment, 36/417
O-Isobutyl *O*-carboxymethyl ethylphosphonothioate intoxication
chemical classification, 37/547
organophosphorus compound, 37/547
Isobutyrate
myopathy, 40/548
Isocitrate dehydrogenase
CSF, *see* CSF isocitrate dehydrogenase
tricarboxylic acid cycle, 27/16
wallerian degeneration, 7/215
Isodemeton intoxication, *see* Demeton-S intoxication
Isoelectric focussing
multiple sclerosis, 47/86, 94, 103
Isoenzyme
creatine kinase, 40/342, 41/410
difference, 66/52
distribution, 66/53
lactate dehydrogenase, 40/356
myophosphorylase deficiency, 40/189, 41/188
phosphorylase, *see* Phosphorylase isoenzyme
property, 66/53
Isoetarine
toxic myopathy, 62/597
Isoetharine, *see* Isoetarine
Isoflurane
animal experiment, 37/412
general anesthetic agent, 37/412
Isoflurane intoxication
inhalation anesthetics, 37/412
spike, 37/412
Isoflurophate, *see* Dyflos
Isolan, *see* Dimethylcarbamic acid 1-isopropyl-3-methylpyrazol-5-yl ester
Isolated angiitis
CNS, *see* Granulomatous CNS vasculitis
Isolated CNS vasculitis, *see* Granulomatous CNS vasculitis
Isolated growth hormone deficiency, *see* Pituitary dwarfism type I
Isolated head perfusion
brain metabolism, 27/28
Isolated trigeminal neuropathy, *see* Sensory trigeminal neuropathy
Isoleucine
vitamin B3 deficiency, 51/333
Isoleucine metabolism
2-methylacetoacetyl-coenzyme A thiolase deficiency, 66/649
Isometheptene
migraine, 48/202
Isometric exercise
autonomic nervous system, 74/625
Isometric myography
physiology, 1/50
Isoniazid
acquired myoglobinuria, 62/577
axonal degeneration, 65/481
behavior disorder, 46/601
chronic axonal neuropathy, 51/531
drug induced polyneuropathy, 51/294
drug induced systemic lupus erythematosus, 16/222
epilepsy, 28/114-116
iatrogenic neurological disease, 65/479
inducing pellagra, 28/97
inducing seizure, 28/114-116
multiple sclerosis, 9/396, 47/188
myopathy, 40/561
nerve conduction, 7/174
neurotoxicity, 51/294, 71/368
neurotoxin, 7/516, 533, 582, 584, 588, 644, 28/97, 115, 41/333, 42/648, 47/565, 51/68, 294, 297, 334, 531, 52/211, 62/601, 65/479
overdose, 70/429
phenytoin intoxication, 52/211
pyridoxal phosphokinase, 51/297
sensory polyneuropathy, 41/333, 52/211
slow acetylation, 51/297
toxic encephalopathy, 65/480
toxic myopathy, 62/601
toxic neuropathy, 65/480
toxic polyneuropathy, 51/68
tuberculous meningitis, 33/226-228, 52/211
vitamin B6, 51/297
vitamin B6 deficiency, 42/468, 47/565, 51/297, 334
wallerian degeneration, 47/565
Isoniazid intoxication
axonal neuropathy, 65/481
dementia, 65/480
epilepsy, 65/480
experimental spongiform leukoencephalopathy, 47/565
leukoencephalopathy, 65/480
mania, 65/480
mental depression, 65/480
myelin ultrastructure, 65/480
optic atrophy, 65/480
optic neuropathy, 65/480
polyneuropathy, 65/480
psychosis, 65/480
secondary demyelination, 65/481
stupor, 65/480

tinnitus, 65/480
vertigo, 65/480
vitamin B6 dependency, 65/480
Isoniazid leukoencephalopathy
features, 47/565
vitamin B6, 47/565
Isoniazid neuropathy, 70/428, 71/382
acetyltransferase, 42/648
animal experiment, 7/588
biochemistry, 7/533
clinical picture, 7/582
CNS, 7/582
deficiency neuropathy, 7/582, 584, 588, 644
iatrogenic neuropathy, 7/533
paresthesia, 7/516, 533, 42/648
peripheral nervous system, 7/582
phenytoin intoxication, 42/641
sensory loss, 42/648
subsarcolemmal myofibril, 7/584
toxic neuropathy, 7/516
tuberculosis, 7/516, 533, 582
vitamin B6 deficiency, 7/516, 533, 582, 42/648
wallerian degeneration, 7/174
Isoniazid polyneuropathy
axonal degeneration, 47/565
optic atrophy, 51/297
sensory neuropathy, 51/297
vitamin B6 deficiency, 51/334
Isonicotinic acid hydrazide, *see* Isoniazid
Iso-Ompa intoxication
chemical classification, 37/549
organophosphorus compound, 37/549
Isoprenaline
blood-brain barrier, 48/64
classification, 74/149
testing, 74/169
Isopropyl alcohol, *see* 2-Propanol
Isopropyl methylphosphonofluoridate intoxication, *see* Sarin intoxication
4-(Isopropylamino)phenazone
migraine, 48/175
Isopropyl-N-dimethyl phosphoramidocyanidate intoxication
chemical classification, 37/544
organophosphorus compound, 37/544
Isopropylmethylphosphonofluoridate, *see* Sarin
Isoprotein abnormality
a-alphalipoproteinemia, 51/396
Isoproterenol, *see* Isoprenaline
Isopyrine, *see* 4-(Isopropylamino)phenazone
Isosorbide
intracranial pressure, 23/212
Iso-Systox sulfoxide intoxication
chemical classification, 37/548
organophosphorus compound, 37/548
Isotachophoresis
multiple sclerosis, 47/91
Isothermognosia
terminology, 1/87
Isotope cisternography
chemical meningitis, 56/130
hydrocephalic dementia, 46/328
traumatic hydrocephalus, 24/236, 238, 248
tuberculous meningitis, 52/209
Isotopic bolus
brain death, 24/772
Isotretinoin
adverse effect, 69/499
benign intracranial hypertension, 67/111
vitamin A intoxication, 65/569
Isovaleric acidemia
acid base metabolism, 28/517
diet, 42/580
feeding disorder, 42/579
heterozygote detection, 42/580
hyperglycinemia, 29/204
hypocalcemia, 42/579
hypoglycin intoxication, 65/131
isovaleryl coenzyme A dehydrogenase, 42/580, 66/652
leucine metabolism, 42/579, 66/651
maple syrup urine disease, 29/55
mental deficiency, 42/580
neutropenia, 42/579
periodic ataxia, 60/647
psychomotor retardation, 42/580
sweaty foot syndrome, 42/579
Isovaleryl coenzyme A dehydrogenase
isovaleric acidemia, 42/580, 66/652
Isovaleryl coenzyme A dehydrogenase deficiency, *see* Isovaleric acidemia
Italy
neurology, 1/10
Itch
coelenterata intoxication, 37/38
laryngeal, *see* Laryngeal itch
terminology, 1/86
Iteration
dialysis encephalopathy, 64/277
Pick disease, 46/234
Ito hypomelanosis
hemihypertrophy, 59/483
Ivory vertebra
carcinoid tumor, 19/171
coccidioidomycosis, 19/171
Ewing sarcoma, 20/35

fibrous dysplasia, 19/172
fluorosis, 19/171
Hodgkin disease, 19/171
hyperparathyroidism, 19/171
osteitis deformans (Paget), 19/172
osteopetrosis, 19/171
osteosarcoma, 19/171
spinal medulloblastoma, 19/171
tuberous sclerosis, 14/375
Iwashi-kujira intoxication, *see* Balaenoptera borealis intoxication
Ixodes dammini
Lyme disease, 47/613
tick-borne disease, 51/199
Ixodes persulcatus
tick-borne encephalitis, 56/139
Ixodes ricinus
tick-borne disease, 51/199
tick-borne encephalitis, 56/139

J receptor
autonomic nervous system, 74/575
sensation, 74/576
stimulation, 74/578
J reflex
exercise, 74/577
reflex bradycardia, 74/578
survey, 74/577
J-shaped sella
optic nerve glioma, 14/151, 18/345
J1-glycoprotein, *see* Tenascin
Jabs and jolts syndrome
chronic paroxysmal hemicrania, 48/260
cluster headache, 48/260
Jaccodo petrosphenoid syndrome, *see* Petrosphenoid syndrome (Jaccodo)
Jaccodo syndrome, *see* Petrosphenoid syndrome (Jaccodo)
Jackfish intoxication
classification, 37/78
Jackson classification
epilepsy, 15/5
Jackson, J.H., 1/4, 6
Jackson syndrome
accessory nerve, 1/189
clinical symptom, 2/100
hypoglossal nerve, 1/189
medulla oblongata syndrome, 2/223
nonrecurrent nonhereditary multiple cranial neuropathy, 51/571
Schmidt syndrome, 1/189
tongue paralysis, 1/189, 2/223
Jacob, A.M., 1/9, 27
Jacob, C., 1/12
Jacobson nerve
anatomy, 5/351
glossopharyngeal neuralgia, 5/351
stimulation, 48/491
Jacobson neuralgia
intermedius neuralgia, 48/490
Jacod syndrome, *see* Orbital apex syndrome
Jactatio capitis nocturna
differential diagnosis, 15/221
hereditary olivopontocerebellar atrophy (Menzel), 21/441
juvenile parasomnia, 45/141
sleep disorder, 3/101-103
Jadassohn nevus phakomatosis, *see* Linear nevus sebaceous syndrome
Jaeken syndrome
adenylosuccinate deficiency, 60/673
deafness, 60/658
pericardial effusion, 60/668
secondary pigmentary retinal degeneration, 60/726
Jaffe-Lichtenstein disease, *see* Fibrous dysplasia
Jake paralysis, *see* Organophosphate induced delayed polyneuropathy
Jakob-Creutzfeldt disease, *see* Creutzfeldt-Jakob disease
Jamaican ginger, *see* Tri-*o*-cresyl phosphate
Jamaican ginger paralysis, *see* Tri-*o*-cresyl phosphate paralysis
Jamaican neuropathy
anterolateral myelopathy, 7/643
ataxic type, 7/480
deficiency neuropathy, 7/643, 51/323
differential diagnosis, 9/205
multiple sclerosis, 9/205
spastic type, 7/480
tropical ataxic neuropathy, 7/640, 51/323
Jamaican polyneuropathy
ataxic neuropathy, 28/26-31
Jamaican vomiting sickness
see also Hypoglycin intoxication *and* Reye syndrome
acidosis, 65/80
acyl glycinuria, 65/93
age, 37/514
animal experiment, 37/513
autopsy finding, 37/514
Blighia sapida, 37/511, 515, 65/79
clinical course, 37/512
clinical symptom, 37/512-514
2 clinical types, 37/512, 65/80
clofibrate feeding, 65/96

coma, 37/512, 533, 65/79
convulsion, 37/512, 65/79
death, 37/512
diagnosis, 37/533
differential diagnosis, 37/533
ethylmalonic aciduria, 37/533, 535
etiology, 37/515-529
fatty acid β-oxidation, 65/87, 89
fulminant course, 65/80
glucose, 37/535
glutaric aciduria type II, 37/533-535, 65/104
glycine, 65/104
histologic change, 37/515
history, 37/511, 65/79
hypoglycemia, 37/511, 532, 65/79
hypoglycin A, 37/515-521
hypoglycin A intoxication, 37/511-539
hypoglycin B, 37/515
hypoglycin metabolism, 65/83
intravenous glucose, 65/105
laboratory finding, 37/529-532
lethargy, 37/533
3-methylenecyclopropylacetic acid, 37/533
mortality, 65/80
organic acidemia, 65/84
organic aciduria, 37/529, 65/93
β-oxidation inhibition, 65/90
pathology, 37/514
peroxisome, 65/96
plasma free fatty acid, 65/84
Reye syndrome, 29/338, 37/533, 65/80
salicylic acid like syndrome, 65/80
toxic encephalopathy, 65/79, 115
treatment, 37/535, 65/104
valproic acid like syndrome, 65/80
vitamin B_2, 65/105
vomiting, 37/512, 65/79
Jamestown Canyon virus
California encephalitis, 56/140
Jannetta mechanism
intermedius neuralgia, 48/492
Jansky-Bielschowsky syndrome, *see* Late infantile neuronal ceroid lipofuscinosis
Janz-Herpin syndrome, *see* Juvenile myoclonic epilepsy
Janz syndrome, *see* Juvenile myoclonic epilepsy
Japanese autumn fever, *see* Leptospirosis
Japanese B encephalitis
acute viral encephalitis, 34/75, 56/134, 138
brain stem, 56/138
Culex tritaeniorhyncus, 56/138
dystonia, 56/138
fatality rate, 56/138
flavivirus, 56/12
geography, 56/138
reticuloendothelial system, 56/30
thalamus, 2/486
Togavirus, 56/12
tremor, 56/138
Japanese encephalitis
epilepsy, 72/157
Japanese puffer fish
tetrodotoxin intoxication, 37/80, 41/292
Japanese sandfish intoxication
vitamin A intoxication, 37/87
Jargon
aphasia, *see* Paraphasia
neologistic, *see* Neologistic jargon
phonemic, *see* Phonemic jargon
phonetic, *see* Phonetic jargon
semantic, *see* Semantic jargon
Jarish-Herxheimer reaction, *see* Herxheimer reaction
Jarretière
amyotrophy, 21/276
Jarvi-Hakola disease, *see* Nasu-Hakola disease
Jasper-Kershman classification
epilepsy, 15/8
Jaundice
Abt-Letterer-Siwe disease, 42/441
acholuric, *see* Acholuric jaundice
acid maltase deficiency, 42/644
arthrogryposis multiplex congenita, 59/73
Crigler-Najjar syndrome, 42/540
cytoside lipidosis, 10/545
fructose intolerance, 42/548
galactosemia, 42/551
glucose-6-phosphate dehydrogenase deficiency, 42/644
glycosaminoglycanosis type VII, 66/311
gram-negative bacillary meningitis, 52/119
hemochromatosis, 42/553
kernicterus, 42/582
neuropathy, 8/18
newborn physiologic, *see* Newborn physiological jaundice
rhodopsin, 13/193
valproic acid induced hepatic necrosis, 65/125
Wilson disease, 49/224
Wolman disease, 10/546
Javelin throwing
rehabilitative sport, 26/527, 529
Jaw
headache, 48/392
osteomyelitis, 5/277
pain, 5/227

small, *see* Micrognathia
tumor, 5/227
Van Buchem disease, 38/407
Jaw cyst
multiple nevoid basal cell carcinoma syndrome, 14/466, 18/172
nevus unius lateris, 43/40
Jaw jerking
reading epilepsy, 2/621, 42/717
Jaw reflex
diagnostic value, 1/240
Jaw tremor
amiodarone intoxication, 65/458
Jaw winking, *see* Marcus Gunn phenomenon
JC virus
acquired immune deficiency syndrome, 71/271
animal experiment, 18/171
BK virus, 71/415
enhancer, 56/37
epidemiology, 71/413
growth, 71/414
human infection, 47/517
immunocompromised host, 56/480
medulloblastoma, 18/171
nervous system, 71/415
neural tumor, 71/418
pathology, 71/414
progressive multifocal leukoencephalopathy, 34/315-317, 319-321, 38/656, 46/388, 47/504-507, 514-518, 56/480, 497, 514, 63/354, 71/412
subtype, 47/516
Jeanné sign
claw hand, 2/36
Jefferson fracture
atlantal arch fracture, 61/30
Jefferson syndrome, *see* Foramen lacerum syndrome (Jefferson)
Jejunoileal bypass
deficiency neuropathy, 51/328
Jejunoileal diverticulitis
sensory radicular neuropathy, 21/82
Jellyfish intoxication, *see* Scyphozoa intoxication
Jendrassik, E., 1/10
Jendrassik maneuver
Achilles reflex, 1/246
diagnostic value, 1/238
gamma 2 motoneuron, 1/67
Jensen chorioretinitis juxtapapillaris, *see* Chorioretinitis juxtapapillaris (Jensen)
Jerk
hemifacial spasm, 49/610
myoclonic, *see* Myoclonic jerk
pendulous knee, *see* Pendulous knee jerk
Jervell-Lange-Nielsen syndrome
anemia, 43/388
birth incidence, 42/389
congenital deafness, 50/218
hereditary deafness, 50/218
LEOPARD syndrome, 14/117
Refsum disease, 21/197
Jet leg
clock function, 74/468
Jeune syndrome, *see* Jeune-Tommasi disease
Jeune-Tommasi disease
amyotrophy, 22/516
cutaneous pigmentation, 22/513, 516
deafness, 60/657
hereditary hearing loss, 22/516
mental deficiency, 22/516
myocardial sclerosis, 22/516
JHM virus
age, 56/444
animal model, 56/439
class II antigen, 56/446
coronavirus, 56/439
demyelinating encephalomyelitis, 56/443
demyelination, 56/442, 444
immune response, 56/445
multiple sclerosis, 9/125, 47/325, 56/447
rat, 56/442
serotype, 56/441
target cell, 56/445
viral encephalitis, 56/441
virus-host relationship, 56/445
JHM virus encephalomyelitis
animal viral disease, 34/293
experimental infection, 56/442
multiple sclerosis, 47/431
Jitter
leg, *see* Restless legs syndrome
Jitter phenomenon
see also Single fiber electromyography
Eaton-Lambert myasthenic syndrome, 41/354
myasthenia gravis, 41/128
Johanson-Blizzard syndrome
microcephaly, 30/508
Johanson vein, *see* Dorsal callosal vein
Joint
see also Neurogenic osteoarthropathy
Charcot, *see* Neurogenic osteoarthropathy
child, 25/184
Clutton, *see* Clutton joint
congenital deformity, 1/483
contracture, 25/184
diarthrodial, *see* Diarthrodial joint

hyperextensible, *see* Hyperextensible joint
occipitoatlantal, *see* Occipitoatlantal joint
pain, 8/12
temporomandibular, *see* Temporomandibular joint
Joint deformity
arthrogryposis multiplex congenita, 1/483
Joint dislocation
multiple, *see* Multiple joint dislocation
Joint dysfunction
temporomandibular, *see* Temporomandibular joint dysfunction
Joint fixation
arthrogryposis, 40/337
contracture, 40/327, 337
cytoarchitectonic change, 40/145
Duchenne muscular dystrophy, 40/361
Emery-Dreifuss muscular dystrophy, 40/390
muscular atrophy, 40/143
Joint hyperlaxity
Down syndrome, 50/519, 529
Jolly, F., 1/13
Jones-Thompson index
vertebral canal stenosis, 50/466
Joseph disease, *see* Joseph-Machado disease
Joseph-Machado disease, 42/261-263
see also Hereditary olivopontocerebellar atrophy
amyotrophic lateral sclerosis, 59/410
athetosis, 42/262
autonomic features, 75/189
autonomic nervous system, 75/35
biochemical finding, 60/472-474
bulbar paralysis, 60/658
choreoathetosis, 60/663
chromosome 1p21, 60/474
chromosome 2p23, 60/474
clinical features, 60/467-471
CSF homovanillic acid, 42/262
dementia, 59/410
distal amyotrophy, 60/659
dominant ataxia, 60/476
dystonia, 42/262, 60/664
dystonia musculorum deformans, 49/524
epidemiology, 60/469
external ophthalmoplegia, 60/656
facial myokymia, 42/262
fasciculation, 60/659
gait disturbance, 42/261
genetics, 42/263, 60/471
glial protein, 60/472
hereditary spinal muscular atrophy, 59/26
hypoesthesia, 60/661
modifier gene, 60/474
molecular genetics, 60/474
multiple neuronal system degeneration, 59/137
neuropathy, 60/660
nigrospinodentate degeneration, 42/262, 60/467
nosology, 60/476
nuclear ophthalmoplegia, 42/262, 60/468
nystagmus, 42/262
olivopontocerebellar atrophy, 60/476
ophthalmoplegia, 42/262
opsoclonus, 42/262
pallidoluysionigral degeneration, 49/459
pallidonigral degeneration, 49/445
progressive dysautonomia, 59/137
progressive external ophthalmoplegia, 60/656
progressive external ophthalmoplegia classification, 22/179, 62/290
ptosis, 60/656
Refsum disease, 60/661
rigidity, 42/261
spasticity, 42/261, 60/661
spinal muscular atrophy, 59/26
striatonigral degeneration, 42/262, 60/468, 543
tremor, 42/262, 60/661
type I, 42/263, 60/469
type II, 42/263, 60/469
type III, 42/263
Joseph syndrome
convulsion, 29/231
CSF protein, 29/231
mental deficiency, 29/231
Joubert-Boltshauser syndrome, *see* Joubert syndrome
Joubert syndrome
apnea, 60/667
ataxia, 50/191, 63/437
ataxia telangiectasia, 60/387
cerebellar agenesis, 50/190
cerebellar aplasia, 63/437
cerebellar hypoplasia, 50/191
cerebellar vermis aplasia, 30/384, 50/190
computer assisted tomography, 60/672
corpus callosum agenesis, 63/437
epilepsy, 50/191, 63/437
episodic apnea, 63/437
episodic hyperpnea, 50/191, 63/437
hemifacial spasm, 63/437
hereditary congenital cerebellar atrophy, 60/291
hyperpnea, 60/667, 63/437
meningocele, 63/437
mental deficiency, 50/191, 63/437
non-REM period, 63/437
occipital meningocele, 63/437
polydactyly, 63/437
retinal pigmentary dystrophy, 63/437

secondary pigmentary retinal degeneration, 60/734
Jowar
millet, *see* Millet jowar
Juberg-Hayward syndrome
cheilopalatoschisis, 30/247, 43/446
digital abnormality, 43/446
hypertelorism, 30/247
microcephaly, 43/446
microcrania, 43/446
Judgment error
migraine, 48/162
Jugular body
paraganglioma nodosum, 8/493
Jugular body tumor, *see* Glomus jugulare tumor
Jugular chemodectoma, *see* Glomus jugulare tumor
Jugular foramen
dura mater, 30/424
skull base metastasis, 69/128
skull base tumor, 68/469
Jugular foramen neurinoma, 68/480
skull base, 68/466
Jugular foramen syndrome
see also Multiple cranial neuropathy *and* Vernet syndrome
anterior condylar canal syndrome, 2/99
brain metastasis, 71/612
clinical picture, 2/97
combined nerves IX, X and XI deficit, 2/20, 18/444
critique, 2/225
multiple cranial neuropathy, 51/569
neurofibromatosis type I, 14/151
signe du rideau de Vernet, 2/97
skull base, 67/147
skull base tumor, 17/182
Jugular receptoma, *see* Glomus jugulare tumor
Jugular vein
external, *see* External jugular vein
internal, *see* Internal jugular vein
Jugular vein compression
Monro-Kellie doctrine, 19/100
Jugular vein compression test, *see* Queckenstedt test
Jumpers of the Maine
see also Gilles de la Tourette syndrome
anorexia, 42/232
Blitzkrämpf, 42/232
differential diagnosis, 49/635
echolalia, 42/231, 49/629
latah, 6/301, 49/621
myoclonus, 49/621
myriachit, 42/231-233, 49/621
startle reaction, 42/231
Junction scotoma
visual field defect, 2/567
Junin virus
arenavirus, 56/355
Argentinian hemorrhagic fever, 56/355, 364
murine encephalitis, 56/358
rodent, 56/367
transmission, 56/368
Junius-Kuhnt disease
macula lutea, 13/265
pseudotumor, 13/265
Jürgens-Von Willebrand hereditary thrombocytopenia
hemorrhage, 9/598
Juvenile absence epilepsy
definition, 72/5, 73/149
EEG, 73/151
epidemiology, 73/150
etiology, 73/149
history, 73/149
onset, 73/150
prognosis, 73/151
seizure, 73/150
survey, 73/143
treatment, 73/151
Juvenile Alzheimer disease
amyotrophic lateral sclerosis, 59/285-287
Juvenile amaurotic idiocy
membranovesicular body, 10/226
neuronal ceroid lipofuscinosis, 10/589
Juvenile amyotrophic lateral sclerosis
β-*N*-acetylhexosaminidase A deficiency, 59/402
β-*N*-acetylhexosaminidase B deficiency, 59/402
differential diagnosis, 59/377
Juvenile angiofibroma
skull base, 68/466
skull base tumor, 68/488
Juvenile brain infarction
sickle cell anemia, 55/464
Juvenile cataract
anhidrotic ectodermal dysplasia, 46/91
Juvenile cerebrovascular accident, *see* Stroke
Juvenile cerebrovascular disease
drug abuse, 55/517
paroxysmal nocturnal hemoglobinuria, 55/467
polycythemia vera, 55/467
pseudoxanthoma elasticum, 55/453
sickle cell anemia, 55/504
sickle cell trait, 55/466
Juvenile chronic myeloid leukemia
neurofibromatosis type I, 50/370
Juvenile diabetes mellitus
aminoaciduria, 42/411

brain atrophy, 42/411
Eaton-Lambert myasthenic syndrome, 62/421
optic nerve agenesis, 50/211
recessive optic atrophy, 22/505
Tunbridge-Paley disease, 22/507
Juvenile dystonic lipidosis
cerebellar ataxia, 42/446
dementia, 42/446
dystonia, 60/664
foam cell, 42/447
sea blue histiocyte, 42/447
seizure, 42/446
symptomatic dystonia, 49/543
Juvenile fucosidosis, *see* Fucosidosis
Juvenile Gaucher disease
acid phosphatase, 10/523
ataxia, 42/438
athetosis, 10/307
behavior disorder, 10/307, 523
bone change, 42/438
ceramide galactose, 10/525
choreoathetoid movement, 29/352
clinical features, 60/157
CNS involvement, 66/125
distractibility, 10/307
dysmetria, 10/307
epilepsy, 10/523
Erlenmeyer flask deformity, 10/522
genetics, 10/309, 522
glucosylceramidase, 42/439
grand mal epilepsy, 42/438
hepatosplenomegaly, 10/307, 29/352, 42/438
laboratory study, 60/158
mental deficiency, 10/307, 523, 42/438
muscle rigidity, 10/307
neuronal lipidosis, 10/524
neuropathology, 60/158
ophthalmoplegia, 10/307
pain, 42/438
restlessness, 10/307
seizure, 10/307, 42/438
splenomegaly, 10/522, 565
strabismus, 29/352, 42/438
treatment, 42/439, 60/159
tremor, 10/307
trismus, 42/438
Juvenile Gaucher disease type A
anemia, 66/125
ataxia, 66/125
brain involvement, 66/125
cognitive impairment, 66/125
EEG, 66/125
horizontal gaze paralysis, 66/125
seizure, 66/125
spasticity, 66/125
thrombocytopenia, 66/125
Juvenile Gaucher disease type B
anemia, 66/125
bone pain, 66/125
brain stem auditory evoked potential, 66/128
glucosylceramidase, 66/128
hepatosplenomegaly, 66/125, 128
horizontal gaze paralysis, 66/125
portal hypertension, 66/125
thrombocytopenia, 66/125
Juvenile GM1 gangliosidosis, *see* GM1 gangliosidosis type II
Juvenile GM2 gangliosidosis, *see* GM2 gangliosidosis type III
Juvenile hereditary benign chorea
age at onset, 49/345
autosomal dominant, 49/346
behavior change, 49/346
case, 49/340-343
clinical features, 49/339, 345
course, 49/345
differential diagnosis, 49/346
dysarthria, 49/345
epidemiology, 49/339, 345
event related potential, 49/339
gait, 49/346
genetics, 49/346
Gordon-Hay sign, 49/345
hereditary essential myoclonus, 49/345
Huntington chorea, 49/292, 335
intelligence, 49/346
milking grip, 49/345
muscular hypotonia, 49/345
nosologic status, 49/346
paramyoclonus, 49/336
personality change, 49/346
pronator sign, 49/345
regressive generalized chorea, 49/346
sex ratio, 49/345
sibling, 49/345
stationary generalized chorea, 49/346
tremor, 49/345
Juvenile hereditary diurnal fluctuating dystonia
Huntington chorea, 49/293
Juvenile Huntington chorea
age at onset, 49/286
clinical features, 49/285
criteria, 6/316-323
disease course, 49/286
juvenile parkinsonism, 49/286
Melanesia, 49/269

Negro, 49/269
neuropathology, 49/321
paternal descent, 42/226, 49/286
paternal inheritance, 49/274
review, 6/322
South Africa, 49/270
Venezuela, 49/270
Juvenile hypereosinophilic temporal arteritis
temporal arteritis, 55/345, 347
Juvenile macula degeneration, *see* Stargardt disease
Juvenile metachromatic leukodystrophy
areflexia, 60/660
arylsulfatase variability, 47/591
behavior disorder, 66/168
bone dysplasia, 60/671
characteristics, 66/93
contracture, 60/669
deafness, 60/657
dementia, 60/665
epilepsy, 60/666
facial dysmorphia, 60/666
hepatosplenomegaly, 60/668
leukocyte inclusion, 60/674
lower motoneuron disease, 66/168
mental symptom, 29/358, 60/125, 66/168
motor symptom, 29/358, 66/168
nerve conduction velocity, 60/670
polyneuropathy, 66/168
visual symptom, 29/358
weakness, 66/168
Juvenile muscular atrophy
male, 42/77
unilateral upper extremity, 42/77
Juvenile myoclonic epilepsy
absence seizure, 73/162
antiepileptic agent, 73/168
chromosome, 72/137
definition, 72/5
EEG, 73/164
generalized tonic clonic seizure, 73/160
genetics, 73/165
hyperventilation, 73/165
myoclonic jerk, 73/160
myoclonus, 49/619
pathology, 72/117
sleep, 73/164
survey, 73/157
Juvenile neuroaxonal dystrophy
see also Progressive myoclonus epilepsy
primary neuroaxonal dystrophy, 49/397
Juvenile neuronal ceroid lipofuscinosis
history, 66/671
prevalence, 42/465
sea blue histiocyte, 60/674
survey, 66/24
vacuolated lymphocyte, 42/465
variant, 66/678
Juvenile Niemann-Pick disease, *see* Niemann-Pick disease type C
Juvenile optic atrophy
dyschromatopsia, 42/410
HLA antigen, 42/410
IgA, 42/410
IgG, 42/410
IgM, 42/410
Juvenile paralysis agitans, *see* Progressive pallidal atrophy (Hunt-Van Bogaert)
Juvenile parasomnia
EEG, 45/141
enuresis, 45/141
jactatio capitis nocturna, 45/141
nightmare, 45/140
non-REM sleep, 45/141
pavor nocturnus, 45/140
somnambulism, 45/140
Juvenile parkinsonism
see also Pallidoluysionigral degeneration
asymmetry, 49/155
autonomic nervous system, 49/156
biopterin, 49/159
clinical features, 49/155
depression, 49/156
dystonia, 49/163
dystonia musculorum deformans, 49/163
familial incidence, 49/159
hereditary progressive diurnal fluctuating dystonia, 42/213, 49/536
hypokinesia, 49/155
juvenile Huntington chorea, 49/286
levodopa, 49/156
levodopa adverse reaction, 49/156
mental deterioration, 49/156
neurochemistry, 49/162
neuropathology, 49/159
pallidal degeneration, 49/454
pallidoluysionigral degeneration, 49/454, 458
pallidonigral degeneration, 49/454
pallidopyramidal degeneration, 49/454
Parkinson disease, 6/635, 49/108, 153, 163
pharmacokinetics, 49/157
prevalence, 49/153
rigidity, 49/155
slow progression, 49/156
subgroup, 49/159
thalamic ventral lateral nucleus surgery, 49/157
thalamic ventralis intermedius nucleus surgery,

49/157
tremor, 49/155
Juvenile periarteritis nodosa, *see* Kawasaki syndrome
Juvenile progressive bulbar atrophy
ptosis, 59/70
Juvenile progressive bulbar palsy, *see* Fazio-Londe disease
Juvenile proximal spinal muscular atrophy, *see* Wohlfart-Kugelberg-Welander disease
Juvenile retinoschisis
sex linked, *see* Sex linked juvenile retinoschisis
Juvenile rheumatoid arthritis
see also CINCA syndrome
ankylosing spondylitis, 26/180
classification, 71/4
Juvenile Sly disease, *see* Glycosaminoglycanosis type VII
Juvenile spinal cord injury, 25/175-194, 61/233-250
achondroplasia, 61/233
age, 25/184
atlantoaxial dislocation, 61/233
bladder management, 61/244
bowel management, 61/245
cause, 25/176
cervical predominance, 61/238
computer assisted tomography, 61/243
deformity, 61/245
delayed deficit, 61/239
difference with adults, 25/175, 179, 61/236
differential diagnosis, 61/233
education, 25/175, 61/250
etiology, 61/234
fall injury, 25/177, 61/237
glycosaminoglycanosis type IV, 61/233
Grisel syndrome, 61/233
hypercalcemia, 61/243
immature skeleton, 61/242
incidence, 25/175, 61/234
initial treatment place, 61/250
injury mechanism, 61/242
joint contracture, 25/184
Klippel-Feil syndrome, 61/233
kyphoscoliosis, 25/175, 183
kyphosis, 25/183
lesion level, 25/184, 61/237, 239
life expectancy, 25/193
literature review, 61/235
lordosis, 25/183
lumbar puncture, 25/179
management, 61/241
mobilization, 61/248
muscular imbalance, 25/183
nuclear magnetic resonance, 61/243
orthopedic measure, 61/243
orthosis, 61/249
orthostatic hypotension, 61/243
osteoporosis, 61/247
pelvic asymmetry, 61/246
pressure sore, 25/188, 61/245
pulmonary function, 61/247
radiologic sign absence, 61/239-241
recurrent deficit, 61/239
rehabilitation, 25/175, 192, 61/247
rheumatoid arthritis, 61/233
scoliosis, 25/183
sex distribution, 61/234
spina bifida aperta, 25/176
spinal deformity, 25/183
sport injury, 61/237
symptom, 61/239
torticollis, 61/233
traffic injury, 25/176, 61/237
treatment, 25/185-187, 191, 61/243
urology, 25/190
Juvenile spinal muscular atrophy, *see* Wohlfart-Kugelberg-Welander disease
Juvenile sulfatidosis
Austin, *see* Austin juvenile sulfatidosis
Juvenile Tay-Sachs disease
β-*N*-acetylhexosaminidase A deficiency, 29/365
Juvenile vertebral column injury
achondroplasia, 61/233
atlantoaxial dislocation, 61/233
cervical predominance, 61/238
differential diagnosis, 61/233
glycosaminoglycanosis type IV, 61/233
Grisel syndrome, 61/233
Klippel-Feil syndrome, 61/233
literature review, 61/235
rheumatoid arthritis, 61/233
torticollis, 61/233
Juvenile xanthogranuloma
neurofibromatosis type I, 50/366
Juxtacarotid chemodectoma
tissue culture, 17/69
Juxtarestiform
dentatorubropallidoluysian atrophy, 49/442

K-1 antigen
Escherichia coli, 33/61, 64, 52/118
Kaguo-kuzira intoxication, *see* Balaenoptera borealis intoxication
Kahler disease, *see* Multiple myeloma
Kahnschädel, *see* Scaphocephaly
Kainate, *see* Kainic acid

Kainic acid
epilepsy, 72/56
excitotoxin, 64/20
Huntington chorea, 49/58, 261
striatal necrosis, 49/511
Kainic acid intoxication
striatal necrosis, 49/500
Kainic acid neurotoxicity
Huntington chorea, 49/297
Kainic acid receptor
domoic acid, 65/151
excitotoxic mechanism, 64/20
lathyrism, 65/14
Kaiser Wilhelm Institute
history of neurology, 1/37
Kala Azar, *see* Visceral leishmaniasis
Kallidinogenase, *see* Kallikrein
Kallikrein
bradykinin, 48/97
multiple sclerosis, 9/397
pain generation, 1/129
Kallikrein-kinin system
brain edema, 67/80, 83
Kallmann syndrome
anosmia, 30/471, 43/418
arhinencephaly, 30/471
holoprosencephaly, 30/471
hypogonadism, 43/418
microcephaly, 30/100
septo-optic dysplasia, 30/471
Kalmuk idiocy, *see* Down syndrome
Kana reading
alexia, 45/446
Kanamycin
acoustic neuropathy, 64/9
neurobrucellosis, 52/596
neurotoxin, 50/219
toxic neuropathy, 64/9
Kanamycin intoxication
symptom, 64/9
Kandori disease
primary pigmentary epithelium degeneration, 13/31
Kanji reading
alexia, 45/446
Kaolin
hydrocephalus, 50/427
hydromyelia, 50/427
Kaposi sarcoma
acquired immune deficiency syndrome, 56/489, 515
epidemiology, 71/340
human immunodeficiency virus, 56/489
human immunodeficiency virus infection, 71/340
immunosuppressive therapy, 65/549
Kaposi syndrome, *see* Hemangiosarcoma
Kappa light chain
antibody, 9/502
viral infection, 56/52
Karyorrhexis
Huntington chorea, 49/325
Karyotype
abnormal, *see* Abnormal karyotype
aqueduct stenosis, 30/613
astrocytoma, 67/43
brain tumor, 67/43
holoprosencephaly, 30/459-466, 42/33
Kasabach-Merritt syndrome
intravascular consumption coagulopathy, 63/318
Kast syndrome
Maffucci syndrome, 14/118
Katayama fever, *see* Acute schistosomiasis
Kaufman valve
technical detail, 30/576
Kawasaki syndrome, 56/637-641
see also Polyarteritis nodosa
acetylsalicylic acid, 52/267
acrodynia, 56/638
angiopathic polyneuropathy, 51/446
arteritis, 52/266
arthralgia, 51/453
arthritis, 52/265
brain infarction, 52/265
cause, 52/264
cervical lymph node, 56/638
child, 52/264
clinical features, 52/264, 56/638
CNS, 51/453, 52/253
coma, 52/265, 56/639
conjunctival congestion, 56/638
coronary artery aneurysm, 56/637
corticosteroid, 52/267
cranial nerve palsy, 52/266, 56/639
cranial neuropathy, 56/639
creatine kinase, 56/639
CSF, 51/453, 52/265, 56/639
CSF cytology, 56/639
desquamation, 56/638
diagnostic criteria, 52/265
differential diagnosis, 52/266
EMG, 56/639
epidemiology, 52/264, 56/637
epilepsy, 52/265
erythroderma, 56/638
facial paralysis, 51/453
fever, 56/638

ganglionitis, 52/266
histopathology, 52/266
immune complex vasculitis, 56/638
iridocyclitis, 56/638
ischemic heart disease, 56/638
keratitis, 56/638
laboratory diagnosis, 52/266
lethargy, 52/265, 56/639
lymphocytic meningitis, 52/265
meningitis, 52/264, 266
meningitis serosa, 51/453
mental symptom, 52/265
mortality, 52/264, 56/637
multiple cranial neuropathy, 51/453
multiple sclerosis, 51/453
muscle biopsy, 56/639
myocardial dysfunction, 52/265
myoclonic epilepsy, 52/265
myositis, 56/639
necrotizing myositis, 56/639
necrotizing vasculitis, 56/641
nerve conduction, 56/639
neuritis, 52/266
neurologic symptom, 52/265
neuropathy, 56/639
polyarteritis nodosa, 48/285, 51/446, 453, 55/358
polymorphous exanthema, 56/638
polymyositis, 62/373
Propionibacterium acnes, 56/638
pyuria, 52/266
relapse, 56/637
retrovirus, 52/264
sensorimotor neuropathy, 56/639
sex ratio, 56/637
strawberry tongue, 52/265, 56/638
thrombocytosis, 52/266
treatment, 52/267, 56/641
uveitis, 56/638
vasculitis, 56/641, 71/8
viral meningitis, 56/639
Kayser-Fleischer ring
primary biliary cirrhosis, 49/228
pseudo, *see* Pseudo-Kayser-Fleischer ring
Wilson disease, 6/267, 270, 274, 27/379, 386, 42/270, 46/58, 49/224, 228
Kearns-Sayre-Daroff-Shy syndrome, 38/221-230
see also Progressive external ophthalmoplegia
acanthocyte, 49/332
Adams-Stokes syndrome, 43/143
ataxia, 62/508
biochemistry, 62/311
cardiac disorder, 62/309
cardiac involvement, 38/229
cardiomyopathy, 22/211, 43/142, 60/667, 62/310
celiac disease, 62/313
cerebellar ataxia, 43/142
cerebellar involvement, 38/225
cerebellar syndrome, 62/308
classification, 40/281
clinical features, 38/222-227, 40/302, 62/307
complex IV, 62/314
computer assisted tomography, 62/306, 311
cranial nerve, 38/225
cranial neuropathy, 62/313
CSF, 38/225, 228, 43/142, 62/311
CSF protein, 62/508
deafness, 62/313
definition, 62/307
dementia, 62/308, 508
diabetes mellitus, 62/309, 508
differential diagnosis, 38/229, 40/298
dwarfism, 60/666
ECG, 62/311
ECG abnormality, 60/670
endocrine disorder, 62/309
epilepsy, 62/308, 72/222
etiology, 38/227-229
external ophthalmoplegia, 60/655
facial paralysis, 43/142
familial spastic paraplegia, 59/334
genetics, 62/312
heart block, 38/224, 62/508
histochemistry, 40/47, 54
histopathology, 62/310
history, 38/221
hypoparathyroid myopathy, 62/541
hypoparathyroidism, 41/215, 43/142, 62/309, 508, 70/121
intestinal pseudo-obstruction, 62/313
lipid, 40/242
mental deficiency, 38/225, 43/142, 62/308
MEPOP syndrome, 62/315
metabolic myopathy, 62/508
mitochondrial abnormality, 22/196, 43/142
mitochondrial change, 40/220
mitochondrial disease, 62/497
mitochondrial DNA, 66/397, 426
mitochondrial DNA defect, 62/507
mitochondrial encephalomyopathy, 62/498
mitochondrial myopathy, 41/215, 62/313
mitochondrial protein, 66/17
MNGIE syndrome, 62/313
molecular genetics, 62/311
muscle biopsy, 62/508
muscle fiber type I, 43/142
muscle histopathology, 62/310

neuromuscular manifestation, 38/225
neuropathy, 62/313
oculopharyngeal muscular dystrophy, 40/302
onset before age 20, 62/508
ophthalmoplegia, 22/179, 38/224, 229, 43/142
OSPOM syndrome, 62/313
pathogenesis, 38/227
pathologic reaction, 40/253
pathology, 38/228
Pearson syndrome, 62/508, 66/427
pigmentary retinopathy, 62/308, 508
POLIP syndrome, 62/313
polyneuropathy, 62/309, 313
progressive external ophthalmoplegia, 22/192, 211, 40/302, 41/215, 43/137, 60/52-57, 62/287, 306, 508
progressive external ophthalmoplegia classification, 22/179, 40/281, 62/290
ptosis, 38/224, 229, 60/655
ragged red fiber, 40/220, 60/54, 62/292, 314, 508
Refsum disease, 21/212
renal dysfunction, 62/309
retina, 38/224, 229
retinopathy, 62/311
secondary pigmentary retinal degeneration, 22/211, 43/142, 60/654, 725
sensorineural deafness, 43/142, 62/508
short stature, 43/142, 62/508
striatopallidodentate calcification, 49/424
treatment, 38/229, 62/312
vitamin B_c deficiency, 51/336
Kearns-Sayre-Daroff syndrome, *see* Kearns-Sayre-Daroff-Shy syndrome
Kearns-Shy syndrome, *see* Kearns-Sayre-Daroff-Shy syndrome
Kegel exercise
bladder, 75/677
stress incontinence, 75/677
Keimzellen of His, *see* Germ cell *and* Matrix cell
Kell blood group
chorea-acanthocytosis, 49/330
genetic linkage, 42/611
McLeod phenotype, 62/127
Kell blood group antigen
McLeod phenotype, 63/288
Keloid
basal cell, 43/146
clinodactyly, 43/146
congenital torticollis, 43/146
hyperpigmentation, 43/146
hypertension, 43/146
renal abnormality, 43/146
spasmodic torticollis, 43/146
Turner syndrome, 50/543
Kemp sign
lumbar intervertebral disc prolapse, 20/586
Kennedy, F., *see* Foster Kennedy
Kennedy syndrome, *see* Foster Kennedy syndrome
Kennmuskel, *see* Segment pointer muscle
Kenyan tick-borne typhus
rickettsial infection, 34/657
Kepone intoxication, *see* Chlordecone intoxication
Kerasin
chemistry, 10/249
globoid cell leukodystrophy, 10/304
Keratan
connective tissue, 10/432
corneal, 27/145
glycosaminoglycanosis, 10/327, 432
glycosaminoglycanosis type IV, 10/342
GM1 gangliosidosis, 10/223, 473
skeletal, 27/144
Keratan sulfate
chemistry, 27/144
degradation, 66/64
glycosaminoglycanosis, 66/281
glycosaminoglycanosis type IV, 66/303, 305
glycosaminoglycanosis type IVB, 66/57
GM1 gangliosidosis, 10/479, 27/155
intervertebral disc, 20/526, 535
Keratansulfaturia
symptomatic dystonia, 49/543
Keratin
CSF, *see* CSF keratin
Keratitis
curvulariosis, 52/484
drechsleriasis, 52/486
hereditary sensory and autonomic neuropathy type V, 60/17
herpes simplex virus, 56/8
interstitial, *see* Interstitial keratitis
Kawasaki syndrome, 56/638
neurosyphilis, 52/274
neurotrophic, *see* Neurotrophic keratitis
Keratitis bullosa
cornea, 8/350
edema, 8/350
Keratitis dystrophica
ectodermal dysplasia, 14/112
Keratoconjunctivitis
endocrine candidiasis syndrome, 42/642
Keratoconus
congenital retinal blindness, 13/272, 22/530
cutis verticis gyrata, 43/244
Down syndrome, 50/526
primary pigmentary retinal degeneration, 13/179

Stargardt disease, 13/133
Keratoderma
multiple hamartoma syndrome, 42/754
Keratopachydermia
hereditary deafness, 50/218
sensorineural deafness, 42/378
Keratopathy
ischemic, *see* Ischemic keratopathy
Keratosis
follicular, *see* Darier disease
hereditary palmoplantar, *see* Hereditary palmoplantar keratosis
palmoplantaris, *see* Palmoplantar keratosis
Keratosis follicularis, *see* Darier disease
Keratosis palmoplantaris, *see* Palmoplantar keratosis
Keratosis pilaris
atypical, 14/791
epilepsy, 14/791
mental deficiency, 14/791
Keratosulfate, *see* Keratan
Keraunoparalysis
electrical injury, 61/192
spinal cord electrotrauma, 61/192
Kern calcinosis, *see* Striatopallidodentate calcification
Kernicterus, 6/491-513
see also Acquired hepatocerebral degeneration, Bilirubin encephalopathy *and* Posticteric encephalopathy
ABO incompatibility, 42/583
acholuric jaundice, 6/492
athetosis, 6/450, 503
bilirubin encephalopathy, 27/415, 419
biochemistry, 6/729
blood-brain barrier, 28/388-391
brain stem nucleus, 6/502
chemistry, 49/218
chorea, 6/438
choreoathetosis, 6/160, 42/583
clinical features, 6/493-496
cor de chasse, 6/494
cortex, 6/503
Crigler-Najjar syndrome, 21/53, 27/420
deafness, 6/507, 513, 49/384
demyelination, 9/638
dentate nucleus, 6/502
dentatorubropallidoluysian atrophy, 21/520
double athetosis, 49/384
dystonia, 6/552
enamel defect, 42/583
episodic dystonia, 6/505
globus pallidus, 6/497
hippocampus, 6/502
jaundice, 42/582
leukoencephalopathy, 9/638, 47/585
maternal vitamin K3, 65/570
mental deficiency, 42/583
Moro reflex, 42/583
neuropathology, 6/512
opisthotonos, 42/583
pallidal lesion, 6/661
pallidal necrosis, 49/465
pallidonigral necrosis, 49/465
pallidostriatal necrosis, 49/465
pathology, 6/496-503
posticteric encephalopathy, 6/452, 491
predisposing factor, 6/493
premature infant, 6/495
prematurity, 6/496
prevention, 6/494
psychomotor retardation, 6/503
Rh incompatibility, 42/583
seizure, 42/583
sensorineural deafness, 42/583
setting sun sign, 6/494, 509, 511
striatum, 6/502
subthalamic body, 6/497, 502
thalamus, 6/502
treatment, 6/496
vertical gaze palsy, 49/384
vitamin K3 intoxication, 65/570
Kernig sign
bacterial meningitis, 33/1
lymphocytic choriomeningitis, 56/359
pneumococcal meningitis, 33/38, 52/43
significance, 1/545
technique, 1/543
Kerosene intoxication, *see* Kerosine intoxication
Kerosine intoxication
headache, 48/420
Kestenbaum syndrome, *see* Aqueduct stenosis
Ketamine intoxication
convulsion, 37/408
dreaming, 37/408
hyperexcitability, 37/408
intravenous, 37/408
psychomotor effect, 37/408
seizure, 37/408
Ketanserin
migraine, 48/193
serotonin 2 receptor, 48/88
Ketoacidosis
alcohol intoxication, 64/112
diabetic, *see* Diabetic ketoacidosis
migraine, 48/158

mucormycosis, 52/474
survey, 70/193
therapy, 70/198
3-Ketoacyl-coenzyme A thiolase, *see* Acetyl-coenzyme A acyltransferase
Ketoconazole
aspergillosis, 52/381
candidiasis, 52/405
coccidioidomycosis, 52/421
cryptococcosis, 52/433
Cushing syndrome, 68/355
drechsleriasis, 52/486
Histoplasma meningitis, 52/442
histoplasmosis, 52/441
neurotoxin, 65/472
North American blastomycosis, 52/391
paracoccidioidomycosis, 52/461
α-Ketoglutaramate, *see* 2-Oxoglutaramic acid
α-Ketoglutarate, *see* 2-Oxoglutaric acid
α-Ketoglutarate dehydrogenase deficiency, *see* Oxoglutarate dehydrogenase deficiency
Ketone
toxic polyneuropathy, 51/68
Ketonemia
diabetes mellitus, 70/195
Ketonuria
ketotic hyperammonemia, 29/106
newborn, 66/640
organic acid metabolism, 66/641
subarachnoid hemorrhage, 12/151
Ketoprofen
chronic paroxysmal hemicrania, 48/264
headache, 48/174
migraine, 48/174
Ketosis
organic acid metabolism, 66/641
17-Ketosteroid
Alzheimer disease, 46/265
basophil adenoma, 2/457
cerebral gigantism, 31/333
migraine, 5/42
spinal cord injury, 26/397, 402, 451
β-Ketothiolase, *see* Acetyl-coenzyme A acyltransferase
Ketotic coma
systemic brain infarction, 11/463
Ketotic hyperammonemia
coma, 29/106
glycine, 29/106
hyperlactic acidemia, 29/106
hyperpyruvic acidemia, 29/106
hypotonia, 29/106
ketonuria, 29/106
tonic clonic seizure, 29/106
Ketotic hyperglycinemia, *see* Propionic acidemia
Ketotic hypoglycemia
idiopathic hypoglycemia, 27/72
Ketotifen
chronic migrainous neuralgia, 48/252
Kety-Schmidt method
brain blood flow, 11/120, 269, 54/139
Khalifeh-Zellweger-Plancherel syndrome
ulceromutilating acropathy, 21/77
Khat leaf
cathine, 65/50
cathinone, 65/50
methcathinone, 65/50
Kidney
allergic granulomatous angiitis, 63/383
amyloidosis, 71/508
bismuth intoxication, 64/336
cystic, *see* Cystic kidney
glycosaminoglycanosis, 10/454
hereditary amyloid polyneuropathy type 2, 60/98
hereditary amyloid polyneuropathy type 3, 60/100
horseshoe, *see* Horseshoe kidney
lead, 64/434
leptospirosis, 33/398
myolipoma, 14/370
polycystic, *see* Polycystic kidney
Refsum disease, 10/345, 13/314, 21/238, 36/348, 42/148, 66/497
Reye syndrome, 29/332, 34/168, 56/157
scleroderma, 39/363
Smith-Lemli-Opitz syndrome, 66/586
thallium intoxication, 36/249
Kidney agenesis, *see* Renal agenesis syndrome
Kidney carcinoma, *see* Renal carcinoma
Kidney malformation
Bardet-Biedl syndrome, 13/402
caudal aplasia, 50/511
multiple nevoid basal cell carcinoma syndrome, 14/468
Patau syndrome, 14/120, 31/508, 43/528, 50/558
trisomy 9, 50/564
Kifa frame
decubitus, 26/222
Kii Peninsula amyotrophic lateral sclerosis, 22/353-416
age at onset, 22/367
2-amino-3-methylaminopropionic acid, 59/274
associated disease, 22/371
blood group, 22/360
cause of death, 22/369
clinical features, 22/369
cycad, 22/130

death rate, 22/355, 359, 361
dementia, 22/369
description, 21/34
diabetes mellitus, 22/371
duration of illness, 22/367
EEG, 22/371
EMG, 22/371
environment, 59/274
epidemiology, 22/354, 359, 367, 59/273
exogenous agent, 22/373
features, 21/26
geography, 22/354
Guam amyotrophic lateral sclerosis, 59/253
Hallervorden-Spatz syndrome, 22/403
hereditary type, 22/360, 371, 373
history, 22/353
Hobara focus, 22/360, 373
investigation, 22/283
Kozagawa focus, 22/357
laboratory data, 22/371
Little disease, 22/371
manganese, 22/359
Marinesco body, 22/414
metal, 22/357-359
neurofibrillary tangle, 22/366, 373, 59/278, 283
neuropathology, 22/373-408, 59/237, 294-296
olivopontocerebellar atrophy, 22/371
3-oxalylaminoalanine, 59/274
Parinaud syndrome, 22/371
Parkinson dementia, 59/237
Parkinson dementia complex, 22/361, 377, 59/274
pathology, 22/129
polyneuropathy, 22/371
prevalence, 22/407
psychotic episode, 59/283
sex ratio, 22/359, 367
spinal cord malformation, 22/371
spinocerebellar degeneration, 22/371
twins, 22/371
Killer cell
multiple sclerosis, 47/352
natural, *see* Natural killer cell
Kiloh-Nevin syndrome
progressive external ophthalmoplegia, 43/136, 60/52
Kimmerle abnormality, *see* Foramen arcuale
Kinase, *see* Phosphotransferase
Kinesia paradoxa, *see* Paradoxical kinesia
Kinesigenic ataxia
hereditary periodic ataxia, 60/433
Kinesigenic choreoathetosis
ballismus, 42/207
basal ganglion degeneration, 27/308, 42/208
dopamine β-mono-oxygenase, 42/208
dystonia, 42/207
EEG, 42/208
epilepsy, 42/208
hereditary paroxysmal, *see* Hereditary paroxysmal kinesigenic choreoathetosis
paroxysmal, *see* Hereditary paroxysmal kinesigenic choreoathetosis
sex ratio, 42/208
speech disorder in children, 42/207
symptomatic paroxysmal, *see* Symptomatic paroxysmal kinesigenic choreoathetosis
Kinesigenic myoclonus
cerebellar influence, 21/513
Kinesigenic pain
migrant sensory neuritis, 51/461
Kinesimetry
Huntington chorea, 49/289
Kinesin
neuroaxonal dystrophy, 49/395
Kinesthetic hallucination
body scheme disorder, 4/218
Kinesthetic neuron
thalamic ventralis intermedius nucleus, 49/600
Kinesthetic reflex epilepsy, *see* Hereditary paroxysmal kinesigenic choreoathetosis
Kinetic sense
proprioception, 1/95
Kinetic tremor, *see* Intention tremor
King-Denborough syndrome
malignant hyperthermia, 62/570
myoglobinuria, 62/558
Kinin
cluster headache, 48/233
pain generation, 1/129
Kininogen
bradykinin, 48/97
hemostasis, 63/303
migraine, 48/97
pain generation, 1/129
Kinky hair disease, *see* Trichopoliodystrophy
Kinky hair syndrome, *see* Trichopoliodystrophy
Kirschner coagulation
gasserian ganglion, 5/306
trigeminal neuralgia, 5/399
Kit test
spinal cord lesion, 2/205
Kjellin syndrome, 22/467-471
see also Hereditary olivopontocerebellar atrophy
Babinski sign, 42/174
deafness, 22/471
EEG, 42/174
familial spastic paraplegia, 59/326

fasciculation, 42/173
hereditary, 22/470, 471
mental deficiency, 22/470, 42/173-175
muscular atrophy, 22/469, 42/173-175
neuropathy, 42/173
retinal degeneration, 22/470, 42/173-175
secondary pigmentary retinal degeneration, 60/733
spastic paraplegia, 22/469-471, 42/173-175
Kjer optic atrophy, *see* Infantile optic atrophy
Klaus index
basilar impression, 32/51
definition, 32/21
Klazomania (Benedek)
postencephalitic shouting tic, 6/788, 795
Klebsiella
adult meningitis, *see* Gram-negative bacillary meningitis
bacterial meningitis, 52/3
brain abscess, 52/149
gram-negative bacillary meningitis, 52/117
infantile enteric bacillary meningitis, 33/62
Klebsiella pneumoniae
bacterial meningitis, 52/7
gram-negative bacillary meningitis, 52/104
Klebsiella type 26
gram-negative bacillary meningitis, 52/118
Kleeblattschädel, *see* Cloverleaf skull
Kleeblattschädel syndrome, *see* Cloverleaf skull syndrome
Klein syndrome, *see* Klein-Waardenburg syndrome
Klein-Waardenburg syndrome
see also Mende syndrome
acoustic nerve agenesis, 50/217
brachycephaly, 14/116
cheilopalatoschisis, 43/52
clinical features, 14/115, 50/218
congenital megacolon, 43/52, 75/635
deaf mutism, 14/115
deafness, 14/115
differential diagnosis, 14/116
dystopia canthorum, 43/51
gene frequency, 43/53
genetics, 43/53, 50/218
glaucoma, 13/462
gyrate atrophy (Fuchs), 14/116
hereditary deafness, 30/408, 50/218
heterochromia iridis, 14/115, 30/408, 43/51, 50/218
hypertelorism, 14/116, 30/408, 50/218
Mende syndrome, 14/116
mutation rate, 43/53
neurocutaneous syndrome, 14/101
neuropathology, 59/162
partial albinism, 14/106, 115, 30/408, 50/218
pathology, 50/218
prevalence, 43/53
progressive dysautonomia, 59/162
sensorineural deafness, 30/408, 43/51, 50/218
sural nerve biopsy, 59/162
synophrys, 14/116, 30/408, 43/51, 50/218
Telfer syndrome, 22/513
type I, 43/51-53
type II, 43/42-44, 52
white hair, 14/115, 30/408, 43/51
Kleine-Levin syndrome
clinical features, 46/418
intermittent daytime hypersomnia, 2/449, 45/136
sleep disorder, 3/97
Whipple disease, 52/138
Kleist, K., 1/8, 24, 28
Kleptomania
dysphrenia hemicranica, 5/79
migraine, 48/162
Klinefelter syndrome
aggression, 43/560
aneuploidy, 50/546
azoospermia, 50/546
behavior, 31/481, 50/549
behavior disorder, 43/557, 50/548
behavior disorder in children, 43/557
birth incidence, 43/558
cardinal symptom, 31/479-481
cerebellar ataxia, 21/467
chorioretinal degeneration, 13/40
clinical features, 31/479-483, 50/548
CNS, 31/482
convulsion, 50/549
corpus callosum agenesis, 50/163
cretinism, 50/549
criminality, 43/205, 46/75
definition, 31/477, 50/546
diagnosis, 31/485
double trisomy, 31/485
dysmorphic sign, 50/550
EEG, 31/482, 50/549
endocrine function, 31/483
epicanthus, 43/560
epidemiology, 31/477-479, 50/547
epilepsy, 31/482
epiphyseal dysplasia, 43/560
foramina parietalia permagna, 50/142
gene penetrance, 13/25
gynecomastia, 43/557, 50/546, 550
heart disease, 43/560
history, 31/477

hypergonadotropic hypogonadism, 50/546, 549
hypertelorism, 43/560
hypogonadism, 43/558, 560, 60/668
intelligence, 50/549
intelligence quotient, 50/550
kyphoscoliosis, 60/669
maternal age, 31/478
mental deficiency, 21/467, 31/477, 479, 481, 43/557, 560, 46/75
mental subnormality, 50/548
microcephaly, 43/560
mosaicism, 31/485
muscular hypotonia, 50/550
optic atrophy, 13/40
parkinsonism, 43/558
pathogenesis, 31/478, 50/547
prevalence, 43/558
rare variant, 50/549
recurrence risk, 43/558
schizophrenia, 43/557
scrotum hyperplasia, 50/550
seizure, 43/558
self-mutilation, 43/560
stature, 50/550
testis, 50/550
thromboembolic disease, 50/549
treatment, 31/485
tremor, 21/467, 43/558, 50/549
XX male, 31/484
49,XXXXY variant, 46/76
48,XXXY variant, 46/75
48,XXYY variant, 46/75
46,XY variant, 46/75
47,XYY variant, 46/75

Klinefelter variant XXXXY
aggression, 43/560
cheilopalatoschisis, 43/560
congenital heart disease, 31/483
cryptorchidism, 31/484
epiphyseal dysplasia, 43/560
growth retardation, 31/483
heart disease, 43/560
hypertelorism, 30/250, 31/483, 43/560
hypoplasia, 31/484
mental deficiency, 31/483, 43/561
microcephaly, 31/484, 43/560
mongoloid eyes, 31/483
palatoschisis, 31/483
radioulnar dysostosis, 31/483
skeletal abnormality, 31/483

Klinefelter variant XXXY
birth incidence, 43/559
epicanthus, 43/559
glycogen storage disease, 43/559
gynecomastia, 43/559
hypogonadism, 43/559
mental deficiency, 31/483, 43/559
prevalence, 43/559
radioulnar dysostosis, 31/483, 43/559
tall stature, 31/483, 43/559

Klinefelter variant XXXYY
aggression, 43/563
mental deficiency, 31/484

Klinefelter variant XXYY
aggression, 31/484, 43/562
birth incidence, 43/563
cryptorchidism, 31/484
dermatoglyphics, 31/484
EEG, 43/563
epilepsy, 31/484
gigantism, 31/484
gynecomastia, 31/484, 43/563
hypogonadism, 43/563
mental deficiency, 31/484, 43/563
motor nerve conduction velocity, 31/484
nerve conduction velocity, 43/563
radioulnar dysostosis, 31/484, 43/563
sensory loss, 43/563
simian crease, 31/484
skeletal abnormality, 31/484
social adjustment, 31/484
tall stature, 43/562

Klintworth syndrome, *see* Acute myoclonic ataxia

Klippel-Feil syndrome, 32/111-121
abducent nerve paralysis, 42/37
Arnold-Chiari malformation type II, 50/407
autosomal dominant inheritance, 50/64
autosomal recessive inheritance, 50/64
birth incidence, 42/37
cervico-oculoacusticus syndrome, 32/123-126
cheilopalatoschisis, 42/37
clinical features, 32/111, 116-120
craniolacunia, 32/187, 50/139
Dandy-Walker syndrome, 42/23
EMG, 32/119
encephalomeningocele, 42/26
epidemiology, 30/150, 162, 32/115
epithelial neurenteric cyst, 32/429
etiology, 32/112-115
genetics, 30/150, 162, 32/112-115
hearing loss, 32/120
heart disease, 42/37
hemangiolipoma, 20/440
hemivertebra, 42/37
history, 32/111
hypertelorism, 30/249

hypogonadism, 42/37
incidence, 30/257, 50/64
iniencephaly, 30/257, 50/131
intraspinal neurenteric cyst, 32/429
juvenile spinal cord injury, 61/233
juvenile vertebral column injury, 61/233
malformation, 32/111, 117, 143
meningocele, 42/37
mirror movement, 1/409, 32/118, 42/233
occipital cephalocele, 50/101
occipital encephalocele, 32/60, 42/26
pathology, 32/115
pterygium colli, 42/37
radiologic appearance, 32/120
rib abnormality, 42/37
scoliosis, 40/337, 42/38
segmentation failure, 32/143
sensorineural deafness, 42/37
skeletal malformation, 30/249
spastic paraplegia, 59/431
spinal lipoma, 20/397
spinopontine degeneration, 21/390
spondylosis, 42/37
Sprengel deformity, 1/483, 42/37, 43/483
Stilling-Türk-Duane syndrome, 50/213
syncope, 7/453
treatment and prognosis, 32/120, 143
Turner syndrome, 14/109
vertebral abnormality, 30/100
vertebral column, 32/111, 143
Wildervanck syndrome, 43/343

Klippel-Trénaunay syndrome, 14/390-401
see also Parkes Weber syndrome
accessory symptom, 14/393
acrocephaly, 14/524
arachnodactyly, 14/395, 524
arteriovenous malformation, 43/25
associated malformation, 14/524
ataxia telangiectasia, 14/252
brain arteriovenous malformation, 31/200
clinical features, 14/116, 391, 395
deafness, 14/523
differential diagnosis, 14/116
epilepsy, 14/395, 43/24
forme fruste, 14/116, 395
genetics, 14/520-524
gynecomastia, 14/524
hemifacial atrophy, 14/524
hemifacial hypertrophy, 14/397-399
history, 14/390
iris coloboma, 14/524
macrencephaly, 14/400
mental deficiency, 14/395, 43/24
micrencephaly, 14/395
microcephaly, 14/395, 524
multiple nevoid basal cell carcinoma syndrome, 14/523
neurocutaneous syndrome, 14/101
neurofibromatosis, 14/397, 532
neurofibromatosis type I, 14/568
neuropathology, 14/398
nevus, 14/392
nevus flammeus, 43/24
osteohypertrophy, 14/13, 67, 116, 393, 43/24
osteoporosis, 43/30
ota nevus, 43/30
Parkes Weber syndrome, 14/395-397
partial gigantism, 14/522
pathogenesis, 14/399
phakomatosis, 31/26, 50/377
polydactyly, 14/524
spina bifida, 14/395, 43/25
Sturge-Weber syndrome, 14/7, 228, 397, 532, 43/49
syndactyly, 14/395, 523
syringomyelia, 14/116
terminology, 14/12
treatment, 14/401
tuberous sclerosis, 14/398
varices, 14/116, 393
Von Hippel-Lindau disease, 14/398, 532

Klippel-Trénaunay-Weber syndrome, *see* Klippel-Trénaunay syndrome

Klumpke paralysis, *see* Brachial paralysis *and* Brachial plexus

Klüver-Bucy syndrome, 45/257-262
affect, 45/45
aggression, 45/262
agnosia, 3/349-356
Alzheimer disease, 45/260
amyotrophic lateral sclerosis, 59/234
animal experiment, 2/602, 708, 716
brain death, 24/723, 745
brain stem, 45/260
disconnection syndrome, 2/603
distractibility, 3/352
emotional disorder, 3/346, 349-356
etiology, 3/350
grasp reflex, 45/258
herpes simplex virus encephalitis, 45/260
Huntington chorea, 49/291
incomplete syndrome, 3/355
lesion, 3/351
localization, 3/353, 45/260
memory disorder, 3/349-356, 45/262
olfactory afferent, 3/356

oral grasping, 3/350, 355, 45/257-260
pathology, 24/704
Pick disease, 45/257, 260, 46/235
placidity, 3/352
prognosis, 24/704
psychic blindness, 2/602, 708
sexual behavior, 3/352, 355, 45/259, 262
sign, 24/723, 45/257
syndrome, 3/350
temporal lobe, 2/602, 24/724
temporal lobe epilepsy, 45/259
temporal lobe infarction, 45/260
temporal lobe lesion, 2/708, 23/11
temporal lobe syndrome, 45/43
temporal lobectomy, 2/708
theory, 3/353
visual agnosia, 45/260
visual association area, 2/602
visual discrimination, 2/603
visual perception disorder, 2/708, 716
visual sensory system, 2/602
Knee
valgum, *see* Genu valgum
varum, *see* Genu varum
Knee fracture
peroneal nerve, 70/38
Kneeling
hereditary multiple recurrent mononeuropathy, 51/554
Knife wound
brachial plexus, 51/144
Knock out
boxing injury, 23/540, 552, 566
Koch phenomenon
viral infection, 56/65
Koch postulate
acoustic neuritis, 56/105
labyrinthitis, 56/105
vestibular neuronitis, 56/105
Koch reaction
hypersensitivity, 9/503
Kocher-Debré-Sémélaigne syndrome
see also Hoffmann syndrome
hypertrophia musculorum vera, 43/84
hypothyroid myopathy, 62/533
hypothyroidism, 40/324, 41/243
Kocher, E.T., 1/17
Koenen fibroma
tuberous sclerosis, 31/7
Koenen tumor
tuberous sclerosis, 14/49, 343
Köhlmeier-Degos disease, 39/435-446, 55/275-280
age, 14/789, 55/275, 277
angiography, 55/277
angiopathic polyneuropathy, 51/446
anterior uveitis, 55/278
aphasia, 55/277
ataxia, 55/276
brain infarction, 14/789, 53/163, 55/276
brain ischemia, 53/163
brain vasculitis, 14/789, 55/277
cause, 55/278
chorea, 55/276
choreoathetosis, 55/276
clinical features, 39/435-437, 55/275
CSF, 55/277
dermatologic lesion, 14/789, 55/275
differential diagnosis, 39/437
duration, 39/436
dysarthria, 55/276
etiology, 39/443-445
facial paralysis, 39/436, 51/454
gastrointestinal lesion, 14/789, 39/437, 55/276
headache, 39/437, 55/276
hemianopia, 39/437
hiccup, 39/437
history, 39/435
immunopathology, 55/279
intermittent headache, 55/276
laboratory finding, 39/437
multi-infarct dementia, 55/277
multiple brain infarction, 14/789, 39/435, 55/276
multiple cranial nerve deficit, 55/277
multiple cranial neuropathy, 55/277
myelomalacia, 39/443, 63/55
neurologic manifestation, 55/276
ophthalmology, 39/440, 55/278
papule size, 39/438, 55/276
paresthesia, 39/437
pathology, 55/278
pigmentary chorioretinitis, 55/278
polyneuropathy, 39/437, 51/454
polyradiculopathy, 51/454
posterior uveitis, 55/278
seizure, 39/437
sex ratio, 39/435, 51/454, 55/275
silent lesion, 55/276
skin, 39/436
superior longitudinal sinus thrombosis, 55/277
systemic brain infarction, 11/464
systemic manifestation, 55/278
treatment, 39/445, 55/279
uveitis, 55/278
venous sinus thrombosis, 55/277
vomiting, 39/437
weakness, 39/437

Kohlu injury
brain injury, 23/597
head injury, 23/597
management, 23/598
mechanics, 23/595
prevention, 23/600
site, 23/595
skull fracture, 23/597
type, 23/597
Kojefnikow epilepsy, *see* Koshevnikoff epilepsy
Kölliker-Fuse nucleus
pneumotaxic center, 63/479
respiratory center, 63/432
respiratory drive, 63/479
Könnecke-Richards-Rundle syndrome, *see* Richards-Rundle syndrome
Konzo, *see* Tropical spastic paraplegia
Korsakoff amnesia
malabsorption syndrome, 70/233
Korsakoff psychosis, *see* Wernicke-Korsakoff syndrome
Korsakoff syndrome, *see* Wernicke-Korsakoff syndrome
Koshevnikoff epilepsy
see also Status epilepticus
classification, 73/118
clinical characteristic, 73/119
definition, 72/9, 73/117
diagnosis, 73/121
EEG, 73/120
epilepsy classification, 15/21-23
far eastern tick-borne encephalitis, 56/140
measles virus, 56/480
myoclonus, 38/584
neuropathology, 73/120
status epilepticus, 15/176
treatment, 73/122
type, 72/9
Kothe-Koenen tumor, *see* Koenen tumor
Kozagawa focus
Kii Peninsula amyotrophic lateral sclerosis, 22/357
Krabbe-Bartels disease, 14/405-412
adiposis dolorosa, 14/406
animal model, 14/412
cerebellar vermis agenesis, 14/410
cheilognathopalatoschisis, 14/410, 529
clinical features, 14/405
corpus callosum agenesis, 14/408
corpus callosum lipoma, 14/12, 409
differential diagnosis, 14/406
dysraphia, 14/12, 407
epilepsy, 14/411
forme fruste, 14/412
genetics, 14/528
Grebe-Myle-Löwenthal syndrome, 14/407
heterotopia, 14/410
HHE syndrome, 14/529
leptomeningeal angiomatosis, 14/82
leptomeningeal lipoma, 14/411
lipodystrophia progressiva (Barraquer-Simons), 14/406
lipomatosis, 14/407, 409-411
Melkersson-Rosenthal syndrome, 14/529
mental deficiency, 14/407, 411, 528
microcephaly, 14/529
myosclerosis, 14/408
neurofibromatosis type I, 14/12, 82
neuropathology, 14/81-83
nosology, 14/406
phakomatosis, 14/12, 31/27
prognosis, 14/412
Riley-Smith syndrome, 14/83
spina bifida, 14/407
spinal lipoma, 14/411
Sturge-Weber syndrome, 14/82, 227
symmetrical pseudolipomatosis, 14/406
syringomyelia, 14/12, 529
treatment, 14/412
twin study, 14/528
Von Hippel-Lindau disease, 14/7
Krabbe classification
diffuse sclerosis, 10/4
infantile diffuse sclerosis, 10/4
Krabbe disease, *see* Globoid cell leukodystrophy
Krabbe, K., 10/67
Krabbe leukodystrophy, *see* Globoid cell leukodystrophy
Krabbe muscular hypoplasia
facial paralysis, 43/110
Kraepelin, E., 1/8, 13, 19
Kraepelin presenile dementia, *see* Presenile dementia (Kraepelin)
Krait intoxication
venom, 37/12, 15
Krause end bulb
sensation, 7/82
Krause, F., 1/8, 17
Krause terminal ventricle, *see* Terminal ventricle (Krause)
Krebs cycle, *see* Tricarboxylic acid cycle
Krebs cycle deficiency, *see* Tricarboxylic acid cycle deficiency
Krebs-Henseleit urea cycle
brain amino acid metabolism, 29/20
disorder, 29/87-106

Kriebelkrankheit, *see* Ergot intoxication
Krimsky-Prine rule
accommodation, 74/423
convergence, 74/423
Krücke classification
nerve disease, 7/201
Kryspin-Exner cell, *see* Opalski cell
Kufs disease, *see* Adult neuronal ceroid lipofuscinosis
Kufs-Hallervorden disease, *see* Adult neuronal ceroid lipofuscinosis
Kugelberg-Welander disease, *see* Wohlfart-Kugelberg-Welander disease
Kugelberg-Welander syndrome, *see* Wohlfart-Kugelberg-Welander disease
Kuhlendahl syndrome
occipital neuralgia, 5/373
Kundrat syndrome
arhinencephaly, 30/384
cerebellar vermis agenesis, 30/384
rhombencephalosynapsis, 30/384
Kunitz type serine protease inhibitor, *see* Serine proteinase inhibitor
Kupffer cell
viral infection, 56/66
Kurland-Hauser classification
epilepsy, 15/23
Kuru
amyloid, 46/256
animal model, 56/558, 586
arthralgia, 56/556
ataxia, 42/657, 56/556
basal ganglion lesion, 6/64
cannibalism, 34/285, 56/556
cerebellar ataxia, 56/556
clinical features, 56/556
Creutzfeldt-Jakob disease, 34/285, 46/289, 292, 294
dementia, 46/389, 56/556
disease duration, 56/557
dysarthria, 56/556
emotional disorder, 42/657
facial expression, 56/556
gait disturbance, 56/556
Gerstmann-Sträussler-Scheinker disease, 56/554
headache, 56/556
hereditary olivopontocerebellar atrophy (Menzel), 21/436
immunology, 56/558
incubation period, 56/557
multiple sclerosis, 9/79, 120
neuropathology, 56/558
origin, 56/558
pain, 42/657
paratonia, 56/557
parkinsonism, 56/557
prion disease, 56/543, 546
pseudodegeneration, 21/59
regression reflex, 56/556
senile plaque, 46/274, 56/551, 555
septal nuclei hypertrophy, 30/311
sex ratio, 42/657, 56/556
slow virus disease, 9/551, 34/275-287
speech disorder, 42/657
speech disorder in children, 42/657
spheroid body, 6/626
spongy dystrophy, 21/68
subacute spongiform encephalopathy, 42/655
transmission, 56/558
tremor, 42/657
truncal ataxia, 56/556
viral mechanism, 56/20
viral origin, 9/126
Kuskokwim arthrogryposis
osteolysis, 43/85
spinal muscular atrophy, 59/373
vertebral abnormality, 43/85
Kveim test
sarcoid neuropathy, 51/196
Kvistad-Dahl-Skre syndrome
hereditary congenital cerebellar atrophy, 60/294
Kwashiorkor
carnitine deficiency, 62/493
deficiency neuropathy, 7/635, 51/322
depigmentation, 7/635, 51/322
hypotonia, 51/322
lipid metabolic disorder, 62/493
mental deficiency, 7/635, 46/18
muscle biopsy, 51/322
myoclonus, 7/635, 51/322
pellagra, 28/76
protein deficiency, 7/635
protein energy malnutrition, 51/322
rigidity, 7/635, 51/322
secondary systemic carnitine deficiency, 62/493
tremor, 7/635, 51/322
Kx blood group antigen
McLeod phenotype, 63/288
Kyasanur Forest disease
acute viral encephalitis, 34/77
Kynureninase
vitamin B6 deficiency, 51/333
xanthurenic aciduria, 42/636
Kyphoscoliosis
see also Scoliosis
anencephaly, 42/12

angiodysgenetic necrotizing myelomalacia, 20/489
anterior spinal meningocele, 42/44
Bassen-Kornzweig syndrome, 13/418, 51/327, 394
Biemond syndrome, 60/669
central core myopathy, 43/81
centrofacial lentiginosis, 43/26
cerebral gigantism, 43/337
Coffin-Lowry syndrome, 43/238
corpus callosum agenesis, 42/101
cri du chat syndrome, 43/501
diastrophic dwarfism, 43/382-384
distal myopathy, 43/114
Dyggve-Melchior-Clausen syndrome, 43/266
Ekbom syndrome type II, 60/669
Flynn-Aird syndrome, 22/519, 42/327
Freeman-Sheldon syndrome, 43/353
Friedreich ataxia, 21/320, 341, 60/669
gait disturbance, 13/414
glycosaminoglycanosis type II, 42/454, 66/298
glycosaminoglycanosis type IV, 66/304
glycosaminoglycanosis type VI, 42/458
glycosaminoglycanosis type VII, 42/459
Goldberg syndrome, 42/439
Goltz-Gorlin syndrome, 43/19
Harrington rod, 25/187
Harvey syndrome, 42/223
hereditary multiple recurrent mononeuropathy, 60/62
hereditary sensory and autonomic neuropathy type III, 43/59, 60/27
homocystinuria, 42/555
hydromyelia, 50/425, 429
hypertrophic interstitial neuropathy, 21/152, 42/317
juvenile spinal cord injury, 25/175, 183
Klinefelter syndrome, 60/669
LEOPARD syndrome, 43/28
Lichtenstein-Knorr disease, 22/516
Marden-Walker syndrome, 43/424
Marinesco-Sjögren syndrome, 42/184, 60/669
mental deficiency, 43/277
microcephaly, 43/427
monosomy 21, 43/540
multiple neurofibromatosis, 43/35
multiple nevoid basal cell carcinoma syndrome, 43/31
nemaline myopathy, 43/122
neurofibromatosis type I, 32/222, 50/369
neuronal intranuclear hyaline inclusion disease, 60/669
olivopontocerebellar atrophy (Wadia-Swami), 60/669
osteogenesis imperfecta, 43/453
4p partial trisomy, 43/500
Pelizaeus-Merzbacher disease, 42/504
21q partial monosomy, 43/542
Richards-Rundle syndrome, 43/265
Roussy-Lévy syndrome, 21/176, 42/108
Ryukyan spinal muscular atrophy, 42/93
Sjögren-Larsson syndrome, 66/615
spinal ependymoma, 20/366
spondyloepiphyseal dysplasia, 43/478
spondylometaphyseal dysplasia, 43/482
Stickler syndrome, 43/484
vertebral hypoplasia, 42/61
Wohlfart-Kugelberg-Welander disease, 42/91
X-linked ataxia, 60/669

Kyphosis
adolescent, *see* Scheuermann disease
angular, *see* Angular kyphosis
ankylosing spondylitis, 50/419
arcual, *see* Arcual kyphosis
clinical features, 32/145, 50/419
congenital, *see* Congenital kyphosis
congenital spondyloepiphyseal dysplasia, 43/478
cranial base, *see* Cranial base kyphosis
diastematomyelia, 50/436
diplomyelia, 50/436
etiology, 32/145
genetics, 32/147
glycosaminoglycanosis type VII, 66/310
Hurler-Hunter disease, 10/436
juvenile spinal cord injury, 25/183
lumbar lordosis, 32/147
operative treatment, 50/421
pathogenesis, 32/147
prognosis, 50/419
radiologic appearance, 32/145
Rubinstein-Taybi syndrome, 31/284, 43/234
scoliosis, 50/418
spina bifida, 32/545
spinal meningioma, 20/214
spinal neurinoma, 20/262-266
spondylotic, *see* Spondylotic kyphosis
treatment, 32/147
vertebral column, 32/147

L protein
measles virus, 56/423

L1-glycoprotein, *see* Lysosome associated membrane protein 1

La Cross virus
bunyavirus, 56/16
California encephalitis, 56/140

La Cross virus encephalitis
acute viral encephalitis, 56/134
Labbé vein
anatomy, 11/47
brain phlebothrombosis, 54/399
radioanatomy, 11/105
temporal lobectomy, 2/708
Labetalol
hypertensive encephalopathy, 54/216
myopathy, 63/87
toxic myopathy, 62/597
Labiobuccal stereognosis
agnosia, 4/35
Labyrinth
arterial supply, 55/129
benign paroxysmal positional nystagmus, 2/370
membranous, *see* Membranous labyrinth
Labyrinth concussion
see also Neuro-otology
dizziness, 24/125, 136
examination, 24/126
hypoxia, 24/125
prognosis, 24/126
sign, 24/125
temporal bone fracture, 24/122, 125, 185
Labyrinthine disease
extended arms test, 1/338
gait ataxia, 1/336
periodic ataxia, 21/570
Von Romberg sign, 1/334
Labyrinthine hemorrhage
clinical features, 55/133
deafness, 55/133
leukemia, 55/133
vertigo, 55/133
Labyrinthine ischemia
anterior inferior cerebellar artery, 55/132
basilar artery aneurysm, 55/131
brain stem auditory evoked potential, 55/131
diagnostic test, 55/131
ENG, 55/131
pathophysiology, 55/130
vertebrobasilar insufficiency, 55/132
Labyrinthitis
Koch postulate, 56/105
viral infection, 56/105
Lacrimal artery
middle meningeal artery, 11/71
radioanatomy, 11/66
topography, 11/66
Lacrimal function
see also Tear
test, 74/428
Lacrimal gland
absence, 42/573
anatomy, 74/425
hereditary sensory and autonomic neuropathy type III, 60/32
parasympathetic innervation, 74/425
phosphoribosyl pyrophosphate synthetase, 42/573
supranuclear innervation, 74/426
sympathetic innervation, 74/426
Lacrimal gland tumor
differential diagnosis, 17/202
Lacrimal system
physiology, 74/427
Lacrimation
carbamate intoxication, 64/190
chronic paroxysmal hemicrania, 48/263, 75/293
chrysaora intoxication, 37/51
cluster headache, 48/9, 221, 237, 270
cluster variant syndrome, 48/270
congenital pain insensitivity, 51/564
disorder, 74/428
headache, 75/284
hereditary sensory and autonomic neuropathy type III, 1/475, 21/109, 111, 51/565, 60/7, 28
hypothalamic syndrome, 2/450
nereistoxin intoxication, 37/56
Lactate dehydrogenase
brain glycolysis, 27/10
central core myopathy, 41/5
CSF, *see* CSF lactate dehydrogenase
dermatomyositis, 62/375
Duchenne muscular dystrophy, 40/376, 41/488
eosinophilia myalgia syndrome, 63/377
ethambutol polyneuropathy, 51/295
facioscapulohumeral muscular dystrophy, 62/166
globoid cell leukodystrophy, 10/303
GM2 gangliosidosis, 10/288-291
histochemical reaction, 40/4
hypertrophia musculorum vera, 43/84
hypothyroid myopathy, 62/533
inclusion body myositis, 62/375
infantile neuroaxonal dystrophy, 42/229
isoenzyme, 40/356
limb girdle syndrome, 62/185
McLeod phenotype, 63/287
myoglobinuria, 62/558
paralysis periodica paramyotonica, 43/167
polymyositis, 22/25, 62/375
Reye syndrome, 29/334, 56/152, 65/118
vertebral fusion, 43/489
wallerian degeneration, 7/215, 9/37
Lactate dehydrogenase deficiency
chromosome 11, 62/490

exercise intolerance, 62/490
inheritance, 62/490
ischemic exercise, 62/490
metabolic myopathy, 62/490
myoglobinuria, 62/490
Lactate dehydrogenase elevating virus
blocking antibody, 56/61
macrophage, 56/28
monocyte, 56/30
reticuloendothelial system, 56/30
Togavirus, 56/28
Lactate dehydrogenase isoenzyme 1
GM2 gangliosidosis, 10/289
Lactate dehydrogenase isoenzyme 2
GM2 gangliosidosis, 10/289
Lactate dehydrogenase isoenzyme 3
GM2 gangliosidosis, 10/289
Lactate dehydrogenase isoenzyme 5
Duchenne muscular dystrophy, 40/356
GM2 gangliosidosis, 10/289
Lactic acid
brain infarction, 53/128
brain injury, 23/116, 119
brain ischemia, 53/128
brain metabolism, 57/70
CSF, *see* CSF lactic acid
hepatic coma, 49/215
hereditary periodic ataxia, 21/565
myophosphorylase deficiency, 27/233, 41/187
respiratory encephalopathy, 63/416
Reye syndrome, 56/152, 238
Lactic acidemia
dyssynergia cerebellaris myoclonica, 60/601
ethylene glycol intoxication, 64/122
mitochondrial disease, 66/16
Lactic acidosis, 42/585-589
acid base metabolism, 28/514-517
alaninuria, 42/516
alcohol intoxication, 64/110
brain death, 24/601, 604, 728
brain ischemia, 63/211
Burkitt lymphoma, 63/352
combined complex I-V deficiency, 62/506
complex I deficiency, 62/503
complex III deficiency, 62/504, 66/423
complex IV deficiency, 62/505, 66/424
CSF, 57/386
cytochrome deficiency, 41/212
diabetes mellitus, 70/200
dihydrolipoamide dehydrogenase, 41/210, 66/419
encephalopathy, 45/122
facioscapulohumeral muscular dystrophy, 43/99
fructose-1,6-diphosphatase deficiency, 41/208, 42/586
glycogen storage disease type I, 41/208, 42/586
infantile, *see* Infantile lactic acidosis
lipoate acetyltransferase, 41/210
malignant hyperthermia, 41/268
marine toxin intoxication, 65/153
MELAS syndrome, 62/509
methylmalonic aciduria, 41/208
mitochondria, 43/188
mitochondrial disease, 66/394
mitochondrial DNA depletion, 62/511, 66/430
mitochondrial myopathy, 41/208, 42/588, 43/188
mitochondrial respiratory chain defect, 42/587
MNGIE syndrome, 62/505
multiple cytochrome deficiency, 41/211
multiple mitochondrial DNA deletion, 62/510
myopathy, 41/219
neuron death, 63/211
occipital lobe epilepsy, 73/110
organic acid metabolism, 66/641
paralytic shellfish poisoning, 65/153
poliodystrophy, 42/588
progressive limb weakness, 41/216
pyruvate carboxylase deficiency, 41/211, 42/516, 586
pyruvate dehydrogenase complex deficiency, 41/211
pyruvate dehydrogenase deficiency, 42/587
Reye syndrome, 49/218
spontaneous, *see* Spontaneous lactic acidosis
subacute necrotizing encephalomyelopathy, 41/212, 42/588, 625
survey, 70/193
trichopoliodystrophy, 41/212
ubidecarenone, 62/504
Lactose malabsorption
symptom, 7/596
Lactoside
chemistry, 10/256
Lactosylceramide
catabolism, 66/60
Fabry disease, 10/544
Lactosylceramidosis, *see* Ceramide lactoside lipidosis
Lactosylsulfatide
enzyme deficiency polyneuropathy, 51/367
Lactrodectus venom
symptom, 64/10
Lacuna sacralis
spinal lipoma, 20/397
Lacunae
chorioretinal, *see* Chorioretinal lacunae
Lacunar infarction

brain, *see* Brain lacunar infarction
brain stem, *see* Brain stem lacunar infarction
Lacunar necrosis
neuroaxonal dystrophy, 49/397
Lacunar skull, *see* Craniolacunia
Lacunar state, *see* Brain lacunar infarction
Lacunar syndrome
brain, *see* Brain lacunar infarction
brain stem, *see* Brain stem lacunar infarction
Laetrile
cyanide intoxication, 65/32
cyanogenic glycoside intoxication, 65/32
Lafora body, 27/171-180, 184-189
see also Lafora progressive myoclonus epilepsy
amyotrophic lateral sclerosis, 42/69
atypical, 15/414
brain cortex, 15/394
chemical analysis, 27/186
clinical features, 73/294
cytoplasmic inclusion body, 1/282
distribution, 27/176
electromicroscopic features, 27/178
GM2 gangliosidosis, 10/411
histochemistry, 27/177-180
histologic features, 27/178
morphology, 27/177
motoneuron disease, 42/69
myoclonic epilepsy, 42/702
nerve fiber, 41/194
polysaccharide type, 21/52, 62
progressive myoclonus epilepsy, 73/293
protein type, 21/62
Purkinje cell, 15/394
retinal ganglion cell, 15/394
ultrastructure, 27/184-186
Unverricht-Lundborg progressive myoclonus epilepsy, 27/176-180, 184-187, 42/706
Lafora body disease, *see* Lafora progressive myoclonus epilepsy
Lafora disease, *see* Lafora progressive myoclonus epilepsy
Lafora progressive myoclonus epilepsy, 15/382-418
see also Lafora body
abnormal storage in muscle, 40/37
age at onset, 15/383, 27/172
autosomal recessive, 15/386, 27/173
Baltic myoclonus epilepsy, 49/617
behavior disorder, 15/383, 385, 27/172, 42/702
bradyphrenia, 15/385
carbohydrate, 15/382
clinical laboratory study, 27/173
clinical syndrome, 15/383, 389
consciousness, 15/384, 27/172
contracture, 60/669
Creutzfeldt-Jakob disease, 56/546
delusion, 27/172
dementia, 15/385, 27/172, 41/194, 42/702, 49/617
diagnosis, 15/389
duration, 15/383
dysarthria, 15/385, 27/173
EEG, 15/383, 387-389, 27/172, 174
epidemiology, 42/706
epilepsy, 72/129
extensor plantar response, 27/173
facial expression, 27/173
Friedreich ataxia, 21/509
hallucination, 15/385, 27/172
hereditary, 15/386
historic aspect, 27/171-173
hyperreflexia, 27/173
hypersalivation, 27/173
hypotonia, 15/386, 27/173
inheritance, 72/133
intention tremor, 27/173
laboratory data, 15/386
lysosome, 40/37, 99, 102
muscle fiber feature, 62/17
muscular twitching, 27/172
myoclonic epilepsy, 60/662
myoclonic jerk, 15/384, 27/173
neuronal storage dystrophy, 21/62
nystagmus, 27/173
pathology, 15/390-418
peroxisome, 40/102
personality change, 15/383, 27/172
polyglucosan, 41/194
polygraphic recording, 15/383
rigidity, 15/386, 27/173
rimmed muscle fiber vacuole content, 62/18
rimmed muscle fiber vacuole type, 62/18
speech disorder in children, 42/207
survey, 73/299
treatment, 15/390
tremor, 27/173
truncal ataxia, 27/173
ultrastructure, 40/99
Unverricht-Lundborg progressive myoclonus epilepsy, 73/293
visual impairment, 15/385, 27/173
Lag in sign
pulse rate, 2/150
Lagophthalmos
Möbius syndrome, 50/215
Lambda light chain
antibody, 9/502
viral infection, 56/52

Lambdoid suture
methotrexate syndrome, 30/122
premature fusion, 30/222
Lambdoid synostosis
craniosynostosis, 50/119
Lambert-Brody syndrome
motor unit hyperactivity, 41/296
Lambert-Eaton syndrome, *see* Eaton-Lambert myasthenic syndrome
Lamella
myelin, *see* Myelin lamella
Lamella widening
myelin, *see* Myelin lamella widening
Lamellar cerebellar sclerosis, *see* Focal cerebellar cortical panatrophy
Lamina basalis
muscle fiber, 62/31
Schwann cell, 47/5
Lamina dissecans
cerebellar cortex, 50/176
Lamina reuniens, *see* Lamina terminalis
Lamina terminalis
development, 30/300
Laminar brain atrophy
tuberous sclerosis, 14/47
Laminar necrosis
brain cortex, *see* Brain cortex laminar necrosis
cortical, *see* Cortical laminar necrosis
Marchiafava-Bignami disease, 28/317
Laminar sclerosis
cortical, *see* Cortical laminar sclerosis
Laminaria
intracranial hypertension, 17/1
Laminated body
concentric, *see* Concentric laminated body
Laminectomy
cervical spine, 25/304
cervical spondylotic myelopathy, 26/109
cervical vertebral column injury, 25/304
droplet metastasis, 20/285
lumbosacral nerve injury, 25/472
multiple myeloma, 20/111
spinal cord blood flow, 12/489
spinal cord injury, 25/304, 420, 459
spinal deformity, 25/462
spinal nerve root injury, 25/420, 424
spondylotic radiculopathy, 26/107
thoracolumbar spine injury, 25/459, 461
Laminin
muscle fiber, 62/31
Laminin receptor
metastasis, 71/645
Lamotrigine
antiepileptic agent, 65/496
epilepsy, 73/360
neurotoxin, 63/496, 65/511
toxic encephalopathy, 65/512
Lamotrigine intoxication
ataxia, 65/512
headache, 65/512
nystagmus, 65/512
vertigo, 65/512
Lance-Adams syndrome, *see* Posthypoxic myoclonus
Lancelet intoxication, *see* Cephalochordata intoxication
Lancinating pain
alcoholic polyneuropathy, 51/316
critical illness polyneuropathy, 51/575
Landau-Kleffner syndrome
aphasia, 46/152, 73/282
behavior, 73/283
definition, 72/9
developmental dysphasia, 46/141
EEG, 73/281, 283, 286
epidemiology, 73/282
epilepsy with continuous spike wave during slow sleep, 73/274
imaging, 72/379
neurophysiology, 73/283
pathophysiology, 73/286
prognosis, 73/287
slow sleep, 73/274
status epilepticus, 73/274
survey, 73/281
treatment, 73/287
verbal auditory agnosia, 72/9, 73/282
Landau reflex
diagnostic value, 1/255
Landing disease, *see* GM1 gangliosidosis
Landouzy-Dejerine disease, *see* Facioscapulohumeral muscular dystrophy
Landouzy-Dejerine dystrophy, *see* Facioscapulohumeral muscular dystrophy
Landouzy-Dejerine syndrome, *see* Facioscapulohumeral muscular dystrophy
Landry-Guillain-Barré-Strohl disease, *see* Guillain-Barré syndrome
Landry-Guillain-Barré syndrome, *see* Guillain-Barré syndrome
Landry paralysis
history, 7/496
polyradiculoneuritis, 7/487
rabies paralysis, 34/254
spinal cord compression, 19/377
Langelüddeke classification

epilepsy, 15/13
Langer-Giedion syndrome
microcephaly, 30/508
Langerhans cell histiocytosis
Abt-Letterer-Siwe disease, 70/56
brain, 70/58
cerebellum, 70/59
Chester-Erdheim syndrome, 70/56
eosinophilic granuloma syndrome, 70/56
Hand-Schüller-Christian disease, 70/56
Hashimoto-Pritzker syndrome, 70/56
history, 70/55
leptomeningeal, 70/58
neurologic symptom, 70/57
optic nerve, 70/60
osteodural, 70/57
self-healing, 70/56
spinal cord, 70/60
survey, 70/55
Langerhans histiocyte
eosinophilic granuloma, 63/422
histiocytosis X, 63/422
Langley ganglion
anatomy, 22/245
brain stem, 22/245
Langley, J.N., 1/7
Language
see also Aphasia
Alzheimer disease, 45/321
amusia, 4/197, 45/488
auditory agnosia, 4/281
auditory evoked potential, 46/148
biological factor, 45/302-306
brain tumor, 16/731, 67/396
cerebral dominance, 45/302
dementia, 45/321
developmental dysphasia, 46/141
dichotic listening, 46/149
disconnection syndrome, 45/103
Down syndrome, 31/404
dyslexia, 46/128, 132
expression, 45/523
head injury outcome, 57/404
hemispherectomy, 45/303
Huntington chorea, 46/308
hyperactivity, 46/181
left thalamus, 45/90
limbic system, 3/33
mental deficiency, 46/30
motor aphasia, 46/152
musical function, 4/197
musical language, 4/197
oriental, *see* Oriental language
parietal lobe epilepsy, 73/102
Parkinson dementia, 46/375
prosody, 4/198
regional brain blood flow, 45/297
rhythmic sense, 4/196
right hemisphere, 45/304
sign, *see* Sign language
subcortical dementia, 46/312
temporal lobe, 46/148
thalamic asymmetry, 45/78
thalamic syndrome (Dejerine-Roussy), 45/90, 94
Language cortex
alexia, 45/12
Language disability
EEG, 73/281
Lanosterol
metabolism, 66/46
Lantermann-Schmidt incisure
Schwann cell, 47/3, 7
Lanthanum
synaptic vesicle, 41/105
Lapemis hardwicki intoxication
antivenene, 37/94
Lapemis semifasciata intoxication
antivenene, 37/94
Laplace law
lower urinary tract, 61/293
Large fiber involvement
hereditary multiple recurrent mononeuropathy, 51/556
Laron dwarfism, *see* Pituitary dwarfism type III
Larsen syndrome
hypertelorism, 30/249
skeletal malformation, 30/249
Larva migrans
helminthiasis, 35/212
parasitic disease, 35/15
Laryngeal abductor paralysis
see also Gerhardt syndrome
Arnold-Chiari malformation type I, 42/319
Arnold-Chiari malformation type II, 50/410
Babinski sign, 42/319
congenital, *see* Congenital laryngeal abductor paralysis
hydrocephalus, 42/319
mental deficiency, 42/318
nucleus ambiguus dysgenesia, 42/319
spina bifida, 42/319
stridor, 42/318
Laryngeal cancer
SBLA syndrome, 42/769
Laryngeal itch
superior laryngeal neuralgia, 48/496

Laryngeal nerve
recurrent, *see* Recurrent laryngeal nerve
superior, *see* Superior laryngeal nerve
Laryngeal nerve block
superior laryngeal neuralgia, 48/495
Laryngeal nerve compression
recurrent, *see* Recurrent laryngeal nerve compression
Laryngeal neuralgia
cough syncope, 48/497
superior, *see* Superior laryngeal neuralgia
Laryngeal pain
superior laryngeal neuralgia, 48/496
Laryngeal spasm
nosologic significance, 1/285
Laryngolabioglossal palsy, *see* Progressive bulbar palsy
Laryngology
autonomic nervous system, 74/387
Larynx
autonomic innervation, 74/387
autonomic nervous system, 74/387
blood vessel, 74/390
cardiovascular response, 74/394
deglutition, 74/392
malfunction, 74/395
multiple neuronal system degeneration, 74/395
nerve supply, 74/391
respiration, 74/392
saliva, 74/392
stimulation, 74/394
stuttering, 46/170
swallow breath, 74/393
sympathetic nerve supply, 74/388
trachea, 74/392
Larynx paresis
diphtheria, 51/183
Larynx stridor
Arnold-Chiari malformation type I, 42/15
congenital, *see* Congenital larynx stridor
infantile Gaucher disease, 42/438
progressive dysautonomia, 59/160
Lasègue, C., 1/543
Lasègue sign
clinical significance, 1/543-545
crossed, *see* Crossed Lasègue test
lumbar intervertebral disc prolapse, 20/586
lumbar vertebral canal stenosis, 20/684, 707
mechanism, 20/587
spinal tumor, 19/72
Laser
surgery, 67/255, 259
Laser injury
arterial vascular injury, 23/666
brain injury, 23/665
experimental injury, 23/665
pathology, 23/668
Lassa fever
arenavirus, 56/16
arenavirus infection, 34/194-196, 203
clinical features, 34/194, 203, 56/370
consciousness, 56/370
convulsion, 56/370
CSF, 34/195
deafness, 56/370
encephalitis, 56/16
epidemiology, 34/194
epilepsy, 56/370
mortality, 56/370
transmission, 56/370
tremor, 56/370
Lassa fever virus
arenavirus, 34/5, 56/16, 355
viral meningitis, 56/16
Lassitude
crab intoxication, 37/58
merostomata intoxication, 37/58
migraine, 48/156, 162
uremia, 63/504
Latah
Gilles de la Tourette syndrome, 49/629, 635
jumpers of the Maine, 6/301, 49/621
myoclonus, 49/621
Late adult muscular dystrophy (Nevin)
see also Limb girdle syndrome
autosomal dominant, 62/181
features, 62/184
late onset, 62/181
limb girdle syndrome, 62/179
Pelger-Huët abnormality, 62/184
Late cerebellar ataxia
cataract, 60/653
diabetes insipidus, 60/668
downbeating nystagmus, 60/656
neurosensory hearing loss, 60/653
Late cerebellar atrophy, 60/581-588
acquired, 60/583-588
alcohol abuse, 60/583
anoxic lesion, 60/587
anti-Purkinje cell antibody, 60/585
ataxia, 42/135
bulbar paralysis, 60/658
chemotherapeutic agent, 60/586
chorea, 60/663
choreoathetosis, 60/663
chromosome 6, 60/583

classification, 60/583
clinical features, 60/585, 587
course, 60/585
Creutzfeldt-Jakob disease, 21/495
cytarabine, 60/586
dementia, 21/495, 60/583, 665
distal amyotrophy, 60/659
dominant cerebello-olivary atrophy, 42/136
drooling, 42/135
electrophysiologic study, 60/583
endocrine disorder, 60/587
exogenous intoxication, 60/586
fluorouracil, 60/586
gait disturbance, 42/135
genetic study, 60/583
hereditary, 60/581-583, 655, 661-663, 665
history, 60/581
Hodgkin disease, 60/585
hypopituitarism, 60/587
hypothyroidism, 60/587
lead intoxication, 60/586
lithium intoxication, 60/587
lymphoma, 60/585
malnutrition, 60/584
mammary carcinoma, 60/585
manganese intoxication, 60/586
Minamata disease, 60/586
muscle stretch reflex, 42/135
muscular hypotonia, 42/135
myoclonus, 60/662
neuropathology, 60/585, 587
nosology, 60/582
oat cell carcinoma, 60/585
ophthalmoplegia, 60/583
optic atrophy, 60/655
organic mercury intoxication, 60/586
ovary carcinoma, 60/585
paraneoplastic, 60/585
phenytoin intoxication, 60/587
Purkinje cell loss, 42/136
rigidity, 60/661
sex chromosomes, 60/586
spasticity, 60/661
speech disorder, 42/135
toxic, 60/587
trimethyltin inhalation, 60/586
vertical nystagmus, 42/135
Von Romberg sign, 42/135
Wernicke encephalopathy, 60/584

Late cord lesion
atlantoaxial dislocation, 25/279
cause, 25/278
cervical, 25/278

Late cortical cerebellar atrophy, 21/477-502
see also Carcinomatous cerebellar degeneration
alcohol, 21/477-479, 483-489
Alpers disease, 21/477
anoxia, 21/477
associated alcoholic disease, 21/479
astrocytosis, 21/483-488
autoimmune disease, 21/495
carcinomatous cerebellar degeneration, 21/477, 489-495
cause, 21/494
cerebellar sign, 21/479
cerebello-olivary atrophy, 60/571
cerebral cholesterosis, 21/477
chorea, 21/495
clinical features, 21/477, 490
clinicopathologic correlation, 21/486, 494
Creutzfeldt-Jakob disease, 21/477
DDT intoxication, 21/497
Down syndrome, 21/477
endocrine disorder, 21/477, 495
focal panatrophy, 21/477, 498
granular layer atrophy, 21/500
Hallervorden-Spatz syndrome, 21/477
heavy metal, 21/477
history, 21/478
Hodgkin disease, 21/495
hyperthermia, 21/477
hypoglycemia, 21/477
industrial solvent, 21/497
inorganic mercury intoxication, 21/497
lead encephalopathy, 21/496
lead intoxication, 21/496
malnutrition, 21/477
manganese intoxication, 21/497
mental deficiency, 21/477
Minamata disease, 21/497
myxedema, 21/496
neurolipidosis, 21/477
neuropathology, 21/480, 491-494
nitrogen chloride intoxication, 21/497
organic mercury intoxication, 21/497
paleocerebellum, 21/480
pathogenesis, 21/487
Pick disease, 21/495
slow virus disease, 21/495
spinopontine degeneration, 21/397
thiophene intoxication, 21/497
tuberous sclerosis, 21/477
vitamin B1 deficiency, 21/489
Wilson disease, 21/477

Late infantile acid maltase deficiency
biochemical aspect, 27/228

calf hypertrophy, 62/480
clinical features, 27/222, 227
clinical laboratory examination, 27/228
delayed walking, 62/480
diagnosis, 27/228
Gowers sign, 27/228
lumbar lordosis, 62/480
muscle biopsy, 27/195, 229
respiratory dysfunction, 27/195, 62/480
toe walker, 27/228, 62/480
treatment, 27/228
waddling gait, 62/480
Late infantile amaurotic idiocy, *see* GM1 gangliosidosis
Late infantile ceroid lipofuscinosis
secondary pigmentary retinal degeneration, 60/654
Late infantile lipidosis
systemic, *see* GM1 gangliosidosis
Late infantile metachromatic leukodystrophy
aphasia, 29/357
bone marrow transplantation, 66/95
ceramide galactose sulfate, 10/311
cerebroside sulfatase, 10/311
classification, 10/17
cortical blindness, 66/167
dihexoside sulfate, 10/141
dysarthria, 10/346, 29/357
extremity involvement, 10/346, 29/357, 66/94
feeding difficulty, 66/167
hypotonia, 29/357
pneumonia, 66/167
seizure, 66/167
spasm, 10/346, 29/357, 66/167
spastic tetraplegia, 66/167
speech disorder, 66/94, 167
stage 1, 60/124
stage 2, 60/124
stage 3, 60/124
stage 4, 60/125
truncal ataxia, 66/167
weakness, 29/357
Late infantile neuroaxonal dystrophy
amyotrophic lateral sclerosis, 49/405
Behr disease, 49/405
clinical features, 49/405
Down syndrome, 49/405
Hallervorden-Spatz syndrome, 42/257, 49/399
Lewy body, 49/405
neurolipidosis, 49/405
neuropathology, 49/400, 405
primary neuroaxonal dystrophy, 49/397
Late infantile neuronal ceroid lipofuscinosis
epidemiology, 66/672
gene, 72/134
survey, 66/24
Late infantile Tay-Sachs disease, *see* GM1 gangliosidosis
Late neuritis, *see* Tardy neuritis
Late onset ataxia
fasciculation, 60/659
nerve conduction velocity, 60/670
neuropathy, 60/660
polyneuropathy, 60/660, 670
rigidity, 60/659-661, 670
striatopallidodentate calcification, 60/671
vestibular function, 60/658
Late onset epilepsy, *see* Epilepsy of late onset
Late onset hereditary cortical cerebellar atrophy, *see* Marie-Foix-Alajouanine disease
Late onset myopathy
multiple mitochondrial DNA deletion, 66/429
neoplasm, 41/82
Latent nystagmus
ENG, 24/130
meningeal irritation, 1/547
Lateral asymmetry
computer assisted tomography, 46/486
positron emission tomography, 46/486
schizophrenia, 46/486
Lateral cerebellar agenesis
clinical picture, 30/382
pathologic finding, 30/380-382
Lateral cerebellar foramina atresia
cerebellar vermis agenesis, 42/5
Lateral cerebellopontine vein
ventro, *see* Ventrolateral cerebellopontine vein
Lateral cutaneous nerve
entrapment, 7/310
meralgia paresthetica, 7/310, 19/86
Lateral femoral cutaneous nerve
compression neuropathy, 51/108
injury, 2/39, 25/475
occupational lesion, 7/34
Lateral gaze
nystagmus, 42/158
Lateral gaze palsy
Cestan-Chenais syndrome, 2/261
progressive supranuclear palsy, 22/219
Lateral geniculate nucleus
anatomy, 2/473
cell type, 2/531
dorsal nucleus, 2/530, 555
inhibition, 2/532
injury, 24/45
lamination, 2/555

macular representation, 2/555
neglect syndrome, 45/156
nerve fiber, 2/530
neuroanatomy, 2/529-532
receptor field organization, 2/532
retinotopic organization, 2/554
rotation, 2/554
synaptic organization, 2/531
vascular supply, 24/31
ventral nucleus, 2/530
visual field defect, 2/577
Lateral habenular nucleus
basal ganglion, 49/29
pallidohabenular fiber, 49/7
Lateral hemiaplasia
cerebellar agenesis, 50/181, 184
Lateral hypothalamic syndrome
aphagia, 1/500
feeding behavior, 46/587
Lateral medullary infarction
ataxia, 53/382
Bonnier syndrome, 2/220
carbon dioxide narcosis, 63/441
clinical type, 2/225
cocaine intoxication, 65/255
corneal reflex, 53/381
cough, 53/382
diplopia, 53/381
dissociated sensation, 53/381
dysautonomia, 53/382
dysphagia, 53/381
facial burning, 53/381
facial pain, 53/381
facial paresthesia, 53/381
forgotten respiration, 53/382, 63/441
headache, 48/281, 53/381
hoarseness, 53/381
Horner syndrome, 2/234, 53/382
nausea, 53/381
nystagmus, 53/382
race, 53/380
Sneddon syndrome, 55/407
symptom, 53/380
temporal arteritis, 55/344
vertebral artery aneurysm, 12/96
vertebral artery occlusion, 53/380
vertebrobasilar system syndrome, 53/380
vertigo, 53/381
vomiting, 53/381
Lateral medullary syndrome
arteriography, 2/235
Babinski-Nageotte syndrome, 2/219
clinical picture, 2/229-236
headache, 5/132
Horner syndrome, 2/232
medial medullary syndrome, 2/228
terminology, 11/31
vascular lesion, 2/229
vertebral artery occlusion, 5/132
Lateral mesencephalic vein
anatomy, 11/51
Lateral nucleus
thalamic ventral, *see* Thalamic ventral lateral nucleus
Lateral occipital artery
radioanatomy, 11/98
topography, 11/98
Lateral pallidal segment
ballismus, 49/372
hemiballismus, 49/372
pallidal lesion, 6/633
Lateral ponticulus
atlas, 32/39
atlas dysplasia, 50/393
Lateral pontine syndrome
classification, 12/22
Foix-Alajouanine disease, 12/17
hemianesthesia cruciata, 12/17
paramedian, 12/18
peripheral facial paralysis, 12/17
pontine hyperpathia, 12/17
Lateral popliteal nerve
leprous neuritis, 51/220
Lateral popliteal nerve palsy
foot drop, 8/153
Lateral reticular nucleus
olivopontocerebellar atrophy (Dejerine-Thomas), 21/418
Lateral sclerosis
amyotrophic, *see* Amyotrophic lateral sclerosis
familial amyotrophic, *see* Familial amyotrophic lateral sclerosis
primary, *see* Primary lateral sclerosis
pseudoamyotrophic, *see* Pseudoamyotrophic lateral sclerosis
Lateral sclerosis (Erb), *see* Primary lateral sclerosis
Lateral sinus, *see* Transverse sinus
Lateral sinus thrombosis, *see* Transverse sinus thrombosis
Lateral spinothalamic tract
pain anatomy, 1/84
Lateral thalamic nuclear group
intermedius nucleus, 2/473
ventralis anterior nucleus, 2/472
ventralis oralis nucleus, 2/472
ventralis posterior nucleus, 2/473

Lateral tract syndrome
see also Pyramidal tract
posterior column syndrome, 2/210
Lateral ventricle tumor
angioma, 17/605
brain angioma, 17/605
brain scanning, 17/600
child, 18/332
choroid plexus papilloma, 18/332
classification, 17/596
clinical features, 17/597
dermoid cyst, 17/606
diagnostic procedure, 17/598
ependymoma, 17/604, 18/332
epidermoid cyst, 17/606
false localization sign, 17/597
frequency, 17/598
lipoma, 17/606
meningioma, 17/597, 603
metastasis, 17/606
radiology, 17/600, 602
sarcoma, 17/606
surgery, 17/603
symptom, 17/598
teratoma, 17/606
Latham-Munro syndrome
hereditary hearing loss, 22/512
progressive myoclonus epilepsy, 22/512
Lathyrism, 36/505-514, 65/1-16
age, 36/506, 509, 65/2, 5
α-amino-3-hydroxy-5-methyl-4-isoxazolepropion
ic acid receptor, 65/14
animal study, 28/37, 65/11
anterior horn cell, 65/9
autonomic nervous system, 65/3
axonal degeneration, 65/4
central motor conduction time, 65/9
chapati, 65/10
corticospinal tract, 65/9
course, 65/7
crawler stage, 65/4
diagnosis, 65/6
differential diagnosis, 65/6
distal axonal degeneration, 65/9
epidemic, 7/644, 36/505, 65/1
epidemiology, 36/506, 65/5
familial spastic paraplegia, 59/306
fasciculation, 65/3
fecal incontinence, 65/3
gangrene, 65/7
geographic distribution, 28/37, 36/505, 65/5
ghontu, 65/10
gracile tract, 65/9
β-(isoxazolin-5-on-2-yl)alanine, 65/13
kainic acid receptor, 65/14
kitta, 65/10
las gachas, 65/8
Lhermitte sign, 65/3
malnutrition, 28/37
N-methyl-D-aspartic acid receptor, 65/14
mitochondrial glutathione, 65/14
molecular mechanism, 65/13
motoneuron disease model, 65/1
muscle spasm, 36/511, 65/3
muscular atrophy, 36/511, 65/6
myoclonus, 65/3
myokymia, 65/3
neurologic manifestation, 28/37
neuron degeneration, 7/644, 65/4
neuronal vacuolation, 65/15
neuropathology, 65/9
neuropathy, 7/520
neurophysiology, 65/9
no stick stage, 65/4
nutritional myelopathy, 7/643
osteolathyrism, 7/644, 36/505, 507, 65/6
pain, 36/509, 65/7
pathogenesis, 65/15
prevention, 65/10
prognosis, 36/511, 65/7
pyramidal tract, 7/644, 36/511, 65/2
reduced nicotinamide adenine dinucleotide
dehydrogenase, 65/14
sensorimotor neuropathy, 65/9
sensory symptom, 65/3
sex ratio, 28/37, 36/506, 65/3, 5
sexual impotence, 36/511, 65/3
spastic paraplegia, 7/520, 643, 28/37, 36/505,
59/438, 64/9, 65/2
spinocerebellar tract, 65/9
stage, 65/4
1 stick stage, 65/4
2 stick stage, 65/4
sudden onset, 28/37, 65/3
symptom, 64/10, 65/3
symptom precipitation, 65/3
tissue necrosis, 65/7
toxic myelopathy, 64/9
treatment, 65/8
urinary incontinence, 65/3
Lathyrus
species, 65/1
use, 65/11
Lathyrus sativus
international network, 65/1
3-oxalylaminoalanine concentration, 65/5, 10

tropical myeloneuropathy, 56/525
Lathyrus sativus intoxication
symptom, 36/505, 64/9
Latissimus dorsi replacement
brachial plexus, 51/152
α-Latrotoxin
arthropod envenomation, 65/196
myasthenic syndrome, 64/16
myoglobinuria, 64/16
neurotoxin, 64/15, 65/194, 196
toxic myopathy, 64/16
toxic neuromuscular junction disorder, 64/15
α-Latrotoxin intoxication
action site, 64/15
pathomechanism, 64/15
symptom, 64/15
Lattice dystrophy
corneal, *see* Corneal lattice dystrophy
Laughing
amyotrophic lateral sclerosis, 22/115
development, 3/360
facial muscle, 3/356
headache, 48/367
muscle contraction, 3/356
narcolepsy, 2/448, 3/94
pathologic, *see* Pathologic laughing
spasmodic, 1/196, 210
Laughing and crying
pathologic, *see* Pathologic laughing and crying
Laughing spell
migraine, 48/162
Laurence-Moon-Bardet-Biedl syndrome
abortive form, 31/326
acrocephalosyndactyly type II, 38/422
Alström-Hallgren syndrome, 13/462
cardinal features, 31/326
chorioretinal degeneration, 13/40
CNS, 31/327
deaf mutism, 13/41, 390
endocrine disorder, 13/133
etiology, 31/326
Grönblad-Strandberg syndrome, 14/114
hydrocephalus, 31/328
macrencephaly, 31/328
mental deficiency, 31/327
metabolic abnormality, 31/328
nomenclature, 13/289, 380, 43/233, 254
optic atrophy, 13/40, 42
pathology, 31/328
prognosis, 31/328
radiologic appearance, 31/327
Refsum disease, 21/215
secondary pigmentary retinal degeneration, 22/536
Stargardt disease, 13/133
treatment, 31/328
Weiss-Alström syndrome, 13/463
Laurence-Moon-Biedl syndrome
acrocephalopolysyndactyly, 46/87
Alström-Hallgren syndrome, 22/503
arachnodactyly, 13/303
Friedreich ataxia, 21/348
mental deficiency, 46/65
nomenclature, 13/289, 380, 43/233, 254
pleiotropy, 30/6
prognostic significance, 30/5
Tunbridge-Paley disease, 22/508
Laurence-Moon syndrome
blindness, 43/254
cataract, 43/254
deafness, 13/302
familial spastic paraplegia, 59/330-332
Friedreich ataxia, 13/300
hypogonadism, 43/253
mental deficiency, 43/253
nomenclature, 13/289, 380, 43/233, 254
Refsum disease, 21/215
secondary pigmentary retinal degeneration, 43/253-255, 60/719, 733
short stature, 43/253
spastic paraplegia, 13/307, 43/254
tetraplegia, 13/307
Lavender
relaxation, 74/365
Law
see also Death
Alexander, *see* Alexander law
anosmia, 24/14
burden of proof, 24/792
civil law, 24/795
codification, 24/793
common law, 24/832
convention on human rights, 24/793
criminal law, 24/796
death, 24/791
death declaration, 24/799, 804, 809, 812
ethics, 24/800
French view, 24/800
German view, 24/800, 804
Harvard report, 24/790, 800, 822
head injury, 24/829
human rights, 24/792, 795
IFSECN view, 24/790, 800
legal model of man, 24/811, 814
legal system, 24/791
Mendel, *see* Mendel law

minor head injury, 24/835
Müller, *see* Müller law
national law, 24/794
physical examination, 24/831
posttraumatic syndrome, 24/838
principle, 24/791
punishable act, 24/794
severe head injury, 24/833
traumatic psychosyndrome, 24/838
Vienna Conference, 24/790, 817, 820
Lawrence syndrome
Berardinelli-Seip syndrome, 42/592
Lawson-Santos sign, *see* Lag in sign
Laxative agent
pain, 69/47
Lazorthes artery
anastomosed anteroposterior spinal artery, 32/500
LCAT, *see* Phosphatidylcholine sterol acyltransferase
LDH, *see* Lactate dehydrogenase
LE cell
systemic lupus erythematosus, 43/420
Le Fort fracture
classification, 23/371
craniofacial injury, 23/371
fixation, 23/375
treatment, 23/375
Lead
absorption, 64/433
air, 64/433
alkyllead, 64/432
amyotrophic lateral sclerosis, 51/268
astrocyte, 64/437
axonal degeneration, 51/263
biokinetic, 64/433
blood level, *see* Blood lead level
body burden, 64/433
brain stem, 64/434
cerebellum, 64/434
chronic axonal neuropathy, 51/531
coproporphyrinogen oxidase, 64/434
daily intake, 64/433
dementia, 64/438
distal axonopathy, 64/13
environment, 64/432
erythrocyte, 64/433
erythrocyte protoporphyrin zinc, 64/434
ferrochelatase, 64/434
food ingestion, 64/433
Guam amyotrophic lateral sclerosis, 59/264
half-life, 64/433
hard water, 64/433
heme synthesis, 64/434
hemorrhagic encephalopathy, 51/264
hippocampus, 64/434
industrial use, 64/432
intestinal tract, 64/434
kidney, 64/434
mobilization test, 36/23
multiple sclerosis, 9/112
myelination inhibition, 51/266
nervous system, 64/434
nervous system development, 64/438
neurotoxicity, 51/265
neurotoxin, 4/330, 6/652, 7/511, 8/366, 9/642, 13/63, 21/496, 22/27, 36/3, 6, 12, 22, 35-61, 65, 46/19, 391, 47/561, 48/420, 49/466, 51/68, 263, 265, 268, 531, 64/10, 13, 431, 437, 443
normal limit, 64/433
oligodendrocyte, 64/437
paint, 64/432
porphobilinogen synthase, 64/434
segmental demyelination, 51/263
sink, 64/437
skeleton, 64/433
tissue distribution, 64/433
toxic neuropathy, 64/10, 13
toxic polyneuropathy, 51/68, 268
water, 64/433
Lead encephalopathy
see also Lead intoxication
acute type, 64/443
adult, 51/268
age, 64/443
animal, 9/684
ataxia, 64/435, 443
blindness, 64/436
blood-brain barrier, 28/392
bradyphrenia, 64/435
brain edema, 64/436
cerebellar calcification, 64/436
child, 51/268
chronic, 36/50-60
chronic type, 64/443
clinical features, 36/6-12
coma, 64/436
computer assisted tomography, 64/436
confusion, 64/435
convulsive seizure, 64/435
death, 64/436
delirium, 64/436
dementia, 46/391, 64/443
dimercaprol, 64/443
dyssomnia, 64/435
edetic acid, 64/443
EEG, 64/436

epilepsy, 64/436, 443
experimental animal, 36/42-50
hallucination, 64/436
headache, 64/435
hemiparesis, 64/436
human, 36/40-42
hydrocephalus, 64/437
insomnia, 64/443
intracranial hypertension, 64/436
late cortical cerebellar atrophy, 21/496
lethargy, 64/435
lucid interval, 64/436
memory disorder, 64/435
α-mercapto-β-(2-furyl) acrylic acid, 64/443
moonshine whiskey, 64/435
myelinoclastic diffuse sclerosis, 9/480
neurofibrillary tangle, 46/274
neuropathology, 64/437
nuclear magnetic resonance, 64/436
optic neuropathy, 64/436
pallidal lesion, 6/664
papilledema, 64/436
parkinsonism, 49/478
pathology, 36/40-42, 50-60
penicillamine, 64/443
polyneuropathy, 64/436
psychosis, 64/436
spongiform encephalopathy, 64/437
striatopallidodentate calcification, 64/436
stupor, 64/435
succimer, 64/443
symptom, 64/443
treatment, 36/25-28, 64/443

Lead intoxication
see also Lead encephalopathy
abdominal colic, 64/435
absorption, 36/3
acetylglucosaminidase, 64/434
action site, 64/10
age incidence, 36/21
amyotrophic lateral sclerosis, 59/200
anemia, 9/642, 64/434
animal experiment, 9/642
anorexia, 64/435
arthralgia, 64/435
artisan, 64/432
astrocyte, 64/437
asymptomatic state, 64/435
ataxia, 64/435
biochemical aspect, 36/65-70
blindness, 64/436
blood lead level, 36/23, 64/434, 443, 445
body burden, 64/445
bone effect, 36/20
brain, 64/445
brain edema, 64/436
chelating agent, 64/439
chelation therapy, 46/391
choroid plexus, 64/437
chronic, *see* Lead encephalopathy
cirrhosis, 9/642
clinical features, 36/1-28
conflicting data, 64/448
convulsion, 9/642
craftsmen, 64/431
crystal decanter, 64/431
culture, 64/447
cytoplasmic inclusion body, 64/437
deciduous teeth, 64/445
delirium, 9/642
dementia, 9/642, 46/391, 64/437
demyelination, 9/641
Devonshire colic, 64/431
diagnostic test, 36/23
diffuse sclerosis, 9/642
dimercaprol, 36/25-28, 64/439
doubtful toxicity, 64/447
dyssomnia, 64/435
economics, 64/447
edetate calcium disodium, 64/439
edetic acid, 36/25-28
EEG, 64/444
electrothermal atomic absorption spectrometry, 64/435
encephalopathy, *see* Lead encephalopathy
endocrine involvement, 36/20
epidemiology, 36/21, 64/432, 448
epilepsy, 64/435
erythrocyte 5-aminolevulinate, 64/433
erythrocyte protoporphyrin, 64/434
ethnic group, 64/447
excretion, 36/4
exposure level, 64/444
fatigue, 64/435
folk medicine, 64/432
free erythrocyte protoporphyrin, 36/23
gingival lead line, 64/435
gout, 64/435
hair, 64/445
hallucination, 4/330
headache, 48/420, 64/435
heart involvement, 36/20
hematologic effect, 36/15-17
history, 36/1, 64/431
incidence, 64/448
inclusion body, 64/436

industrial activity, 64/432
inorganic, 36/6
intelligence quotient, 64/445
interstitial nephritis, 9/642
intracerebral calcification, 49/417
intracranial hypertension, 64/436
laboratory diagnosis, 64/434
late cerebellar atrophy, 60/586
late cortical cerebellar atrophy, 21/496
lead mobilization test, 36/23
leukoencephalopathy, 9/641, 47/560
longitudinal study, 64/446
low lead exposure, 64/438
manifestation, 36/4-6
meningism, 9/642
mental deficiency, 46/19
mental impairment, 64/435
meta-analysis, 64/446
metabolism, 36/3
method critique, 64/445
methodology, 64/444
minimal brain dysfunction, 64/433
misdiagnosis, 18/129
motoneuron disease, 22/27
multiple tumors, 9/641
myalgia, 64/435
myelinoclastic diffuse sclerosis, 9/480
myopathy, 9/642
nail, 64/445
nerve conduction, 7/173
nerve conduction velocity, 36/23
neurobehavior aspect, 64/443
neuropathy, 7/511, 8/366, 9/642, 36/12-15, 38-40
neuropsychologic, 64/433
neuropsychologic deficit, 64/446
occupational disease, 64/431
occupational therapy, 64/432
oligodendrocyte, 64/437
onion bulb formation, 21/252
optic atrophy, 13/63
optic neuropathy, 64/436
organic, *see* Organic lead intoxication
pallidal necrosis, 6/652, 49/466
pallidoluysionigral degeneration, 49/460
pallor, 64/435
papilledema, 9/642, 64/436
penicillamine, 36/27, 64/439
poisoning source, 51/269
polyneuropathy, *see* Lead polyneuropathy
prevention, 36/28
psychometric method, 64/445
race, 64/447
radiologic finding, 36/23
radiologic technique, 64/445
Raynaud phenomenon, 8/337
renal effect, 36/17-20
seasonal incidence, 36/28
segmental demyelination, 9/642, 51/71
skeleton, 64/445
social group, 64/447
statistical significance, 64/445
striatopallidodentate calcification, 49/417
study flaw, 64/446
study sample, 64/445
succimer, 64/439
sudanophilic leukodystrophy, 9/642
susceptibility, 36/21
symptom, 36/4-6, 64/431, 435
teeth, 64/445
treatment, 36/25-28, 64/439
tremor, 64/435
umbilical cord blood, 64/447
unithiol, 64/439
urate retention, 64/434
urinary coproporphyria, 64/434
visceral autonomic nerve, 75/503
visuomoter integration, 64/446
wine conserving, 64/431
wine sweetening, 64/431
Lead myelopathy
amyotrophic lateral sclerosis, 64/436
Lead neuropathy
intestinal pseudo-obstruction, 64/437
pseudo-obstruction, 64/437
Lead pipe rigidity, *see* Plastic rigidity
Lead polyneuropathy
Achilles reflex, 64/437
amyotrophic lateral sclerosis, 64/436
anemia, 51/269
asymmetrical neuropathy, 51/268
autonomic polyneuropathy, 64/437
brain stem auditory evoked potential, 64/437
clinical features, 51/268
criteria, 51/268
CSF, 51/270
dimercaprol, 51/270
distal latency, 64/437
distal motor polyneuropathy, 51/268
ECG, 64/437
edetic acid, 51/270
EMG, 51/270
endoneurial edema, 51/271, 64/436
foot drop, 64/436
frequency, 64/435
heme synthetase, 51/270
intestinal pseudo-obstruction, 64/437

lead colic, 51/269
measurement, 51/269
mechanism, 51/271
nerve conduction velocity, 64/435, 437
neuropathology, 64/437
paresthesia, 64/436
poliomyelitis, 64/436
porphobilinogen synthase, 51/270
porphyrin metabolism, 51/270
segmental demyelination, 51/271, 64/436
somatosensory evoked potential, 64/437
visual evoked response, 64/437
wrist drop, 64/436
Leao spreading depression, *see* Spreading depression
Leaping ague
involuntary movement, 6/301
Learning
see also Memory
aging, 27/470
anesthesia, 27/467
attention, 3/156
biochemistry, 27/459-471
brain injury, 4/9
cerebral localization, 3/312
cholinergic receptor blocking agent, 27/467
conditioned, 3/156
definition, 27/459-461
developmental dyslexia, 46/108-113
electroconvulsive treatment, 27/466
headache, 48/34
historic aspect, 27/459
Huntington chorea, 46/307
idiopathic developmental dyslexia, 4/436
intelligence, 3/302, 305, 312
memory, 3/261-263
ontogeny, 3/360
posttraumatic amnesia, 27/468, 45/189
prefrontal syndrome, 2/748
protein synthesis inhibition, 27/468
sleep, 3/87
treatment, 27/471
Wernicke-Korsakoff syndrome, 27/469
Learning disability, 46/123-134
aptitude, 46/124
brain injury, 46/124
cerebral dominance, 46/125, 132-134
cognition, 46/123
constructional apraxia, 46/125
definition, 46/123
dyscalculia, 46/125
dyslexia, 46/125
dysphasia, 46/125
epilepsy, 73/461
genetics, 46/130
hyperactivity, 46/181
intelligence test, 46/125
maturity, 46/124
minimal brain dysfunction, 46/124
perception, 46/125
psychometry, 46/124-126
Leatherjacket intoxication
tetrodotoxin, 37/79
Leber congenital amaurosis, *see* Congenital retinal blindness
Leber disease, *see* Leber optic atrophy
Leber hereditary optic atrophy, *see* Congenital retinal blindness
Leber hereditary optic neuropathy
age at onset, 62/509
ataxia, 62/509
Babinski sign, 66/428
cardiac conduction, 62/509
clinical features, 62/509
hyperreflexia, 62/509, 66/428
mitochondrial disease, 62/497
mitochondrial disorder, 66/393
mitochondrial DNA mutation, 66/428
mitochondrial DNA point mutation, 62/509
neuropathy, 62/509, 66/428
optic atrophy, 62/509
sex ratio, 62/509
Wolff-Parkinson-White syndrome, 66/428
Leber-Mooren amaurosis congenita, *see* Congenital retinal blindness
Leber optic atrophy
ataxia, 42/109
athetosis, 42/401
Babinski sign, 42/401
Behr disease, 42/109
Bruyn-Went disease, 13/99, 22/454
congenital retinal blindness, 13/67
dystonia musculorum deformans, 49/524
familial spastic paraplegia, 13/97, 22/454
Hallervorden-Spatz syndrome, 13/99
hereditary cerebellar ataxia, 13/97
hereditary motor and sensory neuropathy type I, 21/284
hereditary motor and sensory neuropathy type II, 21/284
hereditary motor and sensory neuropathy variant, 60/246
hereditary spastic ataxia, 21/372
hyperreflexia, 42/401
hypertrophic polyneuropathy, 13/97
Lyon hypothesis, 13/96

misdiagnosis, 13/102
multiple sclerosis, 9/103, 176, 13/102
olivopontocerebellar atrophy variant, 21/454
sex ratio, 42/399
speech disorder in children, 42/401
tobacco alcoholic amblyopia, 13/67
vitamin B12, 28/175
Leber retinal blindness, *see* Congenital retinal blindness
Lecithin, *see* Phosphatidylcholine
Lecithin:cholesterol acetyltransferase, *see* Phosphatidylcholine sterol acyltransferase
Lecithinase, *see* Phospholipase
Lecithinase C, *see* Phospholipase C
Left handedness, *see* Handedness
Left hemisphere
apraxia, 45/428
Gerstmann syndrome, 2/687
motor apraxia, 45/428
schizophrenia, 46/489
Left temporal lobe
schizophrenia, 46/504
Left thalamus
language, 45/90
Leg abnormality
diastematomyelia, 50/436
diplomyelia, 50/436
Leg crossing
orthostatic hypotension, 75/719
Leg jitter, *see* Restless legs syndrome
Leg shortening
diastematomyelia, 50/436
diplomyelia, 50/436
Legal aspects, *see* Brain death, Death *and* Law
Legg-Perthes disease
glycosaminoglycanosis type VI, 66/307
glycosaminoglycanosis type VII, 66/311
Legionella feelii, *see* Legionellosis
Legionella micdadei, *see* Legionellosis
Legionella pneumophila
endotoxin, 52/257
pneumonia, 52/253
polymyositis, 62/373
9 serogroups, 52/253
Legionellosis
acquired myoglobinuria, 62/577
acute cerebellar ataxia, 52/254
ataxia, 52/254
choreoathetosis, 52/254
clinical features, 52/254
CNS, 52/253
computer assisted tomography, 52/256
cranial nerve palsy, 52/254
CSF, 52/254-256
delirium, 46/544
differential diagnosis, 52/257
EEG, 52/256
EMG, 52/256
encephalopathy, 52/254
epidemiology, 52/254
epilepsy, 52/254
erythromycin, 52/257
headache, 52/254
hyponatremia, 52/254
laboratory finding, 52/255
mental change, 52/255
mortality, 52/254
neurologic symptom, 52/254
neuropathology, 52/256
neuropathy, 52/254
polymyositis, 62/373
psychiatric symptom, 52/254
radionuclide scan, 52/256
status epilepticus, 52/254
treatment, 52/257
vertigo, 52/254
Legionnaires disease, *see* Legionellosis
Legs
crossed, *see* Crossed legs
restless, *see* Restless legs syndrome
Leigh disease, *see* Subacute necrotizing encephalomyelopathy
Leigh necrotizing encephalomyelopathy, *see* Subacute necrotizing encephalomyelopathy
Leigh syndrome, *see* Subacute necrotizing encephalomyelopathy
Leiomyoma
cutaneous, *see* Cutaneous leiomyoma
multiple nevoid basal cell carcinoma syndrome, 14/468
skin, *see* Cutaneous leiomyoma
uterine, *see* Uterine leiomyoma
Leipzig, Germany
neurology, 1/19
Leishmaniasis
visceral, *see* Visceral leishmaniasis
Lemieux-Neemeh syndrome
Alport syndrome, 22/522
amyotrophy, 22/521
distal muscle weakness, 42/370
foam cell, 42/371
hereditary hearing loss, 22/520
hereditary motor and sensory neuropathy type I, 22/521
high frequency hearing loss, 42/371
nephritis, 42/370

nerve conduction velocity, 42/370
neuropathy, 22/520
renal disease, 22/522
sensorineural deafness, 42/370
Lemmocytoma
tissue culture, 17/83
Lemnisci diagonalis
nucleus, *see* Nucleus lemnisci diagonalis
Leningrad, USSR
neurology, 1/11
Lennox classification
epilepsy, 15/8
Lennox-Gastaut syndrome
absence, 73/215
age at onset, 73/213
child, 72/7
definition, 72/7
EEG, 72/305, 73/214
epilepsy, 72/118
epilepsy with continuous spike wave during slow sleep, 73/274
etiology, 73/215
generalized tonic clonic seizure, 72/7
imaging, 72/379
myoclonic epilepsy, 73/263
myoclonus, 49/619
pathology, 72/117
petit mal epilepsy, 15/471
polymicrogyria, 73/218
seizure, 73/224
sleep, 73/216
slow sleep, 73/274
status epilepticus, 73/274
sudden drop, 73/213
survey, 73/211
tonic seizure, 15/116
treatment, 73/219
Lens
anatomy, 74/420
Lens luxation
Sturge-Weber syndrome, 14/512
Lenticonus
Alport syndrome, 60/770
Lenticular degeneration
progressive, *see* Wilson disease
Lenticular fasciculus
anatomy, 6/5, 20
Lenticular opacity
galactosemia, 42/551
Lenticulo-optic syndrome
thalamic connection, 2/486
Lenticulostriate artery
brain infarction, 53/365
brain lacunar infarction, 54/236
globus pallidus, 49/467
middle cerebral artery syndrome, 53/365
Parkinson disease, 49/97
radioanatomy, 11/82
striatal necrosis, 49/501
symptomatic dystonia, 49/542
Lentigines syndrome
multiple, *see* LEOPARD syndrome
Lentiginosis
centrofacial, *see* Centrofacial lentiginosis
diffuse, *see* Diffuse lentiginosis
Lentiginosis periorofacialis, *see* Peutz-Jeghers syndrome
Lentiginosis profusa (Darier)
differential diagnosis, 14/600
LEOPARD syndrome, 14/600
Scheidt melanism, 14/600
Lentiginosis syndrome
corticotropin, 70/210
Lentigo
Capute-Rimoin-Konigsmark syndrome, 14/110
LEOPARD syndrome, 14/117
neurocutaneous melanosis, 14/78
Lentivirus
caprine arthritis encephalitis, 56/460
classification, 56/460
human immunodeficiency virus, 56/460, 584
multiple sclerosis, 47/323
retrovirus, 56/460, 490, 529
simian T-cell tropic retrovirus, 56/460
visna, 56/16
visna-maedi disease, 56/16, 460
Lenz syndrome
anophthalmia, 43/419
dermatoglyphics, 43/419
digital abnormality, 43/419
EEG, 43/419
mental deficiency, 43/419
microcephaly, 43/419
microphthalmia, 43/419
renal abnormality, 43/419
Leonardo Da Vinci, 2/6
Leonine face, *see* Leontiasis
Leonine-mouse syndrome
cluster headache, 48/219
Leontiasis
leprous neuritis, 51/217
Leontiasis ossea, *see* Fibrous dysplasia
LEOPARD syndrome
congenital heart disease, 14/117
cryptorchidism, 43/28
deafness, 14/117, 600

differential diagnosis, 14/117, 604
ECG, 43/28
Forney syndrome, 14/117
growth retardation, 43/27
heart disease, 43/27
hereditary deafness, 50/218
hyperextensible joint, 43/28
hypertelorism, 14/117, 30/248, 43/27
hypospadias, 43/28
Jervell-Lange-Nielsen syndrome, 14/117
kyphoscoliosis, 43/28
lentiginosis profusa (Darier), 14/600
lentigo, 14/117
mental deficiency, 30/248
neurocutaneous syndrome, 14/101
pectus excavatum, 43/28
pterygium colli, 43/27
ptosis, 43/27
sensorineural deafness, 43/27
skin manifestation, 30/248
Lepine, J., 1/21
Leprechaunism
aminoaciduria, 42/590
corpus callosum agenesis, 50/163
elfin face, 42/589
endocrine disorder, 42/589-591
hirsutism, 42/589
hypertelorism, 42/590
mental deficiency, 42/590
microcephaly, 42/590
muscular atrophy, 42/589
pancreatic islet cell hyperplasia, 42/590
psychomotor retardation, 42/590
Lepromatous leprosy
classification, 51/216
Raynaud phenomenon, 51/218
Lepromatous mononeuritis
leprous neuritis, 51/220
Lepromatous mononeuritis multiplex
leprous neuritis, 51/220
Lepromatous sensory polyneuritis
ulnar nerve, 51/220
Lepromin skin test
leprosy, 42/650, 51/217
Leprosy, 33/421-462
alopecia, 51/218
autonomic nerve, 51/217
autonomic nervous system, 75/18
bacteriology, 33/423-426
borderline, *see* Borderline leprosy
carpal tunnel syndrome, 7/296
classification, 33/423-426, 51/216
claw hand, 51/218
clinical features, 33/428-437
cutaneous nerve, 33/428
dapsone, 51/217
ECG, 51/218
epidemiology, 51/216
erythema nodosum leprosum, 51/216
etiology, 51/216
facial nerve, 33/436
fever, 51/216
fluorescent leprosy antibody absorption, 51/216
hepatosplenomegaly, 51/216
hereditary amyloid polyneuropathy, 21/128
history, 33/421-423
HLA antigen, 42/650
Huntington chorea, 49/291
hyperhidrosis, 51/218
hypertrophic interstitial neuropathy, 21/145
immune response, 42/649
immunology, 33/426-428
indeterminate, *see* Indeterminate leprosy
infectious disease, 51/215
iridocyclitis, 51/216
lepromatous, *see* Lepromatous leprosy
lepromin skin test, 42/650, 51/217
Lucio phenomenon, 51/217
lymphadenopathy, 51/216
mononeuritis multiplex, 51/220
muscle pathology, 33/456-460
muscular atrophy, 51/218
Mycobacterium leprae transmission, 51/216
nasal involvement, 33/435
nerve, 33/441-448
nerve lesion, 7/477
neuropathology, 33/437-455
neuropathy, 33/428-433
neurotrophic osteoarthropathy, 38/459
ocular involvement, 33/435
orthostatic hypotension, 51/218, 63/154
pandysautonomia, 63/154
pathology, 33/423-426
perforating skin ulcer, 51/218
portal of entry, 51/216
Renaut body, 51/231
Ridley-Jopling classification, 51/217
rifampicin, 51/217
skeletal pathology, 33/460-462
symptomatology, 51/216
syringomyelia, 32/293
treatment, 33/433-435
trigeminal nerve, 33/436
trigeminal neuralgia, 51/220
tuberculoid, *see* Tuberculoid leprosy
tuberculoid nerve, 33/437-441

vasculitis, 51/216
Leprous neuritis
acoustic nerve, 51/220
anhidrosis, 51/217, 219
auricularis magnus nerve, 51/219
autonomic dysfunction, 51/219
blindness, 51/217
claw hand, 51/220
clofazimine, 51/232
dapsone, 51/232
dermal nerve, 51/229
EMG, 51/221
facial nerve, 51/220
foam cell, 51/226
glucocorticoid, 51/233
H-reflex, 51/222
histopathology, 51/222
hypertrophic nerve, 51/225, 231
lateral popliteal nerve, 51/220
leontiasis, 51/217
lepromatous mononeuritis, 51/220
lepromatous mononeuritis multiplex, 51/220
lepromatous type, 51/9, 223
mixed nerve, 51/227
mononeuritis, 51/218
mononeuritis multiplex, 51/219
motor impairment, 51/230
motor nerve conduction velocity, 51/221
nerve conduction velocity, 51/221
neuralgia, 51/221, 234
neurogenic muscular atrophy, 51/221
patchy anesthesia, 51/219
pathogenesis, 51/228
polyneuritis, 51/218
portal of entry, 51/229
radial nerve, 51/219
Renaut body, 51/231
rifampicin, 51/232
rifamycin, 51/232
Schwann cell, 51/226
segmental demyelination, 51/224, 231
sensory conduction velocity, 51/221
sensory impairment, 51/230
sural nerve, 51/224
thermography, 51/218
treatment, 51/232-235
trigeminal nerve, 51/220
trigeminal neuralgia, 51/220
tuberculoid type, 51/223, 225, 229
ultrastructure, 51/224
Leprous neuropathy
differential diagnosis, 60/19
hereditary sensory and autonomic neuropathy, 60/19
hereditary sensory and autonomic neuropathy type I, 60/19
Leptomeningeal angiomatosis
brain atrophy, 18/274
Cobb syndrome, 14/14
Divry-Van Bogaert syndrome, 14/14, 71, 512, 55/320
histopathology, 14/62
Krabbe-Bartels disease, 14/82
leukodystrophy, 10/24, 120
linear nevus sebaceous syndrome, 43/39
orthochromatic leukodystrophy, 10/113
primary pigmentary retinal degeneration, 13/339
Sturge-Weber syndrome, 14/228, 232, 43/47, 50/376
Leptomeningeal cyst, *see* Arachnoid cyst
Leptomeningeal gliomatosis
diffuse, *see* Diffuse leptomeningeal gliomatosis
Leptomeningeal hemangiomatosis, *see* Leptomeningeal angiomatosis
Leptomeningeal lipoma
Krabbe-Bartels disease, 14/411
Leptomeningeal melanism
definition, 14/415
Leptomeningeal melanoblastic hyperplasia
terminology, 14/415
Leptomeningeal melanoma
neurocutaneous melanosis, 14/593, 43/33
Leptomeningeal melanomatosis
terminology, 14/416
Leptomeningeal melanosis
malignant change, 14/420
neurofibromatosis type I, 14/87, 133
terminology, 14/415
Leptomeningeal metastasis, 69/76
biochemical marker, 71/652
biopsy, 71/652
breast cancer, 71/657
breast carcinoma, 71/617
chemotherapy, 71/655
choroid plexus tumor, 71/663
cranial nerve, 69/153
CSF, 69/154, 71/649
germ cell tumor, 71/662
histology, 69/152
Hodgkin disease, 39/36-38
incidence, 69/151, 71/647
intrathecal treatment, 71/656
lymphoma, 71/662
malignant lymphoma, 39/36-38
meningeal lymphoma, 69/266
neoplastic meningitis, 69/158

neuroimaging, 71/652
pinealoma, 71/662
radiotherapy, 71/654
spinal symptom, 69/154
survey, 69/151, 71/641
symptom, 69/152, 71/648
systemic solid tumor, 71/657
therapy, 69/158, 71/654
tumor type, 71/658
Leptomeningeal venous angioma
hemorrhage, 18/272-274
Leptomeninges
cerebellar medulloblastoma, 18/187
desmoplastic cerebellar medulloblastoma, 18/184-186
globoid cell leukodystrophy, 10/77
Huntington chorea, 49/316
verruga peruana, 34/665
Leptomeningitis
brain biopsy, 10/685
spinal cord injury, 25/60
traumatic, 25/60
Leptospira
acute viral encephalitis, 56/126
viral meningitis, 56/126, 128
Leptospira interrogans
epidemiology, 55/415
leptospirosis, 55/415
meningitis, 55/415
temporal arteritis, 55/415
zoonosis, 55/415
Leptospirosis, 33/395-414
acquired myoglobinuria, 62/577
chronic meningitis, 33/402, 56/645
classification, 33/396
demyelination, 9/682
encephalitis, 33/403
encephalomyelitis, 33/402
epidemiology, 33/398
general symptom, 33/395
icterohaemorrhagiae, 33/409
kidney, 33/398
laboratory diagnosis, 33/399-420
Leptospira australis, 33/409
Leptospira autumnalis, 33/409
Leptospira bataviae, 33/409
Leptospira canicola, 33/408
Leptospira grippotyphosa, 33/407
Leptospira hebdomadis, 33/409
Leptospira hyos, 33/407
Leptospira interrogans, 55/415
Leptospira pomona, 33/407
Leptospira pyrogenes, 33/409
Leptospira sejroe, 33/408
liver, 33/398
lymphocytic meningitis, 33/400-402
meningism, 33/397
meningitis, 33/400-402
meningitis rash, 52/25
microbiology, 33/369
mode of infection, 33/395
moyamoya disease, 53/164
nerve lesion, 7/489
neurasthenia, 33/404
neurologic manifestation, 33/396-399
optic neuritis, 7/489
pathology, 33/405-407
polyradiculitis, 7/489
postencephalitis defect, 33/404
postinfection neurasthenia, 33/404
psychosis, 33/404
serologic classification, 33/397
serologic detection, 33/407
subarachnoid hemorrhage, 12/111
treatment, 33/400, 55/415
viral meningitis, 56/130
Lergotrile
chemical structure, 37/236
delirium, 46/539
Parkinson disease, 37/236
Léri pleonosteosis
carpal tunnel syndrome, 7/293
compression neuropathy, 7/293
Léri sign
diagnostic value, 1/245
lumbar intervertebral disc prolapse, 20/588
Leriche syndrome
diagnostic pitfalls, 12/538
Takayasu disease, 63/53
Lesch-Nyhan hyperuricemia, *see* Lesch-Nyhan syndrome
Lesch-Nyhan syndrome, 29/263-275, 42/152-154
see also Hyperuricemia
aggression, 42/152, 63/256
amniocentesis, 29/273
biochemical features, 29/266-268
birth incidence, 42/153
childhood psychosis, 46/194
choreoathetosis, 6/456, 22/514, 29/263, 42/153, 49/385, 59/360, 60/19
clinical features, 29/263-266, 63/256
crystalluria, 29/265, 42/153
dystonia musculorum deformans, 49/524
epilepsy, 63/256, 72/222
gene localization, 42/154
genetics, 29/270-273

glycine conversion, 6/456
hematuria, 6/456, 29/265, 42/153
hereditary sensory and autonomic neuropathy, 60/19
hereditary sensory and autonomic neuropathy type IV, 60/19
heterozygote detection, 42/154
hyperreflexia, 42/153
hyperuricemia, 6/456, 22/514, 29/265, 42/153, 46/194, 49/385, 59/360, 60/19
hypoxanthine phosphoribosyltransferase, 42/154, 59/360, 60/19
hypoxanthine phosphoribosyltransferase deficiency, 22/514, 29/268-270, 59/360
intrauterine diagnosis, 29/273
Lyon principle, 50/542
mental deficiency, 6/456, 22/514, 42/153, 46/66, 60/19, 63/256
molecular basis, 29/268-270
opisthotonos, 42/153
orotic aciduria, 63/256
prenatal diagnosis, 42/154
psychomotor retardation, 42/153
purine metabolism, 29/267
renal calculi, 42/153, 59/360
self-mutilation, 6/456, 22/514, 29/264, 42/153, 46/66, 59/360, 60/19, 63/256
sex ratio, 42/154
symptomatic dystonia, 49/543
tetraparesis, 59/360
tophus, 29/266
treatment, 6/457, 29/273-275, 59/360
uric acid, 6/456, 29/266, 46/66, 60/19
Leschke-Ullmann syndrome
clinical features, 14/117
differential diagnosis, 14/117
incontinentia pigmenti, 14/117
mental deficiency, 14/117
neurocutaneous syndrome, 14/101
neurofibromatosis, 14/741
Lesionectomy
epilepsy, 73/390
Lesions
adrenocortical, *see* Adrenocortical lesion
amygdala, *see* Amygdala lesion
aorta, *see* Aorta lesion
avulsion, *see* Avulsion lesion
brachial plexus, *see* Brachial plexus lesion
brachium conjunctivum, *see* Brachium conjunctivum lesion
brain stem, *see* Brain stem lesion
capsula interna, *see* Capsula interna lesion
cardiac, *see* Cardiac lesion
cerebellar, *see* Cerebellar lesion
cerebellar mass, *see* Cerebellar mass lesion
cerebellopontine, *see* Cerebellopontine lesion
cervical cord, *see* Cervical cord lesion
cingulate gyrus, *see* Cingulate gyrus lesion
CNS, *see* CNS lesion
cord, *see* Cord lesion
coup, *see* Coup lesion
delta, *see* Delta lesion
dorsal root entry zone, *see* Dorsal spinal root entry zone lesion
ectopic, *see* Ectopic lesion
epidural, *see* Epidural lesion
epileptogenic, *see* Epileptogenic lesion
eye, *see* Eye lesion
frontal lobe, *see* Frontal lobe lesion
hemorrhagic, *see* Hemorrhagic lesion
hippocampus, *see* Hippocampus lesion
iatrogenic, *see* Iatrogenic lesion
lumbar plexus, *see* Lumbar plexus lesion
microwave, *see* Microwave lesion
multifocal pontine, *see* Multifocal pontine lesion
multiple brain mass, *see* Multiple brain mass lesion
multiple white matter, *see* Multiple white matter lesion
myelin, *see* Myelin lesion
nerve, *see* Nerve lesion
nucleus ruber, *see* Nucleus ruber lesion
occipital lobe, *see* Occipital lobe lesion
optic tract, *see* Optic tract lesion
oral, *see* Oral lesion
otolith, *see* Otolith lesion
pallidal, *see* Pallidal lesion
parasellar space occupying, *see* Parasellar space occupying lesion
parietal lobe, *see* Parietal lobe lesion
peripheral nervous system, *see* Peripheral nervous system lesion
peroneal nerve, *see* Peroneal nerve lesion
phrenic nerve, *see* Phrenic nerve lesion
pontine, *see* Pontine lesion
posterior fossa, *see* Posterior fossa lesion
posterior lobe, *see* Posterior lobe lesion
primary melanocytic, *see* Primary melanocytic lesion
principal spinal cord, *see* Principal spinal cord lesion
putaminal, *see* Putaminal lesion
pyramidal tract, *see* Pyramidal tract lesion
radio, *see* Radiolesion
recurrent laryngeal nerve, *see* Recurrent laryngeal nerve lesion

rostral tegmental, *see* Rostral tegmental lesion
silent, *see* Silent lesion
skeletal, *see* Skeletal lesion
skin, *see* Skin lesion
space occupying, *see* Space occupying lesion
spinal cord, *see* Spinal cord lesion
spinal nerve root, *see* Spinal nerve root lesion
stereotactic, *see* Stereotactic lesion
striatal, *see* Striatal lesion
subcortical, *see* Subcortical lesion
superior cerebellar peduncle, *see* Superior cerebellar peduncle lesion
temporal lobe, *see* Temporal lobe lesion
thalamic, *see* Thalamic lesion
thermo, *see* Thermolesion
transverse spinal cord, *see* Transverse spinal cord lesion
trigeminal nerve, *see* Trigeminal nerve lesion
tumor growth, *see* Tumor growth lesion
ulnar nerve, *see* Ulnar nerve lesion
vertebral, *see* Vertebral lesion
visceral, *see* Visceral lesion
Lesser occipital nerve
neurofibromatosis, 2/134
Lethal granuloma, *see* Midline granuloma
Lethal multiple pterygia syndrome
infantile spinal muscular atrophy, 59/72
Lethargy
see also Hypersomnia *and* Sleep
angiostrongyliasis, 52/554
ascariasis, 35/359
aspergillosis, 52/378
bacterial meningitis, 52/1
barbiturate intoxication, 65/508
brain abscess, 52/152
chelonia intoxication, 37/91
complex II deficiency, 66/422
digoxin intoxication, 65/450
encephalitis, *see* Encephalitis lethargica
ethylene glycol intoxication, 64/123
glutaric aciduria type II, 62/495
gram-negative bacillary meningitis, 52/119
headache, 48/32
homocystinuria, 55/329
hyperbeta-alaninemia, 29/230
hyponatremia, 63/545
intravascular consumption coagulopathy, 63/316
Jamaican vomiting sickness, 37/533
Kawasaki syndrome, 52/265, 56/639
lead encephalopathy, 64/435
leukemia, 63/344
lymphocytic choriomeningitis, 9/197
meningeal lymphoma, 63/347
mesuximide intoxication, 37/202
metharbital intoxication, 37/202
methotrexate, 63/357
multiple acyl-coenzyme A dehydrogenase deficiency, 62/495
organotin intoxication, 64/139
oxoglutarate dehydrogenase deficiency, 62/503
pellagra, 28/71
petit mal status, 15/170
phenobarbital intoxication, 37/202
pyruvate decarboxylase deficiency, 62/502
pyruvate dehydrogenase complex deficiency, 66/416
reserpine intoxication, 37/438
Shapiro syndrome, 50/167
valproic acid induced hepatic necrosis, 65/125
vincristine, 63/359
Waldenström macroglobulinemia, 63/396
zinc intoxication, 36/337
Letter inversion
migraine, 48/162
Letter recognition
agnosia, 45/344
Letterer-Siwe disease, *see* Abt-Letterer-Siwe disease
Leucine
GM2 gangliosidosis, 10/295
pellagra, 28/72
vitamin B3 deficiency, 51/333
Leucine aminopeptidase
wallerian degeneration, 9/36
Leucine enkephalin
basal ganglion, 49/33
endogenous opioid, 57/83, 65/349
Parkinson disease, 49/120
proenkephalin, 65/350
Leucine metabolism
infantile spinal muscular atrophy, 59/74
isovaleric acidemia, 42/579, 66/651
maple syrup urine disease, 66/651
β-methylcrotonylglycinuria, 66/652
Leucine sensitivity
idiopathic hypoglycemia, 27/73, 42/576
Leu-enkephalin, *see* Leucine enkephalin
Leukapheresis
dermatomyositis, 62/386, 71/143
inclusion body myositis, 62/386
polymyositis, 62/386, 71/143
vasculitis, 71/161
Leukemia
acute, *see* Acute leukemia
acute lymphocytic, *see* Acute lymphocytic leukemia
acute myeloid, *see* Acute myeloid leukemia

adult T-cell, *see* Adult T-cell leukemia
aspergillosis, 35/396, 52/378, 63/345
aspergillus infection, 63/353
ataxia, 63/342, 344
ataxia telangiectasia, 14/271, 46/58, 60/674
autonomic nervous system, 39/12
bacterial CNS infection, 63/353
basal ganglion, 39/12
biochemical marker, 69/241
Bloom syndrome, 14/775, 43/10
brain, 18/246, 251, 321
brain cortex, 39/11
brain embolism, 63/345
brain hemorrhage, 39/6-8, 53/34, 54/293, 63/343
brain microinfarction, 63/345
brain stem, 39/12
brain venous infarction, 63/344
Brown-Séquard syndrome, 63/342
carmustine, 39/18
cauda equina syndrome, 63/343
cerebellum, 39/11
Chédiak-Higashi syndrome, 60/674
chemotherapy, 39/15, 63/356, 69/233-235
child, 67/400
chlormethine, 39/18, 63/358
choroid plexus xanthoma, 18/322
chronic lymphocytic, *see* Chronic lymphocytic leukemia
chronic meningitis, 56/645
chronic myelogenous, *see* Chronic myelogenous leukemia
CNS, *see* CNS leukemia
combined modality therapy, 63/355
cortical blindness, 63/342
corticosteroid, 39/15
cranial nerve, 39/9, 69/238
cranial neuropathy, 63/354
cranial vault tumor, 17/122
cryptococcosis, 52/433
CSF, 39/5, 69/239, 241
cytarabine, 39/17, 63/356
cytostatic treatment, 18/251
diabetes insipidus, 39/10, 63/342
disseminated necrotizing leukoencephalopathy, 63/355
dissemination, 39/1
Down syndrome, 31/424-426, 43/545, 50/519
encephalitis, 63/352
encephalopathy, 63/345, 354
epidemiology, 69/234
epidural tumor, 69/239
epilepsy, 63/345
eye, 39/13
Fanconi syndrome, 31/319
fluorouracil, 39/17
gaze paralysis, 63/342
gene amplification, 69/241
granulocytic, *see* Myeloid leukemia
hairy cell, *see* Hairy cell leukemia
headache, 48/33, 63/344
hemiplegia, 63/342, 345
history, 39/1
Hodgkin lymphoma, 63/354
hyperviscosity syndrome, 55/471, 484
hypothalamus, 39/9
immunocompromised host, 56/469
immunocytochemistry, 69/241
immunotherapy, 18/251
incidence, 39/2-4, 63/340
infection, 39/19
intrathecal methotrexate, 63/355, 357
intravascular consumption coagulopathy, 39/8, 63/315, 345
irradiation encephalopathy, 63/354
irradiation myelopathy, 63/356
juvenile chronic myeloid, *see* Juvenile chronic myeloid leukemia
labyrinthine hemorrhage, 55/133
lethargy, 63/344
leukoencephalopathy, 9/594
listeriosis, 52/91
lymphatic, *see* Lymphatic leukemia
lymphoblastic, *see* Lymphoblastic leukemia
lymphocytic, *see* Lymphocytic leukemia
lymphosarcoma, 18/247
meningeal, *see* Meningeal leukemia
meningeal irritation, 69/238
meningitis, *see* Meningeal leukemia
metastasis, 71/659
metastatic cardiac tumor, 63/93
methotrexate, 39/15
methotrexate effect, 63/357
monocytic, *see* Monocytic leukemia
multiple brain infarction, 63/345
mycotic brain aneurysm, 63/344
myeloblastic, *see* Myeloblastic leukemia
myelocytic, *see* Myeloid leukemia
myeloid, *see* Myeloid leukemia
myelopathy, 63/354
myopathy, 39/13
neuroimaging, 69/241
neurologic chemotherapy complication, 63/356
neurologic features, 18/249
neurologic treatment complication, 63/354
neurology, 69/233
neuropathology, 18/251

neuropathy, 8/145, 63/354
nitrosourea, 63/358
non-Hodgkin lymphoma, 63/354
nonseptic thrombotic endocarditis, 63/345
optic nerve tumor, 17/371
paraneoplastic polyneuropathy, 51/468
paraplegia, 63/343
paraproteinemia, 69/308
parenchymal degeneration, 39/20
pathogenesis, 39/1
pathology, 69/235
peripheral nerve, 39/13
peripheral neurolymphomatosis, 56/180
pituitary, 39/10
plasma cell, *see* Plasma cell leukemia
polyneuropathy, 63/343
pontine myelinolysis, 39/12
procarbazine, 39/18
progressive multifocal leukoencephalopathy, 39/19, 63/354, 71/400
promyelocytic, *see* Promyelocytic leukemia
radiation, 39/14
radiculopathy, 39/13, 63/343
radiotherapy, 18/251, 63/354
RES tumor, 18/247-254
retrovirus, 56/16
SBLA syndrome, 42/769
skull base tumor, 17/174
spinal, *see* Spinal leukemia
spinal cord, 39/13, 67/198
spinal cord compression, 63/343
spinal epidural tumor, 20/103, 112
spinal nerve root, 63/343
spinal subarachnoid hemorrhage, 63/344
spinal subdural hemorrhage, 63/344
subarachnoid hemorrhage, 12/110, 55/3, 63/344
subdural hematoma, 39/8, 63/344
survey, 69/233
systemic brain infarction, 11/461
T-lymphocyte, *see* T-lymphocyte leukemia
thiotepa, 63/359
thrombocytopenia, 63/321, 343
toxoplasmosis, 63/354
transverse spinal cord lesion, 63/343
treatment, 18/251, 39/5, 14-18
tremor, 63/345
venous sinus thrombosis, 63/345
vinblastine, 39/17
vincristine, 39/16, 63/358
viral CNS infection, 63/353
xeroderma pigmentosum, 60/674
Leukemic meningitis, *see* Meningeal leukemia
Leukemoid reaction
Down syndrome, 31/424-426
dwarfism, 43/385
Leukism
partial, 14/105
Leukoaraiosis
see also Binswanger disease
arterial hypertension, 63/76
episodic hypotension, 63/76
hypertension, 63/76
hypertensive arteriolar disease, 63/76
white matter lesion, 54/224
Leukocyte
brain metastasis, 69/112
CSF, *see* CSF leukocyte
glycosaminoglycanosis type VI, 66/308
vasculitis, 71/150
Leukocyte count
CSF, *see* CSF leukocyte count
multiple sclerosis, 47/340
Leukocyte granulation
neuronal ceroid lipofuscinosis, 10/578
Leukocyte inclusion
Chédiak-Higashi syndrome, 60/674
juvenile metachromatic leukodystrophy, 60/674
Leukocytoclastic vasculitis
classification, 71/4
paraneoplastic syndrome, 71/678
Leukocytosis
Abt-Letterer-Siwe disease, 42/441
brain infarction, 53/165
brain ischemia, 53/165
critical illness polyneuropathy, 51/576
CSF, *see* CSF leukocytosis
globoid cell leukodystrophy, 10/73
iatrogenic brain infarction, 53/165
metastatic cardiac tumor, 63/96
primary cardiac tumor, 63/96
zinc intoxication, 64/362
Leukodystrophia cerebri progressiva hereditaria, *see* Metachromatic leukodystrophy
Leukodystrophy
see also Diffuse sclerosis
Addison disease, 10/36, 129, 39/482
adrenal, *see* Adrenoleukodystrophy
adult, *see* Adult leukodystrophy
adult onset spongiform, *see* Adult onset spongiform leukodystrophy
angiomatosis, 10/124
biochemistry, 10/134
catecholamine, 10/48
central visual disturbance, 60/655
cholesterol metabolism, 66/3
cutis marmorata, 10/120

dementia, 46/130
Diezel-Richardson classification, 10/24
dog, 9/678, 10/88
dysmyelinogenic, *see* Alexander disease
epilepsy, 15/373, 376, 425
familial amaurotic idiocy, 10/53
familial spastic paraplegia, 22/433, 441
familial type, 10/12-15
ganglioside, 10/24
globoid cell, *see* Globoid cell leukodystrophy
hereditary, *see* Hereditary leukodystrophy
hereditary progressive, *see* Hereditary progressive leukodystrophy
history, 10/1, 105
history of concept, 10/13
idiocy, 10/24
infantile metachromatic, *see* Infantile metachromatic leukodystrophy
inositol phospholipid, 10/30
juvenile metachromatic, *see* Juvenile metachromatic leukodystrophy
Krabbe-Collier-Greenfield type, 10/19
late infantile metachromatic, *see* Late infantile metachromatic leukodystrophy
leptomeningeal angiomatosis, 10/24, 120
lipid hexosamine, 10/24
lipidosis, 10/22
Löwenberg-Hill, *see* Löwenberg-Hill leukodystrophy
macrencephaly, 10/24
megalencephalic type, *see* Alexander disease
megalobarencephaly, 10/36
melanodermic type, *see* Adrenoleukodystrophy
meningeal angiomatosis, 10/120
metabolism, 66/2
metachromatic, *see* Metachromatic leukodystrophy
microcephalic familial type, *see* Norman microcephalic familial leukodystrophy
MNGIE syndrome, 62/505
myelin, 10/134
myelin lipid, 10/135
myoclonia, 6/764
neuroaxonal, *see* Neuroaxonal leukodystrophy
neuropathy, 8/16
neutral fat, *see* Neutral fat leukodystrophy
Norman, *see* Norman leukodystrophy
nosology, 10/1-37, 66/2
optic atrophy, 60/655
organic acid metabolism, 66/641
orthochromatic, *see* Orthochromatic leukodystrophy
Peiffer classification, 10/106
Pelizaeus-Merzbacher disease, 10/19
prelipid, 10/12, 14
Rosenthal fiber, 42/484-486
spinal muscular atrophy, 59/371
spongy degeneration, 10/19
subacute sclerosing panencephalitis, 10/2
sudanophilic, *see* Sudanophilic leukodystrophy
survey, 66/1
terminology, 10/43
thromboangiitis obliterans, 10/127
vitamin A, 10/58
Wicke classification, 10/14
Leukodystrophy classification, 10/1-37
inclusion body encephalitis, 10/20
panencephalitis, 10/20
subacute encephalitis, 10/20
Leukodystrophy with meningeal angiomatosis, *see* Divry-Van Bogaert syndrome
Leukoencephalitis
acute hemorrhagic, *see* Acute hemorrhagic leukoencephalitis
hemorrhagic, *see* Hemorrhagic leukoencephalitis
orthomyxovirus, 56/15
progressive multifocal, *see* Progressive multifocal leukoencephalopathy
rubella virus, 56/410
subacute sclerosing, *see* Subacute sclerosing panencephalitis
viral, *see* Viral leukoencephalitis
Leukoencephalitis periaxialis concentrica, *see* Concentric sclerosis
Leukoencephalomyelitis
infectious, *see* Infectious leukoencephalomyelitis
Leukoencephalopathy
see also Encephalopathy
acute hemorrhagic, *see* Acute hemorrhagic leukoencephalitis
acute hepatic failure, 47/584
acute necrotizing hemorrhagic, *see* Acute necrotizing hemorrhagic leukoencephalitis
adrenoleukodystrophy, 47/593-598
alcohol, 47/585
allergic angiopathy, 9/609
Alzheimer type II astrocyte, 47/584
amphotericin B, 47/570, 71/384
anemia, 9/613
anoxia, 9/572
anoxic ischemic, *see* Anoxic ischemic leukoencephalopathy
antimetabolite, 9/627
arsenic intoxication, 9/647, 650, 47/559
arsphenamine, 9/647
Binswanger disease, 9/589, 54/221

blood disease, 9/592
blood dyscrasia, 9/572, 593
bone marrow aplasia, 9/599
brain amyloid angiopathy, 54/341
brain thromboangiitis obliterans, 9/613
burn, 47/574-576
burn demyelination, 47/576
burn toxin, 47/576
carbon monoxide intoxication, 9/578, 628, 47/553-555
carbon tetrachloride intoxication, 47/556
cavitating, *see* Cystic leukoencephalopathy
central pontine dystrophy, 9/636
central pontine myelinolysis, 9/636, 47/585-589
chronic, *see* Chronic leukoencephalopathy
chronic ischemic, *see* Chronic ischemic leukoencephalopathy
ciclosporin intoxication, 63/537
copper, 9/645
cuprizone intoxication, 47/563
cyanide intoxication, 9/632, 47/552
cystic, *see* Cystic leukoencephalopathy
demyelinating, *see* Demyelinating leukoencephalopathy
diffuse, *see* Diffuse leukoencephalopathy
diphtheria toxin, 47/569
disseminated necrotizing, *see* Disseminated necrotizing leukoencephalopathy
dog, 9/676
ergot derivative, 9/652, 47/573
experimental spongiform, *see* Experimental spongiform leukoencephalopathy
fludarabine intoxication, 65/534
fluorouracil intoxication, 65/532
gasoline, 66/721
Gerstmann-Sträussler-Scheinker disease, 56/555
globoid cell leukodystrophy, 47/598
hemoblastosis, 9/594
hemorrhage, 9/597
hemorrhagic, *see* Hemorrhagic leukoencephalopathy
Henoch-Schönlein purpura, 9/599, 607
hepatic coma, 9/634-636
hereditary hemorrhagic telangiectasia (Osler), 9/607
Hodgkin disease, 9/594
homocystine, 47/585
hypereosinophilic syndrome, 63/372
hypertensive encephalopathy, 9/613
hyperthyroidism, 9/627
hypoglycemia, 9/626
hyponatremia, 47/586
immunothrombocytopenia, 9/599
inclusion body myositis, 62/373
interleukin-2 intoxication, 65/539
intoxication category, 9/617-620
intoxication diagnosis, 47/556
intoxication neuropathy, 47/554, 556
intoxication treatment, 47/553, 555-557, 565
ischemia, 9/572
isoniazid, *see* Isoniazid leukoencephalopathy
isoniazid intoxication, 65/480
kernicterus, 9/638, 47/585
lead intoxication, 9/641, 47/560
leukemia, 9/594
liver disease, 47/584
lymphoma, 9/594
lysolecithin, 47/570
manganese intoxication, 9/643
Marchiafava-Bignami disease, 9/653, 47/557
metabolic disorder, 47/583-599
metachromatic leukodystrophy, 47/589-593
methanol intoxication, 47/555
methotrexate, 47/551, 570-573, 69/484
methotrexate intoxication, 39/100, 65/531
methylmalonic acid, 47/585
microangiopathy, 9/592
multifocal necrotizing, *see* Multifocal necrotizing leukoencephalopathy
multiple, 54/96
multiple myeloma, 9/594
myelin sheath, 47/551, 564
necrotizing, *see* Necrotizing leukoencephalopathy
neuropathology, 66/726
nuclear magnetic resonance, 54/96
organic mercury intoxication, 9/643, 47/559
organic solvent, 66/721
organic solvent intoxication, 47/555
pathologic autointoxication, 9/620
phosphorus, 9/652
physiologic autointoxication, 9/620
pleomorphism, 67/223
polyarteritis nodosa, 9/609
polycythemia, 9/594, 598
polymicrocavitation, 47/584
porphyria, 9/639, 47/585
progressive multifocal, *see* Progressive multifocal leukoencephalopathy
radiation effect, 63/355
rheumatic disease, 9/609
rickettsial infection, 9/609
scurvy, 9/607
serum sickness, 9/608
sickle cell anemia, 9/614
splenogenic purpura, 9/598
spongiform, *see* Spongiform leukoencephalopathy

Streptococcus viridans, 9/609
subacute combined spinal cord degeneration, 9/614, 47/584
tellurium intoxication, 9/645
temporal arteritis, 9/609
thallium intoxication, 9/646
thrombocytopenic purpura, 9/598
thrombotic thrombocytopenic purpura, 9/599
toluene, 66/721
toxi-infective purpura, 9/598
triethyltin intoxication, 47/564
uremic, *see* Uremic leukoencephalopathy
vitamin B12 deficiency, 47/584
Wegener granulomatosis, 9/610-613, 39/344
Werlhof essential thrombocytopenic purpura, 9/598
Wernicke encephalopathy, 47/586
Wilson disease, 9/634
Leukoencephalopathy with meningeal angiomatosis, *see* Divry-Van Bogaert syndrome
Leukoencephalopathy with meningocerebral angiomatosis, *see* Divry-Van Bogaert syndrome
Leukofuchsin
Schiff, *see* Schiff leukofuchsin
Leukomalacia
anoxic, *see* Anoxic leukomalacia
cocaine intoxication, 65/255
infantile periventricular, *see* Infantile periventricular leukomalacia
periventricular, *see* Periventricular leukomalacia
Leukopenia
adult Gaucher disease, 66/123
Argentinian hemorrhagic fever, 56/364
aspergillosis, 52/378
hyperdibasic aminoaciduria, 42/564
mephenytoin intoxication, 37/202
orotic aciduria, 42/607
penicillamine intoxication, 27/405
renal transplantation, 63/532
Leukoplakia
dyskeratosis congenita, 14/784, 43/392
Leukotomy
prefrontal, *see* Prefrontal leukotomy
transorbital, *see* Transorbital leukotomy
Leukotriene
brain infarction, 53/131
brain ischemia, 53/131
eosinophil, 63/369
Leukotriene C4
eosinophil, 63/370
Leukovorin, *see* Folinic acid
Leuprolide, *see* Leuprorelin
Leuprorelin
benign intracranial hypertension, 67/111
CNS toxicity, 69/500
Levamisole
ataxia telangiectasia, 60/390
multiple sclerosis, 47/203
Levi-Lorain dwarfism
myxedema, 2/454
Peter Pan syndrome, 18/545
pituitary gland, 2/455
thyrotropic hormone, 2/454
Levine-Critchley syndrome, *see* Chorea-acanthocytosis
Levine experimental anoxia ischemia
brain edema, 9/582
complex IV, 9/583
MAO, 9/583
succinate dehydrogenase, 9/583
Levodopa
albinism, 43/3
Alzheimer disease, 46/266
behavior disorder, 46/592-594
brain death, 24/743
chemical structure, 37/235
chorea, 49/190
confusional state, 46/379
deafness, 43/3
delirium, 46/539
dementia, 46/379
depression, 46/474, 593
dystonia, 42/213, 253
dystonia deafness syndrome, 42/372
dystonia musculorum deformans, 42/215-218
hallucination, 46/593
hallucinogenic agent, 37/235
hepatic coma, 27/360, 372
hepatic encephalopathy, 27/360, 372
induced myoclonus, 38/582
juvenile parkinsonism, 49/156
manganese intoxication, 36/225-231, 64/316
mania, 46/462, 593
1-methyl-4-phenyl-1,2,3,6-tetrahydropyridine intoxication, 65/373
micrographia, 45/465
multiple neurofibromatosis, 43/35
multiple neuronal system degeneration, 59/136, 138
myoclonus, 38/582
myotonic syndrome, 40/561
nigrospinodentate degeneration, 22/171
Parkinson dementia, 46/373, 379
Parkinson disease, 42/246, 60/736
parkinsonism, 42/253, 256
pharmacology, 74/144

postherpetic neuralgia, 51/181
presynaptic, 74/144
progressive dysautonomia, 59/136, 138, 140
progressive pallidal atrophy (Hunt-Van Bogaert), 42/248
progressive supranuclear palsy, 22/227, 49/247
psychosis, 46/593
pure autonomic failure, 22/236, 240
sleep disorder, 46/593
striatonigral degeneration, 42/263, 60/541
taurine deficiency, 42/256
uvea melanoma, 60/736
vitamin B6, 28/124-128
Levodopa induced dyskinesia
age, 49/197
animal model, 49/197
blepharospasm, 49/195
chorea, 49/195, 197
clinical features, 49/195
drug holiday, 49/199
dystonia, 49/196
epidemiology, 49/196
hemiballismus, 49/374
linguobuccofacial dyskinesia, 49/195
movement disorder, 49/185
myoclonic jerk, 49/195
myoclonus, 49/196, 199
orofacial dyskinesia, 49/195
oromandibular dystonia, 49/195
Parkinson disease, 49/196
parkinsonism, 46/592, 49/195
pathophysiology, 49/197
peak-dose dyskinesia, 49/198
treatment, 49/198
Levodopa intoxication
depression, 46/593
emotion, 46/593
hypomania, 46/593
nightmare, 46/593
psychosis, 46/593
visual hallucination, 46/593
Levodopa provocation test
Huntington chorea, 49/290
Levonorgestrel
benign intracranial hypertension, 67/111
Levorphanol
chemistry, 37/366
opiate, 37/366, 65/350
Levosan, *see* Fructose
Lewandowsky-Lutz syndrome
multiple nevoid basal cell carcinoma syndrome, 14/114
Lewandowsky, M., 1/5, 23
Lewis triple response
autonomic nervous system, 1/485
Lewy body
Alzheimer disease, 42/276
characteristics, 6/215
diagnostic value, 6/215
Hallervorden-Spatz syndrome, 6/624, 49/401
late infantile neuroaxonal dystrophy, 49/405
neuronal atrophic dystrophy, 21/62
Parkinson dementia, 46/377, 49/178
Parkinson disease, 6/177, 215, 21/54, 42/246, 49/108, 113
pure autonomic failure, 2/120, 22/235, 239, 265
striatonigral degeneration, 6/667, 698
structure, 49/113
Lewy body disease
amyotrophic lateral sclerosis, 59/410
dementia, 59/410
diffuse, *see* Diffuse Lewy body disease
Parkinson disease, 49/110
Leyden, E. Von, *see* Von Leyden, E.
Leyden-Möbius dystrophy, *see* Limb girdle syndrome
Leyden-Möbius pelvifemoral muscular dystrophy, *see* Pelvifemoral muscular dystrophy (Leyden-Möbius)
Leyden-Westphal ataxia, *see* Acute cerebellar ataxia
Leydig cell testicular tumor
benign intracranial hypertension, 67/111
LGB syndrome, *see* Guillain-Barré syndrome
LH, *see* Luteinizing hormone
Lhermitte-Duclos disease, *see* Cerebellar granular cell hypertrophy
Lhermitte, J., 1/7, 27, 31
Lhermitte sign
cisplatin intoxication, 64/358
cisplatin neurotoxicity, 61/200
clinical significance, 1/109
demyelination, 47/32
glucocorticosteroid, *see* Glucocorticoid
intraspinal tumor, 19/35
lathyrism, 65/3
multiple sclerosis, 9/171, 47/55
neuromyelitis optica, 47/401
radiation myelopathy, 26/85, 91, 61/199
rheumatoid arthritis, 71/23
Sjögren syndrome, 71/75
spastic paraplegia, 59/432
vitamin B6 deficiency, 51/335
Li-Fraumeni syndrome
brain tumor, 67/54
CNS tumor, 67/54
survey, 68/301

Libido
brain embolism, 53/207
brain infarction, 53/207
brain injury, 24/593
cerebrovascular disease, 53/207
intracranial hemorrhage, 53/207
sexual function, 26/454
transient ischemic attack, 53/207
Libman-Sacks endocarditis
see also Systemic lupus erythematosus endocarditis
brain embolism, 53/34, 156, 63/29
brain hemorrhagic infarction, 63/29
brain infarction, 53/156, 63/29
brain ischemia, 53/156
cardiac valvular disease, 63/29
systemic lupus erythematosus, 63/29
Lichtenberg figure
electrical injury, 61/192
Lichtenstein classification
nerve disease, 7/201
Lichtenstein-Knorr disease
cataract, 22/515
Friedreich ataxia, 22/515
hereditary hearing loss, 22/515
kyphoscoliosis, 22/516
Licorice, *see* Glycyrrhiza
Lid lag
myotonic hyperkalemic periodic paralysis, 41/165, 62/462
paramyotonia congenita, 40/544
thyroid associated ophthalmopathy, 62/530
Lid twitch sign
myasthenia gravis, 22/205
Liddell, E.G.T., 1/46, 50
Lidocaine
action mechanism, 37/449
antiarrhythmic agent, 37/449
antiarrhythmic agent intoxication, 37/449-451
asterixis, 63/192
brain infarction, 53/116
brain ischemia, 53/116
cardiac pharmacotherapy, 63/190, 192
chemical structure, 37/449
confusional state, 63/192
decompression illness, 61/225
decompression myelopathy, 61/225
encephalopathy, 63/192
epilepsy, 63/192
historic aspect, 37/449
iatrogenic neurological disease, 63/190, 192, 65/454
neonatal seizure, 73/258
tremor, 63/192
Lidocaine intoxication
agitation, 37/450, 65/454
anxiety, 37/450
clinical value, 37/405
CNS, 37/450
color blindness, 37/451
coma, 65/454
confusion, 37/450, 65/454
convulsion, 37/450
cranial nerve, 37/451
delirium, 65/454
depression, 37/450
diplopia, 37/451
dizziness, 37/450
dysarthria, 37/450, 65/454
EEG, 37/451
epilepsy, 37/449, 451, 65/454
euphoria, 37/450
nystagmus, 37/451
psychosis, 37/450
respiratory arrest, 37/451
respiratory depression, 65/454
tinnitus, 65/454
tremor, 65/454
vertigo, 65/454
visual impairment, 37/451
Liepmann, H.C., 1/8
Liesegang ring
concentric sclerosis, 9/438, 47/409
neuromyelitis optica, 47/416
Life
definition, 24/790, 814
family, *see* Family life
Life expectancy
Edwards syndrome, 50/560
epilepsy, 15/794
hypokalemic periodic paralysis, 41/151, 62/459
juvenile spinal cord injury, 25/193
missile injury, 24/457
multiple sclerosis, 47/62
solitary myeloma, 20/13
spinal cord injury, 25/193, 26/310, 61/499
spinal cord injury prognosis, 26/310
trisomy 22, 50/565
Life table method
spinal cord injury, 61/507
Ligament
anterior longitudinal, *see* Anterior longitudinal ligament
clinical lumbar stability, 61/16
dentate, *see* Dentate ligament
experimental injury, 25/2, 6

Gruber petrosphenoid, *see* Gruber petrosphenoid ligament
occipitoatlantoaxial complex, 61/6
serrated, *see* Serrated ligament
spine experimental injury, 25/2
stylomandibular, *see* Stylomandibular ligament
supracondylar, *see* Supracondylar ligament
vertebral experimental injury, 25/2
Ligamentotaxis
thoracolumbar vertebral column injury, 61/97
Ligamentum flavum hypertrophy, 20/809-815
anatomy, 20/809
biomechanics, 20/810
clinical features, 20/814
myelography, 20/812, 814
pathogenesis, 20/811
pathology, 20/811
spondylosis, 20/812
surgery, 20/814
Light
depression, 74/507
melatonin, 74/499
myelin, 9/311
pineal gland, 74/499, 502
seasonal rhythm, 74/490
vampire hormone, 74/499
Light chain Ig
primary amyloidosis, 51/417
Light-near dissociation
Argyll Robertson pupil, 74/416
pupil, 74/416
Light stimulation
epilepsy, 15/441
intermittent, *see* Intermittent light stimulation
Light therapy
Alzheimer disease, 74/496
circadian disorder, 74/496
depression, 74/507
Lightning eye movements, *see* Opsoclonus
Lightning injury
see also Electrical injury
clinical features, 23/694, 701, 703, 719
coma, 7/355, 360
cranial nerve, 51/139
EEG, 23/696
electrophysics, 23/683
lipid, 7/41, 51
nerve injury, 7/344, 347-350, 355, 51/134
paralysis, 7/355, 23/696
pathology, 23/691, 719
pathophysiology, 23/689
spinal cord injury, 23/703, 721
stroke, 7/41, 51, 362, 23/691, 719
vasomotor disorder, 23/695
wallerian degeneration, 7/212
Lightning pain
hereditary sensory and autonomic neuropathy type I, 21/73, 77, 60/6
Roussy-Lévy syndrome, 21/173
tabes dorsalis, 8/201, 33/363, 52/280
Lignocaine, *see* Lidocaine
Lignoceric acid
metachromatic leukodystrophy, 10/49
Lilliputian hallucination
Bonnet syndrome, 45/358
hallucinosis, 46/562
schizophrenia, 45/358
temporal lobe, 45/358
toxicity, 45/358
visual hallucination, 2/617
Limb
anencephaly, 50/84
ataxia, 1/313
phantom, *see* Phantom limb
phantom third, *see* Phantom third limb
Limb apraxia
classification, 45/425
Limb asymmetry
Conradi-Hünermann syndrome, 43/344
Limb atrophy
hydromyelia, 50/429
Limb dystonia
hereditary progressive diurnal fluctuating dystonia, 49/530
Limb fracture
spinal injury, 26/200, 226-232
surgery, 26/226-232
treatment, 26/226-232
Limb girdle dystrophy, *see* Limb girdle syndrome
Limb girdle myasthenia
familial, *see* Familial limb girdle myasthenia
Limb girdle syndrome, 40/433-467, 62/179-191
see also Bethlem-Van Wijngaarden syndrome, Late adult muscular dystrophy (Nevin), Muscular dystrophy, Pelvifemoral muscular dystrophy (Leyden-Möbius), Quadriceps myopathy *and* Scapulohumeral muscular dystrophy (Erb)
adenosine monophosphate deaminase deficiency, 62/190
adult acid maltase deficiency, 62/190
autosomal recessive, 62/179
back kneeing, 40/314
birth incidence, 43/104
carcinoma, 62/188
cardiac involvement, 62/185

cervical neurofibroma, 62/188
classification, 40/277, 279
classification history, 40/433, 62/179
clinical features, 40/306, 312, 316, 322, 342, 449, 62/182
connectin, 62/187
contracture, 43/104
creatine kinase, 40/449, 62/185
differential diagnosis, 40/437-441, 41/177, 62/187-190
dystrophin, 62/187
Emery-Dreifuss muscular dystrophy, 62/189
EMG, 40/453, 62/186
etiology, 62/190
etiology and treatment, 40/466
facioscapulohumeral muscular dystrophy, 40/438, 441
fructose-bisphosphate aldolase, 62/185
genetics, 41/419, 486, 62/181
glycogen storage disease, 62/190
histochemistry, 40/50
history, 40/448
incidence, 62/182
inclusion body myositis, 62/187
infantile acid maltase deficiency, 62/190
infectious cause, 62/187
lactate dehydrogenase, 62/185
late adult muscular dystrophy (Nevin), 62/179
lordosis, 40/436
lung, 62/185
Medical Research Council Scale, 40/447
mitochondrial myopathy, 40/280
motor unit, 40/453
multivariant analysis, 40/441
muscle biopsy, 62/186
muscle histopathology, 62/186
muscle necrosis, 40/252, 453
muscle pseudohypertrophy, 62/182
myopathy, 40/279
myopathy pattern, 62/182
myophosphorylase deficiency, 62/190
myotonic dystrophy, 62/188, 240
necrosis, 40/252, 453
neuromuscular disease, 41/419
non-REM, 62/185
outline, 40/435
pathologic reaction, 40/252, 453
pelvifemoral muscular dystrophy (Leyden-Möbius), 62/179
poliomyelitis, 40/435
polymyositis, 40/457, 62/187
prevalence, 43/104, 62/182
progressive muscular dystrophy, 43/104
proximal muscle weakness, 43/104
pseudohypertrophy, 40/436
pseudomyopathic spinal muscular atrophy, 62/188
pyruvate kinase, 62/185
quadriceps myopathy, 40/438, 440, 43/128
Raynaud phenomenon, 40/438
recent classification, 62/181
REM, 62/185
respiratory dysfunction, 40/437
scapulohumeral muscular dystrophy (Erb), 62/179
secondary type, 62/181
severity index, 40/447
single fiber electromyography, 62/186
sporadic, 62/179
titin, 62/187
toxic cause, 62/187
treatment, 62/190
vitamin D deficiency, 40/438, 62/188
vitamin E deficiency, 62/188
winged scapula, 40/436
Wohlfart-Kugelberg-Welander disease, 40/433, 62/188

Limbic encephalitis
amygdaloid nucleus, 8/135
bronchial carcinoma, 8/137
cingulate gyrus, 8/135
claustrum, 8/135
dementia, 46/387
frontal lobe, 8/135
hippocampus, 8/135
Hodgkin disease, 39/54
insula, 8/135
malignant lymphoma, 39/54
paraneoplastic encephalomyelitis, 71/682
paraneoplastic syndrome, 69/355
progressive bulbar palsy, 22/143

Limbic system
affective disorder, 46/473
akinesia, 45/162
Alzheimer disease, 46/268
autonomic function, 3/16
autonomic nervous system, 2/720, 74/2
basal ganglion, 49/19, 25
Broca, 2/700
consciousness, 2/719
depression, 46/427-429
descending efference, 73/60
dreaming, 3/89
emotion, 3/320, 329, 45/272, 276-278
emotional disorder, 3/343-363
function, 2/440
hallucination, 45/355
hippocampus, 73/60

language, 3/33
mood change, 45/277
organic affective syndrome, 46/427-429
organotin intoxication, 64/140
Papez circuit, 73/57
pharmacology, 74/158
rabies virus, 56/392
rhinencephalon, 73/57
schizophrenia, 46/454, 497
septum pellucidum agenesis, 30/329
sorrow, 45/276
stress, 45/249-251
temporal lobe, 2/719
temporal lobe epilepsy, 73/56
Limit dextrin type glycogen
glycogen storage disease, *see* Debrancher deficiency
Limit dextrinosis, *see* Debrancher deficiency
Limulus lysate test
Hemophilus influenzae meningitis, 33/55
infantile enteric bacillary meningitis, 33/64
pneumococcal meningitis, 33/41
Linamarase
cyanogenic glycoside intoxication, 65/27
neurotoxin, 65/27
Linamarin intoxication, *see* Cyanogenic glycoside intoxication
Lindane
chemical formula, 64/203
epilepsy, 64/202
iatrogenic neurological disease, 65/479
neurotoxin, 62/601, 64/202, 65/478
organochlorine, 64/197
organochlorine insecticide intoxication, 64/197, 202
toxic myopathy, 62/601
Lindane intoxication
benign intracranial hypertension, 65/479
brain vascular necrosis, 65/479
confusion, 65/479
dysarthria, 65/479
EEG, 65/479
epilepsy, 65/479
headache, 65/479
optic neuritis, 65/479
optic neuropathy, 65/479
Lindau disease, *see* Von Hippel-Lindau disease
Lindau tumor, *see* Hemangioblastoma
Lindenov-Hallgren syndrome, 13/452-460
deaf mutism, 13/453
deafness, 13/452
gait disturbance, 13/454
mental deficiency, 13/454
pleiotropism, 13/454
Refsum disease, 21/216, 60/235, 66/497
schizophrenia, 13/455, 458
secondary pigmentary retinal degeneration, 13/451-460
vestibular feature, 13/454
Line orientation
space perception, 45/412
Linear nevus sebaceous syndrome
alopecia, 43/38
anhidrotic ectodermal dysplasia, 14/788
brain atrophy, 43/38
CNS, 31/289, 43/38
cranial nerve, 43/38
CSF, 43/39
cutaneous abnormality, 31/288, 43/38
EEG, 43/39
extrapyramidal sign, 43/38
gene defect, 72/138
hyperkeratosis, 43/38
hypsarrhythmia, 43/39
leptomeningeal angiomatosis, 43/39
mental deficiency, 31/288-290, 43/38, 46/94
multiple pigmented nevi, 43/38
pathology, 31/289
seizure, 31/289, 43/38
Lines
Boogaard, *see* Boogaard line
Chamberlain, *see* Chamberlain line
clinoparietal, *see* Clinoparietal line
digastric, *see* Digastric line
Hultkranz, *see* Hultkranz line
McGregor, *see* McGregor line
McRae, *see* McRae line
Mees, *see* Mees line
Prouzet, *see* Prouzet line
Ranawat, *see* Ranawat line
siphon incisivum, *see* Siphon incisivum line
Twining, *see* Twining line
Lingua plicata
Melkersson-Rosenthal syndrome, 2/72, 8/344, 14/790, 16/222, 42/322
Linguistics
aphasia, 45/306-309
Lingula
boundary, 2/399
Linguobuccofacial dyskinesia
see also Orofacial dyskinesia
levodopa induced dyskinesia, 49/195
Linkage
genetic, *see* Genetic linkage
Linoleate, *see* Linoleic acid
Linoleic acid

Bassen-Kornzweig syndrome, 13/423
chemistry, 10/266
erythrocyte, 13/423
linolenic acid, 10/235
lipid metabolic disorder, 10/280
migraine, 48/95
multiple sclerosis, 9/132, 315, 321, 47/68, 154
plaque, 9/315
plasma, 13/423, 48/96
serum, 9/321

Linolenic acid
linoleic acid, 10/235
multiple sclerosis, 9/315

Lion d'or
characteristics, 64/95
methanol, 64/95
use, 64/95

Lionfish intoxication
picture, 37/73

Lioresal, *see* Baclofen

Lip
cleft, *see* Cheiloschisis
hypoplastic upper, *see* Upper lip hypoplasia
median cleft, *see* Median cleft lip
swelling, 8/205

Lip reflex
diagnostic value, 1/241
newborn behavior, 4/345

Lip retraction
Wilson disease, 49/225

Lipase
myopathy, 40/380

Lipid
see also Triglyceride
abnormal metabolism, 10/265
acidic, *see* Acidic lipid
Batten disease, 40/243
biochemistry, 9/1, 66/33
brain, 10/233
brain arteriosclerosis, 63/77
brain development, 9/17
brain edema, 23/152
composition, 66/198
CSF, *see* CSF lipid
Down syndrome, 31/433
Duchenne muscular dystrophy, 40/380
excessive lipid accumulation, 40/95
fingerprint body, 21/52
glia, 9/6
glial membrane, 9/18
histochemistry, 9/25, 40/36
Kearns-Sayre-Daroff-Shy syndrome, 40/242
lightning injury, 7/41, 51
lipid storage disorder, 40/242
migraine, 48/85, 95
multiple sclerosis, 9/321
myelin, *see* Myelin lipid
myelin chemistry, 10/136
myelin *vs* glia, 9/18
myelination, 9/10, 17
myotonic dystrophy, 40/523, 548
nerve myelin, 51/25
nervous system tumor, 27/504
neutral, *see* Neutral lipid
normal metabolism, 10/233
oil red O staining, 9/25, 28
OsO4 staining, 9/31
OTAN staining, 9/25, 28
polyphenol, 10/24
primary pigmentary retinal degeneration, 13/196
progressive dysautonomia, 59/157
rat brain development, 9/17
serum, 9/321, 10/265
structure, 66/34
Sudan III staining, 9/25, 28
Sudan IV staining, 9/25, 28
synthesis, 9/7
total lipid, 10/265
ultrastructure, 40/94
virology, 34/4
wallerian degeneration, 7/212
white matter, 9/1, 6, 10

Lipid degeneration, *see* Neuronal ceroid lipofuscinosis

Lipid dopaquinone
melanin precursor, 10/24

Lipid ester
hydroxylamine reaction, 9/25

Lipid hexosamine
globoid cell, 10/89
leukodystrophy, 10/24
metachromatic leukodystrophy, 10/47
orthochromatic leukodystrophy, 42/501
sudanophilic leukodystrophy, 10/34

Lipid inclusion body
epilepsy, 42/692
yellow teeth, 42/692

Lipid metabolic disorder
beta oxidation defect, 62/494
biochemistry, 62/491
cardiomyopathy, 62/493
carnitine deficiency, 41/196-201, 62/492
carnitine palmitoyltransferase deficiency, 41/204-207, 62/494
childhood myoglobinuria, 62/566
cholesterol, 10/265

chylomicron, 10/267
dicarboxylic aciduria, 62/494
differential diagnosis, 41/60
Fanconi syndrome, 62/493
fatty acid oxidation, 62/491
fatty acyl-carnitine, 41/195
fatty acyl-coenzyme A, 41/194
globoid cell leukodystrophy, 10/302-306
hepatic encephalopathy, 62/493
history, 62/490
hypoglycemia, 62/494
icosisphingosine, 10/293, 295
kwashiorkor, 62/493
linoleic acid, 10/280
long chain acyl-coenzyme A dehydrogenase deficiency, 62/494
long chain 3-hydroxyacyl-coenzyme A dehydrogenase deficiency, 62/495
macrencephaly, 10/291
medium chain acyl-coenzyme A dehydrogenase deficiency, 62/495
metabolic encephalopathy, 62/494
metabolic myopathy, 41/194-207, 62/490
multiple acyl-coenzyme A dehydrogenase deficiency, 62/495
myoglobinuria, 41/207
Niemann-Pick disease, 10/280-287
other lipid storage myopathy, 41/201-204
short chain acyl-coenzyme A dehydrogenase deficiency, 62/495
short chain 3-hydroxyacyl-coenzyme A dehydrogenase deficiency, 62/495
systemic types, 62/493
Tangier disease, 10/278, 21/12
triglyceride storage disease, 62/496
valproic acid intoxication, 62/493

Lipid metabolism
biochemistry, 66/39
brain infarction, 53/130
brain ischemia, 53/130
brain metabolism, 57/81
Friedreich ataxia, 60/306
globoid cell leukodystrophy, 10/84-87
head injury, 57/81
Huntington chorea, 6/350
lysoglycerophospholipid, 10/244
mitochondrial myopathy, 41/214
multiple sclerosis, 9/134
myelin, 10/248

Lipid peroxidation
neuronal ceroid lipofuscinosis, 66/685
pathogenesis, 40/171
vitamin E deficiency, 40/171
vitamin E intoxication, 65/570

Lipid storage
chemical pathology, 66/140
enzyme defect, 66/140
mitochondrial myopathy, 41/202

Lipid storage disease
rimmed muscle fiber vacuole content, 62/18
rimmed muscle fiber vacuole type, 62/18

Lipid storage disorder, 29/345-378
animal model, 29/378
biochemistry, 29/349
classification, 29/346
detection, 29/375-377
historic aspect, 29/345-349
lipid, 40/242
pathologic reaction, 40/258
treatment, 29/377

Lipid storage myopathy
carnitine, 66/403
carnitine deficiency, 40/95, 258, 280, 308, 317, 422, 43/175, 184, 62/493
carnitine palmitoyltransferase deficiency, 40/95, 258, 339, 342, 438, 43/177
congenital ichthyosis, 41/203
ichthyosis, 43/184
lipofuscin, 40/108
muscle cramp, 41/207-214
muscle fiber type I, 43/184
neuromuscular disease, 41/426
proximal muscle weakness, 43/184

Lipidosis
see also Neurolipidosis
abnormal lipid, 10/265
adult onset, 49/543
Alexander disease, 10/351
amiodarone intoxication, 65/459
ceramide lactoside, *see* Ceramide lactoside lipidosis
cerebral, 15/423
chemistry, 10/325
classification, 10/44
cytoside, *see* Cytoside lipidosis
epilepsy, 15/423
galactosylceramide, *see* Globoid cell leukodystrophy
juvenile dystonic, *see* Juvenile dystonic lipidosis
leukodystrophy, 10/22
mental deficiency, 46/19
metachromatic leukodystrophy, 10/59
myoclonia, 6/765
myoclonus, 38/579
phosphatidylethanolamine, *see* Phosphatidylethanolamine lipidosis

polyneuropathy, 60/165-177
pseudo, *see* Pseudolipidosis
sphingomyelin, *see* Niemann-Pick disease
sulfatide, *see* Metachromatic leukodystrophy
Wolman disease, 66/16
Lipiduria
metachromatic, *see* Metachromatic lipiduria
Lipoamide dehydrogenase, *see* Dihydrolipoamide dehydrogenase
Lipoamine dehydrogenase
Friedreich ataxia, 51/388
Lipoate acetyltransferase
lactic acidosis, 41/210
Lipoblast lipoma, *see* Hibernoma
Lipocalcinogranulomatosis
brachial plexus, 43/185
hyperphosphatemia, 43/185, 187
sciatic nerve, 43/185
Lipochondrodystrophy, *see* Glycosaminoglycanosis type IH *and* Glycosaminoglycanosis type II
Lipodystrophia progressiva (Barraquer-Simons)
differential diagnosis, 42/592
Krabbe-Bartels disease, 14/406
Lipodystrophy
acquired generalized, *see* Lawrence syndrome
athetotic membranous, *see* Athetotic membranous lipodystrophy
bone membrane, *see* Bone membranous lipodystrophy
generalized, *see* Berardinelli-Seip syndrome
GM3 hematoside, *see* GM3 hematoside lipodystrophy
hemiatrophy, 59/480
hemihypertrophy, 59/483
hepatomegaly, 8/346
hirsutism, 8/346
hyperglycemia, 8/346
pigmentation, 8/346
Lipofuscin
cell reaction, 21/48
ceroid, *see* Ceroid lipofuscin
dyssynergia cerebellaris myoclonica, 10/550
dystonia musculorum deformans, 6/535
histogenesis, 10/30
lipid storage myopathy, 40/108
Parkinson disease, 49/117, 129
pigmentary degeneration, 22/45
spinal muscular atrophy, 59/45
striatopallidodentate calcification, 6/715
Lipofuscinosis
adult neuronal ceroid, *see* Adult neuronal ceroid lipofuscinosis
infantile ceroid, *see* Infantile ceroid lipofuscinosis
intention tremor, 42/230
juvenile neuronal ceroid, *see* Juvenile neuronal ceroid lipofuscinosis
late infantile neuronal ceroid, *see* Late infantile neuronal ceroid lipofuscinosis
neuronal ceroid, *see* Neuronal ceroid lipofuscinosis
poliosis, 60/647, 667
progressive spinal muscular atrophy, 22/17
Lipogranulomatosis, *see* Farber disease
Lipohyalinosis, *see* Arteriolar fibrinohyalinoid degeneration
Lipoic acid, *see* Thioctic acid
Lipoid cell splenomegaly, *see* Niemann-Pick disease
Lipoid granulomatosis
see also Histiocytosis X
histiocytosis X, 38/95
history, 38/95
xanthomatosis, 38/96
Lipoid proteinosis, *see* Urbach-Wiethe disease
Lipoma
brain, *see* Brain lipoma
congenital spinal cord tumor, 32/358-386
corpus callosum, *see* Corpus callosum lipoma
corpus callosum agenesis, 2/764, 14/410, 18/356, 30/285, 288
corpus callosum tumor, 14/409, 17/501, 18/355-357
cranial vault tumor, 17/112
diastematomyelia, 50/436, 438
diplomyelia, 50/436, 439
Ekbom syndrome type II, 60/667
epidural spinal, *see* Epidural spinal lipoma
hypothalamic tumor, 68/71
intradural spinal, *see* Intradural spinal lipoma
intraspinal, *see* Intraspinal lipoma
intraspinal neurenteric cyst, 20/65
lateral ventricle tumor, 17/606
leptomeningeal, *see* Leptomeningeal lipoma
lumbosacral, *see* Lumbosacral lipoma
MERRF syndrome, 60/663, 667, 66/427
multiple epidural spinal, *see* Multiple epidural spinal lipoma
muscle tumor, 41/382
neurofibromatosis type I, 14/568
pseudo, *see* Hibernoma
skull base tumor, 17/159
spinal, *see* Spinal lipoma
spinal cord compression, 19/361
spinal epidural, *see* Spinal epidural lipoma
spinal epidural tumor, 20/109
spinal site, 19/17
subcutaneous, *see* Subcutaneous lipoma

syringomyelia, 61/377
Lipoma fetalocellulare, *see* Hibernoma
Lipomatosis
brain, 14/529
epidural, *see* Epidural lipomatosis
Krabbe-Bartels disease, 14/407, 409-411
leptomeningeal, 14/529
macrocephaly, 14/83
multiple, *see* Krabbe-Bartels disease
multiple symmetric, *see* Multiple symmetric lipomatosis
neurocutaneous, *see* Neurocutaneous lipomatosis
perineural, *see* Perineural lipomatosis
pilonidal sinus, 42/47
spinal, *see* Spinal lipomatosis
Von Hippel-Lindau disease, 14/532
Lipomembranous polycystic osteodysplasia
dementia, 42/279-281
Nasu-Hakola disease, 49/406
progressive dementia, 42/279-281
striatopallidodentate calcification, 49/424
Lipomeningocele
spina bifida, 50/499
Lipomucopolysaccharidosis, *see* Mucolipidosis type I
Lipomyoma
liver, 14/51
Lipopigment
histochemistry, 10/30
melanin, 10/24
neuroaxonal dystrophy, 21/56
neuronal ceroid lipofuscinosis, 10/589, 672, 21/58
synthesis, 10/24
vitamin E deficiency, 21/58
Lipopigment dystrophy
features, 21/57
neuronal, *see* Neuronal lipopigment dystrophy
vitamin E, 21/67
vitamin E deficiency, 21/67
Lipoprotein
Alzheimer disease, 46/265
antihypertensive agent, 63/82
apolipoprotein B containing, *see* Apolipoprotein B containing lipoprotein
arterial hypertension, 63/77
ataxia, 21/581
Bassen-Kornzweig syndrome, 13/423-425, 29/419-421
characteristics, 10/267
classification, 66/535
function, 10/268
high density, *see* High density lipoprotein
low density, *see* Low density lipoprotein
multiple sclerosis, 9/321
myelin, *see* Myelin lipoprotein
prebeta, *see* Prebetalipoprotein
serum, *see* Serum lipoprotein
structure, 66/535
very low density, *see* Very low density lipoprotein
α-Lipoprotein
enzymopathy, 21/52
neuropathy, 8/20
β-Lipoprotein
Bassen-Kornzweig syndrome, 8/19
chorea-acanthocytosis, 49/328
enzymopathy, 21/52
headache, 48/100
hyperlipoproteinemia type II, 10/272
low density lipoprotein, 10/267
migraine, 48/99
Parkinson disease, 49/123
Lipoprotein A
process, 66/549
α-Lipoprotein deficiency, *see* Tangier disease
Lipoprotein lipase
hyperlipoproteinemia type I, 42/443
Lipoprotein lipase deficiency, *see* Hyperlipoproteinemia type I
Lipoprotein lipase gene
hepatosplenomegaly, 66/541
Lipoprotein metabolism
cholesterol ester transfer protein gene, 66/552
disorder, 66/535
hyperlipoproteinemia type II, 66/545
molecular biology, 66/537
phosphatidylcholine sterol acyltransferase, 66/551
phosphatidylcholine sterol acyltransferase gene, 66/551
Lipoprotein phenotype
classification, 66/543
Lipoproteinosis
enzyme deficiency polyneuropathy, 51/367
Liposarcoma
neurofibromatosis type I, 50/369
Lipotropin
endorphin precursor, 65/351
β-Lipotropin
cluster headache, 48/233
Liquefaction necrosis
myelomalacia, 12/589
posttraumatic syringomyelia, 25/60
radiation injury, 23/650
spinal cord injury, 25/80, 84
Liquor tympanum
temporal bone fracture, 24/121
Liquorice, *see* Glycyrrhiza

Liquorrhea, *see* CSF fistula
Lisch nodule
  neurofibromatosis, 50/365
  neurofibromatosis type I, 50/366
Lissauer tract
  congenital pain insensitivity, 51/564
Lissauer tract absence
  anhidrosis, 8/199, 42/346
  hereditary sensory and autonomic neuropathy type IV, 8/199, 42/346, 60/13
  indifference to pain, 1/110, 8/199
  universal insensitivity, 1/110
Lissencephalocele
  hydrocephalus, 50/286
Lissencephaly
  see also Agyria, Brain cortex heterotopia *and* Miller-Dieker syndrome
  autosomal recessive, *see* Autosomal recessive lissencephaly
  brain cortex heterotopia, 50/248
  cerebellar atrophy, 42/39
  cerebrohepatorenal syndrome, 50/250
  chromosome deletion syndrome, 50/591, 593
  claustrum absence, 42/40
  clinical features, 30/486
  computer assisted tomography, 50/268
  convulsion, 42/40
  corpus callosum agenesis, 30/485, 42/39, 50/151, 153, 163, 168
  cortex layer type, 30/486
  decerebration, 30/486, 42/40
  dentate nucleus, 42/40
  Edwards syndrome, 50/250
  EEG, 30/486
  epilepsy, 72/342
  falx cerebri, 30/485, 42/39
  features, 50/269
  frequency, 30/485
  gene, 72/132
  genetics, 30/486
  HARD syndrome, 50/593
  HARDE syndrome, 50/593
  heterotopia, 30/485
  hydranencephaly, 30/667, 50/337, 340, 342, 347
  hydrocephalus, 30/485, 50/258
  inferior olivary nucleus heterotopia, 50/250, 259
  Lowe syndrome, 42/607
  macroscopic appearance, 30/485
  malformation, 30/487
  mental deficiency, 30/486, 46/11, 48
  micrencephaly, 50/258
  microcephaly, 42/39, 50/260, 263
  microgyria, 30/485
  microscopic appearance, 30/485
  midline calcification, 50/259
  Miller-Dieker syndrome, 50/260, 593
  morphology, 50/247
  neuronal heterotopia, 42/40
  neuronal migration, 42/40
  Norman-Roberts syndrome, 50/593
  opisthotonos, 30/486, 42/40
  optic nerve agenesis, 50/211
  oxycephaly, 50/260
  17p partial monosomy, 50/591
  pachygyria, 30/485, 42/39-41, 50/258
  reeler mutant mouse, 50/40
  rigidity, 30/486
  seizure, 30/486
  spina bifida, 30/487
  status epilepticus, 42/40
  syndactyly, 42/40
  thalamus, 42/40
  X-linked, *see* X-linked lissencephaly
Lissencephaly syndrome
  autosomal recessive, 30/487, 42/40
  EEG, 42/40
  gastrointestinal disorder, 42/40
  heart disease, 42/40
  polydactyly, 42/40
Listening
  dichotic, *see* Dichotic listening
Listeria
  acquired immune deficiency syndrome, 71/291
  CNS infection, 69/441
  nonrecurrent nonhereditary multiple cranial neuropathy, 51/571
Listeria cerebri, *see* Listeria monocytogenes encephalitis
Listeria monocytogenes, 33/77-92
  see also Listeriosis
  acquired immune deficiency syndrome, 56/495
  ameningitic encephalitis, *see* Listeria monocytogenes encephalitis
  Anton test, 33/82
  bacteriology, 33/80-82
  brain stem encephalitis, 33/77, 52/93
  Burkitt lymphoma, 63/352
  characteristics, 33/80
  epilepsy, 63/531
  headache, 63/531
  Hodgkin lymphoma, 63/353
  L-form, 33/80, 52/95
  laboratory examination, 33/81
  macrophage, 52/90
  microbiology, 52/89
  renal insufficiency, 27/343

renal transplantation, 63/531
serodiagnosis, 33/82, 52/89
serotyping, 52/89
subarachnoid hemorrhage, 12/111
T-lymphocyte, 52/90
terminology, 33/78
virulence, 52/97
Listeria monocytogenes encephalitis
ataxia, 52/93
brain abscess, 33/86, 52/94, 150
brain parenchyma, 63/532
brain stem encephalitis, 52/93
cerebellar atrophy, 52/95
clinical features, 33/86-88
computer assisted tomography, 52/94
headache, 33/87
hydrocephalus, 52/95
somnolence, 33/87
Listeria monocytogenes encephalomyelitis
spinal cord abscess, 52/191
Listeria monocytogenes infection
clinical features, 31/217
CSF, 33/85
epidemiology, 31/217
treatment, 33/91
Listeria monocytogenes meningitis, 33/77-92
acquired immune deficiency syndrome, 56/515
ataxia, 52/93
bacteremia, 33/83, 52/91
bacterial meningitis, 52/2-4
brain granuloma, 52/93
brain stem, 33/85
case cluster, 52/91
cephalosporin, 52/96
chronic, 52/94
consciousness, 63/531
co-trimoxazole, 52/95
cranial nerve palsy, 33/85, 52/93, 63/531
CSF, 33/85, 52/93, 63/531
CSF listeria antigen, 52/98
dementia, 33/88
diagnosis, 52/95
endocarditis, 33/83
ependymitis, 52/93
epilepsy, 52/93
gram-negative bacillary meningitis, 52/110
headache, 33/85, 52/94, 63/531
hemiparesis, 63/531
hemiplegia, 63/531
Hodgkin disease, 33/90
incidence, 52/91
lymphadenopathy, 33/82
lymphoma, 33/90
mortality, 33/84, 52/91, 96, 63/531
nystagmus, 52/93
prevention, 52/95
psychosis, 33/88
rifampicin, 52/95
seizure, 63/531
subarachnoid hemorrhage, 33/85
symptom, 52/93, 63/531
treatment, 33/85, 52/95, 63/531
Listeria psychosis, *see* Listeria monocytogenes meningitis
Listeriosis
see also Listeria monocytogenes
acquired immune deficiency syndrome, 52/90
alcoholism, 52/91
blood culture, 52/94
cancer, 52/91
chronic liver disease, 52/91
chronic renal failure, 52/91
cirrhosis, 52/91
diabetes mellitus, 52/91
dialysis, 52/91
encephalitis, 52/90
epidemiology, 33/79
food-borne outbreak, 52/92
glucocorticoid, 52/91
granulomatosis infantiseptica, *see* Listeria monocytogenes meningitis
hemochromatosis, 52/91
immunocompromised host, 52/91
immunosuppressive agent, 52/91
leukemia, 52/91
lymphoma, 52/91
meningoencephalitis, 52/91
milk, 52/92
newborn, 33/78, 84, 52/92
organ involvement, 52/94
pathogenesis, 33/80, 52/90
pericarditis, 52/90
renal transplantation, 52/91
sarcoidosis, 52/91
sheep manure, 52/92
subarachnoid hemorrhage, 33/86
thrombocytopenia, 52/91
zoonosis, 33/79
Lisuride
behavior disorder, 46/594
cluster headache, 48/230
migraine, 48/192
Litchi chinensis
hypoglycin, 65/81
Literacy
cerebral dominance, 45/304

Lithium
aggression, 46/180
amyotrophy, 41/333
behavior disorder, 46/597
chronic migrainous neuralgia, 48/252
cluster headache, 48/229, 239
delirium, 46/539
digitalis intoxication, 46/538
drug induced polyneuropathy, 51/304
Duchenne muscular dystrophy, 40/378
dyskinesia, 51/304
hypnic headache syndrome, 74/509
mania, 46/450
migraine, 48/97, 198
neurotoxicity, 51/304
neurotoxin, 46/539, 48/420, 49/385, 51/304, 60/587, 62/601, 64/293, 295
potassium imbalance, 28/484
serum creatine kinase, 40/344
toxic myopathy, 62/597, 601
tremor, 51/304
Lithium carbonate
benign intracranial hypertension, 67/111
lithium intoxication, 64/293
Lithium intoxication
acute mania, 64/293
acute neurotoxic effect, 64/295
age, 64/297
amyotrophy, 22/28
antidepressant, 64/299
antiepileptic agent, 64/293
antipsychotic agent, 64/298
ataxia, 64/296
axonal neuropathy, 64/296
bipolar affective disorder, 64/293
carbamazepine, 64/299
cerebellar damage, 64/295
chorea, 64/295
choreoathetosis, 49/385, 64/296
chronic neurotoxic effect, 64/296
clearance, 64/294
coma, 64/295
Creutzfeldt-Jakob disease, 64/297
dementia, 64/296
diuretic agent, 64/299
dosage range, 64/294
downbeating nystagmus, 64/296
drug interaction, 64/298
EEG, 64/296
EMG, 64/296
epilepsy, 64/296
general toxicity, 64/294
half-life, 64/294
headache, 48/420
history, 64/293
hyperthermia, 64/295
late cerebellar atrophy, 60/587
lithium bromide, 64/293
lithium carbonate, 64/293
lithium chloride, 64/293
mania, 64/293
monitoring, 64/294
mortality, 64/297
muscle relaxant, 64/299
myasthenic syndrome, 64/296
myoclonus, 64/296
myopathy, 64/296
nephrotoxicity, 64/294
neuropathy, 64/296
nystagmus, 64/296
optic neuritis, 64/296
papilledema, 64/296
parkinsonism, 64/296
permanent damage, 64/296
pharmacokinetics, 64/293
polyneuropathy, 64/295
precipitating factor, 64/297
pregnancy, 64/298
prevalence, 64/297
prevention, 64/299
renal function, 64/298
safety margin, 64/294
secondary neurotoxicity, 64/294
self-intoxication, 64/295
severity grades, 64/297
somatic illness, 64/297
therapeutic range, 64/294
treatment, 64/300
tremor, 64/295
Little disease
anoxia, 1/209
central motor disorder, 1/209
clinical features, 1/195
incontinentia pigmenti, 14/108
Kii Peninsula amyotrophic lateral sclerosis, 22/371
newborn, 1/209
Livedo
Divry-Van Bogaert syndrome, 14/14
Livedo racemosa
associated disease, 55/402
brain infarction, 55/401
semantic confusion, 55/401
Sneddon syndrome, 55/401
Livedo reticularis
antiphospholipid antibody, 63/122, 330

computer assisted tomography, 54/64
description, 8/340
Divry-Van Bogaert syndrome, 55/317-319
polyarteritis nodosa, 55/362
reflex sympathetic dystrophy, 61/121
semantic confusion, 55/401
Sneddon syndrome, 55/317, 319, 401
Liver cell carcinoma
hepatitis B virus, 56/10
Liver cirrhosis
acanthocyte, 49/332
alcoholic polyneuropathy, 7/613
ascites, 27/351
brancher deficiency, 62/483
deficiency neuropathy, 7/609, 613, 51/324
gastrointestinal bleeding, 27/351
hepatic encephalopathy, 27/351, 49/384
hepatic polyneuropathy, 7/609
pseudoathetosis, 49/384
spheroid body, 6/626
splenomegaly, 27/351
Liver disease
see also Acute hepatic necrosis
acanthocytosis, 63/271
acquired hepatocerebral degeneration, 27/352, 366
alcoholic, 7/613
Alzheimer type II astrocyte, 9/635, 47/584
ammonia level, 27/368
asterixis, 6/821
autonomic nervous system, 75/32
brain hemorrhage, 55/476, 63/313
central pontine myelinolysis, 28/293, 312, 47/585
chronic liver failure, 27/364-366
clinical chemistry, 27/368
coagulopathy, 63/309, 313
debrancher deficiency, 62/482
encephalopathy, *see* Hepatic encephalopathy
epilepsy, 27/362, 73/377
galactosemia, 42/552
glycogen storage disease, 43/179
hemochromatosis, 42/553
hypoglycemia, 27/71
hyponatremia, 28/499
intravascular consumption coagulopathy, 63/314
laboratory diagnosis, 27/368-370
leukoencephalopathy, 47/584
mitochondrial DNA depletion, 62/511
myelopathy, 27/367
nervous system, 27/351-353
neurologic complication, 27/349-372
neuropathy, 7/554, 609, 612, 614, 27/353
obstructive, *see* Obstructive liver disease
pathogenesis, 27/353-361
pathophysiology, 27/351-353
phosphorylase kinase deficiency, 62/486
portacaval shunting, 27/351, 359, 367
rhodopsin, 13/193
toxic factor, 27/365
visceral, 27/350
Wilson disease, 27/383, 42/270
Liver glia
central pontine myelinolysis, 9/635
tellurium intoxication, 9/646
Liver metastasis
brain metastasis, 69/107
Liver phosphorylase
hepatophosphorylase deficiency, 27/222, 41/188
Liver toxicity
copper intoxication, 37/623
iproniazid intoxication, 37/321
manganese intoxication, 37/623
MAO inhibitor intoxication, 37/321
methyldopa intoxication, 37/440
papaverine intoxication, 37/456
Liver transplantation
see also Hepatobiliary transplantation
amanitin, 65/37
complication, 70/252
pathology, 70/256
survey, 70/249
Liverpool Adverse Events Profile
epilepsy, 72/419
The Liverpool Assessment Battery
epilepsy, 72/422
Liverpool Quality of Life Battery
epilepsy, 72/432
Liverpool Seizure Severity Scale
epilepsy, 72/419, 430
Livona intoxication
ingestion, 37/64
Livona pica intoxication
staple food, 37/82
Lobar atrophy
Pick disease, 27/489, 46/236, 241
Lobar holoprosencephaly
classification, 30/438, 50/227
definition, 30/433, 438
gross anatomy, 30/438, 50/227
nomenclature, 30/438, 50/227
Lobe hypoplasia
superior temporal, *see* Superior temporal lobe hypoplasia
Lobectomy
anterior temporal, *see* Anterior temporal lobectomy
epilepsy, 73/390

frontal, *see* Frontal lobectomy
occipital, *see* Occipital lobectomy
temporal, *see* Temporal lobectomy
Lobotomy
anxiety, 1/110, 45/27
frontal lobe syndrome, 45/26
intractable pain, 45/27
obsessional neurosis, 45/27
pain, 1/110
Lobster eye syndrome, *see* Gerhardt syndrome
Lobular sclerosis
epilepsy, 72/109, 112
Lobulus ansoparamedianus
ablation, 2/402
animal experiment, 2/402
hypermetria, 2/403
Lobulus paramedianus
somatotopic localization, 2/407
Lobulus simplex
ablation, 2/402
animal experiment, 2/402
somatotopic localization, 2/407
LOC stimulus
Purkinje cell, 2/410, 416
Local adaptation syndrome
strain, 45/248
Local anesthetic agent
abscess, 65/421
adrenergic nerve ending, 65/426
anesthetic agent intoxication, 37/401-407
anesthetic agent potency, 65/421
apnea, 65/420
arachnoiditis, 65/421
axonal degeneration, 65/423
brain metabolic rate, 65/420
calcium influx, 65/426
chemical character, 65/419
complication, 65/419
complication frequency, 65/421
complication rate, 65/422
demyelination, 65/423
diabetes mellitus, 65/421
dysarthria, 65/420
epidural anesthesia, 65/420
epilepsy, 65/420
hematoma, 65/421
infarction, 65/421
intoxication, 65/419
local administration, 65/420
myoclonus, 65/420
nerve blood flow, 65/428
nerve conduction, 65/419
nerve conduction block, 65/419
nerve injury, 65/419-421
neuropathology, 65/423
neuropathy, 65/421
neurotoxicity, 65/422
nitric oxide, 65/426
orolingual numbness, 65/420
phospholipase, 65/427
phospholipase A2, 65/427
prilocaine intoxication, 37/405
prostacyclin, 65/428
prostaglandin metabolite, 65/427
sodium channel, 65/419, 426
spinal anesthesia, 65/420
systemic administration, 65/420
thromboxane A2, 65/428
tinnitus, 65/420
toxic mechanism, 65/420
toxicity mechanism, 65/423
trauma, 65/421
ultrastructure, 65/424
vasoconstriction, 65/427
Local injection
lumbosacral plexus, 51/159
Localizations
cerebral, *see* Cerebral localization
object, *see* Object localization
spatial, *see* Spatial localization
Locked-in syndrome
bilateral lacunar infarction, 54/241
brain death, 24/702, 708
brain injury, 24/702
brain stem auditory evoked potential, 54/162
central pontine myelinolysis, 63/549
cytarabine intoxication, 65/533
EEG, 53/210
insomnia, 49/245
pontine infarction, 53/389
sign, 24/708
somatosensory evoked potential, 54/163
symptom, 53/210, 389
yawning, 63/493
Locomotor ataxia
paleocerebellar syndrome, 2/418
progressive, *see* Tabes dorsalis
Locura manganica, *see* Manganese intoxication
Locus
esterase D, *see* Esterase D locus
gene, *see* Gene locus
Locus ceruleus
alcohol intoxication, 64/114
autonomic nervous system, 74/155
blood pressure, 74/155
noradrenalin, 74/161

phenylketonuria, 29/34
progressive bulbar palsy, 59/224
stress, 74/307, 310
thalamus, 74/155
vessel permeability, 48/110
Locus ceruleus degeneration
Parkinson dementia complex, 42/250
Parkinson disease, 6/214, 42/246, 49/111
progressive pallidal atrophy (Hunt-Van Bogaert), 42/247
Loeffler endocarditis, *see* Idiopathic hypereosinophilic syndrome
Loeffler eosinophilic fibroblastic endocarditis, *see* Idiopathic hypereosinophilic syndrome
Loeffler fibroblastic endocarditis, *see* Idiopathic hypereosinophilic syndrome
Loeffler syndrome, *see* Idiopathic hypereosinophilic syndrome
Lofepramine
tricyclic antidepressant, 65/312
Logic
death, 24/803
general, 24/802
Logorrhea
dysphrenia hemicranica, 5/79
Pick disease, 46/234, 240
Loiasis
encephalopathy, 52/515, 517
filariasis, 35/161, 163, 52/514, 517
Loken-Senior syndrome
cerebellar ataxia, 60/735
secondary pigmentary retinal degeneration, 60/735
skeletal abnormality, 60/735
Lomustine
antineoplastic agent, 65/528
chemotherapy, 39/115, 67/282, 284
neurologic toxicity, 39/115
neurotoxin, 65/528, 67/356
optic neuropathy, 65/528
London Conference on Chromosomes, 31/343, 376, 549, 590
London, United Kingdom
National Hospital, *see* National Hospital, London, United Kingdom
neurology, 1/6, 19
Long chain acyl-coenzyme A dehydrogenase
inner mitochondrial membrane system, 66/407
metabolic myopathy, 62/492
myoglobinuria, 62/558
Long chain acyl-coenzyme A dehydrogenase cDNA
inner mitochondrial membrane system, 66/407
Long chain acyl-coenzyme A dehydrogenase deficiency
cardiomegaly, 62/495
dicarboxylic aciduria, 62/495
hepatomegaly, 62/495
infantile hypotonia, 62/495
lipid metabolic disorder, 62/494
metabolic encephalopathy, 62/495
metabolic myopathy, 62/495
myalgia, 62/495
myoglobinuria, 62/495
myopathy, 62/495
Long chain acyl-coenzyme A dehydrogenase gene
mitochondrial matrix system, 66/409
Long chain acyl-coenzyme A dehydrogenase protein
inner mitochondrial membrane system, 66/407
Long chain alcohol intoxication
headache, 48/420
Long chain 3-hydroxyacyl-coenzyme A dehydrogenase
inner mitochondrial membrane system, 66/406
Long chain 3-hydroxyacyl-coenzyme A dehydrogenase coding
inner mitochondrial membrane system, 66/408
Long chain 3-hydroxyacyl-coenzyme A dehydrogenase deficiency
cardiomyopathy, 62/495
hypoglycemic encephalopathy, 62/495
inner mitochondrial membrane system, 66/408
lipid metabolic disorder, 62/495
metabolic encephalopathy, 62/495
metabolic myopathy, 62/495
muscle biopsy, 62/495
Long chain polyisoprenol alcohol
neuronal ceroid lipofuscinosis, 51/382
Long circumflex artery
radioanatomy, 11/98
topography, 11/98
Long latency reflex
hypotonia, 49/70
rigidity, 49/70
Long loop reflex
Parkinson disease, 49/70, 102
Long QT syndrome
cardiac dysrhythmia, 39/266
hereditary, *see* Hereditary long QT syndrome
idiopathic, *see* Idiopathic long QT syndrome
Long thoracic nerve
compression neuropathy, 51/102
occupational lesion, 7/331
origin, 2/137
topographical diagnosis, 2/20
Longitudinal fasciculus
medial, *see* Medial longitudinal fasciculus

Schütz, 2/472
Longitudinal sinus
inferior, *see* Inferior sagittal sinus
Longitudinal sinus thrombophlebitis, *see* Superior longitudinal sinus thrombophlebitis
Longitudinal sinus thrombosis
superior, *see* Superior longitudinal sinus thrombosis
Lonidamine
antineoplastic agent, 65/529
neurotoxin, 65/529, 538
Lonsdale-Blass syndrome
periodic ataxia, 21/578
Look-through phenomenon
brain blood flow, 54/149
Lop ear
sensorineural deafness, 42/379
Townes-Brocks syndrome, 42/379
Loquacity
migraine, 48/162
Lorazepam
antiepileptic agent, 37/358, 65/497
benzodiazepine intoxication, 65/334
chemical structure, 37/356
neuroleptic akathisia, 65/290
status epilepticus, 73/328
Lordosis
see also Scoliosis
congenital, *see* Congenital lordosis
dystonia, 42/253
facioscapulohumeral muscular dystrophy, 40/418
juvenile spinal cord injury, 25/183
limb girdle syndrome, 40/436
lumbar, *see* Lumbar lordosis
MASA syndrome, 43/255
myopathy, 40/418, 436
neuroleptic dystonia, 65/281
parkinsonism, 42/253
Rubinstein-Taybi syndrome, 31/284, 43/234
scoliosis, 50/414, 416
spina bifida, 32/545
subsarcolemmal myofibril, 42/253
Loss of consciousness, 72/259
brain concussion, 23/420, 427, 432, 440
brain contusion, 23/427
classification, 23/7
commotio cerebri, 23/420
history, 23/7
intracranial pressure, 23/207
Lotaustralin
cyanogenic glycoside intoxication, 36/515, 65/26
neurotoxin, 36/515, 65/26
Loudness balance
alternate binaural, *see* Alternate binaural loudness balance
Loudness balance test
acoustic nerve tumor, 17/676
equal, *see* Equal loudness balance test
Loudness discomfort level test
audiogram, 16/311
description, 16/309
Loudness recruitment
acoustic neuroma, 16/308
lesion localization, 2/74
Meniere disease, 2/74, 16/309
Louis-Bar syndrome, *see* Ataxia telangiectasia
Louping ill
acute viral encephalitis, 34/77
Louping ill encephalitis
acute viral encephalitis, 56/134
Louse-borne typhus
Brill-Zinsser disease, 34/645
rickettsial infection, 34/644-646
Lovastatin, *see* Mevinolin
Low back pain
see also Intervertebral disc prolapse
infectious endocarditis, 63/117
lumbar intervertebral disc prolapse, 20/578
pseudoradicular syndrome, 20/578
referred pain, 20/578
Low density lipoprotein
Bassen-Kornzweig syndrome, 42/511, 63/275
hyperlipoproteinemia type II, 10/272, 42/445
β-lipoprotein, 10/267
prebetalipoprotein, 10/267
xanthelasma, 10/272
xanthoma tendinosum, 10/272
Low grade fibrillary astrocytoma
chemotherapy, 68/48
Low grade glioma
astroblastoma, 68/54
classification, 68/34
clinical finding, 68/36
epidemiology, 68/35
imaging, 68/37
irradiation, 68/43
pleomorphic xanthoastrocytoma, 68/55
seizure, 68/51
subependymal giant cell astrocytoma, 68/55
surgery, 68/40
symptom, 68/35
treatment, 68/44
tumor cyst, 68/50
Low set ears
acrocephalosyndactyly type I, 43/318
acrocephalosyndactyly type II, 38/422, 43/324

cerebrohepatorenal syndrome, 43/339
Coffin-Lowry syndrome, 43/238
Edwards syndrome, 43/535, 50/275
9p partial trisomy, 43/516
Patau syndrome, 31/507, 43/528
Pena-Shokeir syndrome type I, 43/438, 59/71
11q partial monosomy, 43/523
21q partial monosomy, 43/541
4q partial trisomy, 43/498
7q partial trisomy, 43/506
Saethre-Chotzen syndrome, 43/322
Saldino-Noonan syndrome, 31/243
trisomy 8 mosaicism, 43/510
trisomy 22, 43/548
Lowe syndrome
acidosis, 42/606
aminoaciduria, 42/606
blindness, 42/606
brain atrophy, 42/607
cataract, 42/606
CNS, 21/49, 31/301
congenital fiber type disproportion, 62/355
enzymopathy, 21/49
eye lesion, 31/301
face, 31/302
Fanconi syndrome, 29/172
frontal bossing, 42/606
genetics, 31/301
glaucoma, 42/606
hydrocephalus, 42/607
hyporeflexia, 42/606
lissencephaly, 42/607
mental deficiency, 31/302, 42/606, 46/66
metabolic abnormality, 31/303
osteoporosis, 42/606
pachygyria, 42/607
pathology, 31/303
porencephaly, 42/607
prognosis, 31/304
radiologic appearance, 31/303
renal tubular dysfunction, 42/606
rickets, 42/606
seizure, 31/302
treatment, 31/304
Löwenberg-Hill leukodystrophy
clinical finding, 10/176
features, 10/151
genetics, 10/176
neuropathology, 10/177
orthochromatic leukodystrophy, 10/106
Lower face headache, *see* Sphenopalatine neuralgia
Lower half headache, *see* Sphenopalatine neuralgia
Lower motoneuron disease
ataxia, 60/657
deafness, 60/657
juvenile metachromatic leukodystrophy, 66/168
mental deficiency, 60/657
radiation myelopathy, 61/200
undetermined significant monoclonal gammopathy, 63/402
Lower motoneuron lesion
EMG, 1/636-638
Lower motoneuron sign
adult polyglucosan body disease, 62/484
Lower urinary tract
anal reflex, 61/295
anocutaneous reflex, 61/295
bladder filling, 61/293
bulbocavernosus reflex, 61/295
continence, 61/293
detrusor dyssynergia, 61/296
detrusor function, 61/296
detrusor muscle, 61/292
EMG, 61/298
female bladder neck, 61/292
function, 61/291
investigation, 61/296
Laplace law, 61/293
low intramural pressure, 61/293
male bladder neck, 61/292
pharmacologic test, 61/297
residual urine volume, 61/296
sacral evoked potential, 61/298
segmental innervation, 61/292
spinal cord injury, 61/291
structure, 61/291
ultrasound scan, 61/296
urethral pressure profilometry, 61/297
urodynamic study, 61/296
uroflow measurement, 61/297
videocystography, 61/296
Lowry syndrome
cryptorchidism, 43/368
secondary pigmentary retinal degeneration, 60/720
talipes, 43/368
Lowry-Wood syndrome, *see* Lowry syndrome
Loxapine
acquired myoglobinuria, 62/573
neuroleptic agent, 65/275
toxic myopathy, 62/601
Loxosceles intoxication
chill, 37/112
cytotoxin, 37/112
erythema, 37/112
fever, 37/112

hemolysis, 37/112
intracranial hemorrhage, 37/112
intravascular consumption coagulopathy, 37/112
nausea, 37/112
pain, 37/112
renal insufficiency, 37/112
tissue necrosis, 37/112
vomiting, 37/112
LSD, *see* Lysergide
Lucid interval
brain fat embolism, 55/178, 183, 185
chronic subdural hematoma, 24/304
epidural hematoma, 24/267, 55/182, 185
intracerebral hematoma, 24/357
lead encephalopathy, 64/436
posterior fossa hematoma, 24/348
subdural hematoma, 55/182
traumatic intracranial hematoma, 57/264
Lucio phenomenon
leprosy, 51/217
Lückenschädel, *see* Craniolacunia
Lüers-Spatz total frontal atrophy, *see* Total frontal atrophy (Lüers-Spatz)
Lues, *see* Syphilis
Luft syndrome
chloramphenicol, 41/210
2,4-dinitrophenol, 41/210
histochemistry, 40/47
hypermetabolism, 66/425
hyperthyroidism, 41/209
mitochondrial abnormality, 41/208
mitochondrial change, 40/219
mitochondrial hypermetabolism, 40/160
mitochondrial myopathy, 41/209
nuclear DNA defect, 66/396
oxidation-phosphorylation, 66/425
oxidative phosphorylation, 41/210
ultrastructure, 40/92
Lumbar disc hernia, *see* Lumbar intervertebral disc prolapse
Lumbar disc protrusion
cauda equina intermittent claudication, 12/520, 63/43
discography, 12/520
Lumbar enlargement artery, *see* Adamkiewicz artery
Lumbar intervertebral disc prolapse, 20/573-600
age, 20/582
bony fusion, 25/503
Bragard sign, 20/587
cauda equina syndrome, 20/594
child, 20/582
clinical features, 20/582
CSF, 20/591
disc-root relation, 20/578
EMG, 20/580
Gowers sign, 20/587
Hyndham sign, 20/586
Kemp sign, 20/586
Lasègue sign, 20/586
Léri sign, 20/588
low back pain, 20/578
motor dysfunction, 20/590
myelography, 20/581
pain, 20/584
pathomechanism, 20/576
posttraumatic, 25/494
posture, 20/588
pregnancy, 8/14, 20/595
provocative test, 20/586
pseudoradicular pain, 20/585
radiography, 20/581
ratio, 20/573
referred pain, 20/578
reflex change, 20/591
scoliosis, 20/588
sensory loss, 20/585, 591
sex ratio, 20/582
side, 20/583
spine rigidity, 20/589
teenage disc prolapse, 20/594
trauma, 20/584, 594
treatment, 20/599
upper (L1-L3) disc, 20/595
vertebral canal stenosis, 20/597
Lumbar lordosis
congenital spondyloepiphyseal dysplasia, 43/478
hereditary progressive diurnal fluctuating dystonia, 49/530
kyphosis, 32/147
late infantile acid maltase deficiency, 62/480
Scheuermann disease, 32/147
Lumbar meningomyelocele
Arnold-Chiari malformation type II, 50/403
Lumbar plexitis
opiate intoxication, 37/388
Lumbar plexus
anatomy, 7/421
diamorphine, 37/388
injury, 7/422
lumbosacral plexus, 51/158
nerve compression, 7/313
Lumbar plexus lesion
ionizing radiation neuropathy, 51/137
Lumbar plexus neuropathy
cisplatin intoxication, 65/537
Lumbar puncture

acute subdural hematoma, 24/282
arachnoiditis, 61/162
bacteremia, 61/162
bacterial meningitis, 33/4-7, 61/162
brain arteriovenous malformation, 12/243
brain hemorrhage, 61/164
brain herniation, 61/163
brain injury, 23/237
cardiac arrest, 61/165
chemical meningitis, 61/162
child, 25/179
complication, 61/147
contraindication, 61/148, 163
cranial nerve palsy, 61/163
CSF, 57/318
CSF fistula, 61/158
dermoid cyst, 20/75, 78
epidermoid cyst, 61/159
epidural abscess, 61/161
epidural hematoma, 24/272, 26/18
epilepsy, 61/166
frontal lobe tumor, 17/261
headache, 5/34, 175, 33/6
hearing loss, 61/165
iatrogenic neurological disease, 55/476
indication, 61/148
intervertebral disc rupture, 61/159
intracranial subdural hematoma, 61/164
ischemic myelopathy, 61/160
juvenile spinal cord injury, 25/179
lumbar epidermoid cyst, 61/159
multiple sclerosis, 9/324
myelitis, 61/162
pneumocephalus, 61/164
radicular syndrome, 61/158
spinal block, 61/161
spinal cord abscess, 52/191, 61/162
spinal cord coning, 61/161
spinal cord hemorrhage, 61/160
spinal cord injury, 25/179, 303, 26/186
spinal epidural empyema, 52/186
spinal epidural hematoma, 61/159
spinal nerve root injury, 25/417
stab wound, 25/201
stylet injury, 61/158
subarachnoid hematoma, 61/160
subarachnoid hemorrhage, 12/178, 61/164
subcutaneous CSF, 61/158
subdural effusion, 24/338
subdural empyema, 57/323
subdural hematoma, 26/31, 35, 37, 61/160
technique, 61/147
traumatic tap, 61/158
unstyleted needle, 61/158
vasovagal syncope, 61/165
Lumbar spinal column injury
mechanism, 25/138
type, 25/139
Lumbar spondylosis
occupational, 7/333
Lumbar stenosis
ap diameter, 50/474
Bragard sign, 20/684
canalography, 20/722
cauda equina compression, 20/627-637
four radiologic signs, 50/472
pathomorphology, 50/472
pseudoradicular pain, 20/757
symptom, 50/471
vertebral canal stenosis, 50/465, 471, 474
Lumbar vertebral canal stenosis
see also Spinal cord intermittent claudication
absolute, 32/332
achondroplasia, 20/637, 777
additional factor, 20/667, 674
age, 20/678, 32/331
area variation, 20/665
ataxia, 20/705
cauda equina intermittent claudication, 20/598, 749, 760, 32/337-339
classification, 20/613, 636
clinical features, 20/678
congenital, 32/331-333
congenital deformity, 20/775
CSF, 32/334
definition, 20/720
discontinuity, 20/646
disharmonic, 20/647
EMG, 20/723
epidemiology, 32/332
extent variation, 20/662
first symptom, 32/334
hereditary, 20/678
history, 20/628-635, 32/329
iatrogenic stenosis, 20/784
idiopathic developmental type, 20/636-761
interpedicular narrowing, 20/636
interpedicular stenosis, 20/774
intervertebral disc prolapse, 20/597
Lasègue sign, 20/684, 707
level variation, 20/661
lumbago, 20/680
lumbar region, 20/597, 613, 627-636
mixed, 20/638, 644
morphology, 32/332
muscle cramp, 20/706

myelography, 20/721
neurogenic bladder, 20/688
orchidalgia, 20/689
osteitis deformans (Paget), 20/783
other than developmental, 20/775-791
pain, 20/756
paraparesis, 20/714
phatomorphology, 20/637
postoperative complication, 20/768
pseudomeralgia paresthetica, 20/705
pudendalgia, 20/689
pure absolute, 20/638, 644
pure relative, 20/638, 644
radiculoneuropathy, 20/683
radiology, 20/718
ratio midsagittal diameter, 20/656-661
relative, 32/332
sex ratio, 20/678
spinal cord intermittent claudication, 20/749, 760, 32/337-339
spondylolisthesis, 20/781
stenosis, 20/780
syndrome, 32/333-337
transition area typology, 20/652
treatment, 20/767, 32/334-337
urinary incontinence, 20/706
variant, 20/724
vertebrogenic syndrome, 20/681
Lumbar vertebral column
facet joint plane, 61/8
Lumbar vertebral column biomechanics
anatomy, 61/14
Lumbodorsal spine injury, *see* Thoracolumbar spine injury
Lumboperitoneal shunt
hydrocephalic dementia, 46/329
Lumbosacral lipoma
see also Spina bifida
pseudo-Babinski sign, 19/37
spina bifida, 50/499
Lumbosacral myelocele
anencephaly, 50/77
Lumbosacral nerve injury
anatomy, 25/467
diagnosis, 25/477
distribution, 25/471
incidence, 25/477
laminectomy, 25/472
mechanism, 25/472
muscle stretch reflex, 25/441, 454
pathology, 25/470
prognosis, 25/478
Lumbosacral plexus
anatomy, 2/152, 25/404, 468, 51/157
aneurysm compression, 51/161
anticoagulant, 51/164
direct injury, 51/160
general anesthesia, 51/161
hemophilia, 51/164
hemorrhagic lesion, 51/164
hip arthroplasty, 51/161
iatrogenic nerve injury, 51/159, 161, 164
impairment evaluation, 8/536
injury, 51/157
local injection, 51/159
lumbar plexus, 51/158
maternal birth injury, 51/163
maternal obstetrical paralysis, 2/153, 20/595
pelvic surgery, 51/161
pregnancy, 51/163
psoas hematoma, 51/165
radiation injury, 51/162
retractor use, 51/160
retroperitoneal hemorrhage, 51/165
sacral plexus, 51/159
surgical damage, 51/160
syndrome, 2/128-156
traction injury, 51/160
tumor growth lesion, 51/165
Lumbosacral plexus injury
see also Lumbosacral trunk injury
pelvic fracture, 25/475
Lumbosacral plexus neuritis
see also Femoral neuropathy *and* Mononeuritis multiplex
ischemia, 51/166
neuralgic amyotrophy, 51/167
Schistosoma japonicum, 51/167
serum sickness, 51/167
Lumbosacral plexus neuropathy
Ehlers-Danlos syndrome, 55/456
Lumbosacral plexus palsy
Ehlers-Danlos syndrome, 55/456
Lumbosacral syringomyelia
inherited, 32/259
Lumbosacral trunk injury
see also Lumbosacral plexus injury
pelvic fracture, 25/475
Lumbriconereis heteropoda intoxication
neurotoxic, 37/55
Luminal, *see* Phenobarbital
Lundberg syndrome
ataxia, 13/91
optic atrophy, 13/91
pyramidal tract sign, 13/91
Lung abnormality

holoprosencephaly, 50/237
Lung cancer
brain metastasis, 69/221
chronic tactile hallucinosis, 46/564
incidence, 69/219
metastasis, 71/619
neurocognitive function, 67/398
neurofibromatosis type I, 50/369
opsoclonus, 60/657
paraneoplastic syndrome, 71/673
prophylaxis, 69/219
SBLA syndrome, 42/769
small cell, *see* Small cell lung cancer
Lung carcinoma
brain metastasis, 71/613
Eaton-Lambert myasthenic syndrome, 62/425
metastatic cardiac tumor, 63/93
spinal cord metastasis, 69/169
Lung cyst, *see* Pulmonary cyst
Lung disease
interstitial, *see* Interstitial lung disease
Lung edema, *see* Pulmonary edema
Lung function, *see* Pulmonary function
Lung growth
anencephaly, 50/84
Lung hypoplasia, *see* Pulmonary hypoplasia
Lung infection, *see* Pneumonitis
Lung tumor
pain, 1/139
Lung volume
see also Respiratory dysfunction *and* Respiratory system
early stage, 26/338
infantile spinal muscular atrophy, 59/56
late stage, 26/338
respiratory system, 26/337, 341
rib cage, 26/340, 346
spinal cord injury, 26/337, 341
Lupin seed
distal axonopathy, 64/13
neurotoxin, 64/13
toxic neuropathy, 64/13
Lupin seed intoxication
symptom, 64/13
Lupus
migraine, 5/56
systemic lupus erythematosus, 71/46
Lupus anticoagulant, *see* Antiphospholipid antibody
Lupus erythematosus
amodiaquine, 13/220
chloroquine, 13/220
infectious endocarditis, 63/111
procainamide intoxication, 37/451
systemic, *see* Systemic lupus erythematosus
Lupus vasculitis, 71/8
Luria-Nebraska neuropsychological test battery
neuropsychologic defect, 45/526
Luschka foramina atresia
Dandy-Walker syndrome, 30/623, 627, 632, 50/323
Luschka sinuvertebral nerve, *see* Sinuvertebral nerve (Luschka)
Lust
brain substrate, 45/277
Luteinizing hormone
adenohypophysis, 2/442
anorexia nervosa, 46/586
brain injury, 23/123
Lutheran blood group
genetic linkage, 40/522, 43/153
Lutz-Splendore-Almeida disease, *see* Paracoccidioidomycosis
Luxation
anterior cervical cord syndrome, 25/335
atlantoaxial, *see* Atlantoaxial dislocation
atlanto-occipital, *see* Atlanto-occipital luxation
lens, *see* Lens luxation
Luxol fast blue staining
phospholipid, 9/25
Luxury perfusion syndrome
brain blood flow, 57/70
head injury, 57/70
vascular tone, 53/113
Luys nucleus
Hallervorden-Spatz syndrome, 66/713
Luysian atrophy
ballismus, 21/531
classification, 21/531
Lyle syndrome
aqueduct stenosis, 2/302
gaze paralysis, 2/302
Lyme disease
see also Borrelia burgdorferi *and* Neuroborreliosis
abducent nerve paralysis, 51/204
acrodermatitis atrophicans, 52/260
Amblyomma americanum, 51/206
animal reservoir, 52/261
animal vector, 52/261
arthralgia, 51/203
arthritis, 51/205, 52/260
Bannwarth syndrome, 47/612-614, 51/208
Borrelia burgdorferi, 52/261
Borrelia duttonii, 51/206
brain biopsy, 52/263
cardiac involvement, 51/205
cerebellar ataxia, 51/205

chorea, 51/205
chronic meningitis, 56/645
clinical features, 47/613, 52/261
CNS, 52/253
course, 51/210
cranial nerve palsy, 52/261
CSF, 51/205
delirium, 52/262
demyelinating disease, 52/262
diagnosis, 51/207
differential diagnosis, 52/263
EEG, 52/262
encephalitis, 51/204, 208, 52/262
epidemiology, 52/261
epilepsy, 52/262
erythema chronicum migrans, 51/203, 52/260
etiology, 51/206
facial paralysis, 52/262
fatigue, 51/204
fever, 51/204
headache, 51/204, 208
hemiparesis, 51/205
Ixodes dammini, 47/613
ixodid tick, 52/261
lymphadenopathy, 51/203
lymphocytic meningitis, 52/262
meningitis, 51/204, 52/261, 263
microglia, 52/263
mononeuritis multiplex, 52/262
myalgia, 51/203
myelitis, 51/205
neuritis, 51/204
neuropathology, 52/263
nuclear magnetic resonance, 52/262
pathogenesis, 51/207
penicillin, 52/263
polymyositis, 62/373
prognosis, 51/210
progressive muscular atrophy, 59/21
psychiatric symptom, 52/262
Sjögren syndrome, 71/74
tertiary, 52/262
treatment, 51/208, 52/263
Lymphadenopathy
angioimmunoblastic, *see* Angioimmunoblastic lymphadenopathy
giant follicular, *see* Giant follicular lymphadenopathy
human T-lymphotropic virus type I associated myelopathy, 56/527
leprosy, 51/216
Listeria monocytogenes meningitis, 33/82
Lyme disease, 51/203
lymphoid interstitial pneumonitis, 39/530
mephenytoin intoxication, 37/202
multiple myeloma polyneuropathy, 51/430
neuroborreliosis, 51/203
POEMS syndrome, 47/619, 51/430, 63/394
Lymphadenopathy associated virus, *see* Human immunodeficiency virus
Lymphangiectasia
intestinal, *see* Intestinal lymphangiectasia
Lymphangioma
orbital tumor, 17/177
Lymphangiomyomatosis
pulmonary, *see* Pulmonary lymphangiomyomatosis
Lymphatic leukemia
acute, *see* Acute lymphatic leukemia
cerebral syndrome, 18/250
chronic type, 18/250
cranial neuropathy syndrome, 18/250
cryoglobulinemia, 63/403
facial paralysis, 18/250
hypothalamic syndrome, 18/250
meningeal syndrome, 18/250
mixed syndrome, 18/250
recurrent meningitis, 52/54
reticulosis, 18/247
Lymphedema
arteriovenous malformation, 42/726
familial pes cavus, 21/264
fever, 42/726
Turner syndrome, 43/550, 50/544
Lymphoblastic leukemia
meningeal leukemia, 63/340
Lymphocytapheresis
multiple sclerosis, 47/201
Lymphocyte
see also Cell mediated immunity
arteritis, 48/310
ataxia telangiectasia, 14/270
B, *see* B-lymphocyte
brain metastasis, 69/112
experimental allergic encephalomyelitis, 9/503, 517, 47/441-445, 449
experimental allergic neuritis, 9/531-536, 47/457
GM2 gangliosidosis, 10/335, 409
human T-lymphotropic virus type I associated myelopathy, 59/450
idiopathic multiple hemorrhagic sarcoma, 18/288
interleukin-1, 56/69
interleukin-2, 56/69
multiple sclerosis, 47/221, 228
natural killer cell, 56/67
Niemann-Pick disease, 10/283

phenotype, 56/67
polymyalgia rheumatica, 41/76, 55/346
rheumatoid arthritis, 41/76
sensitized, 9/521, 523
T, *see* T-lymphocyte
T-helper, *see* T-helper lymphocyte
transformation, 9/523, 532
tumor infiltrating, *see* Tumor infiltrating lymphocyte
vacuolated, *see* Vacuolated lymphocyte
Lymphocyte antibody
multiple sclerosis, 47/197-199
Lymphocyte transformation test
multiple sclerosis, 47/358
Lymphocyte vacuolation
Batten disease, 10/567
Fabry disease, 10/283
glycosaminoglycanosis type IH, 10/335, 342
GM1 gangliosidosis, 10/466
GM2 gangliosidosis, 10/335, 409
neuronal ceroid lipofuscinosis, 10/224, 576
Niemann-Pick disease, 10/335, 487, 528, 576
Lymphocytic choriomeningitis
allergic inflammation, 9/503
antibody, 9/544
antibody test, 56/364
aqueduct stenosis, 50/308
arenavirus, 9/193-196, 56/16
arenavirus infection, 34/194-200
biphasic type, 56/130
Brudzinski sign, 56/359
bulbar paralysis, 56/361
carrier mouse, 56/362
chorioretinitis, 9/198, 50/308, 56/360
chronic meningitis, 56/645
CNS autoantigen, 9/536
confusion, 9/197
CSF, 9/544, 34/195, 197, 56/360
deafness, 56/361
diagnosis, 56/364
drowsiness, 56/359
encephalomyelitis, 9/194, 56/360
epidemiology, 56/362
features, 56/355
foreign antigen, 9/536, 544
golden hamster, 56/362
grippal type, 9/194, 56/359
growth hormone deficiency, 56/35
hallucination, 9/197
headache, 9/197, 56/359
hydrocephalus, 9/198, 34/198, 50/308, 56/360
hypersensitivity, 9/501, 544
immunofluorescence test, 56/364
incubation period, 56/359
Kernig sign, 56/359
lethality, 56/361
lethargy, 9/197
meningeal type, 56/359
meningitis serosa, 9/194, 196, 56/129
meningoencephalomyelitic type, 9/194, 197, 56/359
myalgia, 56/359
nuchal rigidity, 9/197, 56/359
papilledema, 9/197
paralysis, 9/197
pathology, 9/199, 56/361
prognosis, 56/130
recurrent disease, 56/360
transverse myelitis, 56/360
treatment, 9/199, 56/363
vaccinia, 9/536
viral encephalitis, 56/360
viral meningitis, 56/129
vomiting, 9/197, 56/359
Lymphocytic choriomeningitis lassa complex
arenavirus, 56/355
Lymphocytic choriomeningitis virus, 56/355-371
acute viral encephalitis, 56/126
arenavirus, 34/5, 56/355
arthralgia, 56/360
cerebellar heterotopia, 56/358
chemical composition, 56/355
deafness, 56/107
EEG, 56/358
epilepsy, 56/358
morphology, 56/355
mouse, 56/358
murine encephalitis, 56/358
myocarditis, 56/360
neurotropism, 56/35
portal of entry, 56/363
retinal hypoplasia, 56/358
T-lymphocyte, 56/358
tolerance, 56/358
transmission, 9/198, 56/363
vertigo, 56/107
viral meningitis, 56/126, 128
Wallgren syndrome, 56/359
zoonosis, 56/359
Lymphocytic cryoglobulinemia, *see* Cryoglobulinemia
Lymphocytic leukemia
acute, *see* Acute lymphocytic leukemia
brain hemorrhage, 63/343
chemotherapy, 63/356
chlormethine, 63/358

chronic, *see* Chronic lymphocytic leukemia
cytarabine, 63/356
encephalitis, 63/352
intrathecal methotrexate, 63/357
meningitis, 63/352
methotrexate effect, 63/357
neurologic chemotherapy complication, 63/356
nitrosourea, 63/358
polyneuropathy, 63/343
radiculopathy, 63/343
spinal nerve root, 63/343
thrombocytopenia, 63/343
vincristine, 63/358
Lymphocytic meningitis
see also Meningitis serosa
African trypanosomiasis, 52/342
brain gumma, 52/278
chronic benign, *see* Chronic benign lymphocytic meningitis
drechsleriasis, 35/565, 52/485
fonsecaeasis, 35/566, 52/487
human T-lymphotropic virus type I associated myelopathy, 56/538
Kawasaki syndrome, 52/265
leptospirosis, 33/400-402
Lyme disease, 52/262
neurobrucellosis, 52/595
neurosyphilis, 52/277
sporotrichosis, 35/569, 52/495
syphilis, 52/275
tabes dorsalis, 52/280
toxoplasmosis, 52/354, 359
tropical spastic paraplegia, 56/538
tuberculous meningitis, 52/205
Lymphocytic meningoencephalitis
acute diffuse, *see* Acute diffuse lymphocytic meningoencephalitis
cytomegalic inclusion body disease, 56/271
Lymphocytic meningoradiculitis, *see* Bannwarth syndrome
Lymphocytoma
congenital spinal cord tumor, 32/355-386
Lymphocytosis
CSF, *see* CSF lymphocytosis
Lymphocytotoxic antibody
disease activity, 47/369
HLA-B7 antigen, 47/369
HLA-B40 antigen, 47/369
HLA-DR2 antigen, 47/369
multiple sclerosis, 47/368-370
Lymphogranulomatosis, *see* Hodgkin disease
Lymphogranulomatous meningoencephalopathy
Hodgkin disease, 18/261
Lymphohistiocytosis
median neuropathy, 56/179
Lymphoid granulomatosis, *see* Lymphomatoid granulomatosis
Lymphoid interstitial pneumonitis
chronic, *see* Chronic lymphoid interstitial pneumonitis
diagnostic problem, 39/530
lymphadenopathy, 39/530
pneumonitis, 39/530
Sjögren syndrome, 39/530
Lymphoid leukemia, *see* Lymphocytic leukemia
Lymphoid organ
parasympathetic innervation, 75/552
Lymphoid radiation
multiple sclerosis, 47/201
Lymphokine
multiple sclerosis, 47/350
viral infection, 56/65, 70
Lymphoma
acquired immune deficiency syndrome, 56/515, 63/346
allergic granulomatous angiitis, 63/350
anticancer agent, 69/275
ataxia telangiectasia, 60/674
B-lymphocyte, *see* B-lymphocyte lymphoma
Bloom syndrome, 43/10
brain, *see* Brain lymphoma
brain hemorrhage, 63/350
brain infarction, 63/350
brain vasculitis, 63/350
brain venous infarction, 63/350
Burkitt, *see* Burkitt lymphoma
cauda equina compression, 63/346
Chédiak-Higashi syndrome, 60/674
chronic axonal neuropathy, 51/531
chronic meningitis, 56/645
ciclosporin, 63/533
classification, 63/345
CNS, *see* CNS lymphoma
cranial neuropathy, 63/350
cryoglobulinemia, 39/181, 63/403
demyelinating neuropathy, 41/343
distribution, 67/130
encephalopathy, 63/350, 69/274
epidemiology, 67/132
epidural, *see* Epidural lymphoma
epidural hematoma, 63/350
epilepsy, 63/350
Epstein-Barr virus, 63/533
headache, 48/285
herpes zoster, 63/353
hoarseness, 63/350

Hodgkin, *see* Hodgkin lymphoma
Hodgkin disease, 63/345
human immunodeficiency virus infection, 71/341, 343
hyperviscosity syndrome, 55/484
hypothalamic tumor, 68/71
immunocompromised host, 56/469
immunosuppressive therapy, 65/549
incidence, 68/530
intravascular consumption coagulopathy, 63/350
late cerebellar atrophy, 60/585
leptomeningeal metastasis, 71/662
leukoencephalopathy, 9/594
Listeria monocytogenes meningitis, 33/90
listeriosis, 52/91
malignant, *see* Malignant lymphoma
Marek disease virus, 56/581
meningeal, *see* Meningeal lymphoma
metastasis, 71/660, 662
metastatic cardiac tumor, 63/93
murine leukemia virus, 56/587
myelomalacia, 63/350
neuropathy, 63/350, 69/278
non-Hodgkin, *see* Non-Hodgkin lymphoma
optic pathway tumor, 68/74
painful neuropathy, 63/350
paraneoplastic syndrome, 71/696
paraproteinemia, 69/308
peripheral neurolymphomatosis, 56/185
phrenic nerve, 63/350
phrenic nerve palsy, 63/350
plexopathy, 63/350
polyarteritis nodosa, 55/359
polyneuropathy, 69/278
primary, *see* Primary lymphoma
primary brain, *see* Primary CNS lymphoma
primary CNS, *see* Primary CNS lymphoma
progressive multifocal leukoencephalopathy, 71/400
recurrent laryngeal nerve, 63/350
recurrent laryngeal nerve palsy, 63/350
renal transplantation, 63/533
retrovirus, 56/16
spinal cord, 19/14, 67/198
spinal cord compression, 63/346, 69/277
spinal epidural, *see* Spinal epidural lymphoma
spinal nerve root compression, 63/346
subarachnoid hemorrhage, 63/350
subdural, *see* Subdural lymphoma
subdural hematoma, 63/350
systemic, *see* Systemic lymphoma
T-cell, *see* T-lymphocyte lymphoma
xeroderma pigmentosum, 60/674

Lymphoma associated paraproteinemia
paraproteinemic polyneuropathy, 51/429
Lymphomatoid granulomatosis
angiopathic polyneuropathy, 51/446
brain infarction, 53/162
brain ischemia, 53/162
brain tumor, 63/424
cerebral, *see* Cerebral lymphomatoid granulomatosis
classification, 71/4
computer assisted tomography, 54/63
corticosteroid, 63/424
cyclophosphamide, 63/424
deafness, 51/451
demyelination, 51/451
facial paralysis, 51/451
granulomatous CNS vasculitis, 55/396
headache, 48/284
internuclear ophthalmoplegia, 51/451
intravascular malignant lymphomatosis, 51/451
malignant lymphoma, 51/451
multifocal necrotizing leukoencephalopathy, 51/451
multiple cranial neuropathy, 51/451
multiple sclerosis, 51/451
neurologic symptom, 55/364
polyarteritis nodosa, 55/364
sex ratio, 51/451
spinal cord tumor, 63/424
treatment, 51/451
Wegener granulomatosis, 39/529
Lymphomatosis
intravascular, 69/280
intravascular malignant, *see* Intravascular malignant lymphomatosis
Lymphomatous meningitis, *see* Meningeal lymphoma
Lymphoproliferative disease
progressive multifocal leukoencephalopathy, 9/491, 34/307, 47/503, 506
Lymphoreticular malignancy
ataxia telangiectasia, 60/349
microcephaly, 50/270
Lymphorrhage
experimental autoimmune myasthenia gravis, 41/110
myasthenia gravis, 41/96
Lymphosarcoma
epilepsy, 72/180
hyperviscosity syndrome, 55/487
leukemia, 18/247
RES tumor, 18/235
spinal epidural tumor, 20/112

Lymphostatic encephalopathy
anoxic ischemic leukoencephalopathy, 47/533
Lyngbya majuscula intoxication
ciguatoxin, 37/97
dermatitis, 37/97
erythema, 37/97
skin lesion, 37/97
Lyon, France
neurology, 1/8
Lyon hypothesis
Leber optic atrophy, 13/96
X-linked genetics, 13/96, 40/353
Lyon principle
aneuploidy, 50/542
Becker muscular dystrophy, 41/408
Duchenne muscular dystrophy, 41/408, 50/542
Lesch-Nyhan syndrome, 50/542
sex chromatin, 50/542
theory, 31/475-477
Lyonization
glucose-6-phosphate dehydrogenase deficiency, 42/645
hemophilia A, 42/738
inheritance, 30/97
orofaciodigital syndrome type I, 43/448
pseudohypoparathyroidism, 42/622
D-Lysergic acid diethylamide, *see* Lysergide
Lysergide
see also Hallucinogenic agent
ataxia, 65/45
biokinetic, 65/43
brain infarction, 53/164
brain ischemia, 53/164
brain vasculitis, 55/522
brain vasospasm, 55/522
carotid artery stenosis, 65/45
chemical structure, 65/43
dopamine receptor stimulating agent, 65/44
drug abuse, 55/522
epilepsy, 65/45
hallucinogenic agent, 37/329, 65/42
hallucinogenic agent intoxication, 37/329, 332-342
hyperreflexia, 65/45
isomer, 65/43
laboratory diagnosis, 65/43
migraine, 48/87
neurotoxin, 36/534, 37/329, 332, 50/219, 62/601, 64/3, 65/44
parasympathomimetic effect, 65/45
paresthesia, 65/45
pharmacology, 65/44
psychosis, 46/594
serotonin 2 receptor, 65/44
status epilepticus, 65/45
toxic myopathy, 62/601
treatment, 65/45
tremor, 65/45
Lysergide intoxication
ataxia, 65/44
color perception, 65/44
emotional change, 37/329
ergot alkaloid, 37/332
flashback, 65/44
hallucination, 4/334, 37/329, 332, 65/44
hereditary deafness, 50/219
hyperthermia, 75/64
macrosomatognosia, 4/231
microsomatognosia, 4/231
physiologic effect, 37/329
symptom, 65/44
tremor, 65/44
Lysine intolerance
congenital, *see* Congenital lysine intolerance
Lysine metabolism
glutaryl-coenzyme A dehydrogenase deficiency, 66/648
Lysine-oxoglutarate reductase deficiency, *see* Hyperlysinemia
Lysinuric protein intolerance
clinical features, 29/210
diagnosis, 29/211
enzyme defect, 29/210
epilepsy, 72/222
features, 29/105, 210
genetics, 29/211
neutropenia, 29/210
osteoporosis, 29/210
pathology, 29/210
thrombocytopenia, 29/210
treatment, 29/211
Lysocephalin
Pelizaeus-Merzbacher disease, 10/195
Lysoglycerophospholipid
biochemistry, 10/244
lipid metabolism, 10/244
Lysolecithin
leukoencephalopathy, 47/570
Lysophosphatidic acid
biochemistry, 9/7
Lysophosphatidyl choline, *see* Lysolecithin
Lysosomal disorder
epilepsy, 72/222
mucolipidosis, 66/64
mucolipidosis type II, 66/64
mucolipidosis type III, 66/64

multiple sulfatase deficiency, 66/64
Lysosomal enzyme deficiency
spinal muscular atrophy, 59/374
Lysosomal protein
mannose residue, 66/83
Lysosomal storage disease
acid maltase deficiency, 41/178
characteristics, 10/670
concept, 27/145
dementia, 46/130
enzyme deficiency polyneuropathy, 51/367
gene therapy, 66/113
genetically engineered, 66/118
germ line therapy, 66/113
glucosylceramidase gene, 66/118
somatic cell therapy, 66/113
spastic paraplegia, 59/439
treatment strategy, 66/117
Lysosome
acid maltase deficiency, 40/153
Batten disease, 40/37, 113
cytosome, 40/37, 113
enzyme, 66/57
intervertebral disc, 20/538
Lafora progressive myoclonus epilepsy, 40/37, 99, 102
metabolic pathway, 66/59
multiple sclerosis, 47/219, 222
muscle fiber, 40/36, 62/22
muscle fiber type, 62/4
muscle tissue culture, 40/190
neuronal ceroid lipofuscinosis, 10/670
peroxisome, 40/37, 99, 102
Lysosome associated membrane protein 1
undetermined significant monoclonal gammopathy, 63/400
Lyssavirus
rabies, 56/13, 383
Lysyl oxidase
Wilson disease, 49/229
Lytic enzyme
brain metastasis, 69/111
Lytic myeloma
paraproteinemia, 69/302
Lytico Bodeg
2-amino-3-methylaminopropionic acid, 64/5
neurotoxicology, 64/5

M band
skeletal muscle, 40/66
M component
serum, *see* Serum M component
M protein, *see* Monoclonal protein
Mabry syndrome
microcephaly, 30/100
MacCarthy sign
technique, 1/183
Macewen sign
see also Cracked pot sign
cerebellar medulloblastoma, 18/173
craniopharyngioma, 18/545
Machado disease
areflexia, 42/155
ataxia, 42/155
cerebellar atrophy, 42/155
CSF, 42/155
diabetes mellitus, 42/155
fasciculation, 42/155
Friedreich ataxia, 21/357
hyperuricemia, 42/155
hyporeflexia, 42/155
muscular atrophy, 42/155
nystagmus, 42/155
sensory loss, 42/155
Machupo, *see* Bolivian hemorrhagic fever
Machupo virus
arenavirus, 56/355
Bolivian hemorrhagic fever, 56/355
natural host, 56/369
transmission, 56/370
Mackerel intoxication
abdominal pain, 37/87
Clostridium, 37/87
diarrhea, 37/87
dizziness, 37/87
Escherichia coli, 37/87
fever, 37/87
headache, 37/87
Proteus morganii, 37/87
Salmonella, 37/87
swelling, 37/87
tachycardia, 37/87
Macrencephaly, 30/647-658
see also Macrocephaly
achondroplasia, 31/271, 42/41, 43/316, 50/263
Alexander disease, 10/101, 30/657, 42/485, 66/702
amaurotic idiocy, 10/101, 50/263
brain cortex heterotopia, 50/263
camptomelic syndrome, 50/263
cerebellar granular cell hypertrophy, 43/252
cerebral gigantism, 31/332, 42/41, 50/263
classification, 30/648
CNS spongy degeneration, 42/506
corpus callosum agenesis, 50/263
definition, 30/647

epidemiology, 30/158, 649
epilepsy, 72/109, 343
etiology, 30/648
external appearance, 30/651-654
genetics, 30/158, 658
glycosaminoglycanosis, 50/263
glycosaminoglycanosis type I, 10/441
GM2 gangliosidosis, 10/394
hemangioma, 42/41
hemi, *see* Hemimegalencephaly
history, 30/647
hydrocephalus, 30/658, 42/41
Klippel-Trénaunay syndrome, 14/400
Laurence-Moon-Bardet-Biedl syndrome, 31/328
leukodystrophy, 10/24
lipid metabolic disorder, 10/291
mental deficiency, 30/650, 42/41, 46/8, 50/263
neurofibromatosis, 42/41
normal body symmetry, 30/648, 657-659
occipitofrontal circumference, 42/41, 43
organic acid metabolism, 66/641
phakomatosis, 14/102
polymicrogyria, 50/257, 263
primary, 30/648-654
psychomotor retardation, 42/41
secondary, 30/648, 654-657
seizure, 30/651, 42/41
sex ratio, 42/42
sphingolipidosis, 50/263
spongy cerebral degeneration, 10/206, 209
tuberous sclerosis, 14/102, 31/2, 6, 42/41
unilateral, 30/648, 657-659
Macrocephaly
see also Macrencephaly
angiomatosis, *see* Riley-Smith syndrome
Bardet-Biedl syndrome, 13/401
cerebral gigantism, 43/336
clinical picture, 14/480
corpus callosum agenesis, 50/152, 162
fetal face syndrome, 30/100
glutaryl-coenzyme A dehydrogenase deficiency, 66/647
glycosaminoglycanosis type VII, 66/310
hemangioma, 14/480, 768
hyperphosphatasia, 30/100
lipomatosis, 14/83
mental deficiency, 46/8
neurofibromatosis type I, 50/369
organic acid metabolism, 66/639, 641
osteopetrosis, 30/100
porencephaly, 30/690, 50/360
pseudopapilledema, 14/480, 768
Riley-Smith syndrome, 14/14, 76, 83, 480, 768, 43/30
Sturge-Weber syndrome, 50/377
Macrocheilia
Melkersson-Rosenthal syndrome, 8/205, 208-211
Macrocrania
mental deficiency, 46/8
subdural hematoma, 57/337
Macrodactyly
hemifacial hypertrophy, 22/550
Macrogenitosomia praecox, *see* Pellazzi syndrome
Macroglia
CNS, 50/7
Macroglobulin
IgM, *see* IgM macroglobulin
α-2-Macroglobulin
Huntington chorea, 6/380, 395
Macroglobulin 19S
primary pigmentary retinal degeneration, 13/195
Macroglobulinemia
brain hemorrhage, 55/485
brain infarction, 55/485
clinical features, 39/190
course, 39/189, 63/396
cryoglobulinemia, 63/403
deafness, 55/132, 485
diagnosis, 39/189
dysproteinemic neuropathy, 8/72-76
encephalopathy, 39/191
history, 39/189
hyperviscosity syndrome, 55/484
myelopathy, 39/192
necrotizing demyelination, 39/532
nerve conduction, 7/167
neuropathy, 7/555, 8/74, 39/193
pathogenesis, 39/194-196
polyneuropathy, 8/146
progressive bulbar palsy, 59/225
Raynaud phenomenon, 55/485
retinopathy, 39/190
stroke, 39/191
subarachnoid hemorrhage, 39/191, 55/485
systemic brain infarction, 11/461
transient ischemic attack, 53/266
treatment, 39/196
type, 63/396
vertebral myeloma, 20/11
Waldenström, *see* Waldenström macroglobulinemia
Macroglossia
infantile acid maltase deficiency, 62/480
Melkersson-Rosenthal syndrome, 8/214
obstructive sleep apnea, 63/454
primary amyloidosis, 51/416

snoring, 63/450
triploidy, 43/564-566
Macrogyria, *see* Pachygyria
Macrophage
autonomic nervous system, 75/556
brain arteriosclerosis, 53/98
CSF, 12/146
eosinophilia myalgia syndrome, 64/253
experimental allergic encephalomyelitis, 9/503, 47/438
experimental allergic neuritis, 47/454-458
lactate dehydrogenase elevating virus, 56/28
Listeria monocytogenes, 52/90
multiple sclerosis, 47/215-225, 228, 232, 234, 337, 355
Mycobacterium leprae, 51/223
segmental demyelination, 51/68
wallerian degeneration, 51/64
Whipple disease, 70/239
Macrophage inclusion
globoid cell leukodystrophy, 51/373
Macropsia
cause, 2/613
hallucination classification, 45/56
higher nervous activity disorder, 3/29
migraine, 48/162
parietal lobe, 45/362
temporal lobe stimulation, 2/646
visual hallucination, 45/356
Macrosomatognosia
body scheme disorder, 4/226, 230, 45/385
epilepsy, 4/230, 45/385, 509
hallucination, *see* Body scheme disorder
lysergide intoxication, 4/231
mescaline, 4/231
migraine, 4/230, 45/385, 510
phantom limb, 4/226, 45/398
psychosis, 4/231, 45/385
schizophrenia, 4/231
Macrosomia, *see* Gigantism
Macrostereognosia
agnosia, 4/35
thalamic lesion, 2/479
Macula
central scotoma, 2/564
microscopic anatomy, 2/515
pseudoatrophic, *see* Pseudoatrophic macula
sparing, *see* Macular sparing
splitting, *see* Macular splitting
unilateral *vs* bilateral representation, 2/579
Macula coloboma, *see* Macula dysplasia
Macula cyst, *see* Vitelliruptive macula degeneration
Macula dysplasia
prenatal infection, 13/134
Macula lutea
cyst, 13/179
hole, 13/179
Junius-Kuhnt disease, 13/265
Macula pseudocyst, *see* Vitelliruptive macula degeneration
Macular corneal dystrophy
hereditary spastic ataxia, 60/464
Macular degeneration
characteristics, 22/536
familial spastic paraplegia, 22/468
Friedreich ataxia, 21/347, 60/310
hereditary cerebellar ataxia, 21/386
hyaloidotapetoretinal degeneration (Goldmann-Favre), 13/37, 276
juvenile, *see* Stargardt disease
olivopontocerebellar atrophy variant, 21/452
olivopontocerebellar atrophy (Wadia-Swami), 60/494
pigmentary, *see* Pigmentary macular degeneration
progressive external ophthalmoplegia, 22/194
retinovestibular hereditary spastic ataxia, 21/386
Sjögren-Larsson syndrome, 66/615
striatopallidodentate calcification, 6/720
vitelliruptive, *see* Vitelliruptive macula degeneration
Macular hypoplasia
albinism, 14/106
Macular sparing
see also Macular splitting
cerebrovascular disease, 55/119
false sparing, 2/666, 671
Grenzstreif, 2/580
occipital lobe syndrome, 45/51
occipital lobe tumor, 17/316, 319
occipital lobectomy, 2/580
visual field, 17/316, 320
Macular splitting
see also Macular sparing
chiasmal lesion, 2/572
occipital lobe tumor, 17/320
occipital lobectomy, 2/580-582, 671
optic chiasm lesion, 2/572
Madelung deformity
dyschondrosteosis, 43/389-391
Huntington chorea, 49/291
Madelung disease
autonomic abnormality, 75/702
Madras disease
amyotrophic lateral sclerosis, 22/323
progressive muscular atrophy, 59/22
Madurella

fungal CNS disease, 35/566, 52/479, 488
Madurelliasis
brain abscess, 35/567, 52/488
epilepsy, 35/567, 52/488
Maduromycosis, *see* Mycetoma
Maedi, *see* Visna-maedi disease
Maffucci-Kast syndrome
bony tumor, 43/29
dyschondromatosis, 43/29
hemangioma, 43/29
hemihypertrophy, 59/483
intracranial tumor, 43/29
Ollier disease, 43/29
vitiligo, 43/29
Maffucci syndrome
angioma, 14/119
angiosarcoma, 14/14
Bean syndrome, 14/106
brain glioma, 14/775
chondrosarcoma, 14/14
differential diagnosis, 14/119
dyschondromatosis, 14/118
Engel syndrome, 14/119
François syndrome, 14/119
Kast syndrome, 14/118
neurocutaneous syndrome, 14/101
nevus, 14/14
vitiligo, 14/119
MAG, *see* Myelin associated glycoprotein
Magee, K.R., 1/23
Magendie, F., 1/7
Magendie foramina atresia
Dandy-Walker syndrome, 30/623, 627, 632, 50/323
Magendie-Hertwig syndrome, *see* Skew deviation
Magendie syndrome, *see* Skew deviation
Magnesium
brain metabolism, 57/74
CNS, 28/434-436
CSF, *see* CSF magnesium
epilepsy, 15/313, 72/85
Guam amyotrophic lateral sclerosis, 59/276
hyperparathyroidism, 27/291
nerve cell, 72/50
neurophysiology, 63/562
pseudocalcium, 6/704, 715, 718
striatopallidodentate calcification, 6/716, 49/425
symptom, 70/113
Magnesium deficiency
Wernicke-Korsakoff syndrome, 45/197
Magnesium imbalance
adult, 28/546, 553
clinical features, 28/569
CSF, 28/552
diagnosis, 28/565
EEG, 28/549
ENG, 28/548
etiology, 28/554
history, 28/545
hypermagnesemia, 28/568-570
hypomagnesemia, 28/546-568
natural history, 28/553
neonatal, 28/569
neurologic form, 28/557-565
neurologic manifestation, 28/545-570
neurologic symptom, 28/546, 568
pathophysiology, 28/555
treatment, 28/566-568, 570
Magnesium intoxication
Eaton-Lambert myasthenic syndrome, 62/425
Magnet reaction
animal experiment, 2/396-398
cerebellar ablation, 2/396-398
Magnetic apraxia
syndrome, 4/50, 54
terminology, 2/760
Magnetic resonance
nuclear, *see* Nuclear magnetic resonance
Magnetic resonance imaging, *see* Nuclear magnetic resonance
Magnetoencephalography
see also EEG
activated, 72/331
epileptic signal, 72/326
evoked activity, 72/324
ictal recording, 72/330
imaging, 72/321
interictal recording, 72/326
localization, 72/329
modelling, 72/320
nuclear magnetic resonance, 72/323
preoperative, 72/330
provocation, 72/330
slow wave, 72/328
source localization, 72/322
spontaneous activity, 72/324
survey, 72/319
temporal lobe, 72/330
Magnus, R., 1/10, 69
Maida Vale hospital, United Kingdom, 1/7
Maier, H.W., 1/10
Main en griffe, *see* Claw hand
Main de singe, *see* Simian hand
Maitotoxin
calcium release, 65/161
ciguatoxin, 65/159

ciguatoxin intoxication, 65/160
marine toxin intoxication, 65/159, 161
neurotoxin, 65/141, 161
sodium release, 65/161
toxic dose, 65/141
Maize
pellagra, 28/64
vitamin B3 deficiency, 28/64, 51/333
Majewski syndrome, *see* Saldino-Noonan syndrome
Major histocompatibility complex
see also HLA system
class restriction, 56/70
cytotoxic T-lymphocyte, 56/68
multiple sclerosis, 47/303
viral infection, 56/70
Major migraine
psychotic migraine, 48/162
Major monosialogangliosidosis, *see* GM1 gangliosidosis
Mal de los rastrojos, *see* Argentinian hemorrhagic fever
Mal de Meleda
palmoplantar keratosis, 14/786
Mal de tête en salves, *see* Cluster headache
Malabsorption
Bassen-Kornzweig syndrome, 7/596, 29/410-412
deficiency neuropathy, 51/321, 324, 328
dibasic amino acid, *see* Dibasic amino acid malabsorption
glucose-galactose, *see* Glucose-galactose malabsorption
hypomagnesemia, 63/562
intestinal, *see* Intestinal malabsorption
lactose, *see* Lactose malabsorption
mesenteric artery occlusion, 7/631
pancreatogenic, *see* Pancreatogenic malabsorption
vagotomy, 75/653
vitamin B12, 7/596
vitamin B12 deficiency, 70/373
Malabsorption state
vitamin E deficiency, 70/442
Malabsorption syndrome
see also Celiac disease *and* Postgastrectomy syndrome
Alzheimer disease, 70/234
brain stem, 70/232
cause, 70/224
cerebellum, 70/232
classification, 7/602
Crohn disease, 28/236, 70/234
deficiency neuropathy, 7/596-599
diphyllobothrium latum myelopathy, 7/631
diverticulosis, 28/235
excitability disorder, 70/227
gastric surgery, 70/230
gastrogenic, 7/598-600
gastrogenous polyneuropathy, 7/598, 600
general symptom, 7/596
hepatic myelopathy, 70/232
intestinal biopsy, 70/226
Korsakoff amnesia, 70/233
laboratory aid, 70/225
methionine, *see* Methionine malabsorption syndrome
motoneuron disease, 70/231
motor end plate, 70/228
muscular atrophy, 70/227
myelopathy, 70/231
myoedema, 70/227
myotonia, 70/227
neurologic manifestation, 28/225-239
neuropathy, 7/595, 70/231
pathophysiology, 7/599
pellagra, 7/572, 38/648
peripheral nerve, 70/229
small intestine diverticulosis, 28/235
spheroid body, 6/628
survey, 70/223
tetany, 70/226
tropical sprue, 7/628, 28/235
visual system, 70/233
vitamin A deficiency, 70/233
vitamin B1 deficiency, 70/229
vitamin B6 deficiency, 70/229
vitamin B12 deficiency, 70/233
vitamin B12 depletion, 70/229
vitamin Bc deficiency, 70/230
vitamin E, 70/230
vitamin E deficiency, 70/228, 444
Wernicke-Korsakoff encephalopathy, 70/233
Whipple disease, 28/237-239, 51/328, 52/137
Malacia
en chapelet, *see* En chapelet malacia
Malacostraca intoxication
allergic reaction, 37/56
bacterial food poisoning, 37/56
eating, 37/56
neurotoxicity, 37/56
symptom, 37/56
Maladie des tics
course, 1/291
semantic, 1/277
Malamud-Cohen syndrome, *see* Infantile sex linked cerebellar ataxia
Malaria, 35/143-158
amodiaquine, 13/220, 35/155, 52/370

antiphospholipid antibody, 63/330
brain metastasis, 69/117
cerebral, *see* Cerebral malaria
chloroquine, 13/220, 35/155, 52/370
clinical features, 52/368
CNS, 31/221
diagnosis, 52/366
diamorphine, 37/393
diamorphine intoxication, 65/355
environmental factor, 35/6, 143
epidemiology, 35/6, 52/365
epilepsy, 72/146, 162
geographic mortality, 35/8
hydroxychloroquine, 52/370
hypoglycemia, 52/369
immunity, 35/144, 52/368
intrauterine infection, 31/221
laboratory test, 52/367
mefloquine, 52/372
myelomalacia, 63/42
nerve lesion, 7/488
neurosyphilis, 33/372, 35/144, 52/282
opiate intoxication, 37/393, 65/355
parasitic disease, 35/6-8
Parkinson disease, 6/244
Plasmodium falciparum, 35/144, 52/365-367
Plasmodium malariae, 35/144, 52/365-367
Plasmodium ovale, 35/144, 52/365-367
Plasmodium vivax, 35/144, 52/365-367
primaquine, 52/372
prophylaxis, 52/371
pyrimethamine, 52/372
quinidine, 52/372
quinine, 52/371
spinal apoplexy, 12/532, 35/148
spinal cord embolism, 63/42
subarachnoid hemorrhage, 12/111
sulfadoxine, 52/372
systemic brain infarction, 11/431
tetracycline, 52/373
transverse myelitis, 52/370
treatment, 35/154, 52/370
vector, 35/6, 52/365
Malate dehydrogenase
GM2 gangliosidosis, 10/291
retinal degeneration, 13/202
wallerian degeneration, 9/37
Malathion intoxication
chemical classification, 37/546
organophosphorus compound, 37/546
Malattia leventinese, *see* Hutchinson-Tay choroidopathy
Maldigestion, *see* Digestive system function disorder
Male sexual function
paraplegia, 61/313
rehabilitation, 61/313
spinal cord injury, 61/313
tetraplegia, 61/313
transverse spinal cord lesion, 61/313
Maley syndrome
brain injury, 23/12
Malherbe epithelioma, *see* Pilomatricoma
Malic enzyme
mitochondrial, *see* Mitochondrial malic enzyme
Malignancy
see also Cancer, Carcinoma *and* Carcinomatous myopathy
amyotrophic myelopathy, 38/671-675
aneuploidy, 50/566
autonomic polyneuropathy, 51/476, 479
blood viscosity, 55/483
brain, 38/625-660
brain embolism, 11/415-420
brain metastasis, 18/227
cavernous sinus thrombosis, 52/173
central pontine myelinolysis, 38/657
cord, 38/669-676
cortical cerebellar degeneration, 38/625-631
delirium, 46/546
dementia, 38/647-651, 657, 46/386
depression, 46/427
dermatomyositis, 62/371, 374, 71/131
ectopic hormone production, 38/651
encephalitis, 8/134
encephalomyelitis, 8/134, 38/631-644
ependymoma, 18/120, 20/363
Foix-Alajouanine disease, 38/669-671
glomus jugulare tumor, 18/449
hereditary multiple exostosis, 19/317
hypercalcemia, 38/652-654, 63/561
hyperhidrosis, 75/529
hyperviscosity syndrome, 55/485
immunocompromised host, 56/469
inclusion body myositis, 62/374
inflammatory myopathy, 41/348-370
intravascular consumption coagulopathy, 55/493, 63/314
Isaacs syndrome, 41/347
lymphoreticular, *see* Lymphoreticular malignancy
muscle fiber type II atrophy, 41/324
mydriasis, 75/529
myotonia, 41/347
neurofibromatosis type I, 50/369
neuromyopathy, 41/347
neuropathy, 7/554, 8/131, 41/347

oligodendroglioma, 16/22, 18/88
optic neuritis, 38/658-660
pellagra, 38/648
phthalazinol, 41/338
pinealoma, 42/766
polymyositis, 41/56, 82, 62/374, 71/131
progressive multifocal leukoencephalopathy, 38/654-657, 47/503
radiation injury, 67/331
skull base tumor, 68/466
spina bifida, 50/494
spinal epidural tumor, 19/63, 67
systemic, *see* Systemic malignancy
tertiary myelopathy, 38/675
vitamin B12 deficiency, 38/650
Wernicke-Korsakoff syndrome, 38/648
Malignant adenoma
pituitary adenoma, 17/383
Malignant angiosarcoma
spinal cord compression, 19/369
Malignant astrocytoma, 67/224
classification, 18/3
congenital spinal cord tumor, 32/355-386
criteria, 18/15-17
differentiation, 18/11
dissemination, 68/91
glioblastoma multiforme, 18/15
histology, 18/5
imaging, 67/169, 176
laboratory diagnosis, 68/97
metastasis, 68/92
morphology, 68/90
optic chiasm compression, 68/75
optic pathway tumor, 68/74
recurrence, 68/110
subarachnoid hemorrhage, 55/2
surgery, 68/100
Malignant atrophic papulosis, *see* Köhlmeier-Degos disease
Malignant cerebellar degeneration
subacute, *see* Subacute malignant cerebellar degeneration
Malignant epithelioma
xeroderma pigmentosum, 60/335
Malignant exophthalmos, *see* Thyroid associated ophthalmopathy
Malignant fibrous histiocytoma
incidence, 68/530
nervous system, 68/380
survey, 68/387
Malignant fibrous histiosarcoma
neurofibromatosis type I, 50/369
Malignant glioma, *see* Malignant astrocytoma
Malignant granuloma, *see* Midline granuloma
Malignant hemangioendothelioma
spinal cord compression, 19/369
Malignant hyperpyrexia, *see* Malignant hyperthermia
Malignant hypertension
arteriolar fibrinohyalinoid degeneration, 12/49
brain hemorrhage, 55/517
diagnostic criteria, 54/212
headache, 48/6
hypertensive encephalopathy, 9/613, 54/211
papilledema, 16/282
prehemorrhagic softening, 12/49
spastic paraplegia, 59/438
vasogenic brain edema, 11/155
Malignant hyperthermia
acidosis, 38/551, 553, 43/117, 63/439
anesthesia, 38/550, 43/117
arrhythmia, 38/551, 41/268, 62/567
ataxia, 63/439
Becker muscular dystrophy, 62/570
caffeine, 40/546, 41/269
calcium channel, 62/567, 569
calcium metabolism, 38/552, 43/118
central core myopathy, 41/3, 43/81, 62/333, 570, 63/440
childhood myoglobinuria, 62/567
chondrodystrophic myotonia, 62/570
chromosome 12q12-13.2, 62/333
chromosome 19q12-13.2, 62/568
clinical course, 62/569
clinical features, 38/550, 40/324, 339, 546
clinical picture, 71/597
cocaine intoxication, 65/252, 256
congenital myopathy, 38/551, 62/335, 63/440
contracture, 41/269
creatine kinase, 38/553, 40/339, 343, 43/117, 63/439
cyclopropane, 41/268
dantrolene, 38/554, 41/270, 298, 62/567
definition, 62/567
diagnosis, 71/597
Duchenne muscular dystrophy, 41/268, 62/570
epidemiology, 62/567
ether, 41/268
excitation contraction coupling, 40/546
experimental myopathy, 41/269
Fukuyama syndrome, 62/570
gallamine, 41/268
general anesthesia, 38/550, 62/567
genetics, 38/550, 41/429
halothane, 38/552, 41/268, 270, 43/118, 62/567
heart failure, 38/551

heat stroke, 38/550, 62/571, 573
histology, 40/54
history, 38/550
hyperkalemia, 38/551, 43/117
hyperpnea, 63/440
hypertension, 43/117
hyperventilation, 38/556, 43/117, 63/439
hypothalamus, 38/554
incomplete dystrophin deficiency, 62/135
inheritance, 41/268, 62/568
intravascular consumption coagulopathy, 38/551
King-Denborough syndrome, 62/570
lactic acidosis, 41/268
metabolic acidosis, 41/268, 62/567
methoxyflurane, 41/268
molecular genetics, 62/569
mortality, 38/550, 41/268, 63/440
muscle rigidity, 38/550, 41/268, 43/117, 63/439
myoglobinuria, 38/551, 40/339, 41/263, 268, 277, 62/567, 63/439
myopathy, 38/554, 43/117
myotonia congenita, 38/551, 62/570
neuroleptic malignant syndrome, 6/258, 62/571
neuromuscular disease, 41/429
pathogenesis, 38/551-554, 63/440, 71/600
pathology, 38/554
pathophysiology, 38/551-554, 62/571
periodic paralysis, 62/570
prevalence, 43/117
prevention, 71/599
procainamide, 37/452, 38/556, 41/270
procaine, 38/556, 41/270
prognosis, 38/551
ryanodine receptor, 62/567, 569
sarcoplasmic reticulum, 38/553, 41/269
sarcoplasmic reticulum adenosine triphosphate deficiency syndrome, 62/570
sex ratio, 38/550, 43/117
succinylcholine, 38/550, 552, 40/546, 41/268, 270, 62/568, 63/439
survey, 71/585, 595
susceptibility, 62/568
suxamethonium induced muscle rigidity, 62/569, 63/439
suxamethonium sensitivity, 43/117
symptom, 71/600
tachycardia, 38/551, 41/268, 43/117, 62/567, 63/439
thermoregulation, 75/73
toxic myopathy, 38/553, 62/601
toxic shock syndrome, 62/574
treatment, 38/555, 62/571, 71/599
D-tubocurarine, 41/268, 63/439

Malignant lymphogranuloma, *see* Hodgkin disease
Malignant lymphoma
see also Hodgkin disease
allergic granulomatous angiitis, 55/488, 490
anticancer agent, 69/275
B-lymphocyte, 18/235
brachial neuritis, 39/56
brain embolism, 55/489
brain hemorrhage, 55/487
brain infarction, 55/487, 489
brain lacunar infarction, 55/488
brain vasculitis, 55/488, 490
bulbar encephalitis, 39/54
Burkitt lymphoma, 18/236
carcinoma, 38/687
centroblastic, 18/235
centrocytic, 18/235
classification, 39/27, 69/262
CNS infection, 39/41-45
complete response, 69/267
cryoglobulinemic polyneuropathy, 51/434
Cryptococcus, 39/43
dermatomyositis, 39/56
encephalopathy, 69/274
endothelial proliferation, 55/488
Guillain-Barré syndrome, 39/56
herpes simplex virus encephalitis, 39/45
herpes zoster, 39/44
hyperviscosity syndrome, 55/487
immunoblastic, 18/236
intracranial metastasis, 39/29-32
intraspinal metastasis, 39/32-38
intravascular consumption coagulopathy, 55/488
leptomeningeal metastasis, 39/36-38
limbic encephalitis, 39/54
lymphangiography, 39/49
lymphoblastic, 18/236
lymphocytic, 18/235
lymphomatoid granulomatosis, 51/451
lymphoplasmacytoid, 18/234
malnutrition, 39/38-40
metabolic disorder, 39/38-40
metastasis, 39/28-38
methotrexate, 69/273
multiple brain infarction, 55/488
myasthenic syndrome, 39/53, 56
neuromyopathy, 39/56, 69/278
nonbacterial thrombotic endocarditis, 55/489
nonmetastatic complication, 39/38-48
optic neuritis, 39/54
paraneoplastic syndrome, 39/53-56
peripheral nerve, 39/38
plexopathy, 69/278

plexus, 39/38
polymyositis, 39/56
progressive multifocal leukoencephalopathy, 39/45
radiation injury, 39/50-53
radiotherapy, 39/50-53
relapse, 69/267
septic embolus, 55/489
splenectomy, 39/50
subacute cerebellar degeneration, 39/54
superior longitudinal sinus thrombosis, 55/489
survey, 69/262
T-lymphocyte, 18/325
toxic disorder, 39/38-40
toxoplasmosis, 39/43
treatment, 69/269
varicella zoster virus encephalitis, 56/235
vascular disorder, 39/45-48
venous sinus thrombosis, 55/489
Malignant lymphomatosis
intravascular, *see* Intravascular malignant lymphomatosis
Malignant melanoma
cell origin, 14/599
metastatic cardiac tumor, 63/93
Parkinson disease, 49/126
scalp tumor, 17/132
treatment, 14/601
type, 14/595
Malignant mesenchymoma, *see* Midline granuloma
Malignant myopia
choroid atrophy, 13/32
Malignant neurinoma, 68/555
features, 14/30
incidence, 14/138
neurofibromatosis, 14/30
neurofibromatosis type I, 8/450, 14/30
Malignant neurofibroma
histopathology, 14/30
incidence, 14/138
Malignant neuroleptic syndrome
amoxapine intoxication, 65/319
heterocyclic antidepressant, 65/319
Malignant peripheral nerve sheath tumor
neurinoma, 68/562
Malignant phakoma
phakoblastoma, 14/201, 562
Malignant purpuric fever, *see* Meningococcal meningitis
Malignant radiculopathy
dysesthesia, 69/73
hyperalgesia, 69/73
paresthesia, 69/73
survey, 69/71
Malignant reticulosis, *see* Midline granuloma
Malignant schwannoma, *see* Malignant neurinoma
Malignant teratoma
sacrococcygeal, 20/112
Malingering
meningeal sign, 1/546
Mallory body myopathy
cardiorespiratory failure, 62/345
congenital myopathy, 62/332
facial weakness, 62/334
Malnutrition
see also Beriberi
acetylcholine, 29/12
acoustic nerve, 28/20
adrenergic neuron, 29/5
alcoholic amblyopia, 28/331
alcoholic polyneuropathy, 51/316
amblyopia, 28/19-22
amino acid, 29/9
ataxia, 28/22
ataxic neuropathy, 28/21-26
body mass, 29/2
brain development, 29/1-13
brain weight, 29/2
burning feet syndrome, 28/9-14
cell proliferation, 29/3
cholinergic system, 29/5
clinical features, 28/23
CNS, 28/1-37
cochlear nerve, 28/21
cranial nerve, 28/20
deafness, 28/20
deficiency neuropathy, 51/321
developing nervous system, 29/1-13
energy metabolism, 29/6
ganglioside, 9/6, 29/6
Hodgkin disease, 39/38-40
late cerebellar atrophy, 60/584
late cortical cerebellar atrophy, 21/477
lathyrism, 28/37
malignant lymphoma, 39/38-40
Marchiafava-Bignami disease, 9/653, 28/324, 46/337, 47/557
microcephaly, 30/519, 50/280
myelination, 29/6-8
neuromotor behavior, 29/11
neuromotor development, 29/11
neuronal process, 29/4-6
neuronal synapse, 29/4-6
neuropathy, 7/633, 28/2-9
ophthalmoplegia, 28/20
pathology, 28/23

pellagra, 28/18, 64, 46/336, 399
protein energy, *see* Protein energy malnutrition
protein metabolism, 29/9
retrobulbar neuropathy, 28/19, 24
ribonucleic acid metabolism, 29/10
serotonin, 29/5
serotoninergic system, 29/5
spastic paraplegia, 28/31-37
spinal ataxia, 28/22
spinal cord syndrome, 28/21
subacute combined spinal cord degeneration, 28/21
syndrome, 2/449
treatment, 28/18
tropical ataxic neuropathy, 7/640, 28/26, 30
tuberculous meningitis, 33/230
vitamin B1 deficiency, 28/2-9, 15-17
vitamin B2 deficiency, 51/332
vitamin B3 deficiency, 28/18
vitamin B12 deficiency, 28/22-24
vitamin Bc deficiency, 28/24
Wernicke encephalopathy, 28/14-19
Malocclusion syndrome
differential diagnosis, 22/112
9p partial monosomy, 50/579
Malonic aciduria, *see* Malonyl-coenzyme A decarboxylase deficiency
Malonyl-coenzyme A decarboxylase deficiency
fatty acid biosynthesis, 66/656
organic acid metabolism, 66/639
Malphigi, M., 2/10
Malta fever, *see* Brucellosis
Mamba intoxication
venom, 37/13, 15
Mamillary body
acetonylacetone intoxication, 64/89
anatomy, 2/436
brain infarction, 55/140
memory disorder, 3/278
Mamillary nucleus
medial, *see* Medial mamillary nucleus
Mamillotegmental tract
anatomy, 2/436
Mamillothalamic tract
Vicq d'Azyr, 2/438
Mammalia intoxication
ingestion, 37/94-97
Ornithorhynchus anatinus, 37/94
vitamin A intoxication, 37/94-97
Mammary carcinoma
late cerebellar atrophy, 60/585
meningioma, 42/744
metastatic cardiac tumor, 63/93
SBLA syndrome, 42/769
spinal tumor, 20/43
vertebral column tumor, 20/43
vertebral metastasis, 20/424
Man in the barrel syndrome, *see* Brachial diplegia
Manchester, United Kingdom
neurology, 1/7
Mandible fracture
cervicomedullary injury, 24/172
Mandibular cyst
multiple nevoid basal cell carcinoma syndrome, 14/79, 114, 456, 466
Mandibular hypoplasia
Pena-Shokeir syndrome type I, 43/438
trigeminal nerve agenesis, 50/214
Mandibular intermittent claudication
temporal arteritis, 48/310
Mandibular joint, *see* Temporomandibular joint
Mandibulofacial dysostosis, *see* Treacher-Collins syndrome
Maneuver
Adson-Coffey, *see* Adson-Coffey maneuver
costoclavicular, *see* Costoclavicular maneuver
head up tilt, *see* Head up tilt maneuver
hyperabduction, *see* Hyperabduction maneuver
Jendrassik, *see* Jendrassik maneuver
Valsalva, *see* Valsalva maneuver
Manganese
basal ganglion, 49/36
biological function, 64/305
CNS, 28/439
content, 28/439
element data, 64/303
Guam amyotrophic lateral sclerosis, 59/264, 275
Kii Peninsula amyotrophic lateral sclerosis, 22/359
neurotoxin, 6/652, 823, 21/497, 36/218-225, 46/392, 49/108, 466, 477, 60/586, 64/6, 303
pseudocalcium, 6/716
source, 64/303
speciation, 64/303
striatopallidodentate calcification, 49/425
toxic encephalopathy, 64/6
Manganese deficiency
ataxia, 64/305
epilepsy, 64/305
free radical, 64/305
manganese intoxication, 64/305
Manganese encephalopathy
pallidal degeneration, 49/477
pallidal lesion, 6/662-664
striatal degeneration, 49/477
Manganese intoxication

alloy manufacturing, 64/313
amyotrophic lateral sclerosis, 59/200, 64/307
anterior horn cell degeneration, 64/307
antiseptic, 64/304
ataxia, 64/305
basal ganglion, 64/307
biological function, 64/305
biological marker, 64/308
blood-brain barrier, 64/304
body burden, 64/304
bradykinesia, 64/306, 308
brain manganese level, 64/305
carbamate intoxication, 64/190
chelation therapy, 64/316
chorea, 49/556, 64/308
chronic, *see* Manganese encephalopathy
clinical features, 36/218-220, 64/308
CNS topography, 64/306
compulsive laughing, 64/308
dementia, 46/392
demyelination, 9/643
differential diagnosis, 49/478, 64/316
dopamine, 64/306
dysarthria, 64/308
dystonia, 64/308
edetate calcium disodium, 64/316
edetic acid, 36/231
elimination, 64/304
emotion, 64/308
epidemiology, 36/217, 232, 64/312
epilepsy, 64/305
experimental study, 49/478
exposure, 36/217, 232
fecal analysis, 64/309
fertilizer, 64/304
food, 64/303
free radical, 64/305
fungicide, 64/304
germicide, 64/304
globus pallidus, 64/307
glutamine synthetase, 64/305
hair, 64/309
half-life, 64/304
hallucination, 64/308
history, 36/217
hyperactivity, 64/306
hypophonia, 64/310
incidence, 36/218
intention tremor, 64/306
intestinal absorption, 64/304
iron deficiency, 64/304
laboratory data, 36/220
late cerebellar atrophy, 60/586
late cortical cerebellar atrophy, 21/497
leukoencephalopathy, 9/643
levodopa, 36/225-231, 64/316
limbic pathway, 64/307
liver toxicity, 37/623
low level exposure, 64/313
manganese deficiency, 64/305
manganese exposure, 64/312
manganese fume, 64/304
mask like face, 64/310
melanin, 64/305
metabolism, 36/222-224
metalloprotein, 64/305
miner, 64/303, 312
Myerson sign, 64/317
neuropathology, 49/477, 64/306
neuropsychology, 64/310, 313
neurotoxicity mechanism, 64/306
nuclear magnetic resonance, 64/309
ore crusher, 64/312
oxitriptan, 36/229
pallidal necrosis, 6/652, 49/466
pallidoluysionigral degeneration, 49/460
Parkinson disease, 6/241, 36/231, 49/94
parkinsonism, 49/108, 64/307
pathogenesis, 64/308
pathology, 36/224
plasma concentration, 64/304
plasma level, 64/305
polution value, 64/304
population study, 64/312
positron emission tomography, 64/310
renal toxicity, 37/623
rigidity, 64/310
serotonin, 64/306
smelting, 64/313
source, 36/217
subthalamic nucleus, 64/307
susceptibility, 36/232, 64/308
symptom, 64/6
treatment, 36/225-231, 64/315
tremor, 6/823, 64/308
urinary concentration, 64/309
water source, 64/315
welding, 64/304, 314
Mania, 46/471-477
Amanita pantherina intoxication, 36/534
beta adrenergic receptor blocking agent intoxication, 65/436
biochemistry, 46/475
brain tumor, 46/430
catatonia, 46/423
classification, 46/444-446

corticosteroid, 46/462, 599
dementia, 46/123
dexamethasone suppression test, 46/450
emotion, 3/338
endocrinology, 46/473-475
epilepsy, 15/596
general paresis, 46/429
genetics, 46/471
glyceryl trinitrate intoxication, 37/454
hyperthyroidism, 46/450
isoniazid intoxication, 65/480
lateralized dysfunction, 46/489
levodopa, 46/462, 593
lithium, 46/450
lithium intoxication, 64/293
MAO inhibitor, 46/462
migraine, 48/162
neurochemistry, 46/449
neuroleptic agent, 46/450
neurologic soft sign, 46/490
organic brain syndrome, 46/476
organic mental disorder, 46/476
physiology, 46/472
pseudodementia, 46/223
psychobiology, 46/471-476
psychopharmacology, 46/449
puerperal psychosis, 46/450
stroke, 46/429
subarachnoid hemorrhage, 46/429
temporal lobe tumor, 46/429
Mania fugax
migraine sine hemicrania, 48/165
Mania transitoria
migraine sine hemicrania, 48/165
Manic depression
hypercalcemia, 63/561
pellagra, 46/336
Manic depressive psychosis
see also Affective psychosis *and* Depression
catecholamine, 43/209
cyclothymia, 43/209
delusion, 43/208
dexamethasone suppression test, 43/208
EEG, 43/208
epilepsy, 15/600
hallucination, 43/208
HLA antigen, 43/209
potassium imbalance, 28/483
prevalence, 43/208
REM sleep, 43/208
semantics, 46/444
sex ratio, 43/208
sleep disorder, 3/93, 43/208
suicide, 43/208
Manihot esculenta
cyanogenic glycoside intoxication, 36/515, 65/27
shrub characteristic, 65/27
Manipulation
cervical vertebral column injury, 25/3
chiropractic, *see* Chiropractic manipulation
spinal, *see* Spine manipulation
Mannitol
brain edema, 16/198, 200, 202, 23/154
brain infarction, 53/427
chronic subdural hematoma, 24/313
CSF, 57/214
head injury, 57/227, 232
intracranial hypertension, 24/616
intracranial pressure, 23/202, 212, 57/227, 232
myoglobinuria, 62/556
neurologic intensive care, 55/213
Reye syndrome, 29/341
Mannitol-fructose
brain edema, 16/202
Mannose oligosaccharide
mannosidosis, 66/63
α-Mannosidase
chemistry, 27/157
degradation, 66/61
enzyme deficiency polyneuropathy, 51/370
genetics, 66/332
glycosaminoglycanosis type IH, 10/430
GM1 gangliosidosis, 10/477
mucolipidosis type II, 42/448
mucolipidosis type III, 42/449
β-Mannosidase
degradation, 66/61
Mannosidase deficiency
bone marrow transplantation, 66/89
α-Mannosidase deficiency
mannosidosis, 10/430, 42/597
Mannosidosis
alpha, *see* Alpha mannosidosis
beta, *see* Beta mannosidosis
biochemistry, 27/157
corneal opacity, 42/597
differential diagnosis, 66/336
enzyme deficiency polyneuropathy, 51/370
epilepsy, 72/222
familial spastic paraplegia, 59/308
foam cell, 42/597
glycosaminoglycanosis type I, 42/597
hepatosplenomegaly, 27/157, 42/597
heterozygote detection, 42/598
mannose oligosaccharide, 66/63
α-mannosidase deficiency, 10/430, 42/597

mental deficiency, 27/157, 42/597
muscle fiber feature, 62/17
prenatal diagnosis, 27/157, 42/598
seizure, 42/597
spastic paraplegia, 59/439
survey, 66/329
vacuolated lymphocyte, 42/597
Mannosyl oligosaccharide, *see* Mannose oligosaccharide
Manometry
anal sphincter, 74/635
esophageal, *see* Esophageal manometry
syringomyelia, 19/92
Manta ray intoxication
elasmobranche, 37/70
Mantakassa, *see* Tropical spastic paraplegia
Manual dexterity
see also Cerebral dominance *and* Handedness
handedness, 4/449, 466
Manual expression
bladder, 26/409
urinary tract, 26/409
Manual reduction, *see* Spine manipulation
Manus cava, *see* Claw hand
MAO
basal ganglion, 6/120
biochemistry, 6/120
depression, 46/448
headache, 48/93, 437
Levine experimental anoxia ischemia, 9/583
migraine, 48/86, 93
noradrenalin, 74/145
platelet, *see* Platelet MAO
schizophrenia, 43/214
sex chromosomes, 48/94
uremic encephalopathy, 63/514
uremic polyneuropathy, 63/514
MAO A
nigrostriatal dopamine system, 49/47
noradrenalin, 74/143
MAO B
Alzheimer disease, 46/266
mesolimbic dopamine system, 49/47
1-methyl-4-phenyl-1,2,3,6-tetrahydropyridine intoxication, 65/381
nigrostriatal dopamine system, 49/53
MAO inhibitor
behavior disorder, 46/597
cheese reaction, 37/320
drug induced extrapyramidal syndrome, 6/262
dyssynergia cerebellaris myoclonica, 60/603
extrapyramidal disorder, 6/262
headache, 48/6, 420
hypotension, 74/145
mania, 46/462
migraine, 5/41, 48/197
neuropathy, 7/519
orthostatic hypotension, 63/152
pharmacology, 74/145
MAO inhibitor intoxication
acute, 37/319
autonomic nervous system, 37/320
CNS, 37/320
liver toxicity, 37/321
Maple syrup urine disease, 29/53-80
acid base metabolism, 28/517
acute metabolic decompensation, 29/79
aminoaciduria, 10/138
benign intracranial hypertension, 67/111
birth incidence, 42/599
brain atrophy, 42/599
branched chain amino acid level, 29/62
branched chain amino acid metabolism, 29/53-57, 42/598
branched chain hydroxy acid level, 29/63
branched chain oxoacid decarboxylase, 42/598
branched chain oxoacid level, 29/63
branched chain oxoacid metabolism, 29/53-57
classic, 29/53, 59, 67, 70-74, 76-79
clinical features, 29/59-62
detection, 29/66, 69
diagnosis, 28/517, 29/66-69
diet, 29/76-79, 42/599
dihydrolipoamide dehydrogenase deficiency, 66/651
2,4-dinitrophenylhydrazine test, 42/599
EEG, 29/64
epilepsy, 72/222
genetics, 29/70
heterozygote, 29/69
heterozygote detection, 29/69
hypervalinemia, 29/55
hypoglycemia, 28/517, 29/64, 42/599
intermittent, 29/53, 60, 68, 80
isovaleric acidemia, 29/55
laboratory finding, 29/62-65
leucine metabolism, 66/651
metabolic error, 29/57-59, 42/598
mild, 29/53, 61
myelin, 10/138
myelin dysgenesis, 42/599
myelination, 29/71
ophthalmoplegia, 42/599
2-oxoisocaproic acid, 42/598
α-oxoisovaleric acid, 42/598
α-oxo-β-methylvaleric acid, 42/598

pathogenesis, 29/74-76
pathology, 29/70-74
prenatal diagnosis, 29/70, 42/599
screening, 29/66-68, 71
treatment, 29/76-81
vitamin B1 response, 29/53, 61, 80
white matter change, 10/138
Maprotiline
heterocyclic antidepressant, 65/312
neurotoxin, 65/311, 319
Maprotiline intoxication
epilepsy, 65/319
myoclonus, 65/319
Marantic endocarditis
brain embolism, 53/156
brain infarction, 53/156
brain ischemia, 53/156
meningitis, 52/297
Marasmus
cavernous sinus thrombosis, 52/173
malnutrition syndrome, 2/449
mental deficiency, 46/5, 18
pellagra, 7/574
progressive myoclonus epilepsy, 15/334
protein energy malnutrition, 51/322
Marathon runner
acquired myoglobinuria, 62/573
Marble bone disease, *see* Albers-Schönberg disease
Marble skin, *see* Cutis marmorata
Marble state, *see* Status marmoratus
Marburg, O., 1/10, 27
Marchesani-Wirz syndrome, *see* Grönblad-Strandberg syndrome
Marchi staining
wallerian degeneration, 9/35
Marchi, V., 1/10
Marchiafava-Bignami disease, 28/317-328
akinetic mutism, 45/33
alcoholic dementia, 17/542, 28/317, 46/337, 47/557
alcoholism, 9/653, 28/317, 46/337, 47/557, 70/353
anoxic ischemic leukoencephalopathy, 47/529
associated condition, 28/325
astasia abasia, 17/542, 28/324
axis cylinder, 9/653
brain white matter, 9/653, 28/321
central pontine dystrophy, 9/638
central pontine myelinolysis, 28/293, 308, 325
cerebellar peduncle, 9/653
classification, 9/472, 10/15
clinical features, 28/324, 47/557
commissura anterior, 9/653, 46/337
complex IV, 9/653, 47/558
corpus callosum, 2/763, 28/317, 46/337, 47/558
corpus callosum tumor, 17/542
cortical laminar sclerosis, 28/317, 325, 342, 46/338, 47/558
course, 28/325
dementia, 28/324, 46/399
demyelination, 9/653, 46/399, 47/558
diagnosis, 28/325
dysarthria, 17/542, 28/317, 324, 46/337
etiology, 28/318
frontal gyrus, 9/653
history, 28/317
laminar necrosis, 28/317
leukoencephalopathy, 9/653, 47/557
malnutrition, 9/653, 28/324, 46/337, 47/557
neuropathology, 9/653, 46/347, 47/558
nutritional deficiency, 17/542, 47/557
oligodendrocyte, 47/557
oligodendroglial lysis, 9/653, 47/558
pallidal necrosis, 49/479
pathogenesis, 28/326-328, 47/558
pathology, 28/318-324, 46/337
porphyrin, 9/653
stupor, 17/542, 28/324
treatment, 28/328
Marchiafava, E., 1/10
Marcus Gunn phenomenon, 42/319-321
see also Abducent nerve abnormality, Oculomotor nerve abnormality *and* Trochlear nerve abnormality
abducent nerve abnormality, 30/403
clinical features, 30/401
cranial nerve, 30/100
definition, 2/62
familial incidence, 30/403, 42/320
features, 50/212
inverted type, *see* Marin Amat syndrome
oculomotor nerve abnormality, 30/401
oculomotor nerve agenesis, 50/212
pathogenesis, 1/409, 30/403
ptosis, 2/326, 42/319
reversal, 2/63
seizure, 30/403
Stilling-Türk-Duane syndrome, 30/403, 50/213
trigemino-oculomotor synkinesis, 42/320
trochlear nerve abnormality, 30/401
type, 30/401
Wolf-Hirschhorn syndrome, 31/552
Marden-Walker syndrome
blepharophimosis, 43/424
cheilopalatoschisis, 43/424
contracture, 43/424
growth retardation, 43/423

heart disease, 43/424
kyphoscoliosis, 43/424
microcephaly, 43/424
micrognathia, 43/424
pectus excavatum, 43/457
renal abnormality, 43/424
short stature, 43/423
Marek disease
acute form, 56/582
avian herpes virus, 47/609, 51/250
chronic type, 56/582
demyelinating polyneuropathy, 51/250, 56/582
demyelination, 47/609, 56/582
Guillain-Barré syndrome, 47/609, 51/250, 56/190, 581
nerve tumor, 56/582
neurolymphomatosis, 56/188
ocular form, 56/189, 582
peripheral neurolymphomatosis, 56/179, 581
polyneuritis, 56/180
transient palsy, 56/582
Marek disease virus
Epstein-Barr virus, 56/249
gammaherpes virus, 56/8
herpes virus, 56/189, 581
lymphoma, 56/581
Marfan like syndrome
sensorineural deafness, 42/381
Marfan syndrome
anterior sacral meningocele, 32/206
aorta aneurysm, 39/389, 63/48
arachnoid cyst, 31/103
autosomal dominant inheritance, 39/393
biochemistry, 39/391
carotid dissecting aneurysm, 39/390, 54/282
carotid system syndrome, 53/309
cerebrovascular disease, 53/33, 43
chorioretinal degeneration, 13/41
concentric laminated body, 40/109
diagnosis, 39/391
differential diagnosis, 39/392
dissecting aorta aneurysm, 39/389, 63/45
ectopia lentis, 39/389
hereditary connective tissue disease, 39/389-393
infantile hypotonia, 40/336
microcoria, 42/403
moyamoya disease, 42/751
myelomalacia, 55/98
neurofibromatosis type I, 32/206
optic atrophy, 13/41
pathology, 39/392
scoliosis, 50/417
subarachnoid hemorrhage, 39/390, 55/29
symptom, 46/51
treatment, 39/393
vertebral dissecting aneurysm, 54/282
Marfan-Weve syndrome, *see* Marfan syndrome
Marginal artery
see also Superior cerebellar artery
radioanatomy, 11/93
topography, 11/93
Marginal gliosis
Huntington chorea, 6/334
Marginal layer
nociceptor, 45/230
nociceptor afferent, 45/230
Marginal sinus
anatomy, 11/57, 61
radioanatomy, 11/112
topography, 11/112
Marginal zone siderosis
subarachnoid hemorrhage, 12/72
Margolis-Ziprkowski syndrome
reproductive fitness, 43/45
sensorineural deafness, 43/44
Margulis vaccine
multiple sclerosis, 9/398
Marie-Charcot-Tooth disease, *see* Hereditary motor and sensory neuropathy type I *and* Hereditary motor and sensory neuropathy type II
Marie-Foix-Alajouanine disease
autosomal dominant, 21/18
hereditary, 21/416
hereditary cerebello-olivary atrophy (Holmes), 21/411, 413
late onset, 21/21
main features, 21/18, 21
olivopontocerebellar atrophy (Dejerine-Thomas), 21/416
pontine infarction, 2/240, 242, 244, 264
sporadic type, 21/416
Marie-Foix classification
synkinesis, 1/408
Marie hereditary cerebellar ataxia, *see* Hereditary cerebellar ataxia (Marie)
Marie heredoataxia, *see* Hereditary cerebellar ataxia (Marie)
Marie-Nonne hereditary ataxia, *see* Hereditary cerebellar ataxia (Marie)
Marie, P., 1/7
Marie-Strümpell disease
back pain, 19/86
plain X-ray, 19/86
rheumatoid arthritis, 19/86
Marihuana, *see* Cannabis
Marihuana intoxication, *see* Cannabis intoxication

Marin Amat syndrome
  definition, 2/327
  EMG, 42/320
  inverse Marcus Gunn symptom, 2/63
Marine toxin, 37/27-98
  algae, 37/97
  sebastapistes mcadamsi intoxication, 37/73
  semaeostomeae intoxication, 37/50-53
  toxic encephalopathy, 65/141
Marine toxin intoxication, 37/27-98, 65/141-168
  see also Amnesic shellfish poisoning, Ciguatoxin intoxication, Diarrheic shellfish poisoning, Neurotoxic shellfish poisoning, Puffer fish intoxication *and* Tetrodotoxin intoxication
  acute gastroenteritis, 65/158
  amnesia, 65/142
  amnesic shellfish poisoning, 65/143
  anuria, 65/153
  ataxia, 37/33, 65/158
  bioconcentration, 65/141
  blue green algae, 65/141
  brevetoxin, 65/147
  brevetoxin A, 65/158
  brevetoxin B, 65/158
  cigua, 65/159
  ciguatoxin, 65/147, 159
  clinical presentation, 65/158
  diagnosis, 65/142, 151, 153, 158
  diarrheic shellfish poisoning, 37/33, 65/143, 148
  diatom, 65/141
  dinoflagellates, 65/141
  domoic acid, 65/150
  dysarthria, 65/153
  EMG, 65/153
  enzyme action, 65/144
  enzyme inhibition, 65/148
  epidemiology, 65/143, 151
  fatality rate, 65/142
  headache, 37/33, 65/153
  ion transport, 65/144
  lactic acidosis, 65/153
  ligand, 65/144
  Livona pica, 65/159
  maitotoxin, 65/159, 161
  management, 65/151, 154
  marine food web, 65/141
  molecular toxicology, 65/143
  mortality, 37/33, 65/153
  myosin light chain phosphate, 65/149
  nausea, 37/33, 65/153
  nerve conduction, 65/161
  neurologic symptom, 65/158
  neurotoxic shellfish poisoning, 65/143
  neurotransmitter stimulating agent, 65/150
  okadaic acid, 65/149
  orphan receptor, 65/144
  orphan receptor site 1, 65/146
  orphan receptor site 5, 65/147
  palytoxin, 65/161
  paralytic shellfish poisoning, 37/33, 65/143, 151
  paresthesia, 65/142, 153, 158
  pathomechanism, 65/142, 158
  phosphorylation, 65/150
  prevention, 65/142
  protein phosphatase 1, 65/150
  protein phosphatase 2a, 65/150
  protein synthesis, 65/144
  Ptychodiscus brevis toxin type I, 65/148
  Ptychodiscus brevis toxin type II, 65/147
  Ptychodiscus brevis toxin type I-X, 65/158
  puffer fish, 65/141
  receptor site interaction, 65/143
  red tide toxin, 65/142
  respiratory failure, 37/33, 65/153
  saxitoxin, 37/33, 65/146, 152
  scaritoxin, 65/159
  sodium channel, 65/146
  sodium channel activation, 65/158
  susceptibility, 65/143
  symptom, 65/153
  temperature reversal, 65/158
  tetrodotoxin, 37/32, 65/146, 152, 155
  tetrodotoxin intoxication, 65/143
  transvector, 65/141
  treatment, 65/142, 154
  vertigo, 65/153
  voltage sensitive sodium channel, 65/145, 149
  vomiting, 37/33, 65/153
Marinesco body
  Kii Peninsula amyotrophic lateral sclerosis, 22/414
Marinesco, G., 1/11
Marinesco-Sjögren-Garland disease, *see* Marinesco-Sjögren syndrome
Marinesco-Sjögren syndrome, 21/555-560, 60/341-344
  alopecia, 60/667
  amyotrophy, 21/556
  anterior horn syndrome, 42/184
  ataxia, 42/184, 60/342
  atypical form, 60/344
  autosomal recessive, 21/555
  Babinski sign, 42/184
  bone dysplasia, 60/669
  carbohydrate metabolism, 42/185
  cataract, 21/555, 42/184, 60/342, 652

cerebellar ataxia, 21/556
cerebellar atrophy, 42/184
cerebellar heterotopia, 21/560
clinical features, 21/555
computer assisted tomography, 60/671
consanguinity, 21/21
course, 21/557
cryptorchidism, 21/556
dementia, 21/556
dwarfism, 60/666
dysarthria, 42/184
ECG abnormality, 60/670
EMG, 21/556
endocrine abnormality, 60/343
epilepsy, 15/329, 42/184
genetics, 21/557
growth retardation, 42/184
hereditary congenital cerebellar atrophy, 60/288
histology, 60/343
hyperreflexia, 42/184
hypogonadism, 42/184, 60/343, 668
infantile, 60/342
inferior olive, 21/560
inheritance, 60/341
juvenile, 60/343
kyphoscoliosis, 42/184, 60/669
laboratory data, 21/556
main features, 21/19
mental deficiency, 42/184-186, 46/63, 60/342, 664
muscular atrophy, 42/184, 60/342
myoclonus, 60/344
myopathy, 21/556, 60/344, 660
neuropathology, 21/557, 60/344
neuropathy, 60/660
nosology, 21/560
nuclear magnetic resonance, 60/673
nucleodegenerative myopathy, 62/351
nystagmus, 21/556, 42/184
optic atrophy, 60/654
pes cavus, 21/556
polyneuropathy, 60/343
positron emission tomography, 60/673
ptosis, 42/184
pyramidal syndrome, 21/556
reflex, 21/556
secondary pigmentary retinal degeneration, 13/320, 21/560, 60/654
seizure, 60/343
sensory neuropathy, 51/398
serum creatine kinase, 60/674
skeletal abnormality, 21/556, 60/343
sleep disorder, 60/344
speech disorder in children, 42/184
spinocerebellar ataxia, 42/184-186
strabismus, 21/556, 42/184
sural nerve, 60/343
vertical gaze palsy, 60/343, 656
Marker
CSF, *see* CSF marker
DNA, *see* DNA marker
muscle fiber type, 62/3
oligodendroglioma, 68/125
Markschattenherde
multiple sclerosis, 47/246
Markscheidenlichtungsherde
multiple sclerosis, 47/246
Maroteaux-Lamy polydystrophy, *see* Glycosaminoglycanosis type VI
Maroteaux-Lamy syndrome, *see* Glycosaminoglycanosis type VI
Marriage
multiple sclerosis, 9/418
sexual function, 26/452
Marrow
bone, *see* Bone marrow
Marshall syndrome
cataract, 42/393
intrauterine growth acceleration, 46/82
mental deficiency, 46/82
myopia, 42/393
Martorell-Fabré syndrome, *see* Takayasu disease
MASA syndrome
adducted thumb syndrome, 43/255
gait, 43/255
lordosis, 43/255
mental deficiency, 43/255
Mask like face
hereditary progressive diurnal fluctuating dystonia, 49/530
manganese intoxication, 64/310
neuroleptic parkinsonism, 65/292
parkinsonism, 6/186
Rett syndrome, 29/311
Mass reflex
central motor disorder, 1/173
description, 1/173
myalgia, 1/139
sexual disturbance, 2/196
spinal cord injury, 61/447
spinal cord lesion, 2/196
tetraplegia, 61/447
transverse spinal cord lesion, 61/447
Massa intermedia
spina bifida, 50/487
Masseter muscle
hypertrophia musculorum vera, 43/84

hypogonadism, 43/138
myotubular myopathy, 43/113
ophthalmoplegia, 43/138
progressive myositis ossificans, 43/191
trigeminal nerve agenesis, 50/214
Masseter muscle reflex
amyotrophic lateral sclerosis, 22/293
lesion localization, 2/57
pseudobulbar paralysis, 2/780
Massive myoclonic jerk
terminology, 15/219
Massive myoclonus
bilateral, *see* Bilateral massive myoclonus
unilateral epileptic seizure, 15/241
Massive spasm
terminology, 15/219
Mast cell
cluster headache, 48/232
Mast syndrome
bradykinesia, 42/282
bulbar paralysis, 22/142
dementia, 22/142, 42/282
dysarthria, 42/282
emotional disorder, 42/282
hyperreflexia, 42/282
hypertonia, 42/282
hypokinesia, 42/282
intention tremor, 42/282
mental deterioration, 42/282
slurred speech, 42/282
Mastadenovirus
41 species, 56/282
Mastication
auriculotemporal syndrome, 2/103
dental headache, 48/392
epileptic, *see* Epileptic mastication
headache, 48/392
pain, 48/392
Pick disease, 46/235
Mastoid emissary foramen
anatomy, 11/55
Mastoid tip
pain, 48/503
pressure, 48/503
Mastoiditis
actinomycosis, 35/386
benign intracranial hypertension, 16/153
brain abscess, 52/144
cranial epidural empyema, 52/170
cranial subdural empyema, 52/170
Gradenigo syndrome, 2/313
gram-negative bacillary meningitis, 52/103
headache, 5/210
Masturbation
sexual function, 26/448, 457
Matas test
carotid angiography, 12/208
carotid artery ligation, 12/219
Maternal age
brain tumor, 16/73
chromosomal aberration, 30/102
Down syndrome, 30/161, 31/373, 379-382, 43/545, 50/523, 525
Klinefelter syndrome, 31/478
Patau syndrome, 31/506, 43/527, 50/558
XXX syndrome, 43/553
Maternal birth injury
lumbosacral plexus, 51/163
Maternal diabetes mellitus
caudal aplasia, 50/512
microcephaly, 50/280
spina bifida, 30/126
teratogenic agent, 30/126
Maternal drug intoxication
microcephaly, 30/125
Maternal hyperphenylalaninemia
microcephaly, 50/280
Maternal hyperthyroidism
teratogenic agent, 30/126
Maternal infection
acoustic nerve agenesis, 50/219
cytomegalic inclusion body disease, 56/265
Maternal obstetrical paralysis
lumbar disc, 20/595
lumbosacral plexus, 2/153, 20/595
Maternal phenylketonuria
embryo, 31/299
microcephaly, 30/508, 511, 50/280
Mathews palate sign, *see* Palate sign (Mathews)
Matrix cell
neural tube, 50/5
Matrix metalloproteinase
metastasis, 71/644
Matthews-Rundle syndrome
areflexia, 60/660
ataxia, 60/578, 652
clinical features, 60/575-578
computer assisted tomography, 60/577, 671
eunuchoidism, 60/575
history, 60/575
hypergonadotropic group, 60/576
hypogonadism, 60/578, 652, 668
hypogonadotrophic group, 60/575
mental deficiency, 60/664
neuropathology, 60/578
normal gonadotropin group, 60/577

secondary pigmentary retinal degeneration, 60/654
tremor, 60/661
uncertain gonadotropin group, 60/577
Maturation
brain, *see* Brain maturation
Plasmodium falciparum, 52/366
Plasmodium malariae, 52/366
Plasmodium ovale, 52/366
Plasmodium vivax, 52/366
Maturity
developmental dyslexia, 46/120
learning disability, 46/124
Max Planck Gesellschaft
history of neurology, 1/26, 37
Maxillary antrum
headache, 48/393
Maxillary artery
external, *see* External maxillary artery
internal, *see* Internal maxillary artery
sphenopalatine neuralgia, 5/328
Maxillary fracture
craniofacial injury, 23/375-377
fixation, 23/375-377
treatment, 23/375-377
Maxillary sinus
headache, 5/213
submento-occipital projection, 32/10
Maxillary sinusitis
headache, 5/207, 213, 218
Maxillary vein
primitive, *see* Primitive maxillary vein
Maxillofacial dysostosis
antimongoloid eyes, 43/425
birth incidence, 43/426
speech disorder, 43/425
speech disorder in children, 43/425
tongue hyperplasia, 43/425
May-White syndrome
ataxia, 22/511
cerebellar ataxia, 42/698
hereditary hearing loss, 22/510, 42/698
myoclonic jerk, 22/511, 42/698
progressive myoclonus epilepsy, 22/512, 42/698
Mayaro virus
alphavirus, 56/12
Mayer sign
diagnostic value, 1/245
MBK, *see* 2-Hexanone
McArdle disease, *see* Myophosphorylase deficiency
McCormick-Lemmi syndrome, *see* Familial pallidoluysionigral degeneration *and* Pallidonigral degeneration
McCune-Albright syndrome, *see* Albright syndrome
McGregor line
basilar impression, 32/17-19, 50/398
definition, 61/7
McLeod phenotype
acanthocyte, 49/330, 62/127
acanthocytosis, 49/332, 63/271, 288
bilirubin, 63/287
carbonate dehydratase, 63/287
cardiomyopathy, 49/332, 63/287
characteristics, 62/127
chorea-acanthocytosis, 49/330, 63/284, 287
chronic granulomatosis, 49/332
chronic granulomatous disease, 63/289
creatine kinase, 49/332, 63/287
Duchenne muscular dystrophy, 62/127, 63/289
EMG, 63/287
Kell blood group, 62/127
Kell blood group antigen, 63/288
Kx blood group antigen, 63/288
lactate dehydrogenase, 63/287
neuropathology, 63/287
neuropathy, 63/287
retinal pigmentary dystrophy, 63/289
secondary pigmentary retinal degeneration, 63/289
serum alanine aminotransferase, 63/287
sex chromosomes, 63/289
McNaughton classification
epilepsy, 15/10
McNicholl syndrome, *see* Cerebrocostomandibular syndrome
McRae line
definition, 61/7
pseudobasilar impression, 19/341
Means syndrome, *see* Oligosymptomatic hyperthyroidism
Measles
animal experiment, 56/116
blastogenesis, 47/357
cancer, 69/450
cellular response, 47/357
chronic meningitis, 56/645
clinical features, 56/115
deafness, 55/131, 56/115
demyelinating antibody, 9/545
diagnosis, 56/115
epilepsy, 72/154
foreign antigen, 9/545
immunocompromise, 56/417
immunosuppression, 56/417
management, 56/115
multiple sclerosis, 9/129, 47/329-331, 357

myxovirus, 9/546
optic nerve agenesis, 50/211
postinfectious encephalomyelitis, 47/326, 476
primary pigmentary retinal degeneration, 13/214
temporal bone pathology, 56/115
teratogenic agent, 34/378
vaccinia, 9/545
Measles antibody
CSF, *see* CSF measles antibody
multiple sclerosis, 9/79, 47/363
Measles encephalitis
epilepsy, 72/149
Hodgkin lymphoma, 63/353
immunocompromise, 56/417
immunosuppression, 56/417
monoclonal protein antibody, 56/426
myoclonus, 56/480
Measles virus
acute measles encephalitis, 56/14
biology, 56/422
culture, 56/425
eosinophilic inclusion body, 56/480
F protein, 56/423
H protein, 56/423
immune mediated encephalomyelitis, 34/408-411
immunocompromised host, 56/480
Koshevnikoff epilepsy, 56/480
L protein, 56/423
monoclonal protein, 56/423
morbillivirus genus, 56/422
multiple sclerosis, 34/435-437
neurotropic virus, 56/417
NP protein, 56/423
P protein, 56/423
paramyxovirus, 56/14
paramyxovirus family, 56/422
paramyxovirus infection, 56/417
persistence, 56/427
postvaccinal measles encephalomyelitis, 56/14
Reye syndrome, 29/332, 34/169, 49/217
slow virus disease, 56/417
spastic paraplegia, 59/436
specific protein, 56/423
subacute sclerosing panencephalitis, 34/344, 346-349, 362-364, 56/14
subacute sclerosing panencephalitis virus, 56/422
teratogenic effect, 34/378
ultrastructure, 56/425
Meat intoxication
meningioma, 27/360
Mebaral intoxication, *see* Methylphenobarbital intoxication
Mechanical injury
nerve injury, 51/134
vasogenic brain edema, 57/261
Mechanisms
brain injury, *see* Brain injury mechanism
DNA repair, *see* DNA repair mechanism
head injury, *see* Head injury mechanism
Jannetta, *see* Jannetta mechanism
Pincers, *see* Pincers mechanism
pincers grip, *see* Pincers grip mechanism
valve, *see* Valve mechanism
Mechlorethamine, *see* Chlormethine
Meckel diverticulum
holoprosencephaly, 50/237
Meckel ganglion neuralgia, *see* Sphenopalatine neuralgia
Meckel-Gruber syndrome, *see* Gruber syndrome
Meckel syndrome, *see* Gruber syndrome
Meclofenamic acid
headache, 48/174
migraine, 48/174
Mecobalamin, *see* Methyl vitamin B_{12}
Medazepam
half-life, 37/357
metabolic pathway, 37/357
psychopharmacology, 37/357
serum level, 15/693
tranquilizing agent, 37/355
Medial aortopathy
idiopathic, *see* Takayasu disease
Medial arteriopathy
idiopathic, *see* Takayasu disease
Medial cutaneous antebrachial nerve
lesion localization, 2/138
topographical diagnosis, 2/38
Medial cutaneous brachial nerve
lesion localization, 2/138
topographical diagnosis, 2/38
Medial forebrain bundle
neurogenic arterial hypertension, 63/231
Medial fronto-orbital artery, *see* Inferior frontal artery
Medial geniculate nucleus
neglect syndrome, 45/156
Medial lemniscus
Benedikt syndrome, 2/298
Cestan-Chenais syndrome, 2/239
opticocochleodentate degeneration, 21/545
Medial longitudinal fasciculus
see also Internuclear ophthalmoplegia *and* Unilateral elevator palsy
Cajal interstitial nucleus, 2/275
eye movement, 1/605
familial amyotrophic lateral sclerosis, 59/246

multiple sclerosis, 42/495
Parinaud syndrome, 55/124
progressive bulbar palsy, 59/224
Medial mamillary nucleus
anatomy, 2/435
Medial medullary syndrome
see also Dejerine syndrome
Babinski-Nageotte syndrome, 2/219
definition, 2/227
Dejerine syndrome, 2/221
lateral medullary syndrome, 2/228
Medial occipital vein
descending, *see* Descending medial occipital vein
Medial pallidal segment
ballismus, 49/372
hemiballismus, 49/372
Medial striate artery, *see* Heubner artery
Medial superior cerebellar vein, *see* Precentral cerebellar vein
Medial temporal lobe
amnesic syndrome, 55/139
Medial thalamic nuclear group
dorsomedial nucleus, 2/472
medial dorsal supranucleus, 2/472
medial nucleus, 2/472
Medial thalamic nucleus
emotion, 2/485
prefrontal leukotomy, 2/491
Medial thalamic syndrome
choroidal artery syndrome, 2/486
dementia, 2/486
thalamic dementia, 2/486
tumor, 2/485
vascular origin, 2/485
Median aplasia
cerebellar agenesis, 50/181
Median cleft lip
holoprosencephaly, 42/32, 50/234
Median cleft lip hypotelorism syndrome
features, 50/234
hydrocephalus, 50/236
Median facial cleft syndrome
see also Holoprosencephaly
cheilopalatoschisis, 30/247
corpus callosum agenesis, 30/243, 46/90, 50/150, 163
encephalocele, 46/90
hypertelorism, 30/247, 46/89
meningocele, 46/90
mental deficiency, 30/243, 46/89
normal intelligence, 30/243
Patau syndrome, 50/558
speech, 46/30
Median longitudinal fasciculus syndrome
diplopia, 42/321
internuclear ophthalmoplegia, 2/278
nystagmus, 42/321
ophthalmoplegia, 42/321
Median nerve
anesthesia, 1/92
carpal tunnel syndrome, 7/286-300, 329, 42/309
compression neuropathy, 51/91
entrapment, 7/286
ganglion, 7/293
lesion site diagram, 2/30
occupational lesion, 7/329
occupational neuropathy, 7/291
plexiform neuroma, 14/26
topographical diagnosis, 2/28-33
upper arm compression, 7/329
wrist fracture, 70/34
Median nerve compression
hand, 7/330
Median nerve lesion
anterior interosseous nerve syndrome, 7/329
brachial plexus, 2/146
carpal tunnel syndrome, 7/329
compression, 7/329
Median nerve palsy
hemophilia, 63/311
Rancho splint, 8/404
Median neuropathy
lymphohistiocytosis, 56/179
neurolymphomatosis, 56/179
reticulosis, 56/179
Mediastinal tumor
neurofibromatosis type I, 14/139
subclavian artery stenosis, 53/372
Medical Research Council Scale
limb girdle syndrome, 40/447
muscle weakness, 40/297
Medication
acquired immune deficiency syndrome, 75/414
pain, 69/37
Medicolegal aspects, *see* Brain death, Death *and* Law
Mediterranean fever, *see* Brucellosis
Mediterranean myoclonus
myoclonic epilepsy, 60/663
Medium chain acyl-coenzyme A dehydrogenase
acylcarnitine moiety, 66/411
metabolic myopathy, 62/492
mitochondrial matrix system, 66/409
Medium chain acyl-coenzyme A dehydrogenase deficiency
hepatomegaly, 62/495

hypoglycemic encephalopathy, 62/495
inheritance, 62/495
inner mitochondrial membrane system, 66/407
lipid metabolic disorder, 62/495
metabolic encephalopathy, 62/495, 66/410
metabolic myopathy, 62/495
mitochondrial matrix system, 66/410
prevalence, 62/495
Reye syndrome, 62/495, 66/410
sudden infant death, 66/410
Medium spiny neuron
Huntington chorea, 49/256
Medroxyprogesterone
benign intracranial hypertension, 67/111
Medulla
adrenal, *see* Adrenal medulla
autonomic nervous system, 74/152
blood pressure, 74/154
brain stem lacunar infarction, 53/394
pharmacology, 74/153
solitary tract, 74/153
vasopressin, 74/154
Medulla oblongata
arterial supply, 2/226-228
brain tumor, 17/693
Cheyne-Stokes respiration, 1/671
commotio, *see* Commotio medullae oblongata
coughing, 63/491
gunshot injury, 25/50
hemangioblastoma, 14/244, 249
hereditary motor and sensory neuropathy type I, 21/302
hereditary motor and sensory neuropathy type II, 21/302
hiccup, 2/228, 235
intrinsic arterial pattern, 11/29
penetrating artery, 11/29
respiration, 1/651
spinal meningioma, 20/206, 278
tumor, 67/160
Medulla oblongata syndrome, 2/217-236
Avellis syndrome, 2/217
Babinski-Nageotte syndrome, 2/218, 330
Barré syndrome, 2/219
Bonnier syndrome, 2/220
Cestan-Chenais syndrome, 2/221
Dejerine syndrome, 2/221
Gasperini syndrome, 2/222
Jackson syndrome, 2/223
Schmidt syndrome, 2/224
Tapia syndrome, 2/224
Medullary cavity
central, *see* Central medullary cavity
Medullary fissure
diastematomyelia, 50/438
diplomyelia, 50/438
Medullary neuroblastoma
adrenal, *see* Adrenal medullary neuroblastoma
Medullary syndrome
dorsal, *see* Dorsal medullary syndrome
lateral, *see* Lateral medullary syndrome
medial, *see* Medial medullary syndrome
paramedian, *see* Paramedian medullary syndrome
Medullary thyroid carcinoma
neurofibromatosis type I, 50/370
Sipple syndrome, 14/750
Medullary tractotomy
trigeminal neuralgia, 48/449
Medulloblast
cerebellum, 18/168
Medulloblastoma
adjunctive treatment, 68/192
adult, 68/201
age, 68/191
astrocytic differentiation, 67/9
cerebellar, *see* Cerebellar medulloblastoma
Chang stage, 68/189
chemotherapy, 68/199
chromosome, 67/51
chromosome abnormality, 67/45
classification, 67/3
clinical investigation, 68/195
clinical staging, 68/188
congenital brain tumor, 31/49
desmoplastic cerebellar, *see* Desmoplastic cerebellar medulloblastoma
distribution, 67/130
epidemiology, 68/182
hemorrhage, 68/95
histology, 68/191
histopathology, 67/9
imaging, 67/183
immunoreactivity, 67/8
JC virus, 18/171
metastasis, 68/190
molecular biology, 67/7
neuroimaging, 68/183
neuropathology, 68/185
neurosurgery, 68/186
pathogenesis, 68/182
pigmented papillary, *see* Pigmented papillary medulloblastoma
pleomorphism, 67/223
prognostic factor, 68/194
spinal, *see* Spinal medulloblastoma
survey, 68/181

Medulloepithelioma
classification, 16/12
definition, 16/22
tissue culture, 17/69
Medullomyoblastoma
cerebellar medulloblastoma, 18/181
Medusae
scalloped, 37/50
Meduso-congestin intoxication
diarrhea, 37/37
dyspnea, 37/37
edema, 37/37
experimental animal, 37/37
Mees line
arsenic intoxication, 36/202, 64/286
thallium intoxication, 64/326
Mefenamic acid
headache, 48/174
migraine, 48/174, 177, 203
Mefentanyl
designer drug, 65/364
potency, 65/368
Mefloquine intoxication
ataxia, 65/485
deafness, 65/485
diplopia, 65/485
epilepsy, 65/485
headache, 65/485
insomnia, 65/485
tinnitus, 65/485
vertigo, 65/485
Megacephaly, *see* Megalocephaly
Megacolon
see also Megadolichocolon
aganglionic, *see* Congenital megacolon
autonomic nervous system, 74/116
Chagas disease, 75/400
congenital, *see* Congenital megacolon
Down syndrome, 50/527
infantile myotonic dystrophy, 40/490
multiple endocrine adenomatosis type III, 42/753
myotonic dystrophy, 40/516
Megaconial myopathy, *see* Mitochondrial myopathy
Megadolichobasilar artery
aneurysm, 12/81
Ekbom other syndrome, 63/85
subarachnoid hemorrhage, 12/81
Megadolichocarotid artery
aneurysm, 12/81
subarachnoid hemorrhage, 12/81
superior orbital fissure syndrome, 12/81
Megadolichocolon
see also Megacolon
hypertrophic interstitial neuropathy, 21/152
Megaduodenum
Melkersson-Rosenthal syndrome, 8/229
Megaesophagus
Chagas disease, 75/400
hereditary motor and sensory neuropathy type I, 21/286
hereditary motor and sensory neuropathy type II, 21/286
hereditary sensory and autonomic neuropathy type III, 21/113
Megalencephalic leukodystrophy, *see* Alexander disease
Megalencephaly, *see* Macrencephaly
Megalobarencephaly
leukodystrophy, 10/36
Megaloblastic anemia
see also Pernicious anemia *and* Vitamin B12 deficiency anemia
biochemistry, 63/253
classification, 38/18
clinical features, 38/18, 24
hereditary vitamin B12 metabolism defect, 63/255
mental deficiency, 43/278
nitrous oxide, 63/256
orotic aciduria, 63/256
thrombocytopenia, 63/322
transcobalamin II deficiency, 63/255
vitamin B_c deficiency, 38/23, 51/336, 63/253
Megaloblastic madness
cyanide intoxication, 38/21
pernicious anemia, 38/21, 42/609
vitamin B12, 38/21
vitamin B_c, 38/21
Megalocephaly
Alexander disease, 10/209, 42/485
GM2 gangliosidosis, 10/206
mental deficiency, 46/8
Megalocornea
ARG syndrome, 50/581
mental deficiency, 43/270
microcephaly, 43/270
N syndrome, 43/283
seizure, 43/270
short stature, 43/270
Megalocorneal mental deficiency syndrome
hypoplasia iridis, 43/270
Megalotestes
heterozygote detection, 43/287
mental deficiency, 43/285-288
sex chromosome fragility, 43/287
Meglumine iocarmate
myelography, 19/186

Meglumine iothalamate
myelography, 19/186
Meiacanthus intoxication
chordata, 37/69
Meige syndrome
see also Dystonia musculorum deformans
age at onset, 1/289
anatomic classification, 1/279
dystonia musculorum deformans, 49/521
hereditary progressive cochleovestibular atrophy, 60/776
Meiosis
chromosome, 30/94
hydrozoa intoxication, 37/40
nereistoxin intoxication, 37/55
occipital lobe tumor, 17/336
pontine infarction, 53/390
vertebrobasilar system syndrome, 53/390
Meiosis type I
Down syndrome, 50/522
Meiosis type II
Down syndrome, 50/522
Meissner corpuscle
morphology, 1/89
neurofibroma, 14/27, 29
sensation, 7/82
Meissner plexus, *see* Submucosal plexus
Meissner submucous plexus, *see* Submucosal plexus
Melancholy, *see* Mental depression
Melanin
albinism, 14/106
lipopigment, 10/24
manganese intoxication, 64/305
metachromatic leukodystrophy, 10/48
pigmented papillary medulloblastoma, 18/180
Sturge-Weber syndrome, 14/102
tuberous sclerosis, 14/102
Melanoblast
terminology, 14/415
Melanoblastosis (Bloch-Sulzberger), *see* Incontinentia pigmenti
Melanoblastosis cutis linearis sive systematisata, *see* Incontinentia pigmenti
Melanocyte
pia, 14/414
terminology, 14/415
Melanocytoma
optic nerve tumor, 17/366
Melanocytosis
oculodermal, *see* Ota nevus
Melanoderma
adrenoleukodystrophy, 10/131, 42/483
demyelination, 10/36
Melanodermic leukodystrophy, *see* Adrenoleukodystrophy
Melanoid pigment
lipid auto-oxidation, 10/29, 31
polymerization, 10/29
Melanoma
brain embolism, 55/163
brain infarction, 55/163
brain metastasis, 18/204, 69/112
breast carcinoma, 71/618
cerebrovascular disease, 55/163
congenital spinal cord tumor, 32/355-386
leptomeningeal, *see* Leptomeningeal melanoma
malignant, *see* Malignant melanoma
metastasis, 71/659
neurinoma, 68/565
spinal cord metastasis, 20/425
subarachnoid hemorrhage, 12/119
uvea, *see* Uvea melanoma
vertebral metastasis, 20/415-417
Melanomalignoma, *see* Malignant melanoma
Melanomatosis
leptomeningeal, *see* Leptomeningeal melanomatosis
meningeal, *see* Leptomeningeal melanomatosis
Melanophakomatosis (Peutz-Touraine-Jeghers), *see* Peutz-Jeghers syndrome
Melanophore
definition, 14/415
Melanosarcoma
CSF cytology, 16/407
Melanosis
leptomeningeal, *see* Leptomeningeal melanosis
meningeal, *see* Leptomeningeal melanosis
neurocutaneous, *see* Neurocutaneous melanosis
Melanosis corii degenerativa (Siemens), *see* Incontinentia pigmenti
Melanosome
definition, 14/415
Melanotic adamantinoma
cranial vault tumor, 17/115
Melanotic neuroectodermal tumor of infancy, *see* Pigmented papillary medulloblastoma
Melanotic progonoma, *see* Pigmented papillary medulloblastoma
MELAS syndrome
see also Progressive myoclonus epilepsy
acanthocyte, 49/332
brain infarction, 62/509
cardiopathy, 62/510
cerebrovascular disease, 53/43
cortical blindness, 62/510
dementia, 62/509

dwarfism, 60/666
epilepsy, 60/666, 62/509, 72/222
episodic vomiting, 60/668
lactic acidosis, 62/509
mitochondrial disease, 62/497, 66/393
mitochondrial DNA point mutation, 62/509, 66/428
mitochondrial encephalomyopathy, 62/498
mitochondrial protein, 66/17
occipital lobe epilepsy, 73/110
progressive external ophthalmoplegia, 60/56
progressive external ophthalmoplegia classification, 62/290
ragged red fiber, 62/292, 509
secondary pigmentary retinal degeneration, 60/725
striatopallidodentate calcification, 49/424, 62/510
Melatonin
age, 74/495, 502
Alzheimer disease, 74/495
biological rhythm, 74/469, 495, 503
circadian rhythm, 59/155, 74/501
cluster headache, 48/233
depression, 74/504
disorder, 74/503
homosexual man, 74/501
immune function, 74/504
light, 74/499
oral contraceptive agent, 74/503
pineal gland, 74/499
progressive dysautonomia, 59/155
sex, 74/502
side effect, 74/504
sleep, 74/501
sleep disorder, 74/501
vampire hormone, 74/499
Melena
dermatomyositis, 62/374
inclusion body myositis, 62/374
polymyositis, 62/374
Melitococcal sciatica
brucellosis, 7/481
Melkersson, E., 8/205
Melkersson-Rosenthal syndrome, 8/205-235
acroparesthesia, 42/322
asialia, 8/221
benign intracranial hypertension, 8/277
blepharitis granulomatosa, 8/212
blepharospasm, 42/322
brain, 16/222
bucca lobata, 8/213
carotid system syndrome, 53/308
cheilitis, 8/213
cheilitis granulomatosa, 8/205-209, 222
congenital megacolon, 8/229
differential diagnosis, 8/227-229
edema, 2/72, 14/790, 42/322
facial edema, 8/208-211, 14/790, 16/222, 42/322
facial paralysis, 2/72, 8/215, 279, 344, 14/790, 16/222, 42/322
gingivitis, 8/213
glossitis granulomatosa, 8/213
headache, 14/790, 42/322
hereditary cranial nerve palsy, 60/39, 42
herpes simplex virus encephalitis, 8/210
history, 8/205
hyperacusis, 14/790, 42/322
Krabbe-Bartels disease, 14/529
lingua plicata, 2/72, 8/344, 14/790, 16/222, 42/322
macrocheilia, 8/205, 208-211
macroglossia, 8/214
megaduodenum, 8/229
mental change, 8/218, 227
narcolepsy, 8/213
neuralgia, 8/209
ophthalmoplegic migraine, 8/228
pachycheilia, 8/209
papilledema, 16/222
pareiitis granulomatosa, 8/216
paresthesia, 42/322
prosopitis granulomatosa, 8/213
psychosis, 8/218, 227
salivation, 8/220
sphenopalatine neuralgia, 8/217
sweating, 8/221
syringomyelia, 14/529
Toxoplasma gondii, 8/226
tuberculosis, 8/227, 229, 233
uranitis granulomatosis, 8/213
vitamin B5, 8/235
vitamin B6, 8/235
Mellontophobia
migrainous social disability, 48/48
Melnick-Needles syndrome
exophthalmos, 43/452
hypertelorism, 30/249, 43/452
micrognathia, 43/452
skeletal malformation, 30/249
Melodic intonation therapy
aphasia, 46/617
Melorheostosis
fibrous dysplasia, 14/199
neurofibromatosis type I, 14/492
Melphalan
antineoplastic agent, 65/528
demyelinating neuropathy, 47/618

dysproteinemic neuropathy, 8/74-78
epilepsy, 65/528
multiple myeloma, 18/256, 20/14, 17
neurotoxin, 39/94, 65/528, 67/357
vertebral myeloma, 20/14, 17
Membrane long chain acyl-coenzyme A dehydrogenase
very long chain acyl-coenzyme A dehydrogenase, 66/409
Membrane protein
translocation, 66/81
Membrane reorganization
axon, 51/52
Membranes
axon, *see* Axolemma
Bruch, *see* Bruch membrane
internode, *see* Internode membrane
neuronal nucleus, *see* Neuronal nucleus membrane
nodal, *see* Nodal membrane
Membranous cytoplasmic inclusion body
chemistry, 21/50
diagram model, 10/371
glycosaminoglycanosis type I, 21/52
glycosaminoglycanosis type IH, 10/445-447
glycosaminoglycanosis type II, 10/445-447
GM2 gangliosidosis, 10/291, 295, 365-372, 408-410, 446, 467
infantile GM2 gangliosidosis, 66/21
metachromatic leukodystrophy, 10/372
sphingomyelin, 10/287
Membranous labyrinth
Bing-Siebenmann type, 30/407
congenital deafness, 30/408
developmental dysgenesis, 30/407-410
genetics, 30/408
infection, 30/410
malformation, 30/409
Michel type, 30/407
Mondini-Alexander type, 30/407
pathology, 30/409
Schiebe type, 30/407
toxic, 30/409
Membranovesicular body
juvenile amaurotic idiocy, 10/226
Memory, 3/258-266
see also Learning *and* Memory disorder
age, 3/265
aging, 27/470, 46/210
alcohol intoxication, 64/115
Alzheimer disease, 46/250
anatomy, 3/37, 258-262, 286
anesthesia, 27/467
antiepileptic treatment, 72/390
autoscopia, 2/714
Binswanger disease, 46/320
biochemistry, 3/262, 27/459-471
brain cortex, 3/260
brain lacunar infarction, 46/355
carbamate intoxication, 64/191
centrencephalic system, 2/450
cerebral coding, 3/263
cerebral dominance, 3/148
cerebral representation of knowledge, 3/265-267
cholinergic receptor blocking agent, 27/467
chronic hepatocerebral degeneration, 49/219
CNS aging, 27/470
consciousness, 3/48
corpus callosum syndrome, 2/762
definition, 27/459-461
delirium, 46/528
depression, 46/123
developmental dyslexia, 46/113, 117
DNA, 3/262
dreaming, 3/87, 175
electroconvulsive treatment, 27/466
elementary amnesic trace, 3/261
emotion, 3/328, 332
enhancement, *see* Memory enhancement
epilepsy, 73/454
final common stimulation, 3/261
fornix, 2/771
frontal lobe lesion, 45/29
hippocampomammillary system, 3/259
histologic study, 3/262
historic aspect, 27/459
Huntington chorea, 46/307
hydrocephalic dementia, 46/327
hysteria, 46/621
immediate memory, 3/259
impairment, *see* Memory disorder
information processing, 3/261
learning, 3/261-263
meaning of stimuli, 3/263
memorization, 3/259
mental deficiency, 46/29
metacircuit, 3/263
migraine, 48/161
mnestic circuit, 3/263
mnestic content, 3/264
molecular theory, 3/263
musical, *see* Musical memory
neuroelectrical activity, 3/262
neuropsychologic defect, 45/522
neuropsychology, 45/516
nonverbal, *see* Nonverbal memory
Papez circuit, 3/259

Parkinson dementia, 46/374
phenobarbital, 72/393
phenytoin intoxication, 65/501
physiology, 3/37, 258-267, 286
Pick disease, 46/234
posttraumatic amnesia, 3/293-295, 27/468, 45/187
protein synthesis inhibition, 27/468
recall, 3/264
recent, 2/712
recognition, 3/264
rehabilitation, 46/620-623
reticular system, 3/259
RNA, 3/262
scopolamine, 46/596
short-term retention, 3/147
spatial, *see* Spatial memory
synapse, 3/262
temporal lobe tumor, 67/152
thalamic syndrome (Dejerine-Roussy), 45/91
toluene intoxication, 64/52
topographical, *see* Topographical memory
treatment, 27/471
tricyclic antidepressant, 65/319
verbal, *see* Verbal memory
visual, *see* Visual memory
Wernicke-Korsakoff syndrome, 27/468-470, 46/621

Memory defect
unilateral, *see* Visuospatial agnosia

Memory deficit
transient global amnesia, 45/209
traumatic, *see* Traumatic amnesia
Wernicke-Korsakoff syndrome, 45/197-199

Memory disorder, 3/268-288
see also Memory, Traumatic amnesia *and* Wernicke-Korsakoff syndrome
acute diffuse brain affection, 3/270
age, 3/274
amusia, 2/599, 691-693, 4/98, 203, 221
amygdala lesion, 2/713
anesthetic agent, 27/463, 467
anomia, 2/599
anomic aphasia, 45/318
anosognosia, 3/284
anterograde amnesia, 3/269
antibiotic blocking agent of macromolecular synthesis, 27/461
aphasia, 2/599, 691-693, 4/98, 221, 302
axial amnesia, 3/259
barbiturate intoxication, 65/509
basilar ischemia, 12/3
beta adrenergic receptor blocking agent intoxication, 65/436
biochemistry, 27/461-464
brain concussion, 23/420, 433
brain injury, 23/12
brain tumor, 16/732, 46/623
carbon dioxide, 27/463
cingulate gyrus lesion, 2/713, 3/286
cingulectomy, 2/713
classification, 3/269, 286-288
color amnesia, 4/17
confabulation, 3/275
confusional state, 3/270
continuing forgetting, 3/269, 283
convulsant, 27/462
cortical amnesia, 3/260
cortical blindness, 3/284
dementia, 3/270-273
diencephalic lesion, 2/713
disorientation, 3/276
dysphrenia hemicranica, 5/78
electroconvulsive treatment, 3/272
epilepsy, 15/96
examination method, 3/268
false memory, 3/269, 275
faulty recognition, 3/275
forgetting, 3/259, 264
frontal lobe lesion, 3/284-286
head injury outcome, 57/397
higher nervous activity disorder, 3/37
hippocampal lesion, 2/713, 3/281-283, 287
hippocampus lesion, 2/713, 46/357
history, 23/7
hyperthermia, 27/463
hypnotic agent, 27/463
hypothalamus, 2/450
immediate forgetting, 3/269, 273
integration amnesia, 3/269
Klüver-Bucy syndrome, 3/349-356, 45/262
lead encephalopathy, 64/435
lesion localization, 3/272, 277-283, 286
loss, 27/467
mamillary body, 3/278
memory defect unawareness, 3/274
memory stock fragmentation, 3/274
mesencephalic diencephalic syndrome, 3/273-281
migraine, 48/162
narcolepsy, 42/711
neuropharmacology, 27/462
neurotic state, 3/265
occipital lobe lesion, 3/283
paramnesia, 3/275
phenytoin intoxication, 65/500
posterior cerebral artery syndrome, 53/412
posttraumatic amnesia, 3/293, 27/468, 57/398

potassium chloride, 27/463
primary amnesia, 3/287
recollection amnesia, 3/269
reduplicative paramnesia, 24/489
registration amnesia, 3/269, 286
registration failure, 3/273
retrograde, 24/479, 489
retrograde amnesia, 2/712, 3/269, 294
secondary amnesia, 3/287
senile dementia, 3/265
short-term memory for spatial location, 3/217
spatial, *see* Spatial memory disorder
spatial orientation disorder, 3/276
stroke, 46/623
temporal lobe abscess, 46/623
temporal lobe electrical stimulation, 2/713
temporal lobe epilepsy, 2/713
temporal lobe lesion, 2/713, 3/147
test, 3/269
thermoregulation, 27/463
time scale, 3/277
topographical memory, 3/219-221, 4/16, 229, 300
traumatic amnesia, 3/293-295
traumatic psychosyndrome, 24/546
visual amnesia, 4/16
vitamin B_1 deficiency, 27/464
vitamin E deficiency, 27/464
Wernicke encephalopathy, 2/450, 3/277
Wernicke-Korsakoff syndrome, 27/469
Memory enhancement
biochemistry, 27/464
Memory impairment, *see* Memory disorder
Memory storage
traumatic amnesia, 24/547
traumatic psychosyndrome, 24/547
Memory training
traumatic psychosyndrome, 24/556
Menadione intoxication, *see* Vitamin K3 intoxication
Menarche
migraine, 48/36
Mende syndrome
see also Klein-Waardenburg syndrome
cheiloschisis, 14/119
Klein-Waardenburg syndrome, 14/116
neurocutaneous syndrome, 14/101
partial albinism, 14/106, 119
Mendel law
see also Genetics *and* Inheritance
dominant inheritance, 30/95
multiple sclerosis, 47/299, 312
Mendel-Von Bechterew reflex
foot, 1/248
hand, 1/244
pyramidal sign, 1/185
technique, 1/185
Mendelejev periodic system
neurotoxicity, 51/265
Mendelian pattern, *see* Mendel law
Mengovirus
Reye syndrome, 56/239
Meniere disease
anterior inferior cerebellar artery, 2/257
audiogram, 16/313
brain stem auditory evoked potential, 47/140
deafness, 2/377
dominant, *see* Dominant Meniere disease
headache, 5/208
hereditary, *see* Hereditary Meniere disease
loudness recruitment, 2/74, 16/309
migraine, 42/402, 48/157
pseudo, *see* Pseudo-Meniere syndrome
recessive, *see* Recessive Meniere disease
reflex neurovascular syndrome, 8/349
Sjögren syndrome, 71/74
statoacoustic nerve, 2/377
tinnitus, 2/377
vertigo, 2/362-364, 42/402
Meningeal angioma
Sturge-Weber syndrome, 14/62
Meningeal angiomatosis
leukodystrophy, 10/120
sudanophilic leukodystrophy, 42/508
Meningeal angiomatosis with leukoencephalopathy, *see* Divry-Van Bogaert syndrome
Meningeal apoplexy, *see* Subarachnoid hemorrhage
Meningeal artery
accessory, *see* Accessory meningeal artery
dorsal, *see* Dorsal meningeal artery
middle, *see* Middle meningeal artery
traumatic aneurysm, 24/392-394
Meningeal carcinomatosis
breast cancer, 69/202
granulomatous CNS vasculitis, 55/391
treatment, 69/202
Meningeal cerebellar sarcoma, *see* Cerebellar medulloblastoma
Meningeal cyst
echinococcosis, 52/524
Meningeal diverticulum
intrasacral meningocele, 20/153
postoperative, *see* Postoperative meningeal diverticulum
spinal nerve root, 20/152
spinal nerve root avulsion, 20/157, 160
Meningeal endarteritis obliterans

neurosyphilis, 52/274
Meningeal gliomatosis
differential diagnosis, 20/189
Meningeal hydrops, *see* Otitic hydrocephalus
Meningeal infection
brain vasospasm, 11/520
venous sinus occlusion, 54/426
Meningeal irritation
Behçet syndrome, 56/598
headache, 48/284
latent nystagmus, 1/547
leukemia, 69/238
migraine, 48/131
spurling test, 1/543
subarachnoid hemorrhage, 12/128, 55/14
supratentorial brain abscess, 33/118
thallium intoxication, 36/248
Meningeal leukemia
adult, 63/340
arachnoid infiltration, 63/341
blindness, 63/341
Brown-Séquard syndrome, 63/343
bulimia, 63/341
cauda equina compression, 63/343
childhood, 63/340
clinical features, 39/4, 63/352
CNS irradiation, 63/340
coma, 16/234, 63/341
convulsion, 63/341
cranial neuropathy, 63/341
CSF, 16/234, 39/4-6, 63/341
CSF contamination, 63/341
CSF diagnosis, 69/240
cytarabine, 18/251, 63/342
diagnosis, 63/341
diplopia, 63/341
epilepsy, 16/234, 63/341
extradural deposit, 63/343
grade I, 63/341
grade II, 63/341
grade III, 63/341
headache, 63/341
hydrocephalus, 63/341
hyperphagia, 63/341
hypothalamic syndrome, 63/341
incidence, 63/340
intrathecal chemotherapy, 63/342, 69/247
intrathecal methotrexate, 63/340
intravenous chemotherapy, 69/247
intraventricular chemotherapy, 69/248
laboratory test, 69/240
leukemia nodule infiltration, 63/342
localized deposit, 63/342
lymphoblastic leukemia, 63/340
methotrexate, 18/251, 39/98, 63/342
neurologic complication, 18/249
neuropathology, 63/341
nuclear magnetic resonance, 63/342
obesity, 63/341
Ommaya reservoir, 63/342
optic chiasm, 63/341
optic nerve, 63/341
papilledema, 16/234, 63/341
paraplegia, 63/343
parenchyma infiltration, 63/341
pathogenesis, 16/234, 63/341
pathology, 39/5
radiotherapy, 69/247
somnolence, 63/341
space occupying lesion, 16/234
spinal cord, 16/234, 63/342
spinal cord compression, 63/343
spinal nerve root, 18/252
spinal nerve root syndrome, 63/341
symptom, 63/341
T-lymphocyte type, 63/340
thiotepa, 63/359
treatment, 39/6, 69/246, 249
Meningeal lymphoma
arabinoside, 63/348
back pain, 63/347
Burkitt lymphoma, 63/346, 351
confusional state, 63/347
cranial nerve palsy, 63/347
CSF cytology, 63/347
CSF eosinophilia, 63/346
cytosine, 63/348
diagnosis, 63/347, 69/265
epilepsy, 63/347
headache, 63/347
intrathecal chemotherapy, 63/348
laboratory test, 63/347
leptomeningeal metastasis, 69/266
lethargy, 63/347
methotrexate, 63/348
β2-microglobulin, 63/347
multiple cranial neuropathy, 63/347
non-Hodgkin lymphoma, 63/346
nuclear magnetic resonance, 63/347
papilledema, 63/347
radiculopathy, 63/347
risk factor, 63/346, 69/266
symptom, 63/347, 69/264
treatment, 63/347, 69/270
Meningeal malformation
Arnold-Chiari malformation type II, 50/405

Meningeal melanomatosis, *see* Leptomeningeal melanomatosis
Meningeal melanosis, *see* Leptomeningeal melanosis
Meningeal meningiomatosis
  differential diagnosis, 20/189
Meningeal oligodendrogliomatosis
  subarachnoid, 18/82
  terminology, 18/82
Meningeal pain
  headache, 1/540
  migraine, 5/371
Meningeal pseudocyst, *see* Postoperative meningeal diverticulum
Meningeal sarcomatosis
  primary, *see* Primary meningeal sarcomatosis
Meningeal sign, 1/536-549
  ascariasis, 35/359
  brain abscess, 33/118
  false, 1/545
  malingering, 1/546
  nuchal rigidity, 1/542, 544, 546
  polyarteritis nodosa, 39/300, 48/285
  true, 1/543
Meningeal syndrome
  cause, 1/547
  lymphatic leukemia, 18/250
  symptom, 1/536
Meningeal system
  middle, *see* Middle meningeal system
Meningeal thickening
  histopathology, 25/40, 59, 89, 94
  nonunited odontoid fracture, 25/280
  spinal cord injury, 25/40, 59, 89, 94, 281
Meningeal tumor
  biopsy, 67/229
Meningeal vein
  anatomy, 11/52
  middle, *see* Middle meningeal vein
  tuberculous meningitis, 52/198
Meningée postérieure
  algie, *see* Algie meningée postérieure
Meninges
  see also Dura mater *and* Meninx
  brain cortex heterotopia, 50/249
  embryo, 50/14
  embryology, 31/80
  hereditary amyloid polyneuropathy type 3, 60/100
  histology, 31/78-80
  morphology, 31/77
  Niemann-Pick disease type A, 10/496
  polymicrogyria, 50/257
Meningioma
  see also Intracranial infection
  acoustic nerve tumor, 17/684
  aged, 68/409
  angioblastic, *see* Hemangiopericytoma
  angiomatous, *see* Angiomatous meningioma
  brain biopsy, 16/722
  brain injury, 24/441
  brain scanning, 16/684
  brain tumor, 18/506
  cauda equina, *see* Cauda equina meningioma
  cavernous sinus, 68/412
  cerebellar, 68/413
  cerebellopontine, 68/412
  chromosome, 67/52, 68/402
  chromosome abnormality, 67/45
  classification, 68/404
  clivus, 68/413
  congenital brain tumor, 31/50
  congenital spinal cord tumor, 32/355-386
  cranial vault tumor, 17/128
  CSF cytology, 16/401
  distribution, 67/130
  droplet metastasis, 20/244
  epidemiology, 68/401
  epilepsy, 15/299, 18/20, 72/181, 359
  falcine, *see* Falcine meningioma
  falx, 68/411
  falx cerebri, 17/542
  foramen magnum, 68/413
  foramen magnum tumor, 17/721, 19/67-69
  Foster Kennedy syndrome, 2/570
  frontal lobe tumor, 17/238
  gasserian ganglion, 5/273
  growth factor, 68/403
  hypervascular, 68/270
  hypothalamus tumor, 18/366
  imaging, 67/182, 187, 189, 68/406
  incidence, 67/132, 68/401
  intraorbital, 18/341
  lateral ventricle tumor, 17/597, 603
  mammary carcinoma, 42/744
  meat intoxication, 27/360
  medical treatment, 68/415
  middle fossa, *see* Middle fossa meningioma
  morbidity, 68/409
  mortality, 68/409
  multiple, *see* Multiple meningioma
  multiple hamartoma syndrome, 42/754
  multiple neurofibromatosis, 43/35
  mummy, 74/187
  nasopharyngeal tumor, 17/205
  neurinoma, 42/744, 68/565
  neurofibroma, 17/176

neurofibromatosis, 14/33, 42/744
neurofibromatosis type I, 14/140, 152, 42/744
neurofibromatosis type II, 50/370
neuropsychology, 45/9
nonrecurrent nonhereditary multiple cranial neuropathy, 51/571
olfactory groove, 68/412
optic chiasm compression, 68/75
optic nerve tumor, 17/362
optic sheath, *see* Optic sheath meningioma
orbital tumor, 17/176
parasagittal, 68/411
parietal lobe tumor, 17/298
progressive supranuclear palsy, 49/246
radiation induced, 68/392
radiology, 68/347
radiotherapy, 18/506, 68/413
recurrence, 68/414
sellar, *see* Sellar meningioma
skull base, 68/411, 466
skull base tumor, 17/176, 68/477
sphenoid wing, 68/410
sphenoidal ridge, *see* Sphenoidal ridge meningioma
sphenoidal sinus syndrome, 2/296
spinal, *see* Spinal meningioma
spinal cord compression, 19/358
spinal epidural tumor, 20/104
spinal site, 19/18
spinal tumor, 19/64
steroid receptor, 68/402
Sturge-Weber syndrome, 50/376
subarachnoid hemorrhage, 12/119
suprasellar, 68/412
supratentorial brain tumor, 18/318
surgery, 68/408, 413
survey, 68/270, 401
tentorial, 68/412
third ventricle tumor, 17/444
thoracic, 67/202
tissue culture, 17/83
traumatic, *see* Traumatic meningioma
treatment, 68/478
trisomy 8 mosaicism, 43/511
tuberous sclerosis, 14/368, 42/744
ventricular, 68/411
vertebral column, *see* Spinal meningioma
Werner syndrome, 43/489
xanthomatosis, 27/246
xanthomatous transformation, 27/246
Meningioma indifferentiale (Globus), *see* Hemangiopericytoma
Meningiomatosis
meningeal, *see* Meningeal meningiomatosis
multiple meningioma, 20/189
neurofibromatosis, 14/36
neurofibromatosis type I, 14/145
spinal meningioma, 20/189
Meningism
Bannwarth syndrome, 51/202
cavernous sinus thrombosis, 52/173
clinical examination, 57/133
CSF, 1/547
gold polyneuropathy, 51/305
lead intoxication, 9/642
leptospirosis, 33/397
migraine coma, 60/651
neuroborreliosis, 51/202
nuchal rigidity, 57/133
OKT3, 63/536
pituitary abscess, 52/152
Meningitis
see also Intracranial infection
actinomycosis, 35/385
acute amebic meningoencephalitis, 52/316, 326
acute aseptic, *see* Meningitis serosa
acute bacterial, *see* Acute bacterial meningitis
acute syphilitic, 52/277
acute viral, *see* Meningitis serosa
adenovirus, *see* Adenovirus meningitis
alternariasis, 35/561, 52/481
angiostrongyliasis, 35/329, 52/545, 548
aseptic, *see* Meningitis serosa
aspergillosis, 35/397, 52/378, 55/432
Aspergillus fumigatus, 63/532
azathioprine intoxication, 65/557
bacillary, *see* Infantile enteric bacillary meningitis
bacterial, *see* Bacterial meningitis
bacterial endocarditis, 52/291, 297
benign recurrent aseptic, *see* Mollaret meningitis
Blastomyces dermatitidis, 52/385, 55/432
blood-brain barrier, 28/380
Borrelia burgdorferi, 52/261
brain abscess, 52/144
brain hemorrhage, 31/166
Brucella, 51/185
Burkitt lymphoma, 63/352
Candida, *see* Candida meningitis
candidiasis, 52/398
carcinomatous, *see* Carcinomatous meningitis
caseous, *see* Caseous meningitis
cephalosporiosis, 35/562, 52/481
cerebral amebiasis, 52/315
chemical, *see* Chemical meningitis
chemotherapy, 69/160
chronic, *see* Chronic meningitis

chronic benign lymphocytic, *see* Chronic benign lymphocytic meningitis
cladosporiosis, 35/563, 52/483
clear fluid, *see* Meningitis serosa
clinical features, 57/139
Clostridium welchii, 33/103
Coccidioides immitis, 55/432
coccidioidomycosis, 52/412
computer assisted tomography, 54/63
congenital spinal cord tumor, 32/366
consciousness, 45/120
cortical blindness, 55/119
cryptococcosis, 35/461, 52/431
Cryptococcus neoformans, *see* Cryptococcal meningitis
CSF fistula, 24/188, 57/311
cytarabine, 61/151
cytomegalovirus, 56/477
depressed skull fracture, 23/407, 411, 24/225
diamorphine intoxication, 65/355
diphtheria, 33/486, 65/240
drechsleriasis, 35/565, 52/485
ECHO virus, *see* ECHO virus meningitidis
encephalopathy, 45/119
enterovirus, 56/326
eosinophilic, *see* Eosinophilic meningitis
epilepsy, 15/306
Epstein-Barr virus, *see* Epstein-Barr virus meningitis
Epstein-Barr virus infection, 56/252
Escherichia coli, *see* Escherichia coli meningitis
exertional headache, 48/374
febrile status epilepticus, 15/179
fibrous, *see* Spinal arachnoiditis
Flavobacterium meningosepticum, 33/103
fonsecaeasis, 35/566, 52/487
fulminant, 57/310
fungal, *see* Fungal meningitis
gonococcal, *see* Gonococcal meningitis
gram-negative, *see* Gram-negative bacillary meningitis
gram-negative bacillary, *see* Gram-negative bacillary meningitis
granulomatous amebic encephalitis, 52/317, 327
granulomatous basal, *see* Granulomatous basal meningitis
head injury, 57/51, 237
headache, 48/31
heart transplantation, 63/179
Hemophilus influenzae, *see* Hemophilus influenzae meningitis
herpes genitalis, 71/267
herpes simplex, 34/85, 156
herpes simplex virus, *see* Herpes simplex virus meningitis
Histoplasma, *see* Histoplasma meningitis
Histoplasma capsulatum, 55/432
hydrocephalus, 50/288
iatrogenic neurological disease, 63/179
immunocompromise, 65/548
inappropriate antidiuretic hormone secretion, 28/499
infantile enteric bacillary, *see* Infantile enteric bacillary meningitis
infection, 57/237
infectious endocarditis, 55/167, 63/23, 112, 116
infectious mononucleosis, 56/252
intravascular consumption coagulopathy, 63/314
intravenous Ig, 65/561
Kawasaki syndrome, 52/264, 266
Leptospira interrogans, 55/415
leptospirosis, 33/400-402
Listeria monocytogenes, *see* Listeria monocytogenes meningitis
listerial, *see* Listeria monocytogenes meningitis
low velocity injury, 24/227
Lyme disease, 51/204, 52/261, 263
lymphocytic, *see* Lymphocytic meningitis *and* Meningitis serosa
lymphocytic leukemia, 63/352
macrophage defense defect, 65/548
management, 24/658, 664
marantic endocarditis, 52/297
meningococcal, *see* Meningococcal meningitis
methotrexate, 61/151
microcephaly, 50/280
migraine, 48/31, 37
Mimae, *see* Mimae meningitis
missile injury, 23/524
Mollaret, *see* Mollaret meningitis
mummy, 74/187
mumps, *see* Mumps meningitis
mycotic brain aneurysm, 52/295
myeloid leukemia, 63/352
Neisseria meningitidis, *see* Meningococcal meningitis
neonatal, *see* Infantile enteric bacillary meningitis
neoplastic, *see* Neoplastic meningitis
neuroborreliosis, 51/204
neurobrucellosis, 33/311, 52/585
neurologic intensive care, 55/204
neurosarcoidosis, 38/521, 63/422
neurosyphilis, 33/355, 52/274, 277, 282
neutropenia, 63/339
Nocardia, *see* Nocardia meningitis
nocardiosis, 16/217, 35/519, 52/447, 449

noninfectious purulent, *see* Noninfectious purulent meningitis
nonpneumococcal gram-positive bacterial, *see* Staphylococcal meningitis *and* Streptococcal meningitis
North American blastomycosis, 35/401, 404, 52/385, 389
OKT3, 63/536
OKT3 intoxication, 65/560
opiate intoxication, 65/355
opisthotonos, 1/542
Paracoccidioides brasiliensis, 55/432
paracoccidioidomycosis, 35/534, 52/457
paragonimiasis, 35/260-262
Paragonimus, *see* Paragonimus meningitis
penetrating head injury, 57/310
phagocyte defense defect, 65/548
pilonidal sinus, 42/47
pneumocephalus, 24/212
pneumococcal, *see* Pneumococcal meningitis
poliomyelitis, 42/652
posttraumatic, *see* Posttraumatic meningitis
progressive external ophthalmoplegia, 22/194
Pseudoallescheria, 52/492
reactive, *see* Meningitis serosa
recurrent, *see* Mollaret meningitis *and* Recurrent meningitis
recurrent aseptic, *see* Mollaret meningitis
renal transplantation, 63/531
rhinorrhea, 24/188, 220
sarcoidosis, 71/488
serosa spinalis, *see* Spinal arachnoiditis
serous, *see* Meningitis serosa
sickle cell anemia, 38/40-42
skull fracture, 23/387, 24/218
spinal, *see* Spinal meningitis
spinal chronic, 32/393
spinal cord abscess, 52/190
spinal dermoid, 19/364
spinal epidermoid, 19/364
sporotrichosis, 35/569, 52/494
staphylococcal, *see* Staphylococcal meningitis
Staphylococcus aureus, *see* Staphylococcal meningitis
Staphylococcus epidermidis, *see* Staphylococcal meningitis
streptococcal, *see* Streptococcal meningitis
Streptococcus pneumoniae, *see* Pneumococcal meningitis
subacute bacterial endocarditis, 33/473
subarachnoid abscess, 16/211
subarachnoid hemorrhage, 12/112, 169, 55/20
subdural effusion, 24/332
superior longitudinal sinus thrombosis, 52/178
sympathetic, *see* Meningitis serosa
syphilitic, *see* Neurosyphilis
T-lymphocyte, 65/548
toxic, *see* Meningitis serosa
toxoplasmosis, 42/662
traumatic intracranial hematoma, 57/281
trichinosis, 52/566
trichophytosis, 35/571, 52/496
tuberculous, *see* Tuberculous meningitis
ustilagomycosis, 35/572, 52/498
varicella zoster virus, *see* Varicella zoster virus meningitis
viral, *see* Viral meningitis
Vogt-Koyanagi-Harada syndrome, 56/611
Whipple disease, 52/138
Zygomycetes, 55/432

Meningitis classification
tuberculous meningitis, 52/196

Meningitis rash
Coxsackie meningitis, 52/25
ECHO virus meningitidis, 52/25
gonococcal bacteremia, 52/25
leptospirosis, 52/25
Moraxella urethralis, 52/25
Rocky Mountain spotted fever, 52/25

Meningitis sarcomatosa
space occupying lesion, 19/354

Meningitis serosa, 34/83-90
see also Lymphocytic meningitis *and* Viral meningitis
arachnoiditis, 20/86
arsenic intoxication, 9/648
bacterial meningitis, 52/6
Behçet syndrome, 71/225
circumscripta, 20/86
Coxsackie virus infection, 34/136
CSF, 34/86
definition, 1/547
drug induced, *see* Chemical meningitis
enterovirus infection, 34/84
Epstein-Barr virus infection, 34/85
herpes simplex, 34/85
infectious mononucleosis, 34/188
Kawasaki syndrome, 51/453
leptospiral arteritis, 55/415
leptospiral infection, 55/415
lymphocytic choriomeningitis, 9/194, 196, 56/129
methotrexate intoxication, 65/530
mumps, 56/129
neuroborreliosis, 51/200
neurosarcoidosis, 63/422
neurosyphilis, 52/277

nonsteroid anti-inflammatory agent intoxication, 65/415
OKT3, 63/536
polyarteritis nodosa, 39/300, 55/356
spinal intradural cyst, 12/425
subarachnoid hemorrhage, 55/7
systemic lupus erythematosus, 39/285, 55/378
tuberculous meningitis, 52/203
Meningitis serosa circumscripta spinalis
arachnoid cyst, 20/86
Meningoangiomatosis, *see* Meningiomatosis
Meningoblastoma
optic nerve, *see* Optic nerve meningoblastoma
optic nerve glioma, 14/630
Meningocele
see also Encephalomeningocele *and* Spina bifida
anterior (lateral) cervical, 32/224
anterior (lateral) lumbar, 32/208-211
anterior (lateral) thoracic, 32/211-224
anterior sacral, *see* Anterior sacral meningocele
anterior spinal, *see* Anterior spinal meningocele
Arnold-Chiari malformation type I, 32/105, 42/15
Arnold-Chiari malformation type II, 32/105
birth incidence, 42/27
brachial plexus, 51/148
caudal aplasia, 32/206
cervical, *see* Cervical meningocele
closed, 50/29
corpus callosum agenesis, 42/7
cranial, *see* Cranial meningocele
epidemiology, 32/206, 211, 223
Gruber syndrome, 14/13
hereditary sensory and autonomic neuropathy type I, 60/9
intrasacral, *see* Intrasacral meningocele
intrathoracic, *see* Intrathoracic meningocele
Joubert syndrome, 63/437
Klippel-Feil syndrome, 42/37
median facial cleft syndrome, 46/90
myelography, 19/187
neurofibromatosis, 14/38, 32/212
neurofibromatosis type I, 14/38, 145, 157
occult intrasacral, *see* Intrasacral meningocele
orbital tumor, 17/202
18p partial monosomy, 43/534
pelvic, 32/193-208
postoperative, *see* Postoperative meningeal diverticulum
pulsating exophthalmos, 14/629
sacral, *see* Sacral meningocele
spina bifida, 19/187, 50/477, 482
spinal arachnoid cyst, 32/394
spinal lipoma, 20/110
treatment, 32/210, 212, 224
vertebral abnormality, 32/193, 206
Meningocele spurius, *see* Postoperative meningeal diverticulum
Meningocerebral angiodysplasia
adenovirus, 56/290
Meningocerebral angiomatosis with leukoencephalopathy, *see* Divry-Van Bogaert syndrome
Meningococcal meningitis, 33/21-30, 52/21-32
see also Neisseria meningitidis
acute cerebellar ataxia, 33/23, 52/26
acute meningoencephalitis, 52/26
age, 33/22, 52/21, 29
allergic arthritis, 52/26
allergic granulomatous angiitis, 52/28
allergic skin vasculitis, 33/23, 52/26
arthralgia, 33/23, 52/28
arthritis, 33/24, 52/26
ataxia, 33/23
bacteremia, 33/23, 52/22
bacterial meningitis, 52/3
brain abscess, 33/23, 52/150
brain edema, 52/25
brain herniation, 52/26, 30
brain infarction, 52/23, 25
C-5,C-9 complement deficiency, 52/29
C-6,C-8 complement deficiency, 52/29
carrier rate, 33/29, 52/22
carrier state, 33/22, 52/22
causalgia, 52/26
cephalosporin, 33/26, 52/30
cerebral palsy, 52/28
chemoprophylaxis, 33/29, 52/31
chloramphenicol, 33/26, 52/30
chronic meningococcemia, 33/23, 52/28
clinical features, 52/24
coma, 33/23
complication, 52/25
conjunctivitis, 33/24, 52/27
conus medullaris syndrome, 52/26
counterimmunoelectrophoresis, 33/25, 52/25
course, 52/24
cranial nerve palsy, 33/23, 52/24, 26
CSF, 33/25, 52/24
CSF endotoxin, 52/25
deafness, 52/24, 26, 28
diagnosis, 33/25, 52/25
differential diagnosis, 33/24, 52/25
encephalitis, 33/27, 52/24
encephalomyelitis, 52/26
endotoxin, 33/23, 25, 28, 52/27
epidemic, 33/22, 52/21

epidemiology, 33/22, 52/21
epilepsy, 52/24, 26, 28
facial paralysis, 52/26
focal metastatic infection, 52/22
fulminant meningococcemia, 52/24
fulminating encephalitis, 52/26
headache, 33/23, 52/24
heart failure, 33/24, 52/30
herpes labialis, 52/24, 27
herpes simplex, 33/23
histopathology, 52/23
hydrocephalus, 33/23, 52/28
IgG2 deficiency, 52/29
immunity, 33/22, 52/22
immunologic reaction, 52/22
immunology, 33/26, 52/23
impaired cerebral perfusion, 52/23
incidence, 52/21
intracranial hypertension, 52/23, 25
intrathecal treatment, 52/30
intravascular consumption coagulopathy, 33/23, 25, 27, 52/25, 27, 30, 63/314
irreversible shock, 52/27
laboratory data, 33/25, 52/25
limulus assay, 33/25, 52/25
management, 52/31
meningeal cry, 52/24
meningococcemia, 33/23, 52/26
meningoencephalitis, 52/24
mental deficiency, 52/28
minimal brain dysfunction, 52/28
minocycline, 33/29, 52/32
mortality, 33/23, 52/29
myelitis, 52/26
myocarditis, 33/24, 27, 52/27, 30
nasopharynx, 33/21, 52/22
nausea, 33/23, 52/24
neurologic complication, 52/26
nitroblue tetrazolium test, 33/25, 52/25
nuchal rigidity, 33/23, 52/24
pathogenesis, 33/27, 52/22
penicillin, 33/26, 52/30
pericarditis, 33/24, 52/27
peritonitis, 52/27
petechia, 33/21, 23, 52/26
pharyngitis, 33/23, 52/27
pneumococcal meningitis, 52/26
pneumonia, 33/24, 52/27
polysaccharide antigen, 33/26
prevalence, 52/22
prevention, 33/28, 52/31
primary meningococcal pneumonia, 52/27
prognosis, 52/29
prognostic factor, 52/29
properdine deficiency, 52/23
psychiatric symptom, 52/24
pulse rate, 52/24
purpura, 33/23, 52/26
rash, 33/21, 23, 52/28
recurrence, 52/23, 29
respiratory rate, 52/24
rifampicin, 33/29, 52/32
season, 33/22, 52/22
septic shock, 33/23, 52/30
septicemia, 52/24, 30
sequela, 52/25, 28
spinal meningitis, 52/190
subarachnoid hemorrhage, 33/23, 52/26
sulfonamide, 52/30
sulfonamide resistance, 33/22, 27, 52/21
systemic lupus erythematosus, 52/23
tetracycline, 33/29
treatment, 33/25, 52/30
vaccination, 33/29, 52/21, 31
vasculitis, 33/23, 52/26
vomiting, 33/23, 52/24
Waterhouse-Friderichsen syndrome, 33/23, 52/27, 30
Meningococcemia
fulminant, *see* Fulminant meningococcemia
meningococcal meningitis, 33/23, 52/26
Meningocortical amyloid angiopathy (Pantelakis), *see* Brain amyloid angiopathy
Meningoencephalic angioneuromatosis, *see* Meningiomatosis
Meningoencephalic gliomatosis
neurofibromatosis, 14/35
Meningoencephalitis
see also Encephalitis
Acanthamoeba, 71/304
acquired immune deficiency syndrome, 56/492, 494
acquired toxoplasmosis, 35/122, 52/352, 354
acute, *see* Acute meningoencephalitis
acute amebic, *see* Acute amebic meningoencephalitis
acute diffuse lymphocytic, *see* Acute diffuse lymphocytic meningoencephalitis
adenovirus, *see* Adenovirus meningoencephalitis
African trypanosomiasis, 52/341
amebic, *see* Amebic meningoencephalitis
angiostrongyliasis, 35/329, 52/554, 556
Argentinian hemorrhagic fever, 56/365
bacterial meningitis, 55/418
brain biopsy, 10/685
brain vasculitis, 55/418

bullous myringitis, 5/210
candidiasis, 55/435
Chagas disease, 52/347
chickenpox, 34/162-167
congenital toxoplasmosis, 52/358
cryptococcosis, 35/461-477, 52/431
Cryptococcus neoformans, *see* Cryptococcal meningoencephalitis
cytomegalovirus, 56/272
dengue fever, 56/12
diphasic, *see* Diphasic meningoencephalitis
Epstein-Barr virus, *see* Epstein-Barr virus meningoencephalitis
experimental virus, *see* Experimental viral meningoencephalitis
heart transplantation, 63/179
herpes simplex virus, 56/478
herpetic, *see* Herpetic meningoencephalitis
histoplasmosis, 52/437
iatrogenic neurological disease, 63/179
immunocompromise, 65/548
immunocompromised host, 56/474
immunodeficiency, 52/347
immunosuppressed host, 52/347
infectious mononucleosis, 56/252
listeriosis, 52/91
lymphocytic, *see* Lymphocytic meningoencephalitis
macrophage defense defect, 65/548
meningococcal meningitis, 52/24
monkey, 56/454
mumps, 34/411, 628, 56/114
neurobrucellosis, 33/311, 313, 52/584, 586
neurosarcoidosis, 38/531
paracoccidioidomycosis, 35/531, 52/457
pertussis, 52/234
phagocyte defense defect, 65/548
primary amebic, *see* Primary amebic meningoencephalitis
sarcoidosis, 38/531
simian T-cell tropic retrovirus, 56/454
Sjögren syndrome, 71/74
subarachnoid hemorrhage, 55/3
T-lymphocyte, 65/548
toxoplasmosis, 52/354, 358
varicella zoster virus, *see* Varicella zoster virus meningoencephalitis
verruga peruana, 34/664
Meningoencephalocele
Arnold-Chiari malformation type III, 50/403, 410
basal, *see* Basal meningoencephalocele
frontoethmoid, *see* Sincipital encephalocele
intracranial, *see* Intracranial meningoencephalocele
Meningoencephalomyelitis
disseminated, *see* Disseminated meningoencephalomyelitis
Meningoencephalopathy
lymphogranulomatous, *see* Lymphogranulomatous meningoencephalopathy
Meningofacial angiomatosis, *see* Sturge-Weber syndrome
Meningohypophyseal artery
anatomy, 11/69
Meningohypophyseal trunk
carotid cavernous fistula, 12/271
Meningomyelitis
Behçet syndrome, 56/597
neurobrucellosis, 52/586
spinal cord compression, 19/377
tuberculous meningitis, 52/202
Meningomyelocele
anencephaly, 50/71
Arnold-Chiari malformation type II, 50/403, 405, 407
craniolacunia, 50/137, 139
diastematomyelia, 50/437, 441
diplomyelia, 50/437, 441
Edwards syndrome, 43/537, 50/562
lumbar, *see* Lumbar meningomyelocele
polyploidy, 50/556
spina bifida, 20/58
spinal muscular atrophy, 59/370
Meningoradiculitis
lymphocytic, *see* Bannwarth syndrome
nerve conduction, 7/178
neuroborreliosis, 51/200
Meningosis leucaemica, *see* Meningeal leukemia
Meningovascular brucellosis
neurobrucellosis, 52/586
Meninx
see also Dura mater *and* Meninges
anatomy, 1/537
innervation, 1/537
referred pain, 1/539
sensitivity, 1/539, 541
sign, 1/536
spinal, 1/539
symptom, 1/536
Menkes disease, *see* Trichopoliodystrophy
Menkes steely hair disease, *see* Trichopoliodystrophy
Menkes syndrome, *see* Trichopoliodystrophy
Menopause
adiposis dolorosa, 8/347
cluster headache, 48/219

headache, 48/433
mental deficiency, 42/341
migraine, 48/433
osteoporosis, 70/19
polyneuritis, 42/341
sleep apnea syndrome, 45/132
sphenopalatine neuralgia, 5/115
Menstrual associated syndrome
daytime hypersomnia, 45/137
Menstrual cycle
body odor, 74/377
olfaction, 74/377
suprachiasmatic nucleus, 74/491
vasopressin, 74/491
Menstrual headache, *see* Catamenial migraine
Menstrual migraine, *see* Catamenial migraine
Menstruation
cluster headache, 48/219
epilepsy, 15/321
spinal cord injury, 61/342
toxic shock syndrome, 52/258
Mental deficiency
Aarskog syndrome, 43/314
abducent nerve paralysis, 43/299
acanthosis nigricans, 30/248
achondroplasia, 31/270, 46/88
acquired hepatocerebral degeneration, 46/18
acrocephalosynanky, 46/87
acrocephalosyndactyly type I, 31/233, 43/320, 46/86
acrocephalosyndactyly type II, 31/236, 38/422, 43/322, 324, 46/87, 50/123
acrocephalosyndactyly type V, 46/87
acrodysostosis, 31/230, 43/291
acrogyria, 14/791
acromicria, 42/341
ADR syndrome, 42/112
adult GM1 gangliosidosis, 60/664
aggression, 43/258, 268
agyria, 46/11, 48
Aicardi syndrome, 31/226, 42/696
Albers-Schönberg disease, 46/64
albinism, 14/781, 30/248
Alexander disease, 10/209, 42/484, 46/49
alexia, 4/134
alopecia, 42/684
Alpers disease, 42/486
aminoaciduria, 42/517
α-amino-n-butyric acid, 13/339
aminopterin syndrome, 46/78
amyotrophic dystonic paraplegia, 42/199
Andermann syndrome, 50/166
anemia, 43/278
anhidrosis, 42/346
anhidrotic ectodermal dysplasia, 14/788, 46/91
aniridia, 42/123, 46/62
anticonvulsant intoxication, 46/24
antimongoloid eyes, 43/257
Apgar score, 46/4
apraxia, 46/27
aqueduct stenosis, 42/57
arachnodactyly, 43/293
areflexia, 43/258
argininosuccinic aciduria, 42/525
Arnold-Chiari malformation type I, 46/8
arteriovenous malformation, 31/188
arthrogryposis multiplex congenita, 46/89
aspartylglucosaminuria, 29/225
aspartylglycosaminuria, 27/159, 42/527
assessment, 46/2-5
associated syndrome, 46/41-96
ataxia, 42/112, 43/278, 46/14, 60/657, 659, 664, 722
ataxia telangiectasia, 14/289, 31/3, 24, 42/120, 46/58
ataxic diplegia, 42/121
athetotic hemiplegia, 42/200
atrophoderma vermiculatum, 43/8
attachment disorder of infancy, 46/25
attention, 46/29
autism, 46/194
autosome, 46/68-73
bacterial meningitis, 52/14
Bardet-Biedl syndrome, 13/337, 387, 22/508, 43/233
Bartter syndrome, 42/529
Bassen-Kornzweig syndrome, 10/548, 42/511
Becker muscular dystrophy, 43/86
Beckwith-Wiedemann syndrome, 43/398, 46/82
behavior disorder in children, 46/26-30
Behr disease, 13/88, 42/415
Berardinelli-Seip syndrome, 42/591
Berlin syndrome, 43/9
Biemond syndrome, 46/65
Biemond syndrome type II, 43/334
Bitter syndrome, 14/107
Bloom syndrome, 14/775, 43/10
body scheme disorder, 4/397
Bonham-Carter syndrome, 30/248
Bonnet-Dechaume-Blanc syndrome, 14/109, 262, 43/36
Börjeson-Forssman-Lehmann syndrome, 31/311, 42/530
boxing injury, 23/544
brachycephaly, 30/226
brachyphalangy, 43/262

bradykinesia, 46/28
bradyphrenia, 46/28
brain abscess, 46/19
brain cortex focal dysplasia, 30/364, 50/208
camptobrachydactyly, 43/236
camptomelic syndrome, 46/63, 88
Candida meningitis, 52/400
cardiac defect, 43/257
carnosinemia, 29/197, 42/532
Carol-Godfried-Prakken-Prick syndrome, 43/8
cataract, 43/123-125, 268, 282, 293
cavum septi pellucidi, 30/325
cavum veli interpositi, 30/328
cavum vergae, 30/325
centrofacial lentiginosis, 14/782, 43/26
cerebellar agenesis, 30/380
cerebellar ataxia, 21/468, 42/123, 130-132, 43/260
cerebellar dysarthria, 42/137
cerebellar granular cell hypertrophy, 43/252
cerebellar hypoplasia, 30/358, 42/19
cerebellar vermis agenesis, 42/4, 43/257
cerebellar vermis atrophy, 60/664
cerebral dominance, 4/465
cerebral gigantism, 31/333, 43/337, 46/83
cerebral hemiatrophy, 30/356
cerebral heterotopia, 30/492
cerebral palsy, 46/13-16
cerebrocostomandibular syndrome, 43/237
cerebrohepatorenal syndrome, 43/339, 46/64
cerebro-oculofacioskeletal syndrome, 43/341
cerebrotendinous xanthomatosis, 10/532, 537, 547, 27/250, 60/166
cerebrovascular disease, 53/31
cervico-oculoacusticus syndrome, 32/127
cheilopalatoschisis, 43/256, 269
cheiloschisis, 43/256, 269
child of phenylketonuric mother, 31/299
childhood psychosis, 46/194
childhood schizophrenia, 46/25
chondrodystrophic myotonia, 62/269
chondrodystrophy, 46/88
chromosomal aberration, 46/19-24, 68-76
chromosomal aneuploidy, 46/19-21
chronic neutropenia, 43/289
citrullinuria, 42/538
classification, 46/1
cloverleaf skull, 46/87
CNS developmental disorder, 46/6-13
Cockayne-Neill-Dingwall disease, 10/182, 13/433, 14/315, 31/236, 43/350, 46/62, 51/399, 60/664
Coffin-Lowry syndrome, 31/245, 43/238, 46/46
Coffin-Siris syndrome, 31/240
Cogan syndrome type I, 42/329
cognition, 46/4
congenital ataxia, 60/664
congenital dyskeratosis, 43/392
congenital heart disease, 38/133, 53/31, 63/3
congenital hip dislocation, 42/75
congenital ichthyosiform erythroderma, 46/59
congenital ichthyosis, 43/276, 46/59
congenital laryngeal abductor paralysis, 42/318
congenital larynx stridor, 30/411
congenital muscular dystrophy, 43/91
congenital myopathy, 62/335
congenital pain insensitivity, 51/564
congenital retinal blindness, 13/272, 22/529, 533
congenital suprabulbar paresis, 59/70
congenital syphilis, 46/76
congenital toxoplasmosis, 35/131, 46/77, 52/358
Conradi-Hünermann syndrome, 43/346, 46/50
convulsion, 42/689, 691
corpus callosum agenesis, 2/765, 30/287, 42/7-10, 100, 102, 46/12, 50/161
craniodigital syndrome, 43/243
craniofacial dysostosis (Crouzon), 31/234, 43/359, 46/85
craniolacunia, 30/273
craniosynostosis, 30/154, 227, 42/20, 43/262, 46/85, 50/126
cretinism, 31/297, 46/18, 83-85
cri du chat syndrome, 31/568, 43/501, 46/21-24, 68, 50/275
Critchley syndrome, 13/427
Crome syndrome, 43/242
cryptophthalmos syndrome, 46/61
cutis laxa, 46/60
cutis verticis gyrata, 14/777, 46/47
cystathioninuria, 42/541
cytomegalovirus, 56/269
cytomegalovirus infection, 31/212, 215, 46/19, 77, 50/277
cytoside lipidosis, 10/545
Dandy-Walker syndrome, 30/636, 46/8, 49
De Lange syndrome, 31/279-281, 43/246, 46/91
deafness, 60/657, 664
definition, 46/1
demyelination, 43/303
dental abnormality, 42/691
dermatoleukodystrophy, 42/488
developmental dyslexia, 4/377
developmental dysphasia, 4/361, 46/141
diabetes mellitus, 42/335, 411
diencephalic syndrome, 46/24
digital abnormality, 43/262, 269
disequilibrium syndrome, 42/140

distal amyotrophy, 60/659
dominant inheritance, 46/42-46
Down syndrome, 31/401-406, 43/544, 46/19-21, 70-73, 50/519, 528-531
Dubowitz syndrome, 30/248, 43/247, 46/59
Duchenne muscular dystrophy, 40/363, 41/479, 46/66, 62/121
Dumon-Radermecker syndrome, 42/477
dwarfism, 43/265, 268, 276, 294, 381, 46/88
Dyggve-Melchior-Clausen syndrome, 43/265
dyskeratosis congenita, 14/784, 43/392
dyskinesia, 46/14
dysmorphism, 60/664
dysostosis, *see* Acrodysostosis
dysraphia, 46/7
dyssynergia cerebellaris progressiva, 42/213
dystonia deafness syndrome, 42/372
dystopic cortical myelinogenesis, 30/362, 50/207
eastern encephalitis, 56/136
ectodermal dysplasia, 43/269
Edwards syndrome, 31/521, 43/535, 46/70, 50/542
EEC syndrome, 43/396, 46/91
EEG, 42/689, 43/268, 46/24
Ehlers-Danlos syndrome, 46/61
elfin face syndrome, 31/318, 43/259
Ellis-Van Creveld syndrome, 43/348
encephalomeningocele, 42/26
endemic cretinism, 46/84
endocrine candidiasis syndrome, 42/642
Engelmann disease, 46/46
environment, 46/76-81
environmental deprivation, 46/5
epidemiology, 46/2
epidermolysis bullosa dystrophica, 14/13
epidermolysis bullosa dystrophica hereditaria, 14/791
epilepsy, 15/586, 42/684, 691, 43/284, 46/24
epileptic seizure, 43/282
Epstein-Barr virus, 56/263
erythrokeratodermia, 46/60
esophageal achalasia, 43/260
eunuchoidism, 60/664
facial asymmetry, 43/296
familial amaurotic idiocy, 46/56
familial ectodermal dysplasia, 60/664
familial hyperglycemia, 46/19
familial hypoglycemia, 46/19
familial spastic paraplegia, 59/312, 374
Fanconi syndrome, 31/320, 43/16
fetal alcohol syndrome, 30/120, 46/81
fetal face syndrome, 43/400
fetal hydantoin syndrome, 30/122, 50/279
fetal Minamata disease, 64/418
fetal trimethadione syndrome, 30/122
fetal warfarin syndrome, 30/123, 46/80
FG syndrome, 43/250
fibrous dysplasia, 14/187
fingerprint body myopathy, 62/347
Flynn-Aird syndrome, 42/327
foramina parietalia permagna, 30/277
formiminotransferase deficiency, 42/546
Förster syndrome, 1/175, 42/202
Freeman-Sheldon syndrome, 43/353
Friedman-Roy syndrome, 43/251
Friedreich ataxia, 21/337, 46/90
frontometaphyseal dysplasia, 31/256
fructose intolerance, 46/19
fucosidosis, 42/550, 66/330
Fukuyama syndrome, 41/37, 43/91
galactosemia, 42/551, 46/19, 52
Gaucher disease, 42/438, 46/55
genetics, 14/528
Gerhardt syndrome, 50/219
Gerstmann syndrome, 4/389
Gillespie syndrome, 42/123
globoid cell leukodystrophy, 46/55
glucose intolerance, 43/299
γ-glutamyl transpeptidase deficiency, 29/122
glycogen storage disease, 46/19, 62/490
glycolipidosis, 10/552
glycosaminoglycanosis, 46/19, 53-55
glycosaminoglycanosis type I, 10/435, 441, 27/147, 42/450, 46/53
glycosaminoglycanosis type II, 10/435, 441, 42/454, 46/54
glycosaminoglycanosis type III, 10/342, 42/455, 46/54, 66/299, 302
glycosaminoglycanosis type IV, 42/457
glycosaminoglycanosis type VI, 66/308
glycosaminoglycanosis type VIII, 42/460
GM1 gangliosidosis type I, 42/431
GM1 gangliosidosis type II, 42/432
GM2 gangliosidosis, 46/56
GM2 gangliosidosis type III, 42/436
Godfried-Prick-Carol-Prakken syndrome, 14/113
Goldberg syndrome, 42/439
Goldenhar syndrome, 31/247, 46/46
Goltz-Gorlin syndrome, 14/787, 31/288, 43/20, 46/61
growth retardation, 43/262, 272
Guérin-Stern syndrome, 14/105
gynecomastia, 43/273-275
hair brain syndrome, 43/297
Hallermann-Streiff syndrome, 43/404, 46/93
happy puppet syndrome, 30/515, 31/309, 46/24, 94

Hartnup disease, 42/146, 46/59
Harvey syndrome, 42/223
Haw River syndrome, 60/664
hearing loss, 60/659
heart disease, 43/257, 262
hemifacial hypertrophy, 59/482
hemifacial hypoplasia, 46/46
hemikeratosis, 46/95
hemivertebra, 43/299
Hemophilus influenzae meningitis, 46/19
hereditary dystonic paraplegia, 59/345
hereditary motor and sensory neuropathy type I, 21/285
hereditary motor and sensory neuropathy type II, 21/285
hereditary paroxysmal kinesigenic choreoathetosis, 42/206
hereditary sensory and autonomic neuropathy type III, 1/477, 30/248, 46/59, 60/27
hereditary sensory and autonomic neuropathy type IV, 60/12
herpes virus hominis syndrome, 31/212, 216
heterozygote detection, 43/287
histidinemia, 42/554
historic note, 46/5
holoprosencephaly, 30/459, 42/32, 50/239
homocystinuria, 42/556, 46/51, 55/327-329
Hooft-Bruens syndrome, 42/508
horizontal gaze paralysis, 42/333
Huntington chorea, 46/45
Hurler-Hunter disease, 10/341
hydranencephaly, 46/11
hydrocephalus, 30/600, 32/552, 46/8, 50/291
hydroxylysinemia, 29/223
hydroxylysinuria, 42/704
hyperammonemia, 42/560
hyperargininemia, 42/562
hyperdibasic aminoaciduria, 42/564
hyperglycinemia, 42/565
hyperlexia syndrome, 46/29
hyperlysinemia, 29/221, 42/567
hyperphenylalaninemia, 42/611
hyperphosphatasia, 42/568
hyperpipecolatemia, 29/222
hyperprolinemia type I, 29/133, 42/569
hyperprolinemia type II, 29/138, 42/569
hyperreflexia, 43/268, 277
hyperserotonemia, 29/225
hypertelorism, 43/257, 277, 417, 439
hypertelorism hypospadias syndrome, 43/413
hypertelorism microtia clefting syndrome, 43/415
hypervalinemia, 42/574
hypogonadism, 43/123-125, 272-275
hypohidrotic ectodermal dysplasia, 30/248
hypoparathyroidism, 46/19
hypophosphatasia, 31/315
hypopigmentation, 43/257
hypoplasia iridis, 43/268
hypothyroid polyneuropathy, 51/515
ichthyosis, 13/476, 43/276
idiopathic hypoglycemia, 42/576
idiopathic scoliosis, 42/51
inborn error of metabolism, 46/17
incontinentia pigmenti, 14/12, 85, 31/242, 43/23, 46/67
indolylacrolyl glycinuria, 29/231
infantile attachment disorder, 46/25
infantile autism, 46/25
infantile cerebrovascular disease, 53/31
infantile ceroid lipofuscinosis, 42/461, 463
infantile epileptogenic encephalopathy, 42/693
infantile Gaucher disease, 42/438
infantile neuroaxonal dystrophy, 6/627, 59/75
infantile neuronal ceroid lipofuscinosis, 42/461, 463
infantile spasm, 15/221, 42/687
infantile spinal muscular atrophy, 59/70
institutionalization, 46/32
intelligence quotient, 46/2-5
intracerebral calcification, 42/534
isovaleric acidemia, 42/580
Jeune-Tommasi disease, 22/516
Joseph syndrome, 29/231
Joubert syndrome, 50/191, 63/437
juvenile Gaucher disease, 10/307, 523, 42/438
Kearns-Sayre-Daroff-Shy syndrome, 38/225, 43/142, 62/308
keratosis pilaris, 14/791
kernicterus, 42/583
Kjellin syndrome, 22/470, 42/173-175
Klinefelter syndrome, 21/467, 31/477, 479, 481, 43/557, 560, 46/75
Klinefelter variant XXXXY, 31/483, 43/561
Klinefelter variant XXXY, 31/483, 43/559
Klinefelter variant XXXYY, 31/484
Klinefelter variant XXYY, 31/484, 43/563
Klippel-Trénaunay syndrome, 14/395, 43/24
Krabbe-Bartels disease, 14/407, 411, 528
kwashiorkor, 7/635, 46/18
kyphoscoliosis, 43/277
language, 46/30
laryngeal abductor paralysis, 42/318
late cortical cerebellar atrophy, 21/477
Laurence-Moon-Bardet-Biedl syndrome, 31/327
Laurence-Moon-Biedl syndrome, 46/65
Laurence-Moon syndrome, 43/253

lead intoxication, 46/19
Lenz syndrome, 43/419
LEOPARD syndrome, 30/248
leprechaunism, 42/590
Lesch-Nyhan syndrome, 6/456, 22/514, 42/153, 46/66, 60/19, 63/256
Leschke-Ullmann syndrome, 14/117
Lindenov-Hallgren syndrome, 13/454
linear nevus sebaceous syndrome, 31/288-290, 43/38, 46/94
lipidosis, 46/19
lissencephaly, 30/486, 46/11, 48
Lowe syndrome, 31/302, 42/606, 46/66
lower motoneuron disease, 60/657
macrencephaly, 30/650, 42/41, 46/8, 50/263
macrocephaly, 46/8
macrocrania, 46/8
management, 46/32
mannosidosis, 27/157, 42/597
marasmus, 46/5, 18
Marinesco-Sjögren syndrome, 42/184-186, 46/63, 60/342, 664
Marshall syndrome, 46/82
MASA syndrome, 43/255
Matthews-Rundle syndrome, 60/664
median facial cleft syndrome, 30/243, 46/89
megaloblastic anemia, 43/278
megalocephaly, 46/8
megalocornea, 43/270
megalotestes, 43/285-288
memory, 46/29
meningococcal meningitis, 52/28
menopause, 42/341
β-mercaptolactate-cysteine disulfiduria, 29/119-121
metabolic disorder, 46/17
metachromatic leukodystrophy, 42/493, 46/57
methionine malabsorption syndrome, 29/127
methylmalonic acidemia, 42/601
metopic suture, 30/228
mevalonate kinase deficiency, 60/664
microcephaly, 30/510, 512-515, 518, 42/45, 81, 43/270, 279, 290, 46/47, 50/269, 277
microgyria, 46/12
microphthalmia, 43/281-283
Mietens-Weber syndrome, 43/304
Minamata disease, 36/135, 46/80, 64/418
Minkowitz disease, 10/552
mirror movement, 42/233
Möbius syndrome, 42/324, 46/90
Mohr syndrome, 43/450, 46/67
monosomy 21, 43/450
monosomy 21 syndrome, 31/598, 605
motor deficit, 4/454
motor disorder, 4/454
motor neuropathy, 42/335
motor stereotypy, 46/27
moyamoya disease, 12/360, 53/33
mucolipidosis type I, 27/161, 42/475, 46/54
mucolipidosis type II, 42/448, 46/55
mucolipidosis type III, 27/162, 42/449
Mulibrey dwarfism, 43/435
multiple mitochondrial DNA deletion, 62/510
multiple neurofibromatosis, 43/35
multiple neuronal system degeneration, 42/75
multiple nevoid basal cell carcinoma syndrome, 14/463, 30/248, 46/44
multiple trichoepithelioma, 14/791
muscle calcification, 43/258
muscular atrophy, 42/81, 43/258
muscular dystrophy, 43/90-92
myelin dysgenesis, 42/498-500
myoclonic epilepsy, 42/704
myoclonus, 42/240
myopathy, 43/123-125, 193
myotonic dystrophy, 40/517, 43/152, 46/44
myotonic myopathy, 43/165
N syndrome, 43/283
Netherton syndrome, 43/305
neurocutaneous melanosis, 14/414, 424
neuroembryology, 46/6-8
neurofibromatosis, 31/3, 32/212-218, 220-222, 46/42
neurofibromatosis type I, 14/156, 50/368
neuronal migration, 43/303
neuropathy, 42/335
neutropenia, 43/289
nevus unius lateris, 14/781, 43/40, 46/95
Niemann-Pick disease type A, 46/57
Niemann-Pick disease type C, 42/470, 46/57, 60/664
Noonan syndrome, 46/74
Norrie disease, 22/508, 31/291, 43/441, 46/66
nutritional deficiency, 46/17
Nijmegen breakage syndrome, 60/426
nystagmus, 42/240
obesity, 43/273-275, 288
oculocerebrocutaneous syndrome, 43/444
oculocerebrofacial syndrome, 43/280
oculocutaneous albinism, 43/3
oculodento-osseous dysplasia, 43/445
oculovertebral syndrome, 31/248
Oguchi disease, 13/262
olfactory bulb agenesis, 43/256
onychotrichodysplasia, 43/289
opticocochleodentate degeneration, 21/540,

22/508, 42/241
orbital hypertelorism, 30/248, 43/257, 439
orofaciodigital syndrome type I, 31/251, 43/447, 46/67
orotic aciduria, 42/607, 63/256
osseous dysplasia, *see* Ruvalcaba syndrome
osteocartilaginous abnormality, *see* Coffin-Lowry syndrome
osteopetrosis, 46/63
osteoporosis, 42/75
otopalatodigital syndrome, 43/457, 46/64
oxycephaly, 30/227
4p partial monosomy, 43/497
8p partial monosomy, 43/508
9p partial monosomy, 43/512, 50/578
17p partial monosomy, 50/591
18p partial monosomy, 43/534, 46/70
4p partial trisomy, 43/500
9p partial trisomy, 43/516, 46/69
10p partial trisomy, 43/519
11p partial trisomy, 43/525, 46/69
20p partial trisomy, 43/539
4p syndrome, 31/559
18p syndrome, 31/579
pachygyria, 46/11, 48
Paine syndrome, 30/514, 43/433
palatoschisis, 43/256, 269, 46/30
pallidocerebello-olivary degeneration, 60/664
pallidopyramidal degeneration, 42/244
palmoplantar keratosis, 42/581
partial trisomy 13, 46/69
Passwell syndrome, 21/216
Patau syndrome, 43/528, 46/69, 50/542, 556, 558
pectus carinatum, 43/257
Pelizaeus-Merzbacher disease, 42/503, 46/57, 59/357
periventricular leukomalacia, 54/42
pertussis, 52/234
phakomatosis, 14/760, 46/42
phenylketonuria, 29/30, 34-37, 42/610, 46/17, 50, 59/75
phosphatidylethanolamine lipidosis, 10/552, 42/429
phosphoglycerate kinase deficiency, 62/489
phosphoribosyl pyrophosphate synthetase, 42/572
photogenic epilepsy, 42/715
pituitary gland, 43/279
plagiocephaly, 30/227
poikilodermatomyositis, 43/258
polydactyly, 43/256
polymicrogyria, 46/11
polyneuritis, 42/341
polyneuropathy, 42/335, 341
porencephaly, 30/681, 685-687, 691, 46/9-11, 50/359
Porot-Filiu syndrome, 42/688
postnatal infection, 46/19
Prader-Labhart-Willi syndrome, 31/322, 324, 43/463, 46/93, 50/590
premature menopause, 42/341
prenatal infection, 46/19
prenatal radiation injury, 46/78
prenatal rubella syndrome, 46/76
prevalence, 43/300
progressive cerebellar dyssynergia, 42/213
progressive external ophthalmoplegia, 62/308
progressive hemifacial atrophy, 22/548, 59/479
progressive muscular atrophy, 42/81
progressive myoclonus epilepsy, 42/704
progressive spinal muscular atrophy, 59/92
pseudo, *see* Pseudomental deficiency
pseudohypoparathyroidism, 46/85, 396, 63/561
pseudoporencephaly, 30/691
pseudopseudohypothyroidism, 46/85
psychiatric disorder, 46/32
psychological adjustment, 46/31
psychological test, 46/4
psychomotor behavior, 46/26-28
psychomotor development, 6/503
psychomotor disorder, 46/26-28
pyknodysostosis, 46/49
Pyle disease, 46/88
pyroglutamic aciduria, 29/230
pyruvate carboxylase deficiency, 29/212, 42/586
pyruvate dehydrogenase deficiency, 29/212, 42/587
4q partial monosomy, 43/496
11q partial monosomy, 43/522
13q partial monosomy, 43/526, 46/69, 50/583
18q partial monosomy, 43/533, 46/70
21q partial monosomy, 43/541
2q partial trisomy, 43/493
4q partial trisomy, 43/498
6q partial trisomy, 43/505
7q partial trisomy, 43/506
9q partial trisomy, 46/69
10q partial trisomy, 43/518
11q partial trisomy, 43/523, 46/69
14q partial trisomy, 43/529
15q partial trisomy, 43/531
19q partial trisomy, 43/538
15q syndrome, 50/590
18q syndrome, 31/589
r(9) syndrome, 43/513, 50/580
r(13) syndrome, 50/585
r(14) syndrome, 50/588

r(15) syndrome, 43/530
r(21) syndrome, 31/494, 43/543
r(22) syndrome, 31/609, 43/547
reaction time, 46/27
recessive inheritance, 46/46-68
recombinant chromosome 3, 43/495
renal abnormality, 43/257, 276
Renpenning type, 43/302
retrolental fibroplasia, 46/28
Rett syndrome, 29/305, 60/636, 63/437
Richards-Rundle syndrome, 22/516, 43/264
Richner-Hanhart syndrome, 42/581
right sided aortic arch, 43/296
Roberts syndrome, 46/79
Robin syndrome, 43/472
Rothmund-Thomson syndrome, 14/777, 43/461, 46/61
Roussy-Lévy syndrome, 21/174
rubella, 46/19
rubella syndrome, 31/212
rubella virus, 56/408
Rubinstein-Taybi syndrome, 13/83, 31/282, 43/234, 46/91, 50/276
Rud syndrome, 13/321, 479, 14/13, 38/9, 43/284, 46/59, 51/398, 60/721
Russell-Silver syndrome, 46/21, 82
saccharopinuria, 29/222
Saethre-Chotzen syndrome, 43/322, 46/87
sagittal suture, 30/225
Salla disease, 43/306, 51/382, 59/360
scaphocephaly, 30/225
scapuloperoneal myopathy, 43/131
Schachter syndrome, 14/792
Schimke syndrome, 62/304
schizencephaly, 30/346, 46/9
scoliosis, 43/124
sea blue histiocyte disease, 42/473
Seckel syndrome, 31/285, 43/378, 46/47
secondary pigmentary retinal degeneration, 13/337-340, 43/293-296, 60/654, 722
secondary speech disorder, 4/420-423
seizure, 42/498, 43/270, 278, 282, 285
self-mutilation encephalopathy, 42/140
sensorineural deafness, 42/131, 385, 389, 43/236, 272, 295, 46/28
sensory disorder, 46/28
sensory neuropathy, 42/346
sensory radicular neuropathy, 42/351
septo-optic dysplasia, 30/320
septum pellucidum agenesis, 30/312, 314-321, 46/13
sex chromosome fragility, 43/287, 303
sex chromosomes, 46/73-76
sex linked recessive aqueduct stenosis, 46/66
sex predominance, 46/66-68
short stature, 43/270, 273-275
sickle cell anemia, 42/623
Sjögren-Larsson syndrome, 13/141, 468, 472, 22/477, 30/248, 43/307, 46/60, 59/358, 66/615
Smith-Lemli-Opitz syndrome, 31/254, 43/308, 46/66
social adjustment, 46/31
spastic cerebral palsy, 42/167
spastic diplegia, 42/715, 46/14
spastic hemiplegia, 46/14
spastic paraplegia, 42/169, 176, 43/277
spastic tetraplegia, 43/295, 46/14
speech, 46/30
speech disorder, 43/257
speech disorder in children, 4/421-425, 43/257
spina bifida, 32/553, 555-557
spinal muscular atrophy, 59/373
spinocerebellar ataxia, 42/181, 60/664
status marmoratus, 46/16
status verrucosus, 30/352
streptococcal meningitis, 52/81
Sturge-Weber syndrome, 31/3, 22, 43/48, 46/95, 50/376, 55/443
succinate semialdehyde dehydrogenase deficiency, 60/664
swayback nose, 43/279, 435
symmetrical spastic cerebral palsy, 42/167
Telfer syndrome, 22/514
temporal lobe agenesis syndrome, 46/13
temporal lobe epilepsy, 42/685
tetraploidy, 43/567
thalidomide syndrome, 46/79
thanatophoric dwarfism, 46/88
thyroid gland dysgenesis, 42/631
Toxoplasma gondii syndrome, 31/212, 219
toxoplasmosis, 35/130, 46/19, 52/358
tranquilizing agent, 30/125
transcobalamin II deficiency, 63/255
traumatic, *see* Traumatic intellectual defect
Treacher-Collins syndrome, 46/46
Treponema pallidum syndrome, 31/212
trichomegaly, 43/294, 46/94
trichopoliodystrophy, 46/61
trigeminal nerve agenesis, 50/214
trigonocephaly, 30/228, 43/486, 50/127
trisomy 8, 31/526, 46/69, 50/563
trisomy 8 mosaicism, 43/511
trisomy 22, 31/525, 43/548, 50/565
true porencephaly, 30/681, 685-687
tryptophanuria, 42/633
tuberculous meningitis, 46/19

tuberous sclerosis, 14/348, 31/6, 8, 43/49, 46/43, 50/371
Turner syndrome, 31/500, 46/74
tyrosinemia type II, 42/635
ubidecarenone deficiency, 66/423
ulegyria, 46/14
undifferentiated, 43/300-302
upper motoneuron disease, 60/657
Urbach-Wiethe disease, 14/777, 42/596
urea cycle defect, 60/664
Usher syndrome, 22/501, 60/664
ventricular dilatation, 43/277
vertebral abnormality, 43/279, 288
viral encephalitis, 46/19
vitamin B1 deficiency, 46/18
vitamin B3 deficiency, 46/18
vitamin B6 deficiency, 42/716
vitamin B6 dependency, 42/716
vitamin B6 dependent xanthurenic aciduria, 29/224
vitamin Bc transport defect, 43/278
Von Hippel-Lindau disease, 46/44, 50/374
W syndrome, 43/271
WARG syndrome, 50/581
Weaver syndrome, 46/82
Wildervanck syndrome, 43/343
Williams syndrome, 43/259, 46/92
Wilson disease, 46/18, 58
Wohlfart-Kugelberg-Welander disease, 59/93
Wolf-Hirschhorn syndrome, 43/497, 46/68
X-linked, *see* X-linked mental deficiency
X-linked optic atrophy, 42/417
xanthinuria, 43/193
xanthurenic aciduria, 42/635
xeroderma pigmentosum, 14/13, 788, 31/290, 43/12, 46/59, 51/399, 60/664
XX male, 43/556
XXX syndrome, 31/493, 43/552, 46/74, 50/553
XXXX syndrome, 31/493, 43/554, 46/74, 50/553, 555
XXXXX syndrome, 31/495, 43/554, 46/74, 50/553, 555
XYY syndrome, 31/487
yellow teeth, 42/691

Mental depression
adiposis dolorosa, 8/347
barbiturate intoxication, 65/509
beta adrenergic receptor blocking agent intoxication, 65/436
calcium antagonist intoxication, 65/441
cardiac arrest, 63/217
chorea-acanthocytosis, 63/281
chrysaora intoxication, 37/51
ethionamide intoxication, 65/481
hypercalcemia, 63/561
β-interferon intoxication, 65/562
isoniazid intoxication, 65/480
Minamata disease, 64/414
myotonic dystrophy, 62/235
organic solvent intoxication, 64/43
paraplegia, 61/537
Parkinson disease, 29/515
reflex sympathetic dystrophy, 61/122
rehabilitation, 55/246
reserpine intoxication, 37/438
sleep disorder, 42/255
speech disorder, 42/255
spinal cord injury, 61/537
striatopallidodentate calcification, 49/423
taurine, 42/255
tetraplegia, 61/537
thalamic syndrome (Dejerine-Roussy), 2/484
transverse spinal cord lesion, 61/537
uremic encephalopathy, 63/506
vigabatrin intoxication, 65/511
vitamin D3 intoxication, 65/569

Mental deterioration
see also Dementia
adiposis dolorosa, 8/347
adult neuronal ceroid lipofuscinosis, 42/466
chronic subdural hematoma, 24/306
constructional apraxia, 45/502
corpus callosum syndrome, 2/761
dominant cerebello-olivary atrophy, 42/136
epilepsy, 15/586
glycosaminoglycanosis type III, 66/299
juvenile parkinsonism, 49/156
Mast syndrome, 42/282
Minamata disease, 36/105, 123
Norman disease, 59/68
phosphatidylethanolamine lipidosis, 10/504, 552
traumatic CSF hypotension, 24/255-257
xeroderma pigmentosum, 51/399

Mental disorder, 46/413-435
see also Psychiatric disorder
behavior change, 46/416
brain tumor, 46/415
corpus callosum tumor, 17/504
definition, 46/413
epilepsy, 15/593, 599, 601, 796
evaluation, 46/415
familial history, 46/415
frontal lobe tumor, 17/241, 244
hemiplegia, 4/219
holoprosencephaly, 50/239
hypertensive encephalopathy, 11/557

neurocysticercosis, 35/300, 46/389, 52/530
occipital lobe tumor, 17/334
organic, *see* Organic mental disorder
parietal lobe tumor, 17/305
patient history, 46/415
precipitating event, 46/415
salicylic acid intoxication, 37/417
subarachnoid hemorrhage, 12/139
temporal lobe tumor, 17/290
thalamic syndrome (Dejerine-Roussy), 2/482-486
Wernicke-Korsakoff syndrome, 28/246
Mental manifestation
Huntington chorea, 49/281
progressive supranuclear palsy, 46/302, 49/239
Mental migraine
symptom schema, 48/156
Mental retardation, *see* Mental deficiency
Mental stress
sleep, 74/651
sweating, 1/455
Mental stress test
autonomic nervous system, 74/624
Mental subnormality
Klinefelter syndrome, 50/548
spinal muscular atrophy, 59/372
Menzel hereditary olivopontocerebellar atrophy, *see* Hereditary olivopontocerebellar atrophy (Menzel)
Meperidine, *see* Pethidine
Mephenesin
anxiety, 37/353
oxidation, 37/353
pyrethroid intoxication, 64/217
Mephenytoin
serum level, 15/689
side effect, 15/711
Mephenytoin intoxication
agranulocytosis, 37/202
aplastic anemia, 37/202
drug induced systemic lupus erythematosus, 37/202
drug interaction, 37/200
fever, 37/202
hepatotoxicity, 37/202
leukopenia, 37/202
lymphadenopathy, 37/202
metabolite, 37/200
normephenytoin, 37/200
pancytopenia, 37/202
rash, 37/202
somnolence, 37/202
stupor, 37/202
thrombocytopenia, 37/202
Mephobarbital intoxication, *see* Methylphenobarbital intoxication
Mepivacaine intoxication
convulsion, 37/404
MEPOP syndrome
Kearns-Sayre-Daroff-Shy syndrome, 62/315
progressive external ophthalmoplegia, 62/315
MEPP, *see* Motor end plate potential
Meprobamate
anxiety, 37/353
neuropathy, 7/519
toxic myopathy, 62/601
Meprobamate intoxication
muscular hypotonia, 65/333
Mepyramine
neuropathy, 7/520
Meralgia paresthetica
cutaneous femoral nerve, 2/40, 19/86, 42/323
lateral cutaneous nerve, 7/310, 19/86
nerve intermittent claudication, 20/794
occupational, 7/334
pain, 19/86, 42/323
pregnancy, 8/15
pseudo, *see* Pseudomeralgia paresthetica
sensory loss, 42/323
terminology, 1/108
Mercaptan
hepatic coma, 27/361, 49/215
hepatic encephalopathy, 27/361
β-Mercaptolactate-cysteine disulfiduria
mental deficiency, 29/119-121
Mercaptopropionyl glycine, *see* Tiopronin
Mercaptopurine
antimetabolite, 39/106
chemotherapy, 39/106
neurologic toxicity, 39/106
polymyositis, 41/80
Mercuria lentis
chronic mercury intoxication, 64/371
Mercuric chloride
toxic myopathy, 62/601
Mercury
analytic method, 64/369
autonomic nervous system, 75/38
CNS content, 64/381
iatrogenic neurological disease, 64/367
inorganic, *see* Inorganic mercury
metabolism, 64/379
metal intoxication, 37/88
molluscs intoxication, 37/88
neurotoxicity, 51/265
neurotoxin, 36/73, 83, 147, 37/2, 88, 46/391, 48/420, 51/68, 265, 479, 63/152, 64/10, 367

tissue distribution, 64/380
toxic encephalopathy, 64/4
toxic neuropathy, 64/10
toxic polyneuropathy, 51/68, 271
Mercury intoxication
see also Minamata disease
abnormal metabolism, 64/383
N-acetylpenicillamine, 64/394
acrodynia, 1/480, 8/337, 51/272, 477
action site, 64/10
acute, *see* Acute mercury intoxication
amalgam filling, 64/368
amyotrophic lateral sclerosis, 59/200
area postrema, 64/381
ataxia, 51/272
autonomic nervous system, 75/503
autonomic polyneuropathy, 51/477
axoplasmic flow, 64/389
blood-brain barrier, 64/388
blood mercury, 64/379, 381
body burden, 64/378
body burden measurement, 64/378
calcium edetic acid, 64/394
carbamate intoxication, 64/187
carrot, 64/368
cell culture, 64/388
cerebellum, 64/381
chelating agent, 46/391, 64/394
cholinergic neurotransmission, 64/389
chorea, 49/559
choroid plexus, 64/381
chronic, *see* Chronic mercury intoxication
cobalt, 64/385
CSF mercury, 64/381
dementia, 46/391
diet, 64/383
dimercaprol, 36/181, 64/394
dysautonomia, 75/503
edetic acid, 36/181-183
embolic mercury, 64/376
enzyme inhibition, 64/393
enzyme polymorphism, 64/385
fish diet, 64/368
glia, 64/381
glutathione, 64/391
glycolysis, 64/391
hair mercury, 64/378
hattery, 64/368
headache, 48/420
hemodialysis, 64/396
history, 64/367
individual susceptibility, 64/377, 382
industrial disease, 64/367
inferior olive, 64/381
inorganic, *see* Inorganic mercury intoxication
mercury biokinetic, 64/380
mercury embolism, 64/367
mercury granuloma, 64/376
mercury vapor, 64/377
mesencephalon, 64/381
methylmercury, 64/368
Morvan fibrillary chorea, 49/293, 559
myoclonus, 64/394
nerve cell, 64/381
nerve conduction, 7/173, 514
nerve pathophysiology, 64/389
neuropathology, 64/381
neuropathy, 7/514
occupational disease, 64/368
organic, *see* Organic mercury intoxication
orthostatic hypotension, 63/152
oxidative metabolism, 64/392
Parkinson disease, 6/243
pathogenesis, 64/387
penicillamine, 36/183
personal hygiene, 64/383
pons, 64/380
prenatal, *see* Minamata disease
protein synthesis, 64/390
Raynaud phenomenon, 8/337
scientific instrument, 64/368
selenium, 64/385
sodium-potassium adenosine triphosphatase, 64/389
subthalamic nucleus, 64/381
succimer, 64/394
symptomatic dystonia, 49/543
trace element, 64/385
treatment, 36/181-184, 64/393
tremor, 6/823, 51/272, 64/394
urine mercury, 64/379
Mercury polyneuropathy
acrodynia, 51/272
anxiety, 51/272
ataxia, 51/272
clinical features, 51/271
CNS change, 51/272
complex organic mercury, 51/271
confusion, 51/272
deafness, 51/272
dorsal spinal nerve root ganglion, 51/271, 273
dysarthria, 51/272
EMG, 51/273
euphoria, 51/272
exposure source, 51/273
granular cell layer, 51/272

hyperhidrosis, 51/272
inorganic mercury, 51/271
irritability, 51/272
mercury vapor, 51/271
Minamata disease, 7/642, 51/272
motor neuropathy, 51/273
nephrotoxicity, 51/272
psychiatric illness, 51/272
Purkinje cell, 51/272
short chain alkylmercury intoxication, 51/271, 273
spasticity, 51/272
stupor, 51/272
tachycardia, 51/273
tissue culture, 51/273
tremor, 51/272
visual field, 51/272
Mercury vapor
mercury intoxication, 64/377
mercury polyneuropathy, 51/271
multifocal ischemic encephalopathy, 64/377
Meridian
vertical, *see* Vertical meridian
Merkel disc
sensation, 7/82
Meroacrania, *see* Meroanencephaly
Meroanencephaly, 42/12-14
see also Anencephaly
acrania subtype, 50/71
foramen magnum, 50/77
iniencephaly, 50/134
neurulation, 50/74
nomenclature, 50/73
skull defect, 42/12
terminology, 30/175, 50/72
Merocrania, *see* Meroanencephaly
Merosin
muscle fiber, 62/31
Merostomata intoxication
arthropoda, 37/58
Carcinoscorpius rotundicauda, 37/58
dizziness, 37/58
headache, 37/58
horseshoe crab, 37/58
ingestion, 37/58
lassitude, 37/58
nausea, 37/58
tachypleus gigas, 37/58
tachypleus tridentatus, 37/58
vomiting, 37/58
xiphosura, 37/58
Merphos, *see* Tributyl phosphorotrithioite
MERRF syndrome
see also Progressive myoclonus epilepsy
ataxia, 62/509
dementia, 62/509
dwarfism, 60/666
dyssynergia cerebellaris myoclonica, 60/599
epilepsy, 62/509, 72/222
gene, 72/136
hearing loss, 62/509
lipoma, 60/663, 667, 66/427
mitochondrial disease, 62/497, 66/393, 73/299
mitochondrial DNA point mutation, 62/509, 66/427
mitochondrial encephalomyopathy, 62/498
mitochondrial protein, 66/17
myoclonic epilepsy, 60/662
myoclonus, 49/618, 62/509
myopathy, 62/509
neuropathy, 62/509
optic atrophy, 62/509
progressive external ophthalmoplegia, 60/56
progressive external ophthalmoplegia classification, 62/290
ragged red fiber, 62/292
secondary pigmentary retinal degeneration, 60/725
spastic paraplegia, 59/429
spasticity, 62/509
striatopallidodentate calcification, 49/424
Merritt, H.H., 1/29
Merzbacher, L., 10/10
Mesantoin, *see* Mephenytoin
Mesaxon
axon, 51/7
definition, 47/2
Guillain-Barré syndrome, 51/251
inner, *see* Inner mesaxon
outer, *see* Outer mesaxon
Mescaline
epilepsy, 72/237
hallucination, 4/333, 45/365
hallucinogenic agent, 65/42
hallucinogenic agent intoxication, 37/330
macrosomatognosia, 4/231
microsomatognosia, 4/231
psychosis, 46/594
toxicity, 37/300
visual perseveration, 45/365
Mescaline intoxication
hallucination, 65/45
mydriasis, 65/46
sympathomimetic effect, 65/45
tremor, 65/46
Mesencephalic artery
anatomy, 11/2, 17, 19

Mesencephalic diencephalic syndrome
memory disorder, 3/273-281
Mesencephalic gray matter
Gilles de la Tourette syndrome, 49/632
Mesencephalic infarction
akinetic mutism, 53/209
amnesia, 53/392
basilar apex syndrome, 53/392
brain stem auditory evoked potential, 53/393
Collier sign, 53/391
computer assisted tomography, 53/392
corectopia iridis, 53/391
hyperconvergence, 53/391
nuclear magnetic resonance, 53/392
peduncular hallucination, 53/392
pseudoabducens nerve sign, 53/391
skew deviation, 53/391
somnolence, 53/391
vertical gaze palsy, 53/391
Mesencephalic pedunculotomy
ballismus, 6/855
chorea, 6/868
Mesencephalic reticular formation
akinesia, 45/162, 166
neglect syndrome, 45/155
Mesencephalic syndrome
Cajal interstitial nucleus, 2/275
central tegmental tract, 2/275
classification, 2/272-283
ophthalmoplegia, 2/276
opisthotonos, 2/283
survey, 2/282
Mesencephalic tectum
Arnold-Chiari malformation type II, 50/406
Mesencephalic tectum beaking
spina bifida, 50/487
Mesencephalic tegmentum syndrome
definition, 2/309
Mesencephalic tumor
symptom, 67/157
Mesencephalic vein
lateral, *see* Lateral mesencephalic vein
Mesencephalitis, *see* Brain stem encephalitis
Mesencephalon
anatomy, 2/272-275
arterial supply, 2/275
astrocytoma, 18/8
dyskinesia, 2/277
EEG, 54/157
hereditary olivopontocerebellar atrophy (Menzel), 21/438
mercury intoxication, 64/381
olivopontocerebellar atrophy (Dejerine-Thomas), 21/425
pathology, 2/275
respiration, 1/653
symptomatology, 2/276
Mesenchymal hypoplasia
iris, *see* Iris mesenchymal hypoplasia
Mesenchymoma
malignant, *see* Midline granuloma
Mesenteric artery
temporal arteritis, 48/317
Mesenteric artery occlusion
chronic superior, *see* Chronic superior mesenteric artery occlusion
deficiency neuropathy, 7/631
malabsorption, 7/631
Mesenteric cyst
multiple nevoid basal cell carcinoma syndrome, 14/468
Mesial cortex
transcortical motor aphasia, 45/20
Mesial occipitotemporal syndrome
migraine, 48/161
Mesial temporal sclerosis
epilepsy, 72/362
temporal lobe epilepsy, 73/69
Mesocephalic anencephaly
cerebellar agenesis, 50/175
Mesocortical pathway
akinesia, 45/165
Mesodermal dysplasia spectrum
axial, *see* Axial mesodermal dysplasia spectrum
Mesodermal insufficiency
cephalocele, 50/98
meso-2,3-Dimercaptosuccinic acid, *see* Succimer
Mesoglia, *see* Oligodendroglioma
Mesolimbic dopamine system
basal ganglion, 49/52
cholecystokinin octapeptide, 49/48
globus pallidus, 49/50
MAO B, 49/47
neurotensin, 49/48
nigrostriatal dopamine receptor, 49/50
noradrenalin, 49/48
prolactin, 49/49
substantia innominata, 49/50
ventral tegmental area, 49/50
Mesolimbic noradrenalin system
acetylcholine, 49/51
GABA, 49/51
Mesolimbic pathway
akinesia, 45/165
dopamine, 29/464
Mesomelic dwarfism

Langer type, 31/272
Nievergelt type, 31/272
Mesopallium
behavior, 2/720
temporal lobe lesion, 2/720
Mesoprosoposchisis
terminology, 30/448
Mesoridazine besylate
chemical formula, 65/275
neuroleptic agent, 65/275
phenothiazine, 65/275
Mesosomatic lipomatosis, *see* Krabbe-Bartels disease
Mestinon, *see* Pyridostigmine
Mesuximide
antiepileptic agent, 37/200
side effect, 15/712
Mesuximide intoxication
drug interaction, 37/200
higher cortical function, 37/203
lethargy, 37/202
metabolite, 37/200
normesuximide, 37/200
reticular activating system, 37/203
somnolence, 37/202
vestibular system, 37/203
Metabolic acidosis
acid base balance, 28/512
Burkitt lymphoma, 63/352
corpus callosum agenesis, 50/167
CSF, 28/512
diabetes mellitus, 70/195
encephalopathy, 45/122
malignant hyperthermia, 41/268, 62/567
methanol intoxication, 36/355, 64/96
organic acid metabolism, 66/641
oxoglutarate dehydrogenase deficiency, 66/419
pyruvate carboxylase deficiency, 29/212, 62/503, 66/418
Reye syndrome, 27/363, 29/333, 335, 56/152, 238
uremic polyneuropathy, 63/512
Metabolic alkalosis
acid base balance, 28/513
CSF, 28/513
hyperventilation syndrome, 63/435
Metabolic ataxia, 21/573-583
aminoaciduria, 21/579
argininosuccinic aciduria, 21/574, 577
ataxia telangiectasia, 21/574, 581, 31/25
Bassen-Kornzweig syndrome, 21/574, 580
Batten disease, 21/575
brain atrophy, 21/577
carbamoylphosphate synthetase deficiency, 21/576
cataract, 21/575
cerebrotendinous xanthomatosis, 21/575
citrullinemia, 21/574, 577
congenital lysine intolerance, 21/578
dementia, 21/575
differential diagnosis, 21/582
eye movement, 21/581
gangliosidosis, 21/575
Hartnup disease, 21/574, 579
hereditary hyperammonemia, 21/574
hyperalaninemia, 21/574, 578
hyperglycemia, 21/578
hyperornithinemia, 21/574, 577
hyperpyruvic acidemia, 21/574
hypothyroidism, 21/575
IgA, 21/581
indicanuria, 21/579
microcephaly, 21/577
neuroblastoma, 21/575
normal betalipoproteinemia, 21/581
nystagmus, 21/580
ornithine carbamoyltransferase deficiency, 21/574, 577
portacaval shunt, 21/576
Refsum disease, 21/575, 581
Reye syndrome, 21/578
subacute necrotizing encephalomyelopathy, 21/578
xanthoma, 21/575
Metabolic disorder
chorea, 49/554
dementia, 46/395-400
drug induced extrapyramidal syndrome, 6/254
epilepsy, 72/221
familial spastic paraplegia, 59/351-361
Hodgkin disease, 39/38-40
inherited, *see* Inborn error of metabolism
leukoencephalopathy, 47/583-599
malignant lymphoma, 39/38-40
mental deficiency, 46/17
myoglobinuria, 62/557
neonatal seizure, 73/255
pellagra, 28/59
progressive muscular atrophy, 59/18
survey, 66/1
Metabolic encephalopathy
acute, *see* Acute metabolic encephalopathy
adenosine triphosphatase 6 mutation syndrome, 62/510
adult polyglucosan body disease, 62/484
agraphia, 45/462
aluminum intoxication, 63/525

anisocoria, 69/400
anoxia, 69/400
beta oxidation defect, 62/494
cancer, 69/396, 399, 401
cardiac arrest, 63/214
carnitine acylcarnitine translocase deficiency, 66/405
central pontine myelinolysis, 69/399
cerebral glycogenosis, 62/490
childhood myoglobinuria, 62/561-563
coma, 45/123
complex V deficiency, 62/506
consciousness, 45/121-123
cytotoxic brain edema, 57/213
delirium, 46/541
dialysis encephalopathy, 63/528
differential diagnosis, 69/407
drug, 69/400
drug overdose, 45/123
eosinophilia myalgia syndrome, 64/257
flaccid paralysis, 69/399
glutaric aciduria type II, 62/495
glycogen storage disease variant, 62/490
heat stroke, 62/573
hepatic encephalopathy, 69/400
hypercalcemia, 69/401
hypernatremia, 63/554, 69/402
hypocalcemia, 63/558
hypoglycemia, 69/404
hyponatremia, 63/547
hypophosphatemia, 63/564
hypothermia, 69/400
hypoxia, 69/400
intoxication, 69/406
lipid metabolic disorder, 62/494
long chain acyl-coenzyme A dehydrogenase deficiency, 62/495
long chain 3-hydroxyacyl-coenzyme A dehydrogenase deficiency, 62/495
medium chain acyl-coenzyme A dehydrogenase deficiency, 62/495, 66/410
MNGIE syndrome, 62/505
motor sign, 69/399
multiple acyl-coenzyme A dehydrogenase deficiency, 62/495
multiple myeloma, 55/470
myoclonus, 49/618, 69/399
nutritional, 69/406
nystagmus, 69/400
ocular motility, 69/400
ocular sign, 69/400
oculomotor disease, 45/123
ophthalmoplegia, 69/400
phosphoglycerate kinase deficiency, 62/489
primary systemic carnitine deficiency, 62/493
pupillary function, 45/122
respiration, 45/121
respiratory encephalopathy, 63/414
salicylic acid, 46/602
seizure, 69/399
septic encephalopathy, 69/405
short chain 3-hydroxyacyl-coenzyme A dehydrogenase deficiency, 62/495
succinylcholine, 69/400
systemic carnitine deficiency, 62/493
toxic shock syndrome, 62/574
vasogenic brain edema, 57/213
water intoxication, 64/240

Metabolic hormone
autonomic nervous system, 74/35

Metabolic myopathy, 41/175-220
acid maltase deficiency, 41/175, 62/480
adenosine triphosphatase 6 mutation syndrome, 62/510
aldosteronism, 40/317
beta oxidation, 62/492
brancher deficiency, 41/184, 62/483
carnitine deficiency, 41/196, 62/492
carnitine palmitoyltransferase deficiency, 41/204, 62/494
complex I deficiency, 62/503
complex II deficiency, 62/504
complex III deficiency, 62/504
complex IV deficiency, 62/505
complex V deficiency, 62/506
debrancher deficiency, 41/181, 62/482
EMG, 62/69
fatty acid oxidation, 62/491
glutaric aciduria type II, 62/495
glycogen storage disease, 41/175-194, 62/479
history, 62/490
Kearns-Sayre-Daroff-Shy syndrome, 62/508
lactate dehydrogenase deficiency, 62/490
lipid metabolic disorder, 41/194-207, 62/490
long chain acyl-coenzyme A dehydrogenase, 62/492
long chain acyl-coenzyme A dehydrogenase deficiency, 62/495
long chain fatty acid, 62/491
long chain 3-hydroxyacyl-coenzyme A dehydrogenase deficiency, 62/495
medium chain acyl-coenzyme A dehydrogenase, 62/492
medium chain acyl-coenzyme A dehydrogenase deficiency, 62/495
mitochondrial myopathy, 41/208-220

MNGIE syndrome, 62/505
multiple acyl-coenzyme A dehydrogenase deficiency, 62/495
myophosphorylase deficiency, 41/185, 62/485
neuromuscular disease, 41/425, 431
oxidation-phosphorylation coupling defect, 62/506
oxidative phosphorylation, 41/216
phosphofructokinase deficiency, 41/190, 62/487
phosphoglycerate kinase deficiency, 62/489
phosphoglycerate mutase deficiency, 62/489
phosphorylase isoenzyme, 41/188
phosphorylase kinase deficiency, 41/192, 62/486
primary systemic carnitine deficiency, 62/493
short chain acyl-coenzyme A dehydrogenase, 62/492
short chain 3-hydroxyacyl-coenzyme A dehydrogenase deficiency, 62/495
systemic carnitine deficiency, 62/493

Metabolic neuropathy
acromegalic neuropathy, 51/68
acute intermittent porphyria, 51/68
carcinoma, 51/68
classification, 7/552
diabetes mellitus, 51/68
hepatic polyneuropathy, 51/68
hypothyroidism, 51/68
intestinal malabsorption, 51/68
multiple myeloma, 51/68
porphyria variegata, 51/68
uremia, 8/1, 51/68
vitamin deficiency, 51/68

Metabolic polyneuropathy
a-alphalipoproteinemia, 51/396
Bassen-Kornzweig syndrome, 51/393
cerebrotendinous xanthomatosis, 51/399
Chédiak-Higashi syndrome, 14/783, 51/397
Cockayne-Neill-Dingwall disease, 51/399
familial hypobetalipoproteinemia, 51/395
hereditary acute porphyria, 51/389
xeroderma pigmentosum, 51/399

Metabolism
acid base, *see* Acid base metabolism
acrylamide intoxication, 64/71
alcohol dehydrogenase, 43/198
alcohol intoxication, 64/109
aluminum, 36/319
amitriptyline, 65/315
aromatic amino acid, 75/235
arsenic intoxication, 64/285
autonomic nervous system, 74/14, 77
basal, *see* Basal metabolic rate
basal ganglion, 49/36
benzodiazepine intoxication, 65/334
bilirubin, 6/491, 27/415
bradykinesia, 1/129
brain, *see* Brain metabolism
brain amino acid, *see* Brain amino acid metabolism
brain death, 24/728, 773
brain glucose, *see* Brain glucose metabolism
brain injury, 23/109, 24/601, 605, 57/211
brain tumor, 69/395
bromide, 36/293
bromide intoxication, 36/293
calcium, *see* Calcium metabolism
cancer, 69/395
carbamate intoxication, 64/188
carbohydrate, *see* Carbohydrate metabolism
ceramide, 66/217
cervical vertebral column injury, 25/288
cholesterol, *see* Cholesterol metabolism
chylomicron, *see* Chylomicron metabolism
consciousness, 3/53, 126
copper, 6/117-123, 27/387-391
cyanide intoxication, 65/28
cyanogenic glycoside intoxication, 65/27
cystine, *see* Cystine metabolism
desipramine, 65/316
energy, *see* Energy metabolism
enzyme deficiency, 66/58
ethylene glycol intoxication, 64/123
fat, *see* Fat metabolism
fatty acid, 66/40
femoral neuropathy, 8/307
GABA, *see* GABA metabolism
galactolipid, 66/3, 199
ganglioside, 10/257, 66/251
genetic gangliosidosis, 66/248
glycerophospholipid, 66/42
glycogen, 27/20-23, 170, 221-224
glycosaminoglycan, *see* Glycosaminoglycan metabolism
glyoxylic acid, *see* Glyoxylic acid metabolism
head injury, 57/233
heterocyclic antidepressant, 65/313
hexachlorophene intoxication, 37/482
holoprosencephaly, 50/238
homocarnosinase, *see* Homocarnosinase metabolism
hydroxyproline, 29/131-133
hypoglycin, 65/83
hypothermia, 71/575
imipramine, 65/316
inborn error, *see* Inborn error of metabolism
inborn error of carbohydrate, *see* Inborn error of

carbohydrate metabolism
intervertebral disc, 20/537
isoleucine, *see* Isoleucine metabolism
lanosterol, 66/46
lead intoxication, 36/3
leucine, *see* Leucine metabolism
leukodystrophy, 66/2
lipid, *see* Lipid metabolism
lipoprotein, *see* Lipoprotein metabolism
lysine, *see* Lysine metabolism
manganese intoxication, 36/222-224
mercury, 64/379
methanol, 64/97
methanol intoxication, 36/354-356
methionine, 29/111
methionine-homocysteine, *see* Methionine-homocysteine metabolism
methylenedioxyphenyl intoxication, 36/440
1-methyl-4-phenyl-1,2,3,6-tetrahydropyridine, 65/380
midbrain syndrome, 24/587, 589
myelin, 66/3
myelination, 51/28
nerve, 7/74
nerve myelin, 51/23, 28
nerve sheath, 7/76
nicotine intoxication, 65/262
noradrenalin, 74/140, 145
nortriptyline, 65/316
organic acid, *see* Organic acid metabolism
organic solvent intoxication, 64/39
organolead intoxication, 64/130
organophosphate, 37/551, 64/155
organophosphate intoxication, 37/551
organotin intoxication, 64/135
parathyroid hormone, *see* Parathyroid hormone metabolism
phenytoin, 15/628
phytanic acid, 27/521, 524-527
pipecolic acid, *see* Pipecolic acid metabolism
porphyria, *see* Porphyrin metabolism
porphyrin, *see* Porphyrin metabolism
proline, 29/129
protein, *see* Protein metabolism
purine, *see* Purine metabolism
pyridoxal, 28/105-109
pyridoxal-5-phosphate, 28/105-109
pyridoxamine, 28/105
pyridoxic acid, 28/105
pyruvate, *see* Pyruvate metabolism
sialic acid, 66/354
sodium, *see* Sodium metabolism
sphingolipid, 66/43
sphingomyelin, 66/139
sphingomyelin phosphodiesterase, 66/139
spinal cord injury, 25/288
sulfur containing amino acid, 29/112
tetrahydrobiopterin, *see* Tetrahydrobiopterin metabolism
transverse spinal cord lesion, 61/278
traumatic vegetative syndrome, 24/581
trichloroethanol intoxication, 65/329
tricyclic antidepressant, 65/313
tryptophan, *see* Tryptophan metabolism
tryptophan nicotinic acid, *see* Tryptophan nicotinic acid metabolism
valine, *see* Valine metabolism
vitamin B_6, *see* Vitamin B_6 metabolism
vitamin B_{12}, 70/369
vitamin B_{12} deficiency, 51/337
vitamin B_W, *see* Vitamin B_W metabolism
Metacarpal
hollow hand sign, 1/184
short, *see* Short metacarpal
Metachromasia
acid phosphatase, 10/53
brown, *see* Brown metachromasia
cause, 10/23-26
experimental allergic encephalomyelitis, 9/38
Ferguson-Critchley syndrome, 22/441
glycosaminoglycanosis, 10/454
glycosaminoglycanosis type I, 10/53
metachromatic leukodystrophy, 10/44, 47, 52
myelin, 9/30
phosphoinositide masking, 10/25
red, 10/22-26
Metachromatic leukodystrophy, 10/43-59, 29/357-363, 60/123-127, 66/163-177
N-acetylgalactosamine-4-sulfatase, 10/28, 568, 29/359
N-acetylglucosamine kinase, 10/48
acid phosphatase, 10/48
adult form, 10/347, 29/358, 42/493, 47/591, 60/125, 66/94, 168
alkaline phosphatase, 10/48
amaurotic idiocy, 10/53
3-amino-2-methylproprionic acid, 10/45
apraxia, 66/168
astrocytosis, 10/51
asymptomatic, 66/93
ataxia, 8/16, 42/491, 493
atypical case, 10/53
Austin urinary test, 10/25
Babinski sign, 42/491-493
behenic acid, 10/49
biochemical aspect, 29/358

biochemical study, 60/125
biochemistry, 10/348, 29/358
biopsy, 66/175
birth incidence, 42/494
blindness, 10/568
bone marrow transplantation, 66/87, 92-95, 176
brain biopsy, 10/680, 685
brown metachromasia, 10/22, 44, 47
Buhot cell, 10/53
carbohydrate deficient glycoprotein syndrome, 66/83
cases reported up to 1969, 10/45
cerebroside, 10/28, 49
cerebroside sulfatase, 10/27, 28, 47, 56, 568, 29/350, 357, 358, 42/491, 494, 47/589, 589-591, 51/368, 60/124, 66/92
cerebroside sulfate ester, 10/25, 44
cerebroside sulfatidase, 10/27, 47
cerebroside sulfatide, 8/16, 10/48
cerebrosulfate ester, 21/51
chemical classification, 10/24
chemical pathology, 10/46, 312
chemical treatment, 10/145
cherry red spot, 66/19
chromosome, 47/590
chromosome 10, 60/126
chromosome 22(q13.31-qter), 60/125
chronic inflammatory demyelinating polyradiculoneuropathy, 51/530
classification, 66/5
clinical features, 10/44-47, 311, 347, 29/357, 47/589-593, 51/368, 59/356, 60/124, 66/166
clinical form, 66/11
clinical presentation, 40/336
CNS, 66/166
congenital, 10/45, 60/124, 66/166
consciousness, 45/121
CSF, 66/175
curvilinear body, 60/126
cytosome, 10/374
definition, 10/37, 60/123
dementia, 10/568, 42/493, 46/401, 59/356
demyelinating neuropathy, 47/622
demyelination, 8/16, 10/19, 50, 29/357, 51/367
detection, 29/359
diagnosis, 66/173
diagnostic procedure, 42/494, 60/127
diagnostic technique, 10/58
diet, 10/58
differential diagnosis, 10/45
dystonia, 66/168
electron microscopy, 10/51, 60/126
emotional disorder, 42/493
encephalopathy, 8/16, 45/119
enzyme abnormality, 66/171
enzyme defect, 29/358, 47/589
enzyme deficiency polyneuropathy, 51/367, 369
epidemiology, 10/568
epilepsy, 10/568, 15/373-418, 72/222
extensor plantar response, 42/493
familial amaurotic idiocy, 10/53
familial diffuse type, 10/17
β-galactosidase, 10/48
galactosyl-3-sulfate ceramide, 42/494
gallbladder peristalsis, 10/58
ganglioside, 9/4
gene defect, 66/171
gene locus, 47/590
gene therapy, 66/176
genetics, 10/46, 314, 568, 29/358, 42/491, 47/592, 60/125
glial dystrophy, 21/65
glucosamine-6-phosphate deaminase, 10/48
glycosaminoglycanosis type I, 10/53
heterozygote detection, 29/359, 42/495
hexagonal subunit, 60/126
Hirch-Peiffer test, 8/16, 10/44
histology, 10/50-55, 347
history, 10/43, 567, 60/123
hydroxylignoceric acid, 10/49
hyporeflexia, 8/16, 42/491
hypotonia, 8/16, 40/336
inclusion body, 10/51
juvenile, *see* Juvenile metachromatic leukodystrophy
kidney change, 10/54
laboratory diagnosis, 47/592
lamellated cytosome, 10/374
late infantile, *see* Late infantile metachromatic leukodystrophy
leukoencephalopathy, 47/589-593
lignoceric acid, 10/49
lipid hexosamine, 10/47
lipidosis, 10/59
long chain fatty acid, 10/49
melanin, 10/48
membranous cytoplasmic inclusion body, 10/372
mental deficiency, 42/493, 46/57
metachromasia, 10/44, 47, 52
metachromatic lipiduria, 10/25
molecular biology, 66/174
multiple sulfatase deficiency, 29/359, 42/493, 47/589, 60/127, 66/65, 169
myelin, 10/140
myelin change, 10/140
myelin dysgenesis, 42/491, 494

nerve biopsy, 8/16, 60/126
nerve conduction, 10/58, 51/368
nerve conduction velocity, 8/16, 59/356, 60/127
neuroglia relationship, 28/410
neuropathology, 10/311
neuropathy, 8/16
neurophysiology, 66/175
nuclear magnetic resonance, 59/356, 66/175
oligodendroglial inclusion, 10/375
ophthalmoplegia, 42/493
overdiagnosis, 59/356
pathogenesis, 10/56
pathology, 10/50, 66/165
pathophysiology, 66/172
peripheral nerve, 66/166
peripheral nerve pathology, 60/126
phosphotungstic acid, 10/47
polyneuropathy, 51/368, 59/356
prelipid, 10/43, 52, 54
prenatal diagnosis, 42/495, 60/127, 66/176
progressive bulbar palsy, 59/375
protagon, 10/47
protease inhibitor, 47/593
protein activator loss, 59/356
pseudoarylsulfatase A deficiency, 47/591
pseudoarylsulfatase deficiency, 59/356, 60/124, 66/176
pseudodeficiency gene, 59/357
psychomotor retardation, 42/491
psychosis, 59/356
Refsum disease, 21/216
schizophrenia, 66/168
Schwann cell, 8/16, 10/372, 42/494, 51/368
secondary pigmentary retinal degeneration, 60/732
segmental demyelination, 51/71
sheep, 10/55
skin fibroblast culture, 47/590
speech disorder in children, 42/493
sphingolipid, 10/25
sphingomyelin, 10/49
sulfatase activator deficiency, 66/170
sulfatide, 10/311-313, 348, 29/357, 42/491
sulfatide activator protein, 47/589, 60/126
sulfolipid, 10/141
sural nerve, 47/593
sural nerve biopsy, 51/372
survey, 66/163
therapy, 60/127
thiosulfate, 47/593
transition, 10/53
treatment, 10/58, 29/360, 47/593, 60/127, 66/176
Tuffstein granule, 21/51, 51/368, 372, 60/126
type, 9/472, 51/368
ultrastructure, 10/372-377, 51/372
uridine diphosphogalactose *N*-acetylglucosamine pyrophosphorylase, 10/48
uridine diphosphoglucose-glycogen transglucosylase, 10/48
urinary arylsulfatase, 10/58
visceral change, 10/54
visceral involvement, 10/54
vitamin A, 10/58, 47/593
in vivo diagnosis, 10/25
zebra body, 51/368, 60/126
Metachromatic leukodystrophy of mink
experimental study, 10/55
human type, 9/670
scrapie, 9/679
Metachromatic leukoencephalopathy, *see* Metachromatic leukodystrophy
Metachromatic lipiduria
metachromatic leukodystrophy, 10/25
Metachromatopsia
hallucination classification, 45/56
Metadrenalin
adrenal medulla, 74/144
noradrenalin, 74/144
pharmacology, 74/144
Metal
Behçet syndrome, 56/604
Guam amyotrophic lateral sclerosis, 59/275-281
heavy, *see* Heavy metal
Kii Peninsula amyotrophic lateral sclerosis, 22/357-359
toxic neuropathy, 7/510
trace, *see* Trace metal
Metal fume fever, *see* Metal intoxication *and* Zinc intoxication
Metal fume intoxication
toxic myopathy, 62/613
Metal intoxication
barium intoxication, 64/353
cisplatin intoxication, 64/353
classification, 37/79
decaborane intoxication, 64/353
diborane intoxication, 64/353
gold intoxication, 64/353
heavy, *see* Heavy metal intoxication
hexaborane intoxication, 64/353
mercury, 37/88
multiple sclerosis, 9/58, 78
pentaborane intoxication, 64/353
platinum intoxication, 64/353
tetraborane intoxication, 64/353
Metaldehyde

toxic myopathy, 62/601
Metallic taste
acute mercury intoxication, 64/375
organolead intoxication, 64/133
toxic shock syndrome, 52/259
uremia, 63/504
Metallothionein
copper, 49/229
Metamidophos, *see* Methamidophos
Metamorphognosia
higher nervous activity disorder, 3/29
Metamorphopsia
central serous retinopathy, 13/142
cerebral asthenopia, 2/662, 670
definition, 2/613
depersonalization, 2/662
Epstein-Barr virus, 56/257
hallucination, 45/361-363
hallucination classification, 45/56
localization, 3/29
migraine, 48/129, 162, 166
occipital lobe epilepsy, 2/647
occipital lobe lesion, 2/662
occipital lobe syndrome, 2/670
occipital lobe tumor, 17/323, 326, 330
parietal lobe, 45/361-363
perception, 45/361
posterior cerebral artery syndrome, 53/411
scotoma, 48/129
thalamic lesion, 2/479
Metamorphotaxia
higher nervous activity disorder, 3/31
Metanephrine, *see* Metadrenalin
Metaphyseal chrondrodysplasia
congenital megacolon, 75/635
Metaphyseal dysplasia, *see* Pyle disease
Metaproterenol, *see* Orciprenaline
Metastasis
astrocytoma, 71/661
biopsy, 71/652
brain, *see* Brain metastasis
brain abscess, 33/114-116, 52/144, 148
brain atrial myxoma, *see* Brain atrial myxoma metastasis
brain tumor headache, 69/23
breast, *see* Breast carcinoma
breast carcinoma, 71/619
cadherin, 71/645
calvarial, *see* Calvarial metastasis
carcinoid syndrome, 70/314
cerebellar medulloblastoma, 18/172, 186-188, 394
cervical, *see* Cervical metastasis
chemotherapy, 71/655
chordoma, 18/160
choriocarcinoma, 63/104
choroid plexus tumor, 71/663
CNS lymphoma, 69/264
colon carcinoma, 71/619
cranial vault tumor, 17/129, 131
CSF, 71/649
droplet, *see* Droplet metastasis
dural, *see* Dural metastasis
ependymoma, 18/125, 71/661
epidural, *see* Epidural metastasis
fibrous dysplasia, 14/198
germ cell tumor, 71/662
glioblastoma, 68/92
glioblastoma multiforme, 18/51
Hodgkin disease, 39/28-38
hourglass tumor, 20/294
hypothalamic tumor, 68/71
idiopathic multiple hemorrhagic sarcoma, 18/288
imaging, 67/172, 179, 185
incidence, 68/530
integrin, 71/645
intracerebral, *see* Brain metastasis
laminin receptor, 71/645
lateral ventricle tumor, 17/606
leptomeningeal, *see* Leptomeningeal metastasis
leukemia, 71/659
liver, *see* Liver metastasis
lung cancer, 71/619
lymphoma, 71/660, 662
malignant astrocytoma, 68/92
malignant lymphoma, 39/28-38
matrix metalloproteinase, 71/644
medulloblastoma, 68/190
melanoma, 71/659
molecule, 71/645
muscle, *see* Muscle metastasis
nervous system, 71/611
neuroblastoma, 8/466, 470, 18/340
neuroimaging, 71/652
neuropathy, 8/131
oligodendroglioma, 72/357
optic chiasm compression, 68/75
optic pathway tumor, 68/74
ovary carcinoma, 71/619
parietal lobe tumor, 17/298
perineural, *see* Perineural metastasis
pineal, *see* Pineal metastasis
pinealoma, 71/662
pituitary, *see* Pituitary metastasis
plexus, *see* Plexus metastasis
primitive neuroectodermal tumor, 68/213, 71/660
protease, 71/645

radiotherapy, 71/654
skull base, *see* Skull base metastasis
skull base tumor, 17/191
spinal cord, *see* Spinal cord metastasis
spinal cord compression, 69/167
spinal cord tumor, *see* Spinal cord metastasis
spinal ependymoma, 20/357
spinal epidural empyema, 52/186
spinal epidural tumor, 20/129
spinal meningioma, 20/189
spinal nerve root, 69/76
spinal neurinoma, 20/244
spine, *see* Spine metastasis
subarachnoid space, 71/660
supratentorial brain tumor, 18/340, 368
thalamic tumor, 68/65
treatment, 71/654
tumor type, 71/658
vertebral, *see* Vertebral metastasis
Metastasis to vertebral column, *see* Vertebral metastasis
Metastatic cardiac tumor
anemia, 63/96
bradyarrhythmia, 63/94
brain embolism, 63/104
brain hypoperfusion, 63/94
brain tumor, 63/93
cachexia, 63/96
conduction disturbance, 63/94
constitutional symptom, 63/94
diagnosis, 63/104
dyspnea, 63/94
echoCG, 63/104
EEG, 63/96
embolic phenomenon, 63/94, 96
endomyocardial biopsy, 63/105
epilepsy, 63/94
fever, 63/96
heart dysfunction, 63/94
increased erythrocyte sedimentation rate, 63/96
inferior vena cava syndrome, 63/94
leukemia, 63/93
leukocytosis, 63/96
lung carcinoma, 63/93
lymphoma, 63/93
malignant melanoma, 63/93
mammary carcinoma, 63/93
nuclear magnetic resonance, 63/105
paroxysmal tachyarrhythmia, 63/94
precordial murmur, 63/94
precordial plop, 63/94
prognosis, 63/105
rash, 63/96
sarcoma, 63/93
serum alanine aminotransferase, 63/96
serum aspartate aminotransferase, 63/96
sudden death, 63/94
superior vena cava syndrome, 63/94
syncope, 63/94, 96
thrombocytosis, 63/96
treatment, 63/105
tumor thrombosis, 63/96
widened pulse pressure, 63/94
Metastatic myopathy, 69/95
pain, 69/72
Metastatic tumor
biopsy, 67/229
Metatarsal
Noack syndrome, 43/321
short, *see* Short metatarsal
Metathalamus
anatomy, 2/529
neural tube, 2/432
Metenkephalin
basal ganglion, 49/21, 33
endogenous opioid, 49/633, 65/349
Gilles de la Tourette syndrome, 49/633
Huntington chorea, 49/258, 298
Parkinson disease, 49/120, 123
proenkephalin, 65/350
Metergoline
migraine, 48/192
Metetoin
serum level, 15/690
side effect, 15/712
Methabarbital, *see* Metharbital
Methacholine
hereditary sensory and autonomic neuropathy type III, 1/478, 21/108, 111, 115, 60/24, 28, 35
tear, 74/427
Methacholine bromide
migraine, 5/98
Methadone
acquired myoglobinuria, 62/577
brain hemorrhage, 55/517
chemistry, 37/366
designer drug, 65/349
opiate, 37/366, 65/350
toxic myopathy, 62/601
Methadone withdrawal syndrome
symptom, 65/354
Methamidophos
organophosphate induced delayed polyneuropathy, 64/176
organophosphate intoxication, 64/155
Methamphetamine

amphetamine intoxication, 65/256
brain vasculitis, 55/520
designer drug, 65/364
epilepsy, 72/234
hallucination, 65/366
phenethylamine, 65/366
Methamphetamine addiction
polyarteritis nodosa, 39/296, 55/360
subarachnoid hemorrhage, 55/360
Methamphetamine intoxication
brain infarction, 65/366
complication, 65/366
epilepsy, 65/366
subarachnoid hemorrhage, 65/366
symptom, 65/366
Methanol
alcoholic amblyopia, 13/62
anion gap, 64/96
bay rum, 64/95
Columbian spirits, 64/95
eagle spirits, 64/95
lion d'or, 64/95
metabolism, 64/97
neurotoxicology, 64/8
neurotoxin, 6/676, 9/500, 36/351, 46/393, 48/420, 64/2, 7, 95
optic atrophy, 13/62
optic neuropathy, 64/8
serum formic acid, 64/97
toxic encephalopathy, 64/2, 7
toxic neuropathy, 64/8
tri-*o*-cresyl phosphate, 64/95
witch hazel, 64/95
Methanol intoxication, 36/351-358, 64/95-103
abdominal pain, 36/353, 64/98
alcohol dehydrogenase, 64/96
alcoholic beverage, 64/98
aldehyde dehydrogenase, 36/354, 357, 64/96
amblyopia, 13/62
amnesia, 36/353, 64/99
anion gap, 64/96, 125
anorexia, 36/353, 64/98
axonal degeneration, 64/101
basal ganglion hemorrhage, 64/100
basal ganglion lesion, 64/99
bicarbonate, 36/353, 357, 64/102
biochemistry, 64/96
blindness, 36/351, 356, 64/99
brain edema, 36/354, 64/99, 101
brain hemorrhage, 64/99
caput nuclei caudati, 64/100
cell asphyxia, 64/98
cellular hypoxia, 64/96
cerebellar infarction, 64/100
clinical features, 64/98
CNS, 36/353
CNS symptom, 64/99
coma, 36/353, 356, 64/99
complex IV, 64/96
confusion, 64/99
dementia, 64/100
differential diagnosis, 64/101
dizziness, 36/353
dysarthria, 64/99
dystonia, 64/100
EEG, 36/354, 64/100
EMG, 64/99
epidemic, 36/357, 64/95
epilepsy, 64/99
ERG, 36/354, 64/99
ethanol administration, 36/357, 64/102
fatality, 64/96
flaccid tetraplegia, 64/99
formaldehyde, 36/354, 357, 64/96
formic acid, 36/354, 357, 64/96
gastrointestinal complication, 36/353
headache, 36/353, 48/420, 64/98
hemodialysis, 36/357, 64/102
hemorrhagic encephalopathy, 64/100
hemorrhagic pancreas necrosis, 36/356, 64/99
hemorrhagic polioleukoencephalopathy, 64/100
hepatic folate system, 64/98
history, 36/351
hypoxic ischemic encephalopathy, 64/102
ipecac, 36/357
laboratory data, 36/354
latent period, 64/98
leukoencephalopathy, 47/555
metabolic acidosis, 36/355, 64/96
metabolism, 36/354-356
methanol exposure, 64/97
methanol TLV, 64/96
4-methylpyrazole, 36/355
myoglobinuria, 64/98
nausea, 36/353, 64/98
neurofilament, 64/101
neuropathology, 64/101
nuclear magnetic resonance, 64/100
optic atrophy, 36/353, 64/99
optic nerve, 36/353, 357, 64/101
optic nerve demyelination, 64/101
optic neuropathy, 64/101
parkinsonism, 64/100
pathology, 36/356
predictive factor, 64/97
prognosis, 36/356

pseudobulbar paralysis, 64/101
pseudomeningitic syndrome, 36/353, 64/99
putamen necrotic infarction, 64/101
putaminal necrosis, 64/99-101
putaminal slit, 36/356, 64/100
reduced nicotinamide adenine dinucleotide/nicotinamide adenine dinucleotide ratio, 64/96
retinal ganglion cell, 36/356, 64/101
salsolinol, 64/98
seizure, 36/353, 356
sensorimotor polyneuropathy, 64/99, 101
serum amylase, 36/354, 64/98
serum creatine kinase, 64/98
serum formic acid, 64/97
sign, 36/352-354
source, 36/351
striatal necrosis, 6/676, 49/500, 503, 64/101
stupor, 36/353, 64/99
subarachnoid hemorrhage, 36/356, 64/102
symptom, 36/352-354
tachypnea, 64/98
tetrahydropapaveroline, 64/98
transverse myelopathy, 64/101
treatment, 36/355, 357, 64/102
visual impairment, 36/352, 64/8, 99
visual manifestation, 36/352
visual symptom, 64/99
Methaqualone
designer drug, 37/355, 65/364
Metharbital
antiepileptic agent, 37/200
serum level, 15/683
side effect, 15/712
Metharbital intoxication
barbital, 37/200
coma, 37/202
drug interaction, 37/200
lethargy, 37/202
metabolite, 37/200
reticular activating system, 37/201
somnolence, 37/202
stupor, 37/202
Methcathinone
designer drug, 65/364
khat leaf, 65/50
Methemoglobinemia
glyceryl trinitrate intoxication, 37/453
Methionine
acquired hepatocerebral degeneration, 49/216
homocystinuria, 55/325
metabolism, 29/111
methyl vitamin B12, 70/378
primary pigmentary retinal degeneration, 13/196
vitamin B12, 70/378
Methionine activating enzyme
tyrosinemia, 42/635
Methionine adenosyltransferase deficiency
endoplasmic reticulum, 29/122
liver biopsy, 29/122
mitochondria, 29/122
Methionine-enkephalin, *see* Metenkephalin
Methionine-homocysteine metabolism
pathway, 55/326
Methionine malabsorption syndrome
convulsion, 29/127, 42/600
diet, 29/127, 42/600
α-hydroxybutyric acid, 29/127, 42/600
mental deficiency, 29/127
Oasthouse urine disease, 29/128, 42/600
Methionine synthesis
vitamin B12, 28/144
Methionylalanine
protein synthesis, 66/77
Methisoprinol
amyotrophic lateral sclerosis, 22/24
Behçet syndrome, 56/605
subacute sclerosing panencephalitis, 56/421
Methodichlorophen, *see* Metodiclorofen
Methohexital intoxication
chemistry, 37/409
convulsant, 37/409
hypnotic, 37/409
Methohexitone intoxication, *see* Methohexital intoxication
Methotrexate
adverse effect, 69/482
antimetabolite, 39/98-104
antineoplastic agent, 65/527
arachnoiditis, 61/151
astrocytoma, 18/39
brain metastasis, 69/201
brain tumor, 18/491, 520, 39/98
cerebellar medulloblastoma, 18/190
chemotherapy, 39/98-104, 65/527
chronic leukoencephalopathy, 47/571, 63/358
CNS leukemia, 69/243, 250
confusional state, 63/357
congenital malformation, 46/79
dermatomyositis, 62/385
encephalomyelopathy, 47/572
encephalopathy, 46/601, 63/355, 357, 65/528
epilepsy, 63/357
glioblastoma multiforme, 18/65
human immunodeficiency virus infection, 71/386
inclusion body myositis, 62/385

intrathecal chemotherapy, 39/100, 61/151
lethargy, 63/357
leukemia, 39/15
leukoencephalopathy, 47/551, 570-573, 69/484
malignant lymphoma, 69/273
meningeal leukemia, 18/251, 39/98, 63/342
meningeal lymphoma, 63/348
meningitis, 61/151
myelopathy, 65/528
neurotoxicity, 39/98-104, 47/551, 63/357, 71/368, 386
neurotoxin, 39/94, 98-104, 119, 46/601, 47/551, 570, 63/357, 65/527, 539, 67/357
non-Hodgkin lymphoma, 69/281
polymyositis, 41/80, 62/385, 71/142
radiculopathy, 63/358
radiosensitizer, 18/491
sarcoidosis, 71/494
systemic lymphoma, 69/273
teratogenic agent, 30/122
transverse myelopathy, 63/358
vasculitis, 71/161
Methotrexate intoxication
acute symptom, 65/529
ataxia, 39/102, 65/530
encephalopathy, 39/101, 65/530
folinic acid, 65/532
headache, 65/529
late symptom, 39/104, 65/531
leukoencephalopathy, 39/100, 65/531
mechanism, 65/531
meningitis serosa, 65/530
myelopathy, 39/100, 65/530
neuropathology, 65/531
pleocytosis, 65/529
radiation influence, 39/100, 65/531
side effect, 65/556
spastic paraplegia, 65/530
stroke like syndrome, 65/530
transverse myelopathy, 65/557
treatment, 65/532
Methotrexate syndrome
bone dysplasia, 30/122
coronal suture, 30/122
fontanelle, 30/122
foot deformity, 30/122
frontal bone, 30/122
growth retardation, 30/122
hypertelorism, 30/122
lambdoid suture, 30/122
micrognathia, 30/122
oxycephaly, 30/122
rib abnormality, 30/122
Methoxychlor
organochlorine insecticide intoxication, 36/405, 64/198
Methoxychlor intoxication
animal study, 36/405
Methoxyflurane
malignant hyperthermia, 41/268
3-Methoxy-4-hydroxyphenylethylene glycol
CSF, *see* CSF 3-methoxy-4-hydroxyphenylethylene glycol
3-Methoxy-4,5-methylenedioxyamphetamine
see also Hallucinogenic agent
amphetamine derivative, 65/50
chemical derivation, 65/50
hallucinogenic agent, 65/42
myristicin, 65/50
phenethylamine derivative, 65/50
3-Methoxy-4,5-methylenedioxyamphetamine intoxication
hallucination, 65/50
5-Methoxy-1-methyl-2-propylaminotetralin
classification, 74/149
2-{2-[4-(2-Methoxyphenyl)-1-piperazinyl]ethyl}-4,4-dimethyl-1,3-(2h,4h)-isoquinolinedione
classification, 74/149
3-Methoxytyramine
Parkinson disease, 49/120
2-Methylacetoacetyl-coenzyme A thiolase deficiency
clinical finding, 66/649
isoleucine metabolism, 66/649
Methyl alcohol intoxication, *see* Methanol intoxication
β-*N*-Methylamino-L-alanine, *see* 2-Amino-3-methylaminopropionic acid
2-Methylamino-1-phenylpropan-1-one, *see* 2-Methylaminopropiophenone
2-Methylaminopropiophenone
designer drug, 65/364
4-Methylaminorex
designer drug, 65/364
N-Methyl-D-aspartic acid
excitotoxin, 64/20
Huntington chorea, 49/58, 261
N-Methylaspartic acid receptor blocker
paraplegia, 61/399, 406
spinal cord injury, 61/399, 406
tetraplegia, 61/399, 406
transverse spinal cord lesion, 61/399, 406
N-Methyl-D-aspartic acid receptor
alcohol intoxication, 64/114
amnesic shellfish poisoning, 65/167
excitotoxic mechanism, 64/20

lathyrism, 65/14
pain, 75/323
N-Methyl-D-aspartic acid receptor blocking agent
anoxic encephalopathy, 63/223
brain ischemia, 63/223
Methylated vitamin B12
Behçet syndrome, 56/605
m-*O*-Methylation
Huntington chorea, 49/285
p-*O*-Methylation
Huntington chorea, 49/285
Methylazoxymethanol glucoside
cycad intoxication, 65/23
Guam amyotrophic lateral sclerosis, 22/130, 344
neurotoxin, 22/344, 64/5
pharmacologic effect, 65/23
toxic encephalopathy, 64/5
Methylazoxymethanol intoxication
experiment, 22/344
symptom, 22/344, 64/5
Methylbenzene, *see* Toluene
Methyl bromide
neurotoxicity, 51/276
Methyl bromide intoxication
action myoclonus, 6/767
acute, 36/311
chronic, 36/311-314
clinical features, 36/311-314
headache, 48/420
industrial use, 36/310
Methyl *n*-butyl ketone, *see* 2-Hexanone
Methyl chloride
chemical formula, 64/82
neurotoxin, 6/652, 36/375, 46/393, 48/420, 49/466, 64/3
organic solvent, 64/82
toxic encephalopathy, 64/3
Methyl chloride intoxication
clinical features, 36/375
headache, 48/420
pallidal necrosis, 6/652, 49/466
symptom, 64/3
3-Methylcholanthrene
craniopharyngioma, 18/537
Methyl-cobalamin, *see* Methyl vitamin B12
β-Methylcrotonic aciduria, *see* β-Methylcrotonylglycinuria
Methylcrotonoyl-coenzyme A carboxylase
mitochondrial disease, 66/417
spinal muscular atrophy, 59/368
3-Methylcrotonoyl-coenzyme A carboxylase, *see* Methylcrotonoyl-coenzyme A carboxylase
Methylcrotonoyl-coenzyme A carboxylase
deficiency, *see* β-Methylcrotonylglycinuria
β-Methylcrotonylglycinuria
acid base metabolism, 28/518
leucine metabolism, 66/652
spinal muscular atrophy, 59/370
α-Methylcyclohexane carboxylic acid
antifibrinolysis, 55/52
brain aneurysm, 55/52
brain aneurysm neurosurgery, 55/52
intravascular consumption coagulopathy, 63/318
subarachnoid hemorrhage, 55/52
4-Methyl-2,5-dimethoxyamphetamine
see also Hallucinogenic agent
amphetamine derivative, 65/50
chemical derivation, 65/50
designer drug, 65/364
hallucinogenic agent, 65/42
phenethylamine derivative, 65/50
4-Methyl-2,5-dimethoxyamphetamine intoxication
symptom, 65/46
Methyl-*N*-dimethyl phosphoramidocyanidate intoxication
chemical classification, 37/544
organophosphorus compound, 37/544
Methyldopa
action mechanism, 37/439
antihypertensive agent, 37/438-440, 63/83, 87
antihypertensive agent intoxication, 37/440
behavior disorder, 46/601
chemical structure, 37/438
choreoathetosis, 37/440
delirium, 46/538
depression, 65/448
fatigue, 65/448
headache, 65/448
historic aspect, 37/438
Huntington chorea, 37/440
hypertensive encephalopathy, 54/217
iatrogenic neurological disease, 65/448
neuroleptic parkinsonism, 6/235
neurologic adverse effect, 65/448
parkinsonism, 37/439, 65/448
side effect, 37/441
tardive dyskinesia, 37/440
vertigo, 65/448
Methyldopa intoxication
carotid endarterectomy, 37/441
choreoathetosis, 37/441
depression, 37/441
dyskinesia, 37/441
dyssomnia, 37/441
facial paralysis, 37/441
galactorrhea, 37/440

headache, 37/440
hemolytic anemia, 37/440
liver toxicity, 37/440
mood change, 37/441
paresthesia, 37/440
parkinsonism, 37/441
sedation, 37/441
sexual impotence, 37/440
speech disorder, 37/441
transient ischemic attack, 37/441
Methylene blue
intrathecal chemotherapy, 61/152
3,4-Methylenedioxyamphetamine
see also Hallucinogenic agent
amphetamine derivative, 65/50
chemical derivation, 65/50
designer drug, 65/364
hallucination, 65/366
hallucinogenic agent, 65/42
phenethylamine, 65/366
phenethylamine derivative, 65/50
popular name, 65/50
3,4-Methylenedioxyamphetamine intoxication
complication, 65/366
epilepsy, 65/366
hallucination, 65/50
opisthotonos, 65/366
symptom, 65/366
3,4-Methylenedioxyethamphetamine
see also Hallucinogenic agent
amphetamine derivative, 65/50
chemical derivation, 65/50
designer drug, 65/364
hallucinogenic agent, 65/42
phenethylamine derivative, 65/50
3,4-Methylenedioxyethamphetamine intoxication
hallucination, 65/50
rhabdomyolysis, 65/367
symptom, 65/367
3,4-Methylenedioxymethamphetamine
see also Hallucinogenic agent
amphetamine derivative, 65/50
chemical derivation, 65/50
designer drug, 65/364
hallucinogenic agent, 65/42
phenethylamine, 65/366
phenethylamine derivative, 65/50
popular name, 65/50
3,4-Methylenedioxymethamphetamine intoxication
brain hemorrhage, 65/367
complication, 65/366
epilepsy, 65/51, 367
hallucination, 65/50
mydriasis, 65/51
rhabdomyolysis, 65/367
subarachnoid hemorrhage, 65/367
symptom, 65/366
Methylenedioxyphenyl intoxication
animal study, 36/439
metabolism, 36/440
symptom, 36/439
5,10-Methylene-*N*-methyltetrahydrofolate reductase deficiency
vitamin B_C deficiency, 51/336
Methylenetetrahydrofolate reductase deficiency
epilepsy, 72/222
homocystinuria, 29/116
Methyl ethyl ketone, *see* 2-Butanone
Methyl ethyl ketone potentiation, *see* 2-Butanone potentiation
3-Methylfentanyl, *see* Mefentanyl
α-Methylfentanyl intoxication
fatality, 65/358
Methylfolate-H4-methyltransferase deficiency
homocystinuria, 29/116
3-Methylglutaconic aciduria
choreoathetosis, 49/385
clinical finding, 66/653
nuclear magnetic resonance, 66/653
phenotype, 66/652
3-Methylglutaconyl coenzyme A hydratase deficiency
organic acid metabolism, 66/641
Methyl green pyronine
skeletal muscle, 40/5
Methylguanidine
uremic encephalopathy, 63/514
uremic polyneuropathy, 51/361, 63/514
3-Methyl histidine
Duchenne muscular dystrophy, 40/380
Methylhydrazine
chemotherapy, 39/118-120
neurologic toxicity, 39/118-120
α-Methyl-β-hydroxybutyric aciduria
acid base metabolism, 28/519
Methyl iodide intoxication
headache, 48/420
2-(1-Methyl-2-isoindolinylmethyl)-2-imidazoline
classification, 74/149
Methyllysergide
Raynaud phenomenon, 8/338
Methylmalonic acid
leukoencephalopathy, 47/585
Methylmalonic acidemia
see also Hyperglycinemia
diagnosis, 29/206

diet, 29/207, 42/603
hepatomegaly, 29/205, 42/601
hypoglycemia, 29/205, 42/601
mental deficiency, 42/601
methylmalonyl CoA carbamyl mutase deficiency, 29/204
methylmalonyl CoA mutase, 29/204, 42/602
methylmalonyl CoA racemase, 29/204, 42/601
microcephaly, 30/509
mitochondrial protein import, 66/396
neutropenia, 29/205, 42/601
type, 66/650
vitamin B12, 42/601, 65/574
vitamin B12 intoxication, 65/574
Methylmalonic aciduria
acid base metabolism, 28/521
ataxia, 55/329
homocystinuria, 55/327, 329
hyperammonemia, 29/105
lactic acidosis, 41/208
osteoporosis, 42/601
subacute combined spinal cord degeneration, 63/255
thrombocytopenia, 42/601
Methylmalonyl CoA carbamyl mutase deficiency
hyperglycinemia, 29/204
methylmalonic acidemia, 29/204
vitamin B12 deficiency, 29/204
Methylmalonyl CoA mutase
methylmalonic acidemia, 29/204, 42/602
vitamin B12 deficiency, 51/338
Methylmalonyl CoA racemase
methylmalonic acidemia, 29/204, 42/601
Methylmercury
cerebellar encephalopathy, 64/5
encephalopathy, 51/265
epilepsy, 64/3
mercury intoxication, 64/368
neurotoxicity, 51/265
neurotoxin, 36/73, 51/265, 64/5, 413
toxic encephalopathy, 64/5
Methylmercury intoxication, *see* Minamata disease
L-α-Methylmetatyrosine
chemistry, 37/234
CNS stimulant, 37/234
Methylnitrosourea
animal experiment, 13/210
brain tumor carcinogenesis, 17/11, 14, 16, 21
α-Methylnorepinephrine, *see* Corbadrine
Methylphenidate
brain vasculitis, 55/520
hyperkinetic child syndrome, 3/197
narcolepsy, 3/95, 15/846
parkinsonism, 6/219
Methylphenidate addiction
brain vasculitis, 55/520
Methylphenobarbital intoxication
drug interaction, 37/200
metabolite, 37/200
phenobarbital, 37/200
1-Methyl-4-phenylpyridinium
analogue, 65/396
analogue structure, 65/396
cell specificity, 65/380
haloperidol, 65/397
1-methyl-4-phenyl-1,2,3,6-tetrahydropyridine, 65/380, 382, 385
1-methyl-4-phenyl-1,2,3,6-tetrahydropyridine intoxication, 65/381, 385
neurotoxin, 65/380
paraquat, 65/397
Parkinson disease, 49/53
pharmacology, 65/380
potency, 65/396
species vulnerability, 65/396
1-Methyl-4-phenyl-1,2,3,6-tetrahydropyridine
aerobic glycolysis, 65/389
akinesia, 49/67
analogue, 65/394
analogue potency, 65/395
analogue structure, 65/395, 397
anatomic selectivity, 65/391
astrocyte, 65/382
astrocyte storage, 65/385
axonal transport, 65/391
complex I inhibition, 65/389
designer drug, 65/364
designer drug intoxication, 64/6
diamorphine intoxication, 65/357
dopamine oxidation, 65/388
dopaminergic neuron uptake, 65/383
enzyme interaction, 65/383
excitotoxic effect, 65/389
free radical, 65/382, 387
history, 65/363
intracellular calcium, 65/390
MAO B conversion, 65/380, 382
MAO concentration, 65/374
metabolism, 65/380
1-methyl-4-phenylpyridinium, 65/380, 382, 385
1-methyl-4-phenylpyridinium clearance, 65/384
1-methyl-4-phenylpyridinium storage, 65/384
mitochondrial inhibition, 65/388
molecular mechanism, 65/379
myoclonia, 65/369
neuromelanin binding, 65/384

neurotoxin, 64/2, 6, 65/363
opiate intoxication, 65/357
oxidative stress, 65/387
Parkinson disease, 49/41, 53, 100, 108, 198
parkinsonism, 65/369
pharmacology, 65/385, 390
primate model, 65/374
rodent model, 65/378
species, 65/373
susceptibility, 65/373
synthesis, 65/369
tissue distribution, 65/380
toxic bioconversion, 65/382
toxic mechanism, 65/387
toxic metabolite, 65/380
tremor, 49/73, 65/369
vesicular storage, 65/385
in vitro study, 65/379
1-Methyl-4-phenyl-1,2,3,6-tetrahydropyridine intoxication
acute symptom, 65/370
aerobic glycolysis, 65/389
age, 65/386
anatomic selectivity, 65/391
animal experiment, 65/373
astrocyte, 65/382
astrocyte storage, 65/385
axonal transport, 65/391
cognitive deficit, 65/371
complex I inhibition, 65/389
computer assisted tomography, 65/372
cytochrome P450, 65/400
dopaminergic neuron, 65/383
EEG, 65/372
EMG, 65/372
eosinophilic inclusion body, 65/371
evoked potential, 65/378
excitotoxic effect, 65/389
extrapyramidal symptom, 65/370
eye movement, 65/372
free radical, 65/382, 387
hallucination, 65/370
history, 65/369
intracellular calcium, 65/390
levodopa, 65/373
MAO B, 65/381
1-methyl-4-phenylpyridinium, 65/381, 385
1-methyl-4-phenylpyridinium clearance, 65/384
1-methyl-4-phenylpyridinium storage, 65/384
molecular mechanism, 65/381
monkey susceptibility, 65/374
neuromelanin binding, 65/384
neuropathology, 65/371
neuroprotection, 65/392
oxidative stress, 65/387
parkinsonism, 65/369
parkinsonism prophylaxis, 65/392
pathobiochemistry, 65/371, 381
pharmacology, 65/385, 390
positron emission tomography, 65/386
postural reflex, 65/372
primate hemiparkinsonism, 65/377
primate inclusion body, 65/376
primate neuropathology, 65/375
primate parkinsonism, 65/374
primate pathochemistry, 65/376
risk factor, 65/370
substantia nigra, 65/371
subtle sign, 65/370
tetrahydroisoquinoline, 65/399
toxic bioconversion, 65/382
toxic mechanism, 65/387
treatment, 65/372
tryptoline, 65/399
vertical gaze palsy, 65/372
vesicular storage, 65/385
Methylprednisolone
cervical spinal cord injury, 61/61
paraplegia, 61/399, 402
spinal cord injury, 61/399, 402
tetraplegia, 61/399, 402
transverse spinal cord lesion, 61/399, 402
Methyl salicylate intoxication
diarrhea, 37/417
drowsiness, 37/417
fever, 37/417
headache, 37/417
hearing loss, 37/417
mental confusion, 37/417
nausea, 37/417
sweating, 37/417
tinnitus, 37/417
vertigo, 37/417
vomiting, 37/417
3-Methylspiperone
Parkinson disease, 49/99
Methylsystox sulfone intoxication
chemical classification, 37/548
organophosphorus compound, 37/548
Methyltetrahydrofolate
vitamin B_c deficiency, 51/335
Methyltetrahydrofolate-homocysteine transferase
vitamin B_{12} deficiency, 51/338
α-Methyl-*p*-tyrosine, *see* Metirosine
Methyl vitamin B_{12}
methionine, 70/378

Methyprylon
clinical use, 37/354
hypnotic agent, 37/354
structural formula, 37/354
Methysergide
carotidynia, 48/339
chronic migrainous neuralgia, 48/252
cluster headache, 5/19, 117, 48/229
intermediate spinal muscular atrophy, 41/326
methylergometrine, 65/66
migraine, 5/18, 56, 100, 381, 48/86, 188, 192
paroxysmal migrainous neuralgia, 5/382
retroperitoneal fibrosis, 65/70
serotonin antagonism, 48/88
serotonin 1 like receptor stimulating agent, 65/66
serotonin 2 receptor blocking agent, 65/66
systemic neoplastic disease, 41/326
Methysergide maleate, *see* Methysergide
Metirosine
akinesia, 49/67
neuroleptic dyskinesia, 65/283
neuroleptic dystonia, 65/283
Metmyoglobin
chemistry, 41/259
Metoclopramide
migraine, 48/125, 197
neuroleptic agent, 65/275
Parkinson disease, 49/108
Metodiclorofen
antineoplastic agent, 65/528
encephalopathy, 65/528
neurotoxin, 65/528
Metolazone
toxic myopathy, 62/597
Metopic suture
see also Craniosynostosis *and* Trigonocephaly
clinical features, 30/227
closure, 50/114
epidemiology, 30/227
external appearance, 30/227
holoprosencephaly, 50/233
malformation, 30/228
mental deficiency, 30/228
r(13) syndrome, 50/584
r(14) syndrome, 50/588
sincipital cephalocele, 50/102
Metopic suture bulge
9p partial monosomy, 50/578
Metopic synostosis
craniosynostosis, 50/117
Metoprolol
beta adrenergic receptor blocking agent, 65/434
classification, 74/149
migraine, 48/180, 203
Metrifonate
organophosphate, 64/152
organophosphate intoxication, 64/152, 155
Metrizamide
adverse effect, 61/149
Arnold-Chiari malformation type II, 50/409
asterixis, 61/166
cauda equina intermittent claudication, 20/722
chemical meningitis, 56/130
delirium, 46/540, 61/165
epilepsy, 61/166
Metrizamide cisternography
bacterial meningitis, 52/9
basal encephalocele, 50/106
basal meningoencephalocele, 50/106
Metrizamide computer assisted tomography
aqueduct stenosis, 50/313
syringomyelia, 50/447
Metrizamide ventriculography
aqueduct stenosis, 50/312
Metronidazole
brain abscess, 52/156
cavernous sinus thrombosis, 52/175
cerebral amebiasis, 52/327
chronic axonal neuropathy, 51/531
cranial epidural empyema, 52/169
drug induced polyneuropathy, 51/301
encephalopathy, 46/602
gram-negative bacillary meningitis, 52/121
iatrogenic neurological disease, 65/487
neuropathy, 69/460
neurotoxin, 46/602, 51/301, 531, 65/487
superior longitudinal sinus thrombosis, 52/179
toxic encephalopathy, 65/487
toxic neuropathy, 51/301
transverse sinus thrombosis, 52/177
Metronidazole intoxication
ataxia, 65/487
axonal degeneration, 65/487
epilepsy, 65/487
polyneuropathy, 65/487
sensory neuropathy, 65/487
sensory polyneuropathy, 65/487
vertigo, 65/487
Metronidazole polyneuropathy
axonopathy, 51/301
sensory neuropathy, 51/301
Mettler, F.A., 1/23, 29
Metyrapone
Cushing syndrome, 68/357
Meuronab CD3, *see* OKT3
Mevalonate kinase deficiency

anemia, 60/674
cataract, 60/653
cholesterol biosynthesis, 66/656
clinical finding, 66/655
facial dysmorphia, 60/666
hepatosplenomegaly, 60/668
mental deficiency, 60/664
nuclear magnetic resonance, 66/655
organic acid metabolism, 66/639
peroxisomal disorder, 66/510
serum creatine kinase, 60/674
Mevalonic aciduria, *see* Mevalonate kinase deficiency
Mevinolin
acquired myoglobinuria, 62/576
nonprimary inflammatory myopathy, 62/372
toxic myopathy, 62/601
Mexico City, 1/12
Neurological Institute, 1/34
Mexiletine
asterixis, 63/192
cardiac pharmacotherapy, 63/192
encephalopathy, 63/192
epilepsy, 63/192
iatrogenic neurological disease, 63/192, 65/454
myotonia, 62/275
myotonic dystrophy, 62/241
tremor, 63/192
Mexiletine intoxication
ataxia, 65/455
headache, 65/455
insomnia, 65/455
nystagmus, 65/455
tinnitus, 65/455
tremor, 65/455
Meyer, A., 1/5
Meyer-Archambault loop
geniculocalcarine tract, 2/534, 557
Meyer-Betz disease
see also Myoglobinuria *and* Rhabdomyolysis
pathologic reaction, 40/263
rhabdomyolysis, 40/263, 340, 41/60
Meyer-Overton rule
alcohol intoxication, 64/113
Meynert decussation
superior colliculus efferent, 2/533
Meynert fasciculus retroflexus
habenular efferent, 2/473
Meynert giant pyramidal cell
visual cortex, 2/537
Meynert nucleus basalis
Down syndrome, 50/530
Parkinson dementia, 46/377
Parkinson disease, 49/110, 112
Meynert solitary cell
visual cortex, 2/537
MGUS, *see* Undetermined significant monoclonal gammopathy
Mianserin
heterocyclic antidepressant, 65/312
migraine, 48/196
Mianserin intoxication
headache, 65/320
tremor, 65/320
Mibelli angiokeratoma syndrome
Bean syndrome, 14/106
xeroderma pigmentosum, 14/529
Michigan Neuropsychological Test Battery
neuropsychologic defect, 45/526
Micky Finn
composition, 65/330
Micrencephaly
congenital heart disease, 63/11
Dandy-Walker syndrome, 42/23
epilepsy, 72/109
GM2 gangliosidosis, 10/415
Klippel-Trénaunay syndrome, 14/395
lissencephaly, 50/258
microcephaly, 30/507, 42/45
occipital encephalocele, 42/26
striatopallidodentate calcification, 49/421
tuberous sclerosis, 14/102, 367
Microabsence
petit mal epilepsy, 15/130
Microadenoma
pituitary, *see* Pituitary microadenoma
Microaneurysm, *see* Charcot-Bouchard aneurysm
Microangioma
subarachnoid hemorrhage, 12/106
Microangiopathy
amyloid, *see* Amyloid microangiopathy
anoxic ischemic leukoencephalopathy, 9/591, 593
brain, *see* Brain microangiopathy
complement mediated, *see* Complement mediated microangiopathy
demyelination, 9/573
diabetes mellitus, 42/544
diabetic polyneuropathy, 51/505
leukoencephalopathy, 9/592
mineralizing, *see* Mineralizing microangiopathy
mineralizing brain, *see* Mineralizing brain microangiopathy
mitral valve prolapse, 9/607
Microcephalic familial leukodystrophy, *see* Norman microcephalic familial leukodystrophy
Microcephaly, 30/507-520

adducted thumb syndrome, 43/331
Aicardi syndrome, 42/696
alaninuria, 42/515
alcohol, 30/511, 50/279
Alpers disease, 30/509, 514, 42/486
aminoaciduria, 42/517
aminopterin syndrome, 30/508
amniotic band syndrome, 30/116
Armendares syndrome, 43/366
autopsy, 50/271
autosomal dominant inheritance, 50/272-274
autosomal recessive inheritance, 50/262, 269-272, 274
axial mesodermal dysplasia spectrum, 50/516
bacterial meningitis, 52/14
Batten disease, 10/614, 42/461
Beckwith-Wiedemann syndrome, 31/329, 43/398
behavior disorder, 30/513
birth incidence, 42/46
Bloom syndrome, 30/100, 508
Börjeson-Forssman-Lehmann syndrome, 30/508
Bowen-Conradi syndrome, 43/335
brain atrophy, 43/427
brain cortex, 50/272
brain cortex disorganization, 50/249
brain cortex heterotopia, 50/248, 261
brain size, 42/46
brain weight, 30/516, 42/46, 50/270
carnosinemia, 42/532
cat eye syndrome, 30/508
cataract, 43/427, 50/278
caudal aplasia, 50/512
cerebro-oculofacioskeletal syndrome, 43/341
cervical spine, 43/428
cervical spine fusion, 43/428
child of phenylketonuric mother, 31/300
chorioretinitis, 30/510, 50/277
chorioretinopathy, 30/512, 43/429
chromosomal aberration, 50/274
classification, 30/508
clinical features, 30/512
CNS malformation, 50/57
Cockayne-Neill-Dingwall disease, 13/432, 30/508, 31/236, 238, 43/350, 60/666
Coffin-Lowry syndrome, 43/239
Coffin-Siris syndrome, 30/508, 43/241
compression neuropathy, 43/428
computer assisted tomography, 50/268, 281, 60/672
congenital heart disease, 30/510, 63/11
congenital heart malformation, 50/272
congenital retinal blindness, 22/533
congenital rubella, 50/278
congenital toxoplasmosis, 35/130, 42/662
Conradi-Hünermann syndrome, 43/346
convulsion, 30/510, 513, 50/272, 277
corpus callosum agenesis, 30/287, 42/7, 43/427, 50/152, 162
craniosynostosis, 50/267
cri du chat syndrome, 30/508, 31/563, 566, 43/501, 50/274
criteria, 30/507-509, 518
Crome syndrome, 43/242
cutis marmorata, 43/429
cutis verticis gyrata, 30/512, 42/45, 43/244, 50/269
cytomegalovirus, 56/268
cytomegalovirus infection, 30/508, 510, 31/212, 215, 50/277, 56/266
De Lange syndrome, 30/508, 31/280, 50/276
deafness, 30/510, 50/277
definition, 30/507, 509
diagnosis, 30/519
dihydrolipoamide dehydrogenase, 41/210
Down syndrome, 31/399, 410, 413, 43/544, 50/274
Dubowitz syndrome, 30/100, 508, 43/247
dwarfism, 43/384
Edwards syndrome, 30/508, 31/517, 521, 50/274, 557
EEC syndrome, 43/396
elfin face syndrome, 43/259
encephalitis, 50/280
epidemiology, 30/156, 515
epiphyseal dysplasia, 43/396-398
etiology, 30/510, 514
experimental, 30/515
face, 50/272
familial type, 50/270
Fanconi syndrome, 30/508, 31/321, 43/16
fetal alcohol syndrome, 30/120, 50/279
fetal hydantoin syndrome, 30/508, 50/279
fetal Minamata disease, 64/419
focal dermal dysplasia, 14/788
foramina parietalia permagna, 42/29
Galloway-Mowat syndrome, 43/431
genetics, 30/156, 507, 512
glycosaminoglycanosis type I, 42/451
Goltz-Gorlin syndrome, 14/113, 788, 30/508, 31/288
growth hormone, 30/513
growth retardation, 30/510, 43/427
Gruber syndrome, 14/115, 505, 30/508, 31/242, 43/391
happy puppet syndrome, 30/509, 515, 31/309
Harper dwarf, 30/100

head circumference, 30/507, 519, 50/270
herpes virus hominis syndrome, 30/508, 510, 31/212
history, 30/507
holoprosencephaly, 30/451, 42/32
hydranencephaly, 50/345
hydrocephalus, 30/510, 50/278
hyperlysinemia, 30/509
hyperphosphatasia, 42/578
hyperreflexia, 43/429
hyperthermia, 50/279
incidence, 50/53, 61
incontinentia pigmenti, 14/85, 216, 30/508, 43/23
infantile neuronal ceroid lipofuscinosis, 60/666, 66/689
infantile spinal muscular atrophy, 59/69
iniencephaly, 50/131
intelligence quotient, 30/519, 50/267
Johanson-Blizzard syndrome, 30/508
Juberg-Hayward syndrome, 43/446
Kallmann syndrome, 30/100
Klinefelter syndrome, 43/560
Klinefelter variant XXXXY, 31/484, 43/560
Klippel-Trénaunay syndrome, 14/395, 524
Krabbe-Bartels disease, 14/529
kyphoscoliosis, 43/427
Langer-Giedion syndrome, 30/508
Lenz syndrome, 43/419
leprechaunism, 42/590
lissencephaly, 42/39, 50/260, 263
lymphoreticular malignancy, 50/270
Mabry syndrome, 30/100
malformation, 30/508, 512
malnutrition, 30/519, 50/280
management, 30/519
Marden-Walker syndrome, 43/424
maternal diabetes mellitus, 50/280
maternal drug intoxication, 30/125
maternal hyperphenylalaninemia, 50/280
maternal phenylketonuria, 30/508, 511, 50/280
measurement, 30/518
megalocornea, 43/270
meningitis, 50/280
mental deficiency, 30/510, 512-515, 518, 42/45, 81, 43/270, 279, 290, 46/47, 50/269, 277
metabolic ataxia, 21/577
methylmalonic acidemia, 30/509
micrencephaly, 30/507, 42/45
microcrania, 30/507, 42/45
microgyria, 30/484, 42/46, 50/277
microphthalmia, 43/429
Miller-Dieker syndrome, 50/592
Minamata disease, 64/419
mitochondrial matrix, 66/410
Möbius syndrome, 30/100
Mohr syndrome, 43/450
morphology, 50/247
muscular atrophy, 42/81
myelin dysgenesis, 42/498-500
Neu syndrome, 43/430
neuronal ceroid lipofuscinosis, 42/461
neuronal migration, 30/518
nonsyndromic, 42/44-46
normal intelligence, 30/519
Norman leukodystrophy, 10/103
Nijmegen breakage syndrome, 60/426
nystagmus, 43/429, 50/269
opticocochleodentate degeneration, 21/538
orthochromatic leukodystrophy, 10/114
oxoglutarate dehydrogenase deficiency, 62/503
4p partial monosomy, 43/497
9p partial monosomy, 50/578
17p partial monosomy, 50/592
4p partial trisomy, 43/500
9p partial trisomy, 43/516
4p syndrome, 31/549-552
pachygyria, 42/46, 50/263
Paine syndrome, 30/509, 514, 43/432, 60/293
parietal cephalocele, 50/108
parietal encephalocele, 50/108
Patau syndrome, 14/120, 31/507, 43/528, 50/556
pathology, 30/515-518
Pelizaeus-Merzbacher disease, 42/503
phenylketonuria, 29/30, 30/509, 42/610
phenytoin, 30/511, 50/279
polymicrogyria, 50/262
porencephaly, 30/685, 50/359
prevalence, 50/267
primary, 30/512
progressive spinal muscular atrophy, 59/92
psychomotor retardation, 42/45
pyruvate dehydrogenase deficiency, 29/212
13q partial monosomy, 43/526, 50/587
18q partial monosomy, 43/532
2q partial trisomy, 43/493
6q partial trisomy, 43/503
10q partial trisomy, 43/517
11q partial trisomy, 43/524
14q partial trisomy, 43/529
15q partial trisomy, 43/531
19q partial trisomy, 43/538
18q syndrome, 31/582
r(13) syndrome, 50/583
r(14) syndrome, 50/588, 590
r(15) syndrome, 43/530
r(21) syndrome, 31/594

r(22) syndrome, 31/607, 43/547
radiation, 50/276
radiologic appearance, 30/519
recombinant chromosome 3, 43/495
renal agenesis syndrome, 50/512
Richner-Hanhart syndrome, 42/581
Robin syndrome, 43/472
Rothmund-Thomson syndrome, 14/777, 43/460
rubella syndrome, 30/508, 510, 31/212
rubella virus, 56/408
Rubinstein-Taybi syndrome, 30/508, 31/282, 43/234, 46/91, 50/276
Ruvalcaba syndrome, 43/290
schizencephaly, 30/346, 50/198
Seckel dwarf, 30/100, 508
Seckel syndrome, 31/285, 43/378
seizure, 42/45, 43/429, 50/269
septum pellucidum, 30/314-321, 325
short stature, 43/429
skull defect, 30/277
Smith-Lemli-Opitz syndrome, 30/508, 31/254, 50/276, 66/582
somniosus, *see* Somniosus microcephalus
spastic cerebral palsy, 42/167
spastic diplegia, 50/269
spastic paraplegia, 50/277
spasticity, 43/427
spinal muscular atrophy, 59/372
strabismus, 43/429, 50/269
symmetrical spastic cerebral palsy, 42/167
Taybi-Linder syndrome, 43/385
Toxoplasma gondii syndrome, 30/510, 31/212
toxoplasmosis, 35/130, 50/277
trichopoliodystrophy, 30/508
trisomy 9, 43/514, 50/564
tuberous sclerosis, 31/2, 6
tyrosinemia type II, 42/635
Williams syndrome, 30/508
X-linked recessive ataxia, 60/636
xeroderma pigmentosum, 30/100, 508, 31/290, 43/12, 51/399, 60/335, 666
XXXX syndrome, 43/554
XXXXX syndrome, 43/554, 50/555
Zackai syndrome, 30/100
Microcoria
Marfan syndrome, 42/403
microcornea, 42/403
myopia, 42/403
spherophakia, 42/403
Microcornea
microcoria, 42/403
Minamata disease, 64/418
oculocerebrofacial syndrome, 43/280
Microcrania
Juberg-Hayward syndrome, 43/446
microcephaly, 30/507, 42/45
Rubinstein-Taybi syndrome, 43/234
Microcytic anemia, *see* Iron deficiency anemia
Microdontia
Down syndrome, 50/526
Microdysgenesis
epilepsy, 72/109, 111
Microelectrode
synapse, 1/52, 55-57
Microembolism
brain, *see* Brain microembolism
brain infarction, 53/108
fat embolism, 9/588
Microencephaly, *see* Micrencephaly
Microfibrillar vitreoretinal degeneration (Favre), *see* Hyaloidoretinal degeneration (Wagner)
Microfilament
nerve, 51/6
Microglia
bone marrow transplantation, 66/89
brain infarction, 11/148
CNS, 50/7
donor bone marrow, 66/90
encephalitis, *see* Acute disseminated encephalomyelitis
globoid cell leukodystrophy, 10/77
GM2 gangliosidosis, 10/398
Lyme disease, 52/263
plaque, 9/272
RES tumor, 18/233
tuberous sclerosis, 14/42
turnover, 66/91
Microglia cell
CNS tumor, 18/233
origin, 30/72
tuberous sclerosis, 14/42
Microglia nodule
acquired immune deficiency syndrome, 56/508
acquired toxoplasmosis, 35/123, 52/353, 357
trichinosis, 52/572
Microglial encephalitis, *see* Acute disseminated encephalomyelitis
Microglioma
brain biopsy, 16/724
clinical features, 18/236
histology, 18/241
RES tumor, 18/236-247
reticulosarcoma, 39/526-529
terminology, 18/234
Microgliomatosis
brain, *see* Primary CNS lymphoma

terminology, 18/234
β2-Microglobulin
back pain, 63/529
camptodactyly, 63/529
discopathy, 63/529
meningeal lymphoma, 63/347
radiculopathy, 63/529
spinal nerve root compression, 63/529
spondylarthropathy, 63/529
structure, 63/528
tetraplegia, 63/529
uremic polyneuropathy, 63/511
β2-Microglobulin amyloidosis
atlantoaxial dislocation, 63/529
camptodactyly, 63/529
carpal tunnel syndrome, 63/528
destructive spondyloarthropathy, 63/529
dialysis, 63/528
dialysis encephalopathy, 63/528
radiculopathy, 63/529
scapulohumeral periarthritis, 63/529
tetraplegia, 63/529
tissue deposition, 63/528
uremia, 63/528
Micrognathia
acrocephalosyndactyly type II, 43/324
Bloom syndrome, 43/10
Bowen-Conradi syndrome, 43/335
cerebrocostomandibular syndrome, 43/237
cerebrohepatorenal syndrome, 43/339
cerebro-oculofacioskeletal syndrome, 43/341
Cockayne-Neill-Dingwall disease, 43/350
corpus callosum agenesis, 42/7
craniodigital syndrome, 43/243
craniofacial dysostosis (Crouzon), 43/361
Dubowitz syndrome, 43/248
dwarfism, 43/380
dysosteosclerosis, 43/393
Edwards syndrome, 50/275
fetal face syndrome, 43/400
Freeman-Sheldon syndrome, 43/353
Herrmann-Opitz syndrome, 43/325
Marden-Walker syndrome, 43/424
Melnick-Needles syndrome, 43/452
methotrexate syndrome, 30/122
Möbius syndrome, 50/215
monosomy 21, 43/540
Neu syndrome, 43/430
oculocerebrofacial syndrome, 43/280
8p partial monosomy, 43/508
Patau syndrome, 43/527
Pena-Shokeir syndrome type I, 43/438
pyruvate dehydrogenase complex deficiency, 66/416
4q partial monosomy, 43/496
13q partial monosomy, 50/587
Robin syndrome, 43/471
Rubinstein-Taybi syndrome, 43/234
Russell-Silver syndrome, 43/477, 46/21
Seckel syndrome, 43/379
sleep apnea syndrome, 45/132
Smith-Lemli-Opitz syndrome, 50/276
Treacher-Collins syndrome, 43/421
trichopoliodystrophy, 29/287
trisomy 9, 43/514
Micrographia
agraphia, 4/168, 45/464
ataxia, 45/464
athetosis, 45/464
bradykinesia, 45/465
chorea, 45/464
dystonia, 45/464
levodopa, 45/465
Parkinson disease, 49/93
parkinsonism, 45/464
Pick disease, 46/234
tremor, 45/464
Microgyria
see also Brain cortex heterotopia
anasarca, 30/483
arhinencephaly, 30/484
Arnold-Chiari malformation type I, 30/484, 50/254
Arnold-Chiari malformation type II, 50/405, 407
biochemical finding, 30/483
brain cortex dysplasia, 50/257
cell migration, 30/479, 481
cerebellar cortical dysplasia, 30/496
congenital syphilis, 30/483
corpus callosum agenesis, 50/150
cortex layer type, 30/480-482
cytomegalovirus infection, 30/483, 50/277
dysraphia, 30/484
dystopic cortical myelinogenesis, 50/206
embryology, 30/482
four layer cortex, 50/257
frequency, 30/485
heterotopia, 30/484
history, 30/479
homocystinuria, 55/330
hydranencephaly, 50/337, 342
incontinentia pigmenti, 14/214
infantile, 1/199
lissencephaly, 30/485
macroscopic appearance, 30/480
malformation, 30/484

mental deficiency, 46/12
microcephaly, 30/484, 42/46, 50/277
microscopic appearance, 30/480, 484
morphology, 50/247
myelin dysgenesis, 42/498
neurofibromatosis, 14/37
orthochromatic leukodystrophy, 10/114
pachygyria, 30/484
pathogenesis, 30/479, 483
periventricular calcification, 50/258
polymicrogyria, 30/484, 50/254
pseudo, *see* Pseudomicrogyria
pseudobulbar paralysis, 1/199
schizencephaly, 30/341, 345, 483, 50/199
sincipital encephalocele, 50/103
spina bifida, 30/484, 42/55, 50/486
status verrucosus deformis, 30/479, 50/202
Sturge-Weber syndrome, 14/66
thanatophoric dwarfism, 43/388
topography, 30/483
toxoplasmosis, 30/483
tuberous sclerosis, 14/42
Turner syndrome, 50/545
ulegyria, 30/479, 50/254
vascular abnormality, 30/482
Microhydrocephaly
hydranencephaly, 50/345
Microneurography
autonomic nervous system, 74/608
Microneuron
thalamic, *see* Thalamic microneuron
Microphthalmia
Aicardi syndrome, 42/696, 50/165
anophthalmia, 42/404, 43/419
basal encephalocele, 50/106
basal meningoencephalocele, 50/106
birth incidence, 42/404
Bitter syndrome, 14/107
cerebro-oculofacioskeletal syndrome, 43/341
corpus callosum agenesis, 50/162
cytomegalovirus, 56/268
definition, 30/469
Edwards syndrome, 50/562
embryology, 30/469
Fanconi syndrome, 43/16
genetics, 30/470
Goltz-Gorlin syndrome, 14/787, 43/20
Gorlin-Chaudry-Moss syndrome, 43/369
Gruber syndrome, 14/505
Hallermann-Streiff syndrome, 43/403
holoprosencephaly, 30/470
hydranencephaly, 30/676, 50/342
Lenz syndrome, 43/419
mental deficiency, 43/281-283
microcephaly, 43/429
neurology, 42/403
Norrie disease, 22/508
oculocerebrocutaneous syndrome, 43/444
oculodento-osseous dysplasia, 43/445
oculovertebral syndrome, 31/248
optic nerve agenesis, 50/211
Patau syndrome, 14/120, 31/504, 43/528
pathologic anatomy, 30/470
polyploidy, 50/555
primary pigmentary retinal degeneration, 13/178
10q partial trisomy, 43/517
r(9) syndrome, 50/584
Stargardt disease, 13/133
Sturge-Weber syndrome, 14/512
thalidomide syndrome, 42/660
trisomy 9, 50/564
Micropolygyria, *see* Polymicrogyria
Micropsia
central *vs* retinal cause, 2/613
central serous retinopathy, 13/142
hallucination classification, 45/56
higher nervous activity disorder, 3/29
migraine, 48/130, 162
parietal lobe, 45/362
visual cortex stimulation, 2/646
visual hallucination, 45/356
Microretrognathia
Edwards syndrome, 50/561
Microscopic polyangiitis
vasculitis, 71/8
Microsomatognosia
body scheme disorder, 4/230, 45/385
epilepsy, 4/230, 45/509
lysergide intoxication, 4/231
mescaline, 4/231
migraine, 4/230, 45/510
psychosis, 4/231
schizophrenia, 4/231
Microsomia
hemifacial, *see* Goldenhar syndrome
Microstomia
Edwards syndrome, 50/275
Freeman-Sheldon syndrome, 38/413, 43/353
Microtia
axial mesodermal dysplasia spectrum, 50/516
Treacher-Collins syndrome, 43/421
Microtubule
acetonylacetone intoxication, 64/90
model, 7/17
neural connection, 30/73
vincristine polyneuropathy, 51/299

Microvilli
Ranvier node, 47/10, 51/12
Schwann cell, *see* Schwann cell microvilli
Microwave lesion
nerve injury, 51/134
Micturition
see also Bladder
autonomic nervous system, 75/665
brain embolism, 53/207
brain infarction, 53/207
central control, 75/668
cerebrovascular disease, 53/207
epilepsy, 15/95, 140
frontal lobe tumor, 17/258
hypothalamus, 2/441
innervation, 61/293
intracranial hemorrhage, 53/207
intrathoracic pressure, 75/716
mechanism, 61/293
physiology, 1/365, 373
pudendal nerve, 1/366
reflex mechanism, 1/375-377, 388
sensation, 1/378
serotonin, 75/670
spinal control, 61/293
spinal cord injury, 26/414
transient ischemic attack, 53/207
vitamin B12 deficiency, 51/341
Micturition reflex
autonomic nervous system, 74/75
Micturition syncope
autonomic polyneuropathy, 51/487
dysautonomia, 22/264
Midazolam
antiepileptic agent, 65/497
brain infarction, 53/419
organophosphate intoxication, 64/173
status epilepticus, 73/328
Midbrain
autonomic nervous system, 74/155
EEG, 72/291
epidural hematoma, 57/256
Foville syndrome, 1/187
Midbrain injury
ocular imbalance, 24/79
Midbrain reticular formation
pain, 45/237
Midbrain syndrome
acute, 24/584
Benedikt syndrome, 2/298, 6/162
blood pressure, 24/585
brain death, 24/587
hyperthermia, 24/585
metabolism, 24/587, 589
oculocephalic reflex, 24/585
pulse, 24/585
pupil, 24/584, 74/416
secondary, 24/587
sign, 24/585
traumatic psychosyndrome, 24/583, 705
tumor, *see* Midbrain tumor
variant, 1/187
Midbrain tremor, *see* Rubral tremor
Midbrain tumor
anatomy, 17/620, 622, 625, 627
angiography, 17/639
astrocytoma, 17/638
blood supply, 17/628
CSF cytology, 17/635
EEG, 17/635
endocrinology, 17/635
incidence, 17/631
multilocular pineal cyst, 17/637
pathology, 17/629
prognosis, 17/643
psychiatric disorder, 17/634
radiology, 17/636
radiotherapy, 17/642
symptom, 17/632
tectum, 17/620
treatment, 17/641
Middle cavernous sinus syndrome
aneurysm, 12/89, 31/147
nasopharyngeal tumor, 2/90
subarachnoid hemorrhage, 12/89, 31/147
Middle cerebellar artery (Foix-Hillemand), *see* Anterior inferior cerebellar artery
Middle cerebral artery
accessory, 11/84
anatomy, 11/9-16, 53/353
aphasia, 4/87
apraxia, 53/363
bacterial endocarditis, 52/293
brain aneurysm, 12/91
brain embolism, 53/355
brain infarction, 11/304, 53/353, 355, 363
branch, 11/12-15, 53/353
carotid angiography, 53/354
confusional state, 55/138
congenital aneurysm, 31/152
delirium, 53/363
diameter, 53/355
dissection, 53/355
headache, 5/129, 48/109
horseradish peroxidase, 48/108
insular segment, 11/84

length, 53/355
migraine, 48/135, 53/355
opercular segment, 11/86
perforating artery, 53/355
pure motor hemiplegia, 53/364
radioanatomy, 11/80-88
radiologic segment, 11/68
Sneddon syndrome, 55/402
sphenoidal segment, 11/82
supply area, 53/353
terminal segment, 11/86
thrombosis, 53/355
transcranial Doppler sonography, 54/12
trigeminal ganglion, 48/109
tuberculous meningitis, 52/198
Middle cerebral artery aneurysm
brain aneurysm neurosurgery, 55/68
surgery, 12/221
Middle cerebral artery infarction
brain stem auditory evoked potential, 54/162
computer assisted tomography, 54/54
EOG, 54/162
optokinetic nystagmus, 54/162
Middle cerebral artery occlusion
brain infarction, 53/107
experimental animal, 53/123
hemiplegia, 53/122
penumbra, 53/119
polymicrogyria, 50/257
striatal necrosis, 49/500
Middle cerebral artery syndrome, 53/353-368
agitated delirium, 53/363
angular gyrus syndrome, 53/363
anisocoria, 53/358
anosognosia, 53/363
aphasia, 53/363
apraxia, 53/362
arteriolar fibrinohyalinoid degeneration, 53/364
brain edema, 53/358
cause, 53/357
Cheyne-Stokes respiration, 53/358
clinical features, 53/357
collateral circulation, 53/355
computer assisted tomography, 53/358
constructional apraxia, 53/363
deep type, 11/305
dominant hemisphere, 53/361, 363
dyspraxia, 53/362
dysprosody, 53/362
embolism, 53/353
external internal carotid bypass, 53/367
fibrinolytic agent, 53/368
Gerstmann syndrome, 53/363
heart disease, 53/356
hemi-inattention, 53/363
ideomotor apraxia, 53/363
infarction, 11/304, 53/353
lenticulostriate artery, 53/365
middle cerebral artery division, 53/361
middle cerebral artery stem occlusion, 53/358
middle cerebral artery stem stenosis, 53/366
misdiagnosis, 53/363
motor aphasia, 53/362
moyamoya disease, 53/355
nondominant hemisphere, 53/362
onset type, 53/357
operculum syndrome, 53/359
perforating artery, 53/364
prognosis, 53/367
race, 53/366
speech, 53/362
streptokinase, 53/368
superficial type, 11/305
tissue plasminogen activator, 53/368
toxic encephalopathy, 53/363
treatment, 53/367
two step middle cerebral artery embolus syndrome, 53/366
variant, 53/366
visual field defect, 53/363
Middle cerebral vein
deep, *see* Deep middle cerebral vein
superficial, *see* Superficial middle cerebral vein
Middle cervical ganglion
cardiac dysrhythmia, 63/238
Middle ear
cholesteatoma, 5/211
Edwards syndrome, 50/562
paratrigeminal syndrome, 48/332
trigeminal ganglion, 48/333
Middle ear carcinoma
skull base tumor, 17/190
Middle ear infection, *see* Otitis media
Middle ear sarcoma
skull base tumor, 17/190
Middle ear tumor
headache, 5/211
Middle fossa
brain stem abscess, 33/139
epidural hematoma, 24/263, 266
skull base metastasis, 69/128
skull fracture, 23/395
Middle fossa arachnoid cyst
temporal lobe agenesis syndrome, 30/357-359
Middle fossa fracture
intracranial infection, 24/224

temporal bone fracture, 23/395
Middle fossa meningioma
nasolacrimal reflex, 2/57
Middle fossa syndrome
brain metastasis, 71/612
Middle frontal vein
anatomy, 11/47
Middle meningeal artery
accessory branch, 11/66
anatomy, 30/418
anterior branch, 11/65
development, 30/417
epidural hematoma, 57/252, 256
lacrimal artery, 11/71
origin variation, 53/299
posterior branch, 11/65
subarachnoid hemorrhage, 12/162
Middle meningeal system
multiple cranial neuropathy, 51/569
Middle meningeal vein
anatomy, 11/52, 54, 61
Middle peduncular artery
radioanatomy, 11/93
topography, 11/93
Middle posterior choroidal artery
radioanatomy, 11/93
topography, 11/93
Middle temporal artery
radioanatomy, 11/86
topography, 11/86
Midface hypoplasia
Gorlin-Chaudry-Moss syndrome, 43/369
Midfrequency hearing loss
optic atrophy, 42/414
Midline calcification
lissencephaly, 50/259
Midline granuloma, 39/347-353
see also Wegener granulomatosis
chronic meningitis, 56/645
clinical features, 39/350
differential diagnosis, 39/351
pathology, 39/349
treatment, 39/352
Midline shift
see also Brain herniation
computer assisted tomography, 57/217
epidural hematoma, 16/102
head injury, 57/217
ventriculography, 57/217
Midline syndrome
cerebral, *see* Cerebral midline syndrome
Midline tumor
supratentorial, *see* Supratentorial midline tumor
Midodrine
blood pressure, 75/724
Miescher syndrome type II
acanthosis, 14/119
acanthosis nigricans, 14/120
cutis verticis gyrata, 14/120
dental abnormality, 14/120
diabetes mellitus, 14/119
differential diagnosis, 14/120
hypertrichosis, 14/120
infantilism, 14/120
neurocutaneous syndrome, 14/101
Sjögren-Larsson syndrome, 14/120
Mietens-Weber syndrome
contracture, 43/304
corneal opacity, 43/304
epiphyseal dysplasia, 43/304
growth retardation, 43/304
mental deficiency, 43/304
strabismus, 43/304
Mifepristone
Cushing syndrome, 68/357
Migraine
see also Headache
abdominal, *see* Abdominal migraine
abdominal pain, 48/3, 8, 38, 159
abortive, *see* Abortive migraine
absenteeism, 48/14
accompagnée, *see* Complicated migraine
acebutolol, 48/180
acetic acid, 48/174
acetylcholine, 5/39
acetylsalicylic acid, 5/98, 48/39, 174, 177
acoustic, 48/162
adenine nucleotides, 48/100
adenosine phosphate, 48/92, 100
adenosine triphosphate, 48/100
adrenalin, 48/91
age, 48/118
aged, 48/19, 120
aggression, 48/161
agitation, 48/162, 165
agnosia, 48/161
agraphia, 48/166
alcohol, 48/85, 122, 146
aldehyde dehydrogenase, 48/86
alexia without agraphia, 48/161
Alice in Wonderland syndrome, 5/244
allergic headache, 5/266
allergy, 5/266, 48/20
alprenolol, 48/180
alternating infantile hemiplegia, 48/146
altitude illness, 48/399

amaurosis fugax, 55/108
amine, 48/61, 85
amino acid, 48/73
aminoaciduria, 5/42
amitriptyline, 5/101, 48/87, 179, 196
amnesia, 48/8, 156, 161
amyl nitrite, 5/379
analgesic agent, 48/39, 173
anesthesia, 48/143
anger, 48/162
angiography, 48/60, 73, 148-150
angioneurotic edema, 48/148
anorexia, 48/125
anterior choroidal artery, 5/85
anterior ischemic optic neuropathy, 55/112
anthranilic acid, 48/174, 176
anticonvulsant, 5/51, 102, 48/202
antidepressant, 48/97
antiemetic agent, 48/197
antihistaminic agent, 48/201
anti-inflammatory agent, 48/173
antiphospholipid antibody, 63/122, 330
anxiety, 5/381, 48/39, 79, 121, 161, 354
aorta coarctation, 63/4
apathy, 48/161
aphasia, 4/86, 5/63, 45/322, 46/150, 48/70
aromatic levo amino acid decarboxylase, 48/86
arterial spasm, 48/27
arteriosclerosis, 48/156
asthenopic scotoma, 48/162
ataxia, 21/570, 48/147
atenolol, 48/180
attack, 48/124
attack profile, 5/49, 52
attack treatment, 5/96, 99, 379
auditory perception, 45/510
aura, 5/46, 48/4, 18, 35, 39, 59, 70, 74, 76, 117, 125, 144, 150, 156
aura status, 48/157
aura treatment, 5/379
autohypnosis, 48/39
autointoxication, 48/143, 149
autokinesis, 48/162
automatism, 48/162
autoscopia, 4/231-233, 45/387, 510, 48/162
autosomal dominant hypoalphalipoproteinemia, 60/136
autosomal dominant inheritance, 48/146
azapropazone, 48/174
barbiturate, 5/98, 101, 48/202
basilar, 48/7, 19, 156
basilar artery, *see* Basilar artery migraine
basilar artery vasoconstriction, 53/388
behavior change, 46/416
benign paroxysmal torticollis, 48/38
benign recurrent vertigo, 48/6, 157
benorilate, 48/174
benzodiazepine, 5/101, 48/202
beta adrenergic receptor blocking agent, 48/151, 179-181, 65/434
bilateral, 48/124
biochemistry, 5/38, 231, 48/85-101
biofeedback, 48/39
biological clock, 48/120
biological rhythm, 74/507
blood-brain barrier, 48/61-63, 79, 93
blood flow, 48/150, 75/295
blood pressure, 48/20, 73, 75/296
blood sodium, 5/231
blood vessel, 48/79
body scheme, 48/129, 161
body scheme disorder, 4/214, 48/162
bradycardia, 48/8
bradykinin, 1/142, 5/39, 48/97
bradyphrenia, 48/161
brain aneurysm, 5/141
brain arteriovenous malformation, 53/35, 54/376
brain artery, 48/61
brain blood flow, 48/60, 63, 70, 73, 79, 54/149
brain infarction, 53/37, 163
brain injury, 23/12
brain ischemia, 53/163
brain stem, 5/83
brain vasospasm, 23/173
bromocriptine, 48/92
caffeine, 5/98
calcium antagonist, 48/200
calcium antagonist intoxication, 65/441
calcium channel blocking agent, 48/199
calcium entry blocker, 48/151
capillary loop of finger, 5/55
carbamazepine, 48/202
carbon dioxide, 48/73, 76, 75/295
carboxylic acid, 48/174
cardiac, *see* Cardiac migraine
cardiac catheterization, 63/177
carotid angiography, 5/19, 48/72
carotid artery, 48/77
catamenial, *see* Catamenial migraine
catecholamine, 5/41, 48/91
cenesthesiopathy, 48/162
cerebellar, *see* Cerebellar migraine
cerebrovascular disease, 45/510, 53/28
cervical, *see* Cervical migraine
cervicogenic headache, 48/405
cheiro-oral, 48/6, 156

cheiro-oral paresthesia, 48/157
chemotherapy, 67/355
chest pain, 48/158
child, 5/239, 241, 48/3, 18, 146
Chinese restaurant syndrome, 48/419
chlorpromazine, 5/380
chocolate, 48/93
chorea, 5/80
chronic migrainous neuralgia, 5/265, 48/250
chronic muscle contraction headache, 48/344
chronic neuralgia, *see* Chronic migrainous neuralgia
ciclosporin, 63/179
circadian rhythm, 48/123
classic, *see* Classic migraine
classification, 5/45
climate, 48/123
clinical features, 5/45, 48, 48/37, 117-132
clivus ridge syndrome, 2/298
clock function, 74/468
clonidine, 37/443, 48/151, 180, 186
cluster headache, 5/265, 48/78, 227, 240, 269
cluster migraine syndrome, 48/268
coca cola, 48/122
cocoa, 48/93
codeine, 5/98
cognition, 48/162
coital, *see* Coital migraine
coitus, 48/377
cold, 48/122
collagen disease, 5/56
coma, 48/144, 148, 156
common, *see* Common migraine
complicated, *see* Complicated migraine
computer assisted tomography, 48/145, 148, 151, 54/64
confusion, 5/79, 48/7, 144, 146, 162
confusional state, 48/6, 8
congenital heart disease, 63/4
consciousness, 48/162
constitution, 5/245, 261
corona, 48/162
cortisone, 5/56
cromoglicate disodium, 48/122
crying, 48/161
crying spell, 48/162
CSF, 48/143
CSF cell, 48/37, 145
CSF protein, 48/145
cutaneous ecchymosis, 48/160
cyclic hypothalamic activity, 48/150
cyclic vomiting, *see* Abdominal migraine
cyclizine, 5/99
cyproheptadine, 5/101, 48/190, 192
cystalgia, 48/160
deafness, 48/27
decompression illness, 61/222
definition, 48/2, 14
déjà vécu, 48/162
déjà vu, 42/246
dejection, 48/162
delirium, 46/545, 48/37, 162, 166
delusion, 48/162, 361
depression, 5/247, 253, 381, 48/73, 162
diarrhea, 48/8, 125, 156
diclofenac, 48/174
dietary, 48/94
differential diagnosis, 48/160
diltiazem, 48/200
dipyridamole, 48/199
disorientation, 48/165
dissociée, 48/6
disturbance, 48/129
diuretic agent, 5/102
diving, 48/402
dominant hemisphere, 48/148
domperidone, 48/198
doom, 48/162
dopamine, 48/92
dopamine β-mono-oxygenase, 48/91, 94
Doppler study, 75/295
dreamy state, 48/162
drowsiness, 48/144, 161
drug treatment, 5/96, 379, 48/173-204, 75/297
duration, 48/4
Duret artery, 5/85
dysarthria, 48/143
dysautonomic, 48/85
dysgraphia, 48/162
dysmorphesthesia, 48/163
dysnociception, 48/79
dysosmia, 48/162
dysphasia, 48/143, 148, 161
dysphoria, 5/79, 48/162
dysphrenia hemicranica, 5/78-80
ebullience, 48/162
echinoidea intoxication, 37/67
echoacousia, 48/162
eclampsia, 39/538
EEG, 5/42, 64, 66, 48/29, 38, 144, 149
elation, 48/162
emotion, 48/60, 156, 161
emotional stress, 48/85, 355
endocrine, 48/431-438
endorphin, 48/100, 151
β-endorphin, 48/100

enkephalin, 48/99
enkephalinergic system, 48/100
enolic acid, 48/174
environmental duplication, 48/162
environmental stimuli, 5/57
enzyme, 48/85, 93
epidemiology, 48/13-20, 35, 118
epilepsy, 5/42, 47, 265, 48/38, 143, 148, 161, 72/216, 255
epileptic aura, 5/47
epileptic focus, 45/510
episodic abdominal pain, 48/18
episodic vertigo, 48/18
ergot derivative, 48/5
ergotamine, 5/56, 96, 379, 48/76, 97, 145, 183-185
ergotamine tartrate, 48/125
erythritol tetranitrate, 5/98
estrogen, 5/381, 48/433, 437
etiology, 48/141
euphoria, 48/162
evoked potential, 48/29
excitation, 48/161
exhilaration, 48/161
extracranial, 48/76
extracranial blood flow, 48/78
extracranial flow index, 48/77
facial paralysis, 48/156
facial thermography, 48/77
facioplegic, *see* Facioplegic migraine
familial hemiplegic, *see* Familial hemiplegic migraine
familial history, 48/3, 25-27, 35
familial periodic paralysis, 5/47
fasting, 48/120, 122
fatigue, 48/122
femoxetine, 48/196
fenamic acid, 48/203
fenclofenac, 48/174
fenoprofen, 48/174
fever, 5/46, 42/746, 48/156, 160
fibrotic disorder, 5/56
fish, 48/122
flickering light, 48/122
flufenamic acid, 48/174, 203
flunarizine, 48/200
Föhn, 48/122
food, 48/20, 36, 60, 121
food intolerance, 5/266, 48/20
footballer, *see* Footballer migraine
forgetfulness, 48/162
fortification spectra, 48/129
free fatty acid, 48/95
frequency, 48/14
gastrointestinal features, 48/4
generalized autonomic dysfunction, 75/296
genetics, 5/42, 55, 258-268, 48/23, 27
gestagen, 48/437
Gilles de la Tourette syndrome, 49/631
global amnesia, 48/162
glyceryl trinitrate, 75/295
gout, 5/47
guilt, 48/354
hallucination, 5/244, 250, 42/746, 48/161, 166, 361
head injury, 48/146, 150
headache, 48/77, 144, 201
heart rate, 75/296
heart transplantation, 63/179
heat, 48/122
hemianopia, 48/131, 143
hemicrania, 48/3
hemiparesis, 48/142
hemiplegic, *see* Hemiplegic migraine
hereditary, 48/27, 35
hereditary cerebral amyloid angiopathy, 54/341
hereditary transmission, 5/262, 265
heterocyclic acetic acid, 48/174
histamine, 5/39, 48/92, 113
history, 48/142
homicide, 48/162
homonymous hemianopia, 48/135, 161, 166
hormone, 48/18, 60, 86, 100, 119
hormone treatment, 5/102
hostility, 48/354
Huntington chorea, 5/80, 49/291
hydralazine, 5/56
hydratropic acid, 48/175
hydrocortisone, 5/380
5-hydroxyindole acetaldehyde, 48/86
5-hydroxyindoleacetic acid, 48/86, 90
hyperperfusion, 48/73
hypertension, 48/19, 85, 156
hypoglycemia, 48/36, 120
hypoperfusion, 48/72
hypothalamus, 74/508
hypothermia, 42/746
hysteria, 46/575, 48/356
ibuprofen, 48/174
ice cream, 48/122
ice pick headache, 48/442
IgE, 48/122
illusionary falsification, 48/162
imipramine, 5/101
inactivity, 48/156
incoherence of thought, 48/162
indecisiveness, 48/162

indoleacetic acid, 48/174
indolence, 48/162
indometacin, 5/56, 102, 48/174, 177
indoramin, 48/187
infant, 48/35, 146
infarction, 48/135
intelligence, 48/19
α-interferon intoxication, 65/562
γ-interferon intoxication, 65/562
International Association for the Study of Pain, 48/60
interstitial cystitis, 48/160
involuntary movement, 5/80
iprazochrome, 48/191
irascibility, 48/162
irritability, 48/162
ischemia, 48/73, 143, 150
isometheptene, 48/202
4-(isopropylamino)phenazone, 48/175
judgment error, 48/162
ketanserin, 48/193
ketoacidosis, 48/158
ketoprofen, 48/174
17-ketosteroid, 5/42
kininogen, 48/97
kleptomania, 48/162
laparotomy, 48/158
lassitude, 48/156, 162
laughing spell, 48/162
letter inversion, 48/162
linoleic acid, 48/95
lipid, 48/85, 95
β-lipoprotein, 48/99
lisuride, 48/192
lithium, 48/97, 198
localization, 48/166
loquacity, 48/162
lupus, 5/56
lysergide, 48/87
macropsia, 48/162
macrosomatognosia, 4/230, 45/385, 510
major, *see* Major migraine
male-female ratio, 48/18
mania, 48/162
MAO, 48/86, 93
MAO inhibitor, 5/41, 48/197
mechanism, 48/113
meclofenamic acid, 48/174
mefenamic acid, 48/174, 177, 203
memory, 48/161
memory disorder, 48/162
menarche, 48/36
Meniere disease, 42/402, 48/157
meningeal irradiation, 48/131
meningeal irritation, 48/131
meningeal pain, 5/371
meningitis, 48/31, 37
menopause, 48/433
menstrual, *see* Catamenial migraine
mental, *see* Mental migraine
mesial occipitotemporal area, 48/166
mesial occipitotemporal syndrome, 48/161
metamorphopsia, 48/129, 162, 166
metergoline, 48/192
methacholine bromide, 5/98
methysergide, 5/18, 56, 100, 381, 48/86, 188, 192
metoclopramide, 48/125, 197
metoprolol, 48/180, 203
mianserin, 48/196
microcirculation, 5/17
micropsia, 48/130, 162
microsomatognosia, 4/230, 45/510
middle cerebral artery, 48/135, 53/355
minor head injury, 48/36
Mistral, 48/122
monoamine, 48/79, 93
monochromatism, 48/161
mosaic vision, 48/129
motion sickness, 5/243, 48/18
motor, 48/6
motor aphasia, 45/510
muscle contraction headache, 48/39, 355, 75/294
mutism, 48/161
naproxen, 48/174
natural history, 5/48
nausea, 48/3, 8, 35, 70, 125, 156
navel colic, *see* Abdominal migraine
nervous tension, 48/79
neurokinin, 5/39
neurotic, 48/119
nifedipine, 48/200
nimodipine, 48/200
nitrate intoxication, 48/419
nitrite intoxication, 48/419
noise, 48/120, 122
noradrenalin, 48/76, 87, 91
nucleotide, 48/86, 100
nystagmus, 48/147
obsessional, 48/354
occipital arteriovenous malformation, 55/118
occipital lobe epilepsy, 73/109
occipital lobe syndrome, 45/57
occipital neurectomy, 5/110
oculomotor nerve, 5/80
oculomotor paralysis, 48/156
oleic acid, 48/95

olfaction, 48/162
onset, 48/18, 118
ophthalmic, *see* Classic migraine
ophthalmic artery, 48/135
ophthalmoplegic, *see* Ophthalmoplegic migraine
opiate peptide, 48/99
optic allesthesia, 48/162
oral contraception, 55/524
oral contraceptive agent, 48/201, 435, 53/36
orbital apex syndrome, 2/322
overstimulation, 48/85
oxicam derivative, 48/174
oxitriptan, 48/86
oxprenolol, 48/180
pain, 1/140, 48/130
pain mechanism, 1/140
pallor, 48/8
palmitic acid, 48/95
papaverine, 37/456
paracetamol, 5/98, 48/175
paranoid state, 48/162
paraphasia, 48/161
paresthesia, 42/746, 48/156
paroxysmal, 48/35
paroxysmal disorder, 5/46, 45/509
paroxysmal tachycardia, 48/157
pathogenesis, 5/37
pathomechanism, 48/166
pathophysiology, 5/37, 48/59-79
pelvic, *see* Pelvic migraine
pelvic pain, 5/46
peptide, 48/61, 85, 99
perception, 45/509, 48/162
periodic ataxia, 21/570
periodic disease, 5/46
periodic paralysis, 41/152
periodic syndrome, 48/38
perivascular nerve, 48/79
perseveration, 48/162
pharmacology, 5/38
phenacetin, 5/98
phenethylamine, 48/93, 97
phenol, 48/94
phenolsulfotransferase, 48/94
phenothiazine, 5/101, 48/197
phenylacetic acid, 48/174
phenylbutazone, 48/174, 177
phobia, 48/28, 161
phonophobia, 42/746, 48/4, 162
phosphene, 45/365, 367
photophobia, 42/746, 48/4, 7, 29, 70, 156, 162
photopsia, 48/128
physiologic disorder, 5/41
physiologic stimuli, 5/57
pial arteriole, 48/61
pigmentary retinopathy, 48/27
pindolol, 48/180
piroxicam, 48/174
pituitary disorder, 48/432
pizotifen, 48/39, 151, 179, 190-192, 201, 203
platelet, 48/89, 93
platelet MAO, 48/92-94
platelet release reaction, 48/90
platelet serotonin, 48/96
polyuria, 48/156
poriomania, 48/162, 164
posterior, *see* Posterior migraine
posterior cerebral artery, 48/165
postheart transplantation, 63/179
posttraumatic, 48/120, 123, 388
posttraumatic dysautonomic cephalalgia, 48/227
posttraumatic headache, 24/506
posttraumatic syndrome, 48/388
potato chip, 48/122
practolol, 48/181
precipitating factor, 48/120
precordial pain, 5/47
precursor, 5/46
pregnancy, 5/54, 56, 48/18, 120, 433
premenstrual, 5/230
prevalence, 48/13-16, 35
prevention, 5/100, 245
prodrome, 5/41
proenkephalin, 48/99
profile, 5/49, 52
progesterone, 5/102
prognosis, 48/39
prolactin, 74/508
pro-opiomelanocortin, 48/99
propanediol, 5/101
prophylactic treatment, 5/100, 48/39
propionic acid, 48/174
propranolol, 37/446, 48/39, 151, 178, 181, 203
prosoplegic, 5/75
prostacyclin, 48/97
prostaglandin, 48/92, 96, 194
pseudoangina, 48/158
psychiatric disorder, 5/78, 244, 247, 250, 253, 381
psychology, 5/42, 54, 57, 48/28
psychotherapy, 48/39
psychotic, *see* Psychotic migraine
pulsation, 48/4, 162
pupil, 75/294
pyrazolones, 48/174
Raynaud phenomenon, 48/85
reading, 48/130

recovery phase, 48/131
recurrence, 48/3
recurrent febrile episode, 48/85
regional brain blood flow, 48/69
related disorder, 5/47
REM sleep, 48/76
renin, 48/98
repeated wave, 48/73
reserpine, 5/40, 37/437, 48/88, 197
restlessness, 48/161
retinal, *see* Retinal migraine
retinal arteriole, 48/135
retinal artery occlusion, 55/112
retrograde amnesia, 48/164
retroperitoneal fibrosis, 5/56
rheumatic disease, 5/56
Roussy-Lévy syndrome, 21/174
salicylic acid, 5/98, 378, 48/174
schizophrenia, 48/361
school, 5/244
scintillating scotoma, 48/4, 65, 155
scotoma, 5/50, 45/510, 48/3, 64, 126, 135
sedation, 5/381, 48/39
seizure, 48/19
sensory, 48/6
sensory symptom, 48/130
serotonin, 5/18, 39, 100, 42/747, 48/38, 86, 88, 90, 97, 113, 182, 188, 192-196
serotonin receptor, 48/87
serotonin release, 48/95
serotonin releasing factor, 42/747, 48/89
sex distribution, 5/259, 48/148
sex hormone, 48/425, 434
shivering, 48/156
shoulder hand syndrome, 48/144
shunting, 48/64
sickle cell trait, 55/466
sinusitis, 48/227
Sirocco, 48/122
skin temperature, 48/77
sleep, 48/76, 121, 156
social group, 48/19
sodium, 5/231
somnambulism, 48/38, 161
sonophobia, 48/7
spasmodic torticollis, 48/38
spectral appearance, 48/128
speech disorder, 42/746, 48/143
spread, 48/73
spreading depression, 5/41, 48/65, 68, 72-74
statistic, 5/258
status migrainosus, 5/380
stearic acid, 48/95
steroid, 5/56
stress, 5/52, 45/253, 48/36, 121
stroke, 5/80-82
strong light, 48/60
stupor, 5/52, 48/6, 161
subarachnoid hemorrhage, 12/110, 112, 55/19
substance P, 48/98, 113
suicide, 48/162
sulindac, 48/174
sulpiride, 48/92
suprachiasmatic nucleus, 74/507
surgical treatment, 5/104
sweating, 48/8, 75/295
syllable mutilation, 48/162
sympathectomy, 5/104
symptomatic, 5/267
syncope, 48/156
syphilis, 5/56
systemic brain infarction, 11/465
tachycardia, 48/156
Takayasu disease, 5/56
teleopsia, 48/162
temperamental change, 48/161
temperature, 48/162
temporal arteritis, 48/77
temporal crescent, 55/118
temporal muscle blood flow, 48/77
temporary paralysis, 48/143
temporomandibular cephalalgia, 48/415
tension, 48/79
terror, 48/162
testosterone, 48/433
thalamogeniculate artery, 5/85
thalamus, 5/85, 48/166
thermography, 48/77, 235
thromboxane, 48/202
timolol, 48/180
tolfenamic acid, 48/174, 203
tolmetin, 48/174
transient ischemic attack, 48/19, 156, 53/234, 263
transient migrainous accompaniment, 48/6
trauma, 48/73, 85
traumatic headache, 24/506
treatment, 5/56, 245, 48/151
tremor, 48/147
trigeminal reflex, 48/113
triggering factor, 48/60
tryptophan, 48/86, 197
tryptophan hydroxylase, 48/86
twilight state, 48/161
twins, 48/19, 23, 28
tyramine, 48/20, 92-94, 97
tyramine sulfate, 48/94

unilateral, 48/3
unusual presentation, 48/37
Valsalva maneuver, 48/419
vascular, 5/19
vascular hypertension, 5/54, 56
vascular mechanism, 5/17, 57
vasoactive intestinal polypeptide, 48/99
vasoconstriction, 48/72, 126
vasovagal syncope, 48/85
verapamil, 48/200
vertebral dissecting aneurysm, 54/277
vertigo, 2/359, 5/47, 48/38, 156, 55/132
vessel ligation, 5/107
vessel wall injury, 48/113
viloxazine, 48/196
violence, 48/161
visceral, 48/157
visual disorder, 5/50, 48/6, 18, 125, 127, 162
visual evoked response, 48/76
visual hallucination, 45/510, 55/146
visual loss, 48/129
visual perseveration, 45/510, 48/162, 166
volition, 48/162
vomiting, 48/3, 8, 35, 117, 125, 156, 158
weather, 48/122
weekend, 74/508
Wernicke field, 5/85
without headache, 48/137
133xenon, 48/70
yawning, 48/59, 124, 156
zimeldine, 48/196
Migraine accompagnée, *see* Complicated migraine
Migraine coma
angiography, 60/651
brain edema, 60/651
contraindication, 60/651
CSF pleocytosis, 60/651
hyperthermia, 60/651
meningism, 60/651
Migraine dissociée
abortive migraine, 48/6, 157
definition, 48/157
Migraine equivalent, 48/155-166
benign paroxysmal vertigo in children, 48/157
cardiac migraine, 48/157
classification, 48/6
history, 48/159
pelvic migraine, 48/160
Migraine sine hemicrania
classification, 48/6, 162
clinical features, 48/160
cortex, 48/165
higher nervous function, 48/162
mania fugax, 48/165
mania transitoria, 48/165
neocortical area, 48/164
symptom schema, 48/156
Migraine personality
characteristics, 5/54, 48/355
description, 48/118
Migraine psychosis
dysphrenia hemicranica, 5/79
pathogenesis, 48/166
site, 48/166
type, 48/161
Migraine variant
main form, 48/155
rare type, 48/26
Migrainous acute confusional state
basilar artery, 48/166
clinical features, 48/163-165
furor transitorius, 48/165
localization, 48/166
spreading depression, 48/166
Migrainous headache
childhood epilepsy with occipital paroxysm, 72/5
epilepsy, 72/5
Migrainous neuralgia
chronic, *see* Chronic migrainous neuralgia
paroxysmal, *see* Paroxysmal migrainous neuralgia
periodic, *see* Cluster headache
Migrainous transient global amnesia
basilar artery, 48/166
cases, 45/211
cause, 55/141
classification, 48/6
clinical features, 48/162-165
EEG, 48/164
form, 48/37
localization, 48/166
migraine type, 45/213
perception disorder, 45/510
posterior cerebral artery, 48/166
spreading depression, 48/166
unknown origin, 45/216
Migrainous trigeminal neuralgia
differential diagnosis, 5/316
Migrant neuritis
ascending neuritis, 8/165
Migrant sensory neuritis
axonal damage, 51/462
cutaneous nerve, 51/461
duration, 51/462
kinesigenic pain, 51/461
nerve biopsy, 51/462
nerve stretch, 51/462

pain, 51/461
sensory loss, 51/461
sensory perineuritis, 51/463
Tinel sign, 51/462
Migration inhibition test
multiple sclerosis, 47/357
Migrenon, *see* Iprazochrome
Miliary aneurysm, *see* Charcot-Bouchard aneurysm
Milk intoxication, *see* Subacute necrotizing encephalomyelopathy
Milking grip
Huntington chorea, 49/280
juvenile hereditary benign chorea, 49/345
Sydenham chorea, 49/360
tardive dyskinesia, 49/187
Milkmaid grip, *see* Milking grip
Millard-Gubler syndrome
Bonnet-Dechaume-Blanc syndrome, 14/109
brain lacunar infarction, 54/244
brain stem lesion, 1/187
facial paralysis, 32/76
Foville syndrome, 2/242, 317
pontine infarction, 2/242, 12/13
subarachnoid hemorrhage, 12/95
symptom, 2/238
Miller-Dieker syndrome
see also Lissencephaly
agyria, 50/592
autosomal recessive inheritance, 50/260
cardiopathy, 50/592
cerebellar heterotopia, 50/592
chromosome deletion syndrome, 50/591
clinical features, 50/592
clinodactyly, 50/592
corpus callosum hypoplasia, 50/592
cryptorchidism, 50/592
cytogenetics, 50/593
dermatoglyphics, 50/593
epilepsy, 50/592, 72/130
episodic apnea, 50/592
four layer cortex, 50/259
lissencephaly, 50/260, 593
microcephaly, 50/592
muscular hypotonia, 50/592
myoclonic epilepsy, 50/592
olivary ectopia, 50/250
17p partial monosomy, 50/591
psychomotor retardation, 50/592
retrognathia, 50/592
short neck, 50/592
syndactyly, 50/592
Miller Fisher syndrome
areflexia, 51/246
ataxia origin, 51/246
brain stem encephalitis, 51/246
cranial nerve, 51/246
Guillain-Barré syndrome, 51/245
neuropathology, 51/246
ophthalmoplegia, 51/246
pupillary involvement, 51/246
symptom, 51/246
triad, 51/245
Miller syndrome
cervical spondylarthrotic myelopathy, 50/470
Millet
multiple sclerosis, 9/103
Millet jowar
vitamin B3 deficiency, 51/333
Mills, A.E., 1/4
Mills syndrome
amyotrophic lateral sclerosis, 22/313
central motor disorder, 1/202
clinical features, 1/203
diagnosis, 1/205
etiology, 1/204
history, 1/202
pathology, 1/205
Milroy disease
edema, 1/483, 8/208-210, 20/142
epidural cyst, 32/395
Milwaukee bracing
spinal deformity, 41/478
Mimae meningitis
chloramphenicol, 33/103
CSF examination, 33/103
petechial rash, 33/103
terminology, 33/102
Mimical apraxia
Huntington chorea, 6/328, 49/280
Mimicking disorder
Parkinson disease, *see* Multiple neuronal system degeneration
Mimicry
alexia, 4/96
Alzheimer disease, 46/252
Minamata disease
see also Mercury intoxication *and* Organic mercury intoxication
acetaldehyde production, 64/414
N-acetylpenicillamine, 64/416
adult, 36/84, 64/413
alkylmercury, 64/425
allowable daily intake, 64/424
animal, 36/127
ataxia, 7/642, 21/583, 36/74, 121, 64/413, 415, 419

athetosis, 64/414, 418, 421
atomic absorption spectrometry, 64/416
autopsy, 36/84
blindness, 36/121, 64/418
blood-hair comparison, 64/417
blood level, 64/416
body burden, 64/424
brain arteriosclerosis, 36/101-105, 123
brain atrophy, 64/415
brain level, 64/416
brain maldevelopment, 64/419
calcarine cortex, 36/84, 107, 121, 64/415
cerebellar atrophy, 36/123, 64/415
cerebral palsy, 36/135, 64/418
chemical diagnosis, 64/417
chorea, 64/414
chronic, 36/115-119
circulatory disturbance, 36/97-101
clinical features, 36/73-80, 89
clinicopathologic correlation, 36/119-127
clumsiness, 36/105
compensation, 64/414
convulsive seizure, 36/105
deafness, 7/642, 36/74, 64/414, 419
deficiency neuropathy, 7/642
dementia, 46/391, 64/414
demyelinating disease, 36/89, 64/416
developmental delay, 64/420
diagnosis, 36/75
dithizone method, 36/105, 117, 64/417
dose-response relationship, 64/416
dysarthria, 7/642, 36/74, 105, 64/413, 420
elimination half-time, 64/416
EMG, 64/415
epidemic, 64/414
epilepsy, 64/419, 421
exchange transfusion, 64/416
experimental, 36/128-135
exposure-symptom ratio, 64/421
extensor plantar response, 64/420
extrapyramidal disorder, 7/642
eye movement, 64/414
fatality rate, 46/391, 64/415
fetal, *see* Fetal Minamata disease
fungicide contamination, 64/415
gait ataxia, 64/413
general population, 64/424
hair level, 64/416
hemodialysis, 64/416
history, 36/73
Hunter-Russell syndrome, 36/74-85, 103
hypersalivation, 36/105, 123
hypotonicity, 64/420
Iraq epidemic, 64/415
Japan epidemic, 64/414
late cerebellar atrophy, 60/586
late cortical cerebellar atrophy, 21/497
latency period, 64/413
ligitation, 64/414
long-term course, 64/422
lowest effect symptom, 64/418
maternal hair level, 64/420
maternal milk level, 64/420
mental deficiency, 36/135, 46/80, 64/418
mental depression, 64/414
mental deterioration, 36/105, 123
mercury polyneuropathy, 7/642, 51/272
methylmercury source, 36/79
microcephaly, 64/419
microcornea, 64/418
molluscs, 37/28, 88
mortality, 64/414, 420
myoclonic epilepsy, 64/418
neonatal blood level, 64/418
neuronal migration disorder, 64/419
neuropathologic aspect, 36/83-141
neuropathology, 36/83, 64/414
neuropathy, 36/109-111
neurotoxicity, 51/264
neurotoxicology, 64/8
neutron activation, 64/416
optic atrophy, 7/642, 64/415
organic mercury intoxication, 37/88
overweight individual, 64/424
paresthesia, 64/413
parkinsonism, 36/121, 64/414
pathology, 37/89
penicillamine, 36/76, 64/416
polydactyly, 64/418
polyneuropathy, 64/415
population survey, 64/420
posterior column deficit, 64/413
pregnancy, 64/419, 424
prenatal, 36/135-139
prevention, 36/79
pyramidal sign, 64/414
rigidity, 36/123, 64/414
scalp hair, 64/416
sensation, 36/74
sensory neuropathy, 64/415
shellfish, 37/28, 88
sign, 36/74
slurred speech, 36/105
sodium selenite, 64/425
spastic tetraplegia, 64/418, 420
spectrometry, 64/420

spontaneous, 36/127
stage, 64/422
symptom, 36/74, 64/5
thiol compound, 64/416
threshold level, 64/417
topography, 7/642, 36/84-93
treatment, 36/76, 37/89, 64/416
tremor, 36/105, 64/418
unithiol, 64/416
in utero intoxication, 36/77
vision, 36/74
visual field, 64/413
visual field constriction, 64/415
visual impairment, 36/105
X-ray fluorescence, 64/420
Mind blindness
agnosia, 4/14
alexia, 4/112
Mind-matter problem
consciousness, 3/56, 175
history, 3/1
Miner experiment
peripheral nervous system, 7/7
Miner nystagmus
oscillation, 1/597, 2/367
Mineralizing brain microangiopathy
blood-brain barrier, 63/355
combined treatment, 63/355
CSF, 63/355
delayed irradiation, 63/355
irradiation encephalopathy, 63/355
radiotherapy, 63/355
Mineralizing microangiopathy
cerebral, *see* Cerebral mineralizing microangiopathy
chemotherapy, 67/361
irradiation, 67/361
Miniature apoplexy
type, 11/592, 596
Miniature motor end plate potential
botulism, 41/290
congenital synaptic vesical paucity syndrome, 62/429
experimental myasthenia gravis, 41/117
features, 62/50
morphology, 1/53
motor unit, 62/50
myasthenia gravis, 41/129, 62/414
myopathy, 62/50
neuromuscular transmission, 40/129
normal, 41/105, 108
Minicore disease, *see* Multicore myopathy
Minicore myopathy, *see* Multicore myopathy
Minimal brain damage
schizophrenia, 46/456
Minimal brain dysfunction
see also Hyperkinetic child syndrome *and* Psychomotor retardation
adrenoleukodystrophy, 47/595
clumsiness, 46/160
developmental dysphasia, 46/141
lead intoxication, 64/433
learning disability, 46/124
meningococcal meningitis, 52/28
Minimal change myopathy
congenital myopathy, 62/332
symptom, 62/357
Mink
metachromatic leukodystrophy, *see* Metachromatic leukodystrophy of mink
Mink encephalopathy
animal viral disease, 34/302
scrapie, 34/287
Minkowitz disease
abdominal pain, 10/552
bilateral ptosis, 10/552
epicanthus, 10/552
glycolipidosis, 10/552
hypertelorism, 10/552
mental deficiency, 10/552
myoclonic epilepsy, 60/662
Minkowski, M., 1/2, 4, 21
Minocycline
ataxia, 65/472
benign intracranial hypertension, 65/472, 67/111
meningococcal meningitis, 33/29, 52/32
neurotoxin, 65/472
Minor benign essential tremor, *see* Benign essential tremor
Minor disease, *see* Benign essential tremor
Minor head injury
compensation neurosis, 46/628
head injury epidemiology, 57/9
law, 24/835
migraine, 48/36
rehabilitation, 46/628
Minor hemisphere syndrome
right thalamus, 45/88-90
Minor sweat test, *see* Sweat test
Minor, V., 1/11
Minsk, USSR
neurology, 1/11
Miosis
à bascules, 2/209
autonomic nervous system, 74/424
chronic paroxysmal hemicrania, 48/263

congenital, *see* Microcoria
cyclic oculomotor paralysis, 42/331
diamorphine intoxication, 65/352
headache, 48/330, 75/281, 283
hereditary sensory and autonomic neuropathy type III, 43/58, 60/28
hereditary spastic ataxia, 60/464
Holmes-Adie syndrome, 43/72
Horner syndrome, 19/59, 43/64
ophthalmoplegia, 43/145
opiate intoxication, 65/353
organophosphate intoxication, 64/167, 170, 228
otitis media, 48/332
paratrigeminal syndrome, 48/330, 334
pharmacologic examination, 2/108
pontine hemorrhage, 2/253
Refsum disease, 13/314, 21/193, 60/229, 66/489
superior orbital fissure syndrome, 2/321
temporal lobe tumor, 17/292
trigeminal ganglion, 48/333
uremic polyneuropathy, 63/512
Mipafox
neurotoxin, 51/68
toxic polyneuropathy, 51/68
Mirex
chemical formula, 64/204
organochlorine, 36/393, 64/197
organochlorine insecticide intoxication, 36/418, 64/197, 203
Mirex intoxication
animal study, 36/418
Mirhosseini-Holmes-Walton syndrome
secondary pigmentary retinal degeneration, 60/720
Mirror focus
epileptic focus, 15/46
Mirror movement
corpus callosum agenesis, 42/233
Friedreich ataxia, 21/330, 42/233
hyperkinetic child syndrome, 43/201
Klippel-Feil syndrome, 1/409, 32/118, 42/233
mental deficiency, 42/233
pyramidal decussation, 42/234
synkinesis, 1/412
Mirror reading
alexia, 4/465
Mirror writing
agraphia, 4/148, 168, 465
benign choreoathetosis, 42/204
Misonidazole
antineoplastic agent, 65/529
drug induced polyneuropathy, 51/301
neuropathy, 69/460
neurotoxicity, 67/364
neurotoxin, 51/301, 65/529, 538, 67/364
toxic neuropathy, 51/299, 301
Misonidazole intoxication
polyneuropathy, 65/529, 538
Misonidazole polyneuropathy
axonal degeneration, 51/301
brain stem encephalocele, 51/301
dorsal spinal nerve root ganglion, 51/301
epilepsy, 51/301
sensory neuropathy, 51/301
symptom, 51/299, 301
Misoplegia
body scheme disorder, 4/219, 45/379
brain infarction, 55/145
cerebrovascular disease, 55/145
hemineglect, 55/145
hemiplegia, 4/219
Misreaching, *see* Ocular ataxia
Missile injury
see also Gunshot injury
age, 24/459
anesthesia, 23/510
antibiotic agent, 23/510, 57/311
anticonvulsant, 57/311
aphasia, 24/468
brain abscess, 23/524, 24/227
brain fungus, 23/524
brain injury, 23/505, 514, 520, 24/455
causalgia, 1/512
complication, 23/522
computer assisted tomography, 57/302
corticosteroid, 57/311
craniectomy, 57/304
craniotomy, 57/304
CSF fistula, 23/523, 57/311
depressed skull fracture, 23/511, 519, 523
dura mater, 23/513, 520
dural grafting, 23/513
dural sinus, 23/517
experimental model, 57/301
gainful employment, 24/462, 464
head injury, 24/455
high velocity, 23/505
history, 23/506
incidence, 24/459, 471, 61/253
incidence armed forces, 24/455
injury type, 24/463
intracerebral hematoma, 23/523
intracranial hematoma, 23/523
intracranial hemorrhage, 23/521
intracranial infection, 24/227, 465
intracranial pressure, 57/300

life expectancy, 24/457
low velocity, 23/505
mechanics, 23/103
meningitis, 23/524
missile track, 23/516, 57/301
mortality, 24/458
neurologic examination, 23/509
neuropathology, 25/49-51
oculomotor nerve injury, 24/61
orbital injury, 23/520
pathology, 23/61
penetrating head injury, 57/299, 302
pneumocephalus, 24/211, 57/312
postoperative care, 23/521
posttraumatic epilepsy, 24/448, 451, 474
posttraumatic meningitis, 57/322
posttraumatic syndrome, 24/460, 474, 48/386
preoperative evaluation, 23/508
radiologic diagnosis, 23/510
retained bone fragment, 23/523
scalp, 23/511, 519
sequela in armed forces, 24/455-475
sex ratio, 61/253
sign, 24/648
site, 24/471
skull fracture, 23/511, 519
spinal cord, *see* Spinal cord missile injury
spinal cord concussion, 25/218
surgery, 24/228, 648, 57/304
surgical procedure, 23/510
tangential wound, 23/519
thoracolumbar spine injury, 25/440
traumatic epilepsy, 23/522, 24/468
treatment, 61/257
type, 23/505, 519
Vietnam war, 24/471
wound infection, 24/458

Mitchell, S.W., 1/4

Mite-borne typhus
rickettsial infection, 34/647

Mithramycin
brain tumor, 18/519
hypercalcemia, 63/562
osteitis deformans (Paget), 38/368, 43/452

Mitochondria
brain edema, 23/157
central core myopathy, 40/91, 43/81
complex IV deficiency, 41/211
Duchenne muscular dystrophy, 40/382
experimental myopathy, 40/138, 158-160
facioscapulohumeral muscular dystrophy, 43/99
fingerprint body myopathy, 43/82
genome, 66/390
histochemistry, 40/2, 31-33
hypoglycin intoxication, 65/86
inclusion body, 40/91
lactic acidosis, 43/188
methionine adenosyltransferase deficiency, 29/122
mitochondrial DNA, 40/188
mitochondrial myopathy, 40/91, 220, 43/119, 127
multicore myopathy, 40/91, 43/121
muscle fiber, 40/31, 90, 216, 62/23
muscle fiber type, 62/3
neurotoxicity, 64/17
normal muscle, 40/212
pathologic reaction, 40/31, 91, 216
respiratory chain defect, 42/587
Reye syndrome, 29/337, 65/119
synapse, 1/53
target fiber, 40/91
troponin, 40/553
ultrastructure, 40/66
valproic acid induced hepatic necrosis, 65/119, 127

Mitochondria jagged Z line myopathy
case, 62/344
congenital myopathy, 62/332
facial weakness, 62/334

Mitochondria lipid glycogen myopathy
see also Mitochondrial myopathy
benign congenital, 41/217
carnitine deficiency, 41/202

Mitochondrial abnormality
see also Oxidative phosphorylation
achondroplasia, 41/214
calcavin, 40/160
carbonylcyanide-meta-chlorophenylhydrazone, 40/158
carnitine deficiency, 41/197
chloramphenicol, 40/158
corticosteroid, 41/251
2,4-dichlorophenoxyacetic acid, 40/160
dimethyl-*p*-phenylenediamine, 40/160
facioscapulohumeral syndrome, 41/218
hypoparathyroidism, 41/248
Kearns-Sayre-Daroff-Shy syndrome, 22/196, 43/142
Luft syndrome, 41/208
plasmocid, 40/160
subacute necrotizing encephalomyelopathy, 41/212

Mitochondrial alpha oxidation
Refsum disease, 51/384

Mitochondrial disease
see also Ragged red fiber
aconitate hydratase deficiency, 62/503

acquired, 66/397
adenosine triphosphatase 6 mutation syndrome, 62/510
Alpers disease, 62/497
biochemical classification, 66/395
biochemistry, 62/498
childhood dermatomyositis, 40/33
childhood myoglobinuria, 62/567
classification, 66/394
clinical features, 62/498
complex I deficiency, 62/503
complex II deficiency, 62/504
complex III deficiency, 62/504
complex IV, 40/31, 62/497
complex IV deficiency, 62/505
complex V deficiency, 62/506
congenital retinal blindness, 59/344
deafness, 66/394
diabetes mellitus, 66/394
dystonia, 60/664
fatal infantile myopathy/cardiopathy, 62/510
fatty acid oxidation, 66/394, 398
fumarate hydratase deficiency, 62/497
genetics, 62/498
history, 62/496
inherited condition, 66/395
Kearns-Sayre-Daroff-Shy syndrome, 62/497
laboratory finding, 62/498
lactic acidemia, 66/16
lactic acidosis, 66/394
Leber hereditary optic neuropathy, 62/497
3 major categories, 66/398
MELAS syndrome, 62/497, 66/393
MERRF syndrome, 62/497, 66/393, 73/299
methylcrotonoyl-coenzyme A carboxylase, 66/417
mitochondrial DNA, 62/500
mitochondrial DNA defect, 62/506
mitochondrial DNA deletion, 62/508
mitochondrial DNA depletion, 62/511
mitochondrial encephalomyopathy, 62/496, 498
mitochondrial metabolism, 62/498
mitochondrial morphology, 62/496
mitochondrial myopathy, 40/33, 62/496
mitochondrial protein transport defect, 62/506
MNGIE syndrome, 62/505
muscle tissue culture, 40/186-188
nuclear DNA disease, 62/502
occipital lobe epilepsy, 73/110
ophthalmoplegia, 40/33, 43/137
oxidation-phosphorylation coupling defect, 62/506
oxoglutarate dehydrogenase deficiency, 62/503
progressive external ophthalmoplegia, 22/189, 196, 60/52-57, 62/497
propionyl-coenzyme A carboxylase, 66/417
pyruvate carboxylase deficiency, 62/503
pyruvate dehydrogenase complex deficiency, 62/496, 502
pyruvate dehydrogenase phosphatase deficiency, 62/503
pyruvate metabolism, 66/398
ragged red fiber, 40/31-33, 187, 62/496
respiratory chain, 66/395
Reye syndrome, 66/395
subacute necrotizing encephalomyelopathy, 41/212, 51/388, 62/497
succinate dehydrogenase, 40/31, 62/497
tetrazolium reductase, 62/497
tricarboxylic acid cycle, 66/395
tricarboxylic acid cycle deficiency, 62/503
trichopoliodystrophy, 62/497

Mitochondrial disorder
epilepsy, 72/222
history, 66/389
inheritance pattern, 66/393
Leber hereditary optic neuropathy, 66/393
mitochondrial myopathy, 66/393
vertical transmission, 66/393

Mitochondrial DNA
complex I, 66/422
complex III, 66/423
complex IV, 66/423
complex V, 66/424
deletion, 66/426
duplication, 66/427
genome, 66/392
Kearns-Sayre-Daroff-Shy syndrome, 66/397, 426
mitochondria, 40/188
mitochondrial disease, 62/500
MNGIE syndrome, 66/430
Pearson syndrome, 66/397
primary defect, 66/426
progressive external ophthalmoplegia, 62/294, 66/426
respiratory chain complex, 66/414

Mitochondrial DNA defect
gene, 66/397
hereditary secondary dystonia, 59/345
Kearns-Sayre-Daroff-Shy syndrome, 62/507
mitochondrial disease, 62/506
mutation, 66/397
Pearson syndrome, 62/508
sporadic progressive external ophthalmoplegia with ragged red fiber, 62/508

Mitochondrial DNA deletion
complex IV deficiency, 62/508

mitochondrial disease, 62/508
multiple, *see* Multiple mitochondrial DNA deletion
progressive external ophthalmoplegia, 60/56
Mitochondrial DNA depletion
DeToni-Fanconi-Debré syndrome, 66/430
hepatomegaly, 66/430
hypotonia, 66/430
iatrogenic neurological disease, 62/511
infantile myopathy, 66/430
inheritance, 62/511
lactic acidosis, 62/511, 66/430
liver disease, 62/511
mitochondrial disease, 62/511
myopathy, 62/511
phenotype, 66/430
ragged red fiber, 62/511
renal dysfunction, 62/511
tetraplegia, 62/511
zidovudine, 62/511
Mitochondrial DNA disease
respiratory chain defect, 66/421
Mitochondrial DNA gene
13 gene products, 66/414
Mitochondrial DNA mutation
congenital retinal blindness, 59/342, 344
Leber hereditary optic neuropathy, 66/428
Mitochondrial DNA point mutation
adenosine triphosphatase 6 mutation syndrome, 62/509
cardiomyopathy, 66/427
diabetes mellitus, 66/427
encephalomyopathy, 66/427
fatal infantile myopathy/cardiopathy, 62/509
Leber hereditary optic neuropathy, 62/509
MELAS syndrome, 62/509, 66/428
MERRF syndrome, 62/509, 66/427
pathogenic, 66/421
Mitochondrial encephalomyopathy
acanthocytosis, 63/271
complex I deficiency, 62/504, 66/422
computer assisted tomography, 60/672
dementia, 60/665
Kearns-Sayre-Daroff-Shy syndrome, 62/498
MELAS syndrome, 62/498
MERRF syndrome, 62/498
microscopic features, 62/25
mitochondrial disease, 62/496, 498
multiple respiratory enzyme defect, 66/425
myoclonic epilepsy, 60/662
myopathy, 60/660
Refsum disease, 60/661
secondary pigmentary retinal degeneration, 60/725
striatopallidodentate calcification, 60/671
Mitochondrial enzyme deficiency
enzyme deficiency polyneuropathy, 51/387
Mitochondrial genome
schematic representation, 66/391
Mitochondrial genomic disease
secondary pigmentary retinal degeneration, 60/725
Mitochondrial malic enzyme
Friedreich ataxia, 51/389
Mitochondrial matrix
acute metabolic encephalopathy, 66/410
carnitine deficiency, 66/410
fatty acid oxidation disorder, 66/410
hypoketotic hypoglycemic, 66/410
microcephaly, 66/410
myoglobinuria, 66/410
myopathy, 66/410
renal dysplasia, 66/410
Mitochondrial matrix system
acetyl-coenzyme A acyltransferase, 66/412
defect, 66/409
2,4-dienoyl-coenzyme A reductase, 66/412
electron transfer flavoprotein, 66/409
flavine adenine dinucleotide, 66/409
long chain acyl-coenzyme A dehydrogenase gene, 66/409
medium chain acyl-coenzyme A dehydrogenase, 66/409
medium chain acyl-coenzyme A dehydrogenase deficiency, 66/410
short chain acyl-coenzyme A dehydrogenase, 66/409
short chain acyl-coenzyme A dehydrogenase deficiency, 66/409
short chain 3-hydroxyacyl-coenzyme A dehydrogenase deficiency, 66/412
Mitochondrial myopathy
see also Mitochondria lipid glycogen myopathy *and* Progressive external ophthalmoplegia
acanthocytosis, 63/290
aminoaciduria, 43/188
areflexia, 43/119
ataxia, 60/649
biochemical classification, 62/498
blindness, 43/188
cardiomyopathy, 43/119, 188, 60/667
cataract, 43/188
classification, 40/288
concentric laminated body, 40/109
congenital myopathy, 41/216
cytoplasmic inclusion body, 43/120

dementia, 43/188
exercise intolerance, 43/188
experimental, 40/158-160
facioscapulohumeral syndrome, 62/168
fatigue, 43/119
fatty acid oxidation, 66/394
genetics, 41/431
growth retardation, 43/188
histochemistry, 40/47
history, 41/208
hyporeflexia, 43/127
IgG, 43/119
infantile hypotonia, 43/127
intestinal pseudo-obstruction, 62/313
Kearns-Sayre-Daroff-Shy syndrome, 41/215, 62/313
lactic acidosis, 41/208, 42/588, 43/188
limb girdle syndrome, 40/280
lipid metabolism, 41/214
lipid storage, 41/202
Luft syndrome, 41/209
metabolic myopathy, 41/208-220
mitochondria, 40/91, 220, 43/119, 127
mitochondrial disease, 40/33, 62/496
mitochondrial disorder, 66/393
muscle tissue culture, 62/100
myasthenia gravis, 62/415
neuromuscular disease, 41/431
ophthalmoplegia, 41/214
oxidative metabolism, 43/119, 128
oxidative phosphorylation, 41/209, 216
pathologic reaction, 40/259
phosphorylation, 41/214
poliodystrophy, 42/588
progressive bulbar palsy, 59/375
progressive external ophthalmoplegia, 22/189, 60/657, 62/289, 313
proximal muscle weakness, 43/119, 188
pyruvic acid metabolism, 41/210
ragged red fiber, 40/33, 220
respiratory chain, 41/211
salt appetite, 43/120, 127
secondary pigmentary retinal degeneration, 60/725
sensorineural deafness, 43/188
striatopallidodentate calcification, 49/421, 424

Mitochondrial protein import
methylmalonic acidemia, 66/396
ornithine carbamoyltransferase, 66/396

Mitochondrial respiratory chain defect
aminoaciduria, 42/587
areflexia, 42/587
complex IV, 42/587
cytochrome b deficiency, 42/587
lactic acidosis, 42/587

Mitochondrial vacuole
neuronal, *see* Neuronal mitochondrial vacuole

Mitogen
CSF cytology, 47/374
multiple sclerosis, 47/374
pokeweed, *see* Pokeweed mitogen

Mitomycin C
antineoplastic agent, 65/528
encephalopathy, 65/528
neurotoxin, 65/528
vitreous body, 13/203

Mitosis
brain tumor, 17/87, 90, 92
chromosome, 30/92

Mitotane
antineoplastic agent, 65/529
ataxia, 65/529
chemotherapy, 39/120
encephalopathy, 65/529
neurologic toxicity, 39/120
neuropathy, 69/461, 472
neurotoxin, 39/95, 113, 119, 65/529, 538, 67/363

Mitotane intoxication
ataxia, 65/538
encephalopathy, 65/538

Mitral annulus calcification
atrial fibrillation, 63/29
brain embolism, 53/156
brain infarction, 53/156, 63/29
brain ischemia, 53/156
cardiac valvular disease, 63/29
infectious endocarditis, 63/23

Mitral stenosis
atrial fibrillation, 63/18
brain embolism, 55/164, 63/17
brain hemorrhage, 55/164
brain infarction, 55/164
iatrogenic neurological disease, 63/177

Mitral valve prolapse
amaurosis fugax, 63/21
atrial fibrillation, 63/22
atrial septal aneurysm, 63/22
auscultatory sign, 63/20
autonomic nervous system, 75/40
autosomal dominant hypoalphalipoproteinemia, 60/136
bacterial endocarditis, 52/289
Bassen-Kornzweig syndrome, 63/274
brain embolism, 53/155, 55/165, 63/20
brain infarction, 53/155, 63/20
brain infarction risk, 63/21

brain ischemia, 53/155
cardiac valvular disease, 38/145, 63/20
cerebrovascular disease, 53/34, 43, 55/165
Charlevoix-Saguenay spastic ataxia, 60/454, 668
echoCG, 63/20
epidemiology, 63/21
epilepsy, 63/21
familial stroke, 63/21
infectious endocarditis, 63/21, 23, 112
microangiopathy, 9/607
monocular blindness, 63/21
myxomatous valve degeneration, 63/20
neurologic complication, 38/143
orthostatic hypotension, 63/155
orthostatic intolerance, 74/326
Parinaud syndrome, 63/21
platelet hyperactivity, 63/22
retinal artery occlusion, 63/21
sex ratio, 55/165, 63/20
stroke, 63/21
stroke risk, 63/22
syncope, 63/21
tachyarrhythmia, 63/22
transient ischemic attack, 55/165, 63/21
treatment, 63/22
ventricular fibrillation, 63/21
vitamin K deficiency, 9/607

Mixed connective tissue disease, 71/121-127
acute pandysautonomia, 51/450
angiopathic polyneuropathy, 51/446
autoimmune disease, 41/78
autonomic polyneuropathy, 51/450, 479
axonal degeneration, 51/450
classification, 71/4
complication, 71/123
dermatomyositis, 62/371, 374
diagnosis, 71/124
epidemiology, 71/122
hepatomegaly, 71/122
inclusion body myositis, 62/374
Mollaret meningitis, 56/630
myelination, 51/450
myositis, 71/122
nasal ulcer, 51/450
neuropathology, 71/126
orthostatic hypotension, 63/152
pathogenesis, 71/123
pathophysiology, 71/127
polymyositis, 62/374
proximal myositis, 51/449
Raynaud disease, 51/449
Raynaud phenomenon, 71/122
sensory trigeminal neuropathy, 51/450
serology, 71/123
symptom, 71/122
treatment, 71/127

Mixed cryoglobulinemia
angiopathic polyneuropathy, 51/446
definition, 63/403

Mixed glioma
biopsy, 67/227
oligodendroglioma, 68/127

Mixed headache
definition, 48/8
features, 48/6
treatment, 48/48

Mixed hepatic porphyria, *see* Porphyria variegata

Mixed lymphocyte reaction
allogeneic, *see* Allogeneic mixed lymphocyte reaction
autologous, *see* Autologous mixed lymphocyte reaction
CSF, *see* CSF mixed lymphocyte reaction
multiple sclerosis, 47/356

Mixed neuronoglial tumor
classification, 67/5

Mixed parotid tumor
skull base tumor, 17/191

Mixed tapetoretinal dystrophy, *see* Centroperipheral tapetoretinal dystrophy

Mixed transcortical aphasia
agraphia, 45/460

Mixoploidy, *see* Mosaicism

Mizuo phenomenon
Oguchi disease, 13/261

MLD, *see* Metachromatic leukodystrophy

MNGIE syndrome
autonomic nervous system, 75/642
diarrhea, 62/505
intestinal pseudo-obstruction, 62/505
Kearns-Sayre-Daroff-Shy syndrome, 62/313
lactic acidosis, 62/505
leukodystrophy, 62/505
metabolic encephalopathy, 62/505
metabolic myopathy, 62/505
mitochondrial disease, 62/505
mitochondrial DNA, 66/430
multiple mitochondrial DNA deletion, 66/429
multiple neuronal system degeneration, 62/505
POLIP syndrome, 62/505
polyneuropathy, 62/505
progressive external ophthalmoplegia, 62/305, 313, 505
progressive external ophthalmoplegia classification, 62/290
ragged red fiber, 62/292, 505, 66/430

MNU, *see* Methylnitrosourea
Moberg ninhydrin test, *see* Sweat test
Mobility
  pressure sore, 61/350
  spina bifida, 50/493
Möbius syndrome
  see also Abducent nerve agenesis *and* Nuclear aplasia
  abducent nerve agenesis, 50/215
  abducent nerve paralysis, 42/324
  Alajouanine syndrome, 2/325
  arthrogryposis, 50/215
  brachydactyly, 50/215
  clinical features, 30/404-406, 40/303, 50/215
  congenital ptosis, 30/401
  convergent strabismus, 50/215
  cranial nerve agenesis, 50/215
  cranial nerve nucleus, 42/325, 50/42
  diabetes insipidus, 50/215
  differential diagnosis, 22/107
  ectropion, 50/215
  epidemiology, 30/156
  epiphora, 50/215
  etiologic controversy, 2/279
  facial paralysis, 42/324
  facioscapulohumeral muscular dystrophy, 40/419, 422, 62/162
  genetics, 30/156, 50/215
  hypogonadotrophic hypogonadism, 59/375
  hypopituitarism, 50/215
  lagophthalmos, 50/215
  malformation, 30/404
  mental deficiency, 42/324, 46/90
  microcephaly, 30/100
  micrognathia, 50/215
  muscle aplasia, 50/215
  myopathy, 43/125, 50/216
  nuclear aplasia, 2/325
  ocular defect, 30/405
  oculomotor nerve agenesis, 50/212
  ophthalmoplegia, 22/179, 43/134, 136
  other cranial nerves palsy, 50/215
  pathogenesis, 30/406
  pathology, 30/406, 50/215
  Poland-Möbius syndrome, 50/215
  polydactyly, 42/324
  polyneuropathy, 50/215, 59/375
  progressive amyotrophy, 59/375
  progressive bulbar palsy, 59/375
  progressive external ophthalmoplegia, 2/279, 43/136
  progressive external ophthalmoplegia classification, 22/179, 62/290
  pseudoprogressive external ophthalmoplegia, 22/204, 208
  Robin syndrome, 43/472
  spinal muscular atrophy, 59/368, 370
  supranuclear mechanism, 22/190
  symphalangism, 50/215
  syndactyly, 42/324, 50/215
  talipes, 42/324
  trigeminal nerve, 30/404
  trigeminal nerve agenesis, 50/213
  3 types, 50/215
  vagus nerve agenesis, 50/219
Mobulidae intoxication
  elasmobranche, 37/70
Models
  animal, *see* Animal model
  computer, *see* Computer model
  experimental, *see* Experimental model
  Hodgkin-Huxley, *see* Hodgkin-Huxley model
  neostriatal, *see* Neostriatal model
  neural, *see* Neural model
Moersch-Woltman syndrome, *see* Stiff-man syndrome
Mogan-Baumgartner-Cogan syndrome, *see* Cogan syndrome type II
Mogigraphia, *see* Writers cramp
Mohr syndrome
  cheilopalatoschisis, 43/449
  Dandy-Walker syndrome, 50/332
  hydrocephalus, 43/450
  hypertelorism, 30/247
  mental deficiency, 43/450, 46/67
  microcephaly, 43/450
  multilobulated tongue, 43/449
  ocular defect, 30/247
  orofaciodigital syndrome type I, 31/252
  polydactyly, 43/449
  porencephaly, 43/450
  tongue hamartoma, 43/450
Mokola virus
  human pathogen, 56/384
  rhabdovirus, 56/384
Molecular genetics
  acid maltase deficiency, 62/481
  ataxia telangiectasia, 60/425
  brain tumor, 67/41
  CNS tumor, 67/41
  debrancher deficiency, 62/483
  Down syndrome, 50/531
  familial amyotrophic lateral sclerosis, 59/248
  hereditary motor and sensory neuropathy type I, 60/200, 202
  hereditary motor and sensory neuropathy type II,

60/200, 202
Joseph-Machado disease, 60/474
Kearns-Sayre-Daroff-Shy syndrome, 62/311
malignant hyperthermia, 62/569
myophosphorylase deficiency, 62/485
Pelizaeus-Merzbacher disease, 66/559
phosphofructokinase deficiency, 62/488
phosphorylase kinase deficiency, 62/487
progressive external ophthalmoplegia, 62/311
X-linked hereditary motor and sensory neuropathy, 60/202
Molindone
chemical formula, 65/276
neuroleptic agent, 65/275
Mollaret meningitis, 34/545-550, 56/627-632
see also Recurrent meningitis
Behçet syndrome, 34/549, 56/629
clinical features, 34/547, 56/627
CSF, 34/547, 56/628
CSF endothelial cell, 34/548, 56/628
dermoid, 56/630
diagnostic criteria, 34/546, 56/629
differential diagnosis, 34/548, 56/629
disease duration, 34/548, 56/628
epidemiology, 34/547, 56/627
epidermoid, 34/548, 56/630
etiology, 34/549, 56/630
familial brucellosis, 34/549, 56/629
history, 34/545
incidence, 56/627
laboratory study, 56/628
Middle East, 56/632
mixed connective tissue disease, 56/630
pituitary abscess, 34/548, 56/630
recurrent meningitis, 34/548, 56/631
Sjögren syndrome, 56/630
systemic lupus erythematosus, 56/629
treatment, 34/548, 56/632
viral meningitis, 56/131
vitiligo, 34/549, 56/631
Vogt-Koyanagi-Harada syndrome, 34/549, 56/629
Möller-Barlow disease
vitamin C deficiency, 28/200, 208
Mollet de coq
amyotrophy, 21/276
Mollusca intoxication
cephalopoda, 37/62
conidae, 37/58
Gastropoda, 37/58
ingestion, 37/64
octopoda, 37/62
Molluscs intoxication
mercury, 37/88
venom, 37/60-62
Molluscum contagiosum virus
neurotropic virus, 56/5
Molluscum pendulum
birth, 8/445
definition, 14/32
description, 8/445
Moloney murine leukemia virus
enhancer, 56/37
gene therapy, 66/114
Molybdenum
striatopallidodentate calcification, 6/716, 49/425
Monakow, K. Von, *see* Von Monakow, K.
Mönckeberg disease, *see* Mönckeberg sclerosis
Mönckeberg sclerosis
brain arteriosclerosis, 53/94
dissecting aorta aneurysm, 63/45
Monday fever, *see* Zinc intoxication
Mondini ear abnormality
congenital deafness, 42/366
Patau syndrome, 50/559
Pendred syndrome, 42/366
Mondrum-Benisty syndrome
differential diagnosis, 5/330
sphenopalatine neuralgia, 5/330
Monensin intoxication
toxic myopathy, 62/612
Mongolism, *see* Down syndrome
Mongoloid eyes
Klinefelter variant XXXXY, 31/483
9p partial monosomy, 50/578
r(9) syndrome, 50/580
Mongoloid idiocy, *see* Down syndrome
Moniliasis, *see* Candidiasis
Moniz arterial axis, *see* Arterial axis (Moniz)
Moniz, E., 1/11, 17
Monkey poxvirus
neurotropic virus, 56/5
Monoalkylphosphoryl enzyme
organophosphate intoxication, 64/224
Monoamine
Alzheimer disease, 46/266
anorexia nervosa, 46/587
autonomic nervous system, 74/139
blood-brain barrier, 48/93
fever, 74/449
headache, 48/79
migraine, 48/79, 93
pharmacology, 74/145
reserpine, 74/145
Monoamine oxidase, *see* MAO
Monoamine oxidase A, *see* MAO A
Monoamine oxidase B, *see* MAO B

Monoamine oxidase deficiency
autonomic nervous system, 75/240
Monoamine oxidase inhibitor, *see* MAO inhibitor
Monoballism
definition, 6/476, 49/369
Monochromatic photography
retina, 13/6
Monochromatism
see also Hemiachromatopsia
agnosia, 45/335
alexia, 45/438
central, *see* Central monochromatism
congenital retinal blindness, 13/121
inferior visual association cortex, 45/15
migraine, 48/161
neuropsychology, 45/14
prosopagnosia, 45/335
temporal lobe epilepsy, 73/71
visual sensory deficit, 45/335
Monoclonal antibody
CSF, *see* CSF monoclonal antibody
multiple sclerosis, 47/85, 232, 234, 342-345
neurotropic virus, 56/26
sensory neuropathy, 71/444
tuberculous meningitis, 52/206
Monoclonal gammopathy
see also Plasma cell dyscrasia
benign, *see* Benign monoclonal gammopathy
chronic axonal neuropathy, 51/531
chronic inflammatory demyelinating polyradiculoneuropathy, 51/530
classification, 69/290
cryoglobulinemic polyneuropathy, 51/434
demyelinating neuropathy, 47/615-618
hereditary amyloid polyneuropathy, 71/446
human T-lymphotropic virus type I associated myelopathy, 56/525
Ig type, 63/398
myeloma, 71/445
neuropathy, 71/441, 445
paraproteinemia, 69/290
polymyositis, 62/373
polyneuropathy, 63/398
segmental demyelination, 51/71
symptom, 71/431
undetermined significant, *see* Undetermined significant monoclonal gammopathy
Waldenström macroglobulinemia, 63/396
Monoclonal IgA
cryoglobulinemia, 63/403
Monoclonal IgG
cryoglobulinemia, 63/403
tropical spastic paraplegia, 56/527
Monoclonal IgM
cryoglobulinemia, 63/403
Monoclonal paraproteinemia
benign, *see* Benign monoclonal paraproteinemia
Monoclonal protein
antibody activity, 69/292
measles virus, 56/423
muscle fiber, 62/26
neuropathy, 69/292
primary amyloidosis, 51/416
specificity, 69/291
symptom, 71/431
Waldenström macroglobulinemia, 63/397
Monoclonal protein antibody
measles encephalitis, 56/426
subacute sclerosing panencephalitis, 56/426
Monoclonal trisomy, *see* Chromosome
Monocular field defect
unpaired retinal fiber, 2/559
Monocular polyopia
occipital lobe lesion, 2/663, 17/331
occipital lobe tumor, 2/663, 17/331
pseudofovea, 2/663, 17/331
Monocular vision loss
brain embolism, 53/207
brain infarction, 53/207
cerebrovascular disease, 53/207
intracranial hemorrhage, 53/207
transient ischemic attack, 53/207
Monocyte
brain arteriosclerosis, 53/98
lactate dehydrogenase elevating virus, 56/30
multiple sclerosis, 47/341, 346, 359
Monocytic leukemia
reticulosis, 18/247
Monomelic spinal muscular atrophy
anterior crural muscle, 59/376
calf, 59/376
compression neuropathy, 59/376
differential diagnosis, 59/376
filum terminale ependymoma, 59/376
Hirayama disease, 59/376
hydromyelia, 59/376
hypertrophic mononeuropathy, 59/376
neuralgic amyotrophic plexopathy, 59/376
neurotoxic enterovirus, 59/376
Poland-Möbius syndrome, 59/376
poliomyelitis, 59/376
quadriceps, 59/376
scapulohumeral distribution, 59/376
syringomyelia, 59/376
traumatic myelopathy, 59/376
traumatic plexopathy, 59/376

traumatic radiculopathy, 59/376
Monomer
dihydrolipoamide acetyltransferase, 66/415
neurofilament, 7/17
Mononeuritis
acquired immune deficiency syndrome, 56/517
carcinoma, 38/686
Epstein-Barr virus, 34/188
infectious mononucleosis, 34/188
lepromatous, *see* Lepromatous mononeuritis
leprous neuritis, 51/218
opiate intoxication, 37/389
terminology, 7/199
tetanus toxoid, 52/240
typhoid fever, 51/187
Mononeuritis multiplex
see also Eosinophilia myalgia syndrome *and* Lumbosacral plexus neuritis
acquired immune deficiency syndrome, 56/493
allergic granulomatous angiitis, 51/453, 63/383, 423
angiopathic polyneuropathy, 51/445
bacterial endocarditis, 52/300
cranial, *see* Cranial mononeuritis multiplex
diabetic, *see* Diabetic neuropathy
dialysis, 63/530
eosinophilia myalgia syndrome, 64/255
hepatitis B virus infection, 56/300
hereditary, *see* Hereditary mononeuritis multiplex
hypereosinophilic syndrome, 63/373
idiopathic hypereosinophilic syndrome, 63/124, 138
lepromatous, *see* Lepromatous mononeuritis multiplex
leprosy, 51/220
leprous neuritis, 51/219
Lyme disease, 52/262
neuralgic amyotrophy, 51/173
plexopathy, 51/166
polyarteritis nodosa, 7/200, 8/125, 9/551, 12/659, 39/296, 305, 51/452, 55/355
polyneuritis, 7/473
Sjögren syndrome, 51/449
Streptococcus viridans, 52/300
systemic lupus erythematosus, 51/447
temporal arteritis, 51/452
terminology, 7/199
Waldenström macroglobulinemia, 63/397
Wegener granulomatosis, 39/344, 51/450, 63/423
Mononeuropathy
Cogan syndrome type II, 51/454
dengue fever, 56/12
diabetic, *see* Diabetic neuropathy
diabetic polyneuropathy, 51/505
hereditary, *see* Hereditary mononeuropathy
hereditary multiple recurrent, *see* Hereditary multiple recurrent mononeuropathy
hereditary neuropathy, 60/254
hypertrophic, *see* Hypertrophic mononeuropathy
hypothyroid polyneuropathy, 51/516
malignant, 69/90
multiple, *see* Multiple mononeuropathy
pain, 69/72
peroneal nerve compression, 7/315
polyneuropathy, 60/254
rheumatoid arthritis, 51/446
silver, *see* Silver mononeuropathy
Sjögren syndrome, 51/449
temporal arteritis, 51/452
Mononeuropathy multiplex
human immunodeficiency virus neuropathy, 71/356
malignant, 69/92
Mononuclear cell
CSF, *see* CSF mononuclear cell
endoneurial, *see* Endoneurial mononuclear cell
multiple sclerosis, 47/340
Mononucleosis
cytomegalovirus, *see* Cytomegalovirus mononucleosis
infectious, *see* Infectious mononucleosis
nerve lesion, 7/489
Mononucleosis infectiosa, *see* Infectious mononucleosis
Monoplegia
brachial, *see* Brachial monoplegia
brain lacunar infarction, 54/243
central, 1/176
central motor disorder, 1/176-178
diastematomyelia, 50/436
diplomyelia, 50/436
venous sinus occlusion, 54/429
Monosialoganglioside
major, *see* GM1 ganglioside
minor, *see* GM2 ganglioside
paraplegia, 61/399, 406
spinal cord injury, 61/399, 406
tetraplegia, 61/399, 406
transverse spinal cord lesion, 61/399, 406
Monosodium glutamate, *see* Sodium glutamate
Monostotic fibrous dysplasia
Albright syndrome, 14/163
differential diagnosis, 19/301
sex ratio, 19/301
Monozygotic twins
foramina parietalia permagna, 30/277, 50/142

restless legs syndrome, 8/315, 42/260, 51/546
Monrad Krohn, G. H., 1/12, 31, 35
Monro-Kellie doctrine
benign intracranial hypertension, 16/152, 67/108
brain blood flow, 11/118, 53/49
jugular vein compression, 19/100
Monstrocellular sarcoma
brain sarcoma, 16/20
glioblastoma, 18/51
Monteggia fracture, *see* Ulnar fracture
Montevideo, Uruguay
Neurological Institute, 1/34
Montreal, Canada
neurology, 1/12
Mood, *see* Emotion
Moon face
pseudohypoparathyroidism, 6/706, 42/621, 46/85, 396, 49/419
r(9) syndrome, 50/584
Moose encephalitis
multiple sclerosis, 9/125
Moraxella
gram-negative bacillary meningitis, 52/104
Moraxella urethralis
meningitis rash, 52/25
Moray intoxication
classification, 37/78
Morbillivirus, *see* Measles virus
Morbus caeruleus
pallidal necrosis, 49/466
Morbus sacer
localization, 2/5
Morbus Weil disease, *see* Leptospirosis
Morel laminar sclerosis, *see* Cortical laminar sclerosis
Morel-Wildi dyshoric angiopathy, *see* Brain amyloid angiopathy
Morgagni-Stewart-Morel syndrome
alkaline phosphatase, 43/412
dementia, 43/411
fibrous dysplasia, 14/199
obesity, 43/411
Pick disease, 46/242
prevalence, 43/412
sex ratio, 43/412
virilization, 43/411
Morgagni syndrome, *see* Morgagni-Stewart-Morel syndrome
Morning headache
brain tumor hemorrhage, 5/141
hypertension, 5/150
Moro reflex
congenital hyperammonemia, 42/560
diagnostic value, 1/255
epilepsy, 42/690
galactosemia, 42/551
hydranencephaly, 30/675, 50/349
hyperammonemia, 42/560
kernicterus, 42/583
newborn behavior, 4/346
prenatal onset, 42/690
Morphea
clinical features, 39/359
differential diagnosis, 39/369
scleroderma, 71/104
symptom, 39/38
Morphine
chemical structure, 37/366, 65/352
chemistry, 37/365
chrysaora intoxication, 37/51
cyanea intoxication, 37/52
half-life, 65/354
intrathecal chemotherapy, 61/151
opiate, 65/350
pain tolerance, 45/228
spastic paraplegia, 61/368
spinal cord injury, 61/368
spinal spasticity, 61/368
subarachnoid hemorrhage, 12/113
toxic myopathy, 62/601
transverse spinal cord lesion, 61/368
Morphine intoxication, 37/365-394
forgotten respiration, 1/654
pallidal necrosis, 6/652, 49/466, 479
Morphopsia
visual hallucination, 45/356
Morquio-Brailsford disease, *see* Glycosaminoglycanosis type IV
Morquio syndrome, *see* Glycosaminoglycanosis type IV
Morquio syndrome type A, *see* Glycosaminoglycanosis type IVA
Morquio syndrome type B, *see* Glycosaminoglycanosis type IVB
Mors subita, *see* Sudden death
Mortality
acquired immune deficiency syndrome, 56/489
acute subdural hematoma, 24/290
angiostrongyliasis, 35/327, 52/556
astrocytoma, 68/103
bacterial endocarditis, 52/304
basilar artery aneurysm, 53/387
Behçet syndrome, 56/598
birth incidence, 25/171
bismuth intoxication, 64/339
Bolivian hemorrhagic fever, 56/369

brain air embolism, 55/193
brain aneurysm, 55/44
brain aneurysm neurosurgery, 55/44
brain fat embolism, 55/182, 192
brain infarction, 54/427
brain injury, 24/671
brain ischemia, 63/210
brain mucormycosis, 35/545, 52/475
brain tumor, 16/59
brain vasospasm, 55/57
carotid artery thrombosis, 26/48
cavernous sinus thrombosis, 52/176
cerebral malaria, 35/146, 52/368
cerebrovascular disease, 53/1
chronic subdural hematoma, 24/315, 317
CNS malformation, 30/151, 50/56
coccidioidomycosis, 52/414
Coma Scale, 57/110
cranial subdural empyema, 52/171
Divry-Van Bogaert syndrome, 55/319
epidermolysis bullosa dystrophica hereditaria, 14/791
epidural hematoma, 26/24, 57/103
epilepsy, 15/495, 812, 72/28
exanthematous encephalomyelitis, 9/547
febrile convulsion, 15/259
gold intoxication, 64/356
gram-negative bacillary meningitis, 52/103
granulomatous CNS vasculitis, 55/392-394
head injury, 57/113
heart transplantation, 63/178
Hemophilus influenzae meningitis, 33/54
hypoglycin intoxication, 65/80
infantile spinal muscular atrophy, 22/82
internal carotid artery syndrome, 53/307
intracerebral hematoma, 24/365
intracranial pressure, 57/68
Jamaican vomiting sickness, 65/80
Kawasaki syndrome, 52/264, 56/637
Lassa fever, 56/370
legionellosis, 52/254
Listeria monocytogenes meningitis, 33/84, 52/91, 96, 63/531
lithium intoxication, 64/297
malignant hyperthermia, 38/550, 41/268, 63/440
marine toxin intoxication, 37/33, 65/153
meningioma, 68/409
meningococcal meningitis, 33/23, 52/29
Minamata disease, 64/414, 420
missile injury, 24/458
motoneuron disease, 21/30
moyamoya disease, 12/366, 42/751, 55/295
multiple sclerosis, 47/260
neurosyphilis, 33/387-389
nocardiosis, 35/519, 52/450
North American blastomycosis, 35/401, 403, 406, 52/385, 390
ostrea gigas intoxication, 37/64
Parkinson disease, 49/125
pediatric head injury outcome, 57/339
pertussis, 52/232, 237
pertussis encephalopathy, 9/551, 33/275, 280, 289
pituitary adenoma, 17/415-417
pituitary apoplexy, 17/421
pneumococcal meningitis, 33/44-46, 52/49
posttraumatic brain abscess, 57/323
posttraumatic syringomyelia, 26/138
rabies, 56/385
Reye syndrome, 31/168, 34/80, 169, 56/164, 238
spinal cord abscess, 52/192
spinal cord injury, 25/142, 281, 26/310, 61/291, 500
spinal cord tumor, 19/2
spinal ependymoma, 20/379
spinal meningioma, 20/237
streptococcal meningitis, 52/81
subarachnoid hemorrhage, 55/44
subdural effusion, 24/339
subdural empyema, 57/324
subdural hematoma, 57/103, 338
subdural hygroma, 57/287
superior longitudinal sinus thrombosis, 52/179
Sydenham chorea, 49/364
thrombotic thrombocytopenic purpura, 55/473
traffic injury, 25/142
traumatic aneurysm, 24/391
traumatic hydrocephalus, 24/247
traumatic venous sinus thrombosis, 24/376
trichinosis, 52/565
tuberculous meningitis, 52/214
valproic acid induced hepatic necrosis, 65/129
varicella zoster virus encephalitis, 56/479
venerupis semidecussata intoxication, 37/64
venous sinus occlusion, 54/427
venous sinus thrombosis, 54/427
water intoxication, 63/547
Wohlfart-Kugelberg-Welander disease, 22/74
Mortimer malady, *see* Sarcoidosis
Morton metatarsalgia, *see* Morton neuralgia
Morton neuralgia
compression neuropathy, 51/106
nerve intermittent claudication, 20/793
Morton neuroma, *see* Morton neuralgia
Morton syndrome, *see* Morton neuralgia
Morvan fibrillary chorea
see also Toxic encephalopathy

anxiety, 6/355
contraction, 6/355
depression, 6/355
gold intoxication, 49/293, 559
Huntington chorea, 49/293
insomnia, 6/355
mercury intoxication, 49/293, 559
myokymia, 49/559
nomenclature, 6/770
pain, 6/355, 49/559
polyneuritis, 49/559
Morvan syndrome, *see* Hereditary sensory and autonomic neuropathy type II
Mosaic vision
hallucination classification, 45/56
migraine, 48/129
Mosaicism
cerebellar agenesis, 50/178
cri du chat syndrome, 31/572
definition, 50/540
Down syndrome, 31/383-386, 46/73, 50/521, 523
Edwards syndrome, 31/523, 46/70, 50/561
Klinefelter syndrome, 31/485
Patau syndrome, 31/506, 514, 50/558
Turner syndrome, 31/502, 50/544
Moschcowitz disease, *see* Thrombotic thrombocytopenic purpura
Moscow, USSR
neurology, 1/11
Mossy fiber
cerebellar stimulation, 2/409
excitatory action, 2/416
Mother milk intoxication, *see* Subacute necrotizing encephalomyelopathy
Motility absence
bowel, *see* Intestinal peristalsis absence
Motility factor
brain metastasis, 69/109
Motility factor receptor
brain metastasis, 69/109
Motion perception
occipital lobe tumor, 17/331
parietal lobe, 45/363
sense modality, 1/95
Motion sickness
alcohol, 74/357
autonomic nervous system, 74/351
balance, 74/354
classic migraine, 5/46
common migraine, 5/46
drug treatment, 74/357
etiology, 74/353
fatigue, 74/357
hypoxia, 74/357
migraine, 5/243, 48/18
perception, 74/358
prevention, 74/356
sleep, 74/357
survey, 74/351
susceptibility, 74/355
symptom, 74/352
treatment, 74/356
Treisman theory, 74/354
vasopressin, 74/358
vestibular function, 74/351
vomiting, 74/354
Motivation
attention, 3/156
emotion, 3/329
Motoneuron
see also Motor unit
definition, 1/49
electrophysiology, 1/55
group 1A activation, 1/56
infantile spinal muscular atrophy, 59/60
organolead intoxication, 64/145
sick, *see* Sick motoneuron
tetanus, 52/231
tonic, *see* Tonic motoneuron
trophic function, 1/59
Motoneuron disease
see also Amyotrophic lateral sclerosis *and* Progressive bulbar palsy
adult onset, 42/67
age, 21/31
agraphia, 45/464
amitriptyline, 59/467
assistive device, 59/469
autonomic nervous system, 75/39
benzatropine, 59/467
cancer, 41/329-338
cause, 21/33
chorea-acanthocytosis, 60/140
clinical features, 40/307, 318, 322, 41/331, 441
Creutzfeldt-Jakob disease, 46/294
dementia, 42/67
differential diagnosis, 40/472, 41/333, 358
drooling, 59/466
Eaton-Lambert myasthenic syndrome, 22/39
electrical injury, 61/194, 196
endocrinopathic, *see* Endocrinopathic motoneuron disease
epidemiology, 21/23
euthanasia, 59/469
fasciculation, 41/300
genetics, 41/441

geography, 21/31
glucose intolerance, 22/30
Guam Parkinson dementia, 42/70
Guam type, *see* Guam motoneuron disease
guanidine, 22/39
hereditary, *see* Hereditary motoneuron disease
histochemistry, 40/45
hydraulic lift, 59/469
juvenile onset, 42/68
Lafora body, 42/69
lead intoxication, 22/27
localized weakness, 40/319, 330, 342
lower, *see* Lower motoneuron disease
malabsorption syndrome, 70/231
mattress, 59/469
morbidity, 21/32
mortality, 21/30
multiple myeloma, 71/440
murine neurotropic retrovirus, 56/586, 59/447
muscle relaxant, 59/468
myoglobinuria, 62/556
neuromuscular disease, 41/441
neuropathy, 8/143
nitrofurantoin, 41/333
nutrition, 59/462
organic mercury intoxication, 22/27
pain, 59/468
paraneoplastic syndrome, 69/363, 71/693, 697
pathogenic mechanism, 41/336
phenytoin, 59/468
phthalazinol, 41/338
poliomyelitis, 22/138, 34/113
progressive muscular atrophy, 59/13
propantheline bromide, 59/467
pseudobulbar paralysis, 40/307, 45/221
quinine sulfate, 59/468
race, 21/31
respiration, 59/463-466
scopolamine patch, 59/467
sex ratio, 21/31
slow virus disease, 22/21
spectrum, 22/10
speech therapy, 59/461
sphincter control loss, 59/468
spinal cord electrotrauma, 61/194
symptomatic, 41/329
systemic neoplastic disease, 41/329
technical aid, 59/469
terminal care, 59/469
therapy, 59/459-471
treatment, 41/335
trichinosis, 52/566, 569
trihexyphenidyl, 59/467
tympanic neurectomy, 59/467
undetermined significant monoclonal gammopathy, 63/402
upper, *see* Upper motoneuron disease
viral infection, *see* Viral infected motoneuron disorder
vitamin E, 22/31
Waldenström macroglobulinemia, 63/397
walking stick, 59/469
weakness spasticity cramp, 59/467
WFN classification, 59/1-7
wheelchair, 59/469
wobbler mouse, 22/32

Motoneuron lesion
lower, *see* Lower motoneuron lesion

Motor activity
developmental dyspraxia, 4/443-446
extensibility, 4/447
extrapyramidal system, 1/299
gesture, 4/445
prehension, 4/447
presence response, 4/448
pyramidal tract, 1/298
reflex, 4/444

Motor aphasia
acalculia, 45/476
age, 45/304
agrammatism, 45/309-311
agraphia, 45/459, 461, 463
alexia, 4/99, 131, 45/436, 447
amusia, 45/487
anarthria, 45/309
anterior language zone, 45/18
apraxia, 45/428
Broca area, 45/35
buccofacial apraxia, 45/428
computer assisted tomography, 45/310
corpus callosum infarction, 2/763
corpus callosum syndrome, 2/762
developmental dyslexia, 4/134
diagnosis, 45/310
disconnection syndrome, 45/463
α-interferon intoxication, 65/562
γ-interferon intoxication, 65/562
language, 46/152
middle cerebral artery syndrome, 53/362
migraine, 45/510
mutism, 45/310
nonfluent agraphia, 45/459
nonverbal ability, 4/107
phonology, 45/308
premotor cortex, 2/734, 45/35
prognosis, 45/323

pure, *see* Pure motor aphasia
radioisotope diagnosis, 45/298, 310
restoration of higher cortical function, 3/400-409
symptom, 2/599
syntaxis, 45/307
transcortical, *see* Transcortical motor aphasia
Motor apraxia
corpus callosum infarction, 2/763
hypokinesia, 6/135
left hemisphere, 45/428
Motor conduction velocity
see also Nerve conduction velocity *and* Sensory conduction velocity
critical illness polyneuropathy, 51/578
galactosialidase deficiency, 51/380
multiple myeloma polyneuropathy, 63/393
POEMS syndrome, 47/619, 51/431, 63/394
Rud syndrome, 51/398
scapuloperoneal spinal muscular atrophy, 62/170
stimulus isolation unit, 7/120
undetermined significant monoclonal gammopathy, 63/399
Motor control
basal ganglion function, 6/91-93
cerebral dominance, 4/283-285
development, 4/443-446
rehabilitation, 46/620
Motor cortex
akinesia, 45/166
area, 1/147-151
Betz cell function, 1/58
frontal lobe syndrome, 2/725, 727
homunculus, 1/151
paracentral lobule, 1/210
primary, *see* Primary motor area
secondary area, 1/151
somatology, 1/303
supplementary, *see* Supplementary motor area
Motor cortex function
disorder of central, 1/169
dissociation, 1/161
history, 1/147
localization, 1/74
paracentral lobule, 1/210
Motor deficit
gainful employment armed forces, 24/466
mental deficiency, 4/454
paracentral lobule syndrome, 53/345
parietal lobe syndrome, 45/67
pediatric head injury, 57/341
spinal ependymoma, 20/367
Motor diaschisis
central motor disorder, 1/174
clinical symptom, 1/174
Motor disorder
Batten disease, 10/611
brain injury, 4/455
central, *see* Central motor disorder
congenital spinal cord tumor, 32/365
drug induced parkinsonism, 6/234
frontal lobe tumor, 17/256, 270
mental deficiency, 4/454
occipital lobe tumor, 17/335
parietal lobe syndrome, 45/67-69
parietal lobe tumor, 17/300
schizophrenia, 46/455
spinal ependymoma, 19/360
spinal meningioma, 19/68, 358
spinal spongioblastoma, 19/360
transverse spinal cord syndrome, 2/180-187
Motor dissociation
eye movement, 1/162
lesion, 1/161
Motor efficiency
developmental dyspraxia, 4/448-450
Motor end plate
see also Motor end plate potential
acetylcholine, 7/86, 40/129
acetylcholine receptor, 40/5, 38
acetylcholinesterase, 40/129, 380
antigen, 69/338
calcium, 69/338
chloroquine intoxication, 65/484
culture, 40/186
curare, 7/87, 40/129, 143
Eaton-Lambert myasthenic syndrome, 41/335
experimental autoimmune myasthenia gravis, 41/115
hereditary motor and sensory neuropathy type I, 21/294
hereditary motor and sensory neuropathy type II, 21/294
histochemistry, 40/3
history, 1/52
intervertebral disc, 20/525
malabsorption syndrome, 70/228
multiple organ failure, 71/536
muscle fiber, 62/28
muscle fiber conductance, 41/108
muscle tissue culture, 62/92, 97
myasthenia gravis, 40/129, 265, 41/97, 102-107, 43/158
myotonic dystrophy, 40/503
neurotoxicology, 64/14
organophosphate intoxication, 64/228
paraneoplastic syndrome, 69/338

Pena-Shokeir syndrome type I, 43/438
poliomyelitis, 56/318
progressive external ophthalmoplegia, 43/136
structure, 40/212
tetanus, 52/231
trophic change, 7/89
Motor end plate dysfunction
organophosphate intoxication, 64/162
Motor end plate noise
features, 62/50
motor unit, 62/50
myopathy, 62/50
Motor end plate potential
see also Motor end plate
arthropod envenomation, 65/195
bulging, 25/125
history, 1/52
miniature, *see* Miniature motor end plate potential
myasthenia gravis, 41/110
neuromuscular transmission, 7/86, 40/127, 129
normal, 41/110
potential, 1/52, 7/86-88
Motor evoked potential
intraoperative monitoring, 67/251
Motor extinction
akinesia, 45/161
Motor function
anatomy, 1/147-168
cerebral dominance, 4/283-285
coordination, 1/305-308
corpus striatum, 1/160
decerebration, 24/714, 716
decortication, 24/714, 716
globus pallidus, 1/160
neuropsychologic defect, 45/525
pathophysiology, 1/147-168
restoration of higher cortical function, 3/387-396
spasmodic torticollis, 6/569
Motor impairment
leprous neuritis, 51/230
posttraumatic syndrome, 24/690
Motor impersistence
akinesia, 45/161
Motor initiation
frontal lobe lesion, 45/32
Motor neglect
frontal lobe lesion, 45/32
Motor nerve conduction
chronic inflammatory demyelinating polyradiculoneuropathy, 51/532
diabetic polyneuropathy, 51/507
dipeptidyl carboxypeptidase I inhibitor intoxication, 65/445
uremic neuropathy, 7/166, 8/1, 27/338, 340
Motor nerve conduction velocity
ataxic neuropathy, 7/641
F wave, 1/57
Klinefelter variant XXYY, 31/484
leprous neuritis, 51/221
phosphofructokinase deficiency, 41/191
Rud syndrome, 51/398
Schut family ataxia, 60/483
uremic neuropathy, 7/166
Wohlfart-Kugelberg-Welander disease, 59/89
Motor neuronopathy
subacute, *see* Subacute motor neuronopathy
Motor neuropathy
antibody, 71/442
autosomal dominant inherited, *see* Autosomal dominant inherited motor neuropathy
cancer, 41/338-347
chorea-acanthocytosis, 51/398
deficiency neuropathy, 51/324
diabetic polyneuropathy, 51/499
gold polyneuropathy, 51/306
hereditary, *see* Hereditary motor neuropathy
hereditary amyloid polyneuropathy, 21/122
mental deficiency, 42/335
mercury polyneuropathy, 51/273
normolipoproteinemic acanthocytosis, 59/402
Sjögren syndrome, 71/85
tri-*o*-cresyl phosphate polyneuropathy, 51/285
Motor performance
clumsiness, 46/163
psychomotor type, 4/457
Motor polyneuropathy
Castleman disease, 63/394
chorea-acanthocytosis, 49/329
neuroborreliosis, 51/201
organophosphate induced delayed polyneuropathy, 64/176
osteosclerotic myeloma, 63/393
POEMS syndrome, 63/394
polyneuritis, 1/230
Motor radiculitis
herpes zoster, 51/181
Motor radiculopathy
pure, *see* Pure motor radiculopathy
Motor retardation
amyotonia congenita, 43/79
congenital ataxic diplegia, 42/121
psycho, *see* Psychomotor retardation
Rubinstein-Taybi syndrome, 31/282, 43/234, 60/92
Motor ritardando
akinesia, 49/67

Motor sensory neuropathy
hereditary, *see* Hereditary motor sensory neuropathy
Motor sequence stimulation
epilepsy, 15/82
right hemisphere, 15/82
Motor stereotypy
mental deficiency, 46/27
Motor syndrome
brachial plexus, 2/146
brachial plexus injury, 2/146
capsula interna, 1/186
peripheral neuron, 1/218
thalamus, 1/186
Motor system
alpha, *see* Alpha motor system
deuteropathic, *see* Deuteropathic motor system
extrapyramidal, *see* Extrapyramidal motor system
Friedreich ataxia, 60/310
protopathic, *see* Protopathic motor system
Motor tic
Gilles de la Tourette syndrome, 6/787, 49/628
Motor unit
see also Motoneuron
acetylcholine, 1/60
amyotrophic lateral sclerosis, 59/186-189
axonal stimulation, 62/59
complex repetitive discharge, 62/62, 67
conventional EMG, 62/49
denervation, 62/62
Duchenne muscular dystrophy, 40/374
electrical response, 7/107
EMG, 19/275, 40/138, 62/49
EMG method, 62/49
extradischarge, 62/50, 67
fasciculation, 62/50
fasciculation discharge, 62/50
fatigue index, 62/54
fiber type, 40/8
fibrillation potential, 62/50, 61, 67
insertional activity, 62/50
intramuscular stimulation, 62/58
limb girdle syndrome, 40/453
macroelectromyography, 62/55
macromotor unit potential, 62/55
miniature motor end plate potential, 62/50
motor end plate noise, 62/50
motor unit counting technique, 62/60
motor unit potential, 62/49, 67
muscle denervation, 62/62
muscle fiber density, 62/54
muscle fiber stimulation, 62/59
muscle reinnervation, 62/62
myogenic jitter, 62/59
myotonic discharge, 62/50, 61, 67
myotonic dystrophy, 40/138, 519
neuromuscular jitter, 62/54
positive sharp wave, 62/50, 67
propagation velocity, 62/53
rate coding, 62/52
recruitment, 62/52
reinnervation, 62/62
scanning electromyography, 62/57
single fiber electromyography, 62/49, 53
size, 62/55
spontaneous activity, 62/50, 67
tremor, 62/52
velocity recovery function, 62/54
voluntary contraction, 62/52
Motor unit action potential
critical illness polyneuropathy, 51/585
Motor unit behavior
tremor, 49/572
Motor unit hyperactivity, 41/295-311
see also Muscle cramp
baclofen, 41/297
black widow spider venom intoxication, 41/310
carbamazepine, 41/296
chondrodystrophic myotonia, 41/296
classification, 40/285
contracture, 41/187, 269, 297, 464
cramp, 40/326
diazepam, 41/296
differential diagnosis, 59/376
fasciculation, 41/299
fibrillation, 41/299
hemifacial spasm, 41/302
hereditary cramp, 41/302
Isaacs syndrome, 40/330, 332, 41/296, 300, 303, 347
Lambert-Brody syndrome, 41/296
muscle disorder, 41/297-299
myelopathy, 41/307
myokymia, 41/299
myopathy, 40/285
myotonia, 40/327, 495, 41/150, 165, 176, 298, 347
nerve, 41/299-303
ordinary muscle cramp, 41/308
painful legs moving toes syndrome, 41/311
phenytoin, 40/286, 546, 41/296
rabies, 59/377
Satoyoshi syndrome, 41/307, 59/377
site, 41/295-297
slow muscle relaxation, 41/297
spinal cord, 41/303-308
spinal myoclonus, 41/308

spinal nerve root compression, 41/303
stiff-man syndrome, 41/305, 348, 59/377
strychnine intoxication, 41/305, 59/377
tetanus, 40/330, 41/303, 59/377
tetany, 40/330, 41/301

Motor unit potential
area, 62/51
complex motor unit potential, 62/51
definition, 62/50
duration, 62/50
motor unit, 62/49, 67
muscle denervation, 62/64
muscle reinnervation, 62/64
myopathy, 62/49, 67
normal value, 62/49
polyphasia, 62/51
satellite potential, 62/51
serrated motor unit potential, 62/51
spike, 62/51
turn, 62/51

Motor weakness
frontal lobe tumor, 17/270
occipital lobe tumor, 17/335
thallium intoxication, 64/325
traumatic intracranial hematoma, 57/265

Mott cell
African trypanosomiasis, 52/340
dysproteinemic neuropathy, 8/75

Mount-Reback disease, *see* Hereditary paroxysmal kinesigenic choreoathetosis

Mountain medicine
autonomic nervous system, 75/259

Mountain sickness
see also Altitude illness
acute, *see* Acute mountain sickness
anorexia, 63/416
ataxia, 63/416
benign intracranial hypertension, 63/417
brain edema, 63/416
chronic, *see* Chronic mountain sickness
coma, 63/416
consciousness, 63/416
course, 63/416
cranial nerve palsy, 63/417
cranial neuropathy, 63/417
dementia, 63/417
dysbarism, 63/416
epilepsy, 63/417
fatigue, 63/416
hallucination, 63/417
headache, 48/397, 419, 63/416
intracranial hypertension, 63/417
nausea, 63/416
papilledema, 63/417
sleep disorder, 63/416
stupor, 63/416
symptom, 63/416

Mouth
carp, *see* Carp mouth
cyclopia, 30/444
inability to open fully, 43/433
syphilis, 33/351

Movement
athetoid, *see* Athetosis
atlantoaxial joint, 61/5
automatic, *see* Automatic movement
choreatic, *see* Chorea
climbing eye, *see* Climbing eye movement
conjugate eye, *see* Conjugate eye movement
contravoluntary, *see* Contravoluntary movement
coordinated associated, *see* Synkinesis
coordination, 1/298
course of impulses, 1/305
decomposition, *see* Decomposition of movement
desautomatization, 2/731
disjunctive, *see* Synkinesis
dysconjugate eye, *see* Dysconjugate eye movement
dystonia, 6/522
eye, *see* Eye movement
fetal, *see* Fetal movement
gaze, *see* Gaze movement
global associated, *see* Synkinesis
identical associated, *see* Synkinesis
intervertebral disc, 20/527
intervertebral disc prolapse, 25/489
involuntary, *see* Involuntary movement
involuntary eye, *see* Involuntary eye movement
mirror, *see* Mirror movement
occipitoatlantal joint, 61/5
periodic, *see* Periodic movement
phantom limb, 4/225
pill rolling, *see* Pill rolling movement
posture, 1/298
pupil, 74/404
sleep periodic, *see* Sleep periodic movement
slow eye, *see* Slow eye movement
visual perception, 2/665

Movement disorder
calcium, 70/120
dementia, 46/130
hyperthyroidism, 70/87
hypoparathyroidism, 70/122
levodopa induced dyskinesia, 49/185
pain, 75/332
paroxysmal, *see* Paroxysmal movement disorder

pyruvate carboxylase deficiency, 41/211
pyruvate decarboxylase deficiency, 41/428
reflex sympathetic dystrophy, 61/128
systemic lupus erythematosus, 39/281, 71/45
Movement induced seizure, *see* Hereditary paroxysmal kinesigenic choreoathetosis
Movement plan
frontal lobe, 3/34
functional system reorganization, 3/381
parietal lobe, 3/34
restoration of higher cortical function, 3/387, 393
Movement reluctance
Parkinson disease, 6/186
Moving toes syndrome
painful legs, *see* Painful legs moving toes syndrome
Moxonidine
blood pressure, 74/171
Moyamoya disease, 12/352-382, 55/293-303
acrocephalosyndactyly type I, 42/751
age, 12/360, 42/751, 53/164, 55/293-295
angiographic stage, 55/296
angiography, 12/366, 42/749, 53/164, 55/297
angiomatous malformation, 55/293
associated disease, 55/295
autopsy finding, 12/373
brain aneurysm, 42/751
brain artery anastomosis, 55/300
brain hemorrhage, 42/749, 53/164, 55/294, 296
brain infarction, 53/164, 55/294-296
brain ischemia, 42/749, 53/164, 55/294
carotid system syndrome, 53/308
cause, 12/374, 53/32
cerebrovascular disease, 12/346, 374, 53/28, 31
clinical features, 53/164
clinical type, 55/295
computer assisted tomography, 54/64, 55/296
consciousness, 55/293
convulsion, 12/360, 42/751
convulsive seizure, 12/360, 374
definition, 55/293
diabetes mellitus, 55/295
diagnostic criteria, 55/294
differential diagnosis, 55/294
Down syndrome, 53/31
durapexy, 55/300
EEG, 12/366, 55/298
encephaloarteriosynangiosis, 55/300
encephaloduroarteriosynangiosis, 55/300
encephalomyosynangiosis, 42/751, 55/300
epidemiology, 55/294
epilepsy, 72/212
epileptic seizure, 55/294
etiology, 12/374
familial occurrence, 42/751, 55/295, 301
follow-up, 12/366
headache, 12/360, 53/32, 55/293
hemiplegia, 12/360, 42/749, 53/33
history, 12/353
hypertension, 55/295
initial symptom, 12/366
internal carotid hypoplasia, 55/293
Japanese occurrence, 12/346, 42/751, 53/32, 164, 55/301
leptospiral arteritis, 55/415
leptospirosis, 53/164
Marfan syndrome, 42/751
mental deficiency, 12/360, 53/33
middle cerebral artery syndrome, 53/355
mortality, 12/366, 42/751, 55/295
neurofibromatosis, 42/751, 53/164, 55/295, 301
neurosurgery, 55/300
nuclear magnetic resonance, 55/296, 298
outcome, 55/296
pathogenesis, 42/751, 55/301
pathology, 55/300
porencephaly, 50/359
positron emission tomography, 55/298
prevalence, 55/301
preventive surgery, 55/300
prognosis, 12/360
radiology, 55/296
recurrence, 55/295
renal artery abnormality, 42/750, 55/295
sex ratio, 12/360, 42/751, 55/293
sickle cell anemia, 53/164, 55/505, 511
subarachnoid hemorrhage, 12/360, 374, 42/749, 53/33, 55/29, 293
symptom, 12/360
telangiectasia, 53/32, 55/293
tonsillitis, 55/295, 302
transient ischemic attack, 53/33, 55/294
treatment, 42/751, 55/298
tuberculous meningitis, 53/164
tuberous sclerosis, 42/751
MPTP, *see* 1-Methyl-4-phenyl-1,2,3,6-tetrahydropyridine
MRI, *see* Nuclear magnetic resonance
mRNA
acid maltase deficiency, 62/482
glycosaminoglycanosis type IVB, 66/269
herpes simplex virus, 56/212
myelin basic protein, *see* Myelin basic protein *m*RNA
polyadenylation, 66/73
prion protein, *see* Prion protein mRNA

RNA splicing, 66/73
structure, 66/74
targeting, 66/80
mRNA stability
sequence motif, 66/78
MS, *see* Multiple sclerosis
Muckle-Wells syndrome
see also CINCA syndrome
amyloidosis, 22/489, 42/396
Corti organ degeneration, 42/397
deafness, 14/113, 22/489
dermatitis, 42/396
dominant urticaria, 22/489, 490
fever, 42/396
hereditary progressive cochleovestibular atrophy, 60/771
nephritis, 22/489, 42/396
pain, 42/396
paresthesia, 42/396
renal disease, 22/490
sensory loss, 42/396
talipes, 42/396
trophic ulcer, 42/396
Mucocele
intrasellar cyst, 17/409, 428
pituitary tumor, 17/428
sphenoidal, 13/69
Mucocutaneous lymph node syndrome, *see* Kawasaki syndrome
Mucocutaneous pigmentation
Peutz-Jeghers syndrome, 14/526, 601, 43/41
Mucoid
histochemistry, 9/24
Mucolipid
GM2 gangliosidosis, 10/390
Mucolipidosis
classification, 66/378
genetics, 66/384
lysosomal disorder, 66/64
sialidase deficiency, 66/356
survey, 66/26, 377
Mucolipidosis type I
N-acetylneuraminidase, 51/380
N-acetylneuraminidase deficiency, 46/55
action myoclonus, 60/662
ataxia, 42/475
biochemistry, 66/353
cataract, 60/653
cherry red spot, 27/162, 42/475, 51/380, 60/654, 66/19
classification, 66/357
corneal opacity, 27/162, 42/475, 51/380
demyelination, 51/380
diagnosis, 66/357
dyssynergia cerebellaris myoclonica, 60/599
enzyme deficiency polyneuropathy, 51/370, 380
epilepsy, 72/222
foam cell, 42/475
genetics, 66/358
GM1 gangliosidosis, 10/466, 470
laboratory finding, 66/357
mental deficiency, 27/161, 42/475, 46/54
miscellaneous case, 66/369
myoclonic epilepsy, 49/617, 60/662
myoclonus, 42/475, 51/380
neuraminidase, 66/60
α-neuraminidase, 42/475
neuraminidase activity, 66/357
pain, 51/380
prenatal diagnosis, 42/476
remyelination, 51/380
seizure, 42/475
sensorineural deafness, 42/475
sialidase deficiency, 66/353
sialidasis type I, 51/380
survey, 66/26, 353
symptom, 66/358
Mucolipidosis type II
see also Fucosidosis
N-acetyl-β-D-hexosaminidase, 42/448
action myoclonus, 60/662
biochemistry, 27/161
carpal tunnel syndrome, 42/309
cerebroside sulfatase, 42/448
cherry red spot, 51/380
clinical features, 27/160
cytoplasm, 66/380
deafness, 51/380
enzyme, 66/381
enzyme deficiency polyneuropathy, 51/370, 380
α-L-fucosidase, 42/448
β-galactosidase, 42/488
genetics, 66/384
β-glucuronidase, 42/448
glycosaminoglycanosis type I, 42/448
GM1 gangliosidosis, 10/466, 470
granular inclusion, 66/379
histopathology, 66/380
α-L-iduronidase, 42/448
inborn error of carbohydrate metabolism, 27/162
lysosomal acid hydrolase, 66/380
lysosomal disorder, 66/64
α-mannosidase, 42/448
mental deficiency, 42/448, 46/55
metabolic defect, 66/381
mucolipidosis type III, 66/382

myoclonic epilepsy, 60/662
myoclonus, 51/380
nuclear magnetic resonance, 66/382
prenatal diagnosis, 27/161
psychomotor retardation, 42/448
radiographic features, 66/379
related disorder, 66/383
Schwann cell vacuole, 51/382
sialyloligosaccharide, 66/380
skeletal deformity, 42/448
survey, 66/377
symptom, 66/378, 382

Mucolipidosis type III
N-acetyl-β-D-hexosaminidase, 42/449
amaurosis, 29/367
amniocentesis, 29/369
biochemical aspect, 29/368
carpal tunnel syndrome, 42/309
cerebroside sulfatase, 42/449
cherry red spot, 29/367
clinical picture, 27/162
corneal opacity, 29/367, 42/449
detection, 29/369
diagnosis, 66/383
dwarfism, 42/449
enzyme deficiency polyneuropathy, 51/370, 381
α-L-fucosidase, 42/449
β-galactosidase, 29/367, 42/449
lysosomal acid hydrolase, 66/383
lysosomal disorder, 66/64
α-mannosidase, 42/449
mental deficiency, 27/162, 42/449
mucolipidosis type II, 66/382
related disorder, 66/383
skeletal deformity, 42/449
treatment, 29/369

Mucolipidosis type IV
athetosis, 51/380
clinical phenotype, 66/384
corneal clouding, 51/380
enzyme deficiency polyneuropathy, 51/380
ganglioside, 66/384
hyperlipidemia type II, 51/380
lysosomal acid hydrolase, 66/384
muscle fiber feature, 62/17
secondary pigmentary retinal degeneration, 60/727

Mucopolysaccharide, *see* Glycosaminoglycan

Mucopolysaccharidosis
type I, *see* Glycosaminoglycanosis type I
type IH, *see* Glycosaminoglycanosis type IH
type IH-S, *see* Glycosaminoglycanosis type IH-S
type II, *see* Glycosaminoglycanosis type II
type IIA, *see* Glycosaminoglycanosis type IIA
type IIB, *see* Glycosaminoglycanosis type IIB
type III, *see* Glycosaminoglycanosis type III
type IIIA, *see* Glycosaminoglycanosis type IIIA
type IIIB, *see* Glycosaminoglycanosis type IIIB
type IIIC, *see* Glycosaminoglycanosis type IIIC
type IIID, *see* Glycosaminoglycanosis type IIID
type IS, *see* Glycosaminoglycanosis type IS
type IV, *see* Glycosaminoglycanosis type IV
type IVA, *see* Glycosaminoglycanosis type IVA
type IVB, *see* Glycosaminoglycanosis type IVB
type V, *see* Glycosaminoglycanosis type IS
type VI, *see* Glycosaminoglycanosis type VI
type VII, *see* Glycosaminoglycanosis type VII
type VIII, *see* Glycosaminoglycanosis type VIII
type F, *see* Fucosidosis

Mucoprotein
multiple sclerosis, 9/320
PAS stain, 9/24

Mucor michai, *see* Mucormycosis

Mucormycosis, 35/541-554
acidosis, 35/545
acquired immune deficiency syndrome, 56/515
amphotericin B, 52/475
associated condition, 52/470
blindness, 35/544, 52/471
brain, *see* Brain mucormycosis
brain abscess, 35/546, 52/151, 471
carotid artery thrombosis, 35/545, 52/471, 473, 55/436
cavernous sinus thrombosis, 35/545, 551, 55/436
cerebrovascular disease, 35/545, 55/435
clinical features, 35/543, 52/470
cranial epidural empyema, 52/168
CSF, 35/551, 52/473
deafness, 52/471
diabetes mellitus, 35/543, 52/469, 474
diabetic ketoacidosis, 35/543, 52/470
diagnosis, 35/551, 52/473
differential diagnosis, 52/473
disseminated, 35/546, 52/472
encephalitis, 52/471
epidemiology, 35/543, 52/468
epilepsy, 52/471
epistaxis, 35/544, 52/471
experimental, 35/552
headache, 35/544, 52/471
histopathology, 52/474
history, 35/541
immunocompromised host, 52/467
immunosuppression, 35/544, 52/474
infant, 35/545, 52/472
ketoacidosis, 52/474

microbiology, 52/467
multiple cranial neuropathy, 52/471
mycosis, 35/375
nasopharyngeal osteitis, 52/471
nonrecurrent nonhereditary multiple cranial neuropathy, 51/571
ophthalmoplegia, 35/551, 52/471
orbital apex syndrome, 35/544, 52/471
panophthalmitis, 35/545
pathology, 35/546-551
predisposing factor, 35/543, 52/472
prognosis, 52/475
radiodiagnosis, 52/473
renal insufficiency, 27/343
rhinocerebral, 35/544, 52/470, 472, 474
spastic paraplegia, 59/436
spinal epidural empyema, 52/187
subarachnoid hemorrhage, 52/471
surgery, 52/475
taxonomy, 52/467
treatment, 35/552-554, 52/474
Mucosal neuroma
multiple, *see* Multiple mucosal neuroma
Mucosal neuroma syndrome, *see* Multiple endocrine adenomatosis type III
Mucoviscidosis
neuroaxonal dystrophy, 6/626
secondary neuroaxonal dystrophy, 49/408
tuberous sclerosis, 14/496
Mulgatoxin a
toxic myopathy, 62/610
Mulibrey dwarfism
hepatomegaly, 43/435
mental deficiency, 43/435
retinal degeneration, 43/435
strabismus, 43/435
swayback nose, 43/435
Müller cell
mitochondrial metabolism, 2/509
origin retinal limiting membrane, 2/507
Müller classification
brain tumor, 16/1
Müller law
specific nerve energy, 7/82
Müllerian aplasia
VACTERL syndrome, 50/514
Müllerian duct
MURCS syndrome, 50/515
Multicore disease, *see* Multicore myopathy
Multicore myopathy
adenosine triphosphatase, 62/339
central core myopathy, 40/158
classification, 40/288
clinical symptom, 62/338
congenital myopathy, 41/16, 62/332
contracture, 43/120
differential diagnosis, 62/338
enzyme histochemistry, 62/339
facial weakness, 62/334
histochemistry, 40/46
hyporeflexia, 43/120
mitochondria, 40/91, 43/121
mitochondria loss, 62/339
muscle fiber type I, 43/121
oxidative metabolism, 43/121
pathologic reaction, 40/262
progressive external ophthalmoplegia, 41/16
proximal muscle weakness, 43/120
ptosis, 43/120
scoliosis, 43/120
ultrastructure, 40/74
Z band, 43/121
Multicystic encephalomalacia
cytomegalic inclusion body disease, 56/268
cytomegalovirus, 56/268
Multicystic necrotizing myelitis
visna, 56/462
Multifocal CNS disease
consciousness, 45/118-121
recurrent, *see* Recurrent multifocal CNS disease
Multifocal cortical epilepsy
stereotaxic operation, 15/769
Multifocal demyelination
encephalitis, 52/342, 357
granulomatous amebic encephalitis, 52/325
Multifocal ischemic encephalopathy
chronic mercury intoxication, 64/372
mercury vapor, 64/377
Multifocal leukoencephalopathy
progressive, *see* Progressive multifocal leukoencephalopathy
Multifocal myoclonus, *see* Myoclonic epilepsy
Multifocal necrosis
acquired toxoplasmosis, 52/357
Multifocal necrotizing encephalopathy
cerebral toxoplasmosis, 52/357
Multifocal necrotizing leukoencephalopathy
lymphomatoid granulomatosis, 51/451
Multifocal neuritis
cytomegalovirus infection, 71/264
Multifocal neuropathy
diabetic polyneuropathy, 51/501
hereditary neuropathy, 60/254
polyneuropathy, 60/254
Multifocal pontine lesion
chemotherapy, 67/365

Multifocal relapsing neuropathy
  a-alphalipoproteinemia, 51/397
Multi-infarct arteriosclerosis
  Parkinson disease, 49/94
Multi-infarct dementia
  age specific incidence rate, 53/14
  Alzheimer disease, 46/265
  atrial myxoma, 63/98
  behavior disorder, 42/285
  Binswanger disease, 54/229, 55/142
  brain atrophy, 42/283-285
  brain blood flow, 46/362, 54/123, 144
  cause, 53/239
  cerebrovascular disease, 53/14, 28, 42, 55/141
  classification, 46/205
  clinical features, 46/354, 55/141
  computer assisted tomography, 46/362, 54/58
  cryptococcal meningitis, 55/435
  detection, 46/201
  Hachinski ischemic score, 46/354, 55/142
  hydrocephalic dementia, 46/326
  Köhlmeier-Degos disease, 55/277
  lacunar stroke, 55/142
  oxygen extraction rate, 54/123
  parkinsonism, 49/471
  perseverative agraphia, 45/467
  positron emission tomography, 54/120
  race, 53/15
  risk factor, 46/354
  sex ratio, 53/14
  single photon emission computer assisted
    tomography, 54/123
  sleep-wake cycle, 46/354
  vascular dementia, 46/353-355
Multilobulated tongue
  Mohr syndrome, 43/449
  orofaciodigital syndrome type I, 43/447
Multilocular pineal cyst
  midbrain tumor, 17/637
Multineuritis, *see* Mononeuritis multiplex
Multinodular stenosing amyloidosis
  apoplexy, 11/581, 613
  arteriosclerosis, 11/613
  cerebrovascular disease, 11/581, 613
  periarteritis, 11/613
Multinucleated giant cell
  acquired immune deficiency syndrome, 56/507,
    509
  acquired immune deficiency syndrome dementia,
    56/507
  Guam amyotrophic lateral sclerosis, 59/294
Multinucleated globoid cell
  globoid cell leukodystrophy, 66/10
Multinucleation
  brain tumor, 17/94
Multiorgan failure, *see* Multiple organ failure
Multiple acyl-coenzyme A dehydrogenase
    deficiency
  see also Glutaric aciduria type II
  coma, 62/495
  episodic encephalopathy, 62/495
  exercise intolerance, 62/495
  fasting, 62/495
  hypoglycemia, 62/495
  infantile hypotonia, 62/495
  lethargy, 62/495
  lipid metabolic disorder, 62/495
  2 main forms, 62/495
  metabolic encephalopathy, 62/495
  metabolic myopathy, 62/495
  myalgia, 62/495
  organic acid metabolism, 66/641
  recurrent coma, 62/495
Multiple ankylosis
  Pena-Shokeir syndrome type I, 43/437
Multiple benign cystic epithelioma (Fordyce), *see*
    Multiple trichoepithelioma
Multiple brain abscess
  candidiasis, 52/398
  nocardiosis, 35/517, 519, 52/447
  Serratia marcescens, 52/298
  toxic encephalopathy, 52/297
Multiple brain infarction
  brain thromboangiitis obliterans, 55/310
  Divry-Van Bogaert syndrome, 55/318
  female, 55/381
  Köhlmeier-Degos disease, 14/789, 39/435, 55/276
  leukemia, 63/345
  malignant lymphoma, 55/488
  multiple myeloma, 55/487
  Parkinson disease, 49/95
  Sneddon syndrome, 55/402
  white matter, 55/488
Multiple brain mass lesion
  acquired toxoplasmosis, 52/354
Multiple brain tumor
  childhood, 18/309
  literature, 16/75
  type, 18/309
Multiple capillary hemangioma
  Patau syndrome, 14/121, 31/504, 507, 50/558
Multiple circumscribed lipomatosis, *see*
    Krabbe-Bartels disease
Multiple cranial nerve
  herpes zoster ganglionitis, 51/181
  pontine infarction, 12/16

Multiple cranial nerve palsy, *see* Multiple cranial neuropathy
Multiple cranial neuritis, *see* Multiple cranial neuropathy
Multiple cranial neuropathy
see also Bulbar nerve syndrome, Cranial mononeuritis multiplex, Jugular foramen syndrome *and* Parapharyngeal space syndrome
acute hemorrhagic conjunctivitis, 56/351
arterial supply, 51/569
ascending pharyngeal system, 51/569
Bannwarth syndrome, 51/201
brain amyloid angiopathy, 54/337
Brucella, 51/185
bulbar nerve syndrome, 51/569
Burkitt lymphoma, 63/351
carotid inferolateral trunk, 51/569
cephalic tetanus, 65/218
chordoma, 18/153
Cogan syndrome type II, 51/454
diphtheria, 65/240
enterovirus 70, 56/329, 351
Epstein-Barr virus, 56/253
ethylene glycol intoxication, 64/124
fonsecaeasis, 35/566, 52/487
Garcin syndrome, 51/570
Guillain-Barré syndrome, 51/245, 247
infraclinoid syndrome, 2/88
jugular foramen syndrome, 51/569
Kawasaki syndrome, 51/453
Köhlmeier-Degos disease, 55/277
lymphomatoid granulomatosis, 51/451
meningeal lymphoma, 63/347
middle meningeal system, 51/569
mucormycosis, 52/471
multiple myeloma, 63/393
neuroborreliosis, 51/201
neurolymphomatosis, 56/182
nonrecurrent nonhereditary, *see* Nonrecurrent nonhereditary multiple cranial neuropathy
parapharyngeal space syndrome, 51/569
peripheral neurolymphomatosis, 56/181
poliomyelitis, 56/329
polyarteritis nodosa, 51/452, 570
posterior condylar canal syndrome, 2/100
recurrent, *see* Recurrent multiple cranial neuropathy
recurrent nonhereditary, *see* Recurrent nonhereditary multiple cranial neuropathy
rhodotorulosis, 52/493
sarcoid neuropathy, 51/196
sarcoidosis, 51/195
schistosomiasis, 52/538
Sicard-Collet syndrome, 2/100, 24/179
Sjögren syndrome, 51/449
superior longitudinal sinus thrombosis, 55/277
Tapia syndrome, 2/99
temporal arteritis, 51/452
thrombotic thrombocytopenic purpura, 55/472
trichinosis, 52/566
trichloroethylene polyneuropathy, 51/280
tuberculosis, 51/186
tuberculous meningitis, 55/426
Villaret syndrome, 2/225
vincristine polyneuropathy, 51/299
Waldenström macroglobulinemia polyneuropathy, 51/432
Wegener granulomatosis, 51/450
Whipple disease, 51/184
Multiple cutaneous hemangioma
Riley-Smith syndrome, 14/14, 76, 480, 768, 43/30
Multiple cystic encephalopathy
see also Brain infarction
animal, 53/28
hydranencephaly, 30/665, 50/341, 346
hydrocephalus, 30/665
neonatal brain, 9/579
phlebostasis, 9/579
prenatal vascular disease, 53/27
Multiple cystic leukoencephalomalacia
brain fat embolism, 55/190
Multiple deficiency syndrome
glycosaminoglycanosis type VI, 66/315
Multiple dentigerous cyst
familial, *see* Multiple nevoid basal cell carcinoma syndrome
Multiple dermal cylindroma
cutis verticis gyrata, 14/591
differential diagnosis, 14/592
histogenesis, 14/591
nosology, 14/589
partial gigantism, 14/589
polydactyly, 14/589
treatment, 14/592
Multiple endocrine adenoma
adrenal gland, 39/503
Multiple endocrine adenomatosis, 42/752-754
APUD cell, 42/753
hyperparathyroidism, 27/286
type I, *see* Wermer syndrome
type II, *see* Sipple syndrome
type IIB, *see* Multiple endocrine adenomatosis type III
Multiple endocrine adenomatosis syndrome
CNS tumor, 30/101
Multiple endocrine adenomatosis type III

café au lait spot, 42/753
diffuse lentiginosis, 42/753
megacolon, 42/753
pheochromocytoma, 39/504, 42/752, 765
thyroid carcinoma, 42/752
Multiple endocrine neoplasia syndrome
bowel, 75/650
neurinoma, 68/538
Multiple endocrine neoplasia type 2b
autonomic nervous system, 75/25
Multiple ependymoma
neurofibromatosis type I, 20/358
Multiple epidural spinal lipoma
associated lesion, 20/406
Multiple exostosis
hereditary, *see* Hereditary multiple exostosis
Multiple hamartoma syndrome
ataxia, 42/755
hirsutism, 42/754
intention tremor, 42/755
keratoderma, 42/754
meningioma, 42/754
neoplasia syndrome, *see* Cowden syndrome
ovarian cyst, 42/754
Von Romberg sign, 42/755
Multiple hemorrhagic sarcoma
idiopathic, *see* Idiopathic multiple hemorrhagic sarcoma
Multiple interstitial telangiectasia
Turner syndrome, 50/544
Multiple intracranial arterial occlusion
cause, 12/347
management, 12/347
onset, 12/346
Multiple joint dislocation
Seckel syndrome, 43/378
Multiple lentigines syndrome, *see* LEOPARD syndrome
Multiple lipomatosis, *see* Krabbe-Bartels disease
Multiple meningioma
meningiomatosis, 20/189
neurofibromatosis, 20/189
vertebral canal, 20/233
Multiple mitochondrial DNA deletion
ataxia, 62/510
cataract, 62/510
EEG, 62/510
exercise intolerance, 62/510
familial myopathy, 66/429
familial recurrent myoglobinuria, 66/429
hearing loss, 62/510
hereditary neuropathy, 66/429
hypoparathyroidism, 62/510
idiopathic cardiomyopathy, 66/429
inclusion body myositis, 66/429
inheritance, 62/510
lactic acidosis, 62/510
late onset myopathy, 66/429
mental deficiency, 62/510
MNGIE syndrome, 66/429
multiple symmetric lipomatosis, 66/429
muscle weakness, 62/510
myoglobinuria, 62/510
neuropathy, 62/510
nystagmus, 62/510
optic atrophy, 62/510
periodic paralysis, 66/429
progressive encephalomyopathy, 66/429
progressive external ophthalmoplegia, 62/510, 66/429
ptosis, 62/510
ragged red fiber, 62/510
tremor, 62/510
Multiple mononeuropathy
diabetic polyneuropathy, 51/499
Tangier disease, 60/135
Multiple mucosal neuroma
autonomic polyneuropathy, 51/476
Multiple myeloma
see also Plasmacytoma
abducent nerve paralysis, 63/393
age, 18/254, 20/11, 15, 39/137
amyloid, 39/132, 141, 156, 47/618
anemia, 20/9, 39/138, 142, 63/392
arthritis, 20/11
ascending motor polyneuropathy, 63/394
ataxia, 55/487
back pain, 39/138, 141, 63/392
Bence Jones protein, 20/9, 11, 39/132, 134, 138
Bence Jones proteinuria, 18/255, 20/9-11, 13-16, 111, 39/133, 136, 138, 148
bone pain, 20/11, 39/138, 146, 63/392
brain lacunar infarction, 55/487
cachexia, 20/9, 39/138
carcinoma, 38/687
carpal tunnel syndrome, 8/7, 39/135, 155, 42/309
cerebrovascular disease, 55/470
chemotherapy, 18/256, 20/14, 111, 63/396
chronic axonal neuropathy, 51/531
chronic inflammatory demyelinating polyneuropathy, 63/393
chronic inflammatory demyelinating polyradiculoneuropathy, 51/530
chronic meningitis, 56/644
classification, 39/135, 69/290
clinical features, 18/254, 39/137-146, 63/392

clinical syndrome, 18/255
confusional state, 63/395
corticosteroid, 18/256, 20/11, 14, 63/396
corticotropin, 18/256
cortisone, 18/256
cranial nerve palsy, 18/255, 39/142, 63/395
cranial neuropathy, 55/470, 63/393, 395
cryoglobulinemia, 39/135, 181, 63/403
cryoglobulinemic polyneuropathy, 51/434
CSF examination, 18/255
cyclophosphamide, 20/14, 111, 63/396
cytotoxic agent, 18/256
dementia, 55/487
demyelinating neuropathy, 47/616, 618
demyelinating polyneuropathy, 55/470
diabetes insipidus, 18/255
diagnosis, 39/158-165
differential diagnosis, 20/12
diffuse plasmacytosis, 18/257
duration, 18/255, 20/11, 39/138
epilepsy, 18/255
exophthalmos, 18/256, 63/393
facial paralysis, 63/393
Fanconi syndrome, 29/172
fever, 20/11
gastrointestinal hemorrhage, 20/11
Guillain-Barré syndrome, 39/153, 63/394
headache, 18/255
hemiplegia, 18/255
hereditary amyloid polyneuropathy, 63/393
history, 18/254, 39/131-135
Hodgkin disease, 71/440
hypercalcemia, 20/11, 14, 39/138, 142, 63/392
hyperviscosity, 39/135, 142, 60/392
hyperviscosity syndrome, 55/484, 486
IgA type, 20/10, 55/486
IgG type, 20/9, 55/486
immunocompromised host, 56/470
incipient, *see* Waldenström macroglobulinemia
intracranial pressure, 18/255
laminectomy, 20/111
leptomeningeal involvement, 39/152
leukoencephalopathy, 9/594
life span, 20/11, 15
melphalan, 18/256, 20/14, 17
metabolic encephalopathy, 55/470
metabolic neuropathy, 51/68
motoneuron disease, 71/440
multiple brain infarction, 55/487
multiple cranial neuropathy, 63/393
neuropathology, 18/257
neuropathy, 18/255, 39/153-156
optic atrophy, 63/393
osteosclerotic, *see* Osteosclerotic myeloma
papilledema, 18/256, 63/393
paraplegia, 20/11, 39/141
paraproteinemia, 69/290, 301
paraproteinemic coma, 18/256, 20/11
paraproteinemic polyneuropathy, 51/429
pathology, 39/146-152, 156
plasma cell, 20/10
plasma cell dyscrasia, 39/132, 63/391
plasmapheresis, 18/256, 63/395
pneumonia, 20/11
POEMS syndrome, 47/619, 51/430, 63/394
polyarteritis nodosa, 55/359
polyneuropathy pathogenesis, 63/395
primary amyloidosis, 51/415, 419
procarbazine, 18/256
prognosis, 18/256, 20/111
progressive multifocal leukoencephalopathy, 55/470
proteinuria, 20/11, 39/138
psychosis, 20/11
radiculopathy, 39/141, 55/470, 63/392
radiotherapy, 18/256, 20/14, 111, 63/395
recurrent meningitis, 52/54
relapsing fever, 20/11
relapsing pneumonia, 20/11
renal disease, 20/11
renal insufficiency, 20/15, 63/392
RES tumor, 18/254-258, 20/9
sclerotic myeloma, 47/618
sensorimotor neuropathy, 38/687
sensorimotor polyneuropathy, 18/255, 51/419, 63/393
sensory action potential, 63/393
sensory neuropathy, 47/618, 63/393
serum M component, 20/14, 39/133, 139, 63/394
sex ratio, 18/254, 39/137
solitary plasmacytoma, 18/256
space occupying lesion, 18/255, 39/135
spinal cord, 67/198
spinal cord compression, 18/255, 19/366, 20/13, 111, 39/135, 141, 63/392, 71/440
spinal epidural tumor, 20/111
spinal nerve root compression, 63/392
spinal tumor, 68/518
status epilepticus, 18/255
survey, 71/434
temporal arteritis, 48/320
treatment, 18/256, 20/14, 39/165-168, 63/395
tumor site, 63/392
uremia, 20/11, 63/392
urethan, 18/256, 20/14, 17
vasculopathy, 63/395

vertebral column, 20/9-18, 39/139
vertebral plasmacytoma, 63/392
vertebral scalloping, 20/13
Multiple myeloma polyneuropathy
see also POEMS syndrome
amyloid deposit, 51/431
amyloid polyneuropathy, 63/393
associated systemic amyloidosis, 63/393
axonal degeneration, 51/431, 63/393
carpal tunnel syndrome, 63/393
CSF protein, 51/430
dissociated sensory loss, 63/393
dorsal column degeneration, 51/431
EMG, 51/431, 63/393
endocrinopathy, 51/430
features, 51/429
heralding symptom, 51/429
hyperglycemia, 51/430
hyperpigmentation, 51/430
hypertrichosis, 51/430
IgG-kappa paraproteinemia, 51/430
incidence, 51/429, 55/470
laboratory finding, 51/430
lymphadenopathy, 51/430
motor conduction velocity, 63/393
nerve conduction velocity, 51/431
orthostatic hypotension, 63/393
osteosclerotic myeloma, 51/430, 63/393
papilledema, 51/430
pathogenesis, 51/431
pathology, 51/431
POEMS syndrome, 51/430
polycythemia, 51/430
relapsing type, 51/430
segmental demyelination, 51/431, 63/393
sensorimotor type, 51/430
sensory type, 51/430
sexual impotence, 63/393
solitary myeloma, 51/430
solitary plasmacytoma, 63/393
systemic amyloidosis, 51/430
treatment, 51/432
visceromegaly, 51/430
Multiple neurinoma syndrome
pheochromocytoma thyroid carcinoma, *see* Pheochromocytoma thyroid carcinoma multiple neurinoma syndrome
Multiple neuritis
neuralgic amyotrophy, 51/172
Multiple neurofibromatosis
acoustic neuroma, 43/35
aqueduct stenosis, 43/34
astrocytoma, 43/35
birth incidence, 43/35
café au lait spot, 43/34
ependymoma, 43/35
exophthalmos, 14/398
glaucoma, 14/398
glioblastoma, 43/35
hereditary sensory and autonomic neuropathy type I, 14/398
hydrocephalus, 43/34
kyphoscoliosis, 43/35
levodopa, 43/35
maternal effect, 43/35
meningioma, 43/35
mental deficiency, 43/35
mutation rate, 43/35
nerve tumor, 8/435-446
oligodendroglioma, 43/35
prevalence, 43/35
seizure, 43/35
Multiple neuronal system degeneration
acetylcholine, 75/192
acid phosphatase, 42/75
adenosine triphosphatase 6 mutation syndrome, 62/510
alcohol, 75/170
alveolar hypoventilation, 63/146
Alzheimer disease, 59/136
amyotrophic lateral sclerosis, 59/411
amyotrophy, 59/138
antecollis, 59/138
anterior horn syndrome, 42/75
apnea, 59/159
ataxia, 75/162
atypical features, 75/163
autonomic dysfunction, 63/146, 75/168
autonomic failure, 75/644
autonomic nervous system, 74/544, 75/32
autonomic symptom, 75/168
blood pressure, 75/170, 174
bradykinesia, 75/162
central autonomic dysfunction, 75/161
cerebellar dysfunction, 75/162
classification, 75/164
complex I deficiency, 62/503
complex III deficiency, 62/504
complex V deficiency, 62/506
congenital hip dislocation, 42/75
congenital pain insensitivity, 51/566
corticospinal tract, 59/143
criteria, 75/166
definition, 59/136
dementia, 59/411
diagnosis, 75/162

diagnostic criteria, 59/137
distal amyotrophy, 60/659
dysarthria, 59/138
eye movement, 59/137, 75/169
familial incidence, 59/143
family history, 75/162
fasciculation, 59/138
forgotten respiration, 63/489
giant axonal neuropathy, 60/82
growth hormone, 75/173
hand deformity, 75/167
history, 75/161
hyperreflexia, 75/162
incidence, 59/135
inspiratory stridor, 59/138
(3-iodobenzyl)guanidine I 123, 75/185
Joseph-Machado disease, 59/137
larynx, 74/395
levodopa, 59/136, 138
mental deficiency, 42/75
MNGIE syndrome, 62/505
nerve conduction velocity, 42/75
nerve lesion, 59/142
neuronal intranuclear hyaline inclusion disease, 60/660
neuropathology, 59/143
noradrenalin, 75/192
nosology, 59/135
nuclear magnetic resonance, 60/673, 75/186
nystagmus, 75/162
olivopontocerebellar atrophy, 59/137, 75/161
olivopontocerebellar involvement, 59/136
orthostatic hypotension, 43/66, 63/145, 232, 75/168
osteoporosis, 42/75
pain, 59/138
pallidonigral degeneration, 49/445
parkinsonism, 49/108, 63/146, 75/162
periodic ataxia, 60/650
peripheral autonomic ganglion, 59/143
peripheral nerve, 59/149
prevalence, 75/163
progressive autonomic failure, *see* Shy-Drager syndrome
progressive dysautonomia, 59/134-138, 141-143
progressive supranuclear palsy, 49/239
pure autonomic failure, 75/166, 172
respiratory dysfunction, 75/171
respiratory failure, 59/159
respiratory pattern, 74/545
rigidity, 75/162
serotonin, 75/192
sexual impotence, 75/168
sleep, 75/171
sleep apnea syndrome, 63/146
somatostatin, 75/193
sporadic type, 59/137
striatonigral degeneration, 22/265, 49/211, 60/540
striatonigral involvement, 59/136
stridor, 63/146
substance P, 75/193
sweating disorder, 75/171
thalamus degeneration, 21/596
treatment, 75/176
tremor, 75/162
ubidecarenone, 62/504
urinary incontinence, 75/162
vocal cord paralysis, 59/138
wakefulness, 59/159

Multiple nevoid basal cell carcinoma syndrome, 14/455-470
autosomal dominant, 14/12
bifid rib, 14/459
blindness, 43/31
brachymetacarpalism, 14/462
cerebellar medulloblastoma, 14/79, 464, 18/172, 42/742
cheilopalatoschisis, 43/32
chromosome, 67/58
CNS lesion, 14/463
CNS tumor, 30/101, 67/54
corpus callosum agenesis, 14/79, 114, 50/163
differential diagnosis, 14/584
dura mater calcification, 14/80, 114, 457
Ellsworth-Howard test, 14/584
endocrine symptom, 14/466
eye lesion, 14/465
face, 14/456
frontal bossing, 43/31
general concept, 14/87
hereditary, 14/468
histogenesis, 14/584
history, 14/580
hydrocephalus, 14/79, 464, 31/26
hypertelorism, 30/248, 43/31
hypogonadism, 14/466, 43/32
intracranial calcification, 14/80, 457, 18/172
intracranial tumor, 43/31
jaw cyst, 14/466, 18/172
kidney malformation, 14/468
Klippel-Trénaunay syndrome, 14/523
kyphoscoliosis, 43/31
leiomyoma, 14/468
Lewandowsky-Lutz syndrome, 14/114
mandibular cyst, 14/79, 114, 456, 466
mental deficiency, 14/463, 30/248, 46/44

mesenteric cyst, 14/468
multiple trichoepithelioma, 14/114
neurocutaneous syndrome, 14/101
neurofibroma, 14/583
neuropathology, 14/79-81
nosology, 14/581
palmar and plantar pit, 14/582, 18/172
parathyroid hormone, 14/80
platybasia, 14/456
prognosis, 14/585
rhabdomyosarcoma, 14/81
rib abnormality, 14/114
rib synostosis, 14/459
sella turcica, 43/31
skeletal lesion, 14/458
skin lesion, 14/80, 457
skin manifestation, 30/248
spina bifida occulta, 14/456, 459
Sprengel deformity, 14/459
syndactyly, 14/456, 462
treatment, 14/585
vertebral abnormality, 43/31
Multiple organ failure
atracurium, 71/536
brain, 71/527
CNS, 71/527
complication, 71/525
corticosteroid, 71/540
critical illness polyneuropathy, 51/575, 584, 63/420
cytokine mediator, 71/540
encephalopathy, 71/527
head injury, 57/235
heart transplantation, 63/180
intensive care, 71/528
motor end plate, 71/536
muscular atrophy, 71/537
myopathy, 71/537, 540
neurologic syndrome, 71/526, 528
neuromuscular block, 71/540
pancuronium, 71/536
peripheral nerve, 71/534
peripheral nervous system, 71/533
polyneuropathy, 71/534, 540
rocuronium, 71/536
septic encephalopathy, 71/540
spinal cord, 71/532
survey, 71/525, 528
vecuronium, 71/536
Multiple pattern test, *see* Harrington-Flocks test
Multiple peripheral circumscribed lipomatosis, *see* Krabbe-Bartels disease
Multiple personality
hysteria, 43/207, 46/576
schizophrenia, 46/576
temporal lobe epilepsy, 46/577
Multiple pigmented nevi
linear nevus sebaceous syndrome, 43/38
Turner syndrome, 50/543
Multiple pituitary hormone deficiency, *see* Pituitary dwarfism type II
Multiple recurrent mononeuropathy
hereditary, *see* Hereditary multiple recurrent mononeuropathy
Multiple respiratory enzyme defect
mitochondrial encephalomyopathy, 66/425
Multiple sclerosis
Abercrombie description, 9/48
N-acetylneuraminic acid, 9/321
acetylsalicylic acid, 9/402
activation therapy, 9/419
active disease phenomena, 47/343
acute, *see* Acute multiple sclerosis
acute disseminated encephalomyelitis, 9/126, 135, 138, 199, 500, 47/72
adenosine phosphate, 9/389
adrenoleukodystrophy, 47/595
adverse effect, 47/201
age, 9/72, 121, 47/268
age at onset, 9/200
aged, 47/63
aggravating factor, 47/64, 149-153, 179
albumin, 47/86
Aleutian mink disease, 9/126
allergic theory, 9/109
allergy, 9/115
allogeneic mixed lymphocyte reaction, 47/356
allotype G1m, 47/363
allotype G3m, 47/363
alprostadil, 47/342, 355
4-aminopyridine, 47/39-41
amyotrophic lateral sclerosis, 22/309, 59/202, 407
Andral description, 9/48
animal, 9/664-687, 47/323
animal model, 56/587
animal neurology, 9/668
ankylosing spondylitis, 38/517, 70/6
antibody, 9/141, 47/274, 339
antibody dependent cytotoxicity, 47/343, 352
antibody response, 47/339
antibrain antibody, 9/141
antidepressant, 47/178
antigen, 47/250
antigen presentation, 47/231
antigen processing, 47/231
aphasia, 9/180

Asia, 47/321
associated disease, 9/115, 232
associated factor, 9/99
astrocytosis, 47/246
ataxia, 9/179, 42/496
Australia, 9/70, 47/321
autoimmune disease, 9/107, 115, 135, 144-148, 539-543, 47/188, 338
autoimmune reaction, 9/316
autologous mixed lymphocyte reaction, 47/356, 361
autonomic nervous system, 47/68, 75/18
autonomic symptom, 9/180
autosomal dominant inheritance, 47/291
autosomal recessive inheritance, 47/290
axis cylinder, 9/264
azathioprine, 47/86, 196
B-lymphocyte, 47/101, 112, 189, 309, 345, 360-363
Babinski drawing, 47/220
baclofen, 47/157
barbiturate, 47/71
behavior change, 46/416
Behçet syndrome, 9/206, 47/71, 56/597
benign type, 9/199, 47/50
benzodiazepine, 47/157
birth order, 47/302
bladder function, 47/160-166
bladder symptom, 9/171
blastogenesis, 47/357
blink reflex, 47/131
blood-brain barrier, 47/79-121
blood chemistry, 9/320, 47/68
blood clotting, 9/131
blood group, 9/99, 47/302
blood transfusion, 9/385
Bobath method, 9/420
Bourneville illustration, 9/50-53
bout frequency, 9/188
bowel dysfunction, 75/646
bowel symptom, 9/171
brain antibody, 47/108, 365
brain antigen, 47/234, 357
brain arteriovenous malformation, 12/248
brain biopsy, 10/685, 47/223
brain blood flow, 9/180
brain IgG, 47/86
brain stem, 9/207
brain stem auditory evoked potential, 47/133, 137-140, 142
brain stem disorder, 47/56, 170
brain stem symptom, 9/166
brain tissue, 47/80-83, 86, 103-106, 111
brain tissue study, 47/86
British Commonwealth, 9/65
C3 receptor, 47/250
C-3 complement, 47/115, 371
C-4 complement, 47/115
cancer, 47/202, 303
canine distemper, 9/668, 47/325
canine distemper encephalomyelitis, 47/430
canine distemper virus, 47/110, 364
capillary fragility, 9/132
carbamazepine, 47/57
carbohydrate metabolism, 9/133
Carswell description, 9/48
case control comparison, 9/73
cassava intake, 9/103
Caucasian patient, 47/399
cause, 9/306
cause of death, 9/200
cell mediated hypersensitivity, 34/440
cell mediated immunity, 47/106
central pontine myelinolysis, 47/72
central scotoma, 9/174
central sleep apnea, 63/463
cerebellar ataxia, 47/55
cerebellar symptom, 9/165
cerebellar tremor, 49/588
cerebral sign, 9/176
Charcot description, 9/49, 47/289, 319
chemical pathology, 9/310-317
chickenpox, 34/437, 47/329
child, 47/62
child infection, 47/430
child onset, 9/202
chloroquine, 9/402
cholesterol, 9/321
chromosome, 9/100, 47/303
chronic, *see* Chronic multiple sclerosis
chronic meningitis, 56/645
chronic relapsing experimental allergic encephalomyelitis, 47/430, 442, 470
chronic silent lesion, 47/449
ciclosporin, 47/359
cigarette smoking, 9/120
classification, 9/327, 47/49-52
climate, 9/111-113
clinical features, 47/49-73
clinical onset, 9/72, 74
clinical silent form, 9/102, 199, 204, 47/131, 136, 190, 487
clofibrate, 9/400
clot retraction, 9/131
CNS distribution, 9/234-242
coated pit, 47/221

cocarboxylase, 9/401
co-dergocrine, 9/390
common exposure hypothesis, 9/100
complement, 47/366
complement component, 47/303
complement system, 47/309
computer assisted tomography, 47/66, 191
concanavalin A, 47/346-349, 374
concentric sclerosis, 9/440, 47/73, 409, 411, 414, 416
concordance, 47/294
congenital retinal blindness, 9/103
conjugal form, 9/98
consanguinity, 47/298
consciousness, 45/119
contraception, 47/153
control patient, 9/410
copolymer I, 47/192
copper, 9/112, 322
corn, 9/103
coronavirus, 47/330, 56/13
corticosteroid, 9/394-396, 411, 47/86, 101, 104
corticotropin, 9/393, 47/101, 104, 107, 116, 192, 356, 373
cortisone, 9/393
cosmic ray, 9/78
course, 9/186-200
course of later bout, 9/187
course of onset bout, 9/187
cranial nerve, 9/167, 47/56
cranial neuralgia, 42/352, 47/56
Creutzfeldt-Jakob disease, 9/79, 120, 126
crossover experiment, 47/361
Cruveilhier description, 9/46
CSF, 9/289, 324-376, 16/369, 42/497, 47/79-82
CSF *N*-acetylneuraminic acid, 9/369
CSF amino acid, 9/368, 16/369
CSF antibody, 9/364
CSF antibody specificity, 47/107-112
CSF autoantigen antibody, 47/108
CSF B-lymphocyte, 47/373-375, 377
CSF brain antibody, 47/378
CSF bromide, 9/369
CSF butyrylcholinesterase, 9/367
CSF C-1 complement, 9/370
CSF C-3 complement, 47/115
CSF C-4 complement, 47/115
CSF cell antigen stimulation, 47/375
CSF cell count, 47/231, 372
CSF cholesterol, 9/368, 47/116
CSF cholesterol esterase, 47/119
CSF corticosteroid, 9/366
CSF 2',3'-cyclic nucleotide 3'-phosphodiesterase, 47/118
CSF cytology, 9/337-340
CSF esterase, 9/368
CSF examination, 9/325, 372-376, 47/79-121, 362, 372-379
CSF glial antibody, 47/109, 119
CSF glial protein, 47/119
CSF β-2-globulin, 9/340
CSF β-glucuronidase, 9/368
CSF glutamic oxalacetic transaminase, 9/367
CSF glycolytic enzyme, 9/368
CSF herpes simplex virus 1 antigen, 47/114
CSF HLA antigen, 47/375
CSF humoral immunity, 47/376
CSF idiotype, 47/379
CSF idiotypic antibody, 47/108, 362, 379
CSF Ig, 9/356, 47/377
CSF Ig subclass, 47/113
CSF IgA, 9/336, 363, 47/103, 113
CSF IgD, 9/363, 47/113
CSF IgE, 47/113
CSF IgG, 9/140, 326, 328, 333, 335, 348-352, 360, 42/497, 47/65, 71, 80
CSF IgG1, 47/104, 377
CSF IgG2, 47/104
CSF IgG3, 47/104
CSF IgG kappa chain, 47/105
CSF IgG lambda chain, 47/105
CSF IgG synthesis, 47/79-82, 87-107
CSF IgM, 9/336, 363, 47/103, 112
CSF immune complex, 47/114, 371
CSF immunoperoxidase staining, 47/97
CSF intrablood-brain barrier IgG synthesis, 47/79-82, 87-107
CSF isocitrate dehydrogenase, 9/367
CSF leukocyte, 47/82-86
CSF leukocyte count, 9/326, 329, 47/83, 100
CSF leukocyte *in vitro* Ig synthesis, 47/84
CSF lipid, 9/368, 47/115
CSF lymphocyte subpopulation, 47/85
CSF measles antibody, 47/377
CSF mixed lymphocyte reaction, 47/375
CSF monoclonal antibody, 47/85
CSF mononuclear cell, 47/83, 231
CSF myelin basic protein, 47/116
CSF myelin basic protein antibody, 47/108, 117
CSF neutral protease, 47/371
CSF nonspecific antibody, 47/378
CSF nonspecific B-lymphocyte activation, 47/105
CSF oligoclonal IgG, 47/65, 94-96, 101, 362, 377
CSF peptidase, 9/368
CSF phosphatidylcholine, 9/368
CSF phosphatidylethanolamine, 9/368

CSF phosphohexose isomerase, 9/367
CSF phospholipid, 9/321, 368, 47/115
CSF plasma cell, 47/83, 112
CSF profile, 9/326, 329, 47/119-121
CSF prostaglandin E, 47/376
CSF protein S100, 47/119
CSF proteinase, 9/368, 47/119
CSF proteolipid, 47/118
CSF sphingomyelin, 9/368
CSF T-helper/T-suppressor cell, 47/85
CSF T-lymphocyte, 47/372
CSF thromboplastic activity, 47/119
CSF total protein, 9/340
CSF trace metal, 9/369
CSF virus antibody, 47/110, 377
CSF vitamin B_{12}, 9/368
CSF vitamin B_c, 9/367
cyclophosphamide, 47/86, 194-196
cytomegalovirus antibody, 47/274
cytomegalovirus infection, 47/330
cytotoxic agent, 9/146, 47/194-197
cytotoxic T-lymphocyte, 47/189
dantrolene, 47/157
Dawson finger, 9/237, 248, 251, 253, 258-260
dazzling test, 9/174
death certificate, 9/64
definite diagnosis, 47/49-52
definition, 9/45, 47/213
delayed hypersensitivity, 9/147, 47/338
dementia, 46/403, 47/58
demyelination, 47/214-223, 226
depression, 46/427, 47/59, 178
diagnosis, 47/65-67
diagnostic criteria, 9/161, 205, 47/50-52, 65-69
diazepam, 47/157
diazepine, 47/157
dicoumarol, 9/384
diet, 9/103, 113, 399, 47/153-155
differential diagnosis, 9/205, 47/71-73, 120
diffuse glioblastosis, 18/77
diffuse sclerosis, 9/138
digestive enzyme, 9/400
dinoprostone, 47/351, 355
diplopia, 42/496, 47/55, 172
Disability Status Scale, 9/162, 409, 421, 47/147
disease duration, 9/72, 191-195
disease severity, 9/189
dissociated nystagmus, 2/367, 16/322
dorsal column stimulation, 47/159, 166
downbeating nystagmus, 60/656
drug treatment, 9/383-407, 47/156-158, 161-168, 178, 187, 191-206
dysarthria, 9/167, 179
dysplastic glial development, 9/57
dystonia, 47/57
early American report, 9/54
EEG, 47/65
electron microscopy, 34/445, 47/213-251
electrophoresis, 47/87, 95, 99, 103
l'emblée hémiplégique (Charcot), 47/56
emotional disorder, 47/178
emotional stress, 9/120, 47/153
empiric risk, 47/312
encephalitogenic myelin basic protein, 9/133
endocrine function, 9/180
ENG, 47/132
environment, 9/86, 111
Environmental Status Scale, 47/149
EOG, 47/132
epidemic, 47/279, 321-323
epidemiology, 9/63-84, 93-96, 47/259-283
epilepsy, 9/178
Epstein-Barr virus, 47/329, 56/256
Epstein-Barr virus antibody, 47/274
Epstein-Barr virus infection, 56/252
erroneous diagnosis, 9/64
erythrocyte count, 47/340
essential fatty acid, 47/199
etiology, 9/7, 57, 77-80, 107-160, 47/290, 311
euphoria, 9/177, 46/429, 47/58, 178
Europe, 9/68, 47/261-263, 320
evoked potential, 47/66, 131-143
exacerbation, 47/149
experimental allergic encephalomyelitis, 9/60, 126, 135, 47/320, 337, 429-432, 447-450, 485, 488
facial myokymia, 9/169, 41/300
facial pain, 5/28
facial paralysis, 9/167, 47/56
familial aggregation, 47/321
familial history, 9/97
familial incidence, 9/94, 47/297, 306
familial nature, 9/75, 85-87, 47/289-295, 306
familial spastic paraplegia, 59/306
Faroe Island epidemics, 47/279-283
fat diet, 9/78
fat intake, 9/114
fatigue, 47/59
fatty acid, 9/321, 399
Fc receptor, 47/250
fecal incontinence, 9/171
Ferguson-Critchley syndrome, 22/442
fibrosis, 9/292
first cousin marriage, 47/298
flexor spasm, 9/412
focalization, 9/71, 75

follow-up, 9/67, 72
food allergy, 47/154
foreign troops, 47/281-283
Frerichs description, 9/48
fucose, 9/321
galactocerebroside, 47/109, 114
galactose, 9/402
gastrointestinal disease, 39/456-459
gaze paralysis, 9/167
gene penetrance, 47/299
genealogy, 47/299, 307-309
general anesthesia, 47/151
genetic epidemiology, 9/93-95
genetic linkage, 42/497, 47/307-309
genetics, 9/85, 97, 47/289-313
geographic clustering, 47/263
geographic correlate, 47/273
geographic distribution, 47/320
geographic latitude, 9/64, 76, 112
geographic pattern, 9/75, 47/260-267
geomagnetism, 9/77, 112
glia degeneration, 9/108
glucocorticoid, 47/192, 353
gluten, 47/155
glycoprotein, 9/321, 47/109
GM1 ganglioside antibody, 47/109
goiter, 9/112
Gowers report, 9/55
granular vacuolar degeneration, 47/215, 222
granulomatous CNS vasculitis, 55/390
Guillain-Barré syndrome, 34/437
Hall description, 9/48
Hammond description, 9/54
haptene, 9/145
haptoglobin, 9/99, 320
headache, 9/182
hearing loss, 9/168
hemiballismus, 49/374
hemiparesis, 47/56
hemiplegia, 47/56
heparin, 9/400
hereditary, *see* Hereditary multiple sclerosis
hereditary paroxysmal kinesigenic choreoathetosis, 49/355
hereditary spastic ataxia, 21/372
heredodegenerative trait, 9/57
herpes simplex, 9/126, 34/437, 47/329-331
herpes simplex virus encephalitis, 9/126
hexosamine, 9/321
hexose, 9/321
high risk area, 9/72
high risk family, 9/87-90, 47/290-295
high to low risk migrant, 47/277
histamine, 9/391
histopathology, 9/124
history, 9/45-62
HLA antigen, 42/497, 47/97, 189, 299-313, 371
HLA-B7 antigen, 47/340
hospital admission ratio, 9/64
human T-lymphotropic virus type I, 56/18
humoral immune response, 34/435-440, 47/363-372
hygiene, 9/123
hyperbaric oxygenation, 47/160, 188
hypercellularity, 47/214-218
hypergammaglobulinemia, 9/326, 354
hyperreflexia, 42/496
hypersomnia, 3/96
hyperthermia, 9/120, 412
hypocalcemia, 47/39
hypothalamic hypophyseal adrenal system, 47/353
hysteria, 9/177, 46/575, 47/72
Iceland epidemics, 47/279
idiotype anti-idiotype theory, 47/368
idiotypic antibody, 47/111
idiotypic IgG, 47/362
Ig, 47/235-239, 309, 360
Ig binding lymphoid cell, 47/81
Ig containing cell, 47/231
Ig Fc fragment, 47/365
Ig injection, 9/411
Ig synthesis, 47/329
IgA, 47/339
IgG, 47/339
IgG allotype, 47/87, 104-106
IgG index, 47/87
IgG synthesis, 47/79-82
immigrant prevalence, 9/70, 47/274
immune complex, 47/114, 239, 370-372
immune disease, 9/80
immune response, 47/83
immune response gene, 47/303
immune system, 47/337
immunization, 47/150
immunocytochemical study, 47/235
immunoelectrophoresis, 47/87, 94, 111
immunofluorescence test, 47/80, 236-239
immunologic aspect, 34/56
immunologic onset, 9/72, 74
immunology, 9/128, 47/189, 337-380
immunopathology, 47/232-235
immunoperoxidase, 47/235
immunoregulation, 47/368
immunosuppression, 9/402, 47/202, 231
incidence rate, 9/67, 47/259-267, 270
incoordination, 47/169

India, 47/398
indometacin, 47/355
infection, 9/58, 117, 121, 47/149
infection source, 9/122
infectious agent, 9/107
infectious mononucleosis, 56/252
inflammation, 9/108, 47/80
inflammatory cell, 47/230
influenza A, 9/129
influenza C, 47/329
inheritance, 47/300
initial symptom, 47/61, 401
injury, 47/152
inoculation, 9/120
interferon, 47/204-206, 347, 350, 353-355
interleukin-1, 47/189
interleukin-2, 47/189, 347, 375
internuclear ophthalmoplegia, 9/167, 47/55, 132, 140
interrelated symptom, 9/183
intestinal function, 47/167
intestinal pathology, 39/456-459
intestinal pseudo-obstruction, 51/493
intrablood-brain barrier IgG synthesis, 47/84, 97-101, 106
intractable hiccup, 63/490
intrathecal IgG synthesis, 47/81, 94, 104
intrathecal phenol, 9/413
intrathecal tuberculin, 9/401
isoelectric focussing, 47/86, 94, 103
isoniazid, 9/396, 47/188
isotachophoresis, 47/91
Jamaican neuropathy, 9/205
Japan, 9/71, 47/398
JHM virus, 9/125, 47/325, 56/447
JHM virus encephalomyelitis, 47/431
Kabat method, 9/420
kallikrein, 9/397
Kawasaki syndrome, 51/453
killer cell, 47/352
kuru, 9/79, 120
labeled IgG, 47/91
laboratory diagnosis, 47/50-52, 65-69
latency period, 9/72
latitude, 9/65
lead, 9/112
Leber optic atrophy, 9/103, 176, 13/102
lentivirus, 47/323
leukocyte count, 47/340
levamisole, 47/203
Lhermitte sign, 9/171, 47/55
life expectancy, 47/62
linoleic acid, 9/132, 315, 321, 47/68, 154
linolenic acid, 9/315
lipid, 9/321
lipid metabolism, 9/134
lipolytic agent, 9/316
lipoprotein, 9/321
local infection, 47/150
low fat diet, 9/386
low to high risk migrants, 47/278
low risk area, 9/72
lumbar puncture, 9/324
lymphocytapheresis, 47/201
lymphocyte, 47/221, 228
lymphocyte antibody, 47/197-199
lymphocyte transformation test, 47/358
lymphocytotoxic antibody, 47/368-370
lymphoid radiation, 47/201
lymphokine, 47/350
lymphomatoid granulomatosis, 51/451
lymphoreticular system, 9/144
lysosome, 47/219, 222
macrophage, 47/215-225, 228, 232, 234, 337, 355
major histocompatibility complex, 47/303
management, 47/147-180
Margulis vaccine, 9/398
Marie description, 47/319
Markschattenherde, 47/246
Markscheidenlichtungsherde, 47/246
marriage, 9/418
measles, 9/129, 47/329-331, 357
measles antibody, 9/79, 47/363
measles exposure, 47/364
measles virus, 34/435-437
medial longitudinal fasciculus, 42/495
Mendel law, 47/299, 312
mental sign, 9/176
metal intoxication, 9/58, 78
metatuberculosis, 9/59
migration, 9/70, 47/274-279, 321
migration age, 47/275-277
migration inhibition test, 47/357
millet, 9/103
misdiagnosis, 13/102
mitogen, 47/374
mitogen induced cell division, 47/353
mitogen stimulation, 47/374
mixed lymphocyte reaction, 47/356
mode of onset, 47/52-54
monoclonal antibody, 47/85, 232, 234, 342-345
monocyte, 47/341, 346, 359
mononuclear cell, 47/340
monosymptomatic onset, 9/186
moose encephalitis, 9/125
morphologic study, 47/330

Morris report, 9/54
mortality, 47/260
mortality rate, 9/65, 47/268
mortality statistics, 9/64, 47/260
Moxon report, 9/56
mucoprotein, 9/320
multiplicity in time, 9/205-207
muscle intermittent claudication, 47/56
muscular atrophy, 9/179
myasthenia gravis, 47/338
myelin associated glycoprotein, 47/235
myelin basic protein, 47/65, 112, 191, 235, 239, 359, 372, 376
myelin sheath, 9/264
myelinated nerve fiber, 47/218-221
myelinoclasis, 9/471
myelinolytic theory, 9/134
myelinolytic toxin, 9/57
myelinotoxic factor, 9/142, 47/366
myokymia, 42/237
myxovirus, 9/123, 129
narcolepsy, 2/448, 9/103, 178
nationality, 9/70
natural history, 47/60
natural killer cell, 47/341, 343, 346, 349-352
natural killer cell receptor, 47/349
nerve conduction block, 47/367
neurobrucellosis, 52/588
neurogenic pulmonary edema, 63/496
neuroleptic agent, 47/178
neuromyelitis optica, 9/230, 235, 426, 428, 431, 47/70, 213, 397-399, 405, 487
neuron-neuroglia interdependence, 9/624
neuropathic trait, 9/57
neuropathology, 9/217-308, 47/213-251
neuropathophysiology, 47/140-143
neurosyphilis, 47/71, 52/278
neurotropic slow virus, 9/118-128
New Zealand, 47/321
nonspecific B-lymphocyte activation, 47/112
Norway, 9/69
Nova Scotia, 9/69
nuclear magnetic resonance, 47/67, 191
nursing care, 47/173
nystagmus, 9/167, 42/496
occupational disease, 9/108
oligodendrocyte, 47/213, 222-230, 241-244
oligodendrocyte antibody, 47/109
oligodendrocyte hyperplasia, 47/224, 244
oligodendroglial lysis, 47/213
olivopontocerebellar atrophy (Dejerine-Thomas), 60/516, 533
onset, 9/72, 74
onset in adult, 9/204
onset bout, 9/183-186
ophthalmoplegia, 9/167
opsoclonus, 47/133
optic atrophy, 13/56-58, 81, 42/495
optic chiasm, 47/55
optic chiasm lesion, 47/55
optic nerve, 47/134-137
optic neuritis, 9/195, 47/54, 69, 171
optic neuropathy, 9/172, 176
optic sign, 9/172
optokinetic nystagmus, 47/133
oral contraceptive agent, 47/153
organic personality syndrome, 46/434
Orientals, 47/399
Orkney, 9/69
orthostatic hypotension, 63/156
osteoporosis, 70/20
pain, 47/170
palatal myoclonus, 9/181, 38/584
pancorphen, 9/388
papillitis, 9/175
parainfluenza virus, 34/438
parainfluenza virus type 1, 9/129, 47/330, 56/431
parainfluenza virus type 2, 9/129
parainfluenza virus type 3, 47/329
paraparesis, 47/56
parental consanguinity, 9/96
paresthesia, 42/496, 47/55
paroxysmal ataxia, 9/179
paroxysmal attack, 47/170
paroxysmal dysarthria, 9/179
paroxysmal symptom, 47/170
pathogenesis, 47/368
pathologic onset, 9/72, 74
pathology, 9/107-160, 47/235-239, 246, 248, 447-450
pathoneurophysiology, 47/131-143
patient management, 9/408-425, 47/179
patient-relative prevalence, 9/93
pattern shift visual evoked response, 47/134-138, 142
pattern visual evoked response, 47/133
Pelizaeus-Merzbacher disease, 47/73
periodic ataxia, 21/570, 60/648
periodic dysarthria, 21/570
peripheral blood cell, 47/346-363
periphlebitis retinae, 9/115, 175
personality, 47/59
phagocytosis, 47/218, 249
phenol block, 47/159, 166
phenylthiocarbamide, 9/99
phosphatidylserine, 3/313

phospholipid, 9/312
physiotherapy, 47/174
phytohemagglutinin, 47/346, 354, 374
piromen A, 9/388
piromen B, 9/388
plaque, *see* Multiple sclerosis plaque
plasma cell, 47/221, 228, 231
plasma exchange, 47/200
plasma fibrinogen, 9/131
plasma infusion, 9/41
plasma transfusion, 9/385
plasmapheresis, 47/142, 367
platelet adhesiveness, 9/131-133
platelet count, 47/340
pleocytosis, 9/330, 333
plurisymptomatic onset, 9/186
pokeweed mitogen, 47/344, 346, 354, 360-362, 374
pokeweed mitogen response, 47/360
poliomyelitis, 9/123
poliovirus, 47/339
polyacrylamide gel electrophoresis, 47/92, 111
polyclonal antibody, 47/232, 234
polyunsaturated fatty acid, 47/154, 187
positional nystagmus, 47/133
possible diagnosis, 47/49-52
postinfectious encephalomyelitis, 9/130, 135, 138, 47/326
postvaccinial encephalomyelitis, 9/135, 138
precipitating factor, 9/116-120
predicting factor, 47/60
predisposing factor, 9/110-116
pregnancy, 9/117, 418, 47/64, 152
pressure sore, 9/415
prevalence, 9/67, 42/496, 47/261-267
primary lateral sclerosis, 9/207
probable diagnosis, 47/49-52
process entity, 9/304
prognosis, 9/199, 47/60
progression, 9/195
progression coefficient, 47/147
progressive course, 9/189
progressive multifocal leukoencephalopathy, 9/120, 126, 47/72, 327
proper-myl, 9/401
proprioceptive impairment, 47/55
prostaglandin, 47/199
protein, 9/320
proteinase, 9/368, 47/214
pseudoneurasthenic prodromal stage, 9/177
psychiatric disorder, 9/204
psychology, 47/178
puerperium, 9/117
pupillary reflex, 47/135
pyramidal sign, 9/163
pyramidal symptom, 47/56
pyruvic acid, 9/321
rabies, 9/138, 47/330
rabies postvaccinial encephalomyelitis, 9/138
rabies vaccine, 9/59
race, 9/75, 99, 47/269
radiation, 47/303
radioisotope exchange, 47/90
rare presentation, 47/59
reagin, 9/501
recessive hypothesis, 9/98
recurrence risk, 47/312
rehabilitation, 9/416, 47/173-178
rehabilitation data, 9/417
Reiber formula, 47/94
Reiber graph, 47/88
Reiter disease, 47/329
relapse rate, 47/149
relative, 47/295-297
remyelination, 47/222, 227, 239-248, 250
respiratory failure, 9/181
respiratory pattern, 63/463
respiratory tract disease, 9/80
retrobulbar neuritis, 9/195-198, 47/52, 54
retrovirus, 47/323, 56/525
rheumatic fever, 9/80
rice, 9/103
Rickettsia, 9/682
risk, 47/271-273
risk factor, 47/273
rubella, 34/438, 47/329
rubella encephalitis, 34/438
rural milieu, 9/99
Salmonella, 47/339
sanitation, 9/123
Scandinavian countries, 47/321
Schuller formula, 47/94
Schwann cell, 47/244
scoring system, 47/148
scrapie, 9/78, 120, 126, 128, 47/330
secretor factor, 9/99
segregation analysis, 47/307
Seguin report, 9/55
seizure, 47/56-58, 170
sensory symptom, 9/169
serum copper, 9/322
serum IgG, 9/326
serum oligodendroglial binding antibody, 9/501
sex distribution, 9/72
sex ratio, 42/496, 47/268, 302
sexual disturbance, 47/168

sexual dysfunction, 75/97
sexual impotence, 9/171
shadow plaque, 9/296
sheep erythrocyte rosetting technique, 47/341, 344, 356, 359
Shetlands, 9/69
sibling, 47/290-292
silver nitrate staining, 47/94
simian virus 5, 47/330
single bout, 9/187
Sjögren syndrome, 51/449, 71/74, 75
sleep, 74/549
smell, 9/180
smooth pursuit eye movement, 47/133
social support, 9/423, 47/177
society, 9/422
socioeconomic class, 9/100, 114
somatosensory evoked potential, 47/140-142
South Africa, 47/321
South Canada, 9/69, 47/321
spastic paraplegia, 42/496, 59/431
spasticity, 9/412, 47/156-158
speech disorder, 42/496
speech therapy, 47/175
spheroid body, 6/626
Spherula insularis, 9/59, 109, 47/319
sphincter disturbance, 47/58
sphingomyelin, 9/311, 368
spinal anesthesia, 47/151
spinal cord, 47/245
spinal cord compression, 19/376
spinal cord hyperthermia, 9/413
spinal glioma, 20/338
spinal injury, 9/206
spinocerebellar degeneration, 9/205
Spirochaeta argentinensis, 9/59, 108
Spirochaeta myelophthora, 9/59, 127, 47/319
spontaneous improvement, 47/175
sporadic, 47/306
stress, 47/179
striatopallidodentate calcification, 49/423
subacute combined spinal cord degeneration, 47/72
subacute myelo-optic neuropathy like virus, 47/330
subacute sclerosing panencephalitis, 9/120, 126, 47/86, 104, 326
subclinical lesion, 47/106
succinic acid, 9/385
sudanophilic leukodystrophy, 10/34
surgery, 9/413, 47/64, 150, 159
survival rate, 47/271
swayback, 9/78, 112, 678
sweating, 9/180
symptomatology, 9/161-208
synovial fluid antibody, 47/112
syphilis, 9/58
syringomyelia, 32/284, 50/452
systemic lupus erythematosus, 47/71, 109, 329, 338, 55/373
T-helper cell, 47/189
T-helper/T-suppressor cell, 47/85
T-lymphocyte, 47/68, 81, 232, 234, 341-345
T-lymphocyte independent response, 47/360
T-suppressor lymphocyte, 47/189, 342, 347
target cell, 47/249
temperature, 47/151
temporal disc pallor, 9/173
tetanus antitoxin, 9/60
tetany, 47/57
tetraethylammonium chloride, 9/387
tetrahydrocannabinol, 47/169
thalamotomy, 47/169
Theiler virus, 47/324
Theiler virus encephalomyelitis, 47/431
thermoregulation, 9/182, 47/136, 142, 150
thrombogenic theory, 9/110, 131
thymectomy, 47/201
thymocyte antibody, 47/197-199
tick-borne encephalitis flavivirus, 47/330
time trend, 9/65
Togavirus, 47/325
tolbutamide, 9/398
tonic seizure, 9/178, 46/170, 47/57, 71
tonsillectomy, 9/123, 47/151
Tourtellotte formula, 47/88-91
Toxoplasma gondii, 47/320
trace metal, 9/113
trace metal deficiency, 9/58, 78
transfer factor, 47/203
transitional sclerosis, 9/473
transmissibility, 9/128
transverse myelitis, 47/72
tranylcypromine, 9/397
trauma, 9/115-117, 47/64, 152
treatment, 47/39-41, 68, 86, 101, 104, 106, 116, 142, 159, 169, 188, 200, 355, 359, 373
treatment criteria, 9/408
treatment evaluation, 9/408
tremor, 9/181, 42/496, 49/604
trigeminal neuralgia, 5/309, 311, 9/168, 42/352, 47/140
truncal ataxia, 47/56
tuberculin, 9/401, 47/354
twin concordance, 9/76, 90
twin study, 9/90-92, 47/292-295

twin zygosity, 47/292-294
twins, 9/101
type, 9/163
United Kingdom, 47/399
urban focalization, 9/75
urban milieu, 9/99
urbanization, 9/78
urethral obstruction, 9/414
urinary incontinence, 9/171, 414, 47/58
urinary symptom, 9/414, 47/58
USA, 47/263-265, 276, 321
use of term, 9/54, 56
uveitis, 9/115, 176, 47/338
vaccinal encephalomyelitis, 9/60
vaccination, 47/339
vaccinia, 34/438
Valentiner description, 9/49
vascular fibrosis, 47/246
vascular origin, 9/59
vascular relationship, 9/257
vascular theory, 9/130-133
venous sheating, 9/175
vertigo, 2/359, 9/168, 47/56
vestibulo-ocular reflex, 47/133
vibratory sense, 9/170
viral antibody, 47/363
viral antigen, 47/356
viral etiology, 47/319
viral infection, 34/56
viral origin, 9/59
virology, 34/435-446, 47/319-332, 357
virus, 9/120, 47/215
virus induced demyelination, 47/327
virus isolation, 34/441-445, 47/330
virus titer, 47/365
visna, 9/124, 47/430
visna like virus, 47/323
visual evoked response, 47/133-135, 142
visual symptom, 9/172
vitamin, 47/155
vitamin B12, 9/322, 368, 400, 47/155
Vogt-Koyanagi-Harada syndrome, 56/623
Von Romberg description, 9/48
wallerian degeneration, 47/222
Wetterwinkel (Steiner), 9/221
wheat, 9/103
Wilson disease, 47/72
world wide, 47/265-267
zygotic twins, 9/91
Multiple sclerosis plaque
acid phosphatase, 9/37
acute, 47/230
alkaline phosphatase, 9/37
axon, 9/38
carbonyl group, 9/37
cholesterol ester, 9/37
chronic, 47/81
demyelination, 9/37
distribution, 9/220
DPN diaphorase, 9/37
glia cell, 9/37
glial cell, 9/37
glial enzyme, 9/37
localization, 9/220
margin, 9/37
neuropathology, 47/213-217
oligodendrocyte, 9/38
PAS procedure, 9/37
plaque forming cell, 47/362
plasma cell, 47/81-83
succinate dehydrogenase, 9/37
Sudan dye, 9/37
sudanophilic lipid, 9/37
Multiple sleep latency test
daytime hypersomnia, 45/131, 135
narcolepsy, 74/562
Multiple spinal ependymoma
CNS, 20/358
visceral malignancy, 20/358
Multiple spinal tumor
neurofibroma, 20/244, 285
Multiple sulfatase deficiency
activator deficiency, 66/172
Alder-Reilly granule, 42/493
corneal dystrophy, 60/652
enzyme deficiency polyneuropathy, 51/369
epilepsy, 72/222
glycosaminoglycanosis type II, 66/314
hepatosplenomegaly, 42/493
lysosomal disorder, 66/64
metachromatic leukodystrophy, 29/359, 42/493, 47/589, 60/127, 66/65, 169
nerve conduction, 51/372
skeletal deformity, 42/493
Multiple symmetric lipomatosis
multiple mitochondrial DNA deletion, 66/429
Multiple system atrophy, *see* Multiple neuronal system degeneration
Multiple system degeneration, *see* Multiple neuronal system degeneration
Multiple tics
childhood chronic, *see* Childhood chronic multiple tic
Gilles de la Tourette syndrome, 1/291
Multiple trauma
critical illness polyneuropathy, 51/575

Multiple trichoepithelioma
clinical features, 14/110
dermal cylindroma, 14/591
differential diagnosis, 14/110
differentiation, 14/588
epilepsy, 14/791
histogenesis, 14/588
history, 14/585
mental deficiency, 14/791
multiple nevoid basal cell carcinoma syndrome, 14/114
neurocutaneous syndrome, 14/101
nosology, 14/586
pathognomonic lesion, 14/586
Potter syndrome type II, 14/110
prognosis, 14/589
sebaceous adenoma, 14/579
symptom, 14/586
treatment, 14/589
tuberous sclerosis, 14/110
Multiple white matter lesion
Binswanger disease, 54/224
classification, 54/96
nuclear magnetic resonance, 54/96
sickle cell anemia, 55/507
Multisynostotic osteodysgenesis
autosomal recessive inheritance, 50/123
Multisystem atrophy, *see* Multiple neuronal system degeneration
Multivitamin deficiency
tropical ataxic neuropathy, 51/323
Mummification process
brain, 74/184
Egypt, 74/184
Mummy
arthritis, 74/187
brain tumor, 74/187
Chagas disease, 74/188
diabetes mellitus, 74/194
enteric nervous system, 74/191
epidural hematoma, 74/187
gout, 74/188
immunoreactivity, 74/189
meningioma, 74/187
meningitis, 74/187
nerve fiber, 74/193
nervous system, 74/181
neurogenic osteoarthropathy, 74/188
neuropathy, 74/189
Parkinson disease, 74/194
rheumatoid arthritis, 74/187
serotonin, 74/192
skin, 74/191
spinal nerve root, 74/190
Mumps
acute cerebellar ataxia, 34/628
animal experiment, 56/115
clinical features, 56/113, 115
cochleitis, 56/113
CSF, 34/628
CSF pleocytosis, 34/411, 56/129
deafness, 55/131, 56/113, 115, 129, 430
demyelinating antibody, 9/545
diagnosis, 56/114, 430
encephalopathy, 31/220
epidemiology, 56/429
facial paralysis, 7/486, 56/430
foreign antigen, 9/545
headache, 56/430
immune mediated encephalomyelitis, 34/411
management, 56/114
mastitis, 56/429
meningitis serosa, 56/129
meningoencephalitis, 34/411, 628, 56/114
nerve lesion, 7/485
neurotropic virus, 9/545, 56/430
orchitis, 9/545, 56/129, 429
parotitis, 7/485, 34/411, 628, 56/113, 129, 429
polyradiculitis, 7/485, 34/411
postinfectious encephalomyelitis, 34/412, 47/326, 56/129, 430
sex ratio, 56/113, 429
spinal muscular atrophy, 59/370
temporal bone pathology, 56/114
treatment, 56/430
Mumps encephalitis
clinical features, 56/430
coma, 34/628
CSF, 56/430
EEG, 56/430
incidence, 56/429
mortality rate, 34/628
prognosis, 34/628
seizure, 34/628, 56/430
Mumps meningitis
age, 9/545
clinical features, 56/430
CSF, 34/411, 56/129, 430
EEG, 56/430
exanthematous encephalomyelitis, 9/546
fatal, 9/545
incidence, 56/429
sex ratio, 9/545
viral meningitis, 56/126, 430
Mumps skin test
delayed hypersensitivity, 47/339

Mumps virus
acute viral encephalitis, 56/126
aqueduct stenosis, 9/551, 50/308
deafness, 56/107
hydrocephalus, 9/551
neurotropism, 56/35
paramyxovirus, 56/14
paramyxovirus infection, 56/430
protein, 56/431
Reye syndrome, 56/150
structure, 56/431
vertigo, 56/107
viral meningitis, 56/126, 128
Mumps virus infection
neurology, 56/430
pathogenesis, 56/431
Munich, Germany
neurology, 1/9
Munro, D., 26/410
MURCS syndrome
see also Caudal aplasia
case history, 50/509
cervicothoracic somite dysplasia, 50/515
müllerian duct, 50/515
renal agenesis syndrome, 50/515
Murex intoxication
ingestion, 37/64
Murine encephalitis
junin virus, 56/358
lymphocytic choriomeningitis virus, 56/358
Murine encephalopathy
Tacaribe virus, 56/358
Murine hepatitis virus, *see* JHM virus
Murine leukemia virus
age acquired resistance, 56/457
amphotropic type, 56/587
animal viral disease, 56/587
ecotropic type, 56/587
immune response, 56/457
lymphoma, 56/587
paraplegia, 56/587
retrovirus, 56/455
slow virus disease, 56/455
spongiform encephalomyelopathy, 56/455
subacute combined spinal cord degeneration, 56/455
Murine neurotropic retrovirus
acquired immune deficiency syndrome, 56/586
animal viral disease, 56/586, 59/447
human T-lymphotropic virus type I associated myelopathy, 59/447
motoneuron disease, 56/586, 59/447
slow virus disease, 56/586
spongiform encephalitis, 56/586
Murine sarcoma virus
acute viral myositis, 56/194
Murine typhus
rickettsial infection, 34/646
Murray Valley encephalitis
acute viral encephalitis, 34/76, 56/134, 139
age, 34/76, 56/139
Culex annulirostris, 56/139
fatality rate, 34/76, 56/139
flavivirus, 34/76, 56/12
geography, 34/76, 56/139
Muscarine
Amanita muscaria, 64/14
Clitocybe, 64/14
Inocybe, 64/14
mushroom intoxication, 36/535
neurotoxin, 36/535, 64/14, 65/39
segmental demyelination, 64/14
synapse, 1/48
toxic neuropathy, 64/14
Muscarine intoxication
action site, 64/14
headache, 65/39
Muscarinic acetylcholine
autonomic nervous system, 75/553
immune response, 75/553
Muscarinic acetylcholine receptor
Huntington chorea, 49/295
Muscarinic cholinergic nerve cell
tardive dyskinesia, 49/189
Muscarinic cholinergic receptor
Parkinson disease, 49/112
Muscarinic receptor
aging, 74/226
cloning, 74/150
pharmacology, 74/150
Muscimol
see also Mushroom toxin
GABA receptor stimulating agent, 65/39
hallucinogenic agent, 65/42
mushroom toxin, 65/39
Muscimol intoxication
delirium, 65/40
epilepsy, 65/40
hallucination, 65/40
symptom, 65/40
tremor, 65/40
Muscle
adenosine triphosphate, 41/264
amyotrophic lateral sclerosis, 59/184
antagonist, 1/297
autonomic nervous system, 74/33

blood flow, 1/465
bulbocavernous, *see* Bulbocavernous muscle
carnitine, 66/403
cerebrotendinous xanthomatosis, 66/604
conductance, *see* Muscle fiber conductance
congenital absence, *see* Congenital muscle absence
congenital weakness, 1/482
connective tissue, 62/40
coordination, 1/297
depressor anguli oris, *see* Depressor anguli oris muscle
developing, 40/113, 146, 183, 199, 260, 439
dystrophin, 62/125
electrical property, 1/631
electrodiagnosis, 1/224
ergoreflex, 75/112
extraocular, *see* Extraocular muscle
eye, *see* Eye muscle
fatigability, 1/231
fiber vacuole, *see* Rimmed muscle fiber vacuole
GM2 gangliosidosis, 10/410
halothane, 40/546
inferior oblique, *see* Inferior oblique muscle
inflammatory cell, 62/42
intramuscular nerve, 62/44
masseter, *see* Masseter muscle
membrane potential, 1/632
neck, *see* Neck muscle
nerve ending, 1/100
neurocysticercosis, 52/529
pectoralis major, *see* Pectoralis major muscle
poliomyelitis, 56/318
postural, 1/264
quadratus labii inferioris, *see* Quadratus labii inferioris muscle
reflex, 2/16
respiratory, *see* Respiratory muscle
ribosome, 40/3, 67
segment pointer, *see* Segment pointer muscle
stapedius, *see* Stapedius muscle
sympathetic activity, 74/250, 75/109
sympathetic nerve, 74/649
trophic change, 7/89
trophic influence, 1/59
vascularization, 62/38
vasoconstriction, 75/110
vitamin E deficiency, 70/448
wasting, *see* Muscular atrophy
X-linked adrenoleukodystrophy, 66/460
Muscle absence
abdominal, *see* Abdominal muscle absence
congenital, *see* Congenital muscle absence
trophic disorder, 1/481
Muscle action potential
acrylamide intoxication, 64/69
botulism, 65/212
cable parameter, 40/550
chemical features, 40/128
compound, *see* Compound muscle action potential
evoked, *see* Evoked muscle action potential
paralytic shellfish poisoning, 65/152
saxitoxin, 65/152
Muscle activity
black widow spider venom intoxication, 40/330
spontaneous electrical, *see* Spontaneous electrical muscle activity
Muscle aging
fiber change, 40/199
Muscle aplasia
Möbius syndrome, 50/215
pectoralis, *see* Pectoralis muscle aplasia
Muscle atrophy, *see* Muscle fiber type I atrophy, Muscle fiber type II atrophy *and* Muscular atrophy
Muscle biopsy
abnormal features, 40/8-43
acid maltase deficiency, 62/481
acquired toxoplasmosis, 52/357
acromegalic myopathy, 62/538
ACTH induced myopathy, 62/537
artefact, 40/8
ataxia telangiectasia, 14/321, 60/376
axonal polyneuropathy, 62/609
Bassen-Kornzweig syndrome, 63/278
Becker muscular dystrophy, 62/123
carnitine palmitoyltransferase deficiency, 62/494
central core myopathy, 62/338
cerebrotendinous xanthomatosis, 66/605
Charlevoix-Saguenay spastic ataxia, 60/456
childhood myoglobinuria, 62/561
chondrodystrophic myotonia, 62/269
chorea-acanthocytosis, 49/331, 60/141
colchicine myopathy, 62/609
congenital end plate acetylcholinesterase deficiency, 62/430
congenital slow channel syndrome, 62/434
cytochemistry, 62/2
debrancher deficiency, 62/483
dermatomyositis, 62/375
diagnostic test, 1/31
disuse *vs* disease atrophy, 1/235
Duchenne muscular dystrophy, 62/123
dyssynergia cerebellaris myoclonica, 60/601
EMG, 62/124
eosinophilia myalgia syndrome, 63/377

eosinophilic polymyositis, 63/385
facioscapulohumeral spinal muscular atrophy, 59/42
familial infantile myasthenia, 62/428
giant axonal neuropathy, 60/79
hereditary amyloid polyneuropathy, 60/104
hereditary brachial plexus neuropathy, 60/72
hereditary periodic ataxia, 60/441
hereditary sensory and autonomic neuropathy type III, 60/31
Hirayama disease, 59/110
histochemical reaction, 40/2, 43-55
hyperaldosteronism myopathy, 62/536
hypereosinophilic syndrome, 63/374
hypothyroid myopathy, 62/533
immunocytochemistry, 62/2
inclusion body myositis, 62/373, 375
indication, 40/1
infantile spinal muscular atrophy, 59/58
interpretation, 40/54, 62/44
Kawasaki syndrome, 56/639
Kearns-Sayre-Daroff-Shy syndrome, 62/508
kwashiorkor, 51/322
late infantile acid maltase deficiency, 27/195, 229
limb girdle syndrome, 62/186
long chain 3-hydroxyacyl-coenzyme A dehydrogenase deficiency, 62/495
microscopy, 62/2
muscle pathology, 62/1
myophosphorylase deficiency, 41/187, 62/485
myotonic dystrophy, 62/225
myotonic syndrome, 62/274
needle biopsy, 62/2
Nelson syndrome, 62/537
oculopharyngeal muscular dystrophy, 60/51
opaque fiber, 62/15
opticocochleodentate degeneration, 60/754
oxidative enzyme, 40/6, 136
oxoglutarate dehydrogenase deficiency, 62/503
periodic paralysis, 62/465
phosphofructokinase deficiency, 41/191, 62/488
phosphoglycerate kinase deficiency, 62/489
phosphoglycerate mutase deficiency, 62/489
phosphorylase kinase deficiency, 62/487
polymyositis, 8/122, 19/87, 38/485, 41/60, 62/375, 65/202
postpoliomyelitic amyotrophy, 59/38
processing method, 62/1
progressive external ophthalmoplegia, 60/54
progressive muscular atrophy, 59/16
removal method, 62/1
Reye syndrome, 31/168, 56/158
scapulohumeral spinal muscular atrophy, 62/171
scapuloperoneal spinal muscular atrophy, 59/46
selection of muscle, 40/2
small caliber fiber, 62/14
spinocerebellar degeneration, 60/677
split fiber, 62/16
sporadic progressive external ophthalmoplegia with ragged red fiber, 62/508
storage disease, 62/17
technique, 40/2
tetrazolium reductase, 40/3
thyrotoxic myopathy, 62/529
toxic oil syndrome, 63/381
trichinosis, 52/568
ubidecarenone, 62/504, 66/423
ultrastructural change, 40/63-114
Wohlfart-Kugelberg-Welander disease, 59/89, 62/200

Muscle calcification
amyotrophy, 43/258
mental deficiency, 43/258

Muscle carnitine deficiency
cardiomyopathy, 66/403
exercise intolerance, 66/403
fatty acid oxidation, 66/404
myalgia, 66/403
myoglobinuria, 66/403
progressive limb weakness, 66/403
short chain acyl-coenzyme A dehydrogenase, 66/404

Muscle cell
see also Sarcolemma
acid phosphatase, 40/5
Golgi apparatus, 40/212
membrane, *see* Muscle cell membrane
normal muscle, 40/63-67
osmotic fragility, 40/386
pathologic change, 40/67-114
whorls of cytomembrane, 40/39

Muscle cell membrane
see also Delta lesion
denervation, 40/127
depolarization, 40/127
diphosphonate, 40/387
Duchenne muscular dystrophy, 40/74, 151, 252, 370, 384, 41/410
experimental autoimmune myasthenia gravis, 41/115
experimental myasthenia gravis, 41/119
hyperpolarization, 40/127
membrane conductance, 41/168
membrane current, 7/65
membrane potential, 7/63, 41/108, 115, 170
myasthenia gravis, 41/109, 125

myotonia, 41/298
myotonia congenita, 40/128, 553
myotonic dystrophy, 40/74, 213, 494, 497, 553
myotonic syndrome, 40/556
periodic paralysis, 28/582, 41/150, 167, 170
potassium channel, 40/128
resting, 40/126, 550, 41/108
sodium channel, 40/126, 558
structure and function, 40/126, 558
Muscle cell nucleus
histologic change, 40/33-36
inclusion body, 40/35, 90
myotubular myopathy, 40/89
ultrastructure, 40/88
Muscle cirrhosis, *see* Myosclerosis
Muscle contraction
adenosine triphosphate, 40/340
arthrogryposis multiplex congenita, 1/483
Bethlem-Van Wijngaarden syndrome, 62/155, 181
brancher deficiency, 62/484
Brissaud-Sicard syndrome, 1/269
denervation, 62/53
Emery-Dreifuss muscular dystrophy, 62/145
EMG, 62/52
Huntington chorea, 49/277
hypothenar dimpling, 41/303
interference pattern, 62/53
iodoacetamide, 41/297
laughing, 3/356
muscle fibrosis, 62/173
myophosphorylase deficiency, 62/485
myotonia, 41/298, 62/262, 273
myotonic syndrome, 62/262, 273
occipital headache, 5/371
pain, 48/349
recruitment pattern, 62/53
Muscle contraction headache
acute, *see* Acute muscle contraction headache
analgesic agent, 48/39
anxiety, 48/39, 355
carisoprodol, 5/168
child, 5/241
chlormezanone, 5/168
chlorzoxazone, 5/168
chronic, *see* Chronic muscle contraction headache
classification, 48/6
cluster headache, 5/115
coital migraine, 48/379
definition, 48/8
depression, 48/355
dopamine β-mono-oxygenase, 48/95
drug treatment, 5/168
EMG, 48/78
emotional stress, 48/355
hypnosis, 5/163
mechanism, 5/20
migraine, 48/39, 355, 75/294
Minnesota Multiphasic Personality Inventory, 48/359
muscle relaxation, 5/168
orphenadrine, 5/169
personality disorder, 48/356
posttraumatic, 5/181
prevention, 5/165
prolotherapy, 5/165-168
psychiatric headache, 48/355
psychodynamics, 5/162
psychotherapy, 5/162
semantics, 5/157
subarachnoid hemorrhage, 55/19
suggestion, 5/163
sweating, 75/295
treatment, 5/168
trigger area, 5/159, 167
vascular headache, 5/20
Muscle contracture
see also Muscle cramp
eosinophilic fasciitis, 63/385
spasticity, 3/389, 26/478-480
Muscle cramp
see also Motor unit hyperactivity *and* Muscle contracture
adenosine monophosphate deaminase deficiency, 62/511
African trypanosomiasis, 52/341
anatomic classification, 1/279
antihypertensive agent, 63/87
bismuth intoxication, 64/337
calcium antagonist intoxication, 65/442
carnitine palmitoyltransferase deficiency, 62/494
childhood myoglobinuria, 62/561
congenital heart disease, 63/4
congenital myopathy, 62/335
continuous muscle fiber activity, 1/290, 40/330, 546
contracture, 40/330
dialysis disequilibrium syndrome, 63/523
differential diagnosis, 40/326
endocrine myopathy, 62/527
eosinophilia myalgia syndrome, 63/377, 64/252
exertion, 40/327
glycogen storage disease, 41/186, 190, 193, 297
hyperventilation syndrome, 63/435
hypomagnesemia, 63/562
hyponatremia, 63/545
hypothyroid myopathy, 41/243, 246, 62/532

lipid storage myopathy, 41/207-214
lumbar vertebral canal stenosis, 20/706
myogelosis, 41/385
myokymia, 40/330, 546
myophosphorylase deficiency, 41/186, 62/485
myotonia, 40/327, 533, 62/261, 273
myotonic syndrome, 62/261, 273
nerve lesion, 41/300
nifedipine intoxication, 65/442
nimodipine intoxication, 65/442
occupational neurosis, 1/290
ordinary cramp, 41/308
phosphofructokinase deficiency, 62/487
phosphoglycerate kinase deficiency, 62/489
phosphoglycerate mutase deficiency, 62/489
progressive spinal muscular atrophy, 22/37
stiff-man syndrome, 41/306
tetanus, 41/306
tetany, 40/330
thyrotoxic myopathy, 62/528
toxic myopathy, 62/596
toxic oil syndrome, 63/381
Trousseau sign, 40/330
uremic encephalopathy, 27/336, 63/505
uremic polyneuropathy, 51/356
water intoxication, 51/356
Wilson disease, 49/224
Wohlfart-Kugelberg-Welander disease, 59/86
writers, *see* Writers cramp
Muscle culture, *see* Muscle tissue culture
Muscle denervation
macroelectromyography, 62/65
motor unit, 62/62
motor unit potential, 62/64
scanning electromyography, 62/66
single fiber, 62/62
Muscle disease, *see* Myopathy
Muscle end plate
transmission, 1/48
Muscle fiber
A band, 62/26
α-actinin, 62/26, 35
aging, 40/199
atrophy, 40/11, 142, 200, 62/13
atrophy cause, 62/13
autophagic vacuole, 62/23
barium, 51/274
C protein, 62/26
cabbage body, 40/192
calcification, 62/18
caps, 62/44
central core, 40/31, 74, 243
change, 62/27
collagen type IV, 62/31
concentric laminated body, 40/39, 109, 154, 249
creatine kinase, 43/116
cytoplasmic inclusion body, 40/26, 107, 245
cytoskeleton, 62/19
cytosome, 40/113
delta lesion, 62/15
denervation, 40/11-14
denervation atrophy, 62/13
dense core tubule, 62/23
development, 40/199
disappearance, 40/203
dystrophin, 62/35
dystrophin related protein, 62/35
entactin, 62/31
exocytosis, 62/18
extralysosomal storage, 62/16
fingerprint body, 40/39, 43, 62/44
focal loss, 40/27
forked fiber, 62/11
forking, 40/206
glycogen, 62/20
glycogen accumulation, 40/240
Golgi system, 62/21
heparan sulfate, 62/31
histochemical type, 40/3-8, 136
honeycomb structure, 40/106, 224
hypercontraction, 62/15
hypertrophy, 62/15
I band, 62/26
lamina basalis, 62/31
laminin, 62/31
lipid accumulation, 40/242
lipid globule, 62/22
lysis, 40/263, 288
lysosomal storage, 62/16
lysosome, 40/36, 62/22
membranous array, 40/154
membranous whorl, 40/154
merosin, 62/31
mitochondria, 40/31, 90, 216, 62/23
monoclonal protein, 62/26
motor end plate, 62/28
muscle fiber type I atrophy, 40/14
muscle fiber type II atrophy, 40/14
myofilamentous degeneration, 40/214
myonucleus, 62/29
nebulin, 62/26
necrosis, *see* Muscle necrosis
nemaline body, 40/24, 87
neural cell adhesion molecule, 62/33
nonnecrotic change, 40/210
nuclear inclusion, 62/31

nucleus, 40/13, 33, 38
opaque fiber, 62/15
partial invasion, 62/9
phagocytosis, 62/6
physiology, 40/8
pigment, 40/108
plasma membrane, 62/32
progressive spinal muscular atrophy, 22/18
ragged red fiber, 62/24
reducing body, 62/44
reducing body myopathy, 40/38, 113
regeneration, 40/20, 89, 62/9
repair, 62/19
ribosome, 40/33
rimmed muscle fiber vacuole type, 62/18
ring fiber, 40/27, 74, 141
ryanodine binding protein, 62/37
sarcolemma, 62/31
sarcoplasmic mass, 40/27, 244, 497, 62/44
sarcoplasmic reticulum, 62/37
sarcotubular change, 40/224
segmental degeneration, 40/205
segmental necrosis, 62/13
small caliber fiber, 62/13
snake coil, 40/141
spectrin, 62/35
spheromembranous body, 40/154
split fiber, 62/16
splitting, 40/15-18
storage material, 40/36, 94-102
striated annulet, 40/27, 74, 141
T tubule, 62/23, 37
target, 40/31, 75, 142
thick filament, 40/26
titin, 62/26
trilaminar, 40/80
tubular aggregate, 40/33, 112, 226, 229
type grouping, 40/8, 18, 139
type predominance, 40/18-20
ultrastructure, 40/66
vacuole, *see* Rimmed muscle fiber vacuole
virus like particle, 40/109
Z band, 62/7, 26
Z disc streaming, 40/24, 81, 151, 157, 215
zebra body, 40/113
Muscle fiber band
A, *see* A band
I, *see* I band
M, *see* M band
Z, *see* Z band
Muscle fiber conductance
depolarization, 40/552
motor end plate, 41/108
myotonia, 40/553
phenytoin, 40/560
procainamide, 40/559
quantal, 41/108
tetrodotoxin, 40/552
Muscle fiber size
aminoaciduria, 43/126
arthrogryposis multiplex congenita, 43/78
benign dominant myopathy, 43/112
congenital muscular dystrophy, 43/93-95
congenital nystagmus, 43/93-95
Duchenne muscular dystrophy, 43/106
fingerprint body myopathy, 43/82
hypertrophia musculorum vera, 43/84
muscular dystrophy, 43/108
myopathy, 43/126
myotonic dystrophy, 43/152
proximal muscular dystrophy, 43/108
Muscle fiber type
adenosine triphosphatase, 62/4
criteria, 62/2
cytoskeleton, 62/3
cytosol, 62/4
determinant molecule, 62/3
display mode, 62/3
fast mature myosin, 62/4
lysosome, 62/4
marker, 62/3
mitochondria, 62/3
myopathy, 62/2
neuromuscular disease, 62/2
nucleus, 62/4
organelle distribution, 62/3
ribosome, 62/4
sarcolemma, 62/3
sarcoplasmic reticulum, 62/3
T tubule, 62/3
type grouping, 62/4
Muscle fiber type I
atrophy, 40/14, 144
benign dominant myopathy, 43/112
carnitine palmitoyltransferase deficiency, 43/177
central core myopathy, 43/81
congenital myopathy, 40/146, 62/349
criteria, 62/4
distal muscular dystrophy, 43/96
etiology, 40/147
experimental myopathy, 40/136
facioscapulohumeral muscular dystrophy, 43/99
familial lysis, *see* Familial lysis muscle fiber type I
fingerprint body myopathy, 43/82
histocytochemistry, 40/3-8
infantile distal myopathy, 43/114, 62/201

Kearns-Sayre-Daroff-Shy syndrome, 43/142
lipid storage myopathy, 43/184
lysis myopathy, 43/116
multicore myopathy, 43/121
muscular dystrophy, 40/149
myopathy, 43/116
myotonic dystrophy, 40/49, 252, 497, 43/152
myotubular myopathy, 43/113
nemaline myopathy, 41/10, 43/122
predilection, 62/5
target fiber, 40/13
ultrastructure, 40/63
Muscle fiber type I atrophy
see also Muscular atrophy
Emery-Dreifuss muscular dystrophy, 62/151
muscle fiber, 40/14
Muscle fiber type I hypertrophy, *see* Myotubular myopathy
Muscle fiber type I predominance
central core myopathy, 41/4, 62/338
congenital fiber type disproportion, 41/15
congenital myopathy, 62/332
myotubular myopathy, 41/12
nemaline myopathy, 41/10
reducing body myopathy, 41/20
uniform muscle fiber type I myopathy, 62/354
Muscle fiber type II
disuse, 40/146, 200
experimental, 40/140, 146
histochemistry, 40/3-8
myotonic dystrophy, 43/153
myotubular myopathy, 40/15, 200, 43/113
neuromuscular disease, 40/14, 140
sarcotubular myopathy, 43/129
steroid myopathy, 40/161
ultrastructure, 40/63
Muscle fiber type II atrophy
see also Muscular atrophy
acquired immune deficiency syndrome, 56/518
corticosteroid, 41/251
Cushing syndrome, 41/251
disuse *vs* disease atrophy, 40/200
endocrine myopathy, 62/527
eosinophilia myalgia syndrome, 63/378
hypereosinophilic syndrome, 63/374
malignancy, 41/324
muscle fiber, 40/14
myasthenia gravis, 41/110
nerve fiber, 41/110, 251, 324
steroid myopathy, 62/605
Muscle fiber type II hypertrophy
size, 40/15
Muscle fiber type II hypoplasia
congenital myopathy, 62/332
Muscle fiber type IIA
criteria, 62/4
predilection, 62/5
Muscle fiber type IIB
criteria, 62/4
predilection, 62/5
Muscle fiber type IIC
criteria, 62/4
Muscle fibrillation
EMG, 1/636
lesion site, 1/639
Muscle fibrosis
arthrogryposis, 62/173
muscle contraction, 62/173
rigid spine syndrome, 62/173
Muscle filament
degeneration, 40/214
histochemistry, 40/2
thick, *see* Thick muscle filament
thin, *see* Thin muscle filament
Muscle filamentous body
atrophy, 40/249
sarcoplasmic mass, 40/249
Muscle hernia
surgery, 41/384
Muscle histopathology
Kearns-Sayre-Daroff-Shy syndrome, 62/310
limb girdle syndrome, 62/186
myotonic dystrophy, 62/219
progressive external ophthalmoplegia, 62/310
Muscle hypertrophy
see also Pseudohypertrophy
autosomal recessive generalized myotonia, 62/267
Berardinelli-Seip syndrome, 42/591
childhood myoglobinuria, 62/561
chondrodystrophic myotonia, 40/284, 62/269
congenital myopathy, 62/335
De Lange syndrome, 40/324
differential diagnosis, 40/322
Duchenne muscular dystrophy, 40/358, 41/408, 461
generalized myotonia (Becker), 43/163
hypertrophia musculorum vera, 43/83-85
hypothyroid myopathy, 41/243
hypothyroidism, 43/84
infantile acid maltase deficiency, 62/480
myofibril increase, 40/200
myotonia congenita, 40/539, 41/298, 43/161, 62/266
myotonic myopathy, 43/165
partial denervation, 40/203
periodic paralysis, 41/151, 160

phenytoin, 40/558
work hypertrophy, 40/200
Muscle innervation
dermatome, 2/132
intensity duration curve, 19/279
nerve, 2/132
Muscle intermittent claudication
ergot intoxication, 65/67
hyperparathyroidism, 20/799
multiple sclerosis, 47/56
myophosphorylase deficiency, 20/799
Muscle ischemia
Duchenne muscular dystrophy, 41/409
nodular myositis, 41/57, 78
polymyositis, 41/57, 78
Muscle mass
evaluation, 41/381
exercise, 74/251
systemic neoplastic disease, 41/381
Muscle membrane dysfunction
critical illness polyneuropathy, 51/584
primary, *see* Primary muscle membrane dysfunction
Muscle metaboreflex
blood pressure, 74/255
exercise, 74/255, 257
spinal cord, 74/257
Muscle metastasis
see also Muscle tumor
skeletal muscle, 41/378
Muscle necrosis
Becker muscular dystrophy, 40/389
Duchenne muscular dystrophy, 40/368
facioscapulohumeral muscular dystrophy, 40/252, 420
generalized, 40/20, 74, 206
histologic features, 62/6
hypokalemia, 41/274
inflammatory myopathy, 40/264
limb girdle syndrome, 40/252, 453
myoglobinuria, 41/263, 62/554
myophosphorylase deficiency, 41/186, 62/485
myotonic dystrophy, 40/252, 497
Muscle pain, *see* Myalgia
Muscle phosphofructokinase deficiency, *see* Phosphofructokinase deficiency
Muscle phosphorylase, *see* Myophosphorylase
Muscle pseudohypertrophy
limb girdle syndrome, 62/182
Muscle regeneration
Duchenne muscular dystrophy, 40/371
myotubule, 40/23
satellite cell, 40/22
target fiber, 40/13
tissue culture, 40/184
ultrastructure, 40/89
Muscle reinnervation
acetylcholine receptor, 40/13, 142
axonal sprouting, 40/140
foreign innervation, 40/139
histochemistry, 40/8
isometric twitch property, 40/139
macroelectromyography, 62/65
motor unit, 62/62
motor unit potential, 62/64
scanning electromyography, 62/66
single fiber, 62/62
target fiber, 40/145
type grouping, 40/139
Muscle relaxation
carisoprodol, 5/168
chlormezanone, 5/168
chlorzoxazone, 5/168
hypothyroidism, 41/298
muscle contraction headache, 5/168
myotonic dystrophy, 62/224
restoration of higher cortical function, 3/390
Muscle response
direct, *see* Direct muscle response
Muscle rigidity
frontal lobe tumor, 17/257
juvenile Gaucher disease, 10/307
malignant hyperthermia, 38/550, 41/268, 43/117, 63/439
Pick disease, 46/234
procaine, 1/66
reserpine, 1/67
scorpion sting intoxication, 37/112
suxamethonium induced, *see* Suxamethonium induced muscle rigidity
Muscle shortening
inability to open mouth, 43/434
muscular dystrophy, 43/109
proximal muscle weakness, 43/109
Muscle spasm
acetonylacetone intoxication, 64/87
Amanita pantherina intoxication, 36/534
calcium antagonist intoxication, 65/442
carbamate intoxication, 64/186
carbon monoxide intoxication, 64/31
chronic muscle contraction headache, 48/349
cisplatin intoxication, 64/358
diborane intoxication, 64/359
eosinophilia myalgia syndrome, 64/252
headache, 5/254
hexachlorophene intoxication, 65/476

hyperventilation syndrome, 63/435
lathyrism, 36/511, 65/3
nifedipine intoxication, 65/442
organophosphate induced delayed polyneuropathy, 64/176
pentaborane intoxication, 64/359
Satoyoshi syndrome, 64/256
snake envenomation, 65/180
tetanus, 52/230
zinc intoxication, 36/337
Muscle spindle
gamma loop, 6/95
hereditary motor and sensory neuropathy type I, 21/294
hereditary motor and sensory neuropathy type II, 21/294
morphology, 1/100
myotonic dystrophy, 40/503
stretch reflex, 1/64, 6/95
Muscle stiffness
painful dominant myotonia, 62/266
Muscle strength
physical rehabilitation, 61/468
Muscle strength scoring
myotonic dystrophy, 62/248
Muscle strength test
brachial plexus, 51/147
Muscle stretch
Parkinson disease, 49/70
scoliosis, 50/414
Muscle stretch reflex
autosomal recessive generalized myotonia, 62/266
brain death, 24/718
brain stem death, 57/451, 476
central motor disorder, 1/171
diastematomyelia, 50/436
diplomyelia, 50/436
erythrokeratodermia ataxia syndrome, 21/216
hereditary sensory and autonomic neuropathy type II, 60/11
late cerebellar atrophy, 42/135
lumbosacral nerve injury, 25/441, 454
neuroacanthocytosis, 42/209
olivopontocerebellar atrophy (Wadia-Swami), 60/493
pallidoluysionigral degeneration, 6/665
paraplegia, 1/206
persistence, 57/476
phenylketonuria, 42/610
physiology, 1/64
posttraumatic syringomyelia, 61/384
presenile dementia, 42/288
renal insufficiency, 27/227
spinal nerve root syndrome, 2/167-170
subarachnoid hemorrhage, 12/131
thoracolumbar spine injury, 25/441, 454
thyrotoxic myopathy, 62/528
Wohlfart-Kugelberg-Welander disease, 22/71
Muscle stretch reflex time
hypothyroid myopathy, 62/532
myxedema neuropathy, 8/58
Muscle tension
headache, 48/34
Muscle tissue culture, 40/183-194, 62/85-111
acetylcholine receptor, 40/186, 62/97
acetylcholine receptor clustering, 62/89
acetylcholinesterase, 62/97
acid maltase deficiency, 40/190, 62/99
aneural culture, 62/86
calcium channel, 62/91
clonal culture, 62/94
cloned satellite cell, 62/88
contractile activity, 62/96
creatine kinase, 62/87
creatine kinase MM, 62/89
debrancher deficiency, 62/99
2,4-dinitrophenol, 40/187
Duchenne muscular dystrophy, 40/191, 375, 62/100
electrical property, 62/96
enzyme histochemistry, 62/94
epidermal growth factor, 62/88
experimental allergic myositis, 40/165
ferritin H, 62/92
fibroblast, 62/87
fibroblast growth factor, 62/88
genetical abnormal muscle, 62/85
glucocorticoid, 62/90
hereditary cylindric spirals myopathy, 62/108
histochemistry, 40/185
HLA antigen, 62/92
inclusion body myositis, 62/104
infantile nemaline myopathy, 62/99, 107
innervated cultured, 62/92
innervation method, 62/92
insulin, 62/88
insulin growth factor 1, 62/90
γ-interferon, 62/92
lysosome, 40/190
methodologic consideration, 62/86
mitochondrial disease, 40/186-188
mitochondrial myopathy, 62/100
monolayer culture, 62/93
motor end plate, 62/92, 97
muscle fiber phenotype, 62/96
muscle phenotype, 62/88

muscle specific isoenzyme, 62/94
myoblast fusion, 62/91
myogenesis, 62/88
myopathy, 62/85
myophosphorylase deficiency, 40/189, 62/109
myotonic dystrophy, 62/105, 222
myotubular myopathy, 40/193
myotubule, 62/93
nerve, 40/186
normal differentiation, 40/184
normal muscle, 62/85
oculopharyngeal muscular dystrophy, 62/102
oxidative phosphorylation, 40/187
phosphofructokinase deficiency, 27/236, 40/190
phosphoglycerate mutase, 62/94
phosphorylase, 62/94, 109
phosphorylase deficiency, 40/189, 62/87
postsynaptic membrane organization, 62/98
preparation, 40/184
primary monolayer culture, 62/87
progressive external ophthalmoplegia, 40/186
ragged red fiber, 40/186-188
ribosomal S6 protein, 62/90
Tomé-Fardeau body, 62/103
tumor necrosis factor-α, 62/92
ultrastructure, 62/94
vascular myopathy, 40/191
virus like particle, 40/192

Muscle tone
action dystonia, 6/523
brain embolism, 53/211
brain infarction, 53/211
central motor disorder, 1/169
cerebellar patient, 1/329
cerebrovascular disease, 53/211
clinical examination, 1/263
CNS, 1/64
coma, 3/65, 75
development, 4/444, 446-448
disorder, 1/169, 261-263
Friedreich ataxia, 21/333
frontalis, *see* Frontalis muscle tension
gamma fiber, 1/258
gamma system, 1/258
Golgi tendon organ, 1/259
history, 1/257
hypertonia, 1/264-269
hypotonia, 1/263, 269-271
intracranial hemorrhage, 53/211
nuchal rigidity, 1/267
physical rehabilitation, 61/465
physiology, 1/257-260
postural reflex, 1/264
r(14) syndrome, 50/588
reflex sympathetic dystrophy, 61/128
Renshaw cell, 1/258
resting tone, 1/263
rigid Huntington chorea, 49/69
sleep, 3/84
spinal deformity, 26/165
transient ischemic attack, 53/211
traumatic spine deformity, 26/165

Muscle tumor
see also Muscle metastasis
desmoid tumor, 41/383
fasciitis, 41/383
hemangioma, 41/382
lipoma, 41/382
rhabdomyoma, 41/381
rhabdomyosarcoma, 41/381

Muscle vasoconstrictor neuron
autonomic nervous system, 74/23

Muscle wasting, *see* Muscular atrophy

Muscle weakness
acromegalic polyneuropathy, 41/249
acrylamide intoxication, 64/63
ACTH induced myopathy, 62/537
acute, *see* Acute muscle weakness
acute intermittent porphyria, 41/167
Addison disease, 41/250
adenosine triphosphatase 6 mutation syndrome, 62/510
autosomal recessive generalized myotonia, 62/266
barium intoxication, 64/354
Bassen-Kornzweig syndrome, 13/417
birgus latro intoxication, 37/58
Cockayne-Neill-Dingwall disease, 13/433
complex III deficiency, 62/504, 66/423
dermatomyositis, 62/371
diabetic proximal neuropathy, 40/318
diborane intoxication, 64/359
distal, *see* Distal muscle weakness
distal muscular dystrophy, 43/96
diuretic agent intoxication, 65/440
dysarthria, 40/306
dysphagia, 40/306
dysphonia, 40/306
eosinophilia myalgia syndrome, 63/376
ethchlorvynol intoxication, 37/354
extraocular, *see* Extraocular muscle weakness
facial weakness, 40/304, 422
globoid cell leukodystrophy, 40/336
heterodontus francisci intoxication, 37/70
hexachlorophene intoxication, 37/486
hollow hand sign, 1/184
hyperkalemia, 63/558

hypermagnesemia, 63/563
hypokalemia, 63/557
hypophosphatemia, 63/565
ichthyohaemotix fish intoxication, 37/87
Medical Research Council Scale, 40/297
multiple mitochondrial DNA deletion, 62/510
muscle testing, 40/297, 447
myxedema neuropathy, 8/57
neck weakness, 40/308
Nelson syndrome, 62/537
ophthalmoplegia, 40/298
organophosphate intoxication, 37/554, 64/151
periodic weakness, 40/297
peroxisomal acetyl-coenzyme A acyltransferase deficiency, 66/518
posttraumatic syringomyelia, 50/454
proximal, *see* Proximal muscle weakness
ptosis, 40/298
quadriceps myopathy, 43/128, 62/184
rhizostomeae intoxication, 37/53
Richards-Rundle syndrome, 22/516, 43/265
sporadic progressive external ophthalmoplegia with ragged red fiber, 62/508
syringobulbia, 32/275
thyrotoxic myopathy, 62/528
toxic oil syndrome, 63/380
toxic shock syndrome, 52/258
ubidecarenone, 62/504
vitamin B6 intoxication, 65/572
Muscular atrophy
see also Muscle fiber type I atrophy *and* Muscle fiber type II atrophy
acetonylacetone intoxication, 64/86
acrylamide intoxication, 64/67
adolescent spinal, *see* Adolescent spinal muscular atrophy
Alföldi sign, 2/15
amyotrophic lateral sclerosis, 22/291, 42/65, 78-81
ankylosing spondylitis, 38/506
anterior horn syndrome, 2/210
areflexia, 42/82, 91
asthma, 63/420
astrocytosis, 42/83
autosomal recessive generalized myotonia, 62/267
beriberi, 7/563-565
birth incidence, 42/84
blennorrhagic polyneuritis, 7/478
brachial neuritis, 42/303
bulbospinal, *see* Bulbospinal muscular atrophy
cardiopathy, 42/343-345
cerebellar ataxia, 42/132
cerebrotendinous xanthomatosis, 60/167
childhood myoglobinuria, 62/561
Cockayne-Neill-Dingwall disease, 13/433
congenital megacolon, 42/3
cordotomy, 40/143
cortical, *see* Cortical muscular atrophy
corticosteroid, 40/160
cranial nerve, 42/91
creatine kinase, 42/85, 91
Creutzfeldt-Jakob disease, 42/657, 46/291
critical illness polyneuropathy, 51/575
cryoglobulinemia, 63/404
CSF, 42/82, 545
CSF paraprotein, 42/603
debrancher deficiency, 62/482
denervation, 22/160, 181, 40/200
diabetic proximal neuropathy, 40/318
diastematomyelia, 50/436
diplomyelia, 50/436
distal, *see* Distal muscular atrophy
distal spinal, *see* Spinal muscular atrophy
disuse, 40/11, 200, 322
eosinophilia myalgia syndrome, 64/253
examination, 40/322
excessive CSF protein syndrome, 42/545
facioscapulohumeral, *see* Facioscapulohumeral muscular atrophy
facioscapulohumeral spinal, *see* Facioscapulohumeral spinal muscular atrophy
fasciculation, 42/78, 82-84
Flynn-Aird syndrome, 42/327
Freeman-Sheldon syndrome, 38/413
Friedreich ataxia, 42/145
frontometaphyseal dysplasia, 31/256, 43/401
gene frequency, 42/85
group atrophy, 40/12
Hallervorden-Spatz syndrome, 6/605, 626
heart disease, 42/343-345
hereditary, *see* Hereditary muscular atrophy
hereditary amyloid polyneuropathy, 42/518, 521
hereditary amyloid polyneuropathy type 1, 51/420
hereditary motor and sensory neuropathy type I, 42/339
hereditary motor and sensory neuropathy type II, 42/339
hereditary multiple recurrent mononeuropathy, 42/316
hereditary progressive spinal, *see* Hereditary progressive spinal muscular atrophy
heterozygote frequency, 42/84
hyperparathyroid myopathy, 62/540
hypertrophic interstitial neuropathy, 42/317
hyperuricemia, 42/571
hyporeflexia, 42/82, 84, 91

infantile spinal, *see* Infantile spinal muscular atrophy
intensive care syndrome, 63/420
intermediate spinal, *see* Intermediate spinal muscular atrophy
intramedullary spinal tumor, 19/32
Isaacs syndrome, 43/160
ischemic neuropathy, 8/152
joint fixation, 40/143
juvenile, *see* Juvenile muscular atrophy
juvenile spinal, *see* Wohlfart-Kugelberg-Welander disease
Kjellin syndrome, 22/469, 42/173-175
late onset, 42/82
lathyrism, 36/511, 65/6
leprechaunism, 42/589
leprosy, 51/218
Machado disease, 42/155
malabsorption syndrome, 70/227
Marinesco-Sjögren syndrome, 42/184, 60/342
mental deficiency, 42/81, 43/258
microcephaly, 42/81
monomelic spinal, *see* Monomelic spinal muscular atrophy
multiple organ failure, 71/537
multiple sclerosis, 9/179
myalgia, 42/84
myasthenia gravis, 43/157
myotonic myopathy, 43/165
neuralgic amyotrophy, 51/171
neuroacanthocytosis, 42/209
neurogenic, *see* Neurogenic muscular atrophy
nonprogressive spinal, *see* Nonprogressive spinal muscular atrophy
olivopontocerebellar atrophy variant, 22/171
organophosphate induced delayed polyneuropathy, 64/176
parietal lobe syndrome, 45/67
parietal wasting, 1/484
Parkinson dementia complex, 42/250
pathologic change, 40/200
peripheral motoneuron, 1/222
postencephalitic parkinsonism, 22/22
posterior column degeneration, 42/80
posttraumatic syringomyelia, 26/127, 61/384
pressure sore, 61/352
prevalence, 42/84
progeria, 43/465
progressive, *see* Progressive muscular atrophy
progressive bulbar palsy, 22/113
progressive spinal, *see* Progressive spinal muscular atrophy
proximal muscle weakness, 42/82
radiation myelopathy, 22/20
Richards-Rundle syndrome, 43/264
Rosenberg-Chutorian syndrome, 22/506, 42/418
Roussy-Lévy syndrome, 42/108
Ryukyan spinal, *see* Ryukyan spinal muscular atrophy
scapulohumeral, *see* Scapulohumeral muscular atrophy
scapulohumeral spinal, *see* Scapulohumeral spinal muscular atrophy
scapulohumeroperoneal, *see* Scapulohumeroperoneal muscular atrophy
scapuloilioperoneal, *see* Scapuloilioperoneal muscular atrophy
scapuloperoneal spinal, *see* Scapuloperoneal spinal muscular atrophy
secondary pigmentary retinal degeneration, 13/306, 308
segmental, 19/26
sensory loss, 42/80
serum paraprotein, 42/603
Shy-Drager syndrome, 1/474
simian hand, 22/291
spastic paraplegia, 42/171-175
spinal, *see* Spinal muscular atrophy
spinal cord tumor, 19/26
spinal tumor, 19/26
spinocerebellar ataxia, 42/182
steroid myopathy, 62/535
striatonigral degeneration, 49/206
tenotomy, 40/143
thalassemia major, 42/629
thoracic outlet syndrome, 51/122
toxic oil syndrome, 63/380
tremor, 42/78
Troyer syndrome, 42/193
unilateral upper extremity, 42/77
vertebral abnormality, 42/82
vertical gaze palsy, 42/188
vitamin B12 deficiency, 51/341
vitamin E deficiency, 70/228
Werner syndrome, 43/489
Wetherbee ail, 42/78-81
Whipple disease, 52/138
Wohlfart-Kugelberg-Welander disease, 22/70
Woods-Schaumburg syndrome, 22/171
X-linked spinocerebellar degeneration, 42/190
xeroderma pigmentosum, 60/336

Muscular dystrophy
see also Limb girdle syndrome
aminoaciduria, 42/517
amniocentesis, 40/357
areflexia, 43/107, 109

Becker, *see* Becker muscular dystrophy
Brossard syndrome, 22/57
cataract, 43/89
cauda equina tumor, 19/87
central sleep apnea, 63/463
collagen disease, 43/103
congenital, *see* Congenital muscular dystrophy
congenital atonic sclerotic, *see* Congenital atonic sclerotic muscular dystrophy
contracture, 43/109
creatine kinase, 43/88, 90, 96, 102
dementia, 46/401
diethylstilbestrol, 40/343, 378
differential diagnosis, 1/233
distal, *see* Distal muscular dystrophy
Duchenne, *see* Duchenne muscular dystrophy
Duchenne like, *see* Duchenne like muscular dystrophy
Emery-Dreifuss, *see* Emery-Dreifuss muscular dystrophy
EMG, 1/641
epilepsy, 43/92
epileptic seizure, 15/326
Erb scapulohumeral, *see* Scapulohumeral muscular dystrophy (Erb)
exercise, 41/463
facioscapulohumeral, *see* Facioscapulohumeral muscular dystrophy
familial spastic paraplegia, 22/433
freeze fracture, 40/371
fructose-bisphosphate aldolase, 40/384
Fukuyama syndrome, 43/91
genetic linkage, 43/458
histochemistry, 40/47-51, 502
hydroxyproline, 40/380
hypogonadism, 43/89, 139
hyporeflexia, 43/109
inflammatory scapuloperoneal, *see* Inflammatory scapuloperoneal muscular dystrophy
late adult Nevin type, *see* Late adult muscular dystrophy (Nevin)
menopausal, 41/54, 58
mental deficiency, 43/90-92
muscle fiber size, 43/108
muscle fiber type I, 40/149
muscle shortening, 43/109
nevus unius lateris, 14/781
ocular myopathy, 1/612
oculopharyngeal, *see* Oculopharyngeal muscular dystrophy
ophthalmoplegia, 22/180, 43/139
orthochromatic leukodystrophy, 10/114
osteoporosis, 41/474
pelvifemoral Leyden-Möbius type, *see* Pelvifemoral muscular dystrophy (Leyden-Möbius)
peripheral, 43/96-98
peroneal, *see* Hereditary motor and sensory neuropathy type I *and* Hereditary motor and sensory neuropathy type II
phakomatosis, 14/492
phosphorylation, 41/413, 489
progressive, *see* Progressive muscular dystrophy
progressive external ophthalmoplegia, 2/279, 22/210, 43/136
proximal muscle weakness, 43/107, 109
pseudohypertrophic, *see* Duchenne muscular dystrophy
recessive, *see* Recessive muscular dystrophy
respiratory pattern, 63/463
Rowley-Rosenberg syndrome, 43/473
scapulohumeral, *see* Scapulohumeral muscular dystrophy
scapulohumeral distal, *see* Scapulohumeral distal muscular dystrophy
scapulohumeral Erb type, *see* Scapulohumeral muscular dystrophy (Erb)
scapuloperoneal, *see* Scapuloperoneal muscular dystrophy
scoliosis, 41/478, 50/417
secondary pigmentary retinal degeneration, 13/319
sick motoneuron, 41/409
tenotomy, 41/471
Welander, *see* Welander muscular dystrophy
X-linked, *see* X-linked muscular dystrophy
Muscular hypertonia
African trypanosomiasis, 52/341
ARG syndrome, 50/581
athetosis, 6/445
neuroleptic syndrome, 6/251
Muscular hypoplasia
abdominal, *see* Abdominal muscle hypoplasia
congenital, *see* Congenital muscular hypoplasia
Krabbe, *see* Krabbe muscular hypoplasia
Muscular hypotonia
alcohol intoxication, 64/115
Argentinian hemorrhagic fever, 56/365
barbiturate intoxication, 65/332
botulism, 65/210
Down syndrome, 50/519, 526, 529
Edwards syndrome, 50/561
elfin face syndrome, 31/318
happy puppet syndrome, 31/309
hereditary sensory and autonomic neuropathy type III, 21/110, 60/27

juvenile hereditary benign chorea, 49/345
Klinefelter syndrome, 50/550
late cerebellar atrophy, 42/135
meprobamate intoxication, 65/333
Miller-Dieker syndrome, 50/592
obstructive sleep apnea, 63/452
9p partial monosomy, 50/579
Prader-Labhart-Willi syndrome, 31/321
progressive pallidal atrophy (Hunt-Van Bogaert), 6/554
sulfite oxidase deficiency, 29/121
trisomy 22, 50/565
Weiss-Alström syndrome, 13/463
Muscular imbalance
child, 25/183
juvenile spinal cord injury, 25/183
spinal deformity, 25/183
Muscular infantilism, *see* Krabbe muscular hypoplasia
Muscular pain fasciculation syndrome
restless legs syndrome, 51/545
Musculocutaneous nerve
compression neuropathy, 51/102
occupational lesion, 7/331
plexus origin, 2/138
topographical diagnosis, 2/22
Mushroom intoxication
Amanita muscaria, *see* Amanita muscaria intoxication
Amanita pantherina, *see* Amanita pantherina intoxication
Amanita phalloides, *see* Amanita phalloides intoxication
chemistry, 36/432
Chlorophyllum molybdites, 36/542
Clitocybe, 36/535
Coprinus atramentarius, 36/535-538
Cortinarius speciosissimus, 36/542
Gyromitra esculenta, 36/536-538
history, 36/429
Hypholoma fasciculare, 36/542
Inocybe, 36/535
muscarine, 36/535
Paxillus involutus, *see* Paxillus involutus intoxication
Psilocybe, 36/535
Raynaud phenomenon, 8/337
taxonomy, 36/330
toxicology, 36/532
Mushroom toxin
see also Amanitin, Coprine, Gyromitrine, Muscimol, Orellanine, Psilocin *and* Psilocybine
acetaldehyde intoxication, 65/39
alcohol, 65/39
Amanita bisporigera, 65/36
Amanita muscaria type, 65/39
Amanita phalloides, 65/36
Amanita tenuifolia, 65/36
Amanita verna, 65/36
Amanita virosa, 65/36
bipyridine congener, 65/36
Clitocybe, 65/39
Coprinus atramentarius type, 65/39
Cortinarius orellanus, 65/36, 38
Cortinarius speciosissimus, 65/38
Cortinarius splendens, 65/38
Cortinarius venenosus, 65/38
delayed effect, 65/35
direct effect, 65/35, 38
epidemiology, 65/35
fatality, 65/36
Gyromitra esculenta, 65/36
Gyromitra gigas, 65/36
ibotenic acid, 65/39
Inocybe, 65/39
laboratory diagnosis, 65/37
Lampteromyces japonicus, 65/35
Lepiota specie, 65/35
muscarinic type, 65/39
muscimol, 65/39
Omphalotus, 65/39
parasympathetic nerve ending, 65/39
phallotoxin, 65/37
Psilocybe type, 65/39
Rhodophyllus rhodopolius, 65/35
taxonomy, 65/35
Tricholoma ustale, 65/35
Music reading
alexia, *see* Musical alexia
Musical agnosia
see also Amusia
temporal lobe lesion, 4/203
Musical agraphia
amusia, 4/203, 45/486
number reading, 4/120
Musical alexia
amusia, 4/203, 45/486
pitch, 45/486
posterior lobe lesion, 45/437
pure alexia, 4/120
rhythm, 45/486
Musical amnesia
amusia, 45/486
Musical aptitude
musical memory, 45/483
musical notation, 45/484

testing, 45/485
tone perception, 45/483
Musical deafness
amusia, 4/203, 45/485
testing, 45/485
Musical hallucination
dreamy state, 2/705
hallucinosis, 46/568
temporal lobe lesion, 45/484
Musical memory
amusia, 45/485
musical aptitude, 45/483
testing, 45/485
Musical perception
amusia, 45/485
Musicianship
amusia, 45/484
Musicogenic epilepsy
amusia, 4/204
EEG, 73/192
emotion, 15/446
perinatal brain injury, 73/192
reflex epilepsy, 15/446
startle epilepsy, 73/192
temporal lobe epilepsy, 15/446, 45/484
Mussels intoxication, *see* Paralytic shellfish poisoning
Mutation
cerebellar hypoplasia, 50/179
CNS malformation, 30/99-101
cytoplasmic DNA, 30/109
definition, 30/97
delayed, 30/105
DNA, *see* DNA mutation
Fabry disease, 66/75
frameshift, 30/98
frequency, 30/97
glycosaminoglycanosis type IVB, 66/306
glycosaminoglycanosis type VII, 66/311
half chromatid, 30/106
Huntington chorea, 49/292
missense, 30/98
mitochondrial DNA, *see* Mitochondrial DNA mutation
mitochondrial DNA defect, 66/397
nosense, 30/98
nuclear base, *see* Nucleotide base mutation
point, 30/98
prealbumin, *see* Prealbumin mutation
premutation, 30/105
proteolipid protein DM-20 gene, 66/565
proteolipid protein gene, 66/570
somatic, 30/97
telomutation, 30/105
Mutism
African trypanosomiasis, 52/341
akinetic, *see* Akinetic mutism
aluminum intoxication, 63/525, 64/279
aluminum neurotoxicity, 63/525
anterior cerebral artery syndrome, 53/346
aphasia, 45/322
brain commissurotomy, 45/104
brain hemorrhage, 54/309
brain infarction, 53/346
brain lacunar infarction, 54/243
carbon monoxide intoxication, 49/475, 64/32
cardiac arrest, 63/215
central pontine myelinolysis, 28/291, 63/549
concentric sclerosis, 47/414
deaf, *see* Deaf mutism
elective, *see* Elective mutism
hyperkinetic, *see* Hyperkinetic mutism
infantile autism, 46/190
migraine, 48/161
motor aphasia, 45/310
Pick disease, 46/233, 240
Shapiro syndrome, 50/167
speech disorder in children, 4/424
supplementary motor area syndrome, 53/346
thalamic aphasia, 54/309
Wilson disease, 49/228
Myalgia
acetonylacetone intoxication, 64/87
adenosine monophosphate deaminase deficiency, 62/511
adrenal insufficiency myopathy, 62/536
adult onset, 42/84
African trypanosomiasis, 52/341
allergic granulomatous angiitis, 63/383
amyotrophic dystonic paraplegia, 42/199
angiostrongyliasis, 52/548, 556
anterior tibial syndrome, 41/385
Aran-Duchenne disease, 42/86
Argentinian hemorrhagic fever, 56/364
azacitidine intoxication, 65/534
Bartter syndrome, 42/528
benzodiazepine withdrawal, 65/340
beta adrenergic receptor blocking agent intoxication, 65/437
Bornholm disease, 40/325
carnitine palmitoyltransferase deficiency, 43/176, 62/494
ciclosporin, 63/190
ciclosporin intoxication, 65/554
ciguatoxin intoxication, 37/83, 65/162
cocaine intoxication, 65/253

Colorado tick fever, 56/141
congenital myopathy, 62/335
corticosteroid withdrawal myopathy, 62/536
dermatomyositis, 62/371
diborane intoxication, 64/359
diuretic agent intoxication, 65/440
enterovirus, 56/11
eosinophilia myalgia syndrome, 63/375, 64/249, 252
eosinophilic fasciitis, 63/384
eosinophilic polymyositis, 63/385
epidemic, *see* Bornholm disease
epidemic benign, *see* Epidemic benign myalgia
ergot intoxication, 65/68
exertional, *see* Exertional myalgia
facioscapulohumeral muscular dystrophy, 40/460
fibrositis, 40/325, 41/385
glutaric aciduria type II, 62/495
glycogen storage disease, 43/181
Haff disease, 41/272
hydrozoa intoxication, 37/38
hypereosinophilic syndrome, 63/371, 374
hypertonia musculorum vera, 43/84
hyperventilation syndrome, 38/319, 63/435
hypoparathyroidism, 42/577
lead intoxication, 64/435
long chain acyl-coenzyme A dehydrogenase deficiency, 62/495
Lyme disease, 51/203
lymphocytic choriomeningitis, 56/359
mass reflex, 1/139
multiple acyl-coenzyme A dehydrogenase deficiency, 62/495
muscle carnitine deficiency, 66/403
muscular atrophy, 42/84
myopathy, 42/660, 43/192
myophosphorylase deficiency, 27/234, 41/185, 62/485
neuralgic amyotrophy, 40/326
neuroborreliosis, 51/203
neurocysticercosis, 40/324
neurolymphomatosis, 56/182
pentaborane intoxication, 64/359
pH, 1/133
phosphohexose isomerase deficiency, 41/192
polyarteritis nodosa, 8/125, 39/305, 55/354, 357
polychlorinated dibenzodioxin intoxication, 64/205
polymyalgia rheumatica, 40/325
polymyositis, 22/25, 38/485, 40/324, 41/56, 58, 62/187, 370
procainamide intoxication, 65/453
rabies, 56/386
Reye syndrome, 65/116
sea snake intoxication, 37/10, 92
spastic, *see* Spastic myalgia
spastic paraplegia, 42/180
Takayasu disease, 55/337
2,3,7,8-tetrachlorodibenzo-*p*-dioxin, 64/205
thalidomide, 42/660
thyrotoxic myopathy, 62/528
toxic myopathy, 62/596
toxic oil syndrome, 63/380
toxic shock syndrome, 52/259
trichinosis, 40/325, 52/565
Venezuelan equine encephalitis, 56/137
viral myositis, 41/53
xanthinuria, 43/192
zimeldine polyneuropathy, 51/304
Myalgia cruras epidemica, *see* Benign acute childhood myositis
Myalgic encephalomyelitis, *see* Postviral fatigue syndrome
Myalgic headache, *see* Muscle contraction headache
Myasthenia
atenolol, 63/195
congenital myopathy, 62/335
familial infantile, *see* Familial infantile myasthenia
familial limb girdle, *see* Familial limb girdle myasthenia
ocular, *see* Ocular myasthenia gravis
progressive muscular atrophy, 59/20
propranolol, 63/195
rheumatoid arthritis, 71/29
Myasthenia gravis, 41/95-136, 62/391-421
see also Myasthenic syndrome
acetylcholine, 41/97, 105, 113, 126
acetylcholine quantum release, 62/392
acetylcholine receptor, 40/129, 41/106, 114, 125, 62/391
acetylcholine receptor antibody, 41/114, 120, 129, 368, 62/408
acetylcholine receptor antibody test, 41/129, 62/414
acetylcholine receptor epitope, 62/407
acetylcholinesterase, 41/107
Addison disease, 41/240
adverse drug reaction, 62/420
age at onset, 41/101, 43/157, 62/399
ambenonium chloride, 41/131
anemia, 62/402
antibody formation, 40/265
anticholinesterase, 41/126, 62/412
anticholinesterase agent, 41/96, 131, 43/158, 62/417

anticholinesterase receptor antibody, 43/158
antigenic modulation, 62/411
Asian race, 62/401
atypical case, 62/79
autoimmune disease, 41/95, 97, 62/391
autoimmune etiology, 62/398
autoimmune experiment, 62/402
azathioprine, 41/96, 133, 136, 62/418
B-lymphocyte response, 62/407
bacterial infection, 62/404
blocking antibody, 62/409
bone marrow transplantation, 62/402
botulism, 41/129, 62/416
α-bungarotoxin, 41/97, 113, 125, 62/409
cancer, 41/361-368
cell mediated immunity, 41/126
central sleep apnea, 63/463
cholinergic crisis, 41/131
chronic ulcerative colitis, 62/402
ciclosporin, 62/418
classification, 41/99, 62/400
clinical course, 62/400
clinical features, 40/297, 303, 306, 308, 317, 323, 330, 335, 41/98, 288, 364-366, 62/399
cobra snake venom intoxication, 41/122
collagen vascular disease, 43/157
complement, 41/119
complement effect, 62/409
Compston classification, 62/401
congenital myasthenic syndrome, 62/415
corticosteroid, 41/93, 132, 62/417
crisis, 43/157, 62/417
cryptococcosis, 52/433
curare, 41/127, 62/413
cyclophosphamide, 41/96, 134, 136, 62/418
decamethonium, 41/113
decremental EMG, 41/128
decremental EMG response, 41/128, 43/158, 62/413
definition, 62/396
diacetate, 41/132
diagnosis, 41/126, 62/412
differential diagnosis, 41/129, 62/415
diplopia, 43/156
dysphagia, 43/157
Eaton-Lambert myasthenic syndrome, 62/391, 402, 415, 421
edrophonium, 41/126, 43/158, 62/412
edrophonium test, 40/298, 303, 321, 41/126, 240, 62/413
elapid snake venom, 62/391
EMG, 1/639, 644, 41/96, 127, 62/398, 413
endocrine myopathy, 41/240, 62/528
ephedrine, 41/132
epidemiology, 41/101, 62/399
experimental, *see* Experimental myasthenia gravis
experimental autoimmune, *see* Experimental autoimmune myasthenia gravis
experimental treatment, 62/420
extraocular muscle weakness, 41/98, 62/399
extrathymic malignancy, 41/366
facioscapulohumeral syndrome, 62/168
false transmitter, 41/113
fatigue, 43/156
germine acetate, 41/132
giant cell polymyositis, 62/402
Hashimoto thyroiditis, 62/534
high dose intravenous Ig, 62/419
history, 41/95, 62/396
HLA antigen, 41/101, 43/158
HLA association, 62/404
HLA haplotype classification, 62/401
HLA loci, 41/101
hyperacusis, 41/98
hypermagnesemia, 63/564
hyperthyroidism, 41/101, 43/157, 62/402, 534, 70/85
hypothyroidism, 62/534, 70/101
immunocytochemistry, 62/415
immunosuppression, 41/132
immunosuppressive agent, 41/96
incidence, 62/399
ion channel effect, 62/409
jitter phenomenon, 41/128
laboratory test, 41/126-219
lid twitch sign, 22/205
lymphoid irradiation, 62/421
lymphorrhage, 41/96
miniature motor end plate potential, 41/129, 62/414
mitochondrial myopathy, 62/415
motor end plate, 40/129, 265, 41/97, 102-107, 43/158
motor end plate potential, 41/110
multiple sclerosis, 47/338
muscle cell membrane, 41/109, 125
muscle fiber type II atrophy, 41/110
muscular atrophy, 43/157
neonatal, *see* Neonatal myasthenia gravis
neostigmine, 41/96, 108, 43/158, 62/417
neostigmine methyl sulfate, 41/127, 131, 62/413
neurasthenia, 62/415
neuroblastoma, 62/415
neurologic intensive care, 55/204
neuromuscular block, 7/109, 40/129, 265, 41/113, 126, 43/158

neuromuscular disease, 43/156
neuromuscular junction disease, 62/392
neuromuscular transmission, 41/95, 107-109, 113, 126, 43/158
nystagmus, 1/598
ocular symptom, 2/323
oculopharyngeal muscular dystrophy, 62/415
Osserman classification, 62/400
paraneoplastic syndrome, 69/374, 382
pathogenesis, 41/111-126, 367, 62/402
pathologic reaction, 40/265
pediatric form, 41/99, 62/400
pemphigus, 62/402
pemphigus vulgaris, 41/102, 240
penicillamine, 41/126, 62/405
pernicious anemia, 41/240
plasmapheresis, 41/96, 136, 62/420
polymyositis, 41/73, 43/157, 62/373, 402
polyneuropathy, 62/413
postactivation exhaustion, 41/356
postactivation facilitation, 41/356
potassium, 41/132
prevalence, 43/157
procainamide, 63/191
progressive bulbar palsy, 59/375, 408
progressive external ophthalmoplegia, 22/181, 188, 204, 62/415
progressive external ophthalmoplegia classification, 22/179, 62/290
progressive spinal muscular atrophy, 59/406
provocative test, 41/127, 62/413
purinethol, 41/133
pyridostigmine, 41/131, 43/158, 62/417
quantum theory, 41/109
quinidine, 63/191
repetitive nerve stimulation, 62/79, 394
respiratory pattern, 63/463
rheumatoid arthritis, 41/240, 62/402
saturating disc model, 62/394
scleroderma, 71/114
sex ratio, 41/101, 43/157, 62/399
single fiber electromyography, 41/128, 62/77, 79, 414
single nerve impulse, 62/393
Sjögren syndrome, 39/430, 41/240, 62/402, 71/74, 89
striated antibody, 62/415
synaptic event, 41/109
synaptic vesicle, 41/97-105, 109
systemic lupus erythematosus, 62/402
T-lymphocyte response, 62/407
thymectomy, 41/96, 134, 43/158, 62/419
thymoma, 41/361, 43/157, 62/402
thymopoietin, 62/406
thymus gland, 43/157, 62/405
thymus hyperplasia, 41/96, 102, 114, 135, 361
thyroid gland, 27/263-267, 41/240, 43/157
thyrotoxic myopathy, 27/263, 41/240, 43/157
transient neonatal, *see* Transient neonatal myasthenia gravis
transmitter release, 62/393
treatment, 41/80, 130-136, 367, 62/416
D-tubocurarine, 41/96, 108, 127
veratrum alkaloid, 41/132
viral infection, 62/404
Myasthenia levis, *see* Ocular myasthenia gravis
Myasthenic myopathic syndrome, *see* Eaton-Lambert myasthenic syndrome
Myasthenic reaction
carbon disulfide intoxication, 64/25
Myasthenic syndrome
see also Eaton-Lambert myasthenic syndrome *and* Myasthenia gravis
acetylcholine, 41/356
acetylcholine quantum release, 62/392
acetylcholine receptor, 62/391
aminoglycoside, 65/483
aminoglycoside intoxication, 65/483
amyotrophic lateral sclerosis, 22/295
anticholinesterase deficiency, 41/130
beta adrenergic receptor blocking agent intoxication, 65/437
botulinum toxin, 64/15
botulism, 51/189
α-bungarotoxin, 64/16
β-bungarotoxin, 64/16
chloroquine, 65/484
chloroquine intoxication, 65/483
congenital, *see* Congenital myasthenic syndrome
curare, 64/16
dipeptidyl carboxypeptidase I inhibitor intoxication, 65/446
Eaton-Lambert, *see* Eaton-Lambert myasthenic syndrome
elapid snake venom, 62/391, 64/16
EMG, 62/78
glycyrrhiza, 64/17
guanidine, 41/356
Hodgkin disease, 39/56
hornet venom, 64/16
α-latrotoxin, 64/16
lithium intoxication, 64/296
malignant lymphoma, 39/53, 56
motor end plate organization, 62/392
neuromuscular block, 40/265, 317, 438
new type, 41/101

organophosphate intoxication, 64/167
phenytoin intoxication, 65/503
repetitive nerve stimulation, 62/394
saturating disc model, 62/394
saxitoxin, 64/15
sea snake venom, 64/16
single nerve impulse, 62/393
snake envenomation, 65/179, 182
tetanospasmin, 64/16
tetrodotoxin, 64/15
tick venom, 64/15
tityustoxin, 64/15
transmitter release, 62/393
Myatrophic ataxia
Friedreich ataxia, 21/355
Mycetoma
cephalosporiosis, 35/562, 52/481
Curvularia, 52/484
curvulariosis, 52/484
fungal CNS disease, 52/479
Pseudoallescheria, 52/491
streptomycosis, 35/570
Mycobacterium
acquired immune deficiency syndrome, 71/285
Mycobacterium avium intracellulare
acquired immune deficiency syndrome, 56/495
Mycobacterium leprae
HLA system, 51/215
immune mechanism, 74/336
macrophage, 51/223
Schwann cell, 51/223
T-lymphocyte, 51/215
Mycobacterium leprae transmission
leprosy, 51/216
Mycobacterium tuberculosis
see also Tuberculous meningitis
acquired immune deficiency syndrome, 56/497
antigen, 52/205
atypical, 52/196
avian type, 52/196
bacterial meningitis, 52/4
brain abscess, 52/150
epidemiology, 52/195
human type, 52/196
neuritis, *see* Tuberculous neuritis
spinal epidural empyema, 52/187
Mycoplasma
acquired myoglobinuria, 62/577
Guillain-Barré syndrome, 47/608
Mycoplasma infection
infantile cerebrovascular disease, 54/40
Mycoplasma pneumoniae
acute cerebellar ataxia, 34/623
bacterial meningitis, 52/4
Guillain-Barré syndrome, 51/240
viral meningitis, 56/130
Mycosis
aspergillosis, 35/374, 395
blastomycosis, 35/380
candidiasis, 35/374
coccidioidomycosis, 35/376-379
cryptococcosis, 35/375
histoplasmosis, 35/376-378
mucormycosis, 35/375
paracoccidioidomycosis, 35/376-379, 531, 52/455
pathogenesis, 35/371-377
pseudo, *see* Pseudomycosis
taxonomy, 35/372
tropical neurology, 35/15-21
Mycosis fungoides
see also Sézary syndrome
brain tumor, 18/235
eosinophilia, 63/371
facial, *see* Midline granuloma
human T-lymphotropic virus type I associated myelopathy, 56/527
neurologic involvement, 38/1-5
spinal epidural tumor, 20/128
Mycotic aneurysm
brain embolism, 11/395, 63/23
brain infarction, 63/23
congenital heart disease, 63/5
diamorphine intoxication, 65/355
infection, 12/80
infectious endocarditis, 63/23, 112
inflammatory response, 11/131
opiate intoxication, 37/391, 65/355
subarachnoid hemorrhage, 12/80, 63/23
Mycotic brain aneurysm
angiography timing, 63/24
antibiotic agent, 52/296
aspergillosis, 52/379
bacterial endocarditis, 52/291
blood culture, 52/295
brain angiography, 52/295
brucella endocarditis, 52/588
candidiasis, 52/401
cardiac valve prosthesis, 52/296
coma, 52/296
computer assisted tomography, 52/295
CSF, 52/295
diagnostic interval, 63/23
headache, 52/294, 296
incidence, 52/295
infectious endocarditis, 55/167, 63/23, 112, 115
late rupture, 63/24

leukemia, 63/344
meningitis, 52/295
neurobrucellosis, 33/308, 314, 52/585
neurosurgery, 52/296
posterior cerebral artery, 52/294
septic embolus, 52/291
site, 52/296
Staphylococcus aureus, 52/295
subarachnoid hemorrhage, 12/112, 52/295
transient ischemic attack, 52/295

Mydriasis
see also Pupil
Amanita pantherina intoxication, 36/534
amphetamine intoxication, 65/51
anticholinergic, *see* Anticholinergic mydriasis
cerebral irritation, 1/620
cyclic oculomotor paralysis, 42/331
gyromitrine intoxication, 65/36
homatropine, 2/108
Horner syndrome, 19/59
hyoscyamine intoxication, 65/48
malignancy, 75/529
mescaline intoxication, 65/46
3,4-methylenedioxymethamphetamine intoxication, 65/51
occipital lobe tumor, 17/336
opiate withdrawal, 65/354
pharmacologic examination, 2/108
phenethylamine intoxication, 65/51
pontine hemorrhage, 2/253
quinidine intoxication, 37/448
scopolamine intoxication, 65/48
scorpion sting intoxication, 37/112
spinocerebellar ataxia, 42/186
superior orbital fissure syndrome, 2/321
temporal lobe epilepsy, 73/73
temporal lobe tumor, 17/292
tricyclic antidepressant intoxication, 65/321
unilateral, 1/622

Myelencephalic vein
ventral, *see* Ventral myelencephalic vein

Myelin
abnormality, 10/138
acid protein, 10/135
acriflavinium chloride, 9/26
aging, 7/47
ali-esterase, 9/33
aminopeptidase, 9/34
anisotropy, 9/27
Baker method, 9/31
bimolecular leaflet, 9/311
biochemistry, 10/134
cerebroside, 7/44, 9/3, 29
chemical aspect, 10/24
chemistry, 7/40
CNS, *see* CNS myelin
composition, 9/311, 66/39
constituent, 9/23-27
decomposition, 10/16
demyelinating disease, *see* Demyelination
desmosterol, 9/18
difference between CNS and nerve, 9/27-30
encephalitogenic proteolipid, 9/13
enzyme, 10/138
enzyme absence, 9/11
fatty acid, 7/45, 9/6
fatty aldehyde, 7/45
fluorochrome, 9/26
formalin fixation, 10/34
α-galactosidase, 9/34
ganglioside, 9/4
globoid cell leukodystrophy, 10/145
β-glucuronidase, 9/34
glycerophospholipid, 9/2
glycoprotein, 9/31
glycosaminoglycan, 9/31
heavy, 9/311
hexosamine, 9/31
histochemistry, 9/31
histophysics, 9/26
immature myelin, 9/40
indicatrix, 9/27
intraperiod line, 9/28
isolation, 9/1
leukodystrophy, 10/134
light, 9/311
lipid fraction, 9/1
lipid metabolism, 10/248
longitudinal arranged phosphoinositide, 9/27
major dense line, 9/28
maple syrup urine disease, 10/138
mass, 9/34
mature myelin, 9/40
metabolic turnover, 66/46
metabolism, 66/3
metachromasia, 9/30
metachromatic leukodystrophy, 10/140
microchemistry, 9/26
molecular model, 9/28
nerve, *see* Nerve myelin
organization, 9/27
organolead intoxication, 64/132
OTAN staining, 9/31
peripheral nervous system, 47/1-5
phenylketonuria, 10/139, 29/34, 42/610
phosphine 3 R, 9/26

phospholipid, 9/30
plasmalogen, 9/2
polysaccharide, 7/52
proteolipid, 9/11, 10/135
proteolysis, 9/11
radially arranged lipid, 9/27
radiation diffraction, 9/27
ribonucleoprotein, 9/34
Schmidt-Lantermann incisure, 9/27
soft X-ray absorption, 9/27
sphingolipid, 7/46, 9/2, 10/136
sphingomyelin, 9/31
stability, 10/49
stripping, 47/437
structure, 9/27
sulfatide, 9/10
tangentially arranged protein, 9/27
unesterified cholesterol, 9/29
X-linked adrenoleukodystrophy, 66/461
Myelin associated glycoprotein
experimental allergic encephalomyelitis, 47/450
IgM paraproteinemic polyneuropathy, 51/437
multiple sclerosis, 47/235
nerve myelin, 51/27
progressive multifocal leukoencephalopathy, 47/512
undetermined significant monoclonal gammopathy, 63/399
Myelin associated protein
nerve myelin, 51/31
neurotropism, 61/479
Myelin basic protein
amino acid, 47/491
anti, *see* Antimyelin basic protein
CSF, *see* CSF myelin basic protein
CSF cytology, 47/376
encephalitogenic, *see* Encephalitogenic myelin basic protein
experimental allergic encephalomyelitis, 10/135, 47/234, 430-433, 442, 446, 450, 468, 474
experimental allergic neuritis, 47/474
immunosuppression, 47/191
multiple sclerosis, 47/65, 112, 191, 235, 239, 359, 372, 376
nerve myelin, 51/25, 31
nuclear histone, 9/514
postinfectious encephalomyelitis, 47/326, 474
progressive multifocal leukoencephalopathy, 47/512
protein P2, 47/492
tryptophan containing fragment, 47/487
Myelin basic protein *m*RNA
nerve myelin, 51/31
Myelin basic protein P1
CNS, 47/5
peripheral nervous system, 47/5
Myelin basic protein P2
Guillain-Barré syndrome, 47/611
peripheral nervous system, 47/5
Myelin breakdown, *see* Demyelination
Myelin constituent
renal insufficiency polyneuropathy, 51/358
wallerian degeneration, 51/33
Myelin decomposition, *see* Demyelination
Myelin deficiency
Pelizaeus-Merzbacher disease, 66/559
Myelin disease
primary, *see* Primary myelin disease
Myelin dysgenesis
see also Demyelination
adducted thumb syndrome, 43/328
anterior horn syndrome, 42/498
argininosuccinic aciduria, 42/525
basal ganglion degeneration, 42/498
Cockayne-Neill-Dingwall disease, 43/350
corpus callosum agenesis, 42/10, 498
Crome syndrome, 43/242
maple syrup urine disease, 42/599
mental deficiency, 42/498-500
metachromatic leukodystrophy, 42/491, 494
microcephaly, 42/498-500
microgyria, 42/498
psychomotor retardation, 42/498
seizure, 42/498-500
spastic cerebral palsy, 42/498
Myelin edema
hereditary multiple recurrent mononeuropathy, 51/558
Myelin figure
enzyme, 9/33
Myelin gene
Pelizaeus-Merzbacher disease, 66/560
Myelin lamella
hereditary multiple recurrent mononeuropathy, 51/556
Myelin lamella widening
paraproteinemia, 63/399
undetermined significant monoclonal gammopathy, 63/399
Waldenström macroglobulinemia, 63/399
Myelin lesion
Guillain-Barré syndrome, 51/251
Myelin like figure
autophagic vacuole, 40/108
drug induction, 40/154
muscle lesion, 40/108

Myelin lipid
biochemistry, 10/135
CNS, 47/5
fatty acid, 10/235
histochemistry, 9/28
leukodystrophy, 10/135
peripheral nervous system, 47/5
Myelin lipoprotein
autoantigen, 9/505
Myelin loop
redundant, *see* Redundant myelin loop
Myelin loss, *see* Demyelination
Myelin poison, 9/656
Myelin protein
chemical composition, 9/31-33
histochemistry, 10/135
organotin intoxication, 64/142
Myelin proteolipid protein
Pelizaeus-Merzbacher disease, 66/561
Myelin remodeling
Schwann cell, 51/18
Myelin sheath
anatomy, 7/40
axon regeneration, 47/21
biochemistry, 9/28
cathepsin, 9/34
chemical composition, 7/40
conductance, 51/41
internode, 47/7
leukoencephalopathy, 47/551, 564
macrophage digestion, 47/226
multiple sclerosis, 9/264
myelin development, 47/21
myelin formation, 47/21
myelin forming cell, 47/21
paraproteinemia, 63/399
plaque, 9/264
protagon hypothesis, 9/30
protein, 7/56, 58
staining, 9/24
structure, 7/18-21, 30, 32, 70, 47/1-5, 66/40
subacute combined spinal cord degeneration, 47/584
toxic disturbance, 47/551, 564
tryptophan, 9/24
undetermined significant monoclonal gammopathy, 63/399
Waldenström macroglobulinemia, 63/399
Myelin storage dystrophy
sphingolipidosis, 21/51
Myelin synthesis
enzyme, 10/138
sulfatide, 10/138
Myelin thickness
nerve, 51/8
nerve conduction velocity, 51/8
Myelin turnover
nerve myelin, 51/32
Myelin vacuolation
organotin intoxication, 64/136
triethyltin, 51/274
vigabatrin intoxication, 65/511
Myelinated fiber loss
amyloid polyneuropathy, 51/418
progressive bulbar palsy, 59/222
Myelinated nerve fiber
diagram, 7/14
function, 47/1-23
multiple sclerosis, 47/218-221
nerve conduction velocity, 47/12, 19
neurofibroma, 14/137
saltatory conduction, 47/12-14
Schwann cell, 7/10
structure, 47/1-23, 51/3
ultrastructure, 7/12, 51/7
Myelinated retinal nerve fiber
Charlevoix-Saguenay spastic ataxia, 60/453
Myelination
antenatal, 9/39
axon, 51/42
Bi-Col procedure, 9/40
biochemistry, 10/136
CNS, 7/31, 41
copper, 9/634
DPN diaphorase, 9/37
familial disorder, 10/101
histochemistry, 9/23-44
lipid, 9/10, 17
malnutrition, 29/6-8
maple syrup urine disease, 29/71
metabolism, 51/28
mixed connective tissue disease, 51/450
myelin composition, 9/40
nerve regeneration, 9/39
neuroglia relationship, 28/407-412
oxidative enzyme, 9/628
phenylketonuria, 29/33, 42/610
physicochemistry, 47/1-9
postnatal, 9/39
satellite cell, 9/40
Schwann cell, 7/24, 51/17
Schwann cell-axon interaction, 51/23
tissue culture, 9/40
ultrastructure, 9/40
white matter, 9/16
Myelination inhibition

lead, 51/266
Myelination ratio
axon, *see* Axon myelination ratio
Myelinoclasis
experimental allergic encephalomyelitis, 9/656
multiple sclerosis, 9/471
postinfectious, 9/650
postinfectious encephalopathy, 9/608
postvaccinal encephalitis, 9/651
Myelinoclastic diffuse sclerosis, 9/469-482, 47/419-426
Addison disease, 9/481
age at onset, 9/474, 47/423
biochemistry, 9/482, 47/426
cholesterol ester, 9/480-482
classification, 10/2, 6
clinical features, 9/474, 47/423
computer assisted tomography, 47/424
CSF examination, 9/479, 47/424
cytoplasmic inclusion body, 9/481
definition, 47/426
differential diagnosis, 9/480
duration, 9/474, 47/423
EEG, 9/475, 47/423
electron microscopy, 9/481, 47/426
ergotamine intoxication, 9/481
evoked potential, 47/424
familial incidence, 9/481
histopathology, 9/479
history, 9/469-471
lead encephalopathy, 9/480
lead intoxication, 9/480
nerve conduction velocity, 47/425
pathology, 47/425
phenylketonuria, 9/481
plasmalogen, 9/482
polysclerotic type, 9/474
progressive type, 9/474
pseudotumoral type, 9/474
psychiatric disorder, 9/474
treatment, 47/425
virus like particle, 9/481
Myelinogenesis
dystopic cortical, *see* Dystopic cortical myelinogenesis
Myelinolysis
central pontine, *see* Central pontine myelinolysis
diuretic agent, 65/439
extra pontine, *see* Extra pontine myelinolysis
hypernatremia, 63/554
hyponatremia, 63/548
neuropathy, 63/548
pontine, *see* Pontine myelinolysis
Myelinopathy
hyperplastic, *see* Hyperplastic myelinopathy
neurotoxicity, 51/263
neurotoxicology, 64/10
Myelinotoxic factor
γ-2-globulin, 9/143
multiple sclerosis, 9/142, 47/366
Myelitis
angiostrongyliasis, 52/555
ascending, *see* Ascending myelitis
bacterial endocarditis, 52/300
cat scratch disease, 52/128
chickenpox, 34/167
chronic, *see* Chronic myelitis
diffuse, *see* Diffuse myelitis
Epstein-Barr virus, *see* Epstein-Barr virus myelitis
Epstein-Barr virus infection, 56/252
herpes genitalis, 71/267
herpes simplex virus, *see* Herpes simplex virus myelitis
herpes zoster, 34/173-175, 71/268
infectious mononucleosis, 56/252
lumbar puncture, 61/162
Lyme disease, 51/205
meningococcal meningitis, 52/26
multicystic necrotizing, *see* Multicystic necrotizing myelitis
necrotizing, *see* Necrotizing myelitis
neuroborreliosis, 51/205
neurobrucellosis, 33/311, 314-316, 52/583, 587
neurologic intensive care, 55/204
opiate intoxication, 37/387
paraneoplastic syndrome, 71/684
paraplegia, 61/161
postpolio, *see* Postpoliomyelitis
postvaccinial transverse, *see* Postvaccinial transverse myelitis
transverse, *see* Transverse myelitis
varicella zoster, 34/167
varicella zoster virus, *see* Varicella zoster virus myelitis
Myelitis necroticans
spinal cord compression, 19/378
Myeloblastic leukemia
acute type, 18/249
age, 18/249
Myeloblastoma
chloroma, 63/343
Myelocele, *see* Spina bifida
Myelocystocele
hydromyelia, 32/235, 50/427
spina bifida, 50/482
Myelocytic leukemia, *see* Myeloid leukemia

Myelocytoma
congenital spinal cord tumor, 32/355-386
Myelodysplasia
caudal aplasia, 32/352-354
neuroacanthocytosis, 63/290
spina bifida, 14/523
thrombocytopenia, 63/322
Myelodysplasia (Fuchs)
dysraphia, 14/111
Myeloencephalopathy
vincristine, 61/151, 63/359
Myelofibrosis
polycythemia, 63/250
polycythemia vera, 63/250
thrombocytopenia, 63/322
Myelography
air myelography, 26/267
angiodysgenetic necrotizing myelomalacia, 20/489-491
arachnoid cyst, 19/189
arachnoiditis, 19/189
Arnold-Chiari malformation type II, 19/188
brachial plexus, 51/148
brain hemorrhage, 61/164
cervical, *see* Cervical myelography
cervical spondylotic myelopathy, 26/104, 107
cervical vertebral column injury, 25/268
coccidioidomycosis, 52/417
complication, 61/147
congenital kyphosis, 32/147
contrast medium choice, 19/182
developmental abnormality, 19/187
diastematomyelia, 19/87, 187, 50/437, 439
diplomyelia, 50/437, 439
epidural ganglion cyst, 20/607
epidural hematoma, 26/22, 25
epilepsy, 61/166
extramedullary spinal tumor, 19/65, 192
foramen magnum tumor, 17/726, 19/193
gamma, *see* Gamma myelography
gas, *see* Gas myelography
hearing loss, 61/165
Hirayama disease, 59/111
hydrosoluble contrast medium, 19/180
intermittent spinal cord ischemia, 12/518
intervertebral disc prolapse, 19/196, 25/499, 506
intracranial subdural hematoma, 61/164
intramedullary spinal tumor, 19/32, 191
ligamentum flavum hypertrophy, 20/812, 814
lumbar intervertebral disc prolapse, 20/581
lumbar vertebral canal stenosis, 20/721
meglumine iocarmate, 19/186
meglumine iothalamate, 19/186
meningocele, 19/187
neurocysticercosis, 52/531
normal appearance, 19/186
oil contrast medium, 19/179
postoperative meningeal diverticulum, 20/162
posttraumatic syringomyelia, 12/607, 26/118, 120, 136, 142-146, 152, 50/454, 61/384
radiation myelopathy, 26/86, 91
spina bifida, 19/87, 187
spinal angioma, 12/632, 19/190, 20/110, 454
spinal arachnoiditis, 19/85
spinal cord abscess, 52/191
spinal cord injury, 25/179, 303, 395, 418, 26/186, 268, 61/515
spinal cord metastasis, 69/179
spinal dermoid, 19/364
spinal ependymoma, 19/360, 20/372
spinal epidermoid, 19/364
spinal epidural empyema, 52/188
spinal epidural hematoma, 26/22, 25
spinal epidural tumor, 19/66, 195, 20/107
spinal lipoma, 20/402
spinal meningioma, 19/68, 358, 20/219
spinal nerve root injury, 25/395, 418
spinal neurinoma, 20/274
spinal perineural cyst, 20/149
spinal subdural empyema, 52/189
spinal tumor, 19/65, 179-203, 67/206
spondylosis, 19/200
spondylotic radiculopathy, 26/104
stab wound, 25/201
subarachnoid hemorrhage, 12/157, 61/164
subdural hematoma, 26/34
syringomyelia, 50/446
technique, 19/182
tethered cord syndrome, 19/189
thoracic intervertebral disc prolapse, 20/568
tuberculous spondylitis, 33/236
vertebral column injury, 61/515
vertebral column tumor, 20/23
vertebral metastasis, 20/420
Myeloid leukemia
acute, *see* Acute myeloid leukemia
age at onset, 18/249
brain hemorrhage, 18/250, 63/343
chemotherapy, 63/356
chlormethine, 63/358
chloroma, 18/247, 63/343
cytarabine, 63/356
encephalitis, 63/352
juvenile chronic, *see* Juvenile chronic myeloid leukemia
meningitis, 63/352

neurologic chemotherapy complication, 63/356
nitrosourea, 63/358
Philadelphia chromosome, 31/435
polyneuropathy, 63/343
radiculopathy, 63/343
spinal nerve root, 63/343
thrombocytopenia, 63/343
vincristine, 63/358
Myeloma
amyloid, 51/413
congenital spinal cord tumor, 32/355-386
demyelinating neuropathy, 41/343
hyperparathyroidism, 20/12
lytic, *see* Lytic myeloma
monoclonal gammopathy, 71/445
multiple, *see* Multiple myeloma
neuropathy, 8/144
osteosclerotic, *see* Osteosclerotic myeloma
paraneoplastic polyneuropathy, 51/468
paraplegia, 20/16
polyneuropathy, 69/304
RES tumor, 18/235
sclerotic, *see* Sclerotic myeloma
solitary, *see* Solitary myeloma
survey, 71/434
systemic brain infarction, 11/461
vertebral, *see* Vertebral myeloma
vertebral compression, 20/11
vertebral pedicle, 20/12
Myeloma polyneuropathy
multiple, *see* Multiple myeloma polyneuropathy
Myelomalacia
see also Central cord necrosis, Ischemic myelopathy *and* Vascular myelopathy
acquired immune deficiency syndrome, 63/55
amyotrophy, 22/19
angiodysgenetic necrotizing, *see* Angiodysgenetic necrotizing myelomalacia
anoxia tolerance, 63/40
anterior spinal artery syndrome, 55/100
aorta arteriosclerosis, 63/41
aorta coarctation, 55/98, 61/112
aorta dissection, 55/98
aorta surgery, 55/99
aortitis, 63/51
arteriosclerosis, 22/20, 55/98, 63/38
arteriovenous malformation, 55/103
arteritis, 63/52
ataxia, 1/319
atrial myxoma, 63/41, 98
bacterial endocarditis, 52/291
blood flow interruption, 63/40
cardiac arrest, 61/114, 63/40, 215
cause, 12/592
central, *see* Central myelomalacia
central hemorrhage, 61/40
central hemorrhagic necrosis, 12/582
central infarction, 12/588
central type, 12/583
chronic venous obstruction, 55/103
course, 61/113
cylindrical infarction, 12/589
cylindrical liquefaction, 12/589
cytomegalovirus, 63/55
decompression illness, 53/160
diabetes mellitus, 63/42
diamorphine myelopathy, 55/519
dissecting aorta aneurysm, 61/112
exercise provocation, 63/40
fibrocartilaginous disc embolus, 63/42
filariasis, 52/516
frequency, 61/111
head injury, 55/100
histiocytic medullary reticulosis, 63/55
iatrogenic lesion, 61/112
iatrogenic neurological disease, 61/112, 63/178
intra-aortic balloon, 63/178
Köhlmeier-Degos disease, 39/443, 63/55
lesion site, 55/98
liquefaction necrosis, 12/589
lumbosacral predilection, 63/40
lymphoma, 63/350
malaria, 63/42
malignant histiocytosis, 63/55
Marfan syndrome, 55/98
myxedema, 55/98
nuclear magnetic resonance, 63/44
paraplegia, 61/111
pathogenesis, 12/584
polyarteritis nodosa, 55/102
posttraumatic syringomyelia, 12/599, 26/130, 154
pregnancy, 55/98
prevalence, 63/40
radiation myelopathy, 12/501, 61/202
recovery, 61/113
schistosomiasis, 63/42
sickle cell trait, 55/466
spastic paraplegia, 59/437
spinal angioma, 55/103
spinal artery thrombosis, 63/38
spinal cord injury, 25/80, 84
spinal cord intermittent claudication, 63/40
spinal cord trauma, 12/557
spinal cord vascular disease, 12/582
sympathectomy, 55/100
symptom, 61/113

systemic hypotension, 61/112
systemic lupus erythematosus, 55/102
treatment, 63/44
vascular lesion, 12/500
vasculitis, 63/51
venous infarction, 55/102
venous infarction symptom, 55/102
venous thrombotic pathology, 55/103
vertebral artery occlusion, 55/100
Wegener granulomatosis, 63/55
Myélomalacia en crayon
artery occlusion, 12/498
Myelomatosis, *see* Multiple myeloma
Myelomeningocele
see also Spina bifida
Arnold-Chiari malformation type II, 50/484
Dandy-Walker syndrome, 42/23
dermoid cyst, 20/76
diastematomyelia, 42/25
double, *see* Spina bifida
epidermoid, 20/76
Gruber syndrome, 14/115, 43/391
infantile enteric bacillary meningitis, 33/61
spina bifida, 50/480, 487, 490
triploidy, 43/566
Myeloneuropathy
tropical, *see* Tropical myeloneuropathy
Myelo-optic neuropathy, *see* Neuromyelitis optica
Myelopathy
acquired immune deficiency syndrome, 56/491, 493
adrenoleukodystrophy, 51/384
alcoholic, *see* Alcoholic myelopathy
amyotrophic, *see* Amyotrophic myelopathy
anterior spinal artery beading, 63/43
anterolateral, *see* Anterolateral myelopathy
aorta arteriosclerosis, 63/41
aortic vasculitis, 63/51
arteriosclerosis, 12/501, 532, 22/20
arteriosclerotic, *see* Arteriosclerotic myelopathy
atlantoaxial dislocation, 12/609
atrial myxoma, 63/41
Babinski sign, 42/156
bronchial arteriography, 63/44
calcium, 70/120
carbon disulfide intoxication, 64/8
cervical, *see* Cervical myelopathy
cervical intervertebral disc prolapse, 7/456
cervical spondylarthrotic, *see* Cervical spondylarthrotic myelopathy
cervical spondylotic, *see* Cervical spondylotic myelopathy
chemotherapy, 67/355
ciclosporin intoxication, 65/552
clinical features, 59/448-450
clioquinol intoxication, 64/8
combined, *see* Combined myelopathies
cyanogenic glycoside intoxication, 64/8
cystic, *see* Posttraumatic syringomyelia
cytarabine, 63/357, 65/528, 67/362
cytarabine intoxication, 65/532
cytidine analog, 67/362
decompression, *see* Decompression myelopathy
delayed, *see* Delayed myelopathy
diabetes mellitus, 27/121-123
diamorphine, *see* Diamorphine myelopathy
diamorphine addiction, 65/356
diphyllobothrium latum, *see* Diphyllobothrium latum myelopathy
disulfiram intoxication, 64/8
familial spastic paraplegia, 59/306
granulomatous CNS vasculitis, 55/390
hepatic, *see* Hepatic myelopathy
hepatic insufficiency, 27/367
Hirayama disease, 59/113
human T-lymphotropic virus type I, 59/306, 65/356, 75/414
human T-lymphotropic virus type II, 65/356
hyperparathyroidism, 27/293, 70/120
hyperreflexia, 42/156
irradiation, *see* Irradiation myelopathy
ischemic, *see* Ischemic myelopathy
lead, *see* Lead myelopathy
leukemia, 63/354
liver disease, 27/367
macroglobulinemia, 39/192
malabsorption syndrome, 70/231
methotrexate, 65/528
methotrexate intoxication, 39/100, 65/530
motor unit hyperactivity, 41/307
necrotic, *see* Necrotic myelopathy
neurobrucellosis, 33/311, 314-316
neurofibromatosis, 14/146
neuromyelitis optica, 9/426
neurotoxicology, 64/10
nitrofurantoin intoxication, 65/487
nitrous oxide, 63/256
nutritional, *see* Nutritional myelopathy
opiate addiction, 65/356
pellagra, 28/59
phosphorus, 70/120
postgastrectomy syndrome, 28/233, 51/327
postlateral, *see* Postlateral myelopathy
progressive necrotic, *see* Foix-Alajouanine disease
progressive senile vascular, *see* Progressive senile vascular myelopathy

radiation, *see* Radiation myelopathy
radiation injury, *see* Radiation myelopathy
radiotherapy, 39/51
sarcoidosis, 71/480
scleroderma, 71/114
Sjögren syndrome, 63/55, 71/74
skeletal fluorosis, 36/476
spinal artery embolism, 63/41
spinal cord atherosclerosis, 63/41
spongiform, *see* Spongiform myelopathy
subacute combined spinal cord degeneration, 28/178
subacute myelo-optic neuropathy, 51/296
systemic lupus erythematosus, 39/282, 55/378, 63/54, 71/45
Takayasu disease, 63/53
temporal arteritis, 63/54
tertiary, *see* Tertiary myelopathy
thallium intoxication, 64/8
thiotepa, 63/359, 65/528
thiotepa intoxication, 65/534
toxic, *see* Toxic myelopathy
transverse, *see* Transverse myelopathy
traumatic, *see* Traumatic myelopathy
tropical ataxic, *see* Tropical ataxic myelopathy
tropical ataxic neuropathy, 7/641, 51/322
tuberculous meningitis, 33/233, 52/216
vascular, *see* Vascular myelopathy
Waldenström macroglobulinemia, 63/396
Waldenström macroglobulinemia polyneuropathy, 39/192, 532
Wegener granulomatosis, 71/179

Myeloradiculopathy
amyotrophic spondylotic, *see* Amyotrophic spondylotic myeloradiculopathy
influenza virus, 51/187

Myeloschisis
Arnold-Chiari malformation type II, 50/405
spina bifida, 32/527, 536

Myeloscintigraphy
abnormal picture, 19/255-258
CSF pressure, 19/118
disadvantage, 19/261
normal picture, 19/255
photoscintigraphy, 19/249
practice, 19/247
RIHSA, 19/247-255
side effect, 19/259
spinal angioma, 20/455
spinal epidural tumor, 20/108
spinal tumor, 19/67, 245-265
styloscintigram, 19/249
technique, 19/251
theory, 19/246

Myelosis
combined funicular, *see* Subacute combined spinal cord degeneration
funicular, *see* Subacute combined spinal cord degeneration

Myelotomy
pain, 26/497
spastic paraplegia, 61/369
spasticity, 26/486
spasticity treatment, 61/369
spinal cord experimental injury, 25/22
spinal cord injury, 25/305, 61/369
spinal pain, 26/497
spinal spasticity, 61/369
syringomyelia, 50/459
transverse spinal cord lesion, 61/369

Myenteric ganglion
anatomy, 75/615

Myenteric plexus
anatomy, 1/503, 75/615
Auerbach, *see* Auerbach myenteric plexus
GM2 gangliosidosis, 10/406
hereditary sensory and autonomic neuropathy type III, 21/113, 60/30
intestinal pseudo-obstruction, 51/491-493, 70/320

Myerson sign
manganese intoxication, 64/317

Myesthesia
terminology, 1/93

Myliobatidae intoxication
elasmobranche, 37/70

Myoadenylate deaminase, *see* Adenosine monophosphate deaminase

Myoadenylate deaminase deficiency, *see* Adenosine monophosphate deaminase deficiency

Myoblast
culture, 40/184
development, 40/199

Myoblastoma
intrasellar abscess, 17/426
pituitary tumor, 17/425

Myocardial disease, 11/436, 441

Myocardial infarction
anticoagulant, 12/458
autonomic nervous system, 74/171, 75/444
brain embolism, 11/388, 53/155
brain infarction, 53/155
brain ischemia, 53/155
cardiac arrest, 63/207
cocaine addiction, 55/522
critical illness polyneuropathy, 51/575
diazoxide intoxication, 37/433

ergot intoxication, 65/69
glyceryl trinitrate, 37/453
myoglobinuria, 62/556
oral contraception, 55/525
polyarteritis nodosa, 11/451, 55/354
pseudoxanthoma elasticum, 55/452
scorpion sting intoxication, 37/112
shoulder hand syndrome, 1/513, 8/330
sleep, 74/553
stroke, 55/161
subarachnoid hemorrhage, 55/9
systemic brain infarction, 11/436
temporal arteritis, 48/316

Myocardial ischemia
antihypertensive agent, 63/74
antihypertensive treatment, 63/74
autonomic nervous system, 74/172
CNS lesion, 63/233

Myocardial sclerosis
cerebellar ataxia, 42/131
Jeune-Tommasi disease, 22/516
sensorineural deafness, 42/131

Myocardium
diphtheria, 51/182
infantile spinal muscular atrophy, 59/67
Reye syndrome, 56/157

Myoclonia, 6/761-778
adrenoleukodystrophy, 60/169
amphetamine intoxication, 65/257
Bardet-Biedl syndrome, 13/401
benzodiazepine intoxication, 65/339
benzodiazepine overdose, 65/339
bismuth intoxication, 64/337
brain stem, 6/775
calcium antagonist intoxication, 65/442
chlorinated cyclodiene, 64/200
chlorinated cyclodiene intoxication, 64/200
chorea electrica, 6/769
Creutzfeldt-Jakob disease, 6/731, 764
dentatorubropallidoluysian atrophy, 49/439
dialysis encephalopathy, 64/277
Divry-Van Bogaert syndrome, 55/322
dyssynergia cerebellaris myoclonica, 6/766, 776
encephalitis, 6/764
encephalitis lethargica, 6/763
essential type, 6/763, 767-772
facial, *see* Facial myoclonia
fetal Minamata disease, 64/418
globoid cell leukodystrophy, 10/72
leukodystrophy, 6/764
lipidosis, 6/765
1-methyl-4-phenyl-1,2,3,6-tetrahydropyridine, 65/369
neuronal ceroid lipofuscinosis, 10/550
nifedipine intoxication, 65/442
ocular, *see* Ocular myoclonia
organochlorine insecticide intoxication, 64/200
organophosphate intoxication, 64/167
pathophysiology, 6/765
pertussis vaccine, 52/242
presenile astroglial dystrophy, 21/67
progressive myoclonus epilepsy, 6/765, 15/334
r(14) syndrome, 50/588
seizure, 73/224
striatopallidodentate calcification, 49/422
subacute sclerosing panencephalitis, 56/419
tremor, 6/772-774
tuberculous meningitis, 55/426
type, 6/761, 763
water intoxication, 64/241

Myoclonia congenita, *see* Hereditary essential myoclonus

Myoclonic akinetic epilepsy
primary generalised epilepsy, 42/682
pyknolepsy, 42/671
theta rhythm abnormality, 42/667

Myoclonic astatic epilepsy
idiopathic generalized, *see* Idiopathic generalized myoclonic astatic epilepsy
pyknolepsy, 42/671

Myoclonic ataxia
acute, *see* Acute myoclonic ataxia

Myoclonic dystonia
hereditary, *see* Hereditary myoclonic dystonia

Myoclonic dystonia musculorum deformans, 49/522, 525, 618

Myoclonic encephalopathy
acute cerebellar ataxia, 34/629-632
early, *see* Early myoclonic encephalopathy
infantile, *see* Infantile myoclonic encephalopathy
postanoxic, *see* Posthypoxic myoclonus

Myoclonic epilepsy
see also Progressive myoclonus epilepsy
adult GM1 gangliosidosis, 60/663
adult neuronal ceroid lipofuscinosis, 38/579, 60/662
ataxia, 42/114
Baltic, *see* Baltic myoclonus epilepsy
behavior disorder, 42/702
behavior disorder in children, 42/702
bilateral massive myoclonus, 15/126
blindness, 42/702
brain atrophy, 42/700
brain edema, 42/701
cardiac arrest, 63/214, 217
cerebellar ataxia, 42/698

cerebello-olivary atrophy, 60/663
choreoathetosis, 42/699
chorioretinal degeneration, 13/39
dentate nucleus, 42/701
dentatorubropallidoluysian atrophy, 49/440, 60/608
Divry-Van Bogaert syndrome, 55/322
dyssynergia cerebellaris myoclonica, 42/211
early childhood, *see* Early childhood myoclonic epilepsy
EEG, 15/513, 27/174, 42/701, 703, 705-707
Farber disease, 60/166
generalized epilepsy, 15/121-128
GM2 gangliosidosis, 49/617
hallucination, 42/702
Hartung type, 42/700
hereditary progressive, *see* Unverricht-Lundborg progressive myoclonus epilepsy
hydroxylysinuria, 42/704
infantile benign, *see* Infantile benign myoclonic epilepsy
juvenile, *see* Juvenile myoclonic epilepsy
Kawasaki syndrome, 52/265
Lafora body, 42/702
Lafora progressive myoclonus epilepsy, 60/662
Lennox-Gastaut syndrome, 73/263
Mediterranean myoclonus, 60/663
mental deficiency, 42/704
MERRF syndrome, 60/662
Miller-Dieker syndrome, 50/592
Minamata disease, 64/418
Minkowitz disease, 60/662
mitochondrial encephalomyopathy, 60/662
mucolipidosis type I, 49/617, 60/662
mucolipidosis type II, 60/662
myoclonic absence, 15/135
neuronal ceroid lipofuscinosis, 27/188, 49/617, 60/662
newborn, 15/195
nonprogressive encephalopathy, 73/264
opsoclonus, 38/582
pallidal degeneration, 6/650
pathophysiology, 6/775
photogenic epilepsy, 22/510
photomyoclonus syndrome, 60/663
primary pigmentary retinal degeneration, 13/338
progressive, *see* Progressive myoclonus epilepsy
Ramsay Hunt syndrome, 42/211, 49/616, 60/593, 663
seizure, 73/224
sensorineural deafness, 42/698
spinocerebellar neuroaxonal dystrophy, 60/663
status epilepticus, 15/160
tuberous sclerosis, 14/347, 496
Unverricht-Lundborg progressive myoclonus epilepsy, 60/662
X-linked hypogammaglobulinemia, 60/663
Myoclonic epilepsy syndrome
classification, 73/262
Myoclonic jerk
aluminum intoxication, 63/525
ataxia telangiectasia, 60/361
calcium antagonist intoxication, 65/442
globoid cell leukodystrophy, 10/303
hypernatremia, 63/554
infantile neuronal ceroid lipofuscinosis, 66/689
juvenile myoclonic epilepsy, 73/160
Lafora progressive myoclonus epilepsy, 15/384, 27/173
levodopa induced dyskinesia, 49/195
massive, *see* Massive myoclonic jerk
May-White syndrome, 22/511, 42/698
nifedipine intoxication, 65/442
nimodipine intoxication, 65/442
respiratory encephalopathy, 63/414
sex ratio, 73/159
Myoclonic spasm
infantile, *see* Infantile myoclonic spasm
Myoclonic spinal neuronitis
subacute, *see* Subacute myoclonic spinal neuronitis
Myoclonic status
generalized, *see* Generalized myoclonic status
Myoclonic torsion dystonia, *see* Myoclonic dystonia musculorum deformans
Myoclonus, 38/575-588
acquired immune deficiency syndrome dementia, 56/491
action, *see* Action myoclonus
acute viral infection, 38/583
adult GM1 gangliosidosis, 60/662
adult neuronal ceroid lipofuscinosis, 42/466
Alpers disease, 42/486
aluminum neurotoxicity, 63/525
Alzheimer disease, 46/252, 49/618
α-amino-n-butyric acid, 13/339
amiodarone intoxication, 65/458
amnesic shellfish poisoning, 65/167
antiepileptic agent intoxication, 60/662
asterixis, 49/622
ataxia, 21/510, 583
benign neonatal sleep, 49/622
bilateral massive, *see* Bilateral massive myoclonus
bilateral striatal necrosis, 21/514
bismuth, 64/4
bone membranous lipodystrophy, 42/279

brain atrophy, 42/700
brain injury, 38/580-582
bulbar, 49/614
C-reflex, 49/611
carbamazepine intoxication, 65/506
cause, 49/611
cerebellar ataxia, 21/510-512, 60/597
cerebellar vermis agenesis, 42/4
child, 49/619
classification, 38/576, 49/611
CNS spongy degeneration, 42/506
complex II deficiency, 66/422
cortical, *see* Cortical myoclonus
Creutzfeldt-Jakob disease, 6/166, 732, 741-745, 38/582, 42/657, 46/291, 49/618, 56/545, 60/662
deafness, 60/657
decompression illness, 49/618
definition, 49/609
dentate nucleus, 21/510, 42/701
dentatorubropallidoluysian atrophy, 60/662
dialysis, 63/525
dialysis disequilibrium syndrome, 63/523
distribution, 49/611
dominant ataxia, 60/660
dominant olivopontocerebellar atrophy, 60/662
dyssynergia cerebellaris myoclonica, 21/511, 38/578, 49/616, 60/597, 599, 662
dystonia, 42/240
EEG, 49/610
Ekbom syndrome type II, 60/662
electric shock, 49/618
EMG, 1/646, 49/610
enflurane intoxication, 37/412
epilepsy, 15/121, 38/579, 73/235
essential, *see* Essential myoclonus
essential startle disease, 38/577
extrapyramidal, *see* Extrapyramidal myoclonus
extrapyramidal disorder, 2/344
facial nerve, 8/274
facial spasm, 42/220
familial, 38/576-579
fasciculation, 42/240
focal, *see* Focal myoclonus
generalized, *see* Myoclonic epilepsy
Gerstmann-Sträussler-Scheinker disease, 56/553, 60/662
Hallervorden-Spatz syndrome, 6/166, 618, 21/514, 49/618
Haw River syndrome, 60/662
heat stroke, 49/618
hereditary, *see* Hereditary myoclonus
hereditary ataxia, *see* Hereditary myoclonus ataxia syndrome
hereditary benign, *see* Hereditary benign myoclonus
hereditary essential, *see* Hereditary essential myoclonus
hereditary olivopontocerebellar atrophy, 60/662
herpes zoster oticus, 38/578
heterocyclic antidepressant, 65/319
Huntington chorea, 21/514, 49/618, 60/662
hydranencephaly, 30/675, 50/349
hypercalcemia, 63/561
hyperexplexia, 38/577
hypernatremia, 63/554
hyperventilation syndrome, 63/435
hypomagnesemia, 63/562
hypoxia, 49/618
inclusion body encephalitis, 6/166
infantile, *see* Infantile myoclonus
infantile myoclonic encephalopathy, 38/583
infantile spasm, 38/579, 49/619
intention, *see* Action myoclonus
jumpers of the Maine, 49/621
juvenile myoclonic epilepsy, 49/619
kinesigenic, *see* Kinesigenic myoclonus
Koshevnikoff epilepsy, 38/584
kwashiorkor, 7/635, 51/322
latah, 49/621
late cerebellar atrophy, 60/662
lathyrism, 65/3
Lennox-Gastaut syndrome, 49/619
levodopa, 38/582
levodopa induced dyskinesia, 49/196, 199
lipidosis, 38/579
lithium intoxication, 64/296
local anesthetic agent, 65/420
maprotiline intoxication, 65/319
Marinesco-Sjögren syndrome, 60/344
massive, *see* Massive myoclonus
measles encephalitis, 56/480
Mediterranean, *see* Mediterranean myoclonus
mental deficiency, 42/240
mercury intoxication, 64/394
MERRF syndrome, 49/618, 62/509
metabolic encephalopathy, 49/618, 69/399
mucolipidosis type I, 42/475, 51/380
mucolipidosis type II, 51/380
multifocal, *see* Myoclonic epilepsy
myokymia, 42/236
myriachit, 49/621
neonatal, *see* Neonatal myoclonus
neuroaxonal dystrophy, 49/618
neuroblastoma, 21/574
neurolipidosis, 21/514
neuropathy, 60/660

nocturnal, *see* Nocturnal myoclonus
noninfantile neuronopathic Gaucher disease, 49/618
non-REM sleep, 49/622
norpethidine intoxication, 65/357
nystagmus, *see* Nystagmus myoclonus
ocular, *see* Ocular myoclonia
olivopontocerebellar atrophy, 60/599
olivopontocerebellar atrophy variant, 21/452
olivopontocerebellar atrophy (Wadia-Swami), 60/662
opsoclonus, *see* Opsoclonus myoclonus
organic acid metabolism, 66/641
origin, 1/279
oscillatory, *see* Oscillatory myoclonus
palatal, *see* Palatal myoclonus
pallidoluysiodentate degeneration, 49/452, 456
paraneoplastic syndrome, 71/691
Parry disease, 42/467
pathophysiology, 6/164-169, 38/585
penicillin, 38/582
peripheral, *see* Peripheral myoclonus
pethidine intoxication, 65/357
phenytoin intoxication, 65/500
pleomorphism, 1/281
polymyoclonia, 38/583
pontine infarction, 53/389
postanoxic, *see* Posthypoxic myoclonus
postanoxic intention, *see* Postanoxic intention myoclonus
posthypoxic, *see* Posthypoxic myoclonus
posture, *see* Postural myoclonus
progressive myoclonus epilepsy, 27/172, 38/577
progressive rubella panencephalitis, 34/338, 56/409
progressive supranuclear palsy, 49/618
Ramsay Hunt syndrome, 60/662
reflex, *see* Reflex myoclonus
renal insufficiency, 27/324-326, 330
respiratory encephalopathy, 63/414
restless legs syndrome, 49/621, 51/356
reticular, *see* Reflex myoclonus
reticular reflex, *see* Reflex myoclonus
rhythmical myoclonus, 38/583
secondary pigmentary retinal degeneration, 13/339
situs inversus, 42/240
sleep, *see* Nocturnal myoclonus
source, 49/611
spinal, *see* Spinal myoclonus
spiromustine intoxication, 65/534
spontaneous, *see* Spontaneous myoclonus
startle, *see* Startle myoclonus
startle syndrome, 49/619
status epilepticus, 15/160
stimulus sensitive, 21/510, 513
subacute sclerosing panencephalitis, 38/584, 49/618
substrate, 21/515
tardive dyskinesia, 49/186
thalamic dementia, 21/591
toxic encephalopathy, 49/618
toxic oil syndrome, 63/381
toxin, 38/581
trauma, 49/618
treatment, 38/586-588
tricyclic antidepressant, 38/582
tricyclic antidepressant intoxication, 65/321
Unverricht-Lundborg progressive myoclonus epilepsy, 15/334, 27/172
uremic encephalopathy, 63/504
uric acid, 21/511
valproic acid intoxication, 65/505
velopalatine, *see* Velopalatine myoclonus
vertical ocular, *see* Vertical ocular myoclonus
vidarabine intoxication, 65/534
viral encephalopathy, 49/618
vitamin Bw deficiency, 49/618
Whipple disease, 52/137
Wilson disease, 21/514, 49/618, 60/662
withdrawal syndrome, 6/166
Myoclonus ataxia syndrome
hereditary, *see* Hereditary myoclonus ataxia syndrome
opsoclonus, *see* Opsoclonus myoclonus ataxia syndrome
Myoclonus body, *see* Lafora body
Myoclonus epilepsy with ragged red fiber, *see* MERRF syndrome
Myoclonus fibrillaris multiplex, *see* Morvan fibrillary chorea
Myoedema
hypothyroid myopathy, 62/532
hypothyroidism, 40/330, 41/245
malabsorption syndrome, 70/227
myxedema neuropathy, 8/59
percussion, *see* Percussion myoedema
Myofascial pain dysfunction syndrome
psychogenesis, 48/414
Myofibril, *see* Muscle fiber
Myofibrillar adenosine triphosphatase
disease specific alteration, 40/8-43
fiber type, 40/3-8
normal histochemistry, 40/3, 136
Myofibrillar lysis myopathy
congenital myopathy, 62/332

enzyme histochemistry, 62/339
histopathology, 62/339
protein synthesis, 62/339
Myogelosis
muscle cramp, 41/385
painful, 41/385
Myoglobin
antibody, 41/262
chemical property, 41/259, 261
function, 41/262
immunochemistry, 41/260
laboratory test, 40/338
preparation, 41/259
sea snake intoxication, 37/10
Myoglobinemia
alcoholic polyneuropathy, 41/272
Duchenne muscular dystrophy, 41/275
Myoglobinuria
see also Meyer-Betz disease *and* Rhabdomyolysis
aconitate hydratase, 62/558
aconitate hydratase deficiency, 62/503
acquired, *see* Acquired myoglobinuria
acute tubular necrosis, 40/338, 41/276
adenosine monophosphate deaminase, 62/558
adenosine monophosphate deaminase deficiency, 62/512
alcohol, 41/272
alcohol intoxication, 40/339
alcoholic polyneuropathy, 40/162, 339
aminocaproic acid intoxication, 40/339
amphotericin B intoxication, 40/339
assay sensitivity, 62/555
asymptomatic creatine kinasemia, 62/554
azacitidine, 41/271
azathioprine, 41/271
barbiturate, 40/339, 41/270
Becker muscular dystrophy, 62/556, 558
biochemical defect, 62/560
α-bungarotoxin, 64/16
β-bungarotoxin, 64/16
carnitine cycle, 66/402
carnitine palmitoyltransferase, 62/558
carnitine palmitoyltransferase deficiency, 40/339, 41/204, 265, 277, 428, 43/176, 62/494
cause, 62/557
central core myopathy, 40/339, 62/558
characteristics, 40/338-340
child, *see* Childhood myoglobinuria
childhood *vs* adult onset, 62/565
chloroquine intoxication, 40/339
chondrodystrophic myotonia, 62/558
classification, 41/277, 62/558
clinical diagnosis, 41/276, 62/554
clinical features, 41/263, 288
coma, 40/338
coma crush, 41/277, 62/556
complex II, 62/558
complex II deficiency, 62/504
complication, 41/276
concomitant, 62/556
consequence, 62/556
Coxsackie virus, 41/275, 56/199
creatine kinase, 40/263, 339, 41/263, 271, 62/554, 556, 560
crush, 40/339, 41/270, 62/557
curare, 64/16
definition, 62/553
dermatomyositis, 40/339, 41/277, 62/556
diagnosis, 41/276
diagnostic criteria, 62/555
diamorphine, 41/271
diamorphine intoxication, 65/357
differential diagnosis, 40/338-340, 41/167, 62/560
drug, 41/271, 62/557
Duchenne muscular dystrophy, 62/555, 558
elapid snake venom, 64/16
electrical injury, 61/192
exercise, *see* Exercise myoglobinuria
exertion, 40/338, 41/266, 62/557, 560
facioscapulohumeral muscular dystrophy, 62/556
familial malignant hyperthermia, 62/558
familial recurrent, *see* Familial recurrent myoglobinuria
fatty acid oxidation, 41/277, 62/558
general anesthesia, 40/339
glucose-6-phosphate dehydrogenase, 62/558
glycogen storage disease, 41/265, 62/558
Haff disease, 40/339, 41/272
headache, 40/338, 65/178
heat, 41/266
heat stroke, 40/339
heme induced nephropathy, 62/556
hereditary cause, 62/558
hereditary disease, 62/559
histochemistry, 40/54
histology, 40/263
history, 62/553
hornet venom, 62/575, 64/16
hypercalcemia, 62/556
hypernatremia, 41/274
hyperphosphatemia, 62/556
hypocalcemia, 62/556
idiopathic, 41/276
incomplete dystrophin deficiency, 62/135
infection, 41/275, 277, 62/557
influenza virus, 56/199

ischemia, 40/338, 41/270, 277, 62/557
King-Denborough syndrome, 62/558
laboratory diagnosis, 41/276
lactate dehydrogenase, 62/558
lactate dehydrogenase deficiency, 62/490
α-latrotoxin, 64/16
lipid metabolic disorder, 41/207
long chain acyl-coenzyme A dehydrogenase, 62/558
long chain acyl-coenzyme A dehydrogenase deficiency, 62/495
long chain fatty acid oxidation, 62/558
malignant hyperthermia, 38/551, 40/339, 41/263, 268, 277, 62/567, 63/439
mannitol, 62/556
metabolic disorder, 62/557
methanol intoxication, 64/98
mitochondrial matrix, 66/410
motoneuron disease, 62/556
multiple mitochondrial DNA deletion, 62/510
muscle carnitine deficiency, 66/403
muscle fasciotomy, 62/556
muscle necrosis, 41/263, 62/554
muscle pathology, 62/556
myocardial infarction, 62/556
myoglobin identification, 40/263, 62/554
myopathy, 41/274
myophosphorylase deficiency, 27/233, 40/339, 41/186, 265, 43/181, 62/485, 558
myotonia congenita, 62/558
myotonic dystrophy, 40/339, 62/556, 558
nomenclature, 62/553
opiate intoxication, 65/357
parathyroid hormone, 62/556
pathogenesis, 62/556
pentose phosphate pathway, 62/558
periodic paralysis, 41/167
phencyclidine intoxication, 41/271
phosphofructokinase, 62/558
phosphofructokinase deficiency, 27/235, 40/327, 339, 41/190, 265, 277, 43/183, 62/487
phosphoglycerate kinase, 62/558
phosphoglycerate kinase deficiency, 62/489
phosphoglycerate mutase, 62/558
phosphorylase kinase, 62/558
plasmapheresis, 62/556
plasmocid, 40/339, 41/271
plasmocid intoxication, 40/339
polymyositis, 40/339, 41/58, 60, 274, 277, 62/556, 577
postexercise, 62/556
precipitating factor, 62/559
progressive muscle disease, 62/557
quail eater disease, 40/339, 41/272
recurrent sporadic, 62/558
renal dialysis, 62/556
rhabdomyolysis, 62/553
salt and water imbalance, 41/273
sea snake intoxication, 37/92
sea snake venom, 62/575, 64/16
sex ratio, 62/559
short chain 3-hydroxyacyl-coenzyme A dehydrogenase, 62/558
short chain 3-hydroxyacyl-coenzyme A dehydrogenase deficiency, 62/495
snake envenomation, 65/179
strychnine, 41/271
succinylcholine intoxication, 40/339
symptom, 62/554
systemic lupus erythematosus, 40/339, 41/275, 277
tetanospasmin, 64/16
thermoregulation, 62/557
thyrotoxic myopathy, 62/529
toxic shock syndrome, 62/574
toxin, 41/271, 62/557
treatment, 62/556
trichinosis, 52/566
type, 43/189
ubidecarenone, 62/504
ubidecarenone deficiency, 66/423
ubiquinone deficiency, 62/558
vincristine intoxication, 40/339
viral myositis, 41/275

Myograph recording
nerve stimulation, 7/106

Myography
isometric, *see* Isometric myography

Myoinositol, *see* Inositol

Myokinesia
EMG, 1/646

Myokymia
anatomic classification, 1/279
black widow spider venom intoxication, 40/330
carbamazepine, 41/300
EMG, 40/330, 546
eosinophilia myalgia syndrome, 64/256
facial, *see* Facial myokymia
facial clonus, 1/280, 285, 2/62
fasciculation, 41/299
gold intoxication, 51/306, 64/13
hereditary, *see* Hereditary myokymia
hereditary periodic ataxia, 60/433
horizontal gaze paralysis, 42/333
hyperhidrosis, 43/160
Isaacs syndrome, 42/237, 43/160

lathyrism, 65/3
Morvan fibrillary chorea, 49/559
motor unit hyperactivity, 41/299
multiple sclerosis, 42/237
muscle cramp, 40/330, 546
myoclonus, 42/236
myotonia, 43/160, 62/273
myotonic syndrome, 62/273
nerve injury, 49/609
periodic ataxia, 60/659
striatonigral degeneration, 42/262
terminology, 1/279
thyrotoxic myopathy, 62/528
toxic myopathy, 62/596
Myokymia (Schultze), *see* Morvan fibrillary chorea
Myokymic discharge
definition, 62/263
Isaacs syndrome, 62/263
myotonia, 62/263
myotonic syndrome, 62/263
Myolipoma
kidney, 14/370
Myomatosis
pulmonary, *see* Pulmonary myomatosis
Myoneural junction, *see* Motor end plate
Myopathy
see also Neuromuscular disease
abnormal myomuscular junction, *see* Abnormal myomuscular junction myopathy
acquired immune deficiency syndrome, 56/518
acromegalic, *see* Acromegalic myopathy
ACTH induced, *see* ACTH induced myopathy
acute necrotizing, *see* Acute necrotizing myopathy
Addison disease, 39/477
adrenal insufficiency, *see* Adrenal insufficiency myopathy
adult acid maltase deficiency, 27/229, 62/480
adult dominant myotubular, *see* Adult dominant myotubular myopathy
alcoholic, *see* Alcoholic polyneuropathy
alkaline phosphatase, 43/126
aminoaciduria, 43/125-127
amiodarone, *see* Amiodarone myopathy
amiodarone intoxication, 65/458
amyloidosis, 71/506
amyotrophic lateral sclerosis, 59/201
ankylosing spondylitis, 70/7
antiepileptic agent, 65/499
antihypertensive agent, 63/87
arthrogryposis, 40/337
arthrogryposis multiplex congenita, 32/512, 515, 40/337, 43/77
artificial respiration, 63/420
asthma, 63/420
ataxia, 43/124
atrophy, 62/13
atrophy cause, 62/13
axonal stimulation, 62/59
Babinski sign, 43/123
Bassen-Kornzweig syndrome, 60/132
Behçet syndrome, 34/486
benign dominant, *see* Benign dominant myopathy
beta adrenergic receptor blocking agent, 63/87
beta adrenergic receptor blocking agent intoxication, 65/437
cachetic, *see* Cachetic myopathy
cancer, 41/317-386
carbamylcholine, 40/150
carcinoid, *see* Carcinoid myopathy
carcinoid syndrome, 70/316
carcinomatous, *see* Carcinomatous myopathy
carnitine cycle, 66/401
carnitine deficiency, *see* Carnitine deficiency myopathy
carnitine palmitoyltransferase deficiency, 62/494
cataract, 43/123-125
cathepsin, 40/380
celiac disease, 28/231
centronuclear, *see* Myotubular myopathy
chloroquine, *see* Chloroquine myopathy
chondrodystrophic myotonia, 31/278
chronic renal failure, *see* Chronic renal failure myopathy
ciclosporin intoxication, 65/552, 554
classification, 40/275-289
clofibrate, 40/155, 509
colchicine, *see* Colchicine myopathy
color blindness, 40/352, 388
combined complex I-V deficiency, 62/506
complex repetitive discharge, 62/62, 67
complex I deficiency, 62/503, 66/422
complex II deficiency, 62/504
complex III deficiency, 62/504, 66/423
complex IV deficiency, 62/505, 66/423
complex V deficiency, 62/506
congenital, *see* Congenital myopathy
congenital hip dislocation, 40/337
congenital muscle absence, 40/323
congenital ophthalmoplegia, 29/232
congenital weakness, 1/482
connective tissue, 40/40-43
contracture, 40/337
conventional EMG, 62/49
corticosteroid, *see* Corticosteroid myopathy
corticosteroid intoxication, 65/555
corticosteroid withdrawal, *see* Corticosteroid

withdrawal myopathy
Coxsackie virus, 56/332
Coxsackie virus infection, 34/138
creatine kinase, 43/116, 126, 193
cricopharyngeal, *see* Cricopharyngeal myotomy
curare sensitive ocular, *see* Curare sensitive ocular myopathy
Cushing syndrome, 41/250
cytoplasmic inclusion body, *see* Cytoplasmic inclusion body myopathy
dementia, 43/132
denervation atrophy, 62/13
diabetic, *see* Diabetic proximal neuropathy
diagnostic electromyography value, 62/71
diagnostic macroelectromyography value, 62/73
diagnostic scanning electromyography value, 62/74
diagnostic single fiber electromyography value, 62/72
diamorphine, 37/390
diazacholesterol, 40/155, 328
differential diagnosis, 1/233
distal, *see* Distal myopathy
dysarthria, 40/306
dysphonia, 40/306
ECHO virus, 56/332
electrodiagnosis, 1/224
EMG, 62/49
EMG method, 62/49
encephalopathy, 43/132
endocrine, *see* Endocrine myopathy
eosinophilia myalgia syndrome, 63/377, 64/252
experimental, *see* Experimental myopathy
experimental ischemic, *see* Experimental ischemic myopathy
extradischarge, 62/50, 67
eye muscle, 1/612
facioscapulohumeral syndrome, 40/280
familial, *see* Familial myopathy
familial visceral, 70/327
fasciculation, 40/330, 62/50
fasciculation discharge, 62/50
fatigue, 40/330
fatigue index, 62/54
female, 43/115
fibrillation potential, 62/50, 61, 67
fingerprint body, *see* Fingerprint body myopathy
flecainide, 63/192
focal, *see* Focal myopathy
forked fiber, 62/11
Friedreich ataxia, 21/355
glycogen storage disease, 22/189, 43/78, 132
Gowers distal, *see* Gowers distal myopathy
granulofilamentous, *see* Granulofilamentous myopathy
granulovacuolar lobular, *see* Granulovacuolar lobular myopathy
hereditary cylindric spirals, *see* Hereditary cylindric spirals myopathy
hereditary motor and sensory neuropathy type I, 21/303
hereditary motor and sensory neuropathy type II, 21/303
honeycomb, *see* Honeycomb myopathy
Horner syndrome, 40/302
human immunodeficiency virus, *see* Human immunodeficiency virus myopathy
hyperaldosteronism, *see* Hyperaldosteronism myopathy
hypercortisolism, 70/206
hyperparathyroid, *see* Hyperparathyroid myopathy
hyperparathyroidism, 27/293-296
hyperreflexia, 43/123-125
hyperthyroid, *see* Thyrotoxic myopathy
hyperthyroidism, *see* Thyrotoxic myopathy
hypogonadism, 43/123-125
hypokalemic periodic paralysis, 43/170
hypoparathyroid, *see* Hypoparathyroid myopathy
hypoparathyroidism, 27/312
hyporeflexia, 43/193
hypothyroid, *see* Hypothyroid myopathy
hypothyroidism, 70/96
imidazole, 40/150
imipramine, 40/169
indometacin, 40/343
infantile, *see* Infantile myopathy
infantile distal, *see* Infantile distal myopathy
infantile nemaline, *see* Infantile nemaline myopathy
inflammatory, *see* Inflammatory myopathy
inner mitochondrial membrane system, 66/406
insertional activity, 62/50
intensive care, 63/420
intensive care complication, 63/420
intestinal pseudo-obstruction, 70/320
intramuscular stimulation, 62/58, 74
ischemic, *see* Ischemic myopathy
isobutyrate, 40/548
isoniazid, 40/561
labetalol, 63/87
lactic acidosis, 41/219
late onset, *see* Late onset myopathy
lead intoxication, 9/642
leukemia, 39/13
limb girdle syndrome, 40/279
lipase, 40/380

lipid storage, *see* Lipid storage myopathy
lithium intoxication, 64/296
long chain acyl-coenzyme A dehydrogenase deficiency, 62/495
lordosis, 40/418, 436
lysis, 43/116
macroelectromyography, 62/55
macromotor unit potential, 62/55
malignant hyperthermia, 38/554, 43/117
Mallory body, *see* Mallory body myopathy
Marinesco-Sjögren syndrome, 21/556, 60/344, 660
McLeod, *see* McLeod phenotype
megaconial, *see* Mitochondrial myopathy
mental deficiency, 43/123-125, 193
MERRF syndrome, 62/509
metabolic, *see* Metabolic myopathy
metastatic, *see* Metastatic myopathy
miniature motor end plate potential, 62/50
minimal change, *see* Minimal change myopathy
mitochondria jagged Z line, *see* Mitochondria jagged Z line myopathy
mitochondria lipid glycogen, *see* Mitochondria lipid glycogen myopathy
mitochondrial, *see* Mitochondrial myopathy
mitochondrial DNA depletion, 62/511
mitochondrial encephalomyopathy, 60/660
mitochondrial matrix, 66/410
Möbius syndrome, 43/125, 50/216
motor end plate noise, 62/50
motor unit counting, 62/76
motor unit counting technique, 62/60
motor unit hyperactivity, 40/285
motor unit potential, 62/49, 67
multicore, *see* Multicore myopathy
multiple organ failure, 71/537, 540
muscle fiber density, 62/54
muscle fiber hypercontraction, 62/15
muscle fiber hypertrophy, 62/15
muscle fiber size, 43/126
muscle fiber stimulation, 62/59
muscle fiber type, 62/2
muscle fiber type I, 43/116
muscle pathology, 62/1
muscle tissue culture, 62/85
myalgia, 42/660, 43/192
myasthenic, *see* Eaton-Lambert myasthenic syndrome
myofibrillar lysis, *see* Myofibrillar lysis myopathy
myogenic jitter, 62/59
myoglobinuria, 41/274
myotonic, *see* Myotonic myopathy
myotonic discharge, 62/50, 61, 67
myotubular, *see* Myotubular myopathy
necrotizing, *see* Necrotizing myopathy
nemaline, *see* Nemaline myopathy
neoplastic, *see* Neoplastic myopathy
neuromuscular jitter, 62/54
neurosarcoidosis, 38/535-539
neurotoxicology, 64/16
nonprimary inflammatory, *see* Nonprimary inflammatory myopathy
nucleodegenerative, *see* Nucleodegenerative myopathy
ocular, *see* Ocular myopathy
oleic acid, 40/158
ophthalmoplegia, 43/125-127
opiate intoxication, 37/390
organic acid metabolism, 66/641
organophosphate intoxication, 64/160, 228
osteoporosis, 43/124
paraneoplastic syndrome, 69/386, 71/681
parathion, 40/150
pargyline, 40/169
partial invasion, 62/9
perhexiline, *see* Perhexiline myopathy
perhexiline maleate polyneuropathy, 51/301
periodic ataxia, 60/660
periodic paralysis, 40/285
pes cavus, 40/321, 337, 427
phenytoin, 40/523, 560
phenytoin intoxication, 65/503
pheochromocytoma, *see* Pheochromocytoma myopathy
phosphodiesterase, 40/150
phosphoglucomutase deficiency, 27/236, 41/192
phosphohexose isomerase deficiency, 27/236
phosphorylase kinase deficiency, 62/487
pituitary gigantism, *see* Pituitary gigantism myopathy
pleoconial, *see* Mitochondrial myopathy
poliodystrophy, 42/588
polyarteritis nodosa, 39/306, 55/355, 357
positive sharp wave, 62/50, 67
potassium depletion, 63/557
primary, *see* Primary muscle disease
primary carnitine deficiency, 66/403
procainamide intoxication, 37/452
progressive external ophthalmoplegia, 13/312, 22/182, 41/215, 43/136
progressive ophthalmoplegia, 40/281
progressive spinal muscular atrophy, 59/407
progressive systemic sclerosis, 51/448
propranolol, 63/87
proximal muscle weakness, 43/115, 132
ptosis, 40/356

quadriceps, *see* Quadriceps myopathy
quail, *see* Quail myopathy
reducing body, *see* Reducing body myopathy
regeneration, 62/9
rheumatoid arthritis, 71/29
rimmed muscle fiber vacuole, 40/153, 229
rimmed vacuole distal, *see* Rimmed vacuole distal myopathy
rod body, *see* Nemaline myopathy
sarcoidosis, 38/535-539, 71/481, 489
sarcoplasmic body, *see* Sarcoplasmic body myopathy
sarcotubular, *see* Sarcotubular myopathy
scanning electromyography, 62/57
scapulohumeral, *see* Scapulohumeral myopathy
scapuloperoneal, *see* Scapuloperoneal myopathy
scleroderma, 71/105, 108
scoliosis, 40/337, 362, 442, 43/124
segmental necrosis, 62/13
sensorineural deafness, 43/126
single fiber electromyography, 62/49, 53
Sjögren syndrome, 39/430, 71/88
skeletal deformity, 40/337
small caliber fiber, 62/13
snake envenomation, 65/179
sotalol, 63/87
speech disorder, 43/124
spheroid body, *see* Spheroid body myopathy
split fiber, 62/16
spontaneous activity, 62/50, 67
sporadic visceral, *see* Sporadic visceral myopathy
steroid, *see* Steroid myopathy
storage disease, 62/16
strychnine intoxication, 40/330
systemic lupus erythematosus, 39/284
talipes, 40/337
talipes equinovarus, 40/337, 427, 480
tardive, *see* Tardive myopathy
thyrotoxic, *see* Thyrotoxic myopathy
toxic, *see* Toxic myopathy
toxic oil syndrome, 63/381
toxic shock syndrome, 52/259
tremor, 43/132
trichinosis, 52/571
trilaminar, *see* Trilaminar myopathy
tryptophan contaminant, 64/252
tubular aggregate, *see* Tubular aggregate myopathy
tubulomembranous inclusion, *see* Tubulomembranous inclusion myopathy
uniform muscle fiber type I, *see* Uniform muscle fiber type I myopathy
vacuolar, 43/96, 108, 129, 171, 178
velocity recovery function, 62/54
vinblastine, 40/555
vincristine, *see* Vincristine myopathy
visceral, *see* Visceral myopathy
Wegener granulomatosis, 51/451, 71/179
Whipple disease, 51/184, 52/138
winged scapula, 40/312, 422
Wohlfart-Kugelberg-Welander disease, 42/92, 59/88, 90
X-linked muscular dystrophy, 40/277
X-linked myotubular, *see* X-linked myotubular myopathy
X-linked neutropenic cardioskeletal, *see* X-linked neutropenic cardioskeletal myopathy
xanthinuria, 43/192
Z band, 62/7
Z band plaque, *see* Z band plaque myopathy
zebra body, *see* Zebra body myopathy
zidovudine, *see* Zidovudine myopathy
Myophosphorylase
central core myopathy, 43/81
debrancher deficiency, 43/179
myophosphorylase deficiency, 27/222, 232, 41/188, 43/181
Myophosphorylase deficiency
adenosine triphosphate, 41/188, 62/486
alanine, 41/189
biochemical aspect, 27/232
biochemical consideration, 41/188
biochemistry, 41/188, 62/485
childhood myoglobinuria, 27/233, 62/564
chromosome 11q13, 62/485
clinical features, 27/233, 40/317, 41/185, 62/485
contracture, 27/234, 40/327, 41/186-188, 297, 43/182, 62/485
creatine kinase, 27/233, 41/185, 187, 62/485
differential diagnosis, 40/438, 547, 41/56, 60, 167, 204, 372
ECG, 27/233, 41/187
EEG, 41/187
EMG, 27/233, 41/187, 62/485
exercise, 27/233
exercise intolerance, 41/185, 62/485
exercise myoglobinuria, 62/485
fatigue, 62/485
fenfluramine, 41/190
genetics, 41/189, 427
glucagon, 41/190
glycogen storage disease, 27/232, 40/4, 53, 189, 257, 317, 327, 339, 342, 438, 62/485
heterozygote detection, 43/181
histochemistry, 40/4, 53, 41/186
histopathology, 41/187, 62/485

inheritance, 62/485
ischemic exercise, 27/233, 41/187, 62/485
ischemic exercise test, 27/233, 41/187, 62/485
isoenzyme, 40/189, 41/188
laboratory, 41/187
lactic acid, 27/233, 41/187
limb girdle syndrome, 62/190
metabolic myopathy, 41/185, 62/485
molecular genetics, 62/485
muscle biopsy, 41/187, 62/485
muscle contraction, 62/485
muscle cramp, 41/186, 62/485
muscle fiber feature, 62/17
muscle intermittent claudication, 20/799
muscle necrosis, 41/186, 62/485
muscle tissue culture, 40/189, 62/109
myalgia, 27/234, 41/185, 62/485
myoglobinuria, 27/233, 40/339, 41/186, 265, 43/181, 62/485, 558
myophosphorylase, 27/222, 232, 41/188, 43/181
pathologic reaction, 40/257
pathology, 27/234, 62/485
phenformin, 41/190
phosphorylase, 41/185-190
phosphorylase isoenzyme, 62/486
phosphorylase limit dextrin, 62/485
physiopathology, 62/486
polymyositis, 41/56, 60
renal insufficiency, 27/233, 41/186, 62/485
rhabdomyolysis, 40/264
rimmed vacuole distal myopathy, 62/485
second wind phenomenon, 41/186, 189, 62/486
secondary pigmentary retinal degeneration, 60/727
seizure, 41/186, 189, 43/181
serum creatine kinase, 27/233, 41/185, 187, 62/485
sex ratio, 41/189, 43/181
treatment, 41/190, 62/486

Myopia
albinoidism, 42/404
amaurosis fugax, 55/108
congenital ophthalmoplegia, 43/134
congenital spondyloepiphyseal dysplasia, 43/478
Down syndrome, 50/526, 529
Flynn-Aird syndrome, 14/112, 22/519, 42/327
Forsius-Eriksson syndrome, 42/397
hereditary deafness, 50/218
hereditary progressive arthro-ophthalmopathy, 13/41
hereditary spastic ataxia, 60/464
homocystinuria, 55/327
horizontal gaze paralysis, 42/333
hyaloidoretinal degeneration (Wagner), 13/37, 274
malignant, *see* Malignant myopia
Marshall syndrome, 42/393
microcoria, 42/403
oculocutaneous albinism, 42/405
Oguchi disease, 13/262
ophthalmoplegia, 43/141
primary pigmentary retinal degeneration, 13/178
progressive external ophthalmoplegia, 43/141
Richner-Hanhart syndrome, 42/581
secondary pigmentary retinal degeneration, 13/218
sensorineural deafness, 42/381, 389, 393
Stargardt disease, 13/140
Stickler syndrome, 43/484

Myosclerosis
arthrogryposis multiplex congenita, 43/133
collagen disease, 43/133
congenital atonic sclerotic muscular dystrophy, 41/45
Emery-Dreifuss muscular dystrophy, 40/392
fibrosing myositis, 41/65
hereditary, *see* Heredofamilial myosclerosis
interstitial fibrosis, 41/65
Krabbe-Bartels disease, 14/408
rigid spine syndrome, 62/173

Myosin
contractile protein, 40/212

Myositis
see also Inflammatory myopathy
acute viral, *see* Acute viral myositis
age, 62/370
autoimmune disease, 62/369
B-lymphocyte, 62/42
bacterial, *see* Bacterial myositis
benign acute childhood, *see* Benign acute childhood myositis
cause, 62/369
CD8+ cytotoxic T-lymphocyte, 62/42
chronic, *see* Chronic myositis
clinical course, 62/369
clinical features, 62/369
complement mediated microangiopathy, 62/369
Coxsackie virus, 56/332
Coxsackie virus infection, 41/20, 53, 73, 377
eosinophilia myalgia syndrome, 64/252
experimental, *see* Experimental myositis
experimental acute viral, *see* Experimental acute viral myositis
experimental allergic, *see* Experimental allergic myositis
fall, 62/369
familial intermittent ophthalmoplegia, 22/190

fibrosing, *see* Fibrosing myositis
granulomatous, *see* Granulomatous myositis
helper inducer lymphocyte, 62/42
immune complex, *see* Immune complex myositis
inclusion body, *see* Inclusion body myositis
Kawasaki syndrome, 56/639
mixed connective tissue disease, 71/122
necrotizing, *see* Necrotizing myositis
nodular, *see* Nodular myositis
ocular, *see* Ocular myositis
orbital, *see* Orbital myositis
ossificans progressiva, *see* Progressive myositis ossificans
postpartum period, 62/370
pregnancy, 62/370
primary inflammatory, *see* Primary inflammatory myositis
proximal, *see* Proximal myositis
rhabdomyolysis, 62/370
rheumatoid arthritis, 38/484
scleroderma, 71/105
sex ratio, 62/370
Sjögren syndrome, 71/74
T-lymphocyte mediated myocytotoxicity, 62/369
thymus hyperplasia, 41/85
Tolosa-Hunt syndrome, 48/302
toxic shock syndrome, 52/259
trichinous, *see* Trichinous myositis
tropical, *see* Tropical polymyositis
viral, *see* Viral myositis
viral disease, 62/369
Myositis ossificans
calcinosis cutis universalis, 41/57
progressive, *see* Progressive myositis ossificans
Myospasm gravis, *see* Satoyoshi syndrome
Myotatic reflex
Friedreich ataxia, 21/331
hereditary motor and sensory neuropathy type I, 21/282
hereditary motor and sensory neuropathy type II, 21/282
Myotome
terminology, 2/159
topographical diagnosis, 2/170
Myotomy
spasticity, 26/485
Myotonia
acid maltase deficiency, 40/539, 41/176
aconitine, 40/557
after-spasm, 40/536
ambient temperature, 62/262
amyotrophy, 1/269
animal model, 40/154, 495, 533, 548
beta adrenergic receptor blocking agent intoxication, 65/437
black widow spider venom intoxication, 40/547
calcium channel blocking agent, 62/276
calcium release, 40/553
cancer, 41/347
cell electrical property, 40/495, 550
chemical induced, 40/496, 548
chloride channel defect, 62/277
chloride conductance, 40/553, 62/276
chondrodystrophic, *see* Chondrodystrophic myotonia
chromosome 17q23.1-25.3, 62/279
classification, 40/538, 41/165, 62/263
clinical features, 40/327, 495, 533-537, 41/298
clomipramine, 62/276
colchicine, 40/555
computer model, 62/277
congenital myopathy, 62/335
cramp, 40/327, 533
cromakalim, 62/276
curare, 41/299
definition, 62/261
denervation, 40/555
depolarization, 41/298
diazacholesterol, 40/155, 514, 548, 555
2,4-dichlorophenoxyacetic acid, 40/548, 555
differential diagnosis, 62/273
diltiazem, 62/276
discharge duration, 62/262
drug induced, *see* Drug induced myotonia
EMG, 1/641-643, 40/327, 533, 535, 41/298
excitation contraction coupling, 40/556
experimental, *see* Experimental myotonia
experimental myopathy, 40/154
fenoterol, 62/276
general, 40/549
generalized (Becker), *see* Generalized myotonia (Becker)
hereditary motor and sensory neuropathy variant, 60/247
Hodgkin-Huxley model, 40/554
hyperhidrosis, 43/160
hypokalemic periodic paralysis, 41/150, 43/170, 62/458
hypothyroidism, 40/547
imipramine, 62/276
Isaacs syndrome, 43/160, 62/263, 273
malabsorption syndrome, 70/227
malignancy, 41/347
membrane conductance, 40/497, 557
mexiletine, 62/275
motor unit hyperactivity, 40/327, 495, 41/150,

165, 176, 298, 347
muscle cell membrane, 41/298
muscle contraction, 41/298, 62/262, 273
muscle cramp, 40/327, 533, 62/261, 273
muscle fiber conductance, 40/553
myokymia, 43/160, 62/273
myokymic discharge, 62/263
myotonia congenita, 40/538, 62/262
myotonic discharge, 62/262
myotonic dystrophy, 40/494, 533-537
neuromyotonic discharge, 62/263
nifedipine, 62/276
orciprenaline, 62/276
painful dominant, *see* Painful dominant myotonia
paradoxical, *see* Paradoxical myotonia
paramyotonia, 62/262
paramyotonia congenita, 40/542, 62/224
paraneoplastic syndrome, 69/374, 380
pathophysiology, 40/549-557, 62/276
percussion, *see* Percussion myotonia
periodic paralysis, 40/283, 544, 41/150, 158, 165
phenytoin, 62/275
pindolol, 62/276
procainamide, 37/452, 62/275
propagation, 40/552
propranolol, 40/547
pseudo, *see* Isaacs syndrome
quinidine, 41/297
quinine, 40/548, 558, 41/297, 62/275
rheobase, 40/553
sodium channel, 62/278
stiff-man syndrome, 62/274
T tubule system, 62/277
taurine, 62/276
tetanus, 40/547
tetany, 40/547
tocainide, 62/275
toxic myopathy, 62/597
treatment, 62/274
triparanol, 40/514
trophic influence, 40/555
vincristine, 40/548
warm up phenomenon, 62/261
Myotonia acquisita
diazacholesterol, 40/328
2,4-dichlorophenoxyacetic acid, 40/328
differential diagnosis, 40/548
myotonic dystrophy, 40/282
Myotonia atrophica, *see* Myotonic dystrophy
Myotonia congenita
ambient temperature, 62/264
case history, 40/562
classification, 40/283, 557
clinical features, 40/324, 328, 539, 41/298, 62/264
differential diagnosis, 40/520, 545, 41/245
EMG, 40/539, 62/70
emotion, 62/264
genetics, 41/421
incidence, 62/266
malignant hyperthermia, 38/551, 62/570
morphology, 40/547
muscle cell membrane, 40/128, 553
muscle hypertrophy, 40/539, 41/298, 43/161, 62/266
myoglobinuria, 62/558
myotonia, 40/538, 62/262
myotonic dystrophy, 62/240
myotonic syndrome, 62/262
pathophysiology, 40/549, 554
pregnancy, 62/264
prevalence, 43/162
progressive external ophthalmoplegia, 62/305
N-propylajmalin, 40/557, 561
sarcolemma, 40/128
sex ratio, 62/266
thyroxine, 27/272
treatment, 40/557
Von Gräfe sign, 2/324
Myotonia fluctuans
dominant inheritance, 62/266
myotonic syndrome, 62/263
pain, 62/266
Myotonia paradoxa
Isaacs syndrome, 43/160
paralysis periodica paramyotonica, 43/167
paramyotonia congenita, 43/169
Myotonic burst
pseudo, *see* Pseudomyotonic burst
Myotonic discharge
features, 62/61
motor unit, 62/50, 61, 67
myopathy, 62/50, 61, 67
myotonia, 62/262
myotonic syndrome, 62/262
Myotonic dystrophy, 40/485-524
ABH secretor gene, 40/541
abortion, 62/228
Adams-Stokes syndrome, 40/507, 522
adenosine triphosphate, 40/497, 523
adult form, 62/216
allelic expansion, 62/212
alopecia, 62/238
alveolar hypoventilation, 40/517, 62/238
anesthesia complication, 62/244
anticipation, 40/488, 62/214
apamin, 62/222, 225

autonomic nervous system, 40/518, 75/641
autosomal recessive generalized myotonia, 62/240
baldness, 43/152
biochemical study, 40/523
brain embolism, 55/163
brain infarction, 55/163
cardiac conduction, 62/217
cardiac dysrhythmia, 41/421, 486
cardiac involvement, 40/509, 523
cardiomyopathy, 43/152
carrier detection, 40/520
case history, 40/562
cataract, 40/512, 41/486, 43/153, 62/217, 233
cerebrovascular disease, 53/43, 55/163
Chamorro people, 22/343
Charcot-Marie-Tooth like syndrome, 62/234
chloride channel, 62/224
chorioretinal degeneration, 13/40
chromosome 19q13.3, 62/210
classification, 40/282, 537, 62/216
clinical features, 40/302-307, 312, 319, 322, 328, 335, 342, 539
clinical and genetic overview, 40/486
clofibrate, 40/548
CNS, 40/517, 62/235
cognitive impairment, 62/235
congenital form, 40/283, 490, 62/216, 218
congestive heart failure, 62/227
contractile protein, 40/523
course, 62/217
cranial hyperostosis, 43/152
creatine kinase, 43/153
demyelination, 62/237
detection, 40/520
diabetes mellitus, 40/511
diaphragm, 40/517
diazacholesterol, 40/548
differential diagnosis, 40/422, 472, 481, 520, 41/13, 245, 62/239
distal myopathy, 62/240
drug treatment, 62/245
Duchenne muscular dystrophy, 40/507, 522
dysphagia, 40/494, 514
early onset, 41/423
ECG, 62/226
echoCG, 62/227
EMG, 40/519, 62/70
endocrine disorder, 40/510
epithelioma, 62/238
erythrocyte, 40/497, 523
esophagus, 40/514
ethnic distribution, 62/214
excitation contraction coupling, 40/523
exercise, 62/249
expressivity, 40/486, 493, 520
facial paralysis, 43/152
facioscapulohumeral muscular dystrophy, 62/240
feeding disorder, 43/152
forgotten respiration, 63/490
Friedreich ataxia, 21/355, 40/508
gastrointestinal symptom, 62/233
gene instability, 62/213
gene lesion, 62/211
gene product, 62/213
genetic linkage, 40/521, 43/153
genetics, 41/421, 486, 62/209
glycogen storage disease, 43/152
gonadal abnormality, 62/229
growth hormone, 62/230, 249
Guam amyotrophic lateral sclerosis, 21/27, 22/342
hearing loss, 62/234
heart disease, 40/507, 522, 62/226
hereditary motor and sensory neuropathy type I, 62/234
hereditary motor and sensory neuropathy type II, 62/234
histochemistry, 40/48
human menopausal gonadotropin secretion, 62/229
hydramnios, 40/490
hyperinsulinism, 62/230
hypersomnia, 40/517, 62/217, 235
hypogonadism, 43/152
hypoventilation, 62/235
hypoxemia, 62/235
IgA, 40/519
IgG, 43/153
immunologic change, 62/239
infantile, *see* Infantile myotonic dystrophy
insulin, 40/511
insulin resistance, 62/230
intellectual function, 41/478
intestinal pseudo-obstruction, 51/493
laboratory data, 62/247
limb girdle syndrome, 62/188, 240
lipid, 40/523, 548
luteinizing hormone secretion, 62/229
management, 62/240-244
maternal complication, 62/229
maternal effect, 43/154
megacolon, 40/516
mental deficiency, 40/517, 43/152, 46/44
mental depression, 62/235
mexiletine, 62/241
motor end plate, 40/503
motor unit, 40/138, 519

muscle biopsy, 62/225
muscle cell membrane, 40/74, 213, 494, 497, 553
muscle fiber size, 43/152
muscle fiber type I, 40/49, 252, 497, 43/152
muscle fiber type II, 43/153
muscle histopathology, 62/219
muscle necrosis, 40/252, 497
muscle pathology, 40/252, 497, 547
muscle relaxation, 62/224
muscle spindle, 40/503
muscle strength scoring, 62/248
muscle tissue culture, 62/105, 222
mutation rate, 43/153
myoglobinuria, 40/339, 62/556, 558
myotonia, 40/494, 533-537
myotonia acquisita, 40/282
myotonia congenita, 62/240
myotonia physiology, 62/224
myotubular myopathy, 62/240
nemaline myopathy, 62/240
neurofibrillary tangle, 62/237
neuromuscular disease, 41/421
neuropathy, 62/234
nuclear magnetic resonance, 62/237
obstetric complication, 62/229
ocular abnormality, 40/512
ocular symptom, 62/233
oculopharyngeal muscular dystrophy, 62/240
ophthalmic aspect, 2/324
ophthalmoplegia, 22/206
optic atrophy, 13/40, 42
orthosis, 62/241
peripheral nervous system disease, 40/518
personality change, 62/235
phakomatosis, 14/492
phenytoin, 40/522, 62/241
pilomatricoma, 43/153, 62/239
pituitary hormone release, 62/229
polyneuritis, 43/152
postoperative care, 62/244
potassium channel, 62/225
pregnancy, 62/228
prenatal diagnosis, 43/153, 62/215
prevalence, 43/153, 62/214
procainamide, 37/452, 40/509, 523
progressive bulbar palsy, 59/375
progressive external ophthalmoplegia, 22/187, 62/302
progressive external ophthalmoplegia classification, 22/179, 62/290
N-propylajmalin, 40/557, 561
quinidine, 40/509, 523
quinine, 40/513
reproductive abnormality, 40/519
reproductive fitness, 43/152
respiratory dysfunction, 40/516
respiratory system, 62/238
retinal degeneration, 40/513
retinography, 40/514
ring fiber, 40/497
sarcolemma, 40/74, 213, 494, 497
sarcoplasmic mass, 40/27, 244, 497
scapuloperoneal muscular dystrophy, 22/59
secondary pigmentary retinal degeneration, 13/318, 60/723
sella turcica, 43/152
sensorineural deafness, 43/152
sleep, 62/235
sleep apnea syndrome, 62/217, 235
smooth muscle abnormality, 40/514
sodium channel, 62/224
spinal muscular atrophy, 59/373
Stargardt disease, 13/140
swan neck deformity, 40/308
syringomyelia, 43/155, 50/454
terminal innervation ratio, 40/519
testosterone, 62/248
testosterone secretion, 62/229
thalamic inclusion, 62/237
thyroid associated ophthalmopathy, 22/179, 204
thyrotropic hormone secretion, 62/230
treatment, 40/522, 557
tubular aggregate, 41/154
ultrastructure, 40/69
ventricular dilatation, 43/152
warm up phenomenon, 62/226
white matter lesion, 62/237

Myotonic hyperkalemic periodic paralysis
ambient temperature, 62/267
attack symptom, 43/152, 62/462
chromosome 17, 62/224, 267
Chvostek sign, 62/462
diplopia, 62/462
features, 62/267
lid lag, 41/165, 62/462
serum potassium, 43/152, 62/267
sodium channel, 62/224
tongue myotonia, 62/462
visual impairment, 62/462

Myotonic muscular dystrophy, *see* Myotonic dystrophy

Myotonic myopathy
chondrodystrophic myotonia, 43/165
classification, 40/282
creatine kinase, 43/166
dwarfism, 43/165

mental deficiency, 43/165
muscle hypertrophy, 43/165
muscular atrophy, 43/165
nasal voice, 43/165
pectus carinatum, 43/165
psychomotor retardation, 43/165
vertebral abnormality, 43/165
Myotonic syndrome, 40/533-563
acquired type, 62/264
ambient temperature, 62/262
animal model, 62/263
9-anthroic acid, 40/548
aromatic carboxylic acid intoxication, 62/264
autosomal recessive generalized myotonia, 62/263
azacosterol intoxication, 62/264
carbamazepine, 40/546, 561
chondrodystrophic myotonia, 62/262
classification, 40/537, 62/263
clofibrate intoxication, 62/264
corticosteroid, 40/561
differential diagnosis, 40/545, 62/273
discharge duration, 62/262
dominant inheritance, 62/263
drug induced, *see* Drug induced myotonic syndrome
hereditary, *see* Hereditary myotonic syndrome
hyperkalemic periodic paralysis, 40/544
incidence, 40/533
insulin, 40/545
Isaacs syndrome, 62/263, 273
levodopa, 40/561
mechanism, 62/261
morphology, 40/547
muscle biopsy, 62/274
muscle cell membrane, 40/556
muscle contraction, 62/262, 273
muscle cramp, 62/261, 273
myokymia, 62/273
myokymic discharge, 62/263
myotonia congenita, 62/262
myotonia fluctuans, 62/263
myotonia model, 40/547
myotonia pathophysiology, 40/549
myotonic discharge, 62/262
neuromyotonic discharge, 62/263
painful dominant myotonia, 62/263
paradoxical myotonia, 62/261
paramyotonia, 62/262
paramyotonia congenita, 40/542, 62/263
phosphorylation, 40/545
sarcolemma, 62/261
stiff-man syndrome, 62/274
treatment, 40/557
triperinol intoxication, 62/264
warm up phenomenon, 62/261
Myotubular myopathy
adult dominant, *see* Adult dominant myotubular myopathy
age at onset, 62/348
chromosome Xq28, 62/348
classification, 40/288
clinical features, 40/335, 337, 41/11
clinical form, 62/30
congenital myopathy, 41/11-14, 62/332
differential diagnosis, 22/99, 40/422
DNA, 43/113
EMG, 62/69
etiology, 40/147
exercise, 40/15, 200
extraocular muscle weakness, 41/11
facial paralysis, 43/113
facioscapulohumeral muscular dystrophy, 62/168
fatality rate, 62/348
genetics, 41/431
histochemistry, 40/46, 62/349
histologic features, 22/188
histopathology, 41/12
masseter muscle, 43/113
muscle cell nucleus, 40/89
muscle fiber type I, 43/113
muscle fiber type I predominance, 41/12
muscle fiber type II, 40/15, 200, 43/113
muscle tissue culture, 40/193
myotonic dystrophy, 62/240
ophthalmoplegia, 43/113
pathologic reaction, 40/260
progressive external ophthalmoplegia, 22/179, 189, 41/11, 43/136, 62/297
progressive external ophthalmoplegia classification, 22/179, 62/290
ptosis, 43/113, 62/297
transmission type, 62/348
Myotubule
acetylcholine receptor, 41/114
culture, 40/113, 184
development, 40/146, 199
muscle regeneration, 40/23
muscle tissue culture, 62/93
Myriachit
Gilles de la Tourette syndrome, 49/629, 635
jumpers of the Maine, 42/231-233, 49/621
myoclonus, 49/621
Myringitis
bullous, *see* Bullous myringitis
Myristicin
3-methoxy-4,5-methylenedioxyamphetamine,

65/50
nutmeg tree, 65/50
Mysoline, *see* Primidone
Mytelase, *see* Ambenonium chloride
Myxedema
see also Hypoparathyroidism
acanthocytosis, 63/271
ataxia, 21/569, 574
carpal tunnel syndrome, 7/296
cerebellar ataxia, 21/496
chronic axonal neuropathy, 51/531
endemic cretinism, 31/297
inappropriate antidiuretic hormone secretion, 28/499
late cortical cerebellar atrophy, 21/496
Levi-Lorain dwarfism, 2/454
myelomalacia, 55/98
obstructive sleep apnea, 63/454
thyroid gland dysgenesis, 42/632
vitamin B12 deficiency, 70/373
Myxedema body
cerebellar ataxia, 21/496
Myxedema neuropathy
acroparesthesia, 7/169, 8/58
ataxia, 8/57, 59
carpal tunnel syndrome, 7/169, 8/57
CNS, 8/57
CSF protein, 8/57, 59
deafness, 8/57
diagnostic test, 8/60
dysarthria, 8/57, 59
muscle stretch reflex time, 8/58
muscle weakness, 8/57
myoedema, 8/59
nerve conduction, 7/169
peripheral nerve, 7/169, 8/57
sensory nerve conduction, 8/58
sensory sign, 8/57
spinal cord, 8/59
treatment, 8/60
Myxoglioma
gelatinoid, *see* Gelatinoid myxoglioma
Myxolipoma
epidural spinal lipoma, 20/405
Myxoma
atrial, *see* Atrial myxoma
cardiac, *see* Cardiac myxoma
primary cardiac tumor, 63/93
Myxosarcoma
cardiac, *see* Cardiac myxosarcoma
primary cardiac tumor, 63/93
vertebral column, 20/33
Myxovirus
encephalitis, 34/79
measles, 9/546
multiple sclerosis, 9/123, 129
subacute sclerosing panencephalitis, 9/551
viral myositis, 41/71

N-CAM, *see* Nerve cell adhesion molecule
N syndrome
growth retardation, 43/283
hypotelorism, 43/283
megalocornea, 43/283
mental deficiency, 43/283
sensorineural deafness, 43/283
spastic tetraplegia, 43/283
NADH-tetrazolium reductase, *see* Tetrazolium reductase
Nadolol
beta adrenergic receptor blocking agent, 65/434
Naegeli syndrome
Divry-Van Bogaert syndrome, 14/108
Goltz-Gorlin syndrome, 14/113
incontinentia pigmenti, 14/215, 31/242
Naegleria
acute amebic meningoencephalitis, 52/313
australiensis, 52/311
culture, 52/318
direct observation, 52/318
gruberi, 52/311
isolation, 52/318
jadini, 52/311
ultrastructure, 52/311
Naegleria fowleri
see also Neuroamebiasis
acute amebic meningoencephalitis, 52/313, 316
amebiasis, 35/25, 48-52
Amoebae, 35/30-32
CNS invasion, 52/323
cytolysis, 52/323
form, 52/311
microscopy, 35/28, 52/310
olfactory nerve, 52/323
primary amebic meningoencephalitis, 35/25
Nageotte nerve
site, 1/228
Nageotte nodule
eosinophilic body, 22/238
hereditary sensory and autonomic neuropathy type I, 60/9
paraneoplastic polyneuropathy, 51/468
pure autonomic failure, 22/238
Nageotte syndrome, *see* Babinski-Nageotte syndrome
NAGLU gene

α-*N*-acetylglucosaminidase, 66/282
glycosaminoglycanosis type IIIB, 66/282
Nail
arsenic intoxication, 36/202, 206
dysplastic, *see* Dysplastic nail
lead intoxication, 64/445
trophic disorder, 1/486
Nail hypoplasia
Coffin-Siris syndrome, 43/241
Edwards syndrome, 43/535
Ellis-Van Creveld syndrome, 43/347
9p partial trisomy, 43/516
Patau syndrome, 31/507, 43/528
Rothmund-Thomson syndrome, 14/777, 43/460, 46/61
sensorineural deafness, 42/384-386
trisomy 8 mosaicism, 43/510
Turner syndrome, 43/550, 50/543
Nairovirus
human pathogen, 56/15
Naja naja α-toxin
experimental myasthenia gravis, 41/113
Nalidixic acid
benign intracranial hypertension, 67/111
iatrogenic neurological disease, 65/486
neurotoxin, 65/486
quinolone intoxication, 65/486
toxic encephalopathy, 65/486
Nalidixic acid intoxication
benign intracranial hypertension, 65/486
epilepsy, 65/486
headache, 65/486
papilledema, 65/486
Nalline intoxication, *see* Nalorphine
Nalorphine
analgesia, 37/366
ataxia, 37/366
chemistry, 37/366
delirium, 37/366
dysphoria, 37/366
euphoria, 37/366
hallucination, 37/366
insomnia, 37/366
opiate, 37/366
Naloxone
brain infarction, 53/430
brain metabolism, 57/83
chemistry, 37/365
congenital pain insensitivity, 51/564, 566
opiate, 37/365, 65/350
opiate intoxication, 65/353
paraplegia, 61/399
spinal cord injury, 61/399
tetraplegia, 61/399
transverse spinal cord lesion, 61/399
Naltrexone
chemistry, 37/365
opiate, 37/365, 65/350
Naming
frontal lobe tumor, 67/150
NANA, *see* *N*-Acetylneuraminic acid
Nanophthalmia, *see* Microphthalmia
Naphthalene intoxication
headache, 48/420
β-Naphthol
alopecia, 13/224
animal experiment, 13/204
antiseptic, 13/224
congenital cataract, 13/204
psoriasis, 13/224
retinal dot, 13/224
scabies, 13/224
typhoid fever, 13/224
Naproxen
behavior disorder, 46/602
headache, 48/174
migraine, 48/174
Narcolepsy, 3/93-98, 42/710-713, 45/147-150
see also Daytime hypersomnia
age at onset, 45/150
amphetamine, 2/448, 3/94, 15/847
automatism, 45/148
awakening disorder, 45/149
body scheme, 45/511
brain injury, 24/593
carotid sinus reflex, 11/546
carotid sinus syndrome, 11/546
cataplexy, 2/448, 3/94, 45/147
catecholamine, 3/96
classic tetrad, 46/418
CNS stimulant, 46/592
depression, 45/149
diplopia, 3/94
dreaming, 3/94
drowsiness, 45/147
EEG, 2/448, 3/94, 45/149
encephalitis, 2/448
epilepsy, 3/96, 72/253
etiology, 3/94
extrapyramidal disorder, 2/344
hallucination, 2/448, 45/354, 511
history, 15/836
hypersomnia, 3/96, 15/838
hypnagogic hallucination, 2/448, 15/845, 45/148, 354
idiopathic, *see* Idiopathic narcolepsy

imipramine, 3/95
incidence, 15/837
laughing, 2/448, 3/94
Melkersson-Rosenthal syndrome, 8/213
memory disorder, 42/711
methylphenidate, 3/95, 15/846
multiple sclerosis, 2/448, 9/103, 178
multiple sleep latency test, 74/562
narcoleptic sleep attack, 2/448, 45/147
narcoleptic syndrome, 42/710
neurobrucellosis, 33/314, 52/587
Niemann-Pick disease type C, 60/151
non-REM sleep, 2/448, 45/150
obesity, 2/448
olivopontocerebellar atrophy (Wadia-Swami), 60/494
onset, 15/837
paranoid symptom, 3/94
paroxysmal disorder, 45/511
pathogenesis, 15/849
pickwickian syndrome, 3/96
polycythemia vera, 2/448
polysomnography, 45/149
prazosin, 65/448
prazosin intoxication, 65/448
prognosis, 3/94
REM sleep, 1/491, 2/448, 42/712, 45/136, 149
secondary, *see* Secondary narcolepsy
sleep, 74/558
sleep paralysis, 2/448, 3/94, 15/844, 45/148
symptomatic, 3/96
temporal lobe epilepsy, 45/148
thalamic lesion, 2/482
treatment, 15/846

Narcoleptic sleep attack
narcolepsy, 2/448, 45/147

Narcoleptic syndrome
cataplexy, 15/842, 42/711
catecholamine, 42/713
hypnagogic hallucination, 42/712
narcolepsy, 42/710
non-REM sleep, 42/712
prevalence, 42/710
REM sleep, 42/710-713
sleep paralysis, 42/712

Narcosis
carbon dioxide, *see* Carbon dioxide narcosis
nitrogen, *see* Nitrogen narcosis

NARP syndrome
adenosine triphosphatase, 66/424
epilepsy, 72/222
subacute necrotizing encephalomyelopathy, 66/425

Narphen, *see* Phenazocine

Nasal bone aplasia
Patau syndrome, 14/121

Nasal deformity
dystonia musculorum deformans, 49/525

Nasal glioma
basal encephalocele, 50/106
basal meningoencephalocele, 50/106
cranial cephalocele, 30/209, 217

Nasal headache
experimental work, 5/212

Nasal and lip defect
miscellaneous, 30/471

Nasal nerve, *see* Nasociliary nerve
Nasal nerve neuralgia, *see* Charlin neuralgia
Nasal nerve syndrome, *see* Charlin neuralgia

Nasal pain
sphenopalatine neuralgia, 5/329

Nasal secretion
Charlin neuralgia, 48/486
chronic paroxysmal hemicrania, 48/262, 75/293
cluster headache, 48/237
headache, 75/284

Nasal septum
abscess, 5/218
headache, 5/212
hematoma, 5/218

Nasal sinus infection
brain abscess, 33/113, 122
toxic hydrocephalus, 12/426, 16/236

Nasal ulcer
mixed connective tissue disease, 51/450

Nasal voice
myotonic myopathy, 43/165

Nasociliary nerve
anatomy, 48/484
Charlin neuralgia, 48/483
Horner syndrome, 2/58
neuralgia, *see* Charlin neuralgia
ocular pain, 48/484

Nasociliary neuralgia, *see* Charlin neuralgia
Nasociliary syndrome, *see* Charlin neuralgia

Nasofacial reflex
Charlin neuralgia, 5/217, 48/485

Nasolacrimal reflex
hysterical anosmia, 2/54
middle fossa meningioma, 2/57

Nasomaxillary fracture
clinical features, 24/98
ocular imbalance, 24/97
orbital fracture, 24/97
radiology, 24/98
symptom, 24/98

Naso-orbital fracture
craniofacial injury, 23/379-381
Nasopalpebral reflex
amyotrophic lateral sclerosis, 22/293
Nasopharyngeal carcinoma
Epstein-Barr virus, 56/8, 249, 255
Epstein-Barr virus infection, 34/190
nonrecurrent nonhereditary multiple cranial neuropathy, 51/571
pituitary adenoma, 17/412
skull base, 68/466
skull base tumor, 68/488
Nasopharyngeal fibroma
skull base tumor, 17/178
Nasopharyngeal tumor
adenopathy, 17/205
chondroma, 17/205
Citelli syndrome, 2/313
differential diagnosis, 17/205
Garcin syndrome, 2/313
meningioma, 17/205
middle cavernous sinus syndrome, 2/90
posterior fossa, 17/206
Nasu-Hakola disease
familial spastic paraplegia, 59/308
lipomembranous polycystic osteodysplasia, 49/406
neuroaxonal leukodystrophy, 49/406
National Acute Spinal Cord Injury Study Scale
spinal cord injury, 61/423
spinal cord injury recovery score, 61/423
transverse spinal cord lesion, 61/423
National Hospital, London, United Kingdom, 1/6, 19, 35
National Institute of Neurological Diseases and Blindness, 1/6, 18, 37
Natriuretic peptide
atrial, *see* Atrial natriuretic peptide
Natural killer cell
herpes simplex virus, 56/209
lymphocyte, 56/67
multiple sclerosis, 47/341, 343, 346, 349-352
Semliki forest virus, 56/73
Sindbis virus, 56/73
Natural killer cell receptor
multiple sclerosis, 47/349
Nausea
alcohol intoxication, 64/112
apistus intoxication, 37/76
Balaenoptera borealis intoxication, 37/94
barotrauma, 63/416
basilar artery migraine, 48/136
brain hypoxia, 63/416
brain tumor, 67/143
buspirone intoxication, 65/344
cardiovascular agent intoxication, 37/426
chelonia intoxication, 37/89
chloral hydrate intoxication, 37/348
cinchonism, 65/452
cluster headache, 48/222
crab intoxication, 37/58
cubomedusae intoxication, 37/43
digoxin intoxication, 65/450
dysbarism, 63/416
echinoidea intoxication, 37/67
epileptic, 15/95
ethchlorvynol intoxication, 37/354
fish roe intoxication, 37/86
glycoside intoxication, 37/426
gymnapistes intoxication, 37/76
headache, 75/286
hexachlorophene intoxication, 37/486
hyponatremia, 63/545
ichthyohaemotix fish intoxication, 37/87
ichthyohepatoxic fish intoxication, 37/87
intracranial hypertension, 16/129
lateral medullary infarction, 53/381
loxosceles intoxication, 37/112
marine toxin intoxication, 37/33, 65/153
meningococcal meningitis, 33/23, 52/24
merostomata intoxication, 37/58
methanol intoxication, 36/353, 64/98
methyl salicylate intoxication, 37/417
migraine, 48/3, 8, 35, 70, 125, 156
mountain sickness, 63/416
nicotine intoxication, 36/427, 65/262
nothesthes intoxication, 37/76
organolead intoxication, 64/133
organotin intoxication, 64/138
ostrea gigas intoxication, 37/64
paraldehyde intoxication, 37/349
paralytic shellfish poisoning, 37/33, 65/153
pheochromocytoma, 39/498, 500, 42/764, 48/427
primary amyloidosis, 51/415
pterois volitans intoxication, 37/75
puffer fish intoxication, 37/80
salicylic acid intoxication, 37/417
scorpaena intoxication, 37/76
scorpaenodes intoxication, 37/76
scorpaenopsis intoxication, 37/76
sea snake intoxication, 37/92
sebastapistes intoxication, 37/76
sodium salicylate intoxication, 37/417
stingray intoxication, 37/72
traumatic vegetative syndrome, 24/577
trichloroethanol intoxication, 65/329

uremia, 63/504
urticaria pigmentosa, 14/790
venerupis semidecussata intoxication, 37/64
vestibular neuronitis, 56/117
vitamin A intoxication, 37/96, 65/568
zinc intoxication, 36/337
zozymus aeneus intoxication, 37/56
Navaja neuropathy
ophthalmic finding, 74/431
Navel colic, *see* Abdominal migraine
Navelbine
neuropathy, 69/460
Nebulin
muscle fiber, 62/26
Neck
bladder, *see* Bladder neck
rigidity, *see* Nuchal rigidity
short, *see* Short neck
webbed, *see* Pterygium colli
wry, *see* Wry neck
Neck chamber
autonomic nervous system, 75/434
blood pressure, 75/434
carotid artery, 75/434
Neck injury
acute infantile hemiplegia, 12/345
carotid dissecting aneurysm, 54/271
cervicogenic headache, 48/124
chiropractic manipulation, 53/377
delayed myelopathy, 12/608
Neck muscle
headache, 5/20
weakness, 40/308
Neck pain
carotid dissecting aneurysm, 53/205
carotidynia, 5/375, 48/338
headache, 48/406
traumatic intracranial hematoma, 57/265
Neck reflex
tonic, *see* Tonic neck reflex
Neck righting reflex
childhood motor development, 4/348
Neck rigidity, *see* Nuchal rigidity
Neck stiffness, *see* Nuchal rigidity
Necrobiosis
nerve cell degeneration, 21/47
selective nucleoperikaryon, *see* Selective nucleoperikaryon necrobiosis
Necrosis
acute hepatic, *see* Acute hepatic necrosis
acute tubular, *see* Acute tubular necrosis
adenohypophyseal postpartum, *see* Adenohypophyseal postpartum necrosis
Becker muscular dystrophy, 40/389
bilateral pallidal, *see* Bilateral pallidal necrosis
bilateral striatal, *see* Bilateral striatal necrosis
boxjelly intoxication, 37/41
brain cortex laminar, *see* Brain cortex laminar necrosis
brain miliary, *see* Brain miliary necrosis
central cord, *see* Central cord necrosis
central hemorrhagic, *see* Central hemorrhagic necrosis
cerebral, *see* Cerebral necrosis
conus medullaris central, *see* Conus medullaris central necrosis
corpus callosum, 17/542
cortical laminar, *see* Cortical laminar necrosis
denervation, 40/23
Duchenne muscular dystrophy, 40/251, 368
familial holotopistic striatal, *see* Hereditary striatal necrosis
fibrinoid, *see* Arteriolar fibrinohyalinoid degeneration
generalized, 40/20, 74, 206
glycogen, 27/210
gray matter, *see* Gray matter necrosis
hemorrhagic, *see* Hemorrhagic necrosis
hereditary putaminal, *see* Hereditary striatal necrosis
hereditary striatal, *see* Hereditary striatal necrosis
inflammatory myopathy, 40/264
ischemic, *see* Ischemic necrosis
lacunar, *see* Lacunar necrosis
laminar, *see* Laminar necrosis
limb girdle syndrome, 40/252, 453
liquefaction, *see* Liquefaction necrosis
multifocal, *see* Multifocal necrosis
muscle, *see* Muscle necrosis
muscle fiber, *see* Muscle necrosis
neuromuscular disease, 62/6
pallidal, *see* Pallidal necrosis
pallidonigral, *see* Pallidonigral necrosis
pallidostriatal, *see* Pallidostriatal necrosis
periventricular, *see* Periventricular necrosis
pituitary hemorrhagic, *see* Pituitary hemorrhagic necrosis
putaminal, *see* Putaminal necrosis
sea wasp intoxication, 37/41
segmental fibrinoid, *see* Segmental fibrinoid necrosis
spinal cord, *see* Myelomalacia
spongy rarefactive, *see* Spongy rarefactive necrosis
striatal, *see* Striatal necrosis
systemic vasculitis, 56/301

tubular, *see* Tubular necrosis
white matter, *see* White matter necrosis
Necrotic myelopathy
diamorphine myelopathy, 65/356
Necrotizing demyelination
macroglobulinemia, 39/532
Necrotizing encephalitis
acute, *see* Hyperacute disseminated encephalomyelitis
aspergillosis, 52/379
cerebral toxoplasmosis, 52/357
Chagas disease, 52/346
cytomegalic inclusion body disease, 56/271
herpes simplex virus, 71/266
multifocal, 52/379
Necrotizing encephalomyelopathy
cocarboxylase, 28/355
subacute, *see* Subacute necrotizing encephalomyelopathy
Necrotizing hemorrhagic leukoencephalitis
acute, *see* Acute necrotizing hemorrhagic leukoencephalitis
Necrotizing leukoencephalopathy
disseminated, *see* Disseminated necrotizing leukoencephalopathy
irradiation, 67/358
multifocal, *see* Multifocal necrotizing leukoencephalopathy
neurotoxicity, 67/358
Necrotizing myelitis
acquired immune deficiency syndrome, 71/263
Necrotizing myelomalacia
angiodysgenetic, *see* Angiodysgenetic necrotizing myelomalacia
Necrotizing myelopathy, *see* Foix-Alajouanine disease
Necrotizing myopathy
paraneoplastic syndrome, 71/681
toxic myopathy, 62/598
vitamin E intoxication, 65/570
Necrotizing myositis
Kawasaki syndrome, 56/639
Necrotizing panarteritis
tuberculous meningitis, 52/200
Necrotizing vasculitis
Kawasaki syndrome, 56/641
Wegener granulomatosis, 71/178
Negative delta sign
venous sinus thrombosis, 54/402
Negative hallucination
hallucinosis, 46/562
Negative symptom syndrome
schizophrenia, 46/501
Neglect
frontal lobe lesion, 45/31
hemiakinesia, 45/16
hemi-inattention, 45/16
hemiplegia, 2/695
hemispatial, 45/167-172, 174
intention, 45/161-167
motor, *see* Motor neglect
neuropathology, 45/175
neuropsychology, 45/16
parietal lobe, 45/16, 172
sensory, *see* Sensory neglect
temporal lobe syndrome, 45/44
thalamic, *see* Thalamic neglect
visuospatial agnosia, 45/16
Neglect syndrome, 45/153-178
acetylcholine, 45/155
akinesia, 45/161-167
akinetic mutism, 45/163
amorphosynthesis, 45/155
anosognosia, 45/178
body scheme disorder, 45/380
brain cortex, 45/157
brain hemorrhage, 45/175
brain infarction, 45/175
brain tumor, 45/175
cerebral dominance, 45/172-175
consciousness, 45/115
contingent negative variation, 45/163
definition, 45/153
EEG, 45/155, 157
epilepsy, 45/175
hemi-inattention, 45/153-158
lateral geniculate nucleus, 45/156
medial geniculate nucleus, 45/156
mesencephalic reticular formation, 45/155
neuropathology, 45/175
parietal lobe, 45/172
prognosis, 45/175-178
reaction time, 45/173
recovery, 45/175-178
rehabilitation, 45/178
right hemisphere, 45/173
sensory extinction, 45/158-161
thalamus, 45/156
visuospatial agnosia, 45/167-172
Negligence
self, *see* Self-negligence
Negri inclusion body
rabies, 34/236, 56/383, 394
Neisseria
bacterial endocarditis, 52/302
childhood myoglobinuria, 62/564

Neisseria meningitidis
  see also Meningococcal meningitis
  bacterial meningitis, 55/416
  bacteriology, 52/23
  brain metastasis, 69/116
  endotoxin, 52/23
  meningitis, *see* Meningococcal meningitis
  serogroup, 52/21, 23
  spastic paraplegia, 59/436
Neisseria meningitidis meningitis, *see* Meningococcal meningitis
Nelson syndrome
  see also Cushing syndrome
  ACTH induced myopathy, 62/537
  adrenalectomy, 68/359
  EMG, 62/537
  fatigue, 62/537
  hyperpigmentation, 62/537
  muscle biopsy, 62/537
  muscle weakness, 62/537
Nemaline myopathy
  see also Rod body
  actin, 41/10
  α-actinin, 41/10
  antidesmin antibody, 62/344
  areflexia, 43/122
  cardiomyopathy, 62/341
  classification, 40/287
  clinical course, 62/342
  clinical features, 40/306, 335, 337
  concentric laminated body, 40/109
  congenital myopathy, 41/8-11, 62/332
  cricopharyngeal dysphagia, 62/342
  differential diagnosis, 22/99, 40/422, 41/372
  facial paralysis, 43/122
  facial weakness, 62/334
  facioscapulohumeral syndrome, 62/168
  fatal form, 62/341
  Fazio-Londe disease, 22/108
  feeding disorder, 43/122
  genetics, 41/432, 62/342
  hereditary motor and sensory neuropathy type I, 60/247
  hereditary motor and sensory neuropathy variant, 60/247
  histochemistry, 40/46
  histopathology, 62/342
  hyporeflexia, 43/122
  inclusion body, 62/342
  kyphoscoliosis, 43/122
  muscle fiber type I, 41/10, 43/122
  muscle fiber type I predominance, 41/10
  myogranule, 62/341
  myotonic dystrophy, 62/240
  ophthalmoplegia, 62/342
  pathologic reaction, 40/259, 287
  pectus carinatum, 43/122
  proximal muscle weakness, 43/122
  rod body, 40/259, 287
  scapuloperoneal syndrome, 40/426
  sex ratio, 43/122
  simulating limb girdle dystrophy, 40/280
  talipes, 43/122
  temporomandibular ankylosis, 62/342
  ultrastructure, 40/87
  Z band, 41/10, 43/122
Nematode
  Angiostrongylus cantonensis, 35/328, 331, 52/545
  Filaria, 52/513
  helminthiasis, 52/505, 509
  Schistosoma, 52/535
  Trichinella, 52/563
Neocerebellar syndrome
  see also Archicerebellar syndrome, Cerebellar syndrome *and* Paleocerebellar syndrome
  adiadochokinesia, 2/422
  archicerebellar syndrome, 2/417
  arrhythmokinesis, 2/422
  associated movement loss, 2/423
  asthenia, 2/420
  clinical symptomatology, 2/417
  corticopontocerebellar deficiency, 2/419-423
  dysarthria, 2/423
  dysmetria, 2/421
  hypotonia, 2/419
  limb deviation, 2/420
  movement delay, 2/420
  movement rate, 2/421
  pastpointing, 2/420
  pendulous knee jerk, 2/420
  rhythmic stabilization test, 2/421
  rombergism, 2/420
  sensation disturbance, 2/423
  static tremor, 2/420
  Stewart-Holmes sign, 2/420
  tremor, 2/420, 423
Neocerebellum
  concept, 1/329
Neocortex
  autonomic nervous system, 74/2
  pharmacology, 74/159
Neologistic jargon
  aphasia, 45/322
  Wernicke aphasia, 45/314
Neonatal adrenoleukodystrophy
  computer assisted tomography, 60/671

deafness, 60/657
distal amyotrophy, 60/659
dysmorphic features, 47/595
epilepsy, 47/595
facial dysmorphia, 60/666
hepatomegaly, 60/668
nuclear magnetic resonance, 60/673
peroxisomal disorder, 66/510
phytanic acid, 60/674
psychomotor retardation, 47/595
retinopathy, 47/595
secondary pigmentary retinal degeneration, 60/654, 729
X-linked adrenoleukodystrophy, 66/512
Neonatal asphyxia, *see* Newborn asphyxia
Neonatal asymptomatic cytomegalovirus
cytomegalovirus infection, 34/217
Neonatal convulsion
benign, *see* Benign neonatal convulsion
epilepsy, 73/235
Neonatal encephalopathy
brain injury, 9/578
developmental disorder, 9/578
Neonatal epilepsy, *see* Newborn epilepsy
Neonatal meningitis, *see* Infantile enteric bacillary meningitis
Neonatal myasthenia gravis
crying, 41/100
facial paralysis, 41/100
ptosis, 41/100
respiratory complication, 41/100
weakness, 41/100
Neonatal myoclonus
epilepsy, 73/242
infantile myoclonus, 73/242
Neonatal seizure
benzodiazepine, 73/257
classification, 72/8
definition, 72/8, 16
differential diagnosis, 73/253
EEG, 73/256
epilepsy, 73/251
etiology, 73/254
inborn error of metabolism, 73/255
infection, 73/255
lidocaine, 73/258
local anesthesia, 73/256
metabolic disorder, 73/255
neuroimaging, 73/257
paraldehyde, 73/258
phenobarbital, 73/257
phenytoin, 73/257
primidone, 73/258
pyridoxine dependency, 73/258
therapy, 73/257
withdrawal, 73/256
Neonatal sleep
benign, *see* Benign neonatal sleep
Neonatal Sly disease, *see* Glycosaminoglycanosis type VII
Neophocaena phocaenoides intoxication
abdominal pain, 37/95
cyanosis, 37/95
death, 37/95
hypersalivation, 37/95
ingestion, 37/95
mortality rate, 37/95
paralysis, 37/95
paresthesia, 37/95
tongue swelling, 37/95
Neoplasia, *see* Cancer *and* Tumor
Neoplasia syndrome
multiple endocrine, *see* Multiple endocrine neoplasia syndrome
multiple hamartoma syndrome, *see* Cowden syndrome
Neoplasm
amyotrophic lateral sclerosis, 22/31
B-lymphocyte, *see* B-lymphocyte neoplasm
brain, *see* Brain tumor
brain hemorrhage, 54/288
brain stem, *see* Brain stem neoplasm
chorea, 49/550
choroid plexus, *see* Choroid plexus neoplasm
downbeating nystagmus, 60/656
Eaton-Lambert myasthenic syndrome, 62/421
epilepsy, 72/352
glossopharyngeal neuralgia, 48/464
human immunodeficiency virus infection, 71/340
intracranial, *see* Intracranial neoplasm
late onset myopathy, 41/82
occipital neuralgia, 19/39
olivopontocerebellar atrophy (Dejerine-Thomas), 60/516
olivopontocerebellar atrophy (Wadia-Swami), 60/497
orbital floor syndrome, 2/88
pneumocephalus, 24/208
progressive multifocal leukoencephalopathy, 71/419
subarachnoid hemorrhage, 55/30
symptomatic dystonia, 6/560
Neoplastic aneurysm
brain aneurysm, 55/80
brain aneurysm neurosurgery, 55/80
subarachnoid hemorrhage, 55/80

Neoplastic angioendotheliomatosis, *see* Intravascular malignant lymphomatosis
Neoplastic meningitis
leptomeningeal metastasis, 69/158
therapy, 69/158
Neoplastic myopathy
azathioprine, 41/374
Neoplastic neuropathy, *see* Paraneoplastic neuropathy
Neoplastic satellitosis
astrocytoma, 10/6
glial cell, 10/6
Neosaxitoxin
neurotoxin, 65/146
paralytic shellfish poisoning, 65/151
Neosebastes nigropunctatus McCulloch intoxication
picture, 37/73
Neospinothalamic tract
pain, 45/234
Neostigmine
cholinesterase inhibition, 37/541
myasthenia gravis, 41/96, 108, 43/158, 62/417
Neostigmine methyl sulfate
myasthenia gravis, 41/127, 131, 62/413
Neostriatal degeneration
carbamate intoxication, 64/188
carbon disulfide intoxication, 64/188
disulfiram, 64/188
Neostriatal GABA
reduced, *see* Reduced neostriatal GABA
Neostriatal glutamic acid decarboxylase
reduced, *see* Reduced neostriatal glutamic acid decarboxylase
Neostriatal model
Parkinson disease, 49/53
Neostriatum
see also Striatum
anatomy, 6/3
aspiny neuron type I, 49/3
aspiny neuron type II, 49/3
aspiny neuron type III, 49/3
cholecystokinin octapeptide, 49/47
connection, 49/3
cytology, 49/2
dorsal, *see* Dorsal neostriatum
GABA, 49/47
glutamic acid, 49/47
hemiballismus, 49/370
limbic part, 49/4
neurotensin, 49/47
neurotransmitter, 49/47
nonlimbic part, 49/4
physiology, 49/22
spiny neuron type I, 49/3
spiny neuron type II, 49/3
striatal afferent, 49/5
substance P, 49/47
ventral, *see* Ventral neostriatum
Nephritis
Alport syndrome, 22/487, 42/376
granulomatous CNS vasculitis, 55/388
hemorrhagic, *see* Hemorrhagic nephritis
hereditary, *see* Hereditary nephritis
hereditary interstitial, *see* Hereditary interstitial nephritis
immune complex, *see* Immune complex nephritis
interstitial, *see* Interstitial nephritis
Lemieux-Neemeh syndrome, 42/370
Muckle-Wells syndrome, 22/489, 42/396
polyarteritis nodosa, 39/296, 55/354
sensorineural deafness, 42/370, 380, 383, 396
systemic lupus erythematosus, 43/420
Nephrolithiasis
cerebrotendinous xanthomatosis, 66/601
choreoathetosis, 49/385
Wilson disease, 49/229
Nephropathy, *see* Renal disease
Nephrotic syndrome
Galloway-Mowat syndrome, 43/431
hyponatremia, 28/499
penicillamine intoxication, 27/405
primary amyloidosis, 51/415
Nephrotoxicity
bismuth intoxication, 64/331, 337
ciclosporin intoxication, 65/551
lithium intoxication, 64/294
mercury polyneuropathy, 51/272
orellanine intoxication, 65/38
Neptunea intoxication
choline, 37/64
choline ester, 37/64
dizziness, 37/64
headache, 37/64
histamine, 37/64
ingestion, 37/64
tetramethylammonium hydroxide, 37/64
urticaria, 37/64
visual disorder, 37/64
vomiting, 37/64
Nereistoxin intoxication
bradycardia, 37/55
chemical structure, 37/55
lacrimation, 37/56
lumbriconereis heteropoda, 37/55
mammal, 37/55
meiosis, 37/55

tachycardia, 37/55
Nernst equation
nerve physiology, 7/63
Nerve
abducent, *see* Abducent nerve
accessory, *see* Accessory nerve
acoustic, *see* Acoustic nerve
actin, 51/6
anterior thoracic, *see* Anterior thoracic nerve
astrocyte, 51/11
auriculotemporal, *see* Auriculotemporal nerve
autonomic, *see* Autonomic nerve
axillary, *see* Axillary nerve
axon myelination ratio, 51/4
axon-neuron ratio, 51/4
axon organelle, 51/6
axon-perikaryon volume ratio, 51/5
axonal cytoskeleton, 51/6
bladder, 74/112
blood-nerve barrier, 51/2
blood supply, 12/644
brachial plexus, 2/135, 137
carotid sinus, *see* Carotid sinus nerve
classification, 7/199
CNS-peripheral nervous system compound axon, 51/1
CNS transitional zone, 51/11
cochlear, *see* Cochlear nerve
cold, 1/449
collagen network, 51/2
common peroneal, *see* Common peroneal nerve
compression neuropathy, 51/2
conduction property, 51/41
conduction velocity, 51/8, 44
cranial, *see* Cranial nerve
cutaneous, *see* Cutaneous nerve
cutaneous femoral, *see* Cutaneous femoral nerve
demyelination, 51/2, 52
dermal, *see* Dermal nerve
development, 7/1, 51/1, 16
digital, *see* Digital nerve
dorsal scapular, *see* Dorsal scapular nerve
electrostimulation, 7/62
embryology, 7/3
endoneurium, 51/2
endoplasmic reticulum, 51/6
epineurium, 51/1
ethmoidal, *see* Ethmoidal nerve
facial, *see* Facial nerve
fasciculus formation, 51/3
femoral, *see* Femoral nerve
fiber branching, 51/4
fibroma, 8/460
function, 7/62
genital tract, 74/113
genitofemoral, *see* Genitofemoral nerve
glossopharyngeal, *see* Glossopharyngeal nerve
gluteal, *see* Gluteal nerve
great auricular, *see* Auricularis magnus nerve
greater occipital, *see* Greater occipital nerve
growth, 51/18
gustatory, *see* Gustatory nerve
hemangioma, 8/459
hemophilia, 38/61-63
Hering, *see* Hering nerve
histology, 7/26
histometric study, 51/3
hypertrophic, *see* Hypertrophic nerve
hypogastric, *see* Hypogastric nerve
hypoglossal, *see* Hypoglossal nerve
iliohypogastric, *see* Iliohypogastric nerve
ilioinguinal, *see* Ilioinguinal nerve
impulse conduction, 51/44
infectious disease, *see* Neuritis
infraorbital, *see* Infraorbital nerve
intercostal, *see* Intercostal nerve
intermediate, *see* Intermediate nerve
intermedius, *see* Intermedius nerve
internode length, 51/8
intrinsic, *see* Intrinsic nerve
iris, 74/400
Jacobson, *see* Jacobson nerve
lateral cutaneous, *see* Lateral cutaneous nerve
lateral cutaneous femoral, *see* Lateral femoral cutaneous nerve
lateral femoral cutaneous, *see* Lateral femoral cutaneous nerve
lateral popliteal, *see* Lateral popliteal nerve
length measurement, 7/130
leprosy, 33/441-448
Lesser occipital, *see* Lesser occipital nerve
long thoracic, *see* Long thoracic nerve
medial cutaneous antebrachial, *see* Medial cutaneous antebrachial nerve
medial cutaneous brachial, *see* Medial cutaneous brachial nerve
median, *see* Median nerve
metabolism, 7/74
microfilament, 51/6
motor unit hyperactivity, 41/299-303
multiple cranial, *see* Multiple cranial nerve
muscle innervation, 2/132
muscle tissue culture, 40/186
musculocutaneous, *see* Musculocutaneous nerve
myelin thickness, 51/8
myelinated, *see* Myelinated nerve fiber

Nageotte, *see* Nageotte nerve
nerve sheath, 51/1
neurite growth, 51/16
neurofilament, 51/5
neurotoxicity, 51/263
neurotubule, 51/5
obturator, *see* Obturator nerve
occipital, *see* Occipital nerve
octavus, *see* Octavus nerve
oculomotor, *see* Oculomotor nerve
olfactory, *see* Olfactory nerve
ophthalmic, *see* Ophthalmic nerve
optic, *see* Optic nerve
osteitis deformans (Paget), 38/367
paleoneurobiology, 74/181
paraneoplastic syndrome, 69/373
parasympathetic, *see* Parasympathetic nerve
pathology, 7/197-233
pathology terminology, 7/199
pelvic, *see* Pelvic nerve
perineurium, 51/2
peripheral, *see* Peripheral nerve
peripheral nervous system, 51/1
peripheral nervous system transitional zone, 51/11
peroneal, *see* Peroneal nerve
petrosal, *see* Petrosal nerve
phrenic, *see* Phrenic nerve
physiology, 7/63
posterior cutaneous femoral, *see* Posterior cutaneous femoral nerve
posterior tibial, *see* Posterior tibial nerve
postganglionic nerve fiber, 51/1
proximal part, 7/218
pudendal, *see* Pudendal nerve
radial, *see* Radial nerve
radiotrauma, *see* Ionizing radiation neuropathy
Ranvier node, 7/22, 51/5, 8, 12
recurrent laryngeal, *see* Recurrent laryngeal nerve
Renaut body, 51/2
sacral, *see* Sacral nerve
saphenous, *see* Saphenous nerve
Schmidt-Lantermann incisure, 51/9, 12
Schwann cell, 51/1, 9, 11
sciatic, *see* Sciatic nerve
sensorimotor, *see* Sensorimotor nerve
sensory, *see* Sensory nerve
sinuvertebral, *see* Sinuvertebral nerve (Luschka)
skin, *see* Cutaneous nerve
spinal, *see* Spinal nerve
spinal accessory, *see* Spinal accessory nerve
spinal nerve root sheath, 51/2
statoacoustic, *see* Statoacoustic nerve
stress, 74/317
structure, 51/1
structure-function relation, 51/45
subclavius, *see* Subclavius nerve
subscapular, *see* Subscapular nerve
superior laryngeal, *see* Superior laryngeal nerve
supraclavicular, *see* Supraclavicular nerve
supraorbital, *see* Supraorbital nerve
suprascapular, *see* Suprascapular nerve
sural, *see* Sural nerve
sympathetic, *see* Sympathetic nerve
thoracodorsal, *see* Thoracodorsal nerve
thrombocytopenia, 38/69
tibial, *see* Tibial nerve
tight junction, 51/2
transport, 51/6
transport mechanism, 7/81
trauma, *see* Nerve injury
trigeminal, *see* Trigeminal nerve
trochlear, *see* Trochlear nerve
trophic function, 7/89
truncal, *see* Truncal nerve
trypsin resistant protein residue, 9/32
tumor, *see* Nerve tumor
ulnar, *see* Ulnar nerve
unmyelinated nerve fiber, 51/3, 7
urethra, 74/112
urinary tract, 74/112
vagus, *see* Vagus nerve
vascular anatomy, 8/154
vascularization, *see* Nerve vascularization
venous drainage, 12/646
Vidian, *see* Vidian nerve
wallerian degeneration, 7/153
Nerve abnormality
hypoglossal, *see* Hypoglossal nerve abnormality
Nerve action potential
acrylamide intoxication, 64/70
hereditary multiple recurrent mononeuropathy, 51/552
tetrodotoxin intoxication, 65/156
Nerve agenesis
abducent, *see* Abducent nerve agenesis
accessory, *see* Accessory nerve agenesis
acoustic, *see* Acoustic nerve agenesis
cranial, *see* Cranial nerve agenesis
facial, *see* Facial nerve agenesis
glossopharyngeal, *see* Glossopharyngeal nerve agenesis
hypoglossal, *see* Hypoglossal nerve agenesis
oculomotor, *see* Oculomotor nerve agenesis
optic, *see* Optic nerve agenesis
trigeminal, *see* Trigeminal nerve agenesis
trochlear, *see* Trochlear nerve agenesis

vagus, *see* Vagus nerve agenesis
Nerve antibody
serum, *see* Serum nerve antibody
Nerve aplasia
olfactory, *see* Olfactory nerve aplasia
optic, *see* Optic nerve aplasia
Nerve atrophy
optic, *see* Optic atrophy
vestibular, *see* Vestibular nerve atrophy
Nerve biopsy
a-alphalipoproteinemia, 51/397
adrenoleukodystrophy, 51/384
amiodarone intoxication, 65/458
arsenic polyneuropathy, 51/267
autosomal dominant pain insensitivity, 60/17
Charlevoix-Saguenay spastic ataxia, 60/456
chorea-acanthocytosis, 49/331
chronic inflammatory demyelinating polyradiculoneuropathy, 51/533
congenital autonomic dysfunction, 60/18
cryoglobulinemia, 63/404
eosinophilia myalgia syndrome, 63/377, 64/254
Friedreich ataxia, 21/81, 325
giant axonal neuropathy, 60/79
Guillain-Barré syndrome, 51/244
hereditary acute porphyria, 51/391
hereditary amyloid polyneuropathy, 21/135-137
hereditary brachial plexus neuropathy, 60/72
hereditary motor and sensory neuropathy type I, 60/195-199
hereditary motor and sensory neuropathy type II, 60/195-199
hereditary multiple recurrent mononeuropathy, 51/552, 60/63
hereditary neuropathy, 60/259
hereditary recurrent neuropathy, 21/90-98
hereditary sensory and autonomic neuropathy type I, 21/81
hereditary sensory and autonomic neuropathy type II, 60/11, 17, 18
hereditary sensory and autonomic neuropathy type III, 21/113, 60/31
hereditary sensory and autonomic neuropathy type IV, 60/13
hereditary sensory and autonomic neuropathy type V, 60/15
hypereosinophilic syndrome, 63/373
hypertrophic interstitial neuropathy, 60/214
metachromatic leukodystrophy, 8/16, 60/126
migrant sensory neuritis, 51/462
olivopontocerebellar atrophy (Wadia-Swami), 60/501
polyneuropathy, 60/259
porphyria, 60/118
progressive muscular atrophy, 59/16
Refsum disease, 21/240
scapuloperoneal spinal muscular atrophy, 62/170
spinocerebellar degeneration, 60/678
sural, *see* Sural nerve biopsy
toxic oil syndrome, 63/381
undetermined significant monoclonal gammopathy, 63/399, 402
vitamin B6 intoxication, 65/572
X-linked hereditary sensory and autonomic neuropathy, 60/17
Nerve block
alpha rigidity, 41/307
laryngeal, *see* Laryngeal nerve block
pain, 26/496, 69/49
spinal pain, 26/496
Nerve cell
adrenergic, *see* Adrenergic neuron
aging, 74/227
aspiny, *see* Aspiny neuron
autonomic, *see* Autonomic neuron
calcium, 72/50
calcium current, 72/45
death, *see* Neuron death
degenerations, *see* Neuron degeneration
dopaminergic, *see* Dopaminergic nerve cell
epilepsy, 15/36
epileptic focus, 15/36
excitability, 72/43
excitation, 72/47
field effect, 72/50
functional connectivity, 72/47
gap junction, 72/50
general, 1/56
globoid cell leukodystrophy, 10/77
inhibition, 72/47
intrinsic membrane stability, 72/45
ionic current, 72/47
kinesthetic, *see* Kinesthetic neuron
magnesium, 72/50
medium spiny, *see* Medium spiny neuron
mercury intoxication, 64/381
muscarinic cholinergic, *see* Muscarinic cholinergic nerve cell
origin, 30/16
potassium, 72/50
potassium current, 72/46
radiation injury, 23/646
resistance, 9/625
retinal ganglion cell, 2/517
Sherrington study, 1/47
sodium current, 72/44

spontaneous activity, 1/60
synaptic connectivity, 72/43
synchronization, 72/49
trophic function, 1/59
Nerve cell adhesion molecule
undetermined significant monoclonal gammopathy, 63/400
Nerve compression
see also Compression neuropathy
acoustic, *see* Acoustic nerve compression
brachial neuritis, 7/430
brachial plexus neuritis, 7/430
chronic, *see* Chronic nerve compression
hereditary multiple recurrent mononeuropathy, 51/559
intermittent, *see* Intermittent nerve compression
lumbar plexus, 7/313
median, *see* Median nerve compression
nerve conduction, 7/164
nerve function, 21/101
nerve injury, 51/133
nerve intermittent claudication, 20/792-795
orbital apex syndrome, 2/570
peroneal, *see* Peroneal nerve compression
peroneal nerve conduction, 7/164
recurrent laryngeal, *see* Recurrent laryngeal nerve compression
tibial, *see* Tibial nerve compression
Nerve conduction
see also Action potential
acute intermittent porphyria, 7/169
age, 7/133, 144
amyotrophic lateral sclerosis, 59/189
antidromic response, 7/148
brachial plexus neuritis, 7/178
bronchogenic carcinoma, 7/167
calcium, 47/39
carbon disulfide intoxication, 7/174
carpal tunnel syndrome, 7/157-160, 42/309
cerebrotendinous xanthomatosis, 66/608
chronic progressive polyneuritis, 7/178
ciguatoxin intoxication, 65/161
compression neuropathy, 7/157
controlling factor, 7/62
deficiency neuropathy, 7/166
diabetic neuropathy, 7/169
diabetic polyneuropathy, 27/103
dimethyl sulfoxide, 7/174
diphtheritic neuropathy, 7/155
electrode, 7/120, 129-131, 139
EMG, 1/640
eosinophilia myalgia syndrome, 63/377
experimental allergic neuritis, 7/178
Fabry disease, 51/376
fast fiber loss, 7/151
fatigue, 7/132
globoid cell leukodystrophy, 51/373
glowing, 7/119
hepatitis, 7/609
hereditary motor and sensory neuropathy type I, 7/179
hereditary motor and sensory neuropathy type II, 7/179
hereditary multiple recurrent mononeuropathy, 42/316, 51/553
hereditary neuropathy, 7/179
hereditary sensory and autonomic neuropathy type I, 60/9
hereditary sensory and autonomic neuropathy type II, 7/187, 60/11
hereditary sensory and autonomic neuropathy type IV, 60/13
human T-lymphotropic virus type I associated myelopathy, 56/537
hyperinsulin neuropathy, 7/166
hyperkalemia, 7/169
hypertrophic polyneuropathy, 7/180
hypocalcemia, 47/39
hypokalemia, 7/169
hypoxia, 7/169
ionic base, 28/463
ischemia, 7/132
isoniazid, 7/174
Kawasaki syndrome, 56/639
lead intoxication, 7/173
local anesthetic agent, 65/419
macroglobulinemia, 7/167
marine toxin intoxication, 65/161
measurement, 1/640
meningoradiculitis, 7/178
mercury intoxication, 7/173, 514
metachromatic leukodystrophy, 10/58, 51/368
motor, *see* Motor nerve conduction
motor nerve, 7/118
multiple sulfatase deficiency, 51/372
myxedema neuropathy, 7/169
nerve compression, 7/164
nerve diameter, 7/137
nerve ischemia, 7/132
paralytic shellfish poisoning, 37/35, 65/152
pellagra, 7/166
phenytoin, 7/174
phosphene, 47/32
polymyositis, 71/137
polyradiculitis, 7/175
pressure, 7/132

proximal distal, 7/135
Refsum disease, 7/181, 21/197, 42/149
renal insufficiency, 27/340
renal insufficiency polyneuropathy, 51/358
rheumatoid neuropathy, 7/169
segmental demyelination, 7/155
sensory, *see* Sensory nerve conduction
sensory neuropathy, 7/167
study, 51/553
subacute polyneuritis, 7/178
syringomyelia, 50/445
temperature, 7/132
tetrodotoxin intoxication, 65/156
thalidomide, 7/174
thoracic outlet syndrome, 51/125
trichloroethylene, 7/174
tri-*o*-cresyl phosphate, 7/173
tropical spastic paraplegia, 56/537
ultrasound, 7/132
uremia, 7/166
vascular neuropathy, 7/165
vitamin E deficiency, 7/166, 70/445
wallerian degeneration, 7/153
xeroderma pigmentosum, 60/337

Nerve conduction block
amyotrophic lateral sclerosis, 47/368
experimental allergic encephalomyelitis, 47/367
local anesthetic agent, 65/419
multiple sclerosis, 47/367
neurapraxia, 7/151
saxitoxin, 65/152
tetrodotoxin intoxication, 65/155

Nerve conduction velocity
see also Motor conduction velocity *and* Sensory conduction velocity
acetonylacetone intoxication, 64/87
acrylamide intoxication, 64/69
ADR syndrome, 42/113
adrenoleukodystrophy, 60/670
amyloid polyneuropathy, 51/417
amyotrophic lateral sclerosis, 22/303
ataxia, 60/670
ataxia telangiectasia, 60/670
Bassen-Kornzweig syndrome, 51/394
benign peroxisomal dysgenesis, 60/670
cerebrotendinous xanthomatosis, 60/670
Charlevoix-Saguenay spastic ataxia, 60/670
chorea-acanthocytosis, 51/398
chronic mercury intoxication, 64/372
ciguatoxin intoxication, 65/163
cisplatin intoxication, 64/358
Cockayne-Neill-Dingwall disease, 51/399
cold neuropathy, 51/136
compression neuropathy, 42/311
critical illness polyneuropathy, 51/575, 584
cryoglobulinemic polyneuropathy, 51/434
per degree centigrade, 51/137
demyelination, 40/128
diabetic polyneuropathy, 51/503
diphtheria, 65/240
facial spasm, 42/315
familial spastic paraplegia, 59/311
Farber disease, 51/379
fiber diameter, 51/44
Friedreich ataxia, 42/144, 51/388, 60/670
Fukuyama syndrome, 43/91
globoid cell leukodystrophy, 42/489
hereditary motor and sensory neuropathy type I, 21/287, 42/339
hereditary motor and sensory neuropathy type II, 21/287, 42/339
hereditary multiple recurrent mononeuropathy, 42/316
hereditary periodic ataxia, 60/439
hereditary recurrent neuropathy, 21/103
hereditary recurrent polyneuropathy, 21/103
hereditary sensory and autonomic neuropathy type III, 75/151
hereditary sensory and autonomic neuropathy type IV, 42/300
IgM paraproteinemic polyneuropathy, 51/437
infantile spinal muscular atrophy, 22/95
internode distance, 47/19
juvenile metachromatic leukodystrophy, 60/670
Klinefelter variant XXYY, 43/563
late onset ataxia, 60/670
lead intoxication, 36/23
lead polyneuropathy, 64/435, 437
Lemieux-Neemeh syndrome, 42/370
leprous neuritis, 51/221
measurement, 7/116-181, 19/273, 278
metachromatic leukodystrophy, 8/16, 59/356, 60/127
motor, *see* Motor nerve conduction velocity
multiple myeloma polyneuropathy, 51/431
multiple neuronal system degeneration, 42/75
myelin thickness, 51/8
myelinated nerve fiber, 47/12, 19
myelinoclastic diffuse sclerosis, 47/425
nerve diameter, 7/151
neuropathy, 60/670
oculopharyngeal muscular dystrophy, 43/101
olivopontocerebellar atrophy (Dejerine-Thomas), 60/670
olivopontocerebellar atrophy (Wadia-Swami), 60/670

organotin intoxication, 64/137
polyneuropathy, 42/315, 60/670
postpoliomyelitic amyotrophy, 59/37
progressive muscular atrophy, 59/16
progressive spinal muscular atrophy, 22/37
Refsum disease, 7/181, 8/23, 21/197, 51/385, 60/230, 66/489
remyelination, 47/20
rigidity, 60/670
Rosenberg-Chutorian syndrome, 22/506
Roussy-Lévy syndrome, 21/177, 42/108
secondary pigmentary retinal degeneration, 60/670
sensory radicular neuropathy, 42/351
spinocerebellar degeneration, 42/190
subacute necrotizing encephalomyelopathy, 51/388, 60/670
thermoregulation, 47/20
uremic neuropathy, 7/554
uremic polyneuropathy, 63/510
value, 40/126
Waldenström macroglobulinemia polyneuropathy, 51/433
X-linked ataxia, 60/670
xeroderma pigmentosum, 60/670

Nerve decompression
conduction, 7/161

Nerve diameter
nerve conduction, 7/137
nerve conduction velocity, 7/151
nerve injury, 51/133

Nerve disease
classification, 7/199-201
enteric, 70/325
Greenfield classification, 7/201
Krücke classification, 7/201
Lichtenstein classification, 7/201
Von Romberg classification, 7/200

Nerve ending
muscle, 1/100
sensation, 7/83
specificity, 7/83
ultrastructure, 1/53

Nerve entrapment, *see* Nerve compression

Nerve fiber
aberrant, 14/36
conductivity, 1/64
delta lesion, 41/410
demyelinated, *see* Demyelinated nerve fiber
diameter, 1/51, 117-119
ghost fiber, 41/67
groups, 1/117
histogenesis, 8/423
inflammatory infiltration, 41/61
Lafora body, 41/194
lateral geniculate nucleus, 2/530
mummy, 74/193
muscle, *see* Muscle fiber
muscle fiber type II atrophy, 41/110, 251, 324
myelinated, *see* Myelinated nerve fiber
myelinated retinal, *see* Myelinated retinal nerve fiber
nonmyelinated, *see* Nonmyelinated nerve fiber
postganglionic, *see* Postganglionic nerve fiber
progressive spinal muscular atrophy, 22/17
retinal, *see* Retinal nerve fiber
single terminal varicose, 74/89
sympathetic, *see* Sympathetic nerve fiber
target, 41/5
trigeminal, *see* Trigeminal nerve fiber
tubular aggregate, 41/154, 161
type I predominance, 41/4, 10, 12, 15, 20
ultrastructure, 7/11-13, 16
unmyelinated, *see* Unmyelinated nerve fiber
unmyelinated fiber, *see* Unmyelinated nerve fiber
vacuole, 41/148, 153, 160, 178, 187, 191, 197

Nerve fiber loss
diabetic polyneuropathy, 51/511
xeroderma pigmentosum, 60/337

Nerve gas, *see* Chlormethine intoxication, Cyanide intoxication *and* Organophosphate intoxication

Nerve grafting
vascularized ulnar, *see* Vascularized ulnar nerve grafting

Nerve growth factor
axon regeneration, 61/484
brain metabolism, 57/84
cobra snake venom, 37/12
congenital pain insensitivity, 51/566
head injury, 57/84
neural induction, 30/50
neural plate formation, 30/50
neuroglia relationship, 28/418
radiation myelopathy, 61/210
Schwann cell development, 51/24
spinal cord repair, 61/484

Nerve hematoma
hemophilia, 55/475

Nerve injury
see also Nerve lesion *and* Neuropathy
abducent, *see* Abducent nerve injury
autonomic paralysis, 7/249
axon reflex, 7/252
axonotmesis, 7/244, 51/75, 134
blood-brain barrier, 65/424
blood-nerve barrier, 65/424

burn, 51/136
calcium influx, 65/425
causalgia, 75/316
classification, 7/244, 51/75
clinical aspect, 7/257
cold, 8/162, 51/136
compression neuropathy, 51/134
conduction block, 51/75
contracture, 7/251
contusion, 51/134
dorsal spinal root ganglion, 75/321
electromagnetic wave, 51/139
electrophysiology, 75/318
etiology, 7/245
evaluation, 8/530
evaluation of impairment, 8/530
facial, *see* Facial nerve injury
general symptomatology, 7/247
hematoma, 51/134
hypoglossal, *see* Hypoglossal nerve injury
iatrogenic, *see* Iatrogenic nerve injury
immersion foot, 51/136
inflammation, 75/317
ionizing radiation, 51/134
lightning injury, 7/344, 347-350, 355, 51/134
local anesthetic agent, 65/419-421
lumbosacral, *see* Lumbosacral nerve injury
mechanical injury, 51/134
microwave lesion, 51/134
myokymia, 49/609
nerve compression, 51/133
nerve diameter, 51/133
nerve ischemia, 51/133
nerve stretch, 51/133
neurapraxia, 7/244, 51/75, 134
neurotmesis, 7/244, 51/75, 134
nitric oxide, 65/425
nosology, 51/79
nutritional disturbance, 7/249
oculomotor, *see* Oculomotor nerve injury
open injury, 7/245
optic, *see* Optic nerve injury
pressure, 51/135
prostaglandin, 65/425
radiolesion, 51/134
reactive hyperemia, 7/271
reflex vasomotor response, 7/271
sensory paralysis, 7/273
sudomotor test, 7/251
superior gluteal, *see* Superior gluteal nerve injury
surgical treatment, 8/513
sympathetic nerve sprouting, 75/315
thermolesion, 51/134
toxicity mechanism, 65/425
traction, 7/245
trench foot, 51/136
trochlear, *see* Trochlear nerve injury
ulnar fracture, 70/33
ultrasound, 51/139
ultrastructure, 65/424
vagus, *see* Vagus nerve injury
vascular change, 7/265
vestibulocochlear, *see* Vestibulocochlear nerve injury
vibration, 51/134

Nerve intermittent claudication
compression neuropathy, 20/792-795
fascia lata syndrome, 20/795
meralgia paresthetica, 20/794
Morton neuralgia, 20/793
nerve compression, 20/792-795

Nerve ischemia
amyloidosis, 12/658
arteriosclerosis, 12/654
carpal tunnel syndrome, 12/660
diabetes mellitus, 12/655
diabetic neuropathy, 12/655
hereditary recurrent neuropathy, 21/102
hereditary recurrent polyneuropathy, 21/102
nerve conduction, 7/132
nerve function, 21/101-103
nerve injury, 51/133
polyarteritis nodosa, 12/659, 39/303
tetany, 63/560
thromboangiitis obliterans, 12/655

Nerve lesion
see also Nerve injury
Alföldi sign, 2/15
anesthesia, 1/89
birth incidence, 25/167
botulism, 7/476
causalgia, 7/433, 8/328
clavicle injury, 70/31
cold, 1/449
cutaneous pigmentation, 7/266
cyclist palsy, 7/334
dengue fever, 7/486
diagnostic principle, 2/15
differentiation nerve root, 2/171
diphtheritic neuropathy, 7/475
electrical injury, 7/344, 349, 351, 358, 365, 375, 51/134
erysipelas, 7/487
fracture, 70/26
Gubler swelling, 2/27
hemophilia, 7/666

hepatitis, 7/486
herpes zoster, 7/483
hypotonia, 1/219
incomplete, *see* Incomplete lesion
influenza, 7/478
ionizing radiation, 7/388
leprosy, 7/477
leptospirosis, 7/489
localization, 2/15
malaria, 7/488
median, *see* Median nerve lesion
mononucleosis, 7/489
multiple neuronal system degeneration, 59/142
mumps, 7/485
muscle cramp, 41/300
neurobrucellosis, 7/481
opiate intoxication, 37/387
peroneal, *see* Peroneal nerve lesion
phantom limb, 4/224, 226, 45/397
phrenic, *see* Phrenic nerve lesion
progressive dysautonomia, 59/142
radiation injury, 7/388
recurrent laryngeal, *see* Recurrent laryngeal nerve lesion
rubella, 7/486
sarcoidosis, 7/479
scarlet fever, 7/487
Streptococcus, 7/487
superficial siderosis, 70/66
syphilis, 7/479
tetanus, 7/476
thermoregulation, 7/252, 266
topographical diagnosis, 2/15-51
toxoplasmosis, 7/488
trichinosis, 7/89
trophic disorder, 7/260
tuberculosis, 7/478
typhoid fever, 7/482
ulnar, *see* Ulnar nerve lesion

Nerve lesion topography
diabetic polyneuropathy, 51/510
ulnar nerve, 2/32-38

Nerve myelin
biochemistry, 51/23
carbonate dehydratase, 51/25
component, 51/25
demyelination, 51/32
dolichol phosphate, 51/30
Golgi route, 51/30
lipid, 51/25
metabolism, 51/23, 28
myelin associated glycoprotein, 51/27
myelin associated protein, 51/31
myelin basic protein, 51/25, 31
myelin basic protein *m*RNA, 51/31
myelin turnover, 51/32
neurokeratin, 9/27-34
P0 protein, 51/27, 29-31
P1 protein, 51/27
P2 protein, 51/27
phosphodiesterase, 51/25
protein, 51/25
protein kinase, 51/25
remyelination, 51/35
Schwann cell, 51/23
sphingomyelin, 9/4
sulfatide, 51/29
wallerian degeneration, 9/36
Wolfgram protein, 51/27

Nerve neuralgia
facial, *see* Intermedius neuralgia

Nerve nucleus
cranial, *see* Cranial nerve nucleus

Nerve paralysis
abducent, *see* Abducent nerve paralysis
cranial, *see* Cranial nerve palsy
edema, 8/382
facial, *see* Facial paralysis
glossopharyngeal, *see* Glossopharyngeal nerve agenesis
lateral popliteal, *see* Lateral popliteal nerve palsy
median, *see* Median nerve palsy
oculomotor, *see* Oculomotor paralysis
phrenic, *see* Phrenic nerve palsy
pseudoabducens, *see* Pseudoabducens nerve palsy
radial, *see* Radial nerve palsy
recurrent laryngeal, *see* Recurrent laryngeal nerve palsy
sciatic, *see* Sciatic nerve palsy
sixth nerve, *see* Abducent nerve paralysis
trigeminal, *see* Trigeminal nerve palsy
trochlear, *see* Trochlear nerve paralysis

Nerve pressure palsy
Ehlers-Danlos syndrome, 55/456

Nerve protein nonenzymatic glycosylation
diabetic polyneuropathy, 51/508

Nerve regeneration
collateral neoformation, 7/226-230
conduction, 7/153
conduction velocity, 51/53
myelination, 9/39
oligodendrocyte, 61/481
spinal cord injury, 61/481
terminal axonic neoformation, 7/226
Tinel sign, 1/97

Nerve repair

suture, 8/513
Nerve root, *see* Spinal nerve root
Nerve root avulsion, *see* Spinal nerve root avulsion
Nerve root cyst, *see* Spinal perineural cyst
Nerve root pain
spinal, *see* Spinal nerve root pain
spinal subarachnoid hemorrhage, 55/32
Nerve sheath
metabolism, 7/76
nerve, 51/1
Nerve sheath tumor
brachial plexopathy, 69/79
Nerve stimulation
antibiotic agent, 7/105
botulism, 7/112
method, 7/105
motor nerve, 7/118
myograph recording, 7/106
orthodromic, *see* Orthodromic stimulation
peripheral, 40/149
sural, *see* Sural nerve stimulation
transcutaneous, *see* Transcutaneous nerve stimulation
Nerve stretch
migrant sensory neuritis, 51/462
nerve injury, 51/133
Rübenzieherlähmung, 7/334
Nerve tissue heterotopia
occipital cephalocele, 50/102
Nerve transitional zone
infantile spinal muscular atrophy, 51/12
Nerve trauma, *see* Nerve injury *and* Nerve lesion
Nerve tumor, 8/412-497
acoustic, *see* Acoustic nerve tumor
carotid body tumor, 8/482-493
ciliary body tumor, 8/496
classification, 8/415
cutaneous leiomyoma, 8/458
ependymoma, 8/454
glomus tumor, 8/455-458
Marek disease, 56/582
multiple neurofibromatosis, 8/435-446
neurinoma, 8/417-435
neuroastrocytoma, 8/453, 462-466
neuroblastoma, 8/466-472
neurocutaneous melanosis, 8/454
neurofibroma, 14/136
neurofibromatosis type I, 8/414, 434, 14/154
neuroma, 8/416
neuroxanthoma, 8/458
optic, *see* Optic nerve tumor
paraganglioma abdominale, 8/495
paraganglioma aorticum, 8/494
paraganglioma nodosum, 8/493
paraganglion, 8/472-496
pheochromocytoma, 8/473-482
sarcoma, 8/446-453
sympathetic ganglion, 8/459-472
terminology, 8/412
3 types, 8/412
vagal body, 8/493
Zuckerkandl organ, 8/496
Nerve vascularization, 8/374, 12/644
Nervi erigentes
anatomy, 1/361
physiology, 1/361
Nervi nevorum
autonomic nervous system, 75/703
Nervon, *see* 15-Tetracosanoic acid
Nervonic acid, *see* 15-Tetracosanoic acid
Nervosa
anorexia, *see* Anorexia nervosa
Nervous headache, *see* Psychiatric headache
Nervous hypertension
intracranial, *see* Intracranial nervous hypertension
Nervous lesion
history, 2/4-14
linear localization, 2/1
localization, 2/1-14
point localization, 2/1
regional localization, 2/2
Nervous system
altitude illness, 48/396
amebiasis, *see* Neuroamebiasis
autonomic, *see* Autonomic nervous system
cancer, 71/611
cancer metabolism, 69/399
central, *see* CNS
cytomegalovirus, 71/261
electrical injury, 61/191
enteric, *see* Enteric nervous system
fibrosarcoma, 68/380
herpes virus, 71/261
hyperglycemia, 27/79-95
JC virus, 71/415
lead, 64/434
liver disease, 27/351-353
malignant fibrous histiocytoma, 68/380
metastasis, 71/611
mummy, 74/181
neurofibromatosis type I, 14/132-158
noradrenergic, *see* Noradrenergic nervous system
opportunistic infection, 71/261
paleoneurobiology, 74/181
parasympathetic, *see* Parasympathetic nervous system

peripheral, *see* Peripheral nervous system
primary melanocytic lesion, 68/380
radiation induced tumor, 68/380
rare tumor, 68/379
rhabdoid tumor, 68/380
rhabdomyosarcoma, 68/380
secondary syphilis, 33/349
superficial siderosis, 70/65
sympathetic, *see* Sympathetic nervous system
systemic malignancy, 71/611
teratoid tumor, 68/380
vegetative, *see* Autonomic nervous system
Nervous system tumor
see also Brain tumor
biochemistry, 27/503-514
carbohydrate content, 27/504
catecholamine, 27/510
enzyme change, 27/508-512
GABA, 27/504, 510
ganglioside, 27/504
glycolytic enzyme, 27/508
glycosphingolipid, 27/504, 508
hydrolytic enzyme, 27/511
lipid, 27/504
mineral content, 27/508
neurotransmitter, 27/509
nucleic acid, 27/506-508
nucleotide, 27/506-508
phospholipid, 27/504
protein, 27/505
serotonin, 27/510
Nervus intermedius neuralgia, *see* Intermedius neuralgia
Nervus intermedius (Wrisberg), *see* Intermedius nerve
Nervus laryngeus recurrence
anatomy, 48/496
Netherlands, *see* The Netherlands
Netherton syndrome
see also Trichorrhexis nodosa
ichthyosiform erythroderma, 43/305
mental deficiency, 43/305
Netsky, M.G., 1/29, 37
Nettleship-Falls albinism
dyschromatopsia, 42/405
genetic linkage, 42/405
Nettleship-Falls syndrome
Forsius-Eriksson syndrome, 42/398
Network
collagen, *see* Collagen network
Neu-Laxova-COFS syndrome
corpus callosum agenesis, 50/163
Neu syndrome
brain atrophy, 43/430
contracture, 43/430
corpus callosum agenesis, 43/430
edema, 43/430
growth retardation, 43/430
hypertelorism, 43/430
microcephaly, 43/430
micrognathia, 43/430
Neubürger classification
diffuse sclerosis, 9/472, 10/6
Neubu6ger, M., 1/3
Neural antigen
autoneuralization, 30/49
NS-4 type, 50/23
Neural arch
segmentation failure, 32/133
Neural cell adhesion molecule
muscle fiber, 62/33
Neural connection
CNS, 30/73
microtubule, 30/73
pathology, 30/74
specificity, 30/73
Neural crest
autonomic nervous system, 74/201
CNS, 30/18
derivative, 14/104, 50/9
embryology, 50/8
Neural crest syndrome, *see* Hereditary sensory and autonomic neuropathy type II
Neural epithelium
primary, *see* Primary neural epithelium
Neural folds
formation, 32/162
fusion, 32/164
Neural induction
archencephalic development, 50/22
autonomic nervous system, 74/203
CNS, 50/24
cyclopia, 50/25
experiment, 50/21
Hensen node, 30/49, 50/22
holocardiac twins, 50/25
holoprosencephaly, 50/25
nerve growth factor, 30/50
prechordal plate, 50/22
Neural model
localization, 1/31
Neural plate
folding, 50/27
Neural plate formation
amphibian, 30/45
basic features, 30/48

cell death, 30/41
chick, 30/46
embryo, 30/46
embryo maldevelopment, 30/51
experimental, 30/46
microtubule development, 30/49
nerve growth factor, 30/50
specific antigen, 30/49
timing, 30/48

Neural tube
caudal neuropore, 50/4
closure defect, 50/28
germ cell, 50/5
matrix cell, 50/5
metathalamus, 2/432
morphologic features, 50/25
rostral neuropore, 50/4

Neural tube closure failure
hydromyelia, 50/426

Neural tube formation
amphibian, 30/53
catecholamine, 30/54
chick, 30/54
CNS, 30/19, 23, 25
CSF, 32/530
embryo, 30/54
embryo maldevelopment, 30/51
hamster, 30/54
morphologic features, 30/52

Neural tumor
BK virus, 71/418
JC virus, 71/418
simian virus 40, 71/418
virus, 71/418

Neuralgia
amyotrophy, *see* Brachial neuralgia
atypical facial, *see* Sphenopalatine neuralgia
auricular, *see* Intermedius neuralgia
brachial, *see* Brachial neuralgia
buccal, *see* Sphenopalatine neuralgia
cervical intervertebral disc prolapse, 20/548, 552-555
cervical rib syndrome, 20/554
Charlin, *see* Charlin neuralgia
chronic migrainous, *see* Chronic migrainous neuralgia
ciliary, *see* Chronic migrainous neuralgia *and* Cluster headache
clinical features, 5/285
cranial, *see* Cranial neuralgia
definition, 5/281
ear, 48/495
facial nerve, *see* Intermedius neuralgia
geniculate, *see* Intermedius neuralgia
glossopharyngeal, *see* Glossopharyngeal neuralgia
Gradenigo syndrome, 2/91, 16/210
greater superficial petrosal, *see* Cluster headache *and* Sphenopalatine neuralgia
headache, 48/5
herpes zoster, 5/207, 209
herpes zoster oticus, 5/509
herpetic geniculate, *see* Herpetic geniculate neuralgia
Hunt, *see* Intermedius neuralgia
Hunt geniculate, *see* Intermedius neuralgia
intermedius, *see* Intermedius neuralgia
Jacobson, *see* Jacobson neuralgia
laryngeal, *see* Laryngeal neuralgia
leprous neuritis, 51/221, 234
major, 5/270
Meckel ganglion, *see* Sphenopalatine neuralgia
Melkersson-Rosenthal syndrome, 8/209
migrainous trigeminal, *see* Migrainous trigeminal neuralgia
minor, 5/270
Morton, *see* Morton neuralgia
nasociliary nerve, *see* Charlin neuralgia
nervus intermedius, *see* Intermedius neuralgia
occipital, *see* Occipital neuralgia
paroxysmal migrainous, *see* Paroxysmal migrainous neuralgia
periodic migrainous, *see* Cluster headache
petrosal, *see* Petrosal neuralgia
postherpetic, *see* Postherpetic neuralgia
pterygopalatine, *see* Sphenopalatine neuralgia
saphenous, *see* Saphenous neuralgia
Sluder, *see* Sphenopalatine neuralgia
Sluder sphenopalatine, *see* Sphenopalatine neuralgia
sphenopalatine, *see* Sphenopalatine neuralgia
superior laryngeal, *see* Superior laryngeal neuralgia
supraorbital, *see* Supraorbital neuralgia
terminology, 1/108, 7/200
trigeminal, *see* Trigeminal neuralgia
tympanic, *see* Tympanic plexus neuralgia
tympanic plexus, *see* Tympanic plexus neuralgia
vagal, *see* Vagal neuralgia
Vail, *see* Sphenopalatine neuralgia
varicella zoster, 5/323-325
Vidian, *see* Sphenopalatine neuralgia

Neuralgic amyotrophy, 51/171-175
acquired immune deficiency syndrome, 56/493
acute brachial radiculitis, 8/77
age, 51/173
allergic neuropathy, 8/82

asthma, 51/174
autosomal dominant, 51/167
axonal degeneration, 51/173
brachial plexus neuritis, 8/76
clinical features, 51/171
compression neuropathy, 51/99, 102
CSF examination, 51/172
differential diagnosis, 51/175
dwarfism, 51/175
Ehlers-Danlos syndrome, 51/175
EMG, 51/172
Epstein-Barr virus, 56/254
etiology, 51/173
facioscapulohumeral syndrome, 51/175
flexion adduction sign, 51/172
hereditary, *see* Hereditary neuralgic amyotrophy
hyperalgesia, 51/172
hypotelorism, 51/175
infectious disease, 51/174
influenza A virus, 51/188
α-interferon intoxication, 65/538
lumbosacral plexus neuritis, 51/167
mononeuritis multiplex, 51/173
multiple neuritis, 51/172
muscular atrophy, 51/171
myalgia, 40/326
pack palsy, 8/81, 51/174
pain stage, 8/78, 51/171
palatoschisis, 51/175
pathogenesis, 51/173
prognosis, 51/173
review, 8/77-84
rifle sling palsy, 51/174
serogenetic neuropathy, 8/99, 51/172
serratus anterior paralysis, 8/77, 51/171
sex ratio, 51/173
shoulder pain, 8/78
stress, 51/174
syndactyly, 51/175
terminology, 8/77
tomacula, 51/559
unilateral paresis, 40/319
vaccinia, 9/550
vaccinogenic neuropathy, 8/86, 51/172
Neuralgic attack
gold polyneuropathy, 51/306
Neuralgic prephase
pseudo, *see* Pseudoneuralgic prephase
Neuralgic shoulder amyotrophy, *see* Brachial plexus neuritis
Neuraminic acid
chemistry, 10/252
Neuraminidase
chemistry, 10/257
mucolipidosis type I, 66/60
α-Neuraminidase
mucolipidosis type I, 42/475
Neuraminidase activity
classification, 66/357
mucolipidosis type I, 66/357
Neuraminidase deficiency, *see* Mucolipidosis type I
Neurapraxia
carpal tunnel syndrome, 51/77
definition, 7/244
histology, 51/76
nerve conduction block, 7/151
nerve injury, 7/244, 51/75, 134
Saturday night palsy, 51/77
Neurasthenia
Epstein-Barr virus, 56/256
iron deficiency anemia, 63/253
leptospirosis, 33/404
myasthenia gravis, 62/415
pellagra, 6/748, 28/86, 46/336
viral, *see* Postviral fatigue syndrome
Neurectomy
auricularis magnus nerve, 5/400
spasticity, 26/485
tympanic, *see* Tympanic neurectomy
Neurenteric canal
spina bifida, 50/501
Neurenteric cyst
area medullovasculosa, 20/57
epithelial, *see* Epithelial neurenteric cyst
features, 31/88
female predominance, 20/59
intra-abdominal, 20/59
intrapelvic, 20/59
intraspinal, *see* Intraspinal neurenteric cyst
intrathoracic, *see* Intraspinal neurenteric cyst
posterior location, 20/60
spina bifida, 50/501
structural group, 20/57
Neurilemmoma, *see* Neurilemoma *and* Neurinoma
Neurilemoma
CSF cytology, 8/417, 16/401
plexiform, 68/564
terminology, 20/238
trigeminal, *see* Trigeminal neurilemoma
Neurinoma
abducent nerve, 68/537, 539
accessory nerve, 68/541
acoustic, *see* Acoustic neuroma
ancient, 68/563
Antoni type A tumor, 14/22, 27, 33, 20/238
Antoni type B tumor, 14/22, 27, 33, 20/238

argyrophil fiber, 14/24
astrocytoma, 68/565
bilateral acoustic, *see* Bilateral acoustic neurinoma
bone, 8/428
brain, 18/321
brain biopsy, 16/723
brain scanning, 16/692
Carney Complex, 68/543
cellular, 68/563
central, 14/145
cerebellopontine angle, 8/417
cervical, 67/202
clinical features, 68/542
collagen fiber, 14/25
congenital, 32/355-386
cranial nerve, 8/417, 68/535
distribution, 67/130, 68/542
droplet metastasis, 20/244, 285
epidemiology, 16/65
facial, *see* Facial neurinoma
facial nerve, 68/481, 538
fallopian canal, 68/538
glossopharyngeal nerve, 68/539
glossopharyngeal neuralgia, 48/464, 68/540
granular cell, 68/564
hourglass, *see* Hourglass neurinoma
hourglass tumor, 19/161
hypoglossal nerve, 14/152, 68/542
imaging, 67/183
intermediate nerve, 68/538
intestinal, *see* Intestinal neurinoma
intracerebral, 68/543
intraparenchymal, 68/543
jugular foramen, *see* Jugular foramen neurinoma
lumbar, 67/202
malignant, *see* Malignant neurinoma
malignant peripheral nerve sheath tumor, 68/562
melanoma, 68/565
melanotic, 68/562
meningioma, 42/744, 68/565
microscopy, 68/559
multiple, 68/564
multiple endocrine neoplasia syndrome, 68/538
nerve tumor, 8/417-435
neurofibroma, 68/535, 564
neurofibromatosis, 14/33, 50/365
neurofibromatosis type I, 14/29, 33, 135, 144, 20/240, 285
neurofibrosarcoma, 68/555
neuropathology, 68/557
oculomotor nerve, 68/536
olfactory nerve, 68/536
palisading, 16/19
perineurinoma, 68/564
phakomatosis, 14/7
primary motor nerve, 68/482
psammomatous, 68/562
pseudoglandular, 68/564
removal method, 68/548
research, 16/37
Schwann cell, 8/461, 14/24
sciatic nerve, 8/428
solitary, 68/535, 555
solitary malignant, 68/544
spinal, *see* Spinal neurinoma
spinal cord, 67/201
spinal cord compression, 19/358
spinal cord tumor, 68/543
spinal meningioma, 20/244
spinal nerve root, 8/417, 68/546
spinal tumor, 19/64, 68
survey, 68/535, 555
terminology, 8/417, 14/135, 67/225
tissue culture, 16/38, 17/82
trigeminal, *see* Trigeminal neurinoma
trochlear nerve, 68/536
tuberous sclerosis, 14/8
vagus nerve, 68/540
vertebral column, *see* Spinal neurinoma
xanthomatosis, 27/246
xanthomatous transformation, 27/246
xeroderma pigmentosum, 43/12
Neurinoma syndrome
pheochromocytoma thyroid carcinoma multiple, *see* Pheochromocytoma thyroid carcinoma multiple neurinoma syndrome
Neurinomatosis centralis (Orzechowski), *see* Tuberous sclerosis
Neurite growth
nerve, 51/16
Neuritis
see also Neuropathy, Polymyalgia rheumatica, Polyneuritis *and* Polyneuropathy
acoustic, *see* Acoustic neuritis
acute paralytic brachial, *see* Acute paralytic brachial neuritis
allergic, *see* Allergic neuritis
angiostrongyliasis, 52/556
ascending, *see* Ascending neuritis
ataxia, 1/318
beriberi, 28/6
botulism, 51/188
brachial, *see* Brachial neuritis
brachial plexus, *see* Brachial plexus neuritis
Brucella, *see* Brucella neuritis
chronic relapsing, *see* Chronic relapsing neuritis

collateral neoformation, 7/226
cranial, *see* Cranial neuritis
diphtheritic, *see* Diphtheritic neuritis
experimental allergic, *see* Experimental allergic neuritis
general aspect, 51/179
headache, 48/5
herpes zoster, *see* Herpes zoster ganglionitis
infectious form, 7/473
influenza virus, 51/187
Kawasaki syndrome, 52/266
leprous, *see* Leprous neuritis
lumbosacral plexus, *see* Lumbosacral plexus neuritis
Lyme disease, 51/204
migrant, *see* Migrant neuritis
migrant sensory, *see* Migrant sensory neuritis
multifocal, *see* Multifocal neuritis
multiple, *see* Multiple neuritis
Mycobacterium tuberculosis, *see* Tuberculous neuritis
neuroborreliosis, 51/204
neurobrucellosis, 33/311, 316, 52/586
optic, *see* Optic neuritis
peripheral, *see* Peripheral neuritis
pseudo, *see* Pseudoneuritis
retrobulbar, *see* Retrobulbar neuritis
retrobulbar optic, *see* Retrobulbar optic neuritis
Salmonella typhosa, *see* Typhoid neuritis
tardy, *see* Tardy neuritis
terminology, 1/108
trigeminal nerve, 5/272
tropical ataxic neuropathy, 70/411
tuberculosis, *see* Tuberculous neuritis
tuberculous, *see* Tuberculous neuritis
typhoid fever, *see* Typhoid neuritis
Whipple disease, 51/184
Neuroacanthocytosis
see also Acanthocytosis *and* Bassen-Kornzweig syndrome
acanthocytosis, 60/674
areflexia, 42/209
ataxia, 21/581
basal ganglion degeneration, 42/209-211
Bassen-Kornzweig syndrome, 7/596, 8/19, 10/280, 549, 13/413, 21/20, 357, 580, 29/412-416, 51/327, 60/673, 63/271
chorea-acanthocytosis, 42/209, 63/271, 280
dementia, 42/209
dystonia, 60/664
epilepsy, 42/511, 63/290
familial hypobetalipoproteinemia, 63/271
Hallervorden-Spatz syndrome, 49/401, 60/673, 63/271
hereditary spinal muscular atrophy, 59/25
hypertriglyceridemia, 60/674
hypobetalipoproteinemia, 29/394, 60/673, 63/271
involuntary movement, 42/209
lipoprotein deficiency, 63/271, 276
muscle stretch reflex, 42/209
muscular atrophy, 42/209
myelodysplasia, 63/290
neurologic dysfunction, 29/394-398
Niemann-Pick disease type B, 60/674
Niemann-Pick disease type C, 60/674
normolipoproteinemic, *see* Normolipoproteinemic acanthocytosis
nosologic group, 49/332
Refsum disease, 21/215
secondary pigmentary retinal degeneration, 13/427
seizure, 42/209
spinal muscular atrophy, 59/374
tic, 42/209
Wolman disease, 60/674, 63/271, 290
Neuroamebiasis
see also Entamoeba histolytica *and* Naegleria fowleri
Acanthamoeba Hartmannella, 52/309
acute amebic meningoencephalitis, 52/309
amebic culture, 52/318
amebic isolation, 52/318
brain abscess, 52/312
brain angiography, 52/321
cerebral amebiasis, 52/309
CNS invasion, 52/314
computer assisted tomography, 52/321
CSF, 52/319
EEG, 52/321
ELISA, 52/320
epidemiology, 52/313
experimental infection, 52/314
geography, 52/312
granulomatous amebic encephalitis, 52/309
Hartmannella, 52/312
immunofluorescence test, 52/318
inappropriate antidiuretic hormone secretion, 52/320
incidence, 52/312
invasion route, 52/314
Iodamoeba bütschlii, 52/310
laboratory diagnosis, 52/317
nasal carriage, 52/314
olfactory nerve, 52/314
pathogenesis, 52/321
primary amebic meningoencephalitis, 52/309

race, 52/312
radioimmunoassay, 52/320
radionucleic scintigraphy, 52/321
wet specimen, 52/317
Neuroanesthesia
brain aneurysm neurosurgery, 55/43
Neuroarthropathy, *see* Neurogenic osteoarthropathy
Neuroastrocytoma
catecholamine, 74/316
central alveolar hypoventilation, 63/464
characteristics, 68/138
classification, 67/5, 7
congenital central alveolar hypoventilation, 63/415, 464
congenital spinal cord tumor, 32/355-386
CSF cytology, 16/401
desmoplastic, 68/138
desmoplastic infantile, *see* Desmoplastic infantile ganglioglioma
dysplastic, *see* Dysplastic gangliocytoma
ganglion cell, 16/10
imaging, 67/174
microscopy, 16/21
nerve tumor, 8/453, 462-466
neurofibromatosis type I, 8/465, 14/138, 145
oligodendroglioma, 72/357
optic chiasm compression, 68/75
optic pathway tumor, 68/74
spinal, *see* Spinal ganglioneuroma
spinal cord compression, 19/363
supratentorial brain tumor, 18/309
sympathocyte, 8/461
third ventricle tumor, 17/444
tissue culture, 17/68, 77
Neuroaxonal degeneration
ataxia telangiectasia, 14/279
Neuroaxonal demyelination
neuroaxonal dystrophy, 49/397
Neuroaxonal dystrophy
see also Axonopathy *and* Spheroid body
α-*N*-acetylgalactosaminidase deficiency, 66/16
adult, *see* Adult neuroaxonal dystrophy
age relation, 49/391
aging, 49/396
alcoholism, 6/626
animal study, 49/393
antikinesin, 49/395
axon terminal, 49/393
axonal content distal stagnation, 49/394
axonal degeneration, 49/393
axonal process, 21/55
boxer dog, 49/394
brown glial pigment, 49/395
central pontine myelinolysis, 6/626
centripetal progression, 49/393
chorea, 60/663
choreoathetosis, 60/663
combined form, 21/63
computer assisted tomography, 60/673
cuneate nucleus, 49/392
definition, 49/391
dementia, 60/665
dorsal spinal nerve root ganglion, 49/393
electron microscopy, 49/392
glial pigment, 49/392
gracile nucleus, 49/392
Guam amyotrophic lateral sclerosis, 21/63
Hallervorden-Spatz syndrome, 6/608, 21/63, 49/395, 397
Herring body, 49/395
hypogonadism, 60/669
hypothalamic hypophyseal tract, 49/395
3,3'-iminodipropionitrile intoxication, 6/626
infantile, *see* Infantile neuroaxonal dystrophy
iron, 49/392
juvenile, *see* Juvenile neuroaxonal dystrophy
kinesin, 49/395
lacunar necrosis, 49/397
late infantile, *see* Late infantile neuroaxonal dystrophy
lipopigment, 21/56
morphology, 49/392
mucoviscidosis, 6/626
myoclonus, 49/618
neuroaxonal demyelination, 49/397
neuroaxonal leukodystrophy, 49/397
neuroaxonal spongy state, 49/396
normal occurrence, 49/392
orthotopic, 21/63
pathogenesis, 49/394
pathologic occurrence, 49/396
pathotopic, 21/63
primary, *see* Primary neuroaxonal dystrophy
pseudohypertrophy, 49/396
renal transplantation, 6/626
retrograde axonal transport, 49/394
Rottweiler dog, 49/394
secondary, *see* Secondary neuroaxonal dystrophy
senile, *see* Senile neuroaxonal dystrophy
spheroid body, 49/391
spinal muscular atrophy, 59/374
spinocerebellar, *see* Spinocerebellar neuroaxonal dystrophy
spongy rarefactive necrosis, 49/396
total white matter, *see* Neuroaxonal leukodystrophy

ventral globus pallidus, 49/392
vitamin E deficiency, 6/626, 49/394
zona reticulata substantia nigra, 49/392
Neuroaxonal leukodystrophy
dementia, 49/406
genodermatosis, 49/406
ichthyosis, 49/406
Nasu-Hakola disease, 49/406
neuroaxonal dystrophy, 49/397
neuropathology, 49/406
primary neuroaxonal dystrophy, 49/397
Neuroaxonal spheroid
dermatoleukodystrophy, 42/487-489
Neuroaxonal spongy state
neuroaxonal dystrophy, 49/396
Neuroaxonal status dysmyelinisatus, *see* Hallervorden-Spatz syndrome
Neurobartonellosis, *see* Verruga peruana
Neurobehavior
brain tumor, 67/395
erythropoietin, 63/535
Neuro-Behçet syndrome, *see* Behçet syndrome
Neuroblast, 14/5
cerebellar medulloblastoma, 18/170
Neuroblastoma
adrenal medullary, *see* Adrenal medullary neuroblastoma
ataxia, 21/574
autonomic nervous system, 42/758, 75/528
birth incidence, 42/759
brain, *see* Brain neuroblastoma
catecholamine, 42/758, 74/315
central, *see* Central neuroblastoma
central alveolar hypoventilation, 63/464
cerebellar medulloblastoma, 18/171, 178
chromosomal aberration, 42/758
classification, 67/4
congenital, 32/355-386
congenital central alveolar hypoventilation, 63/415, 464
congenital megacolon, 42/3, 758
cyclophosphamide, 8/471
cystathionine, 8/471
cystathioninuria, 8/471, 29/118
exophthalmos, 42/758
Horner syndrome, 43/64
metabolic ataxia, 21/575
metastasis, 8/466, 470, 18/340
myasthenia gravis, 62/415
myoclonus, 21/574
nerve tumor, 8/466-472
olfactory, *see* Olfactory neuroblastoma
opsoclonus, 21/574, 34/613, 60/657
optic atrophy, 42/758
posterior fossa tumor, 18/407
precocious puberty, 42/758
prevalence, 42/759
prognosis, 8/471
spinal epidural tumor, 20/103
stress, 74/315
supratentorial brain tumor, 18/340
sympathoblast, 8/461
sympathogonia, 8/461
tissue culture, 17/67
treatment, 8/471
tumor cell, 67/21, 34
vanilmandelic acid, 8/478
vincristine, 8/472
Neuroborreliosis
see also Bannwarth syndrome, Borrelia burgdorferi *and* Lyme disease
abducent nerve paralysis, 51/204
Amblyomma americanum, 51/206
arthralgia, 51/203
arthritis, 51/205
Bannwarth syndrome, 51/199, 52/260-264
Borrelia burgdorferi, 51/199
Borrelia duttonii, 51/206
cardiac involvement, 51/205
cerebellar ataxia, 51/205
chorea, 51/205
course, 51/210
CSF, 51/200, 202, 205
diagnosis, 51/207
difference Lyme disease-Bannwarth syndrome, 51/208
encephalitis, 51/204, 208
epidemiology, 51/200
erythema chronicum migrans, 51/203
erythema migrans, 51/200
etiology, 51/206
facial paralysis, 51/199
fatigue, 51/204
fever, 51/204
headache, 51/204, 208
hemiparesis, 51/205
lymphadenopathy, 51/203
meningism, 51/202
meningitis, 51/204
meningitis serosa, 51/200
meningoradiculitis, 51/200
motor polyneuropathy, 51/201
multiple cranial neuropathy, 51/201
myalgia, 51/203
myelitis, 51/205
neuritis, 51/204

pain, 51/201
pathogenesis, 51/207
prognosis, 51/210
sensory symptom, 51/201
treatment, 51/208
Neurobrucellosis
acute meningoencephalitis, 52/583
antihuman globulin Coombs test, 33/310, 319, 52/585
arachnoiditis, 33/312, 316, 52/587
ataxic quadriparesis, 52/594
Behçet syndrome, 52/594
brachial neuritis, 52/590
brain aneurysm, 52/585
brain edema, 33/313, 52/585, 587
brain hematoma, 52/588
cataplexy, 33/314, 52/587
child, 33/318, 52/595
chorea, 33/314, 52/587
chronic, 33/309, 317, 55/583
clinical category, 52/583
CNS demyelination, 52/586, 588
communicating hydrocephalus, 52/587
complement fixation test, 52/585
compressive myelopathy, 52/585, 590
co-trimoxazole, 52/596
cranial nerve, 33/316
cranial neuritis, 33/312, 316, 52/585, 589
CSF, 33/310, 52/593
deafness, 33/317, 52/589
demyelination, 33/309, 52/584
dexamethasone, 33/321, 52/596
diagnosis, 33/319, 52/593
differential diagnosis, 33/320, 52/593
doxycycline, 52/596
ELISA, 52/585
encephalitis, 33/312-314, 52/587
epilepsy, 52/587
gentamicin, 52/596
granuloma, 52/585
Guillain-Barré syndrome, 33/316, 52/594
headache, 33/309, 312, 52/587
IgA, 33/310
IgG, 33/310, 52/584
IgM, 33/310, 52/584
immunology, 52/584
incidence, 52/583
intervertebral disc prolapse, 33/316, 52/594
intracranial hypertension, 33/312, 52/587
kanamycin, 52/596
lymphocytic meningitis, 52/595
meningitis, 33/311, 52/585
meningoencephalitis, 33/311, 313, 52/584, 586
meningomyelitis, 52/586
meningovascular, 52/584, 588
meningovascular brucellosis, 52/586
2-mercaptoethanol test, 52/584
multiple sclerosis, 52/588
mycotic brain aneurysm, 33/308, 314, 52/585
myelitis, 33/311, 314-316, 52/583, 587
myelopathy, 33/311, 314-316
narcolepsy, 33/314, 52/587
nerve lesion, 7/481
netilmicin, 52/596
neuritis, 33/311, 316, 52/586
neuromyelitis optica, 33/316, 52/588
neuropathology, 33/307, 52/585
no gibbus, 52/595
nuclear magnetic resonance, 52/595
optic neuritis, 33/316, 52/588
papilledema, 33/316, 52/584, 586, 594
paraparesis, 52/588
paraplegia, 33/316, 52/587, 590
parkinsonism, 33/314, 52/587
pathogenesis, 52/585
peripheral neuritis, 52/584, 589
polyarteritis nodosa, 52/588
polyneuritis, 33/316, 52/584
polyradiculitis, 7/481, 52/594
prevention, 52/597
psychiatric manifestation, 33/317
radiculitis, 33/311, 316, 52/585
radiculopathy, 52/585, 590
radiology, 52/595
rifampicin, 52/596
sacral neuritis, 52/590
skin test, 52/585
spastic paraparesis, 33/314, 52/584
spinal nerve root demyelination, 52/590
spondylitis, 33/310, 52/585, 590, 595
standard tube agglutination test, 33/319, 52/584
streptomycin, 33/317, 52/596
subacute disseminated encephalomyelitis, 52/584
subarachnoid hemorrhage, 33/311, 314, 52/584, 588
surgery, 52/597
tetracycline, 33/318, 52/596
transient ischemic attack, 33/314, 52/588
treatment, 33/320, 52/595
tuberculous meningitis, 52/594
vascular syndrome, 33/314
vasculitis, 52/585
Neurocardiac disease
Friedreich ataxia, 21/344
Neurocentral joint, *see* Uncovertebral joint
Neurochemical coding

autonomic nervous system, 74/10, 98
neuropeptide, 74/11
Neurochemistry
affective psychosis, 46/448-450
Binswanger disease, 27/478
congenital Pelizaeus-Merzbacher disease, 10/167
dementia, 27/477-499
depression, 46/448
emotion, 45/278
epilepsy, 15/60, 72/83
epileptic focus, 15/61, 66
familial amaurotic idiocy, 10/332-341
Friedreich ataxia, 60/304
functional psychosis, 46/446-450
head injury, 57/385
head injury outcome, 57/385
juvenile parkinsonism, 49/162
mania, 46/449
Pick disease, 27/489
progressive dysautonomia, 59/145-149
schizophrenia, 46/446-448
status epilepticus, 73/318
Neurocirculatory asthenia, *see* Hyperventilation syndrome
Neurocirculatory syndrome
autonomic nervous system, 74/173
pharmacology, 74/173
Neurocognitive function
assessment, 67/392
lung cancer, 67/398
quality of life, 67/389
Neurocutaneous angiomatosis
phakomatosis, 68/296
Neurocutaneous lipomatosis
phakomatosis, 31/27
Neurocutaneous melanoblastosis, *see* Neurocutaneous melanosis
Neurocutaneous melanosis, 14/414-425
Bamberger-Marie syndrome, 1/488
brain, 68/390
cell type, 14/76
classification, 14/415
clinical features, 14/421
CNS infiltration, 14/78, 43/33
convulsion, 14/422
CSF, 14/422, 43/33
diagnosis, 14/423, 43/34
differential diagnosis, 14/119
differentiation, 14/600
epilepsy, 14/422
family, 14/417, 43/33
forme fruste, 14/424, 43/33
genetics, 14/524-526
Hallervorden-Spatz syndrome, 14/425
histogenesis, 14/425, 597
history, 14/593
hydrocephalus, 14/119, 414, 418, 31/26, 43/33
hypertelorism, 14/600
hypoglycorrhachia, 43/33
lentigo, 14/78
leptomeningeal infiltration, 14/76
leptomeningeal melanoma, 14/593, 43/33
malignant change, 14/416, 428, 596, 43/33
mental deficiency, 14/414, 424
nerve tumor, 8/454
neurocutaneous syndrome, 14/101, 43/33
neurofibromatosis, 14/424, 532, 593, 43/33
neuroid nevus cell, 14/595
neuronevus, 14/78
neuropathology, 14/76-79, 418
nevocellular nevus, 14/595, 597
nosology, 14/593
pathognomonic lesion, 14/594
pigmented nevus, 14/11, 119, 43/33
prognosis, 14/423, 600
psychiatric disorder, 14/768
sebaceous gland agenesis, 14/597
seizure, 43/33
skin lesion, 14/416, 43/33
Sturge-Weber syndrome, 14/8, 424, 532
subarachnoid hemorrhage, 43/33
telangiectasia, 14/424, 597
treatment, 14/601, 43/34
twin study, 14/525
Neurocutaneous melanosis (Van Bogaert), *see* Neurocutaneous melanosis
Neurocutaneous syndrome
albinism, 14/101
ataxia telangiectasia, 14/101
atrophoderma idiopathica (Pasini-Pierini), 14/791
Bean syndrome, 14/101, 774
Bitter syndrome, 14/101
Bonnet-Dechaume-Blanc syndrome, 14/101
Capute-Rimoin-Konigsmark syndrome, 14/101
cerebellar agenesis, 50/191
cerebellar vermis aplasia, 50/191
Chédiak-Higashi syndrome, 14/783
Curtius syndrome type V, 14/101
Divry-Van Bogaert syndrome, 14/101
dysraphia, 14/101
Ehlers-Danlos syndrome, 14/101
epilepsy, 72/181
erythrokeratodermia ataxia syndrome, 21/216
Fabry disease, 14/780
Flynn-Aird syndrome, 14/101
Godfried-Prick-Carol-Prakken syndrome, 14/101

Goltz-Gorlin syndrome, 14/101
Grönblad-Strandberg syndrome, 14/101
hereditary hemorrhagic telangiectasia (Osler), 14/101
incontinentia pigmenti, 14/101
Klein-Waardenburg syndrome, 14/101
Klippel-Trénaunay syndrome, 14/101
LEOPARD syndrome, 14/101
Leschke-Ullmann syndrome, 14/101
Maffucci syndrome, 14/101
Mende syndrome, 14/101
Miescher syndrome type II, 14/101
multiple nevoid basal cell carcinoma syndrome, 14/101
multiple trichoepithelioma, 14/101
neurocutaneous melanosis, 14/101, 43/33
neurofibromatosis type I, 14/101
nevus flammeus, 14/774, 50/191
Ota syndrome, 14/101
Peutz-Jeghers syndrome, 14/101
progressive familial choreoathetosis, 14/101
sebaceous adenoma, 14/101
Sjögren-Larsson syndrome, 14/101
Sturge-Weber syndrome, 14/101
tuberous sclerosis, 14/101
Turner syndrome, 14/101
Von Hippel-Lindau disease, 14/101
xeroderma pigmentosum, 14/101
Neurocysticercosis, 35/291-315, 52/529-533
see also Taenia solium
acquired immune deficiency syndrome, 71/307
age distribution, 16/226, 35/298, 52/529
albendazole, 52/533
angiography, 35/307
arachnoiditis, 52/530
ataxia, 35/299, 52/530
azathioprine, 52/533
brain edema, 35/300, 52/529-531
brain infarction, 52/530, 53/162
brain ischemia, 53/162
brain tumor, 16/225
calcification, 16/225, 52/529
cerebellar tumor, 17/716
chronic meningitis, 56/644
clinical features, 16/225, 35/301-304, 52/529
computer assisted tomography, 52/531
convulsion, 35/300
CSF, 16/226, 20/92, 35/305, 52/531
cyclophosphamide, 52/533
dementia, 46/389, 52/530
diagnosis, 16/226, 35/314, 52/531
differential diagnosis, 52/531
distribution, 35/293-295
EEG, 35/307
epidemiology, 35/13, 293-295, 52/529
epilepsy, 16/225, 52/530, 72/146, 165
etiology, 35/296-299
evolution, 35/315
fatal course, 52/531
frequency, 52/529
gasserian ganglion, 5/273
geographic distribution, 16/225, 20/92, 35/14, 46/389, 52/529
global form, 35/303
headache, 16/225, 52/530
helminthiasis, 35/218, 52/529
history, 35/291-293
hypoglycorrhachia, 52/531
immunosuppression, 52/533
incidence, 35/293-295, 52/529
incubation period, 35/299
infectious striatopallidodentate calcification, 49/417
infestation route, 35/297
intracranial hypertension, 35/300-302, 52/530, 533
laboratory data, 35/304
mental disorder, 35/300, 46/389, 52/530
mode of infection, 35/297, 52/529
muscle, 52/529
myalgia, 40/324
myelography, 52/531
neurosurgery, 52/533
obstructive hydrocephalus, 16/225, 35/315, 52/530
parasitic disease, 35/13, 52/529
parasitology, 35/296-299
pathogenesis, 35/314
pathology, 35/308-314, 52/529
PEG, 35/307
praziquantel, 52/533
prevalence, 35/293-295
prognosis, 35/315
psychiatric disorder, 35/302, 52/530
radiologic examination, 35/306
seizure, 46/389, 52/530
serology, 52/531
skull radiation, 35/307
spastic paraplegia, 59/436
spinal, 20/92, 35/303
spinal cord compression, 35/303, 52/530
spinal cyst, 20/92, 52/530
spinal type, 52/531
swine, 35/293, 52/529
symptomatology, 35/299-301
systemic neoplastic disease, 41/373

Taenia solium, 16/225, 35/296-298, 46/389, 52/529
treatment, 35/315, 52/533
trigeminal neuralgia, 5/273
tropical myeloneuropathy, 56/526
ventriculography, 35/307
Neurocytology
principle, 17/43
Neurocytoma
central, *see* Central neurocytoma
hypothalamic tumor, 68/71
tissue culture, 17/67
Neurodegenerative disease
see also Degenerative disease
dementia, 21/3
epidemiology, 21/3-42
neuropathology, 21/43-71
xanthomatous transformation, 27/246
Neurodevelopmental status
child, 67/394
Neuroectodermal dysplasia
class, 14/4
incontinentia pigmenti, 14/213
Neuroectodermal tumor
classic primitive, *see* Classic primitive neuroectodermal tumor
primitive, *see* Primitive neuroectodermal tumor
Neuroectodermosis, *see* Neuroectodermal dysplasia
Neuroectomesodermosis, *see* Neuroectodermal dysplasia
Neuroeffector transmission
mechanism, 74/42
Neuroembryology
mental deficiency, 46/6-8
Neuroencephalopathy
allergic, *see* Allergic neuroencephalopathy
Neuroendocrine function
autonomic nervous system, 74/626
Neuroendoscope
brain tumor, 67/249
Neuroepidemiology
brain tumor, 16/56, 60
diagnostic function, 1/30
human T-lymphotropic virus type I associated myelopathy, 56/531
tropical spastic paraplegia, 56/531
Neuroepithelial cyst
histology, 68/314
treatment, 68/314
Neuroepithelial tumor
dysembryoplastic, *see* Dysembryoplastic neuroepithelial tumor
Neuroepithelioma, *see* Ependymoma
Neuroepithelium
cell detachment, 50/32
cell migration, 50/32-35
CNS, 30/17, 50/29
life span, 50/29
overgrowth malformation, 50/30
proliferation pattern, 50/30
tissue culture, 17/44
Neurofibrillary disease
argyrophilic dystrophy, 21/57
Neurofibrillary tangle
adult neuronal ceroid lipofuscinosis, 46/274
agnosia, 46/260-263
aluminum, 46/271
aluminum intoxication, 36/320-324, 64/276
Alzheimer disease, 46/247, 258-260
amnesia, 46/260-263
amyotrophic lateral sclerosis, 22/399, 408, 42/70, 46/274, 49/249, 59/282-284
aphasia, 46/260-263
apraxia, 46/260-263
biochemical aspect, 27/486
boxer dementia, 23/557, 46/274
brain amyloid angiopathy, 54/334
chemistry, 27/486, 42/276
choline acetyltransferase, 46/268
clinicopathologic correlation, 46/260-264
colchicine, 46/271
dementia, 27/486, 489-491
Down syndrome, 49/249, 50/530
electron microscopy, 46/258
experimental induction, 46/271
GABA, 46/269
Gerstmann-Sträussler-Scheinker disease, 56/555, 60/623
glutamic acid, 46/269
Guam amyotrophic lateral sclerosis, 22/377-380, 59/257-259, 278, 282-284
Guam amyotrophic lateral sclerosis-Parkinson dementia complex, 65/22
Guam motoneuron disease, 65/22
Guam Parkinson dementia, 27/486
hippocampus, 46/258
Hirano flame shaped type, 49/248
Hirano globose type, 49/248
incidence, 46/274
Kii Peninsula amyotrophic lateral sclerosis, 22/366, 373, 59/278, 283
lead encephalopathy, 46/274
microscopy, 46/258
myotonic dystrophy, 62/237
neurotransmitter change, 46/268
Parkinson dementia, 46/377, 49/401

Parkinson dementia complex, 22/223, 42/250, 49/249, 59/282-284
Parkinson disease, 49/110
parkinsonism, 46/274
pathogenesis, 46/259
Pick disease, 27/489, 42/286
progressive pallidal atrophy (Hunt-Van Bogaert), 42/247
progressive spinal muscular atrophy, 22/17
progressive supranuclear palsy, 22/222-224, 42/259, 46/274, 302, 49/248-250
subacute sclerosing panencephalitis, 46/274
vinblastine, 46/271
vincristine, 46/271
Neurofibroma
see also Perineural fibroblast
benign, 14/138
congenital spinal cord tumor, 32/355-386
dermal, 8/445
differentiation, 14/571
droplet metastasis, 20/285
foramen magnum tumor, 17/721
glossopharyngeal neuralgia, 48/464
hourglass tumor, 8/440, 19/161
intracranial tumor, 14/136
intraorbital, 18/341
intraspinal tumor, 14/136
malignant, *see* Malignant neurofibroma
Meissner corpuscle, 14/27, 29
meningioma, 17/176
multiple nevoid basal cell carcinoma syndrome, 14/583
multiple spinal tumor, 20/244, 285
myelinated nerve fiber, 14/137
nerve tumor, 14/136
neurinoma, 68/535, 564
neurofibromatosis, 14/565-567, 50/365
neurofibromatosis type I, 8/417, 14/21, 25-27, 29, 135-138, 565-567, 20/240, 285, 41/384
nodular, 8/445, 14/32
nonmyelinated nerve fiber, 14/137
orbital, *see* Orbital neurofibroma
orbital tumor, 17/176, 202
paraspinal, *see* Paraspinal neurofibroma
periorbital, *see* Periorbital neurofibroma
peripheral, *see* Peripheral neurofibroma
plexiform, *see* Plexiform neurofibroma
skull base tumor, 17/176
spinal, *see* Spinal neurofibroma
spinal cord, 67/201
spinal epidural tumor, 20/104
spinal meningioma, 20/184
terminology, 8/417, 440, 14/135
vertebral column, *see* Spinal neurofibroma
visceral, 8/438
Wagner-Meissner corpuscle, 8/444
Neurofibromatosis
aberrant nerve fiber, 14/36
acoustic neuroma, 14/33, 151, 492, 18/402, 42/756
acromegaly, 14/701, 740
Addison disease, 14/732
adrenocortical lesion, 14/732
amyotrophic lateral sclerosis, 14/492
angioma, 14/8
angiomatosis, 14/36
animal, 14/492
anterior spinal meningocele, 42/44
aqueduct stenosis, 14/37, 50/302, 309
associated disease, 14/492
astrocytoma, 20/330, 42/732, 68/289
ataxia telangiectasia, 68/298
bilateral acoustic, *see* Neurofibromatosis type II
Bonnet-Dechaume-Blanc syndrome, 68/298
brachial plexus syndrome, 2/143
café au lait spot, 1/484, 50/365
carcinoid syndrome, 70/315
clinical features, 14/669-682, 31/14
CNS tumor, 30/101
congenital kyphosis, 32/145
conjunctiva, 14/632
corneal, 14/633
cortical heterotopia, 14/37
cranial vault tumor, 17/109
cutaneous manifestation, 31/14
diagnosis, 31/15
diagnostic features, 50/365
diencephalic syndrome, 14/741-746
differential diagnosis, 14/570
diffuse glioblastosis, 18/75
droplet metastasis, 20/244
Ehlers-Danlos syndrome, 14/568
elephantiasis, 14/185
Elsberg sign, 20/260-263
ependymoma, 14/35
epidemiology, 30/150, 158, 31/1
epilepsy, 14/155, 72/129
fibrous dysplasia, 14/83, 177, 182, 185, 187, 193, 198
forme fruste, 14/490
general concept, 14/86
genetics, 30/150, 158, 50/366
giant cell astrocytoma, 68/293
glaucoma, 14/398, 635
glial heterotopia, 14/37
glioma polyposis syndrome, 68/299
gliomatosis, 14/35, 102

glucose-6-phosphate dehydrogenase deficiency, 14/492
Hallervorden-Spatz syndrome, 6/625, 49/401
hamartoma, 14/36, 68/291
hemifacial hypertrophy, 14/635
hemihypertrophy, 59/483
hereditary, 14/489, 20/177
hereditary hemorrhagic telangiectasia (Osler), 68/297
hereditary spinal muscular atrophy, 59/26
heterogeneity, 50/365
heterotopia, 14/145
histogenesis, 14/568-570
history, 14/563
hourglass tumor, 20/289
Huntington chorea, 6/315, 14/492, 49/291
hypermelanosis, 1/484
incidence, 68/287
intelligence, 14/776
intramedullary schwannosis, 14/36
intrathoracic meningocele, 32/212-218, 220-222
iris, 14/634
isolated case, 14/490
Klippel-Trénaunay syndrome, 14/397, 532
laboratory aspect, 50/366
Leschke-Ullmann syndrome, 14/741
Lesser occipital nerve, 2/134
Lisch nodule, 50/365
macrencephaly, 42/41
malignant neurinoma, 14/30
meningioma, 14/33, 42/744
meningiomatosis, 14/36
meningocele, 14/38, 32/212
meningoencephalic gliomatosis, 14/35
mental deficiency, 31/3, 32/212-218, 220-222, 46/42
microgyria, 14/37
moyamoya disease, 42/751, 53/164, 55/295, 301
mucous membrane, 14/567
multiple, *see* Multiple neurofibromatosis
multiple meningioma, 20/189
mutation rate, 14/491
myelopathy, 14/146
neurinoma, 14/33, 50/365
neurocutaneous melanosis, 14/424, 532, 593, 43/33
neurofibroma, 14/565-567, 50/365
neuropathology, 14/20-38
ocular lesion, 14/638
ocular manifestation, 31/15
onion bulb formation, 21/252
optic, *see* Optic neurofibromatosis
optic atrophy, 14/151
optic disc, 14/638
optic nerve, 14/630
optic nerve glioma, 50/365, 367
optic pathway tumor, 68/75
orbital, 14/627
osseous manifestation, 31/15
perineural fibroblast, 14/20, 135
phakomatosis, 68/287
pheochromocytoma, 8/475, 14/134, 252, 492, 702, 733, 750, 39/501, 503, 42/765, 50/366
pigmentation, 14/182
polar spongioblastoma, 14/33
pontine glaucoma, 18/397
prognosis, 14/572
psychiatric disorder, 14/762-765
retrovirus, 56/525
Sachsalber syndrome, 14/398
scoliosis, 50/417
Sipple syndrome, 14/736, 738, 750
sphenoid wing aplasia, 30/278, 281, 50/144
spinal meningioma, 20/240
spinal neurinoma, 20/244
spinal site, 20/238
Sturge-Weber syndrome, 14/6, 532
surgical treatment, 20/285
survey, 68/287
syringomyelia, 32/287
treatment, 31/15
tuberous sclerosis, 14/6, 368
tumeur royale, 14/30
tumor type, 31/11-14
vascular abnormality, 14/568
visceral lesion, 14/669-682
Vogt triad, 68/291
Von Hippel-Lindau disease, 14/6, 251, 532
Von Recklinghausen, *see* Neurofibromatosis type I
Neurofibromatosis type I
see also Optic nerve glioma *and* Sphenoid wing dysplasia
acoustic neuroma, 18/402, 42/756
acute intermittent porphyria, 14/492
adrenal medullary neuroblastoma, 50/370
adrenogenital syndrome, 14/492
age at onset, 14/147
anatomic form, 14/31-38
angioneuromatosis, 14/36
anterior spinal meningocele, 42/44
Antoni type A tumor, 14/137
Antoni type B tumor, 14/137
associated disease, 14/133
astrocytoma, 14/140, 50/367
atrophoderma vermiculatum, 14/113
autosomal dominant, 50/366, 370

bilateral acoustic neurinoma, 14/145, 151, 42/756
brain glioma, 50/367
brain tumor, 14/136
buphthalmos, 14/635
café au lait spot, 14/564, 31/14, 50/366, 68/288
central type, 14/145, 150, 153
cerebellopontine angle syndrome, 19/39
chromosome, 67/55
classification, 14/11, 136
cleidocranial dysostosis, 14/492
clinical features, 14/146-158, 32/141
CNS tumor, 67/54
Cockayne-Neill-Dingwall disease, 14/492
congenital glaucoma, 50/367
congenital heart disease, 14/113
cranioectodermal dysplasia, 43/35
cranium bifidum, 14/158
cutaneous dysplasia, 14/101
cutaneous manifestation, 31/14
cutis laxa, 14/32
diagnosis, 68/288
differential diagnosis, 14/570
diffuse glioblastosis, 18/75-77
droplet metastasis, 20/244, 285
dysraphia, 14/146
ependymoma, 14/140
ephelis, 14/102
epidemiology, 30/158, 31/1
epidermolysis bullosa dystrophica, 14/568
epilepsy, 72/131
eyelid, 14/624
familial spinal arachnoiditis, 42/107
fibroma molluscum, 14/135, 149
fibrous dysplasia, 14/83, 134, 165, 492
forme fruste, 14/134, 143, 145, 151, 490
François syndrome, 14/627, 635, 638
freckling, 50/366
fur nevus, 14/102
Gardner disease, 22/517
gelatinoid myxoglioma, 14/32
gene, 67/32
general concept, 14/86
general features, 14/2-4
genetics, 14/488, 16/30, 30/150, 158, 31/10, 32/141, 68/287
genu valgum, 50/369
genu varum, 50/369
glaucoma, 14/635
gliomatosis cerebri, 14/143, 151
glycosaminoglycanosis type I, 14/492
hamartoma, 14/36, 136
hemihypertrophy, 22/549
hereditary, 14/134, 489
hereditary motor and sensory neuropathy variant, 60/247
heterochromia iridis, 14/492
heterotopia, 14/145
history, 14/132, 20/238
hourglass tumor, 14/139, 154, 20/289
hydrocephalus, 14/146
hypopigmentation, 50/366
ichthyosis, 14/492
intestinal neurinoma, 14/604
intracranial tumor, 14/151
intraspinal tumor, 14/153, 19/39
intrathoracic meningocele, 14/38, 157
iris, 14/634
jugular foramen syndrome, 14/151
juvenile chronic myeloid leukemia, 50/370
juvenile xanthogranuloma, 50/366
Klippel-Trénaunay syndrome, 14/568
Krabbe-Bartels disease, 14/12, 82
kyphoscoliosis, 32/222, 50/369
leptomeningeal melanosis, 14/87, 133
lipoma, 14/568
liposarcoma, 50/369
Lisch nodule, 50/366
lung cancer, 50/369
macrocephaly, 50/369
malignancy, 50/369
malignant degeneration, 20/244
malignant fibrous histiosarcoma, 50/369
malignant neurinoma, 8/450, 14/30
Marfan syndrome, 32/206
mediastinal tumor, 14/139
medullary thyroid carcinoma, 50/370
melorheostosis, 14/492
meningioma, 14/140, 152, 42/744
meningiomatosis, 14/145
meningocele, 14/38, 145, 157
mental deficiency, 14/156, 50/368
multiple ependymoma, 20/358
nerve tumor, 8/414, 434, 14/154
nervous system, 14/132-158
neurinoma, 14/29, 33, 135, 144, 20/240, 285
neuroastrocytoma, 8/465, 14/138, 145
neurocutaneous syndrome, 14/101
neurofibroma, 8/417, 14/21, 25-27, 29, 135-138, 565-567, 20/240, 285, 41/384
neurofibrosarcoma, 50/368
neurologic lesion, 14/150
neuronal heterotopia, 14/145
neuropathology, 14/20-38, 134-146
nosology, 14/563
ophthalmology, 14/624-639
optic atrophy, 13/69

optic chiasm glioma, 14/33
optic nerve glioma, 14/134, 142, 147, 151, 18/344
orbital neurofibroma, 50/367
pachygyria, 14/37
parathyroid adenoma, 14/492
partial gigantism, 14/397
Passow syndrome, 14/492
pathology, 31/10-14, 32/141, 143
pectus excavatum, 50/369
periorbital neurofibroma, 50/367
peripheral neurofibroma, 50/368
Peutz-Jeghers syndrome, 14/602, 604
phakomatosis, 14/19, 25, 29-39
phenotype variation, 14/491
pheochromocytoma thyroid carcinoma multiple neurinoma syndrome, 14/134
pituitary adenoma, 17/412
plexiform neurofibroma, 8/438, 440, 50/366
plexiform neuroma, 14/32
polar spongioblastoma, 14/33
polymicrogyria, 14/37
posterior fossa astrocytoma, 50/368
precocious puberty, 14/156
pretumor stage, 14/155
prevalence, 14/134, 488, 50/366
prognosis, 31/15
pseudoarthrosis, 50/369
pseudoatrophic macula, 50/366
psychiatric disorder, 14/762-764
renovascular hypertension, 50/369
retina, 14/638
rhabdomyosarcoma, 50/369
rib notching, 14/158
Rosenthal fiber, 14/33, 37
sacral meningocele, 14/158
scalp tumor, 17/124
scoliosis, 50/369
seizure, 31/3
sensory crest, 14/86
sex ratio, 14/147
skeletal lesion, 14/157
skeletal malformation, 50/369
skin lesion, 14/148
soft tissue sarcoma, 50/369
speech disorder, 50/368
sphenoid wing dysplasia, 50/367, 369
spinal cord malformation, 14/38
spinal cord tumor, 14/136
spinal deformity, 32/141-143
spongioblastoma, 14/33, 102, 142, 50/367
striatopallidodentate calcification, 14/152
Sturge-Weber syndrome, 14/568
syringomyelia, 14/37, 133, 146, 31/12, 14, 32/287
thyroid carcinoma, 14/134
treatment, 14/572, 31/15
tuberous sclerosis, 14/133
tumeur royale, 14/30, 32
tumor cell, 67/31
Turner syndrome, 14/492
twin study, 14/134, 489
uvea, 14/634
vascular abnormality, 14/568
vertebral column, 32/141-143
vertebral scalloping, 14/158
visceral cancer, 50/369
Von Hippel-Lindau disease, 14/133, 568
Wilms tumor, 50/370
Neurofibromatosis type II
autosomal dominant, 50/366
bilateral acoustic neurinoma, 50/370
café au lait spot, 50/370
chromosome, 67/56
CNS tumor, 67/54
diagnosis, 68/288
gene, 67/33
meningioma, 50/370
oral contraception, 50/370
paraspinal neurofibroma, 50/370
tumor cell, 67/33
Neurofibromin
signal transduction, 67/32
Neurofibrosarcoma
neurinoma, 68/555
neurofibromatosis type I, 50/368
tissue culture, 17/83
Neurofilament
acetonylacetone intoxication, 64/89
aluminum intoxication, 64/276
hexane polyneuropathy, 51/277
2-hexanone polyneuropathy, 51/277
methanol intoxication, 64/101
monomer, 7/17
nerve, 51/5
neurotoxicity, 51/264
organization, 7/17
subunit strand, 7/17
Neurofilament axonopathy
disulfiram polyneuropathy, 51/309
Neurofilament neuropathy
neurotoxicity, 51/276
Neurofilament polyneuropathy
axonal transport, 51/278
Neurofilament protein
acetonylacetone intoxication, 64/89
Parkinson disease, 49/114
Neurofilament segregation

acrylamide polyneuropathy, 51/282
Neurogenic acro-osteolysis
areflexia, 42/293
foot deformity, 42/294
hereditary sensory and autonomic neuropathy type I, 42/293-295
hypertelorism, 42/294
hyporeflexia, 42/293
osteolysis, 42/294
osteomyelitis, 42/294
sclerosis, 42/294
sensorineural deafness, 42/293
sensory loss, 42/293-295
stump, 42/294
trophic ulcer, 42/293-295
ulceration, 42/293
Neurogenic amyotrophy, *see* Neurogenic muscular atrophy
Neurogenic arterial hypertension
amygdaloid nucleus, 63/231
genetic animal model, 63/231
hypothalamus, 63/231
medial forebrain bundle, 63/231
medulla oblongata distortion, 63/230
normal pressure hydrocephalus, 63/231
nucleus tractus solitarii, 63/229
posterior fossa tumor, 63/230
vasopressin, 63/230
Neurogenic arthropathy, *see* Neurogenic osteoarthropathy
Neurogenic bladder
see also Autonomous bladder
adult polyglucosan body disease, 62/484
automatic cord bladder, 2/189
bladder neck resection, 26/425
bladder training, 26/210, 416
cauda equina tumor, 19/87
cause, 1/394
diastematomyelia, 42/24
intravenous pyelography, 26/424
lumbar vertebral canal stenosis, 20/688
pathophysiology, 1/370
sacral root, 19/87
spinal cord tumor, 19/34
spinal tumor, 19/59
stage, 2/188-193
Neurogenic diarrhea
diabetic autonomic visceral neuropathy, 27/115
Neurogenic muscular atrophy
ataxia, 60/674
chorea-acanthocytosis, 63/281
diabetes mellitus, 60/674
facioscapulohumeral muscular dystrophy, 22/75
galactosialidase deficiency, 51/380
hereditary, *see* Hereditary motor and sensory neuropathy type I *and* Hereditary motor and sensory neuropathy type II
Huntington chorea, 6/315
leprous neuritis, 51/221
neuropathy, 60/659
scapulohumeral muscular dystrophy, 21/29
secondary pigmentary retinal degeneration, 13/308, 60/674
Wegener granulomatosis, 51/451
Neurogenic orthostatic hypotension
autonomic nervous system, 74/613
management, 75/714
Neurogenic osteoarthropathy
see also Joint
anatomy, 26/510
autonomic nervous system, 1/487
bone scan, 26/517
brain death, 24/722
brain injury, 24/722
calcium, 26/517
clinical features, 26/507
congenital pain insensitivity, 51/564
definition, 26/501
diabetes mellitus, 8/31, 27/115
evolution, 26/509
hemiplegia, 1/183
hereditary sensory and autonomic neuropathy type II, 42/349
hereditary sensory and autonomic neuropathy type III, 21/110, 60/27
hereditary sensory and autonomic neuropathy type IV, 42/346, 60/12
incidence, 26/501
indifference to pain, 8/195
location, 26/503
mummy, 74/188
pathogenesis, 26/506, 510
phosphatase, 26/385, 405, 517
posttraumatic syringomyelia, 26/124
prevention, 26/511
radiodiagnosis, 26/275, 277-281
radiology, 26/502-516
segmental level, 26/503
serum level, 26/385
spinal cord injury, 26/277, 501-519
syringomyelia, 26/124, 32/270
tabes dorsalis, 19/86, 52/280
treatment, 26/511
Neurogenic pulmonary edema, 75/373
brain aneurysm, 55/53
brain aneurysm neurosurgery, 55/53

brain embolism, 55/157
brain hemorrhage, 55/157
brain infarction, 55/157
brain stem infarction, 63/496
brain stem tumor, 63/496
cardiac dysrhythmia, 55/157
definition, 63/493
epilepsy, 63/495
intracranial hypertension, 63/441, 495
medullar obex lesion, 63/485
multiple sclerosis, 63/496
neurologic intensive care, 55/221
nucleus tractus solitarii, 63/495
pathomechanism, 63/495
poliomyelitis, 63/496
status epilepticus, 63/495
stroke, 55/157
subarachnoid hemorrhage, 55/53
sudden death, 63/496
Neurogenic sarcoma
CNS, 14/30
connective tissue, 8/446
Schwann cell, 8/446
spinal cord compression, 19/369
terminology, 14/30
Neurogenic ulcer
diabetes mellitus, 27/114
Neuroglia
CNS, 30/16
Niemann-Pick disease type A, 10/495
origin, 30/16
tissue culture, 17/54
Neuroglia relationship, 28/401-419
brain injury, 28/417
demyelination, 28/410-412
dysmyelination, 28/410
historic aspect, 28/401
metabolic relationship, 28/413-417
metachromatic leukodystrophy, 28/410
morphologic relationship, 28/403-406
myelination, 28/407-412
nerve growth factor, 28/418
Pelizaeus-Merzbacher disease, 28/412
phenylketonuria, 28/410
Refsum disease, 28/410
trophic relationship, 28/412
vitamin B_{12} deficiency, 28/412
Neurohypophysis
anatomy, 2/436
Neuroid nevus cell
neurocutaneous melanosis, 14/595
Neuroimmunology
brain tumor, 67/293
Neurokeratin
chemistry, 9/13
CNS myelin, 9/31, 33
histochemistry, 9/13
nerve myelin, 9/27-34
proteolipid dissimilarity, 9/32
Neurokinin
headache, 48/97
migraine, 5/39
reflex sympathetic dystrophy, 61/126
Neurolathyrism, *see* Lathyrism
Neuroleptic agent
acetylcholine level, 65/278
adenylate cyclase, 65/279
akathisia, 49/191
aliphatics, 65/274
apomorphine antagonism, 65/278
behavior change, 65/279
behavior disorder, 46/595-597
benzamide, 65/275
butaperazine, 65/284
butyrophenone, 65/274
calmodulin, 65/279
chemical structure, 65/274
childhood psychosis, 46/193
chlorpromazine, 13/223, 65/275
chlorprothixene, 65/276
cholinergic function, 65/281
chorea, 49/191
classification, 65/274
clozapine, 65/275
cyclic adenosine monophosphate, 65/279
dibenzoxazepine, 65/275
dopamine, 46/446
dopamine metabolite, 65/278
dopamine receptor type, 65/277
dopaminergic system, 65/277
dystonia, 49/191
epilepsy, 72/233
fluphenazine, 65/275
GABA-ergic function, 65/281
glycogen, 27/208
half-life, 65/277
haloperidol, 65/276
hemiballismus, 6/261, 49/375
histamine H1 receptor antagonist, 65/280
infantile autism, 46/192
loxapine, 65/275
mania, 46/450
mesoridazine besylate, 65/275
metabolite, 65/277
metoclopramide, 65/275
molindone, 65/275

multiple sclerosis, 47/178
noradrenergic function, 65/280
parkinsonism, 6/261, 49/191
perphenazine, 65/275
pharmacology, 65/278
phenothiazine, 65/274
phosphodiesterase, 65/279
Pick disease, 46/244
pimozide, 65/275
piperazine, 65/274
piperidine, 65/274
reserpine, 65/275
schizophrenia, 46/446
serotonergic function, 65/280
sulpiride, 65/275
symptomatic dystonia, 6/560, 49/543
tardive dyskinesia, 49/186
tetrabenazine, 65/275
thioridazine, 65/275
thioxanthene, 65/274
tiotixene, 65/276
tricyclic structure, 65/275
trifluoperazine, 65/275
triflupromazine, 65/275
vitamin B2, 28/206
Neuroleptic agent intoxication
catatonia, 46/423
pallidal necrosis, 49/466
rabbit sign, 49/187
Neuroleptic akathisia
acute type, 65/285
age, 65/287
amantadine, 65/290
anticholinergic effect, 65/289
Barnes Akathisia Scale, 65/286
beta noradrenergic receptor blocking agent, 65/290
Chouinard Scale, 65/286
clinical features, 65/285
clonazepam, 65/290
clonidine, 65/290
diazepam, 65/290
dopaminergic system, 65/288
dyskinesia, 65/286
dysphoria, 65/287
epidemiology, 65/286
fluoxetine, 65/289
Hillside-LIJ modification, 65/286
history, 65/285
lorazepam, 65/290
noradrenergic function, 65/288
pathophysiology, 65/287
prevalence, 65/286
reserpine, 65/287
serotonergic function, 65/289
serum ion status, 65/287
sex ratio, 65/287
Simpson-Angus EPS Scale, 65/286
spinal cord, 65/288
stereotypy, 65/286
suicide, 65/285
symptom, 6/253
transferrin, 65/287
treatment, 65/289
Neuroleptic catalepsy
acute effect, 65/278
anticholinergic basis, 65/277
Neuroleptic dyskinesia
Abnormal Involuntary Movement Scale, 65/282
akathisia, 65/274
baboon, 65/283
catalepsy, 65/277
dopamine depletion, 65/274
dystonia, 65/274
incidence, 65/274
metirosine, 65/283
monkey, 65/283
parkinsonism, 65/274
pathobiochemistry, 65/278
pathophysiology, 65/283
pharmacology, 65/283
squirrel monkey, 65/283
stereotypy, 65/278
Toronto Western Spasmodic Torticollis Rating Scale, 65/282
Neuroleptic dystonia
acute, 65/281
anticholinergic abolition, 65/284
baboon, 65/283
blepharospasm, 65/281
butaperazine, 65/284
characteristic features, 65/281
drug dosage, 65/282
drug potency, 65/282
dysarthria, 65/282
dysphagia, 65/282
dystonia musculorum deformans, 65/281
epidemiology, 65/282
grimacing, 65/281
jaw opening/closing, 65/281
lordosis, 65/281
metirosine, 65/283
monkey, 65/283
oculogyric crisis, 65/281
opisthotonos, 65/281
pathophysiology, 65/283

pharmacology, 65/283
retrocollis, 65/281
risk factor, 65/282
scoliosis, 65/281
squirrel monkey, 65/283
stridor, 65/282
tardive, 65/281
tongue protrusion, 65/281
torticollis, 65/281
treatment, 65/284

Neuroleptic malignant syndrome
acquired myoglobinuria, 62/572
autonomic nervous system, 75/13, 497
clinical picture, 71/593
computer assisted tomography, 71/592
diagnosis, 71/594
drug induced extrapyramidal syndrome, 6/258-260
Hallervorden-Spatz syndrome, 66/715
hypothyroidism, 70/101
lateral, 66/715
malignant hyperthermia, 6/258, 62/571
pathogenesis, 71/600
prevention, 71/595
survey, 71/585, 591, 75/74
symptom, 6/258, 62/573, 71/600
thermoregulation, 75/74
treatment, 71/594

Neuroleptic parkinsonism
acute type, 65/273
age, 65/294
akinesia, 65/292
amantadine, 65/296
aprosody, 65/292
bradykinesia, 65/292
classification, 6/234
clinical characteristic, 65/292
depression, 65/294
description, 6/234
differential diagnosis, 65/293
drug potency, 65/294
EMG, 65/293
epidemiology, 65/294
hereditary factor, 65/295
HLA, 65/295
incidence, 65/294
individual predisposition, 65/294
mask like face, 65/292
methyldopa, 6/235
pathophysiology, 65/295
permanent lesion, 6/238
phenothiazine, 42/651
phenothiazine derivative, 6/236-238
positron emission tomography, 65/295
postural instability, 65/293
postural reflex, 65/292
rabbit syndrome, 65/292
reserpine, 6/235
rigidity, 65/292
sex ratio, 42/651, 65/294
subacute type, 65/273
treatment, 65/296
tremor, 65/292
Unified Parkinson Disease Rating Scale, 65/293

Neuroleptic syndrome
akinesia, 6/250
butyrophenone derivative, 6/257
chorea, 6/257
clinical features, 6/249-254
criteria, 6/248
EEG, 6/259
history, 6/249
hyperkinesia, 6/251
hypokinesia, 6/250
malignant, *see* Malignant neuroleptic syndrome
mental change, 6/254
muscular hypertonia, 6/251
neuroautonomic syndrome, 6/261
neuropathology, 6/259
pathophysiology, 6/259
permanent syndrome, 6/253, 257
piperazine derivative, 6/256
precipitation, 6/254
reserpine, 6/256
treatment, 6/261

Neurolinguistic analysis
aphasia, 46/616

Neurolipidosis
see also Lipidosis
amiodarone induced, 65/459
amiodarone intoxication, 65/458
chloroquine, 51/302
classification, 10/327, 353
dystonia musculorum deformans, 49/524
enzymology, 66/51
epilepsy, 15/423
gangliosidosis, 66/19
glycosaminoglycanosis type IH, 66/19
GM2 gangliosidosis, 66/20
histology, 66/18
late cortical cerebellar atrophy, 21/477
late infantile neuroaxonal dystrophy, 49/405
major type, 66/15
metabolic manipulation therapy, 66/107
molecular biology, 66/69
myoclonic variant, 10/221, 21/58, 62

myoclonus, 21/514
Niemann-Pick disease, 66/20
Niemann-Pick disease type A, 66/20
pigment variant, 21/56
polyneuropathy, 60/165-177
rare type, 10/542-555
Refsum disease, 8/23, 10/345, 13/314, 21/181, 231, 27/520
secondary neuroaxonal dystrophy, 49/398
spastic paraplegia, 59/439
sphingolipid activator protein deficiency, 66/61
sphingolipidosis, 66/20
Stargardt disease, 13/132, 134
survey, 66/1
ultrastructure, 10/362-380
Wolman disease, 10/546
Neurolipoma
phakomatosis, 14/81-83
Neurologic diagnosis
faulty diagnosis, 2/2
introduction, 2/1-3
late theory of localization, 2/14
Neurologic intensive care, 55/203-222
acute stroke treatment, 55/210
admission diagnosis, 55/203
age, 55/205, 213
anticipatory admission, 55/203
barbiturate, 55/209, 211, 215
blood flow, 55/220
blood pressure, 55/217
brain abscess, 55/204
brain blood flow, 55/220
brain death, 55/218
brain death rate, 55/209
brain edema, 55/212, 214
brain hemorrhage, 55/204, 212, 234
brain infarction, 55/234
brain stem auditory evoked potential, 55/217-219
calcium channel blocking agent, 55/217
carotid endarterectomy, 55/222
coma, 55/204
compressed spectral array, 55/220
corticosteroid, 55/216
craniotomy, 55/216
CSF drainage, 55/216
decreasing metabolic demand, 55/210
dextran, 55/210
diuretic agent, 55/209, 212
drug intoxication, 55/220
economics, 55/205
EEG, 55/219
encephalitis, 55/215
etacrynic acid, 55/213
evoked potential, 55/217
excitotoxin, 55/211
fever, 55/221
fibrinolytic treatment, 55/211
fluorocarbon, 55/210
free radical scavenger, 55/210
furosemide, 55/213
general care, 55/220
glucocorticoid, 55/209
glutamic acid, 55/211
glycerol, 55/213
Guillain-Barré syndrome, 55/204, 219
head injury, 55/212, 215
head trauma, 55/204
heparin, 55/210
high dose barbiturate, 55/214
hyperbaric oxygenation, 55/210
hyperosmolar agent, 55/212
hyperventilation, 55/214
hypocapnia, 55/214
hypothermia, 55/217
increasing blood delivery, 55/210
increasing substrate delivery, 55/210
intracranial hypertension, 55/204
intracranial hypertension treatment, 55/212
mannitol, 55/213
meningitis, 55/204
monitoring, 55/217
myasthenia gravis, 55/204
myelitis, 55/204
near drowning, 55/210
neurogenic pulmonary edema, 55/221
opiate antagonist, 55/211
osmotic agent, 55/209
patient body position, 55/211
PCO_2, 55/209
PO_2, 55/209
postoperative care, 55/222
poststroke brain edema, 55/216
prognosis, 55/218
respiratory failure, 55/204
resuscitation, 55/209
Reye syndrome, 55/215
Sellick maneuver, 55/221
serum osmolarity, 55/213
somatosensory evoked potential, 55/218
spinal cord trauma, 55/204
status epilepticus, 55/204
stroke, 55/204
subarachnoid hemorrhage, 55/204
thiopental, 55/209
tumor, 55/204
urinary care, 55/221

vasospasm, 55/211
xylocaine, 55/214
Neurologic syndrome
see also Neurology
animal, 9/666
Arnold-Chiari malformation type I, 32/106
calcium, 70/111
diastematomyelia, 50/436
diplomyelia, 50/436
electrolyte disorder, 63/545
intensive care, 71/528
multiple organ failure, 71/526, 528
phytanic acid, 21/182
Sjögren syndrome, 71/62
Neurologic treatment
basic aspect, 1/34
Neurologists
training, 1/22-27
Neurology
see also Neurologic syndrome
acquired immune deficiency syndrome, 56/489
animal viral disease, 56/581
Argentina, 1/12
Austria, 1/10
Basle, Switzerland, 1/10
Belgium, 1/10
Berlin, Germany, 1/8, 13, 19
Breslau, Germany, 1/8
Canada, 1/12
Chile, 1/12
clinical, *see* Clinical neurology
Columbia, Bogota, 1/12
Columbia University, 1/5
comparative, *see* Comparative neurology
coronavirus infection, 56/439
Czechoslovakia, 1/11
Danzig, Germany, 1/9
definition, 1/1-38
diagnosis, *see* Neurologic diagnosis
diphtheria, 52/229
disease classification, 7/201
Dortmund, Germany, 1/9
Dundee, United Kingdom, 1/7
echoEG, 1/31
Edinburgh, United Kingdom, 1/7
enterovirus infection, 56/349
Epstein-Barr virus, 34/185-191
Epstein-Barr virus infection, 56/249-258
Europe, 1/1-38
fibrous dysplasia, 14/187-190
financial support, 1/37
France, 1/7
Frankfurt, Germany, 1/9
future, 1/23, 35
gastrointestinal disease, 39/449-464
Germany, 1/8, 13
Ghent, Belgium, 1/10
Glasgow, United Kingdom, 1/7
glycosaminoglycanosis, 10/441
Hamburg, Germany, 1/9
Hammersmith, United Kingdom, 1/7
Heidelberg, Germany, 1/9, 19
hepatobiliary disorder, 70/249
herpes simplex virus, 34/145-147
history, 1/1-23
Hungary, 1/11
hypothyroidism, 27/257-260
incontinentia pigmenti, 14/214
infantile spasm, 15/221
intelligence test, 3/296-303
internal medicine, 1/16
intestinal pseudo-obstruction, 70/322
Italy, 1/10
Leipzig, Germany, 1/19
Leningrad, USSR, 1/11
leukemia, 69/233
London, United Kingdom, 1/6, 19
Lyon, France, 1/8
Manchester, United Kingdom, 1/7
meaning of term, 1/2
microphthalmia, 42/403
Minsk, USSR, 1/11
Montreal, Canada, 1/12
Moscow, USSR, 1/11
mumps virus infection, 56/430
Munich, Germany, 1/9
optic atrophy, 13/44-88
Oslo, Norway, 1/12
Oxford, United Kingdom, 1/7
Padua, Italy, 1/11
pancreas transplantation, 70/249
paramyxovirus infection, 56/417
Paris, France, 1/7
Pennsylvania, USA, 1/5
Philadelphia, USA, 1/5
Phoenix, Arizona, USA, 1/35
place in medicine, 1/1-38
Poland, 1/11
Portugal, 1/11
Prague, Czechoslovakia, 1/10
psychiatric disorder, 1/8, 19-21
restorative, *see* Restorative neurology
Rome, Italy, 1/10
Rumania, 1/11
Sao Paulo, Brazil, 1/12
Scandinavia, 1/11

Spain, 1/11
specialization, 1/21
spinal cord missile injury, 25/211, 218
Strasbourg, France, 1/8
Switzerland, 1/10
The Netherlands, 1/10, 24
training, 1/21
tropical, *see* Tropical neurology
tuberous sclerosis, 14/349
Tübingen, Germany, 1/19
Turin, Italy, 1/9
turkey, 1/11
United Kingdom, 1/6
United States, 1/5
Uruguay, 1/12
varicella zoster virus infection, 56/229-243
Vienna, Austria, 1/10
Vogt-Koyanagi-Harada syndrome, 56/619
Warsaw, Poland, 1/11
Wegener granulomatosis, 71/175
Whipple disease, 70/244
Zürich, Switzerland, 1/10
Neurolymphoma
nonrecurrent nonhereditary multiple cranial neuropathy, 51/571
Neurolymphomatosis
cranial nerve palsy, 56/182
human T-lymphotropic virus type I, 56/179
hypertrophic nerve, 56/179
hypertrophic polyneuropathy, 56/179
Marek disease, 56/188
median neuropathy, 56/179
multiple cranial neuropathy, 56/182
myalgia, 56/182
non-Hodgkin lymphoma, 69/279
peripheral, *see* Peripheral neurolymphomatosis
Neuroma
definition, 7/244
history, 20/238
multiple mucosal, *see* Multiple mucosal neuroma
nerve tumor, 8/416
phantom limb, 4/227
plexiform, *see* Plexiform neuroma
Schwann cell, 8/416
synonym, 8/416
tissue culture, 17/82
traumatic, *see* Traumatic neuroma
Neuroma sign
brachial plexus, 51/148
Neuromacular dystrophy, *see* Neurolipidosis
Neuromedin B
Huntington chorea, 49/296
Neuromedin K
Huntington chorea, 49/296
Neuromelanin
1-methyl-4-phenylpyridinium affinity, 65/384
1-methyl-4-phenyl-1,2,3,6-tetrahydropyridine affinity, 65/384
Parkinson disease, 49/116, 124, 129
Neuromery
CNS, 50/2
embryo, 50/3
Neuromodulation
autonomic nervous system, 74/98
Neuromuscular block
aminoglycoside intoxication, 65/483
botulinum toxin, 40/129, 143
botulism, 41/290
α-bungarotoxin, 40/129, 143
curare, 40/129, 143
exercise, 74/251
experimental myasthenia gravis, 41/120
multiple organ failure, 71/540
myasthenia gravis, 7/109, 40/129, 265, 41/113, 126, 43/158
myasthenic syndrome, 40/265, 317, 438
organophosphate intoxication, 37/555, 558, 40/297, 304
regional, 41/296
stiff-man syndrome, 41/306
suxamethonium, 40/129
tetanus, 41/305
Neuromuscular disease
see also Myopathy
atrophy, 62/13
atrophy cause, 62/13
Becker muscular dystrophy, 41/411
cancer, 41/317-386
congenital muscular dystrophy, 41/420
congenital myopathy, 41/3-15, 429
denervation atrophy, 62/13
distal muscular dystrophy, 41/421
Duchenne muscular dystrophy, 41/408
facioscapulohumeral muscular dystrophy, 41/418
forked fiber, 62/11
gastrointestinal tract, 70/325
genetic counseling, 41/484
glycogen storage disease, 41/189, 426
hereditary neuropathy, 41/432
hyperparathyroidism, 70/118
limb girdle syndrome, 41/419
lipid storage myopathy, 41/426
malignant hyperthermia, 41/429
metabolic myopathy, 41/425, 431
mitochondrial myopathy, 41/431
motoneuron disease, 41/441

muscle fiber hypercontraction, 62/15
muscle fiber hypertrophy, 62/15
muscle fiber type, 62/2
muscle fiber type II, 40/14, 140
muscle pathology, 62/1
myasthenia gravis, 43/156
myotonic dystrophy, 41/421
necrosis, 62/6
oculopharyngeal, *see* Oculopharyngeal muscular dystrophy
partial invasion, 62/9
periodic paralysis, 41/151, 425
phagocytosis, 62/6
principle, 41/405
progressive external ophthalmoplegia, 41/420
regeneration, 62/9
scapuloperoneal muscular dystrophy, 41/411
segmental necrosis, 62/13
small caliber fiber, 62/13
spinal muscular atrophy, 41/435
Z band, 62/7

Neuromuscular disorder
antiretroviral therapy, 71/367
intestinal pseudo-obstruction, 70/320
sleep, 74/550

Neuromuscular junction, *see* Motor end plate

Neuromuscular junction disease
abnormal acetylcholine-acetylcholine receptor interaction syndrome, 62/392
autoimmune, 62/392
classification, 62/392
congenital acetylcholine receptor deficiency/short channel opentime syndrome, 62/392
congenital acetylcholine receptor deficiency/synaptic cleft syndrome, 62/392
congenital ε-acetylcholine receptor subunit mutation syndrome, 62/392
congenital end plate acetylcholinesterase deficiency, 62/392
congenital familial limb girdle myasthenia, 62/392
congenital high conductance fast channel syndrome, 62/392
congenital partial characterized, 62/392
congenital slow channel syndrome, 62/392
congenital synaptic vesical paucity syndrome, 62/392
Eaton-Lambert myasthenic syndrome, 62/392
familial infantile myasthenia, 62/392
myasthenia gravis, 62/392

Neuromuscular transmission
see also Trophic interaction
acetylcholine, 40/129
acetylcholine receptor, 40/129
acetylcholine transmission, 7/85
anticholinesterase, 41/131
botulinum toxin, 40/143
α-bungarotoxin, 40/129
β-bungarotoxin, 40/143
corticosteroid, 41/133
curare, 40/143
depolarization, 40/129
drug, 41/134
Eaton-Lambert myasthenic syndrome, 7/112, 41/352
EMG, 1/632, 644
experimental myasthenia gravis, 41/120
guanidine, 7/112
hemicholinium, 40/143
impulse generation, 40/126
miniature motor end plate potential, 40/129
motor end plate potential, 7/86, 40/127, 129
myasthenia gravis, 41/95, 107-109, 113, 126, 43/158
progressive muscular atrophy, 59/16
pyridostigmine, 7/112
testing, 7/104
transmitter release, 7/85

Neuromuscular transmission blocking
critical illness polyneuropathy, 51/584

Neuromyelitis optica, 9/426-435, 47/397-406
abortive case, 9/429
age, 9/428, 47/400
blindness, 9/426, 428, 47/401
carbamazepine, 47/406
cerebral symptom, 9/429
classification, 10/22
clinical features, 9/428, 47/400
computer assisted tomography, 47/402
corticosteroid, 47/406
course, 9/429, 47/402
CSF examination, 9/429, 47/401
definition, 9/426
demyelinating disease, 9/428, 431, 47/398
diagnosis, 47/401
diagnostic criteria, 47/400
differential diagnosis, 9/430, 47/405
disseminated encephalomyelitis, 9/426, 431, 47/473
EEG, 47/402
encephalomyelitis, 9/426, 431
frequency, 47/398-400
gait disturbance, 47/401
heralding sign, 9/429
hereditary spastic ataxia, 21/372
herpes zoster, 47/406
history, 9/426, 47/397

Horner syndrome, 9/428
inheritance, 47/400
initial symptom, 47/401
Japan, 47/398-400
Lhermitte sign, 47/401
Liesegang ring, 47/416
mode of onset, 9/428, 47/401
multiple sclerosis, 9/230, 235, 426, 428, 431, 47/70, 213, 397-399, 405, 487
myelopathy, 9/426
neurobrucellosis, 33/316, 52/588
oculomotor paralysis, 9/428
paraplegia, 9/426, 429, 47/401
pathogenesis, 9/434, 47/406
pathology, 9/430, 47/402-405
postinfectious encephalomyelitis, 9/427, 47/405
prognosis, 9/430
respiratory infection, 9/428
sensory loss, 9/429, 47/401
sex ratio, 9/428, 47/400
subacute, *see* Subacute myelo-optic neuropathy
subacute myelo-optic neuropathy, 9/435
swine influenza vaccination, 47/478, 480, 485
synonym, 9/428, 47/397
transverse myelitis, 47/401
treatment, 47/406
visual field, 9/428
visual impairment, 9/428, 47/397

Neuromyopathy
cancer, 41/322
carcinoma, *see* Carcinomatous neuromyopathy
chloroquine intoxication, 65/483
Hodgkin disease, 39/56
malignancy, 41/347
malignant lymphoma, 39/56, 69/278

Neuromyotonia, *see* Isaacs syndrome

Neuromyotonic discharge
definition, 62/263
Isaacs syndrome, 62/263
myotonia, 62/263
myotonic syndrome, 62/263

Neuron, *see* Nerve cell

Neuron body damage, *see* Neuronopathy

Neuron death
acidosis, 63/211
anaerobic glycolysis, 63/211
arachidonic acid, 63/211
aspartic acid, 63/211
autonomic nervous system, 74/206
blood glucose level, 63/211
brain ischemia, 63/212
calcium, 63/211
cardiac arrest, 63/210, 212
cell population, 63/212
cell vulnerability, 63/212
free radical, 63/212
glutamic acid, 63/211
lactic acidosis, 63/211
neuropathology, 63/212
seizure, 72/108
sodium-potassium adenosine triphosphatase pump, 63/211
synapse number, 51/17

Neuron degeneration
granulovacuolar, *see* Granulovacuolar neuronal degeneration
lathyrism, 7/644, 65/4
organophosphate intoxication, 64/228
progressive spinal muscular atrophy, 22/14
trichloroethylene polyneuropathy, 51/280

Neuron-neuroglia interdependence
multiple sclerosis, 9/624

Neuron ratio
axon, *see* Axon-neuron ratio

Neuronal argyrophilic dystrophy
Alzheimer neurofibrillary change, 21/63
amyloid angiopathy, 21/64
Gerstmann-Sträussler-Scheinker disease, 21/64
Guam amyotrophic lateral sclerosis, 21/64
main senile morphology type, 21/54
progressive supranuclear palsy, 21/63
senile plaque, 21/64

Neuronal atrophic dystrophy
achromatic form, 21/62
amyotrophic lateral sclerosis, 21/62
atypical form, 21/62
Lewy body, 21/62
main senile morphology type, 21/54
simple Spatz type, 21/62

Neuronal ceroid lipofuscinosis, 66/671-695
see also Batten disease *and* Progressive myoclonus epilepsy
adult, *see* Adult neuronal ceroid lipofuscinosis *and* Parry disease
Alzheimer disease, 42/276
Aran-Duchenne disease, 42/87
astroglial dystrophy, 21/67
ataxia, 10/550, 42/230
atypical, 66/24
Batten disease, 66/671
brain atrophy, 42/461-464
cerebellar ataxia, 42/466
cerebellar change, 10/656
cherry red spot, 10/550
classification, 10/224
clinical features, 10/612, 51/382

curvilinear body, 10/226
dementia, 46/402
dermatoglyphics, 10/226
differential diagnosis, 10/620
dolichol, 51/382, 66/685
dolichol diphosphate oligosaccharide, 66/685
dyssynergia cerebellaris myoclonica, 60/599
dystonia, 60/664
EEG, 42/461-468
enzyme deficiency polyneuropathy, 51/371
epidemiology, 10/576
familial amaurotic idiocy, 10/215
form, 10/588, 21/65, 66/24
genetics, 72/134
glia, 10/659
GM2 gangliosidosis, 10/221, 390
histochemistry, 10/640
history, 10/550
inclusion body, 51/382
infantile, *see* Infantile neuronal ceroid lipofuscinosis
juvenile, *see* Juvenile neuronal ceroid lipofuscinosis
juvenile amaurotic idiocy, 10/589
late infantile, *see* Late infantile neuronal ceroid lipofuscinosis
leukocyte granulation, 10/578
lipid peroxidation, 66/685
lipopigment, 10/589, 672, 21/58
long chain polyisoprenol alcohol, 51/382
lymphocyte vacuolation, 10/224, 576
lysosome, 10/670
microcephaly, 42/461
myoclonia, 10/550
myoclonic epilepsy, 27/188, 49/617, 60/662
neuronal dystrophy, 21/62
neuropathology, 66/24
nomenclature, 66/672
optic atrophy, 42/461, 463
pathology, 66/25
pigment chemistry, 10/669
polyunsaturated fatty acid, 10/671
protease inhibitor, 66/685
psychomotor retardation, 42/461-463
Purkinje cell loss, 42/461, 466, 468
retinal change, 10/607
retinal degeneration, 42/461, 463
Schaffer-Spielmeyer cell change, 10/652
secondary pigmentary retinal degeneration, 10/299, 42/464, 60/732
seizure, 42/461, 463, 466
sphingolipid, 10/589
substantia nigra degeneration, 42/468
subtype, 10/225
symptomatic dystonia, 49/543
ultrastructure, 10/636-638, 51/382
vacuolated lymphocyte, 60/674
Neuronal dystrophy
classification, 21/61
definition, 21/47
neuronal ceroid lipofuscinosis, 21/62
Neuronal heterotopia
Aicardi syndrome, 42/696
Berardinelli-Seip syndrome, 42/592
Börjeson-Forssman-Lehmann syndrome, 42/530
cerebellar hypoplasia, 42/18
cerebellar vermis agenesis, 42/5
cerebrohepatorenal syndrome, 43/340
corpus callosum agenesis, 42/7, 10
Dandy-Walker syndrome, 42/23
encephalomeningocele, 42/26
lissencephaly, 42/40
neurofibromatosis type I, 14/145
Patau syndrome, 43/527
peroxisomal acetyl-coenzyme A acyltransferase deficiency, 66/518
spina bifida, 42/55
Taybi-Linder syndrome, 43/386
thanatophoric dwarfism, 43/388
trichopoliodystrophy, 42/584
Neuronal intranuclear hyaline inclusion disease
areflexia, 60/660
Biemond syndrome, 60/449
bulbar paralysis, 60/658
cardiac atrophy, 60/668
caudate nucleus atrophy, 60/671
computer assisted tomography, 60/671
contracture, 60/669
deafness, 60/657
dementia, 60/665
distal amyotrophy, 60/659
ERG, 60/670
external ophthalmoplegia, 60/655
fecal incontinence, 60/664
intestinal pseudo-obstruction, 60/650, 668
kyphoscoliosis, 60/669
multiple neuronal system degeneration, 60/660
neuropathy, 60/660
periodic ataxia, 60/648
ptosis, 60/655
pupillary disorder, 60/655
Refsum disease, 60/661
spasticity, 60/661
telangiectasia, 60/666
trigeminal sensory deficit, 60/657
urinary incontinence, 60/664

Neuronal lipopigment dystrophy
main senile morphology type, 21/54
type, 21/64
Neuronal migration
cerebrohepatorenal syndrome, 43/340
congenital megacolon, 42/3
lissencephaly, 42/40
mental deficiency, 43/303
microcephaly, 30/518
Neuronal migration disorder
cortical dysplasia, 73/400
fetal Minamata disease, 64/419
hemimegalencephaly, 73/400
Minamata disease, 64/419
porencephaly, 42/48
Neuronal mitochondrial vacuole
thallium, 51/266
Neuronal nucleus membrane
Huntington chorea, 49/323
Neuronal storage dystrophy
acid maltase deficiency, 21/61
combined form, 21/62
GM2 gangliosidosis, 21/61
granular form, 21/61
Lafora progressive myoclonus epilepsy, 21/62
neuropathology, 21/51
polymorphic form, 21/62
Schaffer-Spielmeyer cell change, 21/61
spherical deposit, 21/61
Neuronal swelling
Pick disease, 6/614, 42/286, 46/233, 236, 238, 243
Neuronal system degeneration
multiple, *see* Multiple neuronal system degeneration
Neuronal tumor
classification, 68/137
demography, 68/139
epilepsy, 72/180
survey, 68/137
treatment, 68/143
Neuronitis
see also Guillain-Barré syndrome *and* Polyradiculitis
subacute myoclonic spinal, *see* Subacute myoclonic spinal neuronitis
vestibular, *see* Vestibular neuronitis
Neuronoglial tumor
demography, 68/139
mixed, *see* Mixed neuronoglial tumor
survey, 68/137
treatment, 68/143
Neuronopathic Gaucher disease
noninfantile, *see* Noninfantile neuronopathic Gaucher disease
Neuronopathy
amnesic shellfish poisoning, 65/167
neurotoxicity, 51/263
Sjögren syndrome, 71/83
subacute motor, *see* Subacute motor neuronopathy
subacute sensory, *see* Subacute sensory neuronopathy
vitamin B6, 51/264
Neuronophagia
poliomyelitis, 34/108, 56/321
Neuro-oculocutaneous angiomatosis, *see* Sturge-Weber syndrome
Neuro-otology
see also ENG, Hearing loss, Labyrinth concussion *and* Temporal bone fracture
head injury, 24/119-140
Neuroparalytic accident
terminology, 9/507, 550
Neuropathic arthropathy
alcoholic polyneuropathy, 51/317
progressive systemic sclerosis, 51/448
Neuropathic gammopathy
hereditary spinal muscular atrophy, 59/26
Neuropathic joint, *see* Neurogenic osteoarthropathy
Neuropathic ulcer
spina bifida, 50/495
Neuropathology
acetonylacetone intoxication, 64/88
β-*N*-acetylhexosaminidase A deficiency, 59/100
acquired hepatocerebral degeneration, 6/285-291, 27/352
acquired immune deficiency syndrome, 56/507
acquired immune deficiency syndrome dementia complex, 71/249
acquired toxoplasmosis, 52/352, 357
acrodynia, 64/374
acute intermittent porphyria, 63/150
adrenomyeloneuropathy, 60/171
African trypanosomiasis, 52/339
alcohol intoxication, 64/116
Alzheimer disease, 27/483, 46/210
Ammon horn, 21/45
amnesic shellfish poisoning, 65/167
amyloid polyneuropathy, 51/417
amyotrophic lateral sclerosis, 22/127, 59/232
angiostrongyliasis, 52/549
anoxic encephalopathy, 46/347
arsenic intoxication, 47/559
aspergillosis, 52/379
astrocytoma, 18/2, 8-11
ataxia telangiectasia, 14/73, 275, 325, 60/353, 376-383

axonal degeneration, 51/63
ballismus, 6/480-483
barium intoxication, 64/354
Bassen-Kornzweig syndrome, 60/133
Batten disease, 10/628, 631, 66/26
Behçet syndrome, 34/493-501, 56/598, 71/218
beriberi, 7/564
Biemond syndrome, 21/378, 60/446-448
Binswanger disease, 54/224
biopsy, 51/74
bismuth intoxication, 64/340
blind loop syndrome, 63/277
Bonnet-Dechaume-Blanc syndrome, 14/69
brain air embolism, 55/193
brain amyloid angiopathy, 54/333
brain death, 9/587, 63/213
brain embolism, 11/401-403
brain fat embolism, 55/187
brain hemorrhage, 54/294
brain infarction, 11/142-154
brain injury mechanism, 57/24
brain lacunar infarction, 46/355, 54/235
bulbospinal muscular atrophy, 59/44
carbon disulfide intoxication, 49/477
carbon monoxide encephalopathy, 9/629
carbon monoxide intoxication, 29/545, 64/35
cat scratch disease, 34/472
cerebellar granular layer atrophy, 21/500
cerebellar hemangioblastoma, 14/57
cerebrotendinous xanthomatosis, 10/534-537, 60/167
Charlevoix-Saguenay spastic ataxia, 60/456
chorea-acanthocytosis, 60/141
chronic mercury intoxication, 64/372
chronic nodular polioencephalitis, 34/558
classification, 7/201
Cobb syndrome, 14/440
coccidioidomycosis, 52/411, 414
Cockayne-Neill-Dingwall disease, 13/434
computer assisted tomography, 54/66
congenital cerebellar atrophy, 21/460
congenital retinal blindness, 22/541
Creutzfeldt-Jakob disease, 6/749-754
cryptococcosis, 52/433
cyanide intoxication, 49/477, 64/232, 65/28
cyanogenic glycoside intoxication, 65/28
cystic fibrosis, 63/277
cytomegalic inclusion body disease, 56/271
cytomegalovirus infection, 56/271
definition, 1/3
delirium tremens, 46/346
dementia, 67/384
dentatorubropallidoluysian atrophy, 21/522-525, 49/440, 60/610-614
diphtheria, 65/241
Divry-Van Bogaert syndrome, 10/124, 14/71, 55/320, 322
Down syndrome, 21/45, 50/530
drug induced extrapyramidal syndrome, 6/259-261
dyssynergia cerebellaris myoclonica, 60/599
dystonia musculorum deformans, 6/534-536
enzyme deficiency polyneuropathy, 51/368
eosinophilia myalgia syndrome, 64/253
epilepsy, 72/107
epileptogenesis, 72/107
Epstein-Barr virus encephalitis, 34/188
Epstein-Barr virus infection, 56/257
erythropoietin, 63/535
ethambutol intoxication, 65/481
ethambutol polyneuropathy, 51/295
Fabry disease, 10/301, 51/377, 60/173, 66/22
familial amyotrophic lateral sclerosis, 22/133, 342, 59/233, 236, 245-248
familial spastic paraplegia, 22/423, 438, 462, 59/303
Fazio-Londe disease, 22/105, 59/124, 128
Ferguson-Critchley syndrome, 22/436
fetal Minamata disease, 64/419
fibrous dysplasia, 14/83
focal cerebellar cortical panatrophy, 21/500
Foix-Alajouanine disease, 55/103
Friedreich ataxia, 21/322-325, 60/301-304
frontal lobe epilepsy, 73/46
fucosidosis, 66/334
functional psychosis, 46/453-455
galactosialidosis, 66/22
Gaucher disease, 10/512, 515-518, 51/379, 66/23
Gerstmann-Sträussler-Scheinker disease, 56/554
giant axonal neuropathy, 60/82
glioblastoma multiforme, 18/50
globoid cell leukodystrophy, 10/37, 21/50, 51/373
glycosaminoglycanosis, 10/442
GM2 gangliosidosis, 10/385-406
Guam amyotrophic lateral sclerosis, 59/233, 237, 256-259, 294-296
Guam motoneuron disease, 65/22
gunshot injury, 25/49-51
Hallervorden-Spatz syndrome, 49/400, 66/22
head injury, 23/35, 57/43-61
hemorrhagic necrosis, 25/64
hepatic coma, 27/351, 49/214
hereditary amyloid polyneuropathy, 21/134-139, 60/100
hereditary amyloid polyneuropathy type 1, 51/422, 424

hereditary amyloid polyneuropathy type 2, 51/423
hereditary cerebello-olivary atrophy (Holmes), 21/406
hereditary dystonic paraplegia, 22/449, 462, 59/345
hereditary motor and sensory neuropathy type I, 21/288-303
hereditary motor and sensory neuropathy type II, 21/288-303
hereditary olivopontocerebellar atrophy (Menzel), 21/436
hereditary sensory and autonomic neuropathy type I, 21/75, 80-82
hereditary sensory and autonomic neuropathy type II, 60/11, 17
hereditary sensory and autonomic neuropathy type III, 21/111-114, 60/29-31
hereditary sensory and autonomic neuropathy type IV, 60/13
hereditary spastic ataxia, 21/368, 60/462
heredodegenerative disease, 21/43-71
hexachlorophene intoxication, 37/499, 65/473, 476
Hirayama disease, 59/114
history, 1/27-29
Hodgkin disease, 18/262
homocystinuria, 55/329
human T-lymphotropic virus type I associated myelopathy, 56/531-533, 536
Huntington chorea, 6/329-342, 49/256, 315
hypertrophic interstitial neuropathy, 21/154-160, 51/63
hypogonadal cerebellar ataxia, 21/469
incontinentia pigmenti, 14/84, 220
infantile neuroaxonal dystrophy, 49/400, 403, 407, 66/22
infantile spinal muscular atrophy, 22/98, 59/60-63
intracranial meningoencephalocele, 31/121-124
juvenile Gaucher disease, 60/158
juvenile Huntington chorea, 49/321
juvenile parkinsonism, 49/159
kernicterus, 6/512
Kii Peninsula amyotrophic lateral sclerosis, 22/373-408, 59/237, 294-296
Klein-Waardenburg syndrome, 59/162
Klippel-Trénaunay syndrome, 14/398
Koshevnikoff epilepsy, 73/120
Krabbe-Bartels disease, 14/81-83
kuru, 56/558
late cerebellar atrophy, 60/585, 587
late cortical cerebellar atrophy, 21/480, 491-494
late infantile neuroaxonal dystrophy, 49/400, 405
lathyrism, 65/9
lead encephalopathy, 64/437
lead polyneuropathy, 64/437
legionellosis, 52/256
leprosy, 33/437-455
leukemia, 18/251
leukoencephalopathy, 66/726
local anesthetic agent, 65/423
Löwenberg-Hill leukodystrophy, 10/177
Lyme disease, 52/263
manganese intoxication, 49/477, 64/306
Marchiafava-Bignami disease, 9/653, 46/347, 47/558
Marinesco-Sjögren syndrome, 21/557, 60/344
Matthews-Rundle syndrome, 60/578
McLeod phenotype, 63/287
medulloblastoma, 68/185
meningeal leukemia, 63/341
mercury intoxication, 64/381
metachromatic leukodystrophy, 10/311
methanol intoxication, 64/101
methotrexate intoxication, 65/531
1-methyl-4-phenyl-1,2,3,6-tetrahydropyridine intoxication, 65/371
Miller Fisher syndrome, 51/246
Minamata disease, 36/83, 64/414
missile injury, 25/49-51
mixed connective tissue disease, 71/126
multiple myeloma, 18/257
multiple neuronal system degeneration, 59/143
multiple nevoid basal cell carcinoma syndrome, 14/79-81
multiple sclerosis, 9/217-308, 47/213-251
multiple sclerosis plaque, 47/213-217
neglect, 45/175
neglect syndrome, 45/175
neurinoma, 68/557
neuroaxonal leukodystrophy, 49/406
neurobrucellosis, 33/307, 52/585
neurocutaneous melanosis, 14/76-79, 418
neurodegenerative disease, 21/43-71
neurofibromatosis, 14/20-38
neurofibromatosis type I, 14/20-38, 134-146
neuroleptic syndrome, 6/259
neuron death, 63/212
neuronal ceroid lipofuscinosis, 66/24
neuronal storage dystrophy, 21/51
neurosyphilis, 52/281
Niemann-Pick disease type A, 10/282, 490, 60/148
Niemann-Pick disease type C, 10/498, 60/151
nigrospinodentate degeneration, 22/160, 165
obstructive liver disease, 63/278
olivopontocerebellar atrophy (Dejerine-Thomas),

21/419-424, 60/512, 516
olivopontocerebellar atrophy (Wadia-Swami), 60/495
opticocochleodentate degeneration, 21/542, 60/754
organochlorine insecticide intoxication, 64/204
organolead intoxication, 64/131
organophosphate induced delayed polyneuropathy, 64/160
organophosphate intoxication, 64/160, 228
organotin intoxication, 64/136, 140, 142
oxygen deficiency, 29/541, 545
oxygen intoxication, 29/536-538
palatal myoclonus, 6/774
pallidoluysian atrophy, 6/636-640
pallidoluysionigral degeneration, 49/447
paraneoplastic polyneuropathy, 51/468
Parkinson dementia, 49/175
Parkinson dementia complex, 22/343, 383
Parkinson disease, 49/108
pathophysiology, 6/259-261
Pelizaeus-Merzbacher disease, 66/566
pellagra, 7/575
peripheral neurolymphomatosis, 56/182
phakomatosis, 14/19-100
Pick disease, 21/64, 27/489
poliomyelitis, 34/106, 56/318-320
porphyria, 60/118
posthemiplegic athetosis, 49/386
posthemiplegic dystonia, 49/386
postpoliomyelitic amyotrophy, 59/38
posttraumatic syringomyelia, 61/385
presenile glial dystrophy, 21/59
primary alcoholic dementia, 46/346
progressive bulbar palsy, 22/126, 59/221-224
progressive dysautonomia, 59/140
progressive external ophthalmoplegia, 22/182, 62/289
progressive multifocal leukoencephalopathy, 71/402
progressive myoclonus epilepsy, 27/175
progressive rubella panencephalitis, 56/410
progressive spinal muscular atrophy, 22/14-18
progressive supranuclear palsy, 22/221, 226, 46/302, 49/248
pure autonomic failure, 22/237
radiation myelopathy, 61/201
Reye syndrome, 27/351, 56/153, 65/120
rheumatoid arthritis, 51/447
Richards-Rundle syndrome, 22/524
rigid Huntington chorea, 49/285, 321
salicylic acid intoxication, 37/420
Salla disease, 59/361
sarcoidosis, 71/472
scapuloperoneal spinal muscular atrophy, 59/45
schizophrenia, 46/453-455
Schut family ataxia, 60/483
segmental demyelination, 51/63
senile dementia, 21/64
Shy-Drager syndrome, 22/265-272, 381, 390, 59/140
sickle cell anemia, 55/508
Sjögren-Larsson syndrome, 22/477, 66/618
Sjögren syndrome, 71/77
spasmodic torticollis, 6/582-585
spinal cord injury, 25/44-104, 61/37
spinal muscular atrophy, 59/114
spinopontine degeneration, 21/390-398
spongiform encephalopathy, 21/60, 68
spongy cerebral degeneration, 10/206, 66/662
sporadic amyotrophic lateral sclerosis, 59/233, 235
status cribrosus, 6/673
status epilepticus, 73/318-320
storage disorder, 21/48-51
striatonigral degeneration, 6/695-697
striatopallidodentate calcification, 6/712-716
Sturge-Weber syndrome, 14/61-68, 228, 55/447
subacute combined spinal cord degeneration, 28/154-158, 56/531, 63/254
subacute sclerosing panencephalitis, 56/421
subarachnoid hemorrhage, 55/4
sudanophilic leukodystrophy, 21/51
superficial siderosis, 70/76
Sydenham chorea, 6/429, 49/364
Tangier disease, 60/135
tardive dyskinesia, 49/190
telangiectasia, 14/51-53
thalamic syndrome (Dejerine-Roussy), 2/486-491
toluene, 66/726
traumatic hydrocephalus, 24/240
trichinosis, 52/570
tropical spastic paraplegia, 56/531-533, 536
tuberculoma, 18/420
tuberous sclerosis, 14/38-53, 353-369
undetermined significant monoclonal gammopathy, 63/399
Unverricht-Lundborg progressive myoclonus epilepsy, 27/175
uremic polyneuropathy, 63/511
varicella zoster virus meningoencephalitis, 56/235
varicella zoster virus myelitis, 56/236
visna, 56/459
vitamin B1 deficiency, 51/332
vitamin B6 deficiency, 51/334
vitamin B12 deficiency, 51/341

vitamin E deficiency, 63/277
Vogt-Koyanagi-Harada syndrome, 34/533
Von Hippel-Lindau disease, 14/53-61, 243
wallerian degeneration, 7/202-219, 51/63
Wernicke-Korsakoff syndrome, 46/347
Whipple disease, 52/139
Wilson disease, 46/18, 347
Wohlfart-Kugelberg-Welander disease, 22/72
Wolman disease, 51/379, 60/160
X-linked adrenoleukodystrophy, 66/459
xeroderma pigmentosum, 60/337
Neuropathy
see also Denervation, Nerve injury, Neuritis, Polyneuritis *and* Polyneuropathy
a-alphalipoproteinemia, 51/396
absinthol, 7/521
acanthosis nigricans, 22/524
β-*N*-acetylhexosaminidase A deficiency, 60/660
acoustic, *see* Acoustic neuropathy
acquired immune deficiency syndrome, 56/490, 71/367
acrodystrophic, *see* Hereditary sensory and autonomic neuropathy type I
acromegalic, *see* Acromegalic neuropathy
acrylamide intoxication, 75/504
acute demyelinating, *see* Acute demyelinating neuropathy
acute inflammatory demyelinating, *see* Guillain-Barré syndrome
acute intermittent porphyria, 8/366, 27/434, 41/434, 60/120
acute ischemic optic, *see* Acute ischemic optic neuropathy
acute sensorimotor, *see* Acute sensorimotor neuropathy
acute sensory, *see* Acute sensory neuropathy
Addison disease, 39/480
adrenoleukodystrophy, 51/384
adrenomyelo, *see* Adrenomyeloneuropathy
adriamycin, 69/460, 471
African trypanosomiasis, 52/341
agraphia, 45/464
alcohol, 75/494
alcoholic, *see* Alcoholic neuropathy
allergic, *see* Allergic neuropathy
allergic headache, 5/234
altretamine, 69/461, 471
δ-aminolevulinic acid dehydratase deficiency porphyria, 60/120
amiodarone, 63/190
amphotericin B, 7/516
amsacrine, 69/461, 472
amyloidosis, 7/556, 8/4-8, 160, 71/517
anemia, 7/664
anterior ischemic optic, *see* Anterior ischemic optic neuropathy
antibiotic agent, 7/515
antibody, 71/442
anticoagulant, 7/520, 8/13
antidiabetic agent, 7/518
apolipoprotein AI, 71/517
arsenic intoxication, 7/513, 36/203, 47/559, 64/286
arteriosclerosis, 8/161
arthrogryposis multiplex congenita, 32/512, 515, 42/73-75
ascending, *see* Ascending neuropathy
ascending sensorimotor, *see* Ascending sensorimotor neuropathy
ascending symmetrical, *see* Ascending symmetrical neuropathy
asymmetrical, *see* Asymmetrical neuropathy
ataxia telangiectasia, 1/318, 7/640, 14/316, 28/26, 36/522, 60/660
ataxic, *see* Ataxic neuropathy
Austin juvenile sulfatidosis, 60/660
autonomic, *see* Autonomic neuropathy
autonomic nervous system, 74/113, 75/681
autonomic poly, *see* Autonomic polyneuropathy
autosomal dominant inherited motor, *see* Autosomal dominant inherited motor neuropathy
axonal, *see* Axonal neuropathy
azacitidine, 69/461, 472
bacterial endocarditis, 52/291, 300
barbiturate, 7/165, 518
barbiturate coma, 7/165
Bassen-Kornzweig syndrome, 8/19, 42/511, 60/131
Behçet syndrome, 34/486
Behr disease, 60/660
benign peroxisomal dysgenesis, 60/660, 670
beriberi, 7/559, 28/4, 49-55
bilateral sensory trigeminal, *see* Bilateral sensory trigeminal neuropathy
bis(*N*-oxopyridine-2-thianato) zinc, 40/141
brachial neuritis, 42/305
brachial plexus, *see* Brachial plexus neuropathy
bronchiectasis, 8/18
bulbar nerve symptom, 59/404
burn, 8/15
cachexia, 8/15
carbon disulfide intoxication, 7/522, 64/8
carbon monoxide intoxication, 7/521, 64/32
carboplatin, 69/460, 464, 467
carcinoid syndrome, 70/316

carcinoma, 8/139
carnitine deficiency, 43/175, 66/406
cat scratch disease, 52/128
celiac disease, 28/227, 229, 232
cerebellar ataxia, 60/660, 69/356
cerebrotendinous xanthomatosis, 60/660
Charlevoix-Saguenay spastic ataxia, 60/660
Chédiak-Higashi syndrome, 60/660
chemotherapy, 69/459
chloramphenicol, 7/515, 530
chlormethine, 7/517, 537
2-chlorodeoxyadenosine, 69/461
chloroquine, 7/518
chloroquine myopathy, 62/606
chlorpropamide, 7/518
chlorprothixene, 7/518
cholesterol, 8/20
chorea-acanthocytosis, 63/283
chronic axonal, *see* Chronic axonal neuropathy
chronic demyelinating, *see* Chronic demyelinating neuropathy
chronic idiopathic demyelinating polyradicular, *see* Chronic idiopathic demyelinating polyradiculoneuropathy
chronic progressive peripheral, *see* Chronic progressive peripheral neuropathy
chronic recurring, *see* Chronic recurring neuropathy
chronic relapsing, *see* Chronic relapsing neuropathy
chronic relapsing inflammatory, *see* Chronic relapsing neuritis
chronic sensorimotor, *see* Chronic sensorimotor neuropathy
chronic sensory, *see* Chronic sensory neuropathy
ciclosporin, 69/461, 472
ciguatoxin intoxication, 65/163
cisplatin, 69/460, 464, 466
clioquinol intoxication, 64/8
Cockayne-Neill-Dingwall disease, 60/660
cold, *see* Cold neuropathy
collagen disease, 8/118
complex I deficiency, 66/422
complex V deficiency, 62/506
compression, *see* Compression neuropathy
congenital, *see* Congenital neuropathy
congenital amyelinating, *see* Congenital amyelinating neuropathy
congenital sensory, *see* Hereditary sensory and autonomic neuropathy type I
Corynebacterium diphtheriae, 47/608
cranial, *see* Cranial neuropathy
cryoglobulinemia, 63/404, 71/441, 445
cyanogenic glycoside intoxication, 64/8
cytarabine, 69/461, 472
cytomegalovirus, 71/358
dapsone, 71/384
DDT intoxication, 7/522
deafness, 60/660
deficiency, *see* Deficiency neuropathy
definition, 8/29
delayed, *see* Delayed neuropathy
demyelinating, *see* Demyelinating neuropathy
diabetes mellitus, 42/544, 60/660, 70/143, 74/118, 75/589
diabetic, *see* Diabetic neuropathy
diabetic autonomic, *see* Diabetic autonomic neuropathy
diabetic autonomic visceral, *see* Diabetic autonomic visceral neuropathy
diabetic peripheral, *see* Diabetic polyneuropathy
diabetic proximal, *see* Diabetic proximal neuropathy
diamorphine intoxication, 65/356
differential diagnosis, 7/245, 8/357, 40/481
1,4-diketone, *see* 1,4-Diketone neuropathy
dimethylaminopropionitrile, 75/506
diphtheritic, *see* Diphtheritic neuropathy
distal muscle weakness, 40/318, 320
distal muscular dystrophy, 43/97
distal sensory, *see* Distal sensory neuropathy
disulfiram, 7/531
disulfiram intoxication, 64/8
dominant ataxia, 60/660
dominant olivopontocerebellar atrophy, 60/660
doxorubicin, 69/471
drug induced demyelinating, *see* Drug induced demyelinating neuropathy
due to drug, 7/515-546
dying-back, *see* Axonal degeneration
dysproteinemic, *see* Dysproteinemic neuropathy
dyssynergia cerebellaris myoclonica, 60/660
electrical injury, 51/139
endocrine, 8/29
enteric, *see* Enteric neuronopathy
entrapment, *see* Compression neuropathy
eosinophilia myalgia syndrome, 64/252
ergotamine, 7/520
ethambutol, 7/517
ethionamide, 7/516, 535
etoglucid, 7/517, 531
etoposide, 69/460
exercise, 8/393
experimental allergic, *see* Experimental allergic neuritis
extensor plantar response, 59/404

familial visceral, *see* Familial visceral neuropathy
fasciculation, 60/659
femoral, *see* Femoral neuropathy
fludarabine, 69/461
Flynn-Aird syndrome, 22/519, 42/327
focal, *see* Focal neuropathy
food intoxication, 7/520
Friedreich ataxia, 60/659
furan, 7/517
ganglioside, 69/298
gasoline, 7/521
gastric surgery, 70/229
gastrointestinal, *see* Gastrointestinal neuropathy
giant axonal, *see* Giant axonal neuropathy
globular, *see* Hereditary multiple recurrent mononeuropathy
glossopharyngeal neuralgia, 48/467
glutethimide, 7/543
glutethimide intoxication, 7/519, 37/354
gold, 7/532
gold intoxication, 36/332
gout, 8/3
Gowers distal myopathy, 40/471-481
Guillain-Barré syndrome, 8/365, 40/297, 303, 308, 318, 326, 336
heavy metal intoxication, 7/511
hemochromatosis, 8/11-13, 42/553
hemophilia, 7/664, 8/13
hepatitis, 7/609
hereditary, *see* Hereditary neuropathy
hereditary amyloid, *see* Hereditary amyloid polyneuropathy
hereditary brachial plexus, *see* Hereditary brachial plexus neuropathy
hereditary bronchial, *see* Hereditary brachial neuropathy
hereditary coproporphyria, 60/120
hereditary cranial, *see* Hereditary cranial neuropathy
hereditary motor and sensory neuropathy type I, 40/319, 472, 60/659
hereditary motor and sensory neuropathy type II, 40/319, 472, 60/659
hereditary motor sensory, *see* Hereditary motor sensory neuropathy
hereditary peripheral, *see* Hereditary polyneuropathy
hereditary recurrent, *see* Hereditary recurrent neuropathy
hereditary sensory, *see* Hereditary sensory and autonomic neuropathy type II
hereditary sensory and autonomic, *see* Hereditary sensory and autonomic neuropathy
hereditary sensory radicular, *see* Hereditary sensory and autonomic neuropathy type I
heterocyclic antidepressant, 65/316
hexacarbon, *see* Hexacarbon neuropathy
hexachlorophene intoxication, 65/472
hexane, 7/522
histochemistry, 40/45
Hodgkin disease, 6/143
Huffer, *see* Huffer neuropathy
human immunodeficiency virus infection, *see* Human immunodeficiency virus neuropathy
human immunodeficiency virus treatment, 71/362
hydralazine, *see* Hydralazine neuropathy
hyperalgesic diabetic, *see* Hyperalgesic diabetic neuropathy
hypercholesterolemia, 8/18
hyperemesis, 8/14
hypereosinophilia, 63/138
hypereosinophilic syndrome, 63/372
hyperinsulin, *see* Hyperinsulin neuropathy
hyperlipemic, *see* Hyperlipemic neuropathy
hyperlipidemic, *see* Hyperlipidemic neuropathy
hyperlipoproteinemia type II, 42/445
hyperparathyroid myopathy, 62/540
hyperparathyroidism, 27/293
hyperthyroidism, 70/86
hypertrophic interstitial, *see* Hypertrophic interstitial neuropathy
hypertrophic sensory, *see* Hypertrophic sensory neuropathy
hypoglycemia, 8/52, 70/157
hypoparathyroidism, 27/312
hypothyroidism, 70/99
iatrogenic, *see* Iatrogenic neuropathy
idiopathic autonomic, *see* Idiopathic autonomic neuropathy
ifosfamide, 69/461, 471
Ig, 8/72
imipramine, 7/518, 533
immunization, 7/555
infantile spinal muscular atrophy, 59/70
infectious mononucleosis, 7/555
insulin, 8/53
α2-interferon, 69/461, 472
interleukin-2, 69/461, 472
interstitial, *see* Interstitial neuropathy
intestinal autonomic, *see* Intestinal autonomic neuropathy
ionizing radiation, *see* Ionizing radiation neuropathy
ischemic, *see* Ischemic neuropathy
ischemic optic, *see* Ischemic optic neuropathy
isolated trigeminal, *see* Sensory trigeminal

neuropathy
isoniazid, *see* Isoniazid neuropathy
Jamaican, *see* Jamaican neuropathy
jaundice, 8/18
Joseph-Machado disease, 60/660
Kawasaki syndrome, 56/639
Kearns-Sayre-Daroff-Shy syndrome, 62/313
Kjellin syndrome, 42/173
late onset ataxia, 60/660
lathyrism, 7/520
lead, *see* Lead neuropathy
lead intoxication, 7/511, 8/366, 9/642, 36/12-15, 38-40
Leber hereditary optic neuropathy, 62/509, 66/428
legionellosis, 52/254
Lemieux-Neemeh syndrome, 22/520
leprosy, 33/428-433
leprous, *see* Leprous neuropathy
leukemia, 8/145, 63/354
leukodystrophy, 8/16
α-lipoprotein, 8/20
lithium intoxication, 64/296
liver disease, 7/554, 609, 612, 614, 27/353
local anesthetic agent, 65/421
lumbar plexus, *see* Lumbar plexus neuropathy
lumbosacral plexus, *see* Lumbosacral plexus neuropathy
lymphoma, 63/350, 69/278
macroglobulinemia, 7/555, 8/74, 39/193
malabsorption syndrome, 7/595, 70/231
malignancy, 7/554, 8/131, 41/347
malnutrition, 7/633, 28/2-9
MAO inhibitor, 7/519
Marinesco-Sjögren syndrome, 60/660
McLeod phenotype, 63/287
median nerve, *see* Median neuropathy
mental deficiency, 42/335
meprobamate, 7/519
mepyramine, 7/520
mercury intoxication, 7/514
MERRF syndrome, 62/509
metabolic, *see* Metabolic neuropathy
metachromatic leukodystrophy, 8/16
metastasis, 8/131
metronidazole, 69/460
Minamata disease, 36/109-111
misonidazole, 69/460
mitotane, 69/461, 472
mono, *see* Mononeuropathy
monoclonal gammopathy, 71/441, 445
monoclonal protein, 69/292
motoneuron disease, 8/143
motor, *see* Motor neuropathy
multifocal, *see* Multifocal neuropathy
multifocal relapsing, *see* Multifocal relapsing neuropathy
multiple cranial, *see* Multiple cranial neuropathy
multiple mitochondrial DNA deletion, 62/510
multiple myeloma, 18/255, 39/153-156
mummy, 74/189
myelinolysis, 63/548
myeloma, 8/144
myoclonus, 60/660
myotonic dystrophy, 62/234
myxedema, *see* Myxedema neuropathy
Navaja, *see* Navaja neuropathy
navelbine, 69/460
neoplastic, *see* Paraneoplastic neuropathy
nerve conduction velocity, 60/670
neurofilament, *see* Neurofilament neuropathy
neurogenic muscular atrophy, 60/659
neuronal intranuclear hyaline inclusion disease, 60/660
neurosarcoidosis, 38/525-529
neurosensory hearing loss, 60/660
neurotoxicology, 64/10
Niemann-Pick disease type C, 51/375
nitrofurantoin, 7/517, 535
nonmalignant IgG, 71/446
nonrecurrent nonhereditary multiple cranial, *see* Nonrecurrent nonhereditary multiple cranial neuropathy
nucleoside analog, 71/367, 371
obturator, *see* Obturator neuropathy
occupational therapy, 8/402
oculomotor, *see* Oculomotor neuropathy
olivopontocerebellar atrophy (Dejerine-Thomas), 60/660
olivopontocerebellar atrophy (Wadia-Swami), 60/660
opiate intoxication, 65/356
optic, *see* Optic neuropathy
optic atrophy, 60/660
oral contraception, 7/520
organic solvent intoxication, 75/505
organophosphate induced delayed, *see* Organophosphate induced delayed neuropathy
painful, *see* Painful neuropathy
pancreas tumor, 7/607
pancreatitis, 7/603
paraneoplastic, *see* Paraneoplastic neuropathy
paraneoplastic sensory, *see* Paraneoplastic sensory neuropathy
paraprotein, 63/402
paraproteinemia, 69/292, 71/441
paraproteinemic, *see* Paraproteinemic neuropathy

pellagra, 7/552, 70/414
penicillin, 7/515
pentostatin, 69/461
peripheral, *see* Polyneuropathy
peritoneal dialysis, 8/1
periwinkle alkaloid, 7/517
pernicious anemia, 42/609
peroneal, *see* Peroneal neuropathy
pertussis vaccination, 8/91
phrenic nerve, 8/388
physiotherapy, 8/397
podophyllin, 69/471, 75/501
POEMS syndrome, 51/430
poly, *see* Polyneuropathy
polyarteritis nodosa, 8/124, 161, 39/304-306
polycythemia vera, 8/145
polymyositis, 8/122
porphyria, 7/554, 8/9, 41/434, 60/117-121
porphyria variegata, 60/120
postgastrectomy syndrome, 28/234, 51/327
pregnancy, 8/13-15
pressure, *see* Compression neuropathy
primary demyelinating, *see* Primary demyelinating neuropathy
primary hyperoxaluria, 51/398
primary systemic amyloidosis, 71/446
procarbazine, 69/461, 471
progressive, *see* Progressive neuropathy
progressive external ophthalmoplegia, 43/136, 62/313
progressive spinal muscular atrophy, 59/404
proximal, *see* Proximal neuropathy
pure cholinergic, 75/125
pyruvate dehydrogenase complex deficiency, 66/416
rabies vaccination, 8/91
radiation induced, 67/342
radiotherapy, 39/52
rapid reversible, *see* Rapid reversible neuropathy
recurrent acute demyelinating, *see* Recurrent acute demyelinating neuropathy
recurrent multiple cranial, *see* Recurrent multiple cranial neuropathy
recurrent nonhereditary multiple cranial, *see* Recurrent nonhereditary multiple cranial neuropathy
recurring, *see* Hereditary recurrent neuropathy
Refsum disease, 8/21-24, 21/197, 27/523, 40/303, 514, 60/228, 230
relapsing peripheral, *see* Relapsing peripheral neuropathy
remitting peripheral, *see* Remitting peripheral neuropathy
renal insufficiency, 27/338-340
reticulosis, 8/143
retrobulbar, *see* Retrobulbar neuropathy
rheumatoid, *see* Rheumatoid neuropathy
rheumatoid arthritis, 7/555, 8/126, 71/15
rigidity, 60/659
Roussy-Lévy syndrome, 60/659
sarcoidosis, *see* Sarcoid neuropathy
sarcoma, 8/144
sciatic, *see* Sciatic neuropathy
scleroderma, 8/121, 39/365, 71/111, 113
secondary pigmentary retinal degeneration, 42/334, 60/660
segmental demyelination, *see* Segmental demyelination neuropathy
selective serotonin reuptake inhibitor, 65/316
sensorimotor, *see* Sensorimotor neuropathy
sensorimotor paraneoplastic, *see* Sensorimotor paraneoplastic neuropathy
sensorineural deafness, 42/373, 60/660
sensory, *see* Sensory neuropathy
sensory radicular, *see* Sensory radicular neuropathy
sensory radicular neuropathy, 42/351
sensory trigeminal, *see* Sensory trigeminal neuropathy
serogenetic, *see* Serogenetic neuropathy
serous inflammation, 9/608
serum sickness, 7/555, 8/82, 96, 9/608
Sjögren syndrome, 8/17, 39/423-426
skeletal fluorosis, 36/476
slow eye movement, 42/187
small fiber, *see* Small fiber neuropathy
small intestine disorder, 28/25
smallpox vaccine, 7/487, 8/88
spastic paraplegia, 42/171
sprue, 28/25
starvation, *see* Starvation neuropathy
Stilling-Türk-Duane syndrome, 42/307
streptomycin, 7/515
subacute combined spinal cord degeneration, 28/179-184
subacute disabling, *see* Subacute disabling neuropathy
subacute myelo-optic, *see* Subacute myelo-optic neuropathy
subacute necrotizing encephalomyelopathy, 42/588, 60/660
subacute sensory, *see* Subacute sensory neuropathy
sulfatide, *see* Metachromatic leukodystrophy
sulfonamide, 7/538
suramin, 69/461, 472

systemic lupus erythematosus, 7/555, 8/119, 11/139, 215, 450, 39/283, 43/420, 71/44
TAB vaccination, 8/88
Tangier disease, 8/20, 10/547, 21/52, 42/627
taxol, 69/460, 467, 469
taxotere, 69/460, 467, 470
terminology, 7/199
tetanus toxoid, 7/555, 8/91
thalidomide, 7/540, 71/388
thallium, *see* Thallium polyneuropathy
thallium intoxication, 7/515, 543, 64/8
thyrotoxic myopathy, 62/528
toluene intoxication, 36/372
tomacular, *see* Hereditary multiple recurrent mononeuropathy
toxic, *see* Toxic neuropathy
toxoid, 8/87
treatment, 8/373
trichinosis, 35/275, 52/566, 571
trichloroethylene, 7/522
tri-*o*-cresyl phosphate, 7/521, 544
tricyclic antidepressant, 65/316
trigeminal, *see* Trigeminal neuropathy
trigeminal sensory, *see* Trigeminal sensory neuropathy
trophic change, 1/486
tropical ataxic, *see* Tropical ataxic neuropathy
tropical sprue, 28/236
tsukubaenolide, 69/461, 472
typhoid vaccination, 7/482, 8/86
ulnar, *see* Ulnar neuropathy
uremic, *see* Uremic neuropathy
urticaria, 5/234
vaccinogenic, *see* Vaccinogenic neuropathy
vascular, *see* Vascular neuropathy
vasculitic, *see* Vasculitic neuropathy
vermouth, 7/521
vesicourethral, *see* Vesicourethral neuropathy
vibration, *see* Vibration neuropathy
vinblastine, 69/460
vinca alkaloid, 69/459, 463
vincristine, 7/545, 69/460, 75/496, 531
vindesine, 69/460
visceral, *see* Visceral neuropathy
vitamin B_1 deficiency, 7/567, 28/4-9, 49-55
vitamin B_3 deficiency, 7/552
vitamin B_5, 7/553
vitamin B_6, 7/553, 70/432
vitamin B_6 deficiency, 7/580, 70/427
vitamin B_6 deficiency polyneuropathy, 51/298
vitamin B_{12} deficiency, 7/587-594, 28/179-184, 70/380
vitamin B_c, 7/553, 28/200
vitamin B_c deficiency, 28/200-202
vitamin E deficiency, 60/660
vomiting, 8/14
Waldenström macroglobulinemia polyneuropathy, 39/193
Wegener granulomatosis, 71/176
Wernicke encephalopathy, 28/17
Wernicke-Korsakoff syndrome, 28/253, 45/195
whipped cream dispenser, *see* Whipped cream dispenser neuropathy
X-linked hereditary motor and sensory, *see* X-linked hereditary motor and sensory neuropathy
X-linked hereditary sensory and autonomic, *see* X-linked hereditary sensory and autonomic neuropathy
xanthomatosis, 8/18
xeroderma pigmentosum, 60/660
zalcitabine, 71/370
zidovudine, 71/373
Neuropathy target esterase
chemistry, 64/161
organophosphate induced delayed neuropathy, 64/229
organophosphate intoxication, 64/151, 229
Neuropeptide
autonomic nervous system, 74/10, 56
brain infarction, 53/133
brain ischemia, 53/133
central autonomic dysfunction, 75/190
epilepsy, 72/84
exercise, 74/11
neurochemical coding, 74/11
neurotransmitter, 74/94
progressive dysautonomia, 59/148
reflex sympathetic dystrophy, 61/126
Neuropeptide Y
brain blood flow, 53/63
brain microcirculation, 53/79
Huntington chorea, 49/256, 260, 319
penis erection, 61/314
reflex sympathetic dystrophy, 61/128
spinal cord injury, 61/450
tetraplegia, 61/450
Neuropeptide Y neuron
Huntington chorea, 49/298
Neuropharmacology
akinesia, 49/58
basal ganglion, 49/47-59
catatonia, 49/56
choreoathetosis, 49/58
dipeptidyl carboxypeptidase I inhibitor, 65/444
dystonia, 49/58

envenomation, 37/13-16
hemiballismus, 49/58
memory disorder, 27/462
orofacial dyskinesia, 49/55
paradoxical kinesia, 49/58
spasmodic torticollis, 49/57
tremor, 49/57
turning behavior, 45/167
venom, 37/13-16
Neurophysiologic test
brain tumor, 67/312
Guillain-Barré syndrome, 47/610
trigemino-oculomotor synkinesis, 42/320
Neurophysiology
chorea, 49/549
cranial neuralgia, 5/281
epilepsy, 15/30
hyperactivity, 46/183
hypertrophic interstitial neuropathy, 21/160
hysteria, 46/577
infantile autism, 46/191
infantile spasm, 73/202
Landau-Kleffner syndrome, 73/283
lathyrism, 65/9
magnesium, 63/562
metachromatic leukodystrophy, 66/175
neuropsychology, 3/13
opiate intoxication, 37/374
pain, 5/282
Pelizaeus-Merzbacher disease, 66/568
pudendal nerve, 1/362
sleep, 3/66, 80-82, 74/526
status epilepticus, 15/54, 73/318
stress, 74/308
tremor, 6/180
trigeminal neuralgia, 5/281
waking, 3/171-181
Neuropil
oligodendrocyte, 47/1, 7, 12
Neuropil dystrophy
glioneuronal composite, 21/67
Neuropore
caudal, *see* Caudal neuropore
CNS, 50/3, 27
rostral, *see* Rostral neuropore
Neuropsychiatry
Germany, 1/8
Neuropsychologic assessment
Alzheimer disease, 45/519
dementia, 46/202-204
functional psychosis, 46/457-460
paranoid schizophrenia, 46/488
rehabilitation, 46/612
schizophrenia, 46/457-460, 487-489
Neuropsychologic assessment contribution
neuropsychologic defect, 45/525
Neuropsychologic defect
acquired immune deficiency syndrome dementia, 56/491
attention, 45/524
cognition, 45/517-521
disorientation, 45/524
executive function, 45/524
expression, 45/523
Halstead-Reitan test battery, 45/525
intellectual efficiency, 45/524
intelligence, 45/517
Luria-Nebraska neuropsychological test battery, 45/526
Luria neuropsychological investigation, 45/526
memory, 45/522
Michigan Neuropsychological Test Battery, 45/526
motor function, 45/525
neuropsychologic assessment contribution, 45/525
orientation, 45/524
perception, 45/521
premorbid functioning, 45/517
psychometry, 45/518, 520
test battery, 45/525
test selection, 45/518
thinking, 45/522
Wechsler test, 45/521
Neuropsychologic syndrome
anatomic factor, 3/23, 309, 369
focal, *see* Focal neuropsychological syndrome
functional factor, 3/23, 369, 379
premorbid personality, 3/23
psychodynamic factor, 3/23
restoration of higher cortical function, 3/368-431
situational factor, 3/23
Neuropsychologic test
acquired immune deficiency syndrome dementia complex, 71/247
aphasia, 45/523
apraxia, 45/523
cerebral dominance, 4/267, 466
dementia, 45/523, 46/127
Huntington chorea, 46/307-309, 49/283
Neuropsychology
age, 45/3
agnosia, 4/13, 29
alexia, 45/11-13
anatomic aspect, 4/1, 45/7
anatomic localization, 4/2-5, 45/7-11
aphasia, 45/17-20

assessment, 45/515-527
astereopsis, 45/16
astrocytoma, 45/9
Balint syndrome, 45/15
behavior, 45/4, 515
brain, 45/4
brain anoxia, 45/9
brain infarction, 45/9
brain surgery, 45/9
brain tumor, 45/9
central monochromatism, 45/14
cerebral dominance, 45/3
cerebrovascular disease, 45/8
cognition, 45/516
color naming defect, 45/13
computer assisted tomography, 45/1, 7, 11
computerized test, 45/526
congenital abnormality, 45/4
control, 45/516
depression, 46/580
development aspect, 4/9, 340-469, 45/10
emotion, 45/516
epilepsy, 45/9, 11
epilepsy with continuous spike wave during slow sleep, 73/269
examination method, 3/5
experimental study, 4/9
frontal lobe, 45/516
frontal lobe lesion, 45/27-29
Gilles de la Tourette syndrome, 49/630
head injury, 45/9
hemiachromatopsia, 45/15
herpes simplex virus encephalitis, 45/9
higher cortical function, 3/382
historic introduction, 3/1-13
hysteria, 46/580
introductory remark, 3/1-21
lesion size, 45/3, 10
manganese intoxication, 64/310, 313
memory, 45/516
meningioma, 45/9
monochromatism, 45/14
nature of function, 3/380
neglect, 45/16
neurophysiology, 3/13
neuroradiology, 45/11
nuclear magnetic resonance, 45/1, 7
optic aphasia, 45/13
parietal lobe syndrome, 45/79-82
perception, 45/516
personality, 45/4, 516
positron emission tomography, 45/1, 7
rehabilitation, 46/609-631
route of investigation, 3/13-15
schizophrenia, 46/570
sex ratio, 45/3
single photon emission computer assisted tomography, 45/7
split brain, 45/2, 10
symptom, 4/1-12, 45/11-20
symptom in children, 4/340-469
symptomatology, 45/1-5
tactile perception, 45/521
terminology, 3/7
thalamus, 45/2
thinking, 45/516
visual agnosia, 45/13
Wernicke-Korsakoff syndrome, 45/200
Neuroradiology
Arnold-Chiari malformation type II, 50/408
brain aneurysm, 12/96-100
brain death, 24/733
brain metastasis, 18/223
classic primitive neuroectodermal tumor, 68/214
congenital brain tumor, 31/41
craniopharyngioma, 18/546
ependymoma, 18/129
functional psychosis, 46/450-452
glomus jugulare tumor, 18/446
headache, 5/33
hereditary sensory and autonomic neuropathy type II, 60/17
hereditary sensory and autonomic neuropathy type V, 60/15
neuropsychology, 45/11
Pelizaeus-Merzbacher disease, 66/568
place in neurology, 1/31
primary alcoholic dementia, 46/343-346
striatopallidodentate calcification, 6/710, 49/426
Sturge-Weber syndrome, 55/445
superficial siderosis, 70/72
supratentorial brain tumor, 18/335
thoracic angioma, 12/639
triethyltin intoxication, 36/282
X-linked adrenoleukodystrophy, 66/458
Neuroreceptor
single photon emission computer assisted tomography, 72/377
Neuroretinal angiomatosis, *see* Bonnet-Dechaume-Blanc syndrome
Neuroretinal degeneration
classification, 13/24-43
EOG, 13/20-22
HRR plate, 13/15
introduction, 13/1-23
ophthalmoscopy, 13/3-9

X-linked inheritance, 13/25
Neuroretinitis
syphilitic, *see* Syphilitic neuroretinitis
Neurosarcoidosis, 38/521-539
see also Sarcoidosis
angiotensin converting enzyme, 63/422
cerebral, 38/530
CNS involvement, 38/529-531
computer assisted tomography, 63/422
cranial nerve involvement, 38/525-529
cranial neuropathy, 63/422
CSF, 38/533
diabetes insipidus, 63/422
diagnosis, 38/531-533
epilepsy, 63/422
facial paralysis, 63/422
gonadotropin failure, 63/422
granulomatous CNS vasculitis, 55/388
granulomatous inflammation, 63/422
hydrocephalus, 38/530, 63/422
hypothalamic dysfunction, 63/422
hypothalamic involvement, 38/522, 530
incidence, 38/521
manifestation, 38/524-531
meningitis, 38/521, 63/422
meningitis serosa, 63/422
meningoencephalitis, 38/531
myopathy, 38/535-539
neuropathy, 38/525-529
pathology, 38/522-524
prognosis, 38/533
sensorimotor neuropathy, 63/422
sensorimotor polyneuropathy, 63/422
spinal cord, 38/529
symptom, 63/421
treatment, 38/533-535
Neuroschisis
Dandy-Walker syndrome, 30/632
Neurosecretion
colloid, 2/437
hypothalamus, 2/439
osmoreceptor, 2/439
tuberoinfundibular nucleus, 2/437
Neurosensory hearing loss
cataract, 60/653
cerebellar ataxia, 60/657, 660
deafness, 60/657
late cerebellar ataxia, 60/653
neuropathy, 60/660
Turner syndrome, 50/545
Neurosis
anxiety, *see* Anxiety neurosis
compensation, *see* Compensation neurosis
consciousness, 3/119
epilepsy, 73/478
filariasis, 35/168, 52/517
Huntington chorea, 46/306
nomenclature, 46/430
obsessional, *see* Obsessional neurosis
occupational, *see* Occupational neurosis
posttraumatic, *see* Posttraumatic neurosis
posttraumatic headache, 24/511
prevalence, 43/211
pupil, 1/624
sphenopalatine, *see* Sphenopalatine neurosis
torsion, *see* Torsion neurosis
traumatic, *see* Traumatic neurosis
traumatic psychosis, 24/526
traumatic psychosyndrome, 24/565, 587
Neurosurgery
anterior communicating artery aneurysm, 12/221
aspergillosis, 35/399, 52/382
bacterial meningitis, 52/4
brachial plexus, 51/147
brain abscess, 52/158
brain aneurysm, *see* Brain aneurysm neurosurgery
brain hemorrhage, 54/321
brain tumor, 67/235
cavernous sinus thrombosis, 52/175
coccidioidomycosis, 52/415, 422
corticothalamic connection, 2/491
cranial epidural empyema, 52/169
cranial subdural empyema, 52/171
cryptococcoma, 52/433
epilepsy, 45/9
Galen vein aneurysm, 12/260
glomus jugulare tumor, 18/450
gram-negative bacillary meningitis, 52/103
histoplasmosis, 52/441
history, 1/17
intracerebral hematoma, 12/211
medulloblastoma, 68/186
moyamoya disease, 55/300
mycotic brain aneurysm, 52/296
neurocysticercosis, 52/533
new technique, 67/235
nocardiosis, 52/450
pneumococcal meningitis, 33/43, 52/49
posttraumatic syringomyelia, 61/388
rigidity, 6/851
schistosomiasis, 52/541
spinal cord abscess, 52/191
spinal epidural empyema, 52/188
spinal subdural empyema, 52/189
tuberculous meningitis, 52/213
Neurosyphilis, 33/353-390

see also Brain gumma, General paresis of the insane, Spinal cord gumma *and* Tabes dorsalis
acquired immune deficiency syndrome, 71/289
amyotrophic lateral sclerosis, 22/309, 52/278
Argyll Robertson pupil, 1/621, 2/110, 280, 22/20, 33/356, 364, 52/280
arteritis, 52/278
asymptomatic, 33/354
benzathine penicillin, 33/373, 52/282
brain gumma, 52/278
cardiovascular disease, 33/384, 52/282
cephalosporin, 52/282
cerebrovascular complication, 55/411
chloramphenicol, 52/282
classification, 33/353
clinical type, 52/277
computer assisted tomography, 52/278
congenital, 33/367-372
course, 33/389
cranial nerve involvement, 52/278
Creutzfeldt-Jakob disease, 56/546
CSF abnormality, 52/277
dementia, 46/390, 52/278
diagnosis, 71/289
early congenital, 33/369
EEG, 52/277
epidemiology, 33/376-382
erythromycin, 52/283
etiology, 33/338
familial spastic paraplegia, 59/306
fever treatment, 33/372
general paresis of the insane, 33/358-362, 46/390, 52/279
gumma, 33/367, 384, 52/274, 277, 282
headache, 52/277
Herxheimer reaction, 52/279
Heubner arteritis, 52/278
history, 33/337
hydrocephalus, 52/277
incidence, 33/379-382
keratitis, 52/274
late congenital, 33/370-372
lymphocytic meningitis, 52/277
malaria, 33/372, 35/144, 52/282
meningeal, 33/354-373
meningeal endarteritis obliterans, 52/274
meningitis, 33/355, 52/274, 277, 282
meningitis serosa, 52/277
meningovascular, 33/357, 52/278
mental change, 52/277
mortality, 33/387-389
multiple sclerosis, 47/71, 52/278
neuropathology, 52/281
Nissl-Alzheimer arteritis, 52/278
optic atrophy, 33/365, 52/277
paretic, 33/358-362, 52/279
penicillin, 33/373, 52/279, 282
personality change, 52/277
pinpoint pupil, 52/280
prevalence, 33/379-382
procaine penicillin, 33/373, 52/283
prognosis, 33/382-387
progressive spinal muscular atrophy, 22/20
Rosahn study, 52/281
schizophrenia, 46/455
spheroid body, 6/626
spinal, 33/366
tabes dorsalis, 33/362-365, 52/277
tetracycline, 52/282
transient ischemic attack, 12/9
treatment, 33/370-376, 52/282-285
Treponema pallidum, 33/338, 52/273, 282
uveitis, 52/274
vasculitis, 52/277
vertigo, 52/278
visual evoked response, 52/278

Neurotensin
Alzheimer disease, 46/268
basal ganglion, 49/33
gastrointestinal disease, 39/452
Huntington chorea, 49/258
mesolimbic dopamine system, 49/48
neostriatum, 49/47
Parkinson disease, 49/120
progressive dysautonomia, 59/157
suprachiasmatic nucleus, 74/481

Neurotic depression
lateralized dysfunction, 46/489

Neuroticism
prevalence, 43/210
sleep disorder, 43/210
speech disorder, 43/210
speech disorder in children, 43/210

Neurotmesis
definition, 7/244
degree, 51/78
nerve injury, 7/244, 51/75, 134

Neurotomy
cervical, *see* Cervical neurotomy
obturator, *see* Obturator neurotomy
trigeminal, *see* Trigeminal neurotomy

Neurotoxic esterase
organophosphate intoxication, 37/475, 51/284
tri-*o*-cresyl phosphate, 37/475, 51/284

Neurotoxic shellfish poisoning
see also Marine toxin intoxication

acute gastroenteritis, 65/158
ataxia, 65/158
brevetoxin A, 65/158
brevetoxin B, 65/158
clinical presentation, 65/158
diagnosis, 65/158
epidemiology, 65/157
history, 65/157
management, 65/159
marine toxin intoxication, 65/143
neurologic symptom, 65/158
paresthesia, 65/158
pathomechanism, 65/158
Ptychodiscus brevis, 65/157
Ptychodiscus brevis toxin type I-X, 65/158
red tide toxin, 65/158
sodium channel activation, 65/158
temperature reversal, 65/158
treatment, 65/159
Neurotoxicity
acetonylacetone, 51/276
aciclovir, 71/368
acquired immune deficiency syndrome treatment, 71/368
acrylamide, 51/276
alcohol, *see* Alcohol neurotoxicity
almitrine dimesilate, 51/294, 308
aluminum, *see* Aluminum neurotoxicity
amiodarone, 51/294
amphotericin B, 71/368
2-amyl-methylphosphonofluoridate, 37/543
animal model, 64/17
arsenic intoxication, 51/264
asparaginase, 63/359
atomic number, 51/265
autonomic neuropathy, 51/294
axonal degeneration, 51/294
axonopathy, 51/263
azethion, 37/546
bis(diethoxyphosphoryl)ethylamine, 37/549
bismuth intoxication, 64/331, 337
bone marrow transplantation, 67/364
O-N-butyl *O*-carbetoxymethyl ethylphosphonothioate, 37/547
calcium carbimide, 51/309
carbamazepine, *see* Carbamazepine neurotoxicity
carbon dioxide, 51/276
carbonyl group cis-configuration, 51/277
cauda equina syndrome, 61/159
chemotherapy, 67/353, 359, 69/481
chloramphenicol, 51/294
chlordecone intoxication, 51/286
chlormethine, 63/358
ciprofloxacin, 71/368
cisplatin intoxication, 64/358
coumafos, 37/546
crab intoxication, 37/56
crustacea intoxication, 37/56
cytarabine, 63/357, 71/368, 386
dapsone, 71/368
deferoxamine, 63/534
dementia, 67/358
demeton, 37/548
demeton-S, 37/547
2,4-dichlorophenoxyacetic acid, 51/286
dichlorvos, 37/545
didanosine, 71/368
diethyl bisdimethyl pyrophosphordiamide, 37/549
diethyl 4-chlorophenyl phosphate, 37/545
diethyl 2-chlorovinyl phosphate, 37/545
O,O-diethyl *S*-2-diethylmethylammoniumethyl phosphonothiolate methylsulfate, 37/550
O,O-diethyl *S*-ethsulfonylmethyl phosphorothioate, 37/547
O,O-diethyl *S*-eththiomethyl phosphorothioate, 37/547
O,O-diethyl *S*-eththionylmethyl phosphorothioate, 37/547
O,O-diethyl *O*-(4-methylumbelliferyl) phosphorothioate, 37/546
O,O-diethyl s-(4-nitrophenyl) phosphorothioate, 37/548
O,S-diethyl *O*-(4-nitrophenyl) phosphorothioate, 37/548
diethyl phosphorocyanidate, 37/544
diethyl phosphorofluoridate, 37/543
O,O-diethyl *O*-(8-quinolyl) phosphorothioate, 37/547
diethyl trichloromethylphosphonate, 37/545
di-2-fluoroethyl phosphorofluoridate, 37/543
diisopropyl phosphoroiodidate, 37/544
dimefox, 37/543
2-dimethylaminoethyl dimethylphosphinate, 37/545
3,4-dimethyl-2,5-hexanedione, 51/278
O,O-dimethyl *S*-(4-nitrophenyl) phosphorothioate, 37/548
dimethyl 1,2,2,2-tetrachloroethyl phosphate, 37/545
dimpylate, 37/546
dioxin, 51/286
diphenyl phosphorofluoridate, 37/543
diphenyl-2-trimethylammoniumethyl phosphate bromide, 37/550
diphtheria, 65/239
diphtheria toxoid, 52/241

di-*N*-propyl phosphorofluoridate, 37/543
disulfiram, 51/308
disystox, 37/547
DNA, 64/17
dose-response relationship, 51/264
ecothiopate iodide, 37/550
ethambutol, 71/368
ethyl-*N*-diethyl phosphoramidocyanidate, 37/544
O-ethyl *S*-(2-dimethylaminoethyl)
methylphosphonothioate, 37/548
ethylene oxide, 51/275
O-ethyl *S*-(2-ethylthioethyl)
ethylphosphonothioate, 37/548
ethyl-4-nitrophenyl ethylphosphonate, 37/545
forstenon, 37/545
foscarnet, 71/368
ganciclovir, 71/368
gold, 51/265, 294, 305
grouting compound, 51/265
halogenated pyrimidine, 67/364
heavy metal, 51/265
herbicide, 51/265
hexane, 51/276
2-hexanone, 51/276
industrial gas, 51/265
industrial solvent, 51/265
α-interferon, 63/359
O-isobutyl *O*-carboxymethyl
ethylphosphonothioate, 37/547
isoniazid, 51/294, 71/368
Iso-Ompa, 37/549
isopropyl-N-dimethyl phosphoramidocyanidate,
37/544
Iso-Systox sulfoxide, 37/548
kainic acid, *see* Kainic acid neurotoxicity
lead, 51/265
lithium, 51/304
local anesthetic agent, 65/422
malacostraca intoxication, 37/56
malathion, 37/546
Mendelejev periodic system, 51/265
mercury, 51/265
methotrexate, 39/98-104, 47/551, 63/357, 71/368,
386
methyl bromide, 51/276
methyl-*N*-dimethyl phosphoramidocyanidate,
37/544
methylmercury, 51/265
methylsystox sulfone, 37/548
Minamata disease, 51/264
misonidazole, 67/364
mitochondria, 64/17
multiple case, 51/265
myelinopathy, 51/263
necrotizing leukoencephalopathy, 67/358
nerve, 51/263
neurofilament, 51/264
neurofilament neuropathy, 51/276
neuronopathy, 51/263
nitrosourea, 63/358
octamethyl pyrophosphoramide, 37/549
organophosphorus compound intoxication, 37/541
paraquat, 51/286
pathoclisis, 9/623
penicillin, 46/602
pentavalent arsenic, 51/265
perhexiline, 51/294
pertussis vaccine, 52/243
pesticide, 51/265
O-phenyl-*N*-2-dimethylaminoethyl-4-phenylphos
phonamidate methiodide, 37/550
phorate sulfone, 37/547
platinum, 51/265
podophyllin, 51/294
polychlorinated biphenyl, 51/286
procarbazine, 63/359
pyrazoxon, 37/545
radiation, 67/359
renal dialysis, 51/294
segmental demyelination, 51/294
silver, 51/307
spinal anesthesia, 61/150
stavudine, 71/368
systemic features, 51/264
tetraethyllead, 51/265
tetraethylphosphorodiamidic fluoride, 37/544
tetraisopropyl dithionopyrophosphate, 37/549
tetraisopropyl pyrophosphate, 37/549
tetramethyl pyrophosphate, 37/549
thalidomide, 51/304, 71/368
thallium, 51/265
thiotepa, 63/359
tissue level, 51/264
trichloroethylene, 51/279
tri-*o*-cresyl phosphate intoxication, 37/475
tricyclic antidepressant, 51/303
trimethoprim, 71/368
2-trimethylammoniumethyl
methylphosphonofluoridate iodide, 37/550
2-trimethylammonium-1-methylethyl
methylphosphonofluoridate iodide, 37/550
3-trimethylammoniumpropyl
methylphosphonofluoridate iodide, 37/550
trivalent arsenic, 51/265
vascular factor, 9/623
vinca alkaloid, 71/368

vincristine, 39/106-110, 51/294, 63/358
vitamin B6 deficiency, 51/294
zalcitabine, 71/368
zidovudine, 71/368
zimeldine, 51/304

Neurotoxicity Scale
epilepsy, 72/419, 431

Neurotoxicology
autonomic neuropathy, 64/14
basal ganglion syndrome, 64/6
cerebellar syndrome, 64/5
cranial nerve syndrome, 64/8
cycad intoxication, 65/22
domain, 64/1
encephalopathy, 64/2
excitotoxin, 64/19
free radical, 64/18
Guam motoneuron disease, 65/22
Hunter-Russell syndrome, 64/8
latent interval, 64/7
Lytico Bodeg, 64/5
methanol, 64/8
Minamata disease, 64/8
motor end plate, 64/14
myelinopathy, 64/10
myelopathy, 64/10
myopathy, 64/16
neuropathy, 64/10
progressive supranuclear palsy, 64/5

Neurotoxin
see also Designer drug
acetonylacetone, 51/68, 276, 64/12, 81, 84
acivicin, 65/529, 538
acrylamide, 7/522, 36/376, 46/393, 51/68, 276, 479, 531, 64/10, 12, 63
alcohol, 4/331, 6/819, 7/166, 553, 613, 9/653, 12/113, 13/53, 61, 17/542, 21/478, 28/317, 331, 341, 40/54, 162, 317, 339, 438, 41/273, 43/197, 45/194, 46/335-348, 393, 540, 563, 47/557, 621, 48/420, 49/556, 590, 51/56, 315-319, 531, 53/18, 54/288, 55/523, 57/130, 259, 61/370, 62/601, 609, 63/151, 564, 64/2, 5, 107, 111-115, 125
aldrin, 36/411, 64/197, 201
alkylating agent, 67/354
allethrin, 64/212
altretamine, 39/95, 113, 117, 119, 46/394, 65/528, 534
aluminum, 36/320, 46/392, 59/278, 63/525, 64/4, 10, 273
amfebutamone, 65/311
2-amino-3-methylaminopropionic acid, 59/263, 274, 64/5, 65/22
amitriptyline, 51/68, 65/314
amoxapine, 62/601, 65/311
amoxicillin, 65/472
amphetamine, 36/232, 48/420, 62/601, 65/251
amphotericin, 65/472
amphotericin B, 67/363
ampicillin, 65/472
amsacrine, 65/529, 538
amygdalin, 36/515, 65/26
Androctonus mauretanicus mauretanicus, 65/198
ant venom, 65/194
apamin, 37/111, 62/222, 225, 65/194, 199
apamin I 125, 65/198
arsenic, 5/299, 9/650, 37/199, 46/391, 48/420, 51/68, 264, 479, 531, 59/370, 62/284, 63/152, 64/10, 12
arthropod venom, 65/193
asparaginase, 39/95, 113, 117, 119, 46/394, 63/345, 359, 65/529, 538
autonomic nervous system dysfunction, 65/196
azacitidine, 39/94, 99, 106, 119, 41/271, 65/528
barbiturate, 6/450, 7/518, 15/713-715, 37/200, 352, 409, 410, 40/339, 41/270, 46/394, 598, 47/71, 49/478, 53/418, 55/216, 220, 522, 62/601, 65/332, 497, 508
barium, 36/325, 51/274, 64/16, 331, 337, 353
bee venom, 37/107, 111, 62/222, 225, 65/194, 199
bioresmethrin, 64/213
bismuth, 36/327, 46/392, 64/4, 10, 331
bleomycin, 39/111, 114, 119, 46/600, 65/528
botulinum toxin, 7/112, 37/88, 40/129, 143, 41/129, 288, 63/152, 64/2, 15, 65/209, 245, 247
botulinum toxin A, 65/213
bradykinin(6-threonine), 65/202
brevetoxin, 65/142
broxuridine, 65/529
buckthorn toxin, 64/10, 14
α-bungarotoxin, 40/129, 41/97, 106, 113, 125, 62/409, 64/16, 65/177, 181
β-bungarotoxin, 37/17, 40/143, 64/15, 65/181
κ-bungarotoxin, 65/183
buthotoxin, 65/194
Buthus tamulus, 65/198
cadmium, 64/10
caffeine, 5/98, 40/546, 41/269, 48/6, 61/156, 65/259
cannabis, 50/219
carbamate, 37/353, 64/15, 160, 183
carbamazepine, 5/383, 15/653, 669, 707, 37/200, 47/38, 56/234, 64/177, 299, 65/496, 505
carbon disulfide, 36/382, 46/393, 48/420, 49/466, 51/68, 279, 531, 64/8, 10, 23, 25, 40, 82, 86, 188
carbon monoxide, 46/393, 48/420, 49/466, 556,

62/601, 64/7, 31
carboplatin, 65/529, 67/356
carmustine, 18/520, 39/115, 65/528, 67/356
caudoxin, 65/182
cefaloridine, 33/43, 65/472
cefalotin, 65/472
cefazolin, 65/472
ceftazidime, 65/472
ceruleotoxin, 65/182, 184
charybdotoxin, 65/194, 198
chlorambucil, 39/97, 46/601, 65/528
chloramphenicol, 7/516, 530, 41/210, 46/602, 51/68, 294, 65/486
chlordane, 36/408, 64/13, 201
chlormethine, 7/517, 537, 39/18, 94, 96, 119, 46/601, 63/358, 65/528
chloroquine, 7/518, 13/149, 220, 38/496, 40/151, 339, 438, 51/302, 62/372, 606, 65/484
ciguatoxin, 37/4, 64, 82, 97, 64/15, 65/141, 147, 159
cilastatin, 65/472
cipermethrin, 64/212
ciprofloxacin, 65/486
cisplatin, 39/95, 112, 115, 119, 51/299, 64/358, 65/529, 67/354
clindamycin, 65/472
clioquinol, 37/115, 45/207, 51/68, 296, 56/526, 64/8, 10, 13
cocaine, 37/236, 55/521, 62/577, 601, 65/41, 50, 251
colchicine, 40/151, 154, 555, 46/271, 51/180, 62/597, 600, 606, 609, 64/13
colistin, 65/472
corticosteroid, 63/359, 65/529, 555
cortisone, 50/219
crotoxin, 37/17, 65/182
crotoxin A, 65/184
cuprizone, 47/563, 64/10, 17
cyanide, 36/518, 49/466, 65/25
cyanoalanine, 65/11
cyanogenic glycoside, 36/515, 518, 520-524, 64/8, 10, 12, 65/25
Cycas circinalis, *see* Cycas circinalis neurotoxin
cyclophosphamide, 39/119, 46/600, 62/418, 64/244, 65/528, 67/357
cycloserine, 28/116, 33/229, 52/212, 65/481
cyfluthrin, 64/212
cyphenothrin, 64/215
cyproterone acetate, 65/529
cytarabine, 39/94, 99, 105, 119, 60/586, 61/151, 63/356, 65/528, 532, 67/362
cytidine analog, 67/362
dactinomycin, 47/568, 64/10, 17, 67/363
dapsone, 51/68, 531, 65/485
delayed symptom, 64/7
deltamethrin, 64/212
dendrotoxin, 65/177, 186
dendrotoxin chemistry, 65/186
diazoxide, 37/432, 65/446
dibenzo-1,4-dioxin, 64/197
dieldrin, 36/393, 411, 64/197, 201
digitoxin, 37/426, 65/449
digoxin, 37/426, 63/195, 65/449
diketone, 64/84
dinophysiotoxin, 65/149
diphtheria toxin, 33/480, 51/68, 64/10, 14, 65/236
disulfiram, 7/531, 37/321, 46/394, 48/420, 51/68, 308, 531, 64/8, 188
domoic acid, 64/3, 65/141, 151, 166
doxorubicin, 51/300, 64/10, 17, 65/528
dyflos, 37/543, 64/152, 223
elmustine, 65/528
endosulfan, 36/418, 64/201
enoxacin, 65/486
enzyme inhibition, 65/148
ephedrine, 55/521, 65/259
ergot alkaloid, 65/61
erythromycin, 65/472
etacrynic acid, 50/219
ethambutol, 7/517, 33/228, 51/294, 52/211, 64/8, 65/481
ethionamide, 7/516, 535, 33/228, 65/481
O-ethyl *S*-[2-(diisopropylamino)ethyl]methylphosphonothionate, 64/223
ethylene glycol, 46/393, 62/601, 64/121
etoposide, 65/528, 67/363
excitotoxin, 55/211, 64/19
fasciculin, 65/177, 186
fasciculin 1, 65/186
fasciculin 2, 65/186
fasciculin chemistry, 65/186
fenpropathrin, 64/213
fenvalerate, 64/212
fludarabine, 65/528
fluorouracil, 39/17, 94, 99, 104, 119, 46/600, 60/586, 64/5, 65/528, 532, 67/363
fluvalinate, 64/212
free radical, 64/18
freon, 64/13
funnel web spider venom, 65/194
general mechanism, 64/18
gentamicin, 50/219
gold, 7/532, 36/332, 38/497, 46/392, 51/265, 274, 294, 305, 531, 62/597, 64/4, 10, 13, 353
gonyautoxin, 37/33, 65/146

griseofulvin, 65/472
gyromitrine, 36/536, 64/3, 65/36
halogenated pyrimidine, 67/364
helminthiasis, 52/509
heptachlor, 36/392, 417, 48/420, 64/197, 201
heterocyclic antidepressant, 65/311
hexacarbon, 51/531, 64/10
hexachlorophene, 37/479, 51/68, 64/10, 14, 65/471
hexane, 7/522, 36/361, 51/68, 276, 64/12, 81
2-hexanone, 36/365, 51/68, 276, 64/12, 49, 81, 91
holocyclotoxin, 65/194
homidium bromide, 47/569, 64/10, 17
hydralazine, 37/430, 42/647, 51/68, 63/74, 83, 87, 65/446
hydrazine, 36/380, 64/3, 82
hydroxyurea, 67/364
hymenoptera kinin, 65/201
hymenoptera venom, 37/107, 65/193
Hyp2-bradykinin, 65/202
hypoglycin, 37/511, 49/217, 65/79, 130
iberiotoxin, 65/194, 198
ifosfamide, 65/528
imidazole derivative, 67/364
3,3'-iminodipropionitrile, 64/10, 82
imipenem, 65/472
imipramine, 7/518, 533, 46/394, 48/420, 65/311
α-interferon intoxication, 63/359, 65/529, 538, 562
β-interferon intoxication, 65/529, 538, 562
interleukin-2, 65/529
ion channel, 65/146
ion channel plugging, 65/148
isoniazid, 7/516, 533, 582, 584, 588, 644, 28/97, 115, 41/333, 42/648, 47/565, 51/68, 294, 297, 334, 531, 52/211, 62/601, 65/479
kanamycin, 50/219
ketoconazole, 65/472
lamotrigine, 63/496, 65/511
latrodectus venom, 64/10
α-latroinsectotoxin, 65/196
α-latrotoxin, 64/15, 65/194, 196
lead, 4/330, 6/652, 7/511, 8/366, 9/642, 13/63, 21/496, 22/27, 36/3, 6, 12, 22, 35-61, 65, 46/19, 391, 47/561, 48/420, 49/466, 51/68, 263, 265, 268, 531, 64/10, 13, 431, 437, 443
Leiurus toxin, 65/194
linamarase, 65/27
linamarin, 36/515, 65/27
lincomycin, 65/472
lindane, 62/601, 64/202, 65/478
lithium, 46/539, 48/420, 49/385, 51/304, 60/587, 62/601, 64/293, 295
lomustine, 65/528, 67/356
lonidamine, 65/529, 538
lotaustralin, 36/515, 65/26
lupin seed, 64/13
lysergide, 36/534, 37/329, 332, 50/219, 62/601, 64/3, 65/44
maitotoxin, 65/141, 161
manganese, 6/652, 823, 21/497, 36/218-225, 46/392, 49/108, 466, 477, 60/586, 64/6, 303
maprotiline, 65/311, 319
mefloquine, 65/485
melphalan, 39/94, 65/528, 67/357
mercury, 36/73, 83, 147, 37/2, 88, 46/391, 48/420, 51/68, 265, 479, 63/152, 64/10, 367
methanol, 6/676, 9/500, 36/351, 46/393, 48/420, 64/2, 7, 95
methotrexate, 39/94, 98-104, 119, 46/601, 47/551, 570, 63/357, 65/527, 539, 67/357
methylazoxymethanol glucoside, 22/344, 64/5
methyl chloride, 6/652, 36/375, 46/393, 48/420, 49/466, 64/3
3-methylenecyclopropylacetic acid, 65/126
methylmercury, 36/73, 51/265, 64/5, 413
1-methyl-4-phenylpyridinium, 65/380
1-methyl-4-phenyl-1,2,3,6-tetrahydropyridine, 64/2, 6, 65/363
1-methyl-4-phenyl-1,2,3,6-tetrahydropyridine analogue, 65/394
metodiclorofen, 65/528
metronidazole, 46/602, 51/301, 531, 65/487
microbial toxin, 65/209
minocycline, 65/472
mipafox, 51/68
misonidazole, 51/301, 65/529, 538, 67/364
mitomycin C, 65/528
mitotane, 39/95, 113, 119, 65/529, 538, 67/363
mitoxantrone, 65/528
Mojave toxin, 65/182
muscarine, 36/535, 64/14, 65/39
mushroom, *see* Mushroom toxin
myotoxin, 65/185
nalidixic acid, 65/486
neomycin, 50/219
neosaxitoxin, 65/146
α-neurotoxin, 65/181
neurotransmitter stimulating agent, 65/150
nicotine, 36/424, 48/420, 65/251
nimustine, 65/528
nitrile, 65/25
nitrofurantoin, 7/517, 535, 41/333, 51/68, 296, 358, 531, 65/487
3-nitropropionic acid, 64/7
nitrosourea, 67/356

nomifensine, 65/311
norfloxacin, 65/486
notechis II-5, 65/182
notexin, 37/16, 65/182, 184
noxiustoxin, 65/194, 198
octamethyl pyrophosphoramide, 37/549, 64/152, 155
okadaic acid, 65/142
opiate, 37/365, 379, 65/349
organic mercury, 36/83, 64/413
organic solvent, 51/479, 64/39
organochlorine, 36/391, 64/197
organolead, 64/129, 133
organophosphate, 37/541, 46/393, 51/531, 64/11, 151, 161, 229
organotin, 64/4, 129, 135
3-oxalylaminoalanine, 36/507, 64/10, 65/1
oxytetracycline, 65/472
palytoxin, 37/55, 65/161
paradoxin, 65/182
parathion, 8/338, 37/546, 64/177
pectenotoxin, 65/149
pefloxacin, 65/486
penicillin, 46/602, 65/472
4-pentenoic acid, 49/217, 65/126
perhexiline, 51/294, 479, 530, 62/606, 608, 64/10, 14
permethrin, 64/13, 212
phenylpropanolamine, 62/601, 65/259
phenytoin, 7/519, 15/715, 37/200, 203, 40/286, 42/641, 46/394, 600, 47/38, 49/365, 385, 51/68, 507, 52/211, 56/234, 60/587, 656, 64/172, 65/500
philanthotoxin, 65/199
δ-philanthotoxin, 65/194
Phoneutria nigriventer, 65/196
phoneutriatoxin, 65/194
phospholipase, 65/185
polychlorinated biphenyl, 51/286, 64/197, 206
polychlorinated dibenzodioxin, 64/204
polymyxin B, 65/472
poneratoxin, 65/194, 199
prazosin, 65/447
presynaptic neuromuscular blocking toxin, 65/183
procarbazine, 39/18, 95, 113, 119, 63/359, 65/528, 534
2-propyl-4-pentenoic acid, 65/126
pseudoephedrine, 65/259
psilocybine, 4/333, 36/535, 46/594, 64/3, 65/40
pyrithione zinc, 65/478
quinidine, 37/447, 46/538, 63/191, 65/451, 485
quinine, 13/63, 224, 41/297, 559, 50/219, 52/372, 64/8, 65/485
radiosensitizing agent, 67/364
robustotoxin, 65/194
salicylic acid, 50/219
sarin, 37/543, 64/152, 223
saxitoxin, 37/1, 33, 56, 41/291, 64/14, 65/142, 146, 152-154
scaritoxin, 65/142
scorpion venom, 37/112, 65/193, 196
serine proteinase inhibitor, 65/186
serrulatotoxin, 65/194, 198
snake venom, 9/39, 37/1-18, 62/601, 610, 65/177-185
sodium isocyanate, 51/68
soman, 37/543, 64/223
sparfosic acid, 39/95, 113, 119, 65/529, 538
spider venom, 37/111, 65/193
spirogermanium, 65/529, 538
spiromustine, 65/528, 534
streptomycin, 7/515, 50/219, 51/295, 64/9, 65/483
styrene, 64/39, 53, 56
sulfamethoxazole, 65/472
suramin, 65/529, 538
tabun, 37/544, 64/152, 223
taipoxin, 37/17, 62/610, 65/182, 184
tamoxifen, 39/95, 65/529, 538, 67/363
taxol, 65/529, 537
taxotere, 65/529, 537
teniposide, 65/528
tetanospasmin, 52/229, 231, 64/10, 16
tetanus toxin, 1/51, 7/476, 33/491, 37/393, 41/303, 52/229, 62/338, 65/215, 226, 245
tetracycline, 65/472, 568
tetraethyl pyrophosphate, 37/541, 549, 64/151, 223
tetramethrin, 64/213
tetrodotoxin, 28/447, 37/1, 68, 79, 40/143, 41/292, 51/263, 64/15, 65/141, 155
textilotoxin, 65/182
thalidomide, 7/174, 540, 42/660, 46/79, 50/219, 51/68, 304, 531, 65/559
thallium, 7/515, 543, 8/337, 9/646, 30/123, 36/239-273, 46/392, 51/68, 265, 274, 332, 479, 531, 63/152, 64/4, 8, 10, 12, 14, 323, 325
thiotepa, 39/94, 97, 119, 63/359, 65/528, 534, 67/357
ticarcillin, 65/472
tick venom, 37/111, 64/15, 65/194
tityustoxin, 64/15, 65/194
toluene, 36/372, 46/393, 62/601, 613, 64/3, 5, 42, 52, 82
trichloroethanol, 65/329
trichloroethylene, 5/299, 7/522, 36/458, 51/68, 279, 64/9, 40, 44

tri-*o*-cresyl phosphate, 37/471, 475, 40/151, 51/68, 285, 64/10, 152, 160
tricyclic antidepressant, 51/303, 65/311
triethyltin, 16/196, 36/280, 284, 286, 47/564, 51/68, 274, 64/10
trimethoprim, 65/472
tryptoline, 65/398
valproic acid, 37/200, 202, 49/217, 65/115, 125, 497, 503
vancomycin, 50/219
venom, 37/13
venom X, 64/174
versutoxin, 65/195
vidarabine, 65/528, 534
vigabatrin, 63/496, 65/511
vinblastine, 39/17, 94, 107, 110, 64/13, 65/528, 535
vincristine, 7/545, 39/16, 94, 106-110, 119, 40/151, 339, 51/68, 294, 299, 478, 531, 61/151, 62/606, 609, 63/152, 358, 64/10, 13, 65/528, 534, 67/363
vindesine, 39/94, 107, 110, 65/528, 535
wasp kinin, 65/194
wasp venom, 62/601, 65/194
yessotoxin, 65/149
zimeldine, 51/304, 65/311
Neurotransmission
autonomic nervous system, 74/136
colocalization, 74/97
epilepsy, 72/83
parasympathetic nerve, 74/101
sensorimotor nerve, 74/102
strain, 45/251
stress, 45/251
sympathetic nerve, 74/100
Neurotransmitter
acetylcholine, 74/92
acquired hepatocerebral degeneration, 49/216
Alzheimer disease, 46/266-269
amino acid, *see* Amino acid neurotransmitter
autonomic nervous system, 74/38, 55, 89, 91
basal ganglion, 49/33
bladder, 75/669
brain death, 24/744
brain edema, 23/151
brain injury, 57/210
brain metabolism, 57/78
carbon monoxide, 74/94, 96
cerebrovascular disease, 75/362
dystonia musculorum deformans, 49/526
enteric nervous system, 75/617
epilepsy, 72/83, 97
GABA, 29/493
glutamic acid, 29/512-514
glycine, 29/508
head injury, 57/78
Huntington chorea, 42/227
hyperactivity, 46/182
interaction, 74/99
intrinsic nerve, 74/104
neostriatum, 49/47
nervous system tumor, 27/509
neuropeptide, 74/94
nitric oxide, 74/94-96
noradrenalin, 74/92
opiate intoxication, 37/368
orthostatic hypotension, 63/144
Parkinson disease, 42/246
penis erection, 61/314
Pick disease, 46/242
purine, 74/93
schizophrenia, 43/214
suprachiasmatic nucleus, 74/489
Neurotrophic disorder
clinical features, 51/481
Neurotrophic keratitis
hereditary sensory and autonomic neuropathy type II, 60/17
hereditary sensory and autonomic neuropathy type III, 60/17, 33
Neurotrophic osteoarthropathy, 38/431-473
classification, 38/432
clinical features, 38/439-450
congenital indifference to pain, 38/467
diabetes mellitus, 38/462-464
etiology, 38/432-434, 451-453
history, 38/431
leprosy, 38/459
neurotrophic features, 38/448-450
osteolysis, 38/447, 468-470
pathology, 38/434-439
reflex sympathetic dystrophy, 22/254, 38/472
spina bifida, 38/459
syringomyelia, 38/456-459
tabes dorsalis, 38/453-456
traumatic nervous system disorder, 38/459-562
Neurotrophic parietal endocarditis
idiopathic hypereosinophilic syndrome, 38/213
Neurotrophin-3 deficiency
autonomic nervous system, 74/218
Neurotropic agent
hydromyelia, 50/427
Neurotropic mouse hepatitis virus, *see* JHM virus
Neurotropic slow virus
multiple sclerosis, 9/118-128
Neurotropic virus

adenovirus, 56/8
alphavirus, 56/12
avidin, 56/26
avidin-vitamin B_W enzyme complex, 56/26
axonal transport, 56/33
blood-brain barrier, 56/31
choroid plexus endothelium, 56/31
classification, 56/1
coronavirus, 56/439
cowpox virus, 56/5
CSF, 56/31
diaminobenzidine, 56/26
DNA virus, 56/5
enterovirus 70, 56/349
Epstein-Barr virus, 56/249
flavivirus, 56/12
herpes simplex virus spread, 56/32
herpes virus, 56/6
horseradish peroxidase, 56/26
host entry, 56/27
immunocytochemistry, 56/26
leukocyte transport, 56/30
lymphocyte transport, 56/30
measles virus, 56/417
microfold cell, 56/29
molecular approach, 56/21
molluscum contagiosum virus, 56/5
monkey poxvirus, 56/5
monoclonal antibody, 56/26
mumps, 9/545, 56/430
nerve spread, 56/31
Orf virus, 56/5
orthopoxvirus, 56/5
papillomavirus, 56/9
papovavirus, 56/9
parapoxvirus, 56/5
pathogenesis, 56/19, 25-39
pathogenic stage, 56/27
Peyer patch, 56/29
plasma transport, 56/30
poliovirus spread, 56/32
polyomavirus, 56/9
poxvirus, 56/5
proteolysis, 56/28
rabies spread, 56/32
reovirus type 3, 56/33
reticuloendothelial system, 56/30
rubella virus, 56/405
rubivirus, 56/12
in situ hybridization, 56/26
Tana poxvirus, 56/5
Togavirus, 56/12
varicella zoster virus, 56/229
variola virus, 56/5
viremia, 56/30
virus spread, 56/29
vitamin B_W, 56/26
X-ray cristallography, 56/26
Yaba poxvirus, 56/5
Neurotropism
autonomic, *see* Autonomic neurotropism
Coccidioides immitis, 52/410
cowpox virus, 56/35
enhancer, 56/36
influenza virus, 56/34
lymphocytic choriomeningitis virus, 56/35
mumps virus, 56/35
myelin associated protein, 61/479
rabies glycoprotein, 56/34
rabies virus, 56/35
reovirus, 56/35
retrovirus, 56/525
sigma 1 protein, 56/35
vesicular stomatitis virus, 56/35
viral infection, 56/34
viral protein HA, 56/34
viral protein VP1, 56/36
viral protein VP3, 56/36
Neurotubule
nerve, 51/5
twisted, *see* Twisted neurotubule
Neurovascular compression syndrome
Adson-Coffey maneuver, 2/150
cause, 2/146-148
costoclavicular syndrome, 2/146, 148
differential diagnosis, 2/148-151
mechanism, 2/149
sign, 2/144-146
symptom, 2/144-146
Neurovascular hila
anatomy, 2/144
syndrome, 2/143
Neurovascular syndrome
reflex, *see* Reflex neurovascular syndrome
Neurovegetative disorder, *see* Autonomic neuropathy
Neurovegetative dystonia
features, 22/251
Neurovirulence
bunyavirus, 56/39
factor, 56/37
herpes simplex virus type 1, 56/38
herpes simplex virus type 2, 56/38
poliomyelitis, 34/101, 42/653, 56/321
poliovirus, 56/38
rabies virus, 56/38

viral protein HA, 56/39
Neurovisceral lipidosis, *see* GM1 gangliosidosis *and* GM2 gangliosidosis
Neurovisceral storage disease
ophthalmoplegia, 42/604-606
vertical supranuclear ophthalmoplegia, 42/604-606
Neuroxanthoma
nerve tumor, 8/458
Neurulation
anencephaly, 50/74
catecholamine, 30/54
CNS, 50/1, 19
craniorachischisis, 50/74
duration, 50/97
meroanencephaly, 50/74
primary, *see* Primary neurulation
rachischisis, 32/162
Neurulation differentiation
basic features, 50/23
Neutral fat leukodystrophy
history, 10/28, 105
Neutral lipid
biochemistry, 10/233-259
brain, 10/233
phosphatidylcholine, 9/30
sphingomyelin, 9/30
Neutral protease
CSF, *see* CSF neutral protease
Neutral proteinase
wallerian degeneration, 7/215
Neutropenia
candidiasis, 52/398
chronic, *see* Chronic neutropenia
Emery-Dreifuss muscular dystrophy, 62/156
encephalitis, 63/339
fumarate hydratase deficiency, 66/420
hyperglycinemia, 42/565
isovaleric acidemia, 42/579
lysinuric protein intolerance, 29/210
meningitis, 63/339
mental deficiency, 43/289
methylmalonic acidemia, 29/205, 42/601
T-lymphocyte, 63/340
Nevin late adult muscular dystrophy, *see* Late adult muscular dystrophy (Nevin)
Nevocellular nevus
cell type, 14/598
controversy, 14/597
histogenesis, 14/597
neurocutaneous melanosis, 14/595, 597
Nevoid basal cell carcinoma syndrome, *see* Multiple nevoid basal cell carcinoma syndrome
Nevus
basal cell, *see* Basal cell nevus
blue, *see* Blue nevus
fur, *see* Fur nevus
giant, 14/416
hemangiomatous, *see* Hemangiomatous nevus
Klippel-Trénaunay syndrome, 14/392
Maffucci syndrome, 14/14
nevocellular, *see* Nevocellular nevus
pigmented, *see* Pigmented nevus
port wine, *see* Nevus flammeus
Nevus cell
neuroid, *see* Neuroid nevus cell
Nevus epitheliomatosus, *see* Multiple trichoepithelioma
Nevus flammeus
Klippel-Trénaunay syndrome, 43/24
neurocutaneous syndrome, 14/774, 50/191
Patau syndrome, 31/504, 507, 50/558
Sturge-Weber syndrome, 14/2-4, 31/18, 43/47, 55/443
Nevus flammeus glaucoma syndrome, *see* Sturge-Weber syndrome
Nevus hypertrophicans (Bor), *see* Klippel-Trénaunay syndrome
Nevus lipomatosus cutaneous superficialis
Goltz-Gorlin syndrome, 31/288
Nevus osteohypertrophicus, *see* Klippel-Trénaunay syndrome
Nevus of Ota, *see* Ota nevus
Nevus pigmentosus systematicus, *see* Incontinentia pigmenti
Nevus sebaceous
Bitter syndrome, 14/107
linear, *see* Linear nevus sebaceous syndrome
Nevus sebaceous syndrome
linear, *see* Linear nevus sebaceous syndrome
Nevus trichoepitheliomatosus adenoides cysticus, *see* Multiple trichoepithelioma
Nevus unius lateris
café au lait spot, 43/40
conjunctival tumor, 43/40
epilepsy, 14/781, 46/95
hemifacial hypertrophy, 59/482
hemihypertrophy, 59/482
hemiplegia, 43/40
ichthyosiform erythroderma, 43/40
ichthyosis hystrix, 43/40
jaw cyst, 43/40
mental deficiency, 14/781, 43/40, 46/95
muscular dystrophy, 14/781
scoliosis, 43/40
seizure, 43/40

tetraplegia, 43/40
Nevus varicosus osteohypertrophicus, *see* Klippel-Trénaunay syndrome
New York
Neurological Society, 1/5
Newborn asphyxia
clinical features, 49/482
hereditary paroxysmal kinesigenic choreoathetosis, 49/355
pallidal necrosis, 49/466
striatal necrosis, 49/500, 506
Newborn behavior
feeding reflex, 4/345
higher nervous activity, 4/344-347
higher nervous activity disorder, 4/344-347
lip reflex, 4/345
Moro reflex, 4/346
rooting reflex, 4/345
stepping reflex, 4/346
swallowing reflex, 4/345
tonic neck reflex, 4/346
Newborn epilepsy
benign, 42/681
birth lesion, 15/201
dysmaturity, 15/205
EEG, 15/195, 206, 216
etiology, 15/200, 205
examination, 15/206
follow-up, 15/215
hypocalcemia, 15/200
hypoglycemia, 15/200, 203
infection etiology, 15/203
pathology, 15/203
prenatal anoxia, 15/203
prognosis, 15/214
seizure pattern, 15/193
seizure state, 15/189
treatment, 15/216
Newborn physiological jaundice
bilirubin encephalopathy, 27/422
Newcastle disease
immune adherence, 47/358
α-interferon, 47/354
Newcastle disease virus
acute viral myositis, 56/193
Newcastle, United Kingdom, 1/7
Niacin, *see* Vitamin PP
Niacin deficiency, *see* Vitamin B3 deficiency
Niceritrol
restless legs syndrome, 51/548
Nick Krampf
infantile spasm, 15/219
Nicotinamide, *see* Vitamin PP
Nicotinamide adenine dinucleotide
enzymology, 66/51
Nicotinamide adenine dinucleotide phosphate diaphorase, *see* Reduced nicotinamide adenine dinucleotide phosphate dehydrogenase
Nicotinamide adenine dinucleotide tetrazolium reductase, *see* Tetrazolium reductase
Nicotinate nucleotide pyrophosphorylase (carboxylating)
Huntington chorea, 49/298
Nicotine
drug addiction, 65/251
neurotoxin, 36/424, 48/420, 65/251
synapse, 1/48
toxic encephalopathy, 65/251
Nicotine intoxication
absorption, 65/262
animal study, 36/424
brain infarction, 65/263
carotid artery stenosis, 65/264
catecholamine release, 36/425, 65/261
chemoreceptor, 65/261
clinical features, 36/425-428
CNS stimulant, 65/251
complication, 65/261
convulsion, 36/425, 65/261
cotidine, 65/262
drug addiction, 65/263
EEG, 36/429, 65/262
epilepsy, 65/261
half-life, 36/430, 65/262
headache, 36/427, 48/420, 65/262
indoleamine release, 65/261
Jean Nicot, 65/261
laboratory finding, 36/428
metabolism, 65/262
nausea, 36/427, 65/262
neuromuscular motor end plate, 65/261
neuronal nicotinic receptor, 65/261
Nicotiana tabacum, 65/261
nicotine-*N*-oxide, 65/262
nicotinic receptor, 65/261
pathology, 36/430
pharmacokinetics, 65/262
respiratory paralysis, 36/427, 65/261
subarachnoid hemorrhage, 65/263
tachycardia, 36/425, 429, 65/262
tobacco alcoholic amblyopia, 13/62, 65/264
tobacco consumption, 65/262
treatment, 36/430, 65/264
tremor, 36/425, 65/261
vomiting, 36/427, 65/261
withdrawal syndrome, 65/263

Nicotinic acid, *see* Vitamin PP
Nicotinic acid deficiency, *see* Pellagra
Nicotinic acid intoxication, *see* Vitamin PP intoxication
Nicotinic receptor
  cloning, 74/150
  nicotine intoxication, 65/261
  pharmacology, 74/150
Nicotinyl alcohol, 5/379
  restless legs syndrome, 8/318, 51/548
NIDR panel
  nomenclature, 30/2, 8
Nielsen syndrome
  dysraphia, 14/109
  hypertelorism, 30/249
  skeletal malformation, 30/249
  Turner syndrome, 14/111
Niemann-Pick disease, 10/484-505, 66/133-152
  acid phosphatase, 10/487
  adult GM1 gangliosidosis, 60/654
  anemia, 60/674
  atypical case, 10/502
  autosomal recessive, 51/373
  biochemistry, 10/331, 501, 29/355
  brain biopsy, 10/681, 685
  ceramide glucose, 10/286
  cherry red spot, 10/470, 503, 66/19
  classification, 10/557, 60/147
  definition, 60/147
  differential diagnosis, 10/503-505
  enzyme defect, 29/355
  epidemiology, 10/568-572
  foam cell, 10/282, 488, 498, 66/21
  gene frequency, 10/569
  genetics, 10/282, 485, 66/134
  α-2-globulin, 10/284
  historic aspect, 10/484, 29/353-356
  incidence, 66/134
  Jewish population, 10/569-571
  juvenile, *see* Niemann-Pick disease type C
  lipid metabolic disorder, 10/280-287
  lymphocyte, 10/283
  lymphocyte vacuolation, 10/335, 487, 528, 576
  neurolipidosis, 66/20
  phosphatidylethanolamine, 10/488
  phosphatidylethanolamine lipidosis, 10/504
  prognosis, 10/505
  recessive inheritance, 10/485
  secondary pigmentary retinal degeneration, 13/330
  sphingomyelin, 10/286, 327, 487, 501, 29/350
  sphingomyelin phosphodiesterase mutation, 66/141
  sphingomyelinase, 10/281, 332, 29/350, 42/469-471, 46/57
  survey, 66/133
  treatment, 10/505, 29/356
  type, 10/572, 51/373
  typical foam cell, 10/282
  vacuolated lymphocyte, 60/674
  visceral pathology, 10/488
  xanthomatosis, 10/504, 27/242, 251
Niemann-Pick disease type A
  amniocentesis, 29/355
  blindness, 42/468, 51/373
  blood vessel, 10/496
  brain atrophy, 42/469
  cerebellum, 10/491
  cherry red spot, 10/572, 42/469, 51/373
  choroid plexus, 10/496
  classic infantile type, 10/219, 281
  clinical features, 10/330, 486, 29/354, 60/148
  clinical pathology, 10/487
  cutaneous pigmentation, 42/468
  ear, 10/496
  enzyme deficiency polyneuropathy, 51/369, 373
  ependyma, 10/496
  eye, 10/496
  foam cell, 42/469
  hepatosplenomegaly, 42/468
  heterozygote detection, 29/355
  heterozygote frequency, 42/469
  histochemistry, 10/493
  infantile, 10/485, 42/468-470
  meninges, 10/496
  mental deficiency, 46/57
  myelin change, 10/493-495
  neuroglia, 10/495
  neurolipidosis, 66/20
  neuronal change, 10/490
  neuropathology, 10/282, 490, 60/148
  phenotype, 66/135
  psychomotor retardation, 42/469
  Purkinje dendrite limit, 10/492
  retinal degeneration, 42/469
  seizure, 42/468
  ultrastructure, 10/492
  white matter, 10/493
  Wolman disease, 10/504, 546
Niemann-Pick disease type B
  adult type, 10/281
  amniocentesis, 29/355
  neuroacanthocytosis, 60/674
  phenotype, 66/135
  sea blue histiocyte, 60/674
  visceral type, 10/219, 281

Niemann-Pick disease type C, 60/148-153
adult type, 10/502
associated disease, 60/149
ataxia, 42/471
cataplexy, 60/151
ceramide lactoside lipidosis, 29/370
cholesterol homeostasis, 60/153
choreoathetosis, 49/385
classification, 42/470
clinical features, 60/149
computer assisted tomography, 60/152
convulsion, 15/426
dementia, 46/402, 60/665
demyelination, 51/375
differential diagnosis, 10/492
dystonia, 60/664
epilepsy, 72/222
foam cell, 42/472
genetics, 66/143
hepatosplenomegaly, 60/668
incidence, 66/143
inclusion body, 51/375
juvenile neurology, 10/497
juvenile neuropathology, 10/498-501
mental deficiency, 42/470, 46/57, 60/664
narcolepsy, 60/151
neuroacanthocytosis, 60/674
neuropathology, 10/498, 60/151
neuropathy, 51/375
Nova Scotia type, 10/219, 281, 42/471
nuclear magnetic resonance, 60/152
polyneuropathy, 21/66
sea blue histiocyte, 60/674
seizure, 42/471
supranuclear ophthalmoplegia, 60/656
symptomatic dystonia, 49/543
systemic late infantile type III, 10/281
treatment, 60/153
ultrastructure, 51/375
vertical supranuclear ophthalmoplegia, 60/657
visceral late CNS type, 10/219
Niemann-Pick disease type D, *see* Niemann-Pick disease type C
Nifedipine
arterial hypertension, 63/74
brain infarction, 63/74
calcium antagonist, 65/440
dystonia, 63/194
encephalopathy, 63/194
hypertensive encephalopathy, 54/217
migraine, 48/200
myotonia, 62/276
oculogyric crisis, 63/194
stupor, 63/194
toxic myopathy, 62/597
Nifedipine intoxication
akathisia, 65/442
carbamazepine neurotoxicity, 65/442
drug interaction, 65/442
dystonia, 65/442
dystonic posture, 65/442
hallucination, 65/442
muscle cramp, 65/442
muscle spasm, 65/442
myoclonia, 65/442
myoclonic jerk, 65/442
parkinsonism, 65/442
phenytoin toxicity, 65/442
respiratory failure, 65/442
Night blindness, *see* Nyctalopia
Nightmare
child, *see* Pavor nocturnus
clonidine intoxication, 37/443
dreaming, 3/99
juvenile parasomnia, 45/140
levodopa intoxication, 46/593
procainamide intoxication, 37/451
sleep disorder, 3/98
Nigrofugal fiber
anatomy, 6/45
Nigropetal fiber
anatomy, 6/46
Nigrospinodentate degeneration
see also Woods-Schaumburg syndrome
ataxia, 22/165, 168-171
dysarthria, 22/170
dysphagia, 22/170
Friedreich ataxia, 22/174
hereditary, 22/158
Joseph-Machado disease, 42/262, 60/467
levodopa, 22/171
neuropathology, 22/160, 165
nosology, 22/172
nuclear ophthalmoplegia, 22/157-175, 60/468
nystagmus, 22/158, 164
ophthalmoplegia, 21/401, 22/158, 164, 170
parkinsonism, 21/401
progressive external ophthalmoplegia, 22/208
striatonigral degeneration, 60/543
Woods-Schaumburg syndrome, 21/401, 575
Nigrostriatal degeneration, *see* Striatonigral degeneration
Nigrostriatal dopamine receptor
mesolimbic dopamine system, 49/50
Nigrostriatal dopamine system
basal ganglion, 49/52

MAO A, 49/47
MAO B, 49/53
Nigrostriatal fiber, *see* Striatonigral fiber
Nigrostriatal pathway
akinesia, 45/165
Nigrotectal fiber
substantia nigra, 49/10
Nigrotegmental fiber
substantia nigra, 49/10
Nigrothalamic fiber
substantia nigra, 49/10
Nimodipine
brain aneurysm neurosurgery, 55/59
brain blood flow, 57/77
brain infarction, 63/193
brain metabolism, 57/77
calcium antagonist intoxication, 65/441
head injury, 57/77
migraine, 48/200
subarachnoid hemorrhage, 55/59
Nimodipine intoxication
akathisia, 65/442
carbamazepine neurotoxicity, 65/442
drug interaction, 65/442
dystonic posture, 65/442
hallucination, 65/442
muscle cramp, 65/442
myoclonic jerk, 65/442
parkinsonism, 65/442
phenytoin toxicity, 65/442
respiratory failure, 65/442
Nimustine
antineoplastic agent, 65/528
chemotherapy, 67/282, 284
encephalopathy, 65/528
neurotoxin, 65/528
NINDB, *see* National Institute of Neurological Diseases and Blindness
Ninhydrin Schiff staining
protein, 9/23
Ninhydrin test, *see* Sweat test
Nippel
widely spaced, *see* Widely spaced nippels
Nisentil, *see* Alphaprodine
Nishimoto disease, *see* Moyamoya disease
Nissl-Alzheimer arteritis
neurosyphilis, 52/278
Nissl, F., 1/8, 27
Nitrate intoxication
brain blood flow, 65/443
brain ischemia, 65/443
glyceryl trinitrate, 65/443
headache, 48/6, 419, 65/443
indication, 65/443
isosorbide dinitrate, 65/443
isosorbide mononitrate, 65/443
migraine, 48/419
neurologic adverse effect, 65/443
pharmacologic action, 65/443
pituitary apoplexy, 65/444
systemic adverse effect, 65/443
Nitrazepam
antiepileptic agent, 65/497
hypnotic agent, 37/355
serum level, 15/693
side effect, 15/713
structural formula, 37/356
Nitric oxide
autonomic nervous system, 74/137, 75/85
enteric nervous system, 75/617
hypertension, 75/465
local anesthetic agent, 65/426
nerve injury, 65/425
neurotransmitter, 74/94-96
Nitric oxide synthase
autonomic nervous system, 74/137
Nitrite intoxication
headache, 48/6, 419, 444
migraine, 48/419
Nitroblue tetrazolium test
bacterial meningitis, 33/9
meningococcal meningitis, 33/25, 52/25
Nitrofural
polyneuropathy, 7/537
Nitrofurantoin
benign intracranial hypertension, 67/111
chronic axonal neuropathy, 51/531
drug induced polyneuropathy, 51/296
iatrogenic neurological disease, 65/487
motoneuron disease, 41/333
neuropathy, 7/517, 535
neurotoxin, 7/517, 535, 41/333, 51/68, 296, 358, 531, 65/487
renal insufficiency polyneuropathy, 51/358
systemic neoplastic disease, 41/333
toxic neuropathy, 65/487
toxic polyneuropathy, 51/68
Nitrofurantoin intoxication
emotional change, 65/487
myelopathy, 65/487
polyneuropathy, 65/487
renal insufficiency, 65/487
Nitrofurantoin polyneuropathy
EMG, 51/296
glutathione synthetase, 51/296
sensorimotor neuropathy, 51/296

Nitrofurazone, *see* Nitrofural
Nitrogen balance
  brain injury, 23/125
Nitrogen chloride intoxication
  late cortical cerebellar atrophy, 21/497
Nitrogen inhalation
  pallidal necrosis, 49/466
Nitrogen mustard, *see* Chlormethine
Nitrogen narcosis
  barotrauma, 63/419
  behavior impairment, 63/419
  cognitive impairment, 63/419
  diving risk, 63/418
  dysbarism, 12/668, 63/419
Nitrogen oxide intoxication
  headache, 48/420
Nitroglycerine, *see* Glyceryl trinitrate
Nitroimidazole
  CNS toxicity, 69/500
3-Nitropropionic acid
  dystonia, 64/7
  neurotoxin, 64/7
  source, 64/7
  succinate dehydrogenase inhibition, 64/7
3-Nitropropionic acid intoxication
  extrapyramidal symptom, 64/7
Nitroprusside, *see* Sodium nitroprusside
Nitroprusside sodium, *see* Sodium nitroprusside
Nitrosamine
  animal experiment, 16/42
  experimental brain tumor, 16/42
Nitrosourea
  adverse effect, 69/490
  chemotherapy, 39/114, 67/282, 284
  Hodgkin lymphoma, 63/358
  leukemia, 63/358
  lymphocytic leukemia, 63/358
  myeloid leukemia, 63/358
  neurologic toxicity, 39/114
  neurotoxicity, 63/358
  neurotoxin, 67/356
  non-Hodgkin lymphoma, 63/358
Nitrous oxide
  megaloblastic anemia, 63/256
  myelopathy, 63/256
  subacute combined spinal cord degeneration, 63/256
  vitamin B12 deficiency, 70/389
  vitamin B12 deficiency anemia, 63/254
Nitrous oxide anesthesia
  postanesthetic encephalopathy, 9/577, 49/478
NMDA antagonist, *see* *N*-Methyl-D-aspartic acid receptor blocking agent
NMR, *see* Nuclear magnetic resonance
Noack syndrome
  classification, 43/326
  metatarsal, 43/321
Nocardia
  acquired immune deficiency syndrome, 71/292
  actinomycosis, 35/383
  bacterial meningitis, 52/4
  fungal CNS disease, 52/479
  microbiology, 35/525, 52/445
  pure motor hemiplegia, 54/242
Nocardia asteroides, *see* Nocardiosis
Nocardia brasiliensis, *see* Nocardiosis
Nocardia caviae, *see* Nocardiosis
Nocardia meningitis
  CSF, 52/449
  symptom, 52/447
Nocardial brain abscess
  diamorphine intoxication, 65/355
Nocardiosis, 35/517-527, 52/445-451
  acquired immune deficiency syndrome, 52/446, 56/515
  alcoholism, 52/446
  antibiotic agent, 16/218, 52/449
  brain abscess, 16/217, 35/519, 52/150, 447-449
  case report, 35/520-523
  cerebellar abscess, 52/448
  chemotherapy, 35/526, 52/445
  chronic meningitis, 56/645
  CNS, 35/518-523, 52/447
  computer assisted tomography, 35/524, 52/447
  CSF, 35/524, 52/447, 449
  diagnosis, 16/218, 35/523
  empyema, 35/524
  epidemiology, 35/517, 52/446
  epidural empyema, 52/447, 449
  granuloma, 35/524, 52/449
  hematogenous spread, 35/518, 52/446
  histopathology, 52/449
  immunosuppression, 52/451
  management, 52/451
  meningitis, 16/217, 35/519, 52/447, 449
  mortality, 35/519, 52/450
  multiple brain abscess, 35/517, 519, 52/447
  neurologic symptom, 52/445
  neurosurgery, 52/450
  organ transplantation, 52/445
  pathology, 35/534, 52/449
  pneumonia, 52/447
  prognosis, 52/451
  pulmonary infection, 16/217, 35/517, 523, 52/445
  renal insufficiency, 16/217, 27/343
  skin abscess, 35/524, 52/447

spastic paraplegia, 59/436
spinal cord compression, 35/519, 52/449
spinal epidural empyema, 52/187
sulfonamide, 16/218, 35/526, 52/450
treatment, 35/526, 52/449
Nociception
flexor reflex, 1/69
reflex sympathetic dystrophy, 61/125
Nociceptive C fiber
test, 74/596
Nociceptor
acetylsalicylic acid, 45/229
bradykinin, 45/229
marginal layer, 45/230
pain perception, 45/227-229
prostaglandin, 45/229
serotonin, 45/229
Nociceptor afferent
marginal layer, 45/230
substantia gelatinosa, 45/230
tabes dorsalis, 45/230
trigeminal neuralgia, 45/230
Nocturnal cramp
chloroquine, 41/310
quinine, 41/310
Nocturnal enuresis
epilepsy, 72/254
Nocturnal myoclonus
baclofen, 45/140
clonazepam, 45/140
EEG, 45/139
EMG, 45/139
insomnia, 51/544
normal symptom, 6/771
Parkinson disease, 49/97
restless legs syndrome, 38/584, 42/260, 45/139, 49/621, 51/544
sleep, 49/622
sleep disorder, 45/139
Nodal axolemma
internodal axolemma, 47/15
Ranvier node, 51/5, 14
Nodal axoplasm
Ranvier node, 51/15
Nodal membrane
potassium channel, 51/49
Nodding tremor
hereditary cerebello-olivary atrophy (Holmes), 21/403, 409, 60/569
Node
Hensen, *see* Hensen node
Parrot, *see* Parrot node
Ranvier, *see* Ranvier node
Nodosum ganglion
anatomy, 5/351
paraganglioma, 42/762
Nodular dysgenesia
frontal cortex, *see* Frontal cortex nodular dysgenesia
striatopallidodentate calcification, 49/424
Nodular headache
head injury, 5/157
Nodular heterotopia
epilepsy, 72/109
Nodular myositis
muscle ischemia, 41/57, 78
Nodular polioencephalitis
chronic, *see* Chronic nodular polioencephalitis
Nodule
amyloid, *see* Amyloid nodule
boundary, 2/399
Dalen-Fuchs, *see* Dalen-Fuchs nodule
Lisch, *see* Lisch nodule
microglia, *see* Microglia nodule
Nageotte, *see* Nageotte nodule
subependymal, *see* Subependymal nodule
Verocay, *see* Verocay nodule
Nodulus caroticus, *see* Carotid body
Nodus caroticus, *see* Carotid body
Nominal aphasia
Alzheimer disease, 4/98
parietal lobe syndrome, 45/74
temporal lobe tumor, 4/98
Nonaphasic misnaming
reduplication phenomenon, 3/254
Nonbacterial regional lymphadenitis, *see* Cat scratch disease
Nonbacterial thrombotic endocarditis, 11/415-419
see also Systemic lupus erythematosus endocarditis
anoxic ischemic leukoencephalopathy, 47/541
autopsy prevalence, 63/119
blood dyscrasia, 63/121
bone marrow transplantation, 63/28
brain embolism, 11/392, 415, 417, 47/541, 55/165, 489, 63/120
brain hemorrhage, 11/419, 55/164
brain infarction, 11/419, 55/165, 489, 63/28, 120
cancer, 11/417, 47/541, 55/489, 63/28
carcinoma, 11/415, 417, 55/165, 63/111, 120
echoCG, 63/120
encephalopathy, 55/165, 489, 63/120
heparin, 63/121
hyperviscosity syndrome, 55/489
intravascular consumption coagulopathy, 11/415, 55/165, 63/28, 111, 120

malignant lymphoma, 55/489
migratory thrombophlebitis, 63/28
multifocal infarction, 55/489, 63/120
neurologic complication, 11/415, 419, 55/489, 63/120
prevalence, 63/120
thrombophlebitis, 11/415, 417, 63/120
toxic encephalopathy, 63/120
treatment, 63/121
Noncongenital rubella encephalitis
rubella virus, 56/405
Nonconscious hemiasomatognosia, *see* Hemidepersonalization
Nonconvulsive seizure
see also Epilepsy *and* Unilateral epileptic seizure
automatism, 15/140
postural reflex, 15/137
type, 15/130
Nondisjunction
Down syndrome, 30/160, 31/378-395
meiotic, 30/101
Turner syndrome, 31/497
Nondopaminergic substantia nigra, 49/20
Nonencephalitogenic protein
nuclear histone, 9/514
Nonfluent agraphia
motor aphasia, 45/459
transcortical motor aphasia, 45/459
writing, 45/459
Nongerminomatous germ cell tumor
chemotherapy, 68/248
CNS, 68/235
management, 68/251
pathology, 68/235
radiotherapy, 68/248
Nonhereditary multiple cranial neuropathy
nonrecurrent, *see* Nonrecurrent nonhereditary multiple cranial neuropathy
recurrent, *see* Recurrent nonhereditary multiple cranial neuropathy
Nonhereditary recurrent polyneuropathy, *see* Recurrent nonhereditary polyneuropathy
Non-Hodgkin lymphoma
brain tumor, 63/348
Burkitt lymphoma, 63/345
chemotherapy, 69/266, 281
chlormethine, 63/358
classification, 63/345, 69/262
corticosteroid, 63/348
diffuse brain infiltration, 63/348
epidural disease, 69/276
epidural lymphoma, 63/349
epilepsy, 63/349
granulomatous CNS vasculitis, 55/388
headache, 63/349
herpes zoster, 69/279
immunocompromised host, 56/470
immunosuppressive therapy, 65/549
International Index, 69/263
intracerebral treatment, 69/271
intrathecal methotrexate, 63/357
leptomeningeal, 69/271
leukemia, 63/354
meningeal infiltration, 63/346
meningeal lymphoma, 63/346
methotrexate, 69/281
methotrexate effect, 63/357
neurolymphomatosis, 69/279
nitrosourea, 63/358
paraneoplastic polyneuropathy, 51/468
progressive multifocal leukoencephalopathy, 63/354
prophylaxis, 69/271
radiotherapy, 63/348
spinal cord compression, 63/349
subdural lymphoma, 63/349
T-lymphocyte lymphoma, 69/267
treatment, 69/271
vincristine, 63/358
Noninfantile neuronopathic Gaucher disease
myoclonus, 49/618
Noninfectious purulent meningitis
CSF examination, 33/5
Nonketotic coma
see also Diabetic coma
diabetes mellitus, 70/196
survey, 70/196
Nonketotic hyperglycemia
choreoathetosis, 49/385
Nonketotic hyperglycinemia, 29/106, 200-203
epilepsy, 72/222
Nonmidline cephalocele
asterion, 50/110
Nonmongoloid trisomy G, *see* Chromosome 22
Nonmyelinated nerve fiber
neurofibroma, 14/137
saltatory conduction, 47/12-14
terminal axonic neoformation, 7/226
Nonne-Froin syndrome
CSF, 7/498, 19/125, 20/459
CSF examination, 20/217
CSF protein, 19/125
proteinocytologic dissociation, 7/498, 19/125
spinal ependymoma, 20/369
xanthochromia, 7/498, 19/125, 20/459
Nonne, M., 1/8, 12, 15, 21

Nonpneumococcal gram-positive bacterial meningitis, *see* Staphylococcal meningitis *and* Streptococcal meningitis
Nonpolio enterovirus infection
spinal muscular atrophy, 59/370
Nonprimary inflammatory myopathy
chloroquine, 62/372
cimetidine, 62/372
emetine, 62/372
ipecac, 62/372
mevinolin, 62/372
Nonprogressive cerebellar ataxia, 42/124-126
ataxia, 42/124
congenital, 42/124
hereditary, 42/124
infantile hypotonia, 42/124
normal intelligence, 42/124
nystagmus, 42/124
physiotherapy, 42/125
speech disorder in children, 42/124
speech therapy, 42/125
Nonprogressive encephalopathy
myoclonic epilepsy, 73/264
Nonprogressive spinal muscular atrophy
nonprogressive mental deficiency, 59/92
Wohlfart-Kugelberg-Welander disease, 59/92
Nonreactive pupil
acute pandysautonomia, 51/475
Nonreceptor kinase
brain metastasis, 69/109
Nonrecurrent nonhereditary multiple cranial neuropathy
acoustic neuroma, 51/571
actinomycosis, 51/571
Albers-Schönberg disease, 51/571
Arnold-Chiari malformation type I, 51/571
aspergillosis, 51/571
Avellis syndrome, 51/571
basal pachymeningitis, 51/571
Behçet syndrome, 51/571
Candida tropicalis, 51/571
carotid artery aneurysm, 51/571
carotid cavernous fistula, 51/571
cavernous sinus syndrome, 51/571
cerebellar hemangioblastoma, 51/571
chordoma, 51/571
Citelli syndrome, 51/571
craniopharyngioma, 51/571
Cryptococcus neoformans, 51/571
fibromuscular dysplasia, 51/571
Foster Kennedy syndrome, 51/571
giant cell tumor, 51/571
glomus tumor, 51/571
Gradenigo syndrome, 51/571
head injury, 51/571
herpes zoster, 51/571
Jackson syndrome, 51/571
Listeria, 51/571
meningioma, 51/571
mucormycosis, 51/571
nasopharyngeal carcinoma, 51/571
neurolymphoma, 51/571
orbital apex syndrome, 51/571
osteitis deformans (Paget), 51/571
paratrigeminal syndrome, 51/571
pituitary tumor, 51/571
Reiter disease, 51/571
Schmidt syndrome, 51/571
Sicard-Collet syndrome, 51/571
superior orbital fissure syndrome, 51/571
syphilis, 51/571
Tapia syndrome, 51/571
Tolosa-Hunt syndrome, 51/571
tuberculosis, 51/571
Van Buchem disease, 51/571
Vernet syndrome, 51/571
Villaret syndrome, 51/571
Wegener granulomatosis, 51/571
Non-REM sleep
adult parasomnia, 45/142
juvenile parasomnia, 45/141
myoclonus, 49/622
narcolepsy, 2/448, 45/150
narcoleptic syndrome, 42/712
obstructive sleep apnea, 63/453
orthodox phase, 3/66, 81
snoring, 63/450
Nonresorptive hydrocephalus, *see* Normal pressure hydrocephalus
Nonseptic thrombotic endocarditis
leukemia, 63/345
Nonspecific antibody
CSF, *see* CSF nonspecific antibody
Nonspecific B-lymphocyte activation
multiple sclerosis, 47/112
Nonsteroid anti-inflammatory agent
arylacetic acid derivative, 65/416
arylpropionic acid derivative, 65/416
chemical structure, 65/415
classification, 65/415
miscellaneous compound, 65/416
pyrazolinone derivative, 65/416
salicylic acid, 65/416
Nonsteroid anti-inflammatory agent intoxication
brain hemorrhage, 65/415
confusional state, 65/416

headache, 65/416, 563
hearing loss, 65/416
meningitis serosa, 65/415
parkinsonism, 65/416
polyneuropathy, 65/416
psychosis, 65/416
Reye syndrome, 65/415
tinnitus, 65/416, 563
vertigo, 65/416, 563
Nonsystematic system degeneration, *see* Progressive supranuclear palsy
Nonthyroidal hypermetabolism, *see* Luft syndrome
Nontropical sprue, *see* Celiac disease
Nonvasoconstrictor neuron
classification, 74/30
Nonverbal memory
temporal lobe, 45/44
temporal lobe epilepsy, 45/44
temporal lobectomy, 45/44
Noonan-Ehmke syndrome
see also Gonadal dysgenesis
Turner syndrome, 31/503, 46/75, 50/546
Noonan syndrome
see also Gonadal dysgenesis *and* Turner syndrome
hypertelorism, 30/249, 46/75
mental deficiency, 46/74
skeletal malformation, 30/249
syringomyelia, 50/454
Noradrenalin
adrenalin, 74/144
adrenergic receptor, 74/163
affective disorder, 46/475
aging, 74/228
Alzheimer disease, 46/266
autonomic nervous system, 74/92, 75/86
basal ganglion, 49/33
biochemical anatomy, 29/465
biological inactivation, 29/462
biosynthesis, 29/459
blood-brain barrier, 48/64
brain edema, 23/152
brain hemorrhage, 63/234
brain microcirculation, 53/79
caffeine intoxication, 65/260
catabolism, 29/460
catecholamine, 74/162
chorea-acanthocytosis, 63/283
cocaine, 74/145, 407
CSF, *see* CSF noradrenalin
desipramine, 74/145
dopamine, 74/145
dopamine β-mono-oxygenase, 74/141
dystonia musculorum deformans, 49/526
enzymatic pathway, 75/234
epilepsy, 72/91
experimental injury, 25/13
genetics, 75/245
glomus jugulare tumor, 18/442
guanethidine intoxication, 37/436
headache, 48/91
hereditary sensory and autonomic neuropathy type III, 1/478, 21/115, 60/27, 35
hypertension, 75/459
locus ceruleus, 74/161
MAO, 74/145
MAO A, 74/143
mesolimbic dopamine system, 49/48
metabolism, 74/140, 145
metadrenalin, 74/144
migraine, 48/76, 87, 91
multiple neuronal system degeneration, 75/192
neurotransmitter, 74/92
normetadrenalin, 74/141
orthostatic hypotension, 26/319, 63/144
Parkinson disease, 29/475-477, 49/119
penis erection, 75/86
pharmacology, 74/143, 145
plasma concentration, 74/228
postsynaptic, 74/146
presynaptic, 74/144
progressive dysautonomia, 59/139, 152
pure autonomic failure, 22/240, 63/147
regional distribution, 29/465
release, 29/462
Reye syndrome, 56/152
schizophrenia, 46/448
spinal cord experimental injury, 25/13
spinal cord injury, 26/319, 325
storage, 29/462
stress, 45/248, 74/310
subarachnoid hemorrhage, 63/234
sympathetic nerve, 74/162
sympathetic nervous system, 74/138
synthesis, 74/140
temporal artery, 48/87
turning behavior, 45/167
turnover, 29/463
vitamin C, 74/145
Noradrenalin spillover
autonomic nervous system, 75/458
Noradrenalin system
mesolimbic, *see* Mesolimbic noradrenalin system
Noradrenalin transporter deficiency
catecholamine, 75/240
Noradrenergic nervous system
anxiety, 74/309

brain blood flow, 53/58
progressive dysautonomia, 59/149
stress, 74/309
11-Nor-9-carboxy-Δ9-tetrahydrocannabinol, *see* Tetrahydrocannabinolic acid
Nordazepam
benzodiazepine intoxication, 65/334
diazepam intoxication, 37/199, 357
No-rebound phenomenon, *see* Stewart-Holmes sign
Norepinephrine, *see* Noradrenalin
Norfloxacin inoxication
epilepsy, 65/486
Normal pressure hydrocephalus
arterial hypertension, 63/231
Binswanger disease, 54/231
biochemical aspect, 27/492
brain scintigraphy, 23/300, 302
dementia, 27/492
hastening phenomenon, 49/68
hydrocephalic dementia, 46/325
neurogenic arterial hypertension, 63/231
Parkinson disease, 49/94-96
progressive supranuclear palsy, 49/247
progressive type, 30/525
striatonigral degeneration, 60/540
subarachnoid hemorrhage, 46/325
temporal arteritis, 55/344
traumatic, 23/300, 302
Norman disease
bulbar paralysis, 60/658
computer assisted tomography, 60/672
leg adductor spasm, 59/68
mental deterioration, 59/68
orthochromatic leukodystrophy, 10/115
Norman-Landing disease, *see* GM1 gangliosidosis
Norman leukodystrophy
dentate fascia aplasia, 10/104
microcephaly, 10/103
Norman microcephalic familial leukodystrophy
classification, 10/114
demyelination, 10/103, 182
dentate fascia, 10/103
hippocampus, 10/103
pachygyria, 10/103, 182
Norman-Roberts syndrome
lissencephaly, 50/593
Normeperidine, *see* Norpethidine
Normetadrenalin
noradrenalin, 74/141
Normetanephrine, *see* Normetadrenalin
Normokalemic periodic paralysis
acetazolamide, 41/163, 43/172
age, 41/162, 62/463
attack, 28/597, 41/162, 43/171, 62/463
carbohydrate, 41/162
familial periodic paralysis, 28/597, 41/162, 62/457
9α-fluorohydrocortisone, 28/598, 41/163
genetics, 41/162, 425
inheritance, 41/162, 62/463
serum potassium, 28/597, 41/162, 425, 43/171, 62/463
sodium chloride, 28/592, 597, 41/162
symptom, 62/463
treatment, 41/162
Normolipoproteinemic acanthocytosis
motor neuropathy, 59/402
progressive spinal muscular atrophy, 59/402
Norpace, *see* Disopyramide phosphate
Norpethidine
pethidine intoxication, 65/357
Norpethidine intoxication
delirium, 65/357
epilepsy, 65/357
hallucination, 65/357
myoclonus, 65/357
tremor, 65/357
Norrie disease
blindness, 22/508, 31/291, 43/441, 46/66
cataract, 22/508, 31/292, 43/441, 46/66
CNS, 22/508, 31/292
deafness, 22/508, 31/292, 43/441
eye lesion, 31/291
genetics, 31/292
hereditary hearing loss, 22/508
male, 22/508, 43/441, 46/66
mental deficiency, 22/508, 31/291, 43/441, 46/66
microphthalmia, 22/508
retinal detachment, 43/441
retinal vasculoglial proliferation, 22/508
sensorineural deafness, 43/441
11-Nor-Δ9-tetrahydrocannabinol-9-carboxylic acid, *see* Tetrahydrocannabinolic acid
North American blastomycosis, 35/401-409, 52/385-393
see also Blastomyces dermatitidis
amphotericin B, 35/406, 52/390
antibiotic agent, 35/407, 52/392
Blastomyces brasiliensis, 52/385
Blastomyces dermatitidis, 35/408, 52/392
blastomycoma, 52/390
brain abscess, 35/401, 52/385, 387, 389
brain biopsy, 35/406, 52/390
cerebellar abscess, 35/404, 52/388
cerebellitis, 35/406, 52/389
chemotherapy, 35/401, 406, 52/390
chronic meningitis, 35/403, 52/387

clinical features, 35/404, 52/388
confusion, 35/406, 52/389
CSF, 35/404, 406, 52/390
cutaneous, 35/401, 403, 52/387
demyelinating encephalitis, 52/387
diagnosis, 35/401, 52/389
differential diagnosis, 35/406, 52/390
drug dosage, 35/407, 52/391
echoEG, 35/406, 52/389
EEG, 35/406, 52/389
epidemiology, 35/16, 380, 402, 52/386
geographic distribution, 35/380, 402, 52/386
hamycin, 35/407, 52/392
headache, 35/404, 406, 52/389
hematogenous dissemination, 35/402, 52/386
hematoxylin-eosin stain, 35/404, 52/390
history, 35/401, 52/385
hydrocephalus, 35/403, 52/388
hydroxystilbamidine, 35/407, 52/392
immunofluorescence test, 52/390
immunology, 35/408, 52/393
intracranial pressure, 35/404, 52/389
intrathecal injection, 35/407, 52/391
ketoconazole, 52/391
meningitis, 35/401, 404, 52/385, 389
mental symptom, 35/406, 52/389
methenamine silver stain, 35/404, 52/390
mortality, 35/401, 403, 406, 52/385, 390
neurologic incidence, 35/403, 52/387
occupation, 35/402, 52/386
PAS stain, 35/404, 52/390
pathogenesis, 35/404, 52/387
pathology, 35/403, 52/387
primary pulmonary infection, 35/403, 52/386
prognosis, 35/401, 406, 52/390
pulmonary form, 35/401, 52/387
racial distribution, 35/402
radiodiagnosis, 35/406, 52/389
saramycetin, 35/407, 52/392
season, 35/402, 52/386
serology, 35/408, 52/393
sex ratio, 35/402, 52/386
spinal granuloma, 35/403, 52/389
systemic, 35/402, 52/387
treatment, 35/406, 52/390
North Asian tick-borne typhus
rickettsial infection, 34/657
Northern epilepsy syndrome
gene, 72/135
Northern Swedish recessive ataxia
cataract, 60/652
tongue fibrillation, 60/659
vestibular function, 60/658
Nortriptyline
debrisoquine hydroxylation, 65/316
metabolism, 65/316
tricyclic antidepressant, 65/312
Nose
cyclopia, 30/443
glycosaminoglycanosis type IV, 66/303
olfactory neuroblastoma, 8/472
Ramses II, 74/185
saddle, *see* Swayback nose
swayback, *see* Swayback nose
Nose test
finger to, *see* Finger to nose test
Nosencephalus
anencephaly, 50/73
Nosoagnosic overestimation
body scheme disorder, 4/219, 45/379
NOSPECS system
thyroid associated ophthalmopathy, 62/530
Notechis II-5
neurotoxin, 65/182
snake envenomation, 65/182
toxic myopathy, 62/610
Notechis scutatus intoxication
antivenene, 37/94
Notencephalus
encephalocele, 50/73
Notexin
autonomic nervous system, 75/512
chemistry, 65/183
neurotoxin, 37/16, 65/182, 184
snake envenomation, 37/16, 65/182
toxic myopathy, 62/610
Nothesthes intoxication
convulsion, 37/76
death, 37/76
delirium, 37/76
diarrhea, 37/76
hypotension, 37/76
hypothermia, 37/76
ischemia, 37/76
local pain, 37/76
nausea, 37/76
pallor, 37/76
radiation, 37/76
respiratory distress, 37/76
vomiting, 37/76
Nothnagel syndrome
chorea, 2/298, 302
definition, 2/277
pineal tumor, 2/298, 18/362
skew deviation, 2/343
Notochord

chordoma, 18/151, 19/18
development, 32/3
differentiation, 32/6
Notochord syndrome
split, *see* Split notochord syndrome
Nottingham Health Profile
epilepsy, 72/422
NOVA family
paraneoplastic syndrome, 69/330
NOVA gene family
paraneoplastic syndrome, 69/333
NP protein
measles virus, 56/423
NP protein antibody
subacute sclerosing panencephalitis, 56/427
3-NPA, *see* 3-Nitropropionic acid
NSAID intoxication, *see* Nonsteroid anti-inflammatory agent intoxication
Nuchal dystonia
progressive supranuclear palsy, 22/219, 49/247
Nuchal rigidity
bacterial meningitis, 52/1
brain embolism, 53/207
brain infarction, 53/207
bromide intoxication, 36/300, 302
cerebellar hemorrhage, 53/383
cerebellar infarction, 53/383
cerebrovascular disease, 53/207
intracranial hemorrhage, 53/207
intracranial hypertension, 16/136
lymphocytic choriomeningitis, 9/197, 56/359
meningeal sign, 1/542, 544, 546
meningism, 57/133
meningococcal meningitis, 33/23, 52/24
muscle tone, 1/267
occipital headache, 5/371, 48/137
pallidoluysionigral degeneration, 49/456
pallidoluysionigrothalamic degeneration, 49/456
progressive supranuclear palsy, 46/301, 49/240, 242
subarachnoid hemorrhage, 12/128
tetanus, 65/217
transient ischemic attack, 53/207
transtentorial herniation, 1/560
traumatic intracranial hematoma, 57/265
tuberculous meningitis, 52/201
Von Hippel-Lindau disease, 14/115
Nuclear aplasia
see also Möbius syndrome
abducent nerve agenesis, 50/212
external ophthalmoplegia, 22/179
Möbius syndrome, 2/325
oculomotor nerve agenesis, 50/212
trochlear nerve agenesis, 50/212
Nuclear DNA
complex II, 66/422
genome, 66/392
Nuclear DNA defect
genetics, 66/395
Luft syndrome, 66/396
Nuclear histone
myelin basic protein, 9/514
nonencephalitogenic protein, 9/514
Nuclear magnetic resonance
acquired toxoplasmosis, 52/356
acute hemorrhage, 57/171
adrenoleukodystrophy, 60/169
Alzheimer disease, 54/96
amnesic shellfish poisoning, 65/167
angiographic technique, 54/99
anoxic encephalopathy, 63/218
aphasia, 45/300-302
aspergillosis, 52/380
ataxia telangiectasia, 60/376
autism, 60/673
bacterial meningitis, 55/421
Behçet syndrome, 56/600, 71/221
Binswanger disease, 54/94, 223
brain abscess, 57/322
brain air embolism, 55/193
brain arteriovenous malformation, 54/91, 93, 184, 379
brain capillary telangiectasia, 54/93
brain cavernous angioma, 54/93, 379
brain contusion, 57/169, 173, 176
brain edema, 54/87, 92, 57/69, 262
brain embolism, 53/231
brain fat embolism, 55/182
brain hemorrhage, 54/88, 317
brain infarction, 53/231, 54/65, 83, 87
brain ischemia, 57/177, 63/218
brain lacunar infarction, 54/255
brain phlebothrombosis, 54/445
brain swelling, 57/177
brain tumor, 67/235
brain vasculitis, 55/421
carbon monoxide intoxication, 64/33
carotid dissecting aneurysm, 54/275, 278
cavernous sinus thrombosis, 52/174
central neurocytoma, 68/153
central pontine myelinolysis, 47/588, 63/552
cerebellar granular cell hypertrophy, 68/146
cerebrotendinous xanthomatosis, 59/358, 60/673, 66/602
cerebrovascular disease, 53/231, 54/81
cervical vertebral column injury, 61/55

chorea-acanthocytosis, 63/281
chronic hemorrhage, 57/173
coccidioidomycosis, 52/419
computer assisted tomography, 54/65, 57/177, 381
congestive heart failure, 63/132
contraindication, 57/168
corpus callosum agenesis, 50/156
cytomegalovirus polyradiculopathy, 71/359
decompression myelopathy, 61/221
dementia, 54/94, 96
dentatorubropallidoluysian atrophy, 60/609, 673
deoxyhemoglobin, 54/88
dermatomyositis, 62/379
differential diagnosis, 54/96
diffuse axonal injury, 57/169
disseminated encephalomyelitis, 47/481
dysembryoplastic neuroepithelial tumor, 68/156
echo time, 57/167
eosinophilia myalgia syndrome, 64/257
ependymitis granularis, 54/97
epidural hematoma, 57/174
epilepsy, 72/337
extra pontine myelinolysis, 63/552
familial paroxysmal ataxia, 60/673
flow void, 54/91
Friedreich ataxia, 60/320
Gerstmann-Sträussler-Scheinker disease, 60/673
globoid cell leukodystrophy, 59/355, 66/193
glutamic aciduria, 60/673
glutaryl-coenzyme A dehydrogenase deficiency, 66/644
glycosaminoglycanosis type IH, 66/99
Hallervorden-Spatz syndrome, 66/716
head injury, 57/167, 222, 381
head injury outcome, 57/379
hematoma pattern, 57/171
hemosiderin, 54/88
hereditary periodic ataxia, 60/441
hereditary secondary dystonia, 59/344
hereditary striatal necrosis, 49/496
herpes simplex virus encephalitis, 56/216
Hirayama disease, 59/111
Huntington chorea, 49/288
hydrocephalus, 50/292
hydromyelia, 50/430
D-2-hydroxyglutaric aciduria, 66/655
L-2-hydroxyglutaric aciduria, 66/655
hypereosinophilic syndrome, 63/372
imaging interpretation, 57/169
inclusion body myositis, 62/379
infectious endocarditis, 63/118
intracerebral calcification, 54/93
intracerebral hematoma, 54/72, 89
intracranial hemorrhage, 53/231, 54/88
intracranial pressure, 57/176
inversion time, 57/167
juvenile spinal cord injury, 61/243
late posttraumatic sequela, 57/177
lead encephalopathy, 64/436
leukoencephalopathy, 54/96
Lyme disease, 52/262
magnetoencephalography, 72/323
manganese intoxication, 64/309
Marinesco-Sjögren syndrome, 60/673
meningeal leukemia, 63/342
meningeal lymphoma, 63/347
mesencephalic infarction, 53/392
metachromatic leukodystrophy, 59/356, 66/175
metastatic cardiac tumor, 63/105
methanol intoxication, 64/100
methemoglobin, 54/88
3-methylglutaconic aciduria, 66/653
mevalonate kinase deficiency, 66/655
moyamoya disease, 55/296, 298
mucolipidosis type II, 66/382
multiple neuronal system degeneration, 60/673, 75/186
multiple sclerosis, 47/67, 191
multiple white matter lesion, 54/96
myelomalacia, 63/44
myotonic dystrophy, 62/237
neonatal adrenoleukodystrophy, 60/673
neurobrucellosis, 52/595
neuropsychology, 45/1, 7
Niemann-Pick disease type C, 60/152
olivopontocerebellar atrophy, 60/673
olivopontocerebellar atrophy (Dejerine-Thomas), 60/519, 522
olivopontocerebellar atrophy (Schut-Haymaker), 60/673
olivopontocerebellar atrophy (Wadia-Swami), 60/499
orbital myositis, 62/299
organic acid metabolism, 66/642
paramagnetic agent, 54/88
Parkinson disease, 49/98
pediatric head injury, 57/329
periventricular hyperintensity, 54/96
phosphorus 32, 54/82
physical principle, 54/81
polymyositis, 62/379
pontine hemorrhage, 54/313
positron emission tomography, 54/86
postinfectious encephalomyelitis, 56/430
posttraumatic syringomyelia, 61/55, 380, 384, 386, 388, 392, 395

precaution, 57/168
primary cardiac tumor, 63/105
progressive dysautonomia, 59/139
progressive multifocal leukoencephalopathy, 47/509
progressive supranuclear palsy, 49/245
propionic acidemia, 66/644
pure oxygen breathing, 54/87
radiation myelopathy, 61/201
relaxation time, 57/167
Rett syndrome, 63/437
Schut family ataxia, 60/484
sex chromosome fragility, 60/673
sickle cell anemia, 55/506, 512
Sjögren syndrome, 71/65
sodium 23, 54/82
spectroscopy, 54/82, 57/71
spinal cord abscess, 52/191
spinal cord injury, 61/515
spinal cord metastasis, 69/179
spinal epidural empyema, 52/188
striatonigral degeneration, 60/543
study, 57/222
subacute hemorrhage, 57/172
subacute sclerosing panencephalitis, 56/420
subarachnoid hemorrhage, 55/21
subdural collection, 57/176
subdural hematoma, 57/174
subfalcine herniation, 57/175
syphilis, 71/287
syringomyelia, 50/447
systemic lupus erythematosus, 71/47
technical consideration, 57/167
thalamic infarction, 53/392
thoracic vertebral column injury, 61/77
thoracolumbar vertebral column injury, 61/92
thyroid associated ophthalmopathy, 62/531
toxoplasmosis, 52/356
transient ischemic attack, 53/231, 313, 54/85
traumatic intracranial hematoma, 57/171, 267
tropical polymyositis, 62/373
venous sinus occlusion, 54/445
venous sinus thrombosis, 54/404, 445
vertebral column injury, 61/515
vertebrobasilar system syndrome, 53/392
viral infection, 56/86
Nuclear ophthalmoplegia
Joseph-Machado disease, 42/262, 60/468
nigrospinodentate degeneration, 22/157-175, 60/468
ptosis, 2/279
pupillary disorder, 2/279
Woods-Schaumburg syndrome, 21/401, 575
Nucleic acid
nervous system tumor, 27/506-508
rabies virus, 34/243
virology, 34/3
wallerian degeneration, 7/213
Nucleocytomegaly
ataxia telangiectasia, 60/353, 382
Nucleodegenerative myopathy
congenital myopathy, 62/332
intranuclear inclusion body, 62/351
Marinesco-Sjögren syndrome, 62/351
membranous whorl, 62/351
Nucleography, *see* Discography
Nucleoprotein
herpes simplex virus, 56/208
Nucleoside analog
chemical structure, 71/369
neuropathy, 71/367, 371
5'-Nucleotidase
wallerian degeneration, 7/215
Nucleotide
migraine, 48/86, 100
nervous system tumor, 27/506-508
Nucleotide base mutation
amyloid, 51/414
Nucleus
amygdaloid, *see* Amygdaloid nucleus
arcuate, *see* Arcuate nucleus
basal, *see* Basal nucleus
brain stem, *see* Brain stem nucleus
brain tumor, 17/95
Cajal interstitial, *see* Cajal interstitial nucleus
caudate, *see* Caudate nucleus
centromedian thalamic, *see* Centromedian thalamic nucleus
cranial nerve, *see* Cranial nerve nucleus
Darkschewitsch, *see* Darkschewitsch nucleus
dentate, *see* Dentate nucleus
dorsal 10th, *see* Dorsal 10th nucleus
dorsal paramedian, *see* Dorsal paramedian nucleus
dorsal vagus, *see* Vagus nerve dorsal nucleus
dorsomedial thalamic, *see* Dorsomedial thalamic nucleus
Edinger-Westphal, *see* Edinger-Westphal nucleus
epithalamic thalamic, *see* Habenular nucleus
extralamellar thalamic, *see* Extralamellar thalamic nucleus
extraocular motor, *see* Extraocular motor nucleus
fastigial, *see* Fastigial nucleus
gracile, *see* Gracile nucleus
habenular, *see* Habenular nucleus
inferior artery to dentate, *see* Inferior artery to dentate nucleus

inferior olivary, *see* Inferior olivary nucleus
intralaminar, *see* Intralaminar nucleus
Kölliker-Fuse, *see* Kölliker-Fuse nucleus
lateral geniculate, *see* Lateral geniculate nucleus
lateral habenular, *see* Lateral habenular nucleus
lateral reticular, *see* Lateral reticular nucleus
Luys, *see* Luys nucleus
medial geniculate, *see* Medial geniculate nucleus
medial mamillary, *see* Medial mamillary nucleus
medial thalamic, *see* Medial thalamic nucleus
Meynert basal, *see* Meynert nucleus basalis
muscle cell, *see* Muscle cell nucleus
muscle fiber, 40/13, 33, 38
muscle fiber type, 62/4
oculomotor, *see* Oculomotor nucleus
olivary, *see* Olivary nucleus
Onuf, *see* Onuf nucleus
Onufrowicz, *see* Onufrowicz nucleus
parafascicular, *see* Parafascicular nucleus
paramedian, *see* Paramedian nucleus
paraventricular, *see* Paraventricular nucleus
pedunculopontine tegmental, *see* Pedunculopontine tegmental nucleus
Perlia, *see* Perlia nucleus
posterior hypothalamic, *see* Posterior hypothalamic nucleus
preoptic, *see* Preoptic nucleus
pulvinaris intergeniculatis, *see* Pulvinaris intergeniculatis nucleus
pulvinaris lateralis, *see* Pulvinaris lateralis nucleus
pulvinaris medialis, *see* Pulvinaris medialis nucleus
pulvinaris superficialis, *see* Pulvinaris superficialis nucleus
red, *see* Nucleus ruber
reticularis polaris, *see* Reticularis polaris nucleus
solitary tract, 74/153
subthalamic, *see* Subthalamic nucleus
superior red, *see* Nucleus ruber superior syndrome
suprachiasmatic, *see* Suprachiasmatic nucleus
supraoptic, *see* Supraoptic nucleus
thalamic, *see* Thalamic nucleus
thalamic ventral anterior, *see* Thalamic ventral anterior nucleus
thalamic ventral lateral, *see* Thalamic ventral lateral nucleus
thalamic ventral oral posterior, *see* Thalamic ventral oral posterior nucleus
thalamic ventralis intermedius, *see* Thalamic ventralis intermedius nucleus
trigeminal spinal, *see* Trigeminal spinal nucleus
tuber cinereum, 2/436
tuberoinfundibular, *see* Tuberoinfundibular nucleus
vagus nerve dorsal, *see* Vagus nerve dorsal nucleus
ventralis oralis, *see* Ventralis oralis nucleus
ventromedial hypothalamic, *see* Ventromedial hypothalamic nucleus
Nucleus accumbens
alcohol intoxication, 64/114
basal ganglion tier II, 49/21
cocaine intoxication, 65/252
functional anatomy, 49/20
Huntington chorea, 49/318, 323
Parkinson disease, 49/67
pedunculopontine tegmental nucleus, 49/12
tegmental subthalamic fiber, 49/11
Nucleus accumbens septi
anatomy, 6/32
Nucleus ambiguus
autonomic nervous system, 74/63
Avellis syndrome, 1/189
congenital laryngeal abductor paralysis, 42/319
expiration, 63/480
respiration, 63/480
respiratory center, 63/431
Nucleus ambiguus dysgenesia
laryngeal abductor paralysis, 42/319
Nucleus ansa peduncularis
anatomy, 6/33
Nucleus basalis
Meynert, *see* Meynert nucleus basalis
Nucleus lemnisci diagonalis
anatomy, 6/32
Nucleus membrane
neuronal, *see* Neuronal nucleus membrane
Nucleus parabrachialis medialis
apneusis, 63/485
pneumotaxic center, 63/479
respiratory center, 63/432
respiratory drive, 63/479
Nucleus pulposus
intervertebral disc, 20/525
Nucleus raphe magnus
analgesia, 45/238-240
pain control, 45/238-240
Nucleus ruber
Benedikt syndrome, 2/298
cell, 6/39, 40
cerebellar tremor, 49/587
cerebello-olivary atrophy, 42/136
constituent, 6/39-42
dentatorubropallidoluysian atrophy, 49/440
function, 6/107
hereditary sensory and autonomic neuropathy type

I, 60/9
magnocellular part, 6/39
parvocellular part, 6/39
relation, 6/38
spinopontine degeneration, 21/391
Nucleus ruber lesion
Foix syndrome, 2/277
Nucleus ruber superior syndrome
ataxia, 1/187
chorea, 1/187
Nucleus ruber syndrome
crossed, *see* Crossed nucleus ruber syndrome
Nucleus striae terminalis
anatomy, 6/33
Nucleus subcaudalis
trigeminal neuralgia, 5/287
Nucleus subthalamicus luysii, *see* Subthalamic nucleus
Nucleus tegmentalis pedunculopontinus, *see* Pedunculopontine tegmental nucleus
Nucleus tractus solitarii
inspiration, 63/480
neurogenic arterial hypertension, 63/229
neurogenic pulmonary edema, 63/495
orthostatic hypotension, 63/143
pulmonary edema, 63/230
respiration, 63/477
respiratory center, 63/431
Number agraphia
constructional, 4/151
ideomotor, 4/151
Number aphasia
arithmetic, 2/600
finger agnosia, 2/600
Number-form
imagery, 43/224
Number reading
alexia, 4/120, 45/437
musical agraphia, 4/120
Numbness
parietal lobe epilepsy, 73/100
Nutrition
see also Diet
anencephaly, 30/197, 200, 50/80
antiepileptic agent, 65/497
brain amino acid metabolism, 29/24-26
brain injury, 24/615
brain tumor carcinogenesis, 17/28
central pontine myelinolysis, 47/588
head injury, 57/233
hereditary sensory and autonomic neuropathy type III, 75/144
motoneuron disease, 59/462
pressure sore, 61/350
Nutritional amblyopia, *see* Tobacco alcoholic amblyopia
Nutritional cerebellar degeneration, 28/271-283, 70/350
clinical features, 28/281
clinicopathologic correlation, 28/282
diagnosis, 28/282
history, 28/271
neuropathologic finding, 28/275-281
prevention, 28/282
retrobulbar neuropathy, 28/272
treatment, 28/282
Wernicke-Korsakoff syndrome, 28/272
Nutritional deficiency
experimental pigmentary retinopathy, 13/203
Marchiafava-Bignami disease, 17/542, 47/557
mental deficiency, 46/17
paraneoplastic polyneuropathy, 51/469
Nutritional deprivation, *see* Malnutrition
Nutritional myelopathy
ataxic spastic symptom, 7/638
lathyrism, 7/643
Nutritional polyneuropathy, *see* Beriberi
Nyctalopia
adaptometer, 13/159
autosomal recessive, *see* Oguchi disease
Bassen-Kornzweig syndrome, 51/327, 394
blindness, 13/224
deferoxamine, 63/534
Hallervorden-Spatz syndrome, 49/401
hyaloidotapetoretinal degeneration (Goldmann-Favre), 13/37, 276
iron, 13/224
piperidylchlorophenothiazine, 13/210
primary pigmentary retinal degeneration, 13/158
quinidine intoxication, 37/448
Refsum disease, 8/21, 13/314, 21/189, 192, 51/385, 60/228, 66/488
rhodopsin, 13/193
Rosenberg-Chutorian syndrome, 22/506
zinc intoxication, 36/335-337
Nyctohemeral sleep rhythm, *see* Circadian rhythm
Nycturia
pure autonomic failure, 22/232, 234
Nijmegen breakage syndrome
ataxia telangiectasia, 60/425
cancer susceptibility, 60/426
chromosomal aberration, 60/426
immunodeficiency, 60/426
mental deficiency, 60/426
microcephaly, 60/426
radiation hypersensitivity, 60/426

Nyssen-Van Bogaert syndrome, *see* Opticocochleodentate degeneration
Nystagmus
see also Eye movement
acoustic neuroma, 16/315
acrylamide intoxication, 64/68
ADR syndrome, 42/112
after, *see* After-nystagmus
albinism, 14/106
alcohol intoxication, 64/115
Alström-Hallgren syndrome, 13/461, 42/391
alternating, 1/592
amyotrophic lateral sclerosis, 22/114, 322
anhidrotic ectodermal dysplasia, 14/788
anticonvulsant, 2/361
aqueduct stenosis, 2/279, 42/57, 50/311
archicerebellar syndrome, 2/418
ARG syndrome, 50/581
arthropod envenomation, 65/196
ataxia, 5/312
ataxia telangiectasia, 42/119
ataxic hemiparesis, 54/248
barbiturate intoxication, 65/332, 509
Bardet-Biedl syndrome, 13/295
basilar artery migraine, 48/138
Bassen-Kornzweig syndrome, 63/273
Behçet syndrome, 56/598
Behr disease, 13/88, 42/416
benign essential tremor, 42/268
benign paroxysmal positional, *see* Benign paroxysmal positional nystagmus
benign paroxysmal positional nystagmus, 2/370
Biemond syndrome, 21/379
brain abscess, 52/152
brain embolism, 53/213
brain infarction, 53/213
brain injury, 1/596
brain stem, 1/590, 595
brain stem injury, 1/590, 595, 57/137
brain stem tumor, 16/322
burst, 1/595
caloric, *see* Caloric nystagmus
Candida meningitis, 52/400
carbamazepine intoxication, 37/202, 65/507
cat scratch disease, 52/130
cerebellar atrophy, 42/135
cerebellar infarction, 53/383
cerebellar tumor, 16/325, 17/711
cerebellar vermis agenesis, 42/4
cerebellum, 1/595
cerebrovascular disease, 53/213
Chédiak-Higashi syndrome, 42/536
classification, 1/585-587
Cockayne-Neill-Dingwall disease, 43/351
complex IV deficiency, 66/424
congenital, *see* Congenital nystagmus
congenital cerebellar ataxia, 42/124
congenital Pelizaeus-Merzbacher disease, 10/158, 42/503
congenital retinal blindness, 13/272, 22/530, 42/407
congenital seesaw, *see* Congenital seesaw nystagmus
congenital with vertigo, 2/361
convergent, 1/594, 16/323
Crome syndrome, 43/242
cutis verticis gyrata, 43/244
cytarabine, 63/356
cytarabine intoxication, 65/533
Dandy-Walker syndrome, 50/328
darkness, 1/593
De Lange syndrome, 43/246
diencephalic syndrome, 18/367, 42/733
directional preponderance, 2/707
dissociated, *see* Dissociated nystagmus
divergence, 1/595
Down syndrome, 50/526, 529
downbeating, *see* Downbeating nystagmus
dysosteosclerosis, 43/393
dystonia, 42/240
epilepsy, 42/690
epiphyseal dysplasia, 43/396-398
erythrokeratodermia ataxia syndrome, 21/216
ethylene glycol intoxication, 64/123
eyelid, *see* Eyelid nystagmus
facial spasm, 42/315
familial spastic paraplegia, 22/427, 458
fasciculation, 42/240
felbamate intoxication, 65/512
Ferguson-Critchley syndrome, 42/142, 59/319
fixation, 14/284
fluorouracil intoxication, 65/532
fluttering, 1/594
Forsius-Eriksson syndrome, 42/397
forward gaze, 1/592
Foville syndrome, 2/316
Friedreich ataxia, 21/335, 348, 42/143, 60/309
gabapentin intoxication, 65/512
gaze paralysis, 2/335
Goltz-Gorlin syndrome, 14/113, 43/20
gyromitrine intoxication, 65/36
Hartnup disease, 7/577, 21/564
head injury, 24/122, 125, 127, 135
head movement, 1/593
hereditary, *see* Hereditary nystagmus
hereditary cerebellar ataxia, 21/369-371

hereditary cerebello-olivary atrophy (Holmes), 21/403, 409, 60/569
hereditary chin tremor, 42/266
hereditary dystonic paraplegia, 22/447, 458
hereditary motor and sensory neuropathy type I, 21/284, 308
hereditary motor and sensory neuropathy type II, 21/284, 308
hereditary olivopontocerebellar atrophy (Menzel), 21/443
heredopathia ophthalmo-otoencephalica, 42/150
hexachlorophene intoxication, 37/486, 65/476
hyperosmolar hyperglycemic nonketotic diabetic coma, 27/90
hyperpipecolatemia, 29/222
hypertrophic interstitial neuropathy, 42/317
hypervalinemia, 42/574
hypomagnesemia, 63/562
impaired vision, 1/596
infantile neuroaxonal dystrophy, 49/402
infantile optic atrophy, 42/408
infantile optic glioma, 42/733
internuclear ophthalmoplegia, 2/329
intracranial hemorrhage, 53/213
intractable hiccup, 63/491
Joseph-Machado disease, 42/262
Lafora progressive myoclonus epilepsy, 27/173
lamotrigine intoxication, 65/512
latent, *see* Latent nystagmus
lateral gaze, 42/158
lateral medullary infarction, 53/382
lidocaine intoxication, 37/451
Listeria monocytogenes meningitis, 52/93
lithium intoxication, 64/296
localization value, 1/589
Machado disease, 42/155
malnutrition syndrome, 2/449
Marinesco-Sjögren syndrome, 21/556, 42/184
median longitudinal fasciculus syndrome, 42/321
mental deficiency, 42/240
mesencephalic lesion, 2/278
metabolic ataxia, 21/580
metabolic encephalopathy, 69/400
mexiletine intoxication, 65/455
microcephaly, 43/429, 50/269
migraine, 48/147
miner, *see* Miner nystagmus
monocular, 1/593
monocular occlusion, 1/595
multiple mitochondrial DNA deletion, 62/510
multiple neuronal system degeneration, 75/162
multiple sclerosis, 9/167, 42/496
myasthenia gravis, 1/598
myopathic, 1/598
nigrospinodentate degeneration, 22/158, 164
nonprogressive cerebellar ataxia, 42/124
nonvestibular, 16/323
normal, 1/579
occipital lobe tumor, 17/335
oculocutaneous albinism, 42/404, 43/3-7
olivopontocerebellar atrophy, 42/161
olivopontocerebellar atrophy variant, 21/455, 42/161
olivopontocerebellar atrophy with retinal degeneration, 60/507
ophthalmoplegia totalis, 43/145
optokinetic, *see* Optokinetic nystagmus
orthochromatic leukodystrophy, 42/500
parietal lobe epilepsy, 73/102
Parkinson disease, 49/94
Pelizaeus-Merzbacher disease, 42/502, 505, 59/357
pellagra, 28/95
periodic, *see* Periodic nystagmus
periodic ataxia, 21/578
periodic vestibulocerebellar ataxia, 42/116
phakomatosis, 14/653
phencyclidine intoxication, 65/54, 367
phenytoin intoxication, 42/641, 65/500
pineal tumor, 17/656
pontine infarction, 53/390, 394
pontine lesion, 2/249
pontine syndrome, 55/125
positional, *see* Positional nystagmus
posterior fossa hematoma, 57/266
prenatal onset, 42/690
presenile dementia, 42/288, 46/248
primary pigmentary retinal degeneration, 13/178
progressive bulbar palsy, 22/114
progressive supranuclear palsy, 22/219, 49/242
rebound, *see* Rebound nystagmus
refractory, 1/594
Refsum disease, 8/22, 10/345, 21/193, 200, 41/434, 60/229, 231, 728, 66/489
retinal, 1/594
retraction, *see* Retraction nystagmus
retractorius, 16/323
Richards-Rundle syndrome, 22/516, 43/264
Richner-Hanhart syndrome, 42/581
rotation, *see* Rotation nystagmus
rotation vestibular, *see* Rotation vestibular nystagmus
scorpion venom, 65/196
secondary pigmentary retinal degeneration, 13/296, 321
seesaw, *see* Seesaw nystagmus

shimmering, 1/594
situs inversus, 42/240
Sjögren syndrome, 71/75
skew deviation, 2/343
spinal cord injury, 25/385
spinal cord tumor, 19/40, 59
spinal tumor, 19/40
spinocerebellar degeneration, 42/190
spinopontine degeneration, 21/371, 390, 42/192
spontaneous, *see* Spontaneous nystagmus
spontaneous vestibular, 16/322
subacute necrotizing encephalomyelopathy, 42/625
Sylvester disease, 22/505
syringobulbia, 42/59, 50/445
syringomyelia, 42/59
temporal bone fracture, 24/122, 125, 127
de la tête, 42/158
tick bite intoxication, 37/111
transient ischemic attack, 53/213
transverse sinus thrombosis, 52/176
traumatic, 24/122, 125, 127, 135
trigeminal neuralgia, 5/312
type, 1/585
upbeating, *see* Upbeating nystagmus
uremic polyneuropathy, 63/512
valproic acid intoxication, 65/504
vertebrobasilar system syndrome, 53/390, 394
vertical, *see* Vertical nystagmus
vestibular, *see* Vestibular nystagmus
vestibular stimulation, 1/577
vestibular system, 16/303
voluntary, 1/598
water intoxication, 64/241
Wernicke-Korsakoff syndrome, 45/194
XXXX syndrome, 50/554
Nystagmus myoclonus
hyperreflexia, 42/240
talipes, 42/240
tremor, 42/240
Nystagmus retractorius, *see* Retraction nystagmus